THE LIFE OF BIRDS
Fourth Edition

The Life of
BIRDS

FOURTH EDITION

Joel Carl Welty

Luis Baptista

California Academy of Sciences

Saunders College Publishing
Harcourt Brace Jovanovich College Publishers
Fort Worth   Philadelphia   San Diego
New York   Orlando   Austin   San Antonio
Toronto   Montreal   London   Sydney   Tokyo

Text Typeface: Electra
Compositor: The Clarinda Company
Acquisitions Editor: Ed Murphy
Project Editor: Sally Kusch
Art Director: Carol Bleistine
Art Assistant: Doris Roessner
Text Designer: Arlene Putterman
Text Artwork: Larry Ward
Production Manager: Tim Frelick, Harry Dean
Assistant Production Manager: JoAnn Melody

Cover Credit: The Great Blue Heron. © Animals, Animals/Lynn M. Stone

Printed in the United States of America

THE LIFE OF BIRDS, 4th edition

0-03-068923-6

Library of Congress Catalog Card Number: 87-32143

4   039   98765

PREFACE

The scientific discipline of ornithology is like a tree with new branches continually growing and reaching out. Unlike a tree, however, "The older it gets, the more vigorous its growth seems to be" (Mayr, 1983a). The revision of a textbook is at once a frustrating and a euphoric experience. It is frustrating in that the exponential growth of the discipline renders it impossible for authors to keep abreast of all new developments. It is a euphoric experience in that it affords us an opportunity to appreciate the vitality of the field and to absorb as many of the latest discoveries as we can. This done, we try to select the best and weave it into a coherent tapestry, and in the process correct errors and eliminate outdated material.

Early ornithologists wrote only for other ornithologists to read. Today, birds have been subjects of numerous studies contributing to the advancement of many areas of biology (see Foreword). A student of ornithology thus learns, sometimes incidentally, the principles of many biological disciplines.

The aim of this book remains that of the first edition: 1. To present, simply and straightforwardly, the basic facts of bird biology. 2. To arouse in the reader a lasting fascination for birds and for all the wonderful things they do. The book is directed towards the general student, not the specialist. Most of the chapters can be read and understood by the reader without special preparation.

Much of this book deals with natural history of birds. Good natural history is like good descriptive anatomy: when well done, both activities provide raw materials for experimentation and hypothesis-testing (Greene, 1986). Although much of this book consists of description, we tried to incorporate whenever possible good experiments testing biological principles. We try throughout to emphasize the fact that birds are products of natural selection, that selection operates on the individual, and that only individuals endowed with a certain complement of genes survive and reproduce to leave progeny. Here below we point out some of the highlights in our new edition.

Avian taxonomy is a dynamic field, and conclusions regarding relationships of birds are still not the final word in many cases. Most birds species have been discovered and described. However, adaptive radiation may mask true evolutionary relationships, and morphological convergences may lead taxonomists to erect phylogenies that may be erroneous. The field of avian taxonomy now calls upon techniques borrowed from sister disciplines such as biochemistry (DNA/DNA hybridization, electrophoresis), karyotypes, and ethology to help solve taxonomic problems. Recent advances in these areas are discussed in Chapter 2.

In Chapter 3 we bring together the latest studies on the secretory functions of the bird epidermis. The epidermis in birds has often been described as aglandular. Recent studies have shown, on the contrary, that the avian epidermis is very much living tissue capable of lipogenesis. One of the most exciting discoveries participated in by the junior author is that nestling Zebra Finches, *Poephila guttata*, are endowed with epidermal lipids organized into lamellar bodies which enable them to decrease cutaneous water loss and thus enable them to survive when heat-stressed. This is the first demonstration of how birds control transpiration through the skin.

Early texts also inform us that birds in general have a poor sense of smell and sense of taste. Only birds with large olfactory bulbs were thought to have a well-developed sense of smell. We bring together in Chapter 5 descriptive and experimental studies indicating that many birds may have good smell and taste capabilities. It has been suggested that large olfactory bulbs are not so much indicative of greater olfactory sensitivity, as of greater discriminatory abilities. Experiments presented herein also indicate that birds may hear infrasounds and see into the near-ultraviolet.

In Chapter 6 we review some new literature on anatomical specializations adapting birds to certain dietary habits, e.g., grains, fruit, and conifer needles. We also introduce the student to the concept of "optimal foraging," one of the more active areas of ecological investigation today.

Chapter 7 deals primarily with physiological adaptations of birds to their environment or microenvironment. Oxygen consumption tends to increase when animals are exposed to high temperatures beyond the zone of thermal neutrality. Latest studies indicate that some hummingbirds do not conform to this principle. We review new literature showing that torpidity is utilized by small birds not only during emergencies but as a routine energy-saving device. We also discuss the latest literature and point out that the Greenwalt/Stein model of how syringes operate is not the last word and that different species vocalize in different ways.

Photoperiod is the primary trigger of gonadal growth in vertebrates. In Chapter 8 we bring together the latest experiments testing the synergistic role of avian vocalizations and photoperiod in effecting gonadal growth, and point out that extent of ovarian development may differ between adult hand-raised experimentals and those caught in the wild. Latest findings regarding other factors (e.g., food temperature, phytoestrogens) as triggers are also treated.

In Chapter 9, Behavior, we develop the themes put forward by Tinbergen as the ultimate goals of ethology, namely, evolution, ontogeny, causation, and function. We emphasize that there is a continuum between nature and nurture in the development of behaviors and bring together experiments indicating that learning may occur on the level of the single neuron. Although earlier texts have sometimes stated that tool-using in animals other than humans is a rare phenomenon, we discuss the literature indicating that tool-using and cultural traditions of various types are commonplace among birds and are often prime examples of the interplay of genome and environment in shaping behaviors.

In, Chapter 10, Social Behavior, we bring up to date the concept of "imprinting" in birds. The early literature on imprinting is replete with studies on the "following" response but few studies on sexual imprinting. We develop Lorenz's four criteria pertaining to sexual imprinting, namely, the sensitive phase, supraindividuality, irreversibility, and manifestation later in life. We call attention to "contact" versus "distant" species in studying avian social behavior, and in the section on avian aggregations, we speculate on the costs and benefits of being the subordinate in a flock.

Modern instrumentation has permitted scientists to make great strides in the field of avian bioacoustics (Chapter 11). Songs appear to be characteristic of males of the species, and indeed females (e.g., of Zebra Finches) may not even have the neural equipment to store and produce songs. We review experiments on hormone manipulation which result in females developing male song centers in their brain and, with testosterone treatment, actually producing male-like songs. We also bring together experiments demonstrating various functions of bird song. In the section on song learning we develop the principles of "the early sensitive phase" and "stimulus filtering" showing that birds are much more plastic than early investigators have led us to believe. Again, the role of nature versus nuture in ontogeny is discussed. Finally, we discuss the controversial topic of assortative mating and avian song dialects.

In the chapter on territories (Chapter 12) we emphasize the fact that territories did not evolve for the purpose of limiting population density. Individuals compete for territories for their own benefit, with the consequence that density is often limited. We distinguish between territories and home ranges, show examples in which the two may or may not overlap, and point out that there may be geographical differences even within a species. We develop Huxley's idea of territory as an "elastic disc," and introduce the controversial concept of the "superterritory" as a manifestation of spiteful behavior.

The mating systems of birds (Chapter 13) comprise a very active area of ornithological research, especially in this age of "sociobiology." We give examples of the different mating systems, presenting them in the order of increasing male involvement: promiscuity, polygamy, monogamy, and polyandry. We discuss the evolution of various mating systems, and review some studies on the role of hormones in promoting polygyny.

Many species of birds are known to have helpers at their nests (Chapter 13). A superficial glance at these data suggests that this phenomenon is a manifestation of "altruism." Recent long-term studies indicate, however, that this phenomenon may be interpreted in terms of "selfishness" and selection on the level of the individual. We also develop the concepts of outbreeding mechanisms, cuckoldry, and intra- versus intersexual selection *sensu* Darwin. We describe and discuss controlled experiments which have provided support of Darwin's idea on intersexual selection.

In Chapter 14, we develop the theme that a bird's nest is a product of natural selection—the shape of the nest, its placement in a particular microenvironment, its proximity to others of its kind—all these factors may effect the survival of the young. Birds may take advantage of local microclimates, so that there may be geographical variation within a species as to where nests are placed. We also discuss the role of nature versus nurture in the nest-building process.

The avian egg has also been the subject of much research in recent years (Chapter 15). We review studies of egg physiology which have advanced our knowledge of development of the egg, changes occurring in the egg during incubation (e.g., calcium absorption), and gaseous exchange through the shell. We also discuss correlations between life history strategies and size of yolk, number of pores in the shell, and clutch size. Finally, we consider the possible adaptive significance of larger clutch sizes at higher latitudes.

In Chapter 16 we review the recent literature on behavioral and physiological aspects of incubation. We bring together the latest studies on hormones and brood-patch production, point out differences in brood patch construction for different avian taxa and review recent literature on incubation patterns in birds. Modern instrumentation has permitted investigators to monitor egg temperatures, demonstrating that eggs survive large fluctuations in temperature range, and that in certain habitats and under certain conditions birds cover eggs to protect them from overheating rather than to raise their temperature. We review the latest literature on brood parasitism and develop the thesis that this breeding strategy evolved as an adaptive behavior for the parasite rather than as a result of deterioration and loss of the nest-building and incubation instincts, as some have previously suggested. We also discuss brood reduction in birds, a topic of much interest recently, and examine its possible adaptive significance.

Chapter 17 covers the care and development of young birds. We discuss the stimuli eliciting the begging response in young hummingbirds: behavior of nestling hummingbirds differs from that in other birds in a number of ways. Altricial versus precocial young represent two ends of a continuum. In the first, much energy is invested by the parents in the care of the young after hatching. In the second, more energy is invested in the developing embryo, as the hatchling may care for itself to various degrees depending on the species.

Chapter 18 deals with population dynamics in birds. An important concept here is the inability of any organism to maximize all forms of reproductive investment simultaneously. Long-lived (K-selected) species tend to produce fewer young per effort than short-lived (r-selected) species, but they must invest more parental care in each offspring. Nestlings compete with each other for resources so that individuals in small broods weigh more than those in large broods. We also review studies indicating the selective advantage of colonial breeding. Finally, we discuss the phenomenon of "floaters" as an alternate strategy to "helpers" or independent breeders.

Ecology (Chapter 19) is an enormous topic, and indeed various aspects of ecology are scattered throughout the book, since ecology merges into other areas of biology such as foraging strategies and helpers. We discuss how birds have adapted to particular habits: as the extremely successful creatures that they are, birds have evolved different methods of surviving harsh environments or seasons. Thus, hummingbirds resort to torpidity to survive winters atop the Andes, jays and titmice cache seeds as food reserves in time of need, to cite some examples. We update the section on competition and competitive release, and stress the point, giving examples, that competitive exclusion cannot be assumed but must be *demonstrated* by removal experiments.

The subject of avian biogeography (Chapter 20) has benefitted greatly in recent years from studies of fossil and subfossil remains and from modern biochemical systematics. For example, recent DNA/DNA hybridization studies by Sibley and his colleagues indicate that several groups of birds thought to have invaded Australia from Asia have apparently evolved *in situ* along with the marsupials and eucalypts. The celebrated McArthur-Wilson model of island biogeography maintains that immigration balances extinction in island avifaunas. We call attention to studies by Olson and colleagues which advocate caution in computing turnover rates based on census of extant avifaunas, as hunting activities of prehistoric humans, predation by introduced predators, or changes in climate during the Pleistocene may cause extinction, which may result in distorted views of turnover rates.

Migration and orientation (Chapter 22) is another active field of research. We review the breeding hybridization studies by Bethold and his associates showing the heritability of wing-length, a feature associated with the migratory habit, and the heritability of the duration of *Zugunruhe*, which is positively correlated with distance of migration. There often is a differential migration distance between the sexes in birds: males tend to winter farther north than females in several passerines, the reverse being true in snowy owls. Several hypotheses put forth to explain the adaptive significance of this phenomenon are reviewed and discussed. We update the section on orientation and migration, reviewing recent studies on orientation using sun, stars, magnetic fields, and polarized light as cues.

In Chapter 23, Origin and Evolution of Birds, we discuss recent studies on *Archaeopteryx* and call attention to data indicating that, contrary to the opinion of earlier scholars, it was probably capable of flapping flight. We discuss the topic of domestication and remind readers of Charles Darwin's studies on this phenomenon which he considered an analog of the speciation process. We review recent taxonomic studies on the nature of isolating mechanisms, and discuss the fascinating story of disassortative mating in White-throated Sparrows and polymorphisms in their chromosomes, morphology, and behavior. The question of "kin selection" is treated and discussed in accordance with neo-Darwinian theories of selection on the level of the individual.

In Chapter 24, Birds and Man, we discuss various activities by humans resulting in environmental pollution and breeding failure in avian populations (e.g., air pollution, pesticides, and oil spills). Finally, we review the literature on conservation measures taken to save some of the more endangered avian species in the world. Some of these efforts have proved more successful than others, and we point out the factors needed for an effective conservation program.

In sum, we offer this new edition in the hope that it will stimulate the next generation of ornithologists to challenge and test the conclusions based on recent studies. Birds are creatures of wondrous beauty, both in themselves and as subjects of study in a diversity of disciplines. We hope that our efforts will satisfy readers who ask with Aristophanes,

My question, answer in the fewest words,
What sort of life is it among the birds?

ACKNOWLEDGEMENTS

Our deepest appreciation is extended to Susan Welty and Dorothy Fuller for preparing the index, and for all their dedicated assistance in editing the manuscript and seeing it to completion. We thank also the following for their editorial assistance: Sharon Chester, James Cunningham, Helen Horblit, Enid Leff, James Oetse, and Verena Schilling.

Our special gratitude goes to Pepper Trail for his trenchant critique of several chapters of the manuscript. We also thank the following for comments on parts of or the entire manuscript: Walter Bock, Robert Bowman, George Hunt, Paul Johnsgard, Stewart Warter, Glen Woolfenden, and John Zimmerman.

For permission to reproduce their illustrations we thank the following: Robert Bowman, Robert Dooling, Abbott and Sandra Gaunt, Klaus Immelmann, Martin Morton, Masakazu Konishi, Robert Raikow, Charles Sibley, Robert Payne, Eileen Zerba, and John Zimmerman. Maria Elena Pereyra and Judy Merrill prepared some of the illustrations. Gerry Stockfleth and Maima Teo typed the manuscript.

Last but not least we thank Acquisitions Editor, Ed Murphy, Project Editor, Sally Kusch, and the staff of Saunders College Publishing for their expert help in so many ways and for their patience on so many occasions.

FOREWORD— WHY STUDY BIRDS?

Most birds are diurnal creatures, as are humans. We cannot help but notice their song, their antics, their spectacular colors and their delicate plumage. Certainly, the first reason we study birds is because we are attracted by their beauty.

Birds have appeared in paintings since earliest times. They have been subjects of poems, and their songs have provided motifs used in musical compositions. Shelley immortalized the skylark with the pen, and Ralph Vaughan Williams set to music his vision of "The lark ascending." In the second movement of Beethoven's Sixth Symphony (the Pastoral), one immediately recognized the call of the European Quail, *Coturnix coturnix*, and the European Cuckoo, *Cuculus canorus*, simulated by oboe and clarinet, respectively. Antics of the Capercaille, *Tetrao urogallus*, are imitated in a folk dance from the Black Forest of Germany, and courtship rituals of the Prairie Chicken, *Tympanuchus cupido*, are mimicked in the dances of the American Plains Indians.

In addition to their esthetic appeal, birds are ideal subjects for scientific studies. Most bird species have been discovered and described. Ornithology has on the whole passed beyond the observation/descriptive stage and has arrived at the experimental/hypothesis-testing stage. Pioneering and significant contributions to many aspects of biology have been based on bird studies. The following has been developed from essays written by Ernst Mayr (1983a, 1983b), the dean of American ornithology:

Systematics

Mayr reminds us that important developments in systematics at the species and population level were often based on ornithological studies. Ornithologists developed the concepts of polytypic species, evaluated critically the meaning of subspecies, and applied the results of systematics to evolutionary theory. Earlier studies were based entirely on the examination of museum specimens. The modern systematist often borrows techniques from other sister disciplines. In addition to morphology, systematists study behavior of birds, examine chromosomes, look for protein polymorphisms, and hybridize DNA (Chapters 2 to 20). These are often important tools helping to elucidate relationships between avian taxa and between populations.

Speciation

Darwin was of the opinion that speciation could occur sympatrically, that is, species could arise from existing species without being isolated from the original population. It was up to ornithologists to convince the world of the prevalence of allopatric speciation, i.e., species arise as a result of the accumulation of genetic changes (micromutations) following geographical isolation. Mayr reminds us that there are two kinds of geographical isolation: (i) A newly arisen geographical barrier (e.g., an ice sheet) might break up a widely ranging species into isolated subpopulations. Speciation may occur as a result of genetic changes arising in these isolated pockets. (ii) Individuals of a contiguous species (founders) may disperse beyond their geographic range (e.g., to islands or isolated mountain tops) and leave descendants which may evolve into new species. Sometimes barriers are removed before two isolated populations have achieved species status; individuals belonging to the contacting populations may pair freely and produce offspring (Chapter 2).

Darwin called attention to the sexual dimorphism that prevails among birds-of-paradise, pheasants, grouse, and other avian species. He argued that certain individuals have greater reproductive success than others and thus leave more offspring. He saw females selecting as sires males in a population with more spectacular plumage or elaborate songs. Recent advances in the study of mate choice in birds-of paradise, bower-birds, European reed warblers, and others are showing that Darwin was correct (Chapters 11, 13, and 23).

Ecology

Mayr points out that ecology has undergone almost revolutionary advances over the last five decades. This field also has passed beyond a descriptive phase into a phase wherein *why* and *how* questions are being asked. Questions are couched in evolutionary terms. Modern ecology may often merge into the realm of behavioral biology—ecologists study habitat selection, foraging strategies, and resource partitioning (Chapters 6, 12, and 19).

It was an ornithologists, Joseph Grinnell, who, on the basis of his studies of chickadees, first noted that two closely related species cannot occupy the same niche in space and time. This concept was later rediscovered by Gauss and is now popularly known as Gauss's principle.

Population Biology

Because most birds are diurnal and easily observed, individuals of a population may be marked with colored bands or other marking techniques and followed through several breeding seasons or throughout their lifetime. Margaret Morse Nice marked individuals in a population of song sparrows, *Melospiza melodia*, and produced the classic which should be required reading for all students of ornithology. (Nice, 1937, 1943).

A marked population of birds enables the investigator to determine the longevity of individuals, rate of population turnover, reproductive success, and distances fledglings disperse before they settle. It is on the basis of studies of marked populations that ornithologists have discovered, for example, that fledgling females generally disperse farther away from their birthplace than males before settling, with the reverse being true for mammals (Chapter 18).

Behavior

Two field naturalists, Konrad Lorenz and Nico Tinbergen, have been recognized as the founders of the field of ethology, and as such have been honored with Nobel prizes. It was they who shaped ethology into an organized discipline. Other students of bird behavior preceded them. Darwin kept a large collection of domestic pigeons and noted breed differences in behavior and voice qualities. His studies inspired the American Charles Whitman, who kept many species of pigeons and doves and noted the homologous nature of pigeon displays. Oskar Heinroth, in Germany, was enthralled with the behavior of duck species and noted the feasibility of using displays as taxonomic characters. Studies of bird behavior in field and laboratory have contributed much to our knowledge of how behavior patterns evolve, the changes that behaviors undergo during ontogeny, and how behaviors affect the survival of the individual and the perpetuation of its genes (Chapter 10).

As far back as 1773, Gaines Barrington put to paper his observations on song-learning in oscines, regional song dialects, and learning confined to an early period of juvenile development (the sensitive phase). Students of bird song are legion today, and song studies have contributed much to our understanding of the role of nature (inheritance) and nurture (learning) in shaping behavior patterns. Bird-song studies have also contributed much to the field of neurophysiology (Chapter 11).

Migration

The phenomenon of bird migration is truly one of the wonders of nature. Consider, for example, the Arctic Tern, *Sterna paradisaea*, which makes a fall trip from the Arctic to the Antarctic, and then returns the following year to its breeding grounds. How does it find its wintering grounds? Does it have one or more routes that it follows? How does it navigate? Mayr points out once more that ornithologists are the leaders in the field of animal orientation. Day migrants often use the sun to orient by; night migrants may use constellations of stars. In the absence of these, some birds use a magnetic compass. Traces of magnitite have been found in various parts of the bird's head region, but how this is used or indeed whether or not these materials are involved in the orientation process is still a mystery. Avian navigation and migration is a dynamic field, and the unanswered questions are many (Chapter 22).

Physiology

Birds are remarkably successful creatures, and certain species inhabit some of the harshest environments on our planet. Hummingbirds, *Oreotrochilus* spp., may be found high atop the Andes. How do they survive the cold? Zebra Finches, *Poephila guttata*, live in some of the driest regions of Australia. How do they make it through drought conditions? Albatrosses and petrels (Procellariformes) may traverse stormy oceans. Young albatrosses may stay in the open ocean for several years at a stretch and not see land until they are mature. Some populations of Song Sparrows, *Melospiza melodia*, and Savannah Sparrows, *Passerculus sandwichensis*, live in salt marshes, often in the absence of fresh water for drinking. How do these pelagic birds and song birds get rid of excess accumulations of salt? These are some of the many questions that students of avian physiology have answered by careful field observation and experimentation (Chapters 7 and 8).

The Life of Birds was Carl Welty's dream-child into which he poured much of his love and energies. Carl passed away on the eve of his eighty-fifth birthday. He did not live to see the fourth edition come to fruition. We managed to complete sixteen chapters together. His wife, Susan, who was involved with Carl in producing the first three editions, was a tremendous help to me in the completion of the last eight chapters.

I visited Susan and Carl at Beloit, Wisconsin, during the planning stages of this book. We sat in the kitchen looking out into their backyard. There had been a light snowfall. Carl swept away the powdered snow, then placed a trickle of seed in a long thin line along the top of the wall behind the garden. From everywhere some 40 Cardinals, *Cardinalis cardinalis*, appeared and congregated to accept this largesse. So many scarlet poppies on a field of white!

Those of us who were fortunate to count Carl among our circle of friends will remember him as one who loved birds and nature deeply. He loved people even more. He was one of a group (the American Friends Service Committee) who went to Germany to bring food and clothing to the needy. Whenever it was possible, he made a special effort to contact German ornithologists and made many lifelong friends after the Second World War. In the preparation of this book, Carl often reminded me to think of the student in a college in some isolated community, without the benefit of a good reference library. The detailed narratives, sometimes containing material extraneous to ornithology, which Carl included in *The Life of Birds* was for those less fortunate souls. In 1980 he endowed an annual award for the student who contributed the most to the Biology Department at Beloit College where he taught for 32 years.

This book is affectionately dedicated to Susan and Carl, friends of nature and friends of humankind.

LUIS F. BAPTISTA

CONTENTS

THE LIFE OF BIRDS
Fourth Edition

Chapter 1-4, 21.

Bird Biology

Birds as Flying Machines

Birds! Birds! Ye are beautiful things,
With your earth-treading feet
  and your cloud-cleaving wings!
—Cook, *Birds*
  From *The Poets' Birds*, 1883

Each animal comes into the world with a specific set of genes which direct it to respond with a certain range of behaviors to a certain range of stimuli and thereby seek an appropriate environment, find food, find a mate, and raise progeny.

The great struggle in most animals' lives is to avoid change. Physiological constancy is the first biological commandment. For any animal too great a change, especially in the internal economy of an animal, means death. If, for example, the proportion of salt in the blood of a man or a bird is increased or decreased by only one-half of 1 per cent, the animal dies. A Chickadee clinging to a piece of suet on a bitter winter day is doing its unconscious best to maintain the internal status quo—to maintain what the physiologists call homeostasis.

The spectacular fact that birds can fly across oceans, deserts, forests, and mountains gives them exceptional opportunities for preserving their internal stability or homeostasis. Through flight, birds can search out the external conditions and substances they need to keep their internal fires burning clean and steady. A bird's wide search for specific foods and habitats makes sense only when considered in the light of this persistent, urgent need for constancy.

The power of flight has opened up to birds an enormous gaseous ocean, the atmosphere, as a means of quick, direct access to almost any spot on earth. As a consequence, they occupy and exploit a greater range of habitats than any other class of organism, plant or animal. Birds can eat in almost any "restaurant"; they can choose to build their homes anywhere among an almost infinite variety of sites, frequently beyond the reach of predators. Perhaps as a result of their supreme mobility, birds are, numerically at least, the most successful terrestrial vertebrates on earth. There are, in round numbers, about 9000 living species of birds (Bock and Farrand, 1980) as compared with 3000 amphibians, 6000 reptiles, and 4100 mammals. Fishes, which live in a much less demanding aqueous environment, number around 20,000 living species. Among the invertebrates, the insects are overwhelmingly the most successful and dominant class, numbering over 800,000 species. They, too, owe much of their dominance to their ability to fly.

POSITION IN THE ANIMAL KINGDOM

Birds make up the Class *Aves* of the subphylum *Vertebrata*, Phylum *Chordata*. They have descended from bipedal, lizard-like reptiles that lived in the Jurassic period some 160 million years ago. Birds still show many reptilian affinities, such as their habit of laying eggs, the possession of scales on their beaks and legs, and the arrangement of many internal structures. This resemblance is so close that birds and reptiles are for convenience grouped together as the *Sauropsida*.

The three highest classes of vertebrates—reptiles, birds, and mammals—have adapted their reproduction to terrestrial life, largely through the evolution of an egg whose embryo is enveloped in a protective membrane called the amnion. Hence these three classes are grouped under the term *Amniota*, or amniotes. Among all animals, only birds and mammals have evolved the high, constant temperature ("warm-bloodedness," homeothermism, or endothermism) that makes energetic activity possible in all habitats and at all seasons. This, added to their highly evolved homeostatic mechanisms, is what makes these two classes the dominant vertebrates.

THE BIRD AS A FLYING MACHINE

At first glance, birds appear to be quite variable. They differ considerably in size, body proportions, color, song, behavior, and ability to fly. But a deeper look shows that in basic architecture they are far more uniform than, say, mammals. The largest living bird, the Ostrich, *Struthio camelus*, weighs about 150 kilograms (330 pounds) and is approximately 66,000 times heavier than the small Scintillant Hummingbird, *Selasphorus scintilla*, which weighs 2.25 g (0.08 oz). However, the largest mammal, a 136,200-kg (299,880-lb) Blue Whale, *Balaenoptera musculus*, weighs 59,000,000 times as much as the Pigmy Shrew, *Microsorex hoyi*, at 2.3 g (0.08 oz). Mammals, there-fore, vary in mass over 850 times as much as do birds.

In body architecture, the comparative uniformity of birds is even more striking. Mammals may be as fat as a walrus or slim as a weasel, furry as a musk ox or hairless as a naked mole-rat, *Heterocephalus glaber*, long as a whale or short as a mole. They may be built to swim, fly, crawl, burrow, run, or climb. But the design of nearly every species of bird is dictated by one pre-eminent activity—flying. Their structure, inside and out, constitutes a solution to the problems imposed by flight. Their uniformity has been thrust upon them by the drastic demands that determine the design of any flying machine. Stringent natural selection has operated to convert a cold-blooded, earth-bound reptile into a light, warm-blooded, air-borne bird. Birds simply dare not deviate widely from sound aerodynamic design. Nature liquidates deviationists much more drastically and consistently than does any authoritarian dictator.

Birds were able to become flying machines largely through the evolutionary gifts of feathers, powerful wings, hollow bones, warm blood, a remarkable respiratory system, and a large, strong heart. These adaptations all boil down to the two prime requirements for any flying machine: high power and low weight. As early as 1679, the Italian physiologist Borelli in his *De Motu Animalium* (Hall, 1951) showed an awareness of this power-weight relationship in birds:

Such excessive power of the pectoral muscles seems to arise, firstly, from their large size . . . forming a dense and compact fleshy structure.

Secondly . . . the body of a Bird is disproportionately lighter than that of man or of any quadruped . . . since the bones of Birds are porous, hollowed out to extreme thinness like the roots of the feathers, and the shoulder-bones, ribs, and wing-bones are of little substance; the breast and abdomen contain large cavities filled with air; while the feathers and the down are of exceeding lightness.

Weight-Reducing Adaptations

To lighten their airships, birds have thrown overboard everything possible that is not concerned with flight. Perhaps the most effective weight reduction has been accomplished in their bones. The skeleton of a pigeon accounts for 4.4 per cent of its total body weight, whereas that of a comparable mammal such as a white rat amounts to 5.6 per cent (Fig. 1–1). This is in spite of the fact that the bird must have larger and stronger pectoral and pelvic girdles and appendages than a mammal, so that the full burden of locomotion

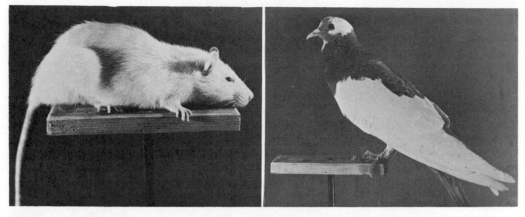

FIGURE 1–1 One of the chief adaptations that permit a bird to fly is its light skeleton. The dried bones of these two animals were weighed and compared. The entire skeleton of the rat weighed 5.61 per cent of its total live body weight; that of the pigeon weighed 4.43 per cent. The rat's skull weighed 1.25 per cent of its body weight; that of the pigeon weighed 0.21 per cent.

may be supported either by the wings alone or by the legs alone—a problem not encountered by the typical quadruped animal.

The skeleton of a Frigate Bird with a 7-foot wingspread is reported by Murphy (1936) to weigh 114 g (4 oz), which is less than the weight of its feathers! Its plumage and breast muscles together make up at least 47 per cent of the total body weight. Although the bird skeleton is very light, it is also very strong and elastic—necessary characteristics in an airframe subject to the great and sudden stresses of aerial acrobatics. This combination of lightness and strength depends mainly on the evolution of hollow, thin bones plus a considerable fusion of bones that are ordinarily separate in other vertebrates. The sacral vertebrae and the bones of the hip girdle, for example, are molded together into a thin, tube-like structure, strong but phenomenally light. The hollow finger bones are fused together, and some of them, in large soaring birds, show internal truss-like reinforcements much like the struts inside airplane wings (Fig. 1–2). Similar struts are commonly seen in the hollow larger bones of the wings and legs. The paper-thin bones of a bird's skull show this type of reinforced construction to a remarkable degree. Coues, the early American ornithologist, called the beautifully adapted avian skull a "poem in bone" (Fig. 1–3).

Birds have reduced the long bony tail of their rep-

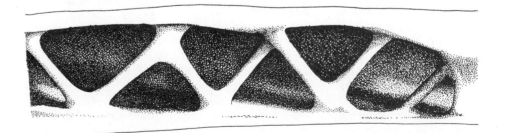

FIGURE 1–2 The metacarpal bone of a vulture's wing in longitudinal section. The internal bone struts are similar to those in a Warren truss used in aircraft and bridge design. After D'Arcy Thompson, 1942.

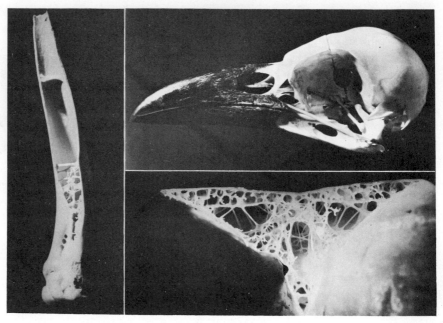

FIGURE 1–3 Why birds' skeletons are light in weight. *(Left)* The upper arm bone (humerus) of a Golden Eagle, cut open to show its hollow interior. *(Right)* The skull of a Common Crow that has been cross-sectioned between *(above)* the eye sockets and *(below)* that part of the section above and between the eye sockets, magnified about three times. The network of bony braces provides rigidity with extreme lightness.

tilian ancestors and of *Archaeopteryx*, the first known bird, to a stunted vestige, the pygostyle. Rump muscles control the steering done by the tail feathers. Further skeletal reduction is found in the bird's hand (wing), where the finger bones are reduced both in size and number; two of them are completely missing, and two of the other three are fused together. Similar fusion and deletion occur in the leg bones.

Ribs are elegantly long, flat, thin, and jointed. They allow extensive movement for breathing and flying, yet are light and strong. Each rib overlaps its neighbor and thus reinforces its own strength with that of the adjoining rib. This arrangement provides the kind of resilient strength found in a woven splint basket. The sternum or breast bone is greatly modified in all flying birds by the addition of a thin, flat keel for the attachment of the large wing muscles.

Feathers, the bird's most distinctive and remarkable acquisition, are magnificently adapted for fanning the air, for insulation against the weather, for streamlining, and for reduction of weight. It has been claimed that for their weight and function they are stronger than any man-made substitute. Their flexibility allows the broad trailing edge of each large wing feather to bend upward with each downstroke of the wing. This produces the equivalent of pitch in a propeller blade, so that each wingbeat provides both lift and forward

propulsion. The heat-insulating value of feathers is so extraordinarily effective that it permits birds to live in parts of the Antarctic too cold for any other animal.

There are no bony plates in the bird's skin like those worn by their saurian ancestors. Teeth have been eliminated in modern birds; consequently their jaw bones and jaw muscles need not be heavy.

In addition to light bones and light feathers, birds possess a weight-reducing system of air sacs that branch throughout the body, often even penetrating the hollow bones. The body of a feathered duck has a specific gravity of 0.6; that of a plucked duck, 0.9. Man and most other vertebrates have a specific gravity of about 1.0.

Further reduction in weight occurs in the urogenital system. Birds possess neither a urinary bladder nor a urethra to hold and discharge urine. Instead of watery urine, the kidneys excrete nitrogenous wastes, concentrated in the form of uric acid. This acid passes directly into the cloaca whose walls effect further concentration by absorbing still more water. The wastes, mixed with those of the intestine, become a concentrated whitish guano that is voided frequently.

Birds are the only class of vertebrates in which there are no species that give birth to live young. Nearly all species possess only one ovary and oviduct (the left). They lay their eggs in the nest soon after the eggs are formed, with the exception of parasitic birds, which seem to be able to control time of laying to coincide with the time the host is absent from the nest (see Chapter 16); development of the young does not occur in the body of the mother as in most mammals, and in some reptiles, amphibians, and fishes. Furthermore, the sharply accented breeding season burdens the bird with heavy sex organs for only part of the year. Both ovaries and testes, as well as their accessory ducts, atrophy during the nonbreeding season when they would be excess baggage. The sex organs of male and female European Starlings, *Sturnus vulgaris*, were found by Bissonnette to weigh about 1500 times as much during the breeding season as they did during the rest of the year.

Even in the foods they select to fuel their engines, birds conserve weight. Just as an airplane cannot be powered by a wood-burning steam engine, so the metabolism of most flying birds is not powered by bulky, low calorie foods such as leaves and grass. Birds eat seeds, fruits, nectar, worms, insects, rodents, fish, and the like—foods rich in calories.

Power-Promoting Adaptations

A basic law of chemistry holds that the speed of any chemical reaction roughly doubles with each rise in temperature of 10 degrees Celsius. In competitive nature, the race often goes to the metabolically swift. Of all the million or so animals on earth, birds and mammals alone are warm-blooded or endothermal, and birds have evolved much the highest operating temperatures. Man with his conservative 37°C (98.6°F) is a metabolic slowpoke compared with some sparrows (42°C, 107°F) or some thrushes (43.5°C, 110.5°F). Birds burn their candles at both ends and live intense lives. Summer or winter, day or night, a bird's metabolic "engine" is, with rare exceptions, always warm and ready for action.

Behind this high temperature in birds lie some interesting anatomical and physiological refinements. Flight itself would be utterly impossible in cool and frigid climates without endothermism. And endothermism, in turn, would be quite impossible without a warm coat of feathers to trap a layer of air over the skin surface (Morris, 1956). It seems probable that feathers evolved originally as an adaptation for heat conservation rather than for flight.

Besides eating an energy-rich diet, birds possess digestive equipment that processes their food rapidly, efficiently, and in large amounts. Fruit fed to young Cedar Waxwings, *Bombycilla cedrorum*, passes through their digestive tracts in as little as 16 minutes (Nice, 1941). Other perching birds may take from one-half to 2 hours to pass food through their bodies. A young growing stork will gain 1 kg in weight for every 3 kg of fish and frogs eaten. Comparable figures for most mammals are 1 kg of growth for 10 kg of food eaten. An examination of the excreta of birds shows that they utilize a very high percentage of the food they eat.

The breast muscles of a bird drive its propeller-like wings. In a strong flier, such as a pigeon, these muscles may account for as much as two-fifths of the total body weight. On the other hand, some species—for example, vultures—fly largely on updrafts of air, as a glider does. In such birds the breast muscles are greatly reduced, and there are well-developed wing tendons and ligaments that enable the bird to hold its wings in the soaring position with little or no effort.

A bird may have strong breast muscles and still be incapable of sustained flight because of an inadequate blood supply to these muscles. This condition is shown in the color of the muscles. The "white meat" of the breast muscles of a chicken or turkey indicates that they have so few blood vessels that they cannot fly far. On the other hand, the dark meat of their legs, in addition to the microscopic structure of the muscles, indicates a good blood supply, and an ability to run a considerable distance without tiring. If a Ruffed Grouse, *Bonasa umbellus*, is flushed four times in rapid succession, its breast muscles will become so fatigued that it can be picked up by hand, unable to fly even a few feet. The blood supply is inadequate to bring fuel and oxygen and to carry away waste products fast enough to keep the muscles functioning. Xenophon's *Anabasis*, written about 400 B.C., relates the capture of bustards by exploiting this weakness:

> But as for the bustards, anyone can catch them by starting them up quickly; for they fly only a short distance like the partridge and soon tire. And their flesh was very sweet.

Birds, like mammals, have four-chambered hearts, which make possible a double circulation; that is, the blood makes a side trip through the lungs for gaseous purification before it is circulated through the body again. The bird's heart is large, powerful, and rapid in its beat. The table on page 6 largely after Quiring (1950), gives the heart sizes and resting pulse rates for various species of animals, both cold- and warm-blooded.

In both birds and mammals the heart rate and the size of the heart in proportion to total body size increase as the animals decrease in size. But the

ANIMAL	HEART AS PER CENT BODY WT	HEART BEATS PER MINUTE
Boa constrictor, *Boa imperator*	0.31	20
Bullfrog, *Rana catesbeiana*	0.32	22
Man, *Homo sapiens*	0.42	78
Dog, *Canis familiaris*	1.05	140
Vulture, *Cathartes aura*	2.07	301
Crow, *Corvus brachyrhynchos*	0.95	342
Sparrow, *Passer domesticus*	1.68	460
Hummingbird, *Archilochus colubris*	2.37	615

increase is significantly greater in birds than in mammals. Any man with a weak heart knows that climbing stairs puts a heavy strain on his pumping system. Birds do a lot of "climbing" and their circulatory systems are built for it. The extra energy required for a bird to become airborne is more than compensated for by the efficiency of aerial locomotion. A migrating swallow expends only 2.5 kilocalories per kilogram of body weight per kilometer flown, as compared to 41 kcal/kg/km for a ground-traveling lemming (Blem, 1980).

The blood of birds is not significantly richer in hemoglobin than that of mammals. The pigeon and the mallard have about 15 g of hemoglobin per 100 cc of blood, the same as man. However, the blood sugar concentration in birds averages about twice that found in mammals. And the blood pressure (systolic), as one would expect, is also somewhat higher; in man it averages 120 millimeters of mercury; in the dog, 110 mm Hg; in the rat, 106 mm Hg; in the pigeon, 135 mm Hg; in the duck, 162 mm Hg; and in the chicken, 180 mm Hg.

In addition to lungs, birds possess an accessory system of usually five pairs of air sacs connected with the lungs, which ramify throughout the body. Branches of these sacs often enter the larger bones of the body to occupy their hollow interiors, some of the sacs even penetrating into the small toe bones. The air sac system supplements the lungs as a supercharger, increasing the utilization of oxygen; and it also serves as a cooling system for the bird's speedy, hot metabolism that accompanies energetic muscle activity. It has been estimated that a flying pigeon uses one-fourth of its air intake for breathing and three-fourths for cooling.

The lungs of man constitute about 5 per cent of his body volume; the respiratory system of a duck, in contrast, makes up 20 per cent of the body volume (2 per cent lungs and 18 per cent air sacs) (Zeuthen, 1942). The anatomical connections of the lungs and air sacs in birds seem to provide a one-way traffic of air through the lungs. Certainly, in man and all other mammals, the mixing of stale with fresh air at each inhalation by the sac-like, dead-end lungs, is inefficient. It seems odd that in the interest of efficient respiration, adaptive evolution has never produced a stale air outlet to animals' lungs, so that their linings could be constantly bathed by fresh air. In birds, the air sac system apparently approaches this ideal more closely than in any other vertebrate. Experiments by Schmidt-Nielsen (1971) indicate that air flows continuously and unidirectionally through a bird's lungs during both inhalation and exhalation.

Balance, Streamlining, and Nervous Adaptations

Any flying machine must be well balanced, with its center of gravity between and somewhat lower than the wings. In order to "trim ship" the weight of a bird's head has been drastically reduced through the elimination of teeth and heavy jaws and jaw muscles. A pigeon's skull weighs about one-sixth as much, proportionately, as that of a rat; its skull accounts for only one-fifth of 1 per cent of its total body weight. In a bird, the function of teeth has been largely taken over by the muscular gizzard (often with ingested grit substituting for teeth), located near the bird's center of gravity. This shifting of weight has eliminated the need of a long, counterbalancing tail such as birds' ancestors possessed.

Because birds must sometimes support their entire weight with their legs, and at other times with their wings, both the pelvic and the pectoral girdles are strikingly adapted. The rigidly fused pelvic girdle is greatly elongated, so that, when a bird alights on its legs, there is a firm support of the entire torso and its contents. There is a redistribution of the abdominal organs: the large gizzard and bulky air sacs have caused a downward and lateral displacement of the liver to the right, matching in position the ovary and oviduct at the left. This lateral broadening of the body, plus the requirements of egg laying, have resulted in a wide separation of the legs.

Rapidly flying birds normally have thoroughly streamlined bodies. The larger vane or contour feathers are responsible for the difference in appearance between the awkward angularity of a plucked bird and the streamlined sleekness of the feathered living bird. A bird has no external projecting ear lobes. It commonly retracts its "landing gear" legs within the belly feathers while it is in flight. The smoothly streamlined Peregrine Falcon, *Falco peregrinus*, is reported to dive on its prey at speeds up to 384 kilometers (256 miles) per hour (Cade, 1982). Some rapid fliers have baffles in their nostrils to protect their lungs and air sacs from excessive air pressures. Even in the water, streamlined birds are among the swiftest of animals.

Skillful flight demands sharp eyesight and quick muscle coordination. The eyes of a bird are relatively large, usually so large that the eye sockets meet in the center of the skull. In keeping with the need to reduce head weight, the eyes have relatively poor mobility as a result of reduced eyeball muscles. This is compensated for, however, by greatly increased neck flexibility.

Weight reduction, however, has not been promoted at the expense of the nervous system. The brain of a small perching bird weighs about ten times that of a lizard of the same body weight. It is about equal in size to that of a comparable rodent, but it is significantly smaller than the brain of an equal-sized carnivore or primate. The relatively large forebrain and cerebellum of a bird are richly supplied with association centers, especially those concerned with muscle coordination. The optic lobes are relatively large. Nerve impulses are rapid. In most birds the sense of smell is apparently almost nonexistent, and the olfactory lobes of the brain are reduced almost to vestiges. Olfactory lobes are largest in swimming birds and medium-sized marsh birds (Bang and Cobb, 1968).

The bird's success in adaptation to flight is best summed up by comparing its efficiency with that of a man-made airplane. The Golden Plover, *Pluvialis dominica*, a strong flier, fattens itself on Labrador bayberries in the autumn and then strikes off across the open Atlantic on a nonstop flight of some 3800 km (2400 miles) to South America. It arrives there weighing about 56 g (2 oz) less than it did on its departure. This is the equivalent of flying a small airplane 256 km (160 miles) on a gallon of gasoline rather than the usual 32 km (20 miles), and the bird accomplishes this fuel economy in spite of the considerable handicaps of greater relative heat loss and greater friction inherent in its smaller body. Man still has far to go to equal such efficiency in flight.

■ SUGGESTED READINGS

There are a number of excellent general works on the biology of birds. Among the older books which still contain much useful information are Newton's pioneering A *Dictionary of Birds*, Pycraft's A *History of Birds*, Beebe's *The Bird*, Thomson's *The Biology of Birds*, and Allen's *Birds and Their Attributes*. Among more recent general texts are those by Pettingill, *Ornithology in Laboratory and Field*, Van Tyne and Berger, *Fundamentals of Ornithology*, and Wallace, *An Introduction to Ornithology*. More extensive and authoritative reference works on birds are Farner and King's 8-volume *Avian Biology*, Grassé's Volume 15 (*Oiseaux*) in *Traité de Zoologie*, Grzimek's lavishly illustrated Volumes 7–9 in *Grzimek's Animal Life Encyclopedia*, Marshall's *Biology and Comparative Physiology of Birds*, and Stresemann's Volume 7 (*Aves*) in Kükenthal and Krumbach's *Handbuch der Zoologie*. Unsurpassed as a one-volume reference in bird biology is Campbell and Lack's A *Dictionary of Birds*. Also recommended is the sumptuously illustrated *The Audubon Society Encyclopedia of North American Birds* by Terres.

The Kinds of Birds
2

Carolus V. Linnaeus, he played Adam with the birds,
With generic and specific nomenclature.
He labeled all the Aves with resounding Latin words—
But some of them cause arguments with Nature.
—Joel Peters (1905–   ),
  What's In a Name?

Any curious person observing the world of birds is struck by the varying degrees of resemblance and difference among them. His intellectual desire to "pigeonhole" diverse objects into systematic order leads him to place the birds of the same general appearance together, and to give them a collective name: wrens, ducks, eagles. The Bible, not especially concerned with natural history, mentions about 40 different varieties of birds. Aristotle, who based his classification more on function than on form, recognized 140 kinds. Even primitive people feel this urge to classify. A tribe of natives in the mountains of New Guinea had 137 specific names for birds classified by scientists today as 138 species. Only one species was confused with another (Mayr *et al.*, 1953).

Closer scrutiny of birds soon shows that there are larger and smaller differences. There are, for example, perching birds, with pointed beaks and separate toes, such as the robin and blackbird, and larger water birds with webbed feet and flat, rounded bills, such as ducks, geese, and swans. But while ducks, geese, and swans have similar feet and bills, they vary, among other things, in length of neck. There are, in short, hierarchies of resemblances and differences: more basic ones that determine the larger, more inclusive groups of birds, and smaller, more superficial ones that determine the smaller groups and eventually the individual "kinds" or species.

NOMENCLATURE

In 1735 the Swedish naturalist Carolus Linnaeus published his famous *Systema Naturae*, which attempted to distinguish and name all the plants and animals.

This seemed no insuperable problem. There were not many species within his geographical horizons, and those he studied had sharply defined characteristics. He named a total of 564 birds and 4235 animals of different kinds.

Linnaeus realized that common or colloquial names were unreliable for purposes of permanent classification. Even today the name "robin" means one bird to an Englishman and a quite different bird to an American. The American Robin is much more closely related to the European Blackbird than it is to the European Robin, or, for that matter, to the American blackbirds. Common names often cause confusion when people talk and write about animals, particularly when such names are used in speaking to an international audience. Many birds, moreover, are too rarely seen to have common names.

To avoid these difficulties, Linnaeus gave each bird two Latin names, a generic name and a specific name. The European Robin, for example, he named *Erithacus rubecula*, which means "a kind of bird inclined to redness"; the European Blackbird he named *Turdus merula*, which means "thrush blackbird." The American Robin belongs to the same genus as the Blackbird, but is a different species, so he named it *Turdus migratorius*, which of course means "thrush which migrates." Originally these Latin names were meant to be descriptive, but in the course of time, relationships unknown in Linnaeus' day have been discovered, so that some names have become meaningless or misleading if considered literally, though they remain convenient designations.

Using two Latin names as he did, Linnaeus established the principle of binomial nomenclature, which is still the basic system for naming all plants and animals. The science of classifying organisms is called taxonomy or systematic biology, and scientists, usually museum workers, who specialize in classifying and naming birds are called bird taxonomists or systematic ornithologists.

Linnaeus also provided Latin names for the larger, more inclusive categories of birds and other animals: kingdom, phylum, class, order, and family. To illustrate: the American Robin and the European Blackbird resemble each other more closely than either resembles a Bluebird, so they are placed in the genus *Turdus*. All three birds, however, resemble each other more closely than any one of them resembles a swallow, so the former are placed in the family Turdi-

dae (thrush family) and the latter in the family Hirundinidae (swallow family). The thrushes and swallows still have enough basic features in common to be grouped into the order Passeriformes (perching birds) as distinct, say, from the order Falconiformes (hawks, eagles, falcons). And because all the 27 orders of living birds possess the irreducible minimum of characteristics that define a bird, such as warm-bloodedness, feathers, and wings, they are placed in the class Aves, the most inclusive bird category. Birds and other animals are thus arranged in a descending series of classifying "pigeonholes" of decreasing inclusiveness from class to species. That is, the more similarities that are shared by a given group, the narrower that group and the lower it is in the taxonomic hierarchy.

In order to provide worldwide uniformity and stability in the classification and naming of animals, the International Congress of Zoology has established an International Code of Zoological Nomenclature, the current edition of which was published in 1985. This code is based on Linnaeus' system of binomial nomenclature. Useful as it is, however, the code must be applied with wisdom and discretion. As Erwin Stresemann, the late dean of German ornithologists, remarked, "Whoever wants to hold firm rules should give up taxonomy. Nature is too disorderly for such a man" (Thomson, 1964).

Shown on page 10 are examples of the use of scientific terminology for four different kinds of animals, the last two both birds.

Species containing populations of slightly different forms are commonly broken down into subspecies, sometimes called races. In this case they are given a trinomial scientific name. The Robin that breeds in northern North America is called *Turdus migratorius migratorius*, and the slightly different form that breeds in the southern United States is called *Turdus migratorius achrusterus*. To the nonscientist the minor distinctions that separate these two races or subspecies are negligible, and he calls both groups Robins.

Just how many different kinds of birds are there? It depends on whom you ask. A recent survey indicates that there are about 9000 species in the world (Bock and Farrand, 1980). Taxonomists can be grouped into "species" too: the "splitters" who delight in classifying birds according to minute differences, and the "lumpers" who differentiate species on the basis of grosser discriminations. This discrepancy in techniques illustrates a central problem. No one has much

Kingdom:	Animalia	Animalia	Animalia	Animalia
Phylum:	Arthropoda	Chordata	Chordata	Chordata
Class:	Insecta	Mammalia	Aves	Aves
Order:	Hymenoptera	Carnivora	Falconiformes	Passeriformes
Family:	Apidae	Felidae	Accipitridae	Turdidae
Genus:*	*Apis*	*Felis*	*Haliaeetus*	*Turdus*
Species:*	*mellifera*	*domestica*	*leucocephalus*	*migratorius*
Common name:	Honeybee	Domestic Cat	Bald Eagle	American Robin

*The genus and species are combined to form the scientific name.

trouble in distinguishing an eagle from a robin. But disagreements arise among observers when species and subspecies are to be differentiated.

On the Aleutian Islands off Alaska there lives a large, dark-colored Song Sparrow that is about twice as large as the light-colored, desert-dwelling Song Sparrow of southern Baja California. Had Linnaeus held these two birds in his hands he would undoubtedly have called them distinct species. Today we know that in North America there are 34 closely intergrading types of Song Sparrows, most of them living up and down the Pacific coast. These birds live as geographical populations, each population only slightly different from its neighbors. While the birds living at the northern and southern extremes of Song Sparrow range are distinctly different in size and coloration, it is quite apparent, as one travels from Alaska to Baja California, that the successive populations in between change gradually from large to small, from dark to light. Consequently the 34 population types are not classed as species but as subspecies of one Song Sparrow: the Aleutian giant is called *Melospiza melodia maxima* and the small Baja California race, *Melospiza melodia rivularis*. The same sort of subspecies geographical populations occur among jays, hawks, larks, and many other birds.

In recent years the trend has been toward lumping slightly divergent species together into a single species. For example, in 1900 the European nuthatch of the genus *Sitta* was split into 19 species and one subspecies. Today these birds are lumped into four "large" or polytypic species (one of which, *Sitta europaea*, alone contains 26 geographic races), and three monotypic, isolated species. Similarly, the Rufous-collared Sparrow, *Zonotrichia capensis*, of Central and South America was once broken up into eight different species but is now classed as one polytypic species with 23 geographical races (Mayr, 1942).

What Is a Species?

So the answer to the question of how many kinds of birds there are depends on the answer to another question: What is a species? Linnaeus thought that all species were created in their different patterns, fixed and unchanging. Since Darwin's epochal publication in 1859, the species concept has itself evolved; it has undergone two fundamental changes. First, we now understand that through evolution one species arises from another pre-existing species. Therefore, birds related through evolutionary descent should show "family resemblances," and they do. Second, a species is no longer defined as a certain idealized type of bird with precisely defined shape, size, structure, and color. Instead, today a species is defined as a living population in nature made up of individuals that have about the same structure, size, color, behavior, and habitat, and breed with each other rather than with members of other similar groups.

This is the working definition of a species. Some distantly related birds may hybridize in captivity if raised together, e.g., domestic duck with domestic goose (Poulson, 1950). Hybrids may even be fertile, e.g., between duck species (Sharpe and Johnsgard, 1966). However, hybrids between species are usually infrequent in the wild and are at a disadvantage. For example, Red-breasted (*Sphyrapicus ruber*) and Red-naped Sapsuckers (*S. nuchalis*) may sometimes hybridize where they occur together. Detailed examination of 145 nesting pairs in their zone of overlap revealed that 75.8 per cent of the birds paired with their own species, 16.6 per cent with backcrosses, 1.4 per cent with F_1 hybrids, and 6.2 per cent formed interspecific pairs (Johnson and Johnson, 1985). Indigo (*Passerina cyanea*) and Lazuli Buntings (*P. amoena*), hybridize in the Great Plains, but backcrosses are rare (Emlen *et al.*, 1975).

Subspecies are geographical races, or fractional populations, of species. Two subspecies represent two populations of a species that have been geographically isolated over time, but have come together before enough genetic differences (micromutations) have accumulated to make them two species. Mating between two "full" species is rare. Mating is random when members of two subspecies meet, and there should be introgression of genes. In the Great Plains one finds interbreeding between eastern Yellow-shafted and western Red-shafted forms of the Northern Flicker *Colaptesawiatus*, between eastern "Baltimore" and western "Bullock" forms of the Northern Oriole, *Icterus galbula*, and between eastern "plain" and western "spotted" forms of the Rufous-sided Towhee, *Pipilo erythrophthalmus* (Rising, 1983). Mating is not assortative (with one's own kind) and hybrids are not selected against. Thus, these eastern and western taxa are considered subspecies of widespread species. Each taxon represents populations once isolated from each other, on the way to becoming species but not yet reaching that stage. Physiological and behavioral differences may accompany the morphological differences (differences in appearance). Thus, a study of numbers of different subspecies may be most instructive in helping us understand the process of evolution.

To sum up, the classification of birds is ideally based on what is known about their evolutionary kinships or their phylogenetic relationships. And this knowledge is by no means complete, nor it is easily obtained. As a consequence, classification is a fluid, inexact science, largely based on subjective judgments.

CRITERIA OF RELATIONSHIPS

The problem of separating one species from another is especially difficult when taxonomists are unable to observe the living populations in nature. Often they must work with only a few birds, perhaps museum specimens, or even fossils, rare as they are. In such cases, they can use only physical characteristics in attempting to describe and delimit a species. Linnaeus and other early taxonomists depended almost entirely on the external and internal structure (morphology) of an animal to determine its species. But morphology alone is often unreliable as a criterion of relationship. Herons, cranes, bustards, Ostriches, and Secretary Birds all have long legs, but are not at all closely related. Contrariwise, some closely related species may occasionally differ strikingly in structure. Delacour (1946) observed that the Eurasian Parrot-bill *Paradoxornis unicolor* has four toes while its congener *Paradoxornis paradoxus* has only three. Yet the two species are otherwise so similiar that they would be classed as two races of a single species.

Problems such as these stem from the fact that two birds can closely resemble each other either because they are descended from a common ancestor, or because, although not closely related, they have each adapted to the same environmental influences or modes of life. This latter principle, known as parallel evolution or adaptive convergence, is seen, for example, in the auks of the Arctic as compared with the penguins of the Antarctic. The two groups are superficially very similar in general appearance and behavior but are not at all closely related (Storer, 1971). Are loons and grebes, for example, similar because of common ancestry or because of adaptive convergence? No one is certain. The great difficulty of distinguishing stable, basic characters due to common descent and resistant to evolutionary change from those more labile, superficial characters based on parallel evolution has created serious problems and frequent disagreement among bird taxonomists.

Because of the many uncertainties inherent in the sources of evidence for bird relationships, it follows that any scheme of bird classification based on fallible human judgment must be an abstraction of their evolutionary history (Bock, 1976). As a consequence it is fairly common for equally competent taxonomists to disagree. In classifying the rails of the world, three different and knowledgeable taxonomists have placed them in 52, 35, and 18 genera, respectively (Keith, 1978). In surveying the criteria used by competent taxonomists in classifying seed-eating finches, Ackermann (1967) discovered that two of them used the work of other taxonomists, another depended on jaw muscles, another used the structure of the hard palate, still another depended on serology and leg muscles, and the last used beak structure, leg muscles, and the mechanism of the vas deferens opening.

How is each different bit of evidence to be weighted? Which of these various measurements are the more stable and diagnostic of degrees of kinship, and which are the more liable to modification by environmental pressures? As examples of the latter, nestling Red-winged Blackbirds, *Agelaius phoeniceus*, in north central Minnesota have higher ratios of wing to tarsus and toe to tarsus than those at Fort Collins, Colorado. James (1983) transported eggs from Colorado and placed them in Minnesota nests. The eggs hatched, and ratio of wing to tarsus in these nestlings shifted towards that of Minnesota nestlings. Local factors, probably diet, must have influenced differential growth of appendages in these nestlings.

Thus, some of the regional differences between nestlings is nongenetic. House Finches, *Carpodacus mexicanus*, from California were released in New York in 1940, and within 9 to 11 years acquired differences in color and size of legs and feet differentiating them from the ancestral stock (Aldrich and Weske, 1978). After only a century of natural selection, North American populations of the introduced House Sparrow, *Passer domesticus*, differed in size and plumage color from the English and German populations (Selander and Johnston, 1967; Johnston and Selander, 1971). Fox Sparrows, *Passerella iliaca*, from two out of seven sites collected in 1978–1980, showed morphometric differences from museum samples taken 50 years earlier (Zink, 1983).

How then does one attempt to determine the true hereditary relationships among birds? Morphological characters still have their use. In fact, until recently classification has been based very largely on morphological features. The bones of a bird's skeleton—especially the structure of the palate, the number and shape of the cervical vertebrae, and the structure of the sternum or of the leg bones—are very useful in determining some of the larger taxonomic categories. For example, the existence of a distinctive canal (coracobrachialis anterioris) in the humerus of the leg characterizes every bird, living or fossil, of the order Charadriiformes (Ballman, 1976). Muscles of the pelvic region have been used in classifying ospreys and vultures, and leg muscles have been used to separate genera of finches. The structure of tongue muscles has aided in classifying parrots, and those of the voice box or syrinx, in classifying perching birds. Even the twistings of the intestines and the arrangement of the large arteries have taxonomic significance. Statistical studies of wing length, body length, and tail length have been helpful in identifying subspecies of birds. The distribution of feathers over the body and

their sequence in molting have been used to differentiate some species. The speckled breasts of young robins and bluebirds help to reveal their membership in the thrush family. Chemical analysis of the waxes from the uropygial gland has clarified relationships between genera of ducks (Jacob, 1977). And of course such external characters as beaks, combs, wattles, leg scales, toes, claws, and spurs have aided in classification.

Besides morphological evidence of taxonomic relationships there is a variety of other evidence, which, while perhaps less compelling, nevertheless provides corroborative clues to avian relationships.

A surprising source of evidence for bird classification comes from their external parasites. Parasites are normally well-adapted to their hosts. As birds evolved, their parasites evolved along with them to such an extent that related species of birds today show related species of parasites. This relationship is known as Fahrenholz' Rule. Although other taxonomists have placed the African Ostrich and the South American Rhea in separate orders, largely on morphological grounds, Rothschild and Clay (1957) have shown that they both possess feather lice of closely related species belonging to the same genus, a genus that occurs on no other birds. The relationship between Ostrich and Rhea is also supported by data on DNA–hybridization (Sibley and Ahlquist, 1985).

The taxonomic validity of using parasites as fairly dependable indicators of avian kinship is shown by the brood-parasitic cuckoos of Eurasia. Rothschild and Clay (1957) point out that cuckoos carry their own genera of lice and not those of their hosts with whom, of course, the young cuckoos are reared. Cuckoos acquire their lice through mating and not from parent–nestling contacts as do other birds.

Since modern birds have evolved from previously existing species, recently derived species should not be very remote either geographically or, in the fossil record, stratigraphically, from their ancestors. Making due allowance for barriers and highways (difficult and easy routes of dispersal), the geographical distribution of birds over the earth provides valuable evidence of relationships. As would be expected, species that resemble each other most closely are generally closest together in space. This is well illustrated in "circles of races" or *Rassenkreise*, a subject considered in Chapter 23.

Avian paleontology has made great strides over the last two decades. Fossil birds (once believed to be rare) and the rich materials accumulated have been instrumental in increasing our understanding of systematics, biogeography, and the evolution of bird groups. They also provide the *only* evidence for lineages now extinct (Olson, 1985).

Microscopic studies of the shapes and numbers of chromosomes have sometimes proved useful in taxonomy. For example, the Congo Peafowl, *Afropavo congensis*, has been regarded variously as a primitive relative of the Asiatic peafowl (*Pavo* spp.), or the African guineafowl (Numididae), or of Asiatic Impeyan pheasants (*Lophophorus* spp.), or placed in a subfamily of its own. Similarities in chromosome morphology have confirmed its relationship with the Blue Peafowl, *Pavo cristatus?* (deBoer and van Bocxstaele, 1981).

To date only 5 per cent of the approximately 9000 extant species of birds have had their chromosomes described (karyotyped) (Shields, 1983). In a survey of the karyotypes of 177 bird species, Shields (1982) found that many unrelated species have identical karyotypes, suggesting that chromosomal change does not accompany avian speciation. On the other hand several genera contain species with identical karyotypes: diving ducks, *Aythya*; gulls, *Larus*; cranes, *Grus*; owls (*Strix, Asio*). Shields suggests that karyotypic differences existing within groups may have developed after reproductive isolation of the lineages. Most ducks have similar karyotypes and may hybridize with each other. An exception is the Mandarin Duck, *Aix galericulata*, which does not form natural hybrids. It has two macrochromosomal characteristics not shared with other ducks (Yamashina, 1952; Shields, 1982).

In recent years new, sophisticated biochemical techniques have been utilized by taxonomists to clarify relationships between avian taxa. Among these are electrophoresis studies utilizing egg-white protein, or proteins from blood and other tissue (Barrowclough *et al.*, 1985). For example, the "Wrenthrush," *Zeledonia coronata*, of Central America was once classified as a thrush or a close relative thereof. More study revealed that *Zeledonia* does not have the "booted" tarsus characteristic of thrushes. It is similar in plumage, life-history, and other characteristics to some Wood Warblers (Parulinae) (Hunt, 1971). Data from starch-gel electrophoresis (Fig. 2–1) of egg-white proteins confirm its relationships with warblers and other "nine-primaried" oscines (Sibley, 1968).

Starch-gel electrophoresis has been used to study allelic variation in proteins from liver, heart, and kidney of galliform birds (Guttierez *et al.*, 1983). They

studied variation in 17 different proteins (loci). Four did not vary across species. Seven were monomorphic within species but differed across species, and the rest were polymorphic, i.e., each locus was represented by several different alleles. These studies indicated that New World quail (Odontophorinae) are distinct from other galliform groups with which they were com-

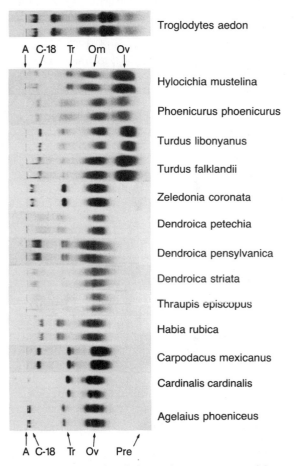

FIGURE 2–1 Starch-gel electrophoretic patterns of the egg-white proteins of a wren *(Troglodytes)*, several thrushes *(Hylocichla, Phoenicurus, Turdus)*, the wren-warbler *(Zeledonia)*, and several nine-primaried oscines *(Dendroica, and all other species below the wren-warbler)*. The wren-warbler was once thought to be related to the thrushes. However, these data support data based on other studies indicating its affinities with the nine-primaried oscines. Note the presence of the ovalbumin (Ov) band in the thrushes, absent in the wren-warbler and the other nine-primaried oscines. Data from Sibley, 1968.

pared. Several authors have proposed that the tube-nosed birds (procellariforms) are most closely related to the penguins (Sibley, 1972). This view has been supported by a detailed analysis of the uropygial gland waxes of 37 species of tube-noses (Jacob and Hoerschelmann, 1982).

Proteins are not an infallible method of determining relationships. Two proteins with similar electrophoretic mobility may reflect either common ancestry or convergent evolution. Conversely, two dissimilar alleles may indicate ancestry or may be a product of adaptive radiation (Diamond, 1983).

A still more recent biochemical technique is that of DNA–DNA hybridization (Sibley and Ahlquist, 1983, 1986). Some authors have suggested, based on evidence from anatomy and behavior, that New World vultures (Cathartidae) traditionally placed with the hawks (Falconiformes) are in fact close relatives of the storks (Ciconiidae) (Ligon, 1967; König, 1982). These conclusions are supported by data on DNA–DNA hybridization (Sibley and Ahlquist, 1985). The New World Wrentit, *Chamaea fasciata*, shares several morphological and behavioral characters with the Old World babblers (Timaliinae) (Brown, 1959; Simmons, 1963) and is now generally considered a member of that group. These conclusions are also supported by studies on DNA–DNA hybridization (Sibley and Ahlquist, 1982). DNA–DNA hybridization has also been used to elucidate relationships between the ratites (Fig. 2–2). The Ostrich, *Struthio camelus*, and the

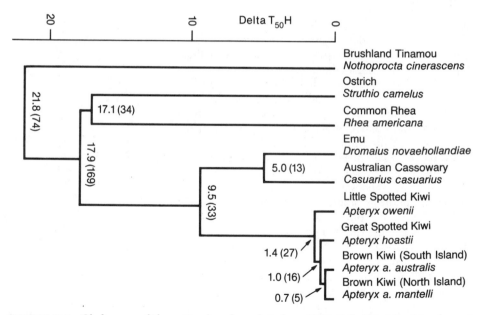

FIGURE 2–2 Cladogram of the ratites based on data from DNA hybridization. Numbers at branching points are the average 50 per cent disassociation temperature [$\Delta T_{50}H(°C)$] for hybrids involving two branches. Numbers in brackets are numbers of hybrids averaged. The value is calculated by taking the difference between the 50 per cent disassociation of a paired strand of cospecific DNA and a paired (hybrid) strand of heterospecific DNA. Note that the Australasian ratites (kiwis, cassowaries, emus) group together, and ostriches and rheas form another group. Courtesy of C. Sibley.

rheas, *Rhea* spp., appear to be closely related and must have originated in Gondwanaland prior to the separation of West Africa and Brazil by the opening of the Atlantic Ocean (Sibley and Ahlquist, 1981).

A recent refinement of the biochemical study of bird relationships involves the analysis of the amino-acid sequences of the macromolecules of proteins. This technique shows promise of revealing more precisely than any previous method the hereditary changes that have been responsible for the phylogenetic relationships of birds (Fitch and Margoliash, 1969).

Partly because of the sometimes inconclusive results obtained by using morphological and biochemical analyses as taxonomic tools, ornithologists are using other criteria with increasing frequency to determine significant genetic relationships.

Behavior patterns may sometimes throw light on phylogenetic relationships. Different species of bower-birds (Ptilinorhynchidae) with almost indistinguishable plumage construct quite different types of courtship bowers. Males belonging to one group build a maypole type bower, and males of another group, a trough-shaped one. Incidentally, the maypole builders lay plain white eggs, the trough- or avenue-builders, heavily speckled ones. Crag Martins, *Ptyonoprogne*, originally classified with the Bank Swallows, *Riparia*, were placed instead with the Barn Swallows, *Hirundo*, because they both build mud nests, agree in voice, and possess concealed white tail spots (Mayr, 1958). Related tyrant flycatcher genera (Tyrannidae) may be grouped according to the type of nest they build: open cups, domed, pendant, or nests in holes (Traylor and Fitzpatrick, 1982).

Based on his studies of pigeons and doves, Whitman (1899, 1919) observed that behavior patterns could be homologous, i.e., could descend from a common ancestral display (see Chapter 9). Heinroth (1911) subsequently noted that homologous displays could be used as taxonomic characters. Courtship displays have since been used in taxonomic studies of many avian taxa, e.g., ducks (Lorenz, 1941, 1958; Delacour and Mayr, 1945; Johnsgard, 1965), gulls (Tinbergen, 1959), pelecaniforms (van Tets, 1965), grebes (Storer, 1963), pigeons (Goodwin, 1966) hummingbirds (Schuchmann, 1978), estrildid finches (Güttinger, 1970, 1976), and others.

Songs and calls have also been extensively used in taxonomic studies (Payne, 1985). They are especially useful in distinguishing locally sympatric populations of sibling species (species that are very difficult to distinguish morphologically), e.g., North American *Empidonax* flycatchers, European treecreepers (*Certhia* spp.), *Phylloscopos* warblers.

Again, caution is required to distinguish behavior due to common descent from convergent behavior due to similar selection pressures. Immelmann (1976) points out that many ecological behavior patterns, for example, feeding and nest-building behavior, are subject to heavy selection pressure and " . . . owing to many convergent adaptations are of comparatively little taxonomic value, whereas aggressive and sexual behavior [are] less directly influenced by environmental factors" and are therefore of more value in tracing phylogenetic relationships.

Other characters of some use to the taxonomist include syrinx morphology, muscle patterns, scutellation of tarsi, form of the stapes, habitat selection, nest construction, feeding and care of the young, nest sanitation habits, wing-flipping, threat behavior, migration, flight patterns, pre- and postcopulatory displays, and many modes of gregarious or social behavior.

Finally, statistics are used in comparing samples from different localities. Birds are collected in the wild and deposited in museums as study skins. Standard measurements, e.g., of wing, tarsus, bill, tail, and many others (Baldwin *et al.*, 1931) are made from the skin and subjected to statistical tests. Skeletons may provide many data points for comparison of samples from different sites (Zink, 1983).

Although birds are the best known and most completely described class of animals, there still are many uncertain systematic relationships to be worked out. A subjective element remains in the determination of any taxonomic group: fallible human judgment is still a necessary tool for the taxonomist. There will probably continue to be controversies over classification in the ornithological journals for years to come. Probably the chief reason for this is the great structural and functional uniformity thrust on birds as they evolved into flying machines in the Cretaceous period over 100 million years ago. This uniformity greatly restricts the diversity, so useful to taxonomists, found in other vertebrate classes.

What diversity exists is sometimes useless for taxonomic purposes. The order Charadriiformes includes long-legged stilts, short-legged seed-snipes, puffins with high, laterally compressed beaks, skuas with hook-shaped beaks, and skimmers with long lower mandibles. Yet, their close relationship, says Niethammer (1972), is clearly seen in the similarity

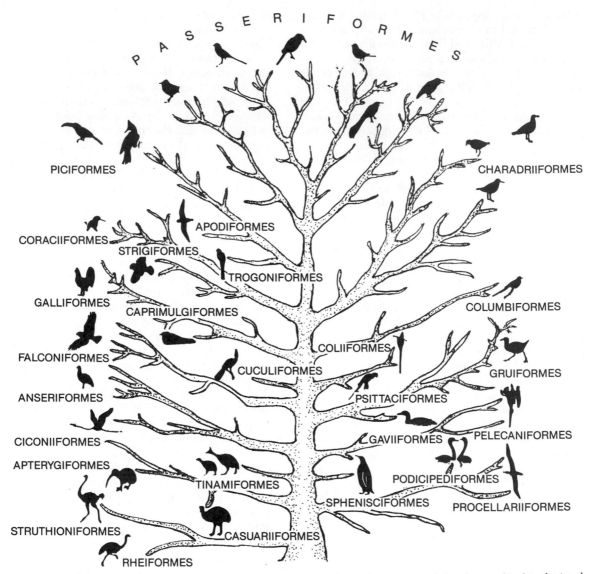

FIGURE 2–3 A hypothetical family tree of birds showing possible relationships. Many of the ideas on kinship depicted here will change as evidence from new and refined taxonomic techniques continue to accumulate. Drawing modified from *Point Reyes Observatory Newsletter*, Autumn, 1986.

of their blood albumin, their internal structure, and the similarity or identity of their external parasites. In a similar fashion Brereton (1963) and Smith (1975) used a great variety of taxonomic criteria in classifying the genera of parrots: skull characters; head-scratching behavior; presence or absence of oil glands, furcu-

la, ambiens muscle; arrangement of carotid arteries; wing shape; tail length; orbital ring development, and others.

To return to the question of how many different kinds of birds there are in the world, Bock and Farrand (1980) note that in Sharpe's (1899–1905) classification 18,939 species were recognized. Under Stresemann's (1927–1934) leadership, groups of birds formerly treated as species were called subspecies of single polytypic species: 8000 species with 30,000 subspecies were recognized. Mayr and Amadon (1951) recognized about 8600 species. Between 1920 and 1940 about 170 new species (by modern definition) of birds were discovered and classified, and between 1940 and 1970, about 90. About three or four new species a year are still being described (Mayr, 1983; Mayr and Vuilleumeier, 1983). Most of these are from the eastern (Amazonian) slope of the Andes. By contrast, between 6000 and 7000 insects are described yearly.

Today there are about 9021 well-defined bird species belonging to 159 families in 28 orders (Bock and Farrand, 1980). Newly discovered species account for only 5 percent of the increase since the 1951 count. The rest of the increment is made up of allopatric forms now recognized as full species, despite small morphological differences.

A Survey of Birds of the World

The list of the birds of the world that follows gives every order of known birds, living or fossil, and every family of living birds. Those American families occurring north of the United States-Mexico boundary are indicated by an asterisk [*]. The arrangement of orders and families mostly follows that of Bock and Farrand (1980). Some groups formerly recognized as families (e.g., Tanagers) are now relegated to subfamilial status.

CLASS AVES

Birds: Possess feathers; forelimbs modified into wings; hindlimbs adapted to walking, swimming, or perching; scales present on feet; mandibles with no teeth (in living species); light skeleton with much fusion; four-chambered heart; extensive airsacs throughout body; warm-blooded; no urinary bladder; oviparous, 9021 living species.

Subclass Archaeornithes

Ancestral birds: Fossil; lizard-like, with true teeth; three clawed-fingers, and a long bony tail of more than 13 vertebrae.

ARCHAEOPTERYGIFORMES: The oldest known birds; five fossils discovered in Jurassic stone quarries in Bavaria; pigeon-sized.
ARCHAEOPTERYGIDAE: *Archaeopteryx*; 1 species.

Subclass Neornithes

True birds: Finger bones fused; tail vertebrae 13 or less; sternum flat or keeled.

SUPERORDER ODONTOGNATHAE.
New World fossil birds.

HESPERORNITHIFORMES: Hesperornithidae:* *Hesperornis*; 4 species; flightless, loon-like swimming and diving bird with toothed jaws, from Upper Cretaceous of Kansas and Montana. About 2 m long. Around 90 million years old.

ICHTHYORNITHIFORMES: Ichthyornithidae:* *Ichthyornis*; 7 species; a tern-like, strong-flying bird with toothed jaws, from Upper Cretaceous of Kansas and Texas. About 20 cm long.

SUPERORDER NEOGNATHAE.
Typical birds.

STRUTHIONIFORMES: Flightless walking birds. Flat sternum; heavy legs with two short toes; few feathers on head, neck, or legs; small brain. Largest living birds, many over 2.2 m tall; weighing 150 kg; precocial young.
STRUTHIONIDAE: Ostriches; 1 species; Africa.

RHEIFORMES: Flightless walking birds. Flat sternum; heavy legs with three toes; 1.5 m tall; precocial young.
RHEIDAE: Rheas; 2 species: South America.

CASUARIIFORMES: Flightless walking birds. Flat sternum; stout legs with three toes; feathers with large aftershafts; 1.5 m tall; precocial young.
CASUARIIDAE: Cassowaries; 3 species; Australia, New Guinea.
DROMICEIDAE: Emus; 2 species; Australia.

FIGURE 2–4 Ostrich. Family Struthionidae.

AEPYORNITHIFORMES: Fossil. Flightless terrestrial birds with flat sternum, rudimentary wings, four toes; varied in size between turkey and ostrich; laid largest known egg: 33 cm by 24 cm.
AEPYORNITHIDAE: Elephant birds; 7 species; Madagascar.

DINORNITHIFORMES: Fossil. Flightless terrestrial birds. Wings rudimentary or absent; flat sternum; three or four toes. Largest known birds; some were over 3 m high and weighed about 450 kg (1000 lb). Became extinct about 300 years ago.
DINORNITHIDAE: Moas; 8 species; New Zealand.
ANOMALOPTERYGIDAE: *Anomalopteryx* and relatives; 20 species; New Zealand.

APTERYGIFORMES: Flightless, terrestrial hen-sized birds. Four toes, long bill with nostrils at tip; degenerative wings; hair-like feathers; nocturnal.
APTERYGIDAE: Kiwis; 3 species; New Zealand.

TINAMIFORMES: Functional wings; keeled sternum; poor fliers; precocial young.
TINAMIDAE: Tinamous; 47 species; Central and South America.

SPHENISCIFORMES: Web-footed marine swimmers.

FIGURE 2–5 Magellanic Penguin. Family Spheniscidae.

Wing modified into a thin, powerful paddle; large keel; scale-like feathers.
SPHENISCIDAE: Penguins; 18 species; Antarctica and cold waters as far north as the Galapagos Islands.

GAVIIFORMES: Short legs located far back on heavy body; webbed feet; good divers; precocial young.
GAVIIDAE:* Loons; 5 species; Northern North America and Eurasia.

PODICIPEDIFORMES: Short legs located posteriorly on body; lobate-webbed toes; rudimentary tail; good divers; precocial young.
PODICIPEDIDAE:* Grebes; 20 species; world-wide.

PROCELLARIIFORMES: Marine birds with tubular nostrils; horny, hooked beak; oily plumage; long

FIGURE 2–6 Red-throated Loon. Family Gaviidae.

narrow wings; feet webbed, with rudimentary hind toe. Excellent fliers.

DIOMEDEIDAE:* Albatrosses; 13 species; mainly on southern oceans. Wandering Albatross, *Diomedea exulans*, has largest wingspread of all living birds: 3.5 m maximum.

PROCELLARIIDAE:* Shearwaters, fulmars; 66 species; world-wide.

HYDROBATIDAE:* Storm petrels; 21 species of small sea birds; world-wide.

PELECANOIDIDAE:* Diving petrels; 4 species of small sea birds; south temperate and south polar seas.

PELECANIFORMES: Four toes united in one web; long beak; nostrils rudimentary or absent; possess a throat pouch (except tropic birds). Fish-eating birds that often nest in large colonies.

PHAETHONTIDAE:* Tropic birds; 3 species; tropical seas.

PELECANIDAE:* Pelicans; 8 species; tropical and warm temperate seas.

SULIDAE:* Boobies, gannets; 9 species; world-wide, except polar seas.

PHALACROCORACIDAE:* Cormorants; 33 species; world-wide in fresh and salt waters; used by Japanese to catch fish.

ANHINGIDAE:* Snake birds; 4 species; tropical and subtropical fresh waters.

FREGATIDAE:* Frigate or man-o'-war birds; 5 species; tropical and subtropical seas.

CICONIFORMES: Long-necked and long-legged waders; toes not webbed (except flamingos); often nest in colonies along shores or in marshes; one family with precocial young.

ARDEIDAE:* Herons, egrets, bitterns; 64 species; world-wide.

BALAENICIPITIDAE: Shoebill Stork; 1 species; Africa.

SCOPIDAE: Hammerhead; 1 species; Africa, Madagascar.

CICONIIDAE:* Storks, jabirus, wood ibises; 17 species; world-wide.

THRESKIORNITHIDAE:* Ibises, spoonbills; 33 species; world-wide.

PHOENICOPTERIDAE:* Flamingoes; 6 species; tropical and temperate regions; precocial young.

ANSERIFORMES. Broadened bills containing many tactile nerve endings and filtering ridges or "teeth" at margins; short legs with webbed feet (except screamers); body well supplied with down and oily feathers; unspotted eggs, precocial young.

ANHIMIDAE: Screamers; 3 species; South America.

ANATIDAE:* Ducks, geese, swans; 147 species; world-wide.

FALCONIFORMES: Diurnal birds of prey; strong bill, fleshy cere (soft skin) at base, hooked at tip and sharp

FIGURE 2–7 Roseate Spoonbill. Family Threskiornithidae.

FIGURE 2–8 European Black Vulture. Family Accipitridae.

on edges; feet with sharp curved talons, opposable hind toe; keen vision; strong fliers as a rule.

CATHARTIDAE:* New world vultures; 7 species; Western Hemisphere.

SAGITTARIIDAE: Secretary Bird; 1 species; Africa.

ACCIPITRIDAE:* Eagles, hawks, kites, Old World vultures, harriers; 217 species; world-wide.

PANDIONIDAE:* Osprey; 1 species; world-wide.

FALCONIDAE:* Caracaras, falcons; 62 species; world-wide.

GALLIFORMES: Vegetarian, hen-like birds with short, stout beaks, short rounded wings, and well-developed tails; heavy feet adapted to scratching ground and running; strong breast muscles; usually a roomy crop and often a caecum; often gregarious; commonly show distinct sex-dimorphism; nest on ground; precocial young; important game and domestic birds.

MEGAPODIIDAE: Megapodes or mound birds; 12 species; Australia and surrounding islands.

CRACIDAE:* Curassows, guans, chachalacas; 44 species; southern United States, Central and South America.

TETRAONIDAE:* Grouse, ptarmigan; 16 species; Northern Hemisphere.

PHASIANIDAE:* Quails, partridge, pheasants, jungle fowl, peafowl; 154 species; mainly Old World.

NUMIDIDAE: Guinea fowl; 7 species; Africa, Madagascar.

MELEAGRIDIDAE:* Turkeys; 2 species; Mexico and United States.

OPISTHOCOMIDAE: Hoatzin; 1 species; young have two clawed-fingers; semiprecocial young; South America.

GRUIFORMES: Some birds long-legged and long-necked, strong fliers (cranes); others smaller with shorter legs and necks, weak fliers (rails and coots); rounded wings; prairie and marsh dwellers; precocial young.

MESITORNITHIDAE: Roatelos (mesites); 3 species; Madagascar.

TURNICIDAE: Bustard quails; 14 species; Africa, Eurasia, Indomalaya, New Guinea, and Australia.

PEDIONOMIDAE: Plains Wanderer; 1 species; Australia.

GRUIDAE:* Cranes; 15 species; world-wide except South America.

ARAMIDAE:* Limpkins; 1 species; South America and southern United States.

PSOPHIIDAE: Trumpeters; 3 species; northern South America.

RALLIDAE:* Rails, coots, gallinules; 142 species; world-wide.

HELIORNITHIDAE: Finfoots (sungrebes); 3 species; wet, tropical Asia, Africa, America.

RHYNOCHETIDAE: Kagu; 1 species; New Caledonia.

EURYPYGIDAE: Sunbitterns; 1 species; southern Mexico to Bolivia.

CARIAMIDAE: Seriemas; 2 species; central and northern South America.

PHORORHACIDAE:* *Phororhacos*; large, flightless, predatory fossil birds; 10 species; Oligocene to Pleistocene; Patagonia, Florida.

DIATRYMIDAE:* *Diatryma*; large, cursorial, preda-

FIGURE 2–9 Prairie Chicken. Family Tetraonidae.

FIGURE 2–10 Sora Rail. Family Rallidae.

tory fossil birds; 4 species; Eocene; North America and Europe.

OTIDIDAE: Bustards; 24 species; Old World.

CHARADRIIFORMES: A large assemblage of 17 living families of shore birds, gulls, terns, auks, and others; toes usually webbed; compact plumage; strong fliers; often colonial; precocial young in most species.

JACANIDAE:* Jaçanas; 8 species; long-toed lily-pad walkers; pantropical; precocial young.

ROSTRATULIDAE: Painted snipe; 2 species; South America, southern Africa, southern Asia, and Australia; precocial young.

HEMATOPODIDAE:* Oystercatcher; 7 species; laterally flattened beaks; world-wide; precocial young.

CHARADRIIDAE:* Plovers, turnstones, lapwings, surfbirds; 64 species; world-wide; precocial young.

SCOLOPACIDAE:* Snipe, woodcock, sandpipers; 86 species; world-wide; precocial young.

RECURVIROSTRIDAE:* Avocets, stilts; 10 species; world-wide; precocial young.

PHALAROPODIDAE:* Phalaropes; 3 species; Northern Hemisphere; precocial young.

DROMADIDAE: Crab-plover; 1 species; northern and western shores of Indian Ocean from Persian Gulf to Madagascar.

BURHINIDAE: Thick-knees; 9 species; world-wide except North America; semiprecocial young.

GLAREOLIDAE: Pratincoles, coursers; 16 species; Old World.

THINOCORIDAE: Seed-snipes; 4 species; western and southern South America.

CHIONIDIDAE: Sheathbills; 2 species; antarctic and subantarctic shores and islands.

PTEROCLIDIDAE: Sandgrouse; 16 species; Eurasia, Africa, Madagascar.

STERCORARIIDAE:* Skuas, jaegers; 5 species; worldwide, especially polar seas; semiprecocial young.

LARIDAE:* Gulls, terns; 88 species; world-wide; semiprecocial young.

RYNCHOPIDAE:* Skimmers; 3 species; North and South America, Africa, southern Asia.

ALCIDAE:* Auks, murres, puffins; 23 species; north temperate and arctic seas; semiprecocial young.

COLUMBIFORMES: Pigeons and doves; short slender bill with cere (soft skin) at base; short neck; short legs; crop produces "pigeon's milk" to feed young.

RAPHIDAE: Dodos (extinct since 1700) and solitaires (extinct since 1800); 3 species; Islands of Mauritius, Reunion, and Rodriguez.

COLUMBIDAE:* Pigeons, doves; 303 species; worldwide.

PSITTACIFORMES: Narrow, hooked beak; upper mandible hinged movably to skull; fleshy tongue; large, rounded head; toes, two front and two rear,

FIGURE 2–11 Herring Gull. Family Laridae.

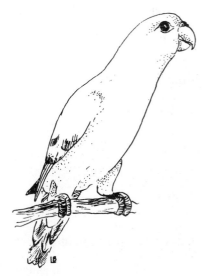

FIGURE 2–12 Parrot. Family Psittacidae.

FIGURE 2–13 Ruby-throated Hummingbird. Family Trochilidae.

FIGURE 2–14 Belted Kingfisher. Family Alcedinidae.

with grasping claws; brilliant plumage; vary in size from 7 cm to 90 cm.

PSITTACIDAE:* Lories, parrots, macaws; 340 species; pantropical and subtropical.

CUCULIFORMES: Toes, two front and two rear; outer hind toe reversible. Many Old and some New World cuckoos are brood parasites.

MUSOPHAGIDAE: Turacos; 18 species; Africa.

CUCULIDAE:* Cuckoos, coucals, roadrunners, anis; 129 species; world-wide.

STRIGIFORMES: Nocturnal predators; large, rounded head with large, forward-directed eyes set in feathered disks; large external ear-openings with flaps; short, powerful, hooked beak; strong feet, sharp talons. Soft, fluffy plumage allows silent flight.

TYTONIDAE:* Barn owls; 11 species; world-wide.

STRIGIDAE:* Owls; 135 species; world-wide.

CAPRIMULGIFORMES: Small to average-sized bill with wide mouth usually surrounded by insect-netting bristles; small legs and feet; twilight insect feeders (Oilbird feeds mainly on palm nuts); majority lay eggs directly on ground.

STEATORNITHIDAE: Oilbird; 1 species; cave dwellers; Ecuador and Peru, east to Trinidad.

PODARGIDAE: Frogmouths; 13 species; India, East Indies, Philippines, Australia.

NYCTIBIIDAE: Potoos; 6 species; Central America and tropical South America.

AEGOTHELIDAE: Owlet frogmouths; 8 species; Australia, New Guinea.

CAPRIMULGIDAE:* Goatsuckers, nightjars; 77 species; world-wide.

APODIFORMES: Small birds with short legs and small feet; bill either small and weak (swifts), or long and slender with tubular or brushy tongue (hummingbirds); wings pointed and having extremely short, stout humeri.

APODIDAE:* Swifts; 83 species (4 in the United States); world-wide.

HEMIPROCNIDAE: Crested swifts; 4 species; southeast Asia, East Indies, Philippines.

TROCHILIDAE:* Hummingbirds; 341 species; includes some of the smallest birds on earth (*Mellisuga helenae* 6.3 cm long); Central and South America; 15 species in the United States.

COLIIFORMES: Small birds of uncertain relationships; marked gregarious habits; first and fourth toes reversible; long tails.

COLIIDAE: Mouse-birds or colies; 6 species; southern Africa.

TROGONIFORMES: Short, stout bill; small weak feet with first and second toes directed backward; richly colored iridescent plumage with green predominant in most species; long tail; nest in cavities.

TROGONIDAE:* Trogons, quetzals; 37 species; pantropical.

CORACIIFORMES: Strong, prominent bill; third and fourth toes joined at base; usually colorful plumage; nest in cavities.

ALCEDINIDAE:* Kingfishers; 91 species; world-wide.

TODIDAE: Todies; 5 species; West Indies.

MOMOTIDAE: Motmots; 9 species; Mexico south to Paraguay and northern Argentina.

FIGURE 2–15 Hairy Woodpecker. Family Picidae.

MEROPIDAE: Bee-eaters; 24 species; Old World tropics and subtropics.

LEPTOSOMATIDAE: Cuckoo-roller; 1 species; Madagascar.

BRACHYPTERACIIDAE: Ground rollers; 5 species; Madagascar.

CORACIIDAE: Rollers; 11 species; Old World.

UPUPIDAE: Hoopoe; 1 species; Eurasia, Africa.

PHOENICULIDAE: Wood-hoopoes; 8 species; southern Africa.

BUCEROTIDAE: Hornbills; 45 species; tropical Africa and Asia to Solomon Islands.

PICIFORMES: Toes, two front and two (occasionally one) rear; highly specialized bill; no down on young or adults (except Galbulidae); nest in cavities.

GALBULIDAE: Jacamars; 17 species; Central and South America.

BUCCONIDAE: Puffbirds; 32 species; Central and South America.

CAPITONIDAE: Barbets, 81 species; pantropical.

INDICATORIDAE: Honey-guides; 16 species; often brood parasites; Africa, India, Malaysia.

RAMPHASTIDAE: Toucans; 33 species; New World tropics.

PICIDAE:* Woodpeckers, piculets, wrynecks; 204 species; strong, gripping toes; long retractile tongue with barbs; world-wide except Madagascar, Australia, and New Zealand.

PASSERIFORMES: Feet adapted to perching on stems or twigs; toes unwebbed, three in front and one behind, joined at the same level; wing with nine or ten primary feathers, tail usually with 12 feathers; characteristic palate; distinctive spermatozoa; young hatched naked, blind, and helpless, able to gape, and reared in the nest. Of all the known birds, over 5000 species, or roughly 60 per cent, are passerines, and of all families, over 40 per cent are passerines. The classification of passerine families is exceptionally difficult because of the high degree of resemblance among them, much of it the result of evolutionary adaptive convergence. As a consequence, there have been differences of opinion among taxonomists in naming and determining the inclusiveness of families and in arranging their sequence.

In the list of families that follows (based largely on Mayr, 1958), the first 15 families (Eurylaimidae through Atrichornithidae) contain the more primitive "suboscine Passeriformes." Nine of these families are restricted to the New World, chiefly South America. None of the 15 occurs in Europe. All other passerine families are known as oscines or songbirds, and generally possess a more highly developed syrinx and hence better vocal powers.

EURYLAIMIDAE: Broadbills; 14 species; Indo-Malaya, Africa.

DENDROCOLAPTIDAE: Woodhewers or woodcreepers; 52 species; Central and South America.

FURNARIIDAE: Ovenbirds; 218 species; Central and South America.

FORMICARIIDAE: Antbirds; 230 species; Central and South America.

CONOPOPHAGIDAE: Ant-pittas; 11 species; New World tropics.

RHINOCRYPTIDAE: Tapaculos; 30 species; Costa Rica to Patagonia.

COTINGIDAE:* Cotingas; 79 species; Texas and Arizona to northern Argentina.

PIPRIDAE: Manakins; 52 species; southern Mexico to Paraguay.

TYRANNIDAE:* Tyrant flycatchers; 375 species; from Canadian tree line to Patagonia.

OXYRUNCIDAE: Sharpbills; 1 species; Costa Rica to southern Brazil.

PHYTOTOMIDAE: Plantcutters; 3 species; western Peru to Patagonia.

PITTIDAE: Pittas; 24 species; Old World tropics.

XENICIDAE: New Zealand wrens; 4 species; New Zealand.

PHILEPITTIDAE: Asities; 4 species; Madagascar.

FIGURE 2–16 Eastern Wood Pewee. Family Tyrannidae.

MENURIDAE: Lyrebirds; 2 species; Australia.

ATRICHORNITHIDAE: Scrub-birds; 2 species; eastern and southwestern Australia.

ALAUDIDAE:* Larks; 78 species; Old World (1 species in America).

HIRUNDINIDAE:* Swallows: 80 species; world-wide except New Zealand and polar regions.

MOTACILLIDAE:* Pipits, wagtails; 54 species; world-wide except Borneo, Moluccas, Philippines, and southern part of North America.

CAMPEPHAGIDAE: Cuckoo-shrikes; 70 species; Africa, Southeast Asia, East Indies, Japan, Australia.

PYCNONOTIDAE: Bulbuls; 123 species; Old World tropics.

IRENIDAE: Fairy bluebirds, leafbirds; 14 species; southeast Asia to Borneo and Philippines.

LANIIDAE:* Shrikes; 74 species; world-wide except Central and South America, New Guinea, Australia.

VANGIDAE: Vanga shrikes; 13 species; Madagascar.

BOMBYCILLIDAE:*

Ptilogonatinae: Phainopepla and other Silky Flycatchers; 4 species; southwestern United States to Panama.

Bombycillinae: Waxwings; 3 species; North America and Eurasia.

Hypocoliinae: Hypocolius; 1 species; Southwestern Asia.

DULIDAE: Palm Chat; 1 species; Hispaniola

CINCLIDAE:* Dippers; 5 species; Northern Hemisphere and South America.

TROGLODYTIDAE:* Wrens; 60 species; New World, 1 species extending to Eurasia and northern Africa.

MIMIDAE:* Thrashers, mockingbirds; 31 species; southern Canada to Patagonia.

PRUNELLIDAE: Accentors, hedge sparrows; 12 species; Eurasia, northern Africa, Japan.

MUSCICAPIDAE:*

Turdinae: Thrushes; 309 species; world wide.

Orthonychinae: Logrunners; 19 species; New Guinea and Australia.

Timaliinae: Babblers, including New World Wrentit; 255 species; Old World tropics except for wrentit.

Panurinae: Reedlings and Parrotbills; 19 species; Old World.

Picathartinae: Rockfowl; 2 species; Africa.

Polioptilinae: Gnatcatchers and Gnatwrens; 12 species; New World.

Sylviinae: Old World warblers, tailor-birds, kinglets; 349 species; world-wide except southern parts of North and South America.

Malurinae: Fairy-wrens; 106 species; Australia and New Guinea.

Muscicapinae: Old-World Flycatchers; 153 species; Old World.

Platysteirinae: Puffback and Wattled Flycatchers; 26 species; Africa.

Monarchinae: Monarch Flycatchers; 91 species; Africa and Australasia.

Rhipidurinae: Fantail Flycatchers; 40 species; Australasia.

Pachycephilinae: 46 species; Africa, Australasia.

AEGITHALIDAE:* Long-tailed tit and Bushtit; 7 species; Eurasia and North America into Mexico (bushtit).

REMIZIDAE:* Penduline tit, Kapok tits, Verdin; 10 species; Europe, Africa, North America (verdin only).

PARIDAE:* Titmice, 47 species; Eurasia, Africa, North America south to Guatemala.

SITTIDAE:* Nuthatches and Wall Creeper; 26

FIGURE 2–17 Wood Thrush. Family Turdidae.

FIGURE 2–18 Chestnut-backed Chickadee. Family Paridae.

FIGURE 2–19 Rose-breasted Grosbeak. Family Fringillidae.

species; Northern Hemisphere, East Indies, Australia, Madagascar.

CERTHIIDAE:* Creepers; 6 species; Eurasia, northern Africa, North America.

RHABDORNITHIDAE: Philippine Creepers; 2 species; Philippine Islands.

CLIMACTERIDAE: Australian Treecreepers; 6 species; Australia, New Guinea.

DICAEIDAE: Flower-peckers; 58 species; southeast Asia, East Indies, New Guinea, Australia.

NECTARINIIDAE: Sunbirds; 117 species; Africa, southeast Asia, East Indies, Philippines, Australia.

ZOSTEROPIDAE: White-eyes; 83 species; Africa, Madagascar, southeast Asia to Japan and Northern Australia.

MELIPHAGIDAE: Honey-eaters; 172 species; southern Africa, Moluccas, New Guinea, Australia, New Zealand.

EMBERIZIDAE:*

Emberizinae: Buntings, New World "Sparrows"; world-wide except Madagascar, Borneo, New Guinea.

Catamblyrhynchinae: Plush-capped Finch; 1 species; Venezuela, Colombia, Ecuador, Peru, Bolivia.

Cardinalinae: Dickcissel, cardinals, grosbeaks, saltators and allies; 39 species; New World.

Thraupinae: Tanagers; 240 species; New World.

Tersininae: Swallow Tanager; 1 species; Panama and Trinidad south to southern Brazil.

PARULIDAE:* Wood warblers; 126 species; Alaska and Labrador south to Paraguay.

DREPANIDIDAE: Hawaiian honeycreepers; 23 species; Hawaiian Islands.

VIREONIDAE:* Vireos; 43 species; New World.

ICTERIDAE:* Blackbirds, troupials; 95 species; New World.

FRINGILLIDAE:

Fringillinae: Chaffinches, Brambling; 3 species; Old World.

Carduelinae: Goldfinches, crossbills, redpolls; 119 species; world-wide except Australia.

ESTRILDIDAE: Waxbills, mannikins; 127 species; Africa, Madagascar, southeast Asia, East Indies, Australia.

PLOCEIDAE: Weaver finches and parasitic viduines; 144 species; Africa, Madagascar, Eurasia, Malaysia; widely introduced elsewhere including North and South America.

STURNIDAE: Starlings; 111 species; Old World (2 species introduced into North America).

ORIOLIDAE: Old World Orioles; 25 species; Eurasia, Africa, East Indies, Philippines, Australia.

DICRURIDAE: Drongos; 20 species; Old World Tropics.

CALLAEIDAE: Wattlebirds; 3 species; New Zealand.

GRALLINIDAE: Mudnest-builders; 4 species; Australia, New Guinea.

ARTAMIDAE: Wood swallows; 10 species; southeast Asia, East Indies, Phillipines, Australia.

CRACTICIDAE: Bellmagpies; 10 species; Australia, New Guinea.

FIGURE 2–20 Blue Jay. Family Corvidae.

PTILONORHYNCHIDAE: Bower-birds; 18 species; Australia and New Guinea.

PARADISAEIDAE: Birds of paradise; 42 species; Moluccas, New Guinea, northern Australia.

CORVIDAE:* Crows, magpies, jays; 106 species; world-wide except New Zealand and Antarctica.

■ SUGGESTED READING

A rich assortment of books is available to anyone interested in the identification and classification of birds. Outstanding for their color plates are the following non-technical works on birds of the world: Austin and Singer, *Birds of the World*; Fisher and Peterson, *The World of Birds*; Gilliard, *Living Birds of the World*; Grzimek, *Grzimek's Animal Life Encyclopedia* (Vol 7–9).

The standard technical reference on classification of world birds is Peters, *Check-list of Birds of the World* (15 vol); and for North America, the American Ornithologists' Union, *Check-list of North American Birds*.

Only a sampling of the numerous regional bird guides can be given here. Among them are: Alexander, *Birds of the Ocean*; Ali and Ripley, *Handbook of the Birds of India and Pakistan* (10 vol); Bent, *Life Histories of North American Birds* (20 vol); Cramp and Simmons, *Handbook of the Birds of Europe, the Middle East and North Africa* (several volumes); Dement'ev, *Birds of the Soviet Union* (6 vol); Glutz von Blotzheim, *et al.*, *Handbuch die Vögel Mitteleuropas* (6 vol); Heinroth, *Die Vögel Mitteleuropas* (4 vol); Mackworth-Praed and Grant, *African Handbook of Birds* (4 vol); Murphy, *Oceanic Birds of South America* (2 vol); Palmer, *Handbook of North American Birds*; de Schauensee, *A Guide to the Birds of South America*; Skutch, *Life Histories of Central American Birds* (3 vol); Smythies, *The Birds of Borneo* and *The Birds of Burma*; Terres, *The Audubon Society Encyclopedia of North American Birds*; Vaurie, *The Birds of the Palearctic Fauna* (2 vol); Voous, *Atlas of European Birds*; Witherby, *et al.*, *The Handbook of British Birds* (5 vol). In addition, there are numerous pocket field guides and many state bird books for the United States, most of them well illustrated.

Among books devoted to special groups of birds are: Brown and Amadon, *Eagles, Hawks and Falcons of the World* (2 vol); Delacour, *The Waterfowl of the World* (4 vol); Goodwin, *Pigeons and Doves of the World*; Swinton, *Fossil Birds*; Wetmore, *A Checklist of the Fossil and Prehistoric Birds of North America and the West Indies*; *Reader's Digest Complete Book of Australian Birds*; Urban, Fry and Stuart, *The Birds of Africa* (several volumes); von Blotzheim and Bauer (eds), *Handbuch der Vögel Mitteleuropas* (10 volumes).

Skin, Scales, Feathers, and Colors

3

From busy beak which cleans and preens
  Each feather gleans its own soft glow:
  The burnished blackness of the crow,
The peacock's blazing blues and greens,
  The white of swan, like moonlit snow.
—Joel Peters (1905-   ).
  Courting Colors

The outer surface of every animal is unavoidably in intimate contact with its life-sustaining—and, at times, life-threatening—external environment. Birds, as creatures of land and air, and sometimes of water, have special needs. As a consequence, natural selection has provided them with a complex, highly adapted exterior.

THE SKIN OF BIRDS

The chief functions of a bird's skin and its derivatives are to protect the bird from harmful forces, substances, or organisms; to prevent the loss of blood and other fluids; to help regulate body temperature; to aid in locomotion and other behavior; and to house sense organs sensitive to pain, temperature, pressure, and vibration.

Like all other vertebrates, birds possess a two-layered skin: epidermis on the outside and dermis underneath. Compared with other vertebrates, birds have a thin skin, extremely thin in owls and goatsuckers. As in man, the outer epidermis is composed of layers of flattened epithelial cells. These are capable of producing large amounts of keratin or horn—a tough, fibrous protein highly resistant to physical or chemical breakdown. Unlike the mammalian epidermis, however, the epidermal cells of birds are also capable of lipogenesis, i.e., of producing fat bodies consisting

27

of neutral lipids. These lipids, along with secretions from the uropygial gland, coat the plumage with a lipid film (Spearman and Hardy, 1985). Lipids may be organized into multigranular bodies in the skin of Zebra Finches during periods of drought and function in controlling cutaneous water loss (Menon *et al.*, ms).

The inner layer of skin, or dermis, a "housekeeping" layer, which supports and nourishes the epidermis, is a tougher, thicker, fibrous layer of connective tissue. It is penetrated by muscles, nerves, and blood vessels that supply the overlying epidermis and help to regulate heat loss. Its innermost layer is often rich in stored fat, valuable for shock absorption, heat insulation, and food storage. Dermal fat storage is particularly great in aquatic birds such as penguins, loons, ducks, albatrosses, petrels, and auks, and, during migration, in many passerine birds. Some species such as emus, boobies, pelicans, and nightjars possess subcutaneous air pockets that the bird may inflate at will. Smooth muscles located in the dermis are attached to the feather sockets. Their action permits the fluffing of feathers to increase the thickness of the insulating layer of trapped air, or the spreading of feathers for purposes of display or flight, or the pressing of feathers against the body for more rapid heat loss in hot weather (Morris, 1956; MacFarland and Baker, 1968). Striated, or voluntary, muscles in and under the dermis produce additional movements of feathers.

Unlike mammals, birds possess no sweat glands.

Sweat would merely plaster the feathers against the skin and hinder their normal functions. The early literature of ornithology stated that with the exception of the uropygial or preen gland, avian skin is completely without glands. However, recent studies have yielded a wealth of information on the secretory functions of avian epidermis. Investigators have identified: (1) anal glands, (2) ceruminous (wax) glands of the external ear canal, (3) and that the whole of body epidermis functions as a lipid gland in addition to the (4) uropygial gland (Menon, 1984).

Anal gland secretions consist primarily of mucopolysaccharides. This gland in male Japanese Quail, *Coturnix coturnix*, produces a frothy substance perhaps associated with the mechanics of internal fertilization. This foam may be left on droppings to act perhaps as a "visual" territorial marker (Schleidt and Shalter, 1972).

The ceruminous glands of the ear canal are present in a row. There are five to seven such glands in the ear canal of the domestic fowl. The functional significance of these glands is as yet unclear (Menon, 1984).

The avian epidermal cells are sebokeratinocytes, which secrete oily lipid substances (Menon *et al.*, 1981; Spearman and Hardy, 1985). These lipoid substances have a function similar to sebum in mammals, protecting the skin and providing an efficient water barrier. Such secretory activity is especially evident on combs of pheasants, wattles of turkeys, bare skin of storks and ibises, and rictus and toe webs of pheasants.

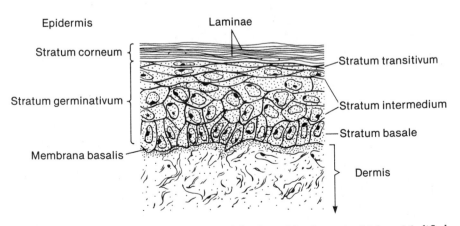

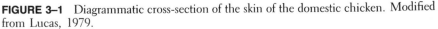

FIGURE 3–1 Diagrammatic cross-section of the skin of the domestic chicken. Modified from Lucas, 1979.

Zebra Finches, *Poephila guttata*, breed in dry regions of Australia. Incubation of eggs or brooding of young stops when temperatures reach 37.5°C (Immelman, 1982). How do these naked nestlings prevent excessive cutaneous water loss? Using an instrument called an Electrolytic Moisture Analyzer, Menon *et al.*, 1987, found that nestlings lose only from 2 to 4 parts per million/square centimeter/hour, whereas adults lose from 60 to 70 ppm/sq cm/hour. Light and electron microscope analyses revealed that the epidermis of nestlings contains many more lipid bodies than that of adults. Many of these lipids are in the form of multigranular bodies, the secreted contents of which fill the extracellular spaces of the lower stratum corneum, preventing transepidermal water loss. The number of lipid bodies is reduced as fledglings mature. In addition, epidermal lipids, especially nonpolar triglycerides, function like an occludant, sealing in the body moisture.

Adult birds have a layer of plumage used in body insulation. Evaporation of water from the skin moistens the feathers and prevents them from becoming too brittle (Spearman, 1980). Pigeons and sand grouse, *Pterocles* spp., also use cutaneous transpiration to cool the body during heat stress (Marder *et al.*, 1986).

As long ago as 1248 A.D., the Emperor Frederick II of Hohenstaufen described the preen gland and its functions in his ornithological landmark, *The Art of Falconry* (Wood and Fyfe, 1943). Frederick described the organ as a double gland on the bird's rump, supplied with a brush-like external opening. The gland secretes an oil that the bird squeezes out with its mandibles and places on its feathers, beak, and claws to keep them waterproof, supple, and in good condition.

Actually, the preen gland may be two-lobed or single. It may or may not have a reservoir; and externally it may have a tuft of feathers at its opening, or a nipple. Most birds possess a preen gland, but it is completely lacking in the Ostrich, *Struthio camelus*, emus, cassowaries, bustards, frogmouths, and in some species of pigeons, parrots, and woodpeckers. Its greatest development occurs in aquatic birds such as ducks, geese, petrels, and pelicans. It is large also in the Osprey, *Pandion haliaetus*, and the Oilbird, *Steatornis caripensis*. Administration of androgens or FSH and LH brought about the hypertrophy of this gland in cockerels and domestic pigeons, respectively (Kusuhara and Ishida, 1984; Deadhikari and Bhattacharyya, 1985). Thyroidectomy leads to atrophy of the gland in pigeons (Salinukul and Menon, 1983).

Studies by Elder (1954) have shown quite clearly that, at least in many waterfowl, the uropygial gland secretes a fluid containing fatty acid, fat, and wax, and that the act of preening automatically stimulates the flow of this secretion. He found further that birds whose glands had been experimentally removed suffered decided deterioration of their feathers, beaks, and leg scales (Fig. 3-2). Interestingly enough, young ducklings from which the oil gland had been removed

FIGURE 3–2 A normal Redhead Duck *(left)*, contrasted with one whose oil gland had been removed. The role of the uropygial gland in feather-care is obvious. Photo from W. H. Elder, 1954.

displayed normal preening behavior but definitely shunned the water.

There is good evidence that feathers supplied with the uropygial secretion have a significant amount of vitamin D, which may serve as a valuable supplement to the diet if swallowed during the preening exercise. Apparently, in some species the secretion of the preen gland contains a vitamin precursor, ergosterol, which, when spread on the feathers, is converted by sunlight into vitamin D. Removal of the preen gland in some species has resulted in rickets, a disease of vitamin D deficiency, but in other species the results have been negative (Thomson, 1964).

Normally ducks and geese oil-preen their feathers about two to seventeen times a day. However, while incubating, the females seem to refrain from oiling their feathers. But when their young hatch, the females begin oiling their belly and flank feathers as often as two or three times an hour (McKinney, 1965). This industrious preening apparently serves to spread oil from the mother's plumage to the downy young, possibly improving the water-repellency of their down. The ducklings themselves perform feather-oiling movements even on the first day of life. Experiments by Fabricius (1959) indicate that at least in the Tufted Duck, *Aythya fuligula*, the uropygial gland is not an absolute requirement for waterproof plumage. Young ducklings whose oil glands had been removed had plumage as water-repellent as that of normal ducklings. The maintenance, by preening, of large amounts of finely distributed air among the ramifications of the feathers seemed to be responsible for their water-repellency. Similar conclusions were arrived at by Van Rhijn (1977) in his studies of water-repellency in various ducks and in the Herring Gull, *Larus argentatus*.

In the Musk Duck, *Biziura lobata*, the Hoopoe, *Upupa epops*, and some petrels, the preen gland gives off an evil-smelling secretion that may ward off predators.

Many species of birds have patches of bare skin, especially on the head and neck: cassowaries, vultures, pelicans, hornbills, Galliformes, and others. These patches may be pigmented as in the turkey, or simply colored red by a rich supply of blood vessels. Many of them are in the form of fleshy lobes and contain a spongy connective tissue which, when inflated with blood, may become erect and turgid. Among such structures are the combs of roosters, the fleshy eyebrows of some grouse, and the wattles of turkeys, pheasants, cotingas, and cassowaries. Some birds possess unfeathered skin-sacs or pouches that serve various functions: capturing food in the pelicans; producing booming calls in some grouse; and providing courtship ornaments in frigate birds and the Marabou Stork, *Leptoptilos crumeniferus*.

BEAKS, SCALES, AND CLAWS

The skin of a bird can produce a variety of structures, depending on its location on the body. In one place it may remain skin, as in the naked head and neck of a vulture. In other places its epidermal cells may produce dense and compact scales, beaks, and claws, or light and complex feathers.

In some birds, including certain grouse and owls, the leg may be feathered as far as the toes, but in most birds the legs and feet are covered only with scales. As a rule, the epidermal cells of the skin build up the scales on their inner side as fast as they are worn away on the outside. Scales may be small and granular or large and plate-like. They provide a wonderfully tough, flexible, and easily cleaned covering for a bird's feet.

Claws are specialized scales protecting the toe-tips. They are well adapted for digging, scratching, clinging, fighting, and cleaning feathers. Each claw possesses a central basal mass of multiplying epidermal cells that become gradually keratinized and pushed outward and forward, resulting in the horny finished claw. Many claws are curved because their upper surface grows more rapidly than their lower. As with other scales, claws are renewed from their bases about as rapidly as normal wear erodes their outsides. But in grouse the claws may be molted and renewed periodically, as are the feathers.

Surprisingly, a few birds still possess claws on their wings, somewhat corresponding to nails on human fingertips. They are commonly found on the thumb in Falconiformes and Anseriformes, and on the second finger in the Ostrich, *Struthio camelus*, Greater Rhea, *Rhea americana*, and young Hoatzin, *Opisthocomus hoazin*. At times claws are found also on the third finger of the Ostrich. The finger claws are well developed and quite functional in the young Hoatzin, which uses them to clamber about branches before it is able to fly.

The cocks of numerous polygamous galliform birds—pheasants, peacocks, turkeys, domestic fowl—have stout spurs on the inner posterior surfaces of the tarsi of their legs. The spurs have bony cores sheathed with a horny keratin covering and are used in courtship battles. Peacock-pheasants, *Polyplectron* spp., may have as many as four spurs on each leg.

Birds have evolved a great variety of beaks adapted to their varied food habits. All of them, however, arise in fundamentally the same way. They are essentially a compact layer of epidermal cells molded around the bony core of each mandible. The epidermis of the skin covering the upper and lower mandibles is stratified into two layers. Outwardly, as in the human skin, there is a horny stratum corneum (Fig. 3-1), which contributes directly to the formation of the beak. Inwardly, next to the dermis, is the stratum germinativum, where the epidermal cells actively multiply. These cells produce the keratin that is concentrated in the stratum corneum to make the beak, or rhamphotheca, proper.

The beak of a bird is normally hard and thick, especially at the tip, where wear is greatest. Frequently the edges, or tomia, are sharp, and useful for cutting food. In ducks the beak is hard only at its tip, and its sides are relatively soft and blunt. Its lateral margins are richly supplied with sensitive tactile nerve endings (Herbst corpuscles), useful in detecting seeds and insects in muddy water. Anyone who has ever held a handful of grain under water to feed a hungry duck can appreciate from the rapid disappearance of the grain the marvelous sensory discrimination of these pressure-sensitive corpuscles in the bird's beak. In a few species, such as the American Woodcock, *Scolopax minor*, the tactile corpuscles occur at the tip of the long sensitive beak, aiding the bird in detecting worms deep in soft soil.

In many species the tough horny material at the base of the upper mandible gives way to bare, thick skin adjoining the forehead. This region, called the cere, is also supplied with touch corpuscles. Frequently the cere is brightly colored, as in parrots and birds of prey.

Nostrils are usually situated in the upper mandible near its base or in the cere, but in the nocturnal kiwis, which find their food largely by scent, the nostrils are located at the tip of the beak. In petrels and albatrosses (Procellariiformes) the nostrils do not open directly outward but are carried forward in a covered tube (Fig. 3-3), while in gannets and cormorants they are entirely covered over.

The Black Skimmer, *Rynchops niger*, is a bird of coastal waters that spends hours at a time flying just above the surface of the water with its upper mandible in the air, its lower mandible cutting the water. One-sided wear on the lower mandible would soon reduce it to a stump but for the remarkable fact that the lower mandible, adapted to the excessive friction, grows more rapidly than the upper. A Black Skimmer reared in captivity, lacking this water friction, will grow a lower mandible over twice as long as its upper mandible.

Ordinarily, each mandible of a bird is covered with a single horny sheath that is renewed from beneath as it is worn down on the outside. However, in some penguins, pelicans, auks, and puffins the keratin sheath may thicken locally into scales, plates, knobs, or other projections that persist only during the breeding season and are shed immediately thereafter. For example, on the vividly colored triangular beak of the Atlantic Puffin, *Fratercula arctica*, grow nine separate superficial scales: six on the upper mandible and three on the lower, and also one above and below each eye. These are apparently courtship ornaments, for they are all shed at the close of the reproductive period in August. Similarly, in certain ptarmigan and grouse the entire sheath is shed at the end of the breeding season. This

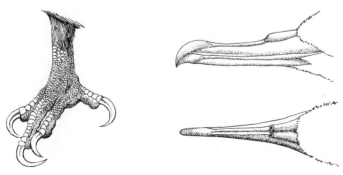

FIGURE 3–3 The horny scales, talons, and beaks of birds are derived from the epidermis of the skin. (*Left*) Scales and talons of the Bald Eagle. (*Right*) Side and top views of the beak of the Pink-footed Shearwater.

periodic molting of beak scales seems to be a primitive characteristic, reminiscent of the shedding of reptilian scales.

From the standpoint of evolutionary selection, beaks seem to be unusually plastic, and as a consequence are of little use taxonomically. For example, the stork, ibis, spoonbill, and flamingo have bills of widely different shapes, yet they are all members of the order Ciconiiformes. The strikingly different diets among these near relatives are obviously the consequence of evolutionary divergence in their eating utensils.

FEATHERS

Types of Feathers

No bird is without feathers, nor is there any other kind of animal that possesses feathers. The horny epidermal substance, keratin, seems to form the scales on the legs of a bird exactly as it forms the scales of a reptile, and in its earliest stages the development of a feather closely resembles that of a reptilian scale. Feathers probably evolved from reptilian scales. Regal

(1975) notes that some reptiles have elongated scales which may be raised or lowered by specialized muscles during thermoregulation. These movable scales may have preadapted the protobird for the evolution of endothermy.

Though modern birds possess both feathers and scales, no intermediate stage between the two has ever been discovered. Highly significant, however, is the fact that transplants of embryonic epidermis may produce either scales or feathers depending on the nature (i.e., induction) of the underlying dermis (Rawles, 1963). Possibly some day a Jurassic fossil animal will be found that reveals the intermediate stage between a scale and a feather. A conjectural reconstruction of such a scale-feather is shown in Figure 23-4.

There are six commonly recognized types of feathers: vaned or contour, down, semiplume, filoplume, bristle, and powder down (Fig. 3-4). The vaned feathers which chiefly cover a bird's body and give it its streamlined form are the contour feathers; those that extend beyond the body and serve for flight are called the flight feathers. Although they exist in great variety (a single feather will often suffice to identify a bird), in basic design they are all quite similar. Once formed,

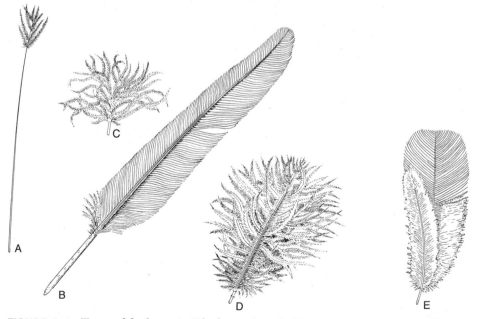

FIGURE 3–4 Types of feathers. A, Filoplume, $3\times$; B, Vane or contour, $1\times$; C, Down, $1\times$; D, Semiplume, $2\times$; E, Pheasant vane feather with aftershaft, $1\times$.

any feather is a dead horny structure without living cells. It receives nothing from the body but physical support.

The typical contour feather (Fig. 3-5) is made up of a central shaft and a vane. The bare proximal (body) end of the shaft is called the calamus or quill. At its base, inserted in the skin, is a tiny opening, the inferior umbilicus, through which the growing feather received nourishment. This part of the shaft is hollow and circular in cross-section. The portion of the shaft between the two webs of the vane is called the rachis. At the point where the calamus ends and the rachis and vane begin is another tiny opening on the under side, the superior umbilicus. At this point in many contour feathers arises a secondary feather called the afterfeather or aftershaft. It is usually small and downy, but sometimes, as in the emu and cassowary, it is as large as the main feather. It is thought that this doubling of the distal (outward) parts of feathers was a device in primitive birds to provide a thicker layer of insulation against heat loss.

From the superior umbilicus to its tip, the rachis is grooved on its inner surface and flattened on its sides, or roughly square in cross-section. It divides the vane of the feather into two webs. These webs exhibit the most highly evolved and precisely adapted epidermal structures known. The flexible flatness of the vane, so valuable in flight, is achieved by means of numerous barbs or rami, small toothpick-like rods or filaments arranged in a closely parallel fashion on both sides of the rachis, running outwardly and diagonally toward the feather tip. They are fastened at their bases to the flat sides of the rachis, and their free ends mark the outer margins of the vane. There are usually several hundred barbs in each web. Those near the tip of the feather are likely to be stiff and flat, while those near the base are often loose and fluffy. Barbs are normally longest near the center of the vane and shortest at its ends. In the primary contour feathers of the wing the vanes are of unequal width. The barbs on the anterior or leading edge of each feather are shorter and often stiffer than those on the inner or trailing edge.

The parallel barbs are held together in the flat webs by means of many tiny barbules or radii. These are, again, parallel filaments set in the two sides of each barb much as the barbs are set in the two sides of the rachis. A single barb of a crane feather has about 600 barbules on each side, which means well over a million barbules for the entire feather. Here, however, the similarity of barbs and barbules ends, for there are two distinctly different types of barbules. Those branching out of the side of the barb toward the feather tip, the distal barbules, bear on their under sides many microscopic hooklets (barbicels or hamuli), while the proximal barbules, on the side of the barb toward the feather base, present rounded ridges or flanges on their upper edges. The hook-bearing distal barbules on one barb overlap at right angles the smooth proximal barbules of an adjoining barb and cling to them with extraordinary tenacity to hold the barbs of a vane parallel and together against the pressure of moving air when the bird is flying. If two adjoining barbs are accidentally separated by some blow (i.e., the web becomes split), the bird needs merely to draw the feather between its mandibles as in preening to lock the barbule hooks and flanges together again and restore the entire web. In addition to hooklets, some distal barbules of flight feathers have on their surfaces lobed barbicels that provide friction against overlapping vanes; this helps control feather movements (Oehme, 1963). Rachis, barbs, and barbules are all hard horny structures on the outside, but more or less hollow or pith-like and filled with air inside.

Down feathers, like the basal portions of many vaned feathers, are downy because their barbules lack

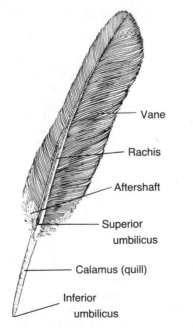

Vane

Rachis

Aftershaft

Superior umbilicus

Calamus (quill)

Inferior umbilicus

FIGURE 3–5 The parts of a typical contour feather.

FIGURE 3–6 A contour feather at increasing magnifications; A, $1\times$; B, $2.5\times$; C, $12.5\times$; D, $75\times$; E, $160\times$. In D and E the tiny hooks that hold the overlapping barbules together to make a flat vane surface are just beginning to show.

hooks. They are generally short fluffy feathers hidden under the contour feathers. They are abundant, especially in ducks and in the downy young of many game birds and waterfowl. Down feathers usually have no rachis. Their chief function is heat conservation.

Semiplume feathers are intermediate between con-

tour and down feathers in general structure. They have a rachis with barbs arranged in two rows as in a vaned feather, but their barbules lack hooks and flanges and therefore are loose and fluffy as in a down feather. Typically they occur under a covering of contour feathers, generally along the sides of the abdomen but also at

FIGURE 3–7 Scanning electron micrographs of a Robin's primary wing feather showing the hooklets on distal barbules and the way they grasp the proximal barbules. Magnifications 160 and 800 diameters. Photos by S. D. Carlson.

times along the neck and midback, and adjoining the large wing and tail feathers. Their function is to provide body insulation, flexibility for the movement of the larger contour feathers, and to increase the buoyancy of water birds.

Filoplumes are fine whip-like shafts with sparse barbs and barbules at the tip. They are more closely associated with the mechanically active (e.g., flight) feathers than others. The follicles surrounding such filoplumes are richly supplied with tactile nerve endings, a fact that stongly suggests a sensory function that aids in controlling feather movements (Borodulina, 1966).

Bristles are modified, vaneless contour feathers, each consisting of a small, stiff rachis with very few or no barbs (Stettenheim, 1973). They occur commonly around the eyes, where they form protective eyelashes as in the Ostrich; around the nostrils, where they filter dust from the air; and around the mouth (rictal bristles), where they form a tactile insect-net for such birds as goatsuckers and tyrant flycatchers. In the Asiatic Little Barbet, *Megalaima australis*, they are twice as long as the bill itself and are extensively movable.

Powder down feathers are unique in that they grow continuously and are never molted. They commonly grow in dense yellowish patches, especially on the breast, belly, or flanks of herons, bitterns, and

tinamous. In other species, such as hawks and parrots, they are more diffusely scattered throughout the plumage. The barbs at their tips constantly disintegrate into a fine, talc-like, water-resistant powder whose particles measure about a micron (1/1000 mm) in thickness. Birds that lack preen glands often have abundant powder down feathers. The powder, scattered by preening or by fluffing the feathers, gives a metallic luster to the plumage of some birds, and in general helps waterproof and preserve the feathers.

The Functions of Feathers

Feathers are adapted to a variety of functions, the chief of which are to protect the body and to promote flight. Feathers protect the body against mechanical damage, rain, the burning actinic rays of the sun, and above all, from disastrous temperature changes in the bird's body. By actively sleeking, fluffing, or ruffling its body feathers, a bird can to a great degree maintain its optimum body temperature despite external temperature changes. Small bodies cool much more rapidly than large ones. Yet the highly effective layer of insulating feathers clothing a tiny Black-capped Chickadee, *Parus atricapillus*, keeps it alive and warm through a long subzero winter night. Many birds carry more feathers in winter than in summer. A Carolina Chickadee, *Parus carolinensis*, taken on February 19, had 1704 contour feathers, while one taken on June 4 had only 1140. Common Redpolls, *Carduelis flammea*, have 31 per cent heavier plumage in November than in July (Brooks, 1968). Close relatives that spend their winters in different climates adapt their plumage to temperature demands. The European Blackcap, *Sylvia atricapilla*, lives in cooler winter quarters than does the congeneric Garden Warbler, *S. borin*. The mean weights of the body plumage and of a single feather of *atricapilla* are 21 and 18 per cent higher, respectively, than those of *borin* (Berthold and Berthold, 1971).

The flat, vaned contour feather is primarily an adaptation to flight. The distal tip of a large wing feather, or primary, has a streamlined cross-section much like that of an airplane propeller. The way the trailing web of a primary bends upward during the downstroke of the wing provides both lift and a forward push to the body. Among smaller birds, Oehme (1959) discovered that the primary wing feathers of strong fliers such as swifts and swallows have firmer, more tightly bound vanes because they have stronger, longer hooklets on their barbules than do weaker fliers like tits or sparrows.

An interesting adaptation in owls' wingfeathers ensures the silent flight so valuable to a nocturnal predator. On the leading edges of exposed primaries are located comb-like projections, while the trailing edges of the feathers are fringed somewhat like a shawl, and often the upper surface of each vane is downy. All these adaptations help to muffle the normal sounds of a blade-like object cutting the air. Significantly, the wing feathers of Asian fishing owls, whose prey lives under water, are of the normal non-silent variety. Further adaptation of feathers to flight will be considered in Chapter 21.

Oily feathers give buoyancy to a water bird by adding greatly to its volume while only slightly increasing its weight. The specific gravity of a feathered duck was determined by Heinroth (1938) to be 0.6, while that of the same bird plucked was 0.9. That is, a plucked duck weighs half again as much per unit volume as a feathered duck. The ventral contour body feathers of water birds, unlike those of land birds, were found by Kennedy (1972) to have shorter, stouter terminal barbules that lacked hooklets and thus did not form flat vanes. The rough-textured tips of these feathers seem to create a water-repellent barrier that preserves the buoyant air trapped within the belly feathers.

In many birds feathers have become strikingly modified in form and color for use as courtship ornaments, recognition marks, social signals, concealing cloaks, and other aids to daily living. Occasionally, feathers are adapted for making sounds, particularly in courtship ceremonies, as when their rachises are clicked together (manakins), or the entire exposed feather hums in the air during flight (snipe). In certain owls, facial feathers are arranged in the shape of a disc and serve as devices for collecting and focusing sound. A common Barn-Owl, *Tyto alba*, was trained to strike at an artificial sound source in complete darkness. Large errors in locating the sound source resulted when the facial discs were removed (Konishi, 1973). Among sandgrouse and a few other species feathers are used to transport water to nestlings, and in some lovebirds, to carry nesting materials. In siskins breast feathers are known to act, with their associated mechanoreceptors, as air-current sense organs used in flight control. All these and other functions of feathers will be considered in greater detail later in the text.

Complex structures that they are, feathers need con-

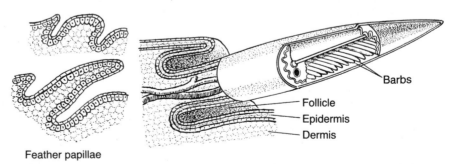

Feather papillae

Barbs
Follicle
Epidermis
Dermis

FIGURE 3–8 A diagram of three successive stages in the development of a contoured feather. After Storer, 1943.

siderably more care than skin or hair. Preening with oil from the uropygial gland, or with powder-down particles, or even with nothing at all, occupies a good share of the waking hours of most birds. Birds of the deserts and steppes where water is scarce take frequent dust baths, while those in more humid regions may take either dust or water baths. Preening commonly takes place after these baths and helps restore oil or powder down to keep the feathers in good condition. Bathing and preening, plus scratching with the claws, help allay itching, remove parasites, and clean the feathers. Preening is such a basic activity that it has achieved sufficient behavior momentum to become ritualized in courtship ceremonies.

The Growth of Feathers

Feathers grow from the base and not from the tip as do most plants. That is, the region of actively growing cells is always at the base of the feather next to the body.

The first evidence of a feather in the skin appears as a localized increase in epidermal and dermal cells to form a small papilla. This tiny pimple appears in an embryo chick of about 6 days incubation. The papilla continues to grow into a cone leaning toward the rear (Fig. 3-8). At the same time its base sinks deeply into the bird's skin, creating a circular moat about the papilla. This moat with the surrounding epidermal cells becomes the feather follicle. Although the feather itself is constructed exclusively of epidermal cells, it is the adjacent dermal cells of the papilla that control the kind of feather produced.

The cells of the papilla next begin to differentiate into several layers. On the outside is a single layer of very thin horny cells that become the epitrichium, a protective sheath. Next, inward, is the malpighian layer of epidermal cells, which will grow into the structure of the feather proper. Enclosed within the malpighian layer is the core of the papilla, made up of dermal cells. This finger of dermal cells, with its rich supply of blood vessels, constitutes an internal pulp that provides nourishment for the growing feather, but contributes no cells to its structure (Fig. 3-9).

As the malpighian layer develops, it differentiates into three layers: (1) a thin outer layer that forms a protective keratin sheath about the developing feather, (2) a thicker middle layer composed of large, rapidly growing epidermal cells that grow into the main structure of the feather, and (3) an internal, thin layer of cells that surround and protect the delicate dermal

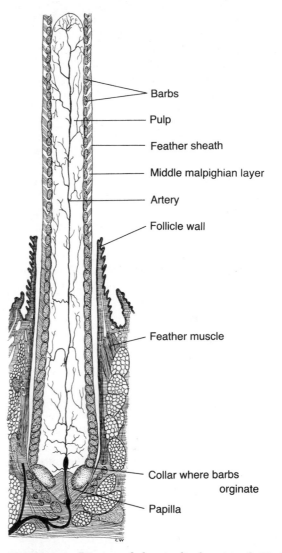

FIGURE 3–9 Diagram of a longitudinal section of a 21-day-old regenerating breast feather of a White Leghorn rooster. After Lillie, 1940.

pulp. These last cells are represented in the completed feather only as a series of hollow pithy caps inside the transparent quill.

The first sign of differentiation into a contour feather appears deep in the follicle in the collar of rapidly growing cells in the middle malpighian layer (Fig. 3-10). These cells begin to grow into a cylinder of parallel longitudinal ridges, something like pick-

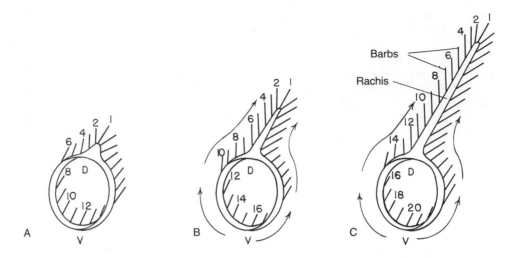

FIGURE 3–10 Diagrams showing successive stages in the growth of a contour feather from the collar of malpighian cells in a feather follicle. Barbs, which first appear in the ventral (V) part of the collar, migrate dorsally (D) and then outward. They are numbered in the order of their appearance. After Lillie and Juhn, 1938.

ets in a cylindrical fence. These pickets will become the barbs of the mature feather. As growth continues, two movements are discernible: the barbs grow in length, and the embryonic malpighian cells from which they originate migrate bilaterally from the ventral to the dorsal side of the collar, where the bases of the barbs unite to form an outward-growing projection, the rachis (Lillie and Juhn, 1938). Every barb originates at this ventral part of the collar. Tension between the surrounding sheath (layer 1 before), and the more rapidly growing middle malpighian layer (layer 2), is thought to effect a secondary splitting of the barbs into the barbules. If the bases of the barbs do not migrate together to form a rachis, the feather will turn into a down feather with only a single circle of barbs on top of the quill. A secretion from the thyroid gland controls the formation of barbules, since its deficiency inhibits their appearance. However, the interaction between hereditary and hormonal influences in feather growth is extremely complex (Voitkevich, 1966). The relationship between the two varies widely with the phase of feather growth and the age and species of the bird. In general, the thyroid gland seems to be the most influential in regulating feather growth, followed by the gonads and lastly, the pituitary gland (Payne, 1972).

With continued growth of the contour feather, the barbs become arranged on the two sides of the rachis, curving diagonally within the cylindrical sheath (Fig. 3-11). The first barbs to grow become the apical barbs of the completed feather; the last, the basal. Ultimately the feather stops growing. The rachis, barbs, and barbules become horny; the nourishing pulp is resorbed; the inferior umbilicus pinches shut; and the protective sheath splits open and is preened off, releasing the mature feather and permitting it to assume the characteristic flat form that makes the feather vane. The feather, now dead, has no more living traffic with the body. A small remnant of the malpighian and pulp cells remains as an embryonic germ at the base of the follicle after the feather has matured.

A typical feather germ grows a feather until it reaches a definite size, then it stops and the cells become dormant until that feather is molted. But in cocks of the Japanese Phoenix Fowl, a domestic breed of *Gallus gallus*, the central tail feathers may grow continuously up to six years without molting, and may reach a length of 6 meters. Usually, however, feather growth is strongly cyclic, and new feathers grow only at certain seasons. However, if a feather is accidentally removed, even in midwinter, the embryonic germ cells in the follicle will promptly grow a new

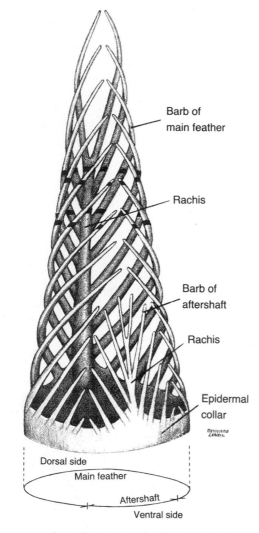

Barb of
main feather

Rachis

Barb of
aftershaft

Rachis

Epidermal
collar

Dorsal side

Main feather

Aftershaft

Ventral side

FIGURE 3–11 A three-dimensional diagram of a developing contour feather along with its after-feather, or aftershaft, which develops from a limited span of the ventral germinal collar. From Lucas and Stettenheim, 1972.

one to replace it. To test the vigor of this regenerative capacity, Pearl and Boring (Stresemann, 1927–1934) pulled out successive feathers from the same follicle on a domestic hen. They found that the follicle would regenerate no more than three times in a given intermolt period. Then the follicle became inactive until the next normal molt period, at which time it resumed activity. It is normally not the loss of a molted feather

that stimulates the growth of a new feather. Rather, at the proper time for molting, the reactivated germinal papilla grows a new feather that pushes out the older, worn feather above it.

Feathers grow at different rates, depending on the species of bird involved, its age, its diet and health, the part of the body concerned, the time of year, and even the time of day, nighttime growth being the slowest. This diurnal rhythm in growth results, in many birds, in feathers with regularly spaced transverse light and dark bands, or growth bars, that vary in width and pigmentation (Wood, 1950).

As Frederick II observed, birds of prey, not subject to as many dangers as "inoffensive" species, can afford to grow their feathers more slowly. However,

In land birds that nest on the ground and to whose young the parents bring no food (because they are able to get it themselves immediately after birth) the growth of feathers is most rapid and soonest completed. . . . Nature has with foresight provided that birds born in nests built on the ground are as early as possible provided with protective plumage and other feathers, because they are exposed to dangers. . . . (Wood and Fyfe, 1943).

Young ground-nesting galliform birds develop their flight feathers most rapidly: some quail are able to fly when only a week old, and mound birds, Megapodiidae, a few hours after hatching. The primaries of the Northern Flicker, *Colaptes auratus*, grow 5 to 7 mm per day. A House Sparrow, *Passer domesticus*, replaced a complete primary in 12 days with an average growth of 4 mm per day; while a Dark-eyed Junco, *Junco hyemalis*, which had a tail feather pulled out in midwinter, grew a new one at the rate of 1.1 mm per day. In general, the feathers of large birds grow in absolute length more rapidly but in relative length less rapidly than the feathers of small birds. For example, the 285-mm wing primary of the Hooded Crow, *Corvus corone*, grows only 1/32 of its length in 24 hours, while the 57-mm primary of the Lesser Whitethroat, *Sylvia curruca*, grows 1/17 of its length in the same time (Stresemann, 1927–1934). Probably the record for rapid feather growth is held by the Wattled Crane, *Bugeranus carunculatus*, whose longest primary grew 480 mm in 54 days, averaging 9 mm per day. Between the 15th and 23rd days it grew almost 13 mm per day (Stresemann and Stresemann, 1966).

Food must be constantly available for a molting bird, "because shortage on only one day is enough to cause the formation on growing feathers of 'fault bars,' where the rachis is pinched and the barbs on the vane are deficient in barbules. The feathers are later liable to break at such points, so that at worst even temporary food shortage may be sufficient to impede the bird's flight for the next year" (Newton, 1972). The nutritional demands of feather growth are reflected in the fact that muscles of a molting Canada Goose, *Branta canadensis*, undergo a decrease in protein content that is not restored until the molt is completed (Hanson, 1963).

Feather Distribution

On primitive birds, feathers were probably rather evenly distributed over the body in a fine checkerboard pattern. Some traces of such an arrangement are still seen in present day species. The bodies of penguins, Ostriches and their relatives (the ratites), screamers, and colies are completely covered by feathers. But on most modern birds, feathers are distributed in scattered patches called pterylae or feather tracts, while the naked intervening regions are called apteria. Within a given pteryla the feathers are usually arranged in definite patterns, often in rows. Among birds with dense plumage such as ducks, the apteria are reduced in size and are often covered with down.

Though species vary, most birds show the following pterylae (Fig. 3–12):

1. Capital tract, which covers the crown of the head.
2. Spinal tract, a quite variable pteryla that runs down the back from the head to the tail.
3. Caudal tract, the pteryla of the tail feathers and the smaller contours that cover them (coverts).
4. Ventral tract, which runs from the throat to the breast, where it divides into two bands (with a naked apterium between them), which in turn continue down the belly to the cloaca, where they unite. Very variable.
5. Humeral tract, which includes feathers of the upper arm and scapular region.
6. Alar tract, which includes the flight feathers and their coverts on the upper arm, forearm, and hand.
7. Femoral tract, which extends diagonally across the thigh.
8. Crural tract, which crosses the remainder of the leg.

The alar and caudal tracts carry the contour feathers responsible for flight. The large feathers of the wings are called the remiges (Latin, *remex*, an oarsman). The largest and most distal remiges are crowded together on the hand, and are called the primaries. Their number varies from 9 to 11. Most passerine birds possess 9 functional primaries on each wing, and most nonpasserines possess 10, but grebes, storks, and flamingos have 11. The secondary remiges are the large quill feathers of the forearm or ulna. They vary in number from 6 in hummingbirds to as many as 40 in albatrosses, but most birds carry between 10 and 20. Many species seem to have a feather missing between the fourth secondary and the next feather toward the body—a situation known as diastataxy. The significance of the feather gap is unknown. Both primary and secondary feathers are anchored firmly to the bones of the hand and forearm of the wing—an arrangement that furthers their control in flight. In addition to these two groups of remiges are groups of smaller contours called coverts or tectrices (Latin, *tegere*, to

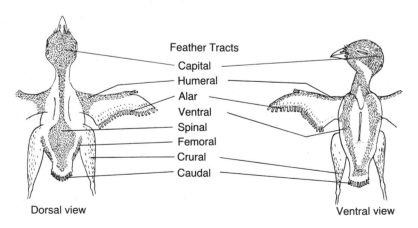

Feather Tracts

Capital
Humeral
Alar
Ventral
Spinal
Femoral
Crural
Caudal

Dorsal view

Ventral view

FIGURE 3–12 Feather tracts of a typical passerine bird.

cover), which act as rows of shingles to cover the quill bases of the larger remiges and to make aerodynamically smooth the upper and lower surfaces of the arm and hand. Attached to the bird's thumb are three or four short vaned feathers that make a separate little wing, the alula or spurious wing. The alula is especially prominent in wings that are short and rounded, such as those of gallinaceous birds. It increases the aerodynamic efficiency of the wing at the slow flight speeds common in ground dwellers.

On the rump are inserted the large tail feathers or rectrices (Latin, *rectrix*, a ruler or leader). These are usually straight, bilaterally paired feathers that range in number from 8 in anis to 24 in pheasants, but most commonly number 10 or 12. They are mainly concerned with steering and maintaining equilibrium in flight. In species that live near the ground they are usually quite short, but in expert flyers that maneuver with great skill they are relatively long.

The Number of Feathers

Birds vary in the number of feathers they carry, according to species, size, sex, age, metabolism, health, season, and geographic distribution. One might expect that adult birds of a given species taken at a given time of year would have a rather stable number of feathers. This was shown not to be the case in Wetmore's (1936) study of the number of contour feathers in birds. For example, two female Song Sparrows, *Melospiza melodia*, on March 5 had 2208 and 2093 feathers, respectively; two female Fox Sparrows, *Passerella iliaca*, on March 26, had 2648 and 2482. As mentioned earlier, birds carry more feathers in winter than in summer: two White-throated Sparrows, *Zonotrichia albicollis*, on February 22 had (male) 2556 and (female) 2710 feathers, respectively, while two females taken October 4 and 8 had 1545 and 1508 feathers, respectively.

Body size is the factor responsible for the greatest variation in feather number. A Ruby-throated Hummingbird, *Archilochus colubris*, had 940 contour feathers (June) and a Tundra Swan, *Cygnus columbianus*, had 25,216 (November). Actually, the hummingbird possesses more feathers (335) per gram of body weight than the swan (ca. 4), which is to be expected since small bodies have relatively more heat-losing surface per unit of weight than large bodies. Perhaps the densest concentration of feathers is to be found in penguins.

A more significant measure of the relationship of body size to plumage was made by Turček (1966). He weighed the plumage of 249 birds of 91 species and found that their mean plumage weight was 6.0 per cent of their mean total body weight (minus stomach contents). For the smaller birds the mean was 7.1 per cent—which suggests a relative increase in their plumage as a heat-conserving adaptation.

Further evidence of the importance of feathers to a bird is seen in their weight. Although they are "light as a feather," the total bird is also light, so that, relatively, feathers make up a sizable part of a bird's substance. A Bald Eagle, *Haliaeetus leucocephalus*, that weighed 4082 g was found by Brodkorb (1955) to possess 7182 contour feathers that weighed 586 g, which was 14 per cent of the bird's total weight. The down feathers weighed an additional 91 g. The contour feathers alone weighed more than twice as much as the bird's skeleton, which was 272 g in weight.

The Molting of Feathers

As feathers become worn, they loosen in their follicles and drop out, pushed by the already growing feathers underneath them. The prime function of molt is, of course, the replacement of worn feathers, but in many species a partial molt, just before the mating season, provides brightly colored courtship plumage even though the old feathers do not need replacement. Molting may also serve a hygienic function. Seaside Sparrows, *Ammospiza maritima*, which molt only once each year, have heavier infestations of bird lice (mallophaga) than Sharp-tailed Sparrows, *A. caudacuta*, which live in the same habitat but have two molts a year (Post and Enders, 1970).

As a rule, the persistence of feathers on a bird, especially the wing primaries, is related to their resistance to wear. Long distance migrants and birds living among bushes, thorns, or coarse grass typically have two molts a year, whereas sedentary or open habitat species are more likely to molt only once a year. In Europe, the Short-toed Lark, *Calandrella cinerea*, has but one molt, in summer; but an Asian race of this species that lives in windy, abrasive deserts has an additional molt of its body feathers in spring (Whistler, 1941).

Adult birds commonly molt and renew all or most of their feathers once a year, usually immediately after the breeding season. However, many species also have a partial molt, mainly of body feathers, just before the breeding season. In young birds, plumage is regularly replaced as they reach adult size. But there are many variations in molting, usually for good reasons. Possibly no other aspect of growth in birds shows such marvelous plasticity and adaptability as molting. This sensitivity of molting to environmental conditions has been widely explored by Erwin and Vesta Stresemann. Much of what follows is based on their monograph (Stresemann and Stresemann, 1966).

Since the two chief functions of feathers—protection and flight—are indispensable, the molting of feathers must occur in such a way as to disturb these functions as little as possible. The energy demands of molting are heavy, especially if the new plumage grows rapidly, but so are the energy demands of reproduction and, for many species, migration. The main problem in molting, then, is chronological: to schedule it and these other peak demands for energy so as to avoid disastrous overlapping, and to fit them into suitable times of the year when food is abundant. Molt-breeding overlap has been documented in many tropical bird species by Foster (1974, 1975). Overlap is rendered energetically possible due to reduced clutch size, and may extend breeding periods and permit more potential renestings.

The heavy energy drain of molting is well illustrated in penguins, which molt all of their short feathers at once. Macaroni Penguins, *Eudyptes chrysolophus*, and Rockhopper Penguins, *E. crestatus*, lose 46 per cent and 43 per cent, respectively, of their body weight during molting when they undergo a period of fast (Brown, 1985).

To prepare for their autumn flights, many migratory species must replace their feathers in a shorter time than sedentary species and, accordingly, use more energy per day in doing so. For example, two European migratory species, the Chaffinch, *Fringilla coelebs*, and Common Redpoll, *Carduelis flammea*, during molt increase their daily energy consumption by 31.8 per cent and 46.4 per cent, respectively; the nonmigratory Bullfinch, *Pyrrhula pyrrhula*, and House Sparrow, *Passer domesticus*, by 19.2 per cent and 24.9 per cent, respectively (Payne, 1972).

The mutual exclusiveness of molting and migration is shown by the fact that long distance migration "is practically never performed as long as a primary or secondary is growing, but nearly always with a complete wing . . ." (Stresemann and Stresemann, 1966). Likewise, molting and reproductive activities generally avoid competing for energy, even in the tropics where food supplies are less seasonal than elsewhere. From 8 to 20.2 per cent of the birds in various tropical avifaunal studies showed overlap (Foster, 1975).

PATTERNS OF MOLT. Feathers do not fall out of the body in a random or haphazard way, or all at once (with a few exceptions), but in accordance with a definite and bilaterally symmetrical pattern. This pattern is "deep-rooted and ancestral" and remarkably constant in any given species, but not necessarily so in families or even genera. The timing and tempo of molting are highly flexible and adaptive.

For purposes of studying and describing molt patterns the important primary and secondary flight feathers of the arm (the remiges), and the tail feathers (rectrices), are given numbers (Fig. 3-13). The primary wing feathers of the hand are numbered from the carpal joint, or wrist, outward: numbers one through six mounted on the fused metacarpal bones; number seven on the third finger or digit; numbers eight through ten (or eleven) on the second digit at the outer tip of the wing. The smaller alula feathers on the thumb are not considered primaries.

The secondary feathers, supported on the ulna of

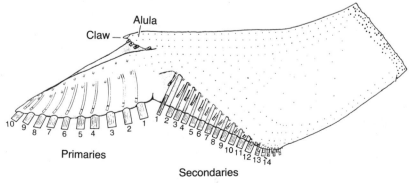

FIGURE 3–13 Dorsal view of the clipped wing of a quail showing the standard numbering of primary and secondary wing feathers. From Ohmart, 1967.

the arm, also begin with number one at the wrist, and run inward along the forearm toward the body for as many secondaries as the bird may have. The innermost tail feathers of each half of the tail begin with number one and proceed outward.

The commonest pattern of primary molt, especially in passerines, is called descending or centrifugal molt, in which feathers drop off beginning with number one at the wrist joint, proceeding in sequence outward. The first secondaries to molt are those at the two ends of the arm—the wrist and the elbow—and the last, those in the center of the arm (Fig. 3-14). Tail feathers commonly molt centrifugally, or from the center outwards, the outer feathers falling when the innermost are completely replaced.

But there are many exceptions to this pattern: ascending molt (rails), centripetal (large owls), synchronous (ducks), and even chaotic (turacos). In the Blue-footed Booby, *Sula nebouxii*, every other tail feather falls during a molt; each new feather then grows beside an old one. In many species replacement of primaries goes so slowly that a new molting cycle begins before the previous cycle ends. Terns molt their first hand feathers (number one at the wrist) at the end of the reproductive period. Molt proceeds as far as number six in about eight months, so that by the end of the next breeding season the outer primaries will be renewed once but the inner primaries twice; or the outer two times to the inner three times. This is called a step-molt.

The genetic plasticity of molt patterns is shown by the fact that the domestic Chicken, *Gallus gallus*, molts its tail feathers centrifugally, while its close relatives the Junglefowl and pheasants molt centripetally (Lucas and Stettenheim, 1972).

ADAPTIVE MODIFICATIONS OF MOLT. Many water birds (ducks, geese, swans, loons, anhingas, flamingos, and most rails, pelicans, jaçanas, auks and grebes)

undergo synchronous molts in which all primaries are lost simultaneously. Synchronous molts never occur in tree-perching birds. Large water birds, such as loons, which have heavy wing-loading (i.e., heavy bodies with relatively small wings), could fly poorly or not at all should one or two wing feathers be missing, as would happen under the standard molting pattern. Consequently, as an adaptive solution to this problem, such birds get the job of molting over with as soon as possible by molting and renewing their primaries simultaneously. They compensate for the hazards of flightlessness by retreating to isolated bodies of water where they can find food and escape enemies by swimming on and under the water. Huge flocks of ducks, geese, and swans commonly undertake special molt migrations whereby they seek sheltered retreats for the safe renewal of their flight feathers. Male ducks, so colorful in their breeding plumage, immediately after the postbreeding molt take on a dull, inconspicuous coat, the "eclipse plumage," which they carry during the period of wing feather renewal. The female alone incubates and cares for the young. Accordingly, she postpones her postbreeding molt until the young are nearly independent.

Temperate zone ducks form new pair bonds each year, whereas geese mate for life. Ducks have two body molts and two plumages a year; geese have only one. The reason seems clear: each year the male duck must court and attract a female anew with his colorful feathers and elaborate courtship rituals. To this end he is provided with an extra prenuptial coat of feathers (Johnsgard, 1965). Unlike the female duck, the female of several African hornbill species loses nearly all her feathers and then quickly replaces them while incubating and brooding, imprisoned in her walled-in nest cavity where she is fed by her mate (Kemp, 1978). The male, however, and the nonbreeding female retain enough feathers to be able to fly. Likewise, the females of the Eurasian Sparrowhawk, *Accipiter nisus*, and

FIGURE 3–14 In most passerines, molting of the flight feathers progresses from the wrist outward to the wing-tip, while secondaries drop off first at the two ends of the forearm, and last in the center.

Osprey, *Pandion haliaëtus*, molt their feathers during the quiet incubation and brooding times, while the males, who continue bringing food to the nest, postpone their molts until the young are flown (Heinroth, 1938). Obviously, natural selection has had a hand in scheduling these sex differences in molting times.

The tempo with which birds replace their feathers varies greatly, often in an ecologically adaptive manner. Birds that molt synchronously usually renew their feathers rapidly: two to six weeks in penguins, for example. At the other extreme, some cranes and eagles require two years for one complete change of feathers, and Barn Owls, *Tyto alba*, even more. The speed of molting is obviously related to the number of molts per year; this may be as many as four in a partridge.

Latitude, or more properly day length and climate, profoundly affect the tempo of molt. The duration of the postnuptial molt in the White-crowned Sparrow, *Zonotrichia leucophrys*, "decreases northward by an average of 2.6 days per degree of latitude between the southernmost (35.2° N, molt duration 83 days) and northernmost (48.9° N, 47 days) limits of the breeding range" (Mewaldt and King, 1978). The Dunlin, *Calidris alpina*, of artic regions, first raises its young and then rapidly molts and replaces its feathers in about two months, whereas the Purple Sandpiper, *Calidris maritima*, which molts in the tropics or subtropics, leisurely occupies four or five months in the renewal of primaries. Similarly, the speed of molt may adapt to the time of year it begins. A study of molting House Sparrows, *Passer domesticus*, in southern Finland showed that molting of primary feathers lasted 83 days if begun before the end of August, and 64 days if begun later (Haukioja and Reponen, 1969).

The following examples, largely from the Stresemanns (1966), show how natural selection has adapted molting in various ingenious ways to the requirements of migration. The European warblers of the genera *Sylvia* and *Phylloscopus* have but one molt a year in sedentary, resident species, but two complete molts per year in those species or races that migrate long distances. Populations of the Peregrine Falcon, *Falco peregrinus*, in southern Europe are relatively sedentary, lay their eggs in April, and molt between then and September or October. Arctic populations of this species are strong migrants. They lay their eggs in June and migrate to the tropics in August or September. Since not enough time

is available for a full wing molt in the arctic, the birds interrupt their molt for the migration period and resume it again in winter quarters, finishing the process in March or April.

A special version of the interrupted molt is found in the Eurasian Yellow-breasted Bunting, *Emberiza aureola*, which breeds across the breadth of Eurasia and winters in southeast Asia and Indonesia. Those southerly populations with the shortest migration routes renew their entire plumage in late summer on the breeding grounds and leave for winter quarters six to nine weeks later than the more northerly populations. The latter leave their breeding grounds in August with their old, worn feathers, interrupt their migration in the lower Yangtse area of China, and there undergo a complete molt in August to September, flying on with new feathers to join the other population in winter quarters. Races of many other species show this same type of interrupted migration for a molting pause, half-way between breeding grounds and winter quarters.

There are, of course, many species that migrate to winter quarters in worn feathers and there undergo a complete plumage replacement at their leisure. Still other northerly breeders rapidly put on a complete set of new wing feathers before migrating south, e.g., the Ruff, *Philomachus pugnax*, and the Thrush Nightingale, *Luscinia luscinia*. For a minority of species in the arctic the breeding season is so short that they are compelled to overlap breeding and molting and begin renewing their wing feathers while laying eggs: for example, the Icelandic populations of the Greater Golden Plover, *Pluvialis apricaria*, and Franklin's Gull, *Larus pipixcan*. In the even shorter antarctic summer, nonmigratory petrels are compelled to grow their new plumage during the breeding cycle. Although molting birds in polar regions must cope with a very short season, compensations in the form of long days, abundant protein-rich food, and the relative absence of predators all promote rapid feather growth.

Another sort of interrupted molt occurs in certain birds known as facultative or opportunistic breeders. Such birds live in regions with unpredictable rainy seasons and are prepared to breed whenever the rains (and food) come. At such times they may interrupt a molt already in progress (Payne, 1972).

Adaptive changes occur also in the molt sequence of feathers. A remarkable example of this, cited by the Stresemanns, is seen in the African Standard-winged Nightjar, *Macrodipteryx longipennis*. During the courtship season the number two primary feather of each wing of the male projects about 35 cm beyond its neighbors in the form of a slender-stemmed flag (Fig. 3–15). When courtship-flight displays cease, the male bites off the projecting parts of these feathers, leaving the remainders in the wings. During the spring molting season the female molts her feathers, as do all other Caprimulgidae, in the descending pattern: number one first and number ten last. But the male *longipennis* begins his molt in the wing's center with numbers five and six and ends it with number ten and the stump of number two, with the result that he begins the next breeding season with brand-new courtship flags.

The time interval between the loss of a feather and the complete growth of its replacement varies according to species, to position on the body, and to environmental conditions. Tail feathers of the Eurasian Kestrel, *Falco tinnunculus*, require 50 days; the more essential primaries, 45 days. Primaries of the no-longer-wild domestic chicken, *Gallus gallus*, require 70 to 90 days; of the Gray Partridge, *Perdix perdix*, 30 days. For water birds with synchronous molts the interval between feather loss and regained flight ability is about three weeks in small ducks, four in larger ducks, five in geese, and six to seven in swans (Heinroth and Heinroth, 1924–1933).

PLUMAGE SEQUENCE. The kinds of feathers that grow after old feathers are cast off commonly vary, especially in color, according to the age of the bird and the season of the year. For this reason descriptive

terms have been proposed by Dwight (1900) for naming the sequence of plumages and molts as applied particularly to birds of the north temperate zone. A domestic chick, for example, hatches out clothed in natal down. When shed, this is succeeded by the juvenal plumage, which is followed in sequence by the first winter (or nonbreeding) plumage, the first nuptial (breeding) plumage, the second winter plumage, and so on.

A substitute terminology for plumages and molts has been proposed by Humphrey and Parkes (1959) in which the term "basic" is used to describe the plumage when it is the only one in the annual cycle or the winter plumage when there are two or three; "alternate" describes the second, or nuptial, plumage, and "supplemental" is used for a third, if such exists.

The second winter plumage (or, in some small species, the first) is essentially the adult plumage in most birds. It follows the first postnuptial molt, which is a complete annual molt in most birds. A few species require more than two years to reach adult plumage: Herring Gull, *Larus argentatus*, four years; albatrosses, seven to eight years; lyrebirds, eight years; Bald Eagle, *Haliaeetus leucocephalus*, five or more years.

It is clear that both innate and environmental stimuli are involved in molting. Different species of birds living under approximately identical conditions have quite different molting patterns and schedules. However, different populations and/or races of Rufous-collared Sparrows, *Zonotrichia capensis* and White-crowned Sparrows, *Zonotrichia leucophrys*, have quite different schedules and degrees of molting, depend-

FIGURE 3–15 The male African Standard-winged Nightjar. The extravagantly long feathers grow and molt at exceptional but highly adaptive times for courtship effectiveness.

TABLE 3–1.
SEQUENCE OF PLUMAGES*

PLUMAGE	LOST BY
Natal down	Postnatal molt
Juvenal plumage	Postjuvenal molt
First winter plumage	First prenuptial molt
First nuptial plumage	First postnuptial molt
Second winter plumage	Second prenuptial molt
Second nuptial plumage	Second postnuptial molt
and so on	and so on

*Cramp and Simmons, pp. 30–31

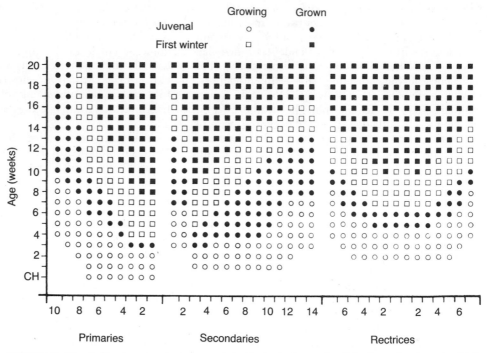

FIGURE 3-16 A diagram showing, for each individual wing and tail feather, the time at which a juvenal feather is replaced with an adult, first winter plumage feather in the California Quail. Note that primary wing feathers are numbered from the wrist outwards; secondaries, from the wrist inwards; tail feathers, from the center outwards. Compare the molt sequence of the secondaries with that of passerines (Fig. 3–14). *CH* = condition at hatching. After Ohmart, 1967.

ing on geographical latitudes (King, 1976; Mewaldt and King, 1978). Tropical ducks typically molt once a year, temperate zone ducks, twice. Hybrids between the two have two molts a year (Stresemann, 1940). These facts also clearly demonstrate a hereditary influence on molting.

THE MECHANISMS OF MOLTING. A variety of experiments have been performed in an attempt to discover the physiological mechanisms that control molting. The results suggest that there are probably a variety of control mechanisms rather than a single one, with different mechanisms in different species. For example, injection of the sex hormone progesterone in fowls caused premature molting (Shaffner, 1954); castration in a sparrow delayed molt (Morton and Mewaldt, 1962); lower air temperatures stimulated more rapid molt in ptarmigan (Salomonsen, 1939); artificially long days in midwinter brought on spring and summer

plumages in ptarmigan (Höst, 1942), but seemed to have no effect on an autonomous molt cycle in sparrows (King, 1968); the hatching of young governed postnuptial molt in quail (Raitt, 1961); alternation of wet and dry seasons seemed to control molting in sparrows (Miller, 1961) and in hummingbirds (Wagner, 1957). Even fright causes some species suddenly to lose their feathers (Dathe, 1955).

Faced with such bewildering data, Juhn (1963) concluded that

molt is essentially an autonomous process, the primary seat of the cyclical renewals being the feather papilla proper. Within limits, one or another environmental factor could well participate as seasonal trigger. This however need not necessarily be because of secretory surges but because of an autonomously changing physiology of the feather papilla which at this time exposes the organ to previously ineffective levels.

For example, less and less thyroid extract is neces-

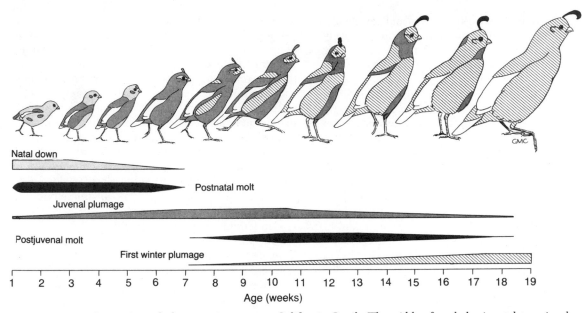

FIGURE 3–17 Succession of plumages in a young California Quail. The width of each horizontal tapering bar indicates the proportion of a given plumage worn by the bird at any given age. From Raitt, 1961.

sary to stimulate a feather papilla, the closer to the date of normal molt the treatment is made. Among various possible molt-controlling mechanisms, Payne (1972) suggests that the high level of sex hormones in breeding birds inhibits the production of thyrox-

in. At the end of the breeding season the inhibition ceases, and a surge of thyroxin stimulates molting and feather renewal. Since for many species molt is under photoperiodic (day-length) control, Murton and Westwood (1977) suggest that it is regulated through pitu-

FIGURE 3–18 Successive plumages in the first two years of a Rose-breasted Grosbeak. After Pettingill, 1970.

itary gland stimulation of the thyroid and sex glands. Their several secretions, in turn, act cooperatively (synergistically) and in phases to carry out the molting process. Much experimental work remains to be done in this intriguing field.

Color in Birds

Birds are more vividly colored than any other class of vertebrates, except, perhaps, coral reef fishes. Since they are able to fly from their enemies and to perch and nest in relatively inaccessible sites, they are somewhat removed from the selective pressures that impel earthbound animals to don inconspicuous coats. Further, their high metabolism produces waste products that may be discarded in the form of feather, skin, or egg pigments.

Colors in birds depend on those wave lengths, or colors, of light that are reflected to the eyes of the observer, and these, in turn, are those "left over" after other wave lengths are removed from the incident white daylight that falls on the bird. Colors are removed in two ways: by pigments or by physical structure.

Pigments are of three main sorts, melanins, carotenoids, and porphyrins (Brush, 1978). Melanins produce black, dull yellow, red, and brown colors. They occur in the form of sharply outlined, microscopic particles. They are nonsoluble in organic solvents, and therefore are not well known chemically. However, they appear to be the result of oxidation of a colorless chromogen, tyrosine, by a copper-containing enzyme, tyrosinase. The amount and kind of melanin formed depends on the amount of various amino acids in the diet, especially tyrosine. Birds that are black require large amounts of riboflavin in their diets.

Melanin granules average about 1 micron in diameter and occur in two forms. The eumelanins, which under the microscope appear as rod-like particles, produce the blacks and grays of a bird's feathers. The phaeomelanins are in the form of oval granules, and they produce the browns, red-browns, and yellow-browns.

Cells known as melanoblasts, derived from the neural crest of the embryo, migrate to the skin and feather papilla where they transform into melanophore cells, or melanocytes, that actively deposit pigment as the feather begins to grow. Melanin granules are formed in the cytoplasm of ameba-like melanophore cells residing in the inner layer of malpighian cells of the feather germ. Long processes from each melanophore reach out to distribute the pigment granules into the developing barbs and barbules. Although both types of melanin granules may be found in a single feather, with the eumelanins characteristically at the tip and the phaeomelanins at the base, a given melanophore will form only one type of pigment. The precise fidelity with which a given melanophore produces its pigment is demonstrated in a fusion papilla, or chimaera. These arise on the junction of a piece of grafted skin and the surrounding host skin of two birds of different pigmentation. Melanin is never distributed into those parts of a "combined" feather that developed from the fraction of the papilla derived from an albino bird (Lillie, 1940).

Carotenoids (related to vitamin A), produce red, yellow, and orange colors. These pigments are derived from dietary sources, but their deposition is under hormonal control (Brush and Power, 1976). Yellow-shafted eastern and Red-shafted western populations of Northern Flickers, *Colaptes auratus*, grew pale feathers in captivity. Raw grated carrots were added to their diet, and after the next molt western birds once again grew red shafts and eastern birds yellow shafts indicating geographical differences in the metabolism of carotenoids (Test, 1969).

A widely distributed carotenoid is zoonerythrin, the red pigment in many birds such as the Northern Cardinal, *Cardinalis cardinalis*. Zooxanthin is the carotenoid that produces the bright yellow colors in canaries, orioles, and many other species, as well as the yellow beaks and feet of many water birds.

Porphyrins have been found in the feathers of at least one-fourth of all birds (Brush, 1978). The copper-containing turacin (red) and turacoverdin (green) found in turacos are examples of these pigments.

Structural colors of birds' feathers depend either upon interference phenomena, which produce shimmering, iridescent colors (as in a soap-bubble) that change hue with the angle of view, or upon the scattering of the shorter wave lengths of light by small vacuoles or alveoli within the feather—vacuoles smaller than the wave length of red light, or less than about 0.6 micron. In this latter type of color production, since only the shorter wave lengths of incident light will be reflected, the apparent color of such feathers will be blue, and the color will remain the same regardless of the angle of vision (Vevers, 1964). This scattering phenomenon is called Tyndall scattering and is responsible for the blue color of the sky. Blue plumage in most birds is based on this principle (Fig. 3–19).

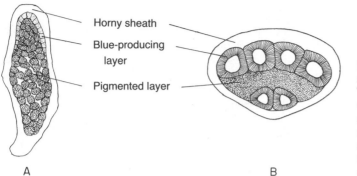

Horny sheath

Blue-producing layer

Pigmented layer

A

B

FIGURE 3–19 Blue colors usually depend on the structure of feathers and rarely on pigments. *A,* The cross-section of a barb of a Blue Jay feather in which a layer of transparent cells containing fine, suspended particles is thought to produce the blue reflected light. After Gower. *B,* A barb of a Tanager in cross-section. Very fine tubes surrounding air chambers are apparently the source of the blue color. After Stresemann, 1927–1934.

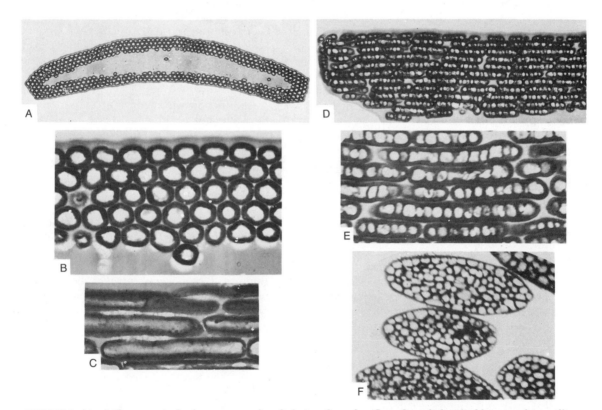

FIGURE 3–20 Iridescence in feathers occurs when light is refracted, reflected, and absorbed by countless millions of tiny particles in the keratin of barbs and barbules. At the left (*A, B, C*) are electron micrographs of the responsible particles in feathers of an African Glossy Starling, *Cinnyricinclus;* at the right (*D, E, F*) are those of a typical Hummingbird. *A,* Cross-section of a starling barbule, ×6000; *B,* Cross- and *C,* Longitudinal-sections of the gas-filled melanin tubules ×30,000. *D,* Cross-section of about one-half of a hummingbird barbule, ×16,000. *E,* Cross-section of six layers of disc-shaped particles, ×45,000; *F,* A surface, or plan, view of several discs, ×45,000. In each bird the multiplicity of the light-bending particles intensifies the brilliant iridescence. *A, B, C,* from Durrer and Villiger, 1970. *D, E, F,* from Greenewalt, 1960.

The underparts of feathers showing physical colors are usually heavily pigmented with melanin that absorbs some wave lengths of light, thus intensifying the other wave lengths reflected from the upper surface.

Interference or iridescent colors are produced when different wave lengths of light are refracted and reflected within the barbs and barbules in such a way that certain wave lengths are mutually reinforced or intensified, while others mutually "interfere" and are nullified. Various forms of fine structure within the feather bring this about. The iridescent barbules of the Common Peacock, *Pavo cristatus*, are covered with three extremely thin layers of horny keratin, each about 0.4 micron, or 1/60,000 inch thick. These reflect colors much as does the thin film of a soap bubble. The iridescent barbules of hummingbird feathers are twisted or angled so that their flat sides face the eyes of the viewer. Near the surface of each barbule are three or more films, each film consisting of a mosaic of oval platelets. Each platelet is about 2.5 microns long, 1 micron wide, and has an optical thickness of one-half the peak wave length of the reflected color. The platelets contain dozens of tiny air bubbles encased in a matrix having a refractive index of two. The color reflected varies with the size of the air bubbles and the thickness of the platelets (Greenewalt, 1960). Still another mechanism produces the intense iridescence of African glossy starlings (Durrer and Villiger, 1970, 1975). For example, in the Violet-backed Starling, *Cinnyricinclus leucogaster*, the barbules contain countless gas-filled tubes of melanin, each about 0.27 by 1.6 microns with walls about 0.066 micron thick (Fig. 3–20). These tubes are packed in about five rather precise layers, much like cigars in a box. The repetitive refraction and reflection of incident light by these layers result in brilliant reflected blue-violet and coppery colors. The specific color reflected seems to be mainly due to the distance separating the tube layers: 0.13 micron produces bluish violet; 0.17 to 0.19 micron, greenish yellow.

The dull velvet appearance of some ornamental feathers is physically the reverse of iridescence. If pigmented feathers, or their barbs or barbules, are vertically directed, they will absorb rather than reflect the light.

A given feather, even a single barbule, may contain different pigments. The various factors responsible for color at times cooperate to produce combination colors. For example, red and yellow pigments commonly produce orange; green is often the result of a yellow sheath of keratin on top of structural blue. This is clearly shown in the tropical Green Magpie, *Cissa chinensis*, whose plumage is green in forest populations but blue among birds living in open country where sunlight fades the superficial yellow pigment. Color intensity is usually a matter of pigment concentration. Eumelanin granules densely packed produce a black feather, and thinly scattered, a blackish-brown feather. A complete lack of pigment results in a white feather.

A feather once formed has no living exchanges with the body and hence can receive no new pigments; nevertheless, its color may change. This is commonly because of abrasion from physical wear and the bleaching effects of sunlight. If the tips of feathers are colored differently from their remaining portions, their wearing away will expose different colors to the viewer (Fig. 3–21). Thus, the nuptial black bib of the House Sparrow, *Passer domesticus*, is the result of wintertime erosion of the lighter-colored tips of the breast feathers. Similar color changes occur in the European Brambling, *Fringilla montifringilla*, the Snow Bunting, *Plectrophenax nivalis*, the Common Redpoll, *Carduelis flammea*, the Bobolink, *Dolichonyx oryzivorus*, the European Starling, *Sturnus vulgaris*, and the Wandering Albatross, *Diomedea exulans*.

FIGURE 3–21 The Snow Bunting changes its color from brown in October to black and white in June because of the wintertime erosion of feather tips. After Chapman, 1912.

October   January   March   June

Pigmented feathers are always more resistant to wear than unpigmented ones. With this fact in mind it is not surprising that the wing-tip primaries of many otherwise white birds are heavily pigmented: for example, storks, gulls, flamingos, herons, and pelicans. It is not the pigment itself that contributes extra strength to feathers. Rather, wherever pigment granules accumulate, it seems "the quantity of viscous keratinized substance present also increases . . . ," and it is this hardened keratin that strengthens feathers (Voitkevich, 1966).

Bleaching of the dark feathers of the South Polar Skua, *Catharacta maccormicki*, occurs rapidly under the actinic rays of 24-hour sunlight in a dustless polar atmosphere. Red porphyrin colors are particularly sensitive to light. The pink breast feathers of certain gulls fade very rapidly when exposed to light, and the pale red down feathers of bustards turn into gray-white in a few minutes when exposed directly to sunlight.

In addition to changes caused by abrasion and bleaching, mature feathers may change coloration as a result of adherent or "cosmetic" colors. Preen oil may change a pelican's white throat to yellow, or the white breast of a gull to pink. The yellowish underparts of the Bearded Vulture, *Gypaetus barbatus*, and many water birds are caused by rust or iron oxide in their environments, and the darker colors of many urban birds are due to soot and other industrial by-products.

Diet undeniably affects feather color in numerous species. Greater Flamingos, *Phoenicopterus ruber*, ordinarily lose their red color in captivity and do not regain it in subsequent molts. But the red can be restored by feeding them certain small crustaceans or even pure dyes, particularly canthaxanthin. Bullfinches, *Pyrrhula pyrrhula*, take on a very dark color if fed a diet of hemp seed. Captive Red Crossbills, *Loxia curvirostra*, that have faded to yellow will grow normal red feathers at their next molt if fed a diet containing rodoxanthin from the yew. Canaries will, in successive molts, gradually change from yellow to intense orange if fed red peppers (Beebe, 1906). Natives in various parts of the world change the colors of parrots by chemical treatment. For example, the green Festive Parrot, *Amazona festiva*, will grow yellow instead of green head feathers if the sprouting feathers are rubbed with the skin secretions of the toad *Bufo tinctorius* (Wallace, 1889). It seems quite possible that variations in the coloration of different geographical races of birds may in part be due to varying amounts of certain foods, or even trace elements, available to them, but the problem needs further study.

Some zoologists believe that humidity is important in determining pigment intensity. It seems a fairly general rule that birds such as pipits, larks, and shrikes are darker when living in moist regions than in dry ones.

Age brings changes in coloration to birds as to other animals, including man. The larger gulls and albatrosses require several years before they assume adult plumage. Full coloration in the Bald Eagle, *Haliaeetus leucocephalus*, does not appear until the fifth year; and in the California Condor, *Gymnogyps californianus*, until the fifth or sixth year. The eyes of accipitrine hawks change from yellow to orange with age; and the leg color of the American Coot, *Fulica americana*, changes in successive years from blue-green to yellow-green to clear yellow, and finally to red-orange in mature adults.

HORMONES AND COLORATION. Adult coloration in many birds coincides with the advent of sexual maturity—a fact that suggests that sex hormones may play a part in plumage coloration. Like many other birds, the Eclectus Parrot, *Eclectus roratus*, shows sexual dimorphism in its coloring: the males are brilliant green, the females, bright red. Castrated males do not lose their colors, but castrated females take on typical male coloring at their next molt. Thus, the female hormone inhibits male coloration. The same situation prevails in the Mallard Duck, *Anas platyrhynchos*. Injection of female hormone into the male changes its plumage to the female type. Different feathers, however, respond differently to such treatment, secondary wing feathers requiring a higher concentration of the female hormone than the primaries.

In phalaropes the nuptial plumage is more colorful in females than in males. Paradoxically, the brighter colors of the female depend on the male hormone, testosterone, which, in most birds, is produced in quantity by the testes of the male. But in, for example, the Red-necked Phalarope, *Phalaropus lobatus*, testosterone occurs in greater concentration in the female's ovaries than in the male's testes (Höhn, 1970). However, in the Ring-necked Pheasant, *Phasianus colchicus*, hormone experiments demonstrated that sex differences in plumage depended both on genetic and hormonal factors (Morejohn and Genelly, 1961; Bird, 1983.) Male plumage in the common House Sparrow, *Passer domesticus*, is determined completely by heredity; castrated males still regenerate normal male plumage.

Precisely how hormones control pigment production has yet to be determined. Some melanophore cells growing in tissue culture are directly sensitive to sex hormones and to antithyroid hormones (Mayaud, 1950). In response to the injection of the gonadotropic or interstitial-cell-stimulating hormone of the pituitary gland, melanophores appeared in regenerating feathers of weaver birds (Ploceidae) within 12 to 50 hours after injection (Ralph *et al.*, 1967).

ABNORMAL COLORATION. Within hereditary limits there may be wide variations in plumage color due to sex, age, diet, disease, temperature, humidity, solar bleaching, abrasion, and other influences. Still other changes occur when the normal hereditary mechanism is disturbed. When this happens, a bird may experience either a lack or an excess of pigmentation. When the lack of pigment is complete, the result is albinism. This may be total albinism (over the entire body, including the eyes), or partial albinism (parts of the body affected). Albinism seems to be mainly due to the genetic absence of the enzyme tyrosinase. Albino birds have been recorded from 163 species of 42 families in Great Britain (Sage, 1963), and 304 species of 54 families of North American birds (Gross, 1965a). It seems most common in those species that are either social or fairly sedentary—both situations conducive to inbreeding. If only a single pigment is lacking, e.g., the melanins, a bird may change its color from, say, a dark to a light yellow, retaining only its carotenoid pigments. This occurs commonly in canaries.

An abnormal excess of pigments is much rarer than albinism. Melanism is said to occur when a bird receives an excess of eumelanin pigments. Gross (1965b) found abnormal melanism in only 29 species of 15 families of North American birds. A normal form of melanism regularly occurs in certain species, especially among the Falconiformes. In this form two phases of plumage, a light and a dark, may appear within a given brood of young. This type of genetic variation in color is known as dichromatism. Melanism may also become established within an isolated population of a species to a degree that completely replaces the normal coloration. This has occurred in the Bananaquit, *Coereba flaveola*, on some islands off Venezuela and on Grenada and St. Vincent in the West Indies. On Grenada the black morph occupies a predominantly wet habitat and the yellow form, dry deciduous areas (Wunderle, 1981).

Erythrism in a bird results from an excess of red pigment. This too is rare. It occurs most frequently in the Galliformes, especially grouse and quail. A normal form of dichromatic erythrism appears in the Eastern Screech Owl, *Otus asio*, and the Tawny Owl, *Strix aluco*; both gray and reddish-brown birds may appear in the same brood.

Xanthochromism in birds refers to excess yellow pigments. It has been reported in parrots, canaries, the Evening Grosbeak, *Coccothraustes vespertinus*, the Eastern Bluebird, *Sialia sialis*, and other species.

Besides the forms of dichromatism mentioned above, there are polymorphic or polychromatic species that exhibit three or more color phases. The Australian Gouldian Finch, *Chloebia gouldiae*, exists in three color phases (red, yellow, and black-headed) which moreover differ in feather structure. The red and yellow feathers lack barbules and possess expanded and flattened barbs absent in the black feathers (Brush and Seifriend, 1968). The male Ruff, *Philomachus pugnax*, varies extremely in the color of its ruff and upper parts; they may be solid, speckled, sandy, buff, chestnut, glossy purple, rufous, or white.

To summarize—in the great majority of birds there are two chief varieties of normal plumage dimorphism: sexual (male and female coloration), and seasonal (nuptial and resting season coloration). On the other hand, many species (e.g., gulls, storks, hawks, parrots, wrens, and others) are monomorphic and show neither sexual nor seasonal changes in plumage coloration. By combining these dichotomous variables in different ways, Udvardy (1972) has classified families of North American birds into seven adult plumage types. Thirty families are almost exclusively monomorphic; 13 are dominantly sexually dimorphic; 13 are primarily seasonally dimorphic; and small minorities of families show various combinations of these variables, particularly the families Parulidae, Icteridae, and Fringillidae.

HEREDITY AND FEATHER COLOR. Heredity, of course, plays a dominant role in controlling feather color. There are no parrot-colored gulls. The heredity genes in the feather germ that determine color are usually very stable. If, for example, melanoblast cells (the forerunners of pigment cells or melanophores) of a barred Plymouth Rock chicken embryo donor are grafted into the base of the wing-bud of a White Leghorn 72-hour embryo host, the resulting wing of the chicken will have barred feathers characteristic of the Plymouth Rock (Fig. 3–22; Willier, 1952). Pigments may be exchanged even between birds

FIGURE 3–22 A young White Leghorn rooster whose barred wing and breast feathers are the result of a graft, during its embryonic life, of pigment cells from a Barred Plymouth Rock embryo donor. After Willier and Rawles, 1940.

of different orders. For example, a White Leghorn chicken developed a black wing whose pigment came from the melanoblasts of an American Crow, *Corvus brachyrhynchos* (Rawles, 1960). Crosses between various species or color strains within species have illustrated such principles as simple dominance, intermediacy or sex-linkage determination of plumage inheritance (Buckley, 1982). The red-headed morph of the Gouldian Finch, for example, is a sex-linked character.

Finally, genetic mutations may occur that affect not only feather color, as in melanism or albinism, but feather structure ("frizzle" in chickens), or crests (as in crested canaries and breeds of domestic pigeons) and mutations may even cause the complete loss of feathers, as in a mutant form of pigeon. In the domestic chicken and pigeon dozens of mutations affecting feathers and their colors have been discovered.

The Uses of Color

Colors in birds are mainly important for their effects on the behavior of other animals, including other birds. Nevertheless, colors have some intrinsic values to the bird. As mentioned, pigmented feathers are stronger and wear better than unpigmented feathers. Pigments absorb radiant heat, which at times is useful to the bird. Experiments by Heppner (1970) showed that black-dyed Zebra Finches, *Poephila guttatta*, absorbed 3.1 calories per minute more than white finches, and that the white birds had a metabolic heat production of 2.7 calories per minute greater than the black birds. This metabolic bonus from pigmentation is particularly advantageous to birds of cool or cold climates. Pigmented feathers stop the harmful ultraviolet rays of the sun, preventing sunburn of the bird's delicate skin. The White Tern, *Gygis alba*, has thin, translucent white feathers, but a black skin, whereas its close relative the Brown Noddy, *Anous stolidus*, has dark, opaque feathers, but a light-colored skin. Apparently the melanin is necessary, whether in skin or feathers, to shield the bird from the harmful actinic rays of the sun (Murphy, 1936).

The evolutionary development of color in birds has resulted in two contrary but equally adaptive general types: cryptic colors, or those promoting concealment, and phaneric colors, or those increasing conspicuousness. The first type generally hinders detection; the second type advertises or facilitates recognition (Cott, 1964).

Cryptic coloration in birds is almost exclusively related to protection against predators. It is most often of a character that more or less matches the bird's habitual surroundings. Because of their wide-ranging and rapid flight, birds have special difficulty in matching their habitats, yet a surprising number of them do just that. The Snowy Owl, *Nyctea scandiaca*, and Greenland Gyrfalcon, *Falco rusticolus*, wear white plumage that resembles their arctic surroundings. The Willow Ptarmigan, *Lagopus lagopus*, even changes its coat from a winter white to a summer brown to match the change in seasons. Many of the ground-dwelling birds such as the Galliformes have an inconspicuous brown color, whereas birds of the treetops like vireos and warblers often resemble the green sun-flecked foliage.

Sometimes the resemblance of a bird to its normal background is remarkably close. This is more apt to be true of small, defenseless species living in relatively bare, featureless habitats. Different races of ground-burrowing birds called miners (Furnariidae) of the deserts and lomas of Peru have plumage that typically matches the color of the soil on which they live (Koepcke, 1963). Larks of the family Alaudidae typically live on steppes where cover is thin and where

FIGURE 3–23 A female Woodcock on its nest, illustrating concealing coloration. Photo by G. R. Austing.

they are quite exposed to predators. Field studies of larks by Meinertzhagen in Syria and Niethammer in Africa (Mayr, 1942) have shown a striking correlation between soil color and plumage color. Blackish subspecies of larks lived on black volcanic soil, reddish subspecies on red soil, and pale sand-colored subspecies on pale sandy soil. Not only did the color of the birds match the soil, but also their color pattern. "The birds will have a smooth, even coloration, if they live on a fine-grained, dusty, or sandy soil. If, on the other hand, they live on a pebble desert, they will have a coarse, disruptive pattern of coloration." Most remarkable of all, when Meinertzhagen attempted to chase reddish larks living on reddish soils to light lime soils only a few yards away, he was unsuccessful. Nor could he force the whitish larks living on light soil to descend on the nonmatching red soils. The birds always alighted on soil matching their plumage, as though they were "conscious of the color of the soil that corresponds to their own coloration." The downy young of the African Spotted Sandgrouse, *Pterocles senegallus*, are gray-brown at hatching, but on about

the fifth day of life change to a sandy-yellow color. Suiting their behavior to their color, they hide among gray-brown stones the first four days, then when sandy-yellow, crouch in windblown patches of yellow sand (George, 1970).

Not only do birds at times match the general background, but in a few instances they match specific objects in it. Cott tells of Saville-Kent's experience with the Australian frog-mouth *Podargus*, which, when disturbed by a predator,

. . . will at once straighten up stiffly and, with its mottled feathers closely pressed to its body, assume so perfectly a resemblance to a portion of the branch upon which it is seated that, even at a short distance, it is almost impossible to recognize it.

The New World Common Potoo, *Nyctibius griseus*, behaves in a similar fashion (Borrero, 1970; Skutch, 1970).

Bitterns, which commonly live among reeds, have streaked breasts which resemble the surrounding vegetation. When threatened by an enemy, a bittern will characteristically stretch its head and neck with the bill pointing skyward and the striped breast presented toward the intruder. W. H. Hudson (1920) tells of encountering the Stripe-backed Bittern, *Ixobrychus involucris*, perched on a reed in an Argentine swamp. Hudson walked in a circle around the bittern, and the bird, while maintaining its mimetic pose, shifted its hold on the rush so as always to present its concealingly colored breast toward him. Even when Hudson several times pushed the bird's head down to its shoulders, it did not fly, but as soon as the hand was removed would resume its rigid vertical stance! Finally, when the bird was forced from its perch, it flew away.

Other cryptic postures are taken by the precocial young of many ground-nesting species. The young of various grouse, gulls, curlews, and other species will, when alarmed, squat low with head and neck stretched out flat on the ground and hold this pose as long as danger threatens. The survival advantage of such a pose is that, in addition to providing immobility, it tends to eliminate tell-tale shadows. Shadow elimination is partly attained in the great majority of adult birds and terrestrial vertebrates by "countershading." This refers to the development of darkest color on the back (where natural illumination is brightest) and lightest color on the belly of the animal (where illumination is weakest). The net effect of the contrary gradients of animal color and lighting is to reduce the three-dimensional roundness of the animal (as ordinarily indicated by shadows) and convert it into an inconspicuous, flat, even-toned object. Experimental evidence of the effectiveness of countershading was obtained by de Ruiter (Tinbergen, 1957) at Oxford. He killed countershaded caterpillars and mounted them

on twigs in a naturally planted aviary, half of them dark side up and half light side up. The European Jays, *Garrulus glandarius*, in the aviary "ate many more of the inverted (light side up) caterpillars than the others."

A further device used by some birds to escape detection is described by Cott as disruptive coloration. Both young and old Ringed Plovers, *Charadrius hiaticula*, possess, as their name suggests, a bold contrasting pattern of rings on the head, neck, and back, which optically break up the body into two pieces; they create a hiatus between head and body. This disruption of the visible form of the bird helps prevent its recognition *as a bird* by a potential predator. Disruptive marks like this are found in the downy chicks of snipe, woodcock, ducks, quail, and other species (Fig. 3–24).

The effectiveness of this cryptic coloration was tested on Sandwich Terns, *Sterna sandvicensis* (Croze, 1970) whose terneries are within gulleries. Eggs and chicks of this tern are of a pale off-white ground color spotted with dark disruptive markings. These terns do not remove eggshells, and defecate all over their nesting area. This leaves a pattern of white splotches all over their territory: the eggs and chicks appear to blend into these surroundings quite well.

Croze trained hand-raised and wild Carrion Crows, *Corvus corone*, to visit a grid full of red mussel shells covering bits of meat as rewards. From a blind he watched Carrion Crows "discover" the first red mussels, and thereafter form a "search image" and seek out red mussel shells to turn over. Croze next splotched white paint in the middle of his grid to simulate tern defecation. He also placed camouflaged black-and-white mussel shells covering meat rewards in the mid-

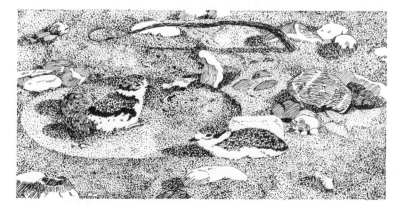

FIGURE 3–24 Concealment by disruptive pattern in the Ringed Plover. The bold markings tend to destroy the outline of the chicks as seen by a predator. The bird at the right shows the characteristic squatting posture of young galliform chicks when alarmed—an instinctive response that aids concealment by eliminating shadow. After Cott, 1940.

FIGURE 3–25   The Dark-eyed Junco, whose white outer tail feathers, conspicuous in flight, may serve as recognition marks to birds of the same species. Photo by G. R. Austing.

dle of his red mussel grid. To entice the crows to visit this new area he placed two familiar red mussels in the middle of his splotched grid. For five days crows found most of the red mussels and none of the camouflaged ones. As the supply of red mussels was depleted, crows found the first pied mussels and thereafter sought out these new mussels for their rewards. The fact that it took five days before crows found the camouflaged mussels is remarkable, and is probably due to the mobbing adults which distracted the predator, which in turn tended to avoid "splotched" areas. Thus two lines of defense are employed by these terns, mobbing and camouflage.

Distinctive coloration can serve as a quick recognition mark for distinguishing friend from foe, or it may hold a flock of birds together. Patterns thought to serve as recognition marks are, for example, the white rump patch of the Northern Flicker, *Colaptes auratus*, and the white outer tail feathers of the Dark-eyed Junco, *Junco hyemalis* (Fig. 3–25). These patterns are conspicuous in flight and may serve to keep the birds of a flock together. Many downy young shorebirds,

although concealingly colored on head and back, possess snowy white wing feathers. When alarmed or lost, the young run about stretching their conspicuous wings high in the air.

Recognition marks also have sexual significance. The male and female Flicker are colored alike except for a pair of black mustache marks on the male's cheeks. In an experiment by Noble (1936), the female of an apparently successfully mated pair was caught and supplied with an artificial mustache. The male immediately attacked his mate as an intruding male rival! When the mustache of the female was removed, she was again accepted by her mate.

A simple experiment by Nekipelov (1969) provided evidence that the bright nuptial coloration of males is a consequence of evolutionary sexual selection. He clipped the showy, iridescent feathers of the head, neck, and wing speculum from male Mallards and then mixed these experimental birds with unclipped control males and introduced both kinds into a flock of females. When females outnumbered the males, the dull-colored males were accepted, but when the males outnumbered the females, the latter selected the brightly colored males. Brightly colored males of many species are the product of sexual selection (e.g., in hummers, ducks, and birds-of-paradise), whereas females tend to be selected for inconspicuous (camouflaged) coloration, and they alone nest-build, incubate, and rear young. The dull-colored female plumage is less likely to attract predators.

To a more limited extent, color may be used for threat or warning, that is, as an energy-conserving device to avoid actual combat. This is called aposematic coloration. Many birds spread their feathers and display their most striking colors when intimidating a competitor or intruder. The broadly spread feathers of a nuthatch before rivals at a feeding tray, or of one turkey cock before another, are common examples. Cott (1940) gives a striking example of an intimidation display in the Turquoise-fronted Parrot, *Amazona aestiva;*

When alarmed or in danger this beautiful species throws its body forward into a horizontal position, partially spreads the wings in a horizontal plane, and widely fans out the tail, at the same time elevating its cobalt blue frontal fringe and green throat ruffle. This attitude effects the display of brilliant red areas on flight and tail feathers, which are normally concealed by a cryptic garment of green.

In addition to this optical onslaught the bird vibrates its feathers, causing a rustling sound, and, at the climax of its display, gives a staccato warning note.

Finally, coloration is used by many birds in courtship ceremonies to attract mates, to stimulate them into sex-readiness, and to synchronize the male and female reproductive time schedules. Such coloration is known as phanerogamic coloration. Lack's (1943) remarks concerning the red breast of the European Robin, *Erithacus rubecula*, as a warning flag, apply also to courtship colors. "When a bird possesses a bright patch of color, one may guess that it plays a part in its life sufficiently important to outweigh the disadvantages of conspicuousness to its enemies."

The colorful male almost invariably presents his showiest plumage directly before the female's eyes (Blume, 1973). Birds-of-paradise will assume quite grotesque poses to display their most colorful feathers, sometimes even hanging upside down from a branch. If a bird lacks gaudy feathers it may possess compensating colorful structures, such as combs, wattles, neck pouches, or even colored legs and feet, for courtship advertising service. The Blue-footed Booby, *Sula nebouxii*, will compete for the attentions of his intended by goose-stepping in front of her "raising his bright blue feet as high as possible and thrusting out his chest" (Murphy, 1936). Among phalaropes and hemipodes (*Turnix* spp.) the female is more brightly colored than the male, and she does the active displaying in courtship. Not only is the male the drab, passive partner in courtship, but he stays home and incubates the eggs and raises the young single-handedly.

◼ SUGGESTED READINGS

The most comprehensive modern references on plumage in birds are Lucas and Stettenheim, *Avian Anatomy: Integument*, and Voitkevich, *The Feathers and Plumage of Birds*. Old, but still valuable for general information on feathers, is Newton's *A Dictionary of Birds*. A somewhat technical account of feather growth and pigmentation is given in Rawles' chapter in Marshall (ed), *Biology and Comparative Physiology of Birds*. Fascinating material on adaptive coloration is to be found in Cott's *Adaptive Coloration in Animals*. The best modern treatise on feather molt is Erwin and Vesta Stresemann's *Die Mauser der Vögel*. An excellent brief treatment of molts is given in Heinroth's paperback, *The Birds*. Thomson's *A New Dictionary of Birds* has excellent, brief, authoritative articles on skin, feathers, plumage, coloration, and molting. Also recommended is R. B. Payne's "Mechanisms and control of molt" *in Avian Biology*, vol. 2.

Bones and Muscles
4

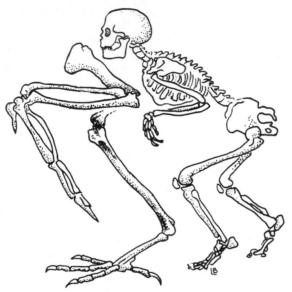

"Oh, for the wings of a dove," men, envious, cry.
What bones, what muscles, have birds and not I,
That give them freedom of the boundless sky?
—Joel Peters (1905–   ),
        Wondering

The three chief functions of the skeleton in birds, as in vertebrates generally, are protection (especially of the vital organs), support (to maintain body shape and posture), and articulation (as of wings and legs in locomotion). Additional functions include the manufacture of blood corpuscles by the bone marrow, and, especially in birds, the storage of an accessible supply of calcium for the secretion of egg shells.

The skeleton of birds has undergone drastic modifications related to the problems of flight. A good way to understand the distinctive modifications of a bird's body is to imagine the difficulties in flying and perching a bird would face if it were built on the plan of the lizard-like reptiles from which it descended. A lizard with feathers instead of scales still could not fly. For flight, extensive remodeling of the body architecture would be necessary. The long, bony tail of the reptile, which could contribute little to flight, would have to go. And with the tail gone, how could the creature maintain its balance while walking on its hind legs? Clearly, a shortening of the body axis would be necessary in order to concentrate the bird's weight over the legs and under the wings. To fly, the front legs must be larger and have powerful muscles, and these, in turn, demand a strong, well-placed anchorage on the ventral side of the body—in other words, a well-developed sternum. The angels one sees in church windows are biological impossibilities—they lack the sternum and breast muscles needed to flap their impressive wings. These modifications birds have developed.

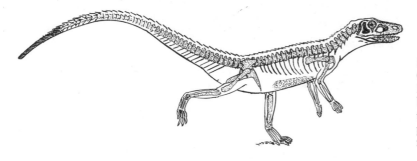

FIGURE 4–1 A restored skeleton of *Ornithosuchus*, a Mesozoic, bipedal, Pseudosuchian reptile thought to have been an ancestor of modern birds. Note the elongated body and the relatively simple hip and shoulder girdles. After Heilmann, 1927.

Then, too, unlike all other vertebrates, a bird must support its weight in two different ways: part of the time by its legs, part of the time by its wings. These alternating requirements have resulted in a skeleton more highly differentiated than that of any other vertebrate. The chief innovations that have met these demands are the two thin but extensive bowl-like structures, the pelvis above and the sternum below. The muscles around these two hip and shoulder "girders" serve to propel the bird and to maintain its balance whether it is on its wings or on its legs. Since these are understandably the strongest and heaviest muscles of the body, it is essential, in a flying machine, that they be located near the center of gravity. As a consequence, the trunk axis of a bird is shorter than that of any other vertebrate of corresponding size except the frog. The rigid pelvis also permits a great reduction in the muscles of the back that, in four-legged vertebrates, or tetrapods, are used to bend the flexible back and to support it and the contents of the abdomen slung below.

As pointed out earlier, the avian skeleton achieves strength-with-lightness by being built with the greatest possible economy of materials. Some bones common to most of the higher vertebrates are completely eliminated; others are fused together. The bones that remain are usually highly mineralized and very strong. Many are "pneumatized" or filled with air spaces instead of bone marrow (see Fig. 1–3). Early in a bird's life the larger bones are filled with marrow, but this is often resorbed (within five weeks in a hen's humerus) and its place taken by outreaching extensions of the air sacs connected with the lungs. In some birds, hornbills for example, the sacs penetrate even into the bones of the toes. Darwin owned a pipe whose stem was contrived from the hollow wing bone of an albatross. Dozens of primitive flutes, made of the long bones of such birds as swans and vultures, have been recovered

from caves and kitchen middens of Europe (Oomen, 1972).

Although pneumatization of bones is a great advantage to flying birds, not all birds show it. As a rule, small birds are less well pneumatized than large birds. Hollow bones are essentially lacking in gulls as well as in the flightless kiwi. In strong divers such as penguins, grebes, and loons, pneumaticity is very poorly developed, and in diving ducks there is less of it than in nondivers.

In general form and structure a bird's skeleton seems to be chiefly molded by heredity, although stresses may increase its mass. Isolated rooster bones subjected to infrasound loads over a six-week period were found to have increased bone-mineral contents (Rubin and Lanyon, 1984). If a limb-bud from a chick embryo is grown in tissue-culture, its cells will differentiate into relatively normal thigh and leg bones with appropriate joints. However, in the absence of muscles and their movements, the joints tend to fuse. This indicates that the finer adaptations of the skeleton depend on movement and mechanical stress—not to mention such influences as diet and hormones (Bellairs, 1964). Osteoclasts increased in size, and spread along bone surfaces within 30 minutes after administration of parathyroid hormone to a female Japanese Quail, *Coturnix coturnix* (Miller *et al.*, 1984). Osteoclasts may effect rapid changes in bone resorption and mineral metabolism.

THE AXIAL SKELETON

To avoid disaster, a rapidly flying animal must see well, possess superior motor coordination, and make quick decisions. This means that birds, as compared with reptiles, must have large eyes and a large brain. The enlargement of these structures has had a profound influence on the form and size of the skull.

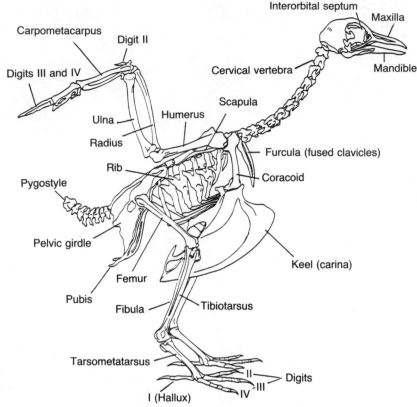

FIGURE 4–2 The skeleton of a typical bird, the pigeon, showing some of the chief functions of its various parts.

The eyes of most birds are so large that they almost touch in the middle of the skull at the interorbital septum. In some instances they occupy so much space that they prevent the formation of part of this septum, so that there is a window between the two eye sockets (see Fig. 1–3). A ring of small, shingle-like bony plates, the sclerotic ring, encircles the eyeball in front—a protective device also found in the eyes of some reptiles. The larger the eyes, the more the brain is forced upward and backward into the bulging cranium. The enlargement of the eyes has been made at the expense of the sense of smell. The reduction in olfactory organs and the substitution of a light, horny beak for a heavy jaw with teeth have resulted in a great reduction in the size of the forepart of the skull.

Pneumatization of the skull occurs in most birds, although the kiwi shows no air spaces at all and the grebe only a few. There are usually two systems of air spaces in the skull: one, in and adjoining the upper mandible, which connects with the nasal cavity; the other, in the roof and base of the cranium and in the interorbital septum, which originates in the tympanic cavity. Some woodpeckers have pneumatized frontal bones which overlap the nasofrontal hinge at the base of the upper joint. This is associated with kinesis (see below).

Since birds have so specialized their front limbs that they are no longer useful for manipulating food, nesting material, and the like, the head and beak have had to take over these functions. Whereas in man and other mammals the skull articulates with the spinal column on two ball-and-socket joints (condyles), in birds and other vertebrates, one suffices. This allows the head greater freedom of movement. In most birds the opening for the spinal cord, the foramen magnum, has shifted from the rear of the skull to its lower side, especially in birds with large eyes and a vertical stance, such as owls and hawks. In more primitive, small-eyed birds, such as the ducks, the opening is still at the rear.

As in reptiles, the lower jaw (mandible) of a bird is composed of several bones, and hinges on two small, movable bones, the quadrates. This allows a double-jointed, wide-gaping type of articulation not found in mammals. In many birds, not only is the lower jaw hinged to the skull, but the upper jaw (maxilla) is also, normally at the forehead (the nasofrontal hinge). By means of separate upper jaw bones (zygomatic arch and pterygopalatal bones) that articulate with the quadrate bones at the angle of the jaws, the upper jaw can be raised or lowered (kinesis). For example, when a woodpecker pecks at a hard surface, the brain case pulls away from the upper jaw at the moment of impact, i.e., the maxillae bend up (Bock, 1974). The overlapping frontal bones prevent the upper jaw from breaking (Bock, personal communication).

The possibility of moving each jaw independently gives birds a much more precise control of the food they manipulate with their beaks. It also makes possible a wider gape, as in insect-catching birds, which helps when feeding on the wing. A similar mechanism allows the raising of only the tip of the upper jaw in the American Woodcock, *Scolopax minor*, and the other long-billed shorebirds. This latter adaptation permits the woodcock to seize an earthworm deep in the ground without expending the energy needed to open its entire beak and push aside the soil for the whole depth that the beak penetrates. In contrast, tyrannid flycatchers have a well-developed jaw ligament (the quadratomandibular) which is stretched when the jaw is opened. The ligament is then loaded in tension and allows the jaw to close rapidly when foraging for insect prey (Bock and Morony, 1972).

A remarkable adaptation, both in skeleton and muscles, is seen in many woodpeckers, hummingbirds, sunbirds, and others that are able to extend their tongues to a greater length than other birds. This permits them to exploit new food niches by probing deep in flower corollas for nectar. Such extreme extension of the tongue is made possible by a lengthening of the hyoid bones that support it, and the development of corresponding sheaths and muscles to house and move these bones. The roots of the hyoid bone are so greatly lengthened in the Green Woodpecker, *Picus viridis*, that they encircle the outside of the skull and their ends come to rest in the nasal cavity (Fig. 4–3). Whereas a sparrow can scarcely extend its tongue the length of its stubby beak, the Green Woodpecker can extend its barbed tongue four times the length of its upper beak (Hess, 1951).

The vertebral column of birds is subdivided, as in reptiles and mammals, into five regions: the cervical (neck), thoracic (chest), lumbar (loins), sacral (hip), and caudal (tail) regions. The number of vertebrae in the spinal column varies greatly, from 39 in a sparrow to 63 in a swan. The most characteristic feature of the avian backbone is its rigidity—a very necessary condition for effective flight as well as for an easy bipedal posture. The separate vertebrae are more or less fused together in all regions but the neck. Rails, however, are exceptional in that they possess very flexible spinal columns.

In the cervical region, as an adaptation to the increased use of the head as a manipulating tool, the vertebrae show extraordinary mobility. This is made possible by their peculiar "heterocoelous" articulating surfaces. These are saddle-shaped, with the anterior face of each vertebra convex up and down, and concave from side to side. The posterior face has corresponding curves to permit proper articulation. These complex vertebrae are supplied with bony processes above, below, and on the sides, for the attachment of ligaments and the complex muscles that move the

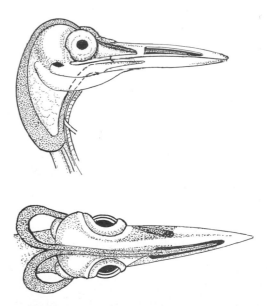

FIGURE 4–3 The highly protrusible tongue of the Green Woodpecker originates in the extremely long hyoid bones. These, sheathed by the muscles that move them, circle the back of the skull, cross its top, and end in the right nostril, which serves as a scabbard to hold them. After Leiber, 1907.

neck. The ligaments that bind these vertebrae together are particularly stout on the upper side to counteract the weight of the long neck and head. An unusual modification of the neck vertebrae occurs in some herons, cormorants, and anhingas. A very long cervical vertebra in the midneck region has an articular surface so directed as to cause a sharp kink in the neck axis, which makes it look dislocated. Special muscles (longus colli anterior and posterior) associated with these vertebrae facilitate the lightning-like stabs these birds use in catching fish. The total number of cervical vertebrae varies from 11 in parakeets to 25 in swans, whereas the number in mammals, whether man or giraffe, is typically seven.

Thoracic vertebrae are those that bear the ribs. They vary in number from three to ten, and usually from three to five of them are fused together in a "dorsal bone." There are normally two or three free thoracic vertebrae just in front of the lumbar region to allow some movement between the dorsal bone and the following synsacrum. The thoracic vertebrae of penguins are not fused, a fact no doubt related to their fish-like swimming movements.

Ribs in birds have a double articulation with the thoracic vertebrae. They may reach only part way, or may extend all the way to the sternum on the lower side of the body. Those that reach all the way occur in two bony sections, an upper and a lower, which are joined almost at right angles to each other on the side of the chest cavity. On each upper rib section occurs a tab-like, backward-directed projection that strengthens the rib cavity by overlapping the adjoining rib to the rear. Interestingly enough, *Sphenodon* and some other reptiles possess such uncinate processes on their ribs. In powerful divers, such as guillemots and loons, there are extra-long uncinate processes that reach across two adjoining ribs, an adaptation that strengthens the rib cage against the pressures encountered in deep dives (see Fig. 21–14). As a further adaptation to diving, the entire rib cage in these birds is compressed and lengthened so that the body offers less resistance to passage through the water. The number of complete ribs in birds varies from three in some pigeons to nine in swans.

The next section of the spinal column, proceeding rearward, is the most extensively fused region of all. What in primitive ancestors were two or three sacral vertebrae have not only fused together in modern birds, but have also fused with lumbar vertebrae in front and a few caudal vertebrae to the rear to

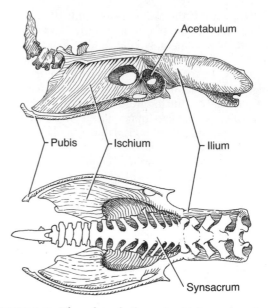

FIGURE 4–4 The pelvis of a domestic chicken as seen from the side and from below. The dome-like structure composed of the spinal column and the pelvic bones provides strength and rigidity with lightness.

make a rigid plate of bone, the synsacrum. The synsacrum consists of between 10 and 23 vertebrae in all. It has also expanded laterally to fuse intimately with the pelvic bones. With its ridges, braces, and projections, the synsacrum makes a wonderfully light and stiff framework for the support of the body by the legs (Fig. 4–4).

The spinal column ends posteriorly in 12 or so caudal vertebrae (there were 20 in *Archaeopteryx*). About half of them, at the end of the tail, are welded together into a single broad bone, the pygostyle, which provides a support for the tail feathers. Movement in the entire backbone itself is possible in only three regions: the cervical vertebrae, between the dorsal bone and the synsacrum, and in some of the caudal vertebrae.

On the ventral side of the body where the ribs come together is found the large, shield-shaped sternum. This bone, more highly developed in the bird than in any other vertebrate, protects the chest and part of the belly against physical blows; more important, it provides a large, ideally placed surface for the attachment of the large wing muscles. In powerful fliers the sternum has a large, thin keel for the attachment of these muscles, but in flightless birds it may bear a very small keel or, as in ostriches (and man), none at all. Although the penguin is flightless, it still has a distinct keel to which are attached the powerful muscles that move its flipper-like wing. The backbone, ribs, and sternum together form a flexible but strong box that houses the heart, lungs, and visceral organs of the bird.

THE APPENDICULAR SKELETON

The energetic thrust of each beating wing is translated to the body through a tripod of bones, one set on each side of the rib cage and collectively called the pectoral or shoulder girdle. These bones are the paired scapula and coracoid, and the single furcula (made up of paired and fused clavicles). This last bone is sometimes called the clavicle or "wishbone." At the place where the three bones come together there is an opening, the foramen triosseum, through which runs the tendon of the supracoracoideus muscle, referred to later. Where the scapula and coracoid unite, there is a hollow depression, the glenoid cavity, with which the chief bone of the wing, the humerus, makes a flexible ball-and-socket joint. A glance at a complete skeleton will show that the wings of the bird are attached above and slightly in front of the body's center of gravity—a necessary location for their function (see Figs. 4–2 and 4–6).

The scapula is a thin, blade-like bone bedded in muscles along the side of the spinal column and tied to the ribs with ligaments. Interposed between the wing muscles and the ribs, the scapula, along with the furcula, braces the shoulders against the displacements caused by the flapping motion of the wings. In strong fliers it is especially long. The stout coracoid bone braces the sternum against the powerful compression created when the chief breast muscles contract to cause the downbeat of the wings. Without this compression support of the coracoid, the muscular stresses of flapping flight would destroy the rib cage. The two halves of the shoulder girdle are united in front by the V-shaped furcula, which may or may not be attached to the sternum. It serves as a strut to brace the two wings apart. In strong fliers it is wide and stout; in ground dwellers it is smaller. The esophagus and windpipe pass between the arms of the furcula. In flightless birds the scapula and coracoid may be greatly reduced, and in gliding birds such as the frigates, all three bones of the pectoral girdle may be rigidly fused

together. In toucans, there is no furcula, the clavicles being unfused and incompletely formed.

To modify a walking leg into a wing requires extensive alterations. For lightness the bones are hollowed, reduced in number and size, and fused together. For compactness, while the wing is at rest, the arm folds closely against the body in a Z-shape. For the attachment of powerful muscles and large feathers, the bones are broadened and provided with prominent heads and crests. To convert the wing into a lifting plane, flat skin membranes—the pre- and postpatagia—are stretched between its front and rear edges and the body. Finally, the arm is equipped with rows of the largest feathers of the body, the primary wing feathers or the remiges. Activated by assorted muscles, the front leg is now ready to propel the bird through the air.

Among the details involved in the development of the wing are the shortening and broadening of the caudally-directed humerus, which, with its large muscle-attachment surfaces, articulates in the glenoid cavity of the pectoral girdle. Near its proximal end, the humerus has a large hole through which its air chambers connect flexibly with the air sacs of the body. The broadened ulna of the forearm provides a base for the attachment of secondary flight feathers. Correlated with the reduced mobility of the hand, the number of wrist bones (carpals) is reduced to two, and the palm bones (metacarpals) to three, the second and third of which are fused together with the second supporting the first six primary flight feathers. Only three fingers appear in the hand of modern birds: digits 2, 3, and 4. Digit 5 is present for a while during embryological development but soon disappears (Hinchliffe and Hecht, 1984). Digit two consists of a single small bone. It supports the three or four small feathers of the alula. The third finger is by far the largest: it consists of two broad segments on which are fastened the distal three or four primaries. The fourth finger is reduced to a single tiny bone that supports primary number 7. In the embryos of many species and the adults of a few (Secretary Bird, Screamer), there are claws on the ends of the fingers. In the African Finfoot, *Podica senegalensis*, the clawed first finger is about 15 mm long and is capable of independent movement. The bird uses it in climbing trees (William, 1963). Claws are also found on the thumb and first finger of nestling Hoatzins, *Opisthocomus hoazin*, which climb around the nest and nesting tree using all four limbs.

The skeletal structure of birds' wings shows wide adaptive responsiveness to different kinds of locomotion. In the penguin's paddle-wing every bone is short, broad, and stout. In birds such as hummingbirds and swifts that depend heavily on flapping flight, the upper arm is short and stout and the hand long. In gliding and soaring birds like albatrosses and vultures, the upper arm is long but the hand is short. Certain soaring birds have small tendonous (sesamoid) bones in the elbow and wrist joints that hold the arms extended in the soaring position without muscular exertion. Muscles pull these bones out of position when the bird folds its wings.

Adapted to running, perching, and occasionally to swimming, the pelvic girdle and the hind legs also show extensive remodeling of ancestral architecture. The bird's pelvis is greatly lengthened so that, welded to the synsacrum, it forms a thin but strong roof that covers about half of the body. The pelvis proper is made of three bones that come together at the leg socket or acetabulum. In the earliest archosaur reptiles from which birds descended, the pelvic bones radiated outward from the acetabulum somewhat like three spokes from a hub, the ilium upward and toward the midline of the back, the pubis downward and forward under the belly, and the ischium downward and toward the rear. In modern birds these bones have shown a more drastic modification than any others in the entire skeleton. The ilium has been greatly broadened and lengthened in an anteroposterior direction and firmly fused to the synsacrum. The ischium and pubis have each evolved into long, thin bars fused to the ilium anteriorly and directed sharply rearward, parallel to the backbone. This exceptional positioning of the pubis is found elsewhere only among ornithischian dinosaurs, a group not ancestral to Aves. In modern birds openings are always left between the ilium and ischium and between the ischium and pubis, through which pass the chief nerves (and one muscle) that supply the legs. These bones do not unite or form symphyses at their distal ends, as they do in primitive reptiles, except in two birds: in the ostrich there is a pubic symphysis and in the rhea an ischial symphysis. The lack of such union leaves the under part of the pelvis open to accommodate the centrally placed abdomen. This openness also facilitates egg laying. Such a ventral location of the abdomen has necessitated spreading the legs apart and widening the pelvic girdle through which the legs support the body. The pelves of some parrots show a pronounced sexual dimorphism, that of the female being the wider. This is evidently an adaptation for egg laying. In running

and climbing birds the pelvis tends to be wide, whereas in diving birds it tends to be narrow.

Since a bird's legs are still used largely for walking, leaping, and perching, they have not undergone as pronounced a modification as its wings. The thighbone, or femur, is generally shortened and at times pneumatic. In most birds, especially those whose bodies are held horizontally, the short femur, buried in the flesh of the body, is directed somewhat forward so that its distal end, from which the leg continues to the ground, will be near the bird's center of gravity. With their legs so wide apart, birds would still have trouble walking were it not for the rotation of the femur and knee joint that results in placing the foot under the bird's center of gravity. Like the humerus, the femur has an enlargement (the trochanter) on its proximal end for muscle attachment. A knob on the trochanter presses against the ilium in such a way that it prevents the bird's body from falling downward and inward when the bird is standing on one leg.

Usually at the knee joint there is, in loons, penguins, and other expert divers, a strong extension of the tibia, the cnemial crest. Leg muscles attached to this crest are provided with an oar-like leverage that greatly increases the thrust of their legs while swimming. This crest illustrates beautifully the way in which different bones may be adapted to the same end. In the giant fossil diver, *Hesperornis*, the cnemial crest consists of the large knee cap, or patella, fused to the end of the tibia. In the diving grebes it consists of contributions from both a conical patella and a prolonged tibia. But in loons, the cnemial crest is almost entirely an extension of the tibia, with but a tiny scale of a patella added to it. All three types of crests are very similar in appearance and serve exactly the same function, in spite of their different origins (see Fig. 21–13).

In the shank or "drumstick" the tibia is the main bone and the fibula is reduced to a splint. This change results in a considerable loss in the ability to rotate the lower leg. When the tibia is rotated at the knee joint, torsion on the fibula causes the leg to resume its normal position when muscle tension ceases. Aside from the inward rotation already mentioned, birds have very little capacity to twist their legs or to step sideways. Their legs move in a rather limited fore-and-aft direction.

It is in the ankle joint that the avian leg shows its most striking changes. Whereas in mammals the ankle joint is usually composed of seven pebble-like tarsal bones, in birds some of these bones are fused to the end of the tibia, which thus becomes the tibiotarsus; and the remaining bones are joined to the three fused metatarsal bones, which all together as one bone are now called the tarsometatarsus (see Fig. 4–2). Instead of a low, many-boned ankle joint, birds have acquired what is essentially an extra bone that greatly lengthens the leg: the tarsometatarsus. This heel bone, now raised high above the ground, adds speed for running, reduces the risk of dislocation, and simplifies leg construction. In the embryo, the separate tarsal bones are still distinguishable, but they soon lose their identity in fusion as the young bird matures. The fused metatarsal end of the tarsometatarsus still shows its three-boned origin in many birds in the three ridges leading to three articular surfaces on which toes are placed. Grooves in and between these ridges serve as channels and pulleys for the tendons that operate the toes.

Modern birds may have two, three, or four toes, but never five. The first toe is generally directed backward and provides the considerable advantages of an opposable toe. The fourth toe in some owls, cuckoos, and plantain-eaters may at will be directed either forward or backward. Parrots and woodpeckers typically have their two outer toes directed backward and the two inner toes forward. In the Ostrich, toes one and two disappear, and only toes three and four, greatly shortened, remain functional. This sort of toe reduction is characteristic of running birds. On the other hand, the toes may be greatly lengthened, as in jaçanas, for walking on floating vegetation.

In general, the bones of a bird, for all their mineral rigidity, have shown an evolutionary plasticity exceeded by no other system of the body. The problems of flight have placed unique and drastic demands on the bird's body, and its skeleton has responded with unique and drastic changes.

MUSCLES

As a result of adaptation to flight, the muscles of birds, like the bones, have become altered in both structure and distribution. The main changes have been in the locomotor muscles of the wings and legs. These massive muscles have been shifted ventrally, resulting in improved aerodynamic balance. Anyone who has carved a fowl knows that the back is as spare in meat as the breast is abundant. The reduction of muscles on a bird's back has been made possible by the rigidity of the synsacrum and dorsal bone. These bones take the place of the fuselage of an airplane and eliminate the need for strong dorsal muscles (as, for example, the loin muscles in a mammal) to hold a flexible backbone against the pull of gravity and the stresses imposed by active flight or running.

Further adaptive modification in the size and distribution of muscles is closely related to the life style of any given species. Studies by Hartman (1961) of the locomotor mechanisms of 360 species of birds of 70 families revealed some striking contrasts in the comparative sizes of locomotor muscle masses. Species like hummingbirds and swallows that are superb fliers but essentially helpless on the ground may devote 25 to 35 per cent of their total body weight to flight muscles but as little as 2 per cent or less to leg muscles. Predatory hawks, falcons, and owls have leg muscles for grasping prey about equally as heavy as wing muscles for flight. Likewise, the proportional weights in coots and grebes, which are strong fliers as well as powerful divers, are divided roughly equally between wing and leg muscles, with 15 per cent each. Ground-dwellers (for example, some rails that depend more on running than flying to escape enemies), have larger leg than wing muscles. In birds of powerful sustained flight, like doves, or of short-range but explosive take-off flight, like tinamous, as much as 42 per cent of the body weight may be devoted to flight muscles (Fig. 4–5).

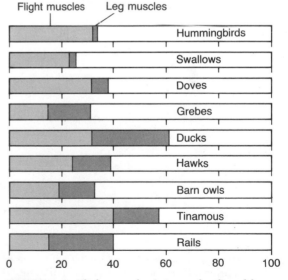

FIGURE 4–5 Flight muscles (wing and tail) and leg muscles shown as percentages of total body weight. Data from Hartman, 1961.

Chief of the flight muscles is the large pectoralis that depresses the wing. It arises on the keel and furcula and is inserted on the under side of the forward-projecting deltoid crest of the humerus at some distance from the shoulder joint. On this muscle falls the main burden of supporting the bird in the air. Acting as its antagonist is the supracoracoideus muscle that raises the wing (Fig. 4–6). Rather than on the backbone where one might expect it, this muscle is also located ventrally, under the pectoralis muscle. It, too, arises on the keel; but whereas the pectoralis extends upward to attach directly on the humerus, the supracoracoideus ends dorsally in a tendon that passes through the foramen triosseum, where the three shoulder-girdle bones come together, and turns outward and downward to attach on the upper side of the humerus. By this kind of rope and pulley arrangement, the supracoracoideus exerts an upward force on the wing by a downward pull, and it keeps its main mass low in the body. These two pairs of muscles may together make up as much as one-fourth of a bird's weight. The supracoracoideus is especially well developed in diving birds that use their wings as paddles and brakes. In flightless birds both of these muscles are weak and poorly developed.

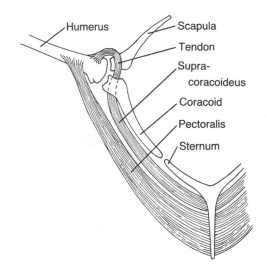

FIGURE 4–6 A cross-section through a typical bird's breast muscles, showing how an ingenious rope-and-pulley mechanism permits the contracting supracoracoideus muscle to raise the wing, although the chief mass of this muscle lies low in the body. This arrangement promotes stable flight. After Storer, 1943.

There are numerous other smaller muscles of the wing and shoulder girdle that pull the wing forward or backward, or that rotate the humerus so as to depress the leading edge of the wing (pronation) or to raise it (supination). Still others may flex or extend the wing, or stretch the skin flaps that occupy the angles of the wing. Even the alula and individual feathers are subject to the control of an elaborate system of small muscles and tendons. All these muscles and their actions are important in making possible the intricate control of bird flight.

As the keel is the chief anchorage for flight muscles, its size and shape are good indices of a bird's capacity for flapping flight. Power fliers have a forward-projecting keel so that the wings are drawn forward on the downstroke. Soarers have a rounded keel so that the outstretched wings are held steady against a vertical tension of the pectoralis muscle during soaring.

Hummingbirds and swifts have relatively enormous keels. They fly with rigid wings rotating at the shoulder and differ from all other fliers in possessing a powered upstroke. Penguins, which fly under water, possess a powered down-stroke. In the Ruby-throated Hummingbird, *Archilochus colubris*, the elevator breast muscles weigh nearly one-half as much as the depressors, whereas in the American Robin, *Turdus migratorius*, they weigh but one-ninth as much (Savile, 1950).

Since the legs of birds still support the weight of the body from below as in reptiles, their muscles have not required as drastic a remodeling as those of the wings. The major muscles of the legs provide forward and backward movement of the legs and very little lateral movement or rotation. Evidently, the ability of birds to escape their enemies by flight has eliminated the necessity to develop the muscles and bone articulations that would make possible the zigzag sort of running characteristic of rabbits. At any rate, most of the leg muscles are concerned with straight-forward walking and running, leaping into the air, cushioning landings, and grasping with the toes. In his detailed study of the Blue Coua, *Coua caerulea,* Berger (1953) listed 38 individual leg muscles. The main mass of these muscles is located in the thigh over the femur. A lesser number are located in the shank or the tibiotarsus, and only six thread-thin miniature muscles are found in the tarsometatarsus, just above the toes. The muscles in the upper parts of the leg flex and extend the lower parts by means of strong, stringy tendons, sliding through sleeve-like sheaths and strategically placed (inside or outside of joints, for example) to provide the proper mechanical actions (Fig. 4–7).

This concentration of the leg muscles on the upper leg has several advantages. First, it places most of their weight near the center of gravity, an important aid in flight. Second, it makes the outer ends of the legs light in weight, enabling them to be moved more quickly and with less exertion than would be the case were they, like human legs, more uniform in structure and thickness from top to bottom. Since the toes and the tarsometatarsus are largely made of tough scales, tendons, and bones, they are, in spite of their exposed location and active use, much less subject to accidental damage and freezing than soft, fleshy extremities

would be. And of course the absence of major muscles at the distal end of the leg permits an economy in the distribution of blood vessels and nerves.

The muscles of the axial skeleton are primarily concentrated in the head, neck, and rump. Although birds have exchanged their teeth for muscular gizzards, they still possess complex muscles that move the beak. Investigations by Beecher (1951) of the jaw musculature of different blackbird species have shown a close correlation between muscle development and the type of diet eaten. Insect eaters such as members of the genus *Euphagus* have relatively weak muscles for closing the jaw. Seed eaters of the genus *Molothrus* have powerful jaw muscles. Meadowlarks of the genus *Sturnella* use the opening or gaping action of their bills in disturbing ground litter for food, and they have unusually strong gaping muscles. Galapagos Large Ground Finches, *Geospiza magnirostris*, feed on large and hard seeds, whereas Small Ground Finches, *G. fuliginosa*, feed on small seeds. There is a positive correlation between the volume of jaw muscle masses and the hardness of the seeds taken (Fig. 4–8, Bowman, 1961).

Probably the most complex muscles in the bird are those that control the elaborately varied movements of its neck. Muscles of the neck are thin and stringy, interwoven and often subdivided and attached to one another. When a certain muscle has several other muscles attached to its fascial sheath and leading in different directions, its motor action will be quite variable, depending on the contractions of its neighbors. A very crucial, so-called hatching muscle is located on the upper neck and back of the head in the unhatched domestic chick. Not only is this muscle strategically located to provide exactly the force needed by the hatching chick to break out of its confining egg shell, but its size, according to careful studies by Fisher (1958), reaches its maximum on the 20th day of incubation, and the chick normally hatches on the 21st day. After the chick hatches, this muscle rapidly decreases in size (see Fig. 16–13).

The rump and tail muscles of a bird are also of considerable complexity. The pygostyle, which supports the tail feathers, is moved in various ways by these muscles. In addition, there are muscles that act directly on the feathers themselves: to fan them out, for example, when a bird alights. Levator caudae and levator coccygis muscles elevate bird tails, and depressor muscles lower them (Fig. 4–9). In the case of Ruddy Ducks, *Oxyura jamaicensis*, these are

FIGURE 4–7 Leg muscles of the common pigeon, showing how the heavier muscles of the leg, placed near the body, control the extremities by means of long, slender tendons. Redrawn after Pettingill, 1970.

FIGURE 4–8 Lateral view of superficial jaw muscles of two of Darwin's Finches. The Large Ground Finch, *Geospiza magnirostris (left)*, feeds on large seeds and fruit stones and is thus equipped with large muscle masses associated with a large heavy bill. The Small Ground Finch, *Geospiza fuliginosa (right)*, feeds mostly on small and soft seeds and thus has much smaller muscle masses associated with its smaller bill. Modified from Bowman, 1961.

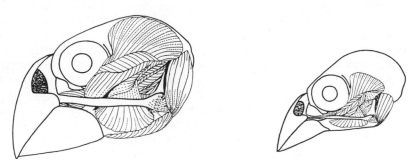

especially important since the tail is raised and fanned during display, and may be lowered beneath the water surface to be used as a rudder (Raikow, 1970).

Throughout the body there are dermal muscles that attach to feather follicles in a systematic fashion. These are smooth, or involuntary, muscles under the control of the sympathetic nervous system. Contour feathers of the domestic Chicken, *Gallus gallus*, are described by Lucas and Stettenheim (1972) as being arranged in rows in such a way that the follicle of any given feather is connected by pairs of antagonistic (i.e., producing contrary motions) muscles to each of four adjacent follicles. There are generally a depressor and an erector muscle in each pair, the former being the lar-

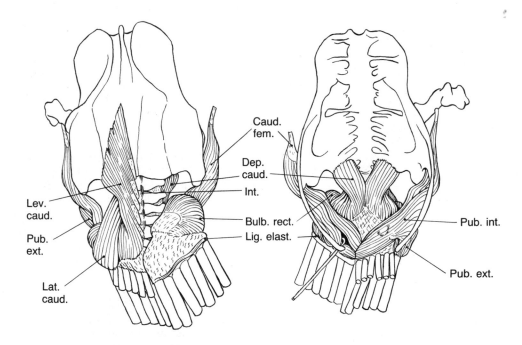

FIGURE 4–9 Dorsal *(left)* and ventral *(right)* tail musculature of the domestic pigeon, *Columba livia*. Levator *(Lev. caud.)* muscles serve to elevate the tail, and depressor muscles *(Dep. caud.* and *caud. fem.)* serve to lower the tail. The Bulbi rectricium *(Bulb. rect.)* and Lateralis caudae *(Lat. caud.)* muscles serve to spread the tail. After Raikow, 1986.

FIGURE 4-10 The Squacco Heron of Europe, showing how muscles may control feathers, either sleeking them against the body or extending them in display. Photo by E. Hosking.

ger. A given muscle is typically attached to the basal end of one follicle and to a higher level on the other. Antagonistic muscles often interdigitate where they cross (Fig. 4–11). Muscles vary in size and number in relation to feather size and activity, and to body location. As many as three or four dozen muscles may attach to a single feather follicle. They are probably activated in response to stimuli from tactile (Herbst) corpuscles located either near the follicles or beside filoplume follicles that accompany contour feathers. Only filoplume feathers lack muscles. In addition to display behavior, feather movements may be variously involved in flight, body insulation control, buoyancy control, brooding, and defecation. Still other skin muscles control the contents of the crop or the tension on inflated air sacs that play a part in the courting antics of some gallinaceous birds.

The skeletal muscles of the body may be either red in color or pale and whitish, as in the "dark" or "light" meat of a roasted turkey. And some muscles are mixtures of both types. These two kinds of muscles are distinctly different in structure and function. The white

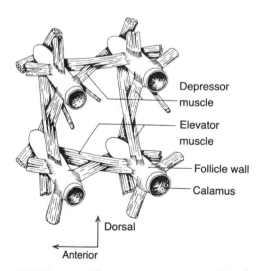

Depressor muscle

Elevator muscle

Follicle wall

Calamus

Dorsal

Anterior

FIGURE 4-11 The typical arrangement of feather muscles. Illustration shows the detail from the left femoral tract of a chicken. From Stettenheim *et al.*, 1963.

muscles normally provide less sustained action than the red muscles. Many of the gallinaceous birds, such as grouse, quail, pheasant, and chickens, have white breast muscles and as a consequence are unable to undertake long flights. In spite of its powerful breast muscles, the Ruffed Grouse, *Bonasa umbellus*, makes an average flight of only 100 or 200 m. If a given bird "is flushed three or four times in quick succession, it can be picked up by the hand, exhausted" (Edminster, 1947). The large pectoralis breast muscle is red in falcons, gulls, crows, sparrows, and other birds of strong, enduring flight.

Muscles are relatively inefficient energy converters: about one-fourth of the fuel they burn produces the work of muscle contraction, the remainder, heat. The heat, however, may be useful in cool or cold weather in maintaining body warmth (Calder and King, 1974).

Microscopically, the red muscles are built of finer fibers than the white, and their nuclei are located at the edges of the fibers, while the nuclei of white muscles are scattered through the fibers. In both muscles the nuclei are remarkably long. The color of the red fibers is largely due to the presence of the oxygen-carrying compound, myoglobin, which is apparently absent or very rare in white muscles. Furthermore, red muscle cells have a higher content of mitochondria (microscopic respiratory bodies) than do white fibers and therefore are better able to carry on sustained oxidative processes (George and Naik, 1960a). In addition to these differences, red muscle fibers are provided with a richer supply of blood capillaries (Fig. 4–12)

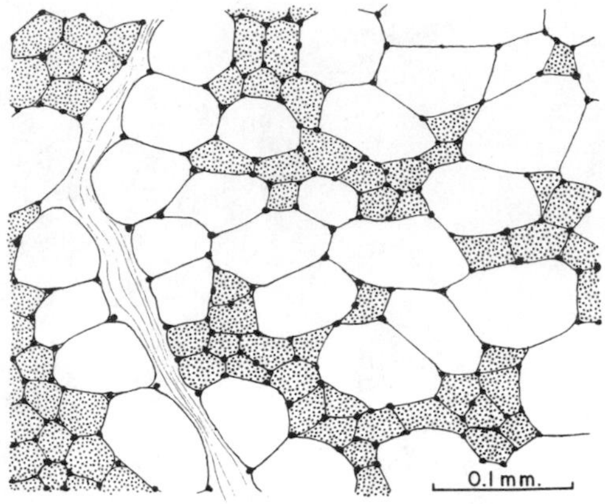

FIGURE 4–12 A magnified cross-section of a pigeon's pectoralis breast muscle, which contains both red and white fibers. The blood capillaries have been injected with India ink, and show as black dots. The narrow red muscle fibers (stippled) are more richly supplied with blood capillaries than the broad white fibers (clear). After George and Naik, 1960a.

and a higher percentage of intracellular fat than white muscles. The amount of fat in the pectoralis breast muscle of the strongly flying Black Drongo, *Dicrurus macrocercus*, of India, was determined by George and Naik (1960b) to be 5.6 per cent, while that of the domestic Chicken, *Gallus gallus*, was 0.98 per cent. In contrast, white fibers are rich in the carbohydrate

TABLE 4–1.
COMPARISON OF WHITE AND RED MUSCLE FIBERS

WHITE FIBERS	RED FIBERS
Thicker in diameter	Thinner in diameter
Nuclei within fibers	Nuclei on sides of fibers
Few mitochondria	Many mitochondria
Few blood capillaries	Many blood capillaries
High in glycogen	Low in glycogen
Little if any myoglobin and cytochrome	High in myoglobin and cytochrome
Low in fat compounds	High in fat compounds
Rich in phosphorylase enzymes for anaerobic breakdown of glycogen	Rich in enzymes for aerobic breakdown of fat and protein compounds
Use little oxygen	Use abundant oxygen
Contract a little more rapidly	Contract more slowly
Capable of strong but short-range flights	Capable of sustained, long-range flights
Fatigue rapidly	Fatigue slowly
Create heat (by shivering) using glycogen	Fat not used for shivering thermogenesis

glycogen. The chief known differences between white and red muscles are summarized in Table 4–1, based on data from George and Naik (1960a, 1960b), and from Hazelwood (1972).

The muscles of birds, like those of humans, respond to exercise or the lack of it by increasing or decreasing in size. In inactive cage birds the red muscles atrophy rapidly, the white, less so. If a domestic pigeon, *Columba livia*, is restrained from flying for nine weeks, there will be a significant reduction in the amount of myoglobin in its heart muscles (Catlett *et al.*, 1978).

In the arctic Snow Goose, *Chen caerulescens*, Hanson (1963) discovered a significant, reciprocal development of wing versus leg muscles. During the complete postnuptial molt the muscles of the legs "became ostrich-like in size and development whereas those of the sternal area (wings) were almost nonexistent." Hanson suggested that the metabolic breakdown of protein in the unused flight muscles contributed needed amino acids for growth of winter plumage. Also, the increase in size of the leg muscles prepared the molting and flightless bird for escape from predators by running and swimming.

All of these pronounced adaptive differences in muscle structure of birds unquestionably play an important role in determining the nature of such life activities as flight, migration, dispersal, habitat preference, size of territory, courtship display, song, diet, and method of feeding.

■ SUGGESTED READINGS

Newton's *A Dictionary of Birds* has much useful material on the skeleton and muscles of birds. Another older reference still worth consulting is Shufeldt's monographic *Osteology of Birds*. Comprehensive treatments of both skeletal and muscular systems are found in Stresemann's volume *Aves*, in Kükenthal and Krumbach's *Handbuch der Zoologie*; and Portmann's chapter on the skeleton, and Oemichen's on the muscles, in Grassé's *Traité de Zoologie, Tome XV, Oiseaux*. Bones and muscles are treated largely from the standpoint of their taxonomic importance in Van Tyne and Berger's *Fundamentals of Ornithology*. Recommended also is the chapter by Bellairs and Jenkin on the skeleton of birds in Marshall's *Biology and Comparative Physiology of Birds*. Very complete descriptions of bird muscles are found in George and Berger's *Avian Myology*.

Brain, Nerves, and Sense Organs
5

O to see like an eagle by day,
To hear like an owl in the dark,
To taste like a gull the salt spray,
To feel the air's lift like a lark!
—Joel Peters (1905–   ),
Wishing

Ever since the advent of bilateral symmetry, hundreds of millions of years ago, animals have been perfecting their organs of locomotion—fins, tails, legs, paddles, wings—largely to escape their enemies or to overtake their prey. Any animal that goes places in a hurry has to know where it is going and has to direct its movements with considerably more alacrity and precision than does a snail. Otherwise, disaster is certain. With the increasing velocity of their headlong flight across the landscape, animals have had to evolve more highly refined nervous mechanisms to handle their increasingly complex problems. An animal sees a moving shape in the distance. Is it friend or foe—or possibly a breakfast? Shall I stand fast, attack, or flee? If I attack, shall it be with stealth or with a rush? These questions, of course, are stated in anthropocentric form, but the problems presented exist for all the higher vertebrates, though they may be solved (or not!) on a much lower intellectual plane than man would solve them.

The terrestrial vertebrates, to whom these problems of locomotion are especially acute, have solved them in a variety of ingenious ways. The rattlesnake, eagle, skunk, lion, and rhinoceros can afford to stand firm in the face of most enemies. The Common Snipe, *Gallinago gallinago*, relies either on its concealing coloration and squats immovably, or seeks escape in zigzag flight. Many vertebrates confine their major activities to times or to places where potential predators are fewer than elsewhere; for example: deermice, to night; prairie dogs, to subterranean runways; breeding birds, to oceanic islands or arctic tundras. Birds, in the main, have so perfected flight through the air as a means of locomotion that even the most fragile finch can wear a coat of gaudy colors and sing its penetrating song on a twig only two leaps away from its inveterate enemy, the house cat. But it must take care to perch two leaps away and not one! All these and other solutions to the problems posed by locomotion depend,

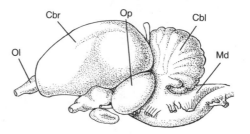

FIGURE 5–2 Brain of a goose, side view. Symbols as in Figure 5–1. After Romer, 1971.

of course, on a variety of nervous mechanisms. The organization and development of an animal's nervous system faithfully reflect the complexities and competence of its behavioral repertoire. It is worth remembering in this connection that behavior itself is based on movement.

THE NERVOUS SYSTEM

As in other vertebrates, the nervous system in birds consists of two major elements: the central nervous system, composed of the brain and spinal cord, and the peripheral nervous system, composed of the cranial and spinal nerves, the autonomic nerves and ganglia, and the sense organs. The peripheral nerves are of two kinds: afferent or sensory, which bring impulses from the sense organs to the central nervous system, and efferent or motor, which conduct impulses from the central nervous system to muscles and glands and thus affect behavior.

The chief functions of the central nervous system are to integrate information received through sensory impulses from the body and outside world, to store this information selectively in the form of memory and learning, and to integrate and coordinate outgoing motor impulses to the viscera and muscles into useful patterns of behavior (Goldby, 1964).

In any moving vertebrate body, it is the anterior end that first comes in contact with the environment, and hence with the associated problems that arise. As a consequence, its chief sense organs and related nerve structures are located here. In the primitive vertebrate the central nervous system, in the form of a neural tube, enlarges into three swollen centers, each associated with one of the three chief sense organs: the forebrain with the nose, midbrain with the eyes, and hindbrain with the ears. As terrestrial vertebrates evolved,

their brains, above all, had to keep pace with the increasingly complex problems brought on by locomotion. Accordingly, specialized layers and outgrowths of the brain developed to accommodate the increased traffic in sensory input and nervous coordination (Fig. 5–4). Cold-blooded reptiles have much smaller, simpler brains than do warm-blooded birds and mammals. For example, a lizard weighing 24 g has a brain weighing 0.134 g (or 0.55 per cent of its body weight), and a Meadow Mouse, *Microtus drummondi*, weighing 23 g, has a brain of 0.64 g (2.8 per cent), while a House Sparrow, *Passer domesticus*, weighing 23 g, has a brain of 1.02 g (4.4 per cent) (Quiring, 1950).

Reptiles, birds, and mammals all have brains based on a common structural plan but varying considerably in internal neural pathways and in the relative proportions of parts. Both birds and mammals have evolved greatly enlarged cerebral hemispheres and cerebella. The olfactory lobes of birds are very small, suggesting poor sense of smell. However, there is considerable evidence that many birds use their sense of smell

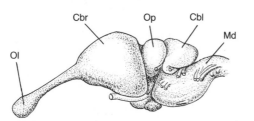

FIGURE 5–1 Brain of an alligator, side view, *Cbl*, Cerebellum; *Cbr*, cerebrum; *Md*, medulla; *Ol*, olfactory bulb; *Op*, optic lobe. After Romer, 1971.

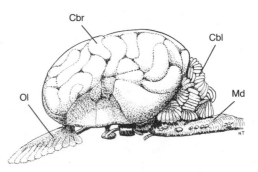

FIGURE 5–3 Brain of a pig, side view. Symbols as in Figure 5–1. After Odlaug, 1965.

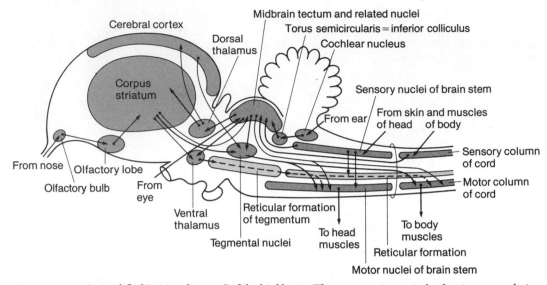

FIGURE 5–4 A simplified "wiring diagram" of the bird brain. The corpus striatum is the dominant correlating center. From Romer, A. S., 1971. *The Vertebrate Body—Shorter Version*. Philadelphia: W. B. Saunders Company.

in various contexts (see beyond). In the midbrain of both reptiles and birds the dorsally placed tectum has become prominent as a correlation center, especially for sight, while in mammals it is small and has lost most of its earlier coordinating functions. In mammals the outer layer or cortex of the cerebral hemispheres is the chief coordinating center of body activities, whereas in reptiles the corpus striatum, which develops from the basal nuclei of the cerebral hemispheres, is an integrating center second only to the tectum in dominance. The cerebral cortex in birds is thin, not fissured, and relatively weakly developed compared to the cortex in mammals; it is probably the seat of conditioned behavior. On the other hand, the bird's corpus striatum swells astonishingly in the floor of the cerebrum to become the dominant coordinating center of the brain. Here in the corpus striatum are located the bird's central controls for sensory perception and most of its instinctive behavior. Extirpation experiments have shown that different parts of the corpus striatum control eating, vocalization, eye movements, locomotion, and those complex instincts related to reproduction, such as copulation, nest construction, incubation, and care of young.

The automatic, built-in nature of these behavioral mechanisms becomes evident when certain brain regions are stimulated directly, culminating in predictable, stereotyped responses. A mild electrical stimulus applied within the anterior part of the upper or hyperstriatum (Fig. 5–5) results in the following complex behavior in a dove: the bird raises its head, erects its crown and neck feathers, then walks about, bowing and cooing (Brown, 1969). Implanting a pellet of the male hormone testosterone in the same region

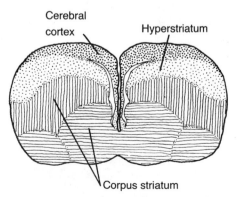

FIGURE 5–5 A diagrammatic cross-section of a bird's forebrain.

induces courtship, copulatory, and aggressive behavior in a dove, but in a chicken it activates only copulatory behavior (Barfield, 1971). The hyperstriatum seems to be particularly sensitive to control by hormones.

Although the cerebral cortex still functions in birds, it lacks the direct motor pathway to the spinal cord, the pyramidal tract, which mammals have. As a consequence it is likely that direct conscious control of body motions is greatly reduced in birds as compared with mammals.

That the cerebral hemispheres of birds are not completely essential to the performance of complex behavior is proved by experiments in which they have been surgically removed. Pigeons whose cerebral cortex has been removed but whose striatum is intact are able to mate and rear young. Removal of large sections of the corpus striatum results in serious disturbances in behavior (Prosser *et al.*, 1950). A falcon without hemispheres is able to capture a mouse, and a pigeon or domestic chick will pick up grain. But here the action stops, sterile and uncoordinated. The falcon blindly holds the mouse in its claws not knowing what to do next; the pigeon will not swallow the grain (Portmann, 1950). The essential difference between mammals and birds in brain coordination is that the brain of the mammal is dominated by the expanded top layer or cortex of the cerebral hemispheres, which has a high capacity for learning; whereas the bird's is dominated by the expanded bottom layer or corpus striatum of the cerebral hemispheres, which seemingly lacks this capacity. Consequently, the behavior of a bird is largely mechanical, stereotyped, and instinctive.

For all their size and importance, the corpora striata and cerebral hemispheres do not completely dominate body activities. Still lingering in the bird brain is a primitive compartmentalization or autonomy of regions that reminds one of brain mechanisms of fishes and amphibians. For that matter, even the spinal cord integrates some forms of reflex behavior, as is evident in the coordinated wing and leg movements of a beheaded chicken. The psychic importance of the cerebrum, however, is indicated by Portmann's studies of the relative size of the cerebrum in relation to other parts of the central nervous system. Birds of superior intelligence, such as parrots, owls, woodpeckers, and crows, have relatively larger cerebral hemispheres than the less intelligent fowls and pigeons.

The thalamus in birds, as in reptiles and mammals, is a visceral brain center. It shares control over visceral activities with the medulla and spinal cord, and acts as an intermediate transfer point between the spinal cord and corpus striatum for incoming sensory and outgoing motor messages. It is apparently the brain center that controls water balance, temperature regulation, sleep, and other vital activities. Lesions in the lateral area of the dorsal hypothalamus produced absence of thirst in Chickens (Lepkovsky and Yasuda, 1967). The hypothalamus is closely associated with the pituitary gland and probably secretes neurohormones that influence pituitary activity.

In birds the largest midbrain structures are the laterally bulging optic lobes. Their large size reflects the size and importance of the eyes in birds. It is chiefly because of the great size and the ventrolateral location of the eyes and optic lobes that the brain has been shifted upward and rearward in the skull.

Each eye of a cat is connected to the thalamic relay nucleus of the brain by contralateral and ipsilateral nerve fibers, i.e., each eye is connected by nerve fibers from the optic nerve to both hemispheres. In contrast, all fibers in the eye of the Barn Owl, *Tyto alba*, connect to the thalamic relay nucleus contralaterally; thus nerve fibers go from the right eye to the left brain hemisphere and from the left eye to the right hemisphere. However, some fibers from each thalamic relay nucleus cross in the supraoptic chiasm of the owl brain and reach the visual wulst in the opposite hemisphere. In this way visual stimuli from each eye reach both hemispheres (see Fig. 5–19). The avian visual wulst thus functions as does the mammalian visual cortex in integrating information from both eyes (Pettigrew and Konishi, 1976)

As one would expect of an animal showing superb muscular coordination in flight, birds have a large, well-developed cerebellum. Its size is related to the large number of spinocerebellar fibers from the spinal cord by which the muscle-sense, or proprioceptive, stimuli arrive from the body. As in other vertebrates, the cerebellum also receives impulses from the inner ear, which, along with the proprioceptive impulses, are assembled and used reflexively in coordinating body movements and in maintaining balance. Birds probably have a greater stake in precision of movement and equilibrium that any other animal, and accordingly the cerebellum is of crucial importance.

The medulla is the hindmost part of the brain. It lies under the cerebellum and gradually tapers into the spinal cord. It seems, as in man, to be a center for the reflex control of breathing. It also serves as a portal for the entrance of eight of the twelve cranial nerves. Only those cranial nerves having to do with smell and vision are located anterior to the medulla. See Figure 5-4 (from Romer, 1971), which shows some of the major nerve pathways within the avian brain.

In other vertebrates the spinal cord is often shorter than the vertebral column, but in birds the two structures agree closely in length. As in all vertebrates, the spinal cord in birds gives off paired spinal nerves, each with two roots, a dorsal sensory and a ventral motor root. As a rule, the sensory tracts within the spinal cord (bundles of nerves that carry sensory impulses to the brain) are smaller than the motor tracts. This condition results from a relative poverty of skin sense receptors, a condition promoted by the covering of feathers. Moreover, the fact that birds need not coordinate the movements of legs with wings as closely as quadrupeds must coordinate the movements of their four legs eliminates the need for considerable neural traffic between these two regions. There are cervical and lumbar enlargements in the spinal cord of birds, each associated with nearby nerve plexi supplying the wings and legs respectively. This is a typical vertebrate situation, but the lumbar enlargement is unusually and (so far) inexplicably large in birds. Some of its cells are richly stored with glycogen and lipids whose function is unknown.

Birds also possess an autonomic nervous system that, like man's, controls reflexes involving particularly the visceral organs and the circulatory system. The autonomic system is made up of efferent or outgoing nerve fibers that are grouped into two antagonistic systems. The sympathetic system, which arises in the thoracic and lumbar regions of the spinal cord, sends out impulses that serve to heighten the activity of the bird: they speed up the heart rate, constrict cutaneous blood vessels, and slow down the digestive activities; in short, they prepare the body for emergency action. Stimuli to the same visceral organs from the parasympathetic system, which arises in the brain and sacral spinal cord, produce exactly opposite results. They slow the heart beat and stimulate digestive processes (Romer, 1971). Secretion of the hormones of the adrenal gland into the blood stream produces effects on visceral organs that are indistinguishable from those caused by sympathetic nerve stimulation. The identical hormones, or neurohumors—adrenalin and noradrenalin—are given off at the ends of sympathetic nerve fibers to produce their effects. The substance

secreted by the parasympathetic nerves is the neurohumor acetylcholine. In birds there are exceptions to this arrangement, and certain autonomic nerve fibers give off other neurohumors than these (Bennett, 1974).

SENSE ORGANS

Sense of Touch

Birds are equipped with the varied sense organs that higher vertebrates typically possess, but with refinements and adaptations appropriate to their way of life. It is likely that a bird's skin possesses sense endings, much like those in man, which pick up stimuli interpreted by the brain as touch, pain, heat, and cold. Such endings are usually more abundant in skin that lacks feathers. The simplest receptors are nerves that zigzag through the stratum germinativum of the skin and end in a disk-like network of fine nerve fibers. These nerve endings are probably the pain and temperature receptors.

More complex are the touch receptors, which lie in the dermis of the skin and elsewhere. These are of two kinds: Grandry corpuscles, and Herbst, or lamellar, corpuscles (Fig. 5–6). Grandry corpuscles occur in the tongue and buccal cavities of ducks and owls. Each corpuscle resembles a sandwich composed of two bun-like sensory cells between which occurs a disk-like "filling" made of a flat terminal network of nerve fibrils. The function of these endings is not certainly known, but is thought to be tactile. Herbst corpuscles are highly developed in birds and are very similar to the Pacinian (touch) corpuscles in mammals. They are found highly concentrated in the tongues of woodpeckers, in the palates and beaks of ducks, and in the mouth flanges of helpless young "nidicole" or nest-dwelling birds.

In his study of the beak anatomy of 17 genera of shorebirds, Bloze (1968) discovered that those species primarily using sight to find food (e.g., Oystercatchers, *Haematopus ostralegus*) have Herbst corpuscles arranged in simple linear order along the length of the bill, but those species depending on touch in probing for food (e.g., snipe, sandpipers) have hundreds of small, bony pockets or alveoli, each containing many corpuscles, concentrated around the tip of the bill. Their effectiveness in locating prey was demonstrated by Kahl and Peacock (1963) through using Wood Storks, *Mycteria americana*, whose eyes were temporarily covered with masks. These birds, used to

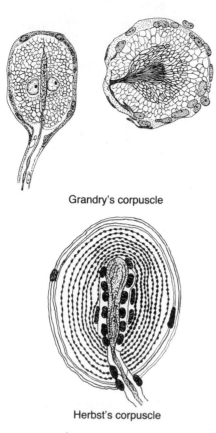

Grandry's corpuscle

Herbst's corpuscle

FIGURE 5–6 Touch, pressure, and vibration receptors from the bill of a duck. Grandry's corpuscle, in longitudinal and horizontal sections; Herbst's corpuscle, in longitudinal section. After Boeke and Clara, in Portmann, 1950.

fishing in turbid water, could close their beaks on a live fish in as little as 0.019 second after initial contact. In response to a loud noise, man requires about 0.040 second to blink his eyes.

By using highly sensitive electrical equipment on anesthetized ducks, Dorward (1970) found that the numerous Herbst corpuscles associated with the follicles of primary wing feathers were extremely sensitive and responsive to slight movements. It is very likely that they play an important sensory role in flight. It seems likely that, in addition to functioning as touch and pressure receptors, they also provide proprioceptive or muscle-sense impulses and are sensitive to low-frequency vibrations. Bullfinches, *Pyrrhula pyrrhula*, whose ears had been removed, were trained by

Schwartzkopff (1949) to respond to vibrations of from 100 to 3200 cycles per second. He found that an aggregation of hundreds of Herbst corpuscles between the tibia and fibula were the receptors. The individual Herbst corpuscle has a club-like core that is the swollen end of a sensory nerve. Arranged along two sides of this club are rows of large nuclei, and the whole core is encased in a capsule of concentric egg-shaped sheaths made of connective tissue.

Sense of Taste

Experiments indicate that birds have an acute sense of taste. For example, Sanderlings, *Calidris alba*, and Dunlins, *C. alpina*, could distinguish between jars containing sand and those containing sand from which worms had been removed before the experiment (van Heezik *et al.*, 1983). Hummingbirds may distinguish between different kinds of sugars (Stiles, 1976). The sense of taste enables birds to discriminate between food items and to reject toxic substances.

Avian taste buds may be of at least three shapes: ovoid as in most songbirds; elongate in ducks, flamingos, and oystercatchers; and round in parrots (Berkhoudt, 1985). Taste buds may be found in the oral mucosa or close to salivary glands. These buds may be in the upper or lower jaws, and on the sides and base of the tongue. Mallards, *Anas platyrhynchos*, have taste buds in the bill tip region.

Birds have few taste buds compared to mammals: man, 9000; rabbit, 17,000; rat 1,265; Mallard 375; parrot 350; Japanese Quail 62; Bullfinch, *Pyrrhula pyrrhula*, 46; Blue Tit, *Parus caeruleus*, 24.

Some species of birds are apparently insensitive to bitter tastes. Bread mixed with quinine and fed to parrots by Heinroth (1938) was eaten without protest, and grain dipped in picric acid was readily eaten by seed eaters and titmice. Many birds eat ants rich in formic acid. In a series of preference tests with pigeons, Duncan (1960, 1964) demonstrated that the birds markedly rejected sour and bitter aqueous solutions, preferred low concentrations of salt and high ones of sucrose, but were indifferent to glucose. Tests using electrophysiological techniques indicated that taste buds of chickens generated impulses when washed with distilled water and with solutions of salt, glycerine, quinine, and acetic acid, but did not respond to sucrose or saccharin solutions. Similar results were obtained with pigeons, except that quinine elicited no response, and about half of the birds responded to saccharin. Inasmuch as certain chemicals (e.g., phenylthiourea) elicit different taste sensations from different humans, different species of birds possess different taste capacities and thresholds. For example, parrots are sensitive to low sodium chloride thresholds (0.35 per cent) as compared to a partridge (20 per cent) (Berkhoudt, 1985).

Sense of Smell

As the small olfactory lobes of the brain indicate, the sense of smell is relatively poorly developed in birds. One need only imagine a winged bloodhound trying to follow a scent-trail in midair to realize in part why this is so. Nevertheless, birds can smell and they vary considerably in this ability. Changes in olfactory bulb electrical activity, and in heart and respiration rate have been noted when odors were presented to any species regardless of olfactory bulb size (Bang and Wenzel, 1985). Electroresponses of olfactory receptor cells in pigeons varied directly with odor concentration (Tonosaki and Shibuya, 1985). In a study of over 108 species of birds, Bang and Cobb (1968) compared the sizes of olfactory bulbs (expressed as the diameter of one olfactory bulb of a given species as a percentage of the greatest diameter of the bird's forebrain). This ratio varied from 37 per cent in the Snow Petrel, *Pagodroma nivea*, to 3 per cent in the Black-capped Chickadee, *Parus atricapillus*. Birds with the largest olfactory bulbs were found among the Apterygiformes, Procellariiformes, Podicipediformes, Caprimulgiformes, and Gruiformes; the smallest, among the Piciformes, Passeriformes, and Psittaciformes. Olfactory bulb size varies also with ecology. It is large in ground-nesting, water-associated, colonial-breeding, carnivorous and piscivorous birds; small in vegetarian, frugivorous, insectivorous, polyphagous and seed-eating birds (Bang and Wenzel, 1985).

In the typical bird two nasal ducts, separated by a thin bony or cartilaginous partition, originate at the external nostrils, run across the palate of the upper beak and end in two large openings or choanae, which connect the tubes with the mouth cavity. Each of these tubes swells, in anteroposterior sequence, into three chambers. In each of these chambers the walls form either a fold or a spiral scroll or concha, which increases the surface over which inhaled air must pass. The first chamber or vestibule, adjoining the nostrils, is usually shaped irregularly, lined with epithelial cells, and provided with a rich blood supply. In the middle chamber the duct develops a prominent scroll of one to two-and-one-half turns—somewhat like a

loosely rolled sheet of paper—and its surfaces are lined with a layer of ciliated epithelial cells. In ducks, the concha of the middle chamber may have up to five turns. The third chamber, which connects with the mouth cavity through the paired choanae, is usually the only one that has olfactory receptors in its lining epithelium. The receptors are slender, columnar cells, each with one or more olfactory hairs extending to the surface of the nasal epithelium. Impulses from these cells are carried to the brain by the olfactory cranial nerve.

Apparently the functions of the first two chambers are to moisten, warm, and clean the air a bird breathes. Because the choanae between the third chamber and the mouth are so large—much larger than the external nostrils in pelicans, for example—it seems probable that they carry odors from the mouth directly to the olfactory receptors. Thus, a bird is able to test food by smelling it while it is held in the mouth. The nocturnal kiwi has stunted eyes but a sharp sense of smell. Its nostrils are located at the tip of its long, probing bill, and the unusually well-developed conchae of the second and third nasal chambers are both covered with an olfactory epithelium. Experimental studies indicate that the kiwi finds its underground prey (largely earthworms) by smelling them (Wenzel, 1972). It has a bulb-to-hemisphere ratio of 34 per cent (Bang and Cobb, 1968).

Birds with olfactory bulb-to-hemisphere ratios of over 28 per cent (e.g., procellariforms and Turkey Vultures) have been shown to have well-developed olfactory abilities (see below). However, Black-billed Magpies, *Pica pica*, with ratios of below 6 per cent are capable of locating buried caches of raisins or suet laced with cod-liver oil (Buitron and Nuechterlein, 1985). The authors suggest that perhaps large olfactory bulbs are not indicative of greater sensitivity to odors, but rather an ability to discriminate more accurately among odors.

Some birds have shown definite abilities to smell potential foodstuffs. It has long been known that albatrosses, petrels, and other sea birds are attracted by fats, especially hot animal oils, spread on sea waters (Murphy, 1936). In experiments with breeding Leach's Storm-Petrels, *Oceanodroma leucorrhoa*, Grubb (1974) showed that captive birds placed in a Y-maze "olfactorium" would choose an air current coming from their own nest material in preference to one from similar materials collected from the forest floor. Petrels taken from their burrows and released did not return if their nostrils were plugged or if their olfactory nerves were cut. Sham-operated controls did return. Carefully controlled experiments under natural conditions at sea by Hutchison and Wenzel (1980) showed that procellariform birds (e.g., albatrosses, shearwaters, fulmars, and storm-petrels) were consistently attracted to sources of food-related odors, approaching them chiefly from down-wind, whereas birds of other orders (e.g., gulls, terns, pelicans, cormorants, and auks) showed no such wind-direction-related approach. Black-chinned Hummingbirds, *Archilochus alexandri*, could distinguish between scented and unscented feeders (Goldsmith and Goldsmith, 1982).

In a series of convincing experiments, Stager (1964, 1967) has demonstrated that the Turkey Vulture, *Cathartes aura*, unlike many other vultures and condors, is definitely capable of locating concealed animal baits entirely by their odors. In the same experiments the birds completely ignored a well-exposed, properly posed, stuffed specimen of a mule deer. The keenness of the birds' sense of smell was further revealed by some enterprising engineers who placed ethyl mercaptan (the odorous substance in carrion) in gas pipe lines to discover leaks. Leaks along a 42-mile-long pipe line were immediately located by observing aggregations of Turkey Vultures circling above them!

In addition to detecting food sources, olfaction may enable some petrels to communicate with pheromones. Bulwer's Petrel, *Bulweria bulweria*, emits a strong, musky odor just before egg-laying and is silent during nocturnal visits. The Tahiti Petrel, *Pterodroma rostrata*, is nocturnal, and characterized by a strong odor. Pheromonal cues have apparently replaced the long, elaborate aerial displays typical of other *Pterodroma* (Thibault and Holyoak, 1978).

Another possible function of the nasal cavity is suggested by the studies of the beaks of "dynamic gliders," such as the albatrosses, petrels, and fulmars, by Mangold and Fürst (Mangold, 1946). Dissections of the nasal chambers of these birds revealed a pair of small forward-opening pockets in the middle chamber which may act as organs for detecting variable pressures produced by differing external air-stream velocities (Fig. 5–7).

Sense of Hearing and Balance

Very probably, the ear arose in the earliest vertebrates not as an instrument for hearing but as an organ of equilibrium. As an organ of balance and motion-perception it reached such a high state of perfection even in the fishes that its basic structure has remained unchanged up the evolutionary ladder all the way to mammals. As an organ of hearing, however, it started out in the fishes as a simple, inefficient device that became progressively refined in the terrestrial reptiles, birds, and mammals.

As in mammals, the bird's ear is divided into three parts: the external, middle, and inner ear. The external ear ordinarily is merely an open tube that carries sound waves from the surrounding air inward to the eardrum at its base. To the rear of the ear opening in many birds is a muscular rim of skin with compact feathers that can be raised to form a sort of hearing trumpet, or, in diving birds, lowered to cover the opening. The middle ear in birds is essentially a cavity in which a rod-like bone, the columella, picks up sound vibrations of the eardrum and transmits them to a membranous oval window in the inner ear. The movements in the oval window, in turn, set up vibrations in the liquids of the inner ear. It is the inner ear that is the sensory receptor for both equilibrium and

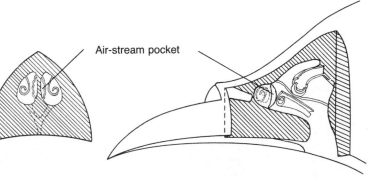

Air-stream pocket

FIGURE 5–7 Cross- and longitudinal sections of the nasal chambers of a fulmar, showing the location of the valve-like pockets, which may serve sea birds as air-velocity sense organs to aid them in exploiting winds of varying speeds during dynamic gliding. After Mangold, 1946.

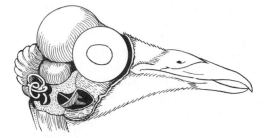

FIGURE 5–8 Dissected inner ear of the pigeon (below and left from the eye) showing the semicircular canals and, curving downward below them, the cochlea. After Krause, from Portmann, 1950.

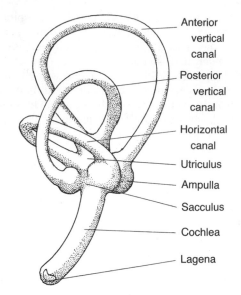

FIGURE 5–9 The right membranous labyrinth of the pigeon.

sound. Impulses of both kinds are carried from the ear to the brain by the eighth cranial, or auditory, nerve. The general construction of the inner ear in birds resembles closely that of crocodiles. It is a complex and delicate structure of bulbs and tubes (somehow reminding one of the contemporary vogue in abstract sculpture) and is called the membranous labyrinth. It is filled with a fluid about two or three times as viscous as water, the endolymph, and encased by the bony labyrinth of the intimately conforming bone of the surrounding skull. The membranous labyrinth does not adhere closely to the bony labyrinth, but is separated from it by a plasma-like fluid, the perilymph.

The central structures of the membranous labyrinth are two connected chambers, the utriculus above, and the sacculus below. Arching out from the utriculus are three semicircular canals (Fig. 5–8), each arranged roughly at right angles to the other two: the anterior, the posterior, and the external or horizontal canals. At its lower end against the utriculus, each canal swells into a bulbous ampulla. Projecting downward from the sacculus like a curved finger is the endolymph-filled cochlear duct, the hearing organ (Fig. 5–9).

In each membranous labyrinth there are six main sensory areas: one in the ampulla of each of the three semicircular canals, and one each in the utriculus, the sacculus, and the cochlea. All sensory areas except that in the cochlea, the organ of Corti, have to do with perception of movement and position. In addition to these six sensory areas, there are two others of uncertain function, one in the utriculus and one at the tip end of the cochlea. The actual receptors inside the ampullae, utriculus, and sacculus are in the form of hair-like sensory cells on whose tips is a gelatinous

membrane. In the utriculus and sacculus this membrane is impregnated with calcium carbonate crystals, the so-called otoliths, which respond to gravitational pull and stimulate the nearest sensory cells to send impulses that the brain interprets as posture or position in space. In the ampullae, with which the semicircular canals connect, the sensory cells respond instead to linear and angular accelerations and send corresponding impressions to the brain. All these receptors collect stimuli from the inner ear, which together with others from the eyes and from proprioceptors in the body, are sent to the brain, where they are integrated to produce compensatory reflexes in the limbs and eyes, and in general maintain posture and balance.

Experiments on live pigeons reported by Coues (1903) revealed that sectioning of the horizontal canal caused the bird to move its head rapidly from side to side in a horizontal plane. The eyes oscillated in nystagmus, and the bird showed a tendency to spin on a vertical axis. Cutting the anterior vertical canals caused the bird to move its head rapidly backward and forward and to turn forward somersaults, while cutting the posterior canals produced similar head motions and backward somersaults.

If two-day-old embryos of the domestic Chicken, *Gallus gallus*, have one ear primordium either

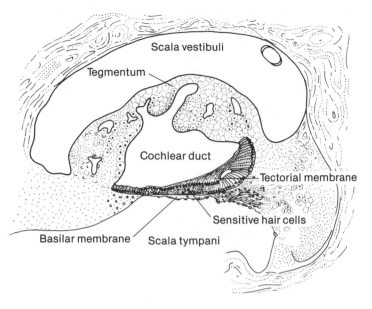

FIGURE 5–10 Diagrammatic cross-section of the cochlea of a duck's ear. Sounds of different frequencies entering the scala vestibuli are thought to cause oscillations at different longitudinal levels of the basilar membrane. Hair cells at these levels send impulses to the brain that are interpreted as sounds of different pitches. Redrawn after Schwartzkopff, 1955.

removed, or rotated 180 degrees and reimplanted, the chick, upon hatching, will show head nystagmus and gross abnormalities in locomotion, balance, and pecking accuracy. However, a few days later its behavior becomes apparently normal (Heaton, 1975).

Bird songs and birds' imitations of sounds in the environment would make little sense unless birds themselves could hear. Both in their ability to make varied sounds and in the numerous ways they respond to them, birds outrank all other vertebrates, man excepted. Hearing in birds depends on the cochlea of the inner ear, which appears to function in birds in essentially the same way as it does in man (Fig. 5–10). The organ of Corti, which is the auditory receptor in the cochlea, rests on a basilar membrane that, stretched from side to side, runs the length of the cochlear duct. On it rest the sensitive hair cells that are in contact with an over-arching tectorial membrane. The basilar membrane varies systematically in a flexibility gradient, being stiffest near the oval window and most flexible at the apex of the cochlea (Konishi, 1974). When the endolymph vibrates in response to movements of the columella against the oval window, vibrations of different frequencies are thought to stimulate the hair cells at different levels along the basilar membrane. Impulses from these cells are then interpreted by the brain as sounds of different pitch. Although the cochlea of the average

bird is approximately only one-tenth the length of the mammalian cochlea, it has about ten times as many hair cells per unit of length. Optimum hearing sensitivity in birds is in the frequency range of 1 to 5 kHz· (Fig. 5–11). Sensitivity declines dramatically for frequencies above this range, reaching a limit of

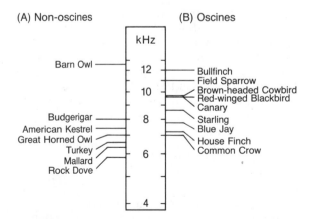

FIGURE 5–11 Comparison of oscines (songbirds) and non-oscines on the basis of high-frequency cutoff (kHz) at 60dB. Note that with the exception of the Barn Owl, most songbirds tend to be able to hear at higher frequencies than nonoscines. From Dooling (1962).

about 10 kHz. Owls do a little better, with a hearing limit of about 12 kHz (Dooling, 1982). Schwartzkopff (1963) reports that songbirds and parakeets can discriminate between tones (frequencies) differing by not more than 0.3 per cent; pigeons, by 5.0 per cent. But probably the chief functional difference between the avian cochlea and the mammalian cochlea is a consequence of the thick, folded glandular tegmentum that in birds practically fills the scala vestibuli chamber (essentially empty in mammals) above the sensory basilar membrane. Sound waves entering the scala vestibuli from the oval window are probably dampened by the tegmentum, and this dampening enables the basilar membrane to "change its pattern of movement within a very short time. This would correspond with recent results of song analysis . . . showing that the song of many birds contains such rapid sequences of notes that the human ear cannot follow them . . . ," but a mimicking Northern Mockingbird, *Mimus polyglottos*, can. In short, birds' ears are capable of exceedingly fine temporal resolution of sound, probably 10 times that of man. Konishi (1969) has determined that single auditory neurons of birds can discriminate discrete sounds separated in time by as little as 0.6- to 2.5-millisecond intervals. Further evidence of extremely short temporal discrimination is seen in the echolocation navigation of the Oilbird, *Steatornis caripensis*. This bird emits short sound pulses (ranging from 1 to 15 kHz, with dominant frequencies of 1.5 to 2.5 kHz), repeated at intervals of 2.3 msec, and uses their echoes to avoid obstacles in dark caves. It may avoid discs as small as 20 cm in diameter while flying in total darkness (Konishi and Knudsen, 1979).

Most birds are unable to hear ultrasonic vibrations, but pigeons and Guinea Fowl, *Numida meleagris*, are able to hear extremely low frequency sounds (infrasounds)—as low as 2 to 10 Hz (Theurich *et al.*, 1984). Below ten cycles per second pigeons "are at least 50 decibels more sensitive than humans" (Kreithen and Quine, 1979). As a rule, the frequency range that a given species can hear is narrower in birds than in man (Schwartzkopff, 1973).

In general, among related species the larger birds have deeper voices and a corresponding trend in hearing sensitivity. But Schwartzkopff points out that there are exceptions to this size-voice relation that have interesting references to behavior. For example, many owls have an unexpected sensitivity to high tones that correspond to the squeak of a mouse. Baby chicks are almost exclusively sensitive to the low clucks of the hen (400 cycles), while the hen shows an extraordinary sensitivity to the high cheeping of her chicks (above 3000 cycles).

In many birds the sensitivity of ears to intensity or volume of sound remains rather constant between 1000 and 6000 cycles per second, but beyond this upper level the sensitivity threshold shoots up abruptly at a rate of over 70 decibels per octave (Whitfield, *in* Pearson, 1972).

Owls possess unique refinements in their hearing equipment that account for their amazing sensitivity to low-intensity sounds. Externally, many species have large, oblong ear-openings bordered in front with a fleshy, erectile flap or operculum, which is framed in small contour feathers. The whole arrangement resembles, and no doubt functions like, a cupped hand held in *front* of one's ear to reflect and concentrate sounds coming from the rear. A further refinement is seen in the striking asymmetry of the external ears. In at least nine different genera of owls the left and right ear openings differ greatly in size and location (Fig. 5–12). This asymmetry of external ears, coupled with an asymmetry of the brain's auditory centers, apparently serves to refine vertical localization of sounds through binaural comparison of their intensity and frequency patterns (Norberg, 1977). It seems reasonable to assume that just as men (and owls) rotate their heads around a vertical axis to localize sounds horizontally, owls commonly rotate their heads around a horizontal axis to localize or "take a fix" on sounds coming from a vertical direction.

Coupled with these external adaptations, owls have very large eardrums, columellae, and cochleae. In addition, nocturnal owls are richly equipped with auditory neurons in the medulla of the brain: a 300-g Barn Owl, *Tyto alba*, has about 95,000 neurons, whereas a 600-g Carrion Crow, *Corvus corone*, has only 27,000 (Schwartzkopff, 1963). The wide head of the owl also helps it locate its prey, since the ear openings are far enough apart to create an appreciable time difference in the arrival of a sound they receive. Owls are said by Schwartzkopff to reach at least the human difference threshold of 0.000,03 second. It is this slight time difference that tells us, for example, that a given sound is at our left instead of at our right. A bird is also able to detect the direction of a sound through the differences in intensity with which it strikes each ear. A sound coming from the left, for example, is louder in the left ear than in the right. Finally, for a minority of species, phase differences in sounds of low

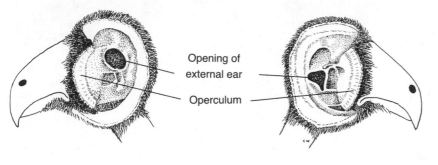

Opening of
external ear

Operculum

FIGURE 5–12 Left and right external ear openings of an owl *(Asio)*, showing their marked asymmetry. After Pycraft, 1910.

pitch as they strike the two ears may give directional clues to the sound's source. However, this method of locating sounds is effective only if the wave lengths of the tones are longer than at least two times the owl's interaural distance (Konishi, 1973). By means of these three clues—differences in timing, loudness, and phase—birds can locate sources of sounds with considerable precision. A Barn Owl, using only its sense of hearing, can locate prey in total darkness with a deviation of only about 1 degree, both in the vertical and horizontal planes (Payne, 1961). Its ears are most sensitively directional with sounds above 9000 cycles per second.

While observing them with infrared illumination, Konishi trained Barn Owls to seize mice in absolute darkness in a soundproof room. When the mice towed pieces of rustling paper by a thread across the floor, the owls consistently struck at the paper, proving that they used neither odor nor infrared radiation from the mice to guide them. Konishi discovered, surprisingly, that the heart-shaped facial disc of densely packed feathers served as a sound-collecting and focusing device. When he removed these feathers the owls made large errors in finding the target. The owl can control the positions of both the disc feathers and the ear flaps—changes that markedly affect the directional sensitivity of the ears.

As a final perfection, velvety feathers muffle the sounds of an owl's flight, so that in addition to hearing its prey very acutely, the owl makes no warning sound in swooping down for the kill. Sounds made by the owl's wings in flight are very faint and lack the higher frequencies (above 1000 cycles per second) to which its ears are most sensitive. Recent experiments indicate that directional hearing of the Northern Harrier, *Circus hudsonius*, is substantially better than a variety of diurnal raptors and comparable to that of owls.

A Northern Harrier could detect prey at a maximum range of 3 to 4 m as compared with 7 m for the Barn Owl (Rice, 1982).

Sense of Vision

A bird can gain more information about its surroundings through its eyes than through all its other sense organs together. An eye can detect the direction, distance, size, shape, brightness, color hue, color intensity, three-dimensional depth, and motion of an object. Combined with the other senses, sight can provide almost perfect information about one's environment. It is not surprising that natural selection has improved on the first dim vertebrate eye, making continuous progress toward sharper acuity, color vision, three-dimensional perception, and other refinements in vision. The eye of the bird has reached a state of perfection found in no other animal.

The avian eye contains numerous legacies from its reptile ancestors, and today resembles in general the eye of living lizards. Variation in the eyes of birds is relatively slight, being no greater throughout the entire class Aves than it is within a single order or suborder of reptiles or amphibians (Walls, 1942).

Birds have enormously large eyes: hawks, eagles, and owls often have eyes actually larger than man's. The eye of the Ostrich, 50 mm in diameter, is the largest of any land vertebrate. Although the weight of the head in both Starlings, *Sturnus vulgaris*, and man is about one-tenth of the total body weight, the ratio of eye weight to head weight in man is less than 1 per cent; in Starlings it is about 15 per cent (Pumphrey, 1961). The value of the large size is, of course, that it provides larger and sharper images—most valuable qualities for rapidly moving animals. In shape the eyeball may be globose or somewhat flattened in the optical axis or, contrariwise, lengthened

FIGURE 5–13 A and B. A Saw-whet Owl striking and carrying off a mouse. These photographs, by G. R. Austing, epitomize adaptations vital to the owl's survival: broad, soft-feathered wings for silent flight, powerful talons and beak for seizing and tearing prey, and hypersensitive eyes with large, light-gathering pupils for locating prey.

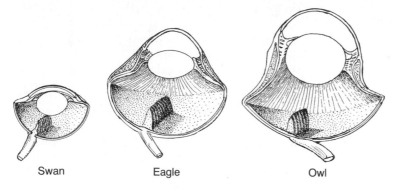

Swan   Eagle   Owl

FIGURE 5–14 Characteristic shapes of birds' eyes. Each figure represents the ventral half of the left eyeball. After Soemmering, from Walls, 1942.

into a somewhat tubular form (Fig. 5–14). As a rule, nocturnal birds such as owls have tubular eyes, and diurnal birds commonly have flattened or occasionally globose eyes. The majority of nocturnal and crepuscular birds (e.g., the Swallow-tailed Gull, *Creagrus furcatus*; Boat-billed Heron, *Cochlearius cochlearius*; Nicobar Dove, *Caloenas nicobarica*; and many others) have larger eyes than their day-living relatives. This eye adaptation to weak illumination is doubtless inherited, but it has been shown that domestic chickens reared under low intensity light developed abnormally large eyes (Bercovitz *et al.*, 1972): after 70 days, the eyeballs of those living in dim blue light were 17 per cent longer and 95 per cent heavier than those of controls reared under normal white light.

In most of its structures the bird's eye resembles the human eye. The eyeball is a three-layered organ with a tough scleroid coat on the outside. Toward the front this coat becomes, with the overlying skin, the trans-

parent cornea. Below the sclera comes the vascular and pigmented choroid coat, and on the inside of the eyeball, the sensitive retina that sends its impulses via the optic nerve to the brain. Just behind the cornea is the anterior chamber, filled with a clear fluid, the aqueous humor. It is followed, in the direction of the retina, by the pigmented iris, then by the crystalline lens, supported at its equator by the annular pad, or *ringwulst*, and the ciliary body. In the main body of the eyeball is the gelatinous vitreous humor. The scleroid coat is reinforced by a circle of usually 11 to 16 small shingle-like bones, the sclerotic ring, surrounding the cornea (Fig. 5–15).

The pigmented iris, containing circular and radial striated muscle fibers, controls the size of the pupil of the eye, much as a diaphragm controls the aperture of a camera lens. It has extraordinary motility in birds, closing down the size of the pupil in bright light and opening it in dim. Many parrot species may contract

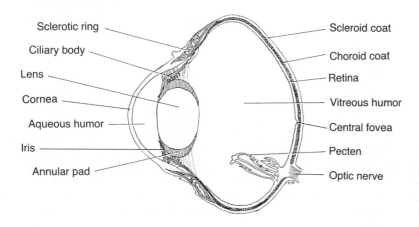

Sclerotic ring — — Scleroid coat
Ciliary body — — Choroid coat
Lens — — Retina
Cornea — — Vitreous humor
Aqueous humor — — Central fovea
Iris — — Pecten
Annular pad — — Optic nerve

FIGURE 5–15 A horizontal section of the eye of the Common Buzzard, *Buteo buteo*, showing its major parts. After Portmann, 1950.

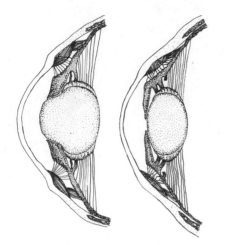

FIGURE 5–16 A bird's eye accommodated for near vision *(left)*, and far vision *(right)*.

the pupil at will and thus use the bright irises in social signalling. The pupil is generally circular in shape but, when constricted, may be oval, slit-like as in the Black Skimmer, *Rhynchops niger*, or even square as in the King Penguin, *Aptenodytes patagonica*. The lens is the highly refractive body in front of the vitreous chamber, which, with major help from the cornea, brings rays of light to a focus on the retina. It is elastic, and when at rest its shape resembles a thick disk in diurnal birds and almost a globe in nocturnal species.

In birds whose eyes have great power of accommodation (focusing) the lens is soft (notably in diving birds such as dippers, cormorants, and diving ducks); however, it is very firm in eyes with low accommodation. The need for extensive and rapid changes in accommodation of a bird that catches insects on the wing, or escapes its enemies by flying through trees and brush, is easily seen. Reflecting this need, most birds have strong powers of near-and-far accommodation (Fig. 5–16), which is measured in diopters (the reciprocal of the focal length of a lens in meters). Whereas a child has an accommodation of 13.5 diopters and a man aged 40 one of 6 diopters, a typical bird has one of 20 diopters; a cormorant possesses the extreme accommodation of 40 to 50 diopters; chickens and doves, 8 to 12 diopters; and night birds such as owls, 2 to 4 diopters (Stresemann, 1927–1934). Owls, unable to focus their eyes on close objects, must back away from food offered them in order to fix it sharply in their eyes

before they pounce. Penguins, formerly thought to be very near-sighted out of water, are now known to be only slightly so, and in water are only moderately far-sighted (Sivak and Millodot, 1977). The relatively flat shape of the cornea causes less change in the penguin's eye-focusing ability, as compared with man's, when changing from vision in air to that in water.

The exact means by which the avian eye accommodates is not known for sure, but there is general agreement that Brücke's muscles of the ciliary body act on the sclerotic ring and annular pad to bring pressure on the lens and thus change its shape. The action is thus different from that in the human eye, in which the ciliary muscles act to release a stretching tension on the lens, allowing the elasticity of the lens itself to determine its shape. In the case of some diving birds, such as cormorants, powerful iris muscles may squeeze the lens into a more rounded, shorter-focusing shape. On the other hand, one ring of ciliary muscles (Crampton's) in hawks and owls acts instead on the cornea to change its curvature and hence the focus of the eye. Still other mechanisms may be employed in eye accommodation in birds. The third eyelid, or nictitans, in diving ducks, loons, and auks has in its center a clear lens-shaped window of high refractive index that serves the bird under water as a "contact lens" (Walls, 1942). Unlike other birds, owls have opaque nictitating membranes. The Black-billed Magpie, *Pica pica*, uses its nictitans in an unusual way: the white membrane has a conspicuous orange spot that the bird displays only during hostile or courtship encounters with other birds. An ingenious adaptation is seen in the machinery on the back of the eyeball that activates this third lid (Fig. 5–17). The highly elastic nictitans is stretched across a bird's eye by means of a greatly lengthened tendon that circles the eyeball to the rear, passes through a sling in another muscle, the quadratus (which holds the tendon away from the optic nerve), and originates in the pyramidalis muscle, which, with the quadratus, does the pulling. The inner surface of the nictitating membrane is covered with epithelial cells that possess brush-like processes, so that the cornea is *brushed* with tears at every flick of this thin, transparent lid.

The eyelids proper are generally moved by smooth muscles, hence slowly; in most species the lids are used to close the eyes only in sleep, while the nictitans is reserved for blinking. Normally the eye is closed by raising the lower lid rather than lowering the upper lid as in man. However, in owls, parrots, toucans,

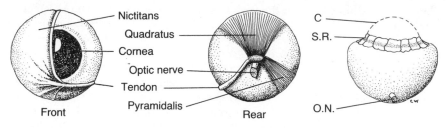

FIGURE 5–17 *(Left)* Front and rear views of the eyeball of a turkey, revealing the nictitating membrane mechanism. After Wolff, from Walls. *(Right)* Eyeball of the Golden Eagle: C, cornea; *S.R.*, sclerotic ring; *O.N.*, optic nerve. After Walls, 1942.

wrens, and ostriches it is the upper lid that is the more movable. Owls use their upper lids to wink and their lower lids to close the eyes in sleep. In birds the lids expose only the cornea and none of the sclera; and since the lids follow the motions of the eyeball, they disguise the large size of the eyes.

Another feature of the bird's eye that links it to the reptiles and sets it apart from the mammals is the pecten, a pigmented, conical, highly vascularized body. It arises near the attachment of the optic nerve and juts out in the vitreous humor toward the lens. Whereas in reptiles the pecten is a simple, cone-like, vascular body, in birds it becomes an elaborate structure of thin folds richly supplied with small blood vessels (not capillaries). The folds either radiate out as spokes from a hub, or snake back and forth in accordion pleats. Inasmuch as the bird's retina is devoid of blood vessels, it seems highly likely that the chief function of the pecten is to supply the eye with oxygen and nourishment and to carry away wastes. As a rule it is smallest in nocturnal birds and largest in diurnal predators.

In its general pattern of construction the bird's retina

is orthodox enough. From the outside (against the choroid coat) inward to the vitreous chamber, it possesses the usual sequence of layers: pigmented epithelium, sensory receptor layer (rods and cones), outer nuclear layer, inner nuclear layer of bipolar cells, ganglion layer, and nerve fiber layer. The explanation of the perfection of the avian retina lies in the abundance and distribution of these parts. Not only are the rods and cones more numerous and tightly packed than in other vertebrates, but the precisely layered conductive cells (Fig. 5–18) are also unusually abundant—so much so that the whole retina is from one-and-one-half to two times as thick as it is in most vertebrates (Walls, 1942). Just as the sharpness or resolving power of a photograph depends on the number and fineness of the silver grains that compose it, so the resolving power of a retina depends largely on the number and size of its sensory cells, especially the cones. Apparently the cone cells connect in a one-to-one fashion with the bipolar cells of the inner nuclear layer, which in turn connect individually with optic nerve-fiber ganglion cells. Therefore each cone cell can have individual representation in the brain—a fact

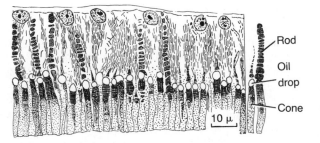

FIGURE 5–18 A cross-section of the rod and cone layer of the retina of the Eurasian Kestrel. In life, the oil droplets above the cones are colored yellow, orange, or red. After Rochon-Duvigneaud, from Grassé, 1950.

ensuring sharp acuity. On the other hand, one bipolar cell ordinarily serves numerous rods—a situation that results in inferior visual acuity in the brain.

As in man, the point of sharpest vision or highest resolution is the fovea, where the overlying nervous tissue is thinned away and the visual cells are packed together in a funnel-shaped pit. Since it is normally the color-perceiving cones that are grouped in the fovea rather than the more sensitive rods, the fovea loses light sensitivity as it gains resolution. Anyone can test this by looking a little to one side of an extremely faint star and then looking directly at it (i.e., fixing it on the fovea) only to have it disappear. A further increase in the acuity of foveal vision is provided by the retinal cells themselves, which have a higher index of refraction than the adjacent vitreous body. Rays of light entering the eye and striking the slanting walls of the foveal pit will therefore be bent outward, magnifying any image that falls on the foveal cones by about 13 per cent linearly, or 30 per cent in area (Walls, 1942). Surrounding the fovea is the so-called "central area" where resolution is higher than in the remaining retina but lower than in the fovea.

Cones are more numerous than rods in the retinas of day birds, and rods more numerous than cones in night birds such as owls and goatsuckers. In the latter the rods are particularly long and thin, are rich in glycogen deposits, and have about 25 times as much scotopic (night vision) pigment as do the rods of diurnal birds (Sillman, 1973). All birds have at least one central area, usually with a fovea, and many diurnal species, especially those that feed on the wing, have a second, temporally placed fovea, useful in binocular vision: e.g., hawks, terns, parrots, swallows, swifts, hummingbirds, and others. Terns and swallows of the genera *Sterna* and *Hirundo* have three foveae in each eye (Chievitz, *in* Stresemann, 1927–1934). Many birds have a horizontal streak or "central" area across the retina, usually with a fovea at each end. These are the hawks, eagles, swallows, ducks, terns, and many shore birds—in short, birds of open country. It does not occur in forest birds. It is significant that this horizontal sensitive streak is parallel not to the long axis of a bird's skull but to the horizon when the bird's head is held in its normal position. Such a device permits a sharp but economical scanning of the horizon without eye or neck movements. The Black-footed Penguin, *Spheniscus demersus*, has only a horizontal central area and no fovea. In many diurnal birds of prey, the sensory cells are more numerous

in the upper hemisphere of the eye (which perceives images from the ground) than in the lower hemisphere (which views the sky). As a consequence the Goshawk, *Accipiter gentilis*, for example, when it wishes to scan the sky more sharply, inverts its head either over its back or down near its belly (Stresemann, 1927–1934).

The "eagle-eyed" sharpness of birds' vision is borne out both by experimental studies on living birds and by microscopic examination of their eyes. Several species of thrushes, finches, and other passerines were trained by Donner (1951) to discriminate between a grating pattern and a solidly gray object of the same brightness to obtain food. By using coarser and finer gratings, he was able to determine the limits of a bird's visual acuity. For these passerines he found that the finest detectable bars were subtended by visual angles of from 20 seconds (20″) to 3 minutes and 50 seconds (3′50″) of arc. This compared very favorably with man's minimum visual angle of about 38 seconds, especially when one considers the small size of the eyes in these birds. A similar study of the gamecock's eyes, which have no fovea, revealed a minimum visual angle of 4 minutes and 4 seconds. By use of choice-reward experiments, Fischer (1969) was able to determine the smallest detectable visual angles for the following species of vultures: Long-billed Vulture, *Gyps indicus*, 13.3 seconds; Griffon Vulture, *G. fulvus*, 17.2 seconds; Egyptian Vulture, *Neophron percnopterus*, 68.9 seconds.

Microscopically, the bird's retina is shown to have even higher acuity. Outside the foveal region, the White Wagtail, *Motacilla alba*, has 120,000 visual cells per square millimeter of retinal surface, while man, even *in* the fovea, has only 200,000. As Walls remarks, the "grand champion of all foveae" is probably that of the Common Buzzard, *Buteo buteo*, which boasts 1,000,000 cones per square millimeter of fovea, which must give it a visual acuity "at least eight times that of man." Other authorities consider avian visual acuity superior to but closer to that of man (Sillman, 1973).

To carry the mosaic perceptions of the retina to the brain where they are assembled into an image requires a great number of nerve fibers. We have seen earlier how some fibers connect each eye to the visual wulst in the opposite hemisphere (Fig. 5–19). The number of nerve fibers in a cross-section of the optic nerve was determined by Bruesch and Arey (*in* Polyak, 1957) for various vertebrates. Here again, birds show their superiority over most other

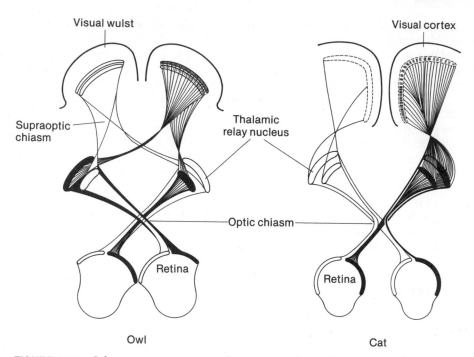

FIGURE 5–19 Schematic representation of the organization of forebrain visual pathways in both the owl and the cat showing how the visual wulst in the owl and the visual cortex in the cat function to integrate information from both eyes. Note that all nerve fibers from the owl's eye cross at the optic chiasm, i.e., they are all contralateral. Some fibers cross once again when they emanate from the thalamic relay nucleus and connect to each visual wulst in the opposite hemisphere. In contrast, the cat's eye is connected to the thalamic relay nucleus by both ipsilateral and contralateral fibers. Modified from Pettigrew and Konishi, 1976.

vertebrates. Man has somewhat over 1,000,000 nerve fibers coursing through each of his optic nerves. Other counts were: canary, 428,000; chicken, 414,000; duck, 408,000; pig, 681,000; dog, 154,000; alligator, 105,000; frog, 29,000; *Necturus maculosus*, 362; goldfish, 53,000; sturgeon, 13,500; brook lamprey, 5217; hagfish, 1579. More recent electron microscope studies by Binggeli and Pauli (1969) have shown not only that the pigeon's optic nerve has more fibers (2,380,000) than formerly thought, but that it has about 10,000 efferent or "centrifugal" fibers carrying impulses from the brain to the eye. Such outgoing fibers are unknown in the optic nerves of other vertebrates. Although their function is unknown, experiments reveal that their electrical stimulation causes the ganglion cells of the retina to "fire" more readily; i.e., it alters their excitability (Miles, 1970).

COLOR VISION. A further refinement of birds' eyes is the abundant presence of colored oil droplets in retinal cone cells, one droplet to a cone. Diurnal birds typically possess red, orange, yellow, and green oil droplets. Nocturnal birds have mostly colorless or pale yellow droplets. Each droplet is a mixture of lipids containing carotenoid pigments. Their function is presently unknown, but various theories have ascribed the following possible roles to them: they increase visual acuity; sharpen contrast; reduce glare; pierce haze; lessen chromatic aberration; and somehow assist in color discrimination. However, when Japanese Quail, *Coturnix coturnix*, were fed a diet free of carotenoids, their chicks developed retinas with colorless oil droplets. These chicks nevertheless were able to distinguish red, yellow, green, blue, and white lights of equal intensities (Dücker and Schulze, 1977).

This indicates that colored oil droplets are not essential to color vision in quail. On the other hand, Bowmaker and Knowles (1977) determined that there are six types of oil droplets in the chicken retina, four of which act as filters with peak transmission of light wavelengths of 575, 520, 497, and 454 millimicrons (mμ) respectively. Besides the oil droplets there are at least four visual pigments in chicken and pigeon cone cells that are thought to be involved, as in other vertebrates, in color vision (Govardovskii and Zueva, 1977). Whatever the basic mechanism involved, microelectrode studies of the ganglion cell layer of the pigeon's retina revealed the apparent presence of three different cellular networks, each responding with peak sensitivity to different regions of the spectrum at wave lengths of about 480 (blue), 540 (green), and 600 (orange) mμ (Donner, 1953).

In view of the colorful plumage of many birds it would seem incredible if they lacked color vision. Whereas man is able to discriminate about 120 different hues of the solar spectrum, a pigeon, under restrictive experimental conditions, can distinguish 20 (Walls, 1942), honeycreepers 12 (Winkel, 1969a), and Emus, *Dromaius novaehollandiae*, 8 (Neumann, 1962). Both honeycreepers and the Emu distinguished colors more easily than shapes or patterns of test objects.

Experiments by Kear (1964) on the naive, or unlearned, color preferences of 41 species of ducklings and goslings revealed a general, innate preference for pecking at spots colored green, and in a few species, yellow. Some goslings tended to avoid blue, others, red and orange. Color preference experiments on chicks of the Laughing Gull, *Larus atricilla*, by Hailman (1966) showed that they pecked most frequently at red and blue stimulus objects. When an achromatic object was placed against colored backgrounds, it received the most pecks when the background was green or yellow. Presumably the achromatic objects assumed in the bird's brain, as in man's, subjective tints of the preferred colors, red and blue, when positioned against their complementary colors of green and yellow. By exploiting the tendency of certain species to remove strange objects, especially fecal pellets, from their nests, Peiponen (1963) found that European Robins, *Erithacus rubecula*, and Bluethroats, *Luscinia svecica*, showed the greatest preference for removing colors similar to their own plumage: orange or blue. This reaction correlates well with the Robin's well-known antipathy for any bird or object of its own color in its territory.

Without doubt, the innate color preferences of many species are adaptive in origin. Newly hatched Black-legged Kittiwakes, *Rissa tridactyla*, which take their food from their parents' mouths, peck less at yellow (the color of the parental beak) than at red (the color of the mouth) (Cullen and Cullen, 1962). Hummingbirds feed predominantly at red flowers. When naive Anna's Hummingbirds, *Calypte anna*, fed at experimental colored feeders, they preferred red over blue, green, yellow, or colorless feeders (Collias and Collias, 1968).

There is now evidence that the domestic pigeon can discriminate polarized light (Delius *et al.*, 1976) and that various songbirds, hummingbirds, and the domestic pigeon can see near-ultraviolet light (Kreithen and Eisner, 1978; Goldsmith, 1980; Parrish *et al.*, 1984). Ability to discriminate polarized and UV light could be of use in migration and orientation (see Chapter 22). Cones with peak sensitivities at the near-ultraviolet spectrum of light have been identified in eyes of the pigeon, Ruby-throated Hummingbird, *Archilochus colubris*, and 13 species of songbirds (Chen *et al.*, 1984).

Wax-blooms of berries are known to reflect ultraviolet light and may contrast against green leaves. Thus the ability to perceive near-ultraviolet light may be important in foraging. Nectar guides of flowers are known to reflect ultraviolet, and may attract hummingbirds to the nectar source.

VISUAL FIELD. The position of the eyes in a bird's head shows close correlation with its life habits. Inoffensive vegetarians, such as ducks, quail, and doves, have eyes laterally placed where they can view possible enemies coming from any quarter. The eyes of hawks and other predators, intent on their next meal, are directed more toward the front. Owls nearly match man with their frontal eyes, but unlike man's their eyes are almost immovably locked in their sockets. In compensation for this rigidity, they have flexible necks that allow them to twist their heads at least 270 degrees. It seems likely that compensatory reflex movements, which in other animals are associated with the eyes, have spilled over into neck movements in some birds.

Eye position may also reflect a bird's special habits. The bittern, which feeds in shallow water, has eyes placed low in the head to facilitate downward vision

FIGURE 5–20 *(Left)* Front and rear views of a Yellow-billed Cuckoo, a species able to converge its eyes on objects either in front of or behind its head. After Polyak. *(Right)* A bittern's head viewed from below, showing its downward-facing eyes. After Berlioz, in Grassé, 1950.

and also to focus directly on an intruder when the bird "freezes" with its head pointed skyward (Fig. 5–20). The mud-probing American Woodcock, *Scolopax minor,* has eyes that are shifted upward and backward to a position effective for preventing surprise attacks. Because of the location of its eyes, the Woodcock has more effective binocular vision of objects to the rear than of objects in front of if! Goatsuckers, cuckoos, and some crows are also able to converge their eyes toward the rear when disturbed by an object approaching from that direction.

The total field of view embraced by a bird's eyes depends on three things: the placement of the eyes, their mobility, and the angle of view that each eye subtends. Pigeons with their laterally disposed eyes (aided by plumage sleeked against the head to the rear) have a total field of view of about 340 degrees. In other words, they can see almost everything in the environment except the space occupied by their own bodies. Owls with their frontal eyes have a total field of 60 to 70 degrees.

When a hawk or a sparrow fixes an item of food before it with both eyes, the natural assumption is that the bird sees it in stereoscopic, three-dimensional vision as we do. There is no guarantee that this is a fact. There are, as Walls (1942) makes clear, numerous clues to distance perception that are available to any bird: size of retinal image—the larger the image, the closer the object; perspective, or the tapering shape of the object; overlap and shadow—near objects hide far objects; vertical nearness to the horizon; aerial perspective—hazy objects are more distant; and parallax, or change in the apparent angular movement of an object produced by lateral movement of the observer's eye.

It is significant that those birds that most need and practice binocular vision (hawks, hummingbirds, swifts, swallows, and others) have a second, temporally placed fovea in each eye. There is good reason to believe that, with the bird's eyes converged nasally, the two images of a single object focused on these foveae fuse in the brain and produce a true stereoscopic image. Even a bird like the pigeon with eyes directed sideways may have an overlap of the two individual eye-fields of from 6 to 25 degrees. The Eurasian Kestrel, *Falco tinnunculus,* which has a 150-

degree field of view for each eye, has a binocular overlap of 50 degrees. It is possible that some birds —for example, penguins and hornbills—may have no binocular overlap whatever. However, most birds are capable of binocular fixation of an object before them and undoubtedly gain a good idea of its distance by unconscious inference from any clues available to them. Their precisely oriented behavior admits no other interpretation.

Newly hatched chicks of the domestic Chicken, *Gallus gallus*, were shown by Hess (1956) to possess innate stereoscopic depth perception. He fitted one-day-old chicks, which had had no previous visual experience, with prism goggles that displaced objects 7 degrees to the right. When these chicks were allowed to peck at a brass nail-head embedded in modeling clay, the marks of their pecks were scattered around the nail, but centered at a point 7 degrees to its right. Similarly scattered marks of the control chicks centered on the target. Tested again at three and four days of age, the control chicks showed increased accuracy: their pecks were less scattered around the nail. The pecks of the prism-wearing chicks were "clustered just as tightly as those of the controls," but were still displaced to the right as much as before.

Subsequent experiments by Rissi (1968) with newly hatched chicks showed progressive compensation for their striking errors. Their goggles displaced the image 8.5 degrees. After four days the chicks missed their target by 3.6 degrees, and after eight days, by only 1.5 degrees. Then the goggle prisms were replaced with flat glass (zero displacement of image), and the chicks over-compensated, missing the target in the opposite direction by about 1.5 degrees.

When chicks were fitted with contact lenses that altered the focal lengths of their eyes, the birds pecked short of their target. However, the chicks adapted in time to their handicap, and after three days errors were greatly reduced. Then, when the lenses were removed, the chicks over-compensated, and pecking errors of the opposite sign appeared (Guiton, 1978). One can assume from this that there is, in newly hatched chicks, an innate depth perception smoothly coordinated with an appropriate-sized pecking reflex.

The survival advantage of depth perception in newly hatched, active, precocial young is made apparent in "visual cliff" experiments. The visual cliff apparatus consists of a horizontal sheet of transparent glass, under the "shallow" half of which is a floor only a centimeter or so away; under the "deep" half the floor

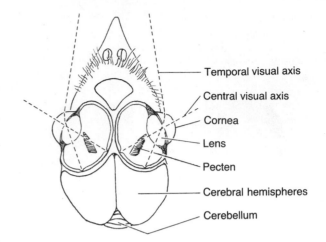

FIGURE 5–21 A horizontal section through the head of a Barn Swallow. With a slight convergence of the eyes toward the beak, a single object in front of the bird may be binocularly focused on the temporal foveae. After Polyak, 1957.

is perhaps a meter away. The sharp drop-off between the two floors is the "cliff." Kear (1967) discovered that domestic chicks and ground-nesting precocial young of other species overwhelmingly chose the shallow side of the cliff when placed on the glass. Young of tree-nesting species showed no preference for either side.

One further adaptation in the eye's structure that promotes binocular use is its twisted asymmetry. Both the cornea and the lens of most birds' eyes are displaced toward the beak so that a ray of light passing through the pupil falls on the retina temporally rather than in its center. This modification reduces the amount of beakward rotation of each eye that is required to cause rays of light from a single object to focus simultaneously on the two temporal foveae. In other words, the eyes need not "toe in" so far to produce binocular fixation (Fig. 5–21).

Even so, many birds very clearly use monocular vision in determining distances. This is the reason many shorebirds such as sandpipers bob their heads vertically. Raising and lowering the head quickly causes the object viewed to shift its relative position against the horizon, which enables the bird to judge its distance. Other birds practice optical fixation or "parallactic localization" of objects by similar means. Coots, *Fulica americana*, while swimming, and fowls, pigeons, and other ground-dwellers, while walking, make rapid movements of the head forward and backward, apparently for object localization. In these birds the backward movement of the head normally compensates precisely for the forward movement of the body, so that, with regard to the immediate surroundings, the head occupies a series of fixed stations from which it not only may take unconscious triangulations of neighboring objects, but, while standing still, may detect more readily any moving object in the landscape. A steady eye can detect a moving fragment in the visible field much more easily than can a moving eye, which causes the whole field to move also. This is no doubt the reason why a bird perching on a moving twig or telephone wire commonly holds its head in a precisely fixed spot even though its body may oscillate several centimeters with the branch. Above all, the eye of the bird must be sensitive to movement of any object in its surroundings, because moving objects mean either danger or food, and both are crucial in survival.

Even owls with their frontal eyes practice parallactic localization. Barn Owls, *Tyto alba* and Tawny Owls, *Strix aluco*, frequently make oscillating horizontal movements of the head when intently observing an object. The Little Owl, *Athene noctua*, makes vertical movements, and still other owls reinforce their depth perception by rotating their heads in rapid vertical circles. These adaptations are quite logical when one considers the poor retinal acuity and limited accommodation found in owls.

Electrophysiological studies of single nerve cells in a monkey's visual cerebral cortex by Bridgeman (1972) revealed the intriguing information that there are cells that are sensitive to absolute motion (as when the entire visual field of view moves across the retina as a result of eye movement), and other cells that respond only to relative motion against the background. Thus, a brain comparison mechanism is available that can apparently distinguish eye movement from object movement. It is unknown whether birds possess such neural equipment, but its potential value for them is obvious and it may be that they possess something like it.

Much experimental work remains to be done to solve some of the problems of acuity, sensitivity, color vision, and depth perception in birds. But from all the evidence available at present, one can at least be certain that birds are equipped with superb eyesight.

■ SUGGESTED READINGS

For a lucid and reliable treatment of the evolutionary relationships of the nervous system and sense organs of birds, see Romer's *The Vertebrate Body*. Excellent and comprehensive works, but in foreign tongues, are Stresemann's volume on *Aves* in Kükenthal and Krumbach's *Handbuch der Zoologie*, and Portmann's chapters on the nervous system and sense organs in Grassé, *Traité de Zoologie*. That sprightly classic on the eye, Wall's *The Vertebrate Eye*, is meaty, and has many references to birds. Contemporary information about the avian eye is well presented by Meyer in Vol. 7 of the *Handbook of Sensory Physiology*, Crescitelli *et al.*, (eds). Much material on the nervous system and the sense organs will be found in the chapters by Portmann, Stingelin, and Pumphrey in Marshall's *Biology and Comparative Physiology of Birds*. Different authorities briefly cover the nervous system and the various sense organs in Campbell and Lack's *A Dictionary of Birds*. In Farner and King's *Avian Biology* are excellent treatments of avian sense organs by Schwartzkopff, Sillman, and Wenzel (Vol. 3) and of the nervous system by Bennett (Vol. 4). Extensive anatomical coverage of the subject will be found in Pearson's *The Avian Brain*. Extensive anatomical coverage of the sense organs is also to be found in Vol. 3 of *Form and Function in Birds*, A.S. King and J. McLelland (eds).

Food, Digestion, and Feeding Habits

6

Where'er there's a thistle to feed a linnet
And linnets are plenty, thistles rife—
Or an acorn-cup to catch dew-drops in it
There's ample promise of further life.
—Tom Hood (1835–1874),
Poets and Linnets

The waking activities of a bird are concerned with fighting, fleeing, feeding and breeding, self preservation, and gene preservation. These mainsprings of behavior are centered in the digestive and reproductive systems, although their workings are of course under nervous and endocrine control. As creatures with intense metabolism, birds require rapid, powerful, and efficient digestion of food. Although their digestive systems are basically similar in structure and function to those of reptiles and mammals, they show numerous adaptations and refinements that support their special needs.

ANATOMY OF THE DIGESTIVE TRACT

While many birds possess a generalized type of digestive system capable of processing a mixed diet, most birds have digestive tracts adapted either to a plant or to an animal diet; this adaptation is evident throughout the entire alimentary canal. More detailed adaptations have also been developed with specific feeding habits.

Starting where the food starts, with the beak, the correlation of food and food-handling machinery is obvious in many species. The main functions of the beak are to expose, seize, kill, and prepare food for swallowing. In birds of prey, the beak is a sharp-edged meat hook; in seed-eaters, a nibbling, crushing forceps; in shorebirds, a long, delicate probe; and in woodpeckers, a heavy, blunt pick or chisel. A multitude of adaptations to feeding habits have molded the bird's beak, showing that, adaptively, it is one of the more plastic parts of the alimentary system. In certain groups, such as birds of prey and gallinaceous birds, the beak shows great uniformity, while in others, such as larks and ground finches (*Geospiza*), it does not. In only a few species does the beak vary between the sexes—for example, in the recently extinct Huias, *Heteralocha acutirostris*, of New Zealand, the bill was long and sickle-shaped in the female, short and conical in the male. The male used its stouter bill to dig holes in dead wood, which the female then probed for wood-boring grubs.

The mouth or buccal cavity is generally roofed with a hard palate. The floor of the mouth is membranous, and in some species, such as pelicans and numerous finches, serves as an extensible pouch for food storage. Among seed-eaters the soft parts of the mouth are well supplied with relatively small mucous glands, but in aquatic species the glands are commonly scanty and may even be absent, as in the pelicans. For birds eating dry foods, the mucus moistens and lubricates the food to be swallowed. Salivary glands are usually present in the pharynx, and in seed-eaters are not only abundant but secrete a starch-digesting enzyme. Many swifts and swallows use the dried saliva to cement their nests. The so-called Edible Swiftlet, *Collocalia*, of the East Indies, builds its nest entirely of saliva. Sometimes as many as three and a half million nests are shipped from Borneo to China in one year to be made into that well known delicacy, birds' nest soup. The glands of these swifts enlarge about 50-fold during the nest-building season and atrophy to the usual size immediately after. Those of the Chimney Swift, *Chaetura pelagica*, enlarge about 12-fold during the nesting season. The European Green Woodpecker, *Picus viridis*, has an unusually large salivary gland, 7 cm long and divided in two sections, one of which secretes a watery solution, the other a sticky solution that coats the remarkably long, protrusible tongue with an insect-holding film. Jays of the genus *Perisoreus* secrete a sticky saliva used to make balls of food that are then attached to twigs and other supports for food-storage (Bock, 1961).

Typically, the tongue of a bird is small, covered with a cornified epithelium, sharply pointed in front and supplied with papillae at the rear. In parrots the tongue is quite muscular, but in most other species it has few if any intrinsic muscles, being moved mainly by muscles of the hyoid framework. It varies greatly in shape and function: tubular, semitubular, or brushy in nectar-eaters; long and barbed in woodpeckers; with backward-directed hooks in penguins; with marginal filtering processes in some ducks and geese; and practically vestigial in the Ostrich, hornbills, and pelicans (Ziswiler and Farner, 1972). The Pesquet's Parrot, *Psittrichas fulgidus*, feeds exclusively on soft fruit, which it swallows without chewing. It bites a large morsel and takes it onto the tip of its tongue, which then retracts into the buccal cavity. The tongue now slips under the morsel, and specialized papillae at the caudal border of the tongue hook onto the food. The tongue retracts again and pushes food into the pharynx (Homberger, 1981).

Because so many birds bolt their food quickly, their tongues and mouths are poorly supplied with taste endings. However, it is probable that all birds can taste foods, at least to a limited extent. Touch endings, on the contrary, are widely distributed in the tongues and mouth parts of many birds. The tips of finches' and woodpeckers' tongues are thickly packed with tactile corpuscles. The beaks of parrots are also well supplied with them, a significant fact in the light of their "nose-rubbing" courtship ceremonies.

The bills of ducks and geese are richly supplied, especially at the edges and tips, with tactile endings. According to Stresemann, these endings are grouped into peg-shaped sensory papillae, about 15 touch endings in a very restricted area. Each papilla contains Grandry corpuscles in its distal half and only Herbst corpuscles in the proximal half. Touch endings occur not only in the papillae, but are also scattered individually in the dermis of the beak, particularly along the edge of the palate, where as many as 27 per square millimeter have been counted in the Mallard, *Anas platyrhynchos*. This compares with 23 touch corpuscles per sq mm in the most sensitive part of man's index finger. In the palatal epithelium of a duck's mouth there are about 18 Herbst and 14 Grandry corpuscles per sq mm, or a total of about 6850 and 6300, respectively, in the entire membrane (Stresemann, 1927—1934).

The esophagus, extending from the pharynx to the stomach, is usually provided with mucous glands and is somewhat muscular. Its size is related to the size

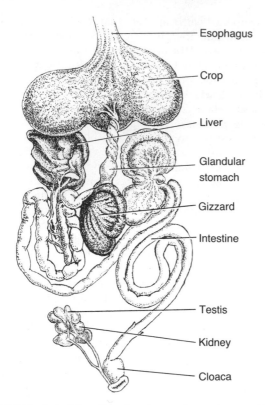

Esophagus

Crop

Liver

Glandular
stomach

Gizzard

Intestine

Testis

Kidney

Cloaca

FIGURE 6–1 The digestive tract of a pigeon. After Schimkewitsch and Stresemann.

of the food particles a species swallows. Insect-eaters, and species that break up their food before swallowing it, have narrow tubes, while species that swallow large items have a wide, distensible esophagus. Unlike mammals, birds seem to suffer no discomfort if food remains lodged in the esophagus. Many sea birds swallow fish so large that the birds go around for hours with the fish heads in their stomachs and the fish tails projecting from their mouths.

In many seed-eating finches there are, either in the mouth cavity or on the sides of the esophagus, paired outpouchings used as storage bags for food. Some carduelid finches develop special pouches in their mouth floors during the breeding season. These open on each side of the tongue and extend back under the jaw as far as the neck (Newton, 1972). The accumulated seeds or insects may later be swallowed or fed to young in the nest by regurgitation. More commonly, the esophagus swells into a larger storage chamber, the crop, where food remains, sometimes a day or more, until the stomach can accommodate it. The crop is usually located over the furcula and widens into a spindle-shaped or globular sac. In pigeons it becomes a large double sac that not only stores grain but also secretes "pigeon's milk" for feeding the young squabs. In species that swallow insects alive and squirming, the esophagus and crop are frequently lined with a heavy epithelium that protects them against physical or chemical damage. Crops are generally prominent in grain-eaters, such as game birds and pigeons. Their great virtue is that they permit their owners to gather and swallow food in a hurry, thus shortening their exposure to enemies while so occupied. Crops also permit hard seeds to be softened with mucus before further digestion in the stomach and intestines. Digestive glands rarely exist in the walls of the crop.

All birds possess two kinds of stomach: an anterior glandular stomach and a posterior muscular stomach or gizzard. The spindle-shaped glandular stomach is an innovation with birds, since only the muscular gizzard is found in reptiles. The inside of the glandular stomach is lined with columnar epithelial cells, the mucosa layer, which is richly supplied with tubular mucous glands. The layer next to this one, the submucosa, is the thickest layer of the stomach and is made up almost entirely of digestive glands. These are tubular glands, either simple or branched, and they secrete a peptic enzyme that attacks protein foods. The secretion is also highly acid (pH 0.7 to 2.5) and is able, in many flesh-eating species, to dissolve large bones. During the breeding season the stomachs of albatrosses and other Procellariiformes contain large amounts of pungent oils that the birds feed to their young. The oils seem to be dietary residues rather than stomach secretions (Lewis, 1969). The oil may also be used as a weapon. When ejected onto another sea bird it causes matting of the feathers, and possibly the waterlogging and death of the bird (Swennen, 1974).

Those species of birds with efficient glandular stomachs, such as loons, pelicans, cormorants, gannets, owls, and accipiters, generally have weak, thin-walled muscular stomachs. It is mainly the herbivores that have powerful muscular stomachs, especially the seed-eaters, such as the parrots, pigeons, gallinaceous birds, and finches, although birds that eat molluscs and crustaceans in their shells may also have strong gizzards.

The gizzard is shaped something like a thick biconvex lens, with striated muscles usually arranged in distinct bands. The mucous epithelium that lines the gizzard secretes a keratinous fluid that hardens into horny plates or ridges serving as millstones for the mechanical grinding of food. This work is furthered by the abrasive action of small bits of grit that many birds swallow, especially the grain-eaters. In time the grit eaten by a bird is worn away and must be replaced. There is evidence that if some grain-eaters are denied grit they will lose weight and eventually die. But experiments on Bobwhite Quail, *Colinus virginianus*, by Nestler (1946) showed that young and also adult breeding quail on a gritless diet survived as well as birds having grit. The gizzard of a domestic goose holds about 30 g of grit; a duck, 10 g; and a turkey, 45 g. Gizzard stones associated with the fossil bones of Moas indicate that they carried as much as 2.3 kg of grit in their gizzards. Many petrels and penguins, and some cormorants, carry gizzard stones, but since they are fish-eaters and do not have very muscular gizzards, their function is problematical. Whalers claim that the penguins use the stones as ballast.

For many carnivorous species the gizzard is a trap that prevents sharp bones and indigestible fragments from proceeding down the alimentary canal. Such resistant items as teeth, feathers, fur, cellulose, or chitin may be rolled up into an elongated "pellet" and regurgitated by mouth. This is a normal occurrence in owls, accipiter hawks, gulls, goat-suckers, swifts, and grouse. The study of a bird's pellets tells much about its diet. The ability to regurgitate is sometimes put to other uses. Boobies and other sea birds often eat so many fish that they must disgorge some of them in order to take off in flight. And, of course, many species regurgitate food that they have collected into the mouths of their young. Grebes are known to pluck their own feathers and eat enough of them to fill their stomachs. It is thought that these feathers either protect the stomach walls against sharp fish bones or plug the pyloric outlet of the stomach long enough for the fish bones to be dissolved before passing on into the intestine. The stomach exit in the Anhinga is covered with a grating of hairlike fibers whose function apparently is to prevent the passage of undissolved fish bones.

Some easily digested foods that do not require rigorous digestion may not even pass through the gizzard. The Thick-billed Flower-pecker, 'Dicaeum agile, of southeast Asia has a gizzard that is located to one side of the alimentary canal and accessible through a sphincter opening. Easily digested mistletoe berries, for example, pass directly from the esophagus to the

intestine and may be evacuated at the cloaca in a very short time, but insects and spiders enter the gizzard for slower, more thorough digestion. Similarly, the nectar eaten by honey-eaters (Meliphagidae) can bypass the gizzard because the end of the esophagus and the beginning of the small intestine are adjacent to one another (Smythies, 1973). Mistletoe specialists, e.g., *Dicaeum* spp., remove the skins with their bills. The Phainopepla, *P. nitens*, of California, swallows mistletoe berries whole. The specialized gizzard extrudes seed and pulp out of the skin. The intestine is then filled with packets of 8 to 16 skins, which are then eliminated separately from the other berry parts (Walsberg, 1975).

In a variety of species a seasonal change in diet parallels a change in stomach structure and function. A study by Spitzer (1972) of the gizzards of Bearded Tits, *Panurus biarmicus*, showed that in winter, when the birds ate largely *Phragmites* seeds, the stomachs were muscular, had hard, keratin linings, contained grit, and weighed 0.88 to 1.21 g. In summer, when the diet was mostly insects, the stomachs were softer, smaller (0.50 to 0.68 g), lacked hard linings, and contained no grit. Mallards, *Anas platyrhynchos*, fed experimental diets high in fiber content, had digestive organs significantly larger than ducks fed normal diets after only 21 days (Miller, 1975).

The two-lobed liver in birds is the largest gland in the body, and is larger than the liver in mammals of equal size. It seems to have the same functions as in mammals: primarily to store excess sugars and fats, synthesize certain proteins, make bile, and excrete waste products from the blood. It is somewhat larger in fish- and insect-eaters than in meat- and grain-eaters. The pancreas likewise is relatively large in birds, and is larger in fish-, insect-, and grain-eaters than in meat-eating species. It secretes digestive enzymes that attack all three types of foods: fats, carbohydrates, and proteins. It is also an endocrine gland that regulates carbohydrate and fat metabolism.

The intestine is the chief organ of digestion and absorption of foods. It shows its reptilian origin in that the circular muscles of the intestine lie outside the longitudinal layer—the reverse of their order in mammals. In birds it is not sharply differentiated into regions as in mammals, but the tube is more or less looped or coiled between the muscular stomach and cloaca. As a rule, the intestine in meat- and fruit-eating birds is short, thin-walled, and broad; that of seed-eaters is long; that of fish-eaters relatively long, thin-walled, and of small diameter (Stresemann, 1927—1934). Among related species, larger birds have a relatively longer intestine than smaller species. This is true because, as solid objects become larger, their volume increases more rapidly than their superficial area.

If the linear dimensions of any object (whether cube, cylinder, sphere, or irregular form) are doubled, its surface area increases four times but its volume increases eight times (Fig. 6–2). It is the mucous epithelium lining the inner surface of the intestine that absorbs the digested food needed for body growth and energy. If the intestine of a small bird is efficiently proportioned to its needs, a simple doubling of its length and diameter will not suffice for a bird twice its size, since the amount of absorbing epithelial surface does not increase as rapidly as the contents of the intestine. Hence, the originally efficient ratio of surface to volume is lost. This is the reason why the intestines of larger birds are proportionately narrower but longer than those of related smaller birds. In this way, more absorbing surface is provided. The absorbing surface is further greatly increased by the growth of hundreds of thousands of microscopic finger-like villi, which project inward toward the lumen from the mucous layer.

Similar adjustments in surface-volume ratios may be seen throughout the animal body. It is for the purpose of providing more surface on which digestive enzymes may act that beaks and gizzards break food into small particles. It is for similar reasons that capillaries, lung tubes, and even cells are very small. Inversely, this surface-volume relation helps us to understand why birds in cold climates are larger than their tropical relatives.

Toward the posterior end of the intestine, at the spot where, in mammals, the small intestine joins the large, there occur in many birds a pair of dead-end sacs or tubes, the caeca. In the more primitive species of birds the caeca are very large, particularly in the Ostrich, cranes, ducks, geese, and Galliformes. In grouse their combined lengths may equal that of the intestine, while in woodpeckers, parrots, swifts, passerines, and certain other groups, they are very small and may even be completely lacking, or converted into a patch of lymphatic tissue. The chief function of the caeca seems to be the absorption of water and digested proteins, and particularly the bacterial decomposition of cellulose in crude fibrous foods. In many species the caecal wall is arranged in spiral ridges, which increase the area of its absorptive surface. The indigestible residue from the caeca is

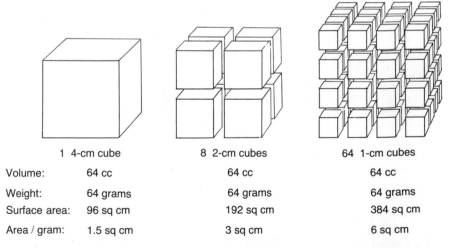

	1 4-cm cube	8 2-cm cubes	64 1-cm cubes
Volume:	64 cc	64 cc	64 cc
Weight:	64 grams	64 grams	64 grams
Surface area:	96 sq cm	192 sq cm	384 sq cm
Area / gram:	1.5 sq cm	3 sq cm	6 sq cm

FIGURE 6–2 A small object has more surface in proportion to its volume than a larger object of similar shape. If this process of successively halving the linear dimensions of each cube were continued ten more times, the end result would be 68,719,476,736 cubes, each one approximately 0.001 cm on a side, with a total surface area for all cubes of 393,216 sq cm, or 6144 sq cm/g. This simple geometric relationship is of far-reaching biological importance.

dark and moist, and is discharged independently of the whitish, drier intestinal feces. In the domestic hen there is one caecal discharge for about every ten intestinal defecations.

At the posterior end of the intestinal canal is the cloaca. It is divided by annular ridges into three regions. The anterior one, the coprodaeum, receives the excrement from the intestines; the middle urodaeum receives the discharges of the kidneys through the ureters, and of the genital organs through the vas deferens or oviduct; and the terminal proctodaeum stores the excrement and is closed posteriorly by the muscular anus. This last chamber is the largest of the three and is supplied with powerful ejection muscles. On its dorsal wall is the bursa of Fabricius, a lymphatic pocket prominent in young birds but generally atrophied in adults. It is involved in the production of the antibody response that protects birds against infections. Farmer and Breitenbach (1966) inoculated two-week-old chickens, one-half of which had had their bursae removed, with the malarial parasite *Plasmodium lophurae*. The normal birds recovered but many of the bursectomized chicks died of the infection—11 of 15 in one experiment.

DIGESTION

Like other animals, birds eat when they are hungry and stop eating when satiated. The mechanism that turns the appetite off and on was once thought to be food or its absence in the stomach, and, later, glucose in the blood. It is now believed that the major control of appetite resides in the hypothalamus of the brain, responding to sensory messages from the eyes (sight of food), stomach or intestine (full or empty), skin (warm or cold), body tissues (dehydrated or not), and other stimuli. Experiments have shown that stimulation of the lateral hypothalamus "produces voracious eating, and lesions in the same area cause loss of appetite." Conversely, stimulation of the ventromedial areas of the hypothalamus results in satiety, whereas lesions there stimulate excessive eating (Sturkie, 1976).

Birds not only eat foods rich in energy and cell-building materials, but they digest them more quickly and more completely than most animals do. As the food is taken into the mouth it is usually torn, cut, nibbled, or ground into small particles, moistened by the saliva and mucus, and swallowed. In some species, such as owls and many sea birds, the prey is swallowed entire, and the glandular stomach takes on the chemical job of dissolving its edible portions. In those species

that possess a crop, the food remains there a variable time and becomes softened while waiting to enter the stomach. The crop is more an organ of storage than of digestion, but it is likely that a small amount of digestion of carbohydrates by saliva may take place there in birds such as finches that have large salivary glands.

It is primarily in the stomach that digestion begins. Here the gastric glands of the glandular stomach secrete a strong peptic enzyme and hydrochloric acid, which together attack proteins and break them down into peptones and proteases. The stomach fluid in which this digestion occurs has a high acidity, commonly varying between pH 0.7 and 2.3 in living birds (Ziswiler and Farner, 1972). In the bearded Vulture or Lammergeier, *Gypaëtus barbatus*, of Mediterranean regions, an entire cow vertebra may be swallowed and digested in a day or two. The gastric juice of some insect-eaters contains two enzymes that digest the tough, chitinous exoskeletons of insects. Domestic chickens with hyperacid gastric juice may develop stomach ulcers.

The grit in the gizzards of seed-eaters may not be indispensable but, at least in the domestic chicken, it increases the digestibility of whole grains and seeds by 10 per cent. Rhythmic contractions of the muscular gizzard vary according to the quantity and hardness of the grain and grit inside. In the domestic chicken, contractions vary between two and three per minute (Sturkie, 1976). The muscles of the gizzard are innervated with autonomic nerves, and, like the heart, will continue their contractions even when the stomach is removed from the body, because the gizzard has its own nerve net.

Calcium from bone or from bone marrow is utilized in the formation of eggs. Calcareous material (grit, mollusc shells) is an important source of calcium for many avian species. Verbeek (1971) documented a variety of female hummingbirds eating grit and suggested this as a means of replenishing calcium utilized in the breeding season.

Considerable force is generated by the kneading walls of the gizzard. Experiments by Mangold (1929) and others revealed the following internal gizzard pressures, expressed in millimeters of mercury: buzzard (with weak carnivore gizzard), 8 to 26 mm Hg; hen, 100 to 150 mm Hg; duck, 180 mm Hg; goose, 265 to 280 mm Hg (Fig. 6–3).

Some early experiments by Réaumur in 1752 and Spallanzani in 1783 have been reported by Stresemann (1927–1934). Réaumur found that a tube of sheet iron that could be dented only under a load of 36 kg (80 lb) was flattened and partly rolled up after being in the stomach of a Turkey, *Meleagris gallopavo*, for 24 hours. A turkey will grind up in its gizzard 24 English walnuts in the shell in four hours. Even more striking are the results of Spallanzani's experiments. He found that a Turkey could grind to pieces 12 steel needles in 36 hours, and 16 surgical lancets in 16 hours!

As the food leaves the stomach and passes into the intestine, it is mixed with digestive juices from the liver and pancreas, and possibly with juices from the walls of the intestine itself. Bile from the liver probably acts to neutralize the acid from the stomach and to emulsify fats in preparation for further digestion. The pancreatic juices digest all three classes of food, but precise investigation into their nature and action remains to be done. Once the food is digested, it is absorbed by the epithelial lining of the intestine, passed on into the blood stream, and distributed through the body. The crude cellulose fibers, digested with the help of bacteria in the caeca, are broken down in varying degrees, corn and wheat fibers being several times more digestible than oat or barley fibers. If the caeca of a hen are removed surgically, the percentage of corn fibers it can digest drops from about 17 per cent to zero; of oat fibers, from about 9 per cent to 1.3 per cent (Sturkie, 1965). In the Willow Ptarmigan, *Lagopus lagopus*, feeding in winter on willow buds and twigs, it was found that the principal bacterial fermentation products were alcohol and acetic, propionic, butyric, and lactic acids (McBee and West, 1969). The energy obtained from this caecal fermentation amounts to about 11 per cent of basal metabolism (Gasaway, 1976).

The high efficiency of avian digestion is shown in the relatively small amount of excrement discharged. But this, too, is variable; birds such as grouse, which live part of the year on a low-grade diet, discharge a greater amount of feces than meat-eaters. A better indication of the efficiency of digestion is the fact that growing young birds, such as storks, can gain 1/6 kg of weight each day from eating 1/2 kg of fishes and frogs (Heinroth, 1938), and commercial broiler chickens can gain 1 kg on 1.9 kg of standard diet (Butz, 1976). An extremely efficient conversion of food into body tissues is seen in the developing embryo. The weight at hatching of a Golden Eagle, *Aquila chrysaëtos*, was reported by Bent (1937) to be 73 per cent of the egg's weight.

The powerful action of a bird's digestive juices is seen in the speed with which foods are digested.

FIGURE 6–3 A turkey gizzard can exert powerful mechanical forces. *(Top row)* At left, a thick-shelled pecan; and at right, a pecan broken after one hour in a turkey's gizzard. *(Bottom row)* At left, a hickory nut beginning to crack after 8 hours in the gizzard; and at right, a broken nut after 31 hours. Experimental pressures required to break similar hickory nuts varied from 56 to 152 kg (124 to 336 lb). After Schorger, 1960.

Although a domestic hen requires 12 to 24 hours to digest a cropful of grain, a shrike can digest a mouse in three hours. Watery fruits pass through the alimentary canal very quickly: the seeds of berries eaten by Blackcaps, *Sylvia atricapilla*, appear in their feces in as little as 12 minutes, and a thrush fed fruits of the elderberry will defecate the seeds 30 minutes later. Caecal digestion proceeds less rapidly. In the male Ring-necked Pheasant, *Phasianus colchicus*, the average maximum time for digestion of a standard diet in the intestine was 8.5 hours; in the caeca, it was 35 hours (Duke *et al.*, 1968).

OPTIMAL FORAGING AND FEEDING HABITS

The basic feeding adaptations of any species lie in such fundamentals as method and speed of locomotion; acuteness of sense organs; psychological bent; diurnal or nocturnal activity; character of beak and tongue; shape and strength of feet; type of plumage; physiological tolerance of such things as high and low temperatures, wet and dry habitats; digestive competence; and tolerance thresholds for certain food ingredients such as toxins, acids, salts, and roughage.

The foods birds eat provide them with the energy

required for survival and reproduction. Birds burn up energy in the act of foraging, however, so that they should generally forage in such a way and seek such foods that they get the maximum energy in return. An Optimal Foraging Theory is based on the rationale that natural selection has shaped certain rules whereby birds make decisions as to (i) what kinds of food to eat; (ii) where to forage; (iii) what types of searching path to take; (iv) when to feed in a new spot (Krebs, 1978, Krebs *et al.*, 1983).

Although birds are said to be selective in their diets, they do not all enjoy the "à la carte" free choice that man does in a restaurant. Natural selection through the years has narrowed the menu for many birds so that they are compelled by their adaptations to dine "table d'hôte." Aesop's fable of the fox and the stork inviting each other in turn to dine out of unsuitable dinnerware illustrates the point.

A bird's bill may place constraints on what foods it can process in the least amount of time, since handling time can also be costly in energy. Bill size varies in individual Darwin's Finches, *Geospiza fortis*. Individuals with larger bills prefer to feed on large *Opuntia* cactus seeds, which they are proficient at handling, and birds with smaller bills are less likely to do so (Grant *et al.*, 1976). In a laboratory study with eight North American finch species, Willson (1971) found that given the opportunity to choose between seeds of high caloric content versus seeds with fewer calories that could be husked more quickly, birds tended to choose the latter. Smaller seeds are more easily handled and more quickly swallowed by smaller finches, permitting the birds to keep moving and suffer less risk of predation.

Sometimes bill shape or size may be different between the sexes within a species and may reflect differences in foraging behavior and food preferences. Selander (1966) has documented such differential niche utilization in woodpeckers, nuthatches, hummingbirds, finches, and a wren. Many female raptors are larger than males, and females tend to take larger prey than males (Mueller and Meyer, 1985; Kennedy and Johnson, 1986). Male Sharp-shinned Hawks, *Accipiter striatus*, average about 100 g in weight; females, 170 g. A study by Storer (1966) showed that male Sharp-shins ate birds averaging 17.6 g., females, 28.4 g. Newton (1979) noted that raptors specializing on fast-moving prey (birds, fish, mammals) tended to be more dimorphic in size than those feeding on slower, less elusive prey (reptiles, insects). Availability may influence the bird's choice of food.

Given the choice between large (profitable) and small (less profitable) prey, Great Tits took equal numbers of both when these were presented to them on a moving belt at low densities. They ignored the smaller prey when the number of larger prey were increased to "optimal" level (Krebs *et al.*, 1977). The larger prey were then kept at a constant optimal level, and the smaller prey increased to twice the density of the large ones. The Tits still ignored the more abundant small prey in favor of the larger profitable items. It would be less costly in energy to reduce searching time for specific prey and behave as a generalist when all prey types are at low densities. Predators can shift to being specialists when profitable prey are in large enough densities so that energy returns outweigh the costs of searching.

Birds finding an abundant food source or "patch" will not restrict their feeding to this one area, but may be continually sampling to acquire new information. Collias and Collias (1968) trained Anna Hummingbirds, *Calypte anna*, to come to feeders filled with sugar solutions dyed red, yellow, green, and blue. Birds would take exploratory sips at each feeder before settling on the solution of highest molarity. Sugar content takes precedent over color. The Colliases pointed out that learning to shift from one flower to another is adaptive since flowers of different colors and different nectar content come into bloom at different seasons. Exploratory sips enable the birds to detect flowers yielding the highest return in energy. Anna Hummingbirds defend territories containing the nectar-poor *Ribes malvaceum* and *Arctostaphylos* in winter when no other flowers are in bloom. They may desert these in favor of nectar-rich *Ribes speciosum* when these come in bloom (Stiles, 1973).

Although theory suggests that birds should forage in manners and in areas yielding the maximum in energy returns, various factors may prevent them from doing so. White-throated Sparrows, *Zonotrichia albicollis*, form linear hierarchies in winter flocks. Artificial feeders containing peanuts were placed at various distances from cover. The feeders contained "nomad matting," a material requiring the birds to search for their food. Dominant individuals fed nearer cover than did subordinates. Dominants nearly depleted an area before moving to the next, however. They sacrificed foraging efficiency in favor of reducing risk from predation (Schneider, 1984).

Pressure from competition may determine a bird's foraging area in a community. Willow Tits, *Parus palustris*, avoided trees used by the Crested Tit, *Parus cristatus*, when the two occurred syntopically. If both species happened to feed together in the same spruce tree, Willow Tits moved to the inner and lower sections, whereas Crested Tits fed on the preferred upper and outer parts of the tree. Great Tits, *Parus major*, preferred foraging in the lower and inner parts, and in their presence Willow Tits fed instead on the outer and upper parts (Alatalo, 1981). The results of experiments (Chapter 19) demonstrate that if a dominant species is removed, the subordinate species will often move into its habitat.

When a species becomes so adapted to its diet that it is restricted to very few food items, or to a single one, it is said to be *stenophagous*. On the other hand, when it eats a wide variety of foods it is described as *euryphagous*. The Snail Kite, *Rostrhamus sociabilis*, of Florida feeds exclusively on fresh-water snails, which it extracts from their shells with its peculiarly hooked beak.

As a stenophagous species increases the perfection of its feeding adaptations, it reduces, or may even eliminate, competition with other species for its special food. That is, it has the restaurant all to itself. But with this considerable advantage comes a price tag. If something should happen to reduce or eliminate the staple food of a stenophagous species, its survival is immediately and drastically threatened. Eel-grass is the staple food of the Brant, *Branta bernicla*, of the Atlantic coast of North America. Between 1931 and 1933 a great blight killed over 90 per cent of the coastal eel-grass, with the predictable result that about 80 per cent of the Brant wintering along that coast disappeared (Moffit and Cottam, 1941). The remnant managed to survive by eating sea-lettuce. This new food changed the flavor of their flesh, so that the species lost favor as a game bird, a fact further enhancing their survival (Barry, 1964). This aspect of diet raises the intriguing possibility that the food eaten by a species may at times be important in limiting the numbers or kinds of predators that prey on it. A euryphagous species such as the Road Runner, *Geococcyx californianus*, of the Southwest, would never be subjected to such a fate. It eats scorpions, tarantulas, snakes, centipedes, mice, rats, horned toads, small birds, eggs, numerous insects, fruits, and seeds (Bent, 1940).

Stenophagy and euryphagy are relative terms. A bird may be stenophagous at a certain stage in its life history or at a certain time of year, and euryphagous at other times. Among birds with stenophagous tendencies are hawks, eagles, doves, pelicans, geese,

mergansers, penguins, hummingbirds, woodpeckers, parrots, swifts, kingfishers, and many passerine species. The more euryphagous species include many shore birds, gulls, some ducks, many gallinaceous birds, herons, crows, starlings, and several passerines (Mayaud, 1950).

Changes in Diet

Seasonal changes in diet are probably the rule, especially for sedentary species living in regions where seasonal climates change extensively. Analysis by McGowan (1973) of the crops of 123 interior Alaskan Ruffed Grouse, *Bonasa umbellus*, in the autumn, showed the following proportions of foods by weight: fruits, 78 per cent; buds, 14 per cent; leaves, 4 per cent. In winter the proportions were 6, 92, and 1 per cent, respectively. The austerity of the winter diet is apparent when one learns that the nutritive value (in calories) of buds of the Quaking Aspen, *Populus tremuloides*—a favorite food—is only one-fifth that of commercial Turkey-growing food (Hill *et al.*, 1968). The diet of Red-winged Blackbirds, *Agelaius phoeniceus*, in Manitoba consisted of up to 70 per cent vegetable matter (mainly seeds) in the spring and autumn, but 70 to 100 per cent animal food (mainly insects) in June and July (Bird and Smith, 1964).

Changes in diet may also be forced by periods of inclement weather. European White Storks, *Ciconia ciconia*, commonly shift their attention from fish and frogs to mice in periods of drought; and Hobbies, *Falco subbuteo*, will eat larger numbers of swifts in cold wet weather than in fair weather, the lack of flying insects in poor weather resulting in weakened swifts.

More than food availability or parental training may be involved in dietary change. Garden Warblers, *Sylvia borin*, and Blackcaps, *S. atricapilla*, were maintained by Berthold (1976) under constant environmental conditions and with constant food supplies. The birds nevertheless showed regular and spontaneous changes in the proportionate amounts of vegetable and animal foods they ate. These changes were thought to be based on some innate physiological periodicity in the birds.

Another likely influence in determining diet is experience. Chicks of Herring Gulls, *Larus argentatus*, and Ring-billed Gulls, *L. delawarensis*, were fed on one kind of experimental food (chopped earthworms, pink- or green-dyed cat food) and then, after five days, given free choice among various kinds of food. They preferred the food they had experienced when it was one of the choices; when it was not, they frequently failed to eat at all (Rabinowitch, 1968). Captive House Finches, *Carpodacus mexicanus*, given one-hour preference tests with untreated canary seed and oats treated with the repellent Methiocarb learned to avoid the oats. Ten- to 12-week-old chicks fledged by conditioned parents also avoided treated oats, although they were never exposed to them. This illustrates the importance of parental contact in influencing food preferences (Avery, 1983). Experiences of this sort met with in nature may equip birds with "search images" or "reward expectancies" that play a large part in determining their diets. Similarly, the frequency with which a bird may encounter, for example, a palatable or distasteful seed or insect, may help shape its "search image." After fortuitously eating a few palatable insects of a certain kind, the bird may show preference for that insect in its future searchings for food (Tinbergen, 1960; Orians, 1971).

It is certain that many species exhibit different diets in different geographical or ecological habitats. The Great Gray Shrike, *Lanius excubitor*, of Europe seems to feed largely on birds in Scandinavia; on insects, spiders, lizards, and shrews in Spain; and primarily on insects in desert regions. Analyses of the pellets of Tawny Owls, *Strix aluco*, showed that the diets of urban London owls consisted of 96 per cent birds, but that owls of nearby suburban woodlands ate 95 per cent mammals and only 5 per cent birds (Harrison, 1960).

Probably the majority of birds change their diets with age. Most young birds eat diets rich in animal food—worms, insects, fish, and the like—even though they may be strict vegetarians as adults. Insects are rich in proteins needed for growth, whereas fruits and greens are richer in calories. The surprising thing is that many newly hatched young that feed themselves instinctively select protein-rich animal food. A study of young Mallards, *Anas platyrhynchos*, by Chura (1961) showed that ducklings 1 to 6 days old consumed animal foods almost exclusively. As the birds matured they took increasing amounts of plant food: between the 19th and 25th days plant foods made up about 50 per cent of the diet, and by 50 days, 100 per cent.

It seems very likely that diet plays a role in the evolution of birds. While it is obvious that specific adaptations may limit a bird in the food it may eat (a hummingbird's beak is unsuitable for gnawing acorns), it is probably no less true that the long-established diet of a species has a selective influence on the structure and function of a bird's feeding equipment. Intestines and caeca of Willow Ptarmigan, *Lagopus lagopus*, and Wood Ducks, *Aix sponsa*, increase in length in the winter when fiber content in their diets increases (Pullainen, 1983; Drobney, 1984). Gizzards and livers of Wood Ducks also increased in length.

Quantity and Kind of Food Eaten

The amount of food eaten by a bird depends on its species, health, experience, age, and sex; on the season and the time of day; on whether the bird is molting, feeding its young, or preparing to migrate; on food availability; and on other conditions. A study of food consumption by the birds and mammals of a 1000-hectare virgin forest in Czechoslovakia indicated that the total bird population consumed food equaling about 25 per cent of its weight daily; the mammals, 20 per cent (Turček, 1952). This fact accords well with the superior rate of metabolism in birds.

Periods of increased demands for energy result in increased food intake. Free-living White-crowned Sparrows, *Zonotrichia leucophrys*, decrease their feeding activities on warm days and increase them on cold cloudy days. Birds warmed with infrared lamps ate less than did unwarmed controls when kept at 7°C. There was no difference in feeding rates of irradiated and nonirradiated groups kept at 20°C. Morton (1967) concluded from this that sunbathing activities alleviate the costs of thermoregulation at low stressful temperatures. Many species, just before a long migratory flight, will eat voraciously and build up a store of fuel in the form of fat that may increase their body weight more than 50 per cent. A female bird will increase its food consumption and also draw on fat reserves in the body during the egg-laying period. Experimental studies by Jordan (1953) on captive wild Mallards, *Anas platyrhynchos*, showed that in the early autumn the birds each consumed about 132 g of grain daily, and in coldest winter about 150 g. During fall and winter months the Mallard drakes ate about 15 per cent more food than the hens, but in the spring the hens consumed an average of 16.6 per cent more food than the drakes. Rapidly growing ducks between 8 and 9 weeks of age ate 44 per cent more food than adults—a statistic that parents with growing children will appreciate. For tropical birds weighing between 100 and 1500 g, Thiollay (1976) found that their daily food ration was only one-half that of comparable birds in the cooler temperate and boreal zones.

The surface-volume relationship referred to earlier imposes a heavy metabolic penalty on small warm-blooded animals. Small birds have relatively more heat-losing surface than large ones. A hummingbird is, in a sense, one fragment of an ostrich cut into 64,000 pieces, and every slice that is made creates more heat-radiating surface without adding a gram of heat-manufacturing tissue. Looking at it another way, an ostrich is a great heap of hummingbirds huddling together to keep warm. Small species must eat relatively more food than large ones to make up their extra heat loss.

A tiny Blue Tit, *Parus caeruleus*, weighing 11 g, will eat around 30 per cent of its body weight in dry food per day; a 100-g Mourning Dove, *Zenaida macroura*, about 11 per cent; and an 1800-g domestic Chicken, *Gallus gallus*, 3.4 per cent (Nice, 1938a). More precisely comparable figures on bird size and appetite were obtained by Marti (1973), who kept four species of owls, each about one year old, in individual outdoor aviaries for one year. He kept accurate records of their daily consumption of live mice. Table 6–1 shows clearly the inverse correlation between body size and food consumption.

The volume of food eaten per day will vary with the type of food. Watery berries will naturally be consumed in greater quantity than dried seeds or insects. A 57-g Bohemian Waxwing, *Bombycilla garrulus*, was estimated to eat about 170 g of the berries of *Cotoneaster horizontalis* in one day (Gibb, 1951).

The work expended in securing food can be considerable. In winter, small birds commonly spend 90 per cent of their waking hours feeding: a Pied Wagtail, *Motacilla alba*, catching a prey item every 4 seconds (Davies, 1976), and a titmouse (*Parus* sp.) one insect every 2½ seconds (Gibb, 1958). Hummingbirds high in the Andes mountains need to visit between 1500 and 2700 flowers each day to meet their energy requirements (Hainsworth and Wolf, 1972). Gray-faced Petrels, *Pterodroma macroptera*, are reported to forage up to 600 km from their breeding colony to secure food for their young (Imber, 1973).

As a general rule, birds eat most heavily early in the morning and again late in the afternoon, logical times in view of the long overnight fast. But here again there is a great variation among different species. Birds that feed on small seeds and insects must eat fairly regularly throughout the day, but those that take in larger items, preying on larger animals or carrion, may feed very irregularly, sometimes going for days without eating. However, it is ordinarily only the larger species that are able to survive a fast of several days or weeks. Probably the world record for long fasting in birds is held by the male Emperor Penguin, *Aptenodytes forsteri*, which, in midwinter in the inhospitable Antarctic, fasts about 115 days preceding and during the incubation of its single egg (Le Maho, 1977).

Experimental starvation studies by Jordan (1953) on wild Mallards, *Anas platyrhynchos*, showed that the birds could live without food under cool air temperatures (average, 14°C) for slightly over three weeks. The hens showed much greater resistance to starvation than the drakes. Similar experiments by Gerstell (1942) indicated that Ring-necked Pheasants, *Phasianus colchicus*, can live without food for two weeks or more during severe winter weather; the wild Turkey, *Meleagris gallopavo*, one week; and the Hungarian Partridge, *Perdix perdix*, four or five days. The Estrildid Finch, *Uraeginthus angolensis*, is said to die of starvation only four or five hours after its alimentary canal is completely empty. For House Sparrows, *Passer domesticus*, denied both food and water and

TABLE 6–1.
OWL WEIGHTS AND DAILY FOOD CONSUMPTION (After Marti, 1973)

SPECIES	MEAN WEIGHT OF OWL (GRAMS)	MEAN WEIGHT MICE EATEN PER DAY (GRAMS)	PER CENT OF BODY WEIGHT EATEN PER DAY
Burrowing Owl, *Athene cunicularia*	166	26.4	15.9
Long-eared Owl, *Asio otus*	295	37.5	12.7
Barn Owl, *Tyto alba*	598	60.5	10.1
Great Horned Owl, *Bubo virginianus*	1333	62.6	4.7

kept in darkened cages, it was discovered by Kendeigh (1945) that survival time was closely related to environmental temperatures. Birds survived the longest (67.5 hours) at 29°C. Below 21° birds died from their inability to maintain sufficient heat production, and above 35° they died from lack of sufficient water for cooling by evaporation.

In fasting birds, there is initially an early decrease in liver glycogen and blood glucose (Hazelhurst, 1972), but with prolonged starvation, tissue proteins are converted into plasma glucose for metabolic needs. In fasting Willow Ptarmigan, *Lagopus lagopus*, more weight is lost from breast muscles alone than from body fat tissues, the muscle protein supplying as much as 40 per cent of the bird's energy needs (Grammeltvedt, 1978). In chickens deprived of food, the level of corticosterone increased markedly. Corticosterone titers declined by 70 per cent to the level of fed birds, within 45 minutes of their being given food *ad lib* (Harvey *et al.*, 1983).

Since small birds have such an intensified hunger problem, it is not surprising that a few species have evolved a special atavistic adaptation to solve it. Hummingbirds living at high altitudes in the Peruvian Andes stretch their slim fuel resources through the cold night by reverting to reptilian dormancy. Their body temperature drops from a daytime high of about 38°C to a lethargic 14°C at night (Pearson, 1953). Without this heat-conserving adaptation, such a small bird probably could not survive the night except in a warm nest. Nestling European Swifts, *Apus apus*, are able to survive as much as 10 days of fasting by reverting to cold-bloodedness. Although the adult Swift is unable to assume dormancy, it is nevertheless able, through other physiological economies, to survive 4.5 days of fasting. Weight losses from fasting average 52.8 per cent in the nestlings and 38 per cent in the adults (Koskimies, 1950). These adaptations in both young and adult Swifts probably arose in response to the common hazard of cold, wet spells during which flying insects, their only food, are not available. The Poor-will, *Phalaenoptilus nuttallii*, of the southwestern states, can endure 88 days without food by a similar dormancy, its body temperature dropping to a low 6°C. Experiments by Marshall (1955) showed that dormancy was not caused directly by low environmental temperatures but by lack of food. The birds did not become torpid until they were denied food and had lost 20 per cent of their weight. Trilling Nighthawks, *Chordeiles acutipennis*, on the contrary, became tor-

pid while very fat. Because they are able to migrate so easily, birds have not needed to adopt hibernation to the extent that mammals have.

Very little experimental work has been done on particular dietary needs of wild birds, but it is known, for example, that a seed-eating Bobwhite Quail, *Colinus virginianus*, must have a minimum of 11 to 12 per cent crude protein in its winter diet to maintain good health (Nestler, 1944). Each of five groups of captive female Ruffed Grouse, *Bonasa umbellus*, were fed the following dietary protein levels: 7.6, 11.5, 13.6, 17.0 and 20.1 per cent dry matter. Associated with an increase in dietary protein was an increase in clutch size, mean egg weight, and hatching success (Beckerton and Middleton, 1982). Newly hatched Wood Ducks, *Aix sponsa*, were fed four different isocaloric diets containing 5, 10, 15, and 20 per cent protein portions until they reached 60 days of age. Both growth rate and survival were significantly reduced in ducklings fed diets low in protein levels (Johnson, 1971). Birds themselves seem to be able to make fine distinctions in selecting protein foods. Bullfinches, *Pyrrhula pyrrhula*, eating the buds of pear trees, did the most damage to those varieties having the highest percentage of bud-protein (Summers and Jones, 1976). Captive Red Grouse, *Lagopus lagopus*, in poor condition preferred heather rich in nitrogen and phosphorus but not calcium. Birds in good condition did not show any preferences. Moreover, grouse in poor condition tended to eat more of the younger, more nutritious heather (Savory, 1983).

Some years ago it was discovered that 70 per cent of the wild Turkeys, *Meleagris gallopavo*, of Missouri lived on a single type of soil, Clarksville stony loam, derived from limestone. Habitats on other soils which appeared similar supported very meager populations or none at all (Allen, 1954). Subsequent nutrition experiments by Dale (1955) have shown that Ring-necked Pheasants, *Phasianus colchicus*, receiving limestone grit produced 10 times as many eggs as those eating a similar diet but with granite grit. That the problem of supplemetary items in the diet is not simple was revealed in an experiment on Bobwhites in which it was found that egg production, fertility, and hatchability "were affected by the calcium and phosphorus levels in the breeding diet of *their parents* a year before" (Allen, 1954). No doubt even minute amounts of trace elements in a bird's diet play a role in its health and survival. Those vitamin nutrition studies that have been done on birds show that they respond much as mammals do to vitamin deficiencies: e.g., vitamin A is important for vision, vitamin D for bone mineralization, vitamin K for blood clotting, and B vitamins prevent polyneuritis (Fisher, 1972). Ducks fed selenium-deficient diets suffered muscle deterioration. There was less collagen in their tendons, collagen fibrils decreased in diameter, and fibroblasts were degenerate (Brown *et al.*, 1982). As more information from nutrition experiments becomes available, it will very likely show that bird distribution and possibly migration habits are linked with the soils and foods characteristic of a given species.

Water Requirements

In addition to food, vitamins, and minerals, birds must have water. Sea birds and most land birds have no difficulty securing water. They either drink it directly or obtain it from the foods they eat. For land birds the common method of drinking is to immerse the beak in water and then hold the head high and allow the water to flow into the throat to be swallowed. A few species drink as many mammals do, by sucking the water continuously while the beak is immersed: doves, certain representatives of the estrildid finches, mousebirds, and some tanagers (Immelmann, 1970; Moermond, 1983). Some pelicans drink by exposing their wide-open beaks to falling rain, and numerous small species drink from dewdrops.

The daily intake and outgo of water varies according to a bird's needs, and needs depend on a wide range of environmental and physiological variables. Under nonstressful conditions, Dark-eyed Juncos, *Junco hyemalis*, kept in cages at 20°C, gained and lost daily water equal to 46.75 per cent of their body weight. Of the water gained, 1.12 per cent (body weight) was in their food, 9.22 per cent came from the metabolic breakdown of foods eaten (so-called metabolic water), and 36.41 per cent the bird drank. Of the water lost, 30.15 per cent was lost by excretion and 16.60 per cent through respiration (Anderson, 1970).

Small birds lose relatively more water daily through evaporation from the respiratory system and skin than do larger birds. Bartholomew and Cade (1963) present the following data for pulmocutaneous water loss as percentages of body weight: a 12-g House Wren, *Troglodytes aëdon*, 37 per cent; a 21-g Dark-eyed Junco, *Junco hyemalis*, 16 per cent; a 40-g Poor-will, *Phalaenoptilus nuttallii*, 7 per cent; a 144-g Bobwhite Quail, *Colinus virginianus*, 4 per cent. Age is another conditioning variable. In young ducks the mean daily intake of water decreased from 43 per cent of body weight at an age of 1 week to 13 per cent at 16 weeks (Fisher, 1972).

Since birds cool themselves largely through the evaporation of body fluids, their water intake is obviously related to the temperature and humidity of the surrounding air. Ground Doves, *Columbina passerina*, which consume about 10 per cent of their body weight in water per day at thermoneutral temperatures, will increase their intake about threefold if the air temperature increases from 30° to 40°C (Willoughby, 1966). The higher the humidity of the air, the less the water lost in respiration.

A reduction in a bird's physical activity also promotes water conservation. Respiratory water loss in a Budgerigar, *Melopsittacus undulatus*, during moderate activity is 150 mg per hour, but while asleep, it is only 44 mg per hour (Cade and Dybos, 1962). Physiological activity likewise takes its toll. A domestic Pigeon, *Columba livia*, denied both food and water, will survive up to 13 days at moderate temperatures, but if fed dried peas and given no water it will survive only 4 or 5 days (Stresemann, 1927–1934). This is because the metabolic water produced by the food is less than that lost through the increased respiration required for the digestion of the food.

Another variable influencing water economy is gender. The mean length of survival time for the desert-dwelling California Quail, *Lophortyx californicus*, denied water and on a dry diet was 28 days for females and 41 days for males (Bartholomew and MacMillen, 1961). During this dehydration the birds lost approximately 50 per cent of their body weight.

Many species of birds adapted to desert life can endure long periods without water, some of them living indefinitely on dry seeds. Budgerigars have survived in good health as long as 150 days on dry seeds; Australian Zebra Finches, *Poëphila guttata*, 1¹⁄₂ years (Sossinka, 1972). Such birds are of course living entirely on metabolic water. Water is conserved by both reducing evaporative pulmocutaneous water loss and excreting very dry feces. A normally watered Budgerigar will excrete feces containing 75 to 80 per cent water; a thirsty one, 60 per cent water (Serventy, 1971).

Since sea water contains about 3 per cent salt and is three times as salty as a bird's body fluids, a severe salt-balance problem exists for pelagic sea birds. Are birds, like man, unable to drink sea water and remain alive? For years this problem has been debated by ornithologists, but only recently has the solution come to light. Most birds' kidneys are less efficient than man's in eliminating salt from the blood. A bird would need to eliminate over 2 liters of urine in order to excrete the salt taken in by drinking 1 liter of sea water. Obviously, some other mechanism is called for.

Years ago Oskar and Magdalena Heinroth (1924–1933) discovered that the enormous supraorbital glands of the Common Eider Duck, *Somateria mollissima*, atrophied sharply after the bird had been kept one year on fresh water. They therefore concluded that the glands "protected the inner nose from the salt irritation of the sea water"—a shrewd observation but the wrong conclusion.

In sea birds, as well as several terrestrial species, these glands, now known as nasal or salt glands, are specialized salt-excreting organs. They are known to occur in 13 orders of birds. The paired, crescent-shaped glands each contain several longitudinal lobes and each lobe contains a central duct from which radiate thousands of tubular glands enmeshed in blood capillaries. It is here that salt excretion occurs. The gland serves as an osmoregulator, responding to osmotic loads in the blood—salt or other substances, e.g., glucose. But it responds to such leads by excreting only sodium chloride and possibly potassium, but no other substances (Schmidt-Nielsen, 1960).

Experiments by Schmidt-Nielsen (1959) proved that gulls, petrels, and other sea birds, as well as sea snakes, excrete from these glands fluids that contain about 5 per cent salt, a fantastically high concentration considering the osmotic problem involved. In petrels the fluid is forcibly ejected through the tubular nostrils (Fig. 6–4), but in other species it dribbles out of the internal or external nostrils. A gull given one-tenth its weight in sea water excreted 90 per cent of the contained salt within three hours. Hughes (1970) gave a Common Puffin, *Fratercula arctica*, 4 ml of 1.0 normal sodium chloride solution. Within 20 minutes the bird's salt glands eliminated 75 per cent of this heavy salt load. The concentration of the salty excretion reached 975 molar equivalents per liter, which is about 7.5 times the salt concentration in normal blood plasma.

Mallard Ducks, *Anas platyrhynchos*, Godwits, *Limosa* spp., and many other birds that live on sea water have larger salt glands than their relatives living on fresh water. Staaland (1967) discovered a strong positive correlation between the size of salt glands, the concentrations of their excretions, and the degree to which shore birds (Charadriiformes) associated with salt-water habitats. Schmidt-Nielsen and Kim (1964) raised domestic Ducklings, *Anas platyrhynchos*, on drinking water composed of 1 to 3 per cent solutions

FIGURE 6–4 A, The salt glands of a gull. After Schmidt-Nielsen, 1960. B, A petrel forcibly ejecting salt droplets from its nostrils. After Schmidt-Nielsen, 1959.

of sodium chloride. The birds all developed larger salt glands that excreted salty fluids more copiously than did the fresh-water control birds. Further experiments revealed that the salt-marsh-dwelling Savannah Sparrow, *Passerculus sandwichensis*, which has no salt gland, uses its exceptionally large kidneys to concentrate salt in its urine to a level 4.5 times that in its blood plasma. This is better than twice the concentration attained by most birds. Some sea birds are so specifically adapted to drinking salt water that they will die if denied access to it (Allen, 1925). Adelie Penguins, *Pygoscelis adeliae*, on the other hand, can change abruptly from drinking sea water to drinking nothing but fresh water (Murphy, 1969).

Although salt may be lethal to some birds, it is necessary in the diet for various metabolic processes such as nerve-impulse transmission, blood clotting, and bone formation. Birds kept on a salt-free diet will eagerly eat pure salt when it is made available. Numerous wild finches of the family Carduelidae have notorious appetites for salt. Crossbills, for example, may be caught in traps baited only with salt.

Adaptations for Taking Animal Foods

There are probably almost as many feeding adaptations, structural and functional, as there are species of birds. Only a few representative examples can be given here. Outstanding among the predominantly carnivorous birds are the diurnal predators (eagles, hawks, and falcons), and the nocturnal predators (owls). They all have strong hooked beaks and sharp powerful talons, the latter useful in grasping their prey. If the feet have become weak, as in the vultures, the bird is compelled to feed on smaller, weaker prey or on carrion. Owls differ from the diurnal predators in that they swallow their prey whole and digest the meat from the bones, fur, and feathers, then cast up this indigestible residue in pellets, generally one pellet per meal (Fig. 6–7). Owls have no crop. Eagles and hawks, on the contrary, tear their food apart, rejecting fur, feathers, and the larger indigestible parts before swallowing the nutritious parts. Because some of the food often remains, as a barrier, in the crop for a few hours, they do not regurgitate stomach pellets as frequently as do owls. Most commonly the prey of eagles and hawks

FIGURE 6–5 A Black Kite, Egyptian Vulture, and Griffon, ready to dine on a dead deer. The bare head and neck of the Griffon permit a more fastidious feeding on the deeper parts of the carcass. Photo by E. Hosking.

FIGURE 6–6 A Screech-Owl eating a house mouse. Owls and parrots are exceptional in that their feet may be used to lift food to the beak. Note the large size of the eyes and their dark-adapted pupils. The mouse is swallowed whole, and later the fur and bones are regurgitated in the form of cylindrical, felted pellets. Photos by G. R. Austing.

FIGURE 6–7 Pellets of the Saw-whet Owl are regurgitated once each day. By dissecting the pellets and studying the contained bones, beetle wing-covers, and other indigestible objects, we can reconstruct a bird's diet. Photo by G. R. Austing.

is made up of smaller birds and rodents, but a few of the larger eagles concentrate on animals as large as themselves. The Harpy Eagle, *Harpia harpyja*, of South America feeds on sloths, peccaries, and monkeys; and the Crowned Eagle, *Stephanoaetus coronatus*, of Africa specializes on monkeys, which it lures to their destruction with a soft whistle. A well-fed, captive Long-eared Owl, *Asio otus*, will refuse to eat meat offered to it, but if it is presented live mice, it will continue to kill them (Räber, 1950). Apparently different behavior centers control killing and eating.

The rich food resources of the sea and fresh waters have been made available to fish-eating birds by a wide variety of adaptations. To hold the slippery fish, the upper mandible may be hooked at the tip as in pelicans, cormorants, frigate-birds, and albatrosses; or the edges of the mandibles may be serrated with "teeth" as in mergansers. Interesting parallel adaptations for the firm grasping of fish are seen in the spiny tubercles under the toes of both the fish-eating Osprey, *Pandion haliaetus*, and the fishing owls of the genus *Ketupa* which, significantly, lack the silent, downy feathers of their relatives that live on terrestrial prey. The Osprey has an enormous uropygial gland, which waterproofs its feathers.

Many fish-eating birds, such as the penguins, possess backward-slanting, horny projections on the tongue, palate, or other mouth parts that push and guide their slippery booty toward the esophagus. Some fish-eaters, like the herons and bitterns, have developed long wading legs, long mobile necks, and forceps-like beaks to seize fish and other aquatic prey. Their silent, immobile stance while awaiting a victim is a profitable behavior adaptation. The African Black Heron, *Egretta ardesiaca*, fishes by holding its two wings out to form two umbrellas and having its back to the sun. The shadow thus created tends to scare off small (10- to 20-mm) fish, but may suppress movement of larger (30 mm and over) fish, which then fall prey to the heron (Winkler, 1982). Many aquatic species use their webbed feet and buoyant feathers to paddle about on the surface of the water to prospect for food. In their manner of foraging, the ducks fit into two categories: the surface-feeders that "tip up" to feed on small aquatic life, and the divers that plunge under the surface completely. These latter usually feed only a meter or two deep, but some Old Squaw Ducks, *Clangula hyemalis*, have been caught in fish nets set over 40 m deep (Scott, 1938).

Among birds that dive for their food, there are those like the penguin or loon that dive from the surface, and those like the terns, gannets, pelicans, and kingfishers that plunge in from some distance in the air. A pelican is so light and buoyant that it cannot get under water without a "running start." Many of the plunging divers are unable to maneuver in pursuit of their prey after they have submerged themselves. However, by virtue of their strong wings or powerful, posteriorly placed legs, the penguins, loons, grebes, cormorants, mergansers, auks, and puffins have become agile underwater swimmers. These birds all seize fish between their mandibles, but the Anhinga uses its darting beak as a stiletto to impale its prey. Most good divers have heavy, vertically compressed bodies. Many fish-eaters such as gulls, petrels, and albatrosses seize their food from the surface of the water. An unusual type of surface feeding is found in the skimmers, *Rynchops* spp., which fly just above the surface of quiet waters with their knife-like lower mandible cutting the water and apparently flipping small fish and crustaceans into the open mouth. The lower mandible, which grows more rapidly than the upper to compensate for the wear caused by water friction, is well supplied with touch corpuscles (Fig. 6–8).

A large variety of birds use their feet either to stir

FIGURE 6–8 A Black Skimmer skimming. As an adaptation to the bird's remarkable mode of feeding, the lower beak grows about twice as rapidly as the upper. Photo by H. J. Lee, courtesy Newspaper National Snapshot Awards.

up or to attract prey. Gulls occasionally "paddle" or move their feet up and down rapidly in shallow water to dislodge small invertebrates, or tramp the soil of wet meadows to bring earthworms to the surface. Plovers, egrets, and herons commonly expose or attract small prey animals by standing on one leg and extending the other forward while they make raking or trembling movements. Cranes often stamp their feet on the ground and startle insects into exposing themselves.

Many birds feed on molluscs. The European Oystercatcher, *Haematopus ostralegus*, preys on small mussels either by stabbing inside the gaping mollusc while it is under water, or hammering through its shell when it is exposed at low tide. Many diving ducks pry small molluscs loose with the hardened nail at the end of their upper mandible and swallow them entire, digesting the shell and all. Many long-billed shore birds probe the mud and sand for burrowing worms, small molluscs, and crustaceans. They frequently regulate their time of feeding not by the sun but by the tides. Gulls, crows, and even eagles are known to fly high over rocky ground with mussels, crabs, turtles, and nuts, and drop them to break them open and expose their edible interiors. Not only does the Northwestern Crow, *Corvus caurinus*, invariably drop whelks over rocks rather than over grass or water, but on successive attempts it usually drops one from increased heights (Zach, 1978). The Lammergeier, *Gypaëtus barbatus*, was given the name Ossifragus ("bone-breaker") in antiquity because of its habit of dropping bones to get at the marrow (Stresemann, 1927–1934). Similarly, eagles crack open tortoises by dropping them on rocks. Pliny says that the Greek poet Aeschylus met his death because an eagle, carrying a tortoise, mistook his bald head for a smooth rock. Aeschylus had remained out of doors all that day, fearful for his life, because an oracle had foretold that he would be killed by a fall of a [tortoise's] house! (Thorndike, 1929).

A number of small birds depend on insects for their chief food. Some, like the flycatchers, perch on an exposed branch and sally out to snap up individual insects. At the angles of their jaws, tyrant flycatchers possess a special pair of ligaments that stretch when the mouth is opened wide. This provides a spring-like tension for the quick snapping-shut of their beaks as the birds snap at insects (Bock and Morony, 1972). Swifts, swallows, and some goatsuckers, with their rapid flight, weak feet, and wide gaping mouths, feed constantly on the wing as animated insect nets. Many small species, such as titmice, nuthatches, and creepers, have strong feet for clinging to bark, and fine pointed beaks for removing insects from the crevices. The woodpecker is adapted in several ways for digging out wood-boring insects and larvae. It has strong grasping feet, stiff tail feathers to brace the body against the hammering head, a heavy skull, a strong pick-like beak, and a tongue that is horny, barbed, richly equipped with tactile corpuscles, and astonishingly extensible. The European Green Woodpecker, *Picus*

FIGURE 6–9 A European Nightjar yawning. The cavernous mouth, fringed with bristles, acts as an insect net for sweeping insects from the air while the bird is in flight. Photo by E. Hosking.

viridis, can stick out its tongue nearly 10 cm (or one-third the length of its body) beyond the tip of its beak (see Fig. 4–3). After an insect gallery has been exposed by the beak, the long flexible tongue can explore its twistings, feel out, impale, and withdraw the larvae encountered. Sapsuckers, which specialize on sap and the insects attracted to it, have a tongue ending in thorny bristles. Industrious sapsuckers will dig about 30 holes a day, usually on the shady side of a tree, to start the sap flowing. At times they drink fermented sap and become too intoxicated to fly well. Lacking a long beak and tongue, the Galapagos Woodpecker-finch, *Cactospiza pallidus*, holds a cactus spine or twig in its beak to pry insects out of cavities in dead wood.

Adaptations for Taking Vegetable Foods

Birds eat all sorts of vegetable foods: roots, bulbs, stems, leaves, flowers, sap, nectar, fruits, seeds. A parrot in New Guinea, *Micropsitta bruijnii*, seems to

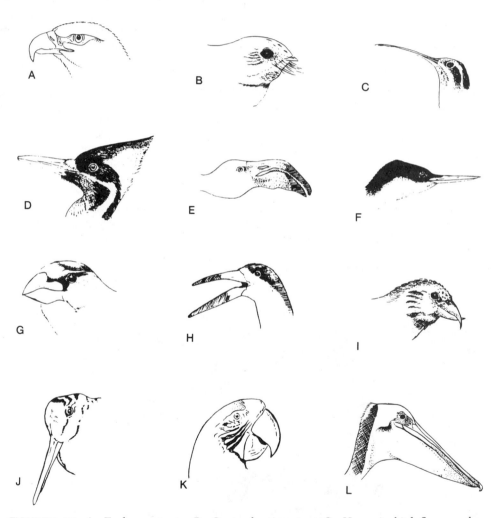

FIGURE 6–10 A., Eagle, meat eater; B., Goatsucker, insect net; C., Hummingbird, flower probe; D., Woodpecker, wood cutter; E., Flamingo, mud sifter; F., Grebe, fish spear; G., Grosbeak, seed cracker; H., Skimmer, water "plow"; I., Crossbill, pine seed extracter; J., Woodcock, earth probe; K., Parrot, nut cracker; L., Pelican, dip-net.

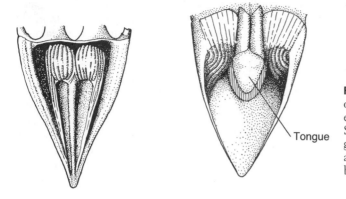

Tongue

FIGURE 6–11 The upper jaw *(left)* and lower jaw *(right)* of the Hawfinch. The two ridged, horny mounds in each jaw serve as "anvils" for cracking nuts and fruit pits. Similar devices are found in the jaws of North American grosbeaks. A Hawfinch can crack olive pits which require a crushing force of 48 to 72 kg (106 to 159 lb). The entire bird weighs only 55 g (2 oz). After Mountfort, 1957.

specialize in eating fungi. Seeds are the preferred food of many vegetarian birds, both because they contain the most concentrated nourishment and because they are available at seasons when other foods are scarce.

A given species of bird will prefer certain kinds of seeds not only because of their size, color, hardness, flavor, and availability, but very likely because of their nutritive value in relation to the bird's needs. Seeds of most conifers, for example, are rich in fats—a highly desirable fuel for cold-climate birds. Oak, birch, and ash seeds are rich in carbohydrates, and seeds of locust, elm, and linden are rich in protein foods (Turček, 1961).

Seed-eaters usually have short, heavy beaks operated by strong jaw muscles. The beaks are commonly sharp at the edges and may have internal ridges against which seeds may be cracked or cut (Figs. 6–11 and 6–12). The tongue is likely to be muscular, scoop-shaped, and horny. Many species, like the doves, swallow the seeds entire; while others, like the finches and grosbeaks, shuck off the hard coat in the mouth. Still others, like the titmice, jays, and ravens, hold the seed with their feet and split it open with blows of the bill (Clark, 1973). Nuthatches, nutcrackers, and some woodpeckers place hard seeds in crevices of bark or rocks and attack them with their beaks. Crossbills, whose beak tips are laterally displaced and move past one another as the mouth is closed, use this remarkable adaptation to get behind conifer seeds in the cone and force them out.

Larger seeds are generally treated in the mouth or gizzard. Parrots, with their powerful beaks, may crack nuts and seeds or may rasp them open by means of many abrasive ridges in the upper mandible opposed

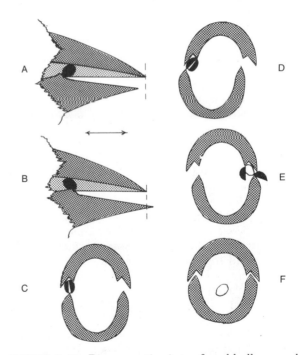

FIGURE 6–12 Diagrammatic views of seed-hull removal by finches. A, B, Side views, and C, cross-section, of beak showing forward and backward movement of lower mandible whose sharp edges slice the seed-coat as it rolls in the groove of the upper mandible. D, E, F, Cross-sections of beak showing lateral movement of lower mandible to remove sliced hull. Redrawn from Ziswiler, 1965.

by a cross-wise ridge on the shorter lower mandible. Both mandibles are hinged to the skull and may be moved freely. The muscular tongue has a horny nail at its tip. Unlike most birds, parrots are able to hold food in one foot and lift it up toward the beak as they eat.

For eating extremely large fruits, the tropical fruit pigeons are described by Stresemann (1927–1934) as possessing three unusual adaptations in the beak, the stomach, and the intestines. The beaks of these birds spread open not only vertically but horizontally, as in snakes, to accommodate the enormous fruits they eat. One species, *Ptilinopus porphyraceus*, swallows fruit pits as large as 30 by 50 mm! This is roughly the size of a small hen's egg. The chief food of many fruit pigeons is the nutmeg, whose pulpy outside is rasped off by numerous horny, conical pegs lining the gizzard, as in *Ducula* spp., or by two opposed pairs of ribbed gizzard plates, as in *Ptilinopus* spp. These rasping surfaces are so hard that the pigeons need no grit in their stomachs. If a domestic pigeon eats a cherry it must regurgitate the seed because its intestines are not large enough to let it pass. But the intestines of fruit pigeons are exceptionally wide and short, so that they can easily pass, undamaged, nutmegs and other large seeds measuring 12.5 by 25 mm.

Smaller fruits and berries are eaten by many species, among them parrots, trogons, colies, toucans, manakins, waxwings, and flower-peckers. Only the pulp is normally digested and the seeds pass on through the alimentary canal. Smaller fruit-eaters (e.g., manakins and tanagers) tend to prefer carbohydrate-rich fruit, whereas larger fruit-eaters (e.g., cotingas and toucans) take a wider range of fruit including small carbohydrate-rich and large lipid-rich fruit. Fruits consist of 50 to 94 per cent water, and tend to have low protein-to-calorie ratios. To meet their maintenance requirements small frigivores must process large quantities of fruit rapidly. They tend to have short intestines, nonmuscular gizzards, and sometimes large livers relative to body size to detoxify secondary compounds found in some fruit (Herrera, 1984; Moermond and Denslow, 1985). In many species, such as thrushes, starlings, and orioles, fruits are eaten only seasonally. Even some primarily carnivorous species, such as shore birds and woodpeckers, occasionally eat fruits. When cold weather prevents their feeding on insects, Tree Swallows, *Iridoprocne bicolor*, may feed in great flocks on bayberries.

Many grouse species browse on buds, catkins, and conifer needles in the winter months (low-grade diet). These galliforms have larger caeca than do seed-eating galliforms (quail, partridges, pheasants, turkeys). For example, a 500-g grouse has a 44-cm caecum, whereas a similar-sized Hungarian Partridge, *Perdix perdix*, has a 17-cm caecum (Leopold, 1953). Bacteria cultured from the caeca of grouse had the ability to decompose cellulose. Coastal greens-eating California Quail, *Lophortyx californicus*, had caeca 19 per cent longer than those of interior seed-eating populations. Geese eat much aquatic vegetation and graze on grass; Brant occasionally eat marine algae as well as eel grass. The aberrant Hoatzin, *Opisthocomus hoazin*, is one of the few birds that eats leaves as a steady diet. This species eats the tough leaves and fruit of arum and mangrove plants, which it grinds, not with its small gizzard, but with its large, muscular crop, which holds 50 times as much as its gizzard and is lined with a tough, horny layer.

Some species, such as orioles and bullfinches, eat flowers and flower buds; and the staple food of the Australian Brush-tongued Lorikeet, *Glossopsitta porphyrocephala*, is pollen of *Eucalyptus* flowers. But a flower's chief attraction for birds is its energy-rich nectar. Some 1600 species, or nearly one-fifth of all birds, are nectar-eaters. Nectar-eaters, such as the honeycreepers, honey-eaters, sunbirds, and hummingbirds, generally have long beaks and long, extensible tongues that enable them to probe the depths of flower corollas for nectar. Often the narrow tongue has grooves on its surface so that, pressed against the palate, they may serve as tubes for sucking in the sweet juices. Or the tongues may have brush-like fringes at the tips or along the sides. In honey-eaters and hummingbirds, as in woodpeckers, the hyoid bones of the tongue circle the skull, thus allowing much greater extensibility.

There is a mutual adaptation between the flowers of "ornithophilous" plants and the beaks of the birds that visit them (Stiles, 1985). Not only do the bills of birds often fit the deep, tubular corollas of the flowers as a curved or straight blade fits its scabbard, but the flowers are typically red and without fragrance. Red colors are peculiarly attractive to hummingbirds and are ignored by at least the honey bee; and birds, with their inferior sense of smell, probably depend slightly if at all on floral fragrance, while many bees are strongly attracted by it. The nectar glands of these flowers flow with a bird-sized rather than an insect-sized abundance, and the nectar pots are so placed that few insects can reach them. Hummingbird flowers usually have nectars rich in sucrose, which hummingbirds prefer, rather than nectars of mixed glucose, fructose, and sucrose, which characterize bee-pollinated flowers (Stiles, 1976). Further, the reproductive organs of the flowers ripen at different times, the sticky pollen usually first, and are so placed that a bird visiting successive flowers will carry the pollen from one flower to another, thus promoting valuable cross-fertilization. While it is true that many nectar-eaters require insects as a protein supplement to their heavy carbohydrate diet, their chief interest in flowers is the nectar. Stresemann tells of one hummingbird kept alive nearly two months on sugar water and flower nectar alone, and of another whose daily intake of nectar was twice the weight of its body.

Since hummingbirds are more likely than bees to seek food and hence to pollinate flowers in foggy or rainy weather, bird-pollinated flowers are more abundant than bee-pollinated flowers in damp, high-elevation habitats. In montane Chiapas, Mexico, Cruden (1972) found that flowers of bee-pollinated *Salvia cacaliaefolia* were 77 per cent pollinated followed by 45 per cent fecundity, whereas bird-pollinated *Salvia chiapensis* flowers were 80 per cent pollinated with 71 per cent fecundity. In Mexican lowlands where flying conditions are more favorable for bees, fecundity figures for four species of labiate flowers were: bee-flowers, 85 per cent; bird-flowers, 69 per cent. Such pronounced differences in reproductive success must play a prominent role in the geographical distribution and evolution of these food plants.

Some plant species depend on birds to distribute their seeds. Many fruit-eating birds either regurgitate the seeds after digesting off the pulp (Moermond and Denslow, 1985), or pass them so rapidly through the alimentary canal that they not only emerge unharmed, but are often better prepared for germination by being softened and provided with dung to hasten their sprouting. Seeds that had passed through the digestive tract of a pheasant were found by Swank (1944) to germinate more quickly than those that had not. Some birds cache seeds in places where later on, if unretrieved, they may germinate. Others transport seeds, like those of the mistletoe, sticking to their beaks, or carried on muddy feet. In his well-known experiment, Darwin (1859) once planted the seeds taken from a ball of mud obtained from the leg of a partridge. The mud had dried for three years, but 82 plants of at least five species germinated from the seeds.

Adaptations to Mixed Diets

Many species are able to eat a variety of both plant and animal foods, often depending on their availability. The ground scratchers, such as the Galliformes, commonly eat seeds, insects, roots, bulbs, or worms as they encounter them. Members of the crow family are well known for their catholic diet. Herring Gulls, *Larus argentatus*, which normally feed on fish, may in the summer move inland and feed largely on grain.

Ducks of the genus *Anas* may eat seeds, leaves, insects, worms, molluscs, and even small vertebrates. They possess a fringed and fluted beak marvellously adapted to sifting small food items from muddy water. Flamingos possess similarly fringed beaks that operate in an inverted position for sifting food from mud. Rather surprisingly, plankton-sifters among the sea birds are rare; but whale birds of the genus *Pachyptila* have laterally placed, parallel plates arranged like teeth in a comb along the inner edge of each upper mandible. These undoubtedly sift food as baleen plates do in a whale (Murphy, 1936) (see Fig. 19–5).

Peculiar Feeding Habits

In their constant struggle to keep alive, birds at times seek out and eat strange foods. As a winter food in Brazil, several species of hummingbirds feed on the sugar-containing excrement, or honeydew, of coccid insect larvae (Reichholf and Reichholf, 1973). In his monograph on the honey-guides, Friedmann (1955) tells of a Portugese missionary in East Africa who wrote, in 1569, of these birds flying into his mission and eating the beeswax altar candles. With genuine symbiotic industry, African honey-guides lead men, ratels, and other animals to the nests of wild bees, where the birds feed on the honeycomb of the plundered nests. Although they eat bee larvae also, the birds may be kept alive on pure beeswax for as long as 32 days. Intestinal bacteria apparently help to break down the wax so that it may be digested and assimilated by the bird.

Other relatively indigestible products sometimes eaten by birds are feathers and wool. As mentioned earlier, grebes often stuff their stomachs with their own feathers, probably for a mechanical effect. Crows and magpies have been observed plucking wool from the backs of sheep in snowy weather near Radolfzell, Germany—perhaps to add fat to their restricted menu (Mühl, 1954). Such a habit may have led to the striking predation of the Kea, *Nestor notabilis*, a New Zealand parrot that attacks sheep in times of winter famine. These birds use their powerful hooked beaks to tear through the wool and to feed on the flesh and fat underneath. The Galapagos Vampire Finch, *Geospiza difficilis*, bites the wings of boobies and drinks their blood as a primary source of food (Bowman and Billeb, 1965). In arctic regions where food is scarce, the Ivory Gull, *Pagophila eburnea*, feeds on the dung of polar bears, walrus, and seals. Puffins and petrels eat great amounts of whale dung (Mayaud, 1950), and at the Kerguelen Islands, Giant Petrels, *Macronectes* spp., catch fish only to eat their stomach contents and not their flesh. In tropical American villages Black Vultures, *Coragyps atratus*, eagerly eat human and dog excrement (Udvardy, personal communication). Postmortem studies of Flamingos, *Phoenicopterus ruber*, in southern France showed that their stomachs held great quantities of mud that contained only 6 to 8 per cent organic matter (Gallet, 1950).

Some of the larger albatrosses regularly eat poisonous, stinging jellyfish, and numerous terrestrial birds eat with impunity berries and fruits poisonous to man. Experimental evidence of birds' capacities in these directions was obtained by Seuter (1970),

who demonstrated that European Blackbirds, *Turdus merula*, and Starlings, *Sturnus vulgaris*, could tolerate doses of up to 100 mg of the poisonous alkaloid atropine; this is 1000 times the human tolerance. Nonetheless, there are numerous foods and substances that are distasteful or poisonous to birds, as is all too evident from the millions of birds killed annually from agricultural pesticides and industrial wastes. Between foods that are wholesome and those that kill are natural foods that contain growth depressants such as toxins, antienzymes, saponins, and others. Soybeans, for example, contain an antitrypsin substance that inhibits growth and causes hypertrophy of the pancreas in chickens (Fisher, 1972).

Many insects are distasteful, if not poisonous, to birds. Whether a given insect is poisonous or not may depend on the food plant on which it grew. Blue Jays, *Cyanocitta cristata*, that eat Monarch Butterflies, *Danaus plexippus*, raised on Asclepiads (milkweeds) containing cardiac glycosides become violently ill. However, the birds quickly learn to recognize the poisonous butterflies and subsequently avoid them (Brower, 1969). Monarchs raised on Asclepiads lacking the glycosides are palatable to Blue Jays. In mixed

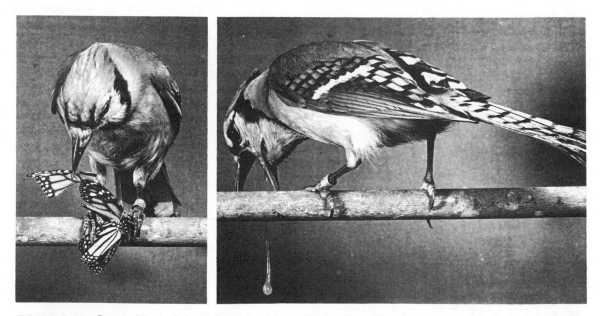

FIGURE 6–13 Certain Monarch butterflies are edible and palatable to Blue Jays; others make them violently ill. The variety of food plant upon which the butterfly caterpillars had been feeding determines their palatability. From Brower, 1969.

populations of Monarchs, some with and some without glycosides, the latter enjoy some immunity from attack by the jays because they of course look exactly like the poisonous variety. In nature a given palatable species of insect, quite unrelated to a distasteful species, may gradually change its outward appearance through the years until it comes to look like the distasteful form. The palatable species, now known as a *mimic*, consequently enjoys the same immunity from avian attack enjoyed by the poisonous species, now called the *model*. For example, the palatable butterfly *Limenitis archippus* closely mimics the distinctive coloration of its model, the distasteful Monarch. This type of sham resemblance occurs widely among insects and is known as Batesian mimicry, named after the 19th century British naturalist, Henry Bates.

The effectiveness of Batesian mimicry was experimentally demonstrated by Schuler (1974), who painted unpalatable mealworm larvae (as "models") with two blue bands and a blue dot. Starlings, *Sturnus vulgaris*, soon learned to reject the model nearly 100 per cent of the time. Then, edible mealworm "mimics" were painted with patterns that deviated progressively from the model: two bars only, one bar and one dot, one bar, etc. The mimics were rejected (i.e., protected) in an amount parallel to their decreasing resemblance to the model. This experiment neatly demonstrates the pathway natural selection may have followed in evolving mimicry among animals.

Feeding Associations

Birds sometimes join forces with other birds or other animals in their endless search for food. Probably the most common feeding association is the intraspecific flocking of a single species: flocks of gulls feeding on a school of fish, or thousands of queleas devastating an African grain field. On occasion such flocks show organized, coordinated behavior, as when White Pelicans, *Pelecanus erythrorhynchos*, form crude arcs and, beating their wings violently, swim toward shallow water where they scoop up the driven fish.

Quite often, however, the feeding flocks are interspecific, i.e., composed of mixed species of birds. In North America, for example, there are troops of Black-capped Chickadees, *Parus atricapillus*, Tufted Titmice, *Parus bicolor*, White-breasted Nuthatches, *Sitta carolinensis*, and perhaps some kinglets or small woodpeckers searching together for food in winter months. Such mixed feeding flocks are common in the tropics and usually are composed of some 25 to 50 birds of 5 to 15 different species. The flocks are often dominated by a few nuclear species who determine the movements of the group. In tropical Andean forests Vuilleumier (1972) found that interspecific feeding flocks were mostly composed of brightly colored, insectivorous passerines. While ornithologists are unsure of the causes of interspecific flocking, Short (1961) suggests these possible advantages: protection against predators, cooperative food-finding, avoidance of duplicate effort in food-finding, and satisfaction of gregarious tendencies.

There are numerous associations in which one species exploits another species as a "beater" to stir up its food. In New Guinea, drongos attach themselves to flocks of babblers and feed on the insects flushed by the latter. In Africa, the Carmine Bee-eater, *Merops nubicus*, perches on the backs of bustards, ostriches, elephants and other animals, to the same end. In ponds, the insects dislodged by diving vegetarian coots are eagerly seized by carnivorous grebes. These are all examples of a *symbiotic* relationship, in which two species eat together, neither at the other's expense. The effectiveness of exploiting beaters was shown by Dinsmore (1973), who determined that Cattle Egrets, *Bubulcus ibis*, are about 3.6 times more efficient in capturing food when foraging with cows or behind moving agricultural machinery than when foraging alone. Such gastronomic profits explain the flocks of crows and gulls that follow the farmer's tractor during spring plowing.

Mutualism is the symbiotic relationship in which two species each benefit from the other's presence, as in the case of tick-birds or ox-peckers that eat ticks on African mammals. Amadon (1967) tells of small Galapagos Finches, *Geospiza fuliginosa*, grooming ticks from the bodies of marine iguanas, *Amblyrhynchus cristatus*.

A very few species of birds have become parasitic on other birds for their food. Jaegers and frigate birds force weaker birds, such as terns, to disgorge the fish they have swallowed, which the parasites immediately seize in midair. The Laughing Gull, *Larus atricilla*, robs pelicans of fish they have caught, at times taking the food directly from the mouth of the larger bird. This kind of interspecific robbery is called food piracy, or *kleptoparasitism*.

FOOD STORAGE. Nonmigratory species of birds living in areas with seasonal fluctuations in food supply prepare for future needs by storing excess food gathered in times of abundance (Smith and Reichman, 1984). Titmice and nuthatches store insects and seeds in bark crevices during the summer and fall for winter consumption. Marsh Tits, *Parus palustris*, were provisioned with radioactively labelled sunflower seeds, and their cache sites located by scintillation counter. Hoarded seeds disappeared more rapidly than control seeds in identical sites, indicating that birds remember storage sites (Cowie *et al.*, 1981).

The Acorn Woodpecker, *Melanerpes formicivorus*, stores great numbers of acorns, each of which fits snugly in a hole that the bird has specially prepared in the bark of pine and oak trees; sometimes as many as 50,000 acorns are placed in one tree (Bent, 1939).

MacGregor's Bowerbird, *Amblyornis macgregoriae*, males store fruit in cavities or on forks of trees near their bowers. The saving in foraging time enables males to spend more time near their bowers, thus increasing interaction time with visiting females and decreasing the rate of marauding by rival males (Pruett-Jones and Pruett-Jones, 1985).

Highly unusual is the altruistic communal food-storing of Clark's Nutcracker, *Nucifraga columbiana*. Late in autumn flocks of these birds harvest pine seeds and hazel nuts and carry them to a communal cache close to their breeding area. The next spring the members of the flock all draw on this food to feed themselves and their offspring (Bock *et al.*, 1973).

Experiments on captive Ravens, *Corvus corax*, by Gwinner (1965) revealed that the hungrier the bird, the more food it would store in caches. Very young Ravens showed this behavior, which suggests that it is innate.

In an attempt to determine the seat of the astonishing visual memory of Nutcrackers, Krushniskaya (1970) performed brain operations on captive birds. Whereas normal control birds were 78 per cent successful in recovering food from caches hidden a day or so before, and birds with neostriatum tissue removed were 91 per cent successful, birds whose hippocampal cortex was removed were only 13 per cent successful. Apparently the cortex plays an essential role in the visual memory of cache locations.

■ SUGGESTED READINGS

Extensive information on the feeding (and other) habits of birds is given in the various volumes of Bent, *Life Histories of North American Birds* and Witherby *et al.*, *The Handbook of British Birds*. Feeding adaptations in birds are well treated by Storer in Marshall's *Biology and Comparative Physiology of Birds*, Vol. 2. In Farner and King's *Avian Biology*, Vol. 2, are well-documented chapters by Ziswiler and Farner on digestion, and by Fisher on nutrition.

Blood, Air, and Heat

7

Darting, hovering helicopter
Fueling at a flower,
Tell me how your engine-heart
Generates such power!
—Joel Peters (1905—   ),
   The Frustrated Engineer

Weight for weight, birds eat more food, consume more oxygen, move more rapidly, and generate more heat than any other vertebrates. In no other vertebrate do the fires of metabolism burn more furiously than in a tiny hummingbird. Two organ systems that are basic to this intense metabolism are the circulatory and the respiratory systems. Each of these systems shows striking adaptations for carrying out its strenuous functions.

THE CIRCULATORY SYSTEM

The chief functions of the circulatory system are to transport digested foods, oxygen, minerals, and hormones to the cells of the body, to remove carbon dioxide and other metabolic wastes, to regulate the water content of tissues and, in part, their hydrogen ion concentration (pH) and temperature. The circulatory system also plays a central role in the prevention and control of disease. It is second only to the nervous system as a powerful homeostatic mechanism for maintaining the body's physiological constancy.

Structurally, the circulatory system of birds is much like that of reptiles and mammals. However, only mammals and birds are *homeotherms*, i.e., they can regulate to maintain a relatively constant body temperature. Only they possess completely separate circulation paths for arterial and venous bloods. This is made possible by a four-chambered heart, the two right chambers of which pump used blood to the lungs for gaseous refreshment while the two left chambers circulate the fresh blood throughout the rest of the body. Reptiles are *poikilotherms*, which do not regulate their body temperatures, and, with their three- or two-chambered hearts, can pump only mixed blood, which is unable to support a high enough level of oxidation to maintain high body temperatures. Birds' hearts are proportionately larger and more powerful

119

than those of mammals. On the average they are 1.4 to 2 times as large as those of mammals of the same weight (Berger and Hart, 1974).

The muscular walls of the left ventricle are more than three times as massive as those of the right ventricle, and the peak blood pressure generated by the left ventricle is five to ten times that of the right. Ventricular muscle fibers are five to ten times smaller than those of mammalian hearts and they are richly supplied with mitochondria (oxidative bodies). Heart sizes of birds vary between the extremes of 0.1 per cent of total body weight in ostriches to 2.7 per cent in some hummingbirds. Small birds generally have relatively larger hearts than larger birds. This relationship undoubtedly stems from their need for higher metabolism. The figures in Table 7–1 illustrate this inverse relationship.

Within a given species, heart size usually increases both with altitude and with latitude. That is, it is larger in alpine races, in response to the need for more rapid circulation in the colder, thinner air; and in polar races, in response to the need for greater heat production. In their study of heart-altitude correlation in passerine species living in highland sites in California and Colorado (2470 to 3670 m) versus lowland sites in California (50 to 210 m), Carey and Morton (1976) found that heart weight of highland populations was 11 per cent higher than those of related taxa in the lowlands. To cite two examples, highland robins have larger hearts than lowland robins and highland chickadees larger hearts than lowland chickadees. Not

unexpectedly, migration seems to affect heart size. The heart of the migratory European Quail, *Coturnix coturnix*, is roughly three times as large as that of the sedentary Bobwhite, *Colinus virginianus* (Hartman, 1961).

As a rule, the heart beats less rapidly in birds than in mammals of comparable size, but its rate depends upon several variables. There is an inverse relationship between body size and heart rate. A Turkey, *Meleagris gallopavo*, at rest, for example, has a heart that beats about 93 times per minute; a Robin, *Turdus migratorius*, 570 times per minute. Exercise, of course, raises the heart rate. A Herring Gull, *Larus argentatus*, resting on the ground or soaring, has a heart rate of 130 beats per minute; in flapping flight, 625 beats per minute. Intensely aggressive behavior accelerates the heart rate nearly to that of flying birds (Kanwisher *et al.*, 1978). There appears to be in many small species a coordination between heart-beat and wing-stroke frequencies, usually in a 1:1 or 2:1 ratio (Aulie, 1971).

The miximum recorded avian heart rate is 1260 beats per minute for the Blue-throated Hummingbird, *Lampornis clemenciae* (Lasiewski and Lasiewski, 1967). Smaller hummingbirds doubtless have higher rates. A drop in air temperature of about 8°C caused an incubating Catbird, *Dumetella carolinensis*, to increase its heart rate 26 per cent. This increase probably reflects heightened metabolism to combat accelerated heat loss at the lower temperature. The temperature of the bird itself has the opposite effect on heart rate: the colder the bird, the slower the heart rate; the

TABLE 7–1.
HEART WEIGHT IN RELATION TO BODY WEIGHT*

SPECIES	BODY WEIGHT (g)	HEART WEIGHT (g/kg BODY WT)
Ostrich, *Struthio camelus*	123,000	0.98
Domestic Goose, *Anser anser*	4,405	8.00
Domestic Chicken, *Gallus gallus*	3,120	4.40
Domestic Duck, *Anas platyrhynchos*	1,685	7.44
Pheasant, *Phasianus* sp.	1,200	4.7
Common Raven, *Corvus corax*	1,200	10.0
Pigeon, *Columba livia*	297	13.80
Bluebird, *Sialia sialis*	29	17.00
Mango Hummingbird, *Anthracothorax nigricollis*	7.7	27.5

*Data based on Portmann (1950), Quiring (1950), Hartman (1961), and Sturkie (1976).

warmer, the faster. As a general rule, the heart rate of any animal will double or treble for every 10°C rise in body temperature within vital limits (Prosser, 1950). At least in chickens, heart rate has little to do with blood pressure. The heart rate of a normal adult chicken may vary from 250 to 400 beats per minute "causing appreciable changes in blood pressure" (Sturkie, 1965). The volume of blood pumped by the heart per minute for the domestic Chicken, *Gallus gallus*, is reported to be 128 to 270 milliliters per kilogram of body weight, and for ducks, 260 to 560 ml/kg (Jones and Johansen, 1972).

Experimental studies by Eliassen (1957) revealed that birds (e.g., guillemots, puffins, eiders) under simulated diving conditions slowed down their heart rates by more than 50 per cent, but with little effect on blood pressure. Gabrielson (1985) reported that the heart rates of Mallards did not slow during voluntary dives lasting 50 to 20 seconds. Heart rates fell 69 per cent if ducks were forced to dive. After 60 such dives, heart rates dropped only 29 per cent.

As a rule, blood pressure is somewhat higher in birds than in mammals, and several times higher in birds than in reptiles, amphibians, fish, or invertebrates. Ordinarily, blood pressure in male birds is higher than in females. Various determinations of arterial blood pressure in adult chickens show a variation of from 130 to 189 millimeters of mercury (mm Hg) for systolic (peak arterial) and 85 to 160 mm Hg for diastolic (lowest arterial) pressure. Systolic pressures for small birds were determined by Woodbury and Hamilton (1937) as follows: domestic Pigeon, *Columba livia*, 135 mm Hg; Starling, *Sturnus vulgaris*, 180 mm Hg; Robin, *Turdus migratorius*, 118 mm Hg; Canary, *Serinus canarius*, 130 and 220 mm Hg. For comparison, among examples given by Prosser of mean arterial blood pressures, expressed in mm Hg, are the following: Man, 100; horse, 175; dog, 110; bat, 50; crocodile, 40; frog, 26; salmon, 64; catfish, 35; shark, 22; lobster, 7; mussel, 0.8; earthworm, 5. Experimental lowering of the body temperature of a chicken decreases both its heart rate and its blood pressure.

Apparently, birds, in their adaptation to respiratory and circulatory efficiency, have pushed their blood pressures close to the margin of mechanical safety. A Cardinal, *Cardinalis cardinalis*, exhausted from repeated territorial battles with another male Cardinal, died from the effects of a 7-mm wound in the heart ventricle, probably caused by the great pressure generated in the heat of battle (Dilger, 1955).

Probably as a result of selecting domestic Turkeys, *Meleagris gallopavo*, for rapid growth and heavy meat production, Turkey breeders have unwittingly selected strains with high blood pressures and weak arteries. A study by Ringer and Rood (1959) revealed that males of the broad-breasted bronze variety of turkey averaged a systolic blood pressure of 296 mm Hg at 22 weeks of age, and some individual birds had pressures as high as 400 mm Hg. (Blood pressures over 150 mm Hg are considered abnormally high for man.) As a consequence, some turkey flocks have suffered high mortalities from aortic rupture followed by internal hemorrhage.

Since the tranquilizing drug reserpine reduces nervous tension and lowers blood pressure in man, it has been used successfully by turkey growers to reduce losses from aortic rupture (Carlson, 1960).

Very little work has been done on the nervous control of heart rate and blood pressure in birds, but evidence suggests that it is similar to that in mammals. Peripheral stimulation of the vagus nerve, a cardioinhibitor, slows the avian heart to one-half or one-third its normal rate and reduces blood pressure from 150 to 100 mm Hg in the chicken. Contrariwise, stimulating vasomotor centers in the medulla may cause constriction in the smaller arteries, which will raise blood pressure as much as 48 per cent (Sturkie, 1965). The regulation of gases in the blood is the function of the carotid body, located on the dorsal carotid artery. As in mammals, it functions to regulate the respiratory reflex in response to changes in the concentrations of oxygen and carbon dioxide in the blood (Jones and Johansen, 1972). Under normal pressure it takes the blood an average of 6 seconds to make a complete circuit of the body of an adult chicken.

In broad outline, the architecture of a bird's circulatory system is very similar to that of reptiles and mammals. Figure 7–1 shows the main paths of blood-flow in a bird. The arterial trunks of birds resemble those of reptiles more than those of mammals. They are derived, as in all vertebrates, from the embryonic aortic arches. Of the six pairs of arches in the embryo, the first, second, and fifth shrink to insignificance or disappear completely; the third pair contributes to the carotid arteries that supply the head; the right arch of the fourth pair becomes the trunk of the aorta, and the left half disappears (just the reverse of the situation in mammals); and the sixth pair becomes the pulmonary arteries. In accord with the heavy demands for blood by the breast and wing muscles, the brachial and

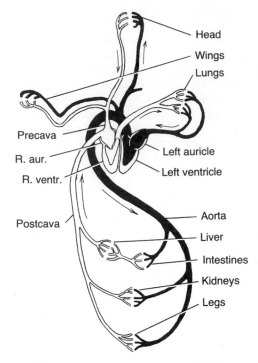

Head

Wings

Lungs

Precava

R. aur.

R. ventr.

Left auricle

Left ventricle

Postcava

Aorta

Liver

Intestines

Kidneys

Legs

FIGURE 7–1 The circulatory system in birds, in simplified diagrammatic form. Black vessels contain "fresh" or oxygenated blood; white vessels contain "used" blood charged with carbon dioxide.

pectoral arteries are relatively huge. Similar regional adaptations in arterial supply are found throughout the body. The venous system shows comparable adjustments to special avian needs. The two jugular veins draining the head are joined by cross vessels near the skull, so that if, for example, a twist of the neck should cramp shut the right vein, the blood could return to the heart via the left jugular. The paired kidneys are supplied with portal veins, a primitive feature not found in mammals. Blood from the posterior regions of the body flows through the iliac (portal) veins to the kidneys, where, by means of prominent valves at the juncture of the iliac and renal veins, the blood may either be deflected into the kidneys, where nitrogenous wastes are removed, or pass directly on to the heart via the renal veins. The hepatic or liver portal system in birds functions as it does in other vertebrates to screen the blood coming from the intestines and to store or transform the food it carries.

Compared with reptiles and mammals, birds have a poorly developed lymphatic system. The dead-end lymph capillaries collect into larger vessels that parallel the larger blood vessels and eventually join the blood system at the thoracic duct. Lymph hearts are found in all bird embryos, usually near the sacral vertebrae, but they ordinarily disappear in adults, although they remain functional in a few species such as ostriches, cassowaries, gulls, storks, and some passerines (Stresemann, 1927–1934).

The spleen in birds is relatively small, has feeble muscles, and apparently is not the important blood storage organ that it is in mammals. In birds it varies seasonally in volume, becoming larger in summer than in winter. It is the site of white blood corpuscle formation and red corpuscle destruction.

Blood

As in other vertebrates, the blood of birds is made up of a fluid plasma and of formed elements. The plasma, which is 80 per cent water, is chemically very complex and in some regards remarkably stable in composition. In contains various salts in solution; digested foods such as glucose, fats, and amino acids; special blood proteins thought to be formed in the liver; waste products (mainly carbon dioxide and urea); gases, hormones, enzymes, vitamins, antibodies, and small amounts of other substances.

The constituents of the blood plasma of a normal, adult chicken are shown by Sturkie (1976) to vary chemically according to age, sex, health, state of egg production, kind of food eaten, and recency of feeding. Laying hens, for example, have about two times as much blood calcium and three to five times as much blood lipids and fatty acids as nonlaying hens. The increase in fats seems due to the secretion of estrogen from the ovary of the laying hen. Glucose level in the blood drops with age. Chickens 18 months old have only 80 per cent of the blood glucose of those 3 months old. The average concentration of glucose in avian blood is about twice that in mammalian blood.

Total blood proteins in birds may vary from 2.82 g to 8.47 g per 100 cc of blood serum. The plasma proteins in males are lower than those in females, a fact suggesting an influence of sex hormones. The chief functions of plasma proteins, as given by Sturkie, are to maintain the normal volume of the blood and water content in tissues by osmotic pressure, and to combat disease. The globulin protein is more concentrated in chicken blood than in mammalian blood. Since globulins manufacture antibodies, this may be the reason chickens are particularly good antibody producers. However, on the debit side of the ledger is the fact that chickens, and probably other birds as well, suffer a remarkably high incidence of atherosclerosis. In a study by Dauber (1945), 45 per cent of chickens over a year old showed visible damage to the aorta alone. The experimental feeding of glucose-soybean meal with 10 per cent fat and 1 per cent cholesterol increased the cholesterol concentrations in blood serum and liver of young European Quail, *Coturnix coturnix*, 7 to 14 times and resulted in heavy accumulations of fatty deposits in their aortas (Morrissey and Donaldson, 1977).

Making up the formed elements of the blood are the erythrocytes or red corpuscles, the leukocytes or white corpuscles, and the tiny, spindle-shaped thrombocytes involved in blood coagulation. The same kinds of granular and non-granular leukocytes are found in birds and in mammals; however, their relative numbers are greater in birds. The chief function of the leukocytes is to protect the body against disease and to assist in fat metabolism.

Erythrocytes are nucleated as in reptiles, biconvex, and oblong in outline. They vary in size from 11 to 16 microns in length and 6 to 10 microns in breadth. (The circular human red corpuscle is about 7.7 microns in diameter.) The largest avian corpuscles are found in the Ostrich and its relatives, the smallest in the hummingbirds. As a rule, the more highly evolved birds have smaller red corpuscles in greater numbers and supplied with richer hemoglobin than do the primitive species. Good fliers have, in relation to their size, smaller erythrocytes than poor fliers. The smaller red corpuscles, with their relatively greater surface for lively gas exchange, illustrate once more the great biological significance of the surface-volume ratio.

In birds the number of red corpuscles varies inversely with body size, from 1.9 million per cu mm of blood in the Ostrich, *Struthio camelus*, to 5.9 million in hummingbirds.

The number of corpuscles (in millions per cu mm) of a few representative lower vertebrates provide another explanation of their metabolic sluggishness (Prosser, 1950): lizard, 1.4; alligator, 0.85; frog 0.5; *Necturus*, 0.05; carp, 1.6; catfish, 2.4; ray, 0.2; lamprey, 0.2.

The centrifuge-packed corpuscle volume (hematocrit) of avian blood averages about 40 per cent. Within a given species, the number of red corpuscles will vary according to sex, hormones, time of day, season of the year, the altitude at which the bird lives, and other factors. Male birds generally have a higher red corpuscle count than females. Mean hematocrits of captive American Kestrels, *Falco sparverius*, varied from winter maxima of 47.9 per cent for females and 50.9 per cent for males to summer minima of 29.1 per cent for females and 30.1 per cent for males (Rehder and Bird, 1983). Experimental studies by Domm and Taber (1946) showed that the red blood corpuscles in adult male chickens dropped from about 3.25 to 2.48 million per cu mm after castration, suggesting that corpuscle count may be sex-hormone-mediated. Thyroid hormone also influences erythrocyte frequency, male chickens with thyroid glands removed having about three-fourths as many red corpuscles as the controls. Females were unaffected by both castration and thyroidectomy.

When Pintail Ducks, *Anas acuta*, were exposed by Cohen (1969) for 20 days to air with the low pressure of 310 mm Hg (equivalent to an altitude of 7.5 km [4.7 miles]), their total blood volume increased by 23 per cent, erythrocyte volume by 68 per cent, and hematocrit by 30 per cent. The reduced oxygen of high altitudes apparently stimulates the production of red corpuscles in ducks as it does in man.

Rather surprisingly, the hemoglobin content in the blood of birds is slightly less than that in mammalian blood. But, what is more important, the hemoglobin of birds is more efficient than mammalian hemoglobin in carrying oxygen to the tissues. The "life" of a red corpuscle in a chicken is thought to be about a month, but a human red corpuscle keeps working for about four months before it is destroyed.

Erythrocytes are formed mainly in the bone marrow of adult birds, although in passerines they are also formed in the spleen and liver (Portmann, 1950). Leukocytes are formed in large numbers early in the life of a bird by the liver, spleen, kidney, pancreas, and bursa of Fabricius. In the adult bird the spleen and caeca are the main organs producing leukocytes.

THE RESPIRATORY SYSTEM

As in all animals, the chief function of the respiratory system in birds is to supply the living cells of the body with oxygen and to rid them of carbon dioxide. In this exchange of gases, the lungs and accessory tubes and sacs suck in fresh air and push out used air, and the blood carries the oxygen of the air from the lungs to the capillaries embracing the various body cells where, by diffusion, the oxygen is given up and the carbon dioxide is absorbed for the return trip to the lungs. The hemoglobin of the red corpuscles is primarily responsible for the transport of oxygen, and the hemoglobin and blood plasma together are responsible for the transport of carbon dioxide. A second vital function of respiration is thermoregulation, and in birds this involves particularly the elimination of excess heat generated in flight.

In order to meet the rigorous demands for air consumption that intense living, flight, and song impose on them, birds must have ample and efficient breathing machinery, as indeed they have. Not only is their respiratory system the most efficient known among all vertebrates, but it is unique in basic structure. Humans and other mammals breathe by means of a cul-de-sac respiratory system in which inhaled fresh air is mixed with residual stale air remaining in the dead-end alveoli of the lungs, which can never be completely emptied. Birds, on the contrary, have a complex system of sacs and interconnecting tubes that

make possible a more thorough bathing of the lung cells with fresh air.

The lungs are the center of the breathing system, and the place where gas exchange between the air and the blood occurs. Lungs are relatively smaller in birds than in mammals. The lungs of man occupy about 5 per cent of his body volume, but those of a duck take up only 2 per cent. In ducks the lungs and air sacs together occupy about 20 per cent of the body volume, but only about 15 per cent in most other birds. In the domestic chicken, *Gallus gallus*, the maximum capacity of the respiratory tract is 550 ml in males and 300 ml in females; the maximum volume of the lungs is 60 ml in males, 30 in females (King and Payne, 1962).

As air is inhaled, it passes from the nasal chambers through the mouth and pharynx into the upper end of the windpipe or trachea via the slit-like glottis. At the top of the trachea is a cylindrical box, the larynx, whose skeletal support is provided by cartilaginous rings that generally become ossified in maturity. The larynx is a reptilian hangover with no vocal cords

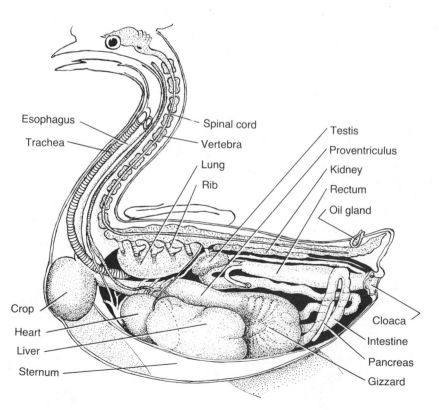

Esophagus
Trachea
Spinal cord
Vertebra
Lung
Rib
Testis
Proventriculus
Kidney
Rectum
Oil gland
Crop
Heart
Liver
Sternum
Cloaca
Intestine
Pancreas
Gizzard

FIGURE 7–2 The internal anatomy of the domestic chicken. The air sacs have been removed. After T. I. Storer, 1943.

and is not as important to birds as it is to mammals. It leads directly to the trachea proper.

The trachea, which conducts air from the mouth cavity toward the lungs, forks into two short tubes, the bronchi, and these connect shortly with the two lungs. The trachea is reinforced with a series of cartilaginous rings, which in most species turn into bone in the adult. In mammals they remain cartilage throughout life. These rings are joined together by tough bands of fibrous connective tissue. Occasionally, in geese, cranes, shorebirds, and others, the trachea is extraordinarily long and serves as a resonating tube to amplify the voice of the bird. This additional length may be provided by tracheal loops or coils located between the breast muscles and the skin, or at the base of the furcula, or in the thorax (see Fig. 11–5). In the Whooping Crane, *Grus americana*, the trachea is nearly as long as the bird itself, and about half its length is coiled in the keel of the sternum. If man had a trachea proportionately as long as that of the Whooping Crane, normal breathing would not suffice to keep him alive—the trachea would merely become filled with stale air shuttling back and forth. This emphasizes the fact that, in birds, the lung and air sac system allows for relatively deep breathing that ensures the presence of oxygen-rich air in the lungs, even though the air has traveled down a long trachea. It also reveals the principle that natural selection may lead to extravagant extremes in adapting to one need (courtship "song," in this case) at the expense of other vital needs (efficient breathing).

At the point, generally above the sternum, where the trachea divides into the two bronchi, there appears a structure found only in birds—the syrinx or voicebox. The syringes show great variety in structure, but generally there are three main types: tracheal, tracheobronchial, and bronchial. A few species have no syrinx whatever (vultures, ostriches, and some storks).

Tracheal syringes are characteristic of primitive groups such as chickens, ducks, and parrots. They have only one pair of membranes on the lateral walls —the lateral tympaniform membranes—which constrict the one air passage (Fig. 7–4). Tracheobroncheal syringes, characteristic of oscines, contain a lateral and a medial tympaniform membrane in each bronchus, resulting in two sound-producing tubes; truly a *syrinx,* the Greek word for a double-flute!

Primitive birds such as ducks and geese have only

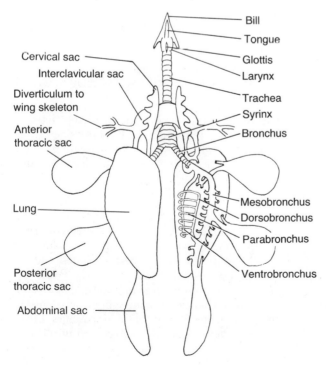

- Bill
- Tongue
- Glottis
- Larynx
- Trachea
- Syrinx
- Bronchus

Cervical sac
Interclavicular sac
Diverticulum to wing skeleton
Anterior thoracic sac
Lung
Posterior thoracic sac
Abdominal sac

Mesobronchus
Dorsobronchus
Parabronchus
Ventrobronchus

FIGURE 7–3 A simplified diagram of the respiratory system in birds. After Hazelhoff, 1951; Brandes and Hirsh.

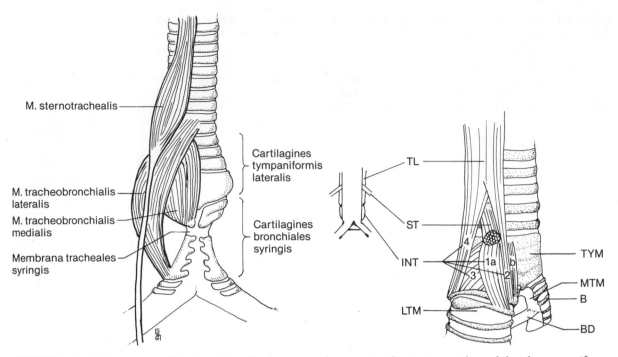

FIGURE 7–4   A, Parrot (psittacid) syrinx. Note that there are only two pairs of intrinsic muscles and that the tympaniform membrane is on the trachea. B, Songbird (oscine) syrinx. Note that in contrast to the parrot syrinx there are four pairs of intrinsic muscles, and the tympanal membranes (*MTM*) are on the bronchi. (Courtesy of A. and S. Gaunt)

two extrinsic muscles associated with the syrinx. Parrots have the same two muscles, plus two intrinsic muscles. Extrinsic muscles arise and insert on the trachea. Intrinsic muscles may arise on the trachea but insert on syringeal elements, or they may arise and insert on the syrinx. Oscine syringes contain extrinsic muscles and more than two pairs of intrinsic muscles (Fig. 7–4).

Extrinsic muscles change the position of the trachea and are thus indirectly involved in sound production. Intrinsic muscles can affect the configuration of the syrinx directly. Acquisition of complex vocal repertoires through learning is to be found only in hummingbirds, parrots, and oscines. These groups have syringes with two or more pairs of intrinsic muscles. Thus syllable imitation is confirmed, but "not distributed throughout" groups with intrinsic muscles (Gaunt and Gaunt, 1985).

The Gaunts inform us that sound may be produced in three ways.

1. The tympaniform membranes vibrate back and forth as air is forced up each bronchus to produce sound. This idea was developed in detail by Greenwalt (1968).

2. The tympaniform membranes are constricted to form a valve. Pressure from the lung side pushes against this valve, forcing it to open and release a burst of air. Each pulse of air produces a sound with a rich frequency spectrum. Pulse-generated sounds are typical of many avian species.

3. The tympaniform membrane constricts and remains motionless, and air forced through the same opening results in a series of vortices producing a whistle. Whistles tend to be pure-toned and are typical of doves (Gaunt *et al.*, 1982). Pitch (frequency) and volume of sound are probably controlled by various combinations of muscle-tension, air pressure, and the bore of the air passages. Loudness in *Streptopelia* dove songs is controlled by

muscles of the body-wall modulating air pressure (Gaunt *et al.*, 1982).

Since each oscine bronchus has its own tympanic membranes, muscles, and nerves it is possible for some birds like the Wood Thrush, *Hylocichla mustelina*, to sing notes of two harmonically unrelated frequencies simultaneously (Borror and Reese, 1956; Greenwalt, 1968). This is popularly known as the two-voiced phenomenon. That each side of the syrinx may function independently has been demonstrated by Nottebohm (1971) who severed either left or right branches of the hypoglossal nerve controlling the syrinx. Severing the left branch resulted in loss of large portions of the song; severing the right branch produced less drastic effects.

There is great variety in the construction of syringes, but the complexity and perfection of a bird's song are not related to the number of syrinx muscles (see Chapter 11). In many species the male does the singing, even though the female possesses a well-formed syrinx. The chief reason that the sexes differ in song is not only because of great structural differences in their syringes (as in ducks, owls), but because song is primarily under the control of nerves and hormones. It is true, however, that in many female birds the syringeal muscles are more weakly developed than in the males.

The bright red lungs of birds are strikingly small, highly vascularized, and relatively inelastic. Above, they are firmly attached to the ribs and thoracic vertebrae. The lower sides are covered with connective tissue on which are inserted several small, weak muscles which originate on the ribs. In the higher reptiles and mammals the bronchi entering the lungs fork again and again into ever smaller tubes, much as the trunk and branches of a tree subdivide until they end in twigs, the lung "twigs" being microscopic air sacs or alveoli. In birds, however, each bronchus passes through the ventral side of the lung and emerges at its posterior end to enter the large, thin-walled air sacs of the abdomen. Inside the lung, each bronchus, now called a mesobronchus, loses its reinforcing rings and gives off four to six secondary tubes called ventrobronchi, which in turn subdivide into the ultimate respiratory units, the parabronchi. The parabronchi are small parallel tubes, each several millimeters long and less than one-half a millimeter in diameter. In weak fliers like the chicken there are

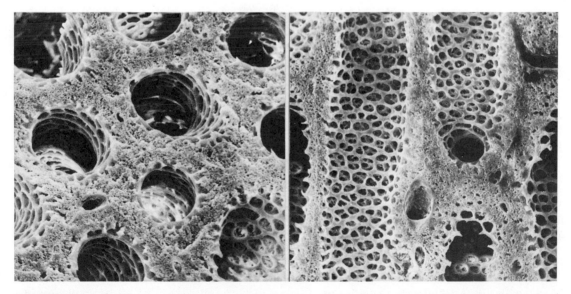

FIGURE 7–5 A cross-section (*left*) and longitudinal section of the tubular parabronchi in which air passes through the spongy lung tissue. The tiny canals in the walls of each parabronchus are the "air capillaries" through whose walls gas exchange between the blood and the inhaled air takes place. Magnifications are about ×105 and ×100. Electron micrographs courtesy of H. R. Duncker. After Schmidt-Nelson, 1971.

about 400 parabronchi per lung; in strong fliers like ducks and pigeons, about 1800. The parabronchial walls are punctured with the openings of hundreds of tiny, branching, and anastomosing "air capillaries" (Fig. 7–5), which extend outward and are intimately surrounded by a profuse network of blood capillaries. It is between these air and blood capillaries that the gas exchange between the lungs and blood takes place. Each parabronchus is lined with smooth muscles and has a massive, sphincter-like band of muscle surrounding the opening where it originates from the ventrobronchus. These muscles may control the caliber of the parabronchi (Lasiewski, 1972).

Unlike the blind alveoli of mammalian lungs, the parabronchi are open at each end. Ventrally, they connect with the ventrobronchi, and dorsally, with the dorsobronchi. In each lung there are two rows of dorsobronchi, six to ten in each row, which join the parabronchi with the mesobronchus. Both anteriorly and posteriorly the mesobronchi and their branches connect with the pulmonary air sacs, of which there are usually eight or nine, but they may vary in number from 6 to 14 (Fig. 7–6). These air sacs act as air reservoirs in the bronchial circuit rather than as air terminals. Thus, the lungs are provided with a system of interconnecting tubes and sacs that allows a continuous flow of air across respiratory surfaces. Air is not drawn *into* the lungs but *through* them.

For many years there has been much conjecture regarding the precise path that a given breath of air follows in ventilating a bird's lungs. Today the details of bird respiration are finally becoming known, thanks largely to the researches of Schmidt-Nielsen and Bretz (Schmidt-Nielsen, 1971; Bretz and Schmidt-Nielsen, 1971, 1972). Through the use of heated thermistor air-flow-indicating probes and a marker gas (argon) monitored by mass spectrometry, the path a breath of air takes in a domestic Duck, *Anas platyrhynchos*, has been traced (Fig. 7–7).

The duck's respiratory system behaves like a two-cycle pump. During the duck's first inspiration the inhaled air passes almost completely into the posterior air sacs. During the first expiration, this air is pumped into the secondary bronchi and the parabronchi of the lungs where it gives up its oxygen and receives carbon dioxide from the lung's blood capillaries. On the bird's secondary inspiration the air is drawn from the lungs into the anterior air sacs, and finally, during the second expiration the air is exhaled from the anterior air sacs through the bronchi and trachea to the outside. In

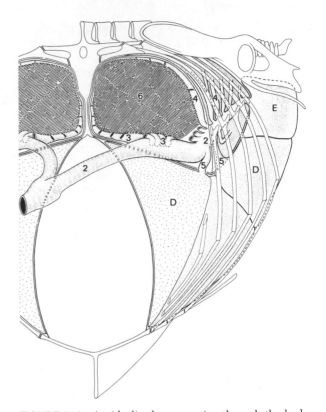

FIGURE 7–6 An idealized cross-section through the body of a White Stork showing the lung and air sac connections. At the middle left is the sectional trachea (dark oval) and the adjacent syrinx. A primary bronchus (2) connects with the lungs (6) by means of the ventrobronchi (3). Dorsobronchi (4) connect the dorsal surface of the lungs with the primary bronchus. The dorsobronchi and ventrobronchi are connected in the lungs by the minute parabronchi (6); this is where gas exchange between the inhaled air and the blood occurs. The abdominal air sac (E) is directly connected with the primary bronchus; the posterior thoracic air sacs (D) join the primary bronchus through the laterobronchi (5). Diagram courtesy of H. R. Duncker (1971) and Springer-Verlag.

short, "the air flows continuously in the same direction through the avian lung during both inhalation and exhalation." Since no valves are known that might control the direction of air flow, fluid dynamics of the rapidly moving air must determine the specific paths taken in the numerous tubes of the respiratory system. A bird's ribcage structure is so arranged that upon inhalation the sternum is lowered, the chest and air sacs are enlarged, but the lungs are diminished in vol-

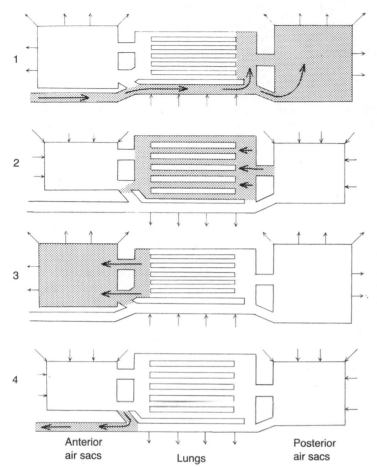

Anterior
air sacs

Lungs

Posterior
air sacs

FIGURE 7–7 A schematic diagram showing the progress of a single breath of air (shaded) through a bird's respiratory tract. Small arrows indicate complementary pressure changes on air sacs and lungs; large arrows show direction of air flow. (1) On first inhalation air flows through trachea and bronchus chiefly into the expanding posterior air sacs. (2) On exhalation air is forced from the diminishing posterior air sacs into the expanding lungs. (3) On the second inhalation the air moves from the shrinking lungs into the enlarging anterior air sacs. (4) Finally, on the second exhalation the air is discharged to the outside. This two-cycle breathing mechanism assures a continuous, unidirectional flow through the lungs. Not shown in the diagrams is a second breath of air that enters the trachea, bronchi and posterior air sacs on the second inhalation of air (identical with step (1) above). Redrawn after Schmidt-Nielsen, 1971.

ume. On exhalation, the reverse occurs. Thus, since lungs expand on exhalation, air flows into them from the shrinking posterior air sacs, and, during the next inhalation, the lungs shrink in volume and force their contained air into the anterior air sacs. The direction of blood flow in the lungs is directly contrary to that of fresh air flowing through parabronchi (i.e., counter-current flow), an anatomical refinement that greatly increases a bird's respiratory efficiency.

There are in the typical bird's body four pairs of air sacs plus one unpaired sac, making a total of nine. Most of them are placed dorsally in the body, a fact that helps stabilize flying by lowering the body's center of gravity. The sacs are all thin-walled, with very little musculature and with a very poor supply of blood vessels, so they cannot be considered as primary res-

piratory organs. The unpaired air sac is the interclavicular, located anteriorly. It sends diverticula into the syrinx and larger pneumatic bones such as the sternum, pectoral girdle bones, and humerus. Associated with the cervical vertebrae and sending lateral branches posteriorly are the paired cervical sacs. Also anteriorly placed are the prethoracic pair of sacs. Posterior in the body are the postthoracic and abdominal pairs of sacs, the latter supplying air to the pneumatic bones of the sacrum, pelvis, and legs. In general, large flying birds have well-developed pneumatic bones, while small fliers have few or none. Birds with pneumatic wing bones are able to breathe through the broken and exposed humerus, even though the trachea may be closed, and apparently this also applies to the femur.

In addition to their essential role in breathing, the air sacs help cool the body during vigorous exercise through the internal evaporation of water. Domestic Pigeons, *Columba livia*, and Budgerigars, *Melopsittacus undulatus*, flying at moderate temperatures, discard 13 to 22 per cent of their heat production through such evaporation, the remainder being lost through radiation and convection (Hart and Roy, 1966; Tucker, 1968b). Certainly, the air sacs are well placed among the body organs for the efficient removal of excess heat.

For aquatic birds the air sacs provide valuable buoyancy. Swimming species have particularly large abdominal and post-thoracic sacs whose volume can be controlled for diving or floating. In aquatic species that plunge from the air into the water for food, the air sacs act as a buffer to protect the body against disabling shocks. It seems probable that the abdominal sacs aid the abdominal muscles in the acts of defecation and egg-laying. The air sacs even play an important role in the courtship of many species, analogous to sartorial finery in the human species. The male Magnificent Frigate-bird, *Fregata magnificens*, and the male Prairie Chicken, *Tympanuchus cupido*, inflate their showy neck bladders through diverticula of the cervical air sacs. The yellow external balloons of the Prairie Chicken also act as resonators for its booming courtship call. Among those species that have long-drawn-out songs, the air sacs may serve as air reservoirs.

It is likely that the rate of breathing is controlled in birds, as it is in mammals, by a respiratory center in the medulla of the brain and the carotid body of the neck. Experimental studies indicate that the breathing center responds to changes in the temperature and pH of the blood, to carbon dioxide inhalation, and even to the position of the body (Sturkie, 1976). Receptors responsive to carbon dioxide in inhaled air were demonstrated in the lungs of the domestic chicken by Peterson and Fedde (1969). A panting center that controls heat-dissipating mechanisms is located in the midbrain. If the brain temperature in a pigeon is raised from 41.7° to 43.6°C, its breathing rate will increase from 46 to 510 times per minute, and the volume of air breathed, from 185 to 610 cu cm per minute (von Saalfeld, 1936). The evaporative cooling that this brings about takes place largely in the air sacs and turbinals of the nasal chambers.

The rate at which a bird breathes varies with species, age, sex, size, activity, air temperature, time of day, and other factors. As a rule, the smaller the bird, the faster its breathing rate.

Because of their large air sacs and because fresh air passes through the lungs during both inhalation and exhalation, birds have a larger capacity of air per breath, and therefore do not need to breathe as rapidly as mammals of comparable size to get the same supply of oxygen. Table 7–2 illustrates this relationship. The animals represented range in size from elephants to shrews, and from ostriches to hummingbirds. The deeper, more ample breaths of birds are reflected in the ratio of heart rate to breathing rate, which is about 7.5 to 1 in birds but only 2.9 to 1 in mammals (Odum, *in litt.*).

As in man, the breathing rate in birds speeds up with vigorous exercise. Miniature telemetry devices have been developed that are small enough to be carried by birds while in otherwise normal flight. These devices are able to broadcast back to recording instruments such information as breathing rates, heart rates, and blood pressure variations in freely flying birds. Using this technique, it was determined that the mean breathing rate of a duck at rest was 14 respirations per minute, but a flying duck had a rate of 96 respirations per minute (Lord *et al.*, 1962).

Several observers are of the opinion that wing-beats in flying birds are synchronized with breathing mechanics, a fact that, if true, would promote the

TABLE 7–2.

APPROXIMATE BREATHING RATES OF BIRDS AND MAMMALS AT REST IN REFERENCE TO BODY WEIGHTS*

WEIGHT OF ANIMAL IN GRAMS	BREATHING RATE IN BREATHS PER MINUTE	
	Mammal	*Bird*
3,600,000	4.5	—
100,000	22	5
10,000	28	10
1000	33	16
100	100	28
10	200	100
3	800	250

*Largely after Calder, 1968.

efficiency both of respiration and of cooling the body. In their observations of nine species of birds in short flights, Berger *et al.* (1970) found that a perfect one-to-one coordination of wing strokes and breathing cycles was rare, but that usually the beginning of inspiration coincided with the end of the upstroke, and the beginning of expiration coincided with the end of the downstroke. Oxygen consumption in flight averaged about 10 times the standard or resting rate.

Electronic studies of the respiratory rate of Black-capped Chickadees, *Parus atricapillus*, by Odum (1943) showed that their basal rate, while asleep at an air temperature of 11°C, was 65 per minute; and at 32°C, 95 per minute—an acceleration related to the need for removing heat. The few species that undergo temporary dormancy or hibernation slow down their breathing rates along with other metabolic activities. Torpid nestlings of the European Swift, *Apus apus*, breathe a minimum of eight cycles per minute during long fasts imposed by cold, wet weather (Koskimies, 1950). In House Wrens, *Troglodytes aëdon*, whose body temperatures were changed experimentally by Baldwin and Kendeigh (1932), birds whose breathing rate at normal body temperatures (38° to 41.5°C) was 92 to 112 per minute, slowed down to a minimum of 28 per minute at 23°C and rose to a rate of 340 per minute at the maximum tolerable body temperature of 47°C (116°F).

A bird that dives deeply in the water faces a special respiratory problem. To reduce buoyancy it must reduce the amount of air in its air sacs. Consequently, any air reserves it may have in them will be curtailed. In the duck, as in most mammals, the heart rate slows during a dive; and in penguins the oxygen consumption by tissues is reduced to 20 to 25 per cent of the normal resting rate (Prosser, 1950). It seems probable that in birds, as in diving mammals, muscle sugar or glycogen is broken down anaerobically to lactic acid in order to provide the needed energy. Such anaerobic glycolysis, while eliminating the immediate need for oxygen, would flood the blood with lactic acid, thus creating an "oxygen debt" that would be repaid when the bird returned to the surface. High tolerance of the breathing center to carbon dioxide, as has been demonstrated in seals and porpoises, also occurs in diving birds to enable them to prolong their time under water. The Mallard Duck, *Anas platyrhynchos*, has been experimentally held alive under water as long as 16 minutes, and even 27 minutes with its trachea closed. Voluntarily, birds stay under water for much shorter times, usually less than a minute; but the Common Loon, *Gavia immer*, has been known to stay under for 15 minutes. Great numbers of Oldsquaws, *Clangula hyemalis*, have been caught and drowned in Lake Superior commercial fishing nets set 21 to 27 meters deep. (As many as 27,000 were so destroyed by one single fishing crew in the spring of 1946!) Undoubtedly these birds required a fairly long immersion to feed at such depths. Loons have been taken in fish nets set 55 meters (180 feet) deep (Schorger, 1947).

The remarkable efficiency of a bird's respiratory system was demonstrated by Tucker (1968a), who exposed House Sparrows, *Passer domesticus*, and mice to a simulated altitude of 6100 m (3.7 miles), which represents air at less than one-half the atmospheric pressure (and oxygen density) of air at sea level. The sparrows were still able to fly, but the mice were comatose. European Quail, *Coturnix coturnix*, kept under similar conditions for six weeks, responded with increases in blood volume of 36 per cent, in hemoglobin concentration of 37 per cent, and in hematocrit ratio of 31 per cent (Weathers and Snyder, 1974). Birds residing permanently at high altitudes adapt to the rarefied air by developing not only larger hearts but also larger lungs and have higher red blood cell counts (Carey and Morton, 1976).

Domestic ducks, *Anas platyrhynchos*, responded to a decline in the partial pressure of arterial oxygen (from 95 to 28 torr) with a fivefold increase in cerebral and coronary blood flow. The highland Bar-headed Goose, *Anser indicus*, responded with only a threefold increase, and oxygen delivery to the heart and brain of the goose was higher than that in the duck under the stressful conditions (Faraci *et al.*, 1984). The goose is clearly better adapted to the thin atmospheres of its native high altitudes than the duck.

ENERGY METABOLISM

Metabolism is the sum total of chemical activities in a living organism that provide energy for heat, movement, irritability, growth, repair, energy storage, and reproduction. This energy is all obtained from the oxidation of foods eaten by the animal: carbohydrates, fats, and proteins. A given quantity of food will provide a precisely specific amount of energy, whether burned in a bomb calorimeter in the laboratory or in the body of a live animal. Fats yield about twice as much energy per gram as carbohydrates or proteins. The energy consumed by an animal may be measured directly, by the caloric value of the food consumed, or indirectly, by the amount of oxygen consumed to oxidize that food and release its energy.

The chemical changes that occur in metabolism are under the control of cellular enzymes that accelerate and direct metabolic reactions. These reactions are of two sorts: anabolic, in which simpler substances are united to form more complex compounds in which energy is stored; and catabolic, in which complex compounds are broken down into simpler substances and energy is released in the form of heat, muscle contraction, nerve impulse transmission, glandular secretion, and other vital activities. Growth is primarily an anabolic reaction in which amino acids from digested foods are united to form the protein molecules of body tissues.

For comparative purposes the metabolic rates of animals are measured under standard conditions of complete rest, nondigestion or absorption of food, and presence of a thermoneutral environment—i.e., under no stress from either heat or cold. Under these conditions a bird's heat production (or its equivalent oxygen consumption) per unit of time is known as its standard or basal metabolic rate. The basal metabolic rate of a 266-g domestic Pigeon, *Columba livia*, for example, can be expressed as 126 kilocalories per kilogram of body weight per hour (126 kcal/kg/hr). Basal metabolism varies roughly in proportion to the surface area of an animal; hence a small bird metabolizes at a higher rate than a larger one. For example, the basal metabolism of a 27-g House Sparrow is 312 kcal/kg/hr, and that of a young adult human weighing 75 kg is about 1600 kilocalories per day, or about 1 kcal/kg/hr. In all animals the basal metabolic rate increases about 5 per cent for each degree Celsius increase in body temperature. In addition, metabolic rate is 10 to 40 per cent higher by day than by night, probably because

of greater muscle tone in day-active birds (Kendeigh *et al.*, 1977), and in passerines is about 1.65 times that of nonpasserines of the same weight (Zar, 1969).

As unusually energetic animals, birds need amounts of energy greatly beyond those required for minimum resting existence. To start with, they have a higher basal metabolism than most mammals because, as a class, they are small animals. Moving in vertical as well as horizontal directions, and at great speed, they require more energy than most other animals, just as hill-climbing requires more energy than walking on a level surface. The tiny hummingbird probably demonstrates the highest metabolic rate and the greatest metabolic range of any vertebrate. Experiments by Pearson (1950) on captive Anna's Hummingbirds, *Calypte anna*, and Allen's Hummingbirds, *Selasphorus sasin*, which weighed between 3.75 and 4.32 g, revealed that their daytime consumption of oxygen while resting ranged from 10.7 to 16.0 cc per gram of body weight per hour. During periods of hovering flight their oxygen consumption rose to 85 cc/g/hr for Allen's Hummingbirds, and 68 cc/g/hr for Anna's Hummingbirds. Similar studies of the metabolic costs of flight have shown that in small passerines the flying rate of metabolism was about 10 to 15 times the resting rate (Teal, 1969; Farner, 1970), in racing pigeons, about 8 times (LeFebvre, 1964), and in hummingbirds, six or seven times (Lasiewski, 1963).

Maintaining a higher body temperature than other animals results in a steeper temperature gradient between body and environment, hence a more rapid heat loss. Undaunted by these handicaps, birds have successfully penetrated the most frigid regions populated by any animal. The Emperor Penguin, *Aptenodytes forsteri*, breeds in antarctic temperatures as low as −62°C (−80°F)—"one of the most remark-

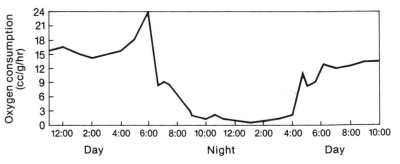

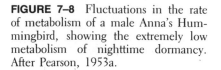

FIGURE 7–8 Fluctuations in the rate of metabolism of a male Anna's Hummingbird, showing the extremely low metabolism of nighttime dormancy. After Pearson, 1953a.

able physiological feats known among warm-blooded animals" (Murphy, 1936). It is not surprising, in view of this fact, that the Emperor is the largest penguin, weighing, with its thick layers of feathers and fat, between 26 and 43 kg (57 to 94 pounds).

In order to achieve their high metabolic rates, birds have had to adapt themselves in various ways. As a rule, they eat foods rich in energy: seeds, fruits, nectar, insects, fish, rodents, and the like—foods from which birds can assimilate 70 to 90 per cent of the contained energy, compared with 30 to 70 per cent from terrestrial vegetation (Ricklefs, 1974). The rich concentration of glucose in their blood (about double that of human blood) promotes higher levels of metabolism. At those times of year when it is cold, or when birds migrate, breed, and molt, their energy requirements are sharply increased. In unconscious anticipation of these extra demands, many species accumulate deposits of fat which tide them over periods of strenuous living. The time spent feeding by a Black-capped Chickadee, *Parus atricapillus*, is about 20 times greater at $-40°C$ than at $10°C$ (Kessel, 1976).

To conserve the heat they generate, they have reduced or eliminated heat-radiating projections such as external ears and fleshy tails and legs; they have covered the body with heat-conserving feathers and, in many aquatic species, with layers of insulating fat. No bats are found in polar regions, for the wings of bats are living membranes with networks of blood vessels, while the main expanse of a bird's wings is made of lifeless feathers that conserve rather than dissipate heat. Exposed skin in many birds is reduced to a minimum. Although air sacs are probably most useful in cooling the body, paradoxically they may also conserve heat. Since air is a poor conductor of heat, the more superficial air sacs, particularly if they contain relatively static air, may act to protect the body from heat loss in cold weather.

The main function of hemoglobin in the blood is to pick up oxygen in the lungs and give it to the oxygen-hungry cells throughout the body. In spite of the fact that avian hemoglobin has a somewhat weaker affinity for oxygen than mammalian hemoglobin, it nevertheless is more efficient in oxygen transport because it gives up oxygen more readily than does mammalian hemoglobin. A study of the difference in oxygen concentration in arterial and venous bloods of the duck and pigeon shows a utilization of 60 per cent, compared with 27 per cent in man (Prosser, 1950). Both the high temperatures and the high carbon dioxide concentrations in body tissues of birds promote a generous dissociation or unloading of oxygen by the red blood cells.

Birds consume energy at variable rates, depending on a variety of conditioning factors. In the winter a Bullfinch, *Pyrrhula pyrrhula*, consumes about 50 per cent of the fat in its body between dusk and the next morning (Newton, 1972). This rapidly cycled fat, at least in pigeons, comes almost exclusively from fatty acids synthesized by the liver (Goodridge and Ball, 1967).

As mentioned earlier, size is an important factor. The resting metabolism of a hummingbird, as measured by its oxygen consumption per gram of body weight, is about 12 times that of a pigeon, 25 times that of a chicken, and 100 times that of an elephant. If an elephant's tissues generated heat as rapidly as a hummingbird's, the elephant, unable to lose heat rapidly enough, would cook to death. If a man metabolized energy at a hummingbird's rate, he would be obliged to eat double his weight in food every day to acquire the needed energy. Apparently hummingbirds and shrews have reached the smallest size theoretically possible for warm-blooded animals. Animals significantly smaller would probably not be able to eat food fast enough to avoid starvation (Pearson, 1953). And, of course, warm-blooded birds and mammals metabolize energy at much faster rates than do cold-blooded animals. A resting canary consumes about 4 cc of oxygen per gram of body weight per hour, but the cold-blooded Goldfish, *Carassius auratus*, consumes only 0.12 cc/g/hr; an earthworm, 0.06 cc/g/hr; and a sea anemone, 0.013 cc/g/hr. With their much lower rates of metabolism, cold-blooded animals do not readily suffocate, which explains why miners keep canaries instead of goldfish in mines to detect foul air.

Birds have higher and more variable temperatures than mammals. Adult mammals range in temperature roughly between 36° and 39°C, whereas birds range from 37.7° to 43.5°C, and the majority of them maintain temperatures of from 40° to 42°C. In their studies of bird temperatures, Baldwin and Kendeigh (1932) found the average resting temperatures of 29 passerine species to be 40.4°C, and Udvardy (1953) determined the mean resting temperatures of 42 species of passerines as 40.6°C and of 13 species of Charadriiformes as 40.09°C. Primitive species tend to have lower temperatures than more advanced species: the kiwi, for example, has a body temperature of 37.8°C. As a rule, the body temperature of small birds fluctuates more than that of large birds: that of a House Wren, *Troglodytes aëdon*, may fluctuate 8° in 24 hours; that of a Robin, *Turdus migratorius*, about 6°, and that of a duck, only about 1°.

There is a rough negative correlation between the weight of a bird and its resting temperature—the smaller the bird, the higher the temperature. With every tenfold increase in body weight there is, roughly, a decrease of 1.5° in body temperature. The same inverse correlation holds for mammals, except for those mammals with weights less than about 5 kilograms. In these, temperatures become lower with decreasing weights rather than higher as in birds (Rodbard, 1950). This contrast is probably due to the fact that fur insulates less well than feathers, so that a small mammal, with its relatively large surface, cannot maintain as high a body temperature as a small bird. Dormancy and hibernation are resorted to much more generally by mammals than by birds, probably because mammals have poorer insulating coats and are less able to escape the rigors of winter by migration.

Sex also influences metabolism. In chickens the metabolic rate of males is higher than that of females. When males are castrated their rate drops about 13 per cent. The thyroid gland has a profound effect on metabolism. Overactivity of the gland elevates, and underactivity depresses metabolic rate (Sturkie, 1976). Removal of the thyroid gland causes heat production in a goose to drop by 15 to 33 per cent.

The daily cycle of activity and rest understandably is accompanied by a temperature cycle. Diurnal birds reach their highest body temperatures late in the afternoon and their lowest temperatures in the early morning. Owls show a major peak in metabolism during the first few hours of darkness and a minor peak just before daybreak—at the times when they are normally most active (Gatehouse and Markham, 1970).

Merely in order to exist in climates with pronounced warm or cold temperatures, birds must generate more energy than that involved in maintaining basal metabolism. This extra heat production is known as *existence metabolism*. And beyond existence metabolism is the *productive metabolism* needed for such activities as flight, migration, courtship and reproduction, molting, singing, and other essential life services (Kendeigh, 1949). As a result, the total metabolic burdens of a bird show seasonal fluctuations. A Rook, *Corvus frugilegus*, requires nearly twice as much food in winter as in summer. The mean exis-

tence energy requirement of a Red-winged Blackbird, *Agelaius phoeniceus*, living under 15 hours of daylight and at an environmental temperature of 21 to 25°C is about 26 kcal per bird per 24 hours. During days with 9 hours of light and temperatures of 11°C, the bird requires 45.5 kcal/bird/24 hr (Brenner, 1966). The differences in metabolic heat production are largely caused by the differences in environmental temperature, but relative daylength (or photoperiod) also plays a role. The net energy cost to a Chaffinch, *Fringilla coelebs*, of growing a new coat of feathers is about 1.1 kcal per day (Dolnik and Gavrilov, 1979).

All animals are popularly considered to be either cold-blooded or warm-blooded, the latter category including only birds and mammals. More precisely, a cold-blooded animal is defined as *poikilothermal*, that is, its body temperature is approximately the same as that of its immediate environment. A warm-blooded animal is called *homeothermal* or *endothermal*—its body temperature is approximately constant regardless of the environmental temperature. Maintaining thermal constancy requires the metabolic production of large amounts of heat. Endothermal animals are called warm-blooded because their bodies are generally (but not always) warmer than their environments.

All endotherms exhibit a slight drop in body temperature at night, a phenomenon known as *nocturnal hypothermia*. However, some endotherms may regularly regulate their body temperatures at two substantially different levels. The large temperature differences may occur daily in *torpor*, or seasonally, in *hibernation*. These organisms are known as *heterotherms*.

Endothermy confers rich benefits on its possessors, but at a heavy metabolic cost. Warm-bloodedness makes possible an internal physiological stability, which in turn makes an animal independent of many environmental restrictions, particularly thermal limitations. Only two species of exothermal reptiles, for example, are found living inside the Arctic Circle; but hordes of endothermal birds live and breed there.

As pointed out in the first chapter, the race in competitive nature usually goes to the metabolically swift, and the constant high temperatures of endotherms make a high level of metabolism possible throughout the year. Endothermism also provides an enormous and obvious psychic advantage to birds and mammals which is nevertheless commonly overlooked. The rate and extent of animal learning depend heavily upon frequency and recency of experience. Even man forgets new mental acquisitions very quickly if they are not soon repeated. Warm-bloodedness makes possible a continuity of experience that powerfully supports the learning process. Most of the little that the slow, chilly brain of a reptile might learn is probably forgotten during the periods of nighttime or winter dormancy.

Temperature Control

To maintain a constant body temperature in spite of environmental temperature changes, a bird requires temperature-regulating mechanisms. These are of two sorts: physical and behavioral mechanisms that control the loss of heat, and chemical or metabolic mechanisms that control the production of heat. Heat may be lost through conduction or convection into the surrounding air or water, by radiation into the environment, and by the evaporation of water from air sacs, lungs, and other respiratory passages and skin. Heat is generated by the catabolism of such energy stores in the body as glucose, glycogen, and fat, largely by way of muscle contraction.

For every species of bird there is a range of environmental temperatures known as the range of *thermoneutrality* in which metabolic heat production is at a minimum. Environmental temperatures above and below this range will result in increased heat production by the body; that is, the bird has to do work, either to keep warm at cold air temperatures or to keep cool at warm temperatures, but throughout both ranges it maintains a relatively constant body temperature.

If, for example, a Cardinal, *Cardinalis cardinalis*, in winter exists at an air temperature of -20°C, it will generate over twice as much heat (as measured by its oxygen consumption) as is required to keep it warm in its zone of thermoneutrality (Fig. 7–9). If the air temperature should increase, the amount of oxygen consumed will decrease, until the thermoneutral zone of 18° to 33°C is reached. There the oxygen consumption levels off. Then, if the air temperature increases still further, oxygen consumption begins to climb, and the bird begins to do work to keep the body from overheating. The temperatures of 18° and 33°C (the turning points in oxygen consumption) are known as the lower and upper *critical temperatures*, respectively (Dawson, 1958). At some point above the upper critical temperature the rate of heat production overtakes the rate of heat loss and the bird's body temperature begins to rise until, at exceptionally high air temperatures, it reaches the lethal limit. Some tropical hummingbirds (*Colibri delphinae* and *Amazilia*

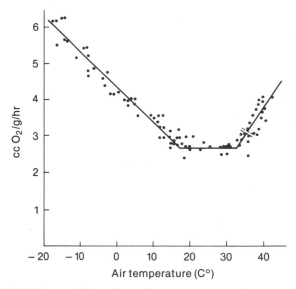

FIGURE 7–9 The relation between oxygen consumption and environmental temperature. The part of the curve between −20°C and 35°C represents data from 22 Cardinals in winter; that between 35° and 43°, from 17 Cardinals in summer. The horizontal portion of the curve represents the zone of thermoneutrality. Redrawn after Dawson, 1958.

tzacatl) do not regulate metabolism according to classical concepts of thermoregulation. Exposed to a range of temperatures from 4° to 40°C, Schuchmann and Schmidt-Marloh (1978) found a linear reduction in oxygen consumption with increased ambient temperature, but there was no increase in oxygen consumption at high temperatures. Instead, oxygen consumption continued to decrease linearly. In most species the upper critical temperature is commonly within the range of 33° to 38°C, but the lower critical temperature varies widely, commonly reflecting the environmental climates of different species. For example, in the antarctic Emperor Penguin, *Aptenodytes forsteri*, it is −10°C; in the boreal Ruffed Grouse, *Bonasa umbellus*, it is −0.3°C; in the temperate zone Horned Lark, *Eremophila alpestris*, it is about 20°; and in the tropical Black-rumped Waxbill, *Estrilda troglodytes*, it is 28°. In the Nighthawk, *Chordeiles minor*, the lower and upper critical temperatures coincide and there is no zone of thermoneutrality.

It is in the zone of thermoneutrality that birds are best able to withstand hunger. Conversely, well-nourished birds can withstand extreme temperatures

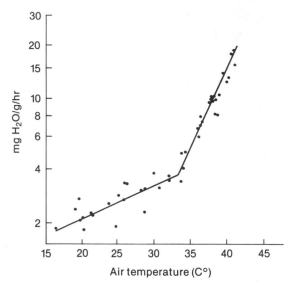

FIGURE 7–10 The relation between evaporative water loss and environmental temperature. Data from 16 Cardinals in spring and early summer. The zone of thermoneutrality ranges from 17°C to 33°C. Redrawn after Dawson, 1958.

better than hungry birds. A sparrow with an empty stomach will die if its body temperature descends to 32.8°C, but if it is well nourished, it will survive a drop to 21°C without harm (Portmann, 1950).

The physical mechanisms that control temperature may operate either to conserve heat in cold weather or to dissipate heat in warm weather. Feathers are remarkably effective heat conservers. Of 25 fasting domestic Pigeons, *Columba livia*, subjected to an air temperature of −40°C, the majority survived 72 hours, and four of them, 144 hours. But Pigeons plucked of their feathers survived less than half an hour (Streicher *et al.*, 1950). The insulating value of feathers is adaptively low in tropical species and high in cold-climate species. In addition, many species living in variable climates carry more feathers in winter than in summer (page 35). Furthermore, in cold weather insulation is increased when the feathers are fluffed so that a thicker layer of stable air surrounds the body. In very hot weather the feathers may be loosely ruffled to allow warm air near the body to escape by convection. In cold weather the desert-dwelling Greater Roadrunner, *Geococcyx californianus*, elevates its feathers and exposes black-pigmented skin of its neck and back to the sun's rays and thereby saves about 550 calories of energy per hour, or about 41 per cent of its basal metabolic rate (Ohmart and Lasiewski, 1971). As a consequence, it can lower its minimum critical temperature to 9°C.

In addition to feathers some species have thick layers of insulating fat to protect their bodies against heat loss. The Emperor Penguin, *Aptenodytes forsteri*, has both a thick layer of downy feathers and a layer of subcutaneous fat 2 to 3 cm thick, the feathers providing about 85 per cent of the barrier to heat loss (Le Maho, 1977). The bird lives in water about 40° colder than its body temperature—an environment in which man can survive only about 10 minutes (Stonehouse, 1967). Penguins in general are so adapted to life in cold water that they face the danger of lethal overheating on land. This explains the nocturnal or crepuscular habits of temperate zone penguins and the burrowing of tropical or subtropical species. Similarly adapted to cold, goslings of the Emperor Goose, *Anser canagica*, in Alaska sometimes die in large numbers during unseasonable heat waves.

Many birds huddle together at night to reduce heat loss (Fig. 7–11). As many as 50 titmice have been found packed together in a ball-like mass on cold winter nights (Tucker, 1949). Experiments by Brenner (1965) on fasting Starlings, *Sturnus vulgaris*, roosting in air temperatures of 2° to 4°C, showed that a single bird consumed 5.83 cc of oxygen per hour, or 92 per cent more than each bird in a huddled group of four. A bird roosting singly at this temperature survived only one day without food as against three days for the grouped birds. By huddling together in groups of thousands, antarctic Emperor Penguins, *Aptenodytes forsteri*, reduce the daily loss of body fat by 25 to 50 per cent (Le Maho, 1977). Many other species conserve heat in cold weather by roosting in tree cavities, or, like some titmice, arctic buntings, and ptarmigan, by plunging into soft snow on bitterly cold nights.

The unfeathered lower extremities of the legs of most birds dissipate large amounts of heat. A domestic chicken can reduce the loss of heat by 40 to 50 per cent by sitting instead of standing, and by tucking its head into its shoulder feathers can further reduce heat loss by 12 per cent (Deighton and Hutchinson, 1940). Many birds, especially shore birds and waders, conserve body heat by sleeping while perched on one leg and with their beaks tucked into their back feathers (Fig. 7–12). Fortunately, birds are often able to turn this apparent liability into an advantage by controlling the amount of warm blood that flows into their legs. In the lower tibiotarsal region of many aquatic and

FIGURE 7–11 A feathery ball of about 15 Tree Creepers, huddled together to keep warm on a cold night in southern Germany. The tails of eight are clearly visible. Photo by Hans Löhrl.

terrestrial birds there is a network of arteries and veins called the *rete mirabile*. This skein of vessels serves as a heat-exchanger to regulate loss of body heat to the environment. Warm, outward-flowing arterial blood is brought into close proximity with colder venous blood returning from the lower leg. This countercurrent mechanism is probably under vasomotor control so that, in very cold weather, the bare lower extremities can be kept from freezing but not so warm as unnecessarily to waste precious heat energy. Cold-acclimatized pheasants exposed to air temperatures of −18°C were found by Ederstrom and Brumleve (1964) to have mean body temperatures of 41°C, with a gradient of temperatures from 35.7° in the feathered lower tibiotarsus to 2.7°C in the toes. Warm-acclimatized pheasants in a temperature of 18°C had toe temperatures of 19.7° (Fig. 7–13).

Not only does the *rete mirabile* restrict heat loss

in cold weather but it promotes heat dissipation in warm weather. If a pigeon, held at an air temperature of about 20°C, has its head and back (but not its feet) warmed with radiant heat, the temperature of its feet will rapidly rise 12° to 15°, although its body temperature remains constant. This is caused by the vasodilation of leg arteries, which allows more blood to bypass the rete and flow to the extremities where it radiates excess heat. Termination of the heating results in vasoconstriction of the arteries and a prompt return of foot temperatures to normal (Bernstein, 1971). The webbed feet of the South Polar Skua *Catharacta maccormicki*, may perform similarly. A variety of birds (hawks, cranes, song birds) fly with their legs exposed during hot weather and tucked under their bodies during cold weather (Nisbert, 1978, Bryant, 1983). Hummingbirds fly with their tarsi exposed in hot weather, but in cold weather they tuck them into special pow-

FIGURE 7–12 A Wilson's Snipe conserves heat by perching on one leg and tucking its long beak into its back feathers.

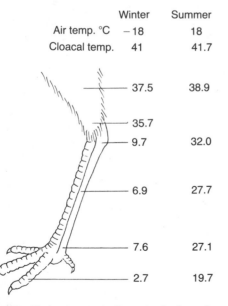

	Winter	Summer
Air temp. °C	−18	18
Cloacal temp.	41	41.7
	37.5	38.9
	35.7	
	9.7	32.0
	6.9	27.7
	7.6	27.1
	2.7	19.7

FIGURE 7–13 Temperature gradients in the legs of warm- and cold-adapted pheasants. Figures are means for 19 cold- and 7 warm-adapted birds. Note the sharp drop in temperature at the heel, just under the feathered tibiotarsus containing the *rete mirabile*. Redrawn after Ederstrom and Brumleve, 1964.

derpuff-like feathers on the distal end of the shank as a mechanism to conserve heat (Udvardy, 1983).

Because of its high capacity for absorbing heat, water will cool a bird's legs more than four times as rapidly as air of the same temperature. At high air temperatures a variety of birds (cormorants, storks, ibises, vultures, and others) take advantage of this principle and cool themselves by squirting liquid excrement on their bare legs (Kahl, 1963; Hatch, 1970). In the Turkey Vulture, *Cathartes aura*, the frequency of excreting on the legs increases with rising air temperatures, and at temperatures below 18°C the excretions are directed away from the legs.

The evaporation of water from body surfaces (skin, lungs, air sacs, buccal cavity, and throat) provides the most effective physical method of controlling heat loss. With their already high body temperatures, birds live perilously close to the lethal body temperature limit of 45° to 46°C. To combat the hazards of high environmental temperatures, birds resort to evaporative cooling mechanisms. A Cardinal, *Cardinalis cardinalis*, loses very little water at thermoneutral temperatures (18° to 33°C), but as air temperatures rise above the upper critical limit of 33°, evaporation of water increases drastically, until at 43° the bird is losing ten times as much water as it did at 18°C (Fig. 7–10).

When birds are under the stress of high air temperatures, evaporative cooling can be greatly increased by more rapid breathing, by panting, and by gular flutter. Gular flutter involves the rapid and sometimes resonant vibration of the upper throat passages, driven by the hyoid apparatus. When heat-stressed, the domestic Chicken, *Gallus gallus*, may pant 300 times per minute, and some caprimulgids, 550 to 690 times per minute (Dawson and Bartholomew, 1968). The effectiveness of such panting is demonstrated by the Ostrich, *Struthio camelus*, which through water evaporation from its respiratory tract can maintain a body temperature of 39.3°C at an air temperature of 51°C (124°F) (Crawford and Schmidt-Nielsen, 1967).

Sandgrouse, *Pterocles senegallus*, while incubating, can withstand air temperatures of 48°C, and soil surface temperatures of 73°C (163.5°F) (Maclean, 1976). Neither gular flutter nor gaping has been observed in incubating sandgrouse. Cooling is achieved by evaporation through the skin, i.e., by cutaneous water-loss (Marder *et al.*, 1986). Cutaneous water loss is an important means of thermoregulation in a number of small birds (Bernstein, 1971; Lasiewski, 1972; Bartholomew, 1972).

Evaporative cooling entails the loss of water. This will vary with the amplitude and frequency of panting (or gular flutter), rate of cutaneous water loss, and the temperature and humidity of the air. Some of this water loss can be regained by metabolic water, and some of it, especially at lower air temperatures, by "countercurrent exchange," whereby moisture-laden exhaled air gives up water by condensation as it passes over the progressively cooler respiratory surfaces from the lungs to the external nostrils (Schmidt-Nielsen *et al.*, 1970). Not only water but also heat can be retrieved in cold climates by this countercurrent mechanism. In antarctic Adelie Penguins, *Pygoscelis adeliae*, at air temperatures of 5°C, about 82 per cent of the water and 83 per cent of the heat added to inhaled air is recovered, largely in the nasal passages, on exhalation (Murrish, 1973).

Evolutionary selection has taken a hand in heat conservation by generally insuring that birds living in colder climates have larger bodies than their relatives in warmer climates. Thus, the larger cold-climate species have proportionately smaller surface areas and consequently lose less heat per unit of mass than their smaller tropical relatives. This principle, known as Bergmann's Rule, has wide ecological and geographical implications (James, 1970).

The other side of temperature control—the production of heat—involves the chemical conversion of foods into heat energy through muscular contractions, especially in the form of shivering. While physical control of body temperature operates at air temperatures above the lower critical temperature, chemical control takes over at temperatures below this point. Experiments on four passerine species by West (1965) disclosed a direct linear correlation between shivering and oxygen consumption. Obviously, the more heat a bird manufactures, the more food it must eat as an energy source. The fat reserves in the bodies of many temperate zone birds correlate roughly with the latitude of each species' winter range. Other species like grouse and arctic finches store large quantities of food in their crops or throat pouches late in the day to carry them through cold winter nights.

A very few species of birds, usually under the combined strains of hunger and cold, are able to effect extraordinary metabolic economies by becoming torpid or dormant. In effect they change temporarily from endotherms to partial exotherms. Perhaps the most famous case of avian dormancy was that of the Poor-will, *Phalaenoptilus nuttallii*, discovered in a state of profound lethargy in a rock niche (Jaeger, 1949). Its eyes were closed, its body temperature varied between 18° and 20°C, and no heart beat or respiration was detectable. Taken into a warm room, the bird opened its eyes and recovered its normal state. The bird, banded and released, was recovered three successive winters in a torpid state in the same spot. Each winter the bird apparently survived three months without eating or drinking. Subsequent experiments with Poor-wills and other caprimulgids have shown that starvation and cold can induce reversible dormancy, or hypothermism, in which the birds' body temperatures may drop as low as 12.8°C (Austin, 1970; Ligon, 1970).

Short-term torpor has been discovered in various other kinds of birds. Torpidity was studied in 18 species of hummingbirds ranging in size from the 2.7-g Racquet-tailed Hummingbird, *Ochreatus underwoodii*, to the 17.5-g Giant Hummingbird, *Patagonas gigas* (Kruger *et al.*, 1982; Schuchman *et al.*, 1983). Torpidity was observed in all species independent of ambient temperature (5° to 25°C), and with a constant food supply. Body temperature at torpidity approached that of ambient temperature but never went below 16°C to 20°C. Metabolism at torpor may be lowered by 60 to 95 per cent of resting metabolic rate. It was concluded that torpor is used not only in a state of emergency, but is absolutely necessary to achieve the energy balance vital to the survival of these diminutive creatures. In contrast to hummingbirds, the 55- to 70-g Red-backed Mousebirds, *Colius castanotus*, are torpid only under conditions of reduced food supply (Prinzinger *et al.*, 1981). The heart rate of the Blue-throated Hummingbird, *Lampornis clemenciae*, at an ambient air temperature of 35° C varied from 480 to 1260 beats per minute, depending on activity; while torpid at an air temperature of 15°C, its heart beat only 36 times per minute, or about 7 per cent of the minimum endothermal rate (Lasiewski and Lasiewski, 1967).

During cool, rainy weather when flying insects are unavailable for food, fasting adult Swifts, *Apus apus*, are able to survive some four days, and nestlings, ten days, by becoming torpid (Koskimies, 1950). Nectar-eating, montane sunbirds (*Nectarinia* spp.) of eastern Africa live at altitudes as high as 4500 meters where hot days may be followed by freezing nights. Body temperatures of individual birds may drop (hypothermia) as much as 17°C at night, but the birds do not become completely torpid (Cheke, 1971). Nocturnal hypothermia has also been documented for several Manakins (Pipridae) (Bartholomew *et al.*, 1983).

A final question remains as to the neural mechanism that directs thermoregulation in birds. While much remains to be discovered, it is known that selective cooling of certain thermosensitive structures in a pigeon's spinal cord will cause constriction of arteries in the feet, erection of plumage, and shivering, whereas heating of such centers will cause dilation of foot arteries, increased carotid and sciatic artery blood flow, thermal panting, and cessation of shivering (Rautenberg, 1972; Bech *et al.*, 1982; Barnas and Rautenberg, 1985). Changes in a bird's skin temperature seem to modify the thermostatic "set point" against which the spinal cord thermoregulators operate. Intracranial local heating of the anterior hypothalamus-preoptic region of the brain of a House Sparrow, *Passer domesticus*, caused the bird to lower its metabolism and decrease its body temperature, whereas cooling the same region caused an increase in metabolism and in body temperature (Mills and Heath, 1970). Current research suggests that nerve centers in the anterior hypothalamus control heat loss by regulating panting and vasodilation, while neurons in the posterior hypothalamus control heat maintenance by regulating shivering and perhaps other forms of heat production (Calder and King, 1974).

As in man, the metabolic rate in birds rises early in life and slows down with age. This general trend follows two different patterns. In birds whose young hatch out naked, blind, and exothermic (altricial species), the nestlings have low metabolic rates compared with adults, but their metabolism (and capacity for thermoregulation) increases during development and usually exceeds that of adults by the time of leaving the nest. The American Robin, *Turdus migratorius*, is an example. Among birds whose young hatch out downy, eyes open, and alert (precocial species), temperature regulation begins developing in the egg during incubation. Such young achieve thermoregulation and reach peak metabolism shortly after hatching and then gradually decline to the adult level as they mature. There are, of course, many variations on this theme. Young of megapodes, or "incubator birds," hatch out completely endothermic. Young of the European Oystercatcher, *Haematopus ostralegus*, while exposed to air temperatures of about 14°C, will maintain body temperatures of 31° at 3 hours of age, 37° at 18 hours of age, and 41° at 30 hours of age (Barth, 1949). The single downy young of Wilson's Petrel, *Oceanites ocean-*

icus, is brooded in an antarctic burrow by its parents only the first day or two after hatching. From then on it remains unattended all day long in an air temperature of 5°C. Its temperature control is established at the age of two days. This seems to be an evolutionary adaptation to the peculiar habits of the adults in an inhospitable environment. The young of the closely brooded Adelie Penguin, *Pygoscelis adeliae*, in the same chilly habitat, do not establish endothermy until about the fifteenth day, at which time chicks leave their parents and gather in nurseries or creches.

American Goldfinches, *Carduelis tristis*, captured in summer, are barely able to tolerate subfreezing temperatures. However, winter-captured birds maintain their body temperatures for up to eight hours when ambient temperatures fall below −60°C. Enzyme activity contributing to catabolism of triglycerides is increased by 50 per cent in winter, thus enabling birds to emphasize fatty acids as an energy source (Marsh and Dawson, 1982). A summer-acclimated House Sparrow, *Passer domesticus*, cannot tolerate ambient temperature much below 0°C, as compared with a winter-acclimated bird that withstands −35°C (Kendeigh, 1969). Fat storage in winter in part enables the bird to sustain the needed metabolism for the longer period, but acclimation to cold involves more than the physical bolstering of fat and feathers.

Because of a variety of physiological limitations, birds are forced to seek out habitats suitable in climate, food availability, water, and other necessities. This may mean different habitats at different seasons of the year—in short, migration. The Yellowhammer, *Emberiza citrinella*, in north and central Europe can withstand a minimum winter temperature of −36°C. It has thick plumage and is sedentary. Its close relative, the Ortolan Bunting, *Emberiza hortulana*, has sparser plumage and can withstand a minimum of only −16°C. Although it breeds in the same regions as the Yellowhammer, it is compelled by its physiological limitations to migrate to Africa for the winter (Wallgren, 1954). Experiments by Kendeigh (1945) have shown that a House Sparrow, *Passer domesticus*, deprived of food will, in an air temperature of −22°C, survive only 4 hours; at −18°, less than 14.5 hours—the length of a temperate zone winter night; and at −14°, 19 hours. Hungry birds survived longest (67.5 hours) at 29°C. With a rise in air temperature from 29° to 46°C, the survival time decreased at the rate of 4 hours per degree. Such considerations alone clearly prescribe the northern geographical limits of a bird's winter residence. Without much doubt, the metabolic idiosyncracies of a species have a great deal to do with its geographical distribution and its migratory or sedentary tendencies.

■ SUGGESTED READINGS

Farner and King's *Avian Biology* contains excellent treatments of the circulatory system by Jones and Johansen, the respiratory system and thermoregulation by Lasiewski, thermal relations by Calder and King, and nutrition and metabolism by Fisher. Much technical information on bird physiology is available in Sturkie's *Avian Physiology*. Marshall's *Biology and Comparative Physiology of Birds* contains much useful material on the respiratory and circulatory systems and on metabolism. The gross and microscopic anatomy of the respiratory system of various species of birds are meticulously described and illustrated in Duncker's *The Lung Air Sac System of Birds*. A series of excellent essays covering various aspects of maintenance and productive energy appear in *Avian Energetics* (R. A. Paynter ed), Publications of the Nuttall Ornithological Club, Vol. 15.

Excretion, Reproduction, and Photoperiodism
8

. . . a cock called Chanticleer.
In all the world for crowing he'd no peer . . . .
Far, far more regular than any clock
Or abbey bell, the crowing of this cock.
The equinoctial wheel and its position
At each ascent he knew by intuition . . . .
—Geoffrey Chaucer (1340?–1400),
    The Nun's Priest's Tale

A seemingly illogical association occurs in vertebrates between the reproductive and excretory systems. In function they have nothing to do with each other, yet in origin and structure they are intimately related. Both systems arise from adjacent ridges of mesodermal cells in the roof of the embryonic body cavity. As an example of their close relationship, the primitive drainage tube of an embryo kidney is taken over by the reproductive system and becomes a sperm duct in the adult bird.

EXCRETION

As in other vertebrates, the kidneys of a bird maintain a homeostatic balance in the concentration of various salt ions in the blood. However, birds' kidneys are generally much less effective than those of mammals in the excretion of electrolytes or salt ions. In compensation, many birds have highly effective salt glands that take over this function (see Chapter 6). Kidneys also remove various waste products from the blood, particularly those that result from nitrogen metabolism. The kidneys are located behind the lungs in a depression against the sacral vertebrae and the pelvis. Because of their rapid metabolism, birds' kidneys are roughly twice as large as those of comparable mammals. Except in rheas and ostriches, there is no urinary bladder, and the urine drains directly by way of the ureters to the cloaca. This arrangement provides one of the weight-saving adaptations that aid flight.

In structure a bird's kidney is much like that of a reptile. It has the two layers—medulla and cortex—of the mammalian kidney, and externally it is usually divided into three lobes, which are again divided into many small lobules. Water-conserving birds that are adapted to living in dry habitats commonly have two to three times as many lobules per unit of kidney cortex as birds of humid habitats (Johnson and Mugaas, 1970). In each lobule is a central vein around which radiate the functional units of the kidney, the nephrons or renal bodies (Fig. 8–1). The nephrons, each composed of a glomerulus and tubule, are of two types: a reptilian type that has no tubule loops (of Henle) and is located in the cortex, and a mammalian type with tubule loops and located in the medulla. From the glomerulus, wastes, salts, glucose, and water filter out into its tubule. In the tubules the water, glucose, and salts are selectively absorbed back into the blood via a network of tubule-clinging capillaries; the wastes, in a concentrated form, pass on down the tubules to the ureter. The renal corpuscles (glomeruli plus their enveloping membranes) of a bird are smaller and more numerous than those of a mammal. In 1 cu mm of a bird's kidney tissue there are between 90 and 500 renal corpuscles, as against only 4 to 15 in a mammal (Benoit, 1950). The total number of renal corpuscles in both kidneys of a bird ranges from 30,000 in small passerines to about two million in ducks and geese. As described earlier, the kidneys of a bird receive both arteries and veins, the latter breaking up into a renal portal system. In this the bird kidneys resemble those of more primitive ancestors rather than those of mammals.

The urine in the renal tubules becomes progressively more concentrated through the absorption of water by the walls of the tubules, the walls of the cloaca, and possibly the walls of the rectum. As the chief concentrators of urine, the tubules absorb between 66 and 99 per cent of the water released by the renal corpuscle. Galliforms can move urine from the cloaca into the lower intestine where water and salts are absorbed. In the desert-living Gambel Quail, *Lophortyx gambelii*, approximately 47 per cent of the water, 62 per cent of the sodium, and 49 per cent of the potassium are absorbed. Additionally, about 68 per cent of the total urate in the ureteral urine is broken down in the gut, allowing further reabsorption of bound ions (Anderson and Braun, 1985).

By the time urine is discharged from the body it has combined with fecal material to form a white or cream-colored paste. The urine of birds differs from that of mammals in its high concentration of uric acid rather than urea, and in the preponderance of creatine over creatinine. The relative amount of nitrogenous wastes in the urine of a domestic chicken are: uric acid, 75 to 80 per cent; ammonia, 10 to 15 per cent; urea, 2 to 10 per cent; creatine, 1 to 5 per cent; and amino acids, 2 per cent (Shoemaker, 1972). Not only does each molecule of uric acid eliminate twice as much nitrogen as a molecule of urea, but the relative insolubility of uric acid makes possible a striking conservation of water. Whereas mammals require about 60 ml of water to excrete 1 g of urea, birds can excrete the equivalent amount of uric acid using only 1.5 to 3 ml of water. The desert-dwelling Kangaroo Rat, *Dipodomys merriami*, has a highly efficient kidney and can excrete urea of a concentration about 20 to 30 times greater than that in its blood, but a bird can excrete uric acid 3000 times more concentrated than that in its blood (Bartholomew and Cade, 1963). It is this water-saving type of excretion of nitrogenous wastes that makes it possible for desert birds, and also reptiles, to exist solely on the water they obtain from the seeds and insects they eat.

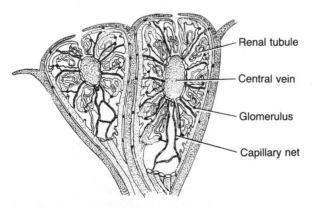

Renal tubule

Central vein

Glomerulus

Capillary net

FIGURE 8–1 Two lobules of a bird's kidney, in section. After Spanner, from Benoit, 1950.

One might assume that the excretion of uric acid instead of watery urine was an invention of birds to reduce weight, but for the fact that reptiles show exactly the same trait. The evolutionary significance of uric acid excretion is made clear by Benoit, who points out that this complex type of excretion is restricted to those two classes of oviparous vertebrates that have adopted an exclusively terrestrial type of reproduction. Reptiles and birds lay eggs that develop in a gaseous rather than an aqueous medium. Only gaseous exchanges can occur between the developing, metabolizing embryo in the egg and the surrounding air. Nitrogenous metabolic wastes cannot be eliminated in a gaseous form; consequently they must accumulate inside the egg shell. In the soluble form of urea such wastes would quickly poison and kill the growing embryo, but in the nearly insoluble form of uric acid they can be stored harmlessly in a large transitory "bladder," the allantois, until the bird or reptile hatches. The allantois of the newly hatched bird contains a heavy load of uric acid crystals, which are then eliminated via the intestines. In the developing fish or amphibian egg this excretory problem does not exist, since the urea of nitrogen metabolism quickly diffuses outward into the surrounding water; in mammals the maternal placenta carries away the urea from the growing fetus. Natural selection, making an adult virtue of embryonic necessity, has preserved the uric-acid-excreting machinery in birds and reptiles to promote their conquest of air and habitation of dry land.

REPRODUCTION

The Origin of Reproductive Organs

The reproductive organs, or gonads, of birds produce the male and female reproductive cells that unite to produce the next generation. The male gonad is the testis, which produces the male sex cells or spermatozoa; the female gonad is the ovary, which produces the female sex cells or ova. The two gonads also secrete hormones that regulate the development and functioning of many body structures. The perennial sex characters that distinguish male from female are determined by heredity, while the seasonal changes in sex characters depend on hormones (Witschi, 1935).

Racial survival for any species depends heavily upon the efficiency of its reproductive system. Animals have evolved in two contrary directions to promote the survival of their kind. Most of the lower vertebrates and most invertebrates produce great numbers of eggs only to abandon them in water; they gamble, in a sense, on the probability that at least one or two eggs will survive the formidable hazards of an inhospitable environment. Reptiles, birds, and mammals, on the other hand, produce very few eggs, but either endow them with protective shells and a generous supply of food for an early and vigorous start in life, or carry and nourish them within the mother's body. Birds provide further care for their eggs by laying them in protective nests, incubating them, and caring for the young when they hatch. An oyster laying millions of eggs each year is probably no more successful in perpetuating its kind than a sea bird which lays only one egg a year.

Normally, the sex of a developing avian embryo is initially determined by the sex chromosomes it receives from its mother. Female birds produce two kinds of eggs that differ from each other in the kinds or numbers of sex chromosomes. In chickens, for example, half of the eggs contain a male-determining sex chromosome and half do not. All sperm, however, are alike, and each contains a male sex chromosome. If an egg containing a sex chromosome is fertilized by a sperm with its sex chromosome, the egg with two male chromosomes will become a rooster. If the egg lacking a sex chromosome is fertilized by a sperm, the result is a hen, because the one male-determining sex chromosome from the sperm is overbalanced by the influence of female-determining genes that reside on other than the sex chromosomes. Occasionally, in certain finches and galliform birds, chromosome aberrations may occur, resulting in sex mosaics or gynanders. Such a bird may have an ovary on its left side and female plumage externally; on its right side a testis and male plumage (Murton and Westwood, 1977).

Regardless of this genetic determination, all embryos start out in life with the essential primordia for forming all the organs of either sex. Which sex a given individual turns into depends, apparently, on enzymes produced by the sex chromosome genes (heredity determiners). These enzymes presumably control the growth and differentiation of cells whose hormones, in turn, control the differentiation of primordial sex cells into either male or female organs. Later on, the sex organs themselves secrete hormones that influence behavior and affect the development of secondary sex characters such as feather shape and color, combs, and spurs. The sex of a bird is normally, but by no means irrevocably, determined at fertilization. Spontaneous and experimental reversals of sex are well-established facts.

In the dorsal wall of the visceral cavity, there occur in the early embryo, according to Domm (1955), sex cords that will develop into the testes if the bird becomes a male, or into the ovary if it becomes a female. Likewise, each embryo starts out with two sets of tubes, the Wolffian and Müllerian ducts. If the animal becomes a male, each of the Wolffian ducts develops into a sperm duct, or vas deferens, and the Müllerian ducts atrophy. The reverse occurs in the developing female, except that as a rule only the left Müllerian duct becomes the oviduct, and only the left ovary develops. Both structures normally atrophy on the right side of the body in females. The origin of sexual differentiation in birds seems to depend on which of two early derivatives of the sex cords attains ascendancy. If the central medullary tissues predominate, they give off substances that suppress the development of the exterior cortex and Müllerian ducts, and the bird becomes a male. If the peripheral cortex predominates, it and the Müllerian ducts enlarge and the bird becomes a female. Which of these two tissues develops first seems to depend on the number of sex chromosomes in the fertilized egg. In the embryos of Chickens and passerines, sexual differentiation begins during the fifth day of development.

The Male Reproductive System

The primary sex organs in the male bird are the paired testes located just ventral to the anterior end of the kidneys. The testes are generally bean-shaped, and their size varies according to season: at their maximum during the breeding season they may be 200 or 300 times larger than at other times of the year. During this seasonal maximum the two testes of a duck may equal one-tenth of its body weight (Benoit, 1950). The left testis is commonly larger than the right. Each testis is composed of numerous seminiferous tubules that give rise to enormous numbers of spermatozoa. As in other vertebrates, the sperm that arise from the tubule cells pass through three stages: (1) Small cells, called spermatogonia, which line the periphery of each tubule, multiply by mitotic cell division until millions are formed. (2) Some of the older spermatogonia move inward toward the central cavity of the tubule and begin a period of growth, increasing about twice in diameter. They are now called spermatocytes. (3) The spermatocytes become spermatozoa through a process of maturation that involves halving the number of chromosomes in each cell (meiosis)—reducing them from two (diploid) of each

kind to one (haploid). The spermatozoa, with vibrating tails and a reduced load of heredity determiners, find their way out of the testis and pass into a wavy, ciliated tube, the vas deferens, where they accumulate, ready to fertilize the female. During the breeding season the vas deferens also increases greatly in size, especially at its caudal end next to the cloaca, and acts as a sperm storage sac. In the Robin, *Turdus migratorius*, this seasonal increase in capacity may be 120-fold. Sexually active male birds may drain sperm into the cloaca. Quay (1985) has found cloacal sperm in some 20 species of migrant birds. Some of these were near their breeding grounds, but others were still some 30 to 900 km away, suggesting that sperm emission may start during spring migration.

Fixed to the ventral wall of the cloaca in primitive birds such as the ratites, tinamous, Galliformes, curassows, storks, flamingos, ducks, and geese, is an erectile, grooved penis that guides the sperm from the male into the cloaca of the female during copulation or "treading." The penis of the Ostrich may reach 20 cm in length. In more advanced species, copulation simply involves the approximation of male and female cloacas, usually while the male stands on the female's back, although some swifts are reported to copulate in

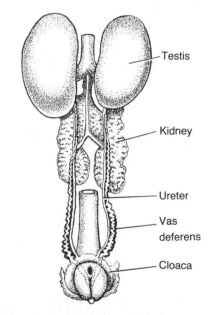

FIGURE 8–2 The reproductive and excretory organs of the male bird. After Jull; and Sturkie, 1965.

flight. A prodigious number of sperm are passed from the male to the female at each coition: 200 million in the pigeon, 8 thousand million in the domestic chicken.

A physiological peculiarity of sperm cells is their inability to develop at high temperatures. Warm-blooded mammals have solved this problem by hanging the testes outside the body in a scrotal sac. Such an arrangement would be obviously unsuitable for birds, both because of heat loss in cold weather and because it would disrupt the aerodynamic efficiency of the bird's smooth streamlined shape. So birds keep their testes inside the body, but solve the problem of the heat vulnerability of sperm in two possible ways. In many birds whose body temperatures are too high for the production of spermatozoa by day, the nocturnal drop in body temperature makes sperm production possible at night (Riley, 1937). Secondly, the caudal storage region of the vas deferens in some species swells into a cloacal protuberance whose temperature, for example, in several finches, may be 4°C cooler than the internal body temperature (Wolfson, 1954).

After insemination of the female takes place, the spermatozoa are intially stored for variable periods in a pouch in the lower oviduct; later they swim up her oviduct and at its upper end fertilize the egg. In the domestic Hen they may travel the length of the oviduct in as little as 26 minutes.

Sperm is stored in the glandular junction of the uterus and vagina for as long as 60 days in the Grey-faced Petrel, *Pterodroma macroptera*, permitting the pelagic male and female to be separated for weeks before laying. A glandular sperm-storage area is also known in other procellariforms and in the Horned Puffin, *Fratercula corniculata* (Hatch, 1983). The time interval between the beginning of copulation and the laying of the first egg varies among species. It may be as short as 20 hours (average 72 hours) in the domestic chicken, about 1 week in some boobies, about 12 days in the Canary, *Serinus canarius*, or as long as several months as in the Common Murre, *Uria aalge*, and the Little Owl, *Athene noctua*. The highest percentage of fertile eggs is found in chickens two or three days after copulation, and good fertility occurs up to five or six days, but then drops rapidly. However, an occasional fertile egg may be laid even 35 days after mating (Sturkie, 1976). Experiments on the hen of a Mallard, *Anas platyrhynchos*, by Elder and Weller (1954) showed a maximum duration of fertility of about two weeks after a separation from the male. There seems little if any relation between frequency of

copulation and the number of eggs laid. In the Griffon Vulture, *Gyps fulvus*, pairs copulate at frequent intervals for an entire month and the female lays only one egg; whereas the hen Turkey, *Meleagris gallopavo*, will lay 12 to 15 eggs after a single copulation. Even when hen Turkeys are inseminated only once every 30 days, they still produce eggs that are 83 per cent fertile (Burrows and Marsden, 1938). Some Turkeys have laid fertile eggs 70 days after copulation.

In numerous species when the proper time comes the drive to mate is urgent. Various species will copulate with stuffed conspecific dummies or even with the human hand. A White-tailed Tropic Bird, *Phaëthon lepturus*, even attempted to copulate with miniature gliders flown by hobbyists along the California coast (Hetrick and McCaskie, 1965). It is evidently the influence of the male hormone on the hypothalamic region of the brain that stimulates copulatory behavior (Barfield, 1967). Wood-Gush (1960) selected domestic Roosters for high and low sexual precocity and was able to establish widely divergent strains in three generations.

Occasionally eggs may develop and hatch without benefit of fertilization. In Turkeys such fatherless chicks are more likely to appear after the hens have been vaccinated for fowl-pox. Parthenogenesis, or the development of embryos in unfertilized eggs, is especially common in the Beltsville Small White strain of Turkeys (Olson, 1960).

Not only do the testes produce sperm that transmit the hereditary traits of the male to the next generation, but they also act as endocrine glands that produce the male hormones or androgens. Between the tubules of each testis are interstitial cells, or cells of Leydig. These are considered the source of the male hormones that are so influential in determining the secondary sex characters. The Leydig cells are unusually rich in lipids and in mitochondria. With the seasonal enlargement of the testes comes a corresponding increase in the activity of interstitial cells and an increase in the secretion of androgens into the blood stream. It is this springtime increase in the concentration of male hormones in the body that excites courtship with its songs and bright colors in the male bird; as Tennyson puts it, "In the spring a livelier iris changes on the burnished dove. In the spring a young man's fancy lightly turns . . . . ."

The Female Reproductive System

Corresponding to the testis in the male is the ovary in the female, located high in the abdominal cavity. This organ, responsible for the manufacture of the eggs, or more properly the ova, is not paired; only the left ovary develops in most birds, and the right ovary and oviduct (Müllerian duct) dwindle to insignificance. If the functional left ovary of a bird is experimentally removed, the vestige of the right ovary will develop into a testis-like organ. The reduction in ovaries from two to one is in part an adaptation to reduce ballast in a flying machine, but also an arrangement that protects the developing egg. If birds had paired ovaries and oviducts, a sudden jolt of the body, as in alighting, might crack mature eggs located side by side in the parallel oviducts. Even so, in some birds of prey, especially in the genera *Accipiter*, *Circus*, and *Falco*, both left and right ovaries may persist and function. However, even in these cases it is usually only the left oviduct that develops and carries the eggs. Abnormal ducks with two functional ovaries have been known to lay two eggs in one day. According to Kinsky (1971), paired ovaries have been recorded in at least 86 species of 16 different orders of birds.

Like the testes, the ovary enlarges greatly during the breeding season, then shrinks almost to invisibility for the rest of the year. At its maximum, it resembles a bunch of various-sized grapes. Each "grape" will become the yolk of a complete egg. In reality the yolk is the ovum, a single giant cell (about 30 mm in diameter in the chicken) greatly enlarged with stored food, and surrounded with a nourishing sphere of connective, epithelial, and vascular cells, the follicle. All the ova develop from much smaller follicles in the cortex or outer layer of the ovary. The central part, or medulla, of the ovary is composed mainly of connective tissues and blood vessels. The follicles may number over 25,000 in the ovary of a Rook, *Corvus frugilegus*, and of these, only five or six follicles mature each year; the great majority never develop into mature ova. Those few destined to mature grow very slowly. In nine months a Rook follicle increases in size from about 0.05 mm to 3.5 mm in diameter. But then it grows very rapidly and reaches the mature size of 14.6 mm in only four days (Benoit, 1950). This rapid increase in size is accounted for largely by the laying down of concentric layers of fat and protein food probably synthesized in the liver. The egg yolk of the domestic hen is composed of 48.7 per cent water, 32.6 per cent fat, 16.6 per cent proteins, 1 per cent carbohydrates, and 1.1 per cent minerals (Romanoff and Romanoff, 1949).

The pattern of growth and maturation of the ovarian follicles is regulated by the follicle-stimulating hormone of the anterior pituitary gland or adenohypophysis. In males the pituitary provides a similar hormone that stimulates testis growth. A second pituitary secretion, the luteinizing hormone, stimulates the growth and activity of the interstitial cells in both the ovary and testes and controls the discharge of the ovum from its follicle. Later in the reproductive cycle, the pituitary gland liberates a third hormone, prolactin, which depresses the production of both the follicle-stimulating and luteinizing hormones and initiates broodiness (nesting and incubation) in a bird. Prolactin also causes milk secretion in mammals and crop secretion ("pigeon's milk") in pigeons and doves. When a follicle reaches maturity, the outer envelope of cells and blood vessels ruptures; the ovum (yolk) breaks out and is swept into the open end of the oviduct. This release of the ovum from the follicle and ovary is called ovulation, and it normally takes place within 15 to 75 minutes after the laying of the preceding egg by the chicken, or four to five hours after laying by the pigeon (Sturkie, 1976). From their microscopic origin to the large mature ova, the female reproductive cells pass through essentially the same cell division stages as the maturing sperm cells. At maturity both types of cells have the reduced or haploid number of chromosomes (single chromosomes), so that fertilization restores the diploid number (paired chromosomes) characteristic of the body cells of the bird.

As in the testes, there are certain cells in the ovary that secrete sex hormones. It is thought that scattered interstitial cells between the follicles, and probably the follicles themselves, are the sources of these hormones. The ovary secretes not only female hormones (estrogens) but also male hormones (androgens) and progesterone.

In the mature female the oviduct is a long, winding tube through which the ovum progresses and in which it acquires layers of albumen, shell membranes, shell, and pigment. The oviduct wall is built up of three layers: an outer connective tissue serosa, a thick layer of circular and longitudinal muscles to move the egg by peristalsis, and an inner glandular epithelial layer that secretes the various substances added to the ovum.

Longitudinally, the oviduct can be divided functionally into five regions (Fig. 8–3). The first is the funnel-shaped infundibulum, whose contractile folds envelop the ovum as it breaks out of the ovarian follicle. Ovulation stimulates the activity of the oviduct. At the time the follicle ruptures and the ovum is released, the infundibulum will actively engulf any suitable

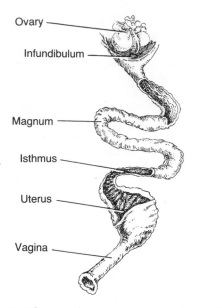

Ovary

Infundibulum

Magnum

Isthmus

Uterus

Vagina

FIGURE 8–3 The reproductive organs of the female bird. After Sturkie, 1965.

object. Cork balls have been experimentally substituted for ova, with the result that the bird laid eggs with centers of cork instead of yolk. If two follicles should release ova stimultaneously, a double-yolked egg will result. Between periods of ovulation the infundibulum is not receptive. The ovum remains in it only a short time—about 18 minutes in the chicken—and then is moved on into the second and largest section of the oviduct, the magnum. This portion, lined with glandular and ciliated cells, secretes layers of albumen (egg-white) around the ovum. The albumen is synthesized by these cells from amino acids removed from the blood. The albumen-secreting glands of the magnum are activated by ovarian hormones, and cease their secretion once the ovary becomes inactive. The developing egg remains in the magnum about four hours in the domestic Hen, and then passes by peristalsis to the third region, the narrow isthmus. Here the egg remains about an hour and a quarter while it receives the keratin shell membranes.

The egg spends most of its time—18 to 20 hours in the hen—in the fourth section of the oviduct, the large muscular uterus, or shell gland, where it acquires some watery albumen and its external limy shell. During the last few hours in the uterus the egg

becomes pigmented. The uterine glands that color the egg secrete pigments that are related to blood and bile pigments. Perhaps this is an adaptive method of getting rid of wastes, for the colors of the eggs of some open-nesting birds are highly concealing. Whether an egg has diffuse coloring, speckles, streaks, or blotches seems to depend on its relative motion while the tiny pigment glands are applying colors (Thomson, 1923). The terminal section of the oviduct is the vagina, whose mucous glands and muscular walls aid in laying the egg. The total time the egg takes to pass through the oviduct varies according to species: it is about 24 hours in the domestic hen, and about 41 hours in the pigeon.

During the breeding season a bird's oviduct weighs from 10 to 50 times as much as during periods of sexual inactivity. Both the size and activity of the oviduct are closely dependent on ovarian hormones. If a laying hen has its ovary removed, the oviduct ceases its secretion and regresses to its inactive size. The injection of various sex hormones, male and female, will powerfully stimulate the growth of the oviduct in immature females of various species (Fig. 8–4). Ovarian hormones injected into a domestic chick will cause its oviduct to increase in size 40-fold. Male hormones, likewise, will vigorously stimulate oviduct growth, and will also stimulate hens to assume male copulatory behavior.

The laying of an egg seems to depend partly on unidentified hormones secreted by the ruptured follicles, and partly on autonomic innervation of the powerful muscles of the uterus and vagina. Hormones and drugs that cause muscle contraction induce premature egg-laying. The intravenous injection of 0.1 to 0.2 cc of obstetrical (posterior) pituitrin will stimulate a hen to lay an egg in 3 to 4 minutes (Sturkie, 1976). Some species of birds, such as warblers, tanagers, and finches, lay their eggs early in the morning, shortly after sunrise. Other species, such as manakins and anis, lay their eggs around noon. That some species possess a psychological control over egg-laying is demonstrated by brood parasites such as cowbirds and cuckoos. These birds can control to within a few seconds the time when they surreptitiously lay their eggs in the nests of their hosts. Most birds seem to be *determinate* layers, in that they lay a certain number of eggs and then stop, regardless of the fate of the eggs. Other birds, *indeterminate* layers, continue to lay eggs until a definite number has accumulated in the nest, replacing any that might have been removed. In some way,

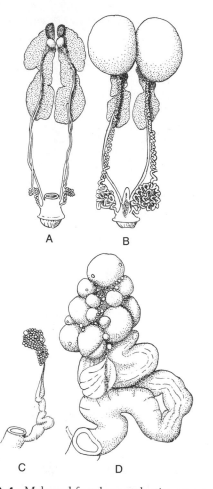

A   B

C   D

FIGURE 8–4 Male and female reproductive organs of the House Sparrow, before and after treatment with hormones. A, Male organs, quiescent phase. Weight of left testis, 0.5 mg. B, Male organs after 17 daily injections of 0.1 mg (20 DRU) of pregnant mare serum. Weight of left testis, 345 mg. C, Female's ovary and oviduct in quiescent phase. Weight of ovary, 10 mg. D, Female organs after 16 daily injections of 2 rat units (2 DRU) of pituitary extract. Weight of ovary, about 500 mg. After Witschi. A and B, $\times 1^{1}/_{2}$; C and D, $\times 2$.

at present unknown, the clutch of eggs then stimulates the anterior pituitary to secrete a prolactin-type hormone that induces regression of the ovary and initiates incubation behavior.

Considerable metabolic expenditure is involved in the laying of a clutch of eggs. The 12 eggs laid by the

Spotted Crake, *Porzana porzana*, weigh one-fourth more than the bird itself. About 35 per cent of the egg mass represents yolk synthesized by the ovary, and 65 per cent represents albumen, membranes, and shell synthesized by the oviduct. Since the eggs of a clutch are commonly laid at intervals of one or two days, rarely more, this means that these materials or their precursors must be made available quickly and copiously to the glands that produce the eggs.

The character of a bird's blood during the egg-laying period clearly reflects this seasonal devotion to egg manufacture. In a laying bird, fatty substances in the blood (phospholipids, fatty acids, and neutral fats, but not cholesterol) increase from 3- to 18-fold over those in a nonlaying or male bird. This great increase in blood lipids, which occurs shortly before and during laying, is stimulated by ovarian hormones. Estrogen injected into immature Hens (and also Roosters) for 10 days will increase the concentration of lipids in the blood sevenfold or more (Sturkie, 1976). Estrogens also stimulate the deposit of fat in body tissues of birds of either sex (Fig. 8–5).

Blood-sugar concentration in a laying Hen is about twice that in a nonlayer or a male. Although experimental studies of blood-sugar mechanisms have given conflicting results, it seems probable that hormones from the pancreas, the thyroid gland, and the adrenal cortex are of chief importance in regulating carbohydrates in the blood.

Blood calcium, important for the formation of the eggshell, is essentially doubled in laying birds. A domestic Chicken that lays 250 eggs a year secretes, as eggshells, about 20 times the calcium content of her entire body (Sturkie, 1976). This calcium comes either directly from the food in the intestine or from the bird's bones. If calcium levels in the food are 3.56 per cent or more, most of the calcium in the eggshell will come from it; if the food contains 1.95 per cent calcium, about 60 to 70 per cent of the shell will come from it and the remainder from the bones. On a calcium-free diet, all of the shell comes from the bones (Hurwitz and Bar, 1969). The large bones of the body act as storage chambers for the extra calcium needed for the shell. As the ovarian follicles grow, they apparently secrete hormones that stimulate a heightened absorption by the intestine of calcium from the food eaten, and also promote a rapid and massive storage of calcium in the hollow bones of the body. Mature female pigeons, for example, have relatively solid bones, while males have hollow bones.

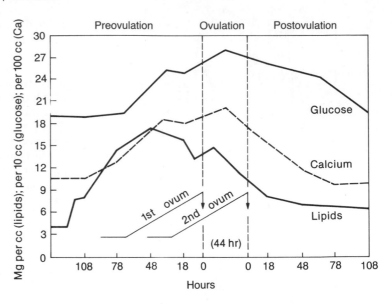

FIGURE 8–5 Changes in the chemical constituents of a pigeon's blood, associated with the ovulation of its two eggs. After Riddle, 1927; and Sturkie, 1965.

Injection of estrogen in various species has increased bone-calcium deposition. Evidence indicates that hormones of both the ovary and parathyroid glands cooperate to regulate the level of blood calcium. Neither one alone is effective, but together they regulate the calcium resources of the body for bone construction or destruction, and for eggshell secretion.

A domestic Hen fed a low-calcium diet will metabolize 8 to 10 per cent of the total calcium in her bones in about 16 hours. The bone absorbed is a special, spongy, spicular kind called medullary bone and is deposited in the bone marrow cavities of laying females but not in those of nonlaying females or males. For calcium storage, specialized cells of the bone marrow called osteoblasts multiply and become active as supervisors of calcium deposition in the bone cavities. Contrariwise, when calcium is needed for making the eggshell, cells known as osteoclasts are mobilized in marrow to extract calcium from the medullary bone and pass it into the blood for transmission to the shell-secreting uterus (Taylor, 1970). The habit that many female birds have of eating the eggshells after their young have hatched may be an instinctive method of restoring depleted calcium reserves. In physiological anticipation of egg laying, a female pheasant will abruptly change her diet in spring and begin eating about 10 times as much calcium as a male.

Not all birds attain sexual maturity with equal speed. As a rule, small species of birds become sexu-

ally mature sooner than large species. Some species lay eggs when very young: some tropical Ploceidae when only eight weeks old; the Budgerigar, *Melopsittacus undulatus*, at three months; button quail, Turnicidae, before they are four months old (Steinbacher, 1936). Several species of Australian grass finches (Estrildidae) begin laying eggs when only 11 weeks old, before the end of their juvenile molt (Immelmann, 1962). Among larger birds that reach sexual maturity after some years, records of birds in the wild are understandably difficult to obtain. White Storks, *Ciconia ciconia*, are reported by Schüz (1936) to reach full sexual maturity ordinarily when four years old, but occasionally not until they are five. Studies of the Satin Bower-bird, *Ptilonorhynchus violaceus*, by Marshall (1954), suggest that the males become sexually mature when they are four or five years old—a remarkable age for a passerine bird. Three individual Mutton Birds or Short-tailed Shearwaters, *Puffinus tenuirostris*, banded as nestlings, were observed by Serventy (1957) to nest for the first time when they were five, six, and seven years old. The Royal Albatross, *Diomedea epomophora*, first breeds when about eight years old. Delayed breeding in such birds may be more than a matter of ripening of reproductive organs. Ainley and Schlatter (1972) point out that Adelie Penguins, *Pygoscelis adeliae*, first breed at the age of four years, and even then only between 3 and 12 per cent of the birds breed.

An interesting characteristic of many species is that all the structures and functions of their reproductive apparatus do not come to full bloom at the same time. Herring gulls, *Larus argentatus*, in immature plumage may pair, build nests, and copulate, but lay no eggs. Their reproductive instincts are apparently out of step with their gonads. Year-old White Storks, *Ciconia ciconia*, do not return to their European birth-nests but remain for the summer in their African winter quarters. Two-year-olds may return and claim nests, but do not breed. Three-year-olds often breed, but either raise fewer young than older birds or none at all (Schüz, 1936). Year-old female Starlings, *Sturnus vulgaris*, Eastern Bluebirds, *Sialia sialis*, and other passerines commonly begin laying later in the season and lay fewer eggs than older birds. In his study of gonad development in 485 specimens of the California Gull, *L. californicus*, Johnston (1956) discovered that the testes of first-year birds showed no significant enlargement; those of second-year birds were about one-fourth adult size; and those of third-year birds about one-half adult size. First- and second-year birds did not breed, but among the three-year-olds a few females and about half of the males were thought to breed.

In a 17-year study of the Yellow-eyed Penguin, *Megadyptes antipodes*, in New Zealand, Richdale (1954) discovered that fertility in females 2 years old was 18 per cent; 3 years old, 82 per cent; 4 years old, 95 per cent; 5 to 15 years old, 93.7 per cent; over 15 years, 91.7 per cent. Sexual maturity also seems to vary with sex in certain species. In Yellow-eyed Penguins, 3-year-old males seemed to be less fertile than 3-year-old females. Only 7 per cent of the 2-year-old males nested, whereas 48 per cent of the 2-year-old females nested. Among Starlings studied in Holland by Kluijver (1935), the majority of the females bred when a year old, but most males not until they were two years old.

Sexual maturity is also a matter of geographical distribution. The Black-headed Gull, *Larus ridibundus*, rarely breeds in the Baltic when a year old, but it commonly breeds in Holland, Switzerland, and Hungary at that age. Starlings seldom breed in Latvia at one year of age, but regularly do in Hungary (Steinbacher, 1936).

Secondary Sex Characters

One of the more exciting discoveries in ornithological history was made by Crew (1923), who found an Orpington Chicken that changed its sex. This bird, a laying hen for three and one-half years, changed its external appearance, became a rooster, and sired two chicks. An autopsy of the bird revealed two functioning testes and a shriveled left ovary, probably destroyed by a tumor. Here was a natural, spontaneous reversal of sex that strikingly revealed the bisexual potentiality of birds. Since Crew's observations, much experimental work has been done on sex and sexual characteristics in birds. The majority of these experiments involve the manipulation of sex hormones through castration, grafting of sex organs, or the injection of hormones.

Birds, more than any other vertebrates, commonly show pronounced somatic differences between the male and the female. Although at hatching both sexes usually look the same, in many species they soon begin to differ in plumage, color, combs and spurs, size, song, and reproductive behavior. These differences are known as secondary sex characters (the primary characters being the gonads themselves), and they are largely due to the actions of sex hormones. The extravagant development of these "badges of sex" in many birds is very likely related to their heightened metabolism, the great velocity of their behavior reactions, and their fugitive mode of living. A male bird, showing off his fine feathers and splendid voice before his intended mate, has to make a persuasive impression in a hurry, before either one of them is frightened away by some predator. And competition with other males of the same species has undoubtedly played a role in intensifying these secondary sex characters. Darwin developed this theme into his principle of sexual selection.

If the two testes are removed from a young male chicken, it will grow into a capon instead of a cock or rooster (Fig. 8–6). The bird will fail to develop a large comb and wattles, but it will retain its spurs. Instead of long glossy tail feathers it will develop dull, hen-like feathers; it will lose its proud posture and pugnacity and will no longer crow, and its metabolic rate will drop about 15 per cent. In short, it loses most of the secondary characters that distinguish a cock from a hen. A castrated hen, or poulard, will develop a smaller comb; it will grow spurs where none existed before; at its next molt it will develop capon-like plumage. If now the capon has an ovary grafted into its belly, it will develop externally into a very close replica of a hen except that it will retain its spurs. If a testis is grafted into a poulard, the bird will take on the external appearance of a cock, spurs and all. Similar results are obtained when appropriate sex hor-

Rooster   Capon   Capon + ovary

Hen   Poulard   Poulard + testes

FIGURE 8–6 Experimental manipulation of secondary sex characters in the domestic chicken. Top row, left to right: normal male; castrated male; castrated male with grafted ovary. Bottom row: normal female; castrated female; castrated female with grafted testes. After Zawadowsky, from Benoit, 1950.

mone injections are substituted for the grafts (Benoit, 1950). If androgens (testosterone and androsterone) are injected into one-day-old chicks and the injections continued from five to ten consecutive days, the chicks will develop large combs and wattles, and some of them will begin crowing when only three days old (Kosin, 1942)! Experiments by Pincus and Hopkins (1958) have shown that merely dipping the pointed end of a fertile hen's egg into a solution of diethylstilbestrol in ethanol (10 mg/100 ml) for 10 seconds caused a feminization of male chicks, as revealed by their external genitalia. Sixteen weeks later, however, these "reversed" chicks developed normal testes. Masculinization of female chicks was not as easily accomplished by this method.

Such experiments as these show that there are two classes of secondary sex characters. In one group are those stimulated by sex hormones: combs, wattles, plumage, song, and sex instincts. The other group includes those characters that are inhibited by sex hormones, such as the spur in chickens, or the penis and male-type syrinx in ducks, whose appearance is prevented by estrogens.

Other endocrine glands than the gonads also play important roles in the development and function of body structures; and some of them, such as the ante-

FIGURE 8–7 Eighteen-day-old domestic chicks. The one on the left was injected with the male hormone testosterone, which caused precocious development of male characteristics. After Selye, from Zuckerman, 1957.

rior pituitary or the thyroid, interact intimately with the gonads. Throughout the life of the bird there continues a complex interplay of endocrine secretions, many details of which are only now being discovered. In broad outline, the endocrine activities in a bird are similar to those in man and other mammals.

Different parts of the body have different thresholds to hormone modification. The feathers on a bird's back, for example, have a lower threshold (i.e., they are more sensitive) to hormone modification than those of the throat. Work by Juhn *et al.* (1931) showed that there was a direct relationship between the speed of feather growth and the height of hormone threshold.

It should be remembered that some somatic characters are irrevocably determined by heredity and cannot be modified by hormones. A crow will remain black no matter what hormones course through its blood. Albino crows are the result of genetic and not endocrine modification. Similarly, the black bib of the male House Sparrow, *Passer domesticus*, is genetically determined. A convincing experiment that illustrates the relative roles of genes and hormones in determining somatic characters was performed by Willier and Rawles (1940). They grafted a piece of head-skin ectoderm from the 72-hour embryo of a Barred Plymouth Rock Chicken to the wing-bud base of a male White Leghorn embryo of the same age. The chicken that resulted had white feathers all over the body except for typical Barred Rock feathers on its wing (see Fig. 3–22). The genes in the graft determined the pigment pattern of the feathers, but the wing-location and hormones of the host (White Leghorn) determined their size, shape, and male character. Female Wilson's Phalaropes, *Phalaropus tricolor*, are more colorful than males. Plumage color is determined by testosterone, and, curiously, their ovaries produce more testosterone than do the testes of males (Höhn and Cheng, 1967).

Secondary sex characters are of course not limited to the outside of the body. Male and female birds show internal sex dimorphism in both structure and function. Some of these differences are unquestionably due to hormones; others are due to heredity. In chickens and pigeons, for example, the blood is richer in red corpuscles in males than in females, and this difference disappears in castrated males. The heart of a capon is only one-half as large as that of a cock. Castration of a cock will lower its respiration rate 30 per cent (Benoit, 1950).

Numerous instinctive behavior patterns are influenced through the effects of sex hormones on the nervous system. A poulard injected with female hormones (estrogens) will crouch and accept copulation from a cock; injected with androgens it will crow and show other forms of male behavior. Apparently some forms of behavior are common to both sexes and may be called forth in either by appropriate hormones (Davis and Domm, 1941).

Not only are psychic activities subject to endocrine control, but endocrine activity itself is subject to psychic control. Experiments by Matthews (1939) with a female pigeon demonstrated that it would lay eggs quite readily if it were placed in a cage with a male pigeon, somewhat less readily with a female pigeon, and not at all if isolated completely from other birds. However, when it was accompanied by a mirror in its cage it began to lay! Free-living female Song Sparrows were treated with estradiol in June and July. Males mated to these females maintained elevated levels of plasma testosterone and defended territories well into the autumn, whereas males mated to untreated (control) females ceased breeding activities and underwent molting. This suggests that the male "fine-tunes" the termination of his breeding behavior to the reproductive state of the female (Runfeldt and Wingfield, 1985).

Among birds that breed in large colonies, such as gulls and terns, communal breeding, in which wave-like spasms of copulation overrun the colony, often occurs. This promotes synchronized breeding. In his studies of gull colonies, Darling (1938) concluded that synchronous breeding was more likely to occur in large than in small colonies. His observations showed that nesting occurred earlier, more eggs hatched, and more young were fledged in large colonies than in small. This social facilitation of breeding success through numbers is known as the "Frazer Darling effect." Studies by Coulson (1968) of Kittiwakes, *Rissa tridactyla*, showed not only that adults in the center of a colony had larger clutches, higher hatching success, and fledged more young per pair than adults on the edge of the colony, but that males from the edge of the colony had a 60 per cent higher mortality rate.

Infectious breeding behavior is not necessarily dependent on sex-ripeness. Pellets of crystalline testosterone were implanted under the skin of wild male California Quail, *Lophortyx californicus*, during the nonbreeding season, by Emlen and Lorenz (1942). The quail were marked for field identification and released. Although the birds did not show complete breeding behavior, they crowed, became pugnacious toward other males, and paired with females. Untreated, wild control males showed contagious behavior and also paired with females, even though this was not during their normal breeding season.

Sex Ratios

Unequal sex ratios commonly occur among birds. In ducks, for example, studies of many species have shown a marked preponderance of adult males (sometimes as high as 3:1) over females. In the Australian Honey-eaters (Meliphagidae), there are genera in which males seem to be ten times as numerous as females (Mayr, 1939). Some examples of sex ratio determinations from relatively large samples are given in Table 8–1. In most instances these birds were adults that had been trapped for banding studies.

The reasons for these unequal ratios are various, and may be real or erroneous. Ornithologists distinguish between a "real" and an "operational" sex ratio. The latter is defined as "the average ratio of fertilizable females to sexually active males at any given time" (Emlen and Oring, 1977). Males of the promiscuous viduine finches (*Vidua* spp.) of Africa sing and display at call sites. An observer counting females vis-

iting males may conclude that females far outnumber males. If displaying males are trapped out, one finds that they are quickly replaced by floaters (Payne and Grotschupf, 1984; Shaw, 1984). The operational sex ratio is observed, but the real sex ratio is masked because males unable to obtain good call-sites may be in hiding.

Differential sex migration may also introduce large errors in determining adult sex ratios of a species. Among Dark-eyed Juncos, *Junco hyemalis*, the proportion of females in wintering populations varies clinally from about 20 per cent in Michigan (42 degrees N. latitude), through 28 per cent in Indiana (39 degrees), and 53 per cent in South Carolina (35 degrees), to 72 per cent in Alabama (33 degrees N. latitude) (Ketterson and Nolan, 1979). Differential sex migration is also known in mallards and diving ducks (Petrides, 1944; Tucker, 1943). A further bias in sex ratios may result from the habit several species have of segregating by sex in flocks. The Redwinged Blackbird, *Agelaius phoeniceus*, and other Icteridae show this trait in their winter quarters. In Nice's (1937) comprehensive study of the Song Sparrow, *Melospiza melodia*, she discovered that although the sex ratio was nearly balanced early in the spring, by the end of the nesting season there were between 8 and 10

TABLE 8–1.
SEX RATIOS IN ADULT WILD BIRDS

SPECIES	TOTAL NUMBER OF BIRDS	PER CENT MALES	PER CENT FEMALES	SOURCE
10 species of ducks	40,904	60	40	Lincoln, 1932
15 species of ducks	10,180	54	46	Beer, 1945
Mallard, *Anas platyrhynchos*	21,723	52	48	Homes, 1942
Pintail, *Anas acuta*	5,707	66	34	McIlhenny, 1937
Blue-winged Teal, *Anas discors*	5,090	59	41	Bennett, 1938
Bob-white, *Colinus virginianus*	45,452	53	47	Leopold, 1945
American Crow, *Corvus brachyrhynchos*	1,000	53	47	Imler and McMurry, 1939
Pied Flycatcher, *Ficedula hypoleuca*	—	57	43	Curio, 1959
House Sparrow, *Passer domesticus*	20,931	55	45	Piechocki, 1958
Red-winged Blackbird, *Agelaius phoeniceus*	6,480	84	16	McIlhenny, 1940
Boat-tailed Grackle, *Quiscalus major*	5,333	33	67	McIlhenny, 1940
Brown-headed Cowbird, *Molothrus ater*	4,281	74	26	McIlhenny, 1940
Giant Cowbird, *Scaphidura oryzivora*	—	80	20	Schafer, 1973
Purple Finch, *Carpodacus purpureus*	1,380	57	43	Magee, 1938

per cent fewer females than males. The incubating females were probably subject to heavier losses than the males.

Actual (rather than apparent) sex ratios for many adult birds may be largely determined by environmental forces that favor one sex above the other. Very probably the sexes are often differentially vulnerable to predation (including hunting by man), diseases, accidents, parasites, malnutrition, temperature extremes, and other hazards. Females are probably much more vulnerable to predators in the many species in which they alone incubate the eggs and care for the young. For unknown reasons the annual mortality rate of the Pied Flycatcher, *Ficedula hypoleuca*, is 65 per cent for females as against 50 per cent for males (Curio, 1959).

Sex ratios of nestlings may be determined by a technique called *laparotomy* (Fiala, 1979) in which a careful incision is made in the body wall, allowing the investigator to note whether testes or ovaries are present in a bird. Howe (1977) found a 1:1 sex ratio in embryos of common grackles, *Quiscalus quiscula*. Nestling mortality, however, reduced the number of males so that 83 females and 52 males fledged. There is a 1:1 sex ratio (with sex determined by laparotomy) in nestling montane White-crowned Sparrows (*Zonotrichia leucophrys oriantha*). Differential mortality after fledging reduced the number of females so that the actual sex ratio is biased towards males (Morton, personal communication). Sex ratios of nestling Redwinged Blackbirds, *Agelaius phoeniceus*, in nests of young and middle-aged females are 50:50, but are biased towards males in nests of old females. Young females fledge twice as many females as males, however (Blank and Nolan, 1983).

The Seasonal Reproductive Cycle

One striking characteristic of bird reproduction is its sharply periodic nature. When the proper season arrives, a bird seems to throw all its physiological resources into an intense and concentrated effort to produce the next generation. This effort usually comes at a time of year when environmental conditions are most benign for the raising of young. The survival value of telescoping the breeding season into a short period becomes apparent when one recalls the added burden a bird must carry when its gonads greatly enlarge to produce reproductive cells and hormones. One consequence of the soft life of domestication is the loss of much of this accentuation in breeding.

Rock Doves, *Columba livia*, ancesters of the domestic Pigeon, nest two or three times a year; domestic breeds nest seven or eight times. A single pair of Budgerigars, *Melopsittacus undulatus*, has raised 23 broods of young within 3.3 years under laboratory conditions (Serventy, 1971). Modifiability of the annual reproductive rhythm has undoubtedly been exploited by natural selection to produce the various breeding timetables exhibited by different birds throughout the world.

With the beginning of the breeding season come not only changes in the size of the gonads but also drastic changes in the bird's physiology and behavior. A typical sequence in the reproductive behavior of the male starts with selecting and defending the nest territory. This may be followed by courting and pairing with the female, and then, in many species, helping her to build the nest and incubate the eggs. The final stage may involve helping to care for the young. The automatic nature of these successive stages may be demonstrated experimentally. If, while a pair of Tricolored Blackbirds, *Agelaius tricolor*, are building their nest, some young from another pair are placed in the nest, the adults will ignore the food calls of the young and continue building, and may even incorporate the young into the structure of the nest (Emlen, 1941). Apparently the nest building and egg laying stage of the sex cycle inhibits the care-of-young stage. Later, when the nest is finished and the eggs laid, the adult blackbirds will accept and feed introduced nestlings. Similar chronological rigidity of instincts has been observed in many other species. After the young have become independent, the parent birds may attempt a second or third brood or may enter a period of sexual quiescence that may last until the beginning of the next breeding season.

In the Northern Hemisphere the breeding seasons of birds generally follow spring in its northward march across the land. The breeding cycle is often timed so that the young birds will hatch out when their standard food is abundantly available and the weather mild. This means that species with long incubation periods must begin their breeding cycle earlier than rapid breeders. The Horned Owl, *Bubo virginianus*, for example, begins breeding in Florida about December 1; in Pennsylvania, mid-February; and in Labrador, late March. Some birds breed normally earlier or later than other species. At the latitude of Washington, D.C., the Eastern Bluebird, *Sialia sialis*, lays its eggs around April 1; the American Goldfinch, *Carduelis tristis*, around July 1. As a general rule, the breeding season of any given temperate zone species begins two or three days later for each degree of latitude as one progresses northward, or 5 degrees of longitude eastward, or 100 m in altitude upward. This principle is known as Hopkin's Bioclimatic Law.

As one travels north from the equator, the summer season becomes progressively shorter. As a consequence, birds breeding in higher latitudes (and also at higher altitudes) may either shorten the time needed to raise a brood or reduce the number of broods per season. Even races within the same species show adaptations of this sort. In coastal California there are two races of the White-crowned Sparrow. One, *Zonotrichia leucophrys nuttalli*, is a year-around resident in the vicinity of San Francisco where it may raise three broods of young in 6.3 months. The other, *Zonotrichia leucophrys pugetensis*, winters in the same locality but migrates north to breed around Puget Sound, some 1200 kilometers away, where it may raise three broods in only four months. Although these two races are exposed to the same environmental influences in winter, their gonads develop at strikingly different times and rates (Blanchard, 1941). White-crowned Sparrows taken from 14 breeding localities from British Columbia to southern California were held in aviaries in San Jose, California. Although exposed to the same (local) photoperiod, each population exhibited its own timing in gonadal development, onset and extent of molt and migratory restlessness. Southern birds achieved maximum testis size in March, midlatitude birds in April, and northern birds in May. These data indicate that differences in reproductive rhythms are genetic (Mewaldt *et al.*, 1968).

Since seasonal fluctuations in temperature and day length dwindle to insignificance as one approaches the equator, it is understandable that wide-ranging species may lose their latitude-breeding correlation in the tropics (Fig. 8–8). The Noddy Tern, *Anoüs stolidus*, has a definite cyclic breeding season in the Dry Tortugas off Florida (24 degrees N. latitude), but breeds throughout the year on St. Paul Rocks (1 degree N. latitude) in the Atlantic (Murphy, 1936). Optimum conditions for breeding are to be found on a circumannual basis on Ascension Island in the South Atlantic (8 degrees S. latitude). Thus, Sooty Terns, *Sterna fuscata*, may be found breeding every month of the year (Fig. 8–9), i.e., there is no fixed breeding season.

More puzzling are breeding schedules like those of the Pied Cormorant, *Phalacrocorax varius*, of southwest Australia, which breeds in the spring (November) on Albrohos Island, but in the autumn on the

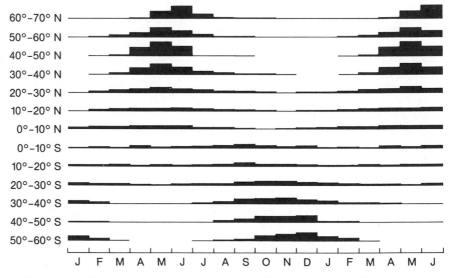

FIGURE 8–8 The relation of breeding seasons to latitude. The height of each bar is proportional to the number of species normally producing eggs during that month of the year at that latitude. After Baker, 1938.

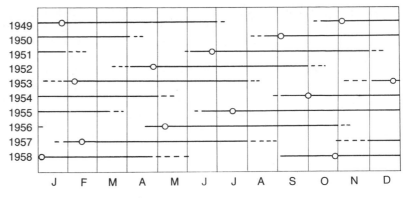

FIGURE 8–9 The nonannual breeding cycle of the Sooty Tern. This species, nesting near the equator, returns to breed on Ascension Island every 9.7 calendar months, or every tenth full moon. Horizontal lines show the approximate length of stay on the breeding grounds, and circles indicate the dates of the first reported eggs in each season. After Chapin and Wing, 1959.

mainland only some 50 kilometers away (Serventy, 1938). Equally enigmatic is the annual cycle of the erratically wandering Red Crossbill, *Loxia curvirostra*, which breeds any month of the year, often in midwinter. And why should large albatrosses breed only once in two years, or many small tropical birds throughout the year, or the White-bellied Swiftlet, *Collocalia esculenta*, three times each year (Immelmann, 1971)? In short, what is it that turns a bird's reproductive activities on and off? Is it an innate physiological rhythm, or some stimulus from the environment—or a combination of the two?

The overwhelming majority of wild birds show a sharply accented breeding season once a year, followed by a period of sexual quiescence. In most species the breeding season occurs during the same months each year. Such facts as these strongly suggest an environmental stimulus. However, when wild species are transplanted from the Southern Hemisphere to the Northern, with the subsequent seasonal displacement of six months, a few of them, such as the Emu, *Dromaius novaehollandiae*, the Australian Black Swan, *Cygnus atrata*, the Short-tailed Shearwater, *Puffinus tenuirostris*, and the Gouldian Finch, *Chloebia gouldiae*, tend to hold on to their Southern-Hemisphere breeding rhythm in spite of the changed seasons. European storks removed to Peru breed there at the same time of year that storks do in Europe. Facts such as these suggest an innate rhythm in breeding cycles. Most species, however, adapt quickly and change their sex rhythm to match the change in seasons.

Since spring and early summer are the typical breeding seasons for most birds, the environmental changes that occur as winter gives way to spring have been widely investigated as possible stimuli triggering reproduction: changes in temperature, day length, light intensity, rainfall, food, exercise. Some evidence has been secured that each of these seasonal changes may play a part in triggering the annual cycle in birds, but only one—the springtime increase in day length—shows a predominant influence, at least among birds of temperate and boreal zones. Warmer temperatures seem to play a secondary role, and in equatorial regions rainfall is of great significance.

It is now abundantly clear that the breeding seasons of birds are determined by two kinds of factors. *Proximate* factors are direct environmental sensory inputs that immediately precede the effect. At middle and high latitudes the most dependable environmental cue is the increasing length of days in spring that foretell the coming of the best weather for rearing young. Increasing day length triggers the endocrinological events that effect gonadal growth.

In tropical regions where seasonal changes in day length are small and not necessarily associated with level of food supply or other conditions of the environment, use of daylight information to trigger gonadal changes is probably minimal or nonexistent (Immelmann, 1971). Rain per se may operate in a proximate manner to initiate gonadal activity in several bird species inhabiting arid regions of Australia (Serventy, 1971; Sossinka, 1975).

Ultimate factors are *environmental factors* operating through natural selection that do not lead immediately to the effect. These establish in a species an inherited seasonal rhythm that roughly matches the environment's seasonal rhythm, ensuring that young are reared at an optimal time of year. These may include such factors as food supply, vegetative cover, breeding site, nesting material, temperature, competition, and predation; and of these, food is probably the most important single factor. There is a considerable interval between the beginning of gonadal recrudescence and hatching of the young; thus, the breeding cycle must be initiated (by a proximate factor) long before the time of peak food availability (an ultimate factor) when young are to be fed.

The physiological demands of reproduction are, for most species, the most rigorous and critical of any in the bird's annual calendar. It is therefore important that they be scheduled at a time of minimum stress on the adults and maximum probablility for the survival of both parents and young (Immelmann, 1971).

Most Northern Hemisphere birds breed in the spring or early summer when the following ultimate factors are at their most beneficent for ensuring reproductive success:

(1) the warming weather promotes the rapid development and survival of both the eggs and the young; it also reduces thermoregulatory energy losses in adults;

(2) growing vegetation provides food, cover, nesting sites, and for some species, nest-building material;

(3) the lengthening days enable parent birds to gather more food per day for the young, especially protein-rich food like insects and worms that become most abundant at this time of year.

The seasonal appearance of different kinds of foods commonly regulates the breeding cycle of a species. The European Robin, *Erithacus rubecula*, feeds its growing young on leaf-eating caterpillars several weeks before the Spotted Flycatcher, *Muscicapa striata*, feeds its young on the later-emerging flying insects. Still later, the Sparrow Hawk, *Accipiter nisus*, feeds its young on the growing crop of young songbirds. Latest of all, in Europe, is the breeding season of Eleonora's Falcon, *Falco eleonorae*, which feeds its nestlings on autumn-migrating small birds (Moreau, 1964; Walter, 1979). It should be noted that in all of these instances the parent birds begin their breeding seasons *before* the food on which they feed their young is in evidence. Therefore the food itself cannot be the proximate stimulus.

The European populations of the Red Crossbill, *Loxia curvirostra*, can breed in any month of the year, even in midwinter, depending on the presence and abundance of crops of conifer seeds on which it feeds its young. Since conifers do not form seeds every year, the Crossbills have become nomads, wandering widely in latitude and altitude in search of seed-bearing trees (Newton, 1972). In this instance, the appearance of food itself probably serves as a proximate factor stimulating the onset of breeding, and the seasonal unpredictability of food the ultimate factor responsible for the non-periodic breeding readiness of the birds.

In temperate zone springs, increasing day length and warming temperatures stimulate the growth of plants and hence of the food required by breeding birds. Warmth itself may play a proximate role in stimulating avian reproduction. Numerous observations show that a warm, early spring is accompanied by earlier nesting in many species. In southern Michigan in 1938 the last 20 days of April were unseasonably warm, and the Prothonotary Warblers, *Protonotaria citrea*, nested two weeks earlier than in cooler 1937 (Walkinshaw, 1939). In England in 1953 the last three weeks of November and all of December were unusually warm. Many resident Blackbirds, *Turdus merula*, Song Thrushes, *Turdus ericetorum*, European Robins, *Erithacus rubecula*, Starlings, *Sturnus vulgaris*, and House Sparrows, *Passer domesticus*, nested. Of 27 nests of the first three species, young were hatched in nine nests, and were fledged from four in December and January (Snow, 1955). Apparently, warm weather at either end of the breeding season may stimulate reproduction.

It seems probable that for some arctic-breeding species temperature may serve both as an ultimate and proximate factor in regulating breeding cycles. The Atlantic Brant, *Branta bernicla*, nests on Southampton Island, Northwest Territories, where the summer season is minimal. Over the years the species' ovaries have become adapted to the brevity of the summer thaw, for if the season is too late in starting, the bird does not attempt to nest, and its ovaries regress in size without ovulating (Barry, 1962). Arctic-nesting Snow Geese, *Chen caerulescens*, respond similarly to shortened summers. As a reproductive stimulus temperature seems to be quite irrelevant to the Band-rumped Storm Petrel, *Oceanodroma castro*, on the Galapagos Island of Plaza. Here, one population of the species breeds annually during the cold season; the other population, using the same burrows, breeds during the hot season (Harris, 1969).

Near the equator, where annual variations in day length are slight, other stimuli must serve as temporal controllers of the reproductive cycle. In regions of lowland tropical rainforests, some birds breed throughout the year, although many species tend to breed during the drier months. In tropical regions with pronounced wet and dry seasons, birds tend to tie their breeding cycles to the rainy seasons. In eastern equatorial Africa, for example, there are two breeding seasons whose peaks coincide with the two rainy seasons (Moreau, 1950). In a similar fashion, the Rufous-collared Sparrow, *Zonotrichia capensis*, of Colombia, breeds twice yearly near the close of the two highland wet seasons (Miller, 1961).

Although many breeding cycles are correlated with wet seasons, it does not follow that it is the rain itself or the food it brings that constitutes either the ultimate or the proximate stimuli that regulate the breeding seasons. In fact, it seems rarely to be true that a single environmental factor determines a bird's annual breeding schedule. The diverse ecological effects of rain may operate differently on different species. The peak breeding season at Cape Town, South Africa, comes at the end of the rains because the rains fall in the cold season and therefore vegetation and insects increase slowly. In warmer parts of Africa breeding seasons are more likely to begin with the rains. Raptors and scavengers breed early in the wet season before tall grass hides many of their prey, and aquatic species breed at its end (Moreau, 1950). Some species depend on the growth of new grass for nest cover or nest-building materials. The African Village Weaver, *Ploceus cucullatus*, instinctively selects flexible green grass-blades to weave its nest; dried grass from the previous season is ignored (Collias and Collias, 1964). Some African savanna species apparently schedule their nesting seasons either before or after the season of grass fires. Eastern Australian mud-nest builders, *Struthidea* and *Corcora*, are said to build their nests immediately after a passing thunder shower makes mud available (Serventy, 1971).

It has been shown that certain hormones found in plants (e.g., 6-MBOA) will stimulate breeding in voles, *Microtus montanus*, (Sanders *et al*, 1981). Field observations (Morton, 1967) revealed that White-crowned Sparrows, *Zonotrichia leucophrys*, switched from eating seeds to grazing prior to breeding. To test for the possible role of phytochemicals in stimulating breeding in White-crowns, Ettinger and King (1981) photostimulated two groups of female White-crowned Sparrows and fed one group green wheat leaves. After 20 days of photostimulation, ovaries were heavier in the green-fed group than in the controls.

Monthly samples of stomach contents indicated that the main diet of the Sharp-tailed Mannikin, *Lonchura striata*, of Malaysia is rice. Just prior to breeding, however, these birds consumed quantities of green filamentous algae, *Spirogyra*, as a source of protein, preparing them physiologically for reproduction (Avery, 1980). Eurasian Tree Sparrows, *Passer montanus*, normally feed on seeds and insects. At the University of Malaysia campus these sparrows also fed on poultry mash, and apparently as a result of this supplemental source of energy, bred for two months more each year than those in other localities (Wong, 1983). Contrariwise, the eating of stunted desert plants in a dry year may inhibit reproduction in California Quail, *Lophortyx californicus*, by means of the concentrated phytoestrogens contained in the leaves (Leopold *et al.*, 1976). The phytoestrogens biochanin A and ferulic acid were fed to captive Japanese Quail, *Coturnix coturnix*. The first inhibited egg-laying, and the second inhibited male copulatory behavior (DeMan and Peeke, 1982).

In desert regions particularly, birds have become physiologically adapted to drought to such an extent that their sexual organs are quickly responsive to rainfall or its effects. Should the rains fail, the birds will not breed and their gonads will remain quiescent, even for several seasons if the drought continues (Keast and Marshall, 1954). The reproductive cycle is therefore erratic and unpredictable. In such habitats it would be disastrous for a species to synchronize its breeding cycle with the astronomical year (e.g., day length or temperature). In the desert region of central Australia, Immelmann (1963) observed the reaction of birds to a heavy May downpour after several months of drought. Wood Swallows, *Artamus melanops*, for example, started courting several minutes after the rain began, copulated two hours later, and began nest-building the next day. Budgerigars, *Melopsittacus undulatus*, and Zebra Finches, *Poephila guttata*, reacted similarly. Changing photoperiods do not affect gonadal growth in Zebra Finches (Sossinka, 1975). To simulate drought conditions Zebra Finches were deprived of water. Their testes were much smaller after three weeks of dehydration, and grew when given unlimited access to water. Spermatogenic activity was high even in dehydrated birds indicating that gonadal tissue remains in a functional state unless birds are severely dehydrated (Vleck and Priedkalns, 1985).

Birds responding in this fashion to rain are called "opportunistic breeders." It appears obvious here that the *unpredictability* of rain is the ultimate factor causing the year-round readiness of birds to begin breeding, while the first rain after a drought is the proximate factor that actually triggers breeding. The Ostrich, *Struthio camelus*, is a facultative opportunistic breeder. In the Namib desert of southwest Africa, exceptional rains fell in March and April, 1969, and the Ostriches responded by breeding four months earlier than usual (Sauer, 1972). The striking synchrony in colonial breeding of the Red-billed Quelea, *Quelea quelea*, of Africa testifies to the sensitive response of this species to rain as a cue. This species lives in enormous colonies of millions of birds. It is "precipitated into mass nesting by prolonged rainfall," with the result that their "eggs all hatch at about the same time and the shells tumbling from the nests resemble falling snowflakes" (Gilliard, 1958).

Jones and Ward (1976) proposed that the immediate proximate control of breeding in Queleas is the body condition of the female, especially the protein reserves in her muscles. In like fashion, the female Canada Goose, *Branta canadensis*, on arrival at her northern nesting grounds, weighs about 46 per cent more than at the end of winter, largely because of fat and protein accumulations in viscera and muscles. The bird fasts during egg-laying and incubation, and by the time the eggs hatch, she has lost about 42 per cent of her body weight, the breast muscles alone atrophying more than 18 per cent (Raveling, 1979).

Rain also controls breeding cycles through changes in water level. The Wood Stork, *Mycteria americana*, of Florida and waterbirds of north-west Argentina breed during the dry season when fish and amphibians are concentrated in dwindling watercourses and food is thus abundantly available to their young (Kahl, 1964; Olrog, 1965). Rising water levels following dry periods provide conditions fostering the reproduction of detritus-feeding invertebrates, which in turn triggers breeding in Australian waterfowl (Crome, 1986). Some species of hole- and ground-nesting birds (e.g., swallows, kingfishers, bee-eaters, skimmers, and terns) depend on dry seasons to expose river banks and sandbars as nesting sites (Preston, 1962).

Even wind seems to control breeding cycles. Cormorants that nest on islands in equatorial African lakes and rivers breed throughout the year except for the months of January through April when windstorms may sweep waves across their flimsy nests (Marshall and Roberts, 1959).

Competition between species undeniably influences the timing of breeding seasons. In tropical Africa and South America some resident species avoid breeding during the period when northern migrants arrive to spend the winter, very probably because of competition for food resources (Immelmann, 1971). And of course the energy requirements of an individual bird at different seasons may be in competition. The European Turtle Dove, *Streptopelia turtur*, usually lays three clutches of eggs in Britain. The last one, in August, is characterized by frequent desertion or by the poor survival of the young, probably because the adult bird then needs to store body fat as a fuel for impending migration (Murton, 1968). Here one can see two ultimate factors competing within the same bird: one concerned with success in reproduction, the other with survival in migration. Even competition within a species affects breeding cycles. Among Sooty Terns, *Sterna fuscata*, aggressiveness of older, established terns in a colony causes younger terns, two to five years old, to nest later in the season (if at all) and in less desirable habitats (Harrington, 1974).

Predators often adjust their reproductive cycle to coincide with the availability of prey. Eagles, hawks, and vultures of southeast Africa commonly adjust their breeding cycles so that the period of greatest food stress (when the young are three to four weeks old) coincides roughly with the August droughts and grass fires when prey animals are more visible and when mortality among the larger ungulates is high because of lack of grazing and water (Benson, 1963). The Mediterranean Eleonora's Falcon, mentioned earlier, feeds during the summer chiefly on insects. In August it nests and raises its young, feeding them almost exclusively on the small passerine birds that are then in full migration to their African winter quarters. This astonishingly late breeding season is undoubtedly the consequence of the rich autumnal availability of food for the young falcons (Walter, 1968).

A contrary influence of predation is shown on the Clay-colored Robin, *Turdus grayi*, of Panama (Morton, 1971). About 70 per cent of the breeding attempts are in the dry season when food resources are low but when there is about a 42 per cent chance of successfully fledging young, even though some nestlings die of starvation. During the rainy season food is more abundant but so is the threat from predators, and only about 15 per cent of the young are successfully fledged. Predation pressure probably also causes many ground-nesting aquatic birds of arid Australia to breed only after sufficient rainfall has created lakes with islands furnishing sanctuary from mammalian predators (Immelmann, 1971).

Finally, the rate at which young birds develop has

an effect on the adults' breeding cycle. Large albatrosses and some vultures and penguins may breed every other year because it requires more than one year to raise their young to independence. The King Penguin, *Aptenodytes patagonica*, raises two young every three years, laying its first egg early in the first antarctic summer, late the second, and not at all the third. In all these ways as well as in others, ultimate and proximate factors mold the reproductive time schedules of birds. Sometimes the factors may operate singly, at other times in combinations; sometimes their effects are sudden and drastic, and at other times, subtle and imperceptible. But whatever their many and varied pressures on birds, the birds respond in ways that tend to promote the survival of their species.

PHOTOPERIODISM

The question now arises: just what is the nature of the physiological machinery that allows birds to respond in so many diverse but adaptive ways to environmental stimuli? In particular, how are they able to respond to ultimate factors (chiefly food supply) that operate toward the close of the breeding season at a time that is much too late to provide a timing clue for the onset of reproduction? And what is it that permits birds, more than any other terrestrial vertebrates, to accclerate, retard, halt, or even completely eliminate breeding in any given season, depending on the conditions in the environment?

In the north temperate zone where most of the experimental work on avian breeding cycles has been done, the annual changing length of daylight is the only absolutely regular, and the most dependable stimulus that fluctuates in synchrony with the breeding cycles of the vast majority of birds. As a consequence, investigators have extensively experimented with day length, or photoperiod, in their attempts to discover in what way physiological mechanisms of reproduction respond to environmental controls.

Man's interest in the relation between day length and breeding has a venerable history. For centuries the Japanese have practiced the art of *Yogai*, which consists of forcing caged birds to sing in midwinter by lengthening their days with candlelight for three or four hours each day after sunset in autumn. The Dutch similarly used to stimulate various finches to sing prematurely in October, and then employed them to entice fall-migrating relatives into traps. Poultry raisers for years have stimulated winter-laying of hens by illuminating their pens for added hours each night. These alterations in the reproductive cycle achieved by manipulating the length of daylight are all manifestations of photoperiodism—the innate response of many plants, animals, and even single-celled organisms to day length.

Pioneer experiments in photoperiodism by Rowan (1929, 1938) demonstrated that Dark-eyed Juncos, *Junco hyemalis*, and other species could be stimulated into sexual activity in subzero winter weather by subjecting them to artifically lengthened days. He believed that the birds were stimulated into out-of-season reproduction by increased wakefulness and exercise. However, later experiments by Riley (1940) and others, involving enforced exercise in darkness, failed to confirm Rowan's hypothesis. Since these early investigations, thousands of experiments involving photoperiodism have been conducted, and they have shown quite clearly that for most temperate zone birds, the late winter and early spring increase in day length stimulates the springtime growth and activity of gonads in birds, and also, that artificial increase in day length at other times of year may stimulate out-of-season sexual activity. In addition to reproductive readiness, photoperiodic stimulation commonly evokes such natural concomitants as molt into breeding plumage, courtship behavior, body fat accumulation, and migratory restlessness.

It was soon discovered that the temporal control of reproduction in birds was more than a simple matter of reaction to day length. At the end of the breeding season, many species enter a physiological phase known as the "refractory" or "regenerative" stage that may endure for days, weeks, or months (Farner *et al.*, 1983). During this stage the bird is totally unresponsive to light stimuli. Further complications soon came to light. For birds living at the equator, how could the insignificant annual variations in day length trigger annual reproductive cycles? And for northern temperate zone migrants spending the winter in the Southern Hemisphere, do shortening days stimulate the onset of reproduction and spring migration? And does the stimulus of aperiodic rain somehow override the effects of periodic changes in day length in desert species, to prevent the disaster of hatching out young in the midst of a prolonged drought?

In an attempt to answer such questions it might be well to examine first a few photoperiod experiments in order to grasp some of the basic mechanisms involved in the control of the reproductive cycle. A typical photoperiod experiment consists in subjecting a sexually quiescent but not photorefractory male bird to an artificially lengthened day. Usually in three or four weeks the bird produces sperm, a sign of mature sexual activity. A normal, quiescent duck kept on a photoperiod of 15 hours of light and 9 hours of darkness each day will develop mature testes in 12 to 15 days. Under optimum conditions the testes will increase in volume 10 times in 10 days, 80 times in 20 days. This maturation of testes includes complete maturity of sperm, interstitial cells, and sperm ducts. Female ducks seem somewhat less sensitive to artificial photoperiodic stimulation, and only rarely reach sufficient sexual maturity to lay eggs, although Bissonnette and Csech (1936) induced pheasants to lay fertile eggs in January in Connecticut. Not only have numerous species of birds been stimulated to lay eggs out of their normal season by "night lighting," but rats, mice, goats, and other mammals have been stimulated to breed precociously; and brook trout have been led to spawn in December instead of March. Photoperiodic control of reproduction is obviously a deep-seated mechanism in many vertebrates.

Continued photoperiod experiments revealed that young male ducks, illuminated by lights of equal intensity but of different wave lengths or colors, show a maximum response in testes development under red light, a very feeble response under blue, and none at all under infrared rays (Benoit, 1950). Further experimentation revealed the surprising fact that radiation of the retina of the eye is not essential to stimulation. If a duck with its eyes removed has a quartz rod inserted in one orbit, oriented so as to carry the light directly to the hypothalamus of the brain, the light will powerfully stimulate the pituitary, and through it, the gonads. In fact, Benoit (1978) found the hypothalamus 100 times more sensitive to photostimulation than the retina (Fig. 8–10). In this case, however, the gonads are not only stimulated by red rays, but even more so by blue. Since the longer red rays penetrate the tissues of the head more effectively than blue rays, this fact may account for the difference in the stimulating effect of blue light through the retina and optic nerve as compared with its more direct effect through the quartz rod.

Further evidence that eyes are not essential to photoperiodic responses was demonstrated by Menaker (1970), who plucked the head feathers from one group of House Sparrows, *Passer domesticus*, to increase the amount of light reaching their brains, and injected india ink under the skin on top of the heads of others for the opposite effect. He then exposed both groups to long days under relatively dim light. Only the plucked birds responded with testicular growth. Long days have stimulated the growth of testes in other species of birds in spite of the removal of their eyes. Even small plates of radioluminous material implanted in the brains of Japanese Quail, *Coturnix coturnix*, caused them to behave as though exposed to long photoperiods, regardless of external light conditions (Kato *et al.*, 1967). When photosensitive sites in a Quail's brain were implanted with bipolar electrodes and then stimulated with mild electric currents for two minutes once daily early in the dark period, testicular growth resulted (Ohta *et al.*, 1984).

While intensity of light has some stimulative effect on birds, it is no substitute for long days. Starlings, *Sturnus vulgaris*, subjected to a constant day length of 10.5 hours, produced no spermatozoa whether the light measured about 5 footcandles or about 190 footcandles (Burger, 1939). On the other hand, a weak red light of only 1.7-footcandle intensity was sufficient

FIGURE 8–10 A, A horizontal section of the head of a Mallard Duck with its right eye removed, the socket filled with cotton, and a quartz rod inserted for conducting light directly to the brain. B, Different responses in the growth of testes in normal Pekin Ducks under different wave lengths of light of equal intensity and equal photoperiod. Top row shows sizes of testes at the beginning, and bottom row, at the end of each experiment. After Benoit, 1950.

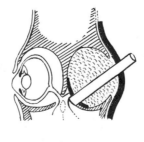

A

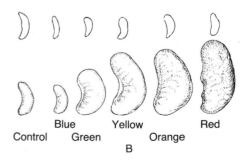

Control   Blue   Green   Yellow   Orange   Red

B

to induce complete spermatogenesis when applied for long days (Bissonnette, 1932). Experiments on House Sparrows by Bartholomew (1949) showed that in winter, 16-hour days of 10-footcandle intensity were as effective in stimulating spermatogenesis as 16-hour days with much higher intensities of illumination; but in the fall, 16-hour, 10-footcandle days were much less effective than 16-hour days with higher intensities. In this experiment, the minimum light intensity to cause full spermatogenic development was 0.7 footcandle. Slight testicular activity was obtained even with 0.04-footcandle illumination.

That an endogenous rhythm may control the breeding cycle was demonstrated by experiments of Marshall and Serventy (1959) on the Short-tailed Shearwater, *Puffinus tenuirostris*, a marine species that spends its winter in the far north Pacific but breeds, with astonishingly rigid regularity and synchrony, during the Southern Hemisphere summer on islands of southern Australia. Captive birds kept for over a year under a steady light-to-dark regime of 12:12 hours (12L:12D) exhibited the same gonad and molt cycles as wild birds that experienced widely different day lengths in their transequatorial migrations. The breeding cycle of these birds can only be logically explained on the basis of some sort of genetic internal clock or calendar. That is, such cycles are something more than passive mechanisms driven by environmental rhythms such as day length, temperature, or rain.

Twenty-nine European Garden Warblers, *Sylvia borin*, and 27 Blackcaps, *S. atricapilla*, all taken as nestlings, were kept under three different constant photoperiods (light:dark ratios of 10L:14D, 12L:12D, and 16L:8D hours) for three years (Berthold *et al.*, 1972). Throughout this time the Garden Warblers exhibited persistent, endogenous, near-annual rhythms in molt, in premigratory restlessness, and in body weight (reflecting fat deposits and gonad growth). The rhythms were somewhat less persistent in the Blackcaps. It is extremely unlikely that these rhythms could have been caused by uncontrolled environmental factors. Roughly 12-month endogenous rhythms in reproduction and other functions have been demonstrated in at least 11 species of birds and also in other animals, vertebrate and invertebrate. For such animals the annual environmental rhythm in day length acts as a cue to adjust the length of the animal's cycle and to synchronize or entrain it with the natural seasons.

These innate cycles are only approximately annual in dimension, varying in length, for example, among warblers of the genus *Sylvia* between 212 and 435 days (Berthold, 1974). They are therefore known as *circannual rhythms*. Not only are these rhythms endogenous, but some birds are able to choose daily light budgets to suit their innate cycles. Dol'nik (1974) trained Chaffinches, *Fringilla coelebs*, to turn lights on and off by hopping on electrical-contact perches. For several months the birds chose photoperiods of 8L:16D and then began to increase them to 14L:10D, and some months later to decrease them again. In parallel to these self-selected daily photoperiods were physiological changes in the birds such as molt, fat deposition, and sexual activity. However, when Berthold (1979) subjected warblers of the genus *Sylvia* to artificial circannual light cycles of 6 months instead of 12 (i.e., two cycles per year), the birds synchronized their molting and migratory restlessness with the artificial cycle, doubling their normal annual pattern, but they retained their once-annual pattern of fat deposition.

The crucial value of innate rhythms is made clear by Aschoff (1967), who pointed out that animals must adapt to niches in time as well as in space. Since different seasons of the year present birds with different opportunities and hazards, and since these periodic variations are usually predictable, it is to a bird's advantage to incorporate in its own body a self-sustained rhythm equivalent to the external rhythm of the seasons, and thus "anticipate in its own organization the respective stages it needs in order to react properly to the environmental conditions which will ensue—it is prepared in advance."

The outward manifestations of these endogenous reproductive cycles are divided by Marshall (1960–1961) into three phases. Immediately after the close of actual breeding, a bird usually enters the *refractory* or regenerative phase during which it is insensitive to photostimulation. In duration, this phase may last about five months as in the Rook, *Corvus frugilegus*, four months in the Starling, *Sturnus vulgaris*, three months in the Mallard, *Anas platyrhynchos*, or it may even be nonexistent as in many doves and pigeons and other birds that can breed continuously throughout the year. The timing mechanism that controls the refractory phase does not seem to be dependent on the annual cycle in the flow of sex hormones but seems to originate in the hypothalamus or perhaps higher brain centers (Immelmann, 1971). That photorefractoriness is a phenomenon of the central nervous system independent of feedback from hormones is borne out by experiments in which castrates or castrates implanted with testosterone were photostimulated, but still became refractory (Farner *et al.*, 1983). The adaptive function of the refractory period is twofold: it allows the bird to devote time and energy to molting and to preparation for fall migration; and it acts as a safety mechanism for many species by preventing breeding during hostile times of the year.

The second phase of the endogenous cycle is the *acceleration* or progressive stage during which the bird again becomes susceptible to environmental stimulation or inhibition. This is the time of territory selection, song, courtship, and the development of secondary sexual characters. The *culminating* or final phase is the period of nest-building, copulation, and ovulation. It too may be accelerated or retarded by various environmental factors, but it seems to be impervious to photoperiodic control. An example of the

meshing together of photoperiodism and the major activities in the life of one species is shown in Figure 8–11.

The actual physiological mechanisms by which reproductive cycles are generated and internally controlled are exceedingly complex, not at all clearly comprehended, and require much further investigation. As presently understood, the way in which photoperiodic control of the breeding cycle begins and runs its course can be described basically as follows (see Eisner, 1960; Farner, 1973; Marshall, 1960–1961; Murton and Westwood, 1977). Light falling either on the retina of the eyes or directly on the brain causes neurosecretory cells in the hypothalamus to secrete various kinds of neurohumors that are carried by portal blood vessels into the anterior pitu-

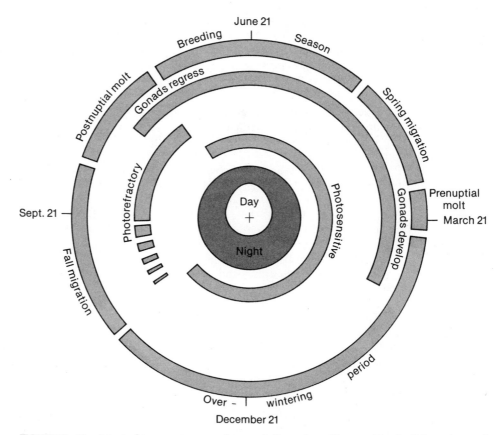

FIGURE 8–11 A typical temperate zone photoperiodic cycle as illustrated by the White-throated Sparrow. Note the avoidance of overlapping in the annual periods of heavy energy demand, shown in the outer circle. Redrawn after Meier, 1976.

itary gland. Here they cause different cells to secrete a variety of hormones or releasing factors that have different targets: two gonadotropins (glycoproteins), the follicle-stimulating, and the luteinizing hormones (which act on the gonads); prolactin (which influences the gonads, feather growth, and behavior); thyroid-stimulating hormone (acts on the thyroid gland); adrenocorticotropic hormone (acts on the adrenal glands); intermedin, or melanophore-stimulating hormone (controls distribution of pigments); and stomatotropic hormone (which controls growth) (Fig. 8–12). Still other hormones arise in this "master gland" of the body. Those mentioned above work together in a bewildering variety of ways to reinforce or inhibit each other. Other external stimuli such as rain, or the sights and sounds of courtship behavior, may activate

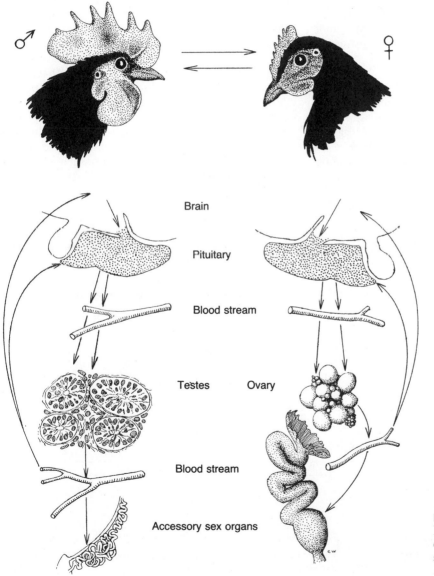

Brain

Pituitary

Blood stream

Testes   Ovary

Blood stream

Accessory sex organs

FIGURE 8–12   The chief glandular mechanisms in birds' reproductive cycles. For explanation see text. Adapted from Marshall, 1960–1961.

the neurosecretory cells of the hypothalamus much as does day length. Whatever the stimuli, they spur the pituitary into seasonal activity.

In the male, the stimulated pituitary releases into the blood stream a gonadotropic hormone that causes both the growth and ripening of testis tubules that produce the spermatozoa, and the development of testis interstitial (or Leydig) cells that, stimulated by the luteinizing hormone, produce the male hormones or androgens. These, in turn, when distributed by the blood, result in stimulation of the development of secondary sex organs such as sperm duct; a regulatory inhibiting feedback effect on the pituitary; and, after acting on the brain, stimulation of instinctive courtship and mating behavior (Fig. 8–13).

In the female, the anterior pituitary releases into the blood a gonadotropic or follicle-stimulating hormone that promotes the growth and maturation of the egg follicles, and also secretes a luteinizing hormone that periodically stimulates ovulation. As its egg follicles develop, the ovary secretes the female sex hormone, estrogen, that causes the growth and activation of accessory sex organs, stimulates courtship and mating behavior, and, as a feedback mechanism, causes the pituitary to reduce its secretion of gonadotropic and luteinizing hormones. The pituitary also secretes prolactin, which, assisted by other hormones, inaugurates broodiness (incubation and care-of-the-young behavior) in the female and promotes the development of the belly brood-patch. In both male and female birds prolactin controls the storage and consumption of premigratory fat deposits, and stimulates migratory restlessness. Whether the pituitary itself is the seat of a bird's innate reproductive cycle is as yet unknown.

Some of the elegantly precise complexities of photoperiodism are only now coming to light, and they remind one of the mechanical complexities of a Swiss calendar watch. A bird's sensitivity to photostimulation may vary greatly according to species, sex, time of year, time of day, and even conditions of freedom. Ovaries of female White-crowned Sparrows captured as adults will grow if photostimulated. Their ovaries grow even more rapidly if females are photostimulated and also exposed to tape recordings of male song (Fig. 8–14). However, ovaries of both groups collapse before ovulation occurs (Morton *et al.*, 1985). Captive female White-crowned Sparrows, *Zonotrichia leucophrys*, captured as adults, do not lay eggs and will develop only one-fourth the pituitary gonadotropin potency that wild females develop under the same stimulating lighting conditions, and females generally are less responsive to photostimulation than are males (Lofts and Murton, 1973). In contrast, a laboratory colony of 36 females captured as nestlings or fledglings laid 134 eggs in one season (Baptista and Petrinovich, 1986). It is suggested that wild females imprint on the habitat and require stimulation from some features of the habitat in order to ovulate. Hand-raised females are "imprinted" on the characteristics of their holding cages, and are thus freed from the constraints of adult wild females confined in cages.

It has long been known that birds, like most other

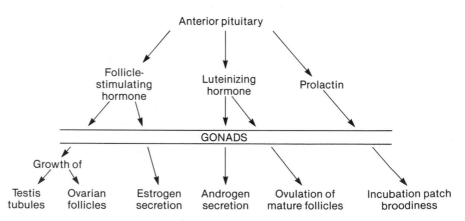

FIGURE 8–13 The role of the pituitary gland in controlling the reproductive cycle in birds. After Eisner, 1960.

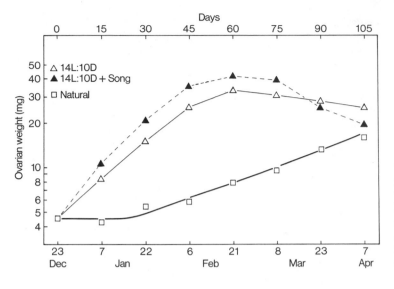

FIGURE 8–14 Mean ovarian growth rates of female White-crowned Sparrows exposed to natural (outdoor) photoperiod, 14 hours of light, and 14 hours of light plus taped male song. Note that females exposed to taped male song exhibited faster ovarian growth rates than females not so exposed. (Data from Morton *et al.*, 1985.)

organisms, possess daily or *circadian* ("circa diem") *rhythms* of approximately 24 hours' length. Even when kept under constant environmental conditions of light, temperature, and the like, birds commonly show a roughly 24-hour periodicity in locomotion, body temperature, metabolism, glucose concentration in the blood, hormone flow, and other physiological variables. The adaptive value of these "internal clocks" becomes apparent when one considers a migrating bird's use of the moving sun as a reference point for orientation during long homing flights (see the section on celestial navigation in Ch. 22).

Even more significant is the circadian cycle as a measuring device to determine the optimum time of year for breeding. Groups of Starlings, *Sturnus vulgaris*, were exposed in winter to photoperiodic cycles of 6 hours of light and between 6 and 66 hours of darkness, the amount varying from group to group. Only those birds experiencing light-dark cycles that occurred at multiples of 24 hours showed sexual development (Gwinner and Eriksson, 1977). These results indicate that an innate circadian rhythm is involved in controlling Starling reproductive cycles. The significance of the 24-hour requirement seems to be that the light must repeatedly coincide with a light-sensitive phase in the bird's circadian cycle to be effective in stimulating reproduction. Similar studies with female Budgerigars, *Melopsittacus undulatus*, indicated that the pituitary was light-sensitive on the twelfth hour of each day (Shellswell *et al.*, 1975). Two groups of

females were kept on long days (LD 14:10), and vocalizations were played to one group during the first seven hours and to the other group during the second seven hours of the light cycle. The group hearing vocalizations in the morning developed larger ovaries than those hearing vocalizations in the afternoon (Gosney and Hinde, 1976). These data suggest that budgerigars are song-sensitive in the morning corresponding to time of peak vocal production (Ferrell and Baptista, 1982), but that the song-sensitive rhythm is not in phase with the light-sensitive rhythm.

It is now becoming clear, at least in some strongly photoperiodic species, that the annual reproductive cycle is intimately linked with a circadian rhythm in hormone flow. In the White-throated Sparrow, *Zonotrichia albicollis*, the peak of the blood plasma concentration of corticosterone, from the adrenal glands, and that of prolactin, from the anterior pituitary, are separated by about 12 hours in the springtime, but by only about 6 hours during the bird's late summer refractory period. The chronological working together (temporal synergism) of these two hormones was experimentally investigated by Meier and his colleagues with fascinating results (Meier, 1973, 1976; Meier and Russo, 1985). When White-throats were kept under an artificial photoperiod of 16 light to 8 dark hours each day, their blood plasma showed a daily rise in concentrations of corticosterone. Different birds were then injected with prolactin at various set times. Injections of prolactin given 12 hours after the daily rise in corticosterone plasma concentration promoted fattening, gonadal growth, and migratory restlessness oriented northward—all springtime characteristics of birds in breeding condition. Injections of prolactin eight hours after the increase in plasma corticosterone induced a summer-like photorefractory response: no fattening, no gonad growth, and no migration restlessness. A four-hour injection pattern promoted fattening, migratory restlessness directed southward, but no gonadal stimulation. White-throats maintained under continuous dim light were treated with both corticosterone and prolactin injections separated by the same time intervals—4, 8, and 12 hours—with similar results. The same orientation results were obtained whether the birds were tested under natural spring skies or fall skies. Meier postulates that in nature the photoperiodic stimulation of the reproductive system involves the synchronization of light cycles with blood plasma corticosterone concentration, which in turn "entrains the photoinducible phases for luteinizing hormone and follicle-stimulating hormone." Similarly, the daily photoperiod entrains the flow of prolactin, responsible for fat deposition, under the influence of adrenal corticosteroids. These results, of course, were obtained for only one species; other species may employ quite different photoperiodic mechanisms.

Experiments by Binkley and her colleagues implicate the pineal gland in circadian rhythms of the House Sparrow, *Passer domesticus*. In this and many other species the circadian rhythm of locomotor activity, body temperature, and various metabolic functions are synchronized (or entrained) with the natural cycle of day and night. A sparrow kept in constant darkness will, for at least 40 days, persist in a steady circadian rhythm of locomotor activity and body temperature change of 3.0° to 4.5°C amplitude. The rhythm, now said to be "free running," may drift out of phase, either gradually lagging behind more and more, or anticipating more and more the natural cycle of day and night, depending on the individual bird (Fig. 8–15). When birds under constant conditions are returned to a natural photoperiod they require a few days to become locked in phase with the daily cycle of light and dark. For probably the same general reasons a human requires a few days to adjust to a new time schedule after flying across several geographical time zones.

If, now, the sparrow's pineal gland is removed (Fig. 8–16), the circadian rhythms are abolished; temper-

FIGURE 8–15 The body temperature of a House Sparrow is about 3.0° to 4.5°C higher by day than by night. When the bird is placed in constant darkness the circadian temperature rhythm persists, although it may drift out of phase with the time of natural daylight. Each horizontal line represents 24 hours of data. The thicker line indicates the higher temperature. *LD* = 12 hours of light:12 hours darkness; *DD* = constant darkness. A sham operation (*SH*) performed as a control against the pineal removal of Figure 8–16, did not disturb the daily rhythm. After Binkley, Kluth, and Menaker, 1971, *Science*, 174:311, Copyright, American Association for the Advancement of Science.

ature fluctuations, for example, become aperiodic (Binkley *et al.*, 1971). The pineal gland may well be essential to these rhythms. Subsequent research has shown daily rhythms in the pineal output of serotonin, melatonin, two enzymes, and norepinephrine. There is, for example, a 10-fold nocturnal rise of melatonin in the pineal glands and a 27-fold increase in the activity of the enzyme (N-acetyltransferase) that

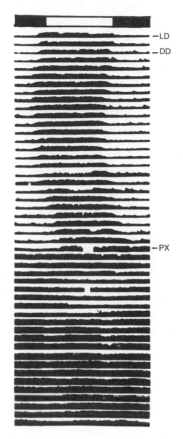

FIGURE 8–16 A House Sparrow kept in constant darkness maintains its circadian temperature rhythm until its pineal gland is removed (*PX*) after which the daily rhythm disappears. After Binkley, Kluth, and Menaker, 1971, *Science*, 174:311, Copyright, American Association for the Advancement of Science.

in vitro in constant darkness, or under continuous illumination, its circadian cycle of N-acetyltransferase production will continue for several days but with gradually diminishing amplitude (Binkley *et al.*, 1978; Kasal *et al.*, 1979). That the pineal gland is not the exclusive self-sustained oscillator responsible for circadian cycles in birds is shown by the fact that circadian rhythms persist in Starlings, *Sturnus vulgaris*, whose pineal glands have been removed (Rutledge and Angle, 1977). Further experiments will undoubtedly reveal additional details in the amazingly intricate biological clockworks of birds.

There still remain many questions, such as those posed by reproductive cycles that are responsive to food, rain, nesting materials, and courtship behavior. Food may act as an ultimate factor in timing breeding seasons, as in the case of Eleonora's Falcon, *Falco eleonorae*, or as a proximate factor, as with California Quail, *Lophortyx californicus*, that ingest chemical inhibitors in food-poor years. Supplemental feeding with protein-rich foods advanced the onset of nesting of Red-winged Blackbirds, *Agelaius phoeniceus* (Ewald and Rohwer, 1982). As a proximate factor, food may affect the breeding cycle through its nutritive effects on metabolism, or through psychoneural effects via the sense organs. Rain is a powerful regulatory factor, but the way it operates is unknown. Possibly for some species the dehydration of body tissues inhibits the pituitary initiation of the breeding cycle. Certainly in species that respond almost immediately to rainfall, the reacting mechanism must be initially psychological.

Low temperatures are inhibitory to testis growth. Black-billed Magpies, *Pica pica*, kept at constant photoperiod (LD 12:12) but separated into two temperature treatments (20°C and 2°C), exhibited significantly larger testes only in the high temperature group (Jones, 1986).

Psychoneural mechanisms must also be involved in species like the European Robin, *Erithacus rubecula*, that refuse to breed unless they own breeding territory (Lack, 1940), or any species that promptly lays a second clutch of eggs if the first one is destroyed. Many experiments have shown that the visual exposure of male dove to a female is sufficient to stimulate egg-laying by the latter (e.g., Barfield, 1966). Courtship songs of male Budgerigars, *Melopsittacus undulatus*, played on a tape recorder to isolated females greatly stimlulated their ovarian development and egg-laying (Brockway, 1965). Taped vocalizations stimulate a

is involved in converting serotonin into melatonin (Binkley *et al.*, 1973). It is known that melatonin injected into mice lowers their body temperature 2° to 3°C. Further significant observations reported by Wolstenholme and Knight (1970) show that the pineal gland, as a result of photoperiod stimulation, responds to neural impulses by synthesizing active compounds. These apparently act selectively either on the hypothalamus or the pituitary to regulate the synthesis and release of gonadotropins and adrenocorticotropins, the latter possibly accounting for the circadian flow of corticosterone. Even if the pineal gland of a young domestic Chicken, *Gallus gallus*, is removed and cultivated male to sing, and hearing his own voice stimulated testicular activity in a male (Brockway, 1967).

A fascinating example of psychological influence on sex rhythm is provided by the synchronous breeding of the White Ibis, *Threskiornis aethiopicus*. In West Java this colonial species builds nests close together on large platforms, each of which holds 20 to 25 nests. The nests on one platform will have eggs or young all of the same stage while those of another platform may be of a very different stage. For example, on August 22 one platform held only eggs, while a neighboring platform held only young that were six weeks old (Hoogerwerk, 1937). Psychologically synchronized breeding of this sort also occurs in gulls, terns, flamingos, weaverbirds, and many other colonial species. In many species of gulls, and no doubt in other birds, a certain threshold of numbers appears to be necessary before the birds will breed at all (Darling, 1938). It may well be that the Heath Hen, *Tympanuchus cupido cupido*, was hurried into extinction when the last remnant on Martha's Vineyard fell below the numbers necessary for successful courtship and breeding.

■ SUGGESTED READINGS

Farner and King's *Avian Biology* is a rich mine of information containing excellent chapters by Gwinner on circannual and circadian rhythms; Immelmann on ultimate and proximate factors controlling reproduction; Lofts and Murton on anatomy and endocrinology of reproduction; Menaker and Oksche on the pineal gland; Serventy on breeding adaptations of desert birds; and Shoemaker on excretion. Sturkie's *Avian Physiology* provides extensive material on reproduction in birds and many references to the journal literature. In *Grassé's Traité de Zoologie* there is good material by Benoit on the urogenital system, endocrine glands, and photoperiodism. Marshall's *Biology and Comparative Physiology of Birds* contains good chapters by Marshall on the anatomy and function of reproductive organs, and by Witschi on sex differentiation. Wolfson's *Recent Studies in Avian Biology* contains dated but still useful chapters by Davis on breeding biology, and by Domm on sex differentiation. Perrin's article in Campbell and Lack's *Dictionary of Birds*, gives a concise introduction to the role of ultimate and proximate factors that control breeding seasons. Murton and Westwood's *Avian Breeding Cycles* and Epple and Stetson's *Avian Endocrinology* are particularly valuable for their treatment of reproductive endocrinology.

Behavior
9

A well-laid scheme doth that small head contain,
At which thou work'st, brave bird, with might
   and main . . .
In truth, I rather take it thou hast got
By instinct wise much sense about thy lot . . .
—Jane Welsh Carlyle (1801–1866),
   To a Swallow Building Under Our Eaves

Birds are popularly considered to be not only intelligent animals, but animals possessing many of the more commendable human traits and emotions. A bird seems to be very clever to find its way in migration over great stretches of trackless forests and seas. It seems grateful when, in return for the privilege of nesting in man's back yard, it sings its cheerful songs. It appears foresighted when, to raise its young, it builds warm sheltering nests; and tenderly devoted when it incubates its eggs, broods, feeds, and defends its young, sometimes at the cost of its life. This pretty, anthropomorphic picture of birds has, through the centuries, become embedded in songs, proverbs, and folktales: Robin Redbreast covering the lost babes with leaves, the birds listening to St. Francis' sermon, Genghis Khan's pet falcon striking the poison cup from his hand, and Noah's dove returning with an olive branch.

Though this anthropomorphic appraisal is misleading, the fact remains that birds are remarkable for many achievements that man, with all his brain power, would find hard to duplicate. Ages of natural selection have provided birds, far more than man, with instincts that are highly adapted to solving routine problems of daily living. As an example, Clark's Nutcracker, *Nucifraga columbiana*, buries pine seeds in the fall to be retrieved as food in the coming winter. By using large objects as remembered visual cues, it is able later to relocate these caches (Vander Wall, 1982). Laboratory experiments have shown that Black-capped Chickadees, *Parus atricapillus*, can accurately relocate cache sites, recall those that have been emptied, and even recall the type of food stored in each cache (Sherry, 1984).

But when environmental changes occur, instinctive behavior cannot always adapt to the new circumstances. The automatic, spur-of-the-moment actions that usually characterize birds as a class then produce strange results. Several species of woodpeckers store acorns in small cavities, usually one acorn to a hole,

171

and feed on them later in the year when those on the ground have become unavailable. Their instinct to store food normally serves, like the Nutcracker's, the same purpose as human foresight. But one Acorn Woodpecker, *Melanerpes formicivorus*, found a small knothole in the wall of a closed cabin and spent the entire fall dropping hundreds of acorns through it to the floor below.

There are two sister disciplines that practice the study of behavior, namely, the fields of comparative psychology and that of ethology. Ethologists are zoologists conducting their studies both in the field and laboratory. As outlined by Tinbergen (1963) questions asked by ethologists tend to be of four different kinds: (1) evolution; (2) ontogeny; (3) causation; (4) function.

Evolution of behavior may be studied because dispositions to perform behavior patterns may be inherited in the same way as color of eyes, hair, and height. Closely related species tend to use similar behavior to achieve certain goals, so that evolutionary relationships may sometimes be elucidated using behavior (e.g., courtship displays) along with morphological characters. The dabbling ducks of the genus *Anas* (e.g., Mallard, Teal, Pintail, Shoveller) tend to feed in shallow water by dabbling motions of the bill; or by "tipping," with only head and foreparts of the body submerged and the hind quarters and tail in the air. The bay-ducks (called pochards in Europe) of the genus *Aythya* (including Scaups, Redheads, Canvas-backs) feed instead by diving. The two groups of ducks may also be distinguished by courtship patterns. Dabbling ducks tend to share elaborate but similar courtship movements (Kaltenhäuser, 1971). One of these movements is called the "grunt-whistle" in which the drake dips its bill in the water and then jumps up and sends a plume of water in the air. Another ritual shared by dabbling duck males is the "headup-tailup" (Fig. 9–1). These are "homologous" displays, i.e., they have evolved from a common "ancestral" display. The above displays are not found in bay-ducks, which share instead a courtship ritual known as a "head-throw" during which the head is thrown quickly backward until the top of the head touches the back. There is a pause, and the head is then thrown abruptly forward, the second movement accompanied by a call (Delacour and Mayr, 1945).

Ontogeny of behavior is the study of individual behavioral development. Some behaviors appear spontaneously, fully perfected. For example, some New World sparrows (*Zonotrichia, Melospiza,*

Passerella) forage with a special movement called a double-scratch in which a bird makes a little jump forward, and then a larger one backwards scratching with both feet at once (Greenslaw, 1977). This behavior first appears in hand-raised Song Sparrows, *Melospiza melodia*, and White-crowned Sparrows, *Zonotrichia leucophrys*, when they are between 16 and 36 days old (Nice, 1943). Some behavior patterns undergo maturational changes. For example, as young domestic ducks, pigeons, cranes, and rails mature, their vocalizations shift from a higher to a lower pitch, a process referred to as the "breaking of the voice" (Abs, 1980).

The study of behavioral ontogeny can be very useful in providing cues as to origins of behavior patterns. Most of us have seen male peacocks (*Pavo* spp.) performing their spectacular courtship displays with shimmering tail (coverts) feathers erected to form a huge fan. The origins of this display may be appreciated by studying the courtship feeding displays of related species of pheasants. Roosters (*Gallus*) and male Ring-necked Pheasants, *Phasianus colchius*, court by scratching the ground and then calling the female. This is the most primitive version of courtship feeding in pheasants. The male Gray Peacock Pheasant, *Polypectron bicalcaratum*, scratches on the ground, then bows and raises its wings and spreads its tail to form a huge fan. He now looks like a miniature peacock. He then moves his head back and forth in the direction of an approaching female. A young peacock raises, fans and shimmers its tail, then scratches on the ground in the manner of roosters or Ring-necked Pheasants. An adult peacock spreads and shimmers the spectacular tail but does not scratch the ground. "Ontogeny has recapitulated phylogeny"; i.e., by studying the developmental stages of courtship feeding in peacocks we have seen the early stages of the behavior probably found in the "ancestral" peacock and extant relatives, but lost in adult modern peacocks (Schenkel, 1958).

Behavioral ontogeny may also be instructive in that it may reveal the interplay between genetic (nature) and environmental influences (nurture) in shaping behavior. Some behaviors are learned (nurtured), but there are often genetic constraints as to what may be learned. For example, Greenfinches, *Carduelis chloris*, may learn syllables from Canary songs. However, the Greenfinch sings this acquired song in the rhythm of Greenfinch song (see Chapter 11). Instinctive behavior may also be improved through practice (see study by Hess, p. 92).

FIGURE 9–1 Courtship displays of Dabbling Ducks and Pochards: A., Grunt-whistle display of Mallard, B., Homologous grunt-whistle display of Pintail, C., Headup-tailup display of Mallard, D., Homologous headup-tailup display of Pintail, E., Head-throw display of European Pochard (*Aythya ferina*), F., Homologous headthrow display of Canvasback (*Aythya valisneria*). A to D redrawn after Lorenz, 1971; E after Blume, 1973; F after Johnsgard, 1965.

The ethologist also asks what proximate internal and external factors cause an animal to behave in a certain way at a certain time. Internal factors could include physiological, hormonal, and neurophysiological mechanisms underlying behavior. External factors could include the photoperiod or the behavior of another bird in the vicinity eliciting a certain behavior. For example, the longer spring day lengths are external factors stimulating the production of testosterone in male White-crowned Sparrows, *Zonotrichia leucophrys*, and estrogen in females. Males sing and females respond with trilling. The high titers of estradiol in females is the internal factor readying females to respond to male song with trilling displays.

The ethologist also asks how a behavior functions in the animal's survival, adaptation to its environment, and reproduction. The detailed studies on the life history of the Kittiwake Gull, *Rissa tridactyla*, by Cullen (1957) (see ch. 23) outlines many specialized behavior patterns adapting the species to living in the relatively predator-free environment of the cliff ledge.

Behavior in birds, as in all animals, is largely directed toward self-survival. It is, in effect, an internally directed system of activities that strives to maintain the physiological stability of the body in the face of many environmental hazards, such as heat and cold, sun and rain, food lack, competition, predators, and parasites.

The way any animal behaves depends in large part on its behavior equipment: the sense organs or receptors, the correlating nervous system, and the effectors or muscles and glands. Since all of these structures are inherited, there exists a solid hereditary base for the behavior patterns exhibited by any given species.

In birds and some other vertebrates, this hereditary component looms so large that Lorenz (1950) considers their behavior not as something that they "may do or not do, or do in different ways, according to the requirements of the occasion, but something which animals of a given species 'have got' exactly in the same manner as they 'have got' claws or teeth of a definite morphological character." Just as the cat with its retractile claws behaves quite differently in fighting and tree-climbing from the dog with its dull, nonretractile claws, or a sniffing dog with its keen sense of smell from a bird with its poor sense of smell, so, to a lesser degree, a hawk with its keen eyesight will differ in its behavior from an owl that possesses eyes of high sensitivity but poor acuity. And just as the inherited quality of its sense organs prescribes limits to an animal's behavior, so does the inherited quality of the correlating central nervous system and of the effectors. Some of the more subtle interrelations of these structural elements of behavior probably account for such patterns of behavior as the innate antagonism many passerine birds show toward owls, or toward the parasitic cuckoos. Behavior is of course influenced by other factors such as hormones, nutrition, periodic physiological cycles, age and maturation, and particularly, the past experiences of the individual. While much of the basic framework of a bird's behavior repertory is hereditary or innate, abundant evidence has also shown that birds are capable of intelligent learning. Recent studies indicate that "innate" and "learned" are opposite ends of a continuum with various combinations in between (Smith, 1983).

INNATE BEHAVIOR

A bird's behavior usually takes place with reference to an environment, internal as well as external, from which the animal is constantly being bombarded with potential stimuli, and toward which it in turn gives off stimuli. The degree to which a given behavior pattern is modifiable by its give and take with the environment—that is, by experience—is broadly a measure of learning or intelligence. Behavior patterns highly resistant to such modification are said to be innate; e.g., the gaping responses of a newly hatched bird, or the cessation of peeping of *unhatched* grouse chicks when their incubating mother gives a warning call as a hawk sails overhead.

There are different grades of inborn behavior. The simplest is the *reflex*, a quick automatic response of an organ to a simple stimulus, as in the blinking of an eye,

or the familiar knee-jerk in the doctor's office. A *kinesis* is the nondirectional active movement of the whole organism in response to a continuing stimulus, like the aimless scurrying of cockroaches when a light is turned on. A *taxis* represents oriented locomotion toward or away from the acting stimulus, as in the moth flying into the candle's flame.

An *instinct*, as the most complex form of innate behavior, is more difficult to define. Its nature has been the subject of perennial argument. Some ornithologists, such as David Lack, think the term should be abandoned. Historically, the word has been freighted with a great variety of meanings, and this makes its modern use at times precarious. However, no generally accepted substitute term is available at the present time.

The following definition of instinct, based largely on Lorenz (1950) and Thorpe (1956) is tentative and no doubt subject to numerous exceptions and modifications. An instinct is an inborn, particulate, stereotyped form of coordinated behavior, characteristic of a given animal species. It is relatively unmodifiable. A general internal "drive" or nervous tension sets instinctive behavior in action and possibly determines its direction. Such behavior is released completely, rather than guided, by environmental stimuli. An instinctive act weakens with repetition (i.e., its stimulus threshold rises), and gets stronger with disuse (its threshold lowers), until in certain situations the instinctive behavior may be performed without apparent stimulus. Inborn in the animal is a disposition to recognize and pay attention to certain stimuli or environmental conditions and to ignore others. Normally, instinctive behavior is directed toward ends that promote survival. Implicit in many definitions of instinct is the idea that an animal does not realize the end of its actions but performs them blindly. This notion, of course, cannot easily be proved or disproved. If, however, one grants that an animal has memory, one can hardly deny that it knows the consequences of repetitions of instinctive acts.

The Nature-Nurture Controversy

There is sometimes great difficulty in distinguishing the relative contributions of heredity (nature) and experience (nurture) to any specific behavior pattern of an animal. The nature-nurture controversy usually involved workers in two disciplines: on the one hand were the ethologists, who largely studied and described stereotyped, species-characteristic behavior of unconfined animals living in natural surroundings;

on the other hand were the more analytical and experimental psychologists, who were more apt to study animals (chiefly white rats and pigeons) confined under such unnatural conditions as mazes and problem boxes. Some contemporary psychologists have even abandoned using the much-abused word *instinct*, either handling it gingerly in quotation marks as "instinct," or with some such circumlocution as "behavior formerly considered innate." In fact, much of the disagreement was more semantic than biological. Nevertheless, no single, suitable, widely comprehended word other than instinct is available for naming behavior whose chief ingredient is innate or hereditary, and we shall so use the word in the following discussions.

A reasonable approach to the nature-nurture or instinct-learning problem can be made by considering first the roles of heredity and environment in determining structural or morphological traits in an animal. No respectable biologist disputes the hereditary control of head-color in, say, a cross between yellow and black-headed Gouldian Finches, *Chloebia gouldiae*, of Australia, in which the second filial generation will appear in a predictable Mendelian ratio of three-fourths black-headed to one-fourth yellow-headed birds (Wrenn, 1978; Immelmann, 1982). This ratio appears regardless of the food the birds eat, or the temperature, humidity, sunlight, exercise, or any other reasonably normal environmental variation to which the animals are exposed. Somewhat different is the inheritance of plumage color in male House Finches, *Carpodacus mexicanus*. The face and breast of male House Finches in the wild is red. A male House Finch brought into captivity and fed a diet of seed alone will turn yellow after its first molt. If carotenoids are added to its diet it will molt into a red of the wild species. Here the final appearance of the animal depends both on heredity and environment. Carotenoids determine the red color but only if it has genes to control the processing of the pigment. A White-crowned Sparrow fed carotenoids would not have a red head. Somewhat similarly, baby chicks reared from hatching on a smooth surface like a varnished table-top, will develop as hopeless cripples because they lack the friction-producing earth to which countless generations of their ancestors became adapted. These examples underscore the principle that the *degree* of normalcy or naturalness of the environment in which an animal develops and lives may play a role in the expression of its inherited traits.

If the great majority of animals of a given species living under the normal environmental variations of their natural habitat exhibit a certain characteristic morphological trait, one is usually justified in saying that the trait is inherited. In saying so, however, he does not deny that various contributions from the environment are essential to the development and survival of the trait, as well as of the animal itself: food, oxygen, minerals, sunlight, warmth, and so on. Naturally, if the animal fails to survive, the trait dies with it.

Exactly the same reasoning applies to the inheritance of functional traits whether they be levels of metabolism, heart rate, blood sugar, critical temperatures—or behavior. If one can say that nerves, brain, sense organs, muscles, and other anatomical components of behavior are inherited, one can with justifiable logic say that their functioning is at least equally under hereditary control. The kind of machinery that is inherited determines in large measure the way it will work. One does not use a sewing machine to type a letter. This principle would quickly become evident to anyone who attempted to train a cat, rather than a dog, to fetch a stick thrown into a pond. The nervous system, of course, plays a crucial role in determining patterns of behavior, and its functioning is well known to be subject to modification by internal and external environmental influences. But the important point is that the *kind* of nervous system that is inherited both directs the development of behavior along certain programmed channels and sets limits to the amount by which behavior can be modified by environmental influences or experience. Besides the nervous system, heredity also controls other mechanisms that may have profound effects on behavior. For example, the innate circannual and circadian rhythms in hormone-flow in birds have definite, predictable effects on reproductive and migratory behavior.

A striking example of hereditary determinism in behavior is demonstrated by intergeneric tissue grafts between amphibian larvae (Roessler, 1976). A small piece of the neural plate (later the medulla oblongata) grafted from *Xenopus laevis* into *Hymenochirus boettgeri* produced a chimaera or graft-hybrid animal whose mouth and throat feeding-behavior matched that of the donor and not that of the host species. That is, both nervous tissue (with its genes) and behavior were grafted. There is no reason to assume that similar grafts between embryo birds would act differently.

One should not assume, however, that the nervous system of an animal is inherited as a rigidly fixed, completely finished mechanism. As it develops in a growing animal it is subject to both structural and functional modification by internal and external environmental influences. For example, Leppel-sack (1983) played computer-simulated songs to naive hand-raised White-crowned Sparrows, *Zonotrichia leucophrys*. One group heard normal songs with syllables that start from a high and end at a lower pitch. A second group was tutored with songs containing syllables that rise in pitch. These songs were played to the birds at maturity, and neural activity in the mediocaudal neostriatum region of the brain was recorded with electronic instruments. It was found that 11.3 per cent of the auditory neurons responded by firing mostly or exclusively to the particular song type the birds were exposed to early in life. Single neurons learned responses early in the bird's development. There tends to be plasticity in behavior during the developmental stage followed by rigidity as a bird matures. If the left hypoglossal nerve (to the syrinx) of an adult Chaffinch, *Fringilla coelebs*, is cut, the bird loses most of its ability to sing. Cutting its right hypoglossus has little, if any, effect on song. If, however, the left hypoglossus in a young bird is cut before the bird learns to sing, the right hypoglossus takes over control of song and the bird, as it matures, will sing normally (Nottebohm, 1975).

The role of heredity in determining behavior is most evident when genetic mutations occur, such as, for example, colorblindness or phenylketonuria (which can cause mental retardation) in man, or congenital tremor in fowls, the aerial tumbling, rolling, ring-beating, and pouting in pigeons (Nicolai, 1976). Widely aberrant forms of inherited behavior are rarely encountered in nature since natural selection soon eliminates such individuals. It seems probable, however, that the following examples of normal patterns of behavior are largely genetically programmed: the building of supernumerary or "cock" nests by wrens, both in the Old and New Worlds; the plastering of mud or resin about their nest openings by various Old and New World nuthatches; the impaling of their prey on thorns and twigs by shrikes; and the de-venoming of bee-stings by bee-eaters. When, as in these cases, a certain kind of behavior is exhibited by the vast majority of individuals of a species that inhabits a wide variety of habitats, ethologists maintain that it is generally safe to assume that the trait in question is due to a preponderance of genetic (innate) as compared to environmental (learned) influences, and therefore, using an excusable shorthand, they call the behavior primarily instinctive. Dobzhansky (1967) in discussing the nature-nurture problem, says

It is biologically significant that different traits . . . occupy different positions in this heredity-environment spectrum. Nor are these positions accidental: by and large, traits appear in those environments in which they are useful; those useful in most or in all environments always appear, and thus seem to be unconditionally 'innate.' To put it another way, the 'innateness' or 'learnedness' of a behavior is fixed by natural selection at a level more or less close to that most favorable to the species.

The fact that young birds, incubator-hatched, hand-raised, and isolated from any possible education by their parents, still perform species-characteristic behavior, further supports the concept of innately ordered behavior. The following young birds, all meeting these qualifications, demonstrate some of the variety found in instinctive behavior. Young White Storks, *Ciconia ciconia*, almost immediately after hatching, invert their heads over their backs and make bill-clapping movements similar to the greeting ceremony of adult birds (Heinroth and Heinroth, 1924–1933). Two-day-old Wilson's Phalaropes, *Phalaropus tricolor*, "spin" in tight circles, like adults, while feeding (Johns, 1969). Even blinded young Black-headed Gulls, *Larus ridibundus*, demonstrate the foot-paddling behavior used by adults to stir up food in shallow water (Rothschild, 1962). Five- to seven-week-old Pied Flycatchers, *Ficedula hypoleuca*, recognize and mob live owls (Curio, 1970), and ducklings five days *before* they hatch from the egg can distinguish the recorded maternal call of their own species from that of a different species in simultaneous discrimination tests (Gottlieb, 1971). Even before its eyes are open, the young European Cuckoo, using elaborately coordinated behavior, tosses its foster-siblings out of the nest. One of the most remarkable instances of innate behavior is provided by the Bronzed Cuckoo, *Chalcites lucidus*, of New Zealand, whose young, without benefit of adult guidance, leave the nests of their foster parents and, about one month after their own parents have departed, migrate some 3500 kilometers, largely across open seas, to the species' winter quarters (Gilliard, 1958).

Further persuasive evidence of the instinctive basis of much stereotyped behavior is seen in the behavior of certain hybrid birds. The Silverbill Finch, *Lonchura cantans*, is capable of manipulating items with its feet when feeding, and drinks by scooping, i.e., it scoops water into its bill, then points it skyward as water drips down its throat, in the manner of chickens. The related Zebra Finch, *Poephila guttata*, does not use its foot to manipulate objects and drinks by sucking like a pigeon. Hybrids between the two species may use their feet while feeding, like a Silverbill, but suck like a Zebra Finch when drinking (Baptista, 1981). As a general rule, an interspecific or intergeneric hybrid displays behavior that is either intermediate between that of its two parents, similar to that of one parent or the other (see Fig. 9–2), or resembles neither (e.g., Wells and Baptista, 1978). Hybrid innate behavior has been observed in hybrids from crosses between different species of ducks, grouse, pheasants, hummingbirds, finches, doves, domestic fowl, and turkeys (Baptista, 1981).

Fixed Action Patterns

In his pioneering study of pigeons, Whitman (1899) discovered that much of their behavior was *particulate*, in that a given action was automatic and relatively independent of the activities of the organism as a whole. It is difficult to believe that Whitman's interpretation of instinctive behavior was written so long ago and not today:

It is quite certain that pigeons are totally blind to the meanings we discover in incubation. They follow the impulse to sit without a thought of consequences; and no matter how many times the act has been performed, no idea of young pigeons ever enters into the act. They sit because

A

B

C

FIGURE 9–2 Courtship displays of A., Zebra Finch, B., African Silverbill, and C., their hybrid. The Zebra Finch courts with mandibles opening and closing and nape and lower belly feathers ruffled. The Silverbill courts with bill almost completely closed and no feather ruffling. The hybrid courts with bill closed, as does the Silverbill, nape and lower belly feathers ruffled, as does the Zebra Finch. The ruffling of feathers in the upper belly is found in neither parent. Drawn from 8-mm cine film, from Baptista (unpublished).

they feel like it, begin when they feel impelled to, and stop when the feeling is satisfied. Their time is generally correct, but they measure it as blindly as a child measures its hours of sleep . . . The same holds true of the feeding instinct. The young are not fed from any desire to do them good, but solely for the relief of the parent.

The modern study of comparative animal behavior or ethology can be said to date from Whitman's work.

As an example of the particulate nature of instinctive behavior, Lorenz describes how a Muscovy Duck, *Cairina moschata*, will respond to the distress cry of a Mallard duckling, *Anas platyrhynchos*, by defending it, but will then respond to its coloration pattern by killing it as it would any strange, small, nest enemy. Similar particulate responses in Sooty Terns, *Sterna fuscata*, were observed by Watson (1908). An adult tern will vigorously drive a strange tern chick from its nesting area. Should the chick, however, avoid the jabs of the adult and touch its breast, a new instinctive response is immediately thrown into gear, and the adult accepts the chick as one of her own and "solicitously" broods it. It reveals the compartmentalized nature of instinctive behavior. The adult tern is a creature of the moment, responding in quick succession to strangeness, to breast contact, or to distress calls, all with sublime impartiality. It is when a bird's behavior "miscarries" in this fashion that its automatic and particulate nature is most evident.

Pondering such nonadaptive examples of instinctive behavior, Craig (1918) came to the conclusion that in much of its behavior an animal aims not at survival but at the *discharge* of these automatic actions which Craig called *consummatory* actions. The general preparatory restlessness or striving which leads toward the consummatory act he called *appetitive behavior*. Appetitive energy may be discharged through specific consummatory acts or may result in spontaneous behavior, such as aimless or exploratory wandering.

In an attempt to explain instinctive behavior in concepts free of teleological purpose, Lorenz (1950) proposed his widely accepted hypothesis of *fixed action patterns*. These fixed action patterns are rigidly stereotyped, predictable, species-characteristic movements that serve as consummatory actions. They arise independently of experience and are essentially complete on their first performance, which once begun, continues independently of environmental cues. They are relatively complex actions that are normally performed only in response to rather simple *sign stimuli*. In brief, they are instinctive acts, the majority of which are

adapted to the individual's reproductive success and survival.

Lorenz's idea of instinctive behavior as rigidly stereotyped has recently been modified by Barlow (1977) who noted that there is variability in fixed action patterns. For example, the duration of 66 head-throw displays of the Goldeneye Duck, *Bucephala clangula*, averaged 1.29 seconds with a standard deviation of 0.08 seconds (Dane *et al.*, 1959). Despite variability, our senses perceive these behaviors as stereotyped because there is a mode. The term "modal action pattern" has been proposed as a more accurate description of stereotyped behavior (Barlow, 1977). Variability in traits is of great evolutionary significance as it provides the "raw material," so to speak, for natural selection to act upon.

Instinctive behavior is further characterized by its dependence not on a simple stimulus-response mechanism as in reflex actions but on the accumulation of a poorly comprehended but probably neurophysiological internal drive, compartmentalized into what Lorenz called reaction-specific energy or *specific action potential*. A specific "packet" of nervous energy regulates the release of a specific consummatory act in preference to all other fixed action patterns. Once a consummatory act is performed, this energy diminishes or is "consumed." In other words, the threshold for the performance of the act rises. If, however, a consummatory act has not occurred for some time, the reaction-specific energy accumulates and the release threshold for the act lowers until, on occasion, the consummatory act may occur without an apparent sign stimulus. In such a case, a bird is said to be performing overflow or *vacuum activity*. The blocking factor that prevents the occurrence of a fixed action pattern before an animal has encountered the appropriate sign stimulus is called an *innate releasing mechanism*—a genetically determined neural mechanism selectively sensitive to specific sign stimuli and capable of regulating the flow of nervous impulses that culminate in a fixed action pattern (Tinbergen and Perdeck, 1951).

In many animals certain structures and patterns of behavior have developed into special, exaggerated displays or sign stimuli, now called *releasers*, whose chief function is the transmission of sign stimuli to other animals. These releasers may be visual (color, postures, movements), auditory (sound), chemical (pheromones), or tactile (Tinbergen, 1948). The evolution of such releasers has probably paralleled that of the associated specific responses until a given releaser is related to a certain innate releasing mechanism as a key is to a lock. Occasionally a releaser may inhibit rather than release a response. Releasers are of great importance in social communication. They are classified by Eibl-Eibesfeldt (1970) as signals that either signify or promote contact between individuals in pairs or in groups; courtship behavior; threat or appeasement behavior; warning against or distraction of predators; mobbing behavior; distress calls; and the like.

The sensitivity of a bird to sign stimuli may change with time or with maturity. If a gull encounters a pipped egg in its nest early in the incubation period, it will nibble at the cracks and probably kill the chick within the egg, but the same egg toward the end of the incubation period will release a pattern of brooding and feeding (Tinbergen, 1960). Sometimes a fixed action pattern is normally evoked by several sign stimuli, the absence of any of which will result in a decrease in the intensity of the consummatory act. This relationship is known as *heterogeneous summation* (Seitz, 1940–1941). Svärdson (1949) is of the opinion that heterogeneous summation is involved in habitat selection by many species of birds.

When releasers are lacking, drive may find expression in exploratory or *appetitive behavior* that represents an early, low-energy stage of a fixed action pattern in which behavior is more variable, less rewarding, and more likely to be influenced by experience than is the consummatory act. Appetitive behavior may be short or long in duration. Udvardy (*in litt*) suggests that spring migration is one long expression of appetitive behavior. Appetitive behavior is usually described in terms of the type of consummatory behavior toward which it leads: e.g., feeding, fighting, flying, sleeping, courting. Sometimes, when motivational energy is low, appetitive behavior is expressed in the form of incipient or *intention movements*. The following examples may help clarify some of these various aspects of fixed action patterns.

In his remarkable studies of the Jackdaw, *Corvus monedula*, Lorenz (1952) discovered that any animal, man included, that carried a dangling or fluttering black object would become the focus of a furious attack by these birds. One's immediate guess might be that these highly gregarious birds mistake the black object for a Jackdaw and rush to its aid, or at least attempt to drive off the dangerous predator. But the releasing stimulus in this case does not need to resemble a bird at all closely. Lorenz's first experience with the

phenomenon came when he withdrew some black swimming trunks from his pocket. Immediately he was "surrounded by a dense cloud of raging, rattling Jackdaws, which hailed agonizing pecks upon my offending hand." Even a Jackdaw carrying to its nest a black wing-feather of a raven is subject to attack by other Jackdaws. On the other hand, when Lorenz held unfeathered, baby Jackdaws in his hand, their parents were not in the least disturbed. But on the day their quill feathers burst open and turned the baby birds black, he was immediately and furiously attacked by the parents. "Dangling black" is the sign stimulus or releaser of an innate predator-attacking mechanism in adult Jackdaws. The behavior is not released by the situation as a whole, nor by those elements in it that a human would consider significant, but instead by some characteristic part of it to which the innate releasing mechanism through the sense organs is "tuned."

As a result of their classic experiments with cardboard silhouettes of various kinds of birds in flight, Lorenz and Tinbergen (Tinbergen, 1948) discovered that both outline and movement were significant visual characteristics in releasing fear reactions in such birds as Willow Ptarmigan, *Lagopus lagopus*. A generalized silhouette (Fig. 9–3) when moved in one direction released fright reactions, but when moved in the opposite direction, did not. Apparently, the silhouette when moving in one direction suggested a hawk; when moving in the other direction, a goose.

However, experiments of a similar nature performed on female Turkeys, *Meleagris gallopavo*, by Schleidt (1961) showed that these birds did not react to the "long-neck versus short-neck" criterion any more than to a circular disk of the same area. More important in eliciting fear responses were apparent size and relative speed of the moving models. The birds, he suggested, were reacting more to novelty than to the configuration of the models. There appear to be species differences in this behavior, for naive Mallard ducklings showed more fear responses to the hawk than to the goose model (Green *et al.*, 1968). Incubator-hatched naive Mallard ducklings and domestic chicks were hooked up to a device to measure heart rates. Heart beating was more irregular (showed a greater variance in response) when birds were shown the hawk model as compared with the goose model (Mueller and Parker, 1980; Moore and Mueller, 1982).

Martin and Melvin (1964) reopened the question. They exposed 18 naive, adult, incubator-raised Bobwhite Quail, *Colinus virginianus*, one at a time,

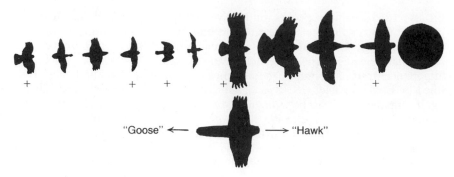

"Goose" ←——— ———→ "Hawk"

FIGURE 9–3 When cardboard models like the above were moved through the air above defenseless birds, the models marked " + " released fright or escape responses. The generalized bird shown below released fright responses only when moved toward the right. The releaser in every case was "short neck." After Tinbergen, 1948.

either to a living Red-tailed Hawk, *Buteo jamaicensis*, trained to glide across the top of their pen, or to a life-sized model silhouette of the hawk, pulled on wires above the pen at the same height and speed as the live hawk's trajectory. Whether the quail were exposed first to a series of trials with the live hawk or with its model, their fear responses to the hawk were always more intense and longer lasting than those to the model. The authors were convinced that the stereotyped fear responses of the quail were innate and specific for the species. An innate fear of the poisonous Coral Snake, *Micrurus fulvius*, was demonstrated by inexperienced, hand-raised Turquoise-browed Motmots, *Eumomota superciliosa*. Models of snakes with green and blue rings or with yellow and red stripes were readily attacked by the birds, but models painted with red and yellow rings to match the venomous snake were avoided (Smith, 1975). Hand-raised, naive herons belonging to three species could distinguish poisonous sea snakes from edible eels: they raised their crests and backed away from the snakes, but gave begging responses when presented with fish (Sullivan and Rubinoff, 1983).

The food-begging response of a newly hatched Herring Gull, *Larus argentatus*, is released by the sight of its parent's red-spotted beak presenting food near by. In an attempt to analyze this stimulus situation, Tinbergen and Perdeck (1951) presented newly hatched chicks with artificial bills of various sizes, shapes, and colors. They discovered that to be an effective releaser the bill must have a definite shape and a red patch near its tip, which is characteristic of the species. Fur-

ther, to be most effective, the bill should be held low, near the chick, pointing downward, in motion, and with something protruding from its outline (food). As a quantitative measure of the relative effectiveness of different kinds of bills and heads, presented under different conditions, the number of pecks made by the chicks toward the experimental models was recorded for comparable exposures. A moving head, for example, received 100 pecks, and a still head only 31; a near object, 81 pecks, a far object, 6 pecks. A bill with a red patch near its tip received four times as many pecks as one with none. A solid red bill released about twice as many pecking responses as a bill of any other solid color. The color or shape of the parent's head had no effect on the peck frequency of the young. A head colored black or green released as many responses as a normally white head. The inborn concern of the chick was for the bill and its general character (Fig. 9–4).

Quite unexpectedly, certain artificial stimuli were found to be more effective releasers than their natural counterparts: a normally shaped bill received 100 pecks, but an abnormally long and thin bill, 174 pecks. Such unusually evocative stimuli were called super-normal releasers—the goal and envy of every advertising executive (Fig. 9–5). Lorenz emphasizes the point that these releasers or "sign stimuli" generally exhibit *relational* properties, and are not simple, absolute qualities or quantities. Neither a limp black rag nor a man standing alone elicits the mobbing attack of Jackdaws, but the rag dangling from a man's hand, or from a dog's mouth, releases the typical antipredator

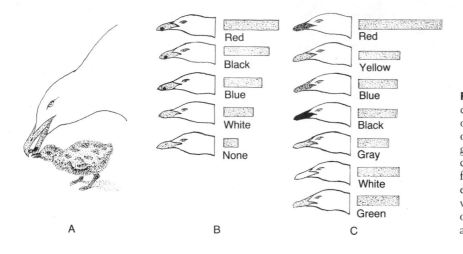

FIGURE 9–4 A, A Herring Gull chick begging for food. B, Models of Gull heads, with varied patches on the bills, used to release begging responses in newly hatched chicks. Bars indicate the relative frequencies of the chicks' responses. C, Bars indicate the releasing values of Gull models with bills of varying solid colors. All figures after Tinbergen, 1951.

assault. Releaser experiments by Tinbergen and Kuenen (1939), using cardboard models on young thrushes, have shown that they will gape toward the nearer, the higher, and the smaller of two objects. All three of these pairs of relations are characteristic of the parent bird's head and body while feeding its young. Each of these relational stimuli alone has the power to release the begging response in the young, and all of them together combine in heterogeneous summation to elicit even more effectively the same consummatory response under natural conditions.

According to Tinbergen, there exists in each animal a hierarchy of functional integrating centers on different levels, one level superimposed on another.

FIGURE 9–5 An example of a super-normal releaser. An Oystercatcher will attempt to incubate a gigantic artificial egg in preference to its own (*foreground*) or that of a gull (*left*). After Tinbergen, 1951.

Each level is activated by nervous potential descending from the level above. Blocks between these levels, each with its innate releasing mechanism, prevent the discharge of motor impulses until an appropriate releaser opens them to allow their passage. If motivational impulses flowing downward from a higher center toward a lower remain blocked, they may be diverted into appetitive behavior or general motivational drive. Other influences that contribute drive to these centers are hormones, sensory stimuli, autonomic impulses from the viscera, and possibly self-generated stimuli from the centers themselves. The blocking or "damming up" of action potential results in lowering the threshold for the stimuli required to release a given action; in other words, it raises the pressure behind the block, sometimes to such a level (threshold zero) that the action goes off without any external stimulus at all (vacuum activity).

Convincing support for the theory of hereditary fixed action patterns is provided by the experiments of von Holst and von Saint Paul (1962) on the domestic Chicken, *Gallus gallus*. By inserting fine electrodes several millimeters deep into the brain of a rooster or hen, they were able to stimulate the brain with weak (0.2 to 1.2 volts) alternating currents. Depending on the region stimulated and the voltage and amperage used they were able to elicit at will a great variety of complex behavior patterns that are normally associated with such everyday activities as feeding, preening, sleeping, courting, and fighting. A rooster, for example, if unstimulated, will ignore a nearby stuffed weasel. But when the appropriate center of its brain is electrically stimulated, it will respond toward the weasel with the following stereotyped sequence of actions: "alertness, visual fixation, approach, attitude of rage, attack with spurs, and triumphant call." This complex train of actions, very different from a simple reflex, is brought on by a single, simple, electrical stimulus.

Subsequent experiments by von Holst and von Saint Paul (1963) on domestic fowl revealed some new facets of fixed action patterns as elicited by direct brain stimulation. They successively stimulated the region of the brain that caused clucking. The first 0.4-volt stimulus required 24 seconds of application to elicit clucking; the second stimulus, 11 seconds; the third, 4 seconds; and the sixth to ninth, 2 seconds. The stimulus threshold, therefore, was progressively lowered. In other experiments they found that the same fixed action pattern could be released by stimulating different regions of the brain, and, contrariwise, that stimulating a given brain region might evoke different behavior patterns at different times. They also discovered that the strong arousal of one response during the performance of another could result in the interruption or suppression of the original behavior. Some sort of inhibitory mechanism was presumably responsible. That the inhibition may have operated against sensory input is suggested by experiments on a cat in which regular impulses from the cochlear (auditory) nerve were recorded every time clicking sounds were played at the ear, but if a mouse were exhibited before the cat, the impulses stopped although the clicks continued (Hernández-Peón et al., 1956).

From analyses of behavior resulting from the weak (less than 1/2 ampere) electrical stimulation of 1482 different brain sites in 87 domestic fowl, Phillips and Youngren (1971) discovered that vocalizations were released at more sites than any other type of behavior except that associated with fear. The following behavior patterns, given in order of descending frequency, were those most commonly elicited: vocalize, circle, crouch, panic, shake head, ruffle plumage, peck, fly, wipe bill, raise hackles, eat, raise tail, attack, hide, scratch head, freeze, stand tall, shake body, gape, fan tail, and others. The authors found that the only complex, natural sequential behaviors evoked were hostile (or agonistic) ones such as fighting, threat, or fleeing. One of the most complex fixed action patterns released was attacking behavior in which a chicken would direct its bill and feet toward objects in the

vicinity: another bird, a stuffed squirrel, a hand, and so on. Stimuli through an electrode fixed in one brain site in a hen caused her to attack 100 times out of 119 trials; 89 times she viciously pecked or kicked the squirrel, but only 11 times did she attack a companion bird. One hen, stimulated strongly, would attack a companion who was even an otherwise dominant bird, but when stimulated weakly, the hen would draw herself tall, raise her hackles and carpals, and spread her tail—all incipient attack reactions. These variable responses may reflect the activity of Tinbergen's hierarchical organization of functional integrating centers.

Often in nature, situations arise when the smooth and routine performance of simple fixed action patterns may be thwarted. A bird, once aroused, may suddenly find an indispensable stimulus lacking, or an obstacle may appear between the bird and an essential stimulus object. A very common situation results from the arousal in the bird of antagonistic or incompatible fixed action patterns as, for example, when an enemy approaches a mother bird brooding her young. The bird may either stay and protect her young or flee and save her own life. Dilemmas like these often result in ambivalent, incongruous, or substitute behavior of one sort or another.

When rising nervous energy is blocked so that it cannot escape into the usual motor channels, it may be diverted into a new path that results in irrelevant or out-of-context *displacement activities*—activities that make no sense according to human standards. A Song Sparrow, *Melospiza melodia*, confined in a cage, tugged strings and attacked straws when his rival alighted outside (Nice, 1943); and grackles, jays, and ravens have been observed attacking boughs and grass when their nests were threatened by man (Skutch, 1946). When thwarted in the performance of an instinctive act, Lapwings, *Vanellus vanellus*, make displacement-feeding movements; crows and jays hammer on a branch; shore birds assume the head-tucked-in-wing sleeping posture; many passerines strop their bills on a twig; and falcons will attack other birds previously unmolested (Armstrong, 1950; 1965). Displacement activity in other birds may take the form of bathing, preening, nest-building, copulation, grass-pulling, drinking, or singing.

A special variety of displacement activities called *redirection movements* occurs when a consummatory act is directed toward an object other than the one releasing the activity, although the original stimulus object remains available. An Argus Pheasant, *Argusianus argus*, for example, courted a stone water-trough when the female to whom he attempted to display refused to stand still (Bierens de Haan, 1926). The term *ritualization* was coined by Julian Huxley (1914) to describe stereotyped communicatory behavior evolved from simple motor habits, or possibly from intention movements or displacement activities. Typically, ritualization involves rhythmically repeated motor acts that serve as releasers to stimulate the opposite sex, intimidate an enemy, strengthen a pair bond, and so on. Displacement preening, used as a courtship ritual, is more conspicuous and more formalized than functional preening.

Related to ritualization is the concept of *emancipation* proposed by Tinbergen (1952). He suggested that during the course of evolution, behavior patterns may be "emancipated" from their original functions to be controlled by new stimuli. For example, fledgling African Mannikins (*Lonchura* spp.) beg for food with their bills wide open, tongue erect and waving, and wing raised. During courtship, adults hold their bills wide open and quiver their tongues, but during aggressive encounters wings are held up like sails (Güttinger, 1970; Baptista, 1973). Thus, different elements of the juvenile begging display have been "emancipated" from the original functions of soliciting for food, and are now ritualized in adult courtship and aggressive contexts.

One of the significant differences between an instinctive response and a reflex movement is the fact that the instinctive act, once begun, rolls on to completion even when the external stimulus that set it off ceases. For example, a goose rolling a displaced egg back into its nest by hooking it in with its lower mandible will carry the motion all the way to the nest if the egg is snatched away in mid-journey. Probably the most famous example of such instinctive *performance momentum* was provided by Lack's (1943) experiments with a mounted European Robin, *Erithacus rubecula*. He wired this stuffed bird on top of a 2-meter stick and placed it in the territory of a female Robin, which vigorously sang and attacked the dummy for 40 minutes. Lack then removed the mounted bird and went off to his breakfast, but chancing to look back he saw the hen Robin return repeatedly to the spot where the stuffed bird had been, and deliver violent pecks at the empty air! In various species, similar instinctive inertia may endure, after a fashion, for several days.

In much of their behavior animals exhibit a directiveness or an apparent purposiveness, since their various behavior patterns seem to promote biologically desirable ends. Intelligence seems chiefly to be displayed in appetitive behavior, while consummatory acts generally follow rigid patterns.

It should be made clear that this theory of fixed action patterns and innate releasing mechanisms is but one of several attempts to explain some of the enormous complexities of inborn behavior in birds. It is the theory that is currently most attractive to students of bird behavior, and it will undoubtedly be subject to alteration as more facts about bird behavior accumulate.

One of the key indications that a given behavior pattern is primarily inherited rather than acquired is provided by its frequency among related birds. If all the members of a given species or genus exhibit a given behavior, it is presumptive but not conclusive evidence that the trait depends in essence on genetic factors. In his behavior studies of a half-dozen species of European titmice, Hinde (1952) found that they agreed remarkably in their acrobatic behavior patterns, and differed mainly in appearance, voice, and ecology. Similarly, the Horned Lark of North America and the Shore Lark of Great Britain, different races of the same species, *Eremophila alpestris*, both have the same peculiar habit of building a sort of flagstone terrace of pebbles or sheep-droppings outside their ground-level nests. These races have very likely been isolated from each other for thousands of years, yet the instinctive pattern continues. Tameness, shyness, and belligerence commonly "run in families" and very likely are based on hereditary behavior patterns. Phalaropes, puffbirds, kinglets, and titmice are relatively tame, confiding birds, while oystercatchers, Roseate Spoonbills, *Ajaia ajaja*, and Redshanks, *Tringa totanus*, are shy, wild species. Sometimes even within a single genus birds demonstrate strikingly different traits. The Ringed Penguin, *Pygoscelis antarctica*, is bold, pugnacious, and quarrelsome, and when approached by man is likely to charge, whereas the Gentoo Penguin, *P. papua*, a "calm philosopher," turns tail, and the Adelie Penguin, *P. adeliae*, stands its ground (Murphy, 1936). There are, of course, differences in temperament between individuals of the same species.

Two strains of Red Junglefowl, *Gallus gallus murghi*, one wild and wary, the other somewhat tamer, were hand-raised from incubator-hatched eggs by Brisbin (1969) to study the source of their wildness. When the birds of the wilder strain were released into a flock of domestic chickens, they soon departed into adjacent woodland where they established themselves and remained very wary. Similarly treated birds from the tamer strain became integrated members of the domestic flock. Chicks of the wild strain, even when hatched and reared by tame-strain hens, persisted in their wild patterns of behavior. Brisbin concluded that behavior patterns of wildness were "integral components of their genetic make-up" and only slightly modified by rearing conditions.

Body-care Behavior

Among forms of behavior that show great stereotypy are those that have to do with care of the body surface: stretching, scratching, bathing, dusting, anting, and preening. Many land birds and all aquatic species bathe in water, while ground-dwelling game birds and dwellers on steppes and deserts bathe in dust. Dust bathing by quail is thought to remove excess lipids that might otherwise cause feather matting (Borchelt and Duncan, 1974). Some birds of prey, doves, kinglets, and Old World sparrows (*Passer* spp.) bathe in both water and dust. Passerines "bathe hurriedly with continuous movement, hawks and pigeons wallow motionlessly between bouts of violent splashing" (Goodwin, 1956). To bathe in water, most species immerse the head, suddenly raise it, and then begin beating the wings. The inborn nature of bathing movements is seen in the fact that four-week-old Goshawks, *Accipiter gentilis*, may go through bathing movements on the bare ground on seeing a brood-mate splashing in water (Bond, 1942), and two different species of young hand-raised hawks attempted to bathe on a shiny piece of plastic food-wrapping with typical movements and postures (Mueller, 1970). Young hand-raised motmots will go through bathing movements even if they only hear the sound of rain on a roof (Smith, 1977). Preening seems invariably to follow bathing.

While sunbathing, most birds adopt a characteristic pose, leaning sideways and extending one wing and half of their tail feathers toward the sun. Kennedy (1969) lists over 170 species of birds in 48 families and subfamilies that have been observed sunbathing. Except in such birds as anhingas and cormorants, which need to dry out their wet feathers after div-

ing, the reasons for sunbathing remain obscure. Heat alone may be the stimulus to sunbathe, according to Lanyon (1958), who observed hand-raised passerines "sunbathing" in the dark while in the path of forced hot air from a space heater. However, Cade (1973) observed many birds in the Los Angeles Zoo sunbathing on a summer day when air temperatures were about 30° to 35°C and most of the birds were probably in the upper range of thermoneutrality. As an experiment he exposed a Bateleur Eagle, *Terathopius ecaudatus*, to the direct rays of four incandescent lamps on a hot summer afternoon. The bird continued to sunbathe in the typical spread-wing fashion even though it became hyperthermic and had to pant to keep cool. One possible function of sunbathing is that it may increase the mobility of ectoparasites, facilitating their removal by preening (Kennedy, 1969). The suggestion has been made that uropygial oil, irradiated by the sun, produces vitamin D, which the bird ingests while preening. The question needs more study.

A widespread but very puzzling instinctive activity of birds called *anting* is a stereotyped behavior pattern in which the bird treats its feathers with ants or substitute materials. Between 200 and 250 species of passerine birds from some 40 families or subfamilies have been observed anting (Simmons, 1985). Anting occurs in either a passive or an active form. In passive anting, seen in crows, the bird spreads its wings, ruffles its plumage, and "sits down" on an active ant hill to let the angry ants crawl through its feathers. In active anting, practiced, for example, by orioles, jays, and starlings, the bird seizes one or more ants, and strokes or jabs them among its feathers. In active anting, the bird usually anoints ventral parts of the body, particularly under the wings and tail. Often, in its contortions to place ants among the under-tail and rump feathers, the bird steps on its own tail and tumbles over backward. Sometimes the bird will eat the ants after anting, and at other times will discard them. The ants used are often those that spray or exude pungent, aromatic, or repugnant fluids. While ants are the standard objects used, birds have been observed "anting" with substitutes such as beetles, bugs, wasps, orange peel, raw onion, hot chocolate, vinegar, hair tonic, moth balls, cigarette butts, burning matches, and smoke. When Whitaker (1957) exposed a hand-raised Orchard Oriole, *Icterus spurius*, to ants on 80 different days, the bird anted on 67 of them. Individual anting sessions commonly lasted 45 minutes.

Ants were offered by Querengässer (1973) to three hand-raised, inexperienced Starlings, *Sturnus vulgaris*, when about a month old. At first they ate them but at 37 days of age the birds began anting instinctively. When one Starling was offered a boiled, acid-free ant, it anted once but no more, whereas dead, acid-yielding ants were used several times.

Various explanations of anting have been proposed by different authors: the bird wipes off the ant's formic acid before eating it; the live ants eat or repel the bird's external parasites; ant fluids produce pleasurable effects, relieve itching, or act as a medicinal tonic on the skin. The metapleural glands of ants produce complex secretions that are known to inhibit fungi and bacterial growth and are spread by the ant over its entire body. The secondary acquisition of these antibiotic secretions may be an important reason for anting (Ehrlich *et al.*, 1986).

Evidence that anting destroys feather mites was presented by V.B. Dubinin (reported by Kelso and Nice, 1963), who found that the principal food of feather mites was the lipids of feathers, and that the number of feather mites on a bird fluctuated according to the amount of fats in the feathers. He further discovered that among heavily infested steppe pipits the mites on birds that had just been anting suffered much higher mortality than mites on nonanting pipits. Until now, however, no single, general explanation of anting has been widely accepted.

Many body-maintenance activities of birds are highly stereotyped and are probably in large part hereditary. Among them are such behavior patterns as preening, scratching, ruffling and shaking plumage, tail-fanning, shaking and rubbing, bill-wiping, bathing (in dust, snow, water, rain, and sunlight), wing- and leg-stretching, foot-pecking, and yawning. Most of these actions are also called "comfort movements" and many of them have become ritualized as social releasers (McKinney, 1965).

In addition to these body-care instincts, birds show numerous other inborn behavior patterns related to feeding, fighting, courtship, nest-building, care of young, sleeping, social relations, and other activities. Many of these forms of instinctive behavior are considered in other chapters.

LEARNED BEHAVIOR

Examples given thus far have perhaps made it appear that birds are capable of little, if any, intelligent behavior. While it is probably true that typically a bird's action is largely stereotyped and instinctive, it does not follow that it is nonadaptive or stupid. Instinctive behavior may be as adaptive as intelligent behavior. A great deal depends on the circumstances in which the behavior occurs. A species living for centuries on an isolated oceanic island with no enemies has no need to maintain instinctive wariness or predator-defense mechanisms. If its instinctive ways of feeding, bathing, courting, and breeding fit its requirements for survival, the species prospers. But once man introduces rats or cats on the island, the bird's "trusting disposition" is no longer adaptive, and the species may become extinct. This has happened all too frequently, for example, with Grayson's Dove, *Zenaida graysoni*, on Socorro Island, Mexico (Jehl and Parkes, 1983; Baptista *et al.*, 1983).

Learning is the adaptive modification of behavior as the result of experience. By "adaptive modification" is usually meant a relatively permanent change that promotes individual and (indirectly) species survival. The use of the word "adaptive" is intended to preclude changes in behavior brought about by fatigue, sensory adaptation, injury, growth, or maturation in an animal. The degree to which an individual animal can adaptively modify its behavior is ordinarily considered a measure of its intelligence. Changes in behavior brought about by learning can occur only if information obtained from the environment is stored in the nervous system and later retrieved in anticipation of similar behavior under similar environmental conditions.

Examples of learning in wild birds are difficult to establish with much certainty, so that most of the observations in this field are the results of experiments on confined birds. European Oystercatchers, *Haematopus ostralegus*, feed on mussels either by prying open the gaping shells and severing the muscles, or by hammering a hole through the shell. Each individual feeds preferentially or exclusively by either method. By cross-fostering eggs Norton-Griffiths (*in* Goss-Custard and Sutherland, 1984) showed that these techniques are learned from the parents or foster parents. An unusual and widespread example of natural learning in birds has appeared in the British Isles. Since 1921 there has been a growing practice among titmice and other birds (at least 11 species) of opening milk-bottle caps and drinking the cream (Fisher and Hinde, 1949). Different methods may be used,

FIGURE 9–6 An example of learning in wild birds. In England, titmice and other birds have learned to open milk bottles and to drink the cream. (*Left*) A Blue Tit removing the cardboard cap from a milk bottle. (*Right*) A Great Tit after puncturing a cap. Photos by D. L. Breeze.

even by the same bird, in getting at the milk. Cardboard and paper tops may be removed, or the paper torn off layer by layer. Metal foil tops are punctured and then torn off in thin strips. The tearing of paper and of paper-like bark is a common trait of tits, which were the first birds reported to open milk bottles. Other species probably picked up the habit by association learning, after watching the tits' success. Similar associative learning has been reported for Greenfinches, *Carduelis chloris*, which in the 1930's began eating fruits of the common garden shrub *Daphne mezereum* in a north British town; the habit spread over Britain by the fifties, and over the continent by the sixties (Newton, 1972).

Associative learning has been demonstrated in a series of experiments with Blue Jays, *Cyanocitta cristata*, in which individual birds were placed in adjacent cages, one bird as a "teacher," the other as a "student" (Brower *et al.*, 1970). In one group of experiments each student watched a teacher attack and eat six butterflies of species A. The student was then offered a simultaneous choice of two butterflies (both edible); one was species A, the other species B, different from A in size and color pattern. Of 21 students that watched A teachers, almost all initially attacked A butterflies rather than B. Reciprocal experiments with B teachers resulted in the majority of students attacking B rather than A butterflies. These results show clearly that a bird's choice of food can be strongly influenced by its companion's food choices.

Striking modifications of behavior to artificial situations are seen in gulls and other birds that follow tractors during spring plowing to feed on the small animals turned up; in Merlins, *Falco columbarius*, which follow slow trains in northern Mexico and prey on the small birds that are stirred up (Kenyon, 1942); or gulls that learned to appear and feed on the dead fish available after each mine-destroying explosion off the coast of Holland following World War II (Vleugel, 1951).

One problem that complicates the study of learning concerns maturation. Does an older bird fly more expertly than a young one because of experience, or because its nerves, muscles, reflexes, and instincts have matured and are capable of more polished action? Herring Gulls, *Larus argentatus*, for example, seem to increase their food-gathering efficiency as follows: first-year birds at a garbage dump ate 0.8 items of food per minute; second-year birds, 1.3; third-year birds, 1.4; and adults, 2.4 (Verbeek, 1977). Here it seems likely that both maturation and the learning of appropriate search-images (p. 103) contribute to the improvement in foraging. Is improvement in song based on practice, on maturation, or on both? Much reproductive behavior is clearly dependent on the maturity of the endocrine glands, particularly the pituitary and sex glands. Without experimental controls, it is often impossible to discriminate maturation from learning. Consequently, caution is required when one interprets behavior that is apparently learned.

Habituation

Learning occurs at different levels of behavior. In his book on learning and instinct, Thorpe (1956) differentiates such types of learning as habituation, latent learning, trial and error, insight, imprinting, and other forms. Much of the discussion that follows is based on Thorpe's contributions.

Perhaps the simplest form of learned behavior is *habituation*, or learning *not* to respond to meaningless stimuli. Thorpe (1964) defines habituation as "the waning of a preexisting response as a result of repeated stimulation when this is not followed by any kind of reward or punishment (reinforcement)." It is a common observation that birds nesting in the vicinity of busy highways and railroads soon learn not to be disturbed by the rush and noise of traffic. Pigeons, Killdeer, *Charadrius vociferus*, and Bobwhite, *Colinus virginianus*, have become habituated to the sound of near-by gunfire.

A quantitative study of habituation in the White-crowned sparrow, *Zonotrichia leucophrys*, was made by Petrinovich and Peeke (1973) who played con-specific songs at a fixed rate to free-living territorial males, and then counted the number of songs sung by the subjects and the number of flights over the speaker. Eight trials were conducted, each separated by a 5-minute silent period. The graph (Fig. 9–7) shows that there is an initial rapid increase in response (the *sensitization* phase), followed by gradual habituation. After a 70-minute silent period, three more trials were conducted, and an increase in response once more obtained (*recovery* of response). The nature

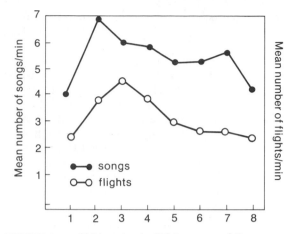

FIGURE 9–7 Habituation by White-crowned Sparrows to playback of con-specific song. Birds responded to playback with song and flights over the speaker. Note that there is an initial rapid response during the first three trials (the sensitization phase), followed by a decrease in response (the habituation phase). Data redrawn from Petrinovich and Peeke, 1973.

of the response varied with the reproductive condition of the bird. For example, in response to playback, females with fledglings fly about and give warning *chinks*. Females with nestlings tend to hide and are less often in view. Females incubating eggs are seen even less (Petrinovich and Patterson, 1979). One other corollary of habituation is the phenomenon of *dishabituation*, which is the restoring of a habituated response by a strong stimulus. A pigeon responds to a clicking noise presented at 10-minute intervals by scanning movements of the head and increased heart rate. The responses weaken at the 10th to 15th trial, but are immediately restored by grabbing the pigeon (strong stimulus) (Delius, 1983).

Conditioned Behavior

This simple form of behavior modification, pioneered by Pavlov, involves the attachment of a response to a new or substitute stimulus. Pavlov's dogs, for example, learned to salivate in response to bell-ringing rather than to the presentation of food after several simultaneous offerings of both food and bell-ringing had occurred.

In Pavlovian or "classical" conditioning, the organism learns to associate *two stimuli*. In a second kind of conditioning behavior known as "operant conditioning" the organism learns to associate between a *specific action* and a *reward*. Confined Jackdaws, *Corvus monedula*, were presented with 10 seconds of a stimulus light that then continued while accompanied by 4 minutes of recorded Jackdaw distress calls. This sequence was alternated with 2-minute intervals of no light, no sound. The Jackdaws soon learned to peck a key during the light-only phase that shut off both the light and the objectionable sound (Morgan and Howse, 1973). Thorpe points out that conditioned behavior is highly important in modifying fixed action patterns to fit environmental circumstances more precisely. Experimental conditioning is commonly used in determining the quantitative ranges of various sensory capacities in birds.

Trial-and-Error Learning

Whereas in conditioned behavior a comparatively passive animal learns to associate a new stimulus with a pre-existing response, in *trial-and-error learning (operant conditioning)* an active animal learns to modify its response or to make a new one. Through the restless, appetitive behavior ("curiosity") that precedes its reinforcement or reward, the animal learns to eliminate nonrewarding actions and to select those actions that achieve reward or avoid punishment. Thorpe points out that this kind of learning is anticipatory in that the motor conditioning that accompanies the performance of the action precedes the reinforcement of a successful trial.

In classical conditioned behavior the animal has no control over the nature of the stimuli presented to it, and the maintenance of the response to the experimental stimulus depends on rather regular rewards or it will wane. In trial and error learning the animal is free to choose among several possible stimuli, and only occasional reinforcement suffices to maintain the learned action. For example, Ferster and Skinner (1957) maintained a learned response in a pigeon although only one out of every 875 pecks was rewarded with food. Since so much trial-and-error experimentation occurs in mazes and problem boxes, this kind of learning is also called *instrumental conditioning*. Considering bird behavior generally, Thorpe believes that "trial-and-error learning is infinitely the most important process involved in adjusting voluntary actions to the circumstances of the environment."

In a typical maze experiment, an animal learns, after several trials, to avoid wrong turns and blind alleys, and is eventually rewarded for achieving the right path. Various passerine species were taught by Sadovnikova (1923) to run the fairly complex Hampton Court maze in from 20 to 50 trials—a performance roughly equivalent to that of rats in spite of the fact that a maze is a highly unnatural situation for a bird. In nature, trial-and-error learning is probably involved when birds learn to discriminate palatable from unpalatable seeds and insects, or when they learn to choose the right kinds of materials for building nests. The refinement in accuracy of the instinctive pecking response of a newly hatched chick provides a good example of trial and error learning in nature.

Not surprisingly, different species of birds vary in their ability to learn. Several White Leghorn Chickens, *Gallus gallus*, Bobwhite Quail, *Colinus virginianus*, Yellowhead Amazon Parrots, *Amazona ochrocephala*, and Red-billed Blue Magpies, *Urocissa occipitalis*, individually were required to learn to discriminate right from left in correctly choosing which of two containers held concealed food. Once success was achieved for a given bird, the food was shifted to the opposite container and the bird required to learn the problem anew. The task continued with successive right-left, left-right reversals for 29 times. The results, shown in Figure 9–8, clearly show the intellectual superiority of the magpies and parrots over quail and chickens (Gossette *et al.*, 1966). Significantly, these differences in intelligence are paralleled by differences in the relative cerebral indices (roughly the ratio of cerebral hemisphere size to brain-stem size) of the bird families represented (Portmann and Stingelin, 1961).

Learning by trial and error does not necessarily require numerous trials nor the close following of a stimulus by reinforcement. The intensity of reinforcement may strongly affect the learning process. Young domestic chickens, allowed to drink colored water containing sickness-inducing lithium chloride, learned after only one experience to avoid the colored water even though the illness followed drinking by as much as 1 hour (Genovese and Browne, 1978).

Play is undoubtedly an activity with an important component of trial-and-error learning. Young birds do not indulge in as much play as young mammals do, probably because so much of a bird's behavior repertory is instinctive. Nevertheless, the young of many birds do play and thereby learn more about their environment, and improve skills that will be of use in adult life. Playful activities by crows have been reported frequently. King (1969) observed a Hooded

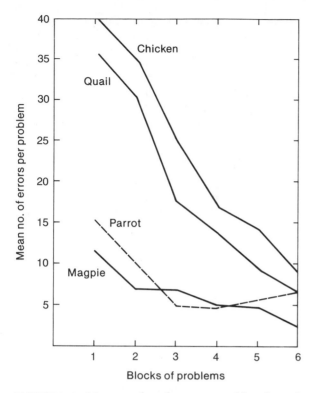

FIGURE 9–8 Mean number of errors per problem for each bird group through 30 reversal problems. Each number below represents a block of five problems. Redrawn after Gossette *et al.*, 1966

Crow, *Corvus cornix*, dropping a small object in midair and then retrieving it after it had fallen about 12 meters; it repeated this performance dozens of times. Similar playing with twigs, leaves, feathers, pine cones, and other inanimate objects has been reported for various species, including frigate birds, gulls, terns, eagles, falcons, kites, vultures, swifts, swallows, rooks, ravens, and other birds. Birds also at times play with each other. Gyrfalcons, *Falco rusticolus*, have been observed in mock aerial combat. Four Common Ravens, *Corvus corax*, were seen taking turns, one by one, sliding down a snow bank feet first, riding on their tail feathers (Bradley, 1978). Jackdaws, *Corvus monedula*, and other Corvidae enjoy a "game" of soaring together in rising air currents, swooping to earth and then repeating the performance over and over. Such activities very likely furnish the satisfaction of discharging consummatory actions, and

conceivably provide something akin to the "fun" children get from skipping rope or playing tag.

A form of learning closely related to trial-and-error learning, and probably one of its ingredients, is known as *latent learning*. It is a form of learning independent of immediate apparent reward. For example, one group of rats is allowed once each day for 10 days to experience random, appetitive exploration of a maze without receiving any reward. Then they are trained to learn the correct path through the maze by being rewarded for each successful run. Rats so treated will learn the maze problem much more quickly than other rats that have been denied this preliminary exploration. In brief, past unrewarded familiarity with a general problem situation facilitates subsequent learning. Play is probably a form of latent learning in that the exercise and polishing of certain instinctive acts expedite their adjustment to adult contingencies later on. Thorpe believes that the unrewarded learning by birds of their home territories, migration routes, and so on may involve latent learning.

Insight Learning

A much higher form of behavior called *insight learning* is defined by Thorpe as an organization of perception that permits the apprehension of relations. Thorpe adds that "insight learning involves the production of a new adaptive response as a result of insight or as the solution of a problem by the sudden adaptive reorganization of experience." In everyday terms, one would say that the animal suddenly "sees through" or understands a problem, and without further ado is able to solve it. In humans this might be called the "Aha!" or "Eureka!" response. A probable example of this occurred in an ingenious experiment by Budgell (1971) with Barbary Doves, *Streptopelia roseogrisea*—birds not considered particularly clever. Doves trained to peck a key in order to gain 5 seconds' access to a reward of grain were run every day until satiation, in a problem box kept at a temperature of 20°C. After this preliminary training the box temperature was lowered to 0°C, and the birds were held in the box for 2 hours before the reward mechanism was turned on. Now, for each peck of the key the dove received not only grain but a 5-second burst of electric heat. This procedure was repeated daily for a month or so. For three weeks or more the birds responded as if they were being rewarded with food only. They would peck until sated, during which time the temperature "incidentally increased." After satiation the birds ceased responding, and the temperature would fall to 0°C. At some time between the 24th and 37th day (depending on the individual), the bird's behavior changed (Eureka!). They continued pecking the key but no longer ate their grain reward. After three days of this regime the birds were given abundant food and water before being placed in the problem box for 2 hours at 0°C, but with the feeding mechanism disconnected. Now their sole reward was a 5-second burst of heat for each peck. Within five weeks' time the birds learned to control the box temperature very nicely within a range between 31.6° and 34.3°C.

It is well-nigh impossible to prove the occurrence of insight behavior in birds living under natural conditions. However, the use of tools in solving a problem, as in the case of apes using a stick to reach a banana, is generally considered a demonstration of insight. Tool-using, once thought to be rare among birds, has been documented in a number of avian taxa (Boswell, 1977, 1983). Colored fruit pulp and masticated charcoal are applied to the walls of its bower by the Satin Bower-bird, *Ptilonorhynchus violaceus*, by means of a "brush" or sponge-like wad of fibrous bark (Marshall, 1954). In searching for food, Brown-headed Nuthatches, *Sitta pusilla*, use scales of bark to pry other bits of bark loose (Morse, 1968); and the Orange-winged Sitella, *Neositta chrysoptera*, of Australia, uses small twigs to probe for wood-boring grubs in eucalyptus trees (Green, 1972).

The classic example of tool-using by birds is that of the Woodpecker Finch, *Cactospiza pallidus*, of the Galapagos Islands. One of "Darwin's Finches," this bird employs a cactus spine or a splint of wood broken to suitable length by the bird to probe holes and crevices too deep for its short bill. When the grub squirms out, the bird drops its tool and eats its prey. The tool may also be used as a spear to impale prey. The Mangrove Finch, *Cactospiza heliobates*, living on one of the Galapagos Islands isolated from the Woodpecker Finches, has a similar tool-using habit (Curio and Kramer, 1964). A captive large Cactus Finch, *Geospiza conirostris*, soon learned to use a tool when allowed to observe a Woodpecker Finch thus performing, demonstrating that this behavior is learned by imitation (Millikan and Bowman, 1967). There have been scattered field observations of other ground finches (*Geospiza* spp.) and the Warbler Finch, *Certhidea olivacea*, using tools (Hundley, 1963). Efforts to teach other manipulative species, e.g., Plain Titmice, *Parus inornatus*, to use tools were unsuccessful. It seems probable that there exists a genetic tendency for all of Darwin's Finches to use tools, although the behavior is usually expressed only in Woodpecker and Mangrove Finches (Fig. 9–9).

Perhaps the most spectacular tool-user is the Egyptian Vulture, *Neophron percnopterus*, of Africa, which throws stones at the eggs of the Ostrich, *Struthio camelus*, to get at their nutritive contents. This was first reported by a sportsman in 1867 (Skead, 1971) and more recently studied by van Lawick-Goodall (1968). Alcock (1970) suggests that rock-throwing at eggs may originally have been redirected behavior of vultures

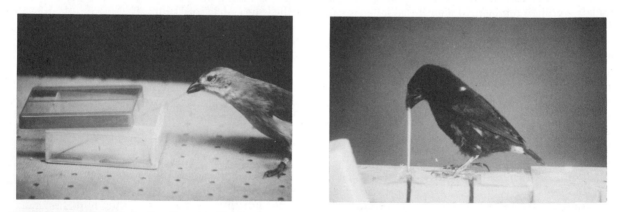

FIGURE 9–9 (*Left*) Tool-using Darwin's Finch, *Cactospiza pallidus* using stick in attempt to secure worm. (*Right*) One of Darwin's Ground Finches, *Geospiza conirostris*, using a tool. This behavior was apparently learned by observing a *Cactospiza* tutor. Courtesy of R. I. Bowman.

accustomed to throwing smaller eggs at the ground to break them. Unable to pick up a large Ostrich egg, a frustrated bird may have redirected its behavior to an available but manageable rock in the vicinity, fortuitously hit an egg, and eventually came to "associate stone-throwing with food and actively seek out stones in encountering an Ostrich egg." Other vultures could then learn the technique through observation.

Various species of tits and other birds can be taught to obtain food tied to the end of a suspended thread by pulling up a loop of the thread and holding it with a foot while the bird reaches with its beak for the next pull. The bird's rapid and unfumbling solution of this problem convinces Thorpe that the act involves insight. Thorpe relates that European Goldfinches, *Carduelis carduelis*,

are so adept at this trick that they have for centuries been kept in special cages so designed that the bird can subsist only by pulling up and holding tight two strings, that on the one side being attached to a little cart containing food and resting on an incline, and that on the other to a thimble containing water. This was so wide-spread in the sixteenth century that the Goldfinch was given the name "draw-water" or its equivalent in two or three European languages.

A similar performance by wild birds is reported from many parts of Norway and Sweden by Homberg (1957). In the early spring Hooded Crows, *Corvus cornix*, pull up fisherman's lines set through holes in the ice to steal the fish or bait. A bird will seize the line with its beak, walk slowly backward with it away from the hole as far as it can, and then walk forward *on top of the line*, thus preventing its slipping back into the water. At the hole it will again grasp the line and repeat the process until the fish or bait is accessible.

Various other examples of possible insight learning in wild birds have been noted. In one instance reported by Lovell (1958), a Green Heron, *Butorides virescens*, repeatedly placed bits of bread in the water, and fed on the small fish that came to nibble on them. The heron did not eat the bread, and drove away other water birds that attempted to. It appeared to be intentionally using the bread as fish bait. Similar fishing behavior has been reported for the Sun Bittern, *Eurypyga helias*, Squacco Heron, *Ardeola ralloides*, and Pied Kingfisher, *Ceryle rudis* (Boswell, 1983).

A very few species of birds have demonstrated that they are able to employ abstract concepts in directing their behavior. For example, eight Great Tits, *Parus major*, were trained by Menne and Curio (1978)

to discriminate between asymmetric and bilaterally symmetric geometric figures. An Indian Hill Mynah, *Gracula religiosa*, was taught to say "hello" to pictures of humans and to bark at pictures of trees (Turney, 1982). Using vocal labels, an African Gray Parrot, *Psittacus erithacus*, could discriminate among more than 40 objects. It also used vocal labels to discriminate between colors and shapes, keys, pieces of wood, and rawhide (Pepperberg, 1983).

In an illuminating series of experiments with Canaries, *Serinus canaria*, Pastore (1954) taught birds to discriminate *uniqueness* in stimuli. Birds were first taught to seek food hidden under the one among nine covering objects that differed from all the others. For example, nine small depressions or wells were covered with eight aspirin tablets and one wood screw. Under the wood screw, the unique or different object, a grain of food was hidden. The bird was taught to push aside all objects until it uncovered the food, its reward. In the second trial there were eight screws and one aspirin tablet, the food this time being hidden under the unique object, the aspirin tablet (Fig. 9–10). In successive trials the reward was placed in different wells, chosen at random, and under either a screw or an aspirin tablet, but always under the unique object among the nine. After about 160 trials, the average Canary succeeded in learning to choose the unique object 15 times out of 20. In a second series of trials, the two sorts of stimulus objects were changed from aspirin tablets and screws to some other pair of

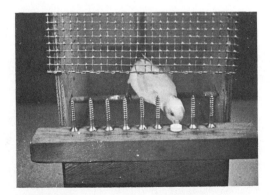

FIGURE 9–10 An example of insight learning. This canary has been taught to discriminate the abstraction "uniqueness." It has learned to seek food hidden under the one different object (here an aspirin tablet) among nine. After Pastore, 1954.

objects: for example, chess pawns and bolts. In all, each Canary was trained to distinguish the unique object in a total of 21 different pairs of objects; and with each successive version of the problem, the birds improved their performance (Fig. 9–11). In short, they learned to respond to the abstract concept of "uniqueness," and not merely to the physical attributes of the stimuli.

After training his smartest Canary, a female named Phyllis, to feed from a food-bin on a post, Pastore elevated the bin beyond the reach of the bird, whose wing feathers had been clipped, and trained her to tug a toy truck on tracks by means of a string until it stopped so near the post that she could perch on the truck to reach the food (Fig. 9–12). Next the string was threaded through a cardboard partition. On one side of the partition the bird could see the truck and post, but could not pull the string. On the other side she could pull the string, but could not see the truck and post. Pastore now began placing the truck at two different distances from the post so that from one position a single tug on the string sufficed to place the truck beside the post, and from the other position two tugs were required. In a series of trials in which the two distances were randomly interspersed, the bird learned after 2600 trials to match the distance of the truck from the post with the appropriate number of tugs from behind the partition. After 6000 trials she learned to correlate four different amounts of tugging with four different distances. This remarkable performance showed clearly that the bird "discriminated among four different dis-

tances, retained a given discrimination while tugging behind the partition, discriminated among four different types of tugging responses and, finally, correlated the four types of tugging responses to the four distances." Obviously, the bird was able to make spatial discriminations, to remember, to form concepts, and to behave with insight.

Insight learning may also be involved in the ability some birds have to "count" objects. For example, a Common Raven, *Corvus corax*, was trained by Köhler (1951) and his students to open one of five boxes. The boxes had two, three, four, five, and six spots on their lids, and the key to the "correct" box was given by a neighboring key card that held the same number of spots. The size, shape, and distribution of spots on both key card and boxes were constantly changed to avoid secondary clues, so that the results indicated clearly that the bird was distinguishing numbers, and numbers alone. Of course, the numbers had no names and probably had no qualitative difference in the bird's mind, any more than one pencil tap in a series is different from any other.

Pastore (1961) trained a Canary to choose the third (to the right) of three aspirin tablets in a row whether they were contiguous (A, A, A) or interspersed with screws and chessmen (S, A, C, A, S, C, A, C) in a total array of nine objects. For each successful choice the bird was rewarded with a seed. Another Canary learned to select the fifth of an identical series of objects in time. One or more aspirin tablets were placed in cubicles so that the bird could not see from

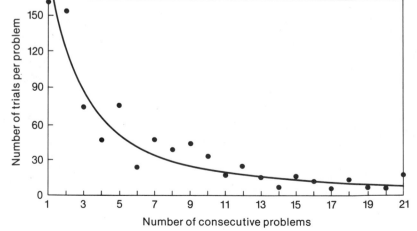

FIGURE 9–11 The learning curve of canaries trained to distinguish "uniqueness." Each point represents the number of trials needed for a bird to select the unique object, in that particular problem, 15 times out of 20 (averaged for four birds). The curve reveals steady improvement on successive problems. After Pastore, 1954.

FIGURE 9–12 In order to reach the food on the top of the post, the canary must perch on the toy truck, which is pulled by a thread. The bird has learned to pull the thread from behind a cardboard screen which hides the truck, the required number of tugs to position the truck beside the post. After Pastore, 1954.

the first cubicle what might be in the second, or third, and so on. Thus, the fifth, or "correct" aspirin, with the reward, might be in the second, third, or some other cubicle. Whether the birds had formed a concept of number was still, according to Pastore, an open question.

The possibility that birds may possess a concept

of number is reinforced by experiments in which birds successfully shift number sense from one sensory modality to another. Lögler (1959) trained a Gray Parrot, *Psittacus erithacus*, to choose the eighth in a line of food trays for its reward. It was then trained to choose the correct tray in the row in accordance with the number of light flashes given in a sequence. Drastic and random changes in the time intervals between flashes did not impair the bird's successful performances. Finally, the bird still performed correctly even when successive notes from a flute were substituted for light flashes.

A most suggestive demonstration of number concept was presented by one of Köhler's (1956) Jackdaws, *Corvus monedula*, that had been trained to open the lids of eight bait boxes standing in a row until it had secured five baits. In this particular trial the baits were distributed in the first five boxes in the order of 1, 2, 1, 0, 1, 0, 0, 0 baits. The Jackdaw raised the lids of the first three boxes and thus secured four baits, and then it returned to its home cage. The investigator was about to record a failure for the bird on this trial, when the bird returned to the boxes and went through an astonishing performance. As it walked along the row of boxes it stopped at the first box, where it had taken one bait, and bowed its head once; it stopped and bowed twice at the second box (two baits), and similarly made one bow at the third box (one bait). Then it proceeded down the row, opening the fourth box (no bait), and finally reached the fifth box where it removed the final bait. Then it ignored the remaining boxes and returned to its cage. The bowing—intention movements in retrospect—seems to give evidence of an (unnamed) number concept as clearly as do the lip movements of a child counting silently.

■ SUGGESTED READINGS

For Suggested Readings see the end of Chapter 10.

Social Behavior
10

Birds learn social ways by imitation
Of parents' acts, and song communication.
—Joel Peters (1905–   )

For practically every bird, as soon as it breaks out of its shell, its first experience is likely to involve social behavior. It must interact with its parents, its nest mates, or both. And later in life it must remain social if for no other reason than to breed. As a social animal, every bird must communicate with others, and this requires chiefly visual and auditory signals.

Early infancy is a time of extreme vulnerability to such environmental hazards as adverse weather, starvation, parasites, or predators. In all probability, natural selection has provided young birds with a remarkable form of learning called *imprinting* that serves as a countervailing adaptation to these hazards. The word imprinting (from the German word *Prägung*) was first suggested by Konrad Lorenz (1935) as a result of experiences he and Oskar Heinroth had had with incubator-raised goslings. Typically, a young gosling, lacking a mother, would react toward Lorenz by following him wherever he went, eventually making him its substitute mother.

Ethologists distinguish between two forms of imprinting. "Filial imprinting" or "imprinting of the following response" refers to the tendency for young of

precocial species to learn to recognize and then to follow their parent or parent substitute. "Sexual imprinting" (see beyond) describes the tendency for young of both *altricial* and *precocial* species (see p. 355) to learn characteristics of their parents or foster parents which will influence subsequent mating preferences (Immelmann, 1972; Immelmann and Suomi, 1981).

The times of peak sensitivity to the various forms of imprinting stimuli occur at different stages. Filial imprinting of precocials usually occurs during the first day of life. This is followed, perhaps on the second day, by habitat imprinting and on the third day by imprinting on food (Hess, 1964). Sexual imprinting may occur in the next few days or weeks.

A rough, practical knowledge of some aspects of filial imprinting has been known, especially among farmers, for many generations. As long ago as 77 A.D. Pliny, in his *Historia Naturalis*, wrote that a goose "was the constant companion of the philosopher Lacydes and would never leave him, either in public or when at the bath, by night or by day" (Hess and Petrovich, 1977). And in 1518 Sir Thomas More in his *Utopia*, describing the artificial incubation of chicken eggs, says, "As soon as the chicks come out of the eggs, they follow men and recognize them as if they were their mothers." Still later, Spalding (1873) wrote, "Chickens as soon as they are able to walk will follow any moving object. . . . they seem to have no more disposition to follow a hen than to follow a duck, or a human being." After keeping three chicks hooded for four days, he unhooded them, upon which "they evinced the greatest terror of me . . . [but] had they been unhooded on the previous day they would have run to me instead of from me. . . ."

Repeated experiments have shown that newly hatched goslings become attached to the first large moving object they see, and promptly follow it as soon as they are able. This object is ordinarily their mother, but in her absence it may be a human being or some inanimate object; the attraction becomes so strong that the goslings will later completely ignore their real mother. A Ruffed Grouse, *Bonasa umbellus*, became imprinted on a farm tractor and not only followed it about the fields but put on courtship displays in front of it for four successive springs (Hassler, 1968).

Sexual imprinting has been demonstrated in a great variety of birds with precocial young—ducks, geese, chickens, turkeys, pheasants, quail, bitterns, and coots—and even in such altricial species as owls, doves, ravens, and finches. Different species of birds show varying degrees of imprintability. The young of wild Mallards, *Anas platyrhynchos*, and Canada Geese, *Branta canadensis*, imprint very readily; those of the Ring-necked Pheasant, *Phasianus colchicus*, very poorly. And, of course, in brood parasites such as cuckoos, whose young are always reared by other species, imprinting would be disastrous for the species should the nestling cuckoo permanently fix its attention and reproductive behavior on the foster-parent species. Family groups or coveys of mixed species have been experimentally established through imprinting (Cushing and Ramsey, 1949). In order to hybridize two different species of pigeons, Craig (1908) found that he had to raise the young of one species under foster parents of the other. Then, as these young grew up, they mated by preference with birds of the same species as their foster parents.

Lorenz (1935) established four important criteria for imprinting. (1) Imprinting is usually an extremely rapid form of learning, ordinarily confined to a very brief critical period (or *sensitive phase*) in the individual's life. (2) It is extraordinarily stable and often irreversible. (3) Imprinting involves learning of "supra-individual," species-specific characteristics. (4) Imprinting is completed long in advance of the adult behavior being performed or expressed. When Oskar Heinroth hand-raised a Ring-necked Pheasant, it not only became imprinted on and courted him but it attacked his wife as a rival. When, as an experiment, the couple exchanged clothing, the bird, at first perplexed, looked closely at their faces and then began again to court Dr. Heinroth and to attack Mrs. Heinroth. On the other hand, Lorenz (1970) found that incubator-raised young Curlews, *Numenius arquata*, would imprint only on adult Curlews.

Imprinting differs from associative learning in that reinforcement, or reward, is not required. For example, Thorpe (1956) tells of Coots between 20 and 60 days old that persisted in following an experimental "parent" as many as 100 consecutive runs, with but 1-minute intervals for rest, without the reaction weakening. And instead of receiving a reward, the birds were mildly punished at the end of each run by being caught and picked up with their legs dangling. Seemingly, the behavior itself of following and proximity to the object followed are sufficient reward.

In a series of experiments aimed at revealing some of the details of the imprinting mechanism, Hess (1959) allowed newly hatched Mallard ducklings to follow a mechanically operated decoy duck, painted to resemble a male, around a circular runway (Fig. 10–1). Each duckling was given a 10-minute imprinting run of about 55 meters behind the male decoy, which uttered man-made sounds, "*Gock, gock, gock, gock*" as it circled the runway. The duckling was then automatically returned to a box and placed in a warm brooder until it was tested for the strength of the imprinting effect. This was done by placing the now imprinted duckling halfway between two model ducks, the male on which it had become imprinted and a similar decoy painted like a female. If the duckling followed the original male decoy the response was recorded as positive; any other response was considered negative. With this technique Hess found that the greatest number of positive responses came from those ducklings imprinted between 13 and 16 hours after hatching (Fig. 10–2). The imprinting effect dropped off very quickly with increased age. At 24 hours after hatching 80 per cent of the ducklings not only failed to follow moving objects but either avoided them or showed signs of fear. At 30 hours all ducklings behaved negatively. As a result of his experiments, Hess concluded that the critical

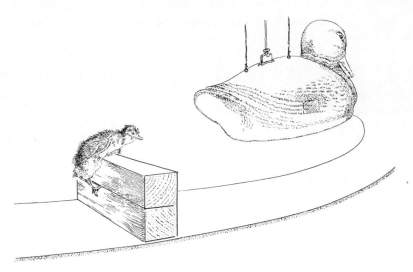

FIGURE 10–1 The greater the effort a duckling makes to follow a moving decoy, the more firmly it becomes imprinted on the decoy. Here a duckling is scrambling over a 10-centimeter barrier to keep pace with the decoy. After Hess, 1958.

period for imprinting begins as soon as the young hatchlings are able to move about on their legs, and rapidly wanes with the onset of fear, since a fearful animal will avoid rather than follow a potential imprinting object. The fact that young coots are wild and difficult to tame if they are more than eight hours old indicates an extremely brief critical period.

Further experiments by Hess showed than the distance a duckling traveled was more significant than the duration of the imprinting exercise. The greater the distance traveled, up to 15 meters, the stronger the imprinting. Ducklings forced to clear 10-cm high hurdles in the runway made the highest imprinting scores. Domestic Chicks, *Gallus gallus*, allowed to follow a moving object, were more strongly imprinted on it than were chicks confined in a transparent plastic box moving in the wake of the moving object (Thompson and Dubonoski, 1964).

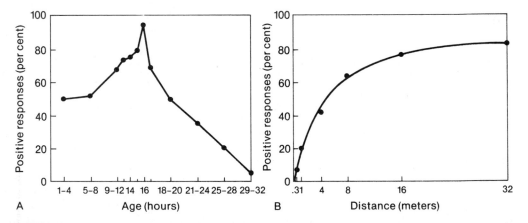

FIGURE 10–2 A, The age at which ducklings are most strongly imprinted by following a decoy. Each dot is the average test score of ducklings imprinted at that age. Mallard ducklings imprint best on a moving object at about 16 hours after hatching. B, The greater the distance traveled by a duckling during imprinting, the stronger the imprinting. After Hess, 1958.

There is evidence that the waning of the critical period for imprinting on an object is in part brought on by the simultaneous growth of fear of the same object (Fabricius, 1964). Hess and Schaefer (1959) exposed 124 domestic chicks, 1 to 44 hours in age, to a stationary, potential imprinting object. The chicks gave off either contentment or distress notes; they either approached or fixated the object. The ratio of chicks showing these contrasting behaviors at different ages completely reversed itself by the time the chicks were 21 hours old. The most pronounced change occurred between the ninth and twentieth hours after hatching. The critical period for imprinting on an object evidently occurs after the beginning of locomotion and before the onset of fear.

The stability, or irreversibility, of imprinting varies with species, with age of bird, with type of imprinting, and perhaps other factors. Newly hatched Mallard ducklings were exposed for 20 continuous hours to humans and followed them wherever they went. The ducklings were then given to a female Mallard that had just hatched a clutch of ducklings. After 1$^{1}/_{2}$ hours' exposure to this female and her own young, the human-imprinted ducklings followed the female when she left the nest and stayed with her. On the other hand, wild ducklings taken from their natural mother 16 hours after hatching could not be imprinted on humans even after two months of trying (Hess, 1972). Quite obviously, there must be some degree of innate preference in the ducklings for their own species. Gottlieb (1971) found that the young of domestic fowl, Mallards, and Wood Ducks, *Aix sponsa*, imprinted selectively on their own species when given a choice.

In numerous species it has been found that visual stimuli alone are much less effective in eliciting following behavior than the combination of visual and auditory stimuli. Here again, some young birds exhibit an innate recognition of their own species. The young of domestic fowl showed a definite preference for the clucking of a broody hen over other sounds (Graves, 1970a). Even more significantly, Wood Duck embryos, still in their shells, respond to a Wood Duck's maternal call with increased bill-clapping, but to a Mallard's call with decreased clapping (Heaton, 1972). In naive, posthatching ducklings this innate species-specific discrimination is less marked, and therefore the young probably require reinforcement by the mother's calls in order to become firmly imprinted on her calls. In nature the Wood Duck mother seems instinctively to realize this. She begins calling, about once every 5 seconds, when the eggs are first pipped, which is about 20 to 36 hours before the ducklings are ready to leave the nest. When exodus draws near she increases her calling to about five calls per second (Gottlieb, 1963). Young Mallard Ducklings begin vocalizing at the time they penetrate the egg air-chamber membrane, which is about two days before hatching. The incubating mother responds with frequent clucks. During the hour-long hatching period the mother vocalizes at a rate between zero and four times per minute, but when the ducklings leave the shell, the mother's vocalizing increases to 45 to 68 calls per minute for about two minutes. This mother-duckling vocal interplay probably facilitates the rapid formation of the posthatching filial bond (Hess, 1972). Similar "conversations" between a parent Laysan Albatross, *Diomedea immutabilis*, and its peeping egg are thought by Fisher (1971) possibly to establish a bond of individual recognition between parent and young.

Peak sensitivities for imprinting to visual and to auditory stimuli do not necessarily coincide in time. Mallard ducklings exposed to silent moving models exhibited a peak critical period for following at between 10 to 20 hours of age, while following sensitivity to auditory stimuli rose to a peak between 40 to 50 hours after hatching (Boyd and Fabricius, 1965). Mallard ducklings were visually imprinted at 24 hours of age on a stuffed female Mallard. Then, at 48 and 72 hours, they were given simultaneous choice tests between the familiar Mallard and an unfamiliar red and white box. When both models were silent, the ducklings followed the Mallard. But when the maternal assembly call was played on a speaker in the box, the ducklings followed it (Johnston and Gottlieb, 1981).

This discrepancy in the times of visual and auditory sensitivity makes adaptive sense when one considers that in nature the ducklings, very soon after hatching, are visually led from the nest by their mother to nearby water where, in time, the young may become separated from their mother in thick vegetation where only auditory stimuli can hold the brood together (Fabricius, 1964). There are also differences between species in their general reliance on either visual or auditory stimuli for initial imprinting. Young poults of the Turkey, *Meleagris gallopavo*, typically grow up in open forests with little undergrowth where visual contact with their mother is almost constant. Experimental studies show that turkey poults will follow and will form stable social bonds when exposed to visual imprinting stimuli alone. For chicks of the Jungle Fowl, *Gallus gallus*, which normally inhabit dense forests, auditory stimuli (clucking) alone are much more potent than visual stimuli alone in imprinting effectiveness (Graves, 1970b).

Because of their thick coats of feathers and poor sense of smell, birds depend very little if at all on tactile or olfactory stimuli for imprinting. Olfactory imprinting is well known in many mammals, e.g., in shrews of the genus *Crocidura*, whose young form "caravans" with their mother by attaching themselves together in a long chain, each young biting and holding on to the rump fur of either its mother or the littermate in front. Young six or seven days old instinctively attach themselves to each other or even to a rag, but from the eighth day on they will attach only to an object to whose odor they have become imprinted. Young shrews reared by nurses of another species become imprinted on the specific odor of the nurse and will participate only with her in a caravan, even ignoring their own mother (Zippelius, 1972).

Fish are known to imprint to odors. Juvenile Rainbow Trout, *Salmo gairdneri*, exposed to the synthetic chemical morpholine, will, as adults, preferentially spawn in a stream to which that chemical has been added (Scholz *et al.*, 1978).

The adaptive survival value of early imprinting of the young on the mother is easy to understand in species like ducks and geese. In these birds the mother gathers her brood of a dozen or so about her and leads them off to a pond on their first day—leads them to food, safety, and education. Within the first 32 hours, as the imprinting machinery runs down, the young are progressively reinforced in their learning to recognize and depend on their mother by her coloration, her characteristic movements, her warmth, quacking, and so on. They also soon learn to distinguish food from nonfood items and to recognize the general appearance of an ecologically suitable environment. By the second or third day of life, the imprinting mechanism not only is unnecessary, but could be a threat to survival should it persist, leading the ducklings into risky adventures with enemies or with "unsympathetic" ducks. Briefly, the adaptive significance of imprinting seems to be to gather the extremely impressionable and vulnerable young around a tutor who immediately begins to teach them the hard facts of survival in an unfriendly world.

Imprinting further enables young birds to learn quickly the characteristics of their own species. This information is of great importance later on, enabling a bird to select as its mate a bird of the same species, and thus to avoid the genetic wastefulness of hybridization. Normally a young bird first experiences the sights and sounds of its own parents and is therefore imprinted on its own species. But in some species imprinting can encompass a wide variety of determining stimuli, including humans or even farm machinery.

Bambridge (1962) showed that young male domestic chickens exposed to a colored ball during days 2 to 9 after hatching and injected daily with testosterone to stimulate precocious sexual development, will attempt to copulate with the ball when they are 19 days old. Control chicks given equal exposure to the ball from days 10 to 17 after hatching (after the critical sexual imprinting period) will ignore or avoid it. A genetic change in appearance does not necessitate a change in the imprinting mechanism (Sonnermann and Sjölander, 1977). A sexually imprintable species would thus have the capacity for rapid adaptation to changed appearances in conspecific animals (Immelmann, 1970).

In cross-fostering experiments in which the young of one species are reared by foster-parents of another species, young are commonly imprinted on their foster-parents and later, when mature, will seek out birds of the foster-species as mates. Eighty doves of four different species were all raised by species other than their own. At sexual maturity, every bird directed his courtship and sexual behavior toward the foster-species and never toward his own genetic species (Brosset, 1971). These imprinted birds also directed their aggressive behavior towards the genetic and not the foster-parent species. Cross-fostering experiments using different species of doves (Craig, 1908), Sparrows (*Passer* spp.) (Cheke, 1969) or grass parakeets (*Neophema* spp.) (Crosby, 1978), resulted in hybrid pairings which produced hybrid young.

The sensitive period for sexual imprinting in ducks comes considerably later than that for visual following. Ducks were kept with their own species from the time of hatching until they were 3 weeks old and then were placed with an imprinting object for the next 5 or 6 weeks. They later showed themselves to be sexually imprinted on that object, and that sexual imprinting lasted many years (Schutz, 1965). Schutz imprinted males, but not females, on ducks of various other species. Female Mallards were raised with females of a different strain by Kruijt *et al.* (1982), and they noted that these females usually chose males of the rearing strain, demonstrating sexual imprinting, but chose a male of the opposite strain if that male was more active. In contrast to Schutz's ducks, Harris (1970) cross-fostered eggs of Herring Gulls, *Larus argentatus*, and Lesser Black-backed Gulls, *L. fuscus*, and found that females, rather than males, were sexually imprinted and formed heterospecific pair bonds.

In an extensive series of experiments in which he cross-fostered finches that exhibited elaborate courtship behavior, Immelmann (1969; 1972a) observed the behavior or 68 male Zebra Finches, *Poephila guttata*, raised by Bengalese Finches, *Lonchura striata*, and 17 male Bengalese Finches raised by Zebra Finches. The males were then isolated until they became sexually mature. In free-choice situations, all males directed their social activities (mutual- or allopreening and contact) and sexual activities (courtship and copulation) toward females of the foster-parent species, and on numerous occasions ignored or even attacked females of their own species. Here again, however, the birds showed a hereditary bias toward their own kind. If a Zebra Finch male is reared by mixed parents (a Zebra Finch male and Bengalese female, or vice versa), it will, in a free-choice situation, nearly always prefer its own species, regardless of the sex of the conspecific parent.

The sensitive period for determination of sexual preferences in these finches lasts from about the 13th to the 40th days of life, and at its completion sexual imprinting is in most cases irreversible. Male Bengalese-imprinted Zebra Finches were forced to pair with female Zebra Finches in sound-proof chambers. Deprived of the opportunity to see and hear Bengalese finches, these male Zebra Finches eventually accepted the conspecific mates and raised young (Immelmann, 1972b). After nine months to seven years of such intraspecific experience, time intervals markedly exceeding the life expectancy of wild Zebra Finches, the experimental males were once again tested in free-choice situations. The males once again chose Bengalese females, demonstrating the *irreversibility* of the imprinting process.

Female Zebra Finches were cross-fostered under Bengalese Finches by Sonnermann and Sjölander (1977) and their sexual preferences tested in free-choice situations. These investigations quantified the amount of time experimental females spent in front of either male Bengalese or male Zebra Finches and

FIGURE 10–3 A, Interspecific sexual imprinting. The male Zebra Finch was raised by two Bengalese Finch parents and is courting a "stuffed" Bengalese Finch female in preference to a stuffed Zebra Finch female. The dummy does not call or move, indicating that the male Zebra Finch is reacting to the appearance of the foster species. B, Intraspecific sexual imprinting. The male grey (wild-type) Zebra Finch at right was raised by two mutant white Zebra Finches and has now paired with a white Zebra Finch female. This demonstrates imprinting on strain color (see text). Photos, courtesy of Klaus Immelmann.

also the number of greeting displays, copulation invitation displays (crouching and tail-quivering), and displacement activities directed at each male. Control Zebra females raised by conspecific parents preferred Zebra Finch males by all these criteria. Experimental females showed the same amount of interest in both species, demonstrating that cross-fostering does influence female sexual preferences.

In contrast to male Zebra Finches, which demonstrate clear-cut preferences, cross-fostered females were attracted to both conspecific and allospecific males. This is probably because females experience a conflict between genetic preferences for their own species and imprinting preferences (Sonnermann and Sjölander, 1977).

The influence of siblings on sexual imprinting has also been demonstrated by cross-fostering Zebra Finches with one, two — up to four siblings, under Bengalese Finches (Kruist *et al.*, 1983). All experimental birds cross-fostered alone, or with one sibling, sang almost exclusively to females of the foster species in free-choice experiments. Sixty-two percent of the experimental birds cross-fostered with two to four siblings sang exclusively or almost exclusively to the foster species. The rest of the cross-fostered birds sang to both Bengalese and Zebra Finches; 17.5 per cent of the birds actually sang more to Zebra Finches.

By switching eggs of pigmented (wild-colored) and white Mallards, Klint (1978) demonstrated that ducklings raised with white families later chose white mates and vice-versa, thus demonstrating the influence of siblings on imprinting in a precocial species.

The sexual sensitive period differs among species. In the Mourning Dove, *Zenaida macroura*, it extends from ages of about 7 to 52 days; Grey Lag Goose, *Anser anser*, 50 to 140 days; and the Bullfinch, *Pyrrhula pyrrhula*, up to as much as two years. Fabricius (1964) defined the sensitive period as the period of maximum sensitivity. Noting that one of the most important attributes of imprinting is the fact that social attachments can be determined only during specific stages in an animal's life cycle, and that the same experience may lead to quite different behavioral results at different ages, Immelmann (1972a) defined the sensitive period more precisely as "the *whole* space of time in the individual's life during which a particular preference can be determined or altered, whereas outside its limits social experience does *not* exert a similar influence." Immelmann concluded that sexual imprinting normally enables birds to recognize and restrict their sexual behavior to birds of their own species and thus helps to maintain sexual isolation between species. A paradoxical "perversion" of sexual imprinting is employed by certain African brood-parasites in duping their hosts. This subject is discussed in Chapter 16.

Ethologists also distinguish between interspecific sexual imprinting (imprinting on alien species), and intraspecific sexual imprinting (imprinting on one's own species). Intraspecific imprinting was demonstrated by Immelman *et al.* (1978) by cross-fostering grey and white morphs of domesticated Zebra Finches. Experimentals of both sexes raised by white parents chose white mates, and individuals raised by grey parents later paired with grey consorts. Wild lesser Snow Geese, *Chen caerulescens*, occur as either white or blue phase morphs. Cooke (1978) and his associates found that most birds with white parents chose white mates and those with blue parents chose blue mates. Only 10 per cent chose mates opposite to the familial color.

Some of the examples of imprinting given above suggest that the basic element in the imprinting mechanism is innate. In only three inbred generations, Graves and Siegel (1966) selected out two strains of domestic chicks that differed significantly in the promptness with which they followed a moving object. Similarly, 18 generations of genetic selection in Japanese Quail, *Coturnix coturnix*, by Kovach (1983) produced chicks with nearly perfect imprinting for early approach to red rather than blue objects, and vice versa, as well as chicks with preferences for grated versus dotted visual patterns.

Several scientists have investigated the neural basis of imprinting by making lesions in a distinct area of the forebrain (encompassing the intermediate and medial hyperstriatum ventrale, or IMHV). Bilateral lesions of this area resulted in loss of filial imprinting; unilateral destruction did not interfere with the process. Thus, at least one IMHV is necessary for both actuation and retention of filial imprinting (McCabe *et al.*, 1981; Horn *et al.*, 1983). Additionally, increased metabolic activity in this brain region was detected as measured by incorporation of labelled amino acids (RNA precursors) during the sensitive phase of filial imprinting (Bateson *et al.*, 1969).

Using the technique of radioimmunoassaying, Pröve (1985) noted increased levels of plasma testosterone and progesterone associated with the beginning and end of sexual imprinting in the Zebra Finch. Nestling Zebra Finches were cross-fostered under Bengalese finches and then implanted with either testosterone or inert material (as a control) between 10 and 35 days of age. Free-choice experiments were conducted when the birds were 100 days old. Testosterone-treated males courted females of the foster species exclusively, whereas control males showed either a preference for the foster species or courted both species with equal fervor. Pröve concluded that although more research is in order, testosterone appears to intensify sexual imprinting.

BIRD AGGREGATIONS

"Birds of a feather flock together," and sometimes flock with other kinds of birds as well. It is a rare bird that leads a solitary life outside the breeding season. Woodpeckers and both diurnal and nocturnal birds of prey are usually examples of this minority group. On the other hand, very few species unite into permanent year-round flocks; weavers are an example. The great majority of birds come together at different times of year, for different reasons, to form a great variety of homo- and heterospecific aggregations, large or small, and tightly or loosely organized. Certain solitary species that show pronounced intolerance of other individuals during the breeding season may form flocks after their gonads decline in activity. Many other species, including most sea birds, come together in colonies for the breeding season.

Usually these colonies are formed of one species of bird, but sometimes they are of mixed groups. For example, mixed breeding colonies on Danish islands were reported by Salomonsen (1947) to contain a nucleus of Black-headed Gulls, *Larus ridibundus*, with about one-half as many Common Gulls, *L. canus*, one-third as many Sandwich Terns, *Sterna sandvicensis*, plus a scattering of Kittiwakes, *Rissa tridactyla*, and Herring Gulls, *L. argentatus*.

Some species of birds that breed in isolated pairs migrate together in flocks—many ducks and geese, for example. Chimney Swifts, *Chaetura pelagica*, spend the night, during migration, clinging to the inside walls of large chimneys, often several thousands of birds in one chimney. Common Goldeneyes, *Bucephala clangula*, form compact, floating rafts of hundreds of birds to spend the winter night on a lake or river. Swallows, finches, crows, and many other species often spend the night roosting in enormous aggregations. During the irruptive invasion of Switzerland by Bramblings, *Fringilla montifringilla*, in the winter of 1950–1951, an estimated 72 million birds roosted in two small (6.8 hectare) pine woods every night (Mühlthalen, 1952).

In cold weather, European tits, creepers, wrens, kinglets, and other small species sometimes gather in compact homospecific masses and spend the night sleeping in a feathery ball. As many as 50 Longtailed Tits, *Aegithalos caudatus*, were observed in one such slumber party (Tucker, 1943). In particularly cold winter weather, over 60 European Wrens, *Troglodytes troglodytes*, were observed nightly, for two weeks, crowding into a nest box whose inside dimensions were about $11.5 \times 14 \times 14.5$ cm (Flower, 1969)! In southern North America the Inca Dove, *Scardafella inca*, in cold weather will sometimes conserve heat by roosting in layers two and three birds deep (Johnson, 1972). Evidence that such nocturnal clumping represents adaptive heat-conserving behavior is presented by Smith (1972), who observed a flock of Common Bushtits, *Psaltriparus minimus*, roosting nightly in a hawthorn tree. In ordinary winter weather the birds roosted on branches evenly spaced with no two birds being closer than 5 cm from each other. But on ten exceptionally cold nights the birds roosted in a row, each bird in snug contact with its neighbor. Starlings, *Sturnus vulgaris*, roosting in wintertime pine woods, are estimated to lower their existence metabolism between 12 and 38 per cent through reduced convective heat loss (Kelty *et al.*, 1977).

That low temperatures are not the only stimulus for such aggregations is suggested by the existence of other sleeping parties among tropical birds. Skutch (1944) observed 16 Prongbilled Barbets, *Semnornis frantzii*, pack into one small tree cavity to spend the night; and Pycraft (1910) writes of wood-swallows (Artamidae) huddling together on tree branches, and colies (Coliidae) hanging upside down on branches in small, compact clusters during the night.

Roosts may also be heterospecific. In both winter and summer, mixed flocks of grackles, *Quiscalus* spp., Brown-headed Cowbirds, *Molothrus ater*, Robins, *Turdus migratorius*, Starlings, *Sturnus vulgaris*, and other species concentrate to spend the night. In one deciduous grove of about 2 hectares near Lexington, Kentucky, a winter roost of starlings, grackles, and cowbirds was estimated to contain between one-half and several million birds. Their numbers caused many branches to break and their guano accumulated 10 cm deep on the ground in some places (Loefer and Patten, 1941). In the Mississippi valley, winter roosts of blackbirds commonly contain a million, and some have been estimated to contain as many as 15 million, birds.

Different birds show great variation in the times at which they go to roost in the evening and leave in the morning. Light intensity is probably the one factor most universally influential in determining these times, but observations have implicated such other factors as circadian rhythms, length of day, season of year, stage of reproductive cycle, hunger, ambient temperature, and type of habitat.

Birds of different species also aggregate to form feeding flocks. In Ghana woodlands, Greig-Smith (1978) found that 34 flocks contained 56 species from 20 families, "including insectivorous, granivorous, and nectarivorous species using a wide range of foraging methods."

Since these and other varieties of bird groups are so common, it is natural to speculate on what survival advantage there may be in bird societies. Without doubt, some birds congregate in breeding colonies simply because suitable nesting spots (e.g., isolated islands) are hard to find. Different species may be thrown together simply because they have similar ecological or other preferences. Some small bird groups may be no more than a family that remains together, as in Canada Geese, *Branta canadensis*. But in many flocks or colonies there are demonstrable advantages (and occasional disadvantages) in group living.

In many colonial species there is strength and security in sheer numbers. Large, closely packed colonies of gulls, for example, are less subject to predation by crows than small colonies. The more eyes and ears there are spread over a given area, the more likely it is that a predator will be detected. Particularly in mixed flocks, the differential alertness of different species, plus possible birds with specialized antipredator behavior, will enhance vigilance for the entire flock (MacDonald, 1977). Cordon Bleus, *Uraeginthus bengalus*,

FIGURE 10–4 Starlings assembling before going to roost for the night. Photo by E. Hosking.

foraging alone spend more time "looking up" and "scanning" than those living in flocks, and single birds also react sooner to approaching danger (Hamed and Evans, 1984). Amount of scanning also decreases as flock size increases in Eurasian Curlews, *Numenius arquata*, Dark-eyed Juncos, *Junco phaeonotus*, House Sparrows, *Passer domestica*, and Brant Geese, *Branta bernicla*, (Abramson, 1979; Caraco, 1982; Studd *et al.*, 1983; Inglis and Lazarus, 1982). In mixed flocks of tits, creepers, nuthatches, and other small passerines, Herrera (1979) estimated that foraging success was roughly twice that of the birds foraging solitarily. This increase in feeding efficiency may arise from the stirring up of insects and the like by mere numbers, or it may depend in part on associative learning by one bird or species from another. As an additional benefit, flocks or colonies of birds may serve as centers for giving out information as to sources of food, especially food distributed in patches. Further group benefits are possible in heat conservation, in the stimulation and synchronizing of breeding, in care of young, and particularly in education, both of adults and of young.

COLLECTIVE BEHAVIOR

Birds vary in the degree to which they practice social behavior. Hediger (1964) recognized two categories of animals which he called "distance" and "contact" species. At one extreme some species will permit the approach of a conspecific to a defined "individual distance." By moving food hoppers progressively closer Marler (1956) found, for instance, that nonbreeding male Chaffinches, *Fringilla coelebs*, would permit another male to approach up to 20 cm before aggressive displays are elicited. Females are permitted to approach to half that distance before aggression occurs. At the other extreme are many species of estrildid finches and parrots which invite bodily contact of others of its kind. Australian wood swallows (*Artamus* spp.) feed together, bathe together, roost in tight clusters, jointly attack flying predators, breed in colonies, preen each other, put on aerial "screaming parties," and feed each others' nestlings (Immelmann, 1966).

Even within a single species pronounced differences may occur. In Costa Rica the Yellow-faced Grassquit, *Tiaris olivacea*, is discontinuously distributed and highly social in its behavior, whereas in Jamaica, it

is continuously distributed and aggressively territorial (Pulliam *et al.*, 1972). Genetic differences are thought to be involved.

Sooner or later in its lifetime, almost every bird will display aggressive, hostile, or agonistic behavior. Aggression is defined by Marler (1964) "as a means of enabling animals to obtain and keep those commodities needed for survival and reproduction"—for example, food, territory, mates, nest sites. Aggression rarely reaches the stage of overt attack but is more commonly expressed in ritualized postures, movements, or calls ("bluff") that serve to repel, intimidate, or appease enemies or competitors without the biological costs of actual combat. Aggressive behavior may result from excitation due to the bird's internal physiological state, especially the level of its male hormones, the surrounding external circumstances (including the proximity of another bird), and past experiences.

When two domestic fowl that had lived together for about six weeks were deprived of food for 24 hours and then frustrated by the presentation of food under a transparent, plastic cover, there was a more than 18-fold increase in aggressive acts between them: threats, pecks, grips of combs or hackles, and chases (Duncan and Wood-Gush, 1971).

Many of the stereotyped actions in aggressive or agonistic behavior are ambivalent in origin and commonly represent a compromise between attack and escape behavior. If two gulls own adjoining territories, one gull will attack the other when it invades the first gull's territory; contrariwise, an invader will retreat under the territory owner's attack. At the boundary between their territories, the gulls cannot attack and retreat simultaneously; instead they engage in a "fierce displacement activity of grass-pulling" (Tinbergen, 1960).

Allopreening (mutual preening) is a widespread, ritualized form of agonistic behavior that bridges the narrow gap between sexual behavior and aggressive attack (Harrison, 1965). This behavior may be found in a wide variety of birds, from tiny estrildid finches to enormous macaws and birds of prey.

Even the most casual observation shows that much bird behavior is socially conditioned; that is, one bird's behavior depends in part on how other birds behave. In colonies of gulls and terns, one alarm cry may set off a contagious panic of cries and upflights by thousands of birds. Perhaps by the same nervous mechanism that makes yawning infectious in man, great waves of instinctive behavior may suddenly sweep a gull colony, and just as suddenly fade away. As one would suspect, there are species differences in these epidemics of mass conformity. In a colony off the coast of Scotland that contained two species (Lesser Black-backed, *Larus fuscus*, and Herring Gulls, *L. argentatus*), Richter (1939) noted that courtship display and copulation spread infectiously only among members of the same species, but that alarm, place-changing, preening, and sleeping spread from one species to the other. Infectious behavior of this sort is by no means restricted to colonial species. When one of Nice's (1939) hand-raised Song Sparrows, *Melospiza melodia*, bathed, ate, preened, or flew to her desk, others were likely to do the same.

A particularly intense form of collective behavior is exhibited in *mobbing*. This ordinarily involves the noisy, massed attack of a predator by a number of smaller birds. Crows and jays, for example, often attack hawks or owls. Experiments by Barash (1976) revealed that a plastic replica of an owl would be attacked by an average of six mobbing Crows, *Corvus brachyrhynchos*, uttering about 18 decibels of "caws"; but if the owl was accompanied by a plastic Crow lying on its side near by (or even a black cloth), the number of crows and the decibels of cawing protest more than doubled. This emphasizes the significance of mobbing as an adaptive response to predation. Probably mobbing is basically innate, but evidence shows that some of it is learned and is even used to transmit information on the identity of predators from generation to generation as a cultural tradition.

In ingenious experiments by Curio and his colleagues (1978), a "tutor" European Blackbird, *Turdus merula*, mobbed a stuffed Little Owl, *Athene noctua*, in an aviary. Barriers were so arranged that an "observer" Blackbird could see the tutor but not the Owl. Instead it saw a novel stuffed Noisy Friarbird, *Philemon corniculatus*, which appeared to be the object of the tutor's mobbing. Observer Blackbirds soon learned to mob the harmless Friarbird during subsequent presentations of the novel bird alone. They even learned to mob a multicolored plastic bottle under similar conditions, but with significantly less vigor than in their response to the Friarbird. Most significantly, observer Blackbirds could become tutors and pass on the information they had received to new observers who, in turn, became knowledgeable tutors for the next observers. In this fashion, cultural transmission of enemy recognition "was passed on in chain-like fashion and without decrement over at least six individuals." Curio believes that such cultural transmission of information occurs in the wild as well. While this is perhaps the chief function of mobbing, other explanations include: alerting others to the presence of a predator; silencing young offspring, confusing the predator through the numbers and erratic flight movements of prey animals; and causing the predator to move on.

Often the concerted action of a group of birds is not merely the simultaneous release of a fixed action pattern, but a continuing performance that requires a constant adjustment of the individual's behavior to that of the group. The group flights of pigeons or shore birds with their split-second coordination and breath-taking precision furnish a common example. Sometimes the concerted behavior seems to make little sense to humans, but in other instances the survival value is apparent. A straggling flock of Coots, *Fulica atra*, paddling on the water, will close ranks and jam together at the approach of an attacking Bald Eagle, *Haliaeetus leucocephalus*. The eagle is able to pick off only isolated stragglers, and the coots instinctively behave as though they knew this (Munro, 1938). In the same manner, starlings close ranks in the air when attacked by a falcon. The group fishing formations of long lines of cormorants and pelicans probably have some of the efficiency of a long seine in catching fish. In the simultaneous diving of many water birds to feed on fish, the advantage may lie in the fact that in escaping one bird a fish plunges into the open beak of another.

SOCIAL FACILITATION

When a hungry domestic hen, *Gallus gallus*, is allowed to feed at a pile of grain, it will eat until satisfied and then stop. If another hungry hen is then introduced, the satisfied hen will resume eating and consume an average 34 per cent more grain; or if three hungry hens enter the scene, she will stuff herself with an average 53 per cent more grain (Bayer, 1929). This is an example of *social facilitation*: the way in which the behavior of one animal may be enhanced by that of its associates. As an indication of the deep-rooted phyletic nature of social facilitation, fishes of various species will also eat more in a small group than they will when isolated from each other; and Goldfishes, *Carassius auratus*, will learn to run a simple maze more quickly in small groups than they will by themselves (Welty, 1934).

In social feeding experiments, Tolman and Wilson (1965) examined the effects of young chicks on each other. A chick deprived of food for six hours and allowed to eat by itself made about 40 food pecks per 5-minute period. Such a chick, when paired with a nondeprived chick, made about 115 pecks every 5 minutes. If paired with a similarly deprived chick (6 hours without food), the chick responded with over 150 pecks per 5 minutes, and when paired with a 24-hour deprived chick, its pecking rate rose to about 260 pecks per 5 minutes! Pecking rate of chicks also rose with increased rate of pecking movements by an artificial hen and nearly doubled with the addition of pecking sounds to the model's movements (Tolman, 1967).

Spontaneous gaping, which appears in young starlings when they are only an hour or two old, disappears after 4 or 5 days in isolated birds, but it will persist up to 11 days in groups of young when mutual stimulation occurs (Holzapfel, 1939). Mutual stimulation varies at times according to the size of the group. Williamson (1949) found that Parasitic Jaegers, *Stercorarius parasiticus*, living in a colony, will often fly at and strike human intruders with their wings, but pairs living alone will not; likewise, Arctic Terns attack intruders more readily in large than in small colonies.

The advantages of synchronized breeding in bird colonies have already been mentioned. Coulson (1968) determined that in colonies of Kittiwakes, *Rissa tridactyla*, birds in the middle of a colony nested earlier, laid and hatched more eggs, fledged more young, and changed mates less often that those that nested on the edges of the colony. Patterson (1965) arrived at essentially the same conclusions from his observations on Blackheaded Gulls, *Larus ridibundus*. Social facilitation in such cases often seems to depend on a numerical threshold. According to Fisher (1952), Fulmars, *Fulmaris glacialis*, not only have less reproductive success in small colonies than in large, but are unlikely to breed at all if their colony contains less than 8 to 12 pairs. This is often referred to as the "Frazer Darling effect," after Frank Frazer Darling (1938), who suggested that the milieu of visual and vocal displays of colonial breeders will tend to accelerate and synchronize the reproductive cycle of the colony. Studies by the MacRoberts (1972), however, failed to demonstrate any intraspecific social facilitation of reproduction in colonies of Herring Gulls, *L. argentatus*, or Lesser Black-backed Gulls, *L. fuscus*.

Observational learning by Blue Jays in eating butterflies was described in the previous chapter. This form of social facilitation has also been demonstrated by Alcock (1969) in the Fork-tailed Flycatcher, *Muscivora tyrannus*. It is evident also in the Muscovy Duck, *Cairina moschata*, which learned to avoid an electrically wired food dish by observing the reactions of two other ducks trained at the apparatus (Klopfer, 1957). Similar "instruction" undoubtedly occurs between parents and young birds under natural conditions.

Whether true imitation enters into these situations is a moot point. Whereas social facilitation is a contagious form of behavior in which the performance of a more or less instinctive pattern of behavior by one animal will act as a releaser for the same behavior in another, imitation is the copying of a novel or otherwise improbable action for which there is no apparent instinctive tendency (Thorpe, 1963). A parrot or mynah reciting human speech is definitely demonstrating imitation. But in some actions by birds it is difficult to distinguish social facilitation from true imitation. At a Louisiana hunting club, free-living decoys and tame crippled geese were called to their evening feeding with beating on a tin pan. Numbers of wild geese often accompanied the tame ones on such occasions, even feeding from the caretaker's hand, but at other times they were unapproachable (Hanson and Smith, 1950). Facilitation (or imitation) of behavior can also work between species. In mixed colonies of Black-headed Gulls, *Larus ridibundus*, and Common Terns, *Sterna hirundo*, the young gulls, which swim a great deal, facilitate swimming in the young terns. Young terns swim much less in colonies that lack gulls (Frederikson, 1940).

It seems probable that social facilitation can work against the interests of birds and become instead what might be called social hindrance. As Huckleberry Finn once remarked, "A body is always doing what he sees somebody else doing, though there mayn't be no sense in it" (Mark Twain, 1935). An adult Greenfinch, *Carduelis chloris*, that had been trained to discriminate between two patterns, was then allowed to observe an untrained bird making errors during its preliminary trials, upon which the original bird began to make errors itself (Klopfer, 1962). A hungry chicken will not eat for a while if placed in a group of satiated birds (Lorenz, 1935). Common Terns and Roseate Spoonbills, *Ajaia ajaja*, sometimes desert their nests in a body as the result of a minor disturbance. Rooks, *Corvus frugilegus*, habitually pilfer sticks from each other's nests, usually slyly and with restraint, but at times contagious waves of wholesale thievery spread through rookeries and cause serious nest destruction (Ogilvie, 1951). In a maze-learning problem with Budgerigars, *Melopsittacus undulatus*, Allee and Masure (1936) found that paired birds learned less readily than single birds.

Patterns of socialized behavior probably become established in the individual bird through initial instinctive reactions plus various types of learning that are mainly shaped and strengthened by experiences with parents and brood mates early in life. In his experiments with domestic chicks only 2 or 3 hours old, Collias (1952) found that pairs of them "were very slow to approach one another or failed to come together, even when only five or six inches apart, until after they had experienced some minutes of bodily contact, following which the members of each pair rapidly came together when separated by short distances." To a cold and hungry chick a broody hen represents a "complex of attractive stimuli including warmth, contact, clucking and movement; and repeated exposure to these stimuli, as well as the food-guidance and protection that a hen gives her chicks, helps to strengthen the family bond" and the chick's social behavior patterns.

SOCIAL ORGANIZATION

Anyone who has raised domestic chickens has very likely wondered at the frequent bullying that goes on, and even more at the refusal of some of the worst-treated hens to fight back. Baby chicks live in pens together quite amicably, but after a few weeks begin to peck and jump at one another. With increased fighting, certain individuals come out on top, and other "hen-pecked" individuals consistently give way. By the time the chicks are about seven weeks old, the flock has become organized into a social hierarchy of dominance and submission called a *peck order*. In Curlews, *Numenius arquata*, this order is achieved only three weeks after hatching. The bird at the top, who wins the greatest number of contests, is called the Alpha bird or despot, and the one at the bottom of the social ladder, the Omega bird. Alpha has earned the right to peck all other birds in the flock; Beta may peck all but Alpha; Gamma may peck all but Alpha and Beta; and so on down to poor Omega, who is able to peck no one with impunity. Flocks of poultry are not always arranged in such a linear order, but may be organized in triangles (A pecks B pecks C pecks A) or polygons. Once a high-ranking

hen achieves her social status she rarely has to fight to maintain it; the merest threatening gesture will cause a subordinate hen to make way for her highness. Both the threatening gesture and the reaction of submission seem to have become symbolic formalities in many species. A submissive Jackdaw, *Corvus monedula*, for example, bends his head low and turns the nape of his neck toward his superior when they meet. He knows his place!

Social hierarchies of this sort are commonly found where animals are thrown into close contact with one another, such as in artificial pens or in natural colonies. They occur in a great variety of birds, in mammals (including man), reptiles, and even fish. In many birds, social hierarchies occur during the breeding season only; but in Jackdaws and certain jays, which live in permanent flocks, the peck order persists throughout the year. The conspicuously social finches of the family Estrildidae apparently thrive without any peck order whatever (Güttinger, 1970). In some species, as in flocks of titmice, doves, or canaries, the peck order is rather loosely organized, but in chickens and jackdaws it is rigorously constructed.

As to rigidity, there seem to be two sorts of peck orders. If, after one or a few belligerent encounters, one bird establishes its superiority over another, and the submissive bird accepts its status from then on, the dominant bird is said to possess a *peck right* over the subordinate bird. If, however, the two birds continue to jockey for dominance, the bird that wins the majority of encounters is said to have *peck dominance* over the other (Allee, 1936).

High status in the peck order carries such rewards as priority at the feeding trough, watering tank, and dust bath, the warmest spot on the roost, and fewest abrasions from combat. Because dominate Rooks, *Corvus frugilegus*, monopolized the best feeding grounds and the warmest positions in winter roosts, the younger, submissive Rooks suffered a higher mortality rate (Swingland, 1977). Among Silvereyes, *Zosterops lateralis*, those individuals that won two-thirds or more of aggressive encounters with other flock members, also tended to survive better over the winter (Kikkawa, 1980). The advantages to the species of flock organization derive mainly from the fact that it prevents incessant fighting and allows time for more constructive pursuits. Two flocks of hens were compared by Guhl (1956, 1968) to see what effect social disorganization had on productivity. One flock was allowed to maintain a stable peck order; the other was disrupted by frequently shifting its members. The birds in the disorganized flock "fought more, ate less food, gained less weight and suffered more wounds."

In a subsequent study of social stability in domestic fowl, Guhl (1968) established two kinds of flocks of female White Leghorns, six birds in each flock. One "constant membership" flock was allowed to remain unchanged for ten weeks; in the other "rotating membership" flock, which had constant members for three weeks, an old bird was removed and a new one, from isolation, was introduced every other day beginning with the fourth week. Every day, when a small bowl of food was placed in each pen, the amount of competitive, aggressive pecking that involved every bird was recorded during a 15-minute period of observation.

During weeks 4 to 10, the stable group pecked at the mean rate of 14 pecks per 15 minutes, and the unstable group, 71.7 pecks. Figure 10–5 shows the mean weekly levels of dominance and subservience for individual birds in each flock: each rank in the stable flock represents a specific individual, while each rank in the rotated flock represents different individuals. A positive number of pecks indicates that a given individual pecked others more than it was pecked, and a negative rank indicates the opposite. The results show clearly the role of social instability on aggressive behavior.

The role of competition in social conflict was ingeniously studied by King (1965) through controlling access to food. Three flocks of White Leghorn roosters were raised from hatching and kept in pens for several months until stable social hierarchies were established. The birds were then fed for only 1 hour each day. Feeding conditions were changed daily and randomly as follows: (1) food was distributed evenly over the floor of the pen so that birds had unrestricted access to it, (2) food was placed in a circular hopper so that all birds could feed at once, but were crowded, (3) food was available at a point source (hole) so that only one bird could feed at a time. Table 10–1 gives the experimental results. In all flocks peck frequencies increased as food accessibility became more restricted. In two

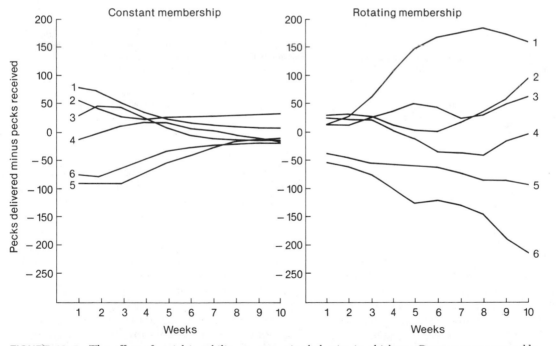

FIGURE 10–5 The effect of social instability on aggressive behavior in chickens. Curves represent weekly total interactions in the form of pecks delivered minus pecks received according to social ranks in one flock having a constant membership and another flock having a regularly rotating membership. Figure redrawn after Guhl, 1968.

TABLE 10–1.
PECK ORDER AMONG CAPTIVE *ZONOTRICHIA LEUCOPHRYS GAMBELII* AS DETERMINED BY AGGRESSIVE ENCOUNTERS*

	LOSER							
Winner	A1 M	A2 M	A3 M	J1 M	J2 M	J3 F	A4 F	J4 F
A1	—	23	10	16	15	13	6	9
A2	1	—	15	20	36	13	2	12
A3	3	4	—	19	19	9	17	4
J1		1	1	—	29	19	17	15
J2		1	4	1	—	15	5	6
J3		1	1			—	14	11
A4	1	24			1	1	—	5
J4								—
A1	—	18	18	20	72	31	9	12
A2		—	33	15	64	33		14
A3			—	24	38	10	12	12
J1				—	109	46	47	20
J2					—	30	23	17
J3						—	15	14
A4		38					—	18
J4								—

*A = adult; J = juvenile; M = male, F = female. Birds are listed vertically and horizontally in order of dominance rank. Top ranking bird located at the top and left, lowest ranking bird at the bottom and right. Numbers in table are actual counts of wins. Gaps in the table indicate no encounters observed. Each aggressive encounter is tallied to the right of the winner (found in the vertical "winners" column) and below the loser (found in the horizontal "losers" column). The term "stabilization" refers to a sequence of observations when dominance reversals (numbers below the diagonal) are rare or have ceased, usually 2–4 weeks following introduction of unfamiliar birds. Upper part of the table shows the hierarchy before stablization (data from 12 to 24 February are pooled). Lower part of the table shows the stablized peck order (26 February–17 March). Note the dominance triangle involving A2 and A4, i.e., A4 scored 38 wins over A2 but lost to all other birds. From Parsons and Baptista, 1980.

flocks peck orders remained linear and undisturbed until food was restricted to a point source, at which stage drastic but temporary disruptions of peck order occurred.

The effect of prolonged social conflict on endocrine glands was investigated by Siegel and Siegel (1961), who found that a rooster continually exposed to the social stress of adjusting his behavior to strange birds had significantly heavier adrenal glands than comparable roosters kept in individual cages. Plasma levels of corticosterone, a hormone released during stressful situations, was found to be higher in low-ranking Harris Sparrows, *Zonotrichia querula*, than in dominant individuals in the flock (Rohwer and Wingfield, 1981).

Attainment of status in a flock does not always depend on strength or belligerence, although they are commonly important factors. In European Jays, *Garrulus glandarius*, the dominant or submissive status is assumed without physical combat and seems to depend on psychological factors. In some species, territory seems to be an important ingredient in attaining dominance. Seneca perceived this nearly twenty centuries ago when he remarked, "The cock is at his best on his own dunghill." Strange Canaries, *Serinus canarius*, introduced into a cage of resident birds, usually take subordinate positions (Shoemaker, 1939); and low-ranking pigeons allowed to remain in a given area will dominate a superior pigeon when it first enters the area (Noble, 1939).

The social hierarchy of 23 natural flocks of Black-capped Chickadees, *Parus atricapillus*, was studied by Glase (1973), who found that within each flock, males dominated females, and within sexes, adults dominated juveniles. Birds attracted to seed bait in California formed a hierarchy with larger species dominating the smaller. Gambel's Quail, *Lophortyx gambelii*, lorded over Scrub Jays, *Aphelocoma coerulescens*; Cactus Wrens, *Campylorhynchus brunneicapillus*, dominated House Finches, *Carpodacus mexicanus*, which dominated Black-throated Sparrows, *Amphispiza bilineata* (Fisler, 1977).

Once a peck order is established in a flock, it can be maintained only if its members recognize and remember the various individuals in the flock. One domestic hen, for example, was able to distinguish 27 others in four different flocks. In order to check the clues a bird uses to recognize other individuals in a flock, Schjelderup-Ebbe (1935) altered the appearances of the heads of hens in a stable flock by placing bonnets over combs, or dyeing their combs or feathers. A transformed hen was treated as a stranger in the flock, and had to fight to reestablish her position in the peck order. In two large flocks of young hens, it was found that the surgical removal of combs and wattles consistently lowered a bird's rank in the social organization (Marks *et al.*, 1960). Apparently head furnishings on a bird can become symbols of status just as does gold braid on a military uniform. In the Marabou Stork, *Leptoptilos crumeniferus*, inflation of the ventral air pouch indicates dominance in an individual, while inflation of the dorsal pouch indicates "apprehension" (Akester *et al.*, 1973). An amusing instance showing how superficial appearance can affect social rank was reported by Kneutgen (1970), who observed that a deformed Chiff-chaff, *Phylloscopus collybita*, whose wings were permanently spread and lowered and the tail cocked upright, as in aggressive display, inadvertently became dominant over its three cagemates. Some birds have eminence thrust upon them.

The recognition of individuals in a flock depends on experience. The male European Robin, *Erithacus rubecula*, is able to recognize his mate at least 27 meters away (Lack, 1943); and Lorenz (1952) tells of Jackdaws, *Corvus monedula*, in a colony, immediately recognizing a member that had been away for seven months. Black-and-white crowned adults dominated brown-crowned juveniles in winter flocks of White-crowned Sparrows, *Zonotrichia leucophrys*, (Parsons and Baptista, 1980). Plucking the brown crown feathers in juveniles induced premature development of black-and-white crowns, giving these birds the appearance of adults. Such juveniles gained high rank when introduced into a flock of unknown juveniles or a new flock consisting of both age classes. Thus

age alone does not ensure high rank. Low-ranking juveniles isolated, disguised by plucking, and then reintroduced into their original flock did not move up in hierarchy, suggesting that individual recognition takes precedence over the status symbol (the bright crown); their flock mates are not fooled by the disguise. Individual birds may be recognized not only by their appearance but also by their calls (review by Falls, 1982). Tinbergen (1939) noticed that Herring Gulls, *Larus argentatus*, reacted differently to the alarm calls of different members in a colony; when bird A called they merely became attentive, but when B called they flew off in alarm. Apparently bird A was psychologically related to the boy who called "Wolf, wolf!" when there was no wolf.

Because of the difficulties of long-continued observations in the field, the stability of peck order in wild birds has received little study. In a flock of domestic hens the despot may retain her dominant position for life, and in a flock of hand-reared Herring Gulls, a straight-line dominance hierarchy was established when the young were 23 days old, remaining unchanged for 18 months (Goethe, 1953).

Sex plays a role in social hierarchy among many species. Roosters are normally dominant over hens. The male Chaffinch, *Fringilla coelebs*, is dominant over the female in winter, but during the nesting season the dominance is reversed. In a captive winter flock, females with under parts dyed red, in imitation of the male, dominated normal females (Marler, 1955). Males of many cardueline finch species tend to be dominant to females in the winter, but females become dominant to males in the spring (Balph, 1977; Smith, 1980). The African Red-billed Quelea, *Quelea quelea*, shows the same seasonal reversal in dominance.

The psychic subtleties of peck order are further revealed by realignments when birds of different status mate. In his semiwild flock of jackdaws, Lorenz (1952) noted that in the union of a high-ranking male and a low-ranking female, the wife took on the status of her husband.

The extraordinary part of the business is not the promotion as such but the amazing speed with which the news spreads that such a little jackdaw lady, who hitherto had been maltreated by eighty per cent of the colony, is, from today, the 'wife of the president' and may no longer receive so much as a black look from any other jackdaw.

In somewhat reverse fashion, a male Steller's Jay, *Cyanocitta stelleri*, ranking in the lower third of his flock's hierarchy, gradually rose to become the despot after mating with the dominant female (Brown, 1963).

Social dominance is also related to breeding success. In a flock of domestic chickens with several roosters, the dominant rooster normally is most successful in mating with hens, and the Omega rooster the least. In one flock, Guhl (1956) found that the rooster lowest in the peck order was completely suppressed sexually and failed to mate with hens he knew even when the other roosters were removed from the flock; he was "psychologically castrated." Hens high in social rank, on the contrary, are less likely to submit to coition than those low in rank, and in adoption experiments are less willing to accept strange chicks than the low-ranking hens (Ramsay, 1953).

In the winter, penned Mourning Doves, *Zenaida macroura*, establish social rank independent of sex. Dominant males pair with dominant females, and such pairs were found by Goforth and Baskett (1971) to be the first birds to pair and to select nesting territories. They were also more successful in raising young than were pairs of lower rank.

The influence of hormones on social hierarchy differs between species. Allee and Collias (1938) injected testosterone propionate in hens of low social standing. The injections slowed down egg-laying and stimulated comb growth and crowing. The treated hens successfully revolted against their superiors and achieved top status, which, once won, was retained even though injections ceased and the other secondary effects of the treatment disappeared. Hens whose ovaries have been removed lose their aggressive behavior and decline in social status. Male Chaffinches, *Fringilla coelebs*, Japanese Quail, *Coturnix coturnix*, and numerous other species respond to injections of male hormones (testosterone) with increased aggressiveness and a rise in social rank (Baptista *et al.*, 1987). The injection of male hormones into female canaries, male and female doves, and even female swordfish, raised their social rank. However, neither castration nor injections of massive doses of testosterone affected the social position of Starlings, *Sturnus vulgaris* (Davis, 1957). Nor did administration of testosterone affect the social rank of Red-billed Queleas, *Quelea quelea*, but the experimental administration of luteinizing hormone raised the social status of both starlings (Davis, 1963) and Red-billed Queleas (Crook and Butterfield, 1968). There was no correlation between either plasma levels of luteinizing hormone or testosterone and social rank in Harris Sparrows, *Zonotrichia querula* (Rohwer and Wingfield, 1981).

The fact that some breeds of domestic fowl not only are routinely dominant over other breeds but are accepted as dominant without challenge by individuals of the subordinate breed (Hale, 1956–1957) suggests strongly that there is a genetic foundation for differences in the dominance-submissiveness spectrum of bird behavior. To test this possibility, selection for both social aggressiveness and submissiveness was pursued through five generations of mature domestic fowl of two breeds, with the result that large strain differences were obtained (Craig *et al.*, 1965). These selected strains were strikingly different in the frequency with which they initiated aggressive contests, in their ability to win decisions, and in physical intensity of their interactions. Heritability of aggressiveness and dominance has also been shown in Japanese Quail, *Coturnix coturnix*, and captive Red Grouse, *Lagopus lagopus*, (Boag, 1982; Moss, *et al.*, 1982).

The aggressive pugnacity associated with hierarchy-building in many birds seems a distressingly wasteful way to achieve social stability. Subordinates in a flock

TABLE 10–2.
TOTAL INTERMEMBER PECK FREQUENCIES IN ROOSTERS UNDER DIFFERING CONDITIONS OF FOOD ACCESS*

	NUMBER OF BIRDS	NO FOOD CONDITIONS	UNRESTRICTED ACCESS	SEMI-RESTRICTED ACCESS	POINT SOURCE ACCESS
Flock 1	9	15	4	41	874
Flock 2	10	41	5	98	831
Flock 3	10	28	6	396	686

*After King, 1965

tend to be bullied by dominants (Craig *et al.*, 1982), or may be forced to occupy poorer winter habitat (Rohwer and Rohwer, 1978). Low-ranking individuals also tend to have poorer chances of survival (Murton *et al.*, 1966; Fretwell, 1968, 1969; Smith, 1967). Why then, are subordinate individuals not selected out of the population? The benefits of flock-living must far outweigh the cost of being relegated to their inferior positions.

Most importantly, subordinates rarely challenge the social order; consequently flocks function with little fighting, and this allows constructive activities to proceed. Wintering Golden-crowned Sparrows, *Zonotrichia coronata*, occupy exclusive home ranges, and each individual within a flock tends to feed at an individual-specific site, and each orients in a particular direction! Fighting within a marked flock is seldom observed (Pearson, 1979).

Rohwer and Ewald (1981) suggest that dominants and subordinates coexist in a manner analogous to shepherds and sheep. Dominants defend a good habitat against other conspecific flocks to the benefit of subordinates in their flock. As if in compensation, subordinates feeding at a new site are observed by the dominants who then usurp the newly found food source. The authors noted that dominant Harris Sparrows, *Zonotrichia querula*, tended to displace other dominants more frequently than they did subordinates when the flock was not feeding. Thus subordinates also benefit by fighting less with birds of a higher rank for a place in good habitat. Even so, in the long run, survival, in most species, is probably biased in favor of the dominant individuals. There is, very likely, still some truth in the old proverb "Only the brave deserve the fair," or, to put it in Darwinian terms, only the vigorous, highly endowed deserve offspring.

■ SUGGESTED READINGS

Very useful works on animal behavior in general are Thorpe's *Learning and Instinct in Animals* and the more recent Marler and Hamilton's *Mechanisms in Animal Behavior*. Recommended for brief introductions to the subject are Dewsbury's *Comparative Animal Behavior*, and Johnsgard's *Animal Behavior*. For a solid introduction to the concept of fixed action patterns see Tinbergen's *The Study of Instinct* and also his brief *Social Behavior in Animals*.

More specifically concerned with bird behavior are the excellent, brief chapters by Hinde in Marshall's *Biology and Comparative Physiology of Birds*, and by Emlen in Wolfson's *Recent Studies in Avian Biology*. Armstrong's *Bird Display and Behaviour* is rich in illustrative examples of reproductive behavior. Avian social organization is well treated in Grassé's *Traité de Zoologie* and Wilson's *Sociobiology: the New Synthesis*. The Heinroths' *Die Vögel Mitteleuropas* contains hundreds of first-hand, perceptive observations of bird behavior. As excellent brief references the following articles in Campbell and Lack's *A Dictionary of Birds* are recommended: Aggression; Behavior, development of; Counting; Displacement Activity; Imprinting; Learning.

Each of the following books contains extensive material on the behavior of a single species: Armstrong's *The Wren*; Kluijver's *The Population Ecology of the Great Tit*; Lack's *The Life of the Robin*; Nice's *Studies in the Life History of the Song Sparrow II*; Val Nolen, Jr.'s *The Ecology and Behavior of the Prairie Warbler*; and Tinbergen's *The Herring Gull's World*. For ducks, geese, and swans, Johnsgard's *Handbook of Waterfowl Behavior* is recommended.

Songs, Calls, and Other Sounds
11

As a man must speak, so a bird must sing,
  To complete a full life-story.
He sings to survive, to mate, to thrive,
  To defend his territory.
—Joel Peters (1905—   )
  Listen to the Birds

Wherever man travels, whether on land or sea, mountain or plain, arctic tundra or tropical jungle, he is rarely beyond the reach of some bird's call. With an uncommonly lavish hand, natural selection has produced in birds an extraordinary variety of songs, calls, and other sounds. Some hint of this variety is seen in the

booming of the Emeu, the harsh cry of the Guillemot, . . . the plaintive wail of the Lapwing, the melodious whistle of the Wigeon, 'the Cock's shrill clarion.' the Cuckow's 'wandering voice,' the scream of the Eagle, the hoot of the Owl, the solemn chime of the Bellbird, the whip-cracking of the Manakin, the Chaffinch's joyous burst, the hoarse croak of the Raven, . . . the bleating of the Snipe, or the drumming of the Ruffed Grouse (Newton, 1896).

A few species of birds have no voice—for example, storks, some pelicans, and some vultures. The vast majority of birds produce vocal sounds of one sort or another. More than one-half of all living birds are "songbirds," members of the order Passeriformes, suborder Oscines, and they all possess specialized voice apparatus. Birds of other orders are less well endowed musically, and are generally restricted to calls instead of songs, although a few species like the Mourning Dove, *Zenaida macroura*, Killdeer, *Charadrius vociferus*, and tinamous have simple and often beautiful songs. It is likely that the superior mobility of birds and their capacity for flight have made increased and complex sound production both useful and relatively safe.

Communication occurs whenever the actions of one animal influence those of another. This influence usually involves the transfer of information which, between birds, most commonly is in the form of either visual displays or vocal sounds, predominantly the latter. Since most bird vocalizations are highly stereotyped or ritualized, their meanings are generally simple, clear, and unambiguous.

Bird vocalizations are of two kinds: calls and songs. Calls, as defined by Thorpe (1956a, 1964), are brief sounds "with relatively simple acoustic structure." They are usually mono- or disyllabic and rarely involve more than four or five notes. In longer bursts of calls there is no clear organization or pattern, and the calls may continue as long as external circumstances require. Calls generally are concerned with coordinating the behavior of other members of the species (young, family groups, flocks) in nonsexual, maintenance behavior such as flocking, feeding, migration, and reactions to predators. Experiments indicate that calls in some groups, e.g., several members of the finch family Estrildidae, are genetically determined (Güttinger and Nicolai, 1973), whereas in others (e.g., the Cardueline finches), they are learned (Mundinger, 1979).

A song is more complex than a call and "consti-

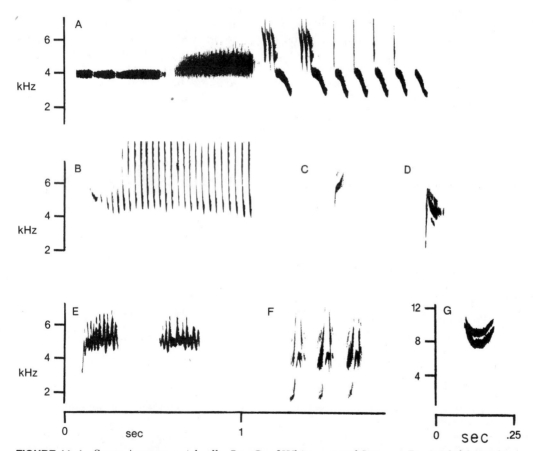

FIGURE 11–1 Song, A, versus social calls, B to G, of White-crowned Sparrow, *Zonotrichia leucophrys.* B, Aggressive trill of male or female. C, Excitement "chink" of both male or female. D, Fledgling location call. E, Distress call of fledglings. F, Fighting calls of adults. G, Seeep call of female, delivered when leaving or approaching nest.

tutes a group of notes separated from another group of notes by a pause longer than the pauses between the notes themselves" (Thielcke, 1969a). Song commonly functions as a substitute for physical combat in defense of territory, and is important in maintenance of the pair bond between mates and in stimulating and synchronizing the reproductive cycle. Song may also act as a factor in promoting reproductive isolation (Payne, 1973). Songs are more often susceptible to modification by learning than are calls.

Some of the advantages of songs and calls over visual displays, scents, and other forms of communication are in the wide spectrum of vocal frequencies and intensities available; the modest energy expenditure required; the fact that sounds carry far, penetrate visual barriers, and are effective in the dark; and that they vanish as quickly as they are uttered, making possible a quick succession of varied communications (Thorpe, 1964).

Considerable variation in song is commonly shown by birds of a given species and also by a single individual bird. Differences in song between birds of the same species are usually not great enough to obscure the basic species pattern. That is, despite their variations, the songs are still recognizable as the songs of a Robin or of a White-throated Sparrow; they are, in effect, "variations on a theme." Some species show much greater variation in song than others. Nearly 900 variations in the song of the Song Sparrow, *Melospiza melodia*, have been recorded by Saunders (1951), with as many as 20 different songs by an individual bird (Mulligan, 1966). On the other hand, the individual Field Sparrow, *Spizella pusilla*, usually sings but one song, which it repeats over and over. The complexity and richness of song in some gifted songbirds is illustrated by the Brown Thrasher, *Toxostoma rufum*, which during 113 minutes of recorded singing, sang a total of 4654 song units (one unit equals a song separated from another song by at least one-fourth second). Of these units about one-half were different from each other and never recurred during the 113 minutes (Kroodsma and Parker, 1977).

In general, it is the male of a species that does the singing, but in a few species such as the Red-winged Blackbird, *Agelaius phoeniceus*, Black-headed Grosbeak, *Pheucticus melanocephalus*, and many tropical species, the female sings also (Beletsky, 1983; Ritchison, 1983; Farabaugh, 1982). And in a very few species such as the African Painted Snipe, *Rostratula bengalensis*, or the European Striped Button Quail,

Turnix sylvatica, in which the sex roles are largely reversed, the females deliver loud, resonant, courtship calls, and the males are practically voiceless.

THE FUNCTION OF VOICE

The importance of song to a bird can be judged partly by the amount of time the bird devotes to it. One of Nice's (1943) unmated Song Sparrows on May 11th spent nine hours singing, nine hours sleeping, and six hours eating and in other occupations. A Red-eyed Vireo, *Vireo olivaceus*, watched by de Kiriline (1954), sang 22,197 songs in the course of one day; and a tropical manakin, *Tyranneutes virescens*, sang nearly 6000 times in one day—a total of 86 percent of his waking hours (Snow, 1961). It is inconceivable that an activity demanding such a large share of a bird's time and energy could survive the winnowing of natural selection unless it had definite survival values.

Songs and other utterances of birds clearly have a variety of functions. They may be grouped tentatively under three major headings: reproductive, social, and individual.

Reproductive Functions

To proclaim the sex of an individual.
To induce another individual to reveal its sex.
To indicate vigor and dominance in an individual.
To advertise for and to allure a mate.
To establish sovereignty over territory.
To stimulate and synchronize sex behavior.
To synchronize the hatching of eggs.
To entice offspring to eat.
To strengthen the pair bond.
To signal changes in domestic duties.
To identify the individual to his mate, parent, or offspring.
To foster reproductive isolation and thereby promote speciation.

Social Functions

To serve as a "password" for species identification.
To rally a flock for collective action.
To hold a flock together, as in dense foliage or during nocturnal migration.
To intimidate or drive away enemies or competitors.
To space out birds of the same species in a given environment.
To convey information, as about enemies and food.

To pass on cultural traditions to the next generation.
To educate offspring and newcomers.
To teach young their species' song.
To announce the emotional state or mood of a bird.

Individual Functions

To discharge nervous energy or provide emotional release.
To identify one individual to another.
To perfect song through practice.
To permit flight in dark caves through echolocation.
And that some birds may sing from a sense of well-being, or simply "for the joy of it," should not arbitrarily be ruled out!

Some of the functions listed are perhaps open to question, and undoubtedly there are other functions that have not been mentioned (review in Thielcke, 1970). The whole subject of sound production in birds is filled with subtle complexities and needs more investigation. High-quality tape recorders and electronic instruments permitting the qualitative and quantitative analysis of sound have helped investigators immensely to answer some of the interesting problems involved in bird song. For example, modern instrumentation may permit one to alter frequency (pitch), rhythm, or tonal quality of recorded bird songs. By playing recorded songs altered in various ways, one may learn which parameters are used by birds in recognizing their species (Dorrscheidt, 1978; Emlen, 1972; Becker, 1982). A great deal could be learned about the functions of bird calls through simple observation. For example, the Pine Siskin, *Carduelis pinus*, gives calls when foraging in dense foliage, but is silent in open fields. Well over two centuries ago the significance of location notes was grasped by that amazingly "modern" ornithologist, Baron von Pernau of Steinach, Austria, whose voluminous writings on bird behavior remained in obscurity until brought to light by Stresemann (1947). In his first book on birds, published in 1702, von Pernau writes,

The Wood-Lark *[Lullula arborea]* eagerly follows the attraction call, in contrast to the Skylark *[Alauda arvensis]* that does not care about it; the reason for the difference probably is that God's inexpressible wisdom, which shines forth from the humblest things, did not implant in Skylarks that method of attracting each other, because they can see their companions on the flat field and can find them without such help, whereas the Wood-Larks, when flying among bushes and over completely wild ground, would often lose each other if they did not utter the attraction call constantly. —(Stresemann's translation.)

The significance of song in defense of territory was clearly demonstrated by Smith (1976) who devocalized Red-winged Blackbirds by puncturing the interclavicular air sac. Birds so devocalized tended to lose large parts of their territories and experienced increased intrusions on their territories. The incisions healed after two or three weeks, and the birds regained their voices and their original territorial boundaries.

Yasukawa (1981) tested the territorial function of song by "speaker-occupation" experiments. Male Red-winged Blackbirds were removed from their territories and replaced with speakers broadcasting Blackbird song. Intrusions by nonneighbors were fewer when songs were played than during silent control periods.

Song in most passerines subserves both a territorial and sexual function (Morton *et al.*, 1985). Some birds (e.g., viduine finches, *Vidua* spp., and Cuban Grassquits, *Tiaris canora*) sing two different songs, one for defense of territory and one for courtship (Payne, 1973; Baptista, 1978).

The use of call notes for sexual and individual recognition is illustrated by the Emperor Penguin, *Aptenodytes forsteri*, of Antarctica. In this species, the trumpeting call of the male is rather musical and the final note long drawn out, whereas that of the female is a cackle with a short final note. Superimposed on these sexual differences in calls are individual differences, so that these similarly colored birds are able not only to recognize the sex of a stranger but to recognize their own mates and young. The young of the King Penguin, *Aptenodytes patagonica*, of South Georgia gather together in nurseries or "crèches" when five or six weeks old. Studies of banded birds by Stonehouse (1956) "showed that parents fed their own chicks exclusively, recognizing them primarily by sound and calling them out of the crèche with distinctive 'call signs,' to which their own chick responds." Two- to three-day old hungry chicks of Black-bellied Gulls, *Larus bulleri*, preferentially recognized tape-recorded calls of their own parents (Evans, 1970); and Guillemot chicks (*Uria aalge*), immediately upon emerging from their shells, already knew and recognized their own parents' luring calls, which, of course, they had heard while still inside the egg (Tschanz, 1968). To study the roles of instinct and learning in such recognition of parental calls, Busse (1977) reciprocally exchanged eggs between nests of Arctic and Common

Terns, *Sterna paradisaea*, and *S. hirundo*, and found that the newly hatched young preferentially chose the calls of the species that had reared them, whether foster-parent species or their own.

In many species of birds the male sings vigorously until he finds a mate; then his songs either become much less frequent or cease altogether (Catchpole, 1973). In his exhaustive studies of the Prairie Warbler, *Dendroica discolor*, Nolan (1978) found that the male sang approximately the following number of songs each day: before the arrival of his mate, 2304 times; during active nest building, 965; the final day of nest building, 192; during the laying period, 497; during incubation, 1052; and during the nestling period, 640. Frankel and Baskett (1961) revealed the courtship significance of cooing in the male Mourning Dove, *Zenaida macroura*, by removing the female of a penned pair of birds, whereupon the cooing of the male increased tenfold. Even the quality of song influences reproductive behavior. Female Canaries, *Serinus canaria*, exposed to large repertories of male songs, built their nests faster and laid more eggs than did females exposed to smaller song repertories (Kroodsma, 1976). Without much doubt, song in such cases serves the function of attracting and stimulating a mate. Some of the larger species of birds, such as certain hawks and crows, have no courtship calls; they depend on their large size and on courtship flights to impress the opposite sex. Owls, on the other hand, as nocturnal birds, have loud voices for amorous communication in the dark.

BIRD VOCABULARIES

For many years most naturalists have agreed that certain birds can communicate with their own kind by uttering different calls. For example, when a Common Crow, *Corvus brachyrhynchos*, discovers an owl perching in the woods, it gives what is called a rallying or assembly call, and other crows within hearing answer the call and fly to join the first crow in "mobbing" the owl. Sound tape recordings of the crow's assembly call were amplified and broadcast with loudspeakers by the Frings (1957) while they themselves remained hidden in woods where crows at the time were neither seen nor heard. Within a few minutes crows were attracted to the locality and came practically to the speaker. Crow alarm calls, on the contrary, quickly repelled the crows. In later experiments, the Frings exchanged sound tapes of crow calls with French ornithologists, and discovered the interesting fact that various species of French crows ignored the alarm calls of American crows but that over one-half of the French crows responded to the broadcast of an American crow assembly call.

At times great roosts of Starlings, *Sturnus vulgaris*, and other birds in trees and buildings become nuisances. A loud broadcast of the distress call of a Starling was found by Frings and Jumber (1954) to be effective in disbanding flocks of Starlings, but it had no appreciable effect on Grackles, *Quiscalus quiscula*, or Robins, *Turdus migratorius*, in the flocks.

The selective recognition by Starlings of the different meanings of different calls was demonstrated by subjecting individual birds, isolated in acoustic chambers, to 10-second bursts of tape-recorded Starling calls signifying distress, escape, feeding, and others (Thompson *et al.*, 1968). During exposure to the different sounds the birds showed clear differences in heart-rate responses, the distress-note causing the greatest acceleration in heart rate.

Many kinds of birds are able to recognize not only the different calls characteristic of their species but also the calls of individual birds (Falls, 1982). When a Gannet, *Sula bassana*, returns from the sea to its colony, it is open to attack if it lands at any nest site other than its own. It is therefore important for the incoming bird to be recognized by its mate. To this end, arriving birds always emit "landing calls." Broadcasts of these tape-recorded calls have demonstrated that they are recognized individually by the mate of the landing Gannet (White, 1971).

Gwinner (1964) kept 18 hand-raised Ravens, *Corvus corax*, in aviaries where they enjoyed free flight. Members of a pair recognized each other "personally" and transmitted information directed only at their partners, even over considerable distances. When a given Raven was absent, its partner frequently uttered sounds that otherwise were emitted exclusively or principally by the absent bird, who, on hearing these sounds, would return to the partner at once. One Raven, as it were, called another by name.

The fact that birds can recognize particular calls has been exploited by man not only to disperse bird pests but also to census wild game-bird populations. Territory-owning males of the Blue Grouse, *Dendragapus obscurus*, were quickly and accurately censused by their response to the recorded precopulatory calls of a female (Stirling and Bendell, 1966). Similar techniques have been employed with different species of quail and even with wolves.

In short, birds possess voice repertories, or vocabularies, that enable them to "talk" to one another. Calls and songs seem generally to be given instinctively and to be understood instinctively. This is indicated in the reaction of many gallinaceous young to their mother's calls. Gross describes how the just-hatched chicks of a Prairie Chicken, *Tympanuchus cupido,* come running to their mother if she gives her "brirrb brirrb" call, but immediately "freeze" into motionless lumps when she gives her shrill warning call (Bent, 1932). Chicks of the Ring-necked Pheasant, *Phasianus colchicus,* respond similarly (Heinz, 1973). The young of the Hazel Grouse, *Bonasa bonasia,* are reported by Pynnönen (1950) to react to four different kinds of parental warning notes that notify them of four different kinds of danger.

Various ornithologists have attempted to "translate" the vocabularies of different species of birds. One of the pioneers in this field, Schjelderup-Ebbe (1923), distinguished ten different calls in the domestic Chicken, *Gallus gallus,* each with its own significance. Coutlee (1971) and Zablotskaya (1984) have described from 12 to 32 social calls in 13 species of Cardueline finches. These included contact, threat, alarm, distress, courtship, precoition, feeding, and territorial calls. Two or more species often flock together when not breeding so that natural selection has placed a premium on species distinctiveness in their contact and courtship calls. Threat, alarm, and distress calls appear to lack species specificity and are used in interspecific communication in mixed species flocks.

Those birds thus far studied have rather modest vocabularies. For example, the Chukar Partridge, *Alectoris graeca,* is said to have a vocabulary of 14 different calls; the Red Grouse, *Lagopus lagopus,* 15; the Ring-necked Pheasant, *Phasianus colchicus,* 16; the Crow, *Corvus brachyrhynchos,* 23; the Great Tit, *Parus major,* 20; the Song Sparrow, *Melospiza melodia,* 16; the House Sparrow, *Passer domesticus,* 11; and the Village Weaverbird, *Ploceus cucullatus,* 15. Passerines seem to have an average of about 15 call notes used in the contexts listed above. Songs, as a rule, tend to be more complex structurally and function primarily in territory defense and in advertising for a mate. Collias (1960) has illustrated several of the most important calls of the domestic Chicken, *Gallus gallus,* in simplified diagrammatic form as shown in Figure 11–2.

Warning calls sometimes indicate distinctions

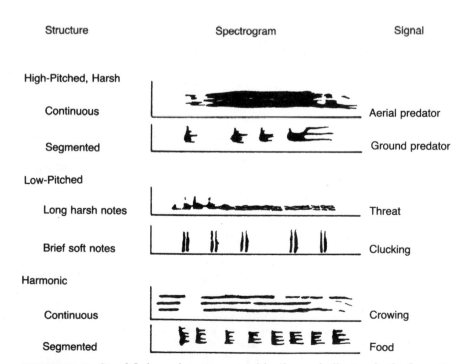

FIGURE 11–2 Simplified sound spectrograms of the classes of calls given by the domestic chicken. Redrawn after Collias, 1960.

between types of danger. A domestic rooster, for example, has two danger calls: one, a "gogógogock," indicates danger approaching on the ground—such as a man or a dog; the other a long-drawn-out "raaaay," warns of danger from the air—such as a hawk (Heinroth, 1938). Similarly, a Field Sparrow, *Spizella pusilla*, gives a "chip-chip-chip" call when a man or a crow approaches; but for a flying hawk it utters a penetrating "zeeeee," which causes all small birds in the vicinity to seek cover (Walkinshaw, 1945). Many small birds throughout the world give essentially the same two warning calls. The European Chaffinch, *Fringilla coelebs*, for example, gives a short, low, "chink-chink-chink" when it sees a static, dangerous enemy, such as a hawk, perched in a tree. This call often attracts other "chinking" small birds, which then proceed to mob the hawk. If, however, the hawk is flying, the Chaffinch gives a "seeet" call and, like the Field Sparrow, flies into the shrubbery along with other species of small birds that heed and repeat the warning call. Careful analyses by Thorpe (1956) using sound oscillograms or sound spectrograms revealed quite unexpected attributes in these two warning calls. Not only do they communicate to other birds the presence of two importantly different kinds of danger—one static and one in motion—but they reveal the location of the caller in the case of the "chink" call, and hide the location of the calling bird in the "seeet" call. Thorpe points out that the chief difference between these two calls is that the "chink" note is easy, the "seeet" call extremely difficult, to locate. The "chink" call is of such low frequency (Fig. 11–3) that the bird's two ears can detect phase differences in the approaching sound waves. Further, the abrupt, click-like nature of this call allows the two ears to detect differences in the time of its arrival, and hence the direction from which it comes. Contrariwise, the source of the high-pitched "seeet" call is difficult to locate because its high frequency probably prevents the detection of phase differences, and its slurred beginning and ending mask differences in the time they reach the two ears. Thus, the "seeet" call warns of danger without revealing to the predator a clue as to the location of the calling bird. Brown (1982) tested this by playing "seeet" calls of American Robins and chink (mobbing) calls of Red-winged Blackbirds to Red-tailed Hawks, *Buteo jamaicensis*, and Great Horned Owls, *Bubo virginianus*, caged outdoors with deciduous woods in the background. Video cameras were used to monitor the orientation of these raptors to playback of the calls. Mean orientation error was 51.5° to playback of mobbing calls but 124.5° to playback of "seeet" calls. When Fleuster (1973) played recorded "chink" and "seeet" calls of the Chaffinch in its natural habitat, Chaffinches responded most strongly to the sounds, but 29 other species also responded, 21 of them merely after hearing the calls, and the rest upon perceiving the calls and the mobbing assemblies.

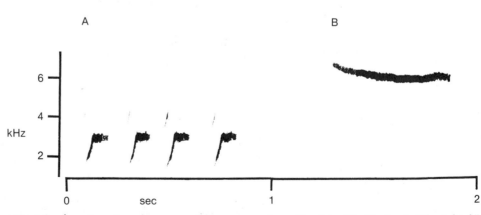

FIGURE 11–3 Sound spectrograms of the two warning calls of the Chaffinch. A, The "chink" note is easily located because of its low frequency (long wave length), short duration, and sharp beginning and end. B, The "seeet" note is difficult to locate because of its high frequency, long duration, and gradual beginning and end. Recordings were made on Lake Constance, Germany, by L. Baptista.

Upon hearing their parents' "seeet" call, nestlings of Great Tits, *Parus major*, react innately by becoming silent and immobile. By coupling recorded playbacks of this warning call with either the feeding (positive experience) of the young, or simulated predation attempts (negative experience), Ryden (1978) showed that their subsequent reactions to the warning calls were either weakened or intensified, respectively.

Most calls are, as a rule, associated with one or another of the following general behavioral situations: social contacts and group movements; finding of food; enemy avoidance; and parent-young relations. Songs are predominantly associated with sexual aggressive situations, including territory defense. These are not hard and fast rules, e.g., California Brown Towhees, *Pipilo fuscus*, utilize a call note in territory defense (Quaintance, 1938).

Once the song of a bird has been recorded on tape, it can be experimentally altered for playback purposes by changing the speed of the tape and thereby the frequency of the notes, by cutting and splicing the tape and thus changing the sequence of elements in the song, by playing the tape backward, and so on. Computer techniques may also be utilized to synthesize songs and alter frequency, rhythm, or syllabic structure (Dorrscheidt, 1978; Margoliash, 1983). By playing back such altered tapes to a bird, investigators are sometimes able to discover which elements in a bird's calls or songs are more significant in conveying information. Using such techniques on male White-throated sparrows, *Zonotrichia albicollis*, Falls (1963) determined that the significant features in its song for species recognition included "pitch, form and arrangement of component sounds, and timing of the sounds and the intervals between them." The first three notes of a Whitethroat's song were sufficient to permit a bird to discriminate its neighbor from a stranger, but a change of more than 10 per cent in the pitch of the song, or even of the first note, interfered with such individual recognition (Brooks and Falls, 1975). Similar studies of the Willow Warbler, *Phylloscopus trochilus*, by Schubert (1971) determined that its song was composed of four major parts, with the essential information contained in parts one and four. The birds seemed to recognize the general form or *gestalt* of the song, for if the order of the parts was changed, the information content of the song was obliterated. Reinert (1965) was also convinced that birds can distinguish gestalts. He trained two Jackdaws, *Corvus monedula*, to distinguish two different rhythms, one with an uneven, rippling tempo, the other of steady beats. The birds also learned to distinguish two series of notes that differed only in the order in which loud and soft notes succeeded one another.

Gestalt is of little importance to the Willow Warbler, *Phylloscopus trochilus*. By replaying artificially constructed territorial songs, Helb (1973) found that many artificial songs released greater reactions than did the species' real song. Interchanging *motifs* in such a way as to emphasize marked changes in pitch increased the reaction among hearers, as did also enriching the song's overtones electronically, or raising or lowering the pitch of notes by 30 per cent. Shortening the song greatly reduced reaction to it. In the Marsh Tit, *Parus palustris*, pitch and time intervals between notes seem essential to the recognition of the species' song. The length of songs, the time interval between songs, and the amplitude modulation of notes, are of little importance (Romanowski, 1979). In the case of the European Chiffchaff, *Phylloscopus collybita*, a single note is sufficient for a species' recognition. A rapid decrease in frequency in the first part of the note is the most important releaser for recognition (Becker *et al.*, 1980).

SOUND-MAKING APPARATUS

The voice-box of the bird is not the larynx with its vocal cords, as in mammals, but the syrinx, a structure that is peculiar to birds. In song birds it is located at the lower end of the trachea, or windpipe, where the two bronchi join. In parrots it is at the lower end of the trachea itself (Gaunt, 1983). In Chapter 7 the basic structure and function of the syrinx were outlined. Here some additional factors involved in voice production will be considered.

During a single crow of a domestic rooster, air sac pressure rose to 60 cm of water, and the amount of air delivered through the respiratory system was over 400 ml, or almost as much as the total volume of the lung-air sac system (Brackenbury, 1978). Rupture of the interclavicular air sac in chickens and Starlings, *Sturnus vulgaris*, rendered them voiceless (Gaunt *et al.*, 1976).

The tracheal air column acts as a resonant filter, at least in geese, by damping out the nonresonant frequencies generated by the tympanic membranes. The resonant frequencies of the trachea depend on its length, its diameter, and the hardness of its tissues. Sutherland and McChesney believe that the dominant frequencies in the calls of geese are those tympanic membrane frequencies that most nearly approach the tracheal resonant frequencies. In small passerines the resonant influence of the small trachea is probably greatly reduced and the dominant vocal frequencies are likely to be those of the tympanic membranes alone. Nevertheless, the tracheal and other air chambers enhance or dampen certain harmonic overtones that provide timbre to a bird's voice. As mentioned earlier, in some gifted songbirds like the Wood Thrush, *Hylocichla mustelina*, Mockingbird, *Mimus polyglottos*, Brown Thrasher, *Toxostoma rufum*, and Reed Warbler, *Acrocephalus scirpaceus*, two membranes may vibrate independently and thus produce two different, harmonically unrelated notes simultaneously.

There is much variation in the anatomy and mode of operation of avian syringes. The tympaniform membranes may be the source of the sound or they may be constricted to form a valve (Fig. 11–4), so that pressure from the lung results in either pulse-generated sound or whistles (Chapter 7). In general, the more primitive birds have few syringeal muscles. Gaunt (1983) noted no correlation, however, between syringeal complexity and vocal virtuosity. Oscines have four to nine pairs of intrinsic syringeal muscles. Consummate mimics such as lyrebirds (Menuridae) have only three pairs, and parrots have only two. Gaunt noted that the three groups known to mimic complex sounds (passerines, hummingbirds, parrots) have intrinsic muscles, whereas nonmimics do not. Galliforms (grouse, quail), with "primitive" syringes containing no intrinsic muscles, may mimic temporal components of conspecifics, but are not known to learn detailed syllables. Gaunt and Gaunt (1985) conducted electromyographic studies on syringeal muscles of various parrots. Intrinsic muscles emitted strings of pulses matched to pulses of sound in amplitude modulated calls, extrinsic muscles only showed simple activation patterns. They concluded that extrinsic muscles functioned as on-off switches, whereas intrinsic muscles were necesssary to produce complex sounds and vocabularies.

Very probably the volume and carrying quality of a bird's voice depend more on resonance than on air pressure from the respiratory tract. As a rule, the longer and wider the trachea, the deeper a bird's voice; the shorter and narrower the trachea, the higher the voice. In the Plain Chachalaca, *Ortalis vetula*, the adult male has a trachea lengthened by a loop that is lacking in young males and females, and its voice is consequently an octave lower than theirs (Sutton, 1951). This length-pitch relationship was experimentally studied by Myers, who cut the trachea of a live hen in the middle and exposed the caudal portion through the skin of the neck. Before the operation the average pitch of the hen's voice was lower (375 cycles

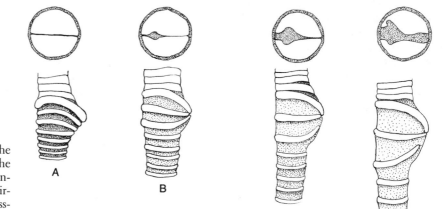

FIGURE 11–4 The syrinx of the Variegated Tinamou, seen from the right side, in four stages of contraction and extension. The circles above show corresponding cross-sections that reveal the condition of the vocal membranes and the opening (shaded) between them. A produces high-pitched, and D, low-pitched tones. After Beebe, 1906.

per second) than afterward (500 cycles per second) (Stresemann, 1927–1934). In hens, the sternotracheal muscles shorten the length of the trachea by one-third or one-fourth by compressing together the ten tracheal rings nearest the syrinx. When the trachea is so shortened, the pitch of the hen's voice rises. Several species of birds, such as the Whooping Crane, *Grus americana*, and Trumpeter Swan, *Cygnus buccinator*, possess enormously lengthened windpipes that provide a deep trombone-like quality to their calls. In these birds, part of the long trachea is coiled within the swollen keel of the sternum. The trachea of the Whooping Crane is 147 cm long, and 71 cm of its length are looped back and forth within the sternal keel. In the Australian Trumpet Manucode, *Manucodia keraudrenii* (a bird of paradise), the male carries its fantastically long windpipe coiled under the skin of its breast (Fig. 11–5).

Other forms of resonance chambers are provided by air sacs in various parts of the body. The paired cervical sacs of many Galliformes are inflated during the breeding season to add both resonance to courtship calls and visual appeal to the male's appearance. The ostrich, rhea, and some bitterns, grouse, bustards, pigeons, and other species can inflate the esophagus with swallowed air to increase the effectiveness of their calls. The male Sage Grouse, *Centrocercus urophasianus*, may increase the volume of its esophagus 25 times by this means. The Emu, *Dromaius novaehollandiae*, makes its booming call through the resonant help of a long inflated air sac in the neck, parallel to the trachea and connected with it through a narrow slit.

Central Control of Song

Attempts to localize those centers in the brain that control voice in birds have been inconclusive, but most experiments in brain surgery or in direct brain stimulation with electricity agree in centering vocal control in the corpus striatum of the forebrain, the tegmentum of the midbrain, and the hypoglossal nucleus of the medulla. In the Chaffinch, *Fringilla coelebs*, each bronchus of the syrinx is supplied with a cranial hypoglossal nerve. If both nerves are cut, the bird loses all voice. If the left nerve of an adult bird is severed, most of the bird's voice is lost. However, if either the left or the right nerve is severed in a young bird before it develops full song, the bird can still develop normal song under the control of the intact nerve (Nottebohm, 1971).

It has long been known that male hormones stimulate singing in a wide variety of birds. For example, castration will abolish song in an adult male Chaffinch, but administration of testosterone will restore it (Zigmond *et al.*, 1973). If radioactive (tritiated) testosterone is administered to castrated birds, it becomes concentrated in certain areas of the brain:

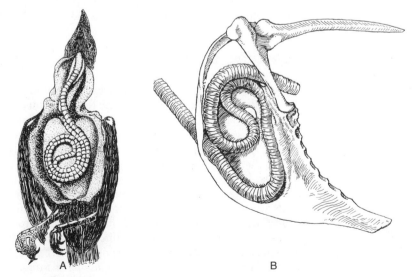

A   B

FIGURE 11–5 A, A dissection of the adult male Trumpeter Manucode showing the long, coiled windpipe between the breast muscles and the skin. After Rüppell, *in* Grassé, 1950. B, Approximately one-half of the trachea of a crane may be looped within the keel of the sternum. After Portmann, 1950.

paired nuclei in the hyperstriatum, and a pair in the medulla at the roots of the hypoglossal nerves (Zigmond *et al.*, 1977; Kelley, 1978). These nuclei, plus another pair in the archistriatum, lie in the neural pathway of song control in songbirds or Oscines. The pathway is formed by the song-producing syrinx innervated by the hypoglossal nerves whose axons arise on the medullary nuclei, which in turn are innervated by the archistriatal nuclei, which receive neural input from the hyperstriatal nuclei (Fig. 11–6). These nuclei seem to be absent from such nonsingers as chickens, doves, and other suboscine birds. Electrical stimulation of the nuclei causes vocalization; their bilateral extirpation abolishes song. Significantly, they are strikingly larger in males of Canaries, *Serinus canaria*, and Zebra Finches, *Poephila guttata*, than in females (Nottebohm and Arnold, 1976).

Gurney (1981) noted that the nuclei of the song system in female Zebra Finches are present, but in a reduced state in numbers and size of neurons. Gurney and Konishi (1980) found that postnatal estrogen-plus-testosterone treatment caused enlargement of the female song system. When silastic tubes filled with estradiol were implanted subcutaneously in nestling females (Pohl-Apel and Sossinka, 1984), the birds, after treatment with testosterone, produced song as adults. Timing and duration of treatment with estradiol affected the quality of female song. Females

treated from days 4–6 after hatching did not sing. Females treated from days 2–18 posthatching produced male-like song. Females treated from days 2–11 produced poor song, similar in quality to that of juvenile males in practice.

The exact role of testosterone in song production is not entirely clear. Sedentary White-crowned Sparrows, *Zonotrichia leucophrys nuttalli*, will sing in the fall and winter months in San Francisco, California, if challenged by other White-crowns or in response to tape-recorded song (DeWolfe, Baptista and Petrinovich, in prep.). During these months the testes are collapsed and testosterone levels are minimal (Blanchard, 1941). Although song level in the population is low, quality of song and singing rate of individuals that sing are comparable to song quality and rate in the spring months. Testosterone thus affects the motivation to sing, but the capacity to sing is still there at low testosterone levels. This is borne out by Pröve's (1974) studies which showed that courtship song in male Zebra Finches disappears following castration. Solitary song, i.e., song not directed at any other bird, is still present, however, at low levels.

A question commonly arises regarding the relationship between breathing and vocalization in birds. By analogy with man, it seems reasonable to suppose that most songs occur during the expiration of a breath. A study by Kneutgen (1969) of the beautiful song of the White-rumped Shama, *Copsychus malabaricus*, indicated that this species produced song during both expiration and inspiration, and that "the rhythmic structure of the *motifs* [was] determined by respiration."

NONVOCAL SOUNDS

Some of the sounds that birds make are not vocal at all but are mechanically produced by the action of their feathers. The "drumming" of the Ruffed Grouse, *Bonasa umbellus*, is produced by the vigorous fanning of its wings while the bird is perched on a log. Among goatsuckers, owls, doves, and larks are several species that make loud clapping noises in flight as their wings strike together. Not only does the Flappet Lark, *Mirafra rufocinnamomea*, of Zambia use the noise of its wings in defending territory, but the flapping "notes" occur in phrases that are shared among birds within a local population but that differ, as do dialects, from the flappings of a neighboring population (Payne, 1978). The Common Nighthawk, *Chordeiles minor*, is known in certain localities as the "bull bat" because

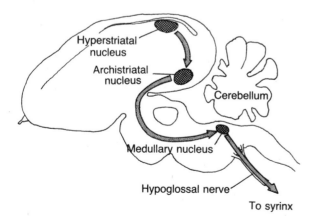

FIGURE 11–6 The neural pathway of song control. A parasagittal section of a bird's brain showing the nuclei in which radioactive testosterone becomes concentrated. A chain of efferent nerve impulses from the hyperstriatal nuclei to the muscles of the syrinx control singing. If testosterone is lacking, song is abolished. After Kelley, 1978.

of the whirring roar of its wing feathers when it pulls up sharply from its steep dives during courtship flights.

In many species certain feathers of the wings are specially modified for sound production. In the Woodcock, *Scolopax minor*, the three outer wing feathers are remarkably narrowed and stiffened, so that, when they are spread apart during the springtime courtship flight, the air rushing between them causes them to vibrate with a high, whistling sound. This same type of feather adaptation is found in the wing feathers of certain ducks, bustards, doves, hummingbirds, cotingas, manakins, and others. The male Broad-tailed Hummingbird, *Salasphorus platycercus*, has narrow first-primary wing feathers that emit shrill whistles during territorial display flights. Experimentally silenced males were less aggressive in territorial defense, and tended to lose their territories to nonsilenced birds (Miller and Inouye, 1983). Two different types of sound-making adaptations are found in the feathers of male manakins (Pipridae) of tropical America. The five outer wing primaries of the White-bearded Manakin, *Manacus manacus*, are greatly narrowed and stiffened, and the bird uses them in flight to make loud sounds. While perching, it can make snapping or growling sounds by striking together the thickened feather shafts of its secondary wing feathers. The quills of these feathers are not firmly inserted on the ulna but have slender muscles attached to them. Quite remarkably modified are the swollen, club-shaped shafts of

the fifth, sixth, and seventh secondaries of the wings of the male Club-winged Manakin *Allocotopterus deliciosus* (Fig. 11–7), which are thought to be somehow snapped together in the folded wing during courtship ceremonies. The wing bones that support these feathers are also abnormally thickened, very likely in correlation with their support and noise-making activity (Newton, 1896). The courtship sounds of this species are described by Willis (1966) as being two clicks followed by an insect-like buzz. The wings flick upward quickly for each click, and then, after a pause, they are again raised and swung conspicuously downward, accompanied by the buzzing sound.

Tail feathers also can be adapted to sound-making. The outer tail feathers of the European Snipe, *Gallinago gallinago*, are much stiffer than the central feathers, and their vanes are held together with more barbule hooklets than normal tail feathers. In courtship flight, these stiffened feathers spread outward into the air stream and set up a rapid, humming vibration. This humming takes on a tremulous character, something like the bleating of a goat, because of air pulses created by the fluttering wings. Reddig (1978) placed these specialized feathers in a wind-tunnel and demonstrated that the spectrograms of sounds thus produced were similar to those obtained from field recordings of displaying birds. This demonstrated that the courtship sounds were mechanical and not vocal. Some species of snipe have extra tail feathers apparently adapted to

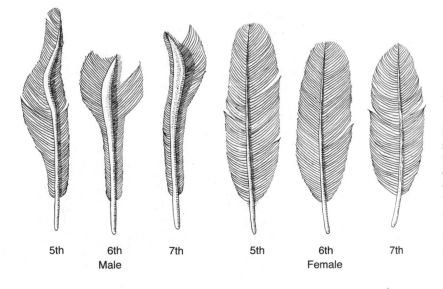

5th	6th	7th	5th	6th	7th
	Male			Female	

FIGURE 11–7 The fifth, sixth, and seventh secondary wing feathers of the South American Manakin, *Allocotopterus deliciosus*, showing the club-like thickened feather shafts of the male, which are thought to act somewhat like castanets in producing sounds. The seventh feather in each case is seen from below; the others from above. After Sclater, in Newton, 1896.

the sole function of making courtship sounds. While the European Snipe has but 14 tail feathers, the Pintail Snipe, *Gallinago stenura*, has 26 feathers, the outer eight pairs of which are sharply narrowed and stiffened for the creation of sound (Fig. 11–8).

Besides using their feathers, birds may make non-vocal sounds with their beaks, either by hammering them on dead limbs or tin roofs as many woodpeckers do, or by clapping their mandibles together as storks and owls do. That a woodpecker's hammerings have courship significance is seen in the male Great Spotted Woodpecker, *Picoides major*, which drums on dead limbs between 500 and 600 times a day during courtship, but only 100 to 200 times a day, and sometimes not at all, after it secures a mate and begins nest-building (Pynnöne, 1939). When tape recordings of tappings of the Lesser Spotted Woodpecker, *Dendrocopus minor*, and Black Woodpecker, *Dryocopus martius*, were played in woods where both species lived, each bird responded only to its own species' sounds (Sielmann, 1958). Several members of the

grouse family make audible noises by stamping their feet on the ground during the courtship period. The vast majority of bird sounds, however, are produced by the syrinx, trachea, and associated vocal organs.

THE NATURE OF SONG

There is a rough correlation between the size of a bird and the pitch of its voice. Large birds, such as owls, cranes, and crows, have low-pitched voices while smaller birds, such as finches and warblers, have high-pitched voices. Ears of oscine birds are best able to hear sounds of about the same frequencies as they themselves produce (Dooling, 1982). This principle does not hold for nocturnal species, e.g., some owls have ears particularly sensitive to high-pitched squeaks, but this is an understandable adaptation to the voices of mice and other small rodents on which they prey. In sound resolution birds are probably superior to all other vertebrates. A bird can probably distinguish 200 discrete sounds per second, whereas for man sounds fuse together into notes or pitch at about 20 per second (Cramp and Simmons, 1977).

Families and genera of birds commonly exhibit similarities in calls and songs, in part because of similarities in basic structure of their vocal equipment and in part due to similarities in the neural equipment controlling bird song. Families of musical instruments—woodwinds, brasses, strings, percussions—share tonal qualities for the same reason. These qualitative differences are most noticeable in the timbre of a bird's voice. The timbre of any sound is based on its fundamental frequency plus a variable number of harmonic frequencies and their relative intensity. Many ornithologists have remarked on the flute-like quality or timbre of the song of various thrushes. Marler (1969) has shown electronically that the song of a Hermit Thrush, *Hylocichla guttata*, and the notes of a clarinet have similar timbre because both emphasize odd-numbered harmonics. The differences between species and individuals in bird songs are more likely to be a matter of sequence and pitch of different notes, and the phrasing, rhythms, and duration of song.

Songs are commonly differentiated, largely on a basis of loudness, into two groups. *Primary songs* are loud songs generally concerned with the daily business of making a living and perpetuating the genes. Primary songs are courtship songs, territory defense songs, special signal songs to other birds as when warning of enemies, and songs representing emotional

FIGURE 11–8 *(Upper)* The Old World Common Snipe; *(lower)* the tail of the Pintail Snipe, showing the narrowed outer tail feathers that make bleating sounds during courtship flight. After Mayaud, *in* Grassé, 1950.

outburst. Ordinarily, primary song is periodic, at its height during territory selection and courtship, and reduced or suspended during the raising of the young. *Secondary songs* are weak, muted, "whisper songs" or "subsongs" that carry only a short distance. They are characteristic of young males, some adult females, and of adult males outside the breeding season. Secondary song may be the means by which a bird "keeps its hand in" and practices singing while the external demands for loud song do not exist. The subsong of the young male Chaffinch, *Fringilla coelebs*, is thought by Thorpe (1956) to be the raw material out of which primary song is constructed, analogous to the babbling a human baby does before it learns to speak. Many of these sounds the infant or young Chaffinch ceases to produce when it does not hear them uttered by its parents or other adults.

A variety of birds practice synchronized singing or duetting, singing simultaneously or alternately, in an antiphonal fashion. Two surveys of duetting in birds revealed that 222 species in 44 families are known to duet (Thorpe, 1972; Kunkel, 1974). Farabaugh (1982) discovered that in eight bird families, 10% or more of the species are known to duet. Notable among these are the following: wrens (Troglodytidae) in which 23 of 59 species duet; the shrikes (Laniidae), in which 21 of 64 species duet; and the honeyeaters (Meliphagidae) in which 22 of 172 species duet.

Duetting is most common in birds of dense forest or scrub vegetation where vision is obscured. A pair of birds may sing a whole song pattern in exact synchrony, or either bird alone may sing a whole duet pattern, particularly when one of a pair is absent. The song repertory of a given pair is usually distinct from that of a neighboring pair. An isolated female will answer reproduced playbacks of her mate's voice, but not those from any neighboring pairs. A common wren of Central America, *Thryophilus modestus*, is called "chinchirigüi" after its loud antiphonal duet. Skutch (1940) relates that one member of a pair calls "chean cherry," while its mate answers "gwee"; the duet is so perfectly synchronized that only when one stands between the two singers does he realize that the calls are not coming from one throat.

A series of 13 consecutive antiphonal duets performed between two mated African Gonoleks, *Laniarius barbarus* (Laniidae) was recorded by Grimes (1965). The mean reaction time between the start of one bird's note and that of the second bird was only 118 milliseconds, and in one instance was a low as 68 msec. Man's comparable minimum reaction time is 444 msec. Birds may change roles, and a duet may be started either by the male or the female of a pair (Baptista, 1978). See Figure 11–9.

Thorpe suggests that duetting may function in helping members of a pair identify and keep in touch with each other in dense vegetation; it helps cement the pair bond; it may be used in common defense of territory; and it may stimulate and synchronize reproductive cycles in a pair. Among Marsh Wrens, *Telmatodytes palustris*, counter-singing may be involved in establishing peck order between neighboring males. If a given male dominates another male in physical encounters, it always precedes it in singing. If, however, the song of a subordinate male is electronically amplified, the leader-follower roles are reversed (Kroodsma, 1979). This may be because a higher intensity song can communicate a higher level of arousal and indicate a greater threat to a receiver (Zahavi, 1979).

THE ECOLOGY OF SONG

Like people who claim they sing their best while taking a shower, most birds are definitely influenced in their song by environmental conditions. Males of many species demand a favorite exposed perch or "singing post" from which they woo their mates or proclaim their sovereignty over real estate. Common examples come quickly to mind: the male Cardinal, *Cardinalis cardinalis*, or Brown Thrasher, *Toxostoma rufum*, singing from the tip-top branch of a tree; the Vesper Sparrow, *Pooecetes gramineus*, or Eastern Meadowlark, *Sturnella magna*, on a fence post; or the field Sparrow, *Spizella pusilla*, atop a mullein stalk. Birds of the prairies, plains, or arctic barrens, lacking suitable song posts, sing on the wing: Bobolinks, *Dolichonyx oryzivorus*, Horned Larks, *Eremophila alpestris*, and Snow Buntings, *Plectrophenax nivalis*.

Birds living in dense vegetation, such as rain forests or thick reed beds, characteristically have loud and persistent voices as a means of keeping in touch with each other. The vegetation not only obstructs vision but absorbs sound.

In an attempt to study this problem systematically, Gish and Morton (1981) recorded songs of Carolina Wrens, *Thryothorus ludovicianus*, in Maryland and Florida. Songs were then played through 50 m of habitat (deciduous forest in Maryland and live oak/palm/palmetto in Florida) and re-recorded. They found that

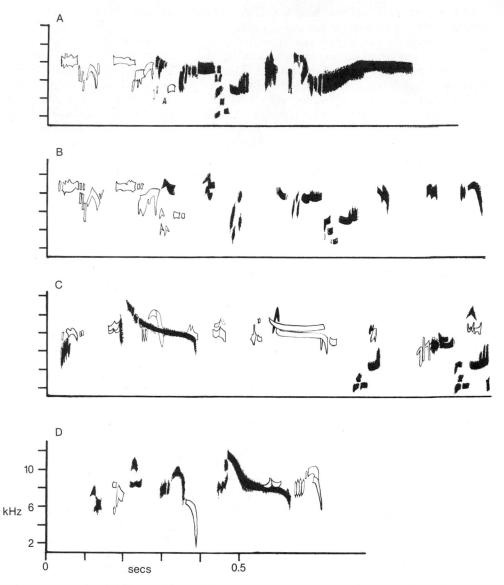

FIGURE 11–9 A to D, Duets of the Cuban Grassquit, *Tiaris canora*. Clear syllables are from songs of the female, and darkened syllables are from songs of the male. In A and B the female begins the duet, in C and D, songs of male and female overlap. From Baptista, 1978.

songs native to the test habitat retained more of the original physical characteristics than songs foreign to the habitat. Bowman (1979, 1983) conducted similar studies with songs of Galapagos Finches recorded on different islands. He also found that songs native to the habitat attenuated with distance at a lower rate than did alien songs.

Analysis of bird songs in an African tropical forest (Chapuis, 1971) and in a Panamanian forest (Morton, 1970) revealed that birds living in areas of dense foliage sang lower-pitched songs whereas those in more open forest or forest edge sang higher-pitched songs. These songs averaged 2200 cycles in the lower levels of forest and 4100–4500 cycles in more open areas (Morton,

1970). In areas of dense vegetation natural selection appears to concentrate sound energy in song in lower frequencies, to increase their carrying power.

The helpless open-nest-living young of many species remain comparatively silent until they are able to fly, but the young of hole-nesting species are well enough protected so that they can, and do, make noise with impunity. The young of colonial nesters are also inclined to be noisy, since the adults in the colony can usually repel invaders through the weight of their numbers.

Weather has a decided influence on bird song. Both cool weather and very hot weather depress the amount of singing in most species, as do rain and wind. A negative correlation between wind velocity and singing has been shown in the Tawny Owl, *Strix aluco*, and European Blackbird, *Turdus merula*. Colquhoun (1939), showed that on a windless evening Blackbirds called 68 times in the period between 30 minutes after sunset and complete darkness; with an 8-km/hr wind there were 22 calls, and with a 24-km/hr wind, no calls. An example of the effect of temperature on song is found in the tallies of calls per minute of Chuck-will's-widow, *Caprimulgus carolinensis*, by Harper (1938). At 16°C (60°F) this species gave 16 to 22 calls per minute, and at 24°C (76°F), 22 to 29 per minute. Latitude has profound effects on bird song. To determine its effects on the Yellowhammer, *Emberiza citrinella*, Rollin (1958) counted the number of songs delivered by an individual per day both in Britain and in arctic Norway. In Britain one bird sang 2279 to 3482 songs per day, whereas in northern Norway one sang only 488 songs. Whether these differences in the amount of song production were caused by day length, light intensity, temperature, or some other factor that varies with latitude remains to be determined.

CYCLES IN SONG

Most birds at higher latitudes show a seasonal variation in song that is mainly correlated with breeding activities and hormone production. The richest, fullest song generally occurs in the spring when birds are busy with territory establishment and courtship. Unmated males commonly sing much more abundantly than mated males. If a male loses his mate he may immediately resume singing to advertise for a new one (Wassermann, 1977). After a pair is mated and the duties of rearing a family begin, song generally wanes or may cease altogether. For obvious reasons, birds rarely sing on or near the nest. An exception that underscores

this protective principle is found in the Central American Gray-capped Flycatcher, *Myiozetetes granadensis*, the female of which sings vigorously while incubating. She can well afford to; she typically builds her nest beside a wasp's nest (Skutch, 1960)! As a rule, birds are completely silent during the period of molt. With the arrival of cold winter weather, most species cease singing until the following spring. The annual singing schedules for numerous species of birds have been worked out by Saunders (1947, 1948), and the relative times for the onset and cessation of song for different species show a striking regularity from year to year (Fig. 11–10). Fall and winter song is known in Bewick's Wren, *Thryomanes bewickii*, Wrentit, *Chamaea fasciata*, Mockingbird, *Mimus polyglottos*, Cardinal, *Cardinalis cardinalis*, and Nuttall's White-crowned Sparrow, *Zonotrichia leucophrys nuttalli*, which defend year-round territories (Kroodsma, 1974; Erickson, 1938; Blanchard, 1938; Baptista, pers. obs.). The male European Robin, *Erithacus rubecula*, sings its best in late fall and winter while establishing winter territory (Fig. 11–11).

For most species, hormones probably play a dominant role in determining the time of year a bird sings. The injection of male hormones into Chaffinches, *Fringilla coelebs*, will bring them into full song in midwinter (Poulsen, 1951). Testosterone injections may also induce song in females of species that sing only rarely in nature, e.g., White-crowned Sparrows (Baptista, 1974; Baptista and Morton, 1982), Chaffinches (Kling and Stevenson-Hinde, 1972), and Canaries, *Serinus canaria*, (Herrick and Harris, 1957).

Most birds that are active by day begin their singing at dawn. At this time of day the most vigorous singing ordinarily occurs. It then gradually tapers off until it reaches low ebb early in the afternoon. Toward evening singing often increases again until gathering dusk stops it for the day. Early in the breeding season many species sing most of the day, but later they confine their singing largely to early morning and evening hours. There are many exceptions to this schedule: wrens and vireos are likely to sing with unabated vigor throughout the long spring day. The Vesper Sparrow, *Pooecetes gramineus*, and Wood Thrush, *Hylocichla mustelina*, although diurnal species, do some of their best singing an hour or more after sunset, and the Eastern Wood Pewee, *Contopus virens*, sings its best before sunrise. A few diurnal species occasionally sing late at night: the domestic Rooster, *Gallus gallus*, Black-billed Cuckoo, *Coccyzus erythrophthalmus*, Nightingale, *Luscinia megarhynchos*, Sedge Warbler, *Acro-

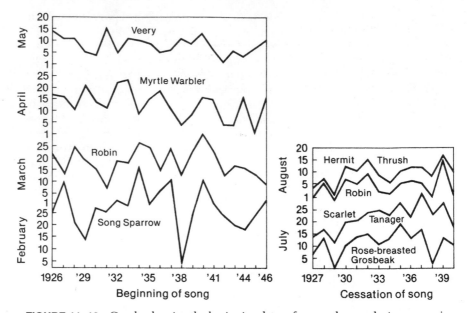

FIGURE 11–10 Graphs showing the beginning dates of seasonal song, during successive years, for four species of birds in southwestern Connecticut; and the dates of ending seasonal song for four species in Allegheny State Park, N.Y. The downward trend of the curves in the spring and their upward trend in the autumn probably are due to the recent gradual warming trend in the climate of the northern hemisphere. After Saunders, 1947.

cephalus schoenobaenus, Skylark, *Alauda arvensis*, and Mockingbird, *Mimus polyglottos*. Contrariwise, a few nocturnal species that normally sing at night may, like the Tawny Owl, *Strix aluco*, occasionally sing by day. Finally, there are a few species that are neither strictly diurnal nor nocturnal but are crepus-

cular and produce their courtship sounds mainly early in the morning or late in the evening: for example, the Woodcock, *Scolopax minor*, and Common Nighthawk, *Chordeiles minor*.

Some birds, like the American Robin, *Turdus migratorius*, sing as soon as they awaken. The Euro-

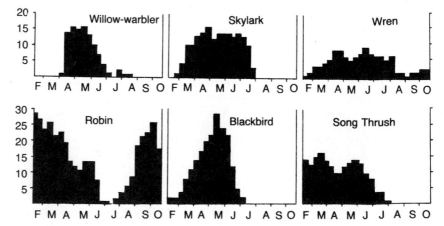

FIGURE 11–11 Histograms showing seasonal variation in the numbers of individuals of different European species heard singing during daily habitat census studies from February to October. After Cox, 1944.

pean Robin, however, begins singing about 3 minutes after awakening; the Blackbird, *Turdus merula*, 5 to 6 minutes; and the Chaffinch, 15 to 20 minutes after awakening (Scheer, 1951). As a rule the time of a species' awakening song is closely related to the time of sunrise so that a bird's first song of the day occurs earlier and earlier as spring progresses and the sun rises earlier each day. The Song Sparrow's first song is given each day at about the time of civil twilight, or approximately one-half hour before sunrise when the sun is 6 degrees below the horizon (Nice, 1938). In January and February the first song is given about 5 minutes after civil twilight, and in March to June, from 3 to 6 minutes before civil twilight. With the lengthening of spring days, many birds rise and sing earlier not only with regard to the clock, but also more minutes before sunrise, so that by early summer in England, the Skylark begins singing at 2 A.M. The Chaffinch, which sings about 15 minutes after sunrise in February, sings 27 minutes before sunrise in June (Armstrong, 1949).

Such facts as these suggest that light intensity is the

chief trigger that sets off awakening song in birds. Several studies have shown this to be true for a variety of species. Observations by Emlen (1937) of the rising time of a Mockingbird on 86 mornings between December 1 and April 1 showed that neither humidity nor temperature correlated with the time of awakening song. Only light intensity showed a consistent positive correlation with the time of the first song. On bright mornings the bird sang about 29 minutes before sunrise, and on cloudy mornings, 13 to 19 minutes.

Using a sensitive photometer, Aldo Leopold studied the relationship between morning song and light intensity by making hundreds of observations on 20 species of birds over a period of four years, in the vicinity of Madison, Wisconsin (Leopold and Eynon, 1961). As a result of these painstaking observations, Leopold confirmed the fact that cloudiness delayed the time of the first morning songs of birds, and that a bright moon, at least for the Robin, caused earlier singing (see also Fig. 11–12). The intensities of incident sky light during the beginning of morning song by the Robin, Cardinal, and Song Sparrow were

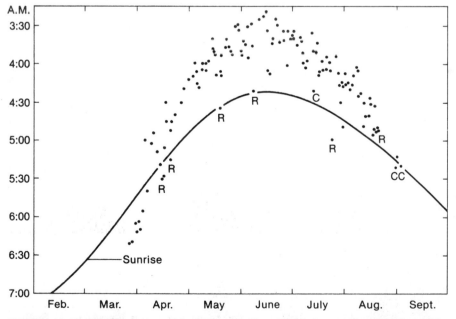

FIGURE 11–12 Daybreak song of the Mourning Dove. Each dot represents the time of the first song given each day at Madison, Wisconsin, 43° N. latitude. Note how the reduced light of rainy *(R)* and cloudy *(C)* mornings delayed the first song on those days, and that the longer the day, the earlier the song with reference to the time of sunrise. After unpublished data of Aldo Leopold, courtesy of J. J. Hickey.

.023, .022, and .015 footcandle, respectively, on clear mornings in April. The intensity of light at the beginning of civil twilight was determined as .04 footcandle. Leopold also found that the light intensity required to stimulate awakening song in the morning was significantly less for nine passerine species than the light intensity coinciding with song cessation in the evening. The Robin, for example, sings in the morning with a light intensity only one fifty-seventh of that in the evening when it ceases singing. This difference perhaps reflects an adaptation to the need of silent preparation for going to roost at the end of the day, and possibly also, as a margin of safety, the need for seeking a second resting place should the first one chosen prove to be unsuitable.

There is an inverse relation between light and singing in nocturnal species. Field observations of the common Poorwill, *Phalaenoptilus nuttallii*, showed that the beginning of its activity at dusk and its cessation at dawn coincided with a light intensity of less than 1 footcandle (Brauner, 1952).

Since distance from the equator affects the length of daylight, there are latitudinal differences in the hours of waking song. The American Robin, *Turdus migratorius*, not only begins its continuous morning song earlier in the higher latitudes (as early as 12:58 A.M. at 60 degrees N. latitude) but sings longer once it starts (89 minutes at 60 degrees N. latitude; 42 minutes at 38 degrees N. latitude) (Miller, 1958). In the continuous June daylight of extreme northern Norway (69°45' N. latitude) Brown (1963) observed that Willow Warblers, *Phylloscopus trochilus*, began singing each day between 10:45 P.M. and 1:45 A.M. and ceased singing for the day at 9:00 P.M. Tinbergen (1939) found that Snow Buntings, *Plectrophenax nivalis*, arose earlier each day as spring progressed in Greenland until about May 1 when they began their activities at 1:00 A.M. Although the days continued lengthening until the summer solstice, the birds refused to rise any earlier. Even at that, they took only two or three hours of sleep each day.

The daily rhythm of activity and song may be inverted in certain species during the breeding and migration seasons. In the antarctic spring, for example, the Snow Petrel, *Pagodroma nivea*, is said to be active by day, but in the summer, only at night. Dur-

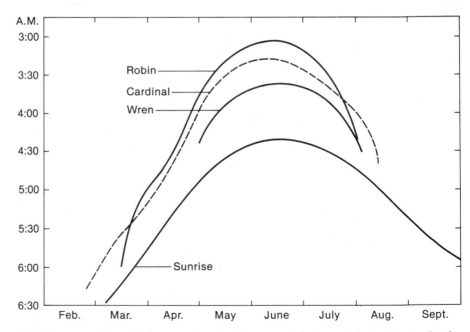

FIGURE 11–13 Smoothed curves showing the mean daily time of daybreak song for the American Robin, Cardinal, and House Wren, for different months of the year at 43° N. latitude. After unpublished data of Aldo Leopold, courtesy of J. J. Hickey.

ing nocturnal migration flights, many diurnal species do not sing, but often they give short chirps or location notes, presumably to hold the flock together.

THE INHERITANCE AND LEARNING OF SONG

Do birds sing the songs they do because they learn them from their parents or because they inherit them? Baron von Pernau observed early in the 18th Century (Stresemann, 1947) that a young bird raised without the company of other young and deprived of an opportunity to hear songs of conspecific adults will never attain its natural song completely and will sing rather poorly. He noted that young Chaffinches, *Fringilla coelebs*, will acquire the Pipit's song. Two centuries ago, Barrington (1773) raised young of one species next to adults of alien species and noted that the young birds acquired the song of the foster species. Modern experiments have confirmed and expanded on the ideas of these insightful early naturalists.

Obviously, the physical structure of a bird's neural and vocal machinery sets limits on the bird's capacity to modify its song, whether by learning or by genetic mutation or cross-breeding. One can no more extract a canary's warble from a goose than a flute's trill from a tuba. But within these rather broad limits, the question still remains: to what extent is a bird's song innate and to what extent acquired? The fact that parrots (including Budgerigars), starlings, and other birds can be taught to repeat words and other sounds is well known. Is this an exceptional trait limited to a few species and to exceptional training conditions, or is it an ability common to many other species and occurring under natural conditions?

The simplest experimental approach to this problem is to raise young birds from incubation to vocal maturity in complete isolation from sounds of their own species. This has been done with a number of species and has shown that the song of some is almost entirely innate—e.g., among the Galliformes and Columbiformes. Vocal learning is known to occur among the Psittaciformes, Apodiformes and 31 families of the Passeriformes (Kroodsma and Baylis, 1982).

Thorpe's (1958, 1961) extensive studies on song development in the European Chaffinch, *Fringilla coelebs*, were a landmark in the history of the field. He hand-raised Chaffinches in lead-lined sound-proof boxes which prevented them from hearing adults of their own species. Their songs the follow-

ing spring were much simpler in structure than those of wild adults. He also played tape-recorded songs of Chaffinches to his hand-raised isolates. He found that these birds would learn Chaffinch songs from tapes during the first 13 months of life but not thereafter. He played songs of other carduelines to his hand-raised birds and found that they would sing songs of alien species during the subsong or practice stage, but as adults sang only Chaffinch songs.

These experiments led Thorpe to conclude that: (1) Songs are learned only during an early "critical period" or "sensitive phase." (2) Chaffinches possess an innate ability to recognize sounds of their own species. A Chaffinch is born with a template or blueprint in its brain of what a Chaffinch song should sound like. Although it hears songs of many species all around it, a naive fledgling matches Chaffinch song with its template and thereby selects conspecific song as a model to be copied.

Thorpe's ideas have been demonstrated and elaborated on or modified for a number of avian species. Marler (1970) demonstrated that White-crowned Sparrows, *Zonotrichia leucophrys*, learn songs during an early sensitive phase. Naive birds would not learn taped songs before 10 days of age. He was successful in teaching them taped White-crowned Sparrow song between ages 10 and 50 days. Fledglings would not learn taped song when exposed to it after 50 days of age. Marler postulated that their sensitive phase for song learning is between the ages of 10 and 50 days.

Baptista and Petrinovich (1984, 1986) also found that 50-day-old White-crowned Sparrows would not copy taped song. They found, however, that about 70 per cent of the 50-day-old experimentals exposed to *live* teachers produced accurate copies of their tutors' songs. These data are comparable to those obtained when naive experimentals were exposed to *taped* song as learning stimuli before 50 days of age (Petrinovich, 1985). Baptista and Petrinovich postulated that a live bird is a more effective learning stimulus, permitting birds to learn beyond the age when tapes are sufficient tutor stimuli. Kroodsma and Pickert (1984a) have obtained similar results with Marsh Wrens, *Cistothorus palustris*. Wrens did not learn songs from tutor tapes during their first spring but did so when exposed to live adults.

Kroodsma and Pickert (1980) demonstrated further that the end of the sensitive phase is not rigidly age-dependent, but that environmental stimuli such as photoperiod and amount of adult song heard during the hatching year may also exert an effect. They raised two groups of Wrens under daylengths simulating June and August hatching dates. These were further divided into one group exposed to many songs types versus a group exposed to few or no song types. Only Wrens raised under August photoperiods and exposed to few or no songs learned songs the next spring. Nottebohm (1969a) castrated a young Chaffinch, and then treated it with testosterone and exposed it to adult song when it was two years old. The Chaffinch successfully learned new songs. Kroodsma and Pickert suggest that Nottebohm may have simulated a late season effect by castrating the chaffinch.

Late-hatching males experience low levels of testosterone and/or reduced singing due to the influence of the shorter photoperiods. A sensitive phase which is partially influenced by environmental conditions gives a late-hatched juvenile some flexibility as to when and where it learns new songs, notably at time of territory establishment.

Using radioimmunoassaying techniques to measure hormonal levels in the blood of canaries, Güttinger *et al.* (1984) found that estradiol levels increased dramatically during the sensitive phase of song learning. These data suggest that estradiol may influence the learning period in some way not yet clear to investigators.

Birds that learn songs may be classified either as critical-period or as open-ended learners (Nottebohm, 1984). A species' breeding ecology may play a role in determining which class it belongs to. Zebra Finches, *Poephila guttata*, learn songs which are exclusively sexual in function during an early sensitive phase ending at about 14 weeks of age (Immelman, 1969). These finches breed in very harsh desert environments where rain may be irregular. A premium is thus placed on early song learning associated with early maturity (as early as 80 days) and pairing, enabling birds to take advantage of the sporadic rain that triggers breeding. Birds such as African Viduine Finches *Vidua* spp.) and Indigo Buntings (*Passerina*) that use song in male/male interaction are open-ended learners (Payne, 1982, 1983a, 1983b). A premium is placed on their being able to learn songs from new neighbors each year because learning their songs gives them some selective advantage such as increased breeding success or greater efficiency in defense of territory.

Canaries are also open-ended learners. New syllable types may be added to their song each year and old types discarded. Syllable repertoires also increase

with age at least up to the second year of life (Nottebohm and Nottebohm, 1978). The authors suggest that song complexity may render a male more attractive to a female. A more complex song may signify an older and therefore more "experienced" individual to an unpaired female. Kroodsma's (1976) laboratory experiments lend some support to the above. He played songs containing either 35 or five syllable types to virgin female canaries placed singly in soundproof boxes. Females that heard the complex songs showed more intense nest-building activity and laid more eggs than females that heard the simpler songs.

Thorpe's template hypothesis has also been demonstrated in other species. When naive hand-raised Swamp Sparrows, *Melospiza georgiana*, White-crowned Sparrows or Marsh Wrens were played tapes of conspecific and alien species songs, experimentals selectively learned conspecific song (Marler, 1970; Marler and Peters, 1977; Kroodsma and Pickert, 1984b; Konishi, 1985). Artificial songs were constructed consisting of Swamp Sparrow and Song Sparrow syllables in various combinations. These were played to hand-raised Swamp Sparrows which then selectively extracted the conspecific syllables with which they constructed songs.

Baptista and Petrinovich (1984, 1986) placed single naive White-crowned Sparrows in special cages which permitted them to see and interact with adult male Strawberry Finches, *Amandava amandava*, Dark-eyed Juncos, *Junco hyemalis*, and Song Sparrows, *Melospiza melodia*, through a screen. The experimentals could hear but could not interact with singing White-crowned Sparrows in the same room. Under these circumstances naive White-crowned Sparrows learned songs of other species instead of their own. The authors suggest that social interaction had in some way cancelled the effects of the auditory template.

Social stimulation may also influence the quality of songs learned. Indigo Buntings and domestic Canaries learn more syllables when exposed to live adult tutors than when exposed only to taped song (Payne, 1981a; Waser and Marler, 1977). Indeed, in some species such as Zebra Finches and European creepers (*Certhia* spp.) social interaction with a live tutor is a prerequisite to learning song (Thielcke, 1970; Price, 1979). Birds exposed to taped song alone sang abnormal songs at maturity.

In some species song development is apparently completely innate, requiring no auditory experience for its normal development. Domestic Fowl, *Gal-*

lus gallus, the domestic Turkey, *Meleagris gallopavo*, and the Ring Dove, *Streptopelia roseogrisea*, if deafened shortly after hatching, will still develop normal vocalizations (Konishi and Nottebohm, 1969). Various New World flycatchers (Tyrannidae) hand-raised away from adults of their kind will produce completely normal songs at maturity (Kroodsma, 1984).

There are often genetic constraints on what a bird learns so that studies of vocal learning in birds provide excellent illustrations of the interplay of heredity and environment in behavioral development. Bailey and Baker (1982) raised day-old Bobwhite Quail, *Colinus virginianus*, in two groups. Their "hoy poo" separation calls were recorded when they were 12 weeks old. Spectrographic analysis indicated that the two groups could be separated on the basis of duration and frequency characteristics of the call. The basic call notes are innate, whereas temporal and frequency components are modifiable by learning. Conversely, Güttinger (1979) and Güttinger *et al.* (1978) have shown that both European Greenfinches, *Carduelis chloris*, and Greenfinch × Canary hybrids are quite capable of learning syllables from Canary songs. The temporal patterning of the learned song, however, is that of the Greenfinch. In this case it is the rhythmic pattern of the song that is innately determined.

Several learning processes are distinguishable during the ontogeny of song in Oscines. The naive fledgling must first listen to and store the model song in its brain; this is the "perceptual learning" phase (Baptista, 1983). Having committed the song to memory, the bird must now learn to use its vocal equipment to produce the song, the "motor learning" phase. Motor learning in songbirds is strongly affected by what they can hear. The song of a surgically deafened adult Chaffinch remains unchanged, indistinguishable from that of intact males. Young Chaffinches deafened at three months old, however, develop simple, unstructured song. Birds deafened between three months of age and adulthood deliver increasingly normal songs the later the deafening occurs (Nottebohm, 1968). That is, audiosensory feedback modifies the motor output. A bird hears what it sings and matches it to its memorized song and thereby improves its motor (singing) performance during the ontogenetic process. Birds deafened as adults sing normal songs because they have heard and memorized their own performance.

Learning to respond to a song need not be coupled with learning to sing it. Richards (1979) discovered several Rufous-sided Towhees, *Pipilo erythrophthalmus*, that had learned songs of Carolina Wrens, *Thyrothorus ludovicianus*. Neighboring Towhees, but not nonneighbors, responded to playback of Wren songs, apparently as a result of associative learning. Similarly, *Acrocephalus* warblers responded to playback song of alien species which they regularly heard and encountered in competitive situations (Catchpole, 1978). Falls and Szijj (1959) obtained similar results in playback studies with Eastern and Western Meadowlarks.

The age at which a young bird first sings varies considerably according to species. Young Song Sparrows begin to sing as early as their 14th day after hatching (Nice, 1943), and the young of many other passerine species, before they are a month old. The recrudescence of adult song that occurs in many birds toward the end of the breeding season may well be a naturally selected device to impress upon the impressionable young the characteristic species song.

Mimicry

Mimicry in bird song is a well-established fact. Mockingbirds, *Mimus polyglottos*, are well known to imitate the calls and songs of other species of birds. One Mockingbird, whose voice was recorded by Cornell University scientists, uttered songs of at least thirty other species of birds. Among some of the better mimics are the Catbird, *Dumetella carolinensis*, Brown Thrasher, *Toxostoma rufum*, Starling, *Sturnus vulgaris*, European Marsh Warbler, *Acrocephalus palustris*, European Red-backed Shrike, *Lanius collurio*, Black Thrush, *Turdus infuscatus*, birds of paradise of New Guinea and Australia, and lyrebirds and bowerbirds of Australia. Some 50 different Australian species have been listed by Marshall (1950) as possessing the powers of mimicry. Tape recordings of the songs of 30 Marsh Warblers by Dowsett-Lemaire (1979) revealed imitations of the songs of 99 European species plus songs of 133 African species acquired by the warblers in winter quarters.

Smith (1968) tells of a Lyrebird, *Menura superba*, that lived for over 19 years around an Australian farm. It could imitate almost any barnyard sound: the noise of a horse and dray; chains rattling; a cross-cut saw; the squeal of a pig; a howling dog; sounds of a violin, piano, or cornet; the calls and songs of most birds in the area, and such human remarks as "Look out, Jack," and "Gee up, Bess." Marshall (1954) also lists some of the sounds mimicked by the Australian

Spotted Bower-bird, *Chlamydera maculata*: the crying of cats, barking of dogs, sheep walking through dead branches, wood-chopping, twanging wire fences, and the calls of many birds. This Bower-Bird is said to mimic the call of a certain eagle so faithfully as to cause a hen and chickens to run for cover. Thorpe (1956) relates the story of a German woman who had a Bullfinch, *Pyrrhula pyrrhula*, trained to sing "God Save the King." A Canary, in an adjoining room, after a year learned the tune from the Bullfinch.

The astonishing fidelity of mimicry, even in a bird not known as a mimic, was demonstrated by Tretzel (1965), who recorded on tape four calls of a Crested Lark, *Galerida cristata*, near Erlangen, Bavaria, which mimicked the four whistled commands that a shepherd gave to his dog: Run ahead! Fast! Halt! Come here! When the taped calls of the Lark were played to the dog, it recognized them and obeyed them correctly. Three other larks in the vicinity produced these whistles with some variations.

It seems somewhat surprising that natural selection has not significantly exploited voice mimicry in birds for the same ends that Batesian mimicry achieves in the realm of animal coloration: that is, the use of sham sounds for ulterior gain. Probably an example of this is seen in the Thick-billed Euphonia, *Euphonia laniirostris*, which uses the alarm calls of other species to promote mobbing of predators (Morton, 1976). Martin (1973) has discovered that female adults and young of the Burrowing Owl, *Athene cunicularia*, when severely distressed, will give a rasping call that closely resembles the rattle of a Prairie Rattlesnake, *Crotalis viridis*, both audibly to human ears and graphically in sound spectrograms. This may serve to warn off predators.

One might anticipate that a good mimic like the Mockingbird could obtain more exclusive sovereignty over its territory by warning away all species whose songs it imitates. Davis (1940) discovered that when a Mockingbird mimicked the alarm note of the Smooth-billed Ani, *Crotophaga ani*, the latter exhibited fright reactions. However, Howard (1974) found that Mockingbirds with larger repertories of other species' songs were more likely to attract mates and to exclude from their territories competitors of their own but not of other species.

Perhaps the closest approach of voice mimicry to Batesian color mimicry is found in the highly precise imitation of their hosts' songs and calls by brood parasites such as the African whydahs (Nicolai, 1964, 1973). The song of the parasitic Straw-tailed Widow Bird, *Vidua fischeri*, matches all eight phrases of the song of its host, the Purple Grenadier, *Uraeginthus ianthinogaster*, and the begging calls of the parasite young match those of the host young. The evolutionary significance of this parasitic relationship is considered in Chapter 23.

SONG POPULATIONS

One result of vocal tradition is the phenomenon of regional dialects which some birds share with man and with some cetaceans (e.g., Humpbacked Whales (Payne and Payne, 1985). Geographical dialects of the English language appear among humans in such variations as the nasal twang of a Brooklyn taxi driver and the drawl of a Kentucky mountaineer. An example of a geographical dialect in birds is found in the "rain-call" of the Chaffinch. Local populations only a few kilometers apart may have distinctly different rain-calls (Sick, 1939; Detert and Bergman, 1984). Not surprisingly, dialects arise in both men and birds in much the same way. Young humans, and many young birds, shape their speech according to the dialect they hear their elders and companions use. They are no more inherited than the cockney accent of "my fair lady," Eliza Doolittle, of Shaw's *Pygmalion*. And, as in the Chaffinch, once a dialect becomes established in the individual, even heroic efforts may fail to change it.

The White-crowned Sparrow, *Zonotrichia leucophrys*, shows distinct dialects in different geographical populations of the San Francisco Bay region (Marler and Tamura, 1964; Baptista, 1975), and in various parts of North America (Baptista, 1977; Baptista and King, 1980). At boundaries between two dialects, individuals may be "bilingual," singing two dialects.

Despite the obvious plasticity of many birds' dialects, they may persist for a long time. For example, a narrowly isolated species of Black-crested Tit, *Parus melanocephalus*, in south central Asia, sings the same song as the more widely distributed Coal Tit, *P. ater*, of Eurasia. This leads Thielcke (1973a) to believe that both species have preserved that same traditional song for thousands of years.

However, like human dialects, some bird dialects undoubtedly change with time. In New England, some 60 years ago, the male Yellowthroat, *Geoth-*

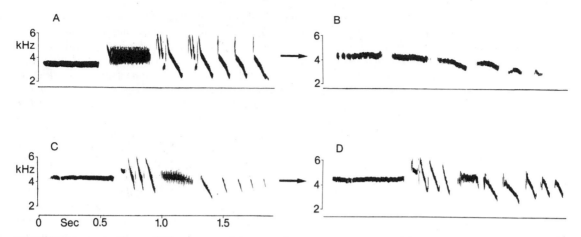

FIGURE 11–14 A, Song of White-crowned Sparrow from San Francisco, California. B, "Kaspar Hauser" song of hand-raised White-crowned Sparrow raised in isolation in a sound-proof box. C, Song of White-crowned Sparrow from Coastal Oregon. D, Song of hand-raised White-crowned Sparrow hatched in California learned from bird C.

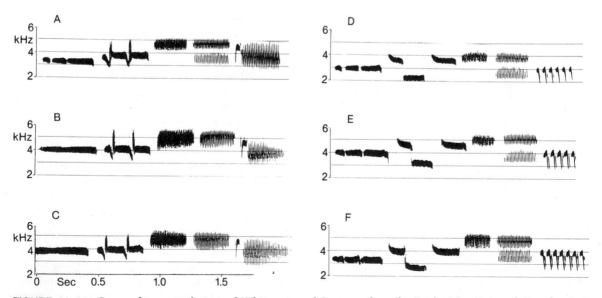

FIGURE 11–15 Songs of two populations of White-crowned Sparrows from the Rocky Mountains of Canada. A to C represent the songs from one dialectal population; D to F represent three songs from a second dialectal population. Because songs are learned, members of a population tend to sing similar songs. Sound spectrograms courtesy of Dr. Ross Lein.

lypis trichas, sang three-note phrases almost exclusively, while today the majority sing four- and five-note phrases (Borror, 1967). New Zealand Chaffinches, *Fringilla coelebs*, introduced from Britain 100 years ago, sing songs with fewer trill phrases and more complex end phrases than their British counterparts (Jenkins and Baker, 1984).

And dialects, of course, change with distance. Wild Cardinals, *Cardinalis cardinalis*, in an Ontario community were exposed to tape recordings of local Cardinals as well as of Cardinals from localities 11 and 3000 kilometers away, and they reacted most vigorously to the local songs and least vigorously to the most distant dialect (Lemon, 1967). Similar results have been obtained from playback studies with White-crowned Sparrows (Petrinovich and Patterson, 1981; Tomback *et al.*, 1983). This was not unexpected. Although they are legatees of the same mother tongue, a Boston Brahmin, a Scottish Highlander, and a London Cockney would undoubtedly have difficulties understanding each other. The dialects of the Great Tits, *Parus major*, of south Germany and of Afghanistan are so widely divergent that tape recordings of the latter's dialect are not recognized at all by the German birds (Thielcke, 1969b).

Are dialects the result of young birds learning songs from adults and not dispersing far from their natal area, or of learning songs from territorial males at sites where they settle? Color-marked sons and daughters of male White-crowned Sparrows singing alien dialects did not sing their father's songs but adopted the local dialects, indicating that White-crowned Sparrows do not sing songs learned directly from fathers (Baptista, 1985). There are also field observations of bilingual birds singing two dialects, then dropping one dialect from their repertoire and using only the dialect of territorial neighbors. These observations indicate that most likely White-crowned Sparrows tend to learn songs after they disperse. Color-marked Bewick's Wrens, *Thryomanes bewickii*, Indigo Buntings, *Passerina cyanea*, and New Zealand Saddlebacks, *Creadion carunculatus*, also learn songs at sites settled (Kroodsma, 1974; Jenkins, 1978; Payne, 1982). On the other hand, field studies with color-marked Galapagos Finches (geospizines) indicate that males learn songs directly from fathers and not from neighbors after dispersal from their birth area (Grant, 1984; Millington and Price, 1985).

Regional dialects then are most likely the result of a "founder" individual, singing a unique song and isolated in an area and then passing his song to future generations of birds either hatched there or moving into the area. Thus bird species isolated on islands often sing songs distinct from those of mainland populations. For example, vocalizations of island populations of the Bearded Bellbird, *Procnias averano*, Dark-eyed Junco, *Junco hyemalis*, Anna Hummingbird, *Calypte anna*, and Brown Creeper, *Certhia americana*, differ from those of their mainland relatives (Snow, 1970; Mirsky, 1976; Baptista and Johnson, 1982).

The rate at which a dialect or some other change in a bird's song spreads geographically is very difficult to study and almost unknown. However, among European Blackcaps, *Sylvia atricapilla*, late in the 19th century, there appeared a new terminal phrase in the species' song called the "Leier song." It first appeared in southeastern Europe and gradually spread west and northward at the rate of about 5 kilometers per year. Now the Leier song is heard in Italy, Spain, and southern France and Germany (Morike, 1953).

Birds can recognize not only dialects but variations in songs of individuals. This was clearly demonstrated by Falls (1969). On the territorial boundary of a White-throated Sparrow, he played tapes of the bird's own song, its neighbor's song, and the song of a stranger—all White-throats. The owner of the territory responded most vigorously (in song-replies per minute) to the stranger's song, about one-half as vigorously to the neighbor's song, and between these two intensities to its own song. When the neighbor's song was played on the "wrong" side of the experimental bird's territory, the owner reacted as to a stranger's song. This almost-human, xenophobic reaction of territory owners to strangers has been demonstrated in at least nine passerine species (Falls and McNicholl, 1979).

Much has been written on song dialects in birds (Krebs and Kroodsma, 1980; Mundinger, 1982). The adaptive significance of song dialects, however, has been a matter of considerable debate. Some authors (Andrew, 1962; Thielcke, 1969a) are of the opinion that song dialects serve no adaptive function and are merely an epiphenomenon of the learning process. Gish and Morton (1981) and Bowman (1979) studied song dialects of Carolina Wrens and Galapagos Finches, respectively. They played tape-recorded songs, measured sound attenuation and found that each dialect maintained its integrity over a greater distance in its home habitat than in a foreign habitat. Dialects in these species therefore are the products of natural selection enabling birds to communicate more effectively over a greater distance in their local area. King (1972) and Nottebohm (1976) studied song dialects of Rufous-collared Sparrows, *Zonotrichia capensis*, and found dialect populations associated with specific habitats or with changing altitude. Although the desired experiments have not yet been conducted, the adaptive significance of these dialects may well be a matter of maintaining song integrity.

Some authors have proposed that dialects, notably in White-crowned Sparrows, may restrict the flow of genes from one population to another (Marler and Tamura, 1962; Nottebohm, 1969b; Baker, 1982). They suggest that both sexes learn the dialect of their birth area, then do not disperse far before settling. A female then chooses her mate by matching his song with that committed to her memory. This has been called the *assortative mating theory*, and may function to fixate genes adapting individuals to a particular habitat.

This theory has been tested with color-marked birds in both sedentary and migratory populations of White-crowned Sparrows. Females normally do not sing, but may be induced to do so with testosterone treatment. Only a few testosterone-induced songs of color-marked females matched those of their mates (Baptista and Morton, 1982; Petrinovich and Baptista, 1984). Moreover, California males singing dialects from Oregon, Washington, Canada, or Alaska were often mated to local females and bred successfully (Baptista, 1974, 1985). Although song is important in mate attraction it does not appear to be the ultimate cue in mate selection in this species. Very few birds banded as nestlings settled in their home populations as breeders (Baptista and Morton, 1982; Petrinovich and Patterson, 1982). Payne (1981b) calculated that the number of White-crowned Sparrows crossing dialect boundaries during dispersal is enough to prevent genetic differentiation. Baker (1982) and associates have published results of biochemical analysis of allelic (genetic) differences in blood proteins between dialect populations. Hafner and Petersen (1985) have re-examined their data using different statistical techniques and found that there were differences between some neighboring dialect pairs but not all. They found moreover that dialect and allelic-change boundaries did not coincide. This suggests that the genetic differences may be the results of "founders" isolated in patchy habitat, and that song dialects do not play a role in maintaining these differences as the assortative mating theory predicts. If dialects are effective in maintaining

genetic differences between local populations (demes), then genetic diversity should be greater in species exhibiting dialects than in species with no dialects. Zink (1985) used an index known as the F_{st} to measure diversity differences based on genetic alleles and found no difference between dialectal and nondialectal species. Why then do these species learn dialects?

One finds that when songs of individuals in a dialect population of White-crowned Sparrows are carefully plotted out on a map, nearest neighbors tend to have songs more similar to each other than nonneighbors (Baptista, 1975). It was shown earlier that this is most likely due to young birds moving into an area and imitating the songs of resident territorial males. Payne (1982, 1983) has shown that yearling Indigo Buntings imitating the dialect of neighbors are more successful (fledge more young) than yearlings that do not imitate. He suggests that nonneighboring birds may avoid an area due to associating a dialect with a previous defeat imposed by a territorial adult. The nonneighbor may also avoid the territory of yearlings that copy the songs of successful (dominant) males. By copying songs of successful males, yearlings become, in an indirect way, successful themselves: energetically costly encounters with strangers are decreased leaving the residents to conduct their breeding activities in peace.

Falls (1982) has reviewed the literature showing that neighbors and strangers may be distinguished by song in a variety of avian species. Responses to neighbors' songs decline as the breeding season progresses due to familiarity and habituation. Responses to strangers' songs remains high, however. Matching neighbors' songs may be one way of distinguishing neighbors from strangers.

Birds with multiple song-matching repertoires often match songs of singing neighbors or songs played from tape recorders with like songs (Krebs *et al.*, 1981; Falls, 1985). Falls and others have suggested that matching directs the signal to a particular individual and may narrow a response to a specific opponent. Neighbors memorize each others songs as a mechanism to better gauge each other's distance after noting the degree of song attenuation from the vocal source (Morton, 1982). This behavior would enhance the value of song as a territory proclaiming tool.

Darwin (1871) suggested that song may be a product of intersexual selection, i.e., females may favor males with more elaborate songs, so that these individuals tend to contribute more progeny to future gene pools. Some evidence is beginning to emerge in support of Darwin's ideas. Catchpole *et al.* (1985) studied polygyny and breeding success in the Great Reed Warbler, *Acrocephalus arundinaceus*. They computed song complexity (measured by diversity in syllable types) of individual males and found that individuals with more complex songs were more successful (attracted more females) than males with simpler repertoires. Kok (1972) found that males of polygynous Boat-tailed Grackles, *Quiscalus major*, with more complex terminal phrases in their songs fledged more young than those with simpler songs.

■ SUGGESTED READINGS

Excellent and extensive material on bird song is provided in Armstrong's A *Study of Bird Song*, Nottebohm's Vocal Behavior in Birds *in* Farner and King's *Avian Biology*, Thorpe's *Bird Song: The Biology of Vocal Communication and Expression in Birds*, Jellis, *Bird Sounds and Their Meaning*, Kroodsma and Miller (eds) *Acoustic Communication in Birds* (2 vols.), and Thielcke, *Bird Sounds*. For books on the identification of North American bird songs, consult Mathews' *Field Book of Wild Birds and Their Music*, and Saunders' A *Guide to Bird Song*. An English calendar for the song periods of many European birds is found in Witherby *et al.*, *The Handbook of British Birds*.

Territory
12

The gull claims only nest-room for her ground,
But if an interloper there be found,
All heaven echoes with protesting sound.
—Joel Peters (1905–   )
   Every Man's House

A common springtime sight on the grassy lawns of yards and parks is the clamorous fighting and chasing that goes on between two American Robins, *Turdus migratorius*. This is the territorial fighting that occurs between two birds, usually males, when both lay claim to the same piece of land for a living site. The same sort of thing occurs in many other species of birds that are landed proprietors and commonly defend their property with all the vigor they can command. Various invertebrates, fishes, lizards, mice and other mammals—man, unfortunately, not excepted—engage in similar battles over land ownership.

Another frequently reported springtime activity of many birds is "reflection fighting," the attack by a bird on its reflected image in a window or a mirror. Brown (1937) documented several species of British birds attacking their images on a mirror set up on an English lawn. A bird apparently mistakes its reflection for an intruder of the same species on its property. Every day between May 2 and July 11 a Brown Towhee, *Pipilo fuscus*, fought its reflection in a university laboratory window at Berkeley, Caifornia, sometimes visiting the window as often as 15 times a day and striking the window as often as 53 times a minute (Ritter and Benson, 1934). That the reflected image is mistaken for another bird of the same species is supported by experiments of Kroodsma (1974). When stuffed dummies of male Rose-breasted Grosbeaks and Black-headed Grosbeaks, *Pheucticus ludovicianus* and *P. melanocephalus*, were placed in the territory of a male Rose-breasted Grosbeak, it attacked the dummy of its own species seven times as often as that of the other species.

HISTORY OF THE TERRITORY CONCEPT
The concept of territory ownership in birds goes back in history at least as far as Aristotle (*ca.* 350 B.C.), who

wrote, "Each pair of eagles needs a large territory and on that account allows no other eagle to settle in the neighborhood" (*in* Nice, 1953). In the next century, Zenodotus observed that "One bush does not shelter two Robins"; and much later, in 1622, Olina wrote of the European Robin, *Erithacus rubecula*, "It has a peculiarity that it cannot abide a companion in the place where it lives and will attack with all its strength any who dispute this claim" (Lack, 1943). Thoughtful observations by later naturalists began to reveal some of the complications of the territory concept. The remarkably perceptive Baron von Pernau wrote in 1707:

On the other hand the Nightingale is forced, for the sake of her feeding requirements, to chase away her own equals, for if many would stay together, they could not possibly find enough worms and would inevitably starve. Nature therefor has given them the drive to flee from each other as much as possible.

Gilbert White of Selborne, writing in 1772, introduced a new notion into the slowly growing concept of territoriality:

During the amorous season such a jealousy prevails amongst male birds that they can hardly bear to be together in the same hedge or field. . . . It is to this spirit of jealousy that I chiefly attribute the equal dispersion of birds in the spring over the face of the country.
(Nice, 1941).

The first comprehensive modern definition of territory came from Bernard Altum in his book *Der Vogel und Sein Leben*, published in 1868. Although his ideas gained currency in Germany, they did not spread to other countries. The following excerpts from Altum are in Mayr's (1935) translation.

It is impossible, among a great many species of birds, for numerous pairs to nest close together, but individual pairs must settle at precisely fixed distances from each other. The reason for this necessity is the amount and kind of food they have to gather . . . together with the methods by which they secure it. All the species of birds which have specialized diets and which . . . limit their wanderings to small areas, cannot and ought not settle close to other pairs because of the danger of starvation. They need a territory of a definite size, which varies according to the productivity of any given locality.

It is, of course, natural that the most suitable localities will be most sought by the species preferring them. Large numbers will gather in such places, overcrowding them, while other available territories would be empty if pairs were not kept apart by force. This force is used by the male as

soon as another gets too close during the breeding season. The interloper is immediately attacked in the most violent manner, and driven to a distance that is determined by the size of the required territory.

Altum goes on to show that song is used both as a proclamation to fix territorial boundaries and as an invitation to females to join the singing male.

Probably unaware of Altum's work, the Irish naturalist Moffat (1903) published a paper on "The spring rivalry of birds" in which he presented similar concepts about territory plus some important new ones. Competition for territory, he believed, produced surplus birds that, lacking territory, failed to breed. Territory was thus a mechanism to protect "their future families against the ills of congestion." Here we find the first clear exposition of density-dependent population regulation, a concept to be neglected by ornithologists for yet another generation. Moffat thought that the bright plumages of quarrelsome males "originally evolved as 'war paint' . . . [or] a sort of 'warning colouration' to rival males rather than attractive colouration to dazzle the females." In other words, he was describing sign stimuli and releasers nearly half a century before Lorenz's presentation of the idea.

A great impetus to the study of bird territory was given by the publication of Howard's (1920) book, *Territory in Bird Life*. Howard emphasized the instinctive intolerance of males for each other during the breeding season, the means by which males advertised their territories, the value of a territory for its exclusive food supply, and the possibility of extremely small, nest-centered territories in the case of birds nesting in colonies.

Numerous definitions of territory have appeared in recent years. They range from very brief statements such as Noble's (1939) "Territory is any defended area," to long discourses covering several paragraphs. Almost every definition, however, embraces the two ideas presented by Craig (1918): (1) an appetite for a place, and (2) an aversion to other members of the species except the mate. Recent studies require a modification of Craig's idea, since some species defend territories as groups or clans (see below). An excellent working definition of territory is offered by Nice (1941):

The theory of territory in bird life is briefly this: that pairs are spaced through the pugnacity of males towards others of their own species and sex; that song and display of plumage and other signals are a warning to other males and an invitation to a female; that males fight primarily for territory and not over mates; that the owner of a territory is nearly invincible in his territory; and finally that birds which fail to obtain territory form a reserve supply from which replacements come in case of death to owners of territory.

In addition, a territory invariably contains a scarce resource such as food, shelter, sex display area, or nest site whose scarcity sets an upper limit on population size. Furthermore, the vital resource must be economically defensible (Brown, 1964; Wilson, 1978).

A distinction should be made between an animal's home range and its territory. The geographical area over which an animal habitually wanders is its home range, but only that portion of it that it defends is its territory.

FUNCTIONS OF TERRITORY

As is already apparent, territory has no single, simple, overall function. The uses of territory may differ for the individual bird, the mated pair or its family, a colony of birds, or a species. The significance of territory for a singing male is at times strikingly different from its significance for the female or for the nestlings. Considered broadly, territory produces its effects, including numerous advantages and some disadvantages, by the isolation of birds, by spacing them apart, by providing geographical stability, and by giving its owners certain psychological advantages.

The first of these functions of territory—the isolation of birds from others of their kind—produces immediately apparent advantages. An isolated male in his territory is unmolested in his courtship of any female that may enter the territory. The ownership of territory in itself is, for many species, a powerful stimulant of the reproductive machinery. Once the breeding cycle begins, territory reduces interference with pairing, copulation, nest-building, and the rearing of young. Ownership of a territory provides more or less a monopoly of the food resources nearby (particularly important in adverse weather) and of nesting materials. Since isolation means fewer contacts with strangers, and possible despots, fighting is reduced and energy thus saved. Also, the hazards of promiscuity are reduced, and family stability thereby promoted. This, of course, is extremely important in many species in which the male assists in the rearing of young. The monopolistic feature of territory probably arose out of the drastic nature of intraspecific competition. A robin's worst enemy—his greatest competitor—is not a hawk or a cat. It is another robin that seeks from the environment exactly those kinds of food, those nesting sites, and that kind of a mate that all Robins seek.

Closely related to, yet different from, the advantages of isolation are those arising from the dispersion of individuals. The mutual antagonism of male birds resembles the restless, mutual repulsion of the molecules of a gas under pressure, seeking always to fill all available space. Such a spacing out of birds of a given species regulates population density and prevents overpopulation. As von Pernau, Altum, and Moffat all pointed out, if birds of a given species gathered only where food is most abundant or most easily obtained, their concentrated appetites might shortly bring on famine, while in other neglected habitats food remained uneaten. Dispersion thus promotes the efficient exploitation of food, nest materials, nest sites, and other requirements. Removal experiments in a Maine forest by Hensley and Cope (1951) demonstrated that only the fittest individuals acquire territories. In a 16-hectare (40-acre) spruce-fir woods there were 154 territory-holding male birds of various species at the beginning of the experiment in early June, 1950. Through the use of firearms this population was reduced to 21 per cent of its original size by June 21, and held at about that level by the continuous shooting of new arrivals in the woods until July 11. By that time a total of 528 adult birds—approximately $3^1/2$ times the original population—had been killed. (Students should realize that an experiment of this kind can be conducted only by special authorization of conservation authorities, and must be aimed at solving a definite scientific problem.) New males moved into the woods and established territories as quickly as territories were vacated by the removed birds. It seems obvious that there were surplus males available—"floaters" without territories and living in less desirable habitats—waiting to rush in and fill the vacuum the moment a given territory became unoccupied. In like manner the territories of Great Tits, *Parus major* (Krebs, 1970), Song Sparrows, *Melospiza melodia* (Knapton and Krebs, 1974), and Cassin's Auklets, *Ptychoramphus aleuticus* (Manuwal, 1974) were all quickly filled with conspecific floaters when the original owners were removed. Even migratory birds, widely removed from their winter quarters, are quickly replaced by others of the same species (Mewaldt, 1964). Removal experiments indicate that territorial behavior may limit a population at a local level. Floaters are prevented from settling by territory holders. It has been proposed that territorial behavior has evolved as a means to limit population density (Wynne-Edwards, 1962). This idea, Davies (1978) points out, implies selection acting at the level of the group or population. The evolution of territoriality is more readily explained if selection is assumed to act on the individual; individuals compete for territories, and one consequence of territorial behavior is the limitation of population density.

A function of territory commonly overlooked is its role in the speciation process by the dispersal of "propagules" which may colonize new habitats. Territorial adults push juveniles to the periphery of a species' range, sometimes forcing them into nontypical habitats where they must eat somewhat different foods, use strange nest materials or nest sites, experience new ranges of climatic extremes, and so on. Some of these marginal birds may be preadapted (due to genetic mutations) to one or more of the new conditions. In general they will, of course, lead the lean, arduous life of pioneers, but like their human counterparts, some small groups of them may occasionally "strike it rich" in the form of evolutionary adaptive divergence encouraged by ecological or geographical isolation. It may well be that ecological divergence, brought on by territorial pressure, has resulted, for example, in altitudinal races of related species such as the European Ptarmigan, *Lagopus mutus*, of the alpine zones, and the Red Grouse, *L. lagopus*, of the lower moors and bogs (Huxley, 1942). If the general principle is true that the less well adapted an individual is to its environment the more drastic the winnowing of natural selection, then certainly the peripheral birds of a given population, exposed to a somewhat unfriendly environment, will be subject to a more intense selection pressure than the better adapted, more comfortably located central birds. Further, small fragments of a peripheral population are more likely to be reproductively isolated from the main population of their species—an important prerequisite to speciation.

Another important aspect of territoriality is that it gives an individual the advantage of perfect familiarity with one small area. After a bird settles down on its territory and explores its surroundings, it soon learns where to go for food, the best escape routes when predators attack, where to find nest materials, and similar facts necessary to survival. Although comparable experiments with birds seem to be lacking, experiments with mice have shown the value of territory possession in the evading of predators. Two Deer

Mice, *Peromyscus leucopus*, were exposed simultaneously in an experimental room to a hungry Screech Owl, *Otus asio*. One of the mice, the "resident," had had several days' familiarity with the room, and the other "transient" mouse had had none. In a series of 20 such confrontations, the owl captured only two of the resident mice but 11 of the transients (Metzgar, 1967). Further, a territory owner comes to know its neighbors "personally" (Falls, 1982) and establishes a practical and more or less pacific *modus vivendi* with them based on their relative aggressiveness. Neighbors have learned territorial boundaries and, posing no threats, are usually spared attacks which are energetically costly and may end in injury and even death. In these various ways a bird in familiar environs can become a creature of habit and thereby save itself the wear and tear of much trial and error learning. The community stability enjoyed by a territory-owning bird is roughly comparable to the law and order of a settled human community (also based in large part on property rights), in contrast to the lawless turmoil of a frontier where every individual is a stranger whose capacity for good or evil must be determined in never-ending face-to-face encounters.

Territoriality thus brings into play two antagonistic forces, one dynamic and centrifugal, the other stable and centripetal. The dynamic, expansive, gas-like force that disperses birds may shove some of them into frontiers where living is rough but where occasionally the species may "make its fortune." The stable, centripetal force that rivets a bird to its territory corresponds to stay-at-home, property-owning conservatism that promotes comfortable living, reproductive success, and the preservation of family life. The poem, "To the Not Impossible Him" by Edna St. Vincent Millay (1922) touches on these two aspects of territorialism in man:

How shall I know, unless I go
    To Cairo or Cathay,
Whether or not this blessed spot
    Is blest in every way?

Now it may be, the flower for me
    Is this beneath my nose;
How shall I tell unless I smell
    The Carthaginian rose?

The fabric of my faithful love
    No power shall dim or ravel
Whilst I stay here,—but oh, my dear,
    If I should ever travel!

Snug in its home territory, the avian family is relatively secure, safe from want, safe from philandering adventurers—but it is just possible that a Carthaginian rose bush would provide a better nest site!

Finally, a territory confers remarkable psychological benefits on its owner. A bird changes its personality drastically when it sets foot off its own territory. The European Robin, *Erithacus rubecula*, defends its territory with song, posturing, and at times actual physical combat. The degree to which it employs these defenses depends on whether the bird is on its own territory or off. One of the male Robins studied by Lack (1943) was caught in a wire-cage trap in his own territory. Shortly after this his neighbor, a male called Double Blue (after the identifying rings on his legs) trespassed on the first bird's territory seeking food. The latter bird "at once postured violently and uttered a vigorous song-phrase from inside the trap, and Double Blue, who was one of the fiercest of all the Robins, promptly retreated to his own territory, although the first male could not, of course, get out to attack him. I now caught the first male and moved him into a trap in the territory of Double blue. The formerly timid Double Blue now came raging over the trap, posturing violently at the first male inside, while the latter, formerly so fierce, made himself as scarce as possible, and did not attempt to fight or posture back."

This principle of belligerent aggressiveness in its own territory and shrinking timidity in its neighbor's territory has been seen in many species of birds. Several examples have been reported of birds that have died from ruptured hearts or aortas, presumably as the result of the emotional excitement of territorial battles (Dilger, 1955). Even baby birds demonstrate the ego-puffing influence of territory possession. Downy chicks of Black-headed Gulls, *Larus ridibundus*, were observed by Kirkman (1937) to drive adult gulls away from their nests. The astonishing part about this is that "outside the nest, the chick is a hunted creature, pecked by every adult it passes and fortunate if it escapes with its life. Yet, inside the nest it opens its stubby wings, raises a small shrill voice, charges heroically, and the adult beats a retreat," A bird is ordinarily invincible in its own territory, or, as Lack puts it, victory "goes not to the strong but to the righteous, the righteous, of course, being the owner of property."

A rare demonstration of the operation of this principle occurred when two male Peregrine Falcons, *Falco peregrinus*, each independently acquired a proprietary interest in the same cliff-side eyrie (Peterson, 1948). Falconers had trapped and removed the original male owner of the territory and, a month later, thinking the eyrie still unoccupied, returned the bird to freedom. However, in its absence another male Peregrine had staked his claim to the cliff.

Hardly had the tercel made a reconnaissance of his home cliff than another male streaked out of nowhere. There had been a replacement after all, . . . each of these birds, no doubt, felt itself to be the rightful owner. The greatest display of flying the falconers had ever seen took place—dog-fights in the air, plunges across the face of the cliff with only inches to spare, barrel rolls, all the maneuvers at their command. Twice the birds grappled on the ledge and tumbled to the wooded foot of the cliff. For three hours the battle lasted until the pale-breasted bird, the latecomer, departed.

When eight male Song Sparrows, *Melospiza melodia*, were captured and held captive, their territories were occupied by floaters. The captives were released 26 days later, and two of them immediately regained some of their previous territory (Smith *et al.*, 1982).

In many species the possession of territory seems to be psychologically necessary for successful breeding. Lack (1940) kept two pairs of European Robins in each of two large aviaries, and in each instance the dominant, territory-owning pair bred, while the subordinate pair showed no traces of breeding behavior. Territory possession, at least in some species, also has a psychological effect on the pair bond of its owners. Lack (1943) tells of one of the rare instances in which a male Robin territory owner was bested in conflict by a vigorous interloper, and eventually chased away. "The hen took no part in this encounter, and after a while began to follow the newcomer about as if he were her mate ('None but the brave,' etc.)." The hen finally reared a family with the victorious male. In some species, the female will join her mate in repelling intruders on their own territory. In such species territory ownership very probably helps to intensify the bond between the mated birds.

Carolina Wrens, *Thryothorus ludovicianus*, form permanent pair bonds, and pairs defend year-long territories as a fighting unit. Males can establish and maintain territories alone, but females cannot. If he dies, she loses her territory to floater males, and eventually she may die also. There is thus division of labor between the sexes in that females assume a predator surveillance role, freeing the male to forage and allot time to singing and patrolling (Morton and Shalter, 1977).

Probably one historical reason for the remarkably emphatic territorialism in birds is their superlative mobility. This brings about frequent contacts between individuals and therefore increases the probability that there will be contests over nest sites, conflicts for food, promiscuity between mated pairs, and unstable family life. Territory ownership tends to put a brake on some of these volatile excesses made possible by flight, and restores some stability to the life of the individual and the family. At the same time, territory is a dynamic force that may be part of the speciation process. The surprising fact is not that birds own real estate, but that, considering the numerous benefits of territory, some species are able to get along without it.

TYPES OF TERRITORY

In different species of birds, territories serve different ends. There have been many classifications of territory based on functions. The following is a modification of those of Mayr (1935), Nice (1941), and Hinde (1956).

The following ten general types of territory may be recognized:

1. Mating, nesting, and feeding territory.
2. Mating and nesting territory.
3. Mating territory.
4. Narrowly restricted nesting territory.
5. Feeding territory.
6. Winter territory.
7. Roosting territory.
8. Group territory.
9. Mobile territory.
10. Superterritory.

1. Mating, nesting, and feeding territories are probably the commonest sort. They are maintained by a great variety of birds, including many woodpeckers, shrikes, thrushes, icterids, warblers, sparrows, and others. When a territory is of this type, courtship, mating, and nest-building all normally occur within it, and after the young hatch, their food comes from it.

2. In mating and nesting territories, reproductive and nesting activities occur on the territory, but the food for the young is obtained elsewhere, often on neutral ground. Examples of such territory are found among some grebes, swans, harriers, and Redwinged Blackbirds and several finches. The male Scarlet Finch, *Carpodacus erythrinus*, which defends its small territory with vigor, will feed amicably with its neighbors on neutral feeding grounds 100 meters to a kilometer distant.

3. Territories restricted exclusively to courtship and mating are held by the males of many promiscuous gallinaceous birds such as the Prairie Chicken, *Tympanuchus cupido*, Sharp-tailed Grouse, *Tympanuchus phasianellus*, and Capercaillie, *Tetrao urogallus*, as well as by males of the Ruff, *Philomachus pugnax*, some birds of paradise, bower-birds, some hummingbirds, some manakins and cotingas. In many of these species the males gather on special display grounds or *leks* where they call, "boom," dance, posture, or—rarely—fight, while the females look on and eventually make their choice of a mate. Males of the Prairie Chicken, for example, strut and boom on their traditional communal booming grounds, each male claiming and defending a particular spot in the area. In one study it was found that fewer than 10 per cent of the males completed 70 per cent of the copulations. Males formed dominance hierarchies, and dominant males held territories in "preferred" central locations within the lek. Thus success in copulations depended on the erotic vigor, social dominance, and position that the cock held in the lek (Wiley, 1973). The male bower-bird builds and decorates his bower, defends it from other males, and may even display in it, months before the breeding season begins. In the breeding season he displays there before a succession of females, and may copulate with them there (Borgia, 1985) (see Fig. 13–10).

4. The fourth type of territory is found among those species that defend only the immediate surroundings of the nest. Many colonial water birds, such as penguins, pelicans, cormorants, shearwaters, gulls, terns, and herons, belong in this category, as well as a few of the solitary-nesting doves, swallows, estrildid finches, and birds of prey. In many colonial birds, the limits of the nest territory are determined by the distance that the sitting bird can jab its beak. The nests of the Peruvian Brown Pelican, *Pelecanus occidentalis*, are spaced about two per square meter, those of the Peruvian Guanay Cormorant, *Phalacrocorax bougainvillii*, about three per square meter. The burrow entrances of the Slender-billed Shearwater, *Puffinus tenuirostris*, may be as crowded as nine per square meter (Murphy, 1936). In a few colonial species such as the Hoatzin, *Opisthocomus hoazin*, anis, *Crotophaga* spp., Galapagos Mockingbird, *Nesomimus macdonaldi*, European Jackdaw, *Corvus monedula*, and the Australian Magpie, *Gymnorhina dorsalis*, the breeding territory is shared in common and defended by all the birds of the colony against intruders of the same species. In addition to cooperating in the defense of the colony, indi-

viduals often must also defend their own nests against other residents of the colony. This is shown by the fact that in many colonial-nesting species and even in some solitary-nesting species, the mutual pilfering of nesting materials is a common practice. Adelie Penguins, *Pygoscelis adeliae*, frequently steal stones from each others' nests.

5. Only a few species are known to have feeding territories separate from their nesting territories. The principal food of the Phainopepla, *Phainopepla nitens*, in the Santa Monica mountains of southern California, is the berry of *Rhamnus crocea* which grows on chaparral hillsides. Trees for nesting sites are found only on canyon bottoms, however, so that pairs defend only small (0.03 hectare) courtship and nesting territories, and commute to the hillsides to forage in their larger feeding territories (Walsberg, 1977). Anna's Hummingbirds, *Calypte anna*, also defend breeding and feeding territories in the breeding season, but only feeding territories in the nonbreeding season (Stiles, 1973).

It is not necessarily the quantity of food that determines territory size; it may be its quality. For the Red Grouse, *Lagopus lagopus*, Lance (1978) discovered that territories became smaller as the nitrogen content of the heather, their main food, became richer.

Since the defense of a feeding territory requires extra energy, birds face the problem of balancing that energy cost against the increased food made available through territory ownership. Hawaiian Honeycreepers, *Vestiaria coccinea*, are primarily insectivorous and nonterritorial in poor flowering years but defend feeding territories in good flowering years. Energy expenditure of territory owners was 17 per cent more than that for nonterritorial individuals (Carpenter and MacMillen, 1976). The benefit of territoriality in this case is not the conservation of foraging time, but the provision of increased resource (flower) availability and predictability. The benefit of defending a predictable renewable resource is illustrated by Gill and Wolf's (1979) study of Golden-winged Sunbirds, *Nectarinia reichenowii*. Individuals owning feeding territories had to forage about 17 per cent more than the rate required for self-maintenance each day to compensate for nectar lost to competing birds; the extra foraging time for non-territory owners, however, amounted to 72 per cent.

Occasionally there are species whose territorial requirements are quite extraordinary. In Nepal, the Orange-rumped Honeyguide, *Indicator xanthonotus*, aggressively defends territory centered around nests of the Giant Honey Bee, *Apis dorsata*, whose honey-

combs he eats (Cronin and Sherman, 1976). When females approach the bee nest to eat the wax, the male copulates with them. In essence, he is trading food for sex.

6. Many species of nonmigratory birds are loosely associated with their territories throughout the year but usually defend them only during the breeding season. Territory may be defended throughout the year by a few species such as the Plain Titmouse, *Parus inornatus*, and Wren-Tit, *Chamaea fasciata*, of California, or the Great Gray Shrike, *Lanius excubitor*, of Europe. True winter territory, however, is an area separate from the mating and nesting territory. Among species known to defend individual winter feeding territories in the United States are the Sanderling, *Calidris alba*, and other shorebirds, the Kestrel, *Falco sparverius*, the Loggerhead Shrike, *Lanius ludovicianus*, and Townsend's Solitaire, *Myadestes townsendi*. In Africa, the following European breeders do the same: the Nightingale, *Luscinia megarhynchos*, European Robin, *Erithacus rubecula*, Pied Wagtail, *Motacilla alba*, Wheatears (*Oenanthe* spp.), and Woodchat Shrike, *Lanius senator*.

In Maryland, Kilham (1958) discovered that the Red-headed Woodpecker, *Melanerpes erythrocephalus*, has a special type of winter territory, in which it conceals stores of acorns in various cavities, and establishes a roost hole. These small storage and roosting territories (Fig. 12–1) are pugnaciously defended, not only against other woodpeckers and such potential acorn robbers as Blue Jays, *Cyanocitta cristata*, and Tufted Titmice, *Parus bicolor*, but also against Starlings, *Sturnus vulgaris*, which attempt to take over the roost holes. The defense is actively maintained from September until early May, when the Red-headed Woodpeckers depart for their mating and nesting territories.

Many arctic-breeding shorebirds wintering in North and South America are also known to defend winter feeding territories (Myers *et al.*, 1979). Arctic-breeding Snowy Owls, *Nyctea scandiaca*, winter in Alberta, Canada, and defend territories. Adult females defend small territories containing a high proportion of stubble field and "edge" habitat rich in rodents. Juveniles defend larger territories with a low proportion of preferred habitat (Boxall and Lein, 1982).

7. Roosting territories are probably the least studied of all territories. Starlings, *Sturnus vulgaris*, show evidence of occupying the same individual perches night after night; and a European Treecreeper, *Certhia*

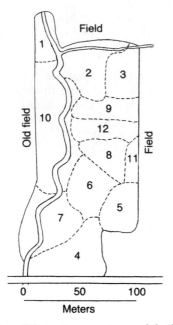

FIGURE 12–1 Winter-time roosting and food-storage territories of 12 individual Red-headed Woodpeckers in a small bottom-land woodlot. Woodpeckers 1, 4, and 10 were immature; the others, adults. After Kilham, 1958.

familiaris, which roosted in a small hole in the bark of a Sequoia tree, vigorously attacked a stuffed creeper placed at the entrance (Rankin, 1940).

8. Occasionally birds will band together and defend a territory occupied by the group. Large colonies of Australian Noisy Miners, *Manorina melanocephala*, will vigorously attack and at times kill birds of their own or any other species that invade their territory, virtually eliminating all competitors (Dow, 1977). Clans of White-fronted Bee-eaters, *Merops bullockoides*, consist of groups made up of breeding pairs and their helpers. Each clan may defend a foraging territory extending up to 7 km away from their (breeding) colony site (Hegner *et al.*, 1982). The Acorn Woodpecker, *Melanerpes formicivorus*, of California deposits enormous numbers of acorns in drilled cavities in tree bark. These caches, representing a group feeding territory, are energetically defended against both intra- and interspecific intruders (MacRoberts and MacRoberts, 1976). Perhaps the acme of group territory development is seen in the multispecies foraging flocks of Peruvian tropical forests. These flocks

are composed of 8 to 12 pairs or families of birds, each representing a different species. They all share a common territory of about 8 hectares that they defend against flocks of similar composition. Defense consists of vigorous singing of each species against conspecifics in the opposing flock (Munn and Terborgh, 1979).

9. When birds make territorial claims on mobile sources of food, the primacy of food as the causal factor in the establishment of their territory seems incontrovertible. South American antbirds (Formicariidae) follow army ants, seizing those insects the ants flush as they march overland. The choicest hunting spots along an ant train are claimed and defended by the largest and strongest species and by the dominant individuals within a species—those high in the peck order. Birds low in the social hierarchy are forced to occupy inferior locations (Willis and Oniki, 1978). In some species a male follows a given female and defends the space around her from conspecific males. The defended space is thus mobile. This behavior is well known in many cardueline species: Rosy Finches, *Leucosticte* spp. (Johnson, 1965), and House Finches, *Carpodacus mexicanus*, (Thompson, 1960), and also in the Brown-headed Cowbird, *Molothrus ater* (Darley, 1983).

10. Superterritories are territories containing more resources than are necessary for an individual's survival and reproduction. They may function to hold a food reserve as insurance against a bad year (Wassermann, 1983). Superterritories may also function to inhibit competitors by preventing their use of vital resources. Verner (1977) postulated that such spiteful behavior may increase an individual's own fitness relative to that of his competitors. The existence of such territories has been demonstrated by Wassermann (1983) for Rufous-sided Towhees, *Pipilo erythrophthalmus*. He mapped out Towhee territories, then used crop-protection netting to cover 10, 20, 30 or 50 per cent of each territory, rendering the arthropods beneath the netting unavailable to the birds. No significant increase in territory size was noted until 50 per cent of the territory was covered with netting. Davis (1982) reported that 28 per cent of nest sites available to Belted Kingfishers, *Megaceryle alcyon*, were unoccupied due to competitive exclusion.

FACILITIES PROVIDED BY TERRITORIES

Although spatial isolation is the irreducible common denominator of territories in general, territories must also be places of biological utility. They cannot be simply blocks of empty space. Therefore the biological uses to which a bird puts its territory determine its composition or general nature. The typical diet of a bird dictates that insects, worms, berry bushes or the like be available in territories that embrace feeding activities. Each pair of Ivory-billed Woodpeckers, *Campephilus principalis*, was estimated by Tanner (1942) to require not less than 15.5 sq km (6 sq mi) of primeval wilderness. This species fed on the wood-boring grubs of trees that had been dead for two or three years. Its disappearance coincided with lumbering operations that eliminated the available habitats of required size. Although possibly extinct from the North American mainland, Ivory-billed Woodpeckers have recently been sighted in Cuba (Horn and Short, personal communication).

For nesting territories, a suitable nest site, nesting materials, and possibly a singing post are required, depending on the species. In one locality in Panama, one-half of the failures of Rieffer's Hummingbird, *Amazilia tzacatl*, to rear broods were attributed by Skutch (1931) to the thievery of nesting materials. Territory defense in this species is either very weak or nonexistent. One female was observed to make 12 fresh attempts to build a nest, each of which was frustrated by the theft of material by hummingbirds of the same species or by birds of other species. Many species require nest cavities that they themselves are unable to make. How the availability of such cavities controlled the number of territories in a given area was revealed in an experiment in which 100 nesting boxes were hung in a German orchard (Creutz, 1949). In the first seven years with increased nesting facilities, the population of Pied Flycatchers, *Ficedula hypoleuca*, rose from two to six pairs, but in the next five years it increased phenomenally to 45 pairs. Presumably something other than simple nest-hole availability operated here, or else the initial increase would have reached the maximum more quickly. A Peregrine Falcon, *Falco peregrinus*, commonly demands for its nest site a nearly vertical cliff with a narrow graveled ledge overlooking a wide span of lowland. Ledges high up on office buildings in large cities sometimes answer this requirement and are consequently chosen by falcons for nesting sites. The territory of the female parasitic European Cuckoo, *Cuculus canorus*, is determined by the location of the nests she parasitizes. Shrike territories must include the thorny shrubs on which the birds may impale their hapless victims.

Scandinavian Black-headed Gulls, *Larus ridibundus*, are reported by Svärdson (1958) to have five requirements for nesting territory: proximity to water; dry nest foundations; unimpeded view in all directions; isolation by water from terrestrial predators; absence of trees. The social Acorn Woodpeckers, *Melanerpes formicivorus*, maintain year-round territories in which they defend interspecifically their acorn caches, acorn-source trees, sap trees, "anvils," insect-hawking perches, and roosts (MacRoberts and MacRoberts, 1976). Some small passerine tropical birds choose nesting territories, possibly for protection, in trees that also hold nests of wasps or ants. If a given habitat has the optimum environmental ingredients required by a certain species for its territory, that habitat becomes magnetically attractive to unsettled birds. This was clearly revealed in Hensley and Cope's (1951) study already mentioned, in which 528 adult birds were removed in 49 days from a 16-hectare forest that originally held 154 territory-owning birds. As the replacement birds moved in, they established territories in the same places that had been occupied by predecessors of the same species.

For some species requirements in the way of a nesting site are very particular. As a consequence there may not be many such nesting sites available and the birds may be forced to accept highly restricted territories. There is an Ecuadorian hummingbird, *Oreotrochilus chimborazo*, whose staple food is the nectar of a flowering shrub, *Chuquiraga acutifolia*, that flourishes on volcanic peaks at elevations of 4000 to 4500 meters. The birds build their nests on walls of ravines under overhangs that protect their young from the powerful noonday sun and the almost daily hailstorms. The rarity of such sites has forced these birds to build as many as five occupied nests within a radius of 2 meters (Smith, 1969).

SHAPES AND SIZES OF TERRITORIES

If space were the only consideration, territories would probably be circular in shape, or perhaps hexagonal like the cells of a honeycomb, since the territory boundaries of one bird would flatten against those of his surrounding neighbors. Hexagonal nesting territories have been described for Royal Terns, *Thalasseus maximus* (Buckley and Buckley, 1977). The radius of each territory would be determined by the time and energy that its owner could devote to defending his area. The geometry of areas tends to keep territories small, since the area that has to be defended increases as the square of its radius. Thus, doubling the radius of any territory quadruples the space to be maintained. The territory of the Chestnut-collared Longspur, *Calcarius ornatus*, approximates an ideal situation in the prairies of Manitoba. There each Longspur territory is approximately circular, with a radius of about 26 to 37 meters (Harris, 1944). However, in environments that are less homogeneous, the size and shape of a territory will vary with the surroundings. If a bird's territory abuts against a natural boundary that needs no defense, such as a river, the territory can be larger since there will be that much less boundary to defend. Should a territory characteristically follow a shoreline, it may be long and narrow, as it is for some coots or for the West Coast Salt Marsh Song Sparrow, *Melospiza melodia*, whose territory averages 40 to 52 meters in length and only 9 meters in width (Johnston, 1956). Such long, narrow plots have more boundary to be defended than circular ones of equal area.

The sizes of avian territories span a tremendous range. As Table 12–1 shows, the Golden Eagle with its 93,000,000 square meters of territory occupies an area roughly 133,000 times as large as that of a Least Flycatcher with its 700 square meters.

Territory size varies for a number of reasons: important variables include: the function of the territory; colonial or solitary habit of life; food; population density; vegetation density; availability of suitable habitat; time of year; sex, age, and size of bird; its individual aggressiveness, and others. The most important variable of all is the species of bird. In a very

rough way there is a direct correlation between the size of a bird and the size of its territory. This is particularly true for predators. Predators have larger territories than herbivores or omnivores of equal size. Feeding-territory size must vary of course with the abundance of suitable available food—the less food, the larger the territory. For predators the total mass of usable food per unit of habitat area decreases as their weight increases (Schoener, 1968). Probably for this reason, feeding-territory size increases more rapidly with body weight for predators than for other birds. The European Woodchat Shrike, *Lanius senator*, feeds chiefly on insects and has a territory size of between 4 and 12 hectares. Its close relative the Great Gray Shrike, *L. excubitor*, feeds largely on small birds and rodents; its territory varies between 46 and 98 hectares (Ulrich, 1971).

Territorial defense, of course, requires the investment of energy. An optimal territory size is one that produces the most benefits to its owner for the least energy invested in its defense. For many species defense strategy depends on the food supply. If food is very abundant, pressure from intruders increases and the owner may relinquish its territory. If, on the other hand, food is so scarce that a bird cannot meet its energy needs, it may abandon its territory. Thus, there are upper and lower thresholds for economical defense (Davies, 1978). Ordinarily, the poorer, or more scattered, the food supply, the larger the territory. Thus, in the high lemming year of 1953, the predatory Pomarine Jaegers, *Stercorarius pomarinus*, at Point Barrow, Alaska, defended territories covering

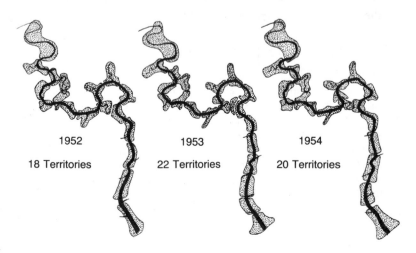

1952
18 Territories

1953
22 Territories

1954
20 Territories

FIGURE 12–2 Territories occupied in successive years by Salt Marsh Song Sparrows along a tidal slough, San Francisco Bay. Individual territories expand or contract in size according to population density. After Johnston, 1956.

TABLE 12–1.
TERRITORY SIZES

SPECIES	LOCALITY	SIZE OF TERRITORY (SQUARE METERS)*	AUTHORITY
Black-headed Gull, *Larus ridibundus*	England	0.3	Kirkman
King Penguin, *Aptenodytes patagonicus*	Antarctica	0.5	Murphy
Least Flycatcher, *Empidonax minimus*	Michigan	700	MacQueen
European Blackbird, *Turdus merula*	England	1200	Snow
American Robin, *Turdus migratorius*	Wisconsin	1200	Young
Willow Warbler, *Phylloscopus trochilus*	England	1500	May
Red-winged Blackbird, *Agelaius phoeniceus*	Wisconsin	3000	Nero
Coot, *Fulica atra*	England	4000	Cramp
House Wren, *Troglodytes aëdon*	Ohio	4000	Kendeigh
American Redstart, *Setophaga ruticilla*	New York	4000	Hickey
Chaffinch, *Fringilla coelebs*	Finland	4000	von Haartman
Song Sparrow, *Melospiza melodia*	Ohio	4000	Nice
European Robin, *Erithacus rubecula*	England	6000	Lack
Oven-bird, *Seiurus aurocapillus*	Michigan	10,000	Hann
Hazel Grouse, *Tetrastes bonasia*	Finland	40,000	Pynnönen
Song Thrush, *Turdus ericetorum*	Finland	40,000	Siivonen
Black-capped Chickadee, *Parus atricapillus*	New York	53,000	Odum
Western Meadowlark, *Sturnella neglecta*	Iowa	90,000	Kendeigh
Great Horned Owl, *Bubo virginianus*	New York	500,000	Baumgartner
Mistle Thrush, *Turdus viscivorus*	Finland	500,000	Siivonen
Red-tailed Hawk, *Buteo jamaicensis*	California	1,300,000	Fitch *et al.*
Bald Eagle, *Haliaeetus leucocephalus*	Florida	2,500,000	Broley
Crowned Hornbill, *Lophoceros melanoleucos*	Africa	5,200,000	Ranger
Ivory-billed Woodpecker, *Campephilus principalis*	U.S.A.	7,000,000	Tanner
Powerful Owl, *Ninox strenua*	Australia	10,000,000	Fleay
Golden Eagle, *Aquila chrysaëtos*	California	93,000,000	Dixon
Bearded Vulture, *Gypaëtus barbatus*	Spain	200,000,000	Terrasse

*One *hectare* = 100 meters square (or 10,000 square meters, or 2.47 acres)
One *square kilometer* = 1,000 meters square (or 1,000,000 square meters, or 0.386 square mile)

6 to 9 hectares. In the previous low lemming year, territories averaged 45 hectares (Pitelka *et al.*, 1955). Similarly, territories of the Snowy Owl, *Nyctea scandiaca*, vary from 120 hectares in peak lemming years to about 5000 hectares in years of scarcity (Pitelka and Schoener, 1968). In the Short-eared Owl, *Asio flammeus*, territory size is negatively correlated with abundance of prey, but Tawny Owl, *Strix aluco*, territories remain constant year after year despite large variations in mouse abundance (Lack, 1966).

Nuthatch pairs make an investment in their year-round territories by storing seeds in crevices in bark,

then defend these territories with vigor, notably in the fall when juveniles are trying to settle (Löhrl, 1957). In years of poor beech mast crop, Nuthatch territories increased threefold in size. By artificially provisioning territories with sunflowers, Enoksson and Nilsson (1983) noted that territory sizes decreased. Thus optimal territory size is larger when food is scarcer. Nuthatch territories were adjusted to ensure an adequate winter food supply.

Individuals of the same species often show variations in territory size even in the same habitat and at the same time of year. Territories of American Redstarts,

Setophaga ruticilla, vary from 0.06 to 0.4 hectare; of Song Sparrows, *Melospiza melodia*, in Ohio, from 0.2 to 0.6 hectare. In Algonquin Park, Ontario, territory held by the Ovenbird, *Seiurus aurocapillus*, varies in size from 0.32 to 1.74 hectares, depending on the type of forest cover in which the bird lives. The area is smallest in stands of aspen, intermediate in conifer-birch and mixed stands, and largest in maple stands. The size of the territory apparently increases with increasing height and density of the forest canopy, and with decreasing ground vegetation (Stenger and Falls, 1959). A study by Weeden (1965) on territoriality in Tree Sparrows, *Spizella arborea*, showed that territories defended by females were almost always smaller than those defended by their male partners. Adjacent territories showed a modest degree of overlap (Fig. 12–3).

Territory size may be a product of sexual selection and may influence a bird's reproductive success. Potter (1972) found that male Savannah Sparrows, *Passerculus sandwichensis*, with territories ranging in size from 601 to 1200 square meters were more successful in attracting females (56 to 89 per cent of the territories with nests) than males holding territories smaller than 601 square meters (11 per cent with nests).

Huxley (1934) has compared a territory with an "elastic disc." As numbers of individuals increase with the season, territorial neighbors may compress territory size, but only to a critical minimum dimension (see Fig. 12–4). A bird is fiercest towards the center of its territory, but the territorial instinct "dies out gradually towards a certain radius from the center." By playing conspecific songs to Redstarts, *Setophaga ruticilla*, or White-throated Sparrows, *Zonotrichia albicollis*, at the center and edge of their territories, investigators have shown a constant level of aggressive response at the center and a significant drop outside, lending support to Huxley's ideas (Ickes and Ficken, 1970; Melemis and Falls, 1982).

Seasonal changes in territory size occur in many species. Early- arriving males of McCown's Longspur, *Calcarius mccownii*, in Wyoming, claim large territories that shrink progressively as other males arrive, until, later in the spring, a minimum is reached beyond which population pressure cannot compress territory size (Mickey, 1943). A number of species show a similar compressibility of territory size in response to population pressure: Little Ringed Plover, *Charadrius dubius*, House Wren, *Troglodytes aëdon*, European Robin, *Erithacus rubecula*, and others.

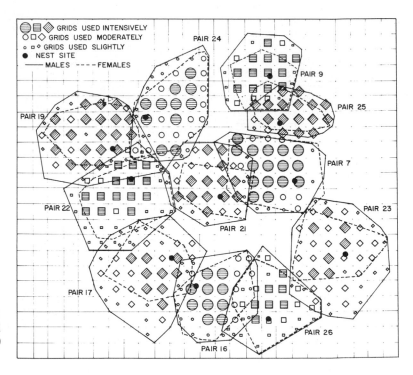

FIGURE 12–3 Territory utilization by 11 pairs of Tree Sparrows in Alaska. Each grid square is approximately 20.1 meters (66 ft) on a side. From Weeden, 1965.

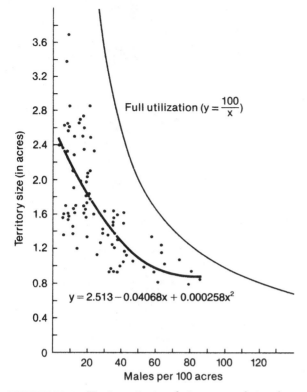

Full utilization $(y = \frac{100}{x})$

$y = 2.513 - 0.04068x + 0.000258x^2$

FIGURE 12–4 Territory size in relation to population density in male Dickcissels, illustrating Huxley's elastic disc theory and compressibility of territories. At low population densities territories may be large, e.g., at about 10 males/100 acres, territory size is about 2.5 acres. At about a density of 60/100 acres, territories are compressed to a size of about 0.9 acres, and decrease little beyond that. From Zimmerman, 1971.

However, the Song Sparrow, *Melospiza melodia*, will not allow itself to be crowded. Territories in Ohio (Fig. 12–5) were found to be as large in a year of surplus population as in years of normal population density (Nice, 1937). Even birds of the same genus may show remarkable differences in regard to territorial compressibility. Territory size among the polygamous bishop birds of Africa was found to be almost indefinitely compressible in the Zanzibar Red Bishop, *Euplectes nigroventris* (from about 1000 to 8 square meters), but remained relatively fixed in the Crimson-crowned Bishop, *E. hordacea* (Moreau and Moreau, 1938). Bishop birds occupying the smallest territories were as successful in reproduction as those having large territories.

Territory size is subject to still other modifying influences, such as food supply. Territories of the Chaffinch, *Fringilla coelebs*, in food-poor spruce forests are six to eight times larger than those in food-rich birch forests (von Haartman, 1971). Similarly, territories of the Blue-throated Hummingbird, *Lampornis clemenciae*, shrink in size from about 3000 to 700 square meters as the abundance of their chief nectar source, *Penstemon kunthii*, increases from about 12 to 110 blossoms per square meter as the summer progresses (Lyon, 1976). Geographical location is another factor: the territory of the Black Woodpecker, *Dryocopus martius*, varies in size from 120 to 180 hectares in Bohemia to 400 hectares in Germany, 850 hectares in the Netherlands, and 800 and 1800 hectares in Finland. A male Field Sparrow, *Spizella pusilla*, in Michigan increased the size of his territory with age and experience. In six consecutive summers, he controlled territories of the following sizes: 0.4, 0.5, 0.8, 0.8, 0.8, 0.8 hectare (Walkinshaw, 1945). Studies of birds nesting on small islands in Minnesota lakes showed that several species could successfully raise families on much smaller pieces of land than they required on the mainland. Territory size of Song Sparrows on Mandarte Island, British Columbia, are from one-third to one-ninth the size of those on the mainland. Here it is because the sparrows are able to find abundant food nearby but not on their territories (Tompa, 1963). This illustrates the fact that, for some birds at least, the various territorial functions are not bound together in a rigid package but may be adjusted to suit local opportunities.

Water surface itself may constitute territory. In three species of Australian ducks observed by Robinson (1945), the males defend certain areas of water, about 3 hectares or less in extent, which may change from day to day. They drive away other males of the same species, but will tolerate both sexes of other species. Facts such as these emphasize the importance in some instances of spatial isolation rather than the physical resources of the environment in determining territory size.

DEFENSE OF TERRITORY

A bird may acquire territory in several ways: through dominance in social contests; through death or emigration of a territory-holding neighbor; through early arrival in migration; and, in family groups, the young (often "nest helpers") may "inherit" a territory released through the senescence or death of an older group

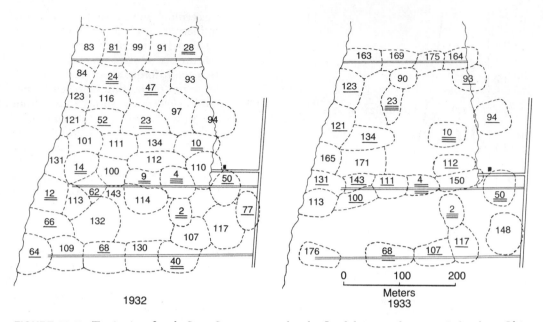

FIGURE 12–5 Territories of male Song Sparrows on a brushy floodplain meadow near Columbus, Ohio. Size of individual territories remains approximately constant regardless of population density. A bird present in the general area the preceding year is underlined, and a line is added for each subsequent year. After Nice, 1937.

member. Once occupied, a territory must be maintained.

Birds may defend their territories by voice, by theatening postures, and, failing these, by pursuit, or by actual physical combat. In territorial defense, song is a particularly economical repellent, effective at a distance and capable of penetrating visual barriers. Visual displays, pursuit, and actual combat are more commonly used in "hand-to-hand" conflicts. Although a bird's aggressive defense of its own territory is an obvious territorial behavior mechanism, the tendency of an invading bird to retreat once it sets foot on a stranger's territory is just as essential a device. What the bird is defending in a particular territory is rarely apparent, and it may not become so even after prolonged study. And the object of a bird's defense may change with the seasons. It seems likely that the chief objects of territorial defense are the mate, the nest, the young, other members of a colony, a song or lookout post, a display ground, and food (Hinde, 1956).

Ordinarily it is the male that arrives first in the spring and establishes and defends the territory while

he waits for the female to arrive. Song Sparrows, Peregrine Falcons, *Falco peregrinus*, and many other species begin the breeding season in this way. Among species in which the birds are already paired when they arrive, for example, the Eastern Kingbird, *Tyrannus tyrannus*, the territory may be established by both sexes. In the polyandrous Chinese Pheasant-tailed Jaçana, *Hydrophasianus chirurgus*, and the Red-necked Phalarope, *Phalaropus lobatus*, of Greenland, it is the more energetic female that establishes and defends the territory while the drab male defends the nest, incubates the eggs, and takes care of the young.

In typical territory establishment the male chooses a territory and begins advertising his ownership of it by singing or by otherwise making himself conspicuous. If, in the course of territory defense, a male should chase an intruder across the territorial boundary into its own area, the roles of owner and intruder become reversed. The vociferous male that just 5 seconds before was burning with righteous wrath now suddenly becomes a silent, cowering intruder himself, and is quickly put to rout by his suddenly dominant neighbor. Territorial boundaries are frequently established by a delicate balancing of these see-saw reactions.

There is a great difference in the vigor with which different species defend their territories. In Finland Pied Flycatchers, *Ficedula hypoleuca*, and Great Tits, *Parus major*, sometimes kill one another in their competition for nesting boxes (Järvi *et al.*, 1978). A few kinds of tyrant flycatchers (Tyrannidae) are particularly bellicose, driving away not only other individuals of their own kind, but also dangerous birds of other species, small mammals, and even humans. Several species of grebes, plovers, woodpeckers, hummingbirds, wrens, shrikes, and thrushes are all known to defend territories against both their own and other species. The European Capercaillie, *Tetrao urogallus*, is famous for attacking humans that may approach its territory, and Sandhill Cranes, *Grus canadensis*, have been observed repelling an adult caribou from their territory (Miller and Broughton, 1971). At the other extreme are the mild-mannered Masked Tityras, *Tityra semifasciata*, which seem to disdain physical combat and defend their nesting territories with long-drawn-out palavers and half-hearted feints (Skutch, 1946). Vigor of territory defense also varies seasonally, often being most intense early in the breeding season when boundaries are being established, and at low ebb after the young have fledged, as well as during the winter.

Birds probably use singing more than any other single means to defend their holdings against intrusion. When male Red-winged Blackbirds, *Agelaius phoeniceus*, were experimentally made mute, their neighbors increased both the frequency and length of their intrusions. But after the experimental birds regained their voices, they successfully regained their territorial boundaries (Smith, 1979). Whereas springtime singing in the European Robin, *Erithacus rubecula*, may serve the two functions of advertising for a mate and defending the territory against trespass, the vigorous singing of the autumn is exclusively territorial (Lack, 1943).

Many species defend their territories with threat displays usually involving stereotyped postures or flights. These may or may not accompany song. In his experiments with stuffed robins, Lack (1934) discovered that the red breast of a trespassing bird had the same effect on a territory's owner as the proverbial red flag has on a bull. In response to the invader's "red flag," the owner postures so as best to display his own red breast to the interloper. If this should not bluff the stranger into retreating, the owner may resort to physical attack. Lack also found that the red breast feathers are themselves a releaser of an antagonistic response in another robin. A mere tuft of red feathers placed in a bird's territory during the fighting season will suffice to bring on violent posturing and even battle, whereas an entire stuffed robin colored brown (like an immature bird) may be ignored! Yellow Wagtails, *Motacilla flava*, similarly display aggressively in their border clashes by puffing out their bright yellow breasts at each other like miniature pouter pigeons (Smith, 1942) (Fig. 12–6). Experiments on various species, using stuffed dummies, indicate that form and color must be accompanied by movement, and perhaps by sound, before lively territorial defense will be released.

FIGURE 12–6 Territorial threat display in male Yellow Wagtails. The birds follow this puffing of the bright, yellow breast feathers by swaying from side to side, then by a pecking and clawing combat or chasing flights. The female does not take part in these disputes. After Smith, 1942.

The significance of color in territory defense involving visual display was nicely verified in experiments by Smith (1972). The red and yellow epaulet (wrist) feathers of 37 male Red-winged Blackbirds, *Agelaius phoeniceus*, were dyed black like the rest of the wing; 32 control males were treated with alcohol minus the dye. Birds were then released near their territories with the result that 64 per cent of the dyed males lost their territories as against only 8 per cent of the controls.

Instead of singing or posturing to display their red breasts, American Robins, *Turdus migratorius*, of both sexes threaten intruders by lifting their tails, crouching, and then making a series of short runs broken by brief pauses. Robin territories frequently overlap (Young, 1951).

If an intruder enters the territory of a Herring Gull, *Larus argentatus*, the owner assumes a threatening attitude and starts walking slowly toward him. If the trespasser holds his ground, the owner bends down and tears out a large mouthful of grass and the intruder does likewise. Sitting opposite each other the birds attempt to pull the material from each others' beaks, and then try to take hold of a wing or beak and begin struggling. Other species, such as some grouse, some hummingbirds, the Great Tit, *Parus major*, the Skylark, *Alauda arvensis*, the Field Sparrow, *Spizella pusilla*, the Lapwing, *Vanellus vanellus*, and Woodcock, *Scolopax minor*, use specifically patterned flights to intimidate territorial competitors. In addition to chasing each other through the air, two male Hazel Grouse, *Tetrastes bonasia*, may contest their territorial boundaries by running side by side on the ground for as long as 45 minutes at a time (Pynnönen, 1950).

Actual physical combat is resorted to in the defense of territory less often than song, posturing, flight, and other nonviolent means of self-assertion. But when fighting does occur it may be lusty and even vicious. White Storks, *Ciconia ciconia*, may fight so vehemently over nest territory that eggs are crushed and sometimes one of the combatants is killed (Schüz, 1944). Penguins often protect their tiny nesting sites with great vigor because of the scarcity of suitable nesting territories. They typically beat one another with their flippers, and jab and bite with their formidable beaks. In describing fights between Humboldt Penguins, *Spheniscus humboldti*, Kearton (1930) remarks, "Combat over possession of burrows is also common between two or more pairs, the fighting going on until all the participants are blood smeared." Mayaud (1950) estimates that territorial combat may be the

chief cause of the high mortality among Emperor Penguins, *Aptenodytes forsteri*. Penguins frequently are seen with one eye missing as a result of either territorial or courtship battles. On rare occasions territorial battles may end in death. Brüll (1972) tells of 20 instances in which adult trespassing Goshawks, *Accipiter gentilis*, were killed either on territories of other Goshawks or of large owls or eagles.

Among the Falkland Flightless Steamer-Ducks, *Tachyeres brachypterus*, both sexes defend their nesting territory. When one pair attempts to invade the territory of another, each bird attempts to seize its opponent by the neck and hold its head under water, meanwhile beating the bird with its hard wings. Murphy (1936) noted that if one succeeds in submerging the head of the other, the beating is redoubled and the water is frequently reddened by blood. Kermadec Petrels, *Pterodroma neglecta*, of South Pacific islands, sometimes nest so close together in their colonies that the adults engage in territorial bickering to such an extent that their young are neglected and become "dwarfed starvelings" (Murphy).

In some species, such as the Herring Gull, *Larus argentatus*, or the European Robin, *Erithacus rubecula*, both the male and female help defend their territory against invaders of either sex, probably because in these species both sexes look alike. In many species with pronounced sex dichromatism, such as the American Goldfinch, *Carduelis tristis*, and the Cardinal, *Cardinalis cardinalis*, males defend their territory against males, females against females. In most wood warblers the male defends the territory proper, and the female defends a smaller area around the nest. Hens of the Hazel Grouse, *Tetrastes bonasia*, neither aid the cocks in defense of the territory nor observe territorial limits when wandering with their broods. In the semicolonial and polygamous Red-winged Blackbird, *Agelaius phoeniceus*, the male defends a sharply defined and stable territory against intruding males but not against females. Each of the females nesting in his territory defends its individual nest against harem mates. In an experiment by Nero and Emlen (1951), males tolerated the introduction of alien nests, moved in gradually with their supporting cattail clumps, and eventually accepted the females that persisted in following their moved nests. But the males would defend neither nests nor females of their harems if these were moved out into neighboring territories. Male Redwings may cooperate, however, in driving a common enemy from the colony. Herring Gulls have been observed protecting not only their own nests but those of their neighbors by chasing away gulls from other colonies.

As a rule, a bird will tolerate birds of other species in its territory and drive out only birds of the same species, especially males. Sometimes as many as six different species of birds may be found nesting within a radius of 10 meters. A Bald Eagle, *Haliaeetus leucocephalus*, and a Great Horned Owl, *Bubo virginianus*, were found in a Florida tree incubating eggs less than 1 meter apart (Peterson, 1948). But occasionally territorial species are intolerant of other species. The European Fieldfare, *Turdus pilaris*, will drive from its territory jays, woodpeckers, and even Ravens, *Corvus corax*. The Wood Thrush, *Hylocichla mustelina*, repels Robins, *Turdus migratorius*, and Veerys, *Hylocichla fuscescens*; and the Black Oystercatcher, *Haematopus bachmani*, will expel from its territory gulls, curlews, eagles, crows, and ravens. Some of these birds are repulsed as egg robbers rather than as territorial intruders. Song Sparrows, *Melospiza melodia*, are known to drive away at least 16 species of birds (Nice, 1937).

FAITHFULNESS TO TERRITORY

A question commonly raised is whether the robins nesting in a particular yard are the same pair that nested there the year before. Numbered aluminum leg-bands have permitted the positive identification of millions of individual birds. A study of nation-wide banding returns by Hickey (1943) indicated that about 74 per cent of the robins in various parts of the United States returned to within 16 kilometers of their original homes. The European Nutcracker, *Nucifraga caryocatactes*, is reported to mate for life and to keep the same territory year after year, one pair for as long as ten years (Swanberg, 1951).

This tendency of many migratory birds to return to their natal territory, or, more commonly, to the site or area of their first breeding season, has been given a variety of names: place faithfulness, site fidelity, site tenacity, philopatry, or (in German) *Ortstreue*. In her study of Song Sparrows, *Melospiza melodia*, Nice found that under favorable conditions over 60 per cent of the breeding males and over 12 per cent of the fledged nestlings returned to the home locality the following year. Birds show a striking site fidelity even for inhospitable antarctic homes. In the ten years between 1948 and 1957, some 7200 penguins, petrels,

and other sea birds, 17 species in all, were banded in antarctic breeding colonies. In spite of wandering thousands of trackless kilometers over the world's oceans between breeding seasons, a great many of these birds returned to their former nest sites, often with their former mates (Sladen and Tickell, 1958). But very few of the birds banded as nestlings were recovered in subsequent seasons at their birthplaces. Carbon dating of frozen mummies of Adelie Penguins, *Pygoscelis adeliae*, found in the ice at Cape Hallet, Victoria Land, indicates the continuous residence of a colony there for some 650 years (Austin, 1961). Carbon dating of peat deposits on Marion Island indicates that Macaroni Penguins, *Eudyptes chrysolophus*, first bred there about 7000 years ago (Lindeboom, 1984). In Germany there is a rookery of Herons, *Ardea cinerea*, near the castle of Morstein, Baden-Württemburg, that has been in existence since well before 1500. An ancient "heron war" was fought over the colony, since in those days herons were game reserved for royalty (Kramer, 1972).

Richdale and Warham (1973) studied the albatross, Buller's Mollymawk, *Diomedea bulleri*, in two New Zealand rookeries. By banding adult birds and by driving numbered pegs near occupied nests in 1948, the authors determined that six birds were still occupying original nest sites as long as 13 to 22 years later, and three others were still in the same general area 19, 21, and 22 years later.

Of 113 female Mallard Ducks, *Anas platyrhynchos*, caught and banded at their artificial nest baskets in North Dakota, 46 per cent returned at least once in subsequent years to the marshes where they were banded; of these, two-thirds returned to the same baskets where originally caught. Notably, the return rate of previously successful nesters (52 per cent) was significantly higher than that of unsuccessful nesters (16 per cent) (Doty and Lee, 1974).

Several studies of faithfulness to breeding territory in the Pied Flycatcher, *Ficedula hypoleuca* (von Haartman, 1949, 1960; Berndt, 1973), have shown that males are much more loyal to their first breeding territory than are the more nomadic females. In successive years a female's breeding sites may be 150 kilometers apart. If she loses her first brood, she seldom returns to the same nest site and may raise a second brood as much as 50 kilometers away that same season—a form of reproductive nomadism. A comparable sexual difference in faithfulness to the natal territory appears in the European White Stork, *Cico-*

nia ciconia. The average distance between birthplace and the site of first breeding as an adult is 33 kilometers in males and 61 kilometers in females (Zink, 1967). Females of many species generally disperse farther than males from birthsite to breeding site (Greenwood, 1980; Greenwood and Harvey, 1982). This bias in dispersal discourages inbreeding.

The attachment that some species have for their nesting territories apparently increases with age and with each additional occupancy of the nest site. Extensive studies of the Common Tern, *Sterna hirundo*, by Austin (1940, 1949) have shown a remarkable tendency toward site fidelity in this species. Of 2964 Cape Cod terns that returned, after being banded, two or more times to the general location of their natal colony, 76.5 per cent returned to the same nest site on their second return; then, as adult birds returned to the colony in three, four, five, and more successive years, increasingly greater percentages of them returned to the same nest sites. One island that originally held a flourishing colony of Arctic Terns, *S. paradisaea*, became so completely overgrown with shrubs that it was unsuitable as nesting terrain. Nevertheless, a few of its oldest tern inhabitants persisted in nesting there under conditions they would never have tolerated elsewhere. The same stubborn site fidelity in the face of a deteriorating ecological habitat has been observed in elderly Blue Grouse, *Dendragapus obscurus* (Zwickel and Bendell, 1972).

A common pattern of site fidelity is observable in the fact that many adult birds return to their nest territories in subsequent years but very few or none of the newly hatched young return. This is due, no doubt, partly to the higher mortality among the young birds, partly to their greater tendency to wander, and partly to the greater attachment of the older birds to the nest territory and to their territorial dominance over, and aversion to, any young that might attempt intrusion. A 14-year study of Tree Swallows, *Iridoprocne bicolor*, in Massachusetts by Chapman (1955) revealed that the percentages of adult swallows banded any given year and returning the next varied from 7 to 55 per cent and averaged 39.6 per cent. The percentage of nestlings banded any year that returned the next year varied from 0 to 5.0 per cent and averaged 2.4 per cent. A similar study of site fidelity in the House Wren, *Troglodytes aëdon*, in Ohio by Kendeigh and Baldwin (1937) showed that of 1831 banded, nesting adults, 631 (34 per cent) returned in subsequent years; and of 7375 banded young, only 152 (2.06 per cent)

returned. The Rose-colored Starling, *Sturnus roseus*, shows essentially no attachment to its birthplace or former breeding site because it feeds, during its breeding season, on swarms of migratory locusts, and these may occur in widely different areas in different years.

Migratory birds become attached to their winter quarters also, but probably not with the fervor that nesting territories command. A 12-year study made by Wharton (1941) in South Carolina revealed some striking inconsistencies between two thrush species in their faithfulness to winter territory. Of 72 American Robins, *Turdus migratorius*, banded there, none ever returned, but of 81 Hermit Thrushes, *Cathanus guttatus*, 12.3 per cent returned in subsequent winters. Of 503 migratory White-crowned Sparrows, *Zonotrichia leucophrys*, banded in their California winter quarters, 34 per cent returned at least once in the next three winters (Mewaldt, 1976).

Each year a flock of about 150,000 Barn Swallows, *Hirundo rustica*, spend their winter roosting nightly in the central business district of Bangkok, Thailand (King, 1969). In 1965 some 20,500 of these birds were banded, and in the following winter about 48 per cent were recaptured. This kind of fidelity to winter quarters appears to be rare. Some birds from ten species of North American wood warblers banded in their winter quarters in Jamaica returned in subsequent winters. The percentage of recaptures varied from 1.4 per cent for the Northern Waterthrush, *Seiurus noveboracensis*, to 17.4 per cent for Swainson's Warbler, *Limnothlypis swainsonii* (Diamond and Smith, 1973).

Moreau (1969) has summarized records of returns of various Eurasian-breeding migrants to their trans-Saharan winter quarters in Africa. In one study, of about 10,000 Yellow Wagtails, *Motacilla flava*, banded in northern Nigeria, 248 were recaptured in the first subsequent winter, 97 in the second, 23 in the third, 6 in the fourth, 2 in the fifth, and 1 in the seventh—a total of about 4 per cent of all birds banded. A variety of other species, in smaller numbers, give parallel records of site fidelity to winter territory. Moreau believed that these heroic journeys to African winter territories began more than 5000 years ago when the Sahara was less of a barren waste than it is today.

The numerous advantages that a territory confers on its owner are reasons enough for a bird to become attached to it. Site fidelity is a learned phenomenon, at least in some birds. Ingenious experiments by Löhrl (1959) with hand-raised Collared Flycatchers, *Ficedula albicollis*, have shown that imprinting early in life determines site faithfulness in this species. A total of 134 juvenile flycatchers, taken from their nests and hand-reared near Stuttgart, southwest Germany, were displaced from their birthplace to a region 90 kilometers south, where the species does not naturally occur. Of those birds displaced two or three weeks before the onset of fall migration, 19 per cent returned to the vicinity of the release site in later years. The return of the males was much better than that of the females. Of 68 other young flycatchers that were displaced and released *after* fall migration had begun, none was ever seen again. Apparently a remarkably short period of about 10 days sufficed to imprint on the first group of juveniles the new "home" range to which they returned and nested.

More recent experiments by Berndt and Winkel (1975, 1979) with Pied Flycatchers, *Ficedula hypoleuca*, banded as wild nestlings reared either in deciduous or pine forests, showed that as returned adults, about 75 per cent preferred deciduous forests to 25 per cent that preferred pine forests regardless of the forest type of

their birthplace. There seems to be no imprinting on the birthplace plant biotope. Nevertheless, birds transferred as eggs, nestlings, or fledglings 250 km away from their original home and released, returned to breed in the release area rather than their birthplace. Faithfulness to birthplace is therefore attributed to an imprinting-like process and not to heredity.

■ SUGGESTED READINGS

Howard's *Territory in Bird Life*, published in 1920, is still the classic in this field. Armstrong's *Bird Display and Bird Behaviour* emphasizes the reproductive aspects of territory. For excellent studies of territorial behavior of specific birds see Nice, *Studies in the Life History of the Song Sparrow*, and Lack, *The Life of the Robin*. For definitions of territory and the history of the territory concept see Mayr, *Bernard Altum and the Territory Theory*, and Nice, *The Role of Territory in Bird Life*. For a modern treatment of the territorial concept, see N.B. Davis's "Ecological questions about territorial behavior" *in* J. R. Krebs and N.B. Davis's *Behavioural Ecology*.

Courtship and Mating Habits

13

Back on budding boughs
  Come birds to court and pair,
Whose rival amorous vows
  Amaze the scented air.
—Robert Bridges (1844–1930)
  Spring

Each year, as the climbing spring sun brightens the landscape with a living coat of green, all nature prepares for a new generation. It is the mating season for birds. The warmth and light of lengthening days affect profoundly not only their plumage and sex organs, but also their behavior—especially behavior that has to do with reproduction. Male songbirds begin to sing, woodpeckers to drum, cock pheasants to strut in their colorful spring garments, and cranes to prance through their grotesque minuets.

Reproductive behavior in most birds can be divided into activities that are commonly rather stereotyped, discrete, and largely innate: territory establishment, courtship, mate selection or betrothal, pair bond establishment, copulation, nest building, laying of eggs, brooding of eggs and young, and the feeding and protection of young. There is, however, great diversity among species in the occurrence, timing, and duration of these breeding activities. Some birds stake no claims to territory; some choose their mates before they establish territory; some birds mate for life, others for two minutes; some do not incubate their eggs or care for their young.

More is known, perhaps, about the life history of the Song Sparrow, *Melospiza melodia*, than about any other American species, thanks to the classic studies of Margaret Morse Nice (1937, 1943). This strongly territorial sparrow may serve as an illustration of the breeding activities of a small passerine. In Ohio, the male acquires territory in late winter or early spring by singing, by posturing, and by fighting other males. He sings almost constantly until he attracts a mate; then he suddenly and almost entirely ceases singing. The male courts his mate by "pouncing" on her—that is, by flying down at her and colliding with her—and then flying away, singing loudly. The male may carry nest materials as a symbolic gesture, but the female constructs the nest by herself. Copulation occurs shortly before nest building begins, and continues at intervals until incubation starts. During nest building and incubation the male again sings, and defends both the territory and his mate, and later he attends the young. Incubation is performed by the female, who alone has incubation patches—bare skin on the belly—for warming the eggs. The eggs usually hatch in 12 or 13 days, and the young in the nest are cared for by both parents for about 10 days. When they leave the nest they are also cared for by both parents, although the female gradually turns her attention to a new nest and begins preparations for a second brood. Between mated Song Sparrows there is a strong pair bond that usually persists throughout the breeding season of two or three broods. At the end of the season this bond dissolves, and each individual goes its own way. It is unlikely that the same pair will reunite the next year, even though both members may come back to the same general locality, chiefly because returning males have an opportunity to mate with resident females before their former mates return.

Sexual maturity is achieved at different ages in different species, usually at an earlier age in small birds than in large. In most small passerines, small owls, most gallinaceous birds, doves, many ducks, and others, sexual maturity is reached within 9 to 12 months after hatching, and as early as 10 weeks in the Australian Zebra Finch, *Poëphila guttata*. Some of the larger passerines, such as crows, many gulls, shore birds, hawks, geese, and the males of some pheasants, first breed when they are two years old. Larger gulls, cormorants, boobies, and loons first breed at three or more years of age, and large eagles at the age of 4, 5, or 6 years. The male ostrich breeds at 4 years of age; the female, at $3\frac{1}{2}$. The majority of Laysan Albatrosses, *Diomedea immutabilis*, breed for the first time when 8 or 9 years old, and some even later (Fisher and Fisher, 1969).

In some species, sexual maturity does not appear fullblown at a given age, but develops by stages. Yearling White Storks, *Ciconia ciconia*, commonly fail to migrate back to Europe, spending the summer in their African winter quarters. Second-year Storks migrate back and may visit nests, but do not breed. Third-year Storks may pair, copulate, and lay small clutches of eggs, but few of them successfully hatch and rear young. Most White Storks first breed successfully when they are 4 or 5 years old (Schüz, 1949). Similarly, first-year European Cormorants, *Phalacrocorax carbo*, show incipient sexual behavior; second-year birds generally pair and build nests but do not lay eggs, while third-year birds generally breed successfully (Kortlandt, 1942). Maturity in breeding behavior, like maturity in the physiology of the gonads, probably depends on the development of both endocrine and nervous systems and their influence on each other. Cumulative experience is also an important factor in achieving reproductive success—experience in acquiring territory, courting and winning mates, building nests, rearing young, and the like. Ashmole (1971) points out that only older, experienced sea birds are able to bring sufficient food to their colonies to allow for the successful raising of young.

The survival of a species on earth depends of course on its success in leaving enough young to replace adult mortality. And this, in turn, depends on a number of complexly interrelated variables, chief among which, for birds, are the number of eggs and clutches a pair lays each year, the minimum breeding age of a bird, and its longevity. Birds are thus faced with two largely conflicting alternatives: to maximize the number of young produced annually or to minimize the mortality of breeding adults. One way of achieving the latter is to raise few young at a time and give them concentrated care for a long period while they mature. This gives them a chance to learn from their parents and from their own sheltered experiences some of the facts necessary for survival. Such a program is largely incompatible with a high annual production of offspring. Some species of birds survive through high productivity of young (r-selected species); others through emphasis on longevity (K-selected species). Accordingly, species vary in the ages at which they reach sexual maturity. Since most of the larger species are more likely to survive such hazards as bad weather and predators, they can better afford to follow the longevity-promoting path. Smaller birds, on the contrary, and many ground-nesting larger species, are more likely to follow the path of high fecundity plus high mortality.

PAIR FORMATION

The coming together of two birds for the procreation of young often originates a more or less durable bond between them, a "marriage" or pair bond that usually results in monogamy. Butterfield (1970) defines a pair bond as "a reciprocal 'mutual attachment' between two heterosexual, sexually mature organisms such that aggressive tendencies are largely suppressed and sexual ones enhanced." The duration of the pair bond varies greatly between species and between individuals. Its significance for all birds, however, is much the same, as Hinde (1964) makes clear. Pair formation "implies that a number of social responses, potentially elicitable by any member of the species, become more or less limited to one individual; at the same time other responses (e.g., aggressive ones) become inhibited towards the partner." Pair formation is facilitated through an exchange of visual and auditory signals between potential mates. It is highly important that these be birds of the same species because hybrids are "usually at a selective disadvantage compared to their parental types." Consequently, to avoid hybridization, display signals involved in pair formation are commonly highly species-specific, whether in plumage coloration, behavior, or song. As an example, several species of similarly colored viduine finches may coexist at the same locality in Africa. Each viduine sings two distinct song types, one used in aggressive and the other in sexual contexts. Females will respond more strongly to playback of their own sexual song both in laboratory and field, supporting the thesis that these songs are used in species discrimination (Payne, 1973a, 1973b).

A mated pair may remain together for life, for several years, for one year, for one brood of young, or for even shorter periods. In birds the pair bond is probably more lasting than in mammals, notably in the tropics where many species have long-lasting or permanent pair bonds (Kunkel, 1974). A long "engagement" period before the physical union seems in some species to promote an enduring conjugal state. According to Lorenz (1952), Bearded Tits, *Panurus biarmicus*, become engaged before their first molt, at about 2$^{1}/_{2}$ months of age, but do not breed until they are 1 year old. Jackdaws, *Corvus monedula*, and

wild geese are betrothed in the spring following their birth, but do not become sexually mature until 12 months later. These species probably remain mated for life. Laysan Albatrosses typically become paired two years before they first breed (Fisher and Fisher, 1969). On the other hand, northern hemisphere ducks commonly form pairs in the autumn and winter in their winter quarters and do not breed until the following spring; yet in spite of this long betrothal the male and female usually separate when the eggs are laid or when they hatch.

It is difficult to determine the nature of the psychological bond that holds a pair together. It may be common attachment to a territory; it may be "personal" recognition of each other or a comfortable familiarity with each other; or it may be something akin to human affection. That there may exist an affectionate bond between two birds is strongly suggested by the observations of Trautman (1947) on Black Ducks, *Anas rubripes*, during autumn courtship and pair formation. On two different occasions the partner of a shot bird refused to flush and leave its dying mate when the flock flew away from gunners. Pairs of Adelie Penguins, *Pygoscelis adeliae*, likewise seem held together by personal bonds. Of 10 pairs of birds banded at their nesting sites in the South Orkneys, 12 birds returned two years later, and five of the original pairs were still intact (Sladen, 1953). Compatibility between pairs is a prerequisite to successful reproduction. Captive female Canvasbacks, *Aythya valisneria*, randomly assigned to specific males, refused to accept them as mates. Captives allowed to choose their own partners soon paired and laid eggs (Bluhm, 1985).

Territory probably plays an important role in the pair bond of numerous species. A White Stork, *Ciconia ciconia*, fights viciously for possession of its nest, and once in possession it forges a stronger bond with the nest than with its mate (Schüz, 1938). Similarly, pairs of Wilson's Storm Petrels, *Oceanites oceanicus*, return year after year to the same burrow, common ownership of the burrow presumably providing the attraction that holds the birds together (Roberts, 1940). Site fidelity is undeniably influential in the formation and maintenance of pairs. However, many species of birds pair before territories are set up; for example, northern hemisphere ducks, Blackcapped Chickadees, *Parus atricapillus*, and American Goldfinches, *Carduelis tristis*.

Pair faithfulness also depends on the outward

appearance of a bird's mate. Perhaps it is a simple matter of recognition. After several pairs of Ring Doves, *Streptopelia roseogrisea*, had completed two breeding cycles, the males and females were separated from each other for as long as seven months. Then, when each female was offered a choice between her former mate and a strange male, 12 out of 15 females chose their former mates (Morris and Erickson, 1971). Ringed Plovers, *Charadrius hiaticula*, establish enduring pair bonds, but two birds that had each lost a foot were rejected by their former mates. Fortunately the two were of opposite sexes; they met, paired, and successfully raised normal offspring (Laven, 1940).

Durability of Pair Bonds

To determine the duration of pair bonds in wild birds is such arduous, time-consuming work that few authentic records are available for species reputed to mate for life. Records of lifetime matings of pairs living in cages or under other artificial conditions are, of course, of little value. Field studies of banded birds living under natural conditions indicate that relatively few species of birds mate for life, or even for several successive years. And promiscuous species like many grouse and hummingbirds form no pair bond at all.

Most albatrosses and petrels are considered to have durable mating ties. From his 16-year study of the Royal Albatross, *Diomedea epomophora*, in New Zealand, Richdale (1952) concluded that this species normally mates for life. Two mated pairs were known to be intact after 15 years. A four-year study of 96 banded Fulmars, *Fulmarus glacialis*, by Macdonald (1977) revealed an adult mortality rate of 2 per cent per year, a divorce rate of 4 per cent per year, and fidelity to the nest site of over 80 per cent per year. Mated pairs of Manx Shearwaters, *Puffinus puffinus*, and British Storm Petrels, *Hydrobates pelagicus*, were found by Lockley (Fisher and Lockley, 1954) to return to the same burrow year after year as long as a given pair remained alive. A life-time pair bond is thought to be the general rule for small petrels.

Many pairs of geese and swans raised as captives have remained mated for life, and many observers believe that the same condition holds among wild birds. In a seven-year study of Mute Swans, *Cygnus olor*, Minton (1968) found that divorce rates were less than 5 per cent per year for breeding pairs and less than 10 per cent per year for non-breeders. Eagles are popularly considered to mate for life and perhaps do, but reliable evidence for wild birds is lacking. Barn Owls, *Tyto alba*, and many parrots are known to establish long-enduring partnerships.

A four-year study of pair-bond persistence in Common Terns, *Sterna hirundo*, by Austin (1947) showed that of 122 mated pairs, the original pair bond persisted in 79.1 per cent from one season to the next. Gulls are believed to possess similarly firm pair bonds.

Enduring nuptial ties are found in many representatives of the Corvidae, including Ravens, *Corvus corax*, Carrion Crows, *Corvus corone*, Jackdaws, *Corvus monedula*, Magpies, *Pica Pica*, Nutcrackers, *Nucifraga caryocatactes*, and others. In several nonmigratory species of the Paridae, pairs hold together for two or more seasons.

From an extensive study of 37,000 individually marked Adelie Penguins, *Pygoscelis adeliae*, at Cape Crozier, Antarctica, Le Resche and Sladen (1970) discovered that 44 per cent of young breeding penguins changed mates as against 16 per cent of old breeders, when both members of a pair returned to the rookery a subsequent year.

That durable pair bonds are the result of more than habit or emotional fixation is suggested by the findings of Coulson (1966) after a 12-year study of Kittiwakes, *Rissa tridactyla*. A female who retained her mate from a previous breeding season bred earlier, laid more eggs, and had greater breeding success than one that had paired with a new male. Natural selection may thus favor enduring pair bonds as a method of enhancing a species' survival. Observing that those Red-billed Gulls, *Larus novaehollandiae*, that had changed mates had had a lower hatching and fledging success the previous year than those that retained mates, Mills (1973) concluded that changing mates was probably advantageous for incompatible pairs, and further, that the delayed breeding and smaller clutch size following a change in partners indicated that a period of adjustment is necessary for efficient breeding.

Types of Mating Systems

Avian species differ with regard to the nature of social bonds between mates (mating systems) and the amount of parental care contributed by each sex in raising the young. Mating systems may be listed in order of increasing male involvement (Mock, 1983) as follows: promiscuity, polygyny, monogamy, polyandry.

The term "promiscuity" denotes a lack of choosiness by sexual partners, and is used to describe polygynous systems in which pair bonds are nonexistent (e.g., in lek species). Although useful, this term is somewhat of a misnomer since females choose certain males over others to fertilize their eggs (see beyond).

A species is called polygynous when each male mates with two or more females. This social system differs from promiscuity in that polygyny implies a social bonding between the male and his females (e.g., many pheasants). The male may help feed the young in some altricial species. In many cases, however, the male's contribution to raising the young is indirect, e.g., in helping defend a territory. Only the female rears the young in most polygynous and in all promiscuous species.

The reverse of polygyny is polyandry, when a female is mated to several males. Davies (1985) has reported on a system he calls "polygynandry" in Dunnocks, *Prunella modularis*, in which several males and several females form a breeding unit.

According to Lack (1968) over nine-tenths of all birds whose young are reared in the nest (nidicoles) are monogamous, as are four-fifths of all birds whose young leave the nest shortly after hatching (nidifuges). He suggested that "each male and each female will, on the average, leave most descendants if they share in raising a brood, particularly when they collect the food for the young."

By removing males from some territories Gowaty (1983) showed, however, that lone female Mountain Bluebirds, *Sialia sialis*, fledged as many young as females with male help. Similar removal experiments with Seaside Sparrows, *Ammospiza maritima*, showed no difference in growth rate of fledglings with and without fathers and that widowed females fledged two-thirds as many young as mated females. This indicates that male help is advantageous but not necessary in enabling females to fledge some young (Greenlaw and Post, 1985). Nestlings of experimentally widowed female Song Sparrows grew more slowly than nestlings of paired females (Smith *et al.*, 1982). Although fledging success was similar in the two groups, they also found that young of experimentally widowed females survived poorly to independence. In removal experiments with male Song Sparrows, Seaside Sparrows and Willow Ptarmigan, *Lagopus lagopus*, polygyny was induced when neighboring males accepted the "widows" as secondary mates. (Smith *et al.*, 1982; Hannon, 1984; Greenlaw and Post, 1985). Smith *et al.* suggest that Song Sparrows are usually monogamous because males with high quality territories outnumber females early in the breeding season. Polygyny occurs when females outnumber males in good territories.

How did polygyny evolve? Some authors (Verner, 1964; Verner and Willson, 1966; Orians, 1969) suggest that polygyny occurs in species with equal sex ratios and in which variance in male territory quality exists. The fittest males own the best territories and are soon paired, and bachelors are relegated to inferior areas. A female might be able to rear more young on the richer territory, unaided by a polygynous mate, than on the poor territory even when assisted by a monogamous mate. Thus as two adjacent territories diverge in their food resources, a "threshold of polygyny" is eventually reached at which such an adaptive choice will be made. Indeed, Holm (1973) discovered that the more females there were in a harem of Red-wing Blackbirds, *Agelaius phoeniceus*, the more young each female raised to fledging.

Studies on the Great Reed Warbler, *Acrocephalus arundinaceus*, (Catchpole *et al.*, 1985) indicate that indeed individual territories vary greatly in size and quality (total area with breeding quality reeds), and that polygynous males have the best territories. Males assisted primary females in feeding young but since secondary females fed their young alone, only one-third survived. The others died of starvation. Polygyny is advantageous to the male as he will leave behind more offspring, but females paired to mated males leave fewer young than primary females. Catchpole *et al.* suggest that males deceive females into pairing with them by concealing their marital status. An evolutionary countermeasure would be for females to spend time checking the paternity of nearby nests, but in delaying her own breeding she would risk even lower breeding success and delayed migration for her and her offspring. It would be better for her to raise some young than not to breed at all.

A study on Pied Flycatchers, *Ficedula hypoleuca*, (Alatalo *et al.*, 1981, 1982) revealed that a male commits a female to breeding in his territory and then establishes a second territory elsewhere and attracts a second mate. The second female is left to raise her young alone, and as a consequence fledges only 3.6 young versus the primary female's 5.5 young. The authors suggest that the male deceives the second female by hiding his own mated state through establishing two or more territories.

Monogamy and polygamy (which may be polygyny or polyandry, see beyond) are not rigidly fixed forms of pair bonds for some species. A number of species that are normally monogamous occasionally shift into polygamous pair bonds: e.g., some swans, hawks, doves, finches, titmice, wrens, flycatchers, warblers, thrushes, icterids, and others (Verner and Willson, 1969).

Unlike most small passerines, such species as the House Wren, *Troglodytes aëdon*, Bank Swallow, *Riparia riparia*, and several other hole-nesters normally remain united for only one brood. Usually in these species the male cares for the fledglings of the first brood while the female seeks another mate and begins a second brood elsewhere. Of 70 pairs of House Wrens that successfully reared one brood and then undertook a second brood the same season, the original pairs remained intact in only 40 per cent of the cases (Kendeigh, 1941).

Although not as common as monogamy, polygamy occurs in a wide variety of birds as either polygyny or polyandry. Polygyny is by far the commoner form. Polygyny increases the reproductive burdens of the female while lightening those of the male; polyandry achieves the reverse (Murton and Westwood, 1977). The first step away from monogamy toward polygyny is the emancipation of the male from family chores (Lack, 1968). Polygyny is found regularly in many gallinaceous species, such as grouse, pheasants, the Green Peacock, *Pavo cristatus*, and in ostriches, and several species of Icteridae—for example, the Red-winged Blackbird, *Agelaius phoeniceus*. In normal polygyny the male inseminates several females, which incubate their eggs in separate nests and rear their young unassisted by him. A remarkable combination of both polygyny and polyandry occurs in the South American Rhea, in which several females lay their eggs, on occasion as many as 50, in one nest, where the male incubates them by himself. The females may then go on to mate with other males (Brunning, 1974). He also is responsible for the care of the young. Another exceptional modification of polygyamy is found in the bigamous Temminck's Stint, *Calidris temminckii*, of northern Europe. Each female pairs in rapid succession with two males on different territories and lays one clutch of eggs on each. The first clutch is incubated by the male, the second by the female, and each bird takes care of its own brood of young (Hildén, 1975).

The bonding between a polygynous male and his harem may be strong. In a colony of Brewer's Blackbirds, *Euphagus cyanocephalus*, Williams (1952) found that of 70 matings, there were 45 cases in which both members of such pairs returned the following year; of these 45 possibilities for rematings, there were 42 (93 per cent) rematings and only three "divorces." In this species polygyny in a colony tends to increase as the ratio of females to males increases.

Observations by Armstrong (1955) on the European Wren, *Troglodytes troglodytes*, underscore the role food plays in determining polygyny. In England and Holland about 50 per cent of all wren matings are polygynous, as against about 6 per cent among American House Wrens, *Troglodytes aëdon*. But a strange state of affairs exists on the isolated Scottish island of St. Kilda. Here the wren is strictly monogamous. On bleak St. Kilda food is scarce, and both parents must scurry to find enough food to feed their brood. In English gardens, on the contrary, food is abundant, the Wrens are commonly polygynous, and each female alone can gather enough food for her young. Similarly, the Savannah Sparrow, *Passerculus sandwichensis*, which is commonly bigamous in warmer climates, is forced into monogamy in arctic tundras where the breeding season is too short for one parent effectively to feed the young (Weatherhead, 1979).

Noting that passerine birds that nest in cavities, crevices, or roofed-over nests are polygynous significantly more often than open-nest birds, von Haartman (1969) expressed the belief that this occurs because their nestlings are better protected from heat-loss and from enemies. A safe, covered-over nest encourages polygyny because it obviates the need for a male to stand guard over the young, and it permits their slow development, which, in turn, reduces the pressure for parental feeding. The other hypothesis is a limitation of nest sites. This may lead to the operational sex ratio deviating from unity in favor of females as Quinney (1983) has found in Tree Swallows, *Tachycineta bicolor*. A combination of a skewed sex ratio and an abundance of food precipitates polygyny.

The reason that species with precocial, downy young are more often polygynous than other species is that both parents are not required to feed such young. Within a day or two after hatching the young feed themselves, and one parent is sufficient to lead the young to suitable feeding places. This probably explains why so many gallinaceous species are polygynous.

Birds that live in rich grasslands and marshes where

solar energy concentrates its production of food in a shallow layer near the earth's surface are more likely to be polygynous than those in woodlands of the same region. Here again it is primarily a matter of food abundance. African weavers (Ploceidae) of grasslands and savannas breed when grass seeds are abundant. Many species in this family are polygynous and the female alone is able to feed her brood. On the other hand, insectivorous and carnivorous species tend toward monogamy, since securing sufficient food for the young is likely to require the efforts of both parents.

Among the 291 passerine birds in North America, 14 species are regularly polygynous. Of these, 13 breed in marshes, and six of the 13 are Icteridae (Verner and Willson, 1966).

A factor favoring polygyny in Red-winged Blackbirds, *Agelaius phoeniceus*, is the fact that the male rarely breeds the first year while the female commonly does. This delay in male sexual maturity—also seen in many polygynous grouse and manakins—makes possible the development of male sexual dominance that allows the insemination of sufficient females to compensate for the loss of earlier deferment (Wilson, 1975). Suggestive in this connection is the fact that in the Orkney Islands the male Hen Harrier, *Circus cyaneus*, is monogamous in its first year but frequently polygynous when 2 years old or older (Balfour and Cadbury, 1979). Male Song Sparrows and White-crowned Sparrows implanted with testosterone became more aggressive, doubled their territory sizes, and became polygynous (Wingfield, 1984). Thus polygyny may be induced by manipulating androgen levels in males.

Two other predisposing factors may influence polygyny in a species. If there is regularly a sexual imbalance in a species resulting in more females than males, polygyny is likely to result. This seems to be the explanation for polygyny in the Ostrich, *Struthio camelus*, the African bishops, *Euplectes* spp., and occasionally in Brewer's Blackbird. Several African weaver finches and whydahs of the family Ploceidae are brood parasites in that they lay their eggs in the nests of other species who raise the ploceid young as foster-nestlings. This of course relieves both the male and female ploceid parents from the duties of rearing their own young, and makes possible the polygyny that characterizes several species.

Polyandry, the reciprocal of polygyny, occurs mainly in those species in which the males incubate the eggs and rear the young: tinamous (Tinamidae), jaçanas (Jacanidae), Spotted Sandpiper, *Actitis macularis*, painted snipe (Rostratulidae), and button quail (Turnicidae). The females are larger, have the brighter nuptial plumage, do the courting, and defend the territory, while the docile, drab males build the nest, incubate the eggs, and rear the young. Among American Jaçanas, *Jacana spinosa*, of Central America, a typical female will have two or three mates. The breeding male, which weighs only about one-half as much as a female, establishes the territory, builds the nest, incubates the eggs, and cares for the young alone (Jenni and Collier, 1972). The female takes a very active role in attacking other species that may potentially prey on their young. She may assist the male in these attacks, or may attack as the male leads the young away to safety (Stephens, 1984). Territories of the flightless Tasmanian Gallinule, *Tribonyx mortierii*, may be held either by pairs or by groups composed of two or three males (often brothers) and one female. Polyandrous groups produce more eggs and more young birds per egg than do pairs. Survival of young per clutch of eggs is 61 per cent for the polyandrous groups versus 45 per cent for pairs (Ridpath, 1972). Polyandrous matings have been reported as occasionally occurring in species normally monogamous such as the White Stork, *Ciconia ciconia*; European Swift, *Apus apus*; Ovenbird, *Seiurus aurocapillus*; and Eastern Bluebird, *Sialia sialis*.

A variety of species are known to practice social breeding habits. The Australian White-winged Chough, *Corcorax melanorhamphos*, lives in groups of from 3 to 20 birds. Group members help build the nest, incubate the eggs (laid by several females), brood the nestlings, and feed them both in and out of the nest. The Acorn Woodpecker, *Melanerpes formicivorus*, of California, lives all year in groups in which all or most members help to incubate the eggs and feed the young (MacRoberts and MacRoberts, 1976). The groups consist of 2 to 15 birds, and may include 1 to 4 sibling males, 1 to 2 sibling females and their offspring (Koenig, 1981). In Colombia the Green Jay, *Cyanocorax yncas*, lives in year-round flocks with only one breeding pair per flock. Flock members defend the group territory and protect and feed the young. The young do not disperse from the parental flock but apparently stay to serve as helpers for the next breeding season (Alvarez, 1975). In 1981, when two-thirds of a marked population of Florida Scrub Jays, *Aphelocoma coerulescens*, died, nonbreeding helpers suddenly became breeders. The population soon recovered. When territories are in short supply, surplus offspring serve as helpers, waiting for adult breeders to die and thus inherit their breeding territories (Woolfenden and Fitzpatrick, 1984). Cooperative breeding occurs in 39 species of Australian birds, including geese, incubator birds, moorhens, kingfishers, babblers, flycatchers, honeyeaters, starlings, wood-swallows, and others (Rowley, 1976). Social breeding also is found in at least 30 of the approximately 80 families of African birds (Grimes, 1976).

Sexual promiscuity occurs in those species that come together only for copulation and then see no more of each other. This form of mating is characteristic of the Ruff, *Philomachus pugnax*, and other waders (Scolopacidae); many grouse such as the Prairie Chicken, *Tympanuchus cupido*, Sage Grouse, *Centrocercus urophasianus*, and Black Grouse, *Lyrurus tetrix*; some pheasants (Phasianidae); cotingas (Cotingidae); manakins (Pipridae); many birds of paradise (Paradisaeidae); parasitic weaver finches (Viduinae); bower-birds (Ptilonorhynchidae); and many hummingbirds (Trochilidae). A biological hazard in the fleeting unions of promiscuous species is the possibility of mating with a bird of another species and producing hybrid offspring—if any. Natural selection has reduced this hazard of errors by a variety of species-recognition mechanisms including spectacular plumes and colors, or strikingly distinctive calls and other sounds, or fantastic postures and dances, or, in the case of the relatively drab-colored bower-birds, the construction of distinctively shaped and decorated bowers. The extreme exaggeration of these characters is a consequence of sexual selection (Diamond, 1981, 1982).

All of these superlative courtship devices are a legacy of the emancipation of the male bird from family duties. Once his reproductive function is reduced to copulation with the female, the pair bond dissolves and he is likely to mate with several females. This situation inaugurates a regime of intensified sexual selection that is responsible for the exaggerated colors and behavior of many polygynous, and especially promiscuous, species (Lack, 1968).

Sporadic promiscuity in otherwise monogamous species, such as swallows, swifts, wrens, and other birds, is probably biologically adaptive in that the casual mate may be available to take the place of a missing parent of half-grown young. Contrariwise, a bird that has lost its mate and young through predation may help a normal pair to rear its hungry brood. Inflexibly monogamous pair bonds would preclude such helpful arrangements. Occasionally, three or more adults of a monogamous species are seen feeding one brood of young. Unmated birds are also often available either as substitutes for a lost mate or as "helpers" to assist a normal pair care for its brood. An experiment demonstrating the availability of such unmated birds is reported by Griscom (1945).

Sporadic promiscuity in otherwise monogamous species has been documented either by observation or experiment. Some fertile eggs occur in nests of Redwinged Blackbirds mated to vasectomized males indicating that some extra-pair copulations had occurred (Bray *et al.*, 1975; Searcy and Yasukawa, 1983). Electrophoretic analysis of the blood proteins of Bobolinks, *Dolichonyx oryzivorus*, revealed genetic traits that could not have come from the presumed pair of parents, proving that extra-pair matings had taken place (Gavin and Bollinger, 1985). During the nest-building and egg-laying stages, one finds that males guard their females very closely to ensure their own paternity and to prevent extra-pair copulations. Mate-guarding behavior has been documented in Acorn Woodpeckers, Herring Gulls, *Larus argentatus*, Murres, *Uria aalge*, Dunnocks, *Prunella modularis*, and other species (Morris and Biodochka, 1982; Mumme *et al.*, 1983; Davies, 1985; Birkhead *et al.*, 1985).

Widowed females may sometimes accept strange males to help them in raising young as documented by the following. The male of a wild pair of Indigo Buntings, *Passerina cyanea*, was shot and removed. The next day the female obtained a new mate. This second male was also shot, and the same process continued until nine different male buntings had been removed; the tenth was left undisturbed to help raise the family (Griscom, 1945). In contrast to this, 33 male Seaside Sparrows were removed. In no case did floater males appear, and no neighboring male fed "non-kin" nestlings. Instead, 32 males enlarged their territories to include those of widowed neighbors (Greenlaw and Post, 1985).

Sexual aberrations of various sorts occur in birds as in many other animals. Inbreeding is often biologically harmful and its rarity in the wild may be accounted for in part by such behavior mechanisms as territory establishment and the dispersal-encouraging intolerance of parent birds toward their grown young. It is significant that brother and sister matings in the Bobwhite Quail, *Colinus virginianus*, reduce the hatchability of their eggs (Nestler and Nelson, 1945). Inbreeding pairs of Great Tits, *Parus major*, have been shown to suffer greater nesting mortality than normal pairs (Greenwood *et al.*, 1978).

Outbreeding mechanisms (incest taboo mechanisms) have developed in some species of birds. Male Japanese Quail, *Coturnix coturnix*, raised in groups tend to approach or copulate with females that they had not previously encountered (Bateson, 1979, 1980). Fledgling Peach-faced Lovebirds, *Agapornis roseicollis*, form pair bonds while still in juvenile plumage before sexual maturation. Pairing preference is for unrelated individuals (Frischdick *et al.*, 1984). Females of group-living Acorn Woodpeckers will not breed so long as their father is still living within the group. These females will breed when the father is replaced or if they emigrate to join a group of unrelated individuals (Koenig and Pitelka, 1979). In contrast, incestuous matings are well known in Splendid Wrens, *Malurus splendida*, which breed in groups consisting of an adult breeding pair and their sons and daughters (Payne *et al.*, 1985).

Homosexual female/female pairing has been reported for a variety of gull species (Conover and Hunt, 1984). By removing males from a colony of Ring-billed Gulls, *Larus delawarensis*, the authors created homosexual female pairings, suggesting that homosexuality results from shortage of males. Some of these homosexual pairs do lay fertile eggs as a result of promiscuous matings with males. Populations of several full species tend to be skewed towards males, possibly because DDT in the environment has caused feminization of male embryos (Fry and Toone, 1981). This may have led to female/female pairing.

Even prostitution and rape seem to occur among birds. Females of the Purple-throated Carib Hummingbird, *Eulampis jugularis*, exploit mating behavior during the non-breeding season to gain access to the food-rich territory being defended by the male (Wolf, 1975). Sperm production is less costly in energy than producing an egg. Males of species that are normally monogamous may indulge in extra-pair copulations or rape to maximize their chances of leaving as many offspring as possible (Trivers, 1972). Forced copulation has been documented in many species of waterfowl (McKinney *et al.*, 1983). Most of these attempts are in the morning hours when eggs are more likely to be fertilized. The rightful husband guards his mate against such attempts, will try to drive off the offending male, and will himself copulate with his mate to better ensure his own paternity (Cheng *et al.*, 1982, 1983).

MEANING OF COURTSHIP DISPLAY

Just as the seasons swing back and forth between the bleak sterility of winter and the green fruitfulness of summer, so the lives of birds oscillate between the relative quiescence of the resting season and the impetuous excitement of the breeding season. The extravagant courtship displays that seem so absurdly exaggerated to human eyes are in part a reflection of the intense living of birds and in part a reflection of special needs resulting from their ways of life. A courting male commonly ceases his singing and sheds his colorful feathers as soon as the female begins incubating the eggs. This reveals the biological costs of advertising either to defend a territory or to win a mate, since every bird has a perpetual need for concealment from predators.

Courtship is ordinarily the province of the male. He shows his wares before the female with an astonishing assortment of tricks, varying according to species. He may posture so as to reveal his gaudiest nuptial plumage; spread his tail and erect his crest or inflate brilliantly colored pouches; parade, dance, fly with dizzying acrobatics; sing his most fetching songs (which to man may be discordant squawks); bring tidbits of food—anything, it seems, to impress his mate-to-be. Very often these courtship displays are presented with dramatic suddenness. Almost always they are uniquely different from those of other species of the same region. That they consume a lot of energy is proved by the fact that the male Great Bustard, *Otis tarda*, weighs one-third more at the beginning than at the end of the courtship period; the male Eurasian Woodcock, *Scolopax rusticola*, sinks from a weight of 320 g in April to 263 g in July (Zedlitz, *in* Stresemann, 1927–1934). Like a lovesick swain who squanders his substance on his intended bride, a male bird almost bankrupts himself biologically in his courtship exertions.

Males usually do the courting and females, the selecting. Pigeons, *Columba livia*, with breeding experience, fledge heavier young than birds breeding for the first time. In double-choice experiments female pigeons selected experienced over inexperienced males for mates (Burley and Moran, 1979; Burley, 1981). Plumage or behavior may broadcast a

FIGURE 13–1 In most birds courtship activities reach their climax in copulation, as shown here in a pair of Black-headed Gulls. Photo by E. Hosking.

male's genetic quality (fitness) and influence a female's choice of a mate (see beyond).

Mate selection, or pair formation, is one of the functions of courtship in most species, but by no means the only one.

Another important function of courtship is the stimulation of ovulation. A virgin dove stroked on the back, or exposed to the sight of another dove in an adjoining cage, will often lay eggs. Taped vocalizations played to Ring Doves, *Streptopelia roseogrisea*, Canaries, *Serinus canaria*, or Budgerigars, *Melopsittacus undulatus*, induce faster gonadal growth rates than those found in control unstimulated birds (review in Morton *et al.*, 1985). In other words, courtship stimulates sexual readiness, not only in the bird being courted but also in the bird doing the courting, through self-stimulation. This reciprocal stimulation may be the chief function of the mutual courtship ceremonies of many colonial birds such as gannets, gulls, and penguins. Such stimulation commonly results in the increase of sex hormones in a bird's body, and the hormones, in a circular fashion, frequently intensify courtship display behavior. Even downy juvenile male Mallards, *Anas platyrhynchos*, will perform adult courtship displays if injected with male hormones (Phillips and McKinney, 1962). In adult Mallards, male hormone levels in the blood correlate directly with intensity of social displays and sexual behavior (Balthazart, 1976).

Courtship activities have the further function of regulating the timing of sexual readiness so that the reproductive physiology of a pair may be synchronized. This is particularly important in a flying animal, which cannot afford to carry indefinitely the extra ballast of greatly enlarged gonads. Simultaneous readiness in both sexes is also important for the successful transfer of sperm in animals lacking a penis. Prolonged courtship displays very likely serve to reinforce the pair bond between mated birds. Finally, courtship displays also function as species-recognition signs. The biological futility of hybridization is commonly expressed in hybrid sterility. In a hybrid between a mallard and a Muscovy Duck, *Cairina moschata*, disruption of normal chromosome activity prevents the formation of spermatocytes (Marchand and Gomot, 1976). And even when hybrids are fertile, their offspring usually exhibit reduced viability, or are selected against (Emlen *et al.*, 1975).

There is much evidence that courtship behavior originates in two conflicting tendencies in the courting male: one, to copulate with the female, and the other, the agonistic tendency to attack or escape (Bastock, 1967). In the singing and "pouncing" behavior of the Song Sparrow, *Melospiza melodia*, cited earlier, the aggressive element initially dominates his courtship actions. In many gulls pair formation begins when the female responds to the hostile displays of an unmated male on his territory by approaching him. Both birds go through apparently hostile displays that often end with the retreat of the female. With successive returns of the female, the intensity of the fights gradually decreases and the two birds eventually become paired (Moynihan, 1958).

Courtship displays in other species undoubtedly assimilate other forms of behavior into their ritualized patterns. In his study of courtship display and fighting in passerines, Andrew (1961) proposed that courtship displays have originated from seven major behavioral elements: sexual, aggressive, fearful, alert, nesting, parental, and juvenile begging for food. The examples that follow illustrate some of these varied roots of courtship displays.

SEXUAL DIMORPHISM AND COURTSHIP

When both sexes of a species are externally similar in appearance, the male and female often take equal parts in courtship activities. In the nuptial display of the Gannet, *Morus bassanus*, the male and female stand facing one another, wings spread, tail depressed, beaks pointing skyward, heads wagging violently from side to side, while each bird gives off hoarse "urrah urrah" calls (see the illustration on the first page of this chapter). Often their beaks clatter together like castanets, or are stroked on each other like knives on whetstones. The display may be varied with elaborate bows of the head and neck or by the opening of mouths, displaying a wide black gape. In general outline this form of mutual display is seen among many sea birds.

Among birds possessing no visible sexual dimorphism, courtship behavior provides a means of sex recognition. Penguins, for example, are apparently unable to determine sex visually, so they adopt a trial and error procedure to solve their problem. In a typical courtship performance, a male may place a pebble at the feet of another bird. If this second bird is a male, he may start a fight. If it is a female, she may be unreceptive and ignore the courting male; but if she is ready for courtship, the pair will go through various courtship ceremonies involving deep bowing, stretching their beaks skyward in the "ecstatic" posture, trumpeting, neck-twining, and other actions that eventually lead to coition. Courtship activities may thus arouse sexual ardor, antagonism, or indifference, depending on the sex and physical state of the partner.

In many species, however, sexual dimorphism provides striking differences between the male and female, especially in their nuptial plumage. The male, with his conspicuous coloration, undertakes the courtship activities. The less conspicuous female assumes the confining domestic duties of the nest, where her drab plumage affords some protection against predators. In those few species (phalaropes, painted snipe, and button quail) in which the male performs the domestic chores, it is, significantly, the female that takes the initiative in courtship, strutting her gaudier plumage, fighting with other females, and defending the home territory.

It seems probable that sex dimorphism first arose in birds as a "badge" for the quick and easy identification of sex. In some species it is still that and no more. To human eyes the only visible difference between a male Northern Flicker, *Colaptes auratus*, and a female is the presence of small black "mustache" or "sideburn" marks on the sides of the male's head. This distinction is also the one the birds themselves use, for when Noble (1936) painted black mustache marks on a female, her mate immediately attacked her as though she were an intruding male.

From such simple beginnings sexual dimorphism presumably evolved into the colorful and even grotesque ornamentations that the males of many species now use for all the various functions of courtship dis-

FIGURE 13–2 Courtship activities do not necessarily stop after coition. Fulvous Tree Ducks perform a stereotyped, postcopulatory dance on the surface of the water, rapidly treading with their feet while holding their bodies erect, breasts puffed out, necks arched, and their outer wings held high. Photo by B. Meanley.

FIGURE 13–3 Visible differences between the sexes may be large, as in peafowl, small as in the Yellow-shafted Flicker, or nonexistent, as in penguins. The chief external evidence of sex in the Flicker is the black mustache mark, seen here in the male, but lacking in the female. If an artificial mustache be painted on a female, other Flickers will treat her like a male. Photo by G. R. Austing.

play, in addition to the primitive function of simply announcing their sex. It is significant that in many species the colorful feathers, and often other structures, used in courtship coincide in their periodic appearance with the appropriate period of the bird's breeding cycle.

Considering the fact that a female commonly invests much more time and energy in reproduction (e.g., eggs, incubation, care of young) than does a male (sperm production), she may forfeit the loss of a complete reproductive cycle should she mistakenly choose a mate of the "wrong" species—a problem not faced by the male. Hence, the evolution of the male's secondary sex characters that help prevent such mistakes (Alcock, 1975). This principle was demonstrated by Klint (1975), who presented individual female Mallards, *Anas platyrhynchos*, with three drakes, one in full nuptial plumage and two with their green head color artifically removed. Nine out of ten hens chose the drake with normal coloration.

Sexual dimorphism in color, or sexual dichromatism, occurs in variable degree. Many woodpeckers differ in sex coloration only in the presence of a small patch of red feathers on the head or nape of the male. Less colorful but more extensive is the dimorphism in the Kelp Goose, *Chloephaga hybrida*, in which the

male is pure white all over and the female brownish black with white barring on her sides. In many species, however, nature has thrown restraint to the winds and developed in the male gorgeous colors, lengthened plumes, distinctive songs, bizarre behavior, and other theatrical effects that he may employ in his pursuit of a bride. Among families showing pronounced sex dimorphism in many of their species are ducks (Anatidae), pheasants (Phasianidae), hummingbirds (Trochilidae), trogons (Trogonidae), manakins (Pipridae), cotingas (Cotingidae), birds of paradise (Paradisaeidae), sunbirds (Nectariniidae), wood warblers (Parulidae), tanagers (Thraupidae), and widow birds (Viduinae).

In his theory of "sexual selection" Darwin (1871) argued that the driving force resulting in extremes of sexual dimorphism are male/male competition and the female choice of a mate. The first is termed "intrasexual selection" and the second, "intersexual selection." The first may have resulted in size dimorphisms between the sexes, with males usually larger, and the second in the often spectacular male adornments or elaborate courtship songs of males.

The male polygynous Long-tailed Widowbird, *Euplectes progne*, possesses a long (0.5 m) tail, which is conspicuous during display flights. Intersexual sexual selection was tested by shortening the tails of some males, and lengthening the tails of others by adding portions of cut tails to the distal ends (Andersson, 1982). Males with artificially lengthened tails acquired additional mates that nested in their territories, whereas males with shortened or unmanipulated tails (controls) did not. Andersson concluded from this that female choice has resulted in selection for long tails in widowbirds.

The role of female choice in selecting for elaborate courtship songs of the Sedge Warbler, *Acrocephalus schoenbaenus*, was studied by Catchpole *et al.* (1984). They implanted naive hand-raised females with estradiol in silastic tubes, then played elaborate and simple songs to single females. Females displayed more to complex songs than to simple songs.

Among the 42 kinds of birds of paradise there are feathers of every hue of the rainbow as well as sparkling whites, dull browns, velvety blacks, and shimmering iridescent shades. Frequently the feather adornments of the gaudy male are modified in shape as well as in color. Some birds have central tail feathers two and three times as long as their bodies. Others, such as the Twelve-wired Bird of Paradise, *Seleucides igno-*

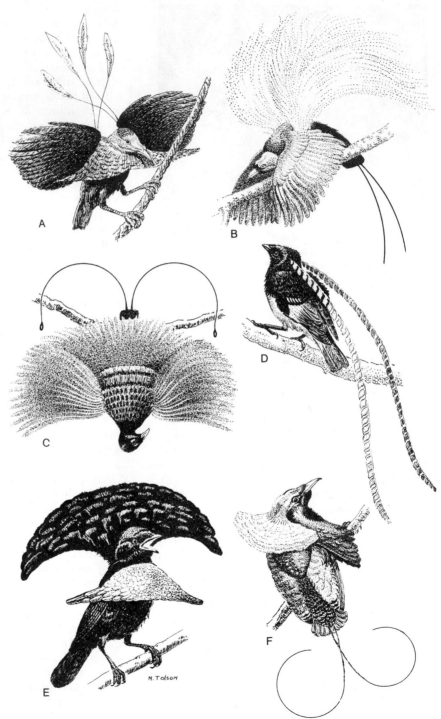

FIGURE 13–4 Male birds of paradise, showing examples of the spectacular plumage which they display in courtship. A, Wallace's Standard-wing Bird of Paradise, *Semioptera wallaceii*. The bib of breast feathers is metallic green, and the crown of the head is violet. The four white wing feathers may be raised or lowered at will. B, The Greater Bird of Paradise, *Paradisaea apoda*, is a brown, crow-sized bird with a yellow crown, glittering green throat, and soft, golden wing plumes which are held erect and vibrated for courtship display. A dozen or more males may display simultaneously and energetically in one tree. C, The Blue Bird of Paradise, *Paradisaea rudolphi*, displays his bright blue plumes while hanging inverted, slowly swinging his body sideways, and singing in a monotone. D, The King of Saxony Bird of Paradise, *Pteridophora alberti*, has two crest feathers in the form of long shafts lined with small blue and white pennants. E, The Lesser Superb Bird of Paradise, *Lophorina superba*, is a small black bird with two sets of erectile plumes: a velvety black cape and an iridescent green bib. In addition, during courtship the male exposes the bright green lining of his mouth cavity. F, The Magnificent Bird of Paradise, *Diphyllodes magnificus*, has bright yellow cape feathers and a metallic green bib, both of which it can erect in display. The two coiled tail "wires" are greatly prolonged feather shafts. After Ripley, and paintings by W. A. Weber, 1950b.

tus, have some feathers reduced to "wires" or simply elongated quills. In the Six-plumed Bird of Paradise, *Parotia* spp., the male possesses six barbless feathers projecting from the top of his head, each tipped with a small oval vane, the whole resembling a tiny racquet. In the excitement of courtship these racquets may be swung directly forward toward the observing female. Perhaps the most extraordinary courtship accessories in the world are the pair of long crown feathers trailing backward from the head of the male King of Saxony Bird of Paradise, *Pteridophora alberti.* Although the bird is no larger than a starling, each feather is about one-half a meter in length, with about three dozen tiny translucent flags or pennants arranged laterally on the shining white rachis. In past years, before the bird was protected, these feathers were also used for human ornamentation and were reputedly worth 280 dollars a pair (Ripley, 1950). Other birds of paradise have feathers coiled into circles, twisted into corkscrews, or gathered into ruffs, capes, crests, false wings or "umbrellas" (Cooper and Forshaw, 1977). If an architect were to classify birds he would almost certainly place birds of paradise in the baroque category.

Males of the Lesser Bird of Paradise, *Paradisaea minor,* and Goldie's Bird of Paradise, *P. decora,* display in leks, calling and waving their beautiful plumes when a female arrives. One male in a lek accounted for 25 of 26 observed copulations in *P. minor* (Beehler, 1983). Dominant plumed adult male *P. decora* would engage in long copulations preceded by display. Unplumed males copulated only briefly with no preliminary displays (LeCroy *et al.,* 1980). These data also show that females select some males over others to sire their young as predicted by Darwin.

Structures other than feathers are used to impress the expectant female. Many species possess lurid colored patches of bare skin, or specialized combs, wattles, lappets, "horns," or pouches, which may be inflated either with blood or air and thus be made more conspicuous. During courtship display, the male Crimson Tragopan, *Tragopan satyra,* of Nepal, stands before the female with his two blood-inflated conical blue wattles erected on the crown of his white-spotted, black head. He makes the display more vivid by rapidly shaking his head and spreading his wings to show their crimson linings. The male Umbrella Bird, *Cephalopterus ornatus,* has, in addition to his erectile "umbrella," a long, feathered, inflatable wattle, nearly as long as the bird himself, hanging in front of his breast. In courtship display the bird spreads his crest so that it overhangs his beak, inflates his dangling wattle with air, and emits a deep roaring call. The Magnificent Frigate Bird, *Fregata magnificens,* possesses during the breeding season a throat pouch of brilliant red skin, which he inflates with air into a bean-shaped balloon practically as large as his body. Similarly inflatable neck pouches appear during the breeding season in several grouse such as the Sage Grouse, *Centrocercus urophasianus.* The pouches function both in display and as sounding boards for calls or "booming." The capacities of the esophageal pouches in two mature Sage Grouse, increase some 40- to 60-fold in volume during the breeding season (Clark, 1942).

Sometimes ordinary structures become brightly colored and are used in courtship display. During the breeding season the male Blue-footed Booby, *Sula nebouxii,* sticks up his tail and goose-steps in front of his prospective mate, lifting each bright blue foot as high as he can. The mouth cavities of many birds are strikingly colored and are prominently displayed in courtship ceremonies, sometimes by the male, sometimes mutually, as in sea birds. Here again, the colors vary widely: the Gannet, *Sula bassana,* black; Razorbill, *Alca torda,* yellow; Common Puffin, *Fratercula arctica,* orange; Black Guillemot, *Cepphus grylle,* red; Lesser Superb bird of Paradise, *Lophorina superba,* bright green; Sickle-bill Bird of

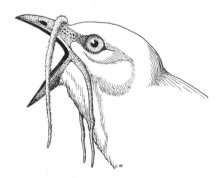

FIGURE 13–5 In addition to brightly colored feathers, some birds possess other accessories used in courtship. The Central American Three-wattled Bellbird, *Procnias tricarunculata,* in the excitement of courtship display, exposes its wide gape. Through increased blood supply, the wattles are increased about three times their length at rest. The bird gets its name from its bell-like call.

Paradise, *Epimachus fastosus*, yellowish green; Rifle Bird, *Craspedophora magnifica*, white. Such gape displays are particularly effective in courtship because they can be made with striking suddenness merely by opening the mouth.

Not all secondary sex characters are externally visible. In ducks—for example, the Rosy-billed Duck, *Metopiana peposaca*, of South America—there may be enlargements in the syrinx and trachea of the male. The Trumpet Manucode, *Manucodia keraudrenii* (Paradisaeidae), gets its name from sounds made possible by the enormously long trachea of the male coiled under the skin of his breast. In addition to this adaptation, which gives him a trumpeting voice, the bird has a collar of narrow, iridescent feathers that he can erect into a fan-like disk through which he thrusts his head while displaying. The trachea of the female is of ordinary length (see Fig. 11–5).

Lekking males of the remarkable nocturnal Kakapo parrot, *Strigops habroptilus*, of New Zealand excavate depressions or bowls 30 to 60 cm in diameter on the ground. They then inflate the thorax to gross proportions and emit resonant booms using the bowls as sound reflectors. Under ideal conditions booming can be heard up to 5 km away (Merton *et al.*, 1984).

VARIETIES OF COURTSHIP DISPLAY

Song may be employed in lovemaking as well as in war or in territorial defense. Its importance in courtship can be judged in some species by the amount of time the male devotes to song before and after he secures a mate (see p. 216). In some species, e.g., Bobolinks, courtship songs directed at females are distinctly different from territorial songs directed at males (Wittenberger, 1983). In some species, like the Prairie Chicken, *Tympanuchus cupido*, certain courtship sounds such as "booming" are restricted to the breeding season because the structures necessary for them exist only at that time of the year. In most species, however, it seems likely that courtship song is largely controlled by hormonal and psychic conditions. Very often the injection of male hormones in a bird, even in a female, will bring on courtship song and behavior.

The lovemaking male serenades his would-be mate not only with calls and songs, but with a great variety of other sounds, according to species. The tom Turkey, *Meleagris gallopavo*, rattles his quills. Hummingbirds, snipe, and nighthawks in flight make assorted humming, buzzing, singing, bleating, or roaring sounds as specialized feathers vibrate in the passing air. Woodpeckers drum on dead limbs and tin roofs. Displaying manakins, probably by striking their specially stiffened wing quills together, produce noises like the snapping of fingers or like a pencil dragged over the teeth of a comb. Voiceless storks clap their mandibles together, and some pigeons, larks, and owls clap their wings over their backs. Nature utilizes whatever resources are at hand to heighten the charms of the wooing male.

Another form of courtship activity involves contact. This type is common among birds lacking sexual dimorphism, particularly sea birds. Such courtship may involve sparring with bills, "kissing," caressing, entwining necks, nibbling at each other's feathers, or simple side-by-side bodily contact. The so-called "love birds" or Budgerigars, *Melopsittacus undulatus*, provide an example of courtship based largely on the sense of touch.

Very rarely does a brightly colored male bird merely stand where the female can see him. Usually he postures or moves about so that his brightest adornments are clearly revealed. In his courtship flights before his intended mate, the male Anna's Hummingbird, *Calypte anna*, actually orients the azimuthal component of his aerial dives with reference to the sun so that his purple-red gorget glows most brilliantly for the female (Hamilton, 1965). There is a high correlation between courtship movements and the particular structures that are the male's showiest ornaments. The Green Peacock, *Pavo cristatus*, for example, spreads his fan-like tail coverts and approaches the female obliquely with the drab rear-side of his fan exposed to her. At just the right distance, he suddenly swings himself around and dazzles her with every one of the hundred or more shimmering "eyes" vibrating and every quill rattling. To climax the performance, he screams with demonic ardor and then settles back to let his theatrics sink in. The male Golden Pheasant, *Chrysolophus pictus*, stands broadside to the female and spreads his beautiful black and gold collar sideways so that she may absorb its full splendor. The Ruff, *Philomachus pugnax*, lowers his head nearly to the ground before the female and extends his large Elizabethan ruff. The male Blue Bird of Paradise, *Paradisaea rudolphi*, hangs upside down on an exposed branch, the better to reveal his magnificent breast plumage. It seems altogether possible that the bird of paradise hanging upside down before his intended mate, and a small boy "skinning the cat" on a tree limb in front of his best girl, are both expressing the same deep ancestral urge. Other paradise birds, some parrots, and some oropendolas also display while inverted. The dramatic effects of courtship displays in nearly all birds are enhanced by their strangeness and frequently by their suddenness.

Courtship display may also take the form of dance rituals. Cranes perform different forms of solemn and stately dances, sometimes in mixed pairs, or in pairs of males with the females looking on, and sometimes in large mixed flocks.

The Black-crowned Night Heron, *Nycticorax nycticorax*, in his courtship overtures lowers his head and wings and

> executes a queer sort of courting dance on the spot of the future nest, treading from one foot to the other with a peculiar weaving action. From time to time he suddenly lowers his head and neck vertically, while his shoulders lift as in a hiccough, and he utters his courting cry. This cry is very deep and quite low, sounding like steam escaping through the safety-valve of a boiler
> (Lorenz, 1938).

It is significant that the legs of the Night Heron develop their bright rosy tinge during the breeding season. Should a female approach a courting male, he lowers his head with one cheek parallel to the ground, at the same time uttering a guttural greeting call.

> The head is then raised and the feathers on the crown, neck and back are raised. At the same time the pupil is contracted and the eyeball actually protruded from its socket, exposing the red iris to its maximum extent. The plumes are erected and may even fall forward over the head as the male bows again to the female and either repeats the greeting or turns his partly open mouth toward her
> (Noble *et al.*, 1938).

The courtship ceremony of the Night Heron illustrates how a variety of resources may be employed in impressing a potential mate: dancing, leg color, plume erection, eyeball protrusion, calls, and others.

Social and Arena Displays

A highly specialized form of communal courtship occurs in specific fixed *arenas* or *leks* where the male birds engage in their display contests. Lek courtship occurs among many species of birds representing 16 different families; among them are snipe, grouse, pheasants, bustards, hummingbirds, cotingas, manakins, bower-birds, birds of paradise, blackbirds, bulbuls, flycatchers, a parrot, and whydahs. In every instance the males of these species are polygynous and most show striking sex dimorphism in plumage and behavior. A given lek represents a collective display ground, typically composed of individually owned territories aggressively defended by each male. The arena of the Sage Grouse, *Centrocercus urophasianus*, of western North America may be about 1 km long by 200 meters wide and accommodate perhaps 400 cocks, each with his own display territory. The females must visit the lek and solicit the attention of a given cock before copulation occurs. There is a definite peck order among the grouse cocks: less than 10 per cent of the males perform more than three-quarters of the copulations. The young, lowest ranking males on the periphery of the lek do not breed. They are thought to move in toward the more active center of the lek as they grow older (and as the dominant cocks presumably die) and may then become the dominant breeders (Wiley, 1973). No pair bonds are formed nor do the males assist in rearing the young. This arena system of reproduction makes it possible for a very small but highly selected percentage of the male population to perpetuate the species.

Leks vary greatly in size and in the number of birds using them. The arenas of some tropical manakins may be only 2 or 3 meters in diameter and involve only two or three displaying males. Others, such as leks of the White-bearded Manakin, *Manacus manacus*, may be 10 to 20 meters in diameter and contain as many as 70 individual territories or display courts (Snow, 1956). In this species, although the females may watch the displaying males, experiments involving the removal of successful males indicate that the females select choice territories rather than superlative mates (Lill, 1974a). Arena display in many species represents a balance between aggressive and social or sexual tendencies of the males, but in some manakins communal displays seem to lack any aggressive element.

The jewel-like manakins (Pipridae) of Central and South America perform astonishing courtship dances and flights. Sometimes the males dance in pairs, sometimes in groups, but frequently they perform on or over dance areas which are floors of the tropical forest which the birds have cleared of all litter. The Long-tailed Manakin, *Chiroxiphia linearis*, of Costa Rica is described by Slud (1957) as performing two kinds of dances. In one dance, two males perch crosswise, facing the same direction, on a horizontal branch or vine near the ground. Each bird rises alternately straight into the air about half a meter, and then descends to the starting point. At the top of each fluttering rise the bird hangs suspended "as though attached to a rubber band" and exposes his conspicuous red cap, fluffed sky-blue back, long arching tail and bright orange legs. Each bird gives a guttural, cat-like "*miaow-raow*" as he rises. Gradually the alternating flights into the air increase in frequency and decrease in height until the two birds seem to be caught in an uncontrolled epileptic frenzy—and suddenly the dance is over. To perform the other dance, two males perch lengthwise on a branch, facing the same direction. As the bird in front gives a cat-call and rises in the air, the second bird hitches forward until under the first bird, and then rises in turn as the first bird descends to the branch diagonally backward. The two birds thus "replace one another . . . like balls in a juggling act." All the while the modestly colored female may be observing them from the sidelines. Stereotyped acrobatic courtship dances such as these are characteristic of many tropical species (Fig. 13–6). The male Standard-wing Bird of Paradise, *Semioptera wallaceii*, even

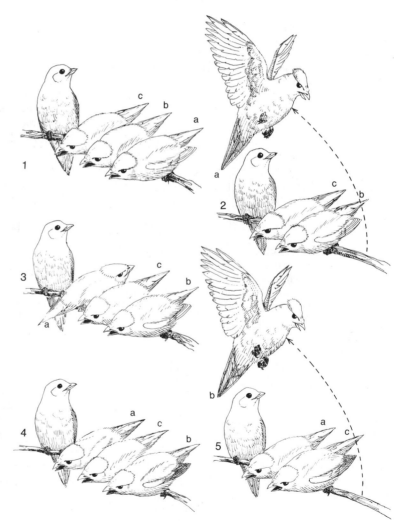

FIGURE 13–6 The courtship display of the Swallow-tailed Manakin, *Chiroxiphia caudata*. Three males, *a*, *b*, and *c*, perform their stereotyped aerial dance while the female watches. Numbers indicate the successive stages in the communal dance, *a*, performing his display flight first, followed by *b*, and then *c*. After Sick, 1967.

performs a backward somersault from his perch, landing on the ground with his wings closed (Armstrong, 1942).

Courtship displays in numerous species are performed in groups. A small group of Yellow-thighed Manakins, *Pipra mentalis*, each male on his own horizontal limb, will act like whirling dervishes whenever a female approaches their assembly in a tropical rain forest. First, a given male stretches high on his legs to expose his bright yellow thighs. He holds his deep black body horizontal, and his scarlet head bent downward. With his wings partly open, each bird then reverses his position on the limb in rapid sequence, facing first forward and then backward, so that his conspicuous head describes fiery circles about the pivot of his gaudy legs. In another type of dance each male, again stretched high on his yellow legs and with his wings held overhead, slides rapidly backward on a horizontal limb for about 30 cm and then jumps forward to the starting place and repeats his stiff-legged, backward fox-trot over and over. Along with this bizarre dance, each manakin gives off loud buzzings and sharp snapping sounds, probably made with the specially modified wing feathers.

In the polygamous Black Grouse, *Tetrao tetrix*, of Europe, a group of males will gather on their lek each spring, and at about sunrise begin calling, dancing, and posturing to defend their individual mating territories and to attract the females. Each male holds his head upright, spreads his bow-shaped tail, droops his wings, inflates his fleshy red "eyebrows" with blood, and begins jumping and fluttering over his chosen spot. Should another male approach his territory, the owner will quickly thrust his head and neck forward and perhaps run a few steps toward the intruder and then retreat. Fighting is highly formalized and is usually restricted to threats. If a female approaches a male, he will circle around her with quick steps, often tilting his body and tail toward her. If she crouches and they copulate, other males may try to interfere.

Village Indigobirds, *Vidua chalybeata*, of Africa practice a form of social reproduction that might be called exploded-lek breeding. During the breeding season individual males defend special trees called "call sites," which are usually several hundred meters apart. All males of a given neighborhood sing the same song dialect. Individual females sample various call sites where they are courted by the resident males. All males successful in winning mates sang more than 18

minutes per hour; unsuccessful males sang less. In one community of 14 males, one male secured 66 per cent of all matings. The fact that females seemed to prefer males that had neighboring males close by suggests a possible method by which lek behavior originated (Payne and Payne, 1977).

Male Guianan Cock-of-the Rocks, *Rupicola rupicola*, not only compete for females visiting a lek but also disrupt copulations by other males or harrass visiting females. This disruptive behavior sometimes succeeds in focusing a visiting female's attention on the attacker, which then results in copulation (Trail, 1985).

Many primitive tribes throughout the world perform dances clearly derived from those of courting or fighting birds. The Blackfoot Indians of the Northwest mimic the foot-stamping, bowing, and strutting of the Sage Grouse, complete with a costume imitating the spread tail of the grouse. According to Armstrong (1942) the Jivaro Indians of South America mimic the dance of the Cock-of-the-Rock; the Chukchee of Siberia, the dance of the Ruff; the Monumbo of Papua New Guinea, that of the cassowary; the Australian aborigines, the Emu; the natives of New Ireland, the hornbill; the Maidu of California, the tree-creeper; and the Tarahumare Indians of Mexico, the turkey. The "Schuhplattler" dance of Bavarian peasants resembles very closely the courtship jumping and fluttering of the native Black Grouse, with the men doing the active dancing and the girls rotating in a fixed position, watching the men. Armstrong was of the opinion that both in birds and in man dances have similar functions for self- and corporate-stimulation (for either love or war) and also for the release of excess energy. Of course, human dances also have intellectual functions such as the propitiation of harmful spirits, the promotion of fecundity in crops, or success in hunting.

Social courtship is practiced not only by numerous species of gallinaceous birds but also by that extraordinary shore bird the Ruff, *Philomachus pugnax*. During the mating season the male has a ruff of prominent neck feathers and exaggerated ear tufts. His courtship dance consists of an excited running about the lek with wings fluttering, head and neck held horizontally and ruff expanded. Suddenly the bird stops, lowers his head until the bill touches the ground, and then freezes into a rigid posture with the spectacular ruff expanded, wings spread, and tail depressed. Shortly

FIGURE 13–7 European male Ruffs in communal courtship display. Note the great individual variation in the coloration of ruff feathers. Photo by C. C. Doncaster.

the bird will shiver his wings and either become quiescent, sometimes even apparently taking a nap, or begin his wild scurrying about once more. If a female approaches a male and perhaps nibbles at his ruff or head feathers, coition ensues. Then the two part, probably never to see each other again. Female Ruffs are most likely to solicit mating in the central part of a lek where male-to-male aggression and male-female courtship activity are the most intense (Shepard, 1971). The unique feature of the system in Ruffs is the occurrence of "satellite" males, which exhibit a white colored ruff and do not hold territories in the lek. Instead, they join territorial males during female visits and the resident and satellite display together (van Rhijn, 1973).

Courtship dancing is not necessarily limited to land. A variety of water birds perform the equivalent of dancing on and in water. In a stereotyped courtship ceremony, the male and female Great Crested Grebe, *Podiceps cristatus*, swim toward each other on the surface of the water and touch beaks. Then they dive and

emerge facing each other, each with a bit of water weed in its beak. Next, in their well-known "penguin dance" the grebes rise upright in the water with their pure white bellies exposed to one another, and while maintaining this position weave their bodies back and forth. In some species of grebes and loons, the pair will actually patter ecstatically and with extraordinary speed over the surface of the water while in this same upright, penguin-like position, sometimes for as far as half a kilometer, only to turn around and skitter back again. Among other forms of display that involve diving, splashing, and pattering over the surface of the water, the Red-throated Diver, *Gavia stellata*, rolls over in the water, belly uppermost, kicks and splashes with its legs in the air, and then dives, emerging upside down and repeating the strange performance.

Many birds conduct their courtship activities in the air. In some species, courtship consists merely in a flight that exposes the more colorful markings of the male to the eyes of the female. Daily, during the nesthole construction period, the Great Spotted

Woodpecker, *Picoides major*, hovers moth-like in the air with his tail raised to expose the fiery red underparts. Just before coition the male Snow Finch, *Montifringilla nivalis*, flies in circles about his mate, exposing the white feathers of his wings and tail. As discussed earlier, many species, such as goatsuckers, snipe, and hummingbirds, combine remarkable aerial acrobatics with unusual sounds caused by the rushing of air across specially adapted feathers.

Among the more spectacular courtship flights are those of some snipe, herons, hawks, doves, hummingbirds, swifts, and Corvidae. In its display flight the male Lapwing, *Vanellus vanellus*, rises slowly from the ground, then speeds up his wing beats and rises at a steep angle. Suddenly he plunges toward earth, turning and twisting and somersaulting "as though out of control" and then "sweeps off with erratic flight tilting from side to side and producing a quite loud humming throb with its wings" (Tucker, 1943). The male Marsh Hawk or Harrier, *Circus cyaneus*, along with numerous other hawks, plunges directly earthward from a great height, turning somersaults and uttering shrill cries during his descent. At other times, he impresses the female by performing an up and down roller-coaster flight, rising only a few meters above the ground and looping the loop and screeching on each downward plunge. Among the ducks, finches, weaver finches, sylviid warblers, and others, there occur reckless courtship flights in which the male pursues the female. In some swifts the chase will climb high in the heavens where it abruptly ends as the pair "lock in a copulatory embrace and fall a thousand feet, their wings flailing the air like a pinwheel" (Peterson, 1948).

Fighting between males is accepted courtship practice in many species. The fighting not only decides which males "deserve the fair" but it also undoubtedly stimulates sexual ardor in many of the combatants. Often, of course, it is difficult to distinguish courtship fighting from territorial fighting or from display to impress the female. Occasionally a male Cock-of-the-Rock will seize the talons of another male and hold on for as long as two hours (Trail, personal communication). One of the chief mating-season activities of the gregarious Avocet, *Recurvirostra avosetta*, is formalized fighting by both sexes. Often, at the height of battle, a bird will put its head under its wing as though going to sleep. Such incongruous action is probably a displacement activity brought on by the intensified emotions aroused by combat. Male penguins engage in courtship fights of impressive vigor, sometimes emerging from conflict minus a tongue or an eye. Rival Adelie Penguins, *Pygoscelis adeliae*, begin fighting with their beaks, "but they soon resort entirely to blows of the flippers and end by leaning against each other in a sidewise position and battering away with the 'outside' wings, like clinched pugilists, raining blows until the battering sounds a tattoo" (Murphy, 1936). The female Adelie seems no less belligerent. When a male suitor approaches her expectantly, she jabs at him with her powerful, sharp beak. This abuse the male meekly accepts, upon which the female refrains from further attack and both birds assume the widespread "ecstatic" posture, crossing their upstretched beaks and weaving their heads from side to side, while uttering their unmusical calls. Unless another male waddles in to interrupt this love making, the pair may copulate and establish a pair bond for the season or for longer. Love and war, eroti-

FIGURE 13–8 Courtship "billing" in the Adelie Penguin. In many species of sea birds the sexes are visually indistinguishable. In such species, the sex of a given bird may be revealed by its response to courtship activities. Photo by Expéditions Polaires Francaises.

cism and aggression, are closely allied partners in the breeding behavior of many birds, and it is likely that both of these seemingly antithetical activities promote the psychological and physiological functions necessary for individual survival.

Bower-bird Courtship

A special category of courtship display is provided by the bower-birds (Ptilonorhynchidae) of Australia and New Guinea. These relatives of crows and birds of paradise are not conspicuously ornamented, but the males make up for their lack of nuptial finery by constructing bowers and display grounds that serve the same functions as secondary sex characters in other birds. This conclusion is supported by the fact that males of the least colorful species of bower-birds build the most elaborate and highly ornamented bowers, and vice versa (Gilliard, 1958; Diamond, 1982).

Among bower-birds a common type of bower is constructed by the male in the form of two parallel hedges of interlaced grasses or twigs stuck in the ground. In the space between the hedges is an avenue in which the male may do some of his displaying. This is the "avenue" type of bower. In another type of construction the bower may be in the form of a stack of twigs erected around a vertical sapling, or arranged in the form of an open-sided tepee or hut whose roof-center is supported by a sapling. These are the "maypole"

FIGURE 13–9 The maypole bower of the Brown Gardener Bower-bird.

bowers, some of which may be as high as 3 meters. The floor under and in front of them is cleared of all litter by the males and decorated with collections of colorful objects neatly piled and, in the case of perishable leaves, flowers, and fruits, often replaced daily, even for months on end. The clear space adjoining each bower is known as the display area, or stage. It seems likely that bower construction originated as a displacement activity for nest-building.

The male Great Grey Bower-bird, *Chlamydera nuchalis*, uses about 1600 twigs, each approximately 2 mm in thickness, to build his two-walled avenue. At each end of the avenue is a stage that he decorates with shells, bleached bone fragments, or shiny metal and glass objects, which, in total, may weigh between 6 and 12 kg (Veselovsky, 1978) (Fig. 13–10).

The Vogelkop Gardener Bower-bird, *Amblyornis inornatus*, builds an open-sided maypole bower that looks very much like a thatched hut. Incorporated in the walls of the hut may be living orchids. In front of the hut the display area may have neatly piled "gardens" of mushrooms, flowers, cartridge shells, colored fruits, or stones of a given color, one kind of object to each pile (Cooper and Forshaw, 1977). Experimenting with the bird's esthetic preference Diamond (in litt.) placed poker chips of different colors in front of each hut. The males selected certain colors which were added to their piles and cast out the others. Each isolated population exhibited preference for certain colors. Diamond suggests that these preferences may have been passed down by tradition as is the case with song dialects.

A few species of bower-birds even paint the walls of their bowers. They make their paint by mixing saliva with chewed-up fruits, grass, rotten wood, or charcoal, and then smear the paste on the sticks of the bower. Two species, of which the Satin Bower-bird *Ptilonorhynchus violaceus*, is one, manufacture wads of fibrous bark which they use as paint brushes to apply the colors (Marshall, 1954).

As the reproductive season opens and hormones begin to flood the tissues of the male bower-bird, he begins to decorate his bower and to display "energetically and noisily," often holding a display object in his beak. The display ground, according to Marshall, is the focal point of the male's territory and the chief agent that attracts the female's interest, keeps rivals away, and helps stimulate and synchronize the male and female reproductive machinery. At least in some species, coition occurs in the bower. After fertiliza-

FIGURE 13–10 The male Great Gray Bower-bird decorates his avenue bower with fragments of bleached bones. Here the male, holding a bone fragment in his beak, displays his neck and bright cerise crest through the bower to the watching female in the background. Photograph courtesy of John Warham.

tion, the female builds her nest at a distance from the bower. There she incubates the eggs and rears the young, unaided by the male. After the young have fledged, the female may bring them to the bower, where the family engages in communal display—possibly an imprinting exposure that may educate the young in the intricacies of display etiquette for that species.

The Satin Bower-bird has intensely blue eyes and favors blue display ornaments: flowers, berries, feathers, bits of paper, all blue. To study the urge of males to assemble blue items for their bowers, Marshall marked

many pieces of broken blue-glass bottles with individual numbers and scattered them near known bowers over an area of 130 sq km (50 sq mi). Within a month 79 per cent of the glass fragments appeared in the bowers. They shifted from one bower to another, however, because of perpetual raiding and counterraiding. To check the influence of bower quality on female solicitation, Borgia (1985) removed the decorations (chiefly blue feathers) from 11 bowers and left 11 others intact as controls; he then monitored them with video cameras. Breeding success decreased in the dispossessed bowers but remained high in the controls, demonstrating that bower ornamentation is a product of sexual selection.

One amazing characteristic of the avenue-bowers of the Satin Bower-bird is their orientation. Of 66 bowers examined by Marshall, the deviation of the axis of the central avenue from true north and south was never more than 30 degrees. When the bower of a male in an aviary was experimentally shifted from its original orientation of 15 degrees (east of north) to 310 degrees (northwest by southeast), the bird reoriented the walls one twig at a time to approximately the original bearing. Avenues of the Yellow-breasted Bower-bird, *Chlamydera cerviniventris*, are oriented roughly east and west.

A bird with a penchant for accumulating bright shiny objects is the Spotted Bower-bird, *Chlamydera maculata*. It will even enter tents and houses to pilfer cutlery, coins, thimbles, nails, screws, lead bullets, and pieces of glass and tin. Marshall relates the story of a car owner who, noting the disappearance of the ignition keys from his car, and familiar with the habits of this bird, walked about a kilometer to the nearest display ground and retrieved his keys!

When alone, the male of this species displays moderately and intermittently, but should a female approach his bower, he immediately throws himself into a frenzied performance. He alternately lifts and lowers his lilac mantle feathers; he contorts his body into strange postures and frequently leaps into the air. He may attack objects on the display floor and fling them about with vigorous abandon. After some 20 minutes of such effervescence, the male may copulate with the passive female outside the bower and then return for additional violent display, after which he calmly tidies up his bower. In this species, as also in the Satin Bower-bird, the male displays around the bower for weeks and even months before coition occurs, and continues to display after the female leaves and begins to incubate. It seems that the long maintenance of sexual activity and display by the male—sometimes four to five months—is an adaptation to ensure the readiness of the male for the sudden, brief, and unpredictable period when the female becomes sexually active. The female's reproductive activity seems to be triggered by some as yet undiscovered environmental factor that precedes that time of year when protein food, mainly insects, will be most abundant for raising young. And in most bower-bird countries this does not occur every year on the same date.

SYMBOLIC COURTSHIP RITES

Many necessary, everyday activities may become formalized and adapted to functions quite different from their original ones. Among these symbolic formalities is courtship feeding, which occurs in species that feed their young by regurgitation, or with their beaks, or with food carried in their talons. It is commonest in species in which the male and female remain together for the breeding season, but rarely occurs in birds that do not form pair bonds or in species like ducks and game birds that do not feed their young. In a review of this subject, Lack (1940) lists 14 orders of birds, including 17 families of the Passeriformes, in which courtship feeding is known to occur. Whenever courtship feeding does occur, the male nearly always feeds the female, with the significant exception of the button quails, in which the male incubates and rears the young. In the usual situation the female may beg her mate for food with much the same posture, wing fluttering, and even infantile calls that the young birds use in importuning their parents for food. In a few species such as terns, either sex may beg food from the other, and in wax-wings the male and female may pass the food back and forth from bill to bill. Courtship feeding and related formalities, such as billing and gaping, are, according to Armstrong (1942), sexual ceremonies that have evolved from the parental habit of feeding young birds—latent activities that reappear in the adult in a new context. Often courtship feeding continues and becomes "real" feeding of the incubating female by the male. The male frequently feeds the female only during the incubation period. (Johnston, 1962).

The innate nature of symbolic courtship feeding was demonstrated by Platz (1964). He raised Red-crested Pochards, *Netta rufina*, from the egg, completely isolated from their parents. The young birds, at maturity, performed the courtship-feeding ceremony exactly as the wild birds did: the males diving for bits of plants that they presented to the females at the surface before copulation.

Nutrition is not ordinarily the reason for courtship feeding. This is easily seen in such species as the Atlantic Common Murre, *Uria aalge*, which carries fish to the nest site and holds them in its beak for several hours. The female European Robin, *Erithacus rubecula*, may beg her mate for food even though she is standing in a dish full of meal worms, or, later in the breeding cycle, beg him for food when her own mouth is full of worms to feed the young (Lack, 1943). However, Nisbet (1977) has shown that courtship feeding by the male Common Tern, *Sterna hirundo*, is an important determinant of the number and size of eggs the female produces. In the case of the female hornbill, imprisoned in her tree cavity nest, "courtship" feeding by the male is an absolute necessity.

In some species, courtship feeding is associated with coition and may function as a releaser for that behavior. Courtship feeding occurs just prior to coition in pigeons and Rooks, *Corvus frugilegus*; it occurs during copulation in the Yellow-billed Cuckoo, *Coccyzus americanus*, and Galapagos finches (Geospizinae); and after copulation in Starlings, *Sturnus vulgaris*, and terns. The male Roadrunner, *Geococcyx californianus*, holds a lizard or mouse in his beak during coition and feeds it to the female immediately afterward. The male Snail Kite, *Rostrhamus sociabilis*, presents a snail to the female who first removes the snail from the shell. As soon as she swallows the snail, the male hops on her back and copulates (Haverschmidt, 1970). Experiments by Mason (1945) on the Corncrake, *Crex crex*, showed that males would display in front of a dummy and attempt copulation with it. After 23 such attempts at coition one bird disappeared, only to return with a green caterpillar, which it offered to the stuffed bird before it resumed its futile attempts. The male Squirrel Cuckoo, *Piaya cayana*, of tropical America first offers the female an insect or caterpillar that both of them then hold on to while they copulate (Skutch, 1966). In some species courtship feeding seems primarily to cement the pair bond: for example, in European Robins.

Food is passed from the male to the female in various ways. Most commonly the male places it in the female's gaping mouth. In many finches, gulls, and doves, the male regurgitates the food for the female. Among hawks, the food may be passed from the talons of the male to the female in midair, the latter turning upside down to accept the symbolic gift. Tengmalm's Owl, *Aegolius funereus*, calls and repeatedly flies to his nest hole, and finally persuades his mate-to-be to enter the nest, where she finds a freshly killed mouse awaiting her. The Adelie Penguin, *Pygoscelis adeliae*, will feed his intended spouse with pebbles and bits of snow.

Dozens of galliform species—partridge, quail, pheasants, guineafowl, and others—perform a courtship feeding ceremony called "tid-bitting" in which the male attracts and appeases the female by calling, "waltzing," and holding or dabbling at food prior to

FIGURE 13–11 Courtship feeding in European Crossbills. The male is passing regurgitated seeds from his crop into the mouth of the female. Photo by E. Hosking.

copulating. According to Stokes and Williams (1971), males display most readily with unusual sources of food (mealworms, live crickets, berries) but rarely with items common in the diet. The male tempts the female with bonbons, as it were, rather than meat and potatoes.

Related to courtship feeding are the stylized ceremonies of billing, fencing, gaping, and similar forms of lovemaking. Billing is common among water birds such as penguins, cormorants, gannets, grebes, puffins, and herons, as well as among some of the parakeets, doves, corvids, and finches. The King Shag, *Phalacrocorax albiventer*, woos his mate by holding "her head gently in his large open bill, and the two will sway from side to side. The birds will then bow, kiss and nibble each other about the head, and utter sundry grunts and coos and a blowing sort of whistle" (Cobb, *in* Murphy, 1936). Courting Ravens, *Corvus corax*, will hold on to each other's beaks in a prolonged kiss. The brightly colored mouth cavities are exposed in ceremonial courtship gaping by the males and sometimes by both sexes of certain cormorants, guillemots, gulls, terns, storks, birds of paradise, thrushes, and other species.

Although in the normal sequence of events courtship precedes nest-building, symbolic or mock nest-building may be employed as a mode of courtship. A great many species handle bits of nesting material as part of the courtship routine. The Black Skimmer, *Rynchops nigra*, gently passes a stick to the female just before coition, and the Roseate Spoonbill, *Ajaia ajaja*, presents sticks to his mate with much head-bobbing and bill-rubbing to reinforce the pair bond. In the pair formation ceremonies of the Yellowhammer, *Emberiza citrinella*, both sexes fly to the ground to pick up and drop pebbles and bits of vegetation (Diesselhorst, 1950). Breeding synchronization in Bullfinches, *Pyrrhula pyrrhula*, is accomplished by the mated pair showing nesting material to each other (Nicolai, 1956). Many estrildid finch species hold a piece of grass or feather, a "nest symbol," as they perform their courtship dance before the female (Goodwin, 1982). (See Fig. 13–12.)

During the mating season, a Gannet, *Morus bassanus*, may "pretend" to receive imaginary nesting material from an invisible mate and build it into a fictitious nest (Lorenz, 1937). Also during the courtship period, female skuas will drop grass before the males; grebes will dive and bring up water weeds in their beaks; and Moorhens, *Gallinula chloropus*,

FIGURE 13–12 Courtship in African Silverbills. Note that the male *(left)* is holding a grass "nesting symbol" as part of his courtship ritual.

will engage in symbolic nest-molding. Possibly an atavistic remnant of nest-display behavior persists in the Blue-diademed Motmot, *Momotus momota*, in which a courting bird approaches its mate holding a leaf or twig in its beak despite the fact that their underground burrow is "quite devoid of lining" (Skutch, 1964).

For some species, the building of the actual nest provides courtship stimulation. The male Phainopepla, *Phainopepla nitens*, begins a nest, and if he acquires a mate, she helps to complete it. If, however, the male finishes the nest without securing a mate, he starts another, dismantling the first to get material for the second (Rand, 1943). Similarly, the male Village Weaver, *Ploceus cucullatus*, weaves the outer shell of the nest and "advertises it to unmated females with a stereotyped wing-flapping display at the entrance. . . ." If females repeatedly reject the nest, "the male tears it down and builds a fresh nest in its place" (Collias and Collias, 1969). There is some evidence that the completed nest built by the male serves in some species to put the female into the mood for breeding. This seems to be the case for some weaver birds (Ploceidae), the Penduline Tit, *Remix pendulinus*, and the Australian Magpie-lark, *Grallina cyanoleuca*. A completed nest or nest site may have a magnetic attraction for an unattached female bird. Males of several species use their nests to "seal the wedding compact": among them are owls, woodpeckers, European flycatchers, wrens, and others. The male Collared Flycatcher, *Ficedula albicollis*, demonstrates his nest hole to passing females with a special call and a slow flight near the hole, exhibiting all his conspicuous plumage characteristics. His urge to display the nest hole is so consuming that he displays to strangers even while his mate is busy carrying nesting materials into the hole. So strongly attractive is the nest hole display that, should the male display at another nest hole while his mate is incubating eggs in the first-chosen nest, she may desert the eggs and start afresh in the new, empty nest hole (Löhrl, 1951).

In all these strange and wonderful fashions male birds court their mates. They use the limited materials and talents they have at hand to win their brides, but they use them in distinctive and striking ways, often with dramatic suddenness. In fundamentals, avian and human courtship patterns are not widely divergent. A young man courting his girl may put on his best suit of clothing (breeding plumage), and set forth with a box of bonbons (courtship feeding) under one arm and a spray of flowers (bower-bird display) under the other. Before taking his girl to a dance (courtship dancing), the two of them may listen to a record of highly rhythmic music and join in on the chorus (courtship singing). During the evening the young man may pull from his pocket a blueprint of their dream home (nest display), while the young lady repairs her make-up with lipstick (gape display). At the end of the evening it is not unlikely that the young couple will indulge in a courtship activity known in both classes of vertebrates as billing and cooing. The hot-springs of love run deep and pervasive in the clay of all vertebrates. It is not surprising that their external bubblings appear to be much the same, whether in a university graduate, an Australian bushman, or a lowly sparrow.

■ SUGGESTED READINGS

A very full and delightfully written treatment of courtship display is found in Armstrong's *Bird Display and Behavior*. For a scholarly and entertaining account of the astonishing courtship displays of bower-birds, see Marshall's *The Bower-Birds*. Lack's *Ecological Adaptations for Breeding in Birds* presents much interesting material on courtship and mating habits. For extensive and thorough treatment of reproductive behavior in two special groups, consult Johnsgard's *Handbook of Waterfowl Behavior* and Richdale's *Sexual Behavior in Penguins*. Many books devoted to individual species or families of birds may be consulted with profit for information on courtship behavior. Modern rigorous treatments of the subjects of courtship and sexual selection may be found it R. B. Payne's "Sexual selection, lek and arena behavior and sexual size dimorphism in birds." Ornithological Monographs, Vol. 33, and L. W. Oring's, "Avian mating systems" *in Avian Biology*, Vol. 6.

Nests
14

With esoteric bills of lading,
For twigs and feathers, mud and grasses,
By searching, seizing, plucking, raiding,
The avian architect amasses
A fine supply of found material
For nest on earth, or nest aerial.
—Joel Peters (1905-   )
The Avian Architect

A great variety of animals build nests, but of them all—insects, fish, amphibia, reptiles, mammals, and others—birds are by far the most expert and most industrious nest-builders. They build nests of many different materials and in a bewildering variety of forms, and locate them in more varied sites than do any other animals.

FUNCTIONS OF NESTS

Birds use their nests chiefly to protect themselves, their eggs, and particularly their developing young from predatory animals and from adverse weather during the breeding season, the most vulnerable period in the life cycle. For protection against predators, birds rely mainly on nests that are inaccessible, armored, camouflaged, or built in colonies that provide the safety of numbers, or in places where they enjoy the protection of aggressive animals.

For inaccessibility, nests are commonly constructed at the tips of tree branches, on ledges of cliffs, in burrows or cavities, on isolated islands, and on or over water. Several species build floating nests, and others construct nests propped up on reeds like the homes of the ancient Swiss lake-dwellers. Each of these varied building sites rules out whole classes of potential predators.

Camouflaged nests are often built of materials found in the immediate vicinity, which are therefore inconspicuous. Nests in grassy habitats are often made of grass; in reed beds, of reeds; in trees and shrubs, of twigs or branches. Sometimes the outsides of nests are covered with such disguising materials as mosses (by kinglets) or lichens (by hummingbirds) which decrease their visibility. The acme of inconspicuousness is achieved in some species (plovers, goatsuckers) by the simple expedient of building no nest at all and laying the eggs (often concealingly colored) directly on the bare ground. Although in such cases the "nest" matches its environment perfectly, other useful functions of a nest may be lacking.

Either through their construction or placement, many nests protect their inhabitants from rain, floods, drifting sand, or the burning rays of the sun, as well as from the eyes of predators. Many ground-dwelling species, such as the Eastern Meadowlark, *Sturnella magna*, and Ovenbird, *Seiurus aurocapillus*, and great numbers of tropical tree dwellers build roofed-over nests. Nests of many desert species are commonly built in the lee of a sheltering rock or in the shade of some shrub.

A second major function of a nest is to maintain the warmth that promotes incubation of eggs and rapid development of the young. The sheltering warmth of the nest, in conjunction with the heat supplied by the brooding parent, parallels the warm protection of the uterus for an embryo mammal. Since warm young birds develop and reach maturity more quickly than those given less warmth, nests reduce the length of their period of highest vulnerability to exposure and predation, and thereby lengthen their life-expectancy. Further, nests may conserve energy for adult birds. Even in the summer, inside the communal nests of the African Sociable Weaver, *Philetarius socius*, air temperatures varied only 7 or 8° and remained within the birds' zone of thermoneutrality, while outside air temperatures ranged from 16 to 33.5°C (Bartholomew *et al.*, 1976). In winter when desert temperatures often dropped to freezing, air temperatures in occupied chambers were 18 to 23°C above outside air temperatures. This nest insulation was estimated to conserve about 40 per cent of the birds' energy expenditures (White *et al.*, 1975).

Nests have still another function—they supply a supporting platform so that the eggs and young may be situated in trees, or on soggy ground, or even floating on water like the nests of grebes. Frequently the base of a nest is composed of coarse sticks or twigs which provide anchorage and support, while the superstructure is made of the finer and warmer materials that provide warmth, protection, and camouflage.

Finally, the construction and possession of a nest satisfies some of the deepest innate urges that a bird has—nest-building, stimulation of hormone-flow and egg-laying, brooding, and the care of the young—all in one package. For species whose young remain in the nest several days or weeks after hatching, nest life enhances family bonds and increases opportunities for educating the young.

EVOLUTION OF NESTS

One can only guess at the evolutionary origin of birds' nests. Much diversity is shown in nest-building behavior, even between close relatives. Among flycatchers of the family Tyrannidae, for example, the Eastern Kingbird, *Tyrannus tyrannus*, builds a nest of weeds, grasses, mosses, and plant down, often at the extremity of a tree branch. The Crested Flycatcher, *Myiarchus crinitus*, lines a tree cavity with grasses, twigs, and rootlets, and often garnishes it with a cast-off snake skin. The Phoebe, *Sayornis phoebe*, constructs a bulky nest of mosses and mud, lined with grass or hair, typically on a beam under a bridge. The Yellow-bellied Flycatcher, *Empidonax flaviventris*, builds its nest of moss and grasses on the ground; the Tody Flycatcher, *Todirostrum cinereum*, builds a hanging, purse-like nest of grasses and other fibers that has a side entrance.

Similar variability is found among the nests of auks, boobies, plovers, gulls, terns, doves, finches, and other birds. Perhaps the group showing the greatest diversity in nest types is the family of ovenbirds or Furnariidae of Central and South America. Within this family there are some species that nest in rock crevices or tree cavities; some drill tunnels in banks or in dead trees; some build nests of twigs or branches in a tree; others make globular two-chambered, oven-shaped nests of mud; and in one species several pairs cooperate to build a communal apartment house of sticks (Gilliard, 1958).

As a general principle, however, passerine birds build the solidest, warmest, and most complex nests. This is only to be expected since the problems of predation and body heat loss focus most drastically on

FIGURE 14–1 A European Nightjar on its nest. The eggs are laid on the bare ground, usually near dead wood. Note that the bird enhances its inconspicuous coloration by flattening itself against the ground and by closing its eyelids. Photo by E. Hosking.

small birds. Nonpasserine birds, as a rule, build either simple nests of vegetation, earth, or stones, or no nest at all.

Nest form and structure may be useful as taxonomic characters distinguishing genera of swallows, swifts, weaverbirds, ovenbirds, and New World and Old World flycatchers (Collias and Collias, 1984).

Components of nest-building behavior may be ritualized and used in courtship. Many passerines hold pieces of grass or feathers (nest materials) during courtship. When a courting male tern, bearing a fish, approaches the female on the ground, she depresses her head and breast while raising her wings and tail, and meanwhile kicks sand rearward with her feet. As the male circles the female, she rotates so as to continue facing him, thus making a saucer-shaped depression or "scrape." She may pick and jerk toward her bits of nesting materials as she keeps rotating on the scrape. Many other species show similar scrape ceremonies during courtship: plovers, curlews, the ostrich, nightjars, grouse, and quail.

The female Red-throated Diver, *Gavia stellata*, lies in a grassy depression near the water's edge during copulation with the male. Immediately after the act, in an instinctive frenzy, she tears out beakfuls of moss and grass which she aimlessly throws behind her without attempting to arrange the material into a nest. Nevertheless, she makes a rude, cup-shaped depression in which she lays her eggs.

Successive steps in the evolution of birds' nests have been postulated by Makatsch (1950) and Collias (1964) to have occurred somewhat as follows. Homeothermy probably evolved gradually so that in their early stages of evolution birds probably depended on the environment to incubate eggs. During this early transitional period they probably buried their eggs as megapodes and the Egyptian Plover, *Pluvialis aegyptius*, still do today. Crocodiles, the close relatives of birds, also bury their eggs in leafy mounds, often in close proximity to megapode eggs. Numerous species of birds (e.g., grebes, ducks, partridges, eagles) cover their eggs, when unattended, with vege-

FIGURE 14–2   One of the chief functions of a nest is protection. The floating nests of grebes, built of aquatic plants, protect them from many terrestrial predators. Shown here is a European Little Grebe and its two young. Photo by E. Hosking.

tation or down feathers as camouflage. From a primitive scraped depression in the ground—a nest form still used by many ground dwellers—there probably evolved a simple nest of scrape-plus-vegetation. By the turning movements of the incubating bird, the twigs or grasses may have become molded into a shallow cup. By adding building materials to the rim (a consequence that might be expected from the position of the incubating bird's bill) the nest cup may have become deepened. It is but a small step from a deep cup-shaped nest to a roofed-over nest. A ground-nester might pull grass tufts together and into the side walls of the nest making an oven-shaped nest. Such a covered nest in a tree might next evolve into the simpler types of hangnests supported from above by a rope of interlaced fibers; this last form might quite naturally develop into the most complex sleeve-shaped hangnests with side entrances and other refinements.

The form of a nest may be influenced by nesting success; for example, young of the Reddish Hermit Hummingbird, *Phaethornis ruber*, are bitten by mosquitoes in shallow nests but not in deep ones (Oniki, 1970). Roofed-over nests and hole-nesting probably evolved primarily as a result of predation pressure. Ground-dwelling Meadowlarks, *Sturnella magna*, and tree-dwelling Magpies, *Pica pica*, build both open-cup and covered-over nests. In each species, losses due to predation were higher in open nests than in covered nests (Roseberry and Klimstra, 1970; Baeyens, 1981). In a study of numerous species, Nice (1957) found that about two-thirds of 94,400 eggs from hole nests survived successfully whereas only about one-half of 22,000 eggs from open nests met with success.

CHOICE OF NEST SITE

Simply because they can fly, birds are practically unlimited in their choice of nest sites even though they may range from oceanic islets to alpine precipices, from polar icefields to "impenetrable" jungles. In this regard, birds have no peers. In combination with their unexcelled mobility, birds seem to exercise an instinctive talent for seeking out sheltered niches in which to build their nests: in crevices, under overhanging ledges, deep in shrubs, even behind waterfalls.

FIGURE 14–3 A pair of Ospreys at their nest. Ospreys feed on fish, and consequently build their nests near water. The nest, which is usually lined with grasses, is often used for many years. Photo by M. D. England.

The keenest competition for nest sites undoubtedly occurs among birds with special requirements, such as those nesting in tree cavities or on cliff ledges. One way in which natural selection discriminates against an inferior nest site is revealed by the mortality rates of the Common Puffin, *Fratercula arctica*, of New-foundland. Pairs of Puffins nesting near a cliff edge had a higher breeding success than those nesting on level ground away the cliff, because those at the cliff edge could more easily escape being robbed by preda-tory gulls of food being brought to their young (Nettle-ship, 1972). Similarly, nest location affects the rate of predation by gulls on fledgling Guillemots, *Uria* spp. When the fledglings can drop directly into the sea from cliff-edge nests, the predation rate varies from 0.6 to 2.2 per cent; but when the young must first traverse difficult terrain to reach the sea, the loss rises to 17.5 per cent (Williams, 1975). Among Northern Gannets, *Sula bassana*, in Newfoundland, more chicks fledge from cliff nest sites (93 per cent) than from slope (77 per

cent) or plateau (70 per cent) sites. Here the predatory attacks come from neighboring adult Gannets (Mon-tevecchi and Wells, 1984).

Both sexes of many species participate in nest-site selection, but in ducks, geese, and gallinaceous birds the male takes little if any part. Among polygynous species such as grouse and pheasants, the female ordinarily chooses the nesting place. Among polyg-ynous Red-winged Blackbirds, *Agelaius phoeniceus*, and weaverbirds of the genus *Euplectes*, the male chooses the territory for his entire harem, while the females select their individual nesting sites within it. However, in many migratory species in which the male arrives first in the spring, the male com-monly establishes the territory and sometimes locates the nest site within it. The male House Wren, *Troglodytes aëdon*, and the male Phainopepla, *Phai-nopepla nitens*, begin building their nests before their mates arrive, and the male Prothonotary Warbler, *Protonotaria citrea*, picks out a cavity in a

FIGURE 14–4 A pair of Reed Warblers at their cup-shaped nest in a reed bed. The nest is constructed by the female alone. The male is shown swallowing a fecal sac from one of the young. Photo by E. Hosking.

dead willow and begins lining it with mosses before his mate appears. In many species the choice of a nesting place becomes a part of the courtship ceremony. This is well illustrated by the Collared Flycatchers, *Ficedula albicollis*, as described in the preceding chapter.

In selecting the nest site, a bird may show anticipatory nest-molding or brooding behavior in a spot completely devoid of nest materials. A female American Robin, *Turdus migratorius*, may crouch and slowly spin around in a suitable tree crotch off and on for several days before she begins building her nest there; and a female European Mistle Thrush, *Turdus viscivorous*, may "brood" a bare nest site for two or three weeks before nest construction begins.

Although the capacity for flight opens up to birds as a class almost unlimited site possibilities, there are anatomical, physiological, and ecological limitations for different species that prevent a completely free choice of nest locations. Swallows with their weak wide bills could not possibly chisel out nest cavities in solid wood as do woodpeckers, and heavy-footed ground nesters like the pheasants could not exploit

soggy marsh habitats for nests as successfully as long-toed jaçanas or web-footed grebes.

An ecological limitation profoundly affects the nesting sites of the Royal Albatross, *Diomedea epomophora*. Like other albatrosses, this species requires strong winds in order to take off from the ground. All of its nests are placed on the protected lee sides of ridges on Campbell Island, New Zealand, where a few short steps will take the bird into the perpetual updrafts of the "roaring forties" (Westerskov, 1963). Similar climatic or topographic requirements restrict the nest-site distribution of many other species. Even nest orientation may be affected by climate. Probably as a temperature-controlling adaptation, the Cactus Wren, *Campylorhynchus brunneicapillus*, orients its nest openings in one direction in the spring to avoid cold prevailing winds, and in another direction in summer to take advantage of cooling breezes (Ricklefs and Hainsworth, 1969). The Piñon Jay, *Gymnorhinus cyanocephalus*, of western North America places over 85 per cent of its nests in the southern half of each tree canopy, apparently to conserve incubation energy during cool spring weather (Balda and Bateman, 1972).

In addition to such limitations as these, there are undoubtedly psychic factors that influence a bird's choice of a nest site. Phoebes, *Sayornis phoebe*, typically build on girders under bridges, and the Prothonotary Warbler, *Protonotaria citrea*, prefers a dead tree stub overhanging water. Old World Sand Grouse, *Pterocles* spp., commonly build their nests among stones approximating their body size—very likely a behavior that helps camouflage the bird.

Foliage density may be important to some species. In studying a nesting colony of American Crows, *Corvus brachyrhynchos*, in a California walnut orchard, Emlen (1942) found that nests built early in the season before the leaves were fully out were located a mean distance of 12.2 trees from the edge of the orchard, while later nests were built a mean distance of 8.8 trees from the edge. Peregrine Falcons, *Falco peregrinus*, are strongly attracted to ledges on cliffs for their eyries, and will use such reasonable substitutes as high office buildings or cathedral spires. For many years, Peregrines have raised young on a ledge at the 20th story of the Sun Life Building in Montreal, and a pair has nested for so many years in the spire of Salisbury Cathedral that they have become special wards of the Dean. The magnetism of a suitable nesting site can be seen in the construction by a Rufous Ovenbird, *Furnarius rufus*, of its globular mud nest on the axle of a windmill. The nest of course went round with each rotation of the axle, but the bird entered its nest to incubate only when the wheel stood still (Makatsch, 1950).

Frequently a species will adapt itself to nontypical nest sites. The Peregrine, for example, will nest in trees where cliffs are unavailable, and on the ground where both cliffs and trees are lacking. In the Baltic countries, about half of its nests are on cliffs, while the remainder are about equally divided between tree and ground sites. The Black Kite, *Milvus migrans*, which typically nests in large trees, is forced by the lack of them near Brienzer See, Switzerland, to become exclusively a cliff nester. Mallards typically nest on the ground, but occasionally one will nest as high as 10 to 15 meters above ground in the branches or hollow cavity of a tree. Abnormal nesting sites may be chosen because of individual variations in instinctive response, or they may be explained by a dearth of normal sites. In eastern Romania, the House Sparrow, *Passer domesticus*, has preempted all nesting places under eaves and in barns. As a consequence, the Eurasian Tree Sparrow, *P. montanus*, has taken to nesting between the stones of village wells, as many as five nests to a well, and from ¹/₂ to 5 meters below ground level (Frank, 1944). Similarly, Chimney Swifts, *Chaetura pelagica*, have nested in wells in Nova Scotia.

That a lack of suitable nest sites may be a critical limiting factor in bird populations is revealed by such studies as that of Creutz (1944), in which the placing of 100 nest boxes in a German orchard increased the number of resident, nesting Pied Flycatchers, *Ficedula hypoleuca*, from 6 to 45 pairs in five years. In like manner, the extensive rubble areas in German cities left by World War II provided excellent nest sites for Black Redstarts, *Phoenicurus ochruros*, whose numbers accordingly increased strikingly.

Birds may become habituated to nest sites that at first would appear quite unsuitable. Wild ducks, turkeys, swallows, and shore birds have all been known to nest within a meter or two of railways where trains thunder by every few minutes. Tree Swallows, *Iridoprocne bicolor*, Barn Swallows, *Hirundo rustica*, Robins, *Turdus migratorius*, and Phoebes, *Sayornis phoebe*, have all raised successful broods on moving ferry boats. A European Robin, *Erithacus rubecula*, once built its nest in a wagon that travelled about 320 km shortly after the young birds hatched. One of the parents accompanied the wagon the entire distance, feeding the young en route (Lack, 1946).

While birds normally nest near their chosen habitat—water birds near water, forest birds in the forest, and so on—this is not always the case. The Peruvian Gray Gull, *Larus modestus*, is an ornithological enigma in that it makes its living in the rich coastal waters of Peru and Chile but nests on flat, rock-strewn deserts from 20 to 100 km inland (Goodall *et al.*, 1945). Although the sterile desert is almost predator-free, the birds must contend with daytime temperatures that may reach 50°C in the sun. An egg or chick exposed to the sun will soon reach the lethal temperature of 45°C. As a consequence, the survival of both

FIGURE 14–5 A nesting colony of Cliff Swallows at Deerfield, Wisconsin. These once-abundant mud-nest builders have been eliminated from most of their former range because House Sparrows evict them from their completed flask-like nests. Photo by J. Emlen.

depends on the sheltering presence of at least one parent (Howell *et al.*, 1974). In the large colony of Lesser Flamingoes, *Phoeniconaias minor*, on the soda flats of Lake Magadi, Kenya, daytime temperatures of the black mud surrounding the nest platforms may reach 75°C (167°F)—a fact that makes the nests inaccessible to mammalian predators. The nest mounds are sufficiently high (23 to 30 cm) to protect eggs and young from lethal temperatures. The White Pelicans, *Pelecanus erythrorhynchos*, of Great Salt Lake nest on isolated predator-free islands where food is lacking. They must fly from 50 to 160 km to obtain food for themselves and their young.

A dearth of normal nesting sites may force birds to inhabit distinctly unsuitable sites. The Magnificent Frigate-bird, *Fregata magnificens*, normally nests atop dense cactus thickets, as at Bimini in the Bahamas, but on Little Swan Island it nests at the tops of trees 10 meters high where the birds lose their balance on the unstable twigs. Often the birds will fall, get caught by the neck in a tree fork and hang there until they die (Murphy, 1936).

HEIGHT OF NESTS ABOVE GROUND

Some species of birds normally nest on the ground: loons, grouse, turkeys, pheasants, most ducks, geese, cranes, rails, shorebirds, gulls, larks, and many more. Other species typically nest at different heights above ground. The Nightingale, *Luscinia megarhynchos*, usually nests about 1 meter above ground level; the American Robin, *Turdus migratorius*, 2 to 3 meters. A species preference for a relatively fixed nesting height was nicely illustrated by observations on the nest sites of Kingbirds, *Tyrannus tyrannus*, by Mayfield (1952). In northern Michigan this species often nests on the intersecting corner braces of steel electric-transmission towers. These braces, which occur at heights above ground of about 2, 7, 9, 11, 13 meters and so on to the top of the tower, appear to human eyes to be equally suitable for nesting sites. Yet, of 16 occupied Kingbird nests found on these towers, 15 were on the 2-meter braces. When nesting in wooded habitats, this species prefers sites 6 to 7 meters above ground. A study by Young (1955) of 202 American Robin nests near Madison, Wisconsin, showed a range in heights of 0.6 to 9 meters, with an average of 2.3 meters. When a suitable nesting site at a normal height is lacking, a given bird may select one at a very different height. Nests of Barn Swallows, *Hirundo rustica*, are commonly built on low rafters in barns, but one was observed by Hickey (1955) in a Minnesota observation tower 32.6 meters (107 feet) above ground. Other examples of nesting-height variations are given in Table 14–1.

The same bird may build its nest at different heights above ground according to the season. Nice (1937) found for the Song Sparrow, *Melospiza melodia*, that nine-tenths of the first nests built each season were constructed on the ground, and two-thirds of the second, but only one-third of the third. The majority of the third nests each season were built higher, generally in bushes. The seasonal rise in nest altitude paralleled and probably was stimulated by the rising growth of grasses and other vegetation. Conversely, early nests of montane White-crowned Sparrows, *Zonotrichia leu-*

TABLE 14–1.
HEIGHTS OF WOOD WARBLER NESTS AT
HOG ISLAND, MAINE*

SPECIES	TOTAL NESTS	LOWEST NEST (METERS)	HIGHEST NEST (METERS)
Northern Parula Warbler, *Parula americana*	71	1.5	16.5
Magnolia Warbler, *Dendroica magnolia*	33	0.3	4.2
Yellow-rumped Warbler, *Dendroica coronata*	44	1.8	13.2
Blackburnian Warbler, *Dendroica fusca*	7	13.2	23.2
American Redstart, *Setophaga ruticilla*	50	0.3	15.9

*After Cruikshank, 1956

cophrys, are placed on branches 21 to 40 cm above the ground, as these are the only available sites protruding above the snow. Nests are built on the ground at the bases of trees when the snow melts (Morton *et al.*, 1972). There are also geographical and ecological variations in nest height. In the eastern United States, the Brown Thrasher, *Toxostoma rufum*, commonly builds in low shrubs, but in its western range it builds on the ground. Predation may also affect nesting height. The Chatham Island Warbler, *Gerygone albofrontata*, generally builds its nests high up in trees. On predator-free islands, however, nests are built low to the ground in dense vegetation (Dennison *et al.*, 1984).

The significance of nest height is difficult to assess. It may be affected by such environmental variables as microclimate, storm damage, density of foliage, food availability, type of substrate, predation, and others. Of six different studies on the relationship of nest-height to reproductive success in the Red-winged Blackbird, *Agelaius phoeniceus*, four found that lower nests were the more productive of young; the two other studies reached the opposite conclusion. Obviously, other factors than nest-height were involved.

COLONIAL NESTING

About 13 per cent of all the birds in the world are colonial nesters, but about 93 per cent of all marine birds nest in colonies. This is probably because they use the safe but limited nesting sites provided by islands and cliffs (Lack, 1967). Among passerine families, only about 16 per cent are composed of some colonial nesters, including those that catch insects in the air in continuous flight (e.g., swifts and swallows). Five-sixths of the passerine families that eat mainly seeds include colonial species, but this is true of only one-sixth of fruit-eating groups (Lack, 1968).

Birds may gather in nesting colonies either because of some kind of gregarious appetite or because community living has been forced upon them by predation pressure, or by lack of suitable nesting sites. This latter reason may account for the dense nesting colonies of African Rosy Bee-eaters, *Merops malumbicus*, which may include as many as 25,000 birds in one colony. The colonies are restricted to sand bars in rivers where they are safe from predators (Fry, 1972). Whatever its origins, colonial nesting has its benefits and its handicaps. Close nesting increases a bird's competition for nest sites and nesting materials and increases opportunities for intraspecific fighting, for infanticide, and for the transmission of parasites and diseases. On the other hand, colonial living improves defense against predators, provides group stimulation and synchronization in breeding, promotes education of the young, and enhances communication about sources of danger or food. The advantage of cooperative defense was demonstrated by Kruuk (1964), who found that chickens' eggs placed inside a colony of Black-headed Gulls, *Larus ridibundus*, suffered lower predation from Herring Gulls, *L. argentatus*, and Carrion Crows, *Corvus corone*, than eggs placed immediately outside the colony. By placing experimental eggs either near colonies of Common Gulls, *Larus canus*, or near nests of solitary gulls, Gotmark and Andersson (1984) found that communal mobbing of predators by the gulls resulted in significantly lower predation of eggs near the colonies. They concluded that nest-predation strongly selects for colonial breeding. Similarly, grebes nesting with Brown-hooded Gulls, *Larus maculipennis*, in Argentina, responded to warning cries of the gulls by covering their eggs and departing from their nests. This behavior resulted in lower predation rates on eggs, chicks, and adults (Burger, 1984).

Colonies in some species are small groups with perhaps a few dozen nests; for example, those of herons, storks, doves, accipiters, swifts, and a few passerines. Other species may nest in colonies totaling hundreds or thousands of birds: penguins, petrels, gannets, pelicans, flamingos, gulls, terns, and auks. On Macquarie Island south of New Zealand, there formerly existed single colonies of King Penguins, *Aptenodytes patagonica*, each including millions of birds, and covering 12 to 16 hectares. Hunters killed hundreds of thousands of these birds each year for their oil, until the total island population was estimated to be 7000 birds. Since 1933, the birds have been protected and their numbers somewhat restored.

Ordinarily, colonies are made up of a single species of bird, but sometimes two or more species may compose a nesting aggregation. This is particularly true of related species that require similar nesting sites: gulls and terns; herons, egrets, spoonbills, and ibises; guillemots, auks, and puffins; Jackdaws and Choughs. Occasionally in a large colony of one species there may be one or a few pairs of birds of another species. A small scattering of ducks, grebes, or shore birds may nest in a large breeding colony of gulls and terns. Heron colonies often attract falcons, kites, or hawks.

Among 13 east European heron colonies, 11 included the nest of a Peregrine Falcon, *Falco peregrinus*, and in one other instance a Falcon nested only 150 meters from the colony. In five cases the herons and Falcons nested in the same tree! (Makatsch, 1950.) In such associations, the Falcons seem to be attracted not to the herons but to their old tree top nests, which the Falcons adopt as their eyries.

Some aggregations of birds may be based on a predator-prey relationship. The Great Skua, *Catharacta skua*, is not primarily a colonial nester, but the easy living made possible by preying on a penguin or gull colony attracts sufficient numbers to make a small adjoining colony of skuas.

In a few colonial species, especially among the Ploceidae, individual nests are built so closely together that they combine to form a large apartment house or arboreal warren. Each nest, however, is independent of the others and has its own entrance. In the case of the Sociable Weaver, *Philetarius socius*, the pairs in a colony work together to build a large thatched dome of grasses in an isolated tree in the African veldt.

Then, on the under side of this roof, the pairs build their individual flask-shaped nests with downward-directed, pike-like entrances. Several hundred individual nests may be accommodated under one thatched roof. At times these community nests weigh enough to cause large branches to break (Fig 14–6).

In the Village Weaverbird, *Ploceus cucullatus*, of Senegal, Collias and Collias (1969) found that large colonies (over 25 nests) generally contained more females than males, whereas in smaller colonies the reverse proportion existed. In the larger colonies each male built, on the average, more than half again as many nests as did each male in small colonies. The authors attributed this difference to social facilitation (see p. 204).

A somewhat different form of community nest is built by babbling thrushes (Timaliidae) and anis (Cuculidae, subfamily Crotophaginae). Among these birds as many as ten pairs may join forces to build a single nest in which several females will then lay their eggs. The eggs are incubated and the young are fed communally. Among the African Malimbe, *Malim-*

FIGURE 14–6 The nest of the Social Weaver, Transvaal, Africa. The underside of a nest with its individual entrances is shown at the right. Photos by H. Friedmann.

bus coronatus (Ploceidae), the elaborate nest is cooperatively built by three to six adults, but only one male and one female incubate, brood, and feed the young (Brosset, 1978).

Among the Magpie-larks (Grallinidae) of Australia, the White-winged Choughs, *Corcorax melanorhamphos*, and the Apostle Birds, *Struthidea cinerea* (so named from their habit of traveling in groups of about 12), are also communal breeders. All members of a group cooperate in nest-building, incubation, and care of the young, but only females high in social rank lay eggs (Serventy, 1973). The White-helmeted Shrikes, *Prionops plumata*, of Africa represent an incipient form of communal breeding. The shrikes roost and travel together in groups of 6 to 12 birds, but build their single-pair nests in groups of five or six, close together. Occasionally two females will lay eggs in one nest. Nest-building, incubation, and the feeding of young seem to be randomly performed by all members of the flock (Gilliard, 1958).

PROTECTIVE NESTING ASSOCIATIONS

There are times when even the cleverest placement of a bird's nest will be unavailing against the designs of a persistent predator. The bird may locate its nest in a quagmire, in a thorny shrub, at the tip of a willowy branch, in a hollow limb, or on the face of a precipitous cliff, but some predatory fish, snake, weasel, or hawk may find it and destroy the eggs or young. Many more eggs and young meet this fate than grow to successful maturity. Because of this high susceptibility to predation, some species, aided by years of natural selection, have adopted extraordinary means of protecting this most vulnerable stage in the bird's life cycle. One of the most striking of these means is the choice, by defenseless birds, of a nesting site in the vicinity of a large aggressive species. In the far north the most dangerous predator of nesting waterfowl is the Arctic Fox, *Alopex lagopus*. In response to predation pressure from the fox, Snow Geese, *Chen caerulescens*, Black Brant, *Branta nigricans*, and Common Eiders, *Somateria mollissima*, all nest in close proximity to the nests of the Snowy Owl, *Nyctea scandiaca*, the chief enemy of foxes. The Red-breasted Goose, *Branta ruficollis*, builds its nest near that of the Peregrine Falcon, *Falco peregrinus*.

Very often small passerine birds, such as sparrows, grackles, weaverbirds, and even birds as large as night herons, will build their nests in the edges and under-

sides of nests occupied by storks, owls, ospreys, hawks, or eagles. In Macedonia, for example, the nests of Imperial Eagles, *Aquila heliaca*, often contain in their fringes the nests of Starlings, *Sturnus vulgaris*, or House Sparrows, *Passer domesticus* (Makatsch, 1950). In such situations the larger, more aggressive bird acts as the "lord protector" of the smaller species by keeping away possible predators, but like a human feudal overlord, it may occasionally exact tribute from the lowly serfs. In the main, however, the larger predatory bird seems to tolerate its smaller tenants, perhaps because it has learned that the quickly darting smaller bird is uncatchable. European Wood Pigeons, *Columba palumbus*, often nest near Hobbies, *Falco subbuteo*, and capitalize on the Hobby's aggression towards predators, which enhances the pigeons' breeding success (Bijlsma, 1984). In British Guiana, small tyrant flycatchers and tanagers characteristically nest around the nest of a large aggressive flycatcher, *Pitangus*, whose noisy intolerance of other predatory birds protects both its own nest and those of its satellites.

A most curious association exists between a shore bird, the Water Thick-knee, *Burhinus vermiculatus*, and crocodiles. Both animals lay their eggs on sandy shores when African rivers are at low ebb, the bird most often in close proximity to the brooding reptile. This arrangement probably protects the bird's eggs from many egg-eating predators (Thomson, 1964).

In a similar way, many birds have become hangers-on of human communities, profiting from the relative absence of hostile predators near man-made dwellings. In the Cameroons, the Village Weaver, *Ploceus cucullatus*, builds its nests along the village streets with the heaviest traffic (Stresemann, 1934). In Europe, the Serin, *Serinum canarius*, seems to nest along congested streets in preference to those with less traffic. Among the species more commonly associated with human communities are storks, pigeons, Jackdaws, *Corvus monedula*, Rooks, *C. frugilegus*, swallows, swifts, weaverbirds, sparrows, robins (both American and European), and starlings. There are, of course, other advantages to living in human communities, such as protected nesting sites and unusual food sources.

Even more curious is the pronounced tendency of many small, inoffensive tropical birds to build their nests in the vicinity of nests of aggressive insects. In Australia, the Black-throated Warbler, *Gerygone palpebrosa*, is called the "hornet-nest bird" because it so commonly nests beside hornets' nests (Chisholm,

1952). The South American Cacique, *Cassicus cela*, commonly builds its flask-shaped hangnest near the nests of arboreal wasps. As many as eight or ten Cacique nests may be clustered around a wasp nest, often so close that the birds' nests rub each other in the wind (Beebe, *in* Stresemann, 1927–1934). Of 61 nests of the Dull-colored Seed-eater, *Sporophila obscura*, found by Contino (1968) in Argentina, 58 were closely associated with nests of carnivorous wasps, 55 of which were wasps of the species *Polistes canadensis*. Frequently the species that associate with ants, bees, and wasps are community nesters. Evidence indicates that in most of these associations, the birds seek the insects' nest and not vice versa (Hindwood, 1955). However, near Belem, Brazil, Oniki (1970) discovered that of 14 nests of the Reddish Hermit Hummingbird, *Phaethornis ruber*, nine contained brood cells of the wasp *Pison* on the underside. Nests of other species in the same locality showed no such association. As a rule, the insects do not bother the nesting birds but do attack humans and other animals that come too close. For example, in eastern Africa the Gray-headed Social Weaver, *Pseudonigrita arnaudi*, nests in small colonies on Acacia trees in whose hollow stems live virulent ants. The ants "take no notice of boughs shaken by birds, but come out in a swarm if they are touched by any other external agent" (Myers, *in* Mackworth-Praed and Grant, 1960).

Most remarkable of all are those tropical birds that make their nests *inside* the nests of stinging or biting insects. The woodpecker *Micropternus brachyurus* regularly nests in the center of the spherical papier-mâché tree-nest of ants of the genus *Cremastogaster* (Stresemann, 1934). Some 49 species of birds in the New and Old World tropics are known to breed in terrestrial and arboreal termite mounds, including kingfishers, parrots, trogons, puff-birds, jacamars, and a cotinga (Hindewood, 1959). The Orange-fronted Parakeet, *Aratinga canicularis*, frequently nests in termite mounds of *Eutermes nigriceps*; the ranges of the two species overlap considerably (Hardy, 1963), suggesting that the parrot is dependent on the termite for success. In many of these odd partnerships the insects neither bother the birds nor seem to benefit from the birds' presence. The birds, on the other hand, are protected, at least by the ants, bees, and wasps, from predators, and in some cases eat the insects and their young.

Tropical birds are so abundantly surrounded by tree-dwelling predators—insects, snakes, lizards, squirrels, weasels, monkeys, toucans, falcons, and other birds—that defenseless species are hard put to lay their eggs and raise their young in security. This no doubt accounts for the fact that whereas in the temperate and subpolar zones most nests are open cups, in the tropics over one-half of all bird nests are covered over on top, with the entrance at the side or bottom where access by predators is difficult. Tree-cavity nests in the tropics offer safety as long as the birds occupy them. Once they are abandoned, bees and wasps soon take them over. In Zimbabwe one ornithologist erected 23 nest boxes to attract hole-nesting birds. in the first year over half of the boxes were occupied by bees and wasps, and in the second year all of them.

FORMS OF NESTS

Birds' nests appear in an almost infinite variety of forms, ranging from a mere depression in the ground to the most intricately woven, multichambered hangnest. Ordinarily the character of a nest is determined by heredity, the raw materials used, the site chosen, the experience and adaptive intelligence of the builder, and possibly by imprinting or imitation. In a large number of species the nest-building instinct has been lost completely, although it may persist among some of their near relatives.

Among birds that build no nest there are those, like many sea birds, shore birds, goatsuckers, and vultures, that lay and incubate their eggs on the bare ground or some similarly unprepared place. The White Tern, *Gygis alba*, lays and incubates its single egg on the bare branch of a tree. The subspecies *candida*, or Cocos Keeling Island, lays its egg in the angle between two leaflets of a horizontal frond of the coconut palm. As the frond matures it droops and withers until it eventually falls. Among settlers discovering a Tern incubating on a withering leaf "it is an object of keen betting . . . whether the young bird will be hatched out before the leaf falls" (Pycraft, 1910). A remarkable substitute for a nest is contrived by the King and Emperor Penguins, *Aptenodytes patagonica*, and *A. forsteri*, which hold their solitary eggs off the antarctic ice on top of their webbed feet and envelop them in a warm fold of belly skin. The penguins are even able to waddle about while thus incubating their eggs.

Other species that build no nests are those that use the abandoned nests or burrows of other birds or mammals, or those that forcibly evict the rightful owner from its nest, or nest parasites, such as many cowbirds and cuckoos, which lay their eggs in the nests of other species, abandoning them to the care of the foster parents. Often hawks, kites, falcons, and owls use abandoned nests of herons, crows, magpies, and rooks. Green Sandpipers, *Tringa ochropus*, lay their eggs in old bird or squirrel nests. Many small passerine species, such as the Tree Swallow, *Iridoprocne bicolor*, the Bluebird, *Sialia sialis*, and Crested Flycatcher, *Myiarchus crinitus*, nest in old woodpecker holes, supplying them, however, with soft lining materials. Species that forcibly expropriate other birds' nests include the House Sparrow, *Passer domesticus*, Chestnut Sparrow, *P. eminibey*, Starling, *Sturnus vulgaris*, and the Piratic Flycatcher, *Legatus leucophaius*. A study of Australian birds by Roberts (1955) revealed 80 species that use the nests of other birds. Shearwaters often use old rabbit burrows, the Burrowing Owls, *Athene cunicularia*, take over the holes of ground squirrels. After campaigns of ground squirrel extermination, the Burrowing Owl population drops accordingly. In southeast Asia, 16 species of birds are said to take over mammal burrows for nest sites.

For many ground nesters, such as ostriches, sand grouse, some petrels, falcons and vultures, and many shore birds, a slight depression, or scrape, in the soil serves as a nest. Some gannets, terns, gulls, or puffins may add a modest amount of vegetation or feathers to the scrape; some penguins, plovers, and terns may collect stones as nesting material. Many ground-nesting birds such as ducks, pheasants, grouse, and sparrows build well-formed, cup-shaped nests of grasses and other vegetation. Some water birds, such as coots, rails, jaçanas, and grebes, build floating or shallow-water nests of reeds, rushes, and other aquatic vegetation.

One of the more secure forms of nest is that made in a cavity. Birds may dig their own holes or use ready-made ones. Excavated cavities may be made in the ground (by most of the smaller Procellariiformes), in vertical banks (by kingfishers, motmots, bee-eaters, some swallows), or in living or dead trees (by woodpeckers, some titmice, some trogons). In North American forests there are at least 85 species of cavity-nesters (Scott *et al.*, 1977). The Rhinoceros

FIGURE 14–7 The nest of the European Stone Curlew is merely a shallow depression on the ground where small stones or rabbit droppings (shown here) may be present. Both the egg and the newly hatched young show obliterative coloration. Photo by E. Hosking.

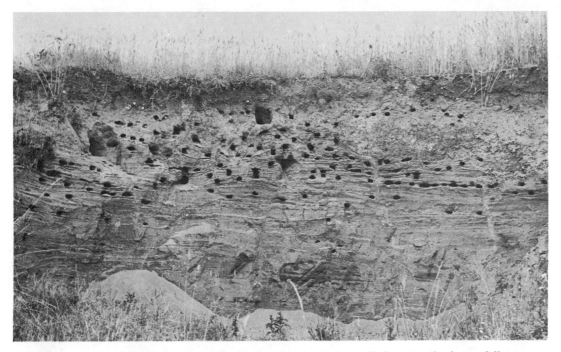

FIGURE 14–8 A Bank Swallow colony. The burrows are dug horizontally from two-thirds to a full meter in depth, ending in a small nest chamber. Photo by E. Hosking.

Auklet, *Cerorhinca monocerata*, of British Columbia, has been observed nesting underground in holes 8 meters long, bee-eaters in holes 3 meters long, and kingfishers in holes 2 meters long—all dug by the birds themselves. The Burrowing Owl, *Athene cunicularia*, uses only its feet in digging its burrow, but the kingfishers, bee-eaters, and swallows dig with their beaks and eject the loose earth with their feet somewhat like a digging dog. A pair of small Asiatic Kingfishers, *Ceyx tridactylus*, dug a hole one-quarter meter deep in 40 minutes. Typically, in these earthen burrows, the eggs are laid in a chamber at the inner end, either on the bare ground or in the midst of some nesting materials. Birds that live in environments lacking in trees are likely to be burrowing species.

Most woodpeckers chisel out their own nest holes, generally in sound wood, whereas the few species of titmice that dig in trees must excavate in pulpy, partially decayed wood because of their weak bills and light skulls. Apparently the illumination within a cavity is significant to some birds. Nest boxes designed for Wood Ducks, *Aix sponsa*, were set up in pairs, one box painted black inside, the other left a natural, light wood color. Of Starlings, *Sturnus vulgaris*, which appropriated some of the boxes, 34 pairs laid eggs in the boxes with black interiors, and three nested in the unpainted boxes (Lumsden, 1976). In contrary fashion, when Eastern Bluebirds, *Sialia sialis*, were presented with paired nest boxes, one painted white inside, the other black, they constructed nests in 33 white boxes but in only three black ones (Pitts, 1977). Some species that use ready-made tree cavities may partially close the nest entrance with pitch or clay as nuthatches do, or with mud and droppings as hornbills do. The opening in the nest of a hornbill is so small that the female must remain a prisoner within the cavity for one to four months, depending on the species, until she and the young break out. In some species, the female leaves before the young, and recloses the entrance. This imprisonment keeps out monkeys and snakes, but it means that the male often must feed the female and young singlehanded. Should the entrance rim be accidentally broken, the unfledged young will instinctively repair it by trowelling into place with their bills fresh "mortar" composed of wet earth, viscous fruits, and dung.

Open nests built above the level of the ground are ordinarily attached to branches or twigs of trees and bushes, and are built of locally available materials such as twigs, leaves, grasses, mosses, mud, plant down, animal hair, feathers, and spider webs. Often the nest material is intercrossed with the surrounding branches to provide firm attachment to the tree.

Birds of most orders build simple, primitive nests that consist of little more than a platform of sticks and twigs. The weight and activities of a bird near the center of its nest cause the nesting material to pack down and be lower than the rim—a fortunate thing for the safety of the eggs. Birds building relatively primitive nests are the cormorants, darters, frigate-birds, herons, bitterns, storks, some accipiters, many pigeons and doves, some cuckoos, and a very few passerine birds—for example, the Rose-breasted Grosbeak, *Pheucticus ludovicianus*. The nests of some of these birds, for example, a frigate bird, or a Mourning Dove, *Zenaida macroura*, are so skimpy that the eggs may be seen from below through the bottom of the nest. Eggs sometimes roll off the nest of a frigate bird when it takes flight. The loose sticks of the frigate bird's nest may become more securely cemented together with droppings after the young hatch. Occasionally some of the birds with primitive nests build up the rims of their nests with twigs or branches, thus improving the security of the nest for eggs or young.

Somewhat more elaborate is the cup-shaped nest characteristic of most passerine birds in the temperate zone. Typically a cup nest is composed of a base or platform of coarse materials, a cup of grasses, moss, or some similar fine materials, and a soft, warm lining of fine grasses, hair, plant down, or feathers. Mud, spider webs, or saliva may be used to cement some of these materials together.

The nest of the American Robin, *Turdus migratorius*, provides an example of a moderately well-finished cup nest, typical of thrushes. The nest is usually straddled on the limb of a tree or braced in a fork. Its base and outside are composed of coarse grasses, leaves, rootlets, and at times shreds of paper or rags. The inner wall is plastered with mud and the cup lined with fine grasses. Craft in nest construction progresses by almost imperceptible degrees from such a nest to the exquisitely fashioned cup nests of such species as the American Goldfinch, *Carduelis tristis*, which builds

FIGURE 14–9 The nest and eggs of a Cooper's Hawk. This species instinctively lines its rather primitive stick nest with flakes of bark. Photo by G. R. Austing.

its nest externally of fine grasses and moss, and lines it inside with thistledown. So firmly are Goldfinch nests put together that at times they hold water and drown the young.

Nests represent an evolutionary compromise between the need to provide a thermally uniform microclimate and protection for the eggs and/or young. In a survey of the insulative properties of various nest-types for 66 nests representing 11 North American songbird species Skowron and Kern (1980) found that well-insulated solid or dense nests were the poorest heat conductors. One should thus expect intraspecific geographical variation in nest structure depending on local climates. Indeed, thermal conductance diminishes with altitude in nests of the Hawaiian Honeycreeper, *Hemignathus virens*. Moreover, nests at lower altitudes are always placed deep in the canopy where they are shielded from rain, whereas those at higher elevations are nearer the canopy to take advantage of the warming solar radiation (Kern and Van Riper, 1984).

Feathers increase nest insulation, but may attract predators. Hole-nesting birds seem to use feather liners more often than open nesters (Moller, 1984). Tropical birds often build thick covered nests to discourage predator entry, but nests built in the canopy are often thin-walled. From below, the latter resemble holes in the forest canopy (Oniki, 1985).

Many species of birds build covered nests. The Magpie, *Pica pica*, both in Eurasia and North America, erects a large dome of thorny sticks over its bulky nest, leaving an opening on one side. A survey of Long-tailed Tits, *Aegithalos caudatus*, by Lack and Lack (1958) disclosed that thorns have a protective value to birds. But of nests built on smooth oak or ash tree branches, every one was destroyed before the young were fledged (Fig. 14–10). Not only does the African White-headed Weaver, *Dinemellia dinemilli*, nest in thorn trees, but it places thorny twigs for a meter or so along boughs leading to its nest (Collias, 1964). One of the ant-thrushes, *Thamnomanes caesius*, fastens over

the top of its nest a large leaf that serves as a sun and rain shield. Wrens and dippers commonly build covered nests with side entrances. A South American ovenbird builds such a large nest of sticks (30 × 30 × 60 cm) that it is called the Firewood Gatherer, *Anumbius annumbi*. The small bird enters its fortress through a crooked passage in the top.

Probably the most elaborate of all nests are the covered, pendent nests of various small passerine birds. These hangnests are especially abundant in the tropics, where they help to outwit tree-dwelling predators. It is here that the building of pensile nests has reached its highest perfection, particularly among the broadbills (Eurylaimidae), and the caciques, oropendolas, troupials, and orioles (members of the blackbird family or Icteridae). Hangnests of the polygamous Montezuma Oropendola, *Gymnostinops montezuma*, of tropical America are normally built in colonies of up to 100 nests, all hanging from the tops of tall trees (Fig. 14–11). Each nest is a tubular sac, often as much as 2 meters in length, with the opening near the top and the nest chamber near the bottom. To leave or enter

FIGURE 14–10 One of the more elaborate covered nests is that of the Long-tailed Tit of Europe. It is made of moss, cobwebs, and hair, and covered externally with lichens. Inside, it is warmly lined with as many as 2000 feathers. Photo by E. Hosking.

FIGURE 14–11 Hangnests of the Central American Oropendola provide security against enemies both by their colonial grouping and by their pendent construction. Photo by A. F. Skutch.

its nest the owner must crawl up or down this long sleeve. Even such apparently secure nests are occasionally robbed by snakes (Skutch, 1954). Some nests have hidden nest chambers and false entrances, apparently evolved as devices to frustrate predators. The Baya Weaverbird, *Ploceus philippinus*, builds a pensile gourd-shaped nest with two chambers, the lower one for eggs, the upper, an entrance tunnel. The tubular side entrance of the felted, spherical nest of the titmouse *Anthoscopus caroli* may be closed each time the bird leaves the nest. Hangnests are frequently suspended at the tips of slender branches, and often they hang over water.

MULTIPLE NESTS

In addition to their breeding nests, some species build extra nests that are variously called incomplete nests,

dummy nests, cock nests, play nests, or sleeping nests. The variety of names given these nests shows that their significance is not completely clear. A common example of this type of nest-building is provided by the House Wren, *Troglodytes aëdon*. Each season before the female arrives, the male builds several globular masses of twigs in sheltered places, some of them nearly complete, some unfinished. When the female arrives, the male sings and displays before a nest, and creeps in and out. If his advertising is persuasive enough, the female will adopt one of the nests, line it with soft materials, and lay her eggs in it.

A field study of the Long-billed Marsh Wren, *Telmatodytes palustris*, by Verner and Engelsen (1970) strongly supports the concept that nest-building has courtship significance in some species. Of 80 males whose territories were mapped, the mean number of nests completed by each unmated male was 17.4; by each monogamous male, 22.1; and by each bigamous male, 24.9. Further evidence of the courtship significance of multiple nests comes from Old World weaverbirds (Ploceidae). In many species of Ploceidae the male constructs the shell of the nest out of green grass stems. He "continues building new nests, at times as many as two dozen in a season, destroying each as it turns brown, until the female signals her acceptance of the nest by starting to line it with soft grass heads" (Collias and Collias, 1964).

The European male Moorhen, *Gallinula chloropus*, sometimes assisted by the female, builds nest platforms or bases on which the young may later be brooded. Extra nests built by the African Cape Weaverbird, *Hyphantornis capensis*, are used as dormitories both by the male during the incubation period and by the female after the young are large enough to crowd her from the brood nest. Probably as a consequence of the vigor of his building instinct, the male may demolish one of the nests he had built only 3 or 4 weeks earlier and use the materials to build a new one. A variation of this activity is seen in the energetic building of the Yellow-tailed Thornbill, *Acanthiza chrysorrhoa*, of Australia. In this species, both sexes build an oblong nest with several chambers inside, one of which is used for rearing the young, and the others possibly as sleeping chambers for the male, or possibly for frustrating predators or nest parasites such as cuckoos. The male may continue adding to the nest long after it is functionally complete. Even the North American Golden Eagle, *Aquila chrysaëtos*, builds supernumerary nests. Of 17 to 19 pairs observed by McGahan (1968), more than half attended more than one nest. Nests averaged 1.8 per pair and ranged from a meter or so to over 6 km apart. In Japan the polygynous Fantail Warbler, *Cisticola juncindis*, builds multiple nests for the females. In one breeding season one male built 20 nests, 8 of which were used by different females. It required an average of 6.6 days to build one nest (Motai, 1973).

For many species an "ideal" nesting site may prove to be irresistibly magnetic. In nature such sites are ordinarily rare, and a given bird is lucky to find even one. But man's repetitious artifacts at times create an embarrassing wealth of desirable nesting sites. The American Robin, *Turdus migratorius*, and the European Blackbird, *T. merula*, have been reported several times to build numerous nests in between the rungs of horizontally hung ladders. In Ohio, a Robin began 26 separate nests on a wooden girder in the spaces between the roof rafters it supported. Work-men erecting this building supplied the bird with pans of wet clay to line its nests, and placed bets on the location of the final, functional nest. After a week of confused effort spread over these delectable multiple niches, the bird settled on one nest, laid eggs in it, and hatched them. A pair of European Robins, *Erithacus rubecula*, started to construct 23 different nests in a stack of pipes laid on their sides, while another started 12 nests in a series of pigeon-holes(!) in a workshop (Lack, 1946). It is not always a matter of mistaken orientation among successively repeated nest sites that prompts multiple nest-building. In his comprehensive study of the Prairie Warbler, *Dendroica discolor*, Nolan (1978) discovered that of 188 females, 70 (37 per cent) built 103 multiple or fragmentary nests, 92 per cent of which were indistinguishable externally from true, finished nests; they all lacked linings and were not used to incubate eggs. Of the 70 females, 44 built one extra nest each; 22 built two each; two built three each; and two built four each. This kind of redundant nest building may represent an "overflow" activity based on the commonly normal construction of a clean, new nest for a second brood as an adaptive method of avoiding nest vermin that often cause nestling mortality.

SIZES OF NESTS

As a rule, birds' nests vary in size with the size of the builder, but this is by no means always true. Hummingbirds probably build the smallest and lightest nests, while eagles and storks build some of the largest and heaviest. Nests of White Storks, *Ciconia ciconia*, have measured over 2 meters deep and 1.7 meters in diameter, and have weighed over a ton. Probably the largest recorded nest of a Bald Eagle, *Haliaëtus leucocephalus*, was one near St. Petersburg, Florida, which was 6.1 meters deep and 2.9 meters wide. These enormous nests are largely the result of years of accumulation of nesting materials at the same site. The average size of Alaskan Bald Eagle nests, as measured by Hensel and Troyer (1964), was 1.6 meters wide by 1 meter deep. Fourteen nests of the Bald Eagle in Florida, that were occupied between 1930 and 1938, were still occupied in 1963 (Howell and Heinzman, 1965). Sea Eagles, *H. albicilla*, have been known to breed annually in the same nest for as long as 80 years. One such nest that crashed down in a storm weighed 2 tons (Makatsch, 1950).

Perhaps the largest nest in the world is that built by

the hen-sized megapodes or "incubator birds" of Australia. The earthen mounds that these birds scratch together with their feet, and in which their eggs are incubated by the heat of decomposing vegetation, may be added to, year after year, and have reached dimensions of 4.2 meters high and 10.7 meters wide (Wetmore, 1931). The nest of the African Hammerhead, *Scopus umbretta*, is remarkable not only for its large size but for its architecture. Although this stork-like bird stands about 30 cm tall, it builds a large ball-shaped, clay-lined nest of sticks up to 2 meters in diameter and firm enough to hold a man on its domed roof. As described by Lydekker (1901), the nest has an entrance on one side so small that the bird must creep in. Inside, the nest contains three chambers: an upper sleeping and incubation room, a middle chamber for the young when they are too large for the upper chamber, and a vestibule or look-out chamber. The nest is usually placed in the fork of a tree near the ground.

Even small passerines may construct massive nests. The Rock Nuthatch, *Sitta neumayer*, of the Balkan peninsula, is a sparrow-sized bird that weighs about 40 g. Both sexes build the nest in a rock cranny, and construct a funnel-shaped entrance over the opening, plastering it with mud mixed with leaves, twigs, and other debris. This earthen nest, including the mortar spread on the rock and in its crevices, may weigh up to 38 kg, or about 950 times the weight of one bird (Mayaud, 1950). American nuthatches have lost this Old World trait of building ramparts around their castles except for a vestige retained by the Red-breasted Nuthatch, *S. canadensis*, which smears pitch about the entrance to its tree-cavity home.

A bird may adapt its nest size to seasonal temperature changes. The Prairie Warbler, *Dendroica discolor*, builds its first nests of the season with an average weight of 4.6 g (dry weight); replacement nests weigh 3.8 g. Since the early season is cooler, warmer nests are desirable because they shorten the time the nestlings are vulnerable to predators (Nolan, 1978).

MATERIALS USED IN NEST-BUILDING

Considering the multitude of materials available for building nests, it is a puzzle how a given species knows which materials to select. What causes the Cliff Swallow, *Petrochelidon pyrrhonota*, to select mud for its nest, and the Wood Pewee, *Contopus virens*, to decorate the outside of its nest with lichens? Undoubtedly the majority of such choices are ordered by instinct, but other factors play important roles. Without its abundant down feathers the duck could not line its nest with warm down, and without copiously flowing salivary glands most swifts would be unable to cement other materials together to make their nests. Obviously, the size of a bird limits the size of the building materials it uses. A hummingbird could not use crow-sized sticks for its nest. Further, the habits and habitat of a species restrict its choice of building materials. Water birds are likely to use aquatic vegetation for their nests; meadow birds, grasses; and woodland birds, forest materials.

Modes of transport also influence the choice of nest materials. Most birds carry nesting materials in their beaks, but accipiters carry their materials in their talons, and various Old World parakeets from separate lineages (*Agapornis, Loriculus, Neophema*) have the remarkable habit of flying to their nests with grass stems, leaves, and woodchips tucked under their rump feathers, which in some species have specially adapted hook-like appendages for this function (Smith, 1979). Although many species carry twigs, grass, or mud from considerable distances, most ducks, geese, and swans have lost the instinct to carry nest materials, and pull in what leaves, grasses, moss, and other plants they can reach with their bills while sitting on the nest site. A short-legged swift, which spends its waking hours in flight, snatches its twigs on the wing.

Without doubt the availability and abundance of nest materials have a bearing on their choice by a bird. Nests of the Gentoo Penguin, *Pygoscelis papua*, are built of woody twigs and stalks in the Falkland Islands; of grass, moss, and bits of seaweed in South Georgia; and, still farther south in the Antarctic Archipelago, where terrestrial vegetation disappears, of molted feathers and the bleached bones of their dead relatives. The Chipping Sparrow, *Spizella passerina*, used to be called the Hairbird because it lined its nest mainly with horsehairs. But with the gradual disappearance of the horse in rural America, it has substituted fine grasses for its nest lining. One Carolina Wren, *Thryothorus ludovicianus*, built its nest of hairpins; and a pair of crows in Bombay once built their nest with 25 pounds worth of gold spectacle frames which they stole from an open shop window (Herrick, 1935).

Materials used in nest construction may be animal, vegetable, or mineral, or various combinations of these. Plant materials are more commonly used than any others. One of the most common types of nests is that made of a cup of interwoven coarser grasses lined with fine grasses, moss, or plant down, and resting on a platform of twigs. Fibers of bark are used in the nests of some vireos, and most woodpeckers use as a nest lining the wood chips left behind from their labors. The tailorbirds of southeast Asia stitch the edges of broad green leaves together to make a funnel-shaped receptacle for their nests. Dippers and hummingbirds often use moss as the main ingredient of their nests. Yellow Warblers, *Dendroica petechia*, and American Goldfinches, *Carduelis tristis*, use large amounts of cottonwood- and thistle-down for their nests, while the European Goldfinch, *Carduelis carduelis*, employs slender fir twigs, fine roots, and wool. The Song Thrush, *Turdus ericetorum*, may line its grassy cup with a plaster made of saliva, rotten wood, or dung. The cave-dwelling Oilbird, *Steatornis caripensis*, constructs its heavy, crater-shaped nest with a plaster made of regurgitated seeds and its own excrement. Many sea birds construct their nests of marine algae, which stiffen upon drying. The Northern Parula, *Parula americana*, typically builds in a hanging clump of the lichen *Usnea*, or the "Spanish moss" *Tillandsia*.

Many birds decorate the outsides of their nests with lichens: several hummingbirds, gnatcatchers, titmice, the Eastern Wood Pewee, *Contopus virens*, Chaffinch, *Fringilla coelebs*, and others. Such a covering, which often matches the surroundings, helps make the nest less conspicuous. Perhaps a latent esthetic sense resides in some species, for they seek brightly colored or shiny objects to decorate their nests. The Moorhen, *Gallinula chloropus*, occasionally places flowers or pieces of paper on the outside of its nest, and various flycatchers, sparrows, finches, and starlings have been observed placing flowers or flower petals in their nests. The fondness of crows and eagles for brightly colored objects in their nests is well known. Gannets, *Sula bassana*, have adorned their nests with such objects as golf balls, blue castor-oil bottles, strings of onions, and even a clockwork toy steamer (Armstrong, 1942). Indian House Crows, *Corvus splendens*, in Calcutta and other areas have been found increasingly to use metallic wire, metal strips, grills, and various other forms of scrap metal instead of twigs and brambles to build nests. Metal nests, which tend to be more stable, are used year after year and are enlarged in successive years (Altevogt and Davis, 1980).

Although the significance of most of these "decorations" remains unknown, the colorful wing-covers of beetles that embellish the nests of some nuthatches may represent a form of nest protection embracing chemical warfare. American White-breasted Nuthatches, *Sitta carolinensis*, have been observed "sweeping" the bark inside and outside their nest holes while holding in their beaks metallic-colored beetles that exude copious oily fluids which may serve to repel squirrels and other competitors for their nest cavities (Kilham, 1971). In Armenia the Rock Nuthatch, *Sitta neumayer*, smears the juices of mashed caterpillars on the outside of its mud-and-saliva nest and lines its interior with rodent fur and bones, and the excrement of carnivorous mammals (Adamyan, 1965). Recalling the use many mammals make of body secretions (pheromones) in marking their territories, it seems reasonable to assume that such chemical treatment of their nests may protect nuthatches from potential predators.

House Sparrows, *Passer domesticus*, breeding near Calcutta, India, incorporate leaves of the Margosa tree, *Azadirachta indica*, in their nests, from the start of nest-building until nestlings fledge. The local people use these same leaves to protect their clothes from insect damage, and recently chemicals have been isolated from the leaves that prevent insects from laying eggs. It is possible that the sparrows were using these leaves to kill parasites (Sengupta, 1981).

Some birds are veritable junk collectors. One Redtail, *Cercomela familiaris*, used in the base of its nest 2 kg of odds and ends, including 361 stones, 15 nails, 146 pieces of bark, 14 bamboo splinters, 3 pieces of tin, 35 old pieces of adhesive tape, 103 pieces of hard dirt, 30 pieces of horse manure, several pieces of rags and bones, 1 piece of glass, and 4 pieces of old inner tubes (Makatsch, 1950). During World War II many species of birds collected for their nests the fine strips of aluminum foil dropped by bombers to confuse ground radar stations.

Animal products used in nest-building are mainly the wool, hair, and feathers used by many species as warm nest linings. However, cast-off snake skins are used by at least 31 species of birds, notably the Crested Flycatcher, *Myiarchus crinitus*, which commonly drapes such a skin half in and half out of its nest cavity. Bones may be used as nesting materials by owls, kingfishers, penguins, petrels, terns, and others. Spider webbing is often used by hummingbirds, titmice, white-eyes, flower-peckers, and spider-hunters in nest-building, either to bind materials together or to suspend the nest from above. The Gabar Goshawk, *Micronisus gabar*, of Africa festoons its nest with spiderwebs, including the live spiders, who then augment the silken coat. Some hummingbird nests are made of a firm felt compounded of plant down and spider silk. Sometimes these felted nests are suspended from leaves or branches or even from the roofs of caves by thin single cords made of spiderwebbing. One such nest built by Hermit Hummingbirds of the genus *Phaethornis* seems to defy gravity in that the nest cup rides in midair in the normal horizontal position, yet its rim is fastened to the supporting cord on only one side. The mystery of its equilibrium lies in an ingenious bit of engineering: the bird fastens counterbalancing lumps of clay on the appropriate side of the cup so that it does not tip and spill its inmates.

Another animal product that some birds use in their nests is their own saliva. Some swifts, hornbills, nuthatches, and hummingbirds secrete unusually abundant amounts of mucilaginous saliva and use it to cement twigs, grass, feathers, plant down, clay, and other ingredients into nest forms, and also to glue their nests to supporting branches or walls. The Great Swallow-tailed Swift, *Panyptila sanctihieronomyi*, builds its tubular pendent nest of floss-borne seed collected in midair and cemented together with saliva. The Cayenne Swift, *P. cayennensis*, makes its tubular, felted nest of feathers and saliva, and the Chimney Swift, *Chaetura pelagica*, cements twigs together to make its half-cup nest and to glue it to the inside of a chimney or hollow tree (Fig. 14–12). Most exceptional is the Palm Swift, *Cypsiurus parvus*, which fastens its simple, bracket-shaped nest of saliva, feathers, and

FIGURE 14–12 A Chimney Swift's nest showing the sticky salivary mucus which cements the twigs together and to the wall. Photo by R. B. Fischer.

plant fluff on the under side of a withered, hanging palm leaf. The two eggs are cemented to the nest shelf with saliva and therefore cannot be rolled around during incubation as is typical procedure in most incubating birds. Probably the swaying of the leaf in the wind substitutes for this necessary movement. One of the smallest and most delicate nests in the world is that of the Crested Tree Swift, *Hemiprocne mystacea*, of southeast Asia, whose thin, parchment-like nest of saliva, bark-strips, and down is glued to one side of a horizontal branch, and is just large enough to hold the bird's solitary egg. While incubating, the bird sits crosswise on the branch so that its weight is supported by the branch and not by the flimsy nest.

Surpassing all of these saliva-containing nests are the nests of the Edible Swiftlets of the genus *Collocalia*. The nests of some of these birds are composed entirely of dried salivary mucus, and are in the form of a small translucent quarter-sphere attached to the face of a rock wall. These are the nests that are prized by the Chinese for making birds'-nest soup. As many as three and one-half million nests (20 to 30 tons) have been exported from Borneo in a single year for this purpose. One Javanese species, *Collocalia gigas*, nests regularly behind waterfalls, and another species, *C. francica*, has achieved what is probably the zenith of nest-site security by building its mucus nests in grottoes than can be reached only through an entrance which, at least during high tide, requires diving under the sea! (Stresemann, 1927–1934).

The salivary glands of some swifts and swallows enlarge greatly during the nesting season. Even so, the Edible Swiftlet, *C. fuciphaga*, that makes its nest exclusively of saliva, requires from 33 to 41 days to make one nest. In one night a single Swiflet deposited a strip of dried salivary mucus 45 mm long, 8 mm high, and 2.5 mm thick (Stresemann, 1927–1934).

Mineral materials used in nest-building are largely restricted to stones and mud, although many species make nest scrapes on rock, earth, or sand, and others burrow in soils of various sorts. Stones to make nest platforms are gathered by several species of penguins, petrels, plovers, terns, and larks. The Black-necked Stilt, *Himantopus mexicanus*, often paves its scrape with mussel shells. Several species of ground-nesting larks—for example, the Horned Lark, *Eremophila alpestris*—typically build at one side of their grass-cup nest a courtyard paved with small flat stones. Desert larks may erect stone walls around their nests; these may shelter the nest from wind or from the eyes

of predators. In Spain, the Black Wheatear, *Oenanthe leucura*, has been known to place as many as 76 small stones under its nest and to arrange 282 stones into a sheltering wall 22 cm long and 6.5 cm high (Makatsch, 1950). Richardson (1965) believes that the accumulation of stones may be the result of a nest-advertising display of the male birds. Around some long-established nests thousands of stones may accumulate. One male was observed carrying 42 stones to a nest in 25 minutes while the female perched nearby.

Mud is used extensively and in a wide variety of nests. It may be used alone or mixed with various sticks, pebbles, or fibers. In Australia, the Mudlarks, *Grallina cyanoleuca*, and Apostle-birds, *Struthidea cinerea*, balance their symmetrical bowls of dried mud high on horizontal branches. The mud that is used to build these pottery-like nests is reinforced with horse-hair and wool, and the nest is lined with soft feathers and grass. Mud nests resembling large inverted pails are built in shallow water by the flamingos (Phoenicopteridae). They scoop together enough mud to erect a mound about 40 cm across the top and 50 cm high. The Yellow-nosed Albatross, *Diomedea chlororhynchos*, builds a similar nest mound about 30 cm high composed of mud mixed with vegetation. Several species of swallows build their nests largely of mud. The Cliff Swallow, *Petrochelidon pyrrhonota*, builds its retort-shaped nests of mud pellets, and locates them in colonies on cliffs or under eaves of barns (Fig. 14–13). The quality of the mud used may vary locally; nests made of clay are much stronger than those made of sandy silt. In Wyoming and Wisconsin, Emlen (1954) found that Cliff Swallows' nests constructed too rapidly or in humid weather often collapsed before completion. Even nests a month old, containing well-developed young, disintegrated during a prolonged rainy spell.

A much more resistant mud nest is that built by the Rufous Ovenbird, *Furnarius rufus*. This species erects its two-chambered, ball-shaped nest on horizontal limbs or on telephone pole cross-arms, where it withstands all kinds of weather. The Ovenbird is destined to become famous as the bird whose formula for mud saved thousands of human lives. In South America, a trypanosome blood infection called Chagas' disease afflicts estimated millions of humans. It causes severe debilitation in adults and is often fatal in children. This incurable disease is transmitted by the bite of a reduviid bug, *Triatoma megista*, the "barbeiro," which flourishes in the cracks of native

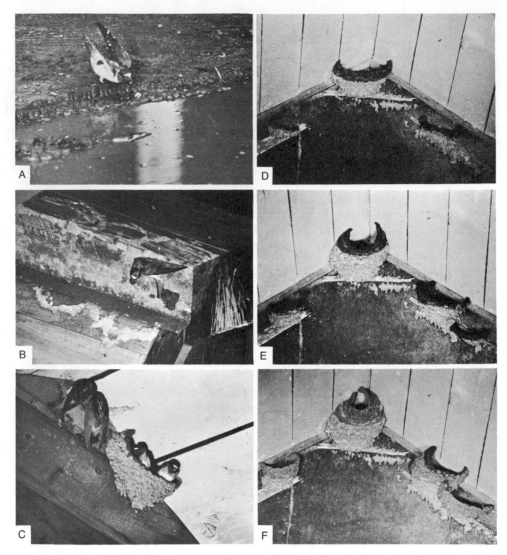

FIGURE 14–13 Cliff Swallow nest construction: *A*, A Swallow gathering a pellet of mud. *B*, Applying a pellet to a horizontal beam. *C*, Two Swallows in a half-finished nest. *D, E, F*, Successive stages in nest construction. Freshly added mud appears dark. Photos by J. Emlen.

mud huts. The simplest way to prevent the disease is to build crack-free houses. Pondering this problem, Dr. Mario Pinotti, head of Brazil's National Department of Endemic Diseases, recalled that as a boy he threw rocks at Ovenbird nests which never cracked. Research showed that the Ovenbird compounded its house mortar of sand and cow dung. Following this lead, government health workers plastered 2000 huts with the Ovenbird's odorless cow-dung formula. After six months, every house was free of cracks and free of the barbeiro, although 98 per cent had been infested previously. In 1958, 200,000 homes were plastered. It is hoped that within a generation 2,500,000 homes will be plastered with the Ovenbird's beneficent mortar, and that Chagas' disease will be eliminated (*Time*, 1958).

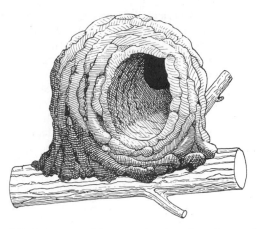 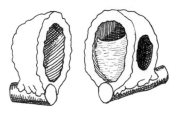

FIGURE 14–14 The two-chambered mud nest of the Rufous Ovenbird, constructed of a mortar made of sand and cow dung. The diagram at the right shows a vertically sectioned nest with the entrance vestibule at the right, separated by a partial partition from the egg chamber at the left. After Pycraft, 1910.

NEST-BUILDING BEHAVIOR

Problems of nest construction do not exist for those birds using abandoned nests of other birds or mammals, or for those that forcibly appropriate occupied nests. By far the majority of species, however, build their own nests. Typically, they build a nest for each brood of young, or, less often, for the successive broods of one season, although some of the larger species, such as storks and eagles, may use the same nest for decades, adding new nesting material each year. Although building a new nest for each generation of young is expensive in labor, it minimizes the chances of disintegration of a nest while it is sheltering eggs or young, and it reduces the possibility of the accumulation of nest parasites. Also, the experience gained in building several nests rather than one may enable the individual bird to do a better job both in site selection and in nest-building.

There can be little doubt that nest-building is primarily an instinctive activity. This interpretation is strongly supported by the simple fact that birds of the same species build nests which conform closely to a given pattern. Over a century ago, Charles Darwin (1845), in his *Journal of Researches*, described the automatic, instinctive nature of nest-building in one of the Ovenbirds, *Geositta cunicularius*, which normally digs a horizontal nesting burrow about 2 meters long in a vertical earthen bank.

Here [at Bahia Blanca] the walls round the houses are built of hardened mud, and I noticed that one, which enclosed a courtyard where I lodged, was bored through by round holes in a score of places. On asking the owner the cause of this, he bitterly complained of the little casarita, several of which I afterward observed at work. It is rather curious to find how incapable these birds must be of acquiring any notion of thickness, for although they were constantly flitting over the low wall, they continued vainly to bore through it, thinking it an excellent bank for their nests. I do not doubt that each bird, as it came to daylight on the opposite side, was greatly surprised at the marvelous fact.

Like the building of multiple nests by robins and other species, this futile digging by the Ovenbird illustrates a stereotyped, instinctive response to a strong releaser situation—in this case, the mud wall.

Instinctive nest-building can sometimes misfire and produce nonadaptive results. When a gull is prevented by its mate or by circumstances from brooding its eggs or young, it commonly indulges in an instinctive displacement activity—in this case, fetching building materials. As a consequence, the size of a gull's nest may be determined by the number of times the bird is prevented from sitting on it (Kirkman, 1937). In Norway, so much material, often stolen from such accumulations, is brought to the nests of Black-headed Gulls, *Larus ridibundus*, that eggs, and at times even newly hatched young, are covered over and perish (Ytreberg, 1956). A similar nonintelligent behavior was observed by Roberts (1940) among Wilson's Petrels, *Oceanites oceanicus*, in Antarctica. When surplus moss was placed in front of the nest burrows of these birds, they carried it inside in such quantities that "there was hardly room for the birds to get in," and Roberts had to remove some of the moss.

Apparently when the time is ripe for nest-building, the activity may proceed regardless of its appropriateness or external adaptability. In his study of the Red-winged Blackbird, *Agelaius phoeniceus*, Emlen (1941) found that eggs or young experimentally introduced into an unfinished nest would not interrupt the normal course of nest-building activity. When Nolan (1978) replaced a beginning, incomplete shell of a nest of the Prairie Warbler, *Dendroica discolor*, with a completed nest, the female continued to build for two more days, "unaffected by the fact that the substituted nest was complete." However, in one of three such experiments, the female's building was influenced by the completed nest. When nestlings were placed in nests under construction, in three instances the females continued construction, but in a fourth nest, after attempting to remove the nestling, the female began feeding it in the normal fashion.

Certainly, hormones play an important role in nest-building behavior. When testosterone was injected into nonbreeding male Village Weaverbirds, *Ploceus cucullatus*, their nest-building activities increased (Collias *et al.*, 1961). Ordinarily the female Quelea, *Quelea quelea*, of Africa assists the male in nest-building very slightly—perhaps 0.5 to 8 per cent of the time. But if she is injected intrapectorally with testosterone, her nest-building activity will essentially equal that of the male (Crook and Butterfield, 1970).

Conversely, nest-building activity in the Ring Dove, *Streptopelia roseogrisea*, appears to stimulate production of the hormone FSH and hence ovulation (Cheng and Balthazart, 1982).

In an attempt to discover to what extent instinct controlled nest-building behavior, Collias and Collias (1984) removed a nestling Village Weaverbird, *Ploceus cucullatus*, from its parents before its eyes were open and hand-raised it. Although it never saw a nest of its own species nor another of its kind building, this isolate built a species-specific nest when it was a year old.

The Colliases also noted that nests of first-year Village Weaverbirds in nature were rather crude in structure as compared to nests of older birds. Suspecting that practice makes perfect, they raised six juveniles of this weaverbird in two groups of three. One group was deprived of opportunity to handle objects, and the second (control) group was given strips of reed grass to manipulate. When they were about a year old, both groups were provided with reed grass and tested for nest-weaving ability. In the first week the deprived birds did not weave a single stitch, whereas the controls wove very well. In the second week the deprived birds wove a few stitches, but only a small percentage of strips were woven as compared with controls. After three months of practice the deprived birds used as many strips as controls; however, controls built 11 nests as compared to the two nests built by deprived birds. One subordinate deprived bird was prevented from manipulating reeds because his dominant cage mates always stole his nesting materials. As a result this bird never learned to build a nest even after eight years. The Colliases concluded from these experiments that although weaving a nest is instinctive, practice is necessary in perfecting the art. Birds deprived of opportunity to manipulate objects in the first two years of their lives never learn to weave.

It seems very unlikely that young megapodes or "incubator birds" (Megapodiidae) learn much about nest-building from their parents. When they burst out of their underground nests they head for a completely independent existence in the brush. Yet, when they reach breeding age, they build nests that are typical of the species.

It is certainly true that birds of many species build better nests as they grow older. Mature female Tree Swallows, *Iridoprocne bicolor*, for example, build better nests, and line them with more feathers, than do year-old females. But whether this is the result of experience or an expression of physiological or psychological maturation, it is at present impossible to say. Experiments by Sargent (1965) with Zebra Finches, *Poëphilia guttata*, showed that experience plays at least a partial role in nest-building behavior. Birds offered a choice of brown, green, and red strands of burlap as nesting material chose brown preponderantly. However, birds reared in green nests selected over twice as many green fibers as did birds reared in brown nests. The diversity of nest-building habits within closely related groups such as the Tyrannidae or Paridae indicates clearly that the instinctive element in their nest-construction behavior is relatively labile and subject to rapid evolution.

That birds are not complete automata in their nest-building is suggested by their adaptability to unusual conditions. When half of the nest of a House Martin, *Delichon urbica*, was destroyed by water, the adults repaired the nest, making roughly alternate trips bearing mud for the nest and insects for the young (Rivière, 1940). Other species show similar adaptability.

In building a nest, the work may be shared about equally by the two sexes, or it may be done entirely or predominantly by one sex or the other.

There are probably almost as many different techniques used in building nests as there are species of birds. Only a few examples of the different methods can be given here. In most species the beak and the feet are the principal tools used. Birds building nests of grasses and weeds more often work hardest early in the morning, partly, perhaps, because at this time of day the building materials are more moist and pliable. The Great Reed Warbler, *Acrocephalus arundinaceus*, will at times soak a mouthful of dry nesting material in water before weaving it into its nest. The male African Weaverbird, *Malimbus rubiceps*, prepares material for building his tubular nest by first breaking off small twigs, 10 to 20 cm long. He then partly removes a thin strip of bark, which, still attached to the twig, he knots into the nest wall (Crook, 1963). Many birds, in constructing the common cup-shaped nest, first build the supporting platform of coarse materials and then build a rim of finer materials to make the bowl.

The shaping of the cup as done by thrushes illustrates a procedure common to many species. The following account is based on the performance of the European Blackbird, *Turdus merula*, as described by Stresemann (1927–1934). The thrush presses its body into the nest mold for about 4 seconds, then rises for about 5 seconds, and then presses again. With each successive down-pressing, the bird rotates its body about 36 degrees from the previous position. After about ten down-pressings, the bird makes two to three complete rotations in the same direction. At its next visit to the nest, it may turn in the opposite direction. Should any loose ends of grass stick out, the bird will pull them out and tuck them back into the walls of the nest.

A 17-day-old fledgling Swainson's Thrush, *Catharus ustulatus*, will go through various movements. One held in cupped hands by Dilger (1956) snuggled down and "simultaneously kicked backward with both feet and forcibly thrust its breast against the side of the cup. The wings were held rather high on the back but not unfolded and the tail was rather depressed. The bird would perform a few rapid thrusts and kicks and then turn slightly in the cup and repeat these acts." Increased pressure by the edge of a hand against the pushing breast provoked an increased pressure by the bird at this point, suggesting that irregularities in the ring-shape of the cup stimulate compensating thrusts by the bird.

In his studies of nest-building in the Canary, *Serinus canaria*, Hinde (1958) determined that various behavior patterns may be integrated into functional sequences. Although hormonal states may control several different behavior activities, one specific activity may lead to a stimulus for the next step. For example, gathering material may stimulate carrying it, and carrying it may lead to placing it in the nest, then weaving it, molding it, and so on. Also, as a bird continues a given activity, its internal tendency to continue or repeat the process decreases, so that a change from nest-weaving back to grass-gathering becomes possible and natural.

The weaving of a hangnest requires a more complex performance, which varies considerably according to the type of nest made. Generally a bird starts such a nest by hanging several long plant fibers over a branch. To ensure that a given fiber will cling to the branch, the Red-billed Quelea, *Quelea quelea*, holds one end of it under its foot and then, with its bill, winds the loose end several times around the branch, as well as around the end of the fiber it is holding with its foot. It then ties a knot in the fiber by drawing the free end through the narrow space between the branch and that portion of the fiber held under its foot. This stereotyped performance is repeated until there are several dangling fibers, each firmly tied to the branch. The bird then flutters or crawls among the hanging fibers while holding one of them in its beak. Repetition of this movement results in the plaited or woven cord that suspends the nest. Other fibers are next woven into the lower ends of this cord by pushing and pulling them through the slowly developing fabric, and occasionally tying knots in them. Gradually the dangling cord is fashioned into a wreath, a hammock, or a cup, and eventually into the hangnest characteristic of the species (Figs. 14–15 and 14–16).

Actual sewing is performed by the famous Tailorbird, *Orthotomus sutorius*, of southeast Asia. With its sharp bill this small bird pierces holes in the edges of large green leaves and then stitches them together by pulling plant fibers through the holes, making a green funnel-shaped receptacle for its nest (Fig. 14–17).

FIGURE 14–15 Hangnest construction by the Baya Weaverbird of India. *(Upper)* A male, on an initial "wad" stage of the nest, is giving the invitation display to a prospecting female arriving in the colony. *(Lower)* Half-built nests at the left. In the lower nest a male is displaying to a female who has just appropriated the nest at the "helmet" stage of construction. When the nest is finished, the lower chamber will be closed to form the egg chamber; the upper chamber will be accessible from the outside only by means of a long, sleeve-like tube directed vertically downward, with the entrance hole at the bottom. Photos by Salim Ali.

The use of mud in forming a nest involves either scooping it together in a mound or carrying it a mouthful at a time to the nest site, where the soft pellets are pressed together into the specific nest form. With much ceremonious bowing and braying, the male Black-browed Albatross, *Diomedea melanophris*, brings mud and moss to the nest where the female, with similar ceremony, accepts it and treads it into a large cylindrical mound. Both sexes of the Cliff Swallow, *Petrochelidon pyrrhonota*, carry in their mouths soft pellets of mud to the nest site, where they deposit them in long superimposed rows to form their flask-shaped nests.

For digging holes in wood, woodpeckers have several serviceable adaptations. Their beaks are often chisel-edged at the tip instead of pointed, and their

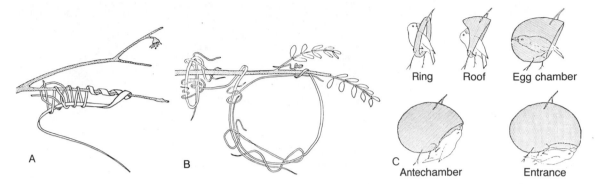

FIGURE 14–16 Stages in the making of hangnests by weaverbirds. A, Appearance of a coiled and threaded strip of grass used to begin a nest by the Village Weaverbird, *Ploceus cucullatus*. B, Details of grass strips in the early ring stage of the nest of *P. jacksoni*. C, Successive stages in nest-building by *P. cucullatus*. From Collias and Collias, 1964.

skulls are unusually heavy. To grip the tree they have strong feet, commonly with two opposed toes (instead of one), and stiff tail feathers to brace the hammering body. In pecking at firm wood they do not strike their blows blindly, but aim them from alternating directions as does a woodchopper. To excavate its nest, the Black Woodpecker, *Dryocopus martius*, delivers about 100,000 hammer blows at the rate of about 5000 per day (Cuisin, 1983).

In digging its nest hole in a vertical bank, a bee-eater, *Merops apiaster*, first flies repeatedly at the wall with a partly opened beak, sometimes from a distance of several meters, sometimes in rapid forward and backward flights from a distance of only 25 to 30 cm. When the hole is deep enough to provide a foothold, the digging pace is accelerated, and when the hole is about 10 cm deep the bird works inside it and kicks out the loose soil with its feet.

Some indication of the amount of labor involved in building a nest is provided by the weights of some of the larger nests mentioned previously. The South American Rufous Ovenbird, *Furnarius rufus*, weighs about 75 g, but its spherical mud nest requires 1200 to 2500 lumps of mortar that together weigh 3.5 to 4.5 kg—about 60 times the weight of the bird. A young, inexperienced Chaffinch, *Fringilla coelebs*, was estimated by Marler (1956) to average 1300 building visits for one nest. Barn Swallows, *Hirundo rustica*, have been estimated to make over 1200 mud-carrying trips to construct one nest. A Long-tailed Tit, *Aegithalos caudatus*, has built into one nest as many as 2457

feathers. The Hammerhead Stork, *Scopus umbretta*, builds a huge nest of sticks, usually in a tree fork, that may be as large as 2 meters wide by 2 meters deep and weigh several hundred kilograms. Kahl (1967) estimated that one pair of these birds made 8000 trips with sticks and grass to build one nest. Probably as a protection against terrestrial predators, the Horned Coot, *Fulica cornuta*, of Andean lakes builds a nest foundation of pebbles about 30 meters from the shore in water 70 to 80 cm deep. It is estimated that this 2- to 4-meter-wide island weighs 1500 kg (Makatsch, 1972).

With so many pieces and so much weight to carry for one nest, it is not surprising that some species have resorted to pilfering and even outright robbery of nesting materials. Such behavior could easily have stemmed from the habit some birds have of tearing apart one of their old nests to build a new. The Puerto Rican Honeycreeper, *Coereba flaveola*, may tear materials out of her nest while it still contains young, to build a new nest for the next brood. She may also steal materials from the nest of a brooding neighbor (Biaggi, 1955). The Mexican Violet-eared Hummingbird, *Colibri thalassinus*, may line its nest with plant down stolen from the nest of the White-eared Hummingbird, *Hylocharis leucotis*, sometimes to such an extent that the latter's nest is completely destroyed (Wagner, 1945). Opportunities for thievery of nest materials are particularly good among colonial nesters. Mutual pilfering of nest materials is quite common among penguins, cormorants, pelicans, and

FIGURE 14–17 The nest of the Long-tailed Tailorbird, showing the stitched leaves which make the container for its fluff-lined nest. The bird first pierces holes in the edges of large leaves with its beak, and then draws plant fibers through, knotting them on the outside so that the stitches will hold. There are nine species of tailorbirds in southeast Asia. Photo by Loke Wan Tho.

storks. At times kleptomania may become infectious, as it sometimes does in colonies of Rooks, *Corvus frugilegus*, and crescendo into wholesale free-for-all plundering. Nest thievery is so universal among the Antarctic Blue-eyed Shags, *Phalacrocorax atriceps*, that one member of a pair stands guard at the nest while the other sets forth to filch bones, feathers, stones, and seaweeds from its neighbors' nests (Murphy, 1936). Skutch (1960) believes that such thievery among oropendolas may be partly beneficial since it discourages careless nest construction.

The time consumed in building a nest depends on a number of variable factors such as the size of the nest and its complexity; the materials used and the distance they are carried; the species building the nest; whether one or both members of the pair do the building; the age and experience of the builder; the time of year and the weather; and the geographical latitude.

Birds that lay their eggs on the bare ground require no time to prepare their nest. The larger and more elaborate the nest, the more time is required to con-

struct it. It is often difficult to say that a given nest requires a certain number of days to build, because its builder may work very energetically at first and then slow down to a leisurely pace. A Kirtland's Warbler, *Dendroica kirtlandii*, observed by Van Tyne (*in* Bent, 1953), made 131 trips the first day, bringing materials to the nest and arranging them, and 59 the second day, completing the body of the nest. The third day she made seven trips, and the fourth day six trips to bring lining for the cup. Several species are persistent builders, adding materials to the nest throughout the incubation period, and in some birds such as Allen's Hummingbird, *Selasphorus sasin*, the Osprey, *Pandion haliaetus*, and Honey Buzzard, *Pernis apivorus*, materials may be added to the nest up to the time the young leave it.

As a rule, small passerines build their nests in a few days. The Corn Bunting, *Emberiza calandra*, requires only 2 days; the Field Sparrow, *Spizella pusilla*, 3 days; Red-eyed Vireo, *Vireo olivaceus*, 5 days; American Robin, *Turdus migratorius*, 6 to 20 days; Carrion

Crow, *Corvus corone*, 9 days; Dipper, *Cinclus mexicanus*, 15 days. The male Long-tailed Tit, *Aegíthalos caudatus*, takes about 9 days to build the nest proper, and the female takes 9 days more to line it. The Golden Eagle, *Aquila chrysaëtos*, may build a complete nest in two months, and the African Hammerhead, *Scopus umbretta*, in four months. Large nests such as these last, however, may be used in successive years with or without annual additions of nesting materials, depending on the species.

The Rufous Ovenbird, *Furnarius rufus*, may require several months to build its nest if dry weather interferes with its mud-gathering. Some tropical hummingbirds need green moss to build their nests and in dry years may omit nest-building altogether. Species that breed two or more times a year generally build their second nests faster than their first. The American Goldfinch, *Carduelis tristis*, takes about 13 days to build a nest in July as against 5 or 6 days in August; and the Prothonotary Warbler, *Protonotaria citrea*, requires about 5 days to build a nest in late May as against 2 days in June. This acceleration of the process of nest-building can probably be accounted for by two influences: first, the experience of having built one nest facilitates the construction of a second; second, more food is commonly available later in the season (insects, seeds) than earlier, hence less time need be taken from nest-building to forage for food. Air temperatures affect the nest building of Prairie Warblers, *Dendroica discolor*. At ambient temperatures 2.5 to 5.2°C below seasonal normal, building activity dropped roughly 50 per cent; at 8.0°C below normal, it ceased entirely (Nolan, 1978).

Geographically, several species have been shown to build their nests with more deliberation in the tropics than do their nearest relatives in temperate regions. Whereas the Great Kiskadee Flycatcher, *Pitangus sulphuratus*, and Vermilion-crowned Flycatcher, *Myiozetetes similis*, of Mexico require about 24 days to build their nests, their tyrant flycatcher relatives in temperate North America take only 3 to 13 days (Pettingill, 1942), in spite of the fact that in Mexico both sexes of *Pitangus* work on the nest, whereas among their northern relatives only the females build. It seems likely that natural selection has accelerated nest-building behavior in birds of temperate regions where the breeding season is shorter.

The survival of any species of animal depends primarily on reproductive success. In the millions of years during which they have been evolving, vertebrates have taken two diametrically opposed paths toward breeding efficiency. One path, followed by the lower vertebrates, has emphasized the production of great numbers of eggs. A fish may lay hundreds of thousands of eggs per year, but may give them absolutely no care. The other path, followed by mammals and birds, emphasizes superlative care of the eggs and young, but of necessity gives that care to very few of them. One of the chief ways in which birds provide that care is by their varied, warm, and wonderfully ingenious nests.

■ SUGGESTED READINGS

Old but still useful books are *Birds' Nests* by Dixon, and *Der Vogel und Sein Nest* by Makatsch. Gilliard's *Living Birds of the World* has much interesting material on nests, including striking photographs. Many regional guides for the identification of birds also contain information on nests. For a good treatment of nest construction see Collias and Collias *Nest Building and Bird Behavior*. The following articles in Campbell and Lack's *A Dictionary of Birds* are recommended for brief introductions to the subjects: *Nest*, by Marchant; *Nest Building*, by Hinde; *Nest Function*, by Davies; *Nest Site Selection*, by Slagsvold; and *Nesting Associations*, by Smith. For identification on North American nests in the field, see Headstrom's *Birds Nests: A Field Guide*, and *Birds Nests of the West: A Field Guide*.

Eggs
15

Yes, Hen, it's a miracle
  To lay a crimson-crested Cock
In a warm almost-lilac shell,
  Quiet as a bit of living rock.
—E. Merrill Root (—1978)
  Lost Eden

A primitive vertebrate, such as the codfish, may lay as many as 10,000,000 tiny, jelly-like eggs at one spawning, and then abandon them in the ocean to an almost certain death. The few eggs that escape the normal hazards of their environment and grow to maturity require four or five years to become full-sized fish about a meter long. Because of this long, dangerous period of immaturity, the codfish *must* lay a great number of eggs or the species will perish.

This "slaughter of the innocents" is greatly reduced among the reptiles by the use of a more highly developed egg. In contrast to the codfish, a somewhat less primitive tortoise lays only about 100 eggs in an exca-

vated hole in the ground, and leaves them to their fate. The greatly increased probability of survival that is reflected in the relatively small number of eggs is mainly due to three evolutionary innovations. First, the reptile eggs have a tough, water-resistant outer membrane that makes possible their development on land, where food and oxygen are more abundant and where the eggs are inaccessible to a great many aquatic predators and diseases. Second, each egg is large and is stored with enough nourishment to give the hatchling tortoise a rapid and vigorous start in life. Third, the poisonous nitrogenous wastes (ammonia and urea) from the growing embryo, which accumulate in the egg and cannot be discarded by diffusion into the surrounding air, are converted into relatively insoluble and harmless uric acid and retained in the egg.

Birds, far less primitive than the tortoise, have carried this principle of a protected, terrestrial development still further. Birds can penetrate even drier habitats than reptiles because 90 per cent of the food oxidized to provide energy for the growing embryo is yolk fat that, when broken down, yields more metabolic water than does the primarily protein food of reptilian eggs. Reptilian eggs consequently are not completely divorced from the need for some external water. Not only do birds lay their large eggs in sheltering nests that are often built in sites inaccessible to most predators, but they hasten the development of the embryo within by warming it with the heat of their own bodies. Moreover, after the young bird hatches, they feed it and protect it until it is able to fend for itself. As a consequence of this painstaking care, a newly hatched bird, in contrast to an infant codfish, may race through the period of vulnerable infancy in a few weeks instead of months or years, and is sooner able to reproduce and leave progeny. Since the incubating bird must be able to warm the eggs with its own body, the number of eggs that can be covered is naturally more limited than the number of fish or even reptile eggs in one clutch. But this limitation is compensated for, both by the accelerated rate of development made possible by incubation and by the superlative care most birds give their newly hatched young. A bird that lays only one or two eggs a year may be more efficient reproductively than a fish that lays millions.

These evolutionary advances in reproduction exact their price from parent birds. In preparation for the metabolic strains of egg-laying, female Blue Tits and Great Tits, *Parus caeruleus* and *P. major*, increase their body weights by 26 and 18 per cent, respectively, while the males simultaneously lose weight owing to the exertions of courtship feeding (Flegg and Cox, 1977). Every day a female domestic Chicken, *Gallus gallus*, deposits about 1.8 per cent of her body weight in the egg; in contrast, the human female deposits about 0.019 per cent of her body weight in the more slowly developing fetus and placenta (Gilbert, 1971). King (1973) estimates that egg production requires about 13 to 16 per cent of daily energy intake (at constant body weights) for Passerine birds, 21 to 30 per cent for Galliformes, and 52 to 70 per cent for Anseriformes. Daily peak energy requirements for egg production vary from about 0.7 to 3.4 times the basal metabolic rate.

FORMATION OF THE EGG

A complete egg is composed of the yolk or ovum, formed in the ovary, and a series of enveloping layers of albumen, shell membranes, and the shell proper, which are added to the yolk as it descends the oviduct. The following account of egg development is based largely on the egg of the domestic hen as described by Sturkie (1965, 1976).

The funnel-like infundibulum of the oviduct grips the ripe ovum shortly before it bursts from the ovary (See Fig. 8–3). The ovum remains in the infundibulum only about 18 minutes and is then passed on either by ciliary movement or by peristalsis to the magnum of the oviduct where, in about three hours, it receives its layer of albumen. The egg next passes to the isthmus where it remains about an hour and receives its shell membranes. Next it passes to the uterus where it may remain about 20 hours while receiving its shell and pigment. Finally, the completed egg quickly passes through the muscular vagina and is laid.

When first secreted by the magnum, the albumen or white of the egg is in the form of a single, dense, jelly-like layer; but by the time the egg is laid, four distinct regions of albumen are visible. Innermost is a thin, watery layer within which the yolk may freely rotate. Outside of this layer is a thicker, middle layer

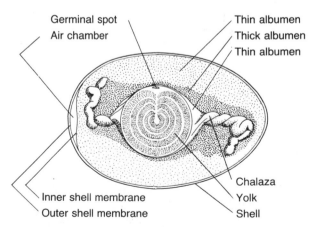

FIGURE 15–1 A diagram of the domestic hen's egg, in longitudinal section.

of albumen, which, in turn, is surrounded by the thin, outermost fluid layer. The fourth region runs through the other three. It is the chalaza, a pair of dense, twisted cords of albumen attached to opposite ends of the yolk and coincident with the long axis of the egg. The yolk is so centered and suspended in the inner layer of thin albumen by these twisted strands of chalaza that it may rotate so that its animal pole is always up and the heavier vegetal pole always down, no matter what the position of the egg in the nest. The chalaza probably arises from protein fibers of mucin of the inner albumen as a consequence of the rotation of the egg as it descends the oviduct.

Egg albumen is composed of three types of protein: mucin and globulin make up about 5 per cent each, and albumin makes up about 90 per cent. Mucin gives the thicker layer of egg white its viscosity. Globulin is most concentrated in the inner thin layer and least concentrated in the outer thin layer. The highest amount of albumin is found in the outer thin layer.

Experiments have shown that the mechanical stimulus of the yolk pressing against the walls of the magnum causes it to secrete albumen. Almost any solid, rounded object inserted in the magnum will become covered with egg white. However, yolkless eggs show that other stimuli must also be involved.

Two fibrous shell membranes are formed by the isthmus—a thick external membrane and an internal one about one-third as thick. The two membranes separate at the blunt end of the egg to form an air chamber. Both membranes are made of the tough protein keratin.

The shell proper is secreted in the uterus and is composed of a light protein framework similar to collagen, and a heavy deposition of inorganic minerals. The minerals, mainly calcium carbonate, are arranged in the protein matrix as vertical crystals separated by minute pores, through which oxygen, carbon dioxide, and small amounts of water vapor may pass. A chicken's egg contains between 6000 and 17,000 pores. The permeability of a shell to air, under a pressure differential of 20 cm of mercury, varies from 3 cc per sq cm per minute in a quail egg to 60 cc per sq cm in an ostrich egg (Romanoff and Romanoff, 1949). The shell of a typical hen's egg is composed of about 94 per cent calcium carbonate, 1 per cent magnesium carbonate, 1 per cent calcium phosphate, and 3 to 4 per cent protein.

A surprisingly rapid and effective transfer of calcium occurs between a mother hen and her growing chicks. During egg-laying, as much as 12 per cent of the bone substance in a hen's body may be released, within a 24-hour period, for use in forming egg shells. Then, as the growing chick within the egg develops, it recycles some of the dissolved calcium carbonate from the egg shell to build its growing bones. Further conservation of calcium occurs when many brooding females eat the broken egg shells after the young hatch.

As a rule, large eggs have thick shells and small eggs, thin shells. Roughly, the thickness of a shell is proportionate to the 0.456 power of the egg weight (Ar *et al.*, 1974). The heavy-footed African Francolins, *Francolinus* spp., have thick-shelled eggs that one "can practically bounce off a wall" (Heinroth and Heinroth, 1958). The heaviest shell known is that of *F. coqui*; the shell amounts to an amazing 28.1 per cent of the total egg weight (Makatsch, 1952). Certain species of Indian woodpeckers, which inhabit ants' nests, are reputed to lay eggs with abnormally soft and transparent shells, perhaps because the woodpeckers' diet of ants is rich in formic acid (Moreau, 1936).

Shell thickness decreases as incubation proceeds. This is because the embryo gets much of its skeletal calcium by removing it from the shell. Eggs of Cedar Waxwings, *Bombycilla cedrorum*, decrease their thickness by 5.6 per cent and eggs of Arctic Terns, *Sterna paradisaea*, by 8 per cent (Rothstein, 1972; Finnlund *et al.*, 1985).

The composition of the entire hen's egg is given by Romanoff and Romanoff (1949) as 65.6 per cent water, 12.1 per cent proteins, 10.5 per cent lipids, 0.9 per cent glucides, and 10.9 per cent minerals. The proteins are made up of 18 amino acids that are used for tissue growth by the developing embryo. The fats, mainly in the yolk, are used chiefly for energy. Excluding the shell, the chemical elements of an egg are carbon, 53 per cent; oxygen, 20 per cent; nitrogen, 15 per cent; hydrogen, 7 per cent; phosphorus, 4 per cent; and sulfur, 1 per cent.

Heinroth and Heinroth (1958) noted that eggs of precocial birds contain more yolk by weight than eggs of altricials (see Ch. 17). A recent survey showed that yolk content in eggs of procochials averaged about 41 per cent and those of altricials about 22 per cent (Carey *et al.*, 1980). Some megapode eggs contain up to 66 per cent (Meyer, 1930). Eggs of altricials contain about 15 per cent more albumin than precocials. Since albumin is mostly water, increased albumin content may compensate for increased water loss in the thinner-shelled more permeable altricials' eggs (Ar and Yom Tov, 1978).

SIZE, SHAPE, TEXTURE, AND POROSITY OF EGGS

An egg is more than a mere package of food to provision the developing embryo for its start in life. The size, shape, surface, and color of an egg often have additional value for a species. Reasonably enough, large birds lay large eggs, small birds, small eggs. The largest known egg is that of the extinct "elephant bird," *Aepyornis*, of Madagascar, whose eggs, known only as fossils, held about 8 liters and measured 34 by 24 cm (13.5 × 9.5 in). Shells of this egg were heavy enough to be fashioned into eating bowls by the former aborigines of Madagascar. At the other extreme is the tiny Vervain Hummingbird, *Mellisuga minima*, of Jamaica, whose egg is less than 10 mm long. One egg of *Aepyornis* could hold the contents of about 50,000 eggs of this small hummingbird.

As a rule, the larger the bird, the smaller its egg is in relation to the parent's size. The big ostrich lays an egg that weighs only 1.7 per cent of its body weight, while the tiny wren lays an egg that equals 13 per cent of its body weight. Among the Procellariiformes, the eggs of the large albatrosses equal about 6 per cent of their body weight, those of the medium-sized fulmars, about 15 per cent, and those of the smallest petrels, about 22 per cent of their body weight. A survey by O'Conner (1984) of the relationship of egg weight to body weight revealed that birds with relatively naked chicks (e.g., passeriforms, pigeons, woodpeckers) lay smaller eggs (4.5 to 7.7 per cent of body weight) than birds with heavily downed young (e.g., owls, waterfowl, charadriiforms, procellariiforms) which lay larger eggs (11.5 to 21.1 per cent of body weight). He noted that downy nidicoles are found in species that leave their young unattended for long periods while they search for food. Probably the species that holds the record for laying the relatively largest egg is the kiwi (*Apteryx* spp.), a hen-sized bird weighing about 1.8 kg. Its enormous egg weighs about 420 g, or about one-fourth of the bird's body weight.

Other weight relationships occur fairly systematically. Precocial species, whose young hatch out alert and active, usually lay larger eggs than do altricial species, whose young are born naked and helpless. A 5-kg precocial crane, for example, lays an egg that weighs 4 per cent of its body weight; while the 5-kg altricial eagle lays an egg that is only 2.8 per cent of its body weight. The egg of the precocial Guillemot, *Uria aalge*, is about five times as large as that of the altricial Raven, *Corvus corax*, although the two birds are of equal size. Birds that lay large clutches of eggs usually lay smaller eggs than do their relatives who lay fewer eggs. Several sandpipers lay four eggs in about four days that together weigh as much as or more than the body of the laying bird. Lastly, young birds commonly lay smaller eggs than older birds of the same species. Eggs of the Razorbill, *Alca torda*, increase in size progressively with age, at least until the bird's fifteenth year (Lloyd, 1979). There are, however, exceptions to all these general rules. A special case of adaptive egg size is found in the European Cuckoo, *Cuculus canorus*, whose unusually small eggs match in size those of the hosts whose nests the Cuckoo parasitizes. Even within a single species, eggs may vary in size. The eggs of the Great Egret, *Casmerodius albus*, are regularly larger in Europe than those of the same species in India. Whether this is caused by an environmental or a hereditary influence has yet to be determined.

FIGURE 15–2 The Greenshank, an Old World sandpiper, whose four large eggs hatch out into precocial young. Photo by E. Hosking.

One important aspect of egg size, especially the amount of yolk fat, was revealed in studies by Parsons (1970, 1975) of the Herring Gull, *Larus argentatus*. He found a positive correlation between egg volume and post-hatching survival of the chicks. In a typical clutch of three eggs, the last egg laid was usually the smallest. Almost every chick that hatched from an egg smaller than 65 cu mm died shortly after hatching. The significant factor was shown to be the amount of residual fat still available to the chick at hatching. The higher mortality of chicks from last-laid eggs was also due in part to their later hatching and subsequent competition with their older siblings.

The shape of a bird's egg is probably acquired while it is in the magnum of the oviduct. The diameter and muscular tension of the walls of the oviduct, as well as the distribution or packing of visceral organs, probably play a part in determining an egg's shape. There is evidence that the shape of the pelvic bones is related to the shape of a bird's eggs: birds with deep pelves are likely to lay eggs that are nearly spherical in shape, while birds with dorsoventrally compressed pelves incline toward elongated eggs (Rensch, 1947) (Fig. 15–3).

Probably the shapes of bird's eggs are of little significance except for the pyriform or top-shaped eggs of shore birds, auks, and guillemots. These eggs, rather pointed at one end and broad at the other, will roll in a small circle on the ground rather than in a straight line—a valuable feature for the guillemots whose one or two eggs are laid on bare rock ledges on ocean cliffs (Ingold, 1980). A further advantage in such eggs is found in the compact way they pack together when there are three or four in a nest—common numbers for many shore birds. If the four eggs in the nest of a Killdeer, *Charadius vociferus*, are disarranged, the bird will rearrange them with pointed ends inward much like the slices of a pie. Not only is the parent better able to cover its eggs, but the heat they receive from its body is dissipated less rapidly, thanks to their compact positioning.

Most birds' eggs have the familiar oval shape of the hen's egg, slightly broader at one end than the other. In some species, such as doves and goatsuckers, the two ends of the egg are equally rounded and the general egg shape approaches an ellipsoid. Tinamou eggs are conical on both ends. Among many hawks and eagles the egg is a short oval approaching the nearly

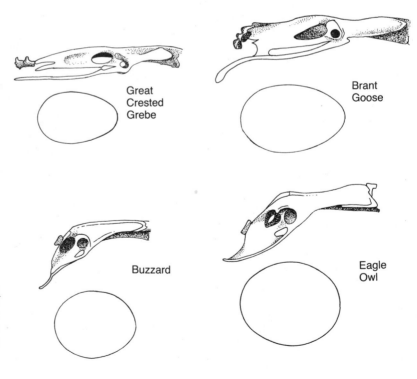

FIGURE 15–3 In some birds the shape of the egg correlates with the shape of the pelvis, especially with the position of the backward-bending ischial bones: the deeper the pelvis, the rounder the eggs. After Rensch, 1947.

spherical egg laid by owls, toucans, kingfishers, and bee-eaters. Long, elliptical eggs are characteristic of the streamlined, rapidly flying swifts, hummingbirds, and swallows (Preston, 1969).

In surface texture most birds' eggs are smooth and have a dull matte finish like that of the domestic hen. Different textures are found, however, in numerous species. The eggs of some ostriches, storks, and toucans are deeply pitted, while those of the emu, cassowaries and Chachalacas, *Ortalis* spp., are rough and corrugated on the surface. Also rough-surfaced are the eggs of grebes, boobies, flamingos, and certain cuckoos. Anis have eggs with a chalky surface layer that easily rubs off and exposes a bluish or greenish deeper layer. In the spectacular egg of the Guira Cuckoo, *Guira guira*, the white chalky layer is deposited in the form of a coarse lattice or network that overlies a deep blue background layer. Another striking variation is found in the tinamous' eggs with their glossy, porcelain-like finish. Woodpeckers also have glossy eggs, and the eggs of many ducks have a greasy, water-repellent surface.

The developing embryo metabolizes and discharges metabolic wastes, which necessitates gaseous exchange (oxygen and carbon dioxide) and water vapor conductance through the eggshell (Carey, 1983). This is achieved through pores in the eggshell, and the rate of gaseous exchange is affected by the ratio of functional pore area to shell thickness (Ar *et al.*, 1974). Eggs of Pied-billed Grebes, *Podilymbus podiceps*, and Black Terns, *Chlidonias niger*, are laid in nests constructed of wet vegetation, creating a highly humid microclimate that may slow down the rate of gas exchange through the egg. To compensate for this, the number of pores per unit area is greater than that in eggs found in drier environments (Davis *et al.*, 1984; Davis and Ackerman, 1985). The Bonin Petrel, *Pterodroma hypoleuca*, with an unusually long incubation period of 48.7 days, is faced with the problem of reducing water loss. By reducing the number of pores in its eggshell, it lowers daily rates of water loss (Grant *et al.*, 1982).

COLORS OF EGGS

Studies of the chemistry of egg pigments indicate that they are mainly derived either from the blood pigment hemoglobin or from bile pigments that are decomposition products of hemoglobin. According to Völker (*in* Makatsch, 1952) there are two chief egg pigments: por-

phyrins, derived from hemoglobin, which are responsible for the brown and olive colors, and cyanin from the bile, which makes blue and green pigments. Pigments may appear singly or together in an egg, or not at all. They may occur in different layers, or penetrate throughout the shell as in tinamou eggs. In the cassowary egg, the outside of the shell is uncolored; the inside layer is green. In certain hawks and passerines, only the very outside of the shell is pigmented. All pigments are secreted by the walls of the oviduct, particularly in the region of the uterus.

In many species, the egg is of a uniform color or tint: blue or greenish-blue in many cormorants, thrushes, herons, and starlings; ivory in the ostrich; pale green in many ducks; dark green in the emus; olive brown in some bitterns, and nearly black in the Chilean Tinamou, *Nothoprocta perdicaria*. Solidly black eggs have been collected in northeastern South America for many years, but the identity of the bird laying them remains a mystery. In Europe, Cetti's Warbler, *Cettia cetti*, lays eggs that are colored deep brick-red. The eggs of many birds are white or nearly so. Some birds occasionally forsake their characteristic egg color and lay reddish eggs, only to return to the usual color later. This inexplicable aberration is known as erythrism and is seen in the eggs of the Herring Gull, *Larus argentatus*, American Crow, *Corvus brachyrhynchos*, Red-backed Shrike, *Lanius collurio*, and other species. Erythrism is common among Australian birds and also among Indian species in regions with red soil. A contrary phenomenon occurs when birds that normally lay pigmented eggs lay white or albino eggs completely lacking in color. In a survey of North American species, Gross (1968) found 34 species that occasionally laid albino eggs.

On these various ground colors—usually white, gray, cream, brown, red, blue, or green—there are arranged in many species blotches, specks, or streaks of brown, red, lavender, gray, black, and other colors. Or, as in the beautiful egg of the Kingbird, *Tyrannus tyrannus*, the spotting may be made of combinations of different colors. The spots may be large or small, many or few. Often, as in the eggs of many sparrows, the spots are concentrated in a wreath around the larger end of the egg. Spots apparently result if the egg stands still while small pigment glands in the wall of the uterus apply the color. If, on the contrary, the egg moves, lines of various shapes—spirals, rays, or random scrawls—result. The blue egg of the Red-winged Blackbird, *Agelaius phoeniceus*, is marked with dark purple or black scrawls that indicate that the egg moved about erratically in the uterus while the pigment glands secreted their color.

Ordinarily, the eggs of a given kind of bird are sufficiently uniform and fixed in coloration so that a student can tell a species by its eggs. In other words, egg pigmentation is largely genetically determined. A study of 810 eggs laid by 35 female Village Weaverbirds, *Ploceus cucullatus*, showed that color and spotting of eggs are inherited. Two pairs of Mendelian alleles were responsible for hue and one pair for spotting (Collias, 1983). However, variations do occur within species and genera that are at times more extreme than the fluctuations between the familiar white and buff eggs of the domestic hen. The highly burnished eggs of the several species of South American tinamous vary "from pale primrose to sage-green or light indigo, or from chocolate-brown to pinkish orange" (Newton, 1896). The Eastern Bluebird, *Sialia sialis*, which normally lays blue eggs, shows a modest variability in color. Of 774 eggs laid in nest boxes in a Nashville, Tennessee, park in 1952, 71 (9.1 per cent) were white; the rest, blue (Laskey, 1943). Probably the most extreme intraspecific variability in egg color is found in the Common Murre, *Uria aalge*. This species lays a large, pyriform egg whose ground color may be "deep blue-green, bright reddish, warm ocherous, pale bluish, creamy or white, usually marked richly or sparingly with blotches, spots, zones or intricate patterns of interlacing lines which vary in color from light yellowish-brown to bright red, rich brown or black, sometimes quite unmarked" (Jourdain, *in* Witherby *et al.*, 1949). Since these birds nest in enormous colonies where each female lays her single egg on an exposed rock ledge, this color variability is apparently an adaptation to help a given bird identify its own egg or nest. Extensive experiments by Tschantz (1959) showed that a bird could be induced to accept a strange egg only if the color and pattern were similar to its own. A different color or pattern caused the egg to be rejected. Contrariwise, some gulls will accept and incubate grotesquely colored and shaped artificial eggs.

The hereditary plasticity of pigment colors in a Chilean race of the domestic Chicken, *Gallus gallus*, was demonstrated in only 20-some years of crossbreeding and genetic selection (Vosburgh, 1948). Individual strains were developed that laid eggs completely covered with tints of blue, green, pink, olive, white, and brown. A given hen laid eggs of only one color.

Even an individual bird may lay eggs of different colors. Sometimes such a change may occur with age, but in some instances eggs of the same clutch may vary in color. A captured bird will at times lay an egg deficient in coloration, indicating either that disturbance and captivity have arrested the secretion of pigment, or that the egg, laid prematurely, did not remain in the uterus long enough to receive its full quota of pigment.

Originally bird eggs were probably white, as reptile eggs now are. With time, natural selection may have favored colored bird eggs because of their inconspicuousness. This explanation gains support from the fact that white eggs are generally restricted to hole nesters (swifts, owls, most petrels, some doves, parrots, woodpeckers, kingfishers, bee-eaters, rollers, and others); or to open nesters that begin incubation after they lay the first egg (some doves, herons, hummingbirds, owls, grebes); or to those open nesters that cover their eggs with down or vegetation when they leave the nest (some ducks, geese, grebes, and many gallinaceous species) (Stresemann, 1927–1934). Most other open nesters, particularly those laying their eggs on the bare ground (excepting those like the Guillemot that nest on inaccessible ledges), lay concealingly colored eggs.

Experimental evidence that concealing coloration actually protects eggs from predators was obtained by Tinbergen and others (1962). On the edges of rookeries of Black-headed Gulls, *Larus ridibundus*, they laid out, in equal numbers, natural spotted gull eggs, uniformly white eggs, and uniformly khaki-colored eggs. The natural eggs suffered less predation than either of the plain-colored eggs from predatory Herring Gulls, *L. argentatus*, and Carrion Crows, *Corvus corone*.

In western India the Yellow-wattled Lapwing, *Vanellus malabaricus*, has eggs that closely match the color of the bare ground on which they are laid. One region along the Malabar Coast has a brick-red, sandy, laterite soil, scattered through which are black nodules of ironstone. In this particular region the Lapwings lay red eggs with brownish specks "exactly like the ground on which they are deposited" (Backer, 1923). Probably the most highly adaptive egg coloration in the world is that found in the European Cuckoo, *Cuculus canorus*. Different races of this species lay eggs whose colors and markings match, often with astonishing fidelity, the eggs of the host species whose nests they parasitize.

THE TIME OF LAYING

The laying of eggs is of course part of the reproductive cycle that, as shown in Chapter 8, is under the close control of the endocrine glands. The activity of the glands, in turn, is responsive to a variety of environmental stimuli: length of day, precipitation, temperature, food availability, psychic stimuli, and perhaps others. Long-continued natural selection has caused each species to lay its eggs at such a time that the young will hatch when conditions are most favorable for rearing them. In species with a long incubation period, this may mean that the eggs must be laid during an inhospitable season, as is the case with the Great Horned Owl, *Bubo virginianus*, which in Florida lays its eggs in mid-December, and in Pennsylvania and Iowa lays them in mid-February. By the time the eggs hatch, the spring crop of rodents is becoming available as food for the hungry young owls. Similarly, the Emperor Penguin, *Aptenodytes forsteri*, lays its single egg in late June or July in the depth of the dark antarctic winter. When the young hatch out 53 days later, it is still bitter winter, with temperatures as low as $-50°C$. In the relatively mild spring and summer days that soon follow, the young are able to fatten up on the abundant food of polar waters, but barely in time to reach independence by the onset of the next winter. That the time of egg-laying and the advance of spring are closely synchronized is seen in the nesting records of the British Trust for Ornithology for the Meadow Pipit, *Anthus pratensis*. At sea level, the bird lays its eggs an average 3.8 days earlier in the south of Britain than in the north, and it begins laying 1 day later for each 40 meters rise in elevation (Coulson, 1956). However, the advance of spring is not rigidly controlled by the calendar, and it may vary from year to year. In their study of the Alpine Swift, *Apus melba*, Lack and Arn (1947) discovered that the species began nesting around May 17 during nine years of good spring weather, and around May 31 in eight years of cold, wet weather.

Much later than other falcons, Eleonora's Falcon, *Falco eleonorae*, begins laying its eggs around the first of August on the islands and shores of the Mediterranean. Here the timing is dictated by food availability. Until late in the summer, small birds, the preferred food of this falcon, are very scarce on these islands, but in the early fall, hordes of small migrating passerines throng the islands and furnish food at just the right time for the growing young falcons (Stresemann, 1927–1934; Walter, 1970). In the northern Sahara the Lanner Falcon, *Falco biarmicus*, feeds its young on north-bound migrants in April and May, probably as an adaptive timing to avoid competition with Eleonora's Falcon (Niethammer, *in* Dawson, 1976).

A remarkable constancy in the annual time of egg-laying is shown by the Short-tailed Shearwater, *Puffinus tenuirostris*, that breeds on the islands of Bass Strait, Australia. After spending their winter season in the far north Pacific, these birds return to their breeding colonies where the great majority lay their single eggs on or within two days of November 25, year after year (Serventy, 1963). This rigid regularity in breeding chronology is apparently caused by some innate rhythm in the bird locked in phase with some as yet unknown astronomical cue.

Numerous population studies have shown that for most species living in regions with contrasting seasons there is usually one limited time of year during which a given species enjoys optimum reproductive success. Studies of the European Oystercatcher, *Haematopus ostralegus*, by Harris (1969) have shown that eggs hatched before May 20 are 88 per cent successful as compared with 38 per cent after June 30. Of the May-hatched chicks, 67 per cent fledged; of the later ones, only 33 per cent fledged. Statistics such as these show one mechanism by which natural selection regulates the times of avian breeding seasons.

Not only may environmental stimuli initiate egg-laying in birds, but they may cause a suspension of laying. If, in the course of laying, the nest of a Serin, *Serinus serinus*, is destroyed, the bird will cease laying until a new nest is constructed (Benoit, 1950).

Like the domestic hen, most passerine species lay their eggs at one-day intervals until the clutch is completed. Usually larger species require longer intervals between eggs than smaller species. Since an ovum will not ordinarily be released by the ovary and enter the oviduct until the previously formed egg has been laid, the egg interval of a species depends chiefly on the time it takes the oviduct to secrete the various egg layers around the ovum. Often the last eggs of a clutch come at greater intervals than the first ones. The European Oystercatcher, *Haematopus ostralegus*, for example, lays its second egg slightly more than 24 hours after the first, but the third and fourth eggs come at intervals of 48 hours or more. Cold, wet weather, as well as psychic influences, may delay the normal pace of egg-laying. The egg interval may on rare occasions be shortened, as was the case with a Northern Flick-

TABLE 15–1.
INTERVALS BETWEEN SUCCESSIVE EGGS IN A CLUTCH

24 hours: most passerine birds, many ducks, some geese, domestic hens, woodpeckers, rollers, the smaller shore birds, and smaller grebes.

38 to 48 hours: ostriches, rheas, larger grebes, some ducks, swans, herons, bitterns, storks, cranes, bustards, doves, some accipiters, owls, some cuckoos, hummingbirds, swifts, kingfishers.

62 hours: some cuckoos and goatsuckers.

3 days: emus, cassowaries, penguins *(Pygoscelis)*.

4 to 5 days: Lammergeier, *Gypaëtus barbatus*, Spotted Eagle, *Aquila clanga*, penguins *(Eudyptes)*.

5 days: condors, kiwis.

5 to 7 days: Booby, *Sula cyanops*, some hornbills.

4 to 8 days: some megapodes.

er, *Colaptes auratus*, which laid two eggs within 13 hours (Sherman, 1910). The examples of egg intervals which follow are taken largely from Stresemann (1927–1934) and Makatsch (1952).

Eggs are laid by many species early in the morning, even before sunrise. As Schifferli (1979) explains, this is advantageous because an egg is most susceptible to damage when the shell is being formed, and the ideal time for shell formation is at night when the bird is inactive. There are, however, many deviations from this schedule. In Swedish Lapland, with its short summer, the Redwing, *Turdus iliacus*, lays five to six eggs in four to five days, at intervals of about 20 hours. This adaptive acceleration in egg-laying of course requires egg-laying at different times of day (Arheimer, 1978). The Ringed Plover, *Charadrius hiaticula*, lays its eggs at all hours of the day and even at night. The European Cuckoo, *Cuculus canorus*, seems to prefer the afternoon. Many pheasants and the Common Gull, *Larus canus*, may lay their eggs in the evening; the American Coot, *Fulica americana*, shortly after midnight. Painstaking studies by Skutch (1952) of Central American birds showed that tanagers, finches, wood warblers, honeycreepers, wrens, and hummingbirds usually lay early in the day, from before sunrise to soon after. Tyrant flycatchers often lay later in the forenoon. Salvin's Manakin, *Manacus aurantiacus*, and two species of Crotophaga lay around midday, and the Pauraque, *Nyctidromus albicollis*, late in the afternoon. For a species whose egg interval does not

correspond with the 24-hour solar day, the time of laying on successive days will vary. The domestic chicken and the Bobwhite, *Colinus virginianus*, lay their eggs somewhat later each day until laying occurs in the evening. Then the birds skip a day and lay the next egg in the early morning.

To lay a single egg requires only a few seconds in the case of brood parasites like the Brown-headed Cowbird, *Molothrus ater*, or the European Cuckoo, but 3 to 10 minutes for the Bobwhite, and an hour or so for geese and turkeys. The Prairie Warbler, *Dendroica discolor*, apparently has no voluntary control over laying. Nolan (1978) was able to capture birds

FIGURE 15–4 A female South American Rhea laying an egg. The Rhea is polyandrous, several females laying eggs in one nest. A single male incubates the eggs and rears the young. Photographs, courtesy of Donald Bruning; from *The Living Bird*.

at the crucial moment and have them lay eggs in his hands. Typically, a female, resting on her heels and with body almost upright and feathers fluffed, would begin to strain rhythmically, moving the anterior part of the body slightly forward every 2 or 3 seconds. With each movement the vent opened progressively larger. After about 5 to 15 straining movements, persisting about 30 seconds, the egg emerged.

CLUTCH SIZE

There is great variation in the number of eggs that different birds lay. Some lay only a single egg: the larger penguins, petrels, albatrosses, guillemots, some larger vultures, most puffins, the larger doves, several swifts and goatsuckers. Two eggs per clutch are laid by the kiwi, most penguins, loons, boobies, many eagles, cranes, some auks, most pigeons and doves, some goatsuckers and hummingbirds, and many passerines in the tropics; three eggs: most gulls and terns; four eggs: most snipe, plovers and sandpipers; four to six eggs: most passerine species at higher latitudes; 8 to 12 eggs: ducks, gallinaceous birds, titmice; 9 to 23 eggs: the European Partridge, *Perdix perdix*. One might expect those species that lay many eggs to be more successful in reproducing their kind than those that lay few eggs, but this does not seem to follow.

Certainly, for most species, clutch size is determined by heredity, probably as a consequence of years of natural selection (Klomp, 1970; Perrins and Jones, 1974). For some species, such as the extinct Passenger Pigeon, *Ectopistes migratorius*, many petrels, shearwaters, and albatrosses, clutch size is constant. In other birds it may be more flexible. A study by Koenig (1984) of 411 egg-sets of the Northern Flicker, *Colaptes auratus*, revealed clutches ranging in size from 3 to 12, with the majority being between 3 and 9.

Certain regular patterns in clutch-size variation suggest a number of factors, chiefly environmental, which may influence the number of eggs a given bird or a given species may lay. Birds in northern latitudes, for example, often lay larger clutches than their relatives in the tropics. The factors in variation may either be ultimate factors that operate through natural selection on the hereditary control of clutch size, or proximate factors that operate more directly on the physiology of egg production. Among factors that have been suggested as having some control over clutch size are the following: heredity; size of bird; age of bird; type of nest, whether open or enclosed; size of nest; size of eggs; the number of eggs the bird can successfully incubate; a "feel" for a certain number of eggs; the number of young that can be successfully raised; the number of broods per season; intensity of predation; population density; geographical distribution; migration; time of year; length of breeding season; length of daylight; climate and weather; type of habitat; food: its nature, abundance, distribution, and availability; the place, time of day, and method of collecting food; experience in collecting food; and the number of parents caring for the young. Certain correlations seem rather obvious and probably represent cause-and-effect relationships, whereas others may be due to subtle mixtures of several variables.

In many instances, larger species lay fewer eggs than their smaller relatives (Moreau, 1944). For example, the Goshawk, *Accipiter gentilis*, lays three or four eggs per clutch while the smaller Sparrow Hawk, *Falco sparverius*, lays five to seven. The large Black Woodpecker, *Dryocopus martius*, lays an average of four eggs, and the Lesser Spotted Woodpecker, *Dendrocopus minor*, six. Possibly this relationship reflects the inability of smaller birds to combat predators as successfully as larger species, and therefore represents an adaptive compensation in clutch-size for a heavier loss of young.

Age also plays a role. Young birds often lay smaller clutches than older birds, perhaps because "in their first adult spring they are inexperienced and find less food than old females" (von Haartman, 1971). Among Lake Ontario Ring-billed Gulls, *Larus delawarensis*, clutch size increases with parental age up to age 4, and then levels off. Hatching success increases up to age 5 (Haymes and Blokpoel, 1980).

Nest size in itself may limit the size of a clutch, and small nests may be completely hidden under the female. Campephagids laying and hatching only one egg build tiny nests that are much more difficult to discover than those of species laying two eggs (Stresemann, 1927–1934).

An experiment that convincingly demonstrated a correlation between nest size and clutch size was performed by Löhrl (1973) in German forests. He put up nest boxes of two different inside diameters: 9 and 20 cm. During a period of two years Great Tits, *Parus major*, laid and hatched more eggs per nest in the larger boxes than in the smaller ones. Nestboxes of two sizes were exchanged after egg-laying had begun. A significant correlation between clutch size and size of cavity was once again noted, indicating that information regarding cavity size was obtained after the onset of laying (Löhrl, 1980). There were no differences in hatching success or nestling mortality between the clutches in the two box types, nor were there differences in the weights of the young. Clearly, the reproductive success of the tits was higher in the larger boxes than in the smaller.

In certain species the temperature regulation of nestlings has important bearings on clutch size. Artificially contrived broods of various numbers of Great Tit nestlings were exposed to cool ambient temperatures (12°C) by Mertens (1969), who reported that broods of one died of chilling, but broods of 2 to 12 produced heat and lost water in proportions correlating with brood size. When van Balen and Cavé (1972) held broods of young Great Tits at an initial air temperature of 30°C, the temperature in a nest box containing 16 young increased to 39.1°C, whereas that in a box with six young increased to only 34.3°C. At an initial warm-summer-day temperature of 35°C, maximum survival of young was obtained in broods of 11. A normal brood has eight to ten young.

Clutch size and egg size seem to be related in a complementary fashion, but the significance of the relationship is not clear. Of two related goatsuckers in India, the larger species, *Batrachostomus javensis*, lays two eggs; the smaller, *B. monileger*, lays one, which, however, is considerably larger than one egg of *javensis* (Stresemann, 1927–1934).

Even the "feel" of eggs against the belly of a bird seems to influence clutch size. The female Herring Gull, *Larus argentatus*, has three bare brood patches and normally lays three eggs. If the number of eggs is experimentally manipulated to become more or less than three, normal incubation behavior becomes dramatically interrupted with other activities, and, probably as a consequence, predation on the eggs increases (Baerends and Drent, 1970). Sandpipers and many other shorebirds lay and incubate four conical eggs that fit snugly together like segments of a pie cut into four pieces. When Hills (1980) added a fifth egg to each of 31 nests of six species of shorebirds, an average of only 2.55 young were hatched from each experimental nest as against 3.86 young from each of 47 normal nests. The incubating birds were apparently unable to cover efficiently the asymmetrically arranged five-egg clutches.

It is obvious that there must be some limit to the number of eggs a given bird can lay, cover, and suc-cessfully incubate. However, this limit does not seem to have been reached, at least in some species. Wagner (1957) demonstrated experimentally that two species of Mexican finches could successfully incubate and rear clutches of four or five, instead of two or three, eggs, suggesting that, in these species, clutch size was determined by other factors. By manipulating the numbers of eggs per clutch in 223 nests of the Coot, *Fulica americana*, Frederickson (1969) demonstrated that Coots could incubate and hatch experimental clutches of up to 21 eggs as successfully as normal ones of about 9 eggs.

In different habitats a variety of environmental factors are combined, any one of which might have some influence (ultimate or proximate) in regulating the clutch sizes of birds living there. The problem is, of course, to determine the amount of influence of each environmental factor. This is an extraordinarily difficult problem and much research remains to be done before solidly conclusive answers can be provided. Currently, multi-variate statistical analyses are being applied to such problems in attempts to discriminate those factors that are most influential in determining clutch size. So far, very few general principles having wide application to this problem have been discovered. In the numerous examples that follow, some idea may be gained of the many and often subtle variables that are involved in clutch size determination.

Among North American blackbirds of the family Icteridae, marsh residents average 4 eggs per clutch; forest residents, 4.25; parkland birds, 4.8; and grassland species, 5.25 (Cody, 1971). Surprisingly, the marsh dwellers suffer the highest predation losses, yet have the smallest clutches. Lack (1968) surveyed some 300 species of tropical African passerine birds of nine different families and noted that in every family but one the clutch size of savanna-dwelling species was larger than that of their evergreen-forest-dwelling relatives. Lack attributes the larger clutches of savanna birds to the severely contrasting seasons of savannas—dry seasons, sometimes with grass fires, alternating with food-rich rainy seasons when the birds breed. The abundant food of the rainy season enables the birds to produce the overpopulation needed to compensate for the heightened mortality caused by the austerities of the dry season. A probable contributing factor is the concentrated availability of food in the essentially two-dimensional savanna compared with the scattered food of the three-dimensional forest.

Island species of birds commonly have smaller

clutches than their closest mainland relatives. Cody (1971) believes this is because islands are likely to have fewer predators, a more equable climate, and larger ecological niches than the mainland. A curious exception to this rule is found in the Scarlet Ibis, *Eudocimus ruber*, which commonly lays three eggs on the island of Trinidad but only two on mainland Surinam.

Geographical variation in clutch size is known to exist in many species. As one moves away from the tropics toward the poles, the clutch size of geographical races of a given species (or of species of a given genus) increases strikingly. This is true for birds of various orders: passerines, gallinaceous birds, owls, hawks, rails, gallinules, herons, and others. Many tropical American tyrant flycatchers, troupials, tanagers, and finches lay only two eggs, but their temperate-zone relatives lay four to six eggs (Skutch, 1949). The European Robin, *Erithacus rubecula*, lays average clutches of 3.5 eggs in the Canary Islands, 4.9 in Spain, 5.8 in Holland, and 6.3 in Finland (Lack, 1953). In another survey of some 300 species of passerine birds belonging to 11 families, Lack (1968) noted that the average clutch size of those family representatives living in central Europe was larger than that of their tropical African relatives—usually by two, and in some families by three or more eggs.

This widespread and striking increase in clutch size with increase in latitude seems to be due, in part, to at least four influential factors: latitudinal differences in mortality, in seasonal food abundance, in day length, and in length of the reproductive season. For birds residing in temperate and arctic zones, the hazards of winter climate must increase the mortality rate; if the birds migrate south for the winter, the sometimes greater hazards of migration have the same effect. Clutch sizes in European titmice seem to bear this out. The sedentary Marsh Tit, *Parus palustris*, Willow Tit, *P. atricapillus*, and Crested Tit, *P. cristatus*, generally have one brood a year and average eight eggs per clutch. The migratory Great Tit, *P. major*, Blue Tit, *P. caeruleus*, Coal Tit, *P. ater*, and Long-tailed Tit, *Aegithalos caudatus*, commonly have two broods per year and average about 12 eggs per clutch (Steinfatt, 1938). In contrast to the dangers of a northern winter, the benefits of a northern summer are equally great. Birds breeding in northern latitudes, their numbers greatly reduced by the rigors of winter, are assured abundant food, particularly protein-rich insect food, and daylight up to 24 hours long for feeding their augmented broods. Birds settled

throughout the year in the tropics are exposed neither to the catastrophes of winter nor to those of migration, and accordingly have less need for large clutches or a high reproductive rate. Paradoxically, the great abundance of predators in the tropics places a premium on small, inconspicuous nests. And small nests require small clutches, which, if destroyed by predators, are replaced at less cost to the breeding bird than are large clutches (Snow, 1978). Even with small nests and only two eggs, tropical manakins commonly suffer over 80 per cent nesting failures from predation (Lill, 1947b).

An unintentional experiment that supports this day-length, clutch-size argument arose from the transplantation of the European Goldfinch, *Carduelis carduelis*, to Australia about 100 years ago. Today, in Australia, this species averages 3.7 eggs per clutch; in ancestral Britain, the species averages 5 eggs. This shrinkage in the clutch size of the Australian Goldfinch is thought by Frith (1957) to be caused at least in part by the shorter day length in Australia for food-collecting, which in turn reduces the number of young that can be raised. From extensive studies of equatorial African birds representing 22 families, Moreau and Moreau (1940) reported that during the breeding season these birds had "a working day more than 30 per cent shorter than the average in the British nesting season" and that "practically without exception African broods run smaller in number than those of allied Temperate Zone birds."

Having noted that clutch size in White-crowned Sparrows, *Zonotrichia leucophrys*, increased latitudinally and altitudinally, Morton (1976) developed the thesis that the short reproductive season is the key force selecting for larger clutches at higher latitudes or altitudes. The short breeding season in these regions permits the successful production of only one brood. Multiple brooding occurs below 49°N among coastal populations, and may occur at times below about 55°N in montane populations. At high altitudes, testicular regression and concomitant lowered male fertility occur once the first brood is fledged, a mechanism ensuring single-broodedness. Breeding events such as nest-building, laying, and fledging age are all telescoped in time in Arctic regions. On the other hand, fledging success is higher in Arctic populations than in populations farther south, e.g., it is 75 per cent in Mountain Village, Alaska, but only 40 per cent in Berkeley, California. In sum, birds at high latitudes and altitudes put more eggs in one basket; they must breed quickly, produce as much as they can and pre-

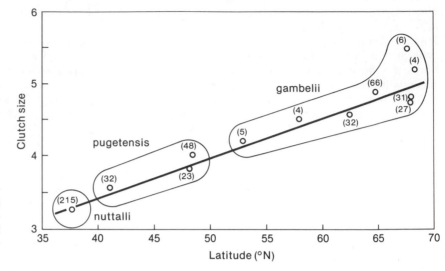

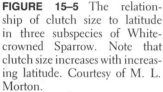

FIGURE 15–5 The relationship of clutch size to latitude in three subspecies of White-crowned Sparrow. Note that clutch size increases with increasing latitude. Courtesy of M. L. Morton.

pare for migration. The milder climates of lower latitudes and altitudes permit smaller clutches, because failure of one clutch may permit the birds to try again and again through the prolonged breeding season. In these regions as many as three broods a year may be raised.

Regional variations in clutch size also occur among populations of the same species living at the same latitude. In southern Africa the Palm Swift, *Cypsiurus parvus*, lays two eggs; but in Madagascar, three. Contrariwise, the average clutch of many passerines, hawks, owls, and gallinaceous birds in central Europe is about one-half an egg larger than the average clutch of the same species in England. This is thought by Lack (1954) to be caused by a difference in food supply, especially in insects, which are probably more abundant in drier central Europe than in rainy England.

Climatic conditions influence clutch size in certain species, probably in most instances through controlling their food supply, but also probably through the direct influence of weather on egg-laying. Cold, humid weather may cause a reduction or even a complete suspension of laying. Swifts and swallows that feed on air-borne insects lay larger clutches in fine sunny seasons than in cold, wet ones. The weather in the canton of Vaud, Switzerland, in 1948, was unfavorable and the mean clutch size of 40 nests of the Barn Swallow, *Hirundo rustica*, was 4.0 eggs. In the favorable summer of 1951, the mean clutch size

for 70 nests was 4.6 eggs (Nicod, 1952). In a contrary way, wet seasons in dry regions may bring more food, and with it, larger clutches. A four-year study of the Horned Lark, *Eremophila alpestris*, in Montana, showed that this species laid larger clutches in wet years than in dry (Dubois, 1936). In Tunisia, a severe drought in 1936 reduced the average clutch size of three species of larks from the normal five or six eggs to two or three (De Guirtchitch, 1937).

Clutch size may also vary seasonally. In many species that lay two or more clutches per year, the first clutch is usually larger than later ones. First clutches of the Blue-winged Teal, *Anas discors*, in Iowa average about 9 eggs; second clutches, about 4 eggs. Early clutches of the Ruffed Grouse, *Bonasa umbellus*, in Ontario average 11.9 eggs; late clutches, 8.5 eggs. First clutches of the Great Tit, *Parus major*, in Germany average 8.6 eggs, replacement clutches average 7.4, and second clutches average 6.8 (Schmidt and Steinbach, 1985). Clutch size may be adapted primarily to day length. Seasonal shrinkage in clutch size may well be caused by different combinations of influences in different species: shorter day length, diminishing food supplies, exhaustion of stored energy (fat) in the laying female, tapering off of the reproductive endocrine cycle, and others.

Among certain species, more food definitely means more eggs. Snowy Owls, *Nyctea scandiaca*, lay twice as many eggs in good lemming years as in poor years. Hawks and owls generally have larger clutches in years when mice are abundant. In Europe, Magpies, *Pica pica*, may increase their clutches by an egg or two in years when June beetles are abundant. In Western Australia there has been an increase in the average clutch size of the Little Eagle, *Hieraëtus morphnoïdes*, Wedge-tailed Eagle, *Uroaëtus audax*, and Australian Goshawk, *Accipiter fasciatus*, ever since the rabbit, introduced in 1859, spread into their range (Serventy and Whittell, 1951). It also seems likely that the type of food a bird eats may influence its clutch size. The corn-eating Scaled Quail, *Callipepla squamata*, of Mexico normally lays 10 to 13 eggs, while the Spotted Wood Quail, *Odontophorus guttatus*, which searches all day for insects, worms, and the like, has a clutch of 4 to 6 eggs (Wagner, 1957).

Clutch size also reflects the way in which a parent bird gathers feed for its young. Crepuscular feeders like the Whip-poor-wills, *Caprimulgus vociferus*, nightjars, and thick-knees, that collect food only during the brief twilight hours of morning and evening, generally lay only two eggs. Sea birds that find food far from their breeding grounds typically lay only one egg—the extensive travel involved in collecting and delivering food to the young requires so much time that feeding two or more young would be an impossibility. In fact, nearly 70 per cent of the single eggs of the Galapagos Red-footed Booby, *Sula sula*, are abandoned before hatching, presumably because the wide-ranging foraging parent is unable to return on schedule to replace its incubating partner (Nelson, 1969). Intertidal and inshore feeders, on the contrary, lay two or more eggs. Their food resources are nearer at hand and probably richer than those of off-shore feeders, so they can feed their young much more frequently (Lack, 1967). In an experiment that supports this interpretation, four Lake Ontario Herring Gulls, *Larus argentatus*, were tracked by radiotelemetry as they foraged for food to feed their young. Two individuals that daily secured their food within 1 km of their nests raised their young successfully; two that foraged more than 30 km away lost their chicks early in the brooding period (Morris and Black, 1980). The European Nutcracker, *Nucifraga caryocatactes*, bases its clutch size on the quantity of nuts and seeds stored from the previous year. When nut stores are poor, the bird will lay a clutch of three eggs; with good stores, whether due to a natural crop or to nuts supplied experimentally, the bird will lay four eggs (Swanberg, 1951).

It seems improbable, in these correlations between clutch size and food supply, that there is necessarily a direct physiological causal connection between the two. In some instances, the larger clutches are already laid before the abundant food is at hand. Most likely the birds respond to anticipatory stimuli that in some manner presage an increase in food abundance. A seven-year study of British titmice, by ornithologists of the Edward Gray Institute of Ornithology at Oxford, has shown that not only do clutches increase in size in those years that caterpillars are more abundant, but that the young titmice appear each year just as the caterpillars emerge in numbers. These two events vary from year to year as much as a month, and yet they always occur synchronously. These parallel happenings both correlate with, and possibly are set in motion by early spring temperatures (Lack, 1955).

Great Tits, *Parus major*, feed their young mainly on leaf-eating caterpillars that are abundant early in the season when first-brood young hatch out, but are scarce later when second-brood young appear. Second broods are attempted only about 7 per cent of the time. First clutches average about ten eggs; second clutches, about seven. Of hatched nestlings 95 per cent flew from first broods, but only 37 per cent from second broods, the majority of the young starving (Lack, 1954). The rarity, small size, and high mortality of second broods in this tit make clear the kind of selective pressures that dictate the size and timing of bird clutches.

Anticipatory weather stimuli are not invariably dependable. Central European Swifts, *Apus apus*, may lay a typical clutch of three eggs early in the summer, but if the weather should turn cold and rainy and reduce the number of flying insects, the parent birds will remove one, two, or three eggs from the nest—"the worse the weather the more eggs"—and drop them to the ground (Koskimies, 1950). This heroic trimming of clutch size is probably an adaptive instinctive act of whose significance the bird is unaware.

A perplexing problem in clutch size is presented by those species that lay more eggs than normally hatch and develop into mature birds. For example, the Brown Booby, *Sula leucogaster*, normally lays two eggs, the second about seven days after the first, but begins incubation with the first egg. As a consequence, the first chick to hatch monopolizes the food and its parents' care, and the second chick, if it ever hatches, dies of starvation, neglect, trampling, or cannibalism. In a large colony of Boobies in the Bahamas,

Chapman (1908) found that less than 1 per cent of the occupied nests contained two live chicks. A similar fate awaits the last-born young of many penguins, pelicans, storks, herons, eagles, hawks, terns, owls, and other species in which the young hatch out at different times. Some observers consider this infanticide a wasteful maladaptation, or at least a Malthusian mechanism to prevent overpopulation. It may, however, be quite the contrary. Lack (1954) makes the point that the staggered hatching of eggs, which results in young of different sizes, may be an adaptation to bring family size into close adjustment with food supply. He says, "In such species the normal clutch tends to be somewhat larger than the parents can raise in an average year, the extra egg or eggs being a reserve that can be utilized in good years."

Still another important factor that probably influences the evolution of clutch size is the intensity of predation that is focused on a species. Young of open-nesting birds are generally subject to heavier predation than those of hole-nesting species. One way to reduce the predation on open nesters is to shorten the highly vulnerable nestling period. In Europe, the average nestling period of open-nesting passerines is around 11 days, while that of the more protected hole nesters is about 19 days—undoubtedly a difference promoted by natural selection. Since a given amount of food per day will permit either a few young to grow rapidly or more young to grow slowly, it is not surprising that open nesters average smaller clutches (5.1 eggs) than hole nesters (6.9 eggs) (Lack, 1954).

A further extension of this principle is illustrated by the open-nesting Anseriformes and Galliformes whose precocial young, even though they leave the nest the day they hatch or the day after, nevertheless have a longer period of dependent infancy than do most altricial, nest-dwelling young. As a consequence they are subject to more hours of predation and other environmental dangers than are altricial young. Accordingly, such birds as ducks, geese, pheasants, and grouse lay unusually large clutches, as if to anticipate the expected losses.

Of 12 European passerine species that have been successfully introduced into New Zealand, 11 have shown a reduction in clutch size. Niethammer (1970) has advanced the theory that this reduction in egg number is caused not by shorter day length but by competition and disturbance caused by greater bird population densities there.

It seems only logical to expect that monogamous species in which both parents feed the young would have clutches about twice as large as polygynous or promiscuous species in which one parent feeds the young. However, this is not generally the case. Monogamous passerines of African savannas have clutches about 40 per cent larger than polygynous Ploceinae of the same habitat. Monogamous Cotingidae have clutches about twice the size of promiscuous species; and 12 monogamous tropical American species average 2.9 eggs per clutch whereas six promiscuous species have 1.2 eggs per clutch (Lack, 1968). However, the difference here may be due to an adaptation to food preference, since the monogamous species eat mostly insects and the promiscuous species mainly fruit.

The influence of type of pair bond on clutch size is minimized by Skutch (1967), who holds that "Among birds of similar size, diet and habitat, there is no correlation between the size of the brood and the number of adults attending it." A survey by von Haartman (1955) showed virtually no difference in clutch size between the monogamous and polygynous species of buntings (Emberizinae), penduline tits (Remizinae), and wrens (Troglodytidae), and only slightly larger clutches in monogamous as compared with polygynous Icteridae. This complex problem requires further study.

Several attempts have been made to discern some general principle that will tie together all these bewildering examples of variable clutch sizes among birds. Lack (1954), while agreeing that various factors may be more or less influential in determining clutch size, proposed that "all factors are related to the maximum number of young that a pair can rear successfully." Experimental efforts to test this theory have yielded equivocal results. Experimentally enlarged clutches of American Coots, *Fulica americana*, have been successfully incubated and hatched (Frederickson, 1969). Morel (1967) found that parents of the Senegal Fire Finch, *Lagonosticta senegala*, could raise a mixed brood parasitized by the Indigobird, *Hypochera chalybeata*, as successfully as a smaller, normal, unparasitized brood. Clutch size was apparently not limited by the number of young the parents could feed. Contrariwise, when Crossner (1977) set up artificial broods of one to ten nestlings of equal-aged Starlings, *Sturnus vulgaris*, their weights progressively decreased as the brood size was increased beyond three nestlings until, in nests with ten chicks, not one survived. If, however, supplemental food was provided, chicks, even in broods of ten, achieved weights equal to those of chicks in normal broods of four or five. These results support Lack's theory of maximized reproduction.

The Gannet, *Sula bassana*, regularly lays one egg and rears one young. Gannets given two eggs reared two young successfully, although the young were slightly underweight and fledged four days late. On the other hand, Brown Boobies, *Sula leucogaster*, and Blue-footed Boobies, *S. nebouxii*, normally lay two eggs but rarely raise more than one young. The Short-tailed Shearwater, *Puffinus tenuirostris*, lays a single egg in its burrow. To 20 occupied burrows, Norman and Gottsch (1969) added one extra egg each. Hatching success in the 20 experimental burrows was only 20 per cent compared to 79.6 per cent in the normal, control burrows.

And, of course, reproductive success is not entirely dependent on the number of young birds fledged per year. Even more important may be their survival to breeding age, and the survival of their parents. In this connection, Lack (1968) tells of a significant experiment in which a male Starling, *Sturnus vulgaris*, worked so hard to feed an artificially enlarged brood of young that it died, apparently of exhaustion.

In considering the entire problem of clutch size, Skutch (1967) points out that the fundamental question is "whether the reproductive rate of animals is always maintained by natural selection at the maximum level that their ability to produce and adequately nourish young permits, or whether it is adjusted to the losses which each species must replace to maintain its population at a favourable level." In short, in Lack's theory of maximum reproduction, mortality is determined by the rate of reproduction, whereas in Skutch's theory of adjusted reproduction, the reproductive rate is determined by the annual mortality. Assuming the latter as true, Skutch suggests the types of mutation that might limit a species' reproductive potential: reduction in clutch size; reduction in the number of broods per year; failure of the male to assist in rearing the young; deferment of reproductive maturity; development of territorialism that limits the number of nests in a region; and others. Whatever the dim origins of clutch size in birds, sufficient problems remain to provoke much productive research in the future.

MULTIPLE AND REPLACEMENT CLUTCHES

If, in the midst of their laying, an egg should be removed from the nest of a Crow, *Corvus brachyrhynchos*, or a Barn Swallow, *Hirundo rustica*, the birds will remain satisfied with the remaining eggs and incubate their clutches of less than normal size. Or, if extra eggs should be added to their nest, the birds will still lay the normal number of eggs for a clutch (five for these species) and try to incubate the abnormally large clutch. Similar experimental manipulations of egg number in the clutches of doves, shore birds, large birds of prey, many passerines, and other species, have failed to change the characteristic number of eggs laid. Even the hatching of previously introduced eggs, or the presence of nestlings being fed by the male, have no effect on the rigidly fixed number of eggs laid by the female Tricolored Blackbird, *Agelaius tricolor* (Emlen, 1941). Species like these that have fixed clutch sizes are known as *determinate* layers. Apparently the number of oocytes in the ovaries that ripen into ova is predetermined for a given season or period of laying, and cannot be modified by external stimlui.

On the other hand, some species are *indeterminate* layers: some penguins, ducks, gallinaceous birds, some woodpeckers, some passerines, and others. In a noted experiment by Phillips (*in* Bent, 1939), a Northern Flicker, *Colaptes auratus*, which normally lays six to eight eggs, was induced to lay 71 eggs in 73 days by removing the eggs as rapidly as they were laid but always leaving one "nest egg." A similarly treated Wryneck, *Jynx torquilla*, whose normal clutch is seven to ten eggs, laid 62 eggs in 62 days; and an ostrich, whose normal clutch is less than 20 eggs, laid 65 eggs in three months in the Marseilles Zoo. The Willow Ptarmigan, *Lagopus lagopus*, normally lays eight or nine eggs, but 11 hens, whose eggs were experimentally removed daily, averaged 16.8 eggs per clutch, and one of them laid 27 eggs before stopping (Höst, 1942). This capacity of indeterminate layers to make up losses in egg number has, of course, been artificially selected by man in domestic fowl. The artificially propagated Japanese Quail, *Coturnix coturnix*, in Japan has been induced to lay 365 eggs in one year (Meise, 1954), and domestic hens, descendants of the Jungle Fowl, *Gallus gallus*, have laid up to 352 eggs in 359 days.

Even though indeterminate layers can lay more eggs than the normal clutch, under natural conditions they all automatically stop laying when the standard number of eggs has been laid. It is thought that the feel of the "proper" number of eggs against the abdominal skin of the bird is somehow relayed to the endocrine glands that cause cessation of ovulation.

This problem was studied in the Black-headed Gull, *Larus ridibundus*, by Weidmann (1956). In this species courtship arouses hormonal states that initiate a sequential development of egg follicles in the ovary. Upon the laying of the first egg the female begins incubating. The act of incubation inhibits follicle development. It does not inhibit full development of the next most advanced follicle, nor possibly the third, but it always causes the degeneration of the fourth follicle. Therefore, the bird lays two eggs, and often three, but not four. If the first and all succeeding eggs are removed as soon as laid, the inhibitory feedback on the ovaries is eliminated, and the bird will continue to lay eggs, at least up to seven in sequence. A bird given a wooden egg to incubate before the follicles reach the critical stage will lay no eggs at all.

The great advantage in indeterminate laying is that it enables the bird to replace lost or stolen eggs quickly and build its clutch up to full size. When an *entire* clutch is destroyed or removed, most birds, with the possible exceptions of large birds of prey and albatrosses, which have long incubation periods, will produce a second replacement or substitute clutch. This, however, may be a slow process and usually involves repeating all the breeding cycle preliminaries: courtship, perhaps nest-building, copulation, and then egg-laying. Even so, pigeons may begin the new cycle half an hour after the loss of their clutch, and the first egg of the replacement set will appear in about five days. This interval between the loss of a clutch and the first egg of its replacement clutch varies with the species. In the Song Sparrow, *Melospiza melodia*, the interval is five days; in the Starling, *Sturnus vulgaris*, eight days.

Feeding of the young is an energetically expensive activity, and female songbirds may lose from 10 to 20 per cent of their body weight during this period (Newton, 1972; Morton, 1976). Indicative of the physiological strain involved in rearing young, McGillivray (1983) found that the interval between fledging and the initiation of the next clutch increases with the number of young fledged in House Sparrows, *Passer domesticus*.

Although many species have only one clutch per year, many others have two, and some have even more. Zebra Finches, *Poephila guttata*, are adapted to living in desert regions of Australia by having their gonads in a state of perpetual readiness, since breeding is primarily stimulated by rainfall. Serventy (1971) reported on one experimental pair raising 23 broods in succession, a fact illustrating this species' ability to initiate breeding rapidly and repeatedly. As a general rule young birds are more likely to lay extra clutches than old birds, and birds living in regions with long summers are apt to lay more clutches than birds of short-summer regions. There is some evidence that extra clutches may be inhibited in birds living under crowded conditions. In temperate latitudes the Mourning Dove, *Zenaida macroura*, commonly attempts five clutches per summer. Sometimes, in doves and in certain passerines, the broods follow one another so closely that the eggs of one brood are laid in a second nest before the young of the preceding brood have flown from the first. In some species, like the Great Reed Warbler, *Acrocephalus arundinaceus*, the male builds a new nest while the female feeds the young, which do not leave the old nest until several days after the second clutch is started. Whatever the method—whether by large clutches, small clutches with better care, numerous clutches or overlapping clutches—natural selection has provided every species, except those on the verge of extinction, with enough replacements to balance mortality.

■ SUGGESTED READINGS

The most exhaustive treatment on birds' eggs is found in Romanoff and Romanoff, *The Avian Egg*. The physiology of egg production is well treated in Murton and Westwood's *Avian Breeding Cycles*, and Sturkie's *Avian Physiology*. For both structure and function of bird eggs, see Carey's chapter in *Current Ornithology, Vol. 1*. Lack's *The Natural Regulation of Animal Numbers* and his *Ecological Adaptations for Breeding in Birds* both contain excellent material on clutch size. The natural history of birds' eggs is well treated in Makatsch, *Der Vogel und Sein Ei*.

Incubation and Brood Parasitism

16

Six white eggs on a bed of hay,
    Flecked with purple, a pretty sight!
There, as the mother sits all day,
    Robert is singing with all his might.
—William Cullen Bryant (1794–1878)
    Robert of Lincoln

Incubation in a bird is the rough equivalent of pregnancy in a mammal. The parent in each case provides food, warmth, and protection for the developing embryo. But whereas in mammals foods are supplied to the embryo and wastes removed from it by way of the blood stream of embryo and mother, in birds, food in the form of a highly specialized "infant formula" is stored in the egg by the ovary and walls of the uterus before the egg is laid, and embryo wastes either accumulate in the embryo's allantois or are given off through the egg shell in gaseous form. Although avian "pregnancy" has its shortcomings, it also has several advantages. For one thing, both parents may share in its duties. This not only reduces the wear and tear on

one parent but it helps hold the pair together so that they both may help raise the young—an arrangement practically unheard of in mammals, except some carnivores like the wolf, and, of course, man. Another advantage is that in birds, the incubating individual, unlike the pregnant mammal, may in cases of dire emergency desert the eggs and escape to live and breed another day.

INCUBATION BEHAVIOR

Many birds begin incubation with the laying of the first egg: loons, grebes, pelicans, herons, storks, eagles, hawks, cranes, many gulls, and other seabirds, cuck-

331

oos, parrots, owls, swifts, hummingbirds, hornbills, and a few passerine species. Early incubation of the clutch provides greater protection for all the eggs from storms and enemies. It results, however, in young that hatch out at different times and therefore are of different ages. In a brood of six young Barn Owls, *Tyto alba*, the first to hatch will be about 15 days older than the last (Fig. 16–1). This has its disadvantages. The parents may start feeding the first-hatched young while eggs or other young need brooding; or the older nestlings may abuse or even eat their younger brothers; or the parents may kill and tear the youngest nestling to bits, as is regularly done by the Lammergeier, *Gypaëtus barbatus*, and several other raptors. The fact that large birds of prey usually lay two or three eggs but rarely raise more than one nestling was known even to Aristotle (Nice, 1954). Hatching asynchrony is adaptive in species faced with irregular or unpredictable food supply in that it enables adults to adjust brood size to food levels prevailing when the eggs hatch (Lack, 1954; O'Conner, 1978). Thus in food-poor years the younger chicks starve, while in good years the entire brood may be raised (Burger, 1979).

The incidences of nests with asynchronous hatching in Great Tits, *Parus major*, and Blue Tits, *P. caeruleus*, increase as the season progresses because incubation starts before clutches are complete (Neub, 1979). Runts are produced if food supply is poor. Late-hatched nestlings beg more feverishly and scramble to a more forward position in the nest to counteract the selective parental feeding of larger nestmates (Ryden and Bengtsson, 1980). When food is ample all nestlings may survive.

Ducks, geese, gallinaceous birds, and most passerines hold off incubating the clutch until the last egg is laid (Fig. 16–2), and even 10 days beyond that in the Rock Partridge, *Alectoris graeca*, or the Bobwhite Quail, *Colinus virginianus*. This makes it possible for all the young to hatch within a short interval, and for the parent bird to shift its behavior completely from incubation to the care of the young. After 28 days of incubation, the dozen or so eggs of the Mallard, *Anas platyrhynchos*, generally all hatch out within a period of about two hours. If any of the ducklings hatch half a day or so later, the female leaves them behind to die as she shepherds her brood to open water. In still other species—ostriches, rheas, rails, and woodpeckers—incubation begins shortly before the completion of the set.

How does a bird know that a clutch is complete and that it is time to change its behavior from laying to brooding? Egg removal experiments show that for many species the number of eggs in a full clutch is not the maximum number the bird can lay. Probably the feel of enough eggs against the belly tells the bird that it is time to incubate. This tactile stimulus seems to cause the pituitary gland to secrete prolactin, which has the two functions of suppressing the output of its follicle-stimulating hormone and therefore suppressing ovulation, and of initiating broodiness or incubating behavior. If Ring Doves, *Streptopelia roseogrisea*, that have never bred are injected with progesterone, they will incubate eggs; if they are injected with both progesterone and prolactin, they will brood eggs and feed young (with pigeon's milk). Birds injected with prolactin alone neither brooded nor fed (Lott and Comerford, 1968). Other stimuli probably bring on broodiness in species which begin incubation with the first egg laid.

A bird's behavior also alters as incubation progresses. A noisy nuthatch becomes secretive and quiet. If a Great Tit, *Parus major*, is removed from her eggs at certain stages of incubation, she will almost invariably desert the nest, but at other stages she can be picked off the nest and returned, and will continue to brood. Brooding owls tend to become more belligerent toward intruders as incubation progresses. When gulls hear the peeps of unhatched young through the egg shell they shift their behavior markedly, and excitedly begin to call, preen, and move about.

FIGURE 16–1 A brood of Snowy Owl young. Since incubation begins with the laying of the first egg, their staggered hatching results in young of different ages and sizes. Photo courtesy of P. S. Taylor.

FIGURE 16–2 Ring-necked Pheasants just hatched. In precocial species with large broods, incubation is delayed until the last egg is laid. As a consequence, all young hatch at about the same time. Photo by E. Hosking.

Hatching of the eggs generally stimulates a rapid shift, especially in precocial parents, from brooding to care of the young. The eggs of European grebes (*Podiceps* spp.) hatch at different times. When the first eggs hatch, the feeding instincts of the parents seem to overpower the brooding instinct, with the result that out of a clutch of four or five eggs only two young hatch and are typically raised (Stresemann, 1927–1934). On the other hand, a female Spotted Flycatcher, *Muscicapa striata*, may at times interrupt incubating a second clutch of eggs in order to feed the young of her first brood.

Adaptive flexibility in timing the incubation period can be demonstrated by experimentally changing nest contents from eggs to young and vice versa. If, for example, four-day-incubated eggs of the American Goldfinch, *Carduelis tristis*, are reciprocally exchanged for two-week-incubated eggs just ready to hatch, each female will adjust her behavior accordingly, one incubating eggs for a total of five days and then shifting to feeding behavior, the other incubating a total of 23 days (Holcomb, 1967).

In the Ring-billed Gull, *Larus delawarensis*, the two parents between them sit on the eggs about 99 per cent of the time. When the eggs hatch, they cover or brood the young about 90 per cent of the time for three days, after which brooding declines to about 13 per cent of the time. This pattern of broodiness was altered by Emlen and Miller (1969) by exchanging eggs and chicks of various ages. The 99 per cent level of sitting could be extended by postponing the introduction of young, or shortened by introducing them early. It could even be restored by replacing chicks with eggs after as much as four days of chick-brooding. The 90 per cent level of the post-hatching period could be induced by replacing eggs with chicks up to eight days before the end of the usual 20 days of incubation, or delayed until eight days after, and regardless of the age of the chicks up to six days of age. The decline of brooding could be delayed by arresting the normal nest progression at early chick stages, or set ahead or hastened by speeding up the apparent chick development. These results suggested a breeding cycle mechanism in which an internal hormonal regulator was replaced early in the incubation phase of the cycle by an external regulator—the perception of the emerging embryo and developing chick.

A veritable passion for incubating eggs is seen in some species, notably penguins, which, lacking eggs, will "incubate" stones and even lumps of ice. Substitutes for eggs are, however, rare. The heightened instinct to incubate is adaptive, especially in the harsh environment of the Antarctic (Simpson, 1976).

For many species, the egg has significance only within the nest, and if it is removed a short distance from the nest, the bird will ignore it. The Laysan Albatross, *Diomedea immutabilis*, will usually brood the empty nest rather than the displaced egg. The Black-footed Albatross, *D. nigripes*, ignores the nest and incubates the egg. Neither species will retrieve either a displaced egg or a chick (Bartholomew and Howell, 1964).

There are a number of species, particularly those that build scrapes of shallow nests on the ground, which will attempt to retrieve eggs displaced short distances from the nest. Among such species are grebes, gulls, terns, shore birds, gallinaceous birds, and those hawks, owls, and doves that nest on the ground. Eggs of Clapper Rails, *Rallus longirostris*, are often displaced by high tide. Kosten (1982) marked eggs from 18 nests and then displaced them 1 m from the base of each nest. Within an hour the rails picked up the eggs with their bills and returned them to their nests. Birds like the passerines, which build deep cupnests, or which nest in trees, will not attempt to retrieve eggs (Poulsen, 1953).

A variety of experiments have thrown some light on the relative releasing values of eggs and nests on incubation behavior. The Arctic Tern, *Sterna paradisaea*, will retrieve one of its eggs from a distance of 70 cm from the nest if the nest is empty, but if one egg remains in the nest the distance will be only 50 cm. When both eggs are displaced together the tendency to brood them outside the nest becomes intensified at a distance of 35 cm (Körner, 1966).

Substitution experiments have shown that many species will incubate almost any object placed in the nest: light bulbs, golf balls, mollusc shells, watches, dice, garishly painted cubes, and other strange objects. Experiments by Tinbergen (1951) have shown that certain objects will act as supernormal releasers of incubating behavior. A European Oystercatcher, *Haematopus ostralegus*, will try to incubate an impossibly large egg in preference to its own (Fig. 9–5). Geese have tried to incubate ostrich eggs and artificial models even larger. When given a choice between its own egg and a larger wooden ball the tiny Painted Quail, *Excalfactoria chinensis*, rolled the ball into its nest (Winkel, 1969).

Such experiments as these raise the question: to what extent do birds recognize their eggs? Obviously those host species that throw out cuckoo's eggs that do not match their own must be able to tell some difference between them. Natural selection has probably heightened the sensitivity for egg recognition in those species commonly parasitized by other birds. However, species vary in this capacity. There are both acceptors and rejectors of such eggs. In aviary experiments Victoria (1972) found that female African Village Weaverbirds, *Ploceus cucullatus*, could generally discriminate between their own and another female's eggs. Even if a nest contained only one of her own

eggs, the female would recognize and reject another's eggs if they were sufficiently distinct from her own in color and spot pattern. This heightened ability to discriminate eggs is presumably a result of selection to combat Didric Cuckoos, *Chrysococcyx caprius*, that often lay eggs in this Weaver's nest. The Cuckoo's eggs are similar to the Weaver's in size and range of colors and spotting, so that presumably only good matches of their host eggs are not rejected.

In field experiments with 35 species of North American passerine birds, Rothstein (1975, 1977, 1978) found that seven species, including the familiar Robin, *Turdus migratorius*, and Blue Jay, *Cyanocitta cristata*, could recognize their own eggs and reject foreign eggs regardless of their number in the nest. When eggs of the Brown-headed Cowbird, *Molothus ater*, were placed in the nest of the Northern (Baltimore) Oriole, *Icterus galbula*, they were ejected or damaged in the nest. Rothstein believes that the ejection behavior is instinctive but that the recognition of its own eggs by the Oriole depends upon an imprinting process whose sensitive period occurs just before and during the onset of laying. A keen ability to discriminate its own eggs from others is found in the Atlantic Common Murre, *Uria aalge*, probably because the species nests in large colonies with similar nest sites and lays eggs with extremely variable markings. An individual Murre not only recognizes its own egg (Tschanz, 1959) but will roll a displaced egg back to the nest from distances of up to 5 meters. So attached are these birds to the appearance of their own eggs that they will brood half eggshells and even broken eggs whose bloody embryos stain their feathers (Johnson, 1941).

For other species the "feel" of the egg is more important. Experiments by McClure (1945) revealed that the Mourning Dove, *Zenaida macroura*, would without hesitation accept eggs of various colors, but if an egg were cracked or punctured the dove would either remove it or abandon the nest. A Herring Gull, *Larus argentatus*, will cease incubating a broken egg and eat it (Tinbergun, 1960). For many species, it is probably the combination egg-plus-nest that evokes normal incubation behavior. The fact that many species will continue to incubate infertile or addled eggs long past their normal incubation period shows that the egg itself is an important stimulus. A female Black-capped Chickadee, *Parus atricapillus*, has been known to incubate its eggs 11 days beyond the normal hatching time; a pair of Eastern Bluebirds, *Sialia sialis*, 34 days; a Great Tit, *Parus major*, 39 days; a Ring-necked

Pheasant, *Phasianus colchicus*, 72 days; and a Herring Gull, *Larus argentatus*, 103 days. Some pigeons, on the other hand, will fail to incubate their eggs even a single day beyond the usual time for their hatching.

BROOD PATCHES

A bird sitting on its eggs is not necessarily incubating them. Feathers are very poor heat conductors. Incubation depends on the copious transfer of heat from the brooding bird to the developing egg, and this requires close contact between the blood stream of the bird and the shell of the egg. Brood or incubation patches achieve this end. These are areas of bare skin on the belly of the bird that develop shortly before egg-laying (in *Zonotrichia leucophrys nuttalli*) or before the complete set of eggs has been laid (Kern, 1979). A bird may have a single median brood patch, as in many passerines, birds of prey, grebes, and pigeons; two lateral brood patches, as in auks, skuas, and shore birds; three brood patches—one median and two lateral—as in waders, gulls, gallinaceous birds, and others; or none at all, as in ducks, geese, cormorants, gannets, and penguins (Tucker, 1943). There is little correlation between the size of a brood patch and the size of a bird's clutch. In general, brood patches arise only in the sex that incubates the eggs; for example, only in the male phalaropes. If both sexes incubate, both usually have brood patches, but a number of male passerines cover the eggs although they lack a brood patch.

A brood patch undergoes two main changes just before incubation starts. It loses its feathers—chiefly down feathers—and the dermis of the skin becomes spongy and richly supplied with large blood vessels so that the skin looks inflamed (Fig. 16–3). Studies by Baldwin and Kendeigh (1932) indicated that a House Wren, *Troglodytes aëdon*, was able, through molting its feathers, to apply to its eggs a temperature 5.6°C higher than would otherwise have been possible.

Hormones initiate the development of the brood patches (Bailey, 1952). Jones (1971) and Kern (1979) have reviewed the advances in this field since Bailey's classic studies. Administration of estrogen and progesterone, or in some species (e.g., *Zonotrichia leucophrys gambelii*) estrogen alone, produces loss of feathers in the ventral brood patch. A combination of estrogen and either progesterone or prolactin, depending on the species, also produces thickening of the epidermis. Number and size of blood vessels (vascularization) in the dermis of the brood patch also increase during incubation. In most songbirds and in California Quail, *Lophortyx californica*, administration of estrogen alone may produce hypervascularity; however, estrogen combined with prolactin, progesterone, or testosterone may increase this vascularization (Jones, 1969). A combination of estrogen, prolactin, and progesterone also induces collagen production and accumulation of fluid (edema) in the dermis. Only small doses of estradiol are necessary to elicit edema in dermal cells. In phalaropes, males incubate the eggs; androgen rather than estrogen plus prolactin induces brood-patch formation (Drent, 1975). Treatment with hormones failed to produce brood

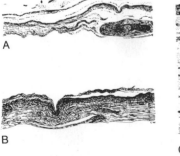

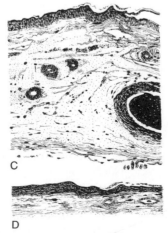

FIGURE 16–3 Cross-sections of the abdominal skin of a Red-winged Blackbird at successive stages in the incubation cycle: *A*, Nonbreeding stage. Down-feather papilla at right. *B*, After laying the first egg. *C*, Early incubation stage, showing the abundant and large blood vessels. *D*, After the young have just left the nest. Magnification 140 ×. After Bailey, 1952.

patches in the Brown-headed Cowbird, *Molothrus ater*, a brood parasite (Selander and Kuich, 1963). In general, estrogen and prolactin, working synergistically, induce the defeathering and vascularization of brood patches, and, combined with progesterone, stimulate their epidermal thickening and tactile sensitivity.

Ducks and geese create a brood patch by plucking down feathers from their breasts and using them to make a warm nest lining. When the female leaves the nest to feed, she covers her eggs with the down. Open-nesting ducks generally have dark colored down that matches the surroundings, while hole nesters, such as the tree ducks, have white down (Stresemann, 1927–1934).

Boobies and gannets warm their eggs by standing on them. The webs of their feet are heavily vascularized with vessels that bypass the capillaries and thus allow a faster, warmer circulation. Murres and penguins have similarly warmed feet but incubate their eggs by holding them on top of the feet. In addition, some penguins have evolved a muscular brood pouch or fold of belly skin that envelops and warms the egg as it rests on top of their feet. Both sexes of the Laysan Albatross, *Diomedea immutabilis*, develop highly vascularized sac-like pouches in the mid-belly region, which accommodate the single egg so snugly that it frequently remains in the pouch when the bird stands (Fisher, 1971).

INCUBATION TEMPERATURES

The chief function of incubation is the transfer of heat from the body of the incubating bird to the embryo within the egg to promote its development. In larger birds (e.g., galliforms), this is performed by a mass of dermal blood vessels which lie near the surface of the brood patch enabling heat to be transferred directly to the egg. In songbirds, blood vessels are deeply situated in the dermis in a large pool of interstitial fluid. Heat is stored in a fluid reservoir and uniformly distributed over the egg (Kern and Coruzzi, 1979).

One of the marvels of bird life is the fact that the Emperor Penguin, *Aptenodytes forsteri*, successfully incubates its egg in the midst of antarctic winter temperatures as low as −60°C (−77°F). This means that the penquin must maintain a temperature in the egg of approximately 34°C (93°F) day and night, without a break, for eight or nine weeks. Actual temperatures of antarctic penguin eggs undergoing incubation were determined by Eklund and Charlton (1959) for the summer-breeding Adelie Penguin, *Pygoscelis adeliae*, by inserting within an egg a miniature electronic telemetering instrument whose broadcast impulses could be received and converted into degrees of temperature. While the average body temperature of the Adelie Penguin was 40°C, fluctuating from 38° to 41.5°C, the average temperature of the incubating egg was 6 degrees less or 34°C, fluctuating from 29° to 37°C. Exceptionally low temperatures occur in the eggs of the Fork-tailed Storm Petrel, *Oceanodroma furcata*, which, during incubation, never exceed 27.5°C (Wheelwright and Boersma, 1978). In contrast, the optimal incubation temperature for eggs of the domestic chicken is 37° to 38°C. No eggs at all survive continuous exposure to temperatures either below 35° or above 40.5°C (Drent, 1975).

In a less stressful climate, the telemetered egg of a Herring Gull, *Larus argentatus*, undergoing natural incubation, varied less than 1 degree from 37°C despite changes in the ambient air temperature of 12° to 27°C (Drent, 1972). Obviously, the incubating parent can somehow control the amount of heat that it delivers to the egg. To investigate this problem, Franks (1967) contrived artificial copper eggs through which was pumped temperature-controlled water. The response of an incubating Ring Dove, *Streptopelia roseogrisea*, to abnormally high egg temperatures was gular flutter and an extended neck; to low temperatures it responded with shivering, fluffed, feathers, and a shortened neck (Fig. 16–4). Incubation persisted at all temperatures between −5°C and 62°C. The doves continued incubating the eggs even when the clutch was kept at warmer than body temperatures for 13 days, or when the eggs were maintained at near-freezing temperatures for 39 hours. Similar experiments with temperature-controlled artificial eggs by Vleck (1981) produced comparable results in the Zebra Finch, *Poephila guttata*. While incubating chilled eggs, the shivering bird's oxygen consumption (i.e., its metabolic rate) increased and the egg temperature rose. It seems likely that the brood patch itself has sensory receptors sensitive to egg temperatures. When White and Kinney (1974) anesthetized brood patches of Village Weaverbirds, *Ploceus cucullatus*, they incubated more intensely than controls, and the eggs reached abnormally high temperatures.

A more direct response to the eggs themselves was reported by Tschanz (1967). Shortly before young

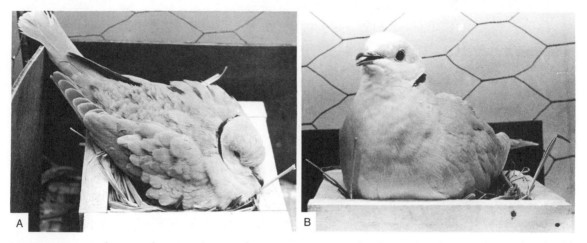

FIGURE 16–4 A dove incubating experimental, temperature-manipulated eggs. A, The typical reaction of cold eggs–feathers elevated, neck pulled in. B, Reaction to hot eggs–feathers unruffled, neck extended, mouth open in gular flutter. After Franks, 1967.

Murres, *Uria aalge* (and other species), pierce the egg shell to break out, they produce a variety of calls, one of which is a distress call when the egg cools. When the incubating parent hears this call persistently, it covers the eggs more closely and raises their temperature. Thus, birds control the temperature of their eggs, either by the amount of time spent on the nest, or by the amount of heat transferred from their bodies to the eggs while incubating them. Both sexes incubate in the Zebra Finch, and nest attendance is about 100 per cent, so that egg temperatures normally do not fluctuate much. Only the female in the White-crowned Sparrow, *Zonotrichia leucophrys*, incubates, and egg temperature is maintained at between 34°C and 40°C. Each time she leaves to forage, however, eggs may be cooled by lower air temperatures or warmed by the sun, and egg temperatures may oscillate between 17.8°C and 43.0°C (Zerba and Morton, 1983). Low or fluctuating temperatures do not prove detrimental to embryos (Fig. 16–5).

Using thermocouples inserted into incubating eggs of 37 different species of birds representing 11 orders, Huggins (1941) determined that the average egg temperature was about 34°C. Eggs in the center of a nest were warmer than the others. A parallel study by Westerkov (1956) of incubation temperatures for the Ring-necked Pheasant, *Phasianus colchicus*, gave a bare brood patch temperature of 39.5°C, an air temperature at the top of the eggs under the sitting hen of 35.1°C, and at the bottom of the eggs, 25°C. In this species, ground temperatures greatly affected egg temperatures.

The female South Polar Skua, *Catharacta skua*, incubates her single egg while holding it between her incubation patch (temperature about 39°C) and the upper surface of her vascularized webbed feet. During the 29-day incubation period, the internal egg temperature gradually rises from about 28° to 39°C (Spellerberg, 1969). That this increase in temperature is due largely to the manufacture of metabolic heat by the growing embryo within the egg is indicated by studies of metabolic heat production in incubating eggs of the Herring Gull, *Larus argentatus*, by Drent (1967, 1970). In this species the embryo is kept at 37.6°C during the first 10 days of incubation; the temperature rises to 39°C at the end of the incubation period owing to embryo-generated heat. This heat comes from combustion of fat stores in the egg yolk, about 60 per cent of which are used up by the time the chick hatches. By the time of hatching, the chick is supplying about 75 per cent of its own heat requirements, the remainder being supplied by the incubating parent. This parental contribution amounts to 8.5 to 14.5 kilocalories per day for the entire clutch.

The metabolic costs of incubation can be high. A female Canada Goose, *Branta canadensis*, may lose 27 per cent of her initial body weight during her 29-day incubation period (Aldrich and Raveling, 1983);

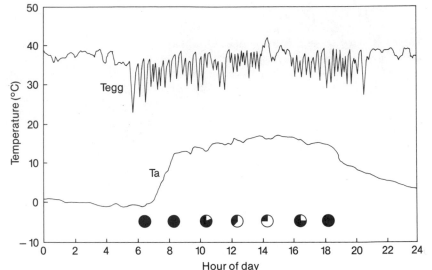

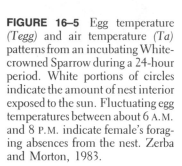

FIGURE 16–5 Egg temperature *(Tegg)* and air temperature *(Ta)* patterns from an incubating White-crowned Sparrow during a 24-hour period. White portions of circles indicate the amount of nest interior exposed to the sun. Fluctuating egg temperatures between about 6 A.M. and 8 P.M. indicate female's foraging absences from the nest. Zerba and Morton, 1983.

the female Common Eider, *Somateria mollissima*, about 32 per cent, much of that loss coming from her pectoral muscles. The blood levels of hematocrit, plasma protein, and free fatty acids all show declines (Korschgen, 1977). Great Tits, *Parus major*, expend about 16 per cent of their available productive energy, and Herring Gulls, about 25 per cent (Drent, 1972). At an air temperature of 10° to 12°C, the metabolic rate of an incubating Starling, *Sturnus vulgaris*, is 25 to 30 per cent higher than that of a nonincubator (Biebach, 1979). When captive Zebra Finches, *Poephila guttata*, were kept at an air temperature of 14.5°C, their eggs failed to hatch because they would have required a productive energy expenditure of 40 per cent from the incubating parent, which was more than it could spare. This probably explains why, in Australia, this bird's geographical range is limited to regions with a mean daily temperature above 12°C (El-Wailly, 1966).

The fact that incubating birds usually turn eggs in the nest is probably an adaptation directed toward providing all the eggs in a nest with more even temperatures. A pheasant will turn her eggs about once every hour; a Mallard, *Anas platyrhynchos*, about once every 40 minutes; the Eurasian Sparrow Hawk, *Accipiter nisus*, about once every 20 minutes; and the American Redstart, *Setophaga ruticilla*, about once every 8 minutes. This periodic moving of the eggs not only provides equal heating throughout the eggs, but prevents embryonic membranes from adhering to the

shell. As the egg rotates, the growing embryo remains uppermost inside the shell. Once the egg is pipped, it is no longer turned. If hens' eggs in an incubator are not turned at least once daily, only about 15 per cent of them will hatch successfully. The eggs of the Palm Swift, *Cypsiurus parvus*, are firmly cemented into its nest with salivary glue, but possibly the swinging of the palm leaf to which the nest is attached provides the equivalent of egg rotation. How megapode eggs, buried in the earth, are able to develop successfully without movement is a problem yet to be solved.

Eggs that rest in spots exposed to full sunlight must sometimes be kept cool. In hot summer weather many incubating birds will protect their eggs from direct sunshine by standing over them, using the body and wings to make a shading canopy. A Common Nighthawk, *Chordeiles minor*, nesting on a flat roof was thus able to hold the temperature of her eggs to 46°C although the temperature of the surrounding roof rose to 61°C (142°F) (Weller, 1958). Birds of at least seven families of the Charadriiformes use wet belly feathers to cool eggs and prevent excessive water loss. Grant (1982) studied incubation behavior of charadriiforms on the shores of the Salton Sea, California, where temperatures regularly approach 38°C and ground temperatures may exceed 50°C (122°F). Parent birds shaded the eggs from the sun and soaked their belly feathers to cool themselves and the eggs. A pair of Black-necked Stilts, *Himantopus mexicanus*, belly-soaked 155 times in a day and brought as much as 3 g of water to the eggs per belly soak. The Egyptian Plover, *Pluvianus aegyptius,* prevents lethally high temperatures by covering both its eggs and newly hatched young with sand and then sprinkling them with water (Howell, 1978).

A variety of ground-nesting birds cover their eggs on leaving the nest. This behavior serves either to keep the eggs warm or to protect them from predators. Grebes cover them with wet vegetation, and the Patagonian Seed-snipe, *Thinocorus rumicivorus,* scrapes dry earth or grass over its eggs. The White-fronted Sandplover, *Charadrius marginatus,* of eastern Africa covers most of its eggs with sand and depends, in part, on solar heat to incubate them. Several species of tinamous cover their eggs mainly with vegetation, but the Ornate Tinamou, *Nothoprocta ornata,* living in the chilly heights of the Peruvian Andes, covers its eggs with feathers. The Eider Duck, *Somateria mollissima,* does the same, but if it is suddenly alarmed by a predator, it flees after squirting excrement over the eggs—a treatment that renders the eggs distasteful to mammalian predators (Swennen, 1968). Northern Shovelers, *Anas clypeata,* do the same. When ferrets and rats were offered a choice between food with or food without Eider excrement added, they rejected the soiled food when the feces came from breeding birds. But when the feces came from nonbreeders, there was no difference in food choice. Eiders apparently have antipredator adaptations in the chemistry of their excrement.

The temperature at which an egg is incubated has a profound effect on the rate of development of the chick embryo within, and therefore on the time of its hatching. Like all metabolic processes, embryonic development, within limits, proceeds more rapidly the higher the temperature. Kendeigh (1940) studied the metabolic rate of the growing embryo of the House Wren, *Troglodytes aëdon,* and showed that at the normal incubation temperature of 35°C the egg required 13 days to hatch, and that at 37.8°C, heightened metabolism so accelerated development (126 per cent) that, theoretically, an egg should hatch in 10 days although no wren's egg is known to do so. Conversely, a subnormal incubation temperature of 32.2°C lowered metabolism sufficiently (to 72 per cent of normal) to require 18 days for hatching. Considering the weighty survival advantages of a short incubation period, it is not to be wondered at that the high normal body temperatures of birds closely approach the lethal limit.

Eggs of many wild birds tolerate a considerable degree of chilling—an adaptation, no doubt, to the absences of the incubating bird for feeding. Crane eggs may be uncovered for an hour at 0°C without killing the embryo. Eggs of the Fork-tailed Storm Petrel, *Oceanodroma furcata,* are frequently deserted in their nest burrows for as many as seven days at a time, or for as many as 28 days during the incubation period, and yet hatch successfully. During these periods of neglect egg temperatures average 10°C. As a consequence of such neglect, incubation periods may range from 37 to 68 days (Boersma *et al.,* 1980). Eggs of Mallards and of domestic fowl, frozen so hard that they have cracked, have nevertheless hatched normal young (Greenwood, 1969). In general, it was found that eggs are more tolerant of low temperatures early in incubation than later, and that occasional chilling of the eggs to 7°C does no great harm, although high mortality occurs among eggs exposed to 0°C for five or six hours. Newly hatched chicks were much more vulnerable to low temperatures than eggs at any stage of incubation.

INCUBATION PATTERNS

Great diversity in incubation patterns is found among birds. Usually both sexes take part in incubating the eggs, particularly in those species lacking pronounced sex dimorphism. A survey by Van Tyne and Berger (1959) of some 160 families, representing the majority of living birds, showed that both sexes usually incubated the eggs in about 54 per cent of the families, the female alone in 25 per cent, the male alone in 6 per cent, and the male, female, or both, in 15 per cent. In those species in which one sex is more conspicuously colored than the other, the drab, less conspicuously marked partner usually takes over the incubation chores, but this is not invariably true. For example, the colorful male Rose-breasted Grosbeak, *Pheucticus ludovicianus,* shares with his mate the brooding of the eggs. Incubation by both sexes is considered the primitive pattern. The following discussion is based on a classification of incubation patterns by Skutch (1957).

Incubation by Both Parents

Although in the majority of species both sexes take part in incubation, the male often takes a minor part. When the female is the chief brooder, as in most passerines, the male often brings food to her. In many water birds the sexes share about equally in incubation. In the Double-crested Cormorant, *Phalacrocorax auritus,* the parents relieve each other at the nest at one- to three-hour intervals. Among fulmars, shearwaters, and many petrels, each parent takes a turn incubating for one to five days at a stretch while its partner is away at sea feeding. Even longer incubation spells are taken by the Short-tailed Shearwater, *Puffinus tenuirostris*, whose male takes the first shift of 12 to 14 days while the females segregate in flocks at sea gathering food. The females return to incubate while the males in turn seek food at sea. This sexual segregation of flocks persists during the period of feeding the young (Serventy, 1967).

Often the two parents will observe a strict incubation schedule. Among doves and pigeons the female sits on the eggs from about 5:00 in the afternoon until about 9:00 the next morning; the male from 9:00 A.M. to 5:00 P.M. A number of sandpipers and plovers follow roughly the same schedule of males by day and females by night: the Dunlin, *Calidris alpina*; Stilt Sandpiper, *C. himantopus*; Lesser Golden Plover, *Pluvialis dominica*, and others. In contrast, the female Sand Grouse, *Pterocles* spp., incubates by day and the male by night. In the Starling, *Sturnus vulgaris*, and Mexican Trogon, *Trogon mexicanus*, both sexes incubate by day but only the female by night. Conversely, among woodpeckers the male incubates by night and both sexes by day. In the Ostrich, *Struthio camelus*, the darker colored male incubates the clutch by night and the gray female by day—an arrangement that aids concealment from foes. Among the Sooty Shearwaters, *Puffinus griseus*, of New Zealand, the sexes take alternate sessions in sitting on the eggs, each session lasting about nine and one-half days. Incubation lasts 53 days (Warham *et al.*, 1982).

There are a few species in which both sexes incubate simultaneously. Simultaneous incubation, but in different nests, is achieved by the European Red-legged Partridge, *Alectoris rufa*. The female builds two nests and then fills the first nest with eggs. These eggs remain unincubated for about 14 days while she lays another 10 eggs in the second nest. She now returns to the first nest to incubate those eggs while the male incubates the second clutch. Each parent cares for its own covey of young; a union of the two groups occasionally occurs (Jenkins, 1957). A similar partnership in incubation occurs in Temminck's Stint, *Calidris temminckii*, except that the male incubates the first clutch of eggs and the female the second.

The instinctive urge to incubate is at times so great that the parent on the nest may be reluctant to be relieved of his or her duties. Often a male goatsucker must gently push his mate off the eggs in order to take her place. If an incubating Magpie-lark, *Grallina*

cyanoleuca, is unwilling to leave the eggs, its mate administeres a peck (Roberts, 1942). In other species, a formalized nest-relief ceremony takes place at the change-over. Chapman (1908) describes such a ceremony among Brown Pelicans, *Pelecanus occidentalis*. A returning bird alights near the nest and walks slowly towards it with its bill directed vertically and its head weaving from side to side. It pauses, and then both birds begin to preen their feathers. A moment later, the bird on duty steps off the nest, and the newcomer takes its place.

Terns, herons, cranes and many other species also use nest-relief rituals.

Incubation by One Parent

While there are disadvantages in reducing the number of egg-caretakers to one, there is the advantage that the fewer birds there are around the nest the less conspicuous it is to predators. In this connection, it is significant that one-parent incubators are common among birds that nest in the open, but not among those nesting in holes, where protection against predators is more assured.

The female alone incubates the eggs in many passerine species (manakins, tyrant flycatchers, buntings, sparrows, most crows and jays), many ducks and geese, hawks, eagles, most owls, hummingbirds, those gallinaceous birds with pronounced sex dimorphism, and, as far as known, in all polygynous species except the rheas and certain tinamous. In many instances, the male feeds the female and guards her and the nest, keeping watch from some inconspicuous nearby perch. Among several species of hawks and harriers, the female rises from the nest and flies out to receive food from the male in midair, either talons to talons, or by catching it in the air as he drops the prey. Female hornbills (Bucerotidae) are imprisoned in tree-cavity nests by the mortaring of most of the entrance with mud, regurigitated food, and droppings, leaving only a small hole through which the males pass fruits, whole or regurgitated. After an imprisonment of 1 1/2 to 4 months the female emerges fat and healthy with her young (Kemp, 1978) but the hard-working male is thin and bedraggled. Food intake of an incubating Streamer-tailed Hummingbird, *Trochilus scitulus*, is 20 per cent less than that of a nonincubating bird. To compensate for this reduced intake, energy expenditure is reduced by lowering body temperature while incubating from 41°C to 32°C. (Schuchmann and Jakob, 1981).

In only a few species does the male alone incubate:

kiwis, tinamous, rheas, emus, cassowaries, painted snipe, button quail, and the Emperor Penguin, *Aptenodytes forsteri*. In the polyandrous phalaropes and button quail, it is the female that is the larger, more brilliantly colored sex and the one that does the courting. The drab males not only stay at home and incubate the eggs, but single-handedly they raise the young. A most unusual incubation pattern is shown by the Chinese Pheasant-tailed Jaçana, *Hydrophasianus chirurgus*. The female lays as many as ten successive clutches of four eggs each, and a different male incubates each of the clutches and rears the young by himself (Hoffman, 1949). Another unusual pattern is demonstrated by the polygynous South American rheas. One male makes a nest in which all the members of his harem will lay their eggs—from 20 to as many as 50—which the male then incubates in solitude. The males of several species of tinamous likewise incubate in one nest, the eggs contributed by several females.

Incubation Constancy

Many kinds of birds incubate their eggs almost constantly, day and night: penguins, woodpeckers, doves, trogons, hornbills, hoopoes, antbirds, and others. On the other hand, several tyrant flycatchers and swallows are inconstant in brooding, sitting on their eggs only half of the daylight hours and spending the rest of their time away from the nest.

The amount of time spent in incubation of eggs and its distribution is largely determined by a bird's hereditary make-up reacting with a variety of external influences. A bird's pattern of incubation, in turn, influences its reproductive success. For example, in studying the European Starling, *Sturnus vulgaris*, introduced in New York around 1890, and the East Asian Crested Mynah, *S. cristatellus*, introduced in Vancouver, British Columbia around 1895, Johnson (*in* Drent, 1972) makes the following comparisons. In their original, native habitats, both species nest in cavities, have the same clutch size (five eggs), brood in April with approximately equal photoperiods (about 13.5 hours), and seem to have equal reproductive success. As is well known, the Starling has prospered mightily in North America, spreading from coast to coast and even into Mexico and Alaska. The Mynah, in contrast, is still restricted to the vicinity of Vancouver. Johnson attributes this discrepancy in success to a genetic difference in incubation habits. The Mynah, adapted to tropical climates, spends only 49 per cent of the daylight hours on its eggs, as against 72 per cent

for the Starling. As a consequence, the Mynah raises only two young per clutch whereas the Starling raises 3.5. When eggs were exchanged between Mynah and Starling nests, their hatching success changed accordingly, showing that it depended not on the genetic quality of the eggs but on the length of the incubation period. When Mynah nest boxes were supplied with artificial heat (28°C), hatching success rose from 64 to 92 per cent. Johnson's conclusion is that the Mynah cannot adapt its tropical incubation schedule to cool Vancouver temperatures, and is thus less successful than the more flexible Starling.

The biological ramifications of incubation patterns have been examined by Skutch (1962), and his conclusions form the basis of the following discussion. For those times of day that a bird spends on the nest incubating its eggs, Skutch proposes the logical term *sessions*, and for those periods away from the nest, *recesses*. These terms are equivalent to the commonly used terms "nest attentiveness" and "nest inattentiveness."

Among those species in which only one parent incubates, the bird usually allocates between 60 and 80 per cent of its daytime activity to brooding. This is considered normal incubation constancy. To achieve higher than 80 per cent constancy, a bird must either be fed on the nest by an attendant, or be able to find food expeditiously during its recesses from the nest. Constancy below 60 per cent is found chiefly in small birds, such as American flycatchers, which feed largely on small flying insects. Birds that fast on the nest for several days on end "require correspondingly long periods for foraging and recuperation, so for them such protracted sitting is feasible only when the sexes alternate on the eggs." For species whose precocious young feed themselves shortly after hatching, the incubating parent is able to fast longer on the nest without losing vital reserves it would otherwise need for feeding its young.

Other influences that may modify the constancy or rhythm of incubation include species, size, sex, and temperament of bird; endocrine levels in the blood, numbers that share in incubation; behavior of the male if he does not incubate; type, abundance, and distribution of food; the size and number of eggs in a clutch (larger masses require more time to recover effective warmth); stimuli from the eggs (visual, thermal, tactile, or vocal); disturbance by other animals; type of nest (open or covered; insulated or not); air temperature; sun exposure; rainfall.

In some species, such as the Palm Swift, *Cypsiu-*

rus parvus, the incubation schedule is very erratic; in others, such as the European Redstart, *Phoenicurus phoenicurus*, attentiveness is regular and almost rhythmic. The female of this species alternates about 15 minutes on her eggs with 8 to 25 minutes off. The number of recesses taken by one female on each of the 12 days of incubation was 23, 24, 27, 32, 28, 27, 28, 29, 30, 29, 32, and 30 (Ruiter, 1941).

Table 16–1 gives examples of incubation constancy for selected small species of birds, mostly passerines. Since the data given represent birds living in different climates and at different stages of incubation, they are valid only for very general comparisons. Most of the figures are rounded off and in some instances represent averages from two different sources.

Knowing that in Finland the incubation period of the Pied Flycatcher, *Ficedula hypoleuca*, was longer in May than in June, von Haartman (1956) investigated the effect that nest-box temperatures might have on the species' brooding rhythm. By using thermostatically controlled nest boxes he discovered that the warmer the box the shorter the individual brooding period. The length of periods off the nest was not affected by outside temperatures. The incubation rhythm of the Great Tit, *Parus major*, was investigated by Kluijver (1950) by placing activity recorders in nest box entrances. He found that as outside air temperatures became warmer, sessions on the nest became shorter and recesses longer (Fig. 16–6).

A hummingbird may leave its eggs 140 times in one day; a robin 20 times; a domestic hen, once or twice. The female Wood Duck, *Aix sponsa*, has an erratic incubating schedule with sessions varying from two to 60 hours in length, and recesses of from 15 minutes to over five hours. The Black-bellied Storm Petrel, *Fregetta tropica*, on the contrary, has an extremely regular incubation schedule. Of 52 sessions on the eggs, 75 per cent of them lasted 72 hours each between change-overs between mates (Beck and Brown, 1972). Some albatrosses incubate steadily, day and night, for three weeks; and a male emu in a zoo refused to leave his eggs for 60 days, fasting all the while. Because the hummingbird is so small it loses body heat rapidly and must therefore leave its nest to seek food many times a day to make up this energy loss. It is more difficult to see why so many large sea birds remain on their eggs for long periods without eating, particularly when in many species both sexes incubate and can

TABLE 16–1.
INCUBATION CONSTANCY OF SMALL BIRD SPECIES IN WHICH ONLY THE FEMALES INCUBATE

SPECIES	AVERAGE TIME ON NEST (MINUTES)	AVERAGE TIME OFF NEST (MINUTES)	ATTENTIVE PERIODS PER DAY	DAYTIME ATTENTIVENESS (PER CENT)
Allen's Hummingbird				
Selasphorus sasin	4.6	1.3	140	78
Barn Swallow				
Hirundo rustica	7.2	2.5	79	70
Black-capped Chickadee				
Parus atricapillus	24	7.8	34	76
Pied Flycatcher				
Ficedula hypoleuca	86.7	4.7	11	95
American Robin				
Turdus migratorius	44	11	20	80
Red-eyed Vireo				
Vireo olivaceus	29	10	28	75
Ovenbird				
Seiurus aurocapillus	105	19	9	82
Bullfinch				
Pyrrhula pyrrhula	111	12	9	90
Song Sparrow				
Melospiza melodia	25	7	34	75

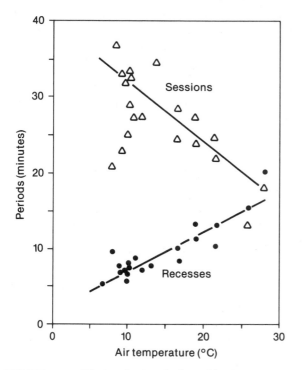

FIGURE 16–6 The incubation rhythm of the Great Tit. In cool weather, sessions on the eggs are longer and recesses shorter; the reverse occurs in warm weather. After Kluijver, 1950.

relieve each other of the duties of incubation. Among petrels and shearwaters the two sexes share incubation, each bird attending the eggs from two to five or more days without relief, while its mate is away at sea feeding. Albatross pairs likewise share incubation, and among the different species the sessions by one parent on the egg vary from 6 to 30 days (Richdale, 1952). Some penguins also refrain from eating during incubation. Within a day after she lays her egg, the female Emperor Penguin, *Aptenodytes forsteri*, turns it over to the male, who incubates it alone for 62 to 64 days, fasting the while (Prevost, 1953). All this time he is living on stored fat, and his weight may drop from 45 to 22 kg. Adelie Penguins, *Pygoscelis adeliae*, may fast as long as four weeks while incubating, with a consequent weight loss of over 25 per cent. Large birds with their low rate of heat loss are able to store sufficient fat in their bodies to tide them over long incubation periods even in cold climates.

Other Patterns of Incubation

As discussed previously, there are species that build community nests and practice communal incubation. The gregarious Anis (*Crotophaga* spp.) cooperate to build a shallow cup of sticks, lined with green leaves, in which several of the females will lay their eggs—as many as five dozen—which several birds then incubate simultaneously. The Babbling Thrush, *Yuhina brunniceps*, of Taiwan is another communal breeder. At one nest, six birds were observed incubating eight eggs; at another, two males and three females fed five young (Yamashina, 1938).

Without question the most remarkable form of incubation is that of the Megapodiidae of Australasia. These birds, known as mound birds, incubator birds, or megapodes (from their big, powerful feet), use other than animal heat to incubate their eggs. For incubation they exploit heat from three sources: the sun, subterranean volcanism, and fermenting vegetation. The account that follows is largely after Frith (1956, 1957).

The Maleo, *Megacephalon maleo*, of the hill forests of the Celebes, migrates on foot as far as 32 km (20 miles) to the nearest ocean beaches having black sand. There the pairs dig holes to bury their large eggs (10 × 6 cm). The black sand absorbs sufficient solar heat to incubate the eggs. Many females may lay their eggs in one communal hole about 30 to 60 cm deep. This same species also lays eggs in the warm soil next to hot springs or volcanic steam fissures.

Another megapode, the Brush Turkey, *Alectura lathami*, lives in the steaming rain forests of Australia and New Guinea. Here, where little sunshine reaches the forest floor, the birds heap up piles of rotting vegetation (Fig. 16–7) and let the heat of fermentation incubate their eggs. These mounds are often 3 meters high; those of a near relative, the Scrub Fowl, *Megapodius freycinet*, are reported to be as high as 5 meters and as wide as 12 meters. The same mound may be used year after year, each female burying in its compost 7 to 12 eggs, usually with the large end up. When the precocial young hatch, they dig their way up to the surface and immediately begin an independent existence, some of them being able to fly the first day.

In the Mallee Fowl, *Leipoa ocellata*, of the semidesert regions of southern Australia, the female lays eggs at four- to eight-day intervals, and may produce as many as 35 in one season. Each time before the female may lay an egg, the male opens the egg

FIGURE 16–7 A Brush Turkey of Australia and its mound. Fermenting vegetation in the mound furnishes the heat to incubate the eggs. Photo from Australian News and Information Service.

chamber in the mound by kicking out 1 or 2 cubic meters of soil. Understandably, the egg-laying period is unusually long, extending from mid-September to late February. For the hard-working male, construction of the mound begins in May (Australian autumn), four months before the first egg is laid, and the eggs require approximately two months to hatch. So for 10 or 11 months of the year he is busy either building or tending the mound to control its temperature (Fig. 16–8).

If, in the spring, the fermenting vegetation raises the temperature of the mound above 33°C (92°F), the male laboriously opens the mound in the cool mornings to let heat escape. When the midsummer sun beats too warmly on the mound, he adds more soil on top to provide an insulating layer to keep the eggs from cooking. In dry autumn weather, when the heat of fermentation is gone, he uncovers the mound enough to let the sun's noon heat penetrate to the

eggs, and then scratches back on top of the mound the sun-warmed earth to keep the eggs warm overnight. Certainly, this substitution for normal incubation is no labor-saving arrangement. By observing birds from blinds for hundreds of hours, Frith learned that the average male Mallee Fowl worked seven hours per day in the spring, 13 in summer, and 10 in the autumn; and throughout all three seasons he averaged 5.3 hours per day of actual digging!

By means of experimental manipulation of mound temperatures (electric heating cables, for instance), Frith proved that the male Mallee Fowl was able to detect internal variation in mound temperatures by sampling beakfuls of earth, and to take proper action to cool or warm the mound as needed. In one mound, for example, which a male had been visiting daily to release heat, Frith removed all the organic matter so that the temperature quickly dropped to 16°C. On the bird's next visit, he detected the lower temperature and began his autumn routine of opening the mound in the heat of the day, while all the other males in the vicinity were busy cooling theirs. In nature, this species is able to maintain the temperature of the egg chamber in each mound at an even 33°C, with not more than a degree or two of fluctuation.

Aside from the megapodes, at least two other species are known to use other than animal heat for incubation. The Egyptian Plover, *Pluvialus aegyptius*, buries its eggs in sandy islands of the Nile and broods them only at night when the sun's heat fails. The White-fronted Sandplover, *Charadrius marginatus*, mentioned earlier, also uses solar heat to incubate its sand-buried eggs. These plovers may also shade the eggs by day to avoid excess heat.

BROOD PARASITISM

Another substitution for the usual pattern of incubation is brood parasitism. This method involves an escape from the chief duties of raising a family—building the nest, incubating the eggs, and rearing the young—through the simple expedient of imposing them on another bird, usually one of another species. This behavior, which occurs in about one per cent of all bird species, is practiced by representatives of five families: Anatidae, Cuculidae, Indicatoridae, Icteridae, and Ploceidae. Brood parasites lay their eggs in nests of other species and abandon them to the care of their foster parents. Among the 130 species of Cuculidae are about 50 known brood parasites; among the

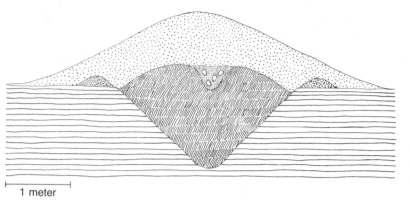

FIGURE 16–8 A diagrammatic cross-section of the mound of a Mallee Fowl. The pit contains fermenting compost, with the eggs laid in a depression on top, and the whole covered with sandy soil. After Frith, 1957.

1 meter

other four families are about 30 more brood parasites, most of them among the honey-guides or Indicatoridae.

How did this behavior originate? Brood parasites have decided advantages over birds providing parental care. Laying in several nests increases the chances that at least some of their offspring will escape predation since predators usually destroy all the young in a nest (Payne, 1977). This strategy, Payne reminds us, is that of Cervantes' Sancho Panza, who counsels Don Quixote that "a wise man. . . [does] not venture all his eggs in one basket."

Thus if a female tends to leave more progeny by laying eggs in several nests, and if this female's progeny behave likewise, then brood parasitism may ultimately be genetically fixed in the population.

Several authors have suggested that brood parasitism very likely arose in the tropics where nest loss to predation is high (Hamilton and Orians, 1965; Payne, 1977; Ricklefs, 1969). A female losing her nest to a predator must find another location to place her eggs. She may choose some other bird's nest. Losing a nest also triggers ovulation; thus, she may leave more young than do other individuals in her population. A mutation may then arise causing a lack of synchronization between nest-building and ovulation. Lacking a nest, the incipient parasite must find another laying site, so again another bird's nest may be chosen.

The habit of some birds of using old or abandoned nests of other species to lay their eggs may have been a precursor to brood parasitism. Owls often nest in old crow nests, sparrows in swallow nests and starlings and flycatchers in woodpecker holes. Several species of ducks may sometimes lay in nests of other

ducks (Giroux, 1981). Some birds may even aggressively expropriate the active nest of another species. This has been observed between House Sparrows, *Passer domesticus*, and Cliff Swallows, *Petrochelidon pyrrhonota*, or between Starlings, *Sturnus vulgaris*, and woodpeckers. This form of eviction has progressed to such a degree in the Chestnut Sparrow, *Passer eminibey*, of Kenya, that it seldom builds its own nests (Payne, 1969).

Whatever its origins, brood parasitism will be promoted in a bird if its eggs have short incubation periods (so its young may hatch before their foster-siblings); if it is an indeterminate layer (to produce more young, and at strategic times); and if it eats a wide variety of foods (so its young may adapt to varied nestling diets). Since a brood parasite is relieved of the necessity of building nests, incubating eggs, and rearing young, the energy thus saved could be used to increase egg production, which, in turn, should improve its chances of success as a brood parasite.

One of the commonest and most highly evolved brood parasites is the European Cuckoo, *Cuculus canorus*, various races of which parasitize over 125 other species of birds. At the onset of the breeding season the female Cuckoo finds suitable nests to parasitize, not by random search, but probably by the intensity of the alarm calls of the potential host species (Seppä, 1969). The Cuckoo is probably helped in this action by its resemblance in flight to a small hawk. Apparently the sight of an appropriate host stimulates the Cuckoo to ripen an egg follicle and lay an egg, because the time interval between such a psychic stimulus and the subsequent laying of the egg is four to five days, an interval corresponding to that in many birds between the time of the destruction of a clutch and the laying of the first egg of the replacement clutch (Stresemann, 1927–1934). The Cuckoo lays usually one egg to a nest at 48-hour intervals. By means of a protrusible cloaca the Cuckoo can lay eggs in covered nests or niches too small to enter. The egg is quickly laid, often in as little time as 5 seconds, and ordinarily between 2:00 and 6:00 P.M. when the owner of the nest is likely to be away. Normally, the Cuckoo will remove one of the host's eggs while visiting the nest. Anatomical studies by Payne (1973) of nine species of parasitic cuckoos show that they lay between 16 and 26 eggs a season and at intervals of about every two days.

Ordinarily, the female Cuckoo lays her egg in an incomplete clutch of the host species, or in a fresly completed clutch. Her sharp discrimination in choosing nests at the proper stage is indicated by Capek's observations (*in* Stresemann, 1927–1934). Of 273 nests containing Cuckoo eggs, 130 contained incomplete host clutches, 68 held fresh, complete clutches, 20 contained incubated complete clutches, 5 were occupied, but contained no host's eggs, and 13 were abandoned.

Host species react to the Cuckoo egg in their nests either by accepting it and incubating it with their own eggs, or by removing it from the nest, or by abandoning the nest completely. Some species are much more sensitive than others to the presence of Cuckoo eggs. In Germany the Wood Warbler, *Phylloscopus sibilatrix*, is very likely to desert its nest if it finds a Cuckoo's egg there, while the Redstart, *Phoenicurus phoenicurus*, is more prone to accept the egg.

A great many birds react to strange objects in their nests by throwing them out. If the Cuckoo egg does not resemble the host's egg in size and color, the host may remove it. This is probably but not necessarily because a given bird knows the color of its own eggs and objects to anything different.

In many instances, the egg of the European Cuckoo strikingly resembles that of its favored hosts (Fig. 16–9). In Finland, the Cuckoo lays bright blue eggs, and its chief hosts are the Redstart and Whinchat, *Saxicola rubetra*, both of which lay blue eggs. In Hungary, the Cuckoo lays greenish eggs boldly blotched with brown and black, and so does its chief dupe, the Great Reed Warbler, *Acrocephalus arundinaceus*. Throughout Europe and Asia, there are races of the Cuckoo whose eggs very closely resemble those of its selected fosterers. In southern England, however, the Cuckoo victimizes a variety of wagtails, pipits, warblers, and other species that lay very dissimilar eggs.

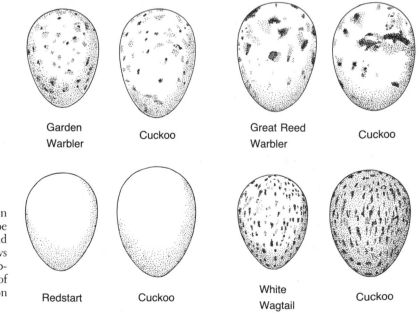

Garden Warbler   Cuckoo   Great Reed Warbler   Cuckoo

Redstart   Cuckoo   White Wagtail   Cuckoo

FIGURE 16–9 Eggs of four European passerines, each paired with the type of Cuckoo egg most frequently found with it in the nest. This mimicry shows clearly the evolutionary ecological adaptation in the eggs of different races of the European Cuckoo. Magnification × 1.2. After Rensch, 1947.

Probably as a consequence, the Cuckoo lays an egg with generalized coloration that matches the eggs of no single host species very closely.

In a study of the Khasia Hills Cuckoo, *Cuculus canorus bakeri*, in India, Baker (1942) found that normal hosts, whose eggs matched the Cuckoo's eggs in coloration, deserted their nests only 8 per cent of the time when parasitized, but abnormal fosterers with discordant eggs deserted 24 per cent of the time. Such selective rejection of the Cuckoo's eggs has doubtless been the factor that produced races of Cuckoos whose eggs match so closely those of their favored hosts.

Cuckoo eggs are adaptive in other ways also. Although the various species of cuckoos are generally larger than the birds they parasitize, their eggs are small, about like those of their hosts. When cuckoos parasitize larger birds, such as crows, their eggs are larger, again matching those of their hosts. The shell of the European Cuckoo egg is thick and strong, weighing about 25 per cent more than shells of other eggs the same size. Above all, the Cuckoo egg hatches in about 12$^{1}/_{2}$ days, whereas the eggs of most host species require 13 or 14 days. This gives the young Cuckoo the enormous advantage of a head start over its foster brothers in growth and in claiming food from the host parents. A part of this temporal advantage is due to the fact that the Cuckoo egg is held in the oviduct long enough for the embryo to begin development (Payne, 1973).

When the young Cuckoo hatches, it is blind, naked, muscular, and able to gape for food like most altricial young. When it is about 10 hours old, an instinct appears that is one of the wonders of the animal kingdom. If any solid object, such as an egg, young bird, or even an acorn, touches the sensitive shallow depression on the little Cuckoo's back, the blind bird manipulates it to the rim of the nest and shoves it overboard. The young Cuckoo persists in this behavior until it has cleared the nest of all objects but itself. This "overboard" instinct disappears in 3$^{1}/_{2}$ to 4 days. The perhaps overly-adaptive urge of many breeding birds to brood and feed almost anything within, and to ignore anything outside the nest, powerfully furthers this ejecting instinct of the young Cuckoo. The host parent will usually ignore her own starving young even when they are just outside the nest cup, and instead of caring for them will feed and brood the rapidly growing, parasitic foster child (Fig 16–10). A further adaptation in some cuckoos relates to their diet. The African Emerald Cuckoo, *Chrysococcyx cupreus*, parasitizes at least five different host species. The diet of its young varies greatly and adapts to that of its specific host (Brosset, 1976).

Almost unbelievable refinements in parasitic adaptations have been reported by Nicolai (1964) and Payne (1973, 1982) in the African widow-birds (subfamily Viduinae), which typically parasitize estrildid finches. The nestlings of the widow-birds do not eject their nest-mates, but they hatch earlier, beg longer and more obstinately, and grow faster than do their host-siblings. They also mimic precisely the plumage and species-specific palate colors of the host young. Not only are their begging calls and behavior identical to those of the host young, but the courtship songs of the adult widow-bird match perfectly those of the host species. The evolutionary significance of these striking parallels will be considered in Chapter 23.

In the Anatidae, only the Black-headed Duck, *Heteronetta atricapilla*, of South America, is a true brood parasite; it regularly lays its eggs in the nests of other species of ducks, coots, gulls, and even a hawk. Its ducklings are remarkably precocious, show little following behavior, and leave their host parent to live an independent existence after the first 24 to 36 hours (Weller, 1968). A strong parasitic tendency is found in the North American Redhead Duck, *Aythya americana*, which frequently lays its eggs in nests of other ducks and leaves them to be incubated by the nest owner (Giroux, 1981).

Brood parasitism is also practiced by the honeyguides, or Indicatoridae, so called because of their famous habit of leading man and other animals to nests of bees where man and bird share spoils. These

FIGURE 16–10 Even after the young Cuckoo has grown nearly to full size, the duped foster-parent still feeds it, so blind is its instinct to place food in a gaping mouth.

FIGURE 16–11 A blind young Cuckoo ejecting the eggs from a Tree Pipit's nest. Young nest-mates are thrown out in the same fashion. Photo by E. Hosking.

birds, largely natives of Africa, all lay their eggs in the nests of their relatives, the barbets and woodpeckers. The nestling of the Greater Honey-guide, *Indicator indicator*, is known to eject its foster brothers from the nest, but probably a more effective means of eliminating its nest-mates is provided by the remarkable, needle-sharp mandibular hooks on the beaks of the blind nestlings of at least two species of Indicatoridae (Fig. 16–12). These weapons are used in a "ferocious, relentless gripping and biting" attack on their foster brothers until they die and are presumably removed from the nest by their uncomprehending parents (Friedmann, 1955). These hooks drop off when the honey-guide nestlings are about two weeks old. The young of some New World cuckoos also have scalpel-sharp hooks in their bills to dispose of their nest-mates (Sick, 1981).

Only one group exists today containing closely related species practicing either parental or parasitic behavior (Friedman, 1929). The nonparasitic Bay-winged Cowbird, *Molothrus badius*, of South America may build its own nest, but usually usurps nests of other species, or occupies abandoned nests. The related Screaming Cowbird, *M. rufoaxillaris*, lays its eggs in nests of *M. badius*, who then rear the parasite's young. The Brown-headed Cowbird, *M. ater*, of North America, is an obligate brood parasite. It lays its eggs in the nests of some 210 species of North American birds (Friedmann and Kiff, 1985).

Compared with the European Cuckoo, the Cowbird is not particularly specialized for parasitism (Chance and Hann, 1942). It has not evolved races that specialize in parasitizing certain host species. Its eggs, unlike the Cuckoo's, are of normal size and shell-thickness, and show no mimicry of the host's eggs. Cowbird's eggs, however, hatch a day or two

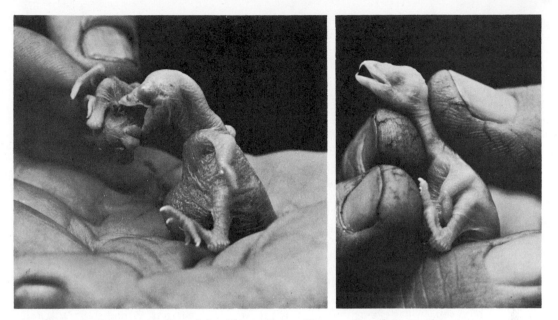

FIGURE 16–12 A young nestling of the African Greater Honey-guide. *(Left)* Attacking its nest-mate, a young barbet. *(Right)* The sharp mandibular hooks drop off after the Honey-guide is about two weeks old and presumably has the nest to itself. Photos by G. Ranger, courtesy of H. Friedmann.

sooner than the typical host's eggs. The young Cowbird does not evict its nest-mates, although in many cases it is larger and stronger than they are, and may crowd some or all of them from the nest, or it may usurp the lion's share of food so that the host's young starve. In a survey of 520 nests of 14 host species in Kansas, Hill (1976) found that 21 per cent were parasitized by the Brown-headed Cowbird. Whereas non-parasitized nests fledged 1.5 young per nest, parasitized nests fledged only 0.5 host young and 0.3 Cowbird young per nest. Cowbird parasitism may cause local extinctions of species. The Kirtland's Warbler, *Dendroica kirtlandii*, in Michigan, and Bell's Vireo, *Vireo belli*, in California, are endangered species due to cowbird parasitism (Friedmann and Kiff, 1985).

A quite unexpected facet of brood parasitism was revealed by Smith (1968) who discovered that the Giant Cowbird, *Scaphidura orizivora*, actually benefitted the host species whose nest it parasitized. Among the colonial hangnest icterids of Panama, Wagler's Oropendola, *Zarhynchus wagleri*, and the Yellow-rumped Cacique, *Cacicus cela*, commonly as many as 90 per cent of their nestlings died from infestations of botflies. However, when their nests were para-

sitized by the Cowbird, only 8 per cent of the nestlings were killed because the Cowbird nestling removed the botfly larvae from the host nest-mates. This beneficial effect, however, applied only to colonies not placed near bee or wasp nests. Since the bees and wasps attack botflies, colonies so protected had no need of Cowbird protection; indeed, they suffered from the Cowbird's competition with their own young. Probably as a consequence, host species enjoying bee and wasp protection throw out Cowbird eggs; those nesting away from bee and wasp nests accept them.

INCUBATION PERIODS

Assuming normal, undisturbed incubation, the average time interval between the laying of an egg and the emergence of the young bird from the shell represents the *incubation period* for that species (Heinroth, 1922). The time that an adult bird sits on the eggs does not necessarily coincide with the incubation period. Studies with thermocouples have shown, for example, that some doves may sit on their eggs without heating them, and at times male goshawks, lacking an incubation patch, will also apparently incubate but not warm the eggs. For practical purposes a species' incubation period may be defined as the time interval between the laying of the last egg of a clutch and the hatching of the last egg (assuming that all eggs hatch).

Great variety is seen in the incubation period of birds, extending from 10 days in the Great Spotted, and Red-cockaded Woodpeckers, *Picoides major*, and *P. borealis*, 11 or 12 days in the Brown-headed Cowbird, *Molothrus ater*, and the White-eyes, *Zosterops* spp., to 81 days in the Royal Albatross, *Diomedea epomophora*. Most small passerine species have incubation periods of 12 to 14 days: the American Robin, *Turdus migratorius*, for example, averages 13 days, but the American Crow, *Corvus brachyrhynchos*, and many other Corividae require 18 to 20 days. The incubation period of the domestic Chicken, *Gallus gallus*, averages 21 days; that of various species of owls, 26 to 36 days; hawks, eagles, and vultures, 29 to 55 days; most petrels, 38 to 56 days; and albatrosses, 63 to 81 days. The land bird with the longest incubation period is probably the Kiwi, *Apteryx* spp., whose eggs require up to 80 days to hatch.

Various naturalists have searched for general principles that might explain the wide diversity found in incubation periods among different species, but so far no universally valid rule has appeared. As a general rule, large birds have longer incubation periods than do small species. There exists a crude relationship between egg weight and incubation period: as egg weight doubles, incubation time increases by 16 per cent (Tullett, 1980). This reflects the fact that, in general, the innate rate of development of the embryo in the egg is slower in large birds than in small. This difference in growth rate continues after hatching. But there are exceptions to this size–growth rate correlation. Tiny hummingbirds require about 16 days, while much larger woodpeckers require only about 12 days to hatch their eggs. This fact reveals another variable in the problem: woodpeckers hatch out at an earlier stage in development than do birds of many other families.

Within families, incubation periods may be directly related to degree of precocity. Ducks in general have incubation periods ranging from 24 to 30 days. The eggs of the Torrent Duck, *Merganetta armata*, require 43 to 44 days to hatch (Moffett, 1970). The young have to be well-developed to survive their rather harsh environment in rushing torrents. Galliform eggs in general require 23 to 25 days to hatch. Eggs of the highly precocial Mallee Fowl, *Leipoa ocellata*, which on hatching are even capable of flight, require 62 days' incubation (Vleck *et al.*, 1984).

Also, as a very rough rule, birds high in evolutionary rank, such as the passerines, have comparatively short incubation periods, but this may be explained largely by their relatively higher body temperatures. Here again, there are exceptions: the Ruddy Quaildove, *Geotrygon montana*, which is low in taxonomic rank, hatches its eggs in 12 days. Undoubtedly a variety of ultimate factors have had a hand in determining, over the years, the rate of embryonic development and the closely associated incubation period.

Ecological factors undeniably have had an effect on the incubation period. Among tyrant flycatchers in the tropics, Skutch (1945) found that those that build pensile, relatively inaccessible nests had longer incubation and nestling periods than did the open-nesters. A comparison of the incubation periods of hole-nesting versus open-nesting birds shows this same relationship (Table 16–2).

Even more striking, once the eggs have hatched, is the variation in duration of the nestling period among tropical American tanagers. Skutch (1954) reported that species that habitually nested low had short periods of 11 to 13 days; those that built open nests higher up had periods averaging 14 to 20 days, while those that built roofed-over nests had nestling periods of 17 to 24 days. A fair assumption is that the shorter periods of open-nesters represent a selective adaptation to higher losses from predation. Heinroth (1938) points out that in Africa, swarming with beasts of prey, the ostrich incubates its 1500-g egg only six weeks, whereas in Australia, with few predators, the much smaller Emu, *Dromaius novaehollandiae*, sits on its 600-g eggs for eight weeks!

Still other environmental influences may affect incubation periods. Among tanagers and warblers, Skutch (1954) found that incubation periods were longer in the tropics than in the temperate zone. In a seasonal study of the European Wren, *Troglodytes troglodytes*, Kluijver *et al.* (1940) discovered that average incubation periods steadily decreased in length as spring progressed: April, 17.5 days; May, 16.3; June, 15.3; and July, 14.5 days. In southern Germany, early clutches of the Great Tit, *Parus major*, take about 16 days to hatch, late clutches take about 11 to 12 days (Zink, 1959). Variation in incubation constancy probably affects incubation time also. Finally, loosely built nests probably mean greater heat loss and therefore slower incubation.

Experiments (Miller, 1972) show that incubation of eggs by the Ring-billed Gull, *Larus delawarenis*, can be terminated after 8 days, but no earlier, by sub-

TABLE 16–2.
INCUBATION PERIODS OF HOLE-NESTING AS COMPARED WITH OPEN-NESTING BIRDS

	PLACE	NUMBER OF SPECIES	AVERAGE DURATION OF INCUBATION	AUTHORITY
Hole nesters	Europe	18	13.8 days	Lack, 1948
Hole nesters	Europe	12	13.8	von Haartman, 1954
Hole nesters	U.S.A.	10	13.8	Nice, 1954
Open nesters	Europe	54	13.1	Lack, 1948
Open nesters	Europe	13	12.2	von Haartman, 1954
Open nesters	U.S.A.	11	12.0	Nice, 1954

stituting young chicks for the eggs. The reluctance of the birds to shift from incubation to parental care any earlier probably means that they are not yet physiologically prepared for the change. But species vary in this regard. Tricolored Blackbirds, *Agelaius tricolor*, will accept nestlings during or shortly following egg-laying (Emlen, 1941), as will some Black-headed Gulls, *Larus ridibundus* (Beer, 1966). The gulls, on the other hand, incubated wooden eggs as long as 75 days.

From her hormone treatments of Bengalese Finches, *Lonchura striata*, Eisner (1969) concluded that estrogen tends to sustain and prolong incubation through stimulating prolactin secretion, and progesterone may interrupt incubation behavior.

Much research remains to be done before all these varied factors can be properly assessed. Without question the dominating factor is heredity, which controls the rate of embryonic development and the time of hatching. All of these other influences are either secondary modifiers of the innate clockwork within each egg, or selective forces that imperceptibly shift the basic hereditary incubation rate.

HATCHING

As the process of incubation goes on, an egg steadily loses weight. Part of this loss represents evaporated moisture, and part is gaseous metabolic waste from the respiring embryo within. The average weight loss of 24 eggs of the Eastern Bluebird, *Sialia sialis*, during its 13-day incubation period was 12.6 per cent (Hamilton, 1943). Much of this weight loss represents metabolized fat from the egg yolk, consumed by the growing embryo for heat production and other energy

uses. Upon hatching, a Herring Gull chick will have consumed 60 per cent of the fat reserves of the fresh egg, leaving a remainder of 40 per cent for use by the chick in the first few days of life (Baerends and Drent, 1970). This remainder consists of the yolk sac in the belly of the newly hatched chick. Records from just-hatched birds of 39 species in the Basel Zoo by Schmekel (1960) showed that for altricial species the residual yolk sac averaged between 5 and 10 per cent of the total body weight, and for precocial species, between 12 and 25 per cent. The weights of the yolk in fresh eggs average about 20 per cent of total egg weight in altricial species, and 35 per cent in precocial species. In young Starlings, *Sturnus vulgaris*, the yolk sac is finally used up in the first four days of life; in the European Quail, *Coturnix coturnix*, in about six days; in the Mallard Duck, *Anas platyrhynchos*, in about seven days; and in the Ostrich, *Struthio camelus*, in eight days or more.

Whether an egg hatches properly or not may depend, among other things, on its treatment during incubation. Eggs of the Mallard, *Anas platyrhynchos*, were incubated under experimentally controlled conditions of moisture by Mayhew (1955). The best hatching success was obtained from eggs dipped in water once a day and incubated at 65 per cent relative humidity. This relationship explains why ducks nesting in fields far from water have poor hatching success in dry summers. Hatchability of eggs from a captive Wood Duck, *Aix sponsa*, fed a diet containing 37 per cent protein, was significantly higher that that of a duck fed 17 per cent protein (Doty, 1972).

Many birds, particularly those that lay eggs at about sunrise, are fairly constant in their hour of laying. The hour at which eggs of a given species hatch, however, is much more variable. Painstaking observations by Skutch (1952) of the time of hatching of eggs of Costa Rican passerines showed that of 26 hatchings of two species of flycatchers of the genus *Myiozetetes*, 11 occurred at night, 14 in the forenoon, and only 1 in the afternoon. This nonrandom distribution of hatching hours was explained by Skutch as apparently due to a diurnal rhythm in the efforts of the young bird to break out of the shell. Records of 93 hatchings among 11 species of finches, tanagers, and wood warblers gave 27 at night, 51 in the forenoon, and 15 in the afternoon. Here the time of hatching was thought possibly to be determined by the hour of laying and a constancy in the rate of embryonic development. It seems more likely that daily cycles of light and dark stimulate coordinated cycles of hatching activity in chicks still within the egg. Young domestic chicks and ducklings, a day or two prior to hatching, clap their beaks within the darkened shell about 20 and 36 times per minute, respectively. After the egg is exposed to light for 1 minute, the rate of bill-clapping in each species roughly doubles (Oppenheim, 1968). In either case, the adaptive "advantage of hatching early in the day is that the nestling can be promptly fed; whereas if it hatched late in the afternoon or after nightfall it might have to wait many hours for its first meal" (Skutch, 1952).

The chief role in hatching is played by the chick itself. In preparation for liberation from its limestone prison, the maturing chick develops two tools. One is a short, pointed, horny "egg-tooth" at the tip of its upper mandible (Fig. 16–13). The other is a set of prominent hatching muscles located largely on the upper side of its neck and head, and used to force the head and egg-tooth upward against the inside of the egg shell. Two days before hatching, the hatching muscle becomes greatly swollen through the absorption of lymph or water. Then, as the muscle contracts repeatedly in forcing the egg-tooth against the shell, triglycerides in the muscle become depleted and the muscle shrinks in size (Hazelwood, 1972). The muscle continues to shrink in size until, in the adult Franklin's Gull, *Larus pipixcan*, it is only one-tenth its relative size at the onset of hatching (Fisher, 1962).

The egg-tooth either drops off or wears away at different times in different species. In megapodes it disappears before the young hatch (Clark, 1961). In terns it may persist for from 6 to 13 days after hatching; in gallinules, about 20 days; in murres, about 35 days; and in some penguins, up to 42 days.

Chicks of the domestic fowl hatch after about 21 days of incubation. Preparatory steps toward hatching are described by Vince (1969) as follows. Two or three days before hatching, the chick's beak punctures the inner shell-membrane that separates it from the air space in the blunt end of the egg. The chick now begins to breathe air from this chamber—air very high in carbon dioxide content and extremely low in oxy-

FIGURE 16–13 A young Whimbrel, *Numenius phaeopus*, two hours after hatching. Note egg-tooth on tip of beak. Photo by E. Hosking.

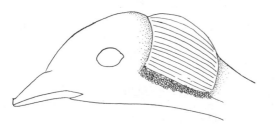

FIGURE 16–14 The hatching muscle in the domestic chick as it appears at the time of hatching. Diagrammatic. After Fisher, 1958.

gen. After a few more hours the chick "pips" the outer shell membrane and the shell, and is now able to breathe fresh air from the outside. It now begins the methodical, laborious behavior of opening the shell. During this activity, blood rapidly drains from the allantoic vessels just within the inner shell membrane. These vessels had until now provided gas exchange for embryonic respiration. During the pipping activity the chick becomes highly responsive to sounds, light, and vibration. It will, for example, stop giving distress calls at the sound of a hen's clucking. In addition to vocal sounds the chick now gives off rather loud, rhythmic clicking sounds that coincide with, and are probably caused by, respiratory movements.

The actual pipping process, as described by Johnson (1969) for the Bobwhite Quail, *Colinus virginianus*, involves spasmodic thrusts of the entire chick at 6- to 8-second intervals. Each thrust begins with a strong shove with the left leg and a weak scratching movement with the right. Then the bird expands its trunk region by the inhalation of air and the contraction of trunk muscles. This puts strong pressure on the egg shell. As the pressure is applied, the hatching muscle forcibly tips up the head so that the egg-tooth penetrates the shell. Then by virtue of a strong depression of the head, the chick rotates its body within the shell in a counterclockwise direction as seen from the blunt end, and becomes positioned for pipping in a new location. Successive punctures of the shell create a ring of breaks that will allow the blunt end of the shell to fall away and the chick to emerge. In the Bobwhite the cutting process requires 10 to 30 minutes and may involve two or three, and rarely, up to eight, complete circlings of the shell. One can see in this performance the significance of rotation of the eggs in the nest during incubation. Neglect of this instinctive

behavior by the parent would result in the adhesion of the chick to the inside of the shell and hence allow no chance for the chick to cut its liberating circle of punctures.

Not all birds hatch with equal facility. Young of the Wood Thrush, *Hylocichla mustelina*, take from 5 to 22 hours between the first chipping and emergence from the shell; the Canada Goose, *Branta canadensis*, about 24 hours; the European Wren, *Troglodytes troglodytes*, two days; the Royal Albatross, *Diomedea epomophora*, three days; Sooty Shearwaters, *Puffinus griseus*, at least four days; and the Laysan Albatross, *Diomedea immutabilis*, two to four, and even six days.

Parent birds rarely assist their young to emerge from the shell, but merely stand by as interested observers. Once the young have freed themselves, the parents either eat the shells or carry them some distance from the nest, where they will not reveal its location to predators. Young flamingos are said to eat their own shells, possibly as a mineral supplement to their first meal.

Precocial birds do not bother about the shells if, like the ducks and pheasants, the parent leads the young from the nest within a day or so after they hatch out. Since such precocial species do not feed their young but merely lead them to food, it is essential that all eggs hatch out at about the same time.

In numerous species as many as a dozen or more precocial young may all hatch out within an hour or two of each other despite the fact that the eggs had been laid over a span of two weeks or more. This remarkable synchrony in hatching has long puzzled ornithologists, and only now are the mechanisms responsible for it coming to light. It is now known that the clicking sounds made by chicks a day or two before they hatch act as signals between young in the same clutch, that serve to accelerate slowly developing chicks or to retard the more advanced ones so that they all hatch at the same time. The frequency of these clicks varies according to species: the Song Thrush, *Turdus ericetorum*, gives off an average of about 35 clicks per minute; the American Tree Sparrow, *Spizella arborea*, about 60; the domestic Fowl, about 80; the Ring-necked Pheasant, *Phasianus colchicus*, about 110; and various quail, about 120 (Vince, 1966). Experiments by Vince (1969, 1972; Vince *et al.*, 1984) have shown that clicks can accelerate or delay (depending on click rate) not only the time of hatching in quail chicks, but also earlier stages of development such as lung ventilation and yolk sac withdrawal into the body.

Even artificial clicks are effective. Eggs in contact with each other in an incubator are much more likely to hatch synchronously than are isolated eggs from the same clutch. Contact apparently transmits the vibrations more effectively. In an incubator tray of quail eggs, hatching may begin on one side and spread in a wave, within an hour or two, across the tray to the other side.

Those young that survive the ordeal of hatching emerge into a world which presents them with two immediate problems: to find food, and at the same time to avoid becoming food for some other animal. A high proportion of young birds meet an unhappy fate before they become independent of parental care.

■ SUGGESTED READINGS

For general treatment of the broad field of eggs and incubation, see Davis, Breeding Biology of Birds, in Wolfson's *Recent Studies in Avian Biology*, and Drent's chapter in Farner and King's *Avian Biology*, V. Good introductions to the problem of brood parasitism are found in Chance, *The Truth about the Cuckoo*, and Freidmann, *The Cowbirds*.

The Care And Development of Young

17

. . . the young have chipped,
Have burst the brittle cage, and gaping bills
Claim all the labour of the parent bird.
—Grahame, Birds of Scotland
(*The Poets' Birds,*1883)

The rearing of young birds follows a general pattern: they hatch out, are fed by the parents, are protected from enemies and from the elements, and finally are feathered and fly off to live their own lives. As regards their maturity at hatching, there are two broadly different types of birds and they fit this pattern in quite different ways.

One type, the *precocial* bird, hatches out covered with down, legs well developed, eyes open and alert, and is soon able to feed itself. Because it is usually able to leave the nest and to run after its parents shortly after hatching, it is also called a *nidifuge* or nest fugitive. The young of the Lesser Golden Plover, *Pluvialis dominica,* leave their arctic nest within 2 hours after hatching. Nidifuges are often ground-nesting species that, as adults, are good runners or good swimmers and feed either on the ground or in the water.

The other type, the *altricial* bird, is born naked, or nearly so, is usually blind, and is too weak to support itself on its legs. About the only thing it can do when newly hatched is to hold up its unsteady, gaping mouth for food. It may be able to do that very soon. A nearly naked young Wood Thrush, *Hylocichla mustelina,* 4 minutes after it frees itself from its shell, can call and gape for food. Altricial parents supply all food until the young are nearly adult in size. Naturally, such birds remain confined to the nest for some days or weeks. They are therefore called *nidicoles* or nest dwellers.

There are, actually, wide variations in the degree of maturity of young birds at the time of hatching. Table 17–1, from Nice (1962) summarizes these variations.

An interesting relationship between egg type and precocity in young birds has been pointed out by Hein-

355

FIGURE 17–1 *(Left)* The one-day-old precocial, or nidifugous, chick of the Ruffed Grouse, eyes open, alert, and ready to run. *(Right)* The one-day-old altricial nestling of the House Sparrow, naked, blind, helpless.

roth (1938). (See Chapter 15.) Altricial birds lay small eggs with low yolk content and thus smaller energy demands on the female. Her energy is expended in the task of rearing her helpless young. Precocials invest energy in a heavier egg with a larger yolk, but benefit subsequently from the relatively independent chicks (Ar and Yom-Tov, 1979). An altricial bird normally uses up its yolk before hatching, and therefore has less

---

TABLE 17–1.
MATURITY OF YOUNG AT HATCHING*

PRECOCIAL
   Eyes open, down-covered, leave nest first day or two.
   1. Completely independent of parents, e.g., megapodes.
   2. Follow parents but find own food, e.g., ducks, shorebirds.
   3. Follow parents and are shown food, e.g., quail, chickens.
   4. Follow parents and are fed by them, e.g., grebes, rails.
SEMIPRECOCIAL
   Eyes open, down-covered, stay at nest although able to walk, fed by parents, e.g., gulls, terns.
SEMIALTRICIAL
   Down-covered, unable to leave nest, fed by parents.
   1. Eyes open, e.g., herons, hawks.
   2. Eyes closed, e.g., owls.
ALTRICIAL
   Eyes closed, little or no down, unable to leave nest, fed by parents, e.g., passerines.

*After Nice, 1962.

---

reserve food to sustain its first days of life. Accordingly, the altricial nidicole is more dependent on parental brooding and feeding. Precocials may retain from one-third to one-seventh of the yolk in their bellies after hatching. The precocial domestic chick, on the other hand, has a generous 6 g of yolk in its belly shortly after hatching. This reserve is gradually discharged over the next seven or eight days into the blood plasma as triglycerides and carbohydrates. This explains why it is that, on a two- or three-day trip from hatchery to farmer, young chicks need not be fed. In general, the weight of a hatched bird is about two-thirds that of the fresh egg from which it came. The evaporation of water and the loss of the shell and of gases and feces from the metabolizing embryo account for the difference.

The most precocious group of birds known are the ones most like the reptiles in hatching from their eggs completely independent young, the Australasian megapodes, or incubator birds, members of the Galliformes. Young megapodes, upon hatching from their buried eggs, scramble to the surface, well feathered and instinctively able to fly and to begin a completely independent life.

A highly significant difference between precocial and altricial birds, as Portmann (1950) makes clear, is one of organ proportion and development (Table 17–2). In the altricial nidicoles the organs of metabolism are enlarged at the expense of the rest of the body, above all at the expense of the nervous system, the sense organs, and the locomotor organs. The young nidicole "is a veritable growth machine permitting prodigious metabolism of an efficiency not found elsewhere among the higher vertebrates." Since altricial nestlings are fed and kept warm by their parents, they need not divert energy toward food-seeking or thermoregulation, and consequently can concentrate their energy expenditures on rapid growth (Ricklefs, 1968a). For example, a cuckoo that weighs but 2 g at hatching will, in only three weeks, attain the adult weight of 100 g—a multiplication of 50 times its original weight. Such fantastically rapid growth is completely unknown in mammals. But some birds, particularly the nidicoles, have achieved it at a heavy price in nervous development, as shown in Table 17–2.

As an example of localized emphasis in precocial development, Sutter (1951) notes that the brain of a newly hatched Quail, *Coturnix coturnix*, "has only to increase its weight three times to reach adult size, while the body weight increases as much as twenty-one

TABLE 17–2.
A COMPARISON OF ORGANS OF PRECOCIAL AND ALTRICIAL YOUNG AT HATCHING*

	PERCENTAGES OF TOTAL BODY WEIGHT		
TYPE AND SPECIES	*Eyes*	*Brain*	*Intestines*
Precocial young			
Snowy Plover, *Charadrius alexandrinus*	10.7	7.2	6.6
Japanese Quail, *Coturnix coturnix*	5.5	6.2	9.7
Ring-necked Pheasant, *Phasianus colchicus*	4.2	4.2	6.5
Water Rail, *Rallus aquaticus*	4.5	6.2	10.5
Altricial young			
Pigeon, *Columba livia*	4.9	2.9	10.3
Alpine Swift, *Apus melba*	6.1	3.1	14.6
Jackdaw, *Corvus monedula*	5.5	3.6	13.1
English Starling, *Sturnus vulgaris*	4.0	3.2	14.1

*After Portman, 1950.

times." On the other hand, the cerebellum of the just-hatched altricial Starling is no further advanced than that of the half-incubated (12-day) domestic chicken embryo still in its shell. It should be made clear, however, that once the nidicole begins its rapid growth, it soon outstrips the average nidifuge in mental endowment.

One further distinction between nidicole and nidifuge relates to the development of nerve fibers in the brain. The growth of myelin sheaths—a sign of maturation—is at hatching much farther along in nidifuge young than in nidicoles. In the evolutionary race for survival, the nidifuge species have gambled on a longer infancy, tempered, however, with alert senses and athletic locomotor equipment. The nidicoles, on the other hand, are born helpless and sluggish, but hurry through the vulnerable period of infancy by eating and growing at a furious tempo.

FEEDING THE YOUNG

As soon as they emerge from the shell, young birds are brooded by the parents to dry their down, if they have any, and to warm and shelter them against the elements. If the young are precocial they are soon led away from the nest to a suitable spot where they shortly learn to feed themselves. In precocial species such as the domestic fowl and many ducks, the female takes complete charge of the brood from this time on, lead-

ing the young to food, but not feeding them directly. The only known exception to this behavior among the Anatidae is found in the somewhat aberrant Magpie Goose, *Anseranas semipalmata*, in which both parents feed their young bill-to-bill (Johnsgard, 1965). In altricial species the young are fed by the parents, normally from the first day on.

Among most hummingbirds, and in other species that are promiscuous, the male does not aid in caring for the young; and of course in brood parasites, such as the cuckoo or cowbird, neither parent ordinarily shows any concern for its offspring. But in many of the precocial shore birds and in the majority of passerines both sexes take some part in rearing the young. For example, female Least Sandpipers, *Calidris minutilla*, attend the young for up to 22 days after hatching, averaging six days. Males attend young for 14 to 27 days, averaging 20 days (Miller, 1985). In many instances the female spends more of her time brooding the delicate young than feeding them, and the male devotes himself to finding food for them. In other cases the female feeds the young birds more assiduously than does the male. The female Red-eyed Vireo, *Vireo olivaceus*, has been observed to do about 75 per cent of the feeding of the young; the female Arctic Warbler, *Phylloscopus borealis*, does about 74 per cent of the feeding, but the male brings more food per trip, as is often the case in such situations.

In some species, such as the English Robin, *Eritha-*

cus rubecula, or the House Wren, *Troglodytes aëdon*, the female may desert her young and begin laying eggs for the second brood, leaving the male alone to finish feeding the first brood. In the Snowy Owl, *Nyctea scandiaca*, the female broods the young constantly while they are small, and the male brings all the food. Most doves run their household duties on a regular shift, the male feeding the young from midmorning to late afternoon, and the female taking charge the remainder of the 24 hours (Goodwin, 1983). In at least 36 species of birds, including grebes, coots, various shorebirds, woodpeckers, and songbirds, the brood is divided, the father feeding part of the young, the mother, the remainder (McLaughlin and Montgomerie, 1985). Individual differences in the begging calls of fledgling Song Sparrows, *Melospiza melodia*, enable the parents to recognize the young divided between them (Smith and Merkt, 1980). In the case of androgynous species such as jaçanas, tinamous, and phalaropes, it is the male parent who feeds the young.

Kinds of Food

The kind of food fed the young varies with species. In nearly all species, whether the adults eat fruits, seeds, insects or other animal food, the young commonly are fed on a diet rich in protein. Such a diet most effectively promotes the rapid multiplication of cells in the growing young bird. An analysis of the stomach contents of Ruffed Grouse, *Bonasa umbellus*, chicks in Virginia, by Stewart (1956) showed a gradual change in diet with age. In late May and early June they ate 91 per cent insect and 9 per cent plant food; in August, 1 per cent insect and 99 per cent plant food. A study of the stomach contents of 4848 grain-eating adult House Sparrows, *Passer domesticus*, by Kalmbach (1958) revealed a diet of 3.4 per cent ani-

mal matter, and 96.6 per cent vegetable; the stomachs of 3156 young, on the contrary, contained 68.1 per cent animal and 31.2 per cent vegetable matter. The fruit-eating Resplendent Quetzal, *Pharomachrus mocinno*, of Central America, Skutch (1945) found, fed its young almost exclusively on insects for the first 10 days. Nestlings of the Wood Pigeon, *Columba palumbus*, are fed a diet of 92 per cent crop milk on day one. The percentage of crop milk is reduced as the amount of seeds and grains in the diet increases, so that at day 19 the percentage of crop milk is only 26 per cent (Murton *et al.*, 1963).

Even young birds seem instinctively aware of the need for high-protein diets early in life. In the first five days after hatching, young Black Ducks, *Anas rubripes*, ate about 90 per cent aquatic invertebrates (versus 10 per cent plant food); this decreased to 43 per cent for fully feathered young (Reinecke, 1979). When Berthold (1977a) fed young European Blackbirds, *Turdus merula*, exclusively, or even 50 per cent, English Ivy *(Hedera helix)* berries, their growth lagged behind that of young fed almost exclusively on an animal diet.

Beyond the need of basic food for energy and tissue-growth, young birds require supplemental nutrients for specific needs, e.g., calcium for bone-growth. Precocial young are relatively advanced at hatching and are often able to run on their first day. After sampling 246 Alaskan sandpipers of four *Calidris* species, Maclean (1974) found, during the egg-laying season, that 38 per cent of the females contained lemming teeth or bones in their stomachs, versus 1.9 per cent of the males. In July 12.2 per cent of the juvenile sandpipers also contained lemming bones or teeth. Calcium demand is similarly met by European Wrynecks, *Jynx torquilla*, by feeding their young egg

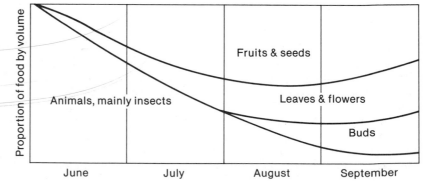

FIGURE 17–2 A generalized diagram showing the seasonal changes in diet for young Ruffed Grouse. After Edminster, 1947.

shells, snail shells, and small bones (Löhrl, 1978). The African Stanley Cranes, *Tetrapteryx paradisea*, feed their young, as their first meal after hatching, bits of their own egg shells (Walkinshaw, 1963).

Experiments by Rabinovitch (1966) with newly hatched, captive Herring Gulls, *Larus argentatus*, and Ring-billed Gulls, *L. delawarensis*, demonstrated that learning is involved in the choice of food by these birds. Different groups of young were fed for six days on a single kind of food (worms, cat food, or dyed cat food). Chicks were then tested to see whether they preferred the training food to an unfamiliar one; 92 per cent of the choices were for the training food.

Methods of Feeding

Great variety is shown in the ways young birds get their food. In many precocial species, such as the domestic hen, the adult merely points out the food to the young, who pick it up off the ground. This technique is followed by the gallinaceous birds, whose young soon become able to distinguish palatable from unpalatable insects. By far the majority of young, particularly the passerines, woodpeckers, and cuckoos, are fed directly from the beak of the parent in the way familiar to anyone who has watched a robin feed its babies. Whereas most altricial young flutter their wings, stretch their necks, and gape vertically for food, nestlings of the Estrildidae beg with their heads turned downward and do not flutter their wings. This peculiar feeding behavior probably stems from the birds' characteristic domed nest with a low entrance hole (Kunkel, 1969).

Many parent birds swallow the food as they find it, and later regurgitate it at the nest. This method of feeding has two advantages. The adult can carry more food per trip, and the digestive juices, regurgitated with the food, aid the digestion of the food in the stomachs of the young. Gulls and storks regurgitate their food in front of the nestlings, who then pick it up from the ground or the nest-rim. Young penguins and pelicans plunge shoulder-deep into the gullets of their parents and feed on the shrimp and fish the parents have caught (Fig. 17–3). The Diving Petrel, *Pelecanoides urinatrix*, has been observed by Richdale (1943) feeding its young at night in its underground burrow. The young petrel places its slightly open beak crosswise within the beak of the parent just in from the sea. The adult regurgitates a red, creamy ribbon of food "as if from a tube of toothpaste," which the young bird seizes with the tip of its rapidly vibrating mandibles. Adult spoonbills and albatrosses likewise

FIGURE 17–3 A young pelican plunges deep into the adult bird's gullet to feed on regurgitated fish.

FIGURE 17–4 A male Little Bittern regurgitating food into the mouth of a nestling. The young bird seizes and tugs the mandible of the parent. Photo by E. Hosking.

take the bills of their young crosswise between their own beaks and direct the vomited food by means of their tongues into the young birds' mouths. Heron young, on the contrary, take their parents' beaks crosswise within their own when feeding. Many finches thrust regurgitated insects and seeds into the gaping beaks of their young, the hummingbird even thrusting its sword-like beak two-thirds its length down the gullet of its young, almost far enough to pierce the stomach.

All pigeons and doves regurgitate "pigeon's milk," a creamy substance with a make-up very similar to rabbit's milk. Typical pigeon's milk contains about 74 per cent water, 12.4 per cent protein, 8.6 per cent fat, 1.3 per cent other nitrogen compounds, 1.4 per cent ash, and no carbohydrates (Fisher, 1972). It is rich in calcium and in vitamins A, B, and B_1. Even more remarkably, as Lewis (1944) has revealed, the same endocrine mechanisms are involved in each animal. The milk of pigeons and mammals comes from fatty cells shed from the epithelial tissues in either the bird's crop or the mammal's mammary glands. If prolactin from the anterior pituitary gland is injected experimentally into these animals, both crop and mammary glands will begin producing milk. Not only will the crop walls of the pigeon begin functioning, but the pigeon will exhibit broodiness and protective care of eggs and young. As little as 0.1 microgram of prolactin injected subcutaneously will cause the crop of a pigeon to begin manufacturing milk.

Oily fluids from the digestive tract are used by other species to feed their newly hatched young (Fisher, 1972). Flamingos, *Phoenicopterus ruber*, initially feed their young exclusively on an esophageal fluid that is 18 per cent fat and 8 per cent protein. The male Emperor Penguin, *Aptenodytes forsteri*, regurgitates a crop fluid whose dry weight is composed of 29 per cent fat, 59 per cent protein, and 5.5 per cent carbohydrate. Ashmole (1971) suggests that the stomach oil of petrels, a secretion of the proventriculus used for feeding the young, is primarily an adaptation to reduce the bulk of collected marine food, and thus the number of wide-ranging food-gathering trips that the parents must make to rear their young.

Predatory birds bring food in their beaks or talons, and feed it, torn to bits, to their small young. They do not tear the food when their offspring are older. In the Eurasian Kestrel, *Falco tinnunculus*, the female takes the prey brought to the nest by the male, and tears it into small pieces and feeds the young. Should the female die (Heinroth, 1938) the young may also die, since they are unable to swallow the entire prey tendered them by the uncomprehending male, who lacks the instinct to tear the meat into bits. When the young are older and stronger, they are able to tear up the prey themselves.

Growing young need water, and if they cannot get it from the food they eat, it must be provided in some other way. In dry habitats this problem becomes acute. The male Namaqua Sandgrouse, *Pterocles namaqua*, of the Kalahari Desert of Africa, each morning, flies (as far as 80 km) to the nearest waterhole. There he squats belly deep in water and rocks his body several minutes until his specially adapted belly feathers have absorbed 25 to 40 ml of water. He then flies back to the precocial seed-eating young, who drink 10 to 18 ml of water by nibbling and stripping the feathers. The young are thus watered daily for about seven weeks until they are able to fly (Cade and Maclean, 1967; Maclean, 1968). On hot days Little Ringed Plovers, *Charadrius dubius*, bring water to their newly hatched young in the same fashion. Adult Common Ravens, *Corvus corax*, have been observed both feeding water to their young, beak-to-beak, and cooling them by applying wet belly feathers to their bodies. Similarly, darters have been observed squirting water into the mouths of their nestlings (Ali and Ripley, 1968).

Helpers at the Nest: Cooperative Breeding

The urge to feed their young is deep-seated in birds. In fact, the instinct appears even in young birds themselves, so that the young of a first brood of the summer may help in feeding their younger brothers and sisters of later broods. This incipient parental care by *helpers* has been observed in over 130 species of birds (Skutch, 1961; Emlen and Vehrencamp, 1983), including coots, hawks, kingfishers, cuckoos, woodpeckers, jays, swallows, tanagers, and others. Helpers may be of almost any age, and they ordinarily help birds of their own species. The Australian White-winged Chough, *Corcorax melanorhamphos*, is a polygynous species that commonly lives in groups of four to eight birds. The young tend to stay in the family group and assist it in rearing subsequent offspring by helping to build the nest, incubate the eggs, and guard, clean, and feed the young (Rowley, 1978). Cooperative breeding in some species seems decidedly beneficial. Among the Australian Kookaburras, *Dacelo gigas*, pairs without helpers raised an average of 1.2 fledglings, while pairs with one or more helpers averaged 2.3 fledglings (Parry, 1973). Similarly, among Florida Scrub Jays, *Aphelocoma*

FIGURE 17–5 The Short-toed Eagle of Europe feeds its nestlings bits of snakes, amphibians, and lizards. When older, the young bird will begin to tear apart the prey brought to it. Photo by E. Hosking.

FIGURE 17–6 The male Namaqua Sandgrouse soaking his belly feathers in a Kalahari Desert pool. The water, carried as far as 80 km in the bird's feathers, will be the young birds' only source of drinking water for about seven weeks. From Cade and Maclean, 1967.

coerulescens, breeding pairs with helpers produced more young than did unassisted pairs (Woolfenden, 1978). Breeding success is higher in Brown Jay, *Psilorhinus morio*, groups containing older, "more experienced," helpers (Lawton and Guidon, 1981). Long-term studies on several species indicate that although helpers appear to be behaving "altruistically," they are in fact waiting for a vacancy in a saturated habitat. Rowley (1981) studied cooperative breeding in the Malurid Wren, *Malurus splendens*. The population was skewed towards males so that most helpers were bachelors unable to find mates. After nine years of study, Rowley encountered a good production year as a result of which females outnumbered males. Helpers "budded off" to occupy new territories, each maintained by a simple pair. In another study disease decimated a breeding population of Florida Scrub Jays, creating territory vacancies. Helpers inserted and became breeders (Woolfenden and Fitzpatrick, 1984). It appears that where territories are in short supply, only dominant individuals actually breed, and subordinates are relegated to the role of helpers. To test this hypothesis, Hannon *et al.* (1983) removed breeders from the cooperatively breeding Acorn Woodpecker, *Melanerpes formicivorous*. Helpers moved in and competed vigorously for breeding vacancies if the reduced group consisted of members of the opposite sex. Strangers were repelled if groups contained members of the same sex as the invaders.

It may be that the numerous cases reported of birds feeding foster children of the same or of other species result from loss of their own young coupled with strong momentum of the feeding instinct.

Perhaps the zenith of interspecific feeding of young is represented by a Cardinal, *Cardinalis cardinalis*, that was observed for several days feeding goldfish in a garden pool. As the goldfish crowded to the edge of the pool with their open mouths, the Cardinal expertly delivered mouthfuls of worms to them! (Fig. 17–7). One can only guess how such a strange association arose, but it seems likely that the Cardinal, bereft of its young, approached the pool to drink, and was met by gaping goldfish accustomed to being fed by humans. The two instinctive appetites, one to feed, the other to be fed, magnetically attracted each other, and a temporary, satisfying bond was set up.

The Timing of Feeding Behavior

When the clockworks of a bird's reproductive cycle reach the proper time, its feeding instincts rise to the

FIGURE 17–7 Sometimes the urge to feed transcends species, and here, even class boundaries. This Cardinal was discovered feeding an adopted "brood" of goldfish. Photo by Paul Lemmons.

surface and will not be denied. But sometimes the reproductive clockworks, like man-made clocks, lose time, and the parents raise their young later than they normally do. In such cases the instinct to care for their young may be superseded by the stronger urge to migrate south, and the hungry young are abandoned and die. This occasionally happens in swallows and martins, when an early autumnal cold spell makes food scarce. Or the urge may appear in an adult before the young are born, in which case anticipatory food-bringing occurs. A male Canada Warbler, *Wilsonia canadensis*, brought food to the nest and attempted to feed the eggs as early as eight days before they hatched (Krause, 1965). Anticipatory "feeding" of this sort has been reported in wood warblers, tanagers, starlings, and other species. Skutch (1942) relates, contrariwise, how the instinct to feed its young dawned tardily in a male Mexican Trogon, *Trogon mexicanus*. Although the female began feeding the young as soon as they hatched, the male would fly to the nest cavity with an insect in its beak, sit there stupidly observing the hungry, clamoring young, and then fly off with the insect undelivered. Two or three days after the young hatched out, the male made five successive trips to the nest with food, but fed the young on only one of them. Two days later, however, he fed them regularly on every trip. In some species, such as the Common Nighthawk, *Chordeiles minor*, the young may not eat until they are two days old (Bent, 1940), living on stored yolk up to that time. If, in such species, the parents are ready to feed the young before they are ready to accept the food, this hereditary disparity in timing could have been the evolutionary beginning of

the habit of regurgitation. Finally, the instinct to feed young seems to achieve in the adult bird a momentum that carries on when the reason for its existence ceases. Common Murres, *Uria aalge*, that have lost their young frequently return and offer food to the empty nest for two or three days following the loss, or may feed the young of other adults (Birkhead and Nettleship, 1984).

The Stimuli for Feeding

Such peculiar happenings raise another question: What stimulus excites the feeding response in young birds? The young of some species will gape spontaneously. Newly hatched Starlings kept in a dark, vibration-free room were observed by Holzapfel (1939) to gape without apparent external stimuli up to four or five days when they were isolated, and 11 days when one bird in a brood stimulated another.

In nature, a variety of stimuli can elicit gaping in the young. The light-colored swollen flanges at the angles of the jaws of many altricial young are highly supplied with tactile nerve-endings (Herbst corpuscles). When these flanges are touched, the nestling's beak suddenly and vigorously springs open as though activated electrically (Fig. 17–8). A slight shaking of a nestful of very young birds will often cause them to gape. When adult Bank Swallows, *Riparia riparia*, enter the nest burrow to feed their young, they usually give a series of high-pitched notes, but if these fail to arouse the proper response, they gently trample on the young until they gape and are fed. Air currents are the tactile stimulus used to elicit the begging response in Chimney Swifts, *Chaetura pelagica*, a response no doubt based on the

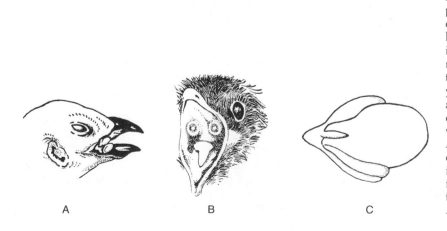

A   B   C

FIGURE 17–8 A, "Reflection pearls" at the angle of the jaws of an oriental parrot-finch. These light-reflecting bodies aid the parent in directing food into the nestling's mouth. After Sarasin. B, The attractive targets on the palate of the young Crested Coua of Madagascar are in the form of a raised bull's-eye and outer ring of glistening white, separated by a bright red ring. After Bluntschli. C, Dorsal view of a young Starling's head, showing the swollen, sensitive mouth flanges. When these are touched, the nestling's mouth springs open. After Portmann, 1950.

wing-flapping descent of the parents in the chimney. Gaping is elicited in nestling hummingbirds during the first five days of life when the mother touches the bulge behind the eye. Thereafter, the hovering activity of the female moves the dorsal down feathers on the nestlings' backs, which then elicits begging. Begging can also be elicited if the investigator blows at the down through a drinking straw (Schuchmann, 1983).

Acoustic stimuli are used by the Chough, *Pyrrhocorax pyrrhocorax*, and even a crude human imitation of their feeding call will arouse the gaping response. Clapping of the hands will cause some young cuckoos to gape. Many altricial young, after several days in the nest, respond to special feeding notes of the parents.

Since altricial young are commonly born blind, visual stimuli are not at first effective, although the young of woodpeckers and other species will often gape when the opening to the nest hole is quietly darkened. When such nestlings grow older and gain their sight, tactile and acoustic stimuli (except feeding-calls) lose their meaning, probably through habituation, and the young respond to visual stimuli. Experiments with crude cardboard models of their parents' beaks show that the pattern and color of the beak are remarkably potent releasers of the begging response. The chick of a Herring Gull, *Larus argentatus*, as Tinbergen (1953) discovered, will respond again and again, up to hundreds of times, to a crude model of its parent's beak, in spite of the fact that the dummy never provides it with food. The chick simply cannot resist the stimulus. In all species, the gaping response of the young probably depends both on the hunger of the young and on the external releasing stimuli.

Just as young birds need signals to set off their begging response, so adults need stimuli to release their feeding behavior. Normally, of course, the hungry young attract and stimulate the feeding adults with their cries and twitters, their vehement head waving and their wide-open mouths. The young of many altricial species also display very remarkable "feeding targets" at which the parents direct their food. Most nidicole young possess gaudy mouth linings, the preferred colors being yellows, reds, or oranges. Moreover, the flanges at the margins of the jaws are bright yellow or white, particularly so in young that live in dark, covered nests (hole-dwelling passerines, Picidae and Coraciidae). In nestlings of the Gouldian Finch, *Chloebia gouldiae* and related Parrot Finches, *Erythrura* spp., there are not only five black spots symmetrically disposed on the roof of the mouth and a black bar crossing the tongue, but also at each angle of the jaws three "reflection pearls," bead-like bodies brilliantly opalescent in emerald green and blue. These reflective spheres, found in many Estrildidae, glow like small lamps in the gloom of the covered nest and help the parents find the open mouths of the young. Not only do these transitory, extravagant structures function to arouse and direct the feeding responses of the parents, but in some birds they seem to identify the species of the young (Steiner, 1960). Among Zebra Finches, *Poephila guttata*, there are white mutants whose young lack these mouth markings. When experimental mixed broods were fed by either wild-colored or white parents, the wild-colored young with mouth targets were fed first, were given more food, grew faster, and had a higher survival rate than the targetless white young (Immelmann *et al.*, 1977).

A question naturally arises regarding the equal sharing of food among the young in a nest. Is it a matter of intelligent judgment on the part of the parent birds? Do they take care to see that all young are fed in turn, or that an underfed weakling gets preferred attention? This, too, seems to be a purely mechanical, instinctive business. When a young bird has just swallowed a worm, its swallowing reflex ceases to function again for some moments. So, when the adult comes to the nest and blindly thrusts some food in the nearest or most attractive gaping mouth, the youngster will swallow the food if it is hungry, but will not if it is sated. If the food is not swallowed, the parent takes it out and thrusts it into other mouths until it goes all the way down. Thus, instinctive responses normally assure an equal distribution of food.

Does instinct explain the apparently intelligent reaction to brood-size: larger and older broods needing and getting more food per day than smaller and younger broods? Or, for example, the remarkable adjustment by a female flycatcher to the emergency needs of her brood when the male parent dies? In such a situation the female, with a great burst of energy, begins feeding her offspring with as much food as the two parents together did previously. In a series of ingenious experiments with the Pied Flycatcher, *Ficedula hypoleuca*, in Finland, von Haartman (1953) analyzed such behavior and determined that here also the adjustment of the parent bird is instinctive and not learned.

By means of mechanical registering devices at 35 nest boxes, von Haartman established the fact that large broods of Pied Flycatchers received food more often than small broods. When he removed five young from a brood of seven, leaving only two in the nest, the number of feeding visits from the adults gradually decreased. If the number was reduced to one, the feeding visits fell off even more. By using young from other nests, and a nest box with a drawer-like bottom, sated young could quickly and unobtrusively be removed and replaced with hungry young in the experimental nest (Fig. 17–9). When five of a brood of seven young were removed from the nest and the two remaining young fed, then shortly replaced with two very hungry young from another nest, and as soon as these were fed, they too were replaced by two other hungry substitutes, and so on, it was found that the parent birds lavished practically as much food on the transient brood of two as they had previously fed to all seven. It was the reaction of the young that stimulated the parents to bring food, and not their number.

To determine whether the parents reacted to the appearance of the gaping young or to their cries, von Haartman arranged a double nest box with one youngster in the normal side with an entrance, and six hungry broodmates in the other side, invisible and inaccessible to the adults and separated from the normal side by a cardboard partition. Thus the parents could hear the hungry six but neither see nor reach them. The hungry, clamoring six in the hidden side stimu-lated the parents to feed the single accessible bird with more than twice its normal budget of food; and when it was so stuffed that it could swallow no more, the parents would urge it to accept more food with their feeding call. These observations, and similar experiments with Great Tits, *Parus major* (Bengtsson and Ryden, 1983), demonstrate that it is the hunger cry of the young that provides the most effective stimulus to feeding behavior in the adults (Figs. 17–10, 17–11).

The essential relation between begging calls and successful feeding of young is further illustrated by experiments with deafened Ring Doves, *Streptopelia roseogrisea*. After being fed for two weeks by deafened parents, young doves weighed only 56 per cent as much as young fed by normal parents (Nottebohm and Nottebohm, 1971). Experiments with deafened Bullfinches, *Pyrrhula pyrrhula*, yielded similar results. More drastic were the consequences of deafening Turkeys, *Meleagris gallopavo*. Unable to hear their young when they hatched out, mother Turkeys treated them as nest enemies and killed them (Schleidt et al., 1960).

Frequency and Amount of Feeding

Precocial young soon learn to feed themselves, usually under parental guidance, but altricial young must be fed for a period of days, weeks, or even months. Food is ordinarily brought to the altricial young fairly regularly, but the frequency of the trips varies according to species, the number of young, the age of the young,

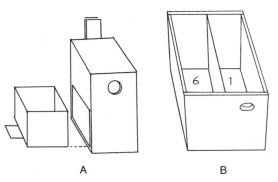

A   **B**   **C**

FIGURE 17–9 Experimental nest boxes used by von Haartman to discover the stimulus that controls the amount of food which adults bring to their broods. A, A nest box with a drawer-like bottom that permits the rapid and unobtrusive changing of nestlings. B, A double nest box (with top removed) showing the thin cardboard partition between chambers. The parent flycatchers could hear the six hungry young in the closed, left-hand chamber, but could see and feed only the single nestling in the right-hand one. C, An adult Pied Flycatcher. A and B after von Haartman, 1949.

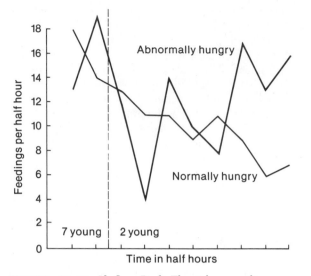

FIGURE 17–10 If five Pied Flycatcher nestlings are removed from a brood of seven, the two remaining birds ("normally hungry" in chart) are fed a decreasing number of times per half-hour by the parents. But if the two that remain in the nest box are continually and surreptitiously replaced with abnormally hungry young, the feeding rate increases sharply. After von Haartman, 1949.

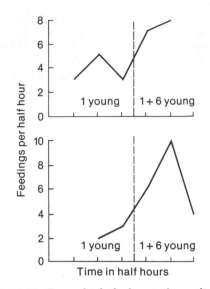

FIGURE 17–11 Parent birds feed a single nestling at an accelerated rate when they are stimulated by the hunger cries of six other invisible, inaccessible young. Left of the dotted line in each curve is shown the rate of feeding for the single nestling by itself; right of the line, the rate for the same bird while six hungry young are clamoring in the adjacent chamber. After von Haartman, 1949.

the method of carrying food, the time of day, the season, and the weather. Young of the Golden Eagle, *Aquila chrysaëtos*, are fed a hare or a grouse twice daily; those of a Bald Eagle, *Haliaeetus leucocephalus*, receive food about four or five times daily. Half-grown young Barn Owls, *Tyto alba*, are brought prey about 10 times a night. Sooty Shearwaters, *Puffinus griseus*, feed their young every other night on the average, but bring single meals that often weigh about as much as the nestling.

At the other extreme are small insectivorous birds such as the Great Tit, *Parus major*, which has been reported making as many as 900 feeding visits to the nest per day, or 60 per hour. These species bring small amounts of food each trip. Species that regurgitate food can bring larger and hence less frequent loads to the nest. An African Swift, *Apus caffer*, that carries food in a sublingual pouch, feeds regurgitated insects to its young about once each hour; but the Pied Flycatcher, *Ficedula hypoleuca*, which brings insects in the tip of its bill, feeds the nearly-fledged young 33 times an hour, or a total of some 6200 feeding trips during the nest life of the young. During cold, wet weather, when insects are not flying, the European Swift, *Apus apus*, may not feed its young for many days. The young live on their stored fat and may lose as much as one-half their weight, but by reverting to reptilian cold-bloodedness, they may survive as long as 21 days without food.

The rate of feeding normally increases with the growth of the young. At first, most nestlings eat *relatively* more food (often as much as their own weight per day) and grow more rapidly than they do later on. As they approach fledging, they eat absolutely more but relatively less per day—perhaps one-fourth of their body weight per day. For a brood of six European Nuthatches, *Sitta europaea*, Bussmann (1943) reported that the parents made 119 feeding trips per day the second day after hatching. This gradually rose to 353 trips per day on the 18th day, then declined to 270 trips on the 22nd day just before the young left the nest. In inclement weather, female House Sparrows, *Passer domestica*, increase their nestling feeding rate (McGillivray, 1983).

These are rather typical figures for a small hole-nesting passerine, and they reveal not only the increase in daily feedings with an increase in age and appetite of the young, but also a decline in the frequency of feedings shortly before the young leave the nest. This, too, is typical of many species and may reflect the fatigue of the overworked parents, or a naturally evolved device to encourage the nearly fledged young to leave the nest and seek their own food. As a possible example of the latter function, the Sooty Shearwater, *Puffinus griseus*, requires about 92 days to fledge, the last 12 of which are spent without food. This starvation period begins when the parent birds depart en masse on their annual migration travels. For the Manx Shearwater, *Puffinus puffinus*, the desertion period is about nine days; during the final few days the fat young come to the mouths of their burrows each night and exercise their wings. At the end of the starvation period the young make their way to the sea, plunge in, and begin an independent life.

In contrast, the parental feeding program of certain gulls and terns is very prolonged. Parents of the single-brooded Royal Tern, *Thalasseus maximus*, for example, continue feeding their young six months after they have hatched, and this occurs thousands of kilometers away from their breeding grounds. Ashmole and Humberto (1968) believe that this extended feeding period is correlated with the need of the young to learn highly skilled feeding methods for the exploitation of active and scarce foods. Adaptation toward reproductive success seems, in this instance, to favor long, subsidized apprenticeship in food gathering for a few young, but with less parental feeding and guidance. In the same way, a pair of Blackfooted Albatrosses, *Diomedea nigripes*, spend as much as 6 months feeding their single youngster.

Work Expended by Parents in Feeding Young

When one considers the energy required to raise a brood of demanding young birds, it is not surprising that the parents show fatigue at the end of the nesting period.

Numbers of Blue Tit, *Parus caeruleus*, were manipulated so that pairs fed 3, 6, 9, 12, or 15 young. Adult weights were examined 8 to 13 days after young hatched. Female weight loss increased linearly with brood size increase, as did female mortality (Nur, 1984). Some species travel great distances to gather food for a single meal for the young; the White Pelicans, *Pelecanus erythrorhynchos*, of Chase Lake, North Dakota, travel as far as 611 km (Johnson and Sloan, 1978). Serventy (1967) believes that the Short-tailed Shearwater, *Puffinus tenuirostris*, may travel as far as 1600 km from the breeding colony in search of food. The Common Swift, *Apus apus*, has been estimated to fly 1000 km per day gathering food for its two or three young. Of course, this search for food requires energy. By using radioactive tracers, Hails and Bryant (1979) determined that while feeding their young, adult House Martins, *Delichon urbica*, expended energy an average of 3.89 times their basal, or resting, metabolic rate. North of the Arctic Circle many species put in a long day caring for their broods. Several 24-hour observations have shown that a female Arctic Warbler, *Phylloscopus borealis*, fed her young steadily for 18 hours; a pair of Pied Flycatchers fed theirs for 19.6 hours, and in Lapland an adult male Bluethroat, *Luscinia svecica*, began feeding his young at 3:00 A.M. and retired for the night at 11:45 P.M.—nearly a 21-hour working day! It is small wonder that the female Pied Flycatcher loses 17 per cent of her weight by the end of the feeding period in Finland.

NEST SANITATION

Most passerine nidicoles will tolerate eggs or young birds in their nests, but no foreign objects. In a small nest crowded with six or eight young birds, the need for scrupulous sanitation seems clear. The nest must be kept dry and warm, and it should not afford a breeding site for insects and other parasites, nor should the careless disposal of feces betray its location to predators. To these ends many altricial species, especially passerines and woodpeckers, instinctively remove the fecal sacs of the young immediately upon their discharge. Since the sacs are enveloped in a clean, tough mucous membrane, the task is simple. In many species the feces are expelled by the young bird as it is fed, and are immediately seized by the parent and either (in the first few days) eaten or (later) carried away from the nest site. In the Prairie Warbler, *Dendroica discolor*, the two parents eat all of the sacs the first day or two, but at the end of the nestling period (about the 11th day), they eat only 5 per cent and carry the rest away (Nolan, 1978). Excreta are taken away from the nest by passerines on an average of about one trip in four.

When adult birds eat the feces of nestlings, they probably make use of the same kind of digestive economy that occurs in rabbits and hares. These animals have a large caecum in which cellulose foods are partly digested by bacteria. Two kinds of feces are formed:

FIGURE 17–12 A Barred Warbler removing a fecal sac from the cloaca of one of its nestlings. Photo by P. O. Swanberg.

the common, small, brown droppings, and a larger, softer, lighter type that comes from the caecum, and that the rabbit eats directly from the anus. By passing this food through its digestive tract a second time, the rabbit brings about its more complete digestion (Schmidt-Nielsen, 1960).

Emus, *Dromaius novaehollandiae*, are known to consume semidigested food particles in their droppings (Davies, 1978). Likewise, among African kudus and other antelopes of the genus *Tragelaphus*, the mothers, regularly, in the first few days, eat the feces and drink the urine of their newborn calves. Apparently the food passing through a very young nestling retains an undigested residue that may provide nourishment for the parent bird who is often too busy feeding the nestlings to seek food for itself. A brood of four nestlings of the White-crowned Sparrow, *Zonotrichia leucophrys*, was found by Morton (1979) to produce from 30 to 67 fecal sacs a day, weighing from 1.69 g (first day) to 24.3 g (seventh day). Those eaten by the two parents were estimated to provide about 10 per cent of their daily energy expenditures.

In disposing of fecal sacs, swallows and martins may drop them over water; wrens and nuthatches deposit them on tree branches away from the nest. The female Lyrebird, *Menura superba*, either submerges the fecal sac under water in a stream or digs a hole and buries it in the ground. As young swallows grow strong enough, they back up to the edge of the nest and defecate in a location more convenient for fecal removal. Some kingfishers possess an unusually muscular cloaca, and the older young turn their bodies around and discharge the fecal mass out of the tunnel opening. Young White-rumped Swifts, *Apus affinis*, defecate through the nest hole even before their eyes are open. Nestlings of the Pied Flycatcher, *Ficedula hypoleuca*, and Wrens, *Troglodytes troglodytes*, have been observed at nest-holes passing droppings from their bills to parents' bills.

Birds whose nests are in no need of concealment, such as eagles, usually shoot the liquid feces over the edge of the nest, marking the ground beneath with a white circle. Nestling hummingbirds also shoot liquid feces out of the nest, but away from the edge so that the nest remains clean (Orr, 1939). Trogons, motmots, hoopoes, and many pigeons are representative of species that practice no nest sanitation. The nest of the Wood Pigeon, *Columba palumbus*, becomes such a culture medium for vermin that in one nest 6573 individual insects and other invertebrates of 58 species were found in one liter of nest-bottom material. However, the Ruddy Quail Dove, *Geotrygon montana*, keeps its nest scrupulously clean by eating the feces.

Many species also remove from the nest the egg shells remaining after the young hatch (Fig. 17–13).

These are often eaten (Tree Sparrow, *Passer montanus*, Yellow Warbler, *Dendroica petechia*); fed to the young (Flamingo, *Phoenicopterus ruber*); or carried away (Golden Plover, *Pluvialis dominica*, Curlew, *Numenius arquata*). Bobwhite Quail, *Colinus virginianus*, however, leave them in the nest. The antipredator effectiveness of eggshell removal was demonstrated by Tinbergen (1963), who scattered eggs of the Black-headed Gull, *Larus ridibundus* (which normally removes its egg shells) about sand dunes. Half of the eggs had empty shells placed 10 cm away; the other half were by themselves. Crows and other gulls preyed on 65 per cent of the eggs accompanied by shells but on only 22 per cent of the unaccompanied eggs. Sometimes Great Horned Owls, *Bubo virginianus*, accumulate so many dead rabbits and other rodents in the nest that the young die from the effects of the insanitary mass. A Brewer's Blackbird, *Euphagus cyanocephalus*, on the other hand, removed its own dead young from the nest and deposited them in the place regularly used for fecal sac disposal.

Probably the most striking witness to the strong urge of most altricial birds to keep their nests clean is the experience many birdbanders have had in banding nestlings. In at least three instances on record, a parent bird tugged vigorously enough to break the leg of a young banded bird. Grinnell and Storer (*in* Bent,

1949) tell of a young American Robin, *Turdus migratorius*, whose parent had fed it a piece of meat too large for it to swallow immediately. The parent bird, obeying the instinct for cleanliness, removed the offending piece of meat with the young bird dangling beneath—a perfect example of throwing the baby out with the bath!

BROODING THE YOUNG

When the infant bird hatches from its shell it is nearly always brooded by the parent. This dries it out and helps to maintain sufficient warmth to promote metabolism and growth, since the young bird is born cold-blooded or exothermal. Different species require different times to establish warm-bloodedness or endothermism. Closely adapted to the gradual development of temperature control in the young is the instinctive reduction in the daily time devoted by the parent to brooding. In Laskey's (1948) study of the Carolina Wren, *Thryothorus ludovicianus*, she found that the female brooded 68 per cent of the daytime the first or hatching day, 65 per cent the second, 15 the third, 10 the fourth, 11 the fifth, and not at all thereafter. At night she brooded constantly. Lack and Lack (1952) determined that the Swift, *Apus apus*, brooded its young 98 per cent of the time when they

FIGURE 17–13 A Greenshank (*left*) and a Bullfinch (*right*) removing eggshells from their nests. Photos by E. Hosking.

were under one week old, 52 per cent the second week, and only 7 per cent after that. For the European Wren, *Troglodytes troglodytes*, Armstrong (1955) found that the average brooding time was reduced daily by 6 to 8 per cent. Not only is the egg of the Emperor Penguin, *Aptenodytes forsteri*, incubated on the upper surface of its feet, but the young chick is constantly brooded there for the first week. After two weeks the chick has developed sufficient body temperature control to remain off its parent's feet for an hour or so (Le Maho, 1977). In nearly all species the amount of brooding given the young increases in cold or wet weather, decreases in warm or dry. In warm habitats, nestlings of many species are protected from excessive heat by a parent standing, sometimes for hours, with spread wings so as to shade the young from direct sunlight.

Mourning Doves, *Zenaida macroura*, brood almost constantly, and nearly up to the time the young leave the nest. Lawrence (1948) observed on different days a nest of Nashville Warblers, *Vermivora ruficapilla*, in Ontario, and noted the time spent brooding the young and the number of feeding trips made to the nest. On the first day (Fig. 17–14) there were in the nest a one-day-old Cowbird, *Molothrus ater*, two just-hatched warblers, and one warbler egg. Five days later a six-day-old Cowbird, and three warblers, five and four days old, occupied the nest. The changes in time spent brooding and in the frequency of feeding trips show very clearly in the graphic record of these observations.

To discover whether the duration of a bird's brooding period was determined by some sort of internal clockwork or by external stimuli from the developing nestlings, Winkel and Berndt (1972) performed a series of nestling-substitution experiments with Pied Flycatchers, *Ficedula hypoleuca*. The normal period of nocturnal brooding by females lasts the first six or seven days of the nestlings' lives, during which they acquire endothermy. When warm, feathered nestlings (over seven days old) were substituted for naked ones, the females' total brooding periods ended earlier than normally. When five-day-old nestlings in one nest box were exchanged with younger ones, the period of brooding was extended up to four times its normal length. The brooding period, therefore, is not completely endogenous but can be shortened or extended by exogenous stimuli from the nestlings.

When birds rear their young in unusually cold environments, they meet difficult problems that require special solutions. One adjustment to the perpetual cold of the Antarctic is the irresistible urge to brood found in parent birds. Murphy (1936) observed a pair of Giant Petrels, *Macronectes giganteus*, that brooded for a week over a chick "which had been crushed as flat as a pancake." Some penguins are such avid brooders that they may fight for the privilege of mothering the chicks—to the detriment of the latter. A special adaptation to antarctic cold is seen in the Emperor Penguin, *Aptenodytes forsteri*. When their downy young are old enough to wander about, they gather in crèches or kindergartens, where they huddle together, surrounded by a watchful "snow fence" of adults. During storms, these crèches of chicks may be completely covered with snow, but the fat young emerge undamaged (Fig. 17–15).

This irresistible instinct to brood even transcends species barriers. Flickers sometimes raise young Starlings; some ducks are notorious for the mixed flotillas

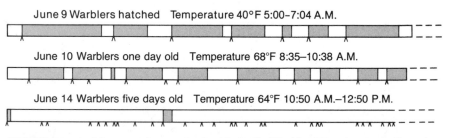

FIGURE 17–14 Nest-attentiveness in the Nashville Warbler. Two-hour samples taken from observations on three different days. Shaded portions of bars represent periods when the female was on the nest brooding the young; clear portions, when she was off the nest. Each dart under a bar indicates a feeding visit by one of the parents. As the young grow older they are brooded less often and fed more. After L. de K. Lawrence (1948).

FIGURE 17–15 A crèche of young Emperor Penguins, in Antarctica, still blanketed with snow from a recent storm. The adults protect the huddled young both from predators and from the sharpest blasts of the wind. Photo from Expéditions Polaires Françaises.

of ducklings in their care; a young eagle slipped into a chicken's nest will be solicitously brooded. Three hen's eggs were substituted by Dubois (1923) for those in the nest of a Short-eared Owl, *Asio flammeus*. The altricial owl hatched the eggs into precocial domestic chicks, brooded the downy young, and quite naturally brought them dead mice and decapitated birds to eat. The owl and the chicks got along amicably together for five days, the chicks being surreptitiously fed corn meal by the observer. Then two of the chicks died from overeating, but the other chick survived the owl's care for five days more and then was abandoned by its foster mother.

DEFENSE OF YOUNG

The commonest defense the young bird itself has against enemies is to crouch in the nest as quietly and inconspicuously as possible, or, if it is able, to flee from the nest and hide in the surroundings. When young Black Skimmers, *Rynchops nigra*, sense danger, they will quickly scratch a depression in the sandy beach and then stretch out in it so that their camouflaged backs are flush with the ground level. Young Hoopoes, *Upupa epops*, have three lines of defense against predators. First, they give a snake-like hiss when any intruder approaches their nest cavity. This may be followed by streams of liquid excrement directed toward the enemy. Lastly, they give off an offensive odor that comes from both their own and their mother's preen glands. Cats and other predators are reputed to avoid Hoopoes because of this odor. Nestlings of the African White-browed Coucal, *Centropus superciliosus*, repel enemies by hissing and vomiting a nauseating black liquid. Parent birds may meet predatory threats by directly attacking the predators, by distracting them from the young, or, less commonly, by leading or moving the young to a safer site. The vigor with which the parent defends its young is, according to Nice (1949), greatest in precocial species at the time the eggs hatch or shortly after, and in altricial species shortly before the young leave the nest. Anyone who has observed young birds extensively has been swooped at and screamed at by outraged parents, and occasionally even struck by their beaks or claws.

Some species, if the threat to their young persists, change their tactics and perform what for many years has been called "injury feigning" or the "broken-wing ruse." Today ornithologists usually call this behavior *distraction display*, placing the emphasis on its demonstrated effectiveness rather than on the conjectured intent of the bird. No one can deny the effectiveness of the performance once he has been its object. Not only humans, but cats, dogs, snakes, and many other predatory animals, including birds, have been "lured" away from nests or young birds by the persuasive drama. The parent bird flutters on the ground as though crippled, and utters piteous cries in its extremity, but it seems always to flutter just beyond reach of the disturbing intruder. In his efforts to catch the apparently incapacitated bird, the intruder is led farther and farther away from the nest, until suddenly the bird "recovers" normal mobility and flies away. Both the stereotyped nature of the performance and its occasional nonadaptive application (for example, by a Ringed Plover, *Charadrius hiaticula*, to a strange egg in its nest) argue against its being interpreted as an intelligent, purposive act. Perhaps the best current interpretation of distraction display is that it originated in the bird's inability to react simultaneously to two great drives: one to protect the nest or young, and the other to flee from the predator. The result is a frenzied compromise refined in its more convincing aspects by years of natural selection into what today appears to be "injury feigning."

Parent birds may also protect their young by carrying them away from danger. This may be done in a variety of ways. Over a dozen species of ducks, geese, swans, and grebes have been seen swimming with young on their backs. At least 16 species of waterfowl have been reported to carry young while in flight, holding them chiefly with the beak (Johnsgard and Kear, 1968). Also reported carrying young in their beaks have been rails, gallinules, harriers, and chickadees. The American Woodcock, *Scolopax minor*, is well known to carry its young, in flight, between its body and its legs; and the Egyptian Nightjar, *Caprimulgus aegyptius*, does the same while walking on the ground. The Chachalaca, *Ortalis vetula*, of Mexico carries her precocial young while they cling to her legs. A number of species, including the Red-tailed Hawk, *Buteo jamaicensis*, and Moorhen, *Gallinula chloropus*, carry their young with their claws or toes. Highly unusual is the method used by the American Finfoot, *Heliornis fulica*. The adult male carries the blind, naked young in pockets of body skin under each wing. The young are securely held whether the parent swims or flies (del Toro, 1971). The parent African Jaçana, *Actophilornis africanus*, carries its young while walking by holding them under its wings against the body.

More effective protection is the instinctive response of nidifugous young to warning cries of their parents. Young Ruffed Grouse, *Bonasa umbellus*, are described by Edminster (1947) as instinctively responding in different ways to four different calls of the mother when a hawk threatens her brood. At one call the young freeze; at a second, more violent call, they scatter and freeze; another call keeps them immobile; and a final "all clear" call brings them back to her. Young Turnstones, *Arenaria interpres*, and other species are reported by Bergman (1946) to freeze at parental warning even before they hatch from the egg. Parent Southern Lapwings, *Vanellus chilensis*, produce different alarm calls, one for reptilian and the other for avian-mammal predators (Walters, 1980).

Infanticide: Cannibalism and Fratricide

Although solicitude for the young is instinctively shown by nearly all species of birds, there are times when protective instincts seem to be imperfectly developed, or perhaps are nullified by competing instincts. In a New Zealand colony of Dominican Gulls, *Larus dominicanus*, Fordham (1970) estimated that 75 per cent of the young were killed (though rarely eaten) by breeding adults. A large proportion of the 23 per cent mortality of young Herring Gulls on the Isle of May, Fife, was attributed by Parsons (1971) to cannibalism by the breeding adults. A parent gull cherishes its own young in the nest, but treats all other young as strangers and trespassers to be driven from its nesting territory. Consequently, many lost and wandering chicks are killed by vicious jabs of nest-defending adults. Once dead, the young become food for adult gulls or predatory and scavenging crows, ravens, and buzzards.

A great variety of birds practice cannibalism of their own young: gulls, terns, storks, ducks, hawks, eagles, cranes, owls, shrikes, and others. Mebs (1964) found evidence that cannibalism was related to food abundance in the Buzzard, *Buteo buteo*. In years of reduced food supplies cannibalism was more frequent in broods of four young than in broods of two or three. Diet-related cannibalism has been observed in Bullfinches, *Pyrrhula pyrrhula*, and other species. Captive Kestrels, *Falco sparverius*, in outdoor aviaries ate their eggs and young until their apparently adequate diet was changed to a slightly different one (Porter and Wiemeyer, 1970).

But factors other than food undeniably enter in. Certain species seem to possess aggressive temperaments that promote infanticide. The most important factor causing mortality among wild Mallards, *Anas platyrhynchos*, in Finland was found by Raitasuo (1964) to be defensive attacks of females against Mallard broods other than their own. Within three days ducklings learn to recognize other ducklings of their brood and thereafter accept no new members to the flock. Ducklings separated from their broods almost inevitably perish.

A significant clue to the origin of infanticide is probably seen in the behavior of young male White Storks, *Ciconia ciconia*, in their first breeding year (Schüz, 1943). They often pick up their own young and shake them as if they were prey, and sometimes even eat them. Here, apparently, the instincts for rearing and protecting the young have not matured sufficiently to supplant or counteract those directed toward protecting the nest against predators. It may well be that the timetable for the seasonal (or lifetime) development of stereotyped behavior patterns must progress in a smooth, integrated sequence for optimum survival of an individual. If either internal or external forces should throw the train of instinctive behavior off its schedule, aberrant behavior may result. It may also result if reciprocal and synchronous forms of programmed stereotyped behavior between parent and young are disturbed. This explanation probably accounts for the killing of newly hatched chicks by deafened mother turkeys, as described earlier. The maternal feeding behavior may be evoked by a heterogeneous summation of both visual and auditory stimuli from the young, and when one of them is lacking, the feeding response fails. Similarly, if a young eagle nestling becomes weakened through competition with an older nestmate, or from dearth of food, and no longer has the strength to beg for food in the normal stereotyped fashion, the parent bird may ignore it and trample it into the nest, or consider it prey and eat it.

A great variety of birds also practice fratricide (siblicide) or "cainism," in which one young bird kills its sibling nestmate. In thus sacrificing close kin (indirect fitness), the individual improves his own chances of survival (direct fitness) (Mock, 1985). Mock (1984) noticed stronger nestling Great Egrets, *Casmerodius albus*, killing weaker siblings in the competition for small fish provided by parents. By contrast, young Great Blue Herons, *Ardea herodias*, which feed on large fish, seldom kill their siblings. Heron chicks, cross-fostered and raised by Egrets on small fish, became siblicidal. Egret chicks raised by Herons, however, did not reduce aggression or siblicide. Fratricide occurs among a large variety of birds and with varying degrees of vigor, such as Brown Boobies, *Sula leucogaster*, White Boobies, *S. dactylatra*, South Polar Skuas, *Catharacta maccormicki*, and White Pelicans, *Pelecanus erythrorhynchos* (Young, 1963; Johnson and Sloan, 1978). In the African Crowned Eagle, *Stephanoaëtus coronatus*, and Verreaux's Eagle, *Aquila verreauxi*, the elder of the two young is reputed invariably to attack and kill the younger (Brown, 1966; Siegfried, 1968). Occasionally nestlings may become cannibals. After a seven-year study of over 1000 nests of Barn Owls, *Tyto alba*, in France, Baudvin (1979) concluded that two-thirds of the deaths of the young were due to cannibalism.

DEVELOPMENT OF YOUNG

As described early in this chapter, there are two basically different types of avian development: precocial and altricial. The essential difference between the two is that the precocial bird undergoes stages of develop-

FIGURE 17–16 A young Arctic Skua, *Stercorarius parasiticus*, kills its brother while an adult stands by. Photo by N. Rankin.

ment within the egg shell that the altricial bird undergoes after hatching. Besides these differences there are other principles of development that apply rather widely. As a general rule, those species that have a short incubation period show rapid cell multiplication and rapid growth in their young; those with a long incubation period show slower growth. Likewise, those young fed by their parents grow rapidly and those that feed themselves grow slowly. Precocials that feed themselves (quail, grouse, rails) burn up extra energy in general activity, food finding, and thermoregulation (Ricklefs, 1968a). The incubation period of the House Wren, *Troglodytes aëdon*, for example, is 14 days, and the young reach their maximum weight in 12 days. The Alpine Swift, *Apus melba*, requires 19 to 20 days for incubation, and reaches adult weight in about 27 days. Generally, large species grow more slowly than small (Ricklefs, 1973): the Field Sparrow, *Spizella pusilla*, reaches adult weight in 8 or 9 days, the Sparrow Hawk, *Falco sparverius*, in about 20 days, the Yellow-billed Tropic-bird, *Phaethon lepturus*, in 41 days, while the California Condor, *Gymnogyps californianus*, reaches only two-thirds its adult weight when 100 days old. Even species of the same adult

size show great variations in growth rate, as is seen in the growth curves of the White Pelican, *Pelecanus erythrorhynchos*, and Mute Swan, *Cygnus olor* (Fig. 17–17).

Before they leave their nests, many young birds weigh more than their parents. Sixty-three days after hatching, the White Pelican weighs 13,850 g, whereas as an adult it will weigh 10,000 g. The chick of a Band-rumped Storm Petrel, *Oceanodroma castro*, weighs 78.4 g at its peak, but only 48.6 g at fledging. The adult weighs 43.5 g (Ricklefs, 1968b). Weight recession is characteristic of penguins, albatrosses, shearwaters, petrels, gannets, pelicans, parrots, owls, kingfishers, swallows, bee-eaters, and other birds. The excess weight represents fat storage that enables the young birds to survive periods of food shortage and also gives them energy to use when they begin to grow feathers and to exercise their muscles.

Quite logically, the asymptote in weight (levelling off of) development in a nestling correlates with its post-fledging care (Cody, 1971). Young of the Common Murre, *Uria aalge*, leave their nests at about one-third the weight of adults and are fed by their parents for several weeks. Young of the Rhinoceros

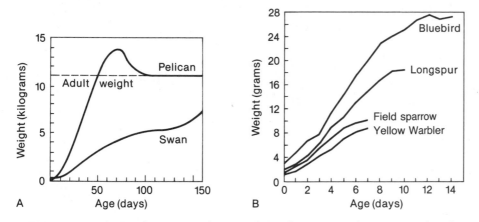

FIGURE 17–17 A, A pelican not only grows faster than a swan, but two months after hatching, it exceeds its own adult weight by nearly 3 kg. After Portmann, 1950. B, Most young passerine birds reach mature weight quickly, and without greatly exceeding their adult weight. Curves end on the day the young typically leave the nest. After data from Hamilton, 1943; Mickey, 1943; Schrantz, 1939; and Walkinshaw, 1943.

Auklet, *Cerorhinca monocerata*, leave after reaching about 80 per cent of adult weight and receive little if any feeding. Cassin's Auklet, *Ptychorhamphus aleuticus*, leaves the nest when it has reached adult weight or more, for a completely independent life.

Most young songbirds, such as the Eastern Bluebird, *Sialia sialis*, Yellow Warbler, *Dendroica petechia*, and Field Sparrow, *Spizella pusilla*, increase in weight rather steadily until they reach adult weight, at which point they level off. The growth weights of all species are subject to decline when adverse weather influences feeding frequency.

Both as an ultimate and as a proximate factor, climate affects the rate of juvenile development. Ricklefs (1976) found that, on average, 30 species of neotropical passerines grew 23 per cent more slowly than did 51 of temperate zone passerines. Reasons suggested for this difference: tropical species have a lower basal metabolic rate, a shorter day for eating and assimilating food, and, especially for fruit eaters, a relatively low content of nitrogen in their diets.

In different species not only does absolute size increase at different rates, but different parts of the body grow and mature at different rates. This has already been pointed out for the growth rates of the alimentary canal and nervous system of altricial and precocial young. Quite often those organs that are needed soonest by the growing young develop first.

Heinroth (1948) has shown that although both the precocial crane and the altricial stork grow up and are capable of flight in about 70 days, the ground-dwelling crane's legs grow most rapidly (7 mm per day) between the 8th and 16th days of life, and those of the nest-dwelling stork most rapidly (5 mm per day) between the 20th and 38th days. It would be hazardous for the roof-top stork to walk too early in life.

Like their reptilian ancestors, birds are born cold-blooded or exothermal. In altricial species, endothermal temperature regulation begins several days after hatching and then develops rapidly (Fig. 17–18). In precocial species it begins in the embryo within the egg and develops more slowly. Reckoned from the beginning of incubation it takes the typical altricial bird 20 to 29 days to develop sufficient endothermy to be independent of parental brooding or shading. The typical precocial bird requires between 37 and 48 days to reach the same status (Nice, 1962). Reckoned from hatching, altricial birds require from about 4.5 to 18 days to achieve endothermy (Dunn, 1975), and precocial birds, zero days to a month or more. The highly precocial megapodes are endothermal on hatching. For example, newly emerged Mallee Fowls, *Leipoa ocellata*, may elevate their body temperature to three times above standard when exposed to cold stress (Booth, 1984). The Bobwhite Quail, *Colinus virginianus*, and American Oystercatcher,

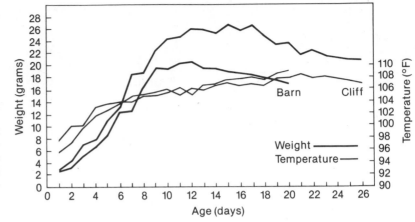

FIGURE 17–18 Growth and temperature development curves of young Barn Swallows and Cliff Swallows. In each species, maximum weight is achieved before complete temperature regulation. Food must first be made available before physiological processes can draw on it for the maturation of structures and functions. After D. Stoner, 1945.

Haematopus palliatus, can maintain thermal independence at moderately cool temperatures when a day or two old. The single downy young of Wilson's Petrel, *Oceanites oceanicus*, is brooded in its antarctic burrow only a day or two after hatching, becoming endothermal at the age of two days. From then on it remains unattended all day long at an air temperature of 5°C. The closely brooded Adelie Penguin, *Pygoscelis adeliae*, in the same chilly habitat, does not establish endothermy until the 15th day. Approximate ages for reaching independent warmbloodedness in a few representative species are: Tree Sparrow, *Spizella arborea*, about 4.5 days; House Sparrow, *Passer domesticus*, 10 days; House Wren, *Troglodytes aëdon*, 15 days; Lapwing, *Vanellus vanellus*, 17 days; domestic Chicken, *Gallus gallus*, 24 days. In most species effective thermoregulation does not develop until the young bird is clothed with a thick coat of feathers.

As young birds develop, in and out of the nest, they achieve increasingly complex behavior, some innate, some learned. Part of the development of the young bird is a simple maturation of structures or of instincts. A bird cannot use its eyes until they are open, or its wings until they are fully feathered. While there are numberless variations of the scheme presented here, the day-by-day development of a typical small, opennesting, altricial bird is shown in Table 17–3.

Some altricial species develop much more slowly than this. The hole-nesting African Lesser Honeyguide, *Indicator minor*, for example, is still blind and nearly naked when 11 days old, and does not leave its tree-cavity nest until it is about 35 days old (Friedmann, 1955). Young toucans do not leave their nests and fly until they are about 50 days old. A precocial bird, on the other hand, is endowed with most of the above accomplishments on the day it hatches. In a sense, it passes its nestling stage in the egg. A three-day-old Lesser Scaup, *Aythya affinis*, for example, can successfully dive, catch a minnow, and return to the surface of the water in less than 5 seconds (Bartonek and Hickey, 1969). The Malau Megapode, *Megapodius pritchardii*, is able to fly at hatching (Weir, 1973), and young of the Hazel Grouse, *Tetrastes bonasia*, can fly when only four days old. At the other extreme, the young Wandering Albatross, *Diomedea exulans*, makes its first flight when about 275 days old.

Obviously, some of these age-related developments are largely genetically controlled, and others are largely environmentally controlled. Juvenile Garden Warblers, *Sylvia borin*, of southern Finland, compared with those of southwest Germany, experience different day-lengths as they develop. Also, the Finnish birds have an earlier and shorter juvenile molt period than the German population, an earlier and shorter migratory restlessness period, and an earlier period of premigratory fattening. To test whether these age differences in development in the two populations were due to the different photoperiods under which they lived, Berthold (1977c) and Gwinner (1979) handreared young warblers under simulated

TABLE 17–3.
DEVELOPMENT OF TYPICAL BEHAVIOR IN ALTRICIAL YOUNG

DAYS AFTER HATCHING	CHARACTERISTIC BEHAVIOR
1	Raises head and gapes at any disturbance. Evacuates in the nest.
2	Eyes begin to open. Primary feathers show through the skin.
3	Eyes open. Uses wings as props. Evacuates at edge or door of nest.
4	Beginning of most rapid growth. Begins to use legs. Voice stronger. Primaries pierce skin.
5	Bird faces opening of nest or the feeding parent. Stands and preens.
6	Shows better muscular coordination. Shows fear and responds to alarm note of parent.
7	Exercises neck, wings, and legs. Crouches, scratches head. Relative growth rate begins to decline. Primary feathers unsheathing.
8	Alert to sights and sounds outside nest. Temperature control nearly established.
10	Preens unsheathed feathers. Stretches wings and legs. Pecks at objects.
13	Acquires fluttering flight. Leaves nest.

photoperiods of each others' natural daylengths. Their results showed that these aspects of development were genetically and not environmentally controlled.

DURATION OF THE NESTLING PERIOD

The length of time that the altricial young stay in the nest is correlated not only with the length of the incubation period (long incubation, long nestling period) but also with the size of the species, larger species having longer nestling periods. Furthermore, birds such as the American Robin, *Turdus migratorius*, which nest in the open and whose eggs and young are accordingly exposed to many environmental hazards, have shorter nestling periods than cavity nesters such as woodpeckers. For 54 European open-nesting species, Lack (1948) determined that the average nestling period lasted 13.2 days, while for 18 hole-nesters it lasted 17.3 days. Nice (1957) found that the corresponding figures for United States species were: 11 species of open-nesters, 11.0 days; 10 species of hole-nesters, 18.8 days. Obviously, natural selection has shortened the nestling period for species most susceptible to disaster.

In addition to the primary factor—an instinctive maturation of behavior that, at a specific time, stimulates the young bird to leave the nest—such factors as the weather, hunger, accidents, and parental nudgings influence the time of departure from the nest. As was observed earlier, the amount of food brought the young is often reduced, or even completely cut off, during the last few days of nest life. Davies (1978) calls this withholding of food parental "meanness," and suggests that its function is to promote independence of offspring. He then points out that the timing of the onset of parental meanness is of survival significance to a species. If food is withheld too soon, the offspring may not survive; if the young are fed too long, the investment in parental care may be at the expense of potential future offspring.

Many birds apparently entice their young from the nest by holding food so far away that the hungry youngster will have to leave the nest to reach it. With drastically reduced diets, the Bald Eagle, *Haliaeetus leucocephalus*, starves its young into leaving the nest. The Japanese Paradise Flycatcher, *Terpsiphone atrocaudata*, has been observed by Jahn (1939) feeding two young that had already left the nest, 30 times in one hour; but it completely ignored the begging nestling left behind until it, too, left the nest. Tree ducks entice their cavity-dwelling children to leave the nest *en masse* by calling them from the ground below. The young tumble out in a torrent, flap their stubby wings ineffectually in their downward plunge, bounce off the earth, and waddle after their mother to the nearest water.

EDUCATION OF THE YOUNG

After leaving the nest the young bird still has much to acquire in the way of behavior: skill in finding and eating suitable food, skill in flight, in song, in social relationships and other functions. Some of these skills are a matter of the maturation of instincts. Many young birds will go through bathing movements the first time they see water, or will perform nest-molding or copulation movements without having observed

FIGURE 17–19 A Chipping Sparrow feeding its young. The eyes of the nestlings are just beginning to open, but they have not yet learned to gape toward the feeding parent. Their beaks still show the touch-sensitive flanges. Gaping upward in response to touch is instinctive. Photo by G. R. Austing.

them in older birds, and before they are themselves old enough to breed. Experience soon teaches a young chick which specks are pebbles and dirt and which are palatable food.

Sometimes enticement from the nest and education of the young in food capture are combined in one operation. Swallows and martins have occasionally been observed feeding their young, late in the nestling period, by flying back and forth in front of the nest with insects in their beaks that the young seize without leaving the nest. A similar technique was reported by Bralliar (*in* Bent, 1940) for the Eastern Belted Kingfisher, *Ceryle alcyon*, which beat a small fish into sluggishness and then dropped it into the water, where the young found it easy prey for their

first inexpert dives for food. The young also practiced fish-catching by diving to retrieve small twigs from the water. Northern Harriers, *Circus cyaneus*, and others give their newly fledged young practice in catching prey by dropping a mouse, for example, in midair; the young bird catches it with its talons after it has fallen only a few feet. Mebs (1972) describes what appears to be well-nigh intentional education of young Peregrine Falcons, *Falco peregrinus*, by their parents. "While one adult flies over the youngster with prey and drops it, the other adult flies lower and catches the prey if the youngster's swoop at it failed. This adult then climbs to repeat the process. Finally the adults also bring living birds which the young catch and kill." Many observers tell of the play of young hawks and eagles that prepares them for the serious duties of making a living. Cade (1953) recounts the play of a young Prairie Falcon, *Falco mexicanus*, with a bit of dried cow manure that it swooped on, picked up with its talons and dropped, only to swoop, pick up and drop, again and again. Play behavior has been noted in a number of avian taxa (Ficken, 1977), and usually involves object manipulation or locomotory play, fighting, or sexual play.

Imitation plays a role in the education of the young. A domestic chick raised with a Swan-goose gosling, *Anser cygnoides*, soon took on some of the gosling's behavior such as grass-eating, and even mimicked the phraseology and modulation of the gosling's voice (Klopfer, 1956). Bildstein (1980) observed that one Northern Harrier, *Circus cyaneus*, playing with a mouse-sized piece of corncob was likely to stimulate others in the vicinity to do the same.

Flight exercises are performed by many young before they leave the nest. Young eagles and hawks will stand on their nest rim and practice flapping their wings with daily increasing skill until they finally take off into the air. Many swifts and swallows, on the contrary, have no opportunity to practice wingstrokes in their cavity nests, and yet are able to fly expertly on

their first attempt, sometimes as far as a half kilometer. Turkey Vultures, *Cathartes aura*, that have been confined in cages too small to allow wing motion are unable to fly even when three months old, because of lack of exercise.

■ SUGGESTED READINGS
Probably the richest source, both for illustrations of developing young birds and for extensive information on their care and development, is Heinroth and Heinroth, *Die Vögel Mitteleuropas*. The Heinroths' small book *The Birds* gives a brief introduction to the subject. Pycraft's *A History of Birds* is still worth consulting for numerous examples of care and development of young. For a theoretical consideration of the evolutionary development of parental care, see Kendeigh's *Parental Care and its Evolution in Birds*. *Parent Birds and their Young*, by Skutch, provides a readable, stimulating introduction to its subject. For an excellent, thoroughly documented treatment of its subject, see Nice's *Development of Behavior in Precocial Birds*. Recent treatises on the development of birds in general are to be found in R. R. O'Conner's *The Growth and Development of Birds* and S. M. Smith's "The Ontogeny of Avian Behavior" in *Avian Biology* Vol. VII.

The Numbers of Birds and Their Regulation

18

Death may come by tooth or claw,
Come by thirst or by starvation,
Come by wind or rain or snow,
By disease, or conflagration—
Risks enough, by Nature's law,
Worse, by man's manipulation.
—Joel Peters (1905–   ),
Forever Death in Life

A favorite pastime of biologists is to compute the total number of descendants left by a single animal or a pair of animals, carried through several generations, assuming that all young live to maturity to reproduce as did the original parents. An oyster, for example, that lays over a million eggs at one spawning would, if all young survived to breed, beget sufficient offspring in only seven generations to make a heap of oysters greater than the mass of the earth. Many organisms are wastefully lavish when it comes to reproducing their own kind. A carp lays two to four million eggs a season; a tapeworm, 120,000 eggs a day. Even the American Robin, *Turdus migratorius*, which commonly produces, in one season, two modest broods of four young each, will leave a spectacular 24,414,060 descendants in 10 years, if one makes the superficially plausible assumption that all young survive at least 10 years and reproduce the same number as the original pair. If one should carry this kind of reckoning to a total of 30 years—not a long time in the evolutionary history of Robins—he will discover that the original pair of birds would have by that time nearly 2000 million million million descendants. These would be sufficient to cover the entire earth with a feathery coat approximately 7.2 kilometers (4.5 miles) deep.

Such intellectual exercises as these merely underscore the reproductive potential of a species. Common sense quickly tells us that this potential is never realized. Nonetheless, certain species of birds sometimes achieve almost incredible populations. In 1871 the Passenger Pigeon, *Ectopistes migratorius*, concentrated in a "great nesting" in central Wisconsin where an estimated 136 million birds bred within an area of 2200 square kilometers (850 square miles) (Schorger, 1937). Another flock of this species was estimated by Alexander Wilson to contain over 2000 million birds. The Mutton Bird, or Short-tailed Shearwater, *Puffinus tenuirostris*, flies at dusk to its home islands in immense river-like streams. One such stream observed in 1909 required the better part of a night to pass a given point, and another flock observed in 1798 by Flinders (Murphy, 1936) took only an hour and a half to fly by, but was estimated to contain more than 1.5 million birds.

Individual breeding colonies of the Red-billed Quelca, *Quelea quelea*, may cover an area of several hundred hectares of African savanna and contain over 10 million nests. Blackbirds (Icteridae) of various species commonly gather in large winter roosts in the United States. Perhaps the largest roost is one in the Dismal Swamp of Virginia that contains an estimated 25 million birds (Meanley and Webb, 1965).

The total numbers of birds in different parts of the world can be only very roughly approximated. From scores of breeding bird censuses taken by the Audubon Society, Peterson (1963) estimates that there are about 6000 million breeding birds in the United States (or about four birds per hectare), and that between two-thirds and three-fourths of them are land birds. For the North American continent north of Mexico he estimates that there are between 12,000 and 20,000 million birds. For the entire world, Fisher (1951) estimates that there are in the neighborhood of 100,000 million birds. This number is only one-third as many as a single pair of robins might leave as descendants in 16 years, if we apply the original assumption that all offspring live to reproduce eight young each year!

Citing various authorities, von Haartman (1971) gives the following estimates of total populations of European birds: Germany 200 million; England, Scotland and Wales 120 million; Finland 64 million.

Obviously, something kills off most of these hypothetical offspring before they reach breeding age. Even so, the reproductive potential of a species is occasionally allowed to express itself with considerable effectiveness. When a species is introduced into a new and favorable habitat, its numbers increase phenomenally at first, and then level off at a population far under theoretical possibilities. This has been the history of the House Sparrow, *Passer domesticus* (Robbins, 1973), and Starling, *Sturnus vulgaris*, in North America, the latter species increasing, in less than 60 years, perhaps a million-fold over the 120 birds introduced into New York City (Peterson, 1948). In 1937, two male and six female Ring-necked Pheasants, *Phasianus colchicus*, were introduced on Protection Island off the coast of Washington. Six years later these eight birds had increased to 1898 (Einarsen, 1942).

Reproductive potential in a given species depends chiefly on the number of eggs laid in a clutch, the number of clutches in a season, the age at which the bird begins breeding, and the bird's longevity. For most species the chief element in breeding potential is the number of eggs laid per year, and in this characteristic there is wide variation.

As seen earlier, different kinds of birds may lay between 1 and 20 eggs per year, and may attain sexual maturity at between 12 weeks and 8 years of age. Obviously, a bird with long-delayed sexual maturity must have a compensating long life-expectancy. Large clutches in themselves are no guarantee of a large population. The parent bird can successfully incubate only so many eggs and feed and protect only so many nestlings. And even these restraints are not as simple and direct as first appears. As brood size increases, heat loss per nestling decreases, so that less food and less heat from the parent are required per nestling. This thermal efficiency of large broods is, however, offset by three counteracting consequences. Larger broods may also mean: (1) greater physiological exhaustion of the parent; (2) higher population densities in the next generation that may have repercussions on food supplies, nesting sites, and territory; and (3) reduced life-expectancy of the young birds. Kluyver (1971), for example, artificially reduced the clutches of a population of Great Tits, *Parus major*, by 40 per cent and observed a near doubling in their subsequent adult survival. Similarly, Murphy (1983) found that 14-day-old Eastern Kingbirds, *Tyrranus tyrranus*, weighed more in broods of three than in broods of four or five. The ability to fledge young successfully in large broods depended on good weather conditions which brought more flying insects for this aerial feeder.

The degree to which a bird will realize its reproductive potential depends on a complex of interrelated external influences: the type of pair bond; the quantity and quality of food fed the young; the age and physiological vigor of the parents; experience in breeding; population density; climate, latitude, and day length; availability and ecological location of nest sites; size and type of nest (open or covered); rate of development of young; tendency toward infanticide; predation, disease, and parasites; brood parasitism; emigration and immigration, and other factors. For a given species certain of these factors will clearly weigh more heavily than others in determining reproductive success. Older birds tend to be more successful breeders than younger birds, partly because foraging efficiency improves with age and practice, and partly because younger birds have poorer access to resources because of competition with older birds (Curio, 1983). For example, a 30-year study of banded Arctic Terns, *Sterna paradisaea*, on the Farne Islands off Britain, revealed that egg volume and clutch size increased with age. Also, eight-year-old and older birds began breeding before younger birds, which were relegated to relatively inferior areas (Coulson and Horobin, 1976). Considering the large variety of factors that influence breeding potential in birds, von Haartman (1971) declared that "at the present time no definite population theory for birds seems possible."

LONGEVITY

Here again, one must distinguish between potential longevity, as realized by captive birds protected from predators, adverse weather, starvation, accidents, and other environmental hazards, and the actual longevity achieved by birds in the wild. To achieve such longevity records, the wild bird must be caught, individually marked—usually with a numbered aluminum leg band—released to live a normal wild life, and then subsequently recovered, usually at the time of its death. Ideally the bird should be banded as a nestling so that its complete life span may be known, but many longevity records are for wild birds

that were caught and banded when already adults. Numerous studies suggest that among the larger sea birds an annual adult survival rate of 90 to 97 per cent may not be uncommon (Ashmole, 1971). This implies that many breeders may survive 50 years or more.

Table 18–1 gives the maximum ages currently attained by banded wild birds as reported chiefly in the journals *Bird-Banding* and *Die Vogelwarte*. About one-half of these records represent the maximum longevities of birds recovered from the approximately 15 million banded wild birds recorded in the files of the U.S. Fish and Wildlife Service. The majority of the remainder represent similar records from European banding programs. It should be emphasized that the table gives the extreme old age attained for each species, usually by only one bird out of thousands banded. For instance, out of 21,715 Purple Finches, *Carpodacus purpureus*, banded by Magee (1940), there were subsequently 1746 recoveries, among which only one bird lived as long as ten years, six as long as eight years, and 18 as long as seven years.

Numerous records of longevity in captive birds have been published. A compilation of such records was made by Flower (1938) from 35 published sources covering over a thousand species of birds. Representative examples from Flower's list include the European White Pelican, *Pelecanus onocrotalus*, 51 years; Canada Goose, *Branta canadensis*, 33; Condor, *Vultur gryphus*, 52; Bateleur Eagle, *Terathopius eucaudatus*, 55; Australian Crane, *Grus rubicunda*, 47; Eagle Owl, *Bubo bubo*, 68; domestic Chicken, *Gallus gallus*, 30; domestic Pigeon, *Columba livia*, 30; Cockatoo, *Cacatua galerita*, 56; Raven, *Corvus corax*, 24; Starling, *Sturnus vulgaris*, 17; Garden Warbler, *Sylvia borin*, 24; House Sparrow, *Passer domesticus*, 23; Cardinal, *Cardinalis cardinalis*, 22; European Goldfinch, *Carduelis carduelis*, 27.

POPULATION STABILITY

Even the most casual observer realizes that the number of robins about his home does not change drastically from year to year. Careful yearly observations of conspicuous species confirm that fact of their relative stability in numbers. This stability can only mean that birth rate equals death rate; that natality matches mortality. This balance between life and death is convincingly demonstrated by investigations of different geographic populations of the Blue Tit, *Parus caeruleus*,

TABLE 18–1.
MAXIMUM KNOWN AGES OF BANDED WILD BIRDS

SPECIES	MAXIMUM AGE (YEARS)	WHERE BANDED	AUTHORITY*
Yellow-eyed Penguin, *Megadyptes antipodes*	18	New Zealand	†
Red-throated Loon, *Gavia stellata*	23	Sweden	BB 24:70
Laysan Albatross, *Diomedea immutabilis*	42	Midway Island	BB 46:3
Short-tailed Shearwater, *Puffinus tenuirostris*	30	Australia	BB 42:222
Brown Pelican, *Pelecanus occidentalis*	31	Florida	BB 46:58
Double-crested Cormorant, *Phalacrocorax auritus*	23	Maine	BB 39:326
Great Frigate-bird, *Fregata minor*	34	Line Islands	BB 40:47
Purple Heron, *Ardea purpurea*	25	Austria	DV 29:230
White Stork, *Ciconia ciconia*	26	Germany	DV 26:355
Trumpeter Swan, *Olor buccinator*	24	Montana	BB 46:59
Canada Goose, *Branta canadensis*	23	Ontario	BB 46:59
Mallard, *Anas platyrhynchos*	29	Iowa	BB 46:59
Teal, *Anas crecca*	20	U.S.S.R.	BB 29:112
Kite, *Milvus milvus*	26	Switzerland	BB 37:128
Golden Eagle, *Aquila chrysaëtos*	25	Germany	‡
Honey Buzzard, *Pernis apivorus*	29	Germany	DV 23:312
Osprey, *Pandion haliaëtus*	32	Delaware	BB 46:61
Peregrine Falcon, *Falco peregrinus*	14	Germany	BB 24:71
Wild Turkey, *Meleagris gallopavo*	12	W. Virginia	BB 46:61
Coot, *Fulica atra*	19	Switzerland	BB 25:114
Oystercatcher, *Haematopus ostralegus*	36	Germany	DV 25:354
Lapwing, *Vanellus vanellus*	25	Netherlands	DV 24:152
Eurasian Curlew, *Numenius arquata*	31	Sweden	DV 23:313
Herring Gull, *Larus argentatus*	32	Netherlands	DV 22:123
Arctic Tern, *Sterna paradisaea*	34	Maine	BB 46:63
Atlantic Puffin, *Fratercula arctica*	21	Norway	BB 31:92
Barn Owl, *Tyto alba*	18	Netherlands	DV 21:329
Swift, *Apus apus*	21	Switzerland	DV 22:282
Ruby-throated Hummingbird, *Archilochus colubris*	5	Oklahoma	BB 42:51
Red-bellied Woodpecker, *Melanerpes carolinensis*	20	N. Carolina	BB 46:65
Eurasian Skylark, *Alauda arvensis*	6	Italy	BB 19:23
Barn Swallow, *Hirundo rustica*	16	England	BB 24:20
American Crow, *Corvus brachyrhynchos*	14	Manitoba	BB 13:112
Rook, *Corvus frugilegus*	20	Netherlands	DV 21:234
European Jay, *Garrulus glandarius*	16	Netherlands	DV 21:234
Black-capped Chickadee, *Parus atricapillus*	12	New Hampshire	BB 46:66
House Wren, *Troglodytes aëdon*	7	U.S.A.	TR 77:94
American Robin, *Turdus migratorius*	11	U.S.A.	TR 77:94
European Robin, *Erithacus rubecula*	11	Ireland	BB 14:150
Reed Warbler, *Acrocephalus scirpaceus*	12	England	BB 35:205
Black and White Warbler, *Mniotilta varia*	11	N. Carolina	BB 46:68
Starling, *Sturnus vulgaris*	20	Belgium	TR 76:70
Red-winged Blackbird, *Agelaius phoeniceus*	14	New York	BB 21:115
Common Grackle, *Quiscalus quiscula*	16	S. Dakota	BB 13:40
Scarlet Tanager, *Piranga olivacea*	9	Pennsylvania	BB 29:43
House Sparrow, *Passer domesticus*	10	England	DV 25:360
Cardinal, *Cardinalis cardinalis*	13	Tennessee	BB 8:128
Dark-eyed Junco, *Junco hyemalis*	11	Massachusetts	BB 14:46
White-crowned Sparrow, *Zonotrichia leucophrys*	13	U.S.A.	TR 77:95
Song Sparrow, *Melospiza melodia*	10	Colorado	BB 46:73

*BB, *Bird-Banding*; DV, *Die Vogelwarte*; TR, *The Ring*.
†Richdale, *A Population Study of Penguins*.
‡*Ornithologische Beobachter*, 73:28.

by Snow (*in* Lack, 1954). In England this bird lays an average clutch of 11 eggs, and its annual adult mortality is 73 per cent; in Spain and Portugal the average clutch is 6 eggs and adult mortality 41 per cent; in the Canary Islands the clutch drops to 4^1/4 eggs and mortality to 36 per cent. Similar studies in other species give similar results. Such a hand-in-glove fitting of birth rate with death rate raises an intriguing question: Has natural selection increased the reproductive potential of a species to compensate for an increased death rate in certain environments, or does a higher breeding rate automatically result in higher mortality? In other words, do species regulate their reproductive potential to the number of young they can successfully rear in their traditional environment, or do they simply rear all the young they can until the geometric increase of their numbers wastefully collides with some density-dependent environmental factor such as food, which trims the population to supportable levels? Is the control of population stability centered primarily in the animal or in the environment? Is mortality dependent on a fixed natality, or is natality dependent on a fixed mortality?

Students of population dynamics do not agree in their answers to this question. David Lack (1954, 1966) has proposed a "theory of maximum reproduction," which holds that a given species has evolved a reproductive potential (based primarily on clutch size) that results in the largest number of surviving young for a given environment. Young from broods above normal size tend to be undernourished, with the result that fewer birds survive per brood than from broods of normal size. The Alpine Swift, *Apus melba*, for example, normally lays three eggs, but sometimes four. From broods of three young, parents successfully raise an average of 2.4 (79 per cent) young to flying independence. From broods of four, they still raise only 2.4 (60 per cent) young to independence. Maximum efficiency therefore is achieved with broods of three young. Although this principle of maximum effective reproduction seems to be true for some species, it is not so for others in which experimentally enlarged broods result in larger numbers of healthy offspring per family.

Alexander Skutch (1967) proposes a theory of "adjusted reproduction," in which a species maintains a rate of reproduction not at the maximum level but at a reduced but favorable rate that corresponds to average annual mortality. The reduction in reproductive potential may be achieved through various kinds of mutations: a reduction in either clutch size or in number of clutches per season; failure of the male to assist in rearing the young; deferment of reproductive maturity; development of territorialism that limits the number of nesting birds, and other means.

There is undoubtedly some validity in both theories. Bumpus (*in* Johnston *et al.*, 1972) noted a large die-off of House Sparrows, *Passer domesticus*, resulting from a snowstorm in 1898. He made specimens of both the moribund survivors and the dead victims. By comparing eight size variables of the two samples, he noted that larger males and intermediate-sized females survived. He concluded that the "unfit" were eliminated. In the temperate zones selection appears to favor high fecundity and rapid development; in the tropics it favors lower fecundity and a slower development, which permits parents to invest more effort in fewer offspring, which, in turn, should increase their fitness (Dobzhansky, 1950).

These two opposing kinds of selection have been termed r-selection (maximal reproduction), and K-selection (carrying capacity) by MacArthur and Wilson (1967). Pianka (1970) summarizes the chief aspects of each form of selection: r-selection is more likely to occur in variable, unpredictable climates where mortality is often catastrophic and independent of population density, and where competition is variable or minimal; K-selection tends to occur in constant, predictable climates where mortality is more selective and density-dependent, and where competition is usually keen. r-selection favors higher fecundity, small bodies, rapid development, and early reproduction. K-selection favors lower fecundity, larger bodies, slower development, and delayed reproductions. In short, r-selection promotes productivity whereas K-selection promotes efficiency. Larger birds (e.g., procellariiforms, anseriforms, falconiforms) tend to be K-selected and longer-lived (Clapp *et al.*, 1982). Smaller birds, e.g., song-birds and hummingbirds, end to be r-selected and shorter-lived (Clapp *et al.*, 1983; Klimkiewicz *et al.*, 1983).

Any given species probably represents a compromise between both kinds of selection. In California, for example, Wilson's Warbler, *Wilsonia pusilla*, lives both at sea level in a rather stable habitat with a predictable climate, and at 3000 meters in the mountains in a relatively unstable habitat with an unpredictable climate. The coastal populations are relatively K-selected: they are monogamous, have smaller clutches, lower reproductive rates, and lower breeding success (33 per cent). The mountain populations are relatively r-selected: they are polygynous, have larger clutches, higher reproductive rates, and higher breeding success (71 per cent) (Stewart *et al.*, 1977).

While more research is needed to solve the problems related to the control of population, current evidence suggests that population stability depends on a variety of controlling factors that undoubtedly vary between species and may even vary at different times for a given species. Food supply and weather are probably the two dominant environmental forces controlling populations, but predation, disease, nesting sites and other restraining influences may at times have profound effects, as subsequent examples may make clear.

POPULATION DENSITIES

Habitat makes a great difference in the numbers of birds found in different parts of the world. Just as plant abundance varies from sterile deserts to lush tropical jungles, so bird abundance varies geographically and regionally. Since all food eventually comes from the chlorophyll of green plants, birds are generally more numerous where vegetation is thick and varied. This relationship exists not only on land but also in the ocean where microscopic plant life, or plankton, uses sunlight to convert nitrates, phosphates, and other nutrients into food. In the northeast Atlantic, where plankton is abundant, one may see from the deck of an ocean liner over 100 sea birds each day, but in the plankton-poor equatorial Atlantic one will see only one or two birds a day (Jespersen, 1924). The teeming flocks of sea birds that nest in dense colonies do not, of course, find their food in the immediate neighborhood of their nests, nor do colonial-nesting swifts and swallows that feed on flying insects.

Although food is by far the most important element

FIGURE 18–1 Sea birds commonly gather in dense nesting colonies, as shown in this aggregation of gannets. Mutual stimulation in breeding activities, and common defense against enemies, probably more than compensate for the disadvantages of congested living. Photo by E. Hosking.

in the habitat that controls population density, it is by no means the only one. Nesting sites, nesting materials, singing posts, the general vegetational appearance of the habitat, and probably other subtler factors play a role. In a wooded area of about 1.25 hectares near Frankfurt, Germany, Pfeifer (1963) was able to increase the bird population density 25-fold by erecting nest-boxes and other nesting facilities. In a larger, 25-hectare woods the nesting population was similarly increased sixfold. The maximum density in the smaller plot rose to 70.4 fledged broods per hectare and in the larger woods to 44.9 per hectare. The effect of cattle-grazing on bird densities was studied in Canada, New York, and Ohio by Dambach (1941), who found that in ungrazed woods there were, on the average, 299 adult birds of 17 species per 40 hectares, while in grazed woods there were only 137 birds of ten species. Even related species with similar nesting and feeding habits may differ markedly in population density in an identical habitat. In a large deciduous forest tract in Louisiana, Tanner (1941) found only one Ivory-billed Woodpecker, *Campephilus principalis*, to 36 Pileated Woodpeckers, *Dryocopus pileatus*, to 126 Red-bellied Woodpeckers, *Melanerpes carolinus*.

Many studies of avian population densities in different habitats have been made in Europe, Africa, and North America. The censuses were made by using different techniques, and may vary in degrees of accuracy. However, the studies agree in broad outline, and show in general that the richer and more varied the plant food resources of an environment, the denser its bird population. They also show that for each habitat there is an optimum number of birds that can be supported, and that if the bird population presses beyond this limit, mortality is increased or fecundity diminished. In a 12-year study of Tawny Owls, *Strix aluco*, in a deciduous woodland, Southern (1970) demonstrated these density-dependent relationships. A population of about 30 pairs of owls would produce about 30 fledged young in a normal year. But when the population of mice and voles (the owl's chief prey) was low, about 25 per cent of the owl young died, and in extremely low years, the owls did not attempt to breed. Supplemental feeding of various tit *(Parus)* species resulted in increased winter survival and a doubling of the breeding population (Jannson *et al.*, 1982). Table 18–2 gives representative examples of population densities in various habitats as reported by different observers.

In his analysis of over 400 population-density studies of birds, Udvardy (1957) found the following ranges in numbers of breeding birds per 40 hectares, for each of the following habitat types: tundra, 100 to 215; grassland, 20 to 120; marsh (omitting colonial species), 50 to 270; brush and scrub, 100 to 300; deciduous forest: 100 to 200 for oak-hickory to 300 to 770 for mixed bottomland forest; coniferous forest, 150 to 500; mixed coniferous-deciduous forest, 150 to 300; edge or ecotone habitats, 250 to 400 birds.

Dense crowding may in itself affect population numbers, either positively or negatively. According to Darling (1938), certain population densities are necessary before some species will breed at all. Contrariwise, dense aggregations of, for example, ducks may, perhaps through endocrine disturbances, increase the amount of fighting, rape, nest parasitism, and abandonment of nests and young (Titman and Lowther, 1975). Shelducks, *Tadorna tadorna*, breeding at isolated sites, fledged more young (0.72 to 1.22) than pairs breeding in colonies (0.04 to 0.32). This was because colony breeders often fought with each other, leaving their broods vulnerable to Herring Gulls, *Larus argentatus* (Pienkowski and Evans, 1982).

To a given bird, death may appear in many forms. It may pounce suddenly with red fangs or talons, or it may creep in slowly and agonizingly in the form of a fungus infection of the lungs, a dwindling food supply, a freezing rain, a drying pond, or a strong headwind during migration. Whatever guise death may take, it increases its toll as an animal population grows. When the breeding potential of a species creates increasing pressure against the carrying capacity of its habitat, environmental resistance in various lethal forms reacts with counterpressure that trims down the excess population. In the great majority of birds, 80 to 93 per cent of the eggs fail to develop into breeding adults. High as this proportion may seem, it is extremely low compared with that of many lower animals. Of the millions of eggs a mackerel lays, 99.9996 per cent will die as eggs or larvae in the first 70 days of life (Sette, *in* Lack, 1954).

PREDATION

Forces in the environment that reduce the numbers of birds not only affect different species in different ways, but act with varying intensity against different stages in the life cycle of any given species. The same air temperature can be fatally warm for a penguin, ideally neutral for a temperature zone thrush, and below

TABLE 18–2.
DENSITIES OF BREEDING BIRDS IN DIFFERENT HABITATS

HABITAT	LOCALITY	NO. ADULT BIRDS PER 40 HECTARES (OR 100 ACRES)	AUTHORITY
Desert, tundra			
Salt marsh fill	New Jersey	6	Hickey '39
Alpine zone	N. Finland	9	Granit '38
Desert steppe	Colorado	10	Bourlière '50
Rock tundra	Canada	44	Soper '40
Grass tundra	Canada	84	Soper '40
Prairie, savanna			
Rough grazing land	England	70	Fisher '51
Grassland	Michigan	112	Kendeigh '49
Prairie	Washington	246	Wing '49
Sagebrush-grassland	Montana	156	Walcheck '70
Dry grass veld	S. Africa	65	Winterbottom '47
Savanna	Tanzania	96	Winterbottom '47
Scrub brush	N. Rhodesia	310	Winterbottom '47
Dense scrub bushveld	E. Cape Prov.	1150	Winterbottom '47
Tropical grassland	Tanzania	4000	Winterbottom '47
Forests			
Burned-over forest (winter)	Finland	2	Lehtonen '43
Dwarf forest	Lapland	36	Bourlière '50
Sub-alpine birch forest	N. Finland	35	Granit '38
Pine forest	N. Finland	120	Granit '38
Aspen-red maple forest	Michigan	118	Kendeigh '49
Cedar-balsam forest	Michigan	292	Kendeigh '49
Pine-juniper forest	Montana	298	Walcheck '70
Bog forest	Ohio	130	Hickey '37
Flood-plain deciduous forest	Illinois	216	Fawver '47
Second-growth hickory forest	Ohio	536	Hickey '37
Young deciduous forest	W. Virginia	522	Audubon F.N. '48*
Mature deciduous forest	W. Virginia	724	Audubon F.N. '48
Young spruce forest	W. Virginia	690	Audubon F.N. '48
Virgin spruce forest	W. Virginia	762	Audubon F.N. '48
Mixed forest	Holland	896	Tinbergen '46
Oak-hornbeam forest	Slovakia	816	Turček '51
Euphorbia-acacia woodland	Ethiopia	1800	Beals '71
Forest bird sanctuary	Germany	5600	Bruns '55
Cultivated lands			
Bare, fallow land	England	200	Fisher '51
Cereals, root crops, clover	England	200	Fisher '51
Scrubby pasture	Hungary	328	Udvardy '47
Orchard, DDT-sprayed and mowed	Maryland	76	Audubon F.N. '48
Orchard, unsprayed, unmowed	Maryland	468	Audubon F.N. '48
Village with many trees	Holland	936	Tinbergen '46
Golf links	England	1000	Fisher '51
Park (zoological garden)	Germany	1170	Steinbacher '42
Farm yard	Germany	1840	Mildenberger '50
Gardens	England	3000	Fisher '51
Bird sanctuary (Whipsnade)	England	5800	Huxley '36
Freshwater lakes			
Oligodystrophic lakes	Finland	7	von Haartman '71
Eutrophic lakes	Finland	60	von Haartman '71

*Audubon Field Notes.

thermoneutrality for a tropical trogon. A spring snow-storm that would kill nestling thrushes is taken in stride by adults. A young shearwater may endure a week or more without food, a young hummingbird less than a day. Predation is no exception to this general principle, and in most species it focuses most intensely on eggs and young birds. In his long-continued banding studies of the Common Tern, *Sterna hirundo*, on Cape Cod, Austin (1946) found that the worst predator was the common rat, which ate the eggs and young. In 1932 the entire crop of several thousand chicks was destroyed by rats that invaded the colony. Owls attacked the adult terns; in a single night one Great Horned Owl, *Bubo virginianus*, might decapitate 15 or 20 adult terns, although it would eat only one of them. Foxes, skunks, and weasels sometimes killed hundreds of tern chicks, and the adult terns themselves frequently pecked and killed stray young that invaded their nesting territories. Continued predation can annihilate large colonies of birds. A colony of 35,000 Noddy Terns, *Anous stolidus*, on the Dry Tortugas, Florida, was reduced by rats to about 400 birds in a span of 19 years (Robertson, 1964).

Nesting success was 45 per cent in an 11-year study involving 1858 Canvasback (*Aythya valisneria*) nests. Predation, mostly by racoons accounted for 86 per cent of nest losses (Stoudt, 1982). In her survey of bird nests in 2 sq km of primary tropical rainforest in Peru, Koepcke (1972) found that the two chief predators of nests were monkeys and toucans, but that a host of other predators despoiled nests, eggs, and young: opossums, ocelots, weasels, wild pigs, snakes, ants, squirrels, raccoons, falconiform birds, crows, and other animals. Predation accounts for about 80 per cent of nestling mortality among Florida Scrub Jays, *Aphelocoma coerulescens* (Woolfenden, 1978), and for 66 per cent among nestlings of six other passerine species (Ricklefs, 1969).

Of course, birds are preyed upon by other birds, especially such raptors as eagles, hawks, owls, jaegers, skuas, and shrikes. In addition, species not particularly adapted to a predatory way of living may at times prey on other birds. The toucan robs the nests of many tropical birds; House Wrens, *Troglodytes aëdon*, frequently destroy the eggs and young of bluebirds and other passerines; some ducks and oystercatchers have been observed eating the eggs and young of terns; and grackles often kill the young and adults of smaller passerine species. Although the Blue-footed Booby,

Sula nebouxii, and Kelp Gull, *Larus dominicanus*, of South America are primarily fish eaters, they are also inveterate enemies. The Gulls stealthily rob the Booby nests of eggs and young. In bloody reprisal the Boobies will attack any wandering Gull chicks that misguidedly approach them, and while "the poor victim cries out in terror, the furious Camanay [Booby] may thrust its saw-edged beak into its open mouth and out through the back of its skull" (Murphy, 1936).

Animals that are normally inconsequential as enemies of birds may, in times of abundance, become serious predators. In 1937 when the wood mouse population in the Rhineland was unusually high, Mildenberger (1940) found that 19 of 26 nests of ground-nesting warblers, especially Chiffchaffs, *Phylloscopus collybita*, were destroyed by mice.

Although birds, mammals, and reptiles are the chief vertebrate predators of birds, fish may at times levy a heavy toll on them. Northern Pike, *Esox estor*, are the major predator on young waterfowl in the Athabaska and Saskatchewan River deltas, killing an estimated 1.5 million ducklings each year, or nearly one-tenth of the young produced (King and Pyle, 1966). Stomachs of the marine goosefish have been found to contain remains of loons, grebes, cormorants, widgeon, scaup, scoters, mergansers, gulls, auks, and guillemots (Bigelow and Welsh, 1924). Crabs in tropical regions are known to be severe predators of young birds, especially ground-nesting terns; and in the southern United States, biting ants prey on quail chicks. Occasionally turtles, frogs, spiders, and praying mantises catch and kill birds, but ordinarily they are not important predators.

The degree to which predators regulate population levels varies from species to species. For example, of 186 nests of the Blue-winged Teal, *Anas discors*, studied in Iowa, 100 (54 per cent) were destroyed, largely by skunk, mink, fox, and raccoon (Glover, 1956). Similar stories have been reported for many other species. On occasion the intensity of predation rises to heights that would seem impossible for a prey species to sustain. On Raoul Island in the Kermadecs (New Zealand) there is a colony of about 80,000 pairs of Sooty Terns, *Sterna fuscata*. Mortality of eggs (from rats) and chicks (from cats), approximated 77 per cent in 1967 (Taylor, 1979). Breeding density of Clay-colored Robins, *Turdus grayi*, was high (50 pairs/10 hectares) at one site and low (15.8 pairs/10 hectares) at another site in lowland Panama. Predation was lower in the overpopulated area, and breeding success was twice as high as that at the other site (Dyrcz, 1983).

Even predatory birds are victims of predation. In England, of 24 nests of the Short-eared Owl, *Asio flammeus*, observed by Lockie (1955) only five survived to hatch young, and only two raised young to fledging age, as a result of predation by foxes and crows. Predation may have been particularly severe in this case because it occurred at a time when meadow mice were scarce.

Sometimes climatic changes throw light on the role of predation in population control. Duebbert (1966) tells of a 2.8-hectare island in a North Dakota lake that supported a very high density of nesting Gadwall Ducks, *Anas strepera*. In 1957, of 109 nests, 92.7 per cent successfully hatched young. Low water the following year allowed access to the island by mammalian predators so that by midsummer only one active nest was found. Quinlan and Lehnhausen (1982) reported that an Arctic fox crossed a tidal flat to Anaulik Island, Alaska, where it buried all 498 eggs of a Common Eider, *Somateria mollissima*, colony. Cached eggs are later recovered by foxes. Fledging success of the White-crowned Sparrow, *Zonotrichia leucophrys*, in Alaska, is 75 per cent, but only 40 per cent in California (Morton, 1976). One may assume that the relative scarcity of predators in Alaska is at least in part responsible for this difference.

In an attempt to determine the effect of predators on wild populations of Ring-necked Pheasants, *Phasianus colchicus*, in southern Minnesota, 434 predators (chiefly skunks, raccoons, and crows) were removed from a 1040-hectare trapped area during three nesting seasons. No predators were removed from an untrapped, control area of 1650 hectares during the same years. There were over 400 pheasant nests on each area. Hatching success increased until it reached 36 per cent in the third year on the trapped area; it remained low, at 16 per cent, on the untrapped area. Chick production was about two times higher on the trapped area (Chesness *et al.*, 1968).

The experimental removal of predators from New York habitats of the Ruffed Grouse, *Bonasa umbellus*, resulted in 24 and 39 per cent destroyed nests, whereas in control areas where predators were not removed, the nest destruction was higher: 51 and 72 per cent. However, mortality of young grouse chicks in the contrasted areas was nearly identical: 57 and 54 per cent in the experimental areas and 67 and 55 per cent

in the control areas. Paradoxically, adult losses were largest in the experimental (predator-removed) areas, and smallest in the control areas, at least during years of grouse abundance (Edminster, 1939).

Such results indicate that predator-prey relations are not simple problems in arithmetic. Undoubtedly numbers play a role in predation. The more animals there are to prey upon in a given area, the more easily they are found by predators and, in all probability, the more intense the predation. In an experiment to test this thesis, Tinbergen and others (1967) placed artificially camouflaged hens' eggs in crowded plots and in more widely scattered plots and found that the crowded eggs suffered higher mortality than the uncrowded from Carrion Crows, *Corvus corone*.

On the other hand, in some species, the denser the population, the more effective its defense against predators and the higher its fecundity. Slagsvold (1980) placed artificial nests containing eggs either within or outside colonies of Fieldfares, *Turdus pilaris*. Predation by Carrion Crows, *Corvus corone*, was higher in nests placed outside the colonies. Indeed, breeding success in Fieldfares varied directly in relation to size of the breeding colony.

Such phenomena illustrate again what are known as density-dependent influences in the life cycles of animals. Population densities, in short, affect intensity of predation, as well as such things as the spread of disease, malnutrition and starvation, competition for territory, the rate of reproduction, infanticide, and other elements that, in complex and often subtle combinations, control the numbers and survival of animals.

One method some species have evolved to reduce predation involves timing. The Antarctic Prion, *Pachyptila desolata*, comes ashore at night to dig its nest burrow and to care for its family while skuas and other predators are asleep. Early the next morning it leaves for the open sea where it feeds in safety. Predation is also lessened through seasonal timing of the life cycle. Normally the Eastern Bluebird, *Sialia sialis*, nests two or three weeks earlier than the House Wren, *Troglodytes aëdon*, but occasionally a late spring freeze will destroy the eggs of the Bluebirds and cause them to lay replacement clutches at about the time the migrating Wrens arrive. This results in competition between the two species and greater than normal destruction of Bluebird eggs by the aggressive Wrens (Musselman, 1946). A species may, of course, have different predators at different stages in its life cycle. The eggs of the Red-footed Falcon, *Falco vespertinus*, are eaten by Rooks, *Corvus frugilegus*, and Tawny Owls, *Strix aluco*, but the young are killed by the Goshawk, *Accipiter gentilis* (Horvath, 1955).

A further chronological defense against predators is found in the synchronized breeding of many species such as gulls. Predators, able to kill only a certain number of prey young each day, are overwhelmed by the synchronous hatching and fledging of large numbers of young, and as a consequence, levy a smaller toll on the total population (Fig. 18–2).

Birds react in a variety of ways to combat losses from predation. First, they must be able to recognize enemies. Many passerine birds instinctively recognize an owl as an enemy (Hartley, 1950). In some species the parent birds apparently educate their young to recognize enemies. Perhaps the most common reaction to the approach of a predator is to fly away or to dive under the water. The next most common reaction is immobility or "freezing," particularly in species with concealing coloration, or in birds brooding their eggs or young. Some of the larger, better-armed birds, such as hawks, owls, and swans, will fight back, as will some passerines of pugnacious disposition like the jays. Many passerines will gather and scold ("mob") the potential predator, or may dive-bomb it. Fieldfares will also defecate with great accuracy on a potential predator (Löhrl, 1983). A few species practice a form of

group defense known as "predator swamping." When a hawk dives toward a flying band of Starlings, *Sturnus vulgaris*, the latter bunch together into a dense flock, and the hawk, unable to single out an individual bird, flies off frustrated. Similarly, fledging Thick-billed Murres, *Uria lomvia*, jump synchronously off the cliffs at certain hours of the day and suffer less predation than asynchronous jumpers at other times of the day (Daan and Tinbergen, 1979). In general, defensive reactions toward predators are heightened during the breeding season. At times these reactions may be detrimental rather than beneficial to the prey species itself. In a colony of Ring-billed Gulls, *Larus delawarensis*, Emlen *et al.* (1966) observed that parental wariness of predators caused neglect of the brooding and care of young, so that hatching success and survival of the young were greatly reduced.

The location of a bird's nest may also affect predation rates. Predation on early nesting Mallards, *Anas platyrhynchos*, and Tufted Ducks, *Aythya fuligula*, was higher than on later nesting ducks because early nesters did not have the benefit of the tall concealing vegetation (Hill, 1984).

A further factor in the complex equation involving predator and prey is the presence or absence of alternative prey. In 1960 an exploding lemming population on Southampton Island, Hudson Bay, produced a larger than normal crop of Arctic Foxes. The following year the lemmings were in a "crash" decline and the foxes turned to nesting tundra birds for food. In some places they consumed 90 per cent of the eggs of Canada Geese, *Branta canadensis* (Hanson and Nelson, 1964). Thus, in years of abundance the rodents acted as *buffers* to absorb some of the predation that might otherwise be directed toward the geese.

Predation is not necessarily an undiluted evil for the species preyed upon. Although exact quantitative studies seem to be lacking, it is probable that predators often act as sanitary police in catching and removing aged, diseased, abnormal, or crippled birds from a prey population. Long-continued investigations by Eutermoser (1961) with three species of trained falcons, (*Falco peregrinus*, *F. rusticolus*, and *F. biarmicus*), showed that when they were flown at flocks of crows they attacked the sick, injured, or underweight birds in preference to sound individuals. Spellerberg (1975) relates that Giant Petrels, *Macronectes giganteus*, prey far more often on weak or wounded chicks

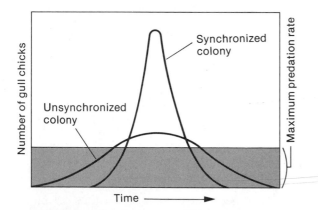

FIGURE 18–2 The mortality of young birds in colonies with synchronized breeding is less than that in unsynchronized colonies, because the predators are unable to kill as large a proportion of young in the limited time available. Redrawn after Wilson, 1975.

of the King Penguin, *Aptenodytes patagonica*, than on healthy chicks. Further, even widespread, indiscriminating predation on a prey species may hold its numbers in check and prevent it from snowballing into destructive overpopulation. This has been demonstrated in the case of the Kaibab Deer of Arizona whose predators, pumas, wolves, coyotes, and bears, were removed in a misguided attempt to increase the deer population. Range cattle were also removed to eliminate those competitors for available food. The experiment succeeded too well, and the deer population rose from about 4000 to 60,000, which exceeded by far the carrying capacity of the environment. As a result, the deer "ate themselves out of house and home" and starved in wholesale numbers until the population dropped to around 10,000 animals. Without much doubt the same sort of checks and balances operate between birds and their predators to establish a relatively stable population equilibrium.

A somewhat unexpected benefit from predation by crows on the eggs and young of ducks has been suggested by Cartwright (1944). He reasoned that such single-brooded birds, undisturbed by predators, would produce synchronized early nestings that would be disastrously vulnerable to late spring freezing, flooding, or similar calamities. Predators provide insurance against such mass catastrophes by destroying many of the first nestings of the ducks, forcing them to lay replacement clutches and thus stagger their nestings through a longer period of time. In the light of present knowledge, it is extremely hazardous to claim that one species is "detrimental" because it preys on another.

To sum up, the role of predation in controlling bird populations seems largely to depend on the number of predators and their selectivity in choice of prey; the number, health, and accessibility of prey animals; their rate and timing of reproduction; their instinctive or learned reactions to predators, including emigration; the presence or absence of buffer populations; and the carrying capacity of the prey species' environment, especially its food and cover.

As a hunter, man is an active predator upon many wild birds. However, his chief influence on bird numbers comes from his use or misuse of the natural environment, which causes tremendous changes in its ecological character. Most of these changes have been detrimental to bird life and have resulted in the decimation and extinction of many species. This topic will be considered more fully in the final chapter.

THE ROLE OF COMPETITION

Most of the things that birds require of an environment are limited in quantity—food, territory, nesting sites, dust baths, singing posts, and so on. Since birds of the same species make the same demands on their habitat, the more birds in a given locality, the less there is of any given requirement for each. Eventually the carrying capacity of the environment acts as a brake on population increase. Usually one element required of the habitat occurs in short supply before others, and this one then becomes the "critical limiting factor" that either holds down the population or causes the surplus birds to emigrate. This phenomenon is particularly evident among species whose numbers fluctuate widely, often in cycles, through the years.

For most birds the ownership of territory has deep and far-reaching significance. The intense competition seen in territorial conflicts becomes more understandable when one recalls the survival benefits that territory ownership confers on a bird or on a species (see Chapter 12). Individuals unable to obtain territories are relegated to the ranks of floaters. For instance, Rufous-collared Sparrow, *Zonotrichia capensis*, floaters were found to move about in flocks; dominant flock members insert into vacancies (Smith, 1978). Young European Nuthatches, *Sitta europaea*, settle at territory borders if these are poorly defended by adults. Here young birds stake out territories waiting for a territory vacancy (Matthijsen and Dhont, 1983).

For a given species it is extremely difficult to determine precisely the functions of territory ownership, but observations by Holmes (1970) of the Dunlin, *Calidris alpina*, clearly emphasize the food-monopoly aspect of territory. In the tundra of arctic Alaska (71 degrees N. latitude) this sandpiper defends territories five times larger (six pairs per 40 hectares) than it does farther south (61 degrees N. latitude) in subarctic Alaska (30 pairs per 40 hectares). This is mainly because the insect food supply in the arctic habitat is less abundant and less predictable and the climate more rigorous than it is farther south. Similarly, territory size varies inversely with insect abundance in the Galapagos Finch, *Geospiza difficilis* (Schluter, 1984).

On Scottish moors the competition for territory among Red Grouse, *Lagopus lagopus*, as described by Watson and Moss (1972), involves much more than competition for food. Each autumn the male grouse compete for territory ownership. Among males that own territory through the winter, mortality is low. Among nonowners, winter mortality is very high, as is also their dispersal into inferior habitats. Only owners of territory will breed in the following spring. Since it appears to be quality of food (heather) that determines survival of grouse, territorial battles are in a sense competition for nutritious food. When the hen joins the cock grouse in his territory in the spring, her breeding success seems to be predetermined by the quality of the eggs she lays, and the quality of her eggs depends on the nutritious value of the heather she eats. In one experiment the nitrogen content of heather was enriched by spreading nitrate fertilizer in an experimental plot. A second plot was not fertilized (the control). All breeding grouse were then removed. Immigrants raised bigger broods in the experimental area than in the control area (Watson *et al.*, 1984).

Changes in territory size appear to be inversely correlated with breeding success. All of these variables, plus those of population density, heredity, possible competition with other species, and predation, show that the control of numbers in Red Grouse is not a simple, single-factor problem.

Competition occurs between birds of different species that live in the same habitat. Hole-nesting birds frequently compete for nesting sites. Starlings, *Sturnus vulgaris*, often steal nests from woodpeckers; and House Sparrows, *Passer domesticus*, often forcibly eject Cliff Swallows, *Petrochelidon pyrrhonota*, from their mud-flask nests. Even the Bald Eagle, *Haliaeetus leucocephalus*, is not immune to nest robbery. Within a period of six years in Florida, Broley (1947) found 31 out of a total of 619 eagle nests usurped by the Great Horned Owl, *Bubo virginianus*. Competition for nest sites does not necessarily embrace competition for food. When artificial nest boxes were placed in birch forests of Swedish Lapland, the population of Pied Flycatchers, *Ficedula hypoleuca*, was increased 15 times, but the population density of 14 other passerine species was not affected (Enemar and Sjöstrand, 1972).

Quantitative studies on competition for food among wild birds are difficult to make. Lack (1954) is of the opinion that closely related species that live in the same habitat have evolved different diets and thus have avoided direct competition for food. However, as a result of studying mixed flocks of African birds, Winterbottom (1950) found evidence of food competition between close relatives. Further study is needed to resolve this problem.

THE ROLES OF CLIMATE, WEATHER, AND ACCIDENTS

Climate has its greatest influence on bird numbers through the indirect means of controlling growth of the plants and plant foods on which birds depend. However, climate does also have a direct influence on bird survival, especially on the tender young, mainly through its extremes in temperature and rainfall. Perhaps the best evidence of climatic control of bird populations is the fact that long-range gradual changes in mean annual temperatures are paralleled by the gradual extension or regression in the geographical ranges of many species. This topic will be considered further in Chapter 20.

However, nature periodically performs a dramatic experiment showing that even moderate climatic changes may have profound effects on bird numbers. About once every seven years the Pacific Equatorial Counter-Current swings southward off the coast of Colombia head-on into the north-flowing Peru Current. It is then called "El Niño," and it sets into play the following events: A band of coastal waters extending about 2000 km in length from Ecuador to southern Peru becomes about 5°C warmer than normal. This causes the death of the usual plankton and the death or disappearance of the teeming fish life characteristic of the Peru Current. Finally, hundreds of thousands of cormorants, boobies, and pelicans, which normally feed on Peru Current fishes, sicken and die (Murphy, 1936; Hughes, 1985). The Niño conditions of 1982–1983 resulted in the disappearance of the entire 18-species sea bird community on Christmas Island in the equatorial central Pacific (Schreiber and Schreiber, 1984).

Weather, as distinct from climate, has short-range but often similarly drastic effects on bird numbers. It, too, may affect a species indirectly through regulating its food supply, as, for example, by a late spring freeze, which kills a crop of wild berries, or a drought, which prevents the maturing of seed crops. More dramatically, weather may kill birds directly by swinging to extremes beyond their limits of tolerance. After the unusually severe winter of 1946–1947, the occupied nests of Herons, *Ardea cinerea*, in Britain, were reduced 40 per cent in numbers (Alexander, 1948). The exceptionally cold winter of 1962–1963 nearly eliminated kingfishers in Germany, and Barn Owls, *Tyto alba*, Little Owls, *Athene noctua*, and Water Rails, *Rallus aquaticus*, in Switzerland. In England dead birds of 46 species, collected during the cold spell, were found to have lost between 55 and 65 per cent of their normal weight (Ash, 1964). In 1941, the combination of a cold wet spring followed by a drought in May caused 55 per cent of the 1830 pairs of White Storks, *Ciconia ciconia*, in Schleswig-Holstein to raise no young, whereas in favorable years the proportion was as low as 13 or 14 per cent (Emeis, 1942). When such adverse weather reduces the food supply so that the young starve, the parent storks generally eat the smaller young and throw the larger ones out of the nest.

Numerous accounts have been published of the disatrous effects on bird life of sudden, brief cold spells or storms. In the spring of 1964, adverse pack-ice conditions along the coast of the Beaufort Sea in North America caused deaths from starvation of about 100,000 migrating Eider Ducks, *Somateria mollissima*, or about 10 per cent of the total population (Barry, 1968). Two July hailstorms in Alberta in 1953 were estimated by Smith and Webster (1955) to have killed over 148,000 waterfowl; a 30-minute October hailstorm in New Mexico killed about 1000 Lesser Sandhill Cranes, *Grus canadensis* (Merrill, 1961); and a heavy, wet snowstorm in March, 1904, killed millions of Lapland Longspurs, *Calcarius lapponicus*, in western Iowa and Minnesota, an estimated 750,000 lying dead on the frozen surfaces of two small lakes alone (Roberts, 1932). In October, 1974, unseasonably cold weather in southern Germany and Switzerland resulted in the near-starvation of Barn Swallows, *Hirundo rustica*, and House Martins, *Delichon urbica*. Thousands, lacking the strength to migrate over the Alps, perished in the ice fields. Conservationists gathered approximately a million of the weakened birds and shipped them by land and air to warmer, insect-rich lands farther south where the majority of them recovered (Ruge, 1974). Facts such as these indicate that many species live dangerously close to their limits of environmental toleration. They apparently possess what an engineer would call a very low safety factor for meeting wider than normal fluctuations in environmental conditions.

The resilience with which a population recovers its numbers after heavy destruction by storms seems to depend chiefly on a species' breeding potential. In the year after the destruction of so many Longspurs, Roberts reported that there was no apparent reduction in their numbers. British Herons require four or five years after a hard winter to regain their former numbers; kingfishers perhaps four years; European Crested Larks, *Galerida cristata*, two years.

At times weather conditions may decimate bird populations in unexpected ways. In 1970, an April storm in the Aleutian Islands persisted for five days. At times the wind velocities reached 135 km per hour. After the storm an airplane survey revealed over 100,000 dead Murres, *Uria aalge*, scattered along 725 km of coastline (Fig. 18–3). The birds were emaciated and had probably died of exhaustion, exposure, and starvation (Bailey and Davenport, 1972). Persistent high winds may blow thousands of sea birds far inland where many of them die of fatigue or starvation. Gentle rains may turn clayey soil into mud that clings to birds' feet, causing "mud-balling," which weighs down the victims (Fig. 18–4). Young gallinaceous chicks often succumb to mud-balling (Yeatter, 1934). In 1945–46 about 500 ducks on a Texas wildlife refuge died from exhaustion by becoming shackled to sticky balls of mud. A typical Mallard, *Anas platyrhynchos*, acquired about 400 g, and a Pintail, *Anas acuta*, about 750 g (1 lb, 11 oz) of mud on its feet (O'Neill, 1947). Fog sometimes causes birds to lose their way and fly into trouble. In 1889, Stevens (*in* Bent, 1932) told of an immense flock of young Passenger Pigeons, *Ectopistes migratorius*, flying across Crooked Lake, Michigan, in a fog, becoming confused, and descending on the water. Thousands drowned and "the shore line for miles was covered a foot or more deep with them." The eggs and young of shore-nesting waterfowl commonly are lost from high flood waters. Catastrophic mortality also occurs in times of drought, sometimes in unusual ways. During an especially severe drought, an Australian farmer removed 5 tons of drowned Budgerigars, *Melopsittacus undulatus*, from one pond. The parched birds, coming to drink, had crowded each other into the water (Immelmann, 1972). During a severe drought in Nigeria, poor feeding conditions caused male Red-billed Queleas, *Quelea quelea*, to abandon their colony after only partly completing their nests. The females laid thousands of eggs through the bottomless nests onto the ground, and many of them died of starvation (Jones and Ward, 1979). In the thornbush savannas of Senegal, there are normally about 6.3 birds per hectare, but in the exceptionally dry year of 1972–1973 there were only 2.9 birds per hectare and no breeding took place (Morel and Morel, 1974).

In addition to weather, birds are subject to a wide variety of natural accidents that injure and kill them. At times they become entangled in, or strangled by, horse hairs or other fibrous nesting materials. Bird banders frequently trap birds with amputated feet or crooked legs that testify to past fractures. Of 6212 bird

FIGURE 18–3 A small part of the more than 100,000 murres killed during a spring storm in the Aleutian Islands. From Bailey and Davenport, 1972.

FIGURE 18–4 Young Hungarian Partridges killed by mud-balling. Wet, clayey soils, under certain conditions, will cling to birds' feet and cause their death through exhaustion, exposure, and starvation. Photo by R. E. Yeatter.

skeletons examined by Tiemeier (1941), 4.5 per cent showed healed injuries, 11.78 per cent of which were the result of gunshot wounds. In passerine birds, 75 per cent of the injuries were of the clavicle.

The eruption of Mt. Pelée exterminated three species of birds on Martinique. Gulls nesting in the vicinity of New Zealand hot springs frequently lose the webbing of their feet from the corrosive action of the water. In one colony less than 3 per cent of the chicks survived to leave the colony, either because of desiccation or because of contact with the hot water (Daniel, 1963). Falling trees and rocks may destroy whole families of birds. At times birds fly into lethal situations as a result of intoxication from eating fermented or narcotic fruits. On all sides and at all times, birds are surrounded by threats to their lives.

PARASITES AND DISEASES

Western civilized man lives such a sheltered antiseptic existence that he finds it difficult to comprehend the high incidence of disease in wild creatures. The probabilities are that any wild bird selected at random can be shown to be infected with one or more forms of parasites or disease organisms. Of 50 Eastern Belted Kingfishers, *Megaceryle alcyon*, collected by Boyd and Fry (1971), 84 per cent had external parasites of various sorts, and 98 per cent had internal worm parasites. A list of parasites and diseases affecting the Bobwhite Quail, *Colinus virginianus*, assembled by Kellogg and Calpin (1971) included 4 viral diseases, 15 bacterial, and 2 fungal. The parasites included 12 species of protozoa, 1 trematode species, 13 cestodes, 3 Acanthocephala, 30 nematodes, and 39 species of arthropods. An examination of 1525 birds of 112 species and subspecies in the American Southwest by Wood and Herman (1943) revealed that 23.4 per cent were infected with blood parasites alone; and a study of Red-winged Blackbirds, *Agelaius phoeniceus*, on Cape Cod by Herman (1938) showed that 60 per cent were infected with avian malaria. Some 45 species of parasites were found infecting about 91 per cent of a sample of 175 scoters of three species (Bourgeois and Threlfall, 1982). The subject of bird parasites and diseases is such a large and complex one that only its bare outlines can be considered here. Birds seem to be subject to the same general types of diseases and functional disorders as man, and probably, if the truth were known, to as many different varieties.

Several investigators have studied the causes of death in wild birds, but their conclusions as to the relative importance of different lethal factors vary so greatly (more than tenfold in three different categories) as to make them highly unreliable. They agree, however, in ascribing dominant influence to physical injury, diseases, parasites, weather, and poisons —but not necessarily in that order. After examining over two million records of nonhunting mortality in fledged North American waterfowl, Stout and Cornwell (1976) declared that diseases and poisons were the major cause (87.7 per cent) of deaths.

ECTOPARASITES. The chief external parasites of birds include biting lice, fleas, calliphorid flies, hippoboscid flies, mosquitoes, black flies, ticks, mites and leeches. With the exception of some of the biting lice (Mallophaga) and mites, which live on skin and feathers, these are all bloodsucking parasites. A list of 198 different species of ectoparasites, taken from 255 species and subspecies of birds in states east of the Mississippi River, was compiled by Peters (1936). Some of these individual parasites may attack a variety of hosts, the tick *Ixodes brunneus*, for example, having been found on at least 64 species of birds (Boyd, 1951). Others, such as the Mallophaga and certain mites, are highly host-specific, a given parasite generally being restricted to a single species or genus of birds. Blowflies of the genera *Protocalliphora* and *Apaulina* lay their eggs in birds' nests, particularly those of hole-nesters, and the maggots feed on nestling birds, sucking the blood from their feet and legs and occasionally entering ear cavities and nostrils, or burrowing under the skin. An examination of 162 cavity-nests by Mason (1944) revealed *Protocalliphora* nest-infestations of 94 per cent for Bluebirds, *Sialia sialis*, 82 per cent for Tree Swallows, *Iridoprocne bicolor*, and 47 per cent for House Wrens, *Troglodytes aëdon*. From one to 200 maggots may infest a single nest. Occasionally the pupae that develop from these maggots are parasitized by the wasp *Nasonia vitripennis* and die. Hippoboscid or louse flies are flat flies that live among a bird's feathers and suck blood from its skin. They, and black flies, mosquitoes, and ticks are known to transmit infectious diseases, particularly of the blood, from one animal to another. The common Bed-bug, *Cimex lectularius*, often infests the nests of falcons and may cause fledglings to leave the nest prematurely or even kill them.

Birds that nest in subterranean burrows of mammals are more likely to carry fleas than are above-ground nesters. From 16 species of burrow-nesting birds in Turkmenskaya, Zagniborodova and Balskaya (1965) obtained 18 different species of fleas, 16 of which were primarily mammalian parasites and 8 of which were known to be transmsitters of the plague.

Ticks are likely to infest ground- or bush-nesting birds, or carnivorous birds that feed on tick-infested rodents. Because of a heavy infestation of ticks on Bird Island, Seychelles, in 1973, about 5000 Sooty Terns, *Sterna fuscata*, deserted their eggs or newly hatched chicks, the latter, of course, starving to death (Feare, 1976). Similar observations were made on sea bird colonies in Peru (Duffy, 1983). Ticks often attack the soft tissues around a bird's eyes, sometimes causing blindness and, in young birds, death. The fact that a nest-dwelling, well-fed tick may survive three or four years on one meal, suggests a possible reason why Cliff Swallows, *Petrochelidon pyrrhonota*, and similar colonial hole-nesters may change nest sites every few years.

Acarine mites commonly infect the nasal cavities and skin of birds. From 100 Starlings, *Sturnus vulgaris*, Mitchell and Turner (1969) obtained 22 species of mites. Heavy mite infestations may reduce the growth rate of nestling birds or may even kill them.

One function of preening must certainly be the removal of ectoparasites. Passerines of three different species, whose upper mandibles were mostly gone and therefore useless for preening, were found to be very heavily infested with Mallophaga. Dust and water baths, and anting, probably reduce the numbers of parasites in a bird's plumage.

It seems quite possible that the rapid development of nestlings is in part an adaptation to ectoparasite infestation, because the less time a young bird remains in the nest, the fewer parasites it is likely to acquire. Because several kinds of ectoparasites accumulate in nest litter, the young birds of second broods are more likely to become heavily infested than those of first broods. Since certain ectoparasites, such as ticks, are transmitted by bodily contact, they are commoner in gregarious species than in non-gregarious.

A striking example of the intimate adaptation of an ectoparasite to its host is provided by the Mallophaga of the Orange-crowned Warbler, *Vermivora celata*. Foster (1969) found that reproduction of the lice coincided with high levels of reproductive hormones in the host's blood and essentially ceased during the host's refractory period. This synchrony in host and parasite reproduction avoided loss of lice and their eggs during periods of feather molt, and prepared large numbers of young Mallophaga to infest the young birds as soon as they appeared.

ENDOPARASITES. Among the commoner internal parasites are the Trematoda or flukes, Cestoda or tapeworms, Nematoda or roundworms, and the Acanthocephala or spiny-headed worms. Although some kinds of parasitic worms may be transmitted directly from one bird to another, more commonly one or more alternate hosts are involved in a worm's life cycle. Thus the adult parasite living in a bird is acquired by eating a fish, tadpole, snail, crustacean, earthworm, mole, or some other animal that contains a larval form of the parasite. Small numbers of parasitic worms in a bird do not ordinarily seem to affect its health or vigor, but large numbers may cause emaciation or death. At times, many young storks die from heavy infestations of intestinal flukes; and Eider Ducks, *Somateria mollissima*, on both sides of the Atlantic, die in great numbers from infestations of spiny-headed worms. Of 162 Eiders found dead in Britain, 86 per cent carried infestations of Acanthocephala. The common shore crab, *Carcinus moenas*, was the intermediate host in this case. As many as 1600 tapeworms of six different species have been found in a single duck, and a survey in Washington of 3400 Mallards, *Anas platyrhynchos*, showed that 94.7 per cent of them contained internal parasites. Hundreds of field studies have been made to determine the incidence of endoparasites in wild birds. As typical findings the following may be given: of 94 wild Turkeys, *Meleagris gallopavo*, 92 per cent were infected with either roundworms or tapeworms; of 106 Prairie Chickens, *Tympanuchus cupido*, 82 per cent had either roundworms or tapeworms; of 406 Common Grackles, *Quiscalus quiscula*, 100 per cent had roundworms and 91 per cent had flukes. In their survey of endoparasites reported to infest Starlings, *Sturnus vulgaris*, Hair and Foster (1970) compiled a list of 81 species of 51 genera that had been reported by various authors around the world. These included 28 species of trematodes, 14 of cestodes, 30 of nematodes, and 9 of Acanthocephala. Of the 81 species, only 11 were common to Starlings of both eastern and western hemispheres.

The variety of internal parasites that may infect birds is not generally appreciated. In his catalog of endoparasitic worms of waterfowl, McDonald (1969) lists approximately 1000 species. In her monograph on nematode parasites of birds, Cram (1927) describes approximately 500 species, not including the numerous microscopic forms.

Specific parasitic worms usually attack certain organs or tissues in the body, such as the intestines, lungs, air sacs, liver, trachea, or blood. A study of pheasants in Nebraska showed that 40 per cent were parasitized with roundworms that infected their eyes. Swans are at times infested with heart worms that may cause death in severe infections. Ten per cent of a flock of wintering Canada Geese, *Branta canadensis*, in North Carolina died of gizzard worms. These nematode worms caused little damage when there were fewer than 150 in each gizzard, but more than that caused denuded gizzard linings and death (Herman and Wehr, 1954). In Delaware, in 1976, large scale mortality occurred in nestling herons and egrets (84 per cent in Snowy Egrets, *Egretta thula*) infected with nematodes that caused perforations in visceral organs. Killifish were the source of the infective larvae (Wiese *et al.*, 1977). The Ruffed Grouse, *Bonasa umbellus*, has harbored at least 20 species of roundworms, 10 of flukes, 9 of tapeworms, and in addition, 10 arthropod and 12 protozoan parasite species (Vanderschaegen, 1972). A sampling of seven of the roundworm parasites of the Ruffed Grouse, after Edminster (1947), is given in Table 18–3.

There is, of course, a great variety of conditioning factors that determine whether, and to what degree, a given bird will be infested with parasites: climate, soil, the presence or absence of alternate hosts, population density, and so on. Cornwell and Cowan (1963) found that worm infestations of ducklings of the Canvasback, *Aythya valisneria*, were five times as numerous as infestations of adults, and that ducklings from marshes were more heavily parasitized than those from prairie potholes. Kellogg and Prestwood (1968) discovered that Bobwhite Quail, *Colinus virginianus*, from high-density populations carried heavier burdens of cestode and nematode parasites than Quail from areas of low populations. It may well be that one of the advantages of territory ownership and of migration is the reduction of potential parasite infection by diminshed exposure to contaminated soil or to alternate hosts.

MICROSCOPIC ORGANISMS. The commoner microorganisms that infect birds are protozoa, fungi, bacteria, and viruses. A listing of only the parasitic blood protozoa of birds by Herman (1944) included 47 different species. Various studies of thousands of birds of scores of species show that from one-fourth to one-third of all wild birds have blood parasites.

Among the protozoa that infect birds, the commonest are probably the sporozoa that cause avian malaria: *Plasmodium*, *Haemoproteus*, and *Leucocytozoon*. These three genera of parasites, all of which destroy red blood corpuscles, are transmitted by the bites of *Culex* mosquitoes, hippoboscid flies, or black flies. A study in Uganda of 1076 birds of 127 species revealed that 404 birds (37 per cent) of 41 species harbored blood sporozoa. *Haemoproteus* represented 95 per cent of all infections detected (Bennett *et al.*, 1977). *Leucocytozoons* are known to parasitize over 150 species of birds and are the most important blood parasite of waterfowl. They are typically transmsitted by the bites of black flies (*Simulium* spp). The life cycle of the parasite is similar to that of the malarial parasite in man. In a given bird these sporozoa may destroy over one-half of the red blood corpuscles and cause marked anemia and great enlargement of the liver and spleen. Mortality rates as high as 90 per cent have occurred among ducks in Michigan and Canada. The incidence of infection in Mallards may be 100 per cent in one region and only 10 per cent in another area only 80 km away (Fallis and Trainer, 1964). Experimental exposure to this parasite of young domestic ducks in northern Michigan by Chernin (1952) revealed a significant seasonal relationship: few birds exposed in late June and early July became infected, and none died of the disease; 90 to 100 per cent of the birds exposed to the sporozoa in late July and early August became infected, and of these, 14 to 83 per cent died; after mid-August the ducks were immune to infection. This type of relationship between season and susceptibility to infection has important bearings on epidemic or, more properly, epizootic infections that sporadically kill great numbers of wild birds. Such epizootics are more common among terrestrial colonial birds than among noncolonial species because of the greater opportunities for contagion in colonies. Likewise, epizootic infection is generally higher in dense populations—among grouse, for example—than in scattered populations. Colonial seabirds seem relatively immune to epizootic outbreaks, probably because of the slight traffic between colony nests and because the ocean, where the birds feed, is a poor culture medium for the infective organisms.

Other protozoan parasites of birds include trypanosomes, coccidia, and trichomonads. Trypanosomes may infect the blood or, more commonly, the bone marrow of a great variety of birds. They are extracellular blood parasites, not particularly pathogenic in birds, and are transmitted by mosquitoes. Nine species of coccidia are known to infect wild birds, the genus

TABLE 18–3.
SEVEN OF THE 20 ROUNDWORMS KNOWN TO PARASITIZE RUFFED GROUSE*

NAME	DESCRIPTION	PRODUCES	NO. PER BIRD	INCIDENCE
Stomach worm *Dispharynx spiralis*	12 mm long White. In proventriculus	Swelling of proventriculus, emaciation, death	Avg. 3–12 Up to 228	Varies greatly from 0–28% Common in domestic Chicken.
Intestinal worm *Ascaridia bonasae*	50–100 mm long Yellowish. In intestines	Emaciation in heavy infections	Avg. 1–15 Up to 200	Ontario—21% Minn.—37% Mich.—20–37% N.Y.—9–71%
Gizzard worm *Cheilospirura spinosa*	30–40 mm long Slender, white. In gizzard lining	Thickens gizzard walls. May cause death	Avg. 12–16 Up to 40	Mich.—42% Minn.—21% Penna.—22% Wis.—19% Ontario—0–14%
Eye worm *Oxyspirura petrowi*	12–18 mm long Lives under nictitating membrane of eye	Conjunctivitis	1–17 per eye	Mich.—15–28%
Blood worm *Microfilaria* spp.	Microscopic In blood	Emaciation	Millions	N.Y.—0–11%
Gape worm *Syngamus trachealis*	15–20 mm long Red. Attacks windpipe	Causes bird to choke and gape	Dozens	N.Y.—1% Common in domestic Chicken
Larval muscle worm *Physaloptera* spp.	2–3 mm long Encysted in muscles of breast and legs	No apparent damage	1 to many dozens	Minn.—3%

*After Edminster, 1947.

Eimeria attacking the lower orders of birds and *Isospora* the higher. These organisms cause the destruction of the intestinal lining, loss of appetite, emaciation, diarrhea, and, in heavy infections, hemorrhage and death. *Eimeria* has caused enormous losses in the poultry industry. In wild birds the incidence of infection averages around 31 per cent in Europe, and is highest in summer and autumn. In wild turkeys of the southeastern United States incidence of infection was found to be 50 per cent in young poults as against 17 per cent in adults.

Trichomonas gallinae, a flagellated protozoon, infects the throats of doves, pigeons, hawks, and domestic fowl. It stimulates a growth that obstructs the esophagus and thus causes starvation. Of 50 feral pigeons collected in the Chicago area in 1966, every one was found to have heavy mouth infestations of *Trichomonas*. An epizootic outbreak of trichomoniasis occurred among the Mourning Doves, *Zenaida macroura*, in Alabama in 1951, killing an estimated 25,000 to 30,000 birds. It has been suggested that the extinction of the Passenger Pigeon, *Ectopistes migratorius*, was hastened by a similar epizootic plague.

A great variety of birds, particularly aquatic species, are occasionally infected with the mold *Aspergillus*, which attacks the windpipe, lungs, and air sacs, but may occur throughout the body. An epizootic outbreak of lung aspergillosis killed about 2000 Canada Geese, *Branta canadensis*, in a Missouri refuge in 1966. Aspergillosis is commonly a fatal disease in penguins kept in zoos in warm climates—a fact that has bearing on their natural geographical distribution.

Birds are responsible for a fungus disease in humans, although the birds themselves do not harbor it. The disease, histoplasmosis, is a pulmonary infection common in the midwestern section of the United States. It affects about 500,000 persons each year, and of these perhaps 5 per cent develop a chronic infection. The disease is caused by the microorganism *Histoplasma capsulatum*, which multiplies in soil under bird roosts and similar sites with heavy fecal deposits. A typical outbreak occurred in Missouri when 64 Boy Scouts worked at clearing and raking the ground in a park where Starlings, *Sturnus vulgaris*, had roosted for nine years. Four of the boys came down with histoplasmosis, and all but two of the 64 developed symptoms of the disease (Murton, 1971).

Bacteria cause relatively few diseases in wild birds, possibly because many of the microorganisms cannot multiply successfully at birds' high body temperatures. Pullorum disease, or bacillary diarrhea, is a common and devastating malady in young chickens, and has been suspected as the cause of death of thousands of young terns. Epizootics of avian cholera caused by the bacterium *Pasteurella multocida* have resulted in high mortality in Eider Ducks, *Somateria mollissima*, and in several species of swans and geese in their winter quarters. In Nebraska in the spring of 1975, about 25,000 waterfowl and 300 Crows, *Corvus brachyrhynchos*, died of avian cholera (Zinkl *et al.*, 1977). Other avian bacterial diseases include tetanus, diphtheria, tuberculosis, tularemia, and botulism. Botulism results from the eating of decaying vegetation or flesh, or of fly larvae feeding on decayed flesh, in which the anaerobic bacterium *Clostridium botulinum* has grown. The virulent toxin given off by the bacterium and consumed by the bird with the food affects the nervous system, causing paralysis and, eventually, death. In Great Salt Lake in 1932, an estimated 250 thousand ducks died of botulism, and in 1910 millions of aquatic birds succumbed to the disease in California and Utah. Chlamydiosis, formerly called ornithosis, is one of the few avian bacterial infections which may be transmitted to man. It has been identified in at least 30 species of parrots and 40 species of other birds. Two-thirds of the pigeons

of Paris are said to be infected with Chlamydiosis. In 1944 about 45 per cent of the pigeons of Chicago carried the disease during an outbreak of ornithosis in humans. It causes in man an infection resembling pneumonia, and is occasionally fatal. Several epidemics of ornithosis have occurred among employees of fowl-processing plants; in one such firm in Pennsylvania, 12 out of 89 employees contracted the disease in 1957 (Boyd, 1958).

Virus diseases include a contagious "foot pox," which causes wart-like growths on the feet and at the base of the bill. It is common in sparrows and may result in the loss of toes or feet. It has been known to reach epizootic proportions in Mockingbirds, *Mimus polyglottos*. Viral hepatitis and Newcastle disease are highly infectious viral diseases that at times cause 100 per cent mortality in flocks of pheasants and chickens.

A more complex mechanism of virus transmission involves both birds and insects. Some 80 species of North American birds act as reservoirs for a virus responsible for encephalitis (inflammation of the brain) in horses, man, and other vertebrates. If mosquitoes or ticks suck blood from an infected bird, the virus multiplies in their tissues and may then be transmitted by a bite to a susceptible vertebrate. This vertebrate may then develop an infection ranging from epidemic fever to fatal encephalitis, depending on the virus strain involved, the resistance of the host, and other factors. Since these are arthropod-borne viruses, they are called arboviruses. The birds themselves rarely suffer from such virus infections. Migrating herons were believed to have spread a virulent strain of encephalitis in Japan and Korea in 1958, causing the deaths of 2800 persons. Two South American strains of equine encephalitis virus have been identified in north-migrating birds captured at the Mississippi delta—a fact that confirms the intercontinental transmission of diseases by birds (Calisher *et al.*, 1971).

A disheartening bit of history that underscores the role insect vectors play in bird distribution and survival began in 1826 with the accidental introduction of the mosquito *Culex pipiens fatigans* into the Hawaiian Islands (Warner, 1968). The spread of the mosquito through the island lowlands was quickly followed, among the native birds, by epizootics of bird pox, malaria, and other diseases. Many bird species, especially among the drepaniids, became extinct, and the susceptible survivors were forced to live at altitudes above 600 meters where the mosquito is relatively scarce. Even today, drepaniids experimentally exposed to lowland environments promptly die from bird pox or malaria.

Still other lethal agents affect birds. Tumors, muscular dystrophy, uremic poisoning, convulsions, arteriosclerosis, aortic rupture, and various genetic structural and functional abnormalities have been found in birds. Arteriosclerosis has a very high incidence in birds: 45 per cent in domestic chickens over one year old, (Dauber, 1944), and 100 per cent in White Carneau Pigeons four years old (Albert *et al.*, 1978). Even "smog" takes its toll. In February, 1959, a persistent smog caused asphyxia in 250 to 300 Starlings, *Sturnus vulgaris*, on the main street of Sleaford, in Lincolnshire, England (Peet, 1959). At times certain marine microorganisms called dinoflagellates multiply sufficiently to produce "red tides" that cause the deaths of great numbers of sea birds. Such red tides killed thousands of marine birds on the coasts of Washington in 1942 (McKernan and Scheffer, 1942), and of Florida in 1974 (Forrester *et al.*, 1977). Following a similar "bloom" of dinoflagellates in the North Sea, 19 species of sea birds were found dead on the northeast coast of Britain in 1968 (Coulson *et al.*, 1968). The birds were apparently killed by eating shell-fish that had filter-fed on the neurotoxin-producing dinoflagellates. Similarly, die-offs of various gulls and terns in Massachusetts followed blooms of toxic dinoflagellates (Nisbet, 1983).

Death in birds may come not only from the presence of lethal agents, such as storms, predators, parasites, and diseases, but also through the absence of certain requirements, such as food, vitamins, trace minerals, oxygen, and others. And, of course, various decimating agents may work in combination. A bird even slightly weakened by parasites is that much less agile in escaping a predator. Since birds apparently live so close to their limits of toleration, even small differences in health may have fatal consequences.

CHANCES OF SURVIVAL

Primarily concerned with disease immunity are the thymus gland and the bursa of Fabricius. The thymus forms lymphocytes that are thought to produce immunoglobulin in the blood responsible for cellular immune reactions. The bursa of Fabricius produces lymphoid cells that produce humoral antibodies that combat bacteria and foreign antigens. Diet very probably affects a bird's disease resistance. Birds fed an excess of good quality protein showed improved

resistance to Newcastle disease (Fisher, 1972). The fact that numerous shore birds, storks, and spoonbills habitually leave water where they are feeding and defecate on land, appears to be a naturally selected behavior pattern to reduce the incidence of parasites and disease organisms in their food (Reynolds, 1965).

Surrounded as birds are by death in so many forms, one begins to understand why so many species have high reproductive potentials. The stable population level that characterizes most wild birds is thus the result of an equilibrium between a species' breeding habits and its living conditions. The agents responsible for this equilibrium have been compared by Leopold (1933) to two antagonistic forces. One of them, which represents the theoretical force of unlimited reproduction, acts like a steel spring always tending to curve upward. This force is counteracted by various strings pulling downward that represent the environmental forces that collectively keep the population leveled off at supportable numbers. These leveling forces are of two kinds: those, like predators, that kill directly, and those, like poor food and water, that reduce the breeding potential. Some forces, like parasites, may do both. Figure 18–5 assumes that one pair of robins, capable of rearing eight young per year, is installed on a natural area capable of supporting about 1000 robins. The actual population curve has been smoothed and does not show the annual saw-teeth of spring and summer rise when young hatch out, nor the fall and winter decline as the young and older birds die.

Since different species of birds enjoy different breeding potentials and live under very different environmental conditions, one would expect them to show different rates of mortality, as they do. Whatever their rate of death, however, most populations remain constant. This means that, ordinarily, deaths in a given species are matched by replacement, but there are three instructive exceptions to this principle. One occurs when a species is introduced into a new and suitable habitat, as the Starling, *Sturnus vulgaris*, was in North America, and the Skylark, *Alauda arvensis*, in New Zealand. Such a species may increase in numbers spectacularly for a few years or decades, but eventually it levels off at a stable population density. A second exception occurs in species that fluctuate in numbers from year to year, sometimes periodically. In either case, these birds increase in numbers until food, predators, disease, or some other density-dependent element in the environment keeps the population in check.

A third exception to the principle of population

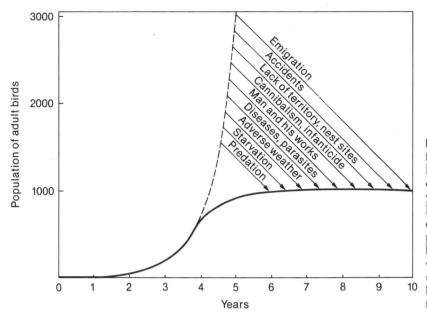

FIGURE 18–5 The counteracting forces that produce population equilibrium. One pair of robins could have over 24 million descendants within ten years if all the young survived that long and all pairs of descendants raised eight young per year. In nature such population explosions are prevented by a variety of forces and influences which cause the "steel spring" of unimpeded population increase to stabilize at supportable levels. Adapted from Leopold, 1933.

stability is difficult to identify and probably occurs rarely. It depends on a genetic change in the population of a species. If, for example, a mutation occurs that changes a species' physiological capacities, this may allow it to live in a warmer or colder climate, to eat different foods, to lay more eggs, and so on. Such a species, so equipped, may exploit new and more productive habitats with a subsequent increase in its numbers. This principle is thought by Mayr (1950) to explain the recent spectacular spread of the Collared Turtle Dove, *Streptopelia decaocto*, over Europe. Years ago the Starling, *Sturnus vulgaris*, was a migratory, single-brooded species. Its recent population explosions both in Europe and in North America may be due in part to genetic changes that now permit it to remain sedentary and to produce two broods a year.

Many difficulties hamper the study of vital statistics of bird populations. Even though a bird may be banded and its identity thus fixed for life, it may not be banded as a young bird of known age. Most banded birds, once released, are not subsequently captured, or if they are, may not be recovered at the time of their death. The great majority of wild birds die only to decay or be eaten, unobserved by human eyes. Moreover, various statistical biases may attach to the small percentages of dead birds that do come to human attention. Nevertheless, many studies of bird survival,

mortality rates, mean longevity and similar vital facts have been undertaken, and much important information obtained.

These studies employ diverse techniques of varying reliability, but most of them rely on population sampling devices similar to those used in public opinion polls. In its simplest form, such a population study, known as the capture, mark, and recapture method, can be compared to a counted handful of black beans (banded birds) inserted into and mixed with a jar full of more numerous but uncounted white beans (an unbanded, wild population of birds). The study of a handful of the mixed black and white beans (the later recovery of banded and unbanded birds) enables one to make quantitative predictions about the whole population, based on the assumption that the same statistics apply to the white beans (unbanded wild birds) as to the black beans (banded and counted birds about whose life span something is known).

In the case of small populations of relatively sedentary birds, more direct and precise studies of population statistics may be undertaken. An example of the facts obtained in such investigations may be furnished by Johnston's (1956) study of the salt marsh Song Sparrows, *Melospiza melodia*, of San Francisco Bay. Of every hundred eggs laid by this species, 26 per cent were lost before hatching, leaving 74 live nestlings. Mortality of nestlings was 30 per cent; as a result only

52 fledglings left the nests. Of these young birds, 80 per cent died the first year, leaving only 10 as adults to breed the following season. During the next year there was 43 per cent mortality among these one-year olds, leaving only six out of each original hundred to carry on the following year; and each subsequent year mortality among the survivors amounted to 43 per cent.

High mortality is the normal fate of all young birds. For a colony of 9000 to 10,000 Emperor Penguins, *Aptenodytes forsteri,* in Adelie Land, Antarctica, Sapin-Jaloustre (1952) estimated that, from the time the egg was laid until the molt of the young, mortality was between 80 and 90 per cent. Quite clearly, the high mortality of eggs and young must be compensated for if a population is to maintain its numbers. High losses can be replaced either by raising large numbers of young, by increasing mean longevity, or both.

Mortality rates vary not only between species, but at different periods in the life cycle. Since the eggs and young are particularly vulnerable to weather and predators, their mortality rate is almost invariably higher than that of adults. Therefore, the longer the duration of the egg and nestling stages of a species, the greater the probable mortality. This is particularly true of altricial species whose eggs and young reside in open nests. A survey of various American and European studies of nesting success of both open- and hole-nesting altricial species has been made by Nice (1957). A few representative results from this survey appear in Table 18–4.

The data in the table show clearly the superior success of cavity-nesting altricial species, both in hatch-

TABLE 18–4.
NESTING SUCCESS OF ALTRICIAL SPECIES*

Species	Eggs Laid (number)	Eggs Hatched (per cent)	Eggs Fledged (per cent)	Authority
OPEN-NESTING BIRDS				
Mourning Dove, *Zenaida macroura*	8018	54.6	46.6	McClure
European Skylark, *Alauda arvensis*	136	68.0	45.0	Delius
American Robin, *Turdus migratorius*	548	57.8	44.9	Young
Redwinged Blackbird, *Agelaius phoeniceus*	1140	72.2	59.2	Smith
Boat-tailed Grackle, *Quiscalus major*	1700	74.0	69.0	Teetor
American Goldfinch, *Carduelis tristis*	696	65.3	48.6	Stokes
Chipping Sparrow, *Spizella passerina*	277	66.8	61.4	Walkinshaw
Field Sparrow, *Spizella pusilla*	1738	51.1	35.7	Walkinshaw
Song Sparrow, *Melospiza melodia*	585	66.5	41.5	Nice
26 studies	21,040	59.8	—	Nice
29 studies	21,951	—	45.9	Nice
HOLE-NESTING BIRDS				
Tree Swallow, *Iridoprocne bicolor*	1123	83.4	61.0	Chapman
Pied Flycatcher, *Ficedula hypoleuca*	3724	70.7	62.2	Creutz
Great Tit, *Parus major*	45,466	—	64.9	Kluijver
House Wren, *Troglodytes aëdon*	6773	82.3	79.0	Kendeigh
Eastern Bluebird, *Sialia sialis*	6260	63.0	44.5	Laskey
Eastern Bluebird, *Sialia sialis*	1290	—	65.0	Musselman
Starling, *Sturnus vulgaris*	10,557	—	75.1	Lack
23 studies (8 species)	34,000	77.0	—	Nice
33 studies (13 species)	94,400	—	66.0	Nice

*Largely after Nice, 1957.

ing eggs and in fledging young. They also show wide variation in the nesting success of such closely related species as sparrows. The great discrepancy in fledging success for the same species (bluebird), as determined by different workers in different parts of the country, suggests that caution should be used in applying such statistics to general principles of population survival. These and similar figures do not, of course, tell what percentage of the fledged young survive to independent breeding existence. Most altricial young after leaving the nest are still dependent on their parents for at least a few weeks.

Precocial or nidifugous young generally leave the nest a few hours after hatching and usually remain in the care of one parent for several days or weeks. It is extremely difficult to follow the fates of either altricial or precocial wild young during this period of youthful wandering, but Lack (1954), in summarizing some 60 studies of survival in various species, both altricial and precocial, estimates that probably "less than one-quarter of the eggs laid give rise to independent young."

One can easily see that this high but variable mortality in young birds plays an important role in shaping the character of a population. The saw-tooth peaks and depressions of the annual population curve have already been mentioned. Age characteristics of any population will naturally depend on the relative gains and losses of young and old birds each year. Species, for example, with pronounced natural longevity or high first-year mortality will tend to show a preponderance of adult birds in their populations, as seems to be the case in many sea birds. Computations by Austin and Austin (1956), based on 6965 recoveries of Common Terns, *Sterna hirundo*, banded as chicks, indicate that 90 per cent of the breeding population on Cape Cod is composed of birds three to ten years old. Birds one and two years old rarely return to the colonies to breed (Fig. 18–6). An example of age structure of a population of California Quail, *Lophortyx californicus*, is shown in Figure 18–7.

Once a bird survives its risk-laden first year, the mortality rate for its remaining years remains more or less constant. That is, the expectation of further life in an adult bird remains the same regardless of its age. Upon analyzing the data from 2690 Mourning Doves, *Zenaida macroura*, banded on Cape Cod over a span of 21 years, Austin (1951) concluded that the annual mortality was about 80 per cent for the first year of life and 55 per cent per year thereafter for the next 10 years. Analysis of 3400 recoveries of Blackheaded Gulls, *Larus ridibundus*, by Flegg and Cox (1975) showed a mortality rate for their first year of about 38 per cent, the second year 27 per cent, and

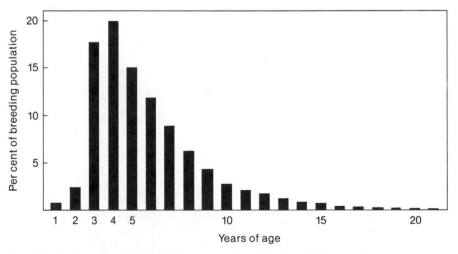

FIGURE 18–6 Age composition of a breeding population of Common Terns on Cape Cod as estimated by Austin and Austin (1956), who calculated the annual adult mortality rate at 25 per cent. More recent work that makes allowances for band losses and other sources of bias suggests a mortality rate of between 10 and 20 per cent (Nisbet, 1978).

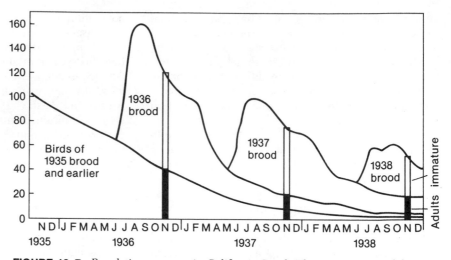

FIGURE 18–7 Population turnover in California Quail. The age structure of the population each November is shown by the vertical bar: black = adult birds; white = immature birds. Eggs and chicks are not shown in the diagram. After Emlen, 1940.

subsequent adult years 24 per cent. For adult Kittiwakes, *Rissa tridactyla*, Coulson and White (1959) estimated adult mortality at 12 per cent. This seems to be a standard pattern for most birds, but there are occasional exceptions. By following the fates of 338 female Pied Flycatchers, *Ficedula hypoleuca*, Berndt and Sternberg (1963) concluded that the relationship of mortality to age was quite similar to that in man. In

the second year of their lives, the birds' average death rate was about 30 per cent; in the third year, 40 per cent; in the fourth year, 65 per cent; in the fifth year, 75 per cent; and in the sixth year, 100 per cent.

Morel (1964) reached opposite conclusions regarding the Senegal Fire-finch, *Lagonosticta senegala*, of Africa whose death rates for the first four years of life were 73, 61, 50, and 40 per cent, respectively. That

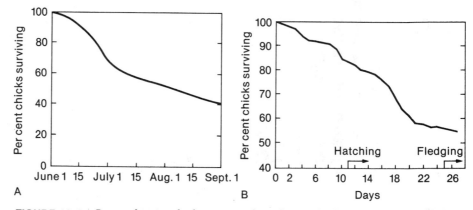

FIGURE 18–8 Curves showing the heavy mortality of young birds in their first summer. A, Survival curve of precocial young Ruffed Grouse between hatching and the first autumn. After Edminster, 1947. B, Survival curve of 261 eggs and the subsequent hatched altricial young of the Bronzed Grackle. After Petersen and Young, 1950.

is, the older the bird the greater its life expectancy. Drost and Hartmann (1949) found the same mortality pattern in the European Oystercatcher, *Haematopus ostralegus*. This apparent absurdity, so different from human experience, probably derives from the fact that, as a bird ages, its growing senility is counterbalanced by its accumulating experience in meeting the problems of survival. Further, in spite of their potential longevity, most birds die long before old age weakens their faculties seriously, so that death from old age is essentially unknown among wild birds.

In addition to such manifest agents of mortality as disease, starvation, and predation, there are other variables that bear significantly on survival. One of these involves seasons of the year. For many species living in cold climates, winter is clearly a time of stress and high mortality. After two unusually cold winters in 1961–1962, the English population of Grey Herons, *Ardea cinerea*, dropped to nearly one-half of its 50-year mean (Reynolds, 1978); and in Turku, Finland, the population of Great Tits, *Parus major*, after a winter with a mean December–February temperature of −10°C, dropped to roughly one-fourth of the size it had after a winter of −2°C (von Haartman, 1971). However, statistics for 817 House Sparrows, *Passer domesticus*, in Britain revealed that 54 per cent of all adult deaths occurred during the breeding season (April through July), with the remaining 46 per cent nearly equally divided between the other four-month periods (Summers-Smith, 1956). This seasonal bias in mortality is probably related to the increased exposure and energy-drain on parent birds that accompany their breeding activities. Breeding season hazards probably account also for a sex-differential in mortality of British Starlings. Among first-year birds females suffer a death rate of 70 per cent as against a 39 per cent rate in males (Coulson, 1960). The explanation here seems to be that females invest more energy than males in reproduction, and also that more females than males breed in their first year. Since energy expended in breeding involves a cost to the individual in reduced survival, birds are faced with a dilemma: whether to breed as soon and as often as possible, or to breed later in life and perhaps intermittently, and thus prolong their breeding careers. Kittiwake Gulls, *Rissa tridactyla*, seem to have evolved the latter strategy (Wooller and Coulson, 1977).

There are also distinct geographical differences in death rates. Adult Barn Owls, *Tyto alba*, for instance, have an annual death rate of 24.8 per cent in the southern United States, 38 per cent in the northeastern United States, and 45.9 per cent in Switzerland (Henny, 1969). Some of this variation may be climate-related, but other factors probably also enter in. In Table 18–5 representative examples, taken from Lack (1954), give adult survival statistics for different kinds of birds. The reciprocal relation between adult annual death rate and life expectancy is obvious. The luxury of a low annual death rate can be enjoyed only by those species demonstrating a high life expectancy.

FLUCTUATIONS IN POPULATIONS

While the great majority of bird populations seem to remain relatively constant in number except for seasonal fluctuations, there are certain species, especially birds of northern regions, whose populations may increase by several hundred per cent over a few years, and then abruptly decrease to their original numbers. These fluctuations may show cyclic, repetitive patterns or may be irregular in nature. When one considers the variety and instability of the factors that cause death in birds, it seems surprising that there are not more disturbances of population equilibrium.

There is not much doubt that food abundance is an important regulator of bird abundance, especially in those species whose diet is narrowly limited in variety. A given species of bird may increase its numbers phenomenally following two or three successive seasons of good food crops. Then, should the crops fail for one year, the birds either die in great numbers or, because of their superior mobility, invade new regions. These mass invasions of exceptional range are called *irruptions*. They have been recorded for certain owls, hawks, woodpeckers, nutcrackers, jays, waxwings, titmice, crossbills, grosbeaks, thrushes, and other species. Svärdson (1957) lists 40 species that show irruptive tendencies in Sweden.

In 1863, 1888, and 1908, the Pallas's Sand Grouse, *Syrrhaptes paradoxus*, emigrated from its normal range in Asia Minor and southern Russia and spread in enormous numbers throughout Western Europe and the British Isles. In Britain in 1888 a special Act of Parliament was passed for their protection, in the hope that the species would establish itself, but by 1892 the birds had completely disappeared. By analyzing data from Christmas counts, Yunick (1984) found irruptions of Boreal Chickadees, *Parus atricapillus*, in New York State, occurring at intervals of six to eight years. Between 1896 and 1939 there were 16 winter invasions

TABLE 18–5.
ANNUAL ADULT SURVIVAL IN VARIOUS SPECIES OF BIRDS*

SPECIES	MEAN ANNUAL ADULT MORTALITY (PER CENT)	MEAN FURTHER LIFE EXPECTANCY (YEARS)	SOURCE
Royal Albatross, *Diomedea epomophora*	3	ca. 36.	Richdale
Fulmar, *Fulmarus glacialis*	6	16.2	Dunnet *et al.*
Sooty Shearwater, *Puffinus griseus*	9	10.6	Richdale
Yellow-eyed Penguin, *Megadyptes antipodes*	10	9.5	Richdale
White-bearded Manakin, *Manacus manacus*	11	8.6	Snow
Canada Goose, *Branta canadensis*	16	5.6	Ratti *et al.*
Common Swift, *Apus apus*	18	5.6	Weitnauer
Rook, *Corvus frugilegus*	25	3.5	Holyoak
Band-tailed Pigeon, *Columba fasciata*	29	3.0	Wight *et al.*
Night Heron, *Nycticorax nycticorax*	30	2.8	Hickey
Herring Gull, *Larus argentatus*	30	2.8	Paynter
Wryneck, *Jynx torquilla*	33	2.5	Peal
Skylark, *Alauda arvensis*	33	2.5	Delius
Blue Jay, *Cyanocitta cristata*	45	1.7	Hickey
American Robin, *Turdus migratorius*	48	1.6	Farner
California Quail, *Lophortyx californicus*	50	1.5	Emlen
Turtle Dove, *Streptopelia turtur*	50	1.5	Murton
Starling, *Sturnus vulgaris*	53	1.4	Coulson
American Coot, *Fulica americana*	60	1.2	Ryder
European Robin, *Erithacus rubecula*	62	1.1	Lack
Song Sparrow, *Melospiza melodia*	70	0.9	Nice

*After Lack, 1954.

of Germany by the Siberian Nutcracker, *Nucifraga caryocatactes*. These invasions seemed always to be correlated with years of poor pine seed crops (the bird's chief food) in the home range of the species.

This same irruptive phenomenon has been observed in mammals, as in squirrels when an acorn crop fails, and even in man, as in the case of the massive Irish emigration that followed the potato famine in 1846.

It is unlikely that birds emigrate from regions of food scarcity because they "recognize" that they will starve if they remain there. In fact, Newton (1970) points out that Swedish Crossbills, *Loxia curvirostra*, put on body fat before mass emigrations just as other migrants do. Therefore it is not imminent starvation that triggers the mass movement. A more probable explanation is that dense populations of birds, built up during years of abundant food crops, are themselves

the immediate stimulus for irruption. Such mass emigrations, set off by sheer numbers, probably represent an adaptive response to sporadic famine, built up through thousands of years of natural selection. In this sense, food lack is the ultimate cause of irruptive behavior, but high population density is the proximate cause. Among species showing sporadic irruption, e.g., the Siberian Nutcrackers, it is likely to be the young birds rather than the old that emigrate. This is probably because the older birds are more attached to the home territory and have peck-order seniority, while the young have been naturally selected as more vigorous exploiters of strange regions. In 1973 an irruption of hundreds of thousands of Long-tailed Tits, *Aegithalos caudatus*, spread from northern Russia into Finland, the Baltic states and Germany, traveling at an average speed of 35 km a day. Hildén (1977) attributed the mass movement to population density and flock territoriality, probably modified by food lack and autumn weather conditions.

This correlation between food-lack and mass movements of birds was clearly shown by Reinikainen (1937). Every March for 11 years he covered the same region of Finland on skis for about 120 km, and recorded both the number of Red Crossbills, *Loxia curvirostra*, he encountered and the abundance of spruce cones, the seeds of which are the staple diet of these birds. Where cones were plentiful, birds were also plentiful, and where the cone crop failed, birds were rare or absent. In the 16 irruptions of Swedish Crossbills that have occurred between 1900 and 1965, none has ever occurred in years when the spruce seed crop was good. However, not all poor crops have been followed by massive emigrations (Newton, 1970). Svärdson (1957) believes that irruptive birds have the capacity, based on endocrine levels and photoperiods, to migrate every year as do other migrants, but that the birds settle down wherever they find abundant food.

From the limited evidence now available, it appears that irruptions depend primarily on three controlling factors: (1) species of birds that, as food specialists, depend essentially on a single type of food, (2) a genetic tendency to emigrate in great numbers from regions of actual or imminent food shortage when populations build to high densities; and (3) irregularity in their staple food crop. This may be brought on either by adverse weather (extreme cold during the trees' flowering period, or extreme drought during the growing season), or by the tendency of many plants to yield crops only in alternate years, or even less frequently. For steppe birds such as the Sand Grouse, or Rosecolored Starling, *Sturnus roseus*, drought is responsible for the lack of seeds or insects that coincides with their mass emigrations.

Only on rare occasions will birds remain to breed in the new region to which they have migrated after an irruption. A very few records have been obtained of irrupting birds returning the subsequent year to their original breeding area. Unlike normal migration movements (roughly north and south), irruptive migrations are likely to be east or west, within the vegetation zone of the species' typical food plant. Recoveries of banded birds have shown irruptive movements of as much as 4000 km.

Cyclic fluctuations in bird populations are largely restricted to predatory and gallinaceous species distributed between 30 degrees and 60 degrees N. latitude, where food supplies are less varied, and climate less regular, than farther south. There are two rather well-defined cyclic patterns in North America: a ten-year cycle characteristic of populations of Ruffed Grouse, *Bonasa umbellus*, Goshawks, *Accipiter gentilis*, and Great Horned Owls, *Bubo virginianus*, as well as the Varying Hare, *Lepus americanus*; and a four-year cycle found in the Willow Ptarmigan, *Lagopus lagopus*, Snowy Owl, *Nyctea scandiaca*, Rough-legged Hawk, *Buteo lagopus*, and Northern Shrike, *Lanius excubitor*, and also the lemmings, *Dicrostonyx* spp. To say that these animals exhibit population cycles of ten and four years does not mean that their population peaks come precisely at those intervals. However, the phenomenon is regular and striking enough to have caused much study and speculation.

To account for these cycles, a great variety of explanations has been suggested, ranging from sunspot and climatic cycles through predators, food-lack, and epizootics, to over-hunting, in-breeding, nervous stress, endocrine exhaustion, territorial spacing behavior, and lack of cover. There is a correspondence between fluctuations in populations of the Varying Hare and those of Great Horned Owls and Goshawks.

Less evident is the basic cause of cycles in gallinaceous birds. Lack (1954) is of the opinion that a producer-consumer, density-dependent oscillation is set up between rodents and their vegetable food. When, for example, the lemmings become too populous, their food plants suffer and decline, and this reduction is followed by a "crash" year of lemming decline. With the lemming population reduced drastically, plants again have a chance to regain their abundance; then the lemming population builds up again until the next crash. When foxes, hawks, and other predators are faced with the disappearance of lemmings, they turn with greater intensity on alternate prey, such as grouse and ptarmigan, thus initiating in these birds a cyclic population oscillation. Nearer the equator, where animal communities are more complex, and where numerous alternative foods are available when a preferred type fails, food relationships are sufficiently elastic to dampen population oscillations to the vanishing point.

Still another explanation of cyclic irruptions is proposed by Svärdson (1957), who points out that 60-year records for the fruiting of Swedish spruce trees show that a heavy seed crop appears every third or fourth year. This may be a plant adaptation which ensures a higher total number of seeds produced and germinated through the years because the staggered years of production prevent the building up of a large, sedentary population of seed-eating birds and other animals. In seedless years the birds must move elsewhere and possibly eat a substitute diet. The spruce trees, in essence, have evolved a mechanism which, to their benefit, drastically reduces the carrying-capacity of the birds' environment, and yet allows the trees sporadic bursts of seed production great enough to surmount the destruction of seeds by a reduced population of birds. Nomadic wanderings and irruptions of birds may represent a complementary adaptation. Galushin (1974) offers the intriguing suggestion that species that oscillate in synchrony with their food source may have evolved the ability to undertake extensive "searching migrations" that enable them to locate areas of adequate food supply—an ability that also may promote the evolution of food specialization. Whatever the ultimate explanation, cycles in avian populations are a well-established fact, and further research will probably clarify the mechanisms responsible.

■ SUGGESTED READINGS

Lack treats the subject of population simply and lucidly in *The Natural Regulation of Animal Numbers* and in *Population Studies of Birds*. For a mathematical scrutiny of population biology, Hutchinson's *An Introduction to Population Ecology* is recommended. Briefer surveys of population biology are found in Gibb's chapter on Bird Populations in Marshall's *Biology and Comparative Physiology of Birds*, and in von Haartman's chapter on Population Dynamics, and Cody's on Ecological Aspects of Reproduction, in Farner and King's *Avian Biology*. Several authors consider the effects of food on bird numbers in Watson's *Animal Populations in Relation to their Food Resources*. Techiques in population research are emphasized in Hickey's chapter in Wolfson's *Recent Studies in Avian Biology*. In the same book will be found a chapter by Herman on Diseases of Birds. Extensive, primarily clinical works on diseases and parasites of domestic and cage birds are Hofstad's *Diseases of Poultry* and Petrak's *Diseases of Cage and Aviary Birds*. Authoritative and highly readable is Rothschild and Clay's *Fleas, Flukes and Cuckoos, A Study of Bird Parasites*.

The Ecology of Birds

19

We have reason to believe that species in a state of nature are limited in their ranges by the competition of other organic beings quite as much as, or more than, by adaptation to particular climates.
—Charles Darwin (1809–1882),
On the Origin of Species

Ecology is the study of the relationships between organisms and their environments, or between organisms and what Darwin called the conditions of their struggle for existence. This is a deceptively simple statement, for ecology covers an unusually broad spectrum of natural phenomena. It is concerned with organisms and their adaptations in structure, physiology, and behavior. It is concerned with all aspects of the environment that may affect organisms: the soil they walk on, the air they breathe, the food they eat, the weather they endure, and all interacting organisms, whether plants or animals. Ecology may be studied from the standpoint of a given organism as related to its environment (autecology), or from that of groups of organisms in an environment (synecology). Emphasis

may be placed on habitat—for example, the ecology of caves—or on the distributional or geographical effects of selected environmental factors—for example, the effects of rainfall on animal distribution. Throughout this book thus far, emphasis has been placed on the adaptations of birds to their conditions of existence. In this chapter an attempt will be made to list and analyze some of these interrelationships between birds and their environments, starting with the physical factors in the environment (soil, water, temperature, and sunlight), progressing through the biological factors (plants and animals), and ending with a study of the more comprehensive fabrics that these threads of influence weave—the ecology of major habitat types such as lakes, forests, and grasslands.

409

Neither the environment nor the genetic composition of a population is entirely static. Mutations constantly emerge in individuals in a population, and the genetic make-up of a species is in a state of flux. As the environment changes, individuals with new genes enabling them to adapt to the new changes will survive and reproduce.

Despite their supreme adaptation, the power of flight, birds often tend to stick closely to the habitat to which they are adapted. Young birds of the year tend to be the pioneers, wandering afar and sometimes colonizing new habitat. "They constitute sort of sensitive tentacles, by which a species keeps aware of the possibilities of aerial expansion. In a world of changing conditions it is necessary that close touch be maintained between a species and its geographical limits . . . " (Grinnell, 1922).

PHYSICAL FACTORS OF ENVIRONMENT
Every bird lives in an environment composed of two fundamentals: matter and energy. Matter provides the medium (air) within which the bird lives, a substrate (ground or water) on which it walks or swims, and the materials on which the bird's body is made. The energy of the environment, which is absorbed by the bird for its own needs, comes ultimately from sunlight, by way of foods eaten by the bird.

Energy Flow
All life on earth is sustained by solar energy which is fixed by green plants in the form of organic compounds by means of photosynthesis. Through photosynthesis, atmospheric carbon dioxide is reduced to carbon compounds that hold potential energy. At the same time, photosynthesis releases molecular oxygen into the atmosphere.

The energy equivalent of all the solar energy that falls annually on the upper atmosphere of the earth is about 156×10^{16} kilowatt hours. Only a very small fraction of this energy—about one-tenth of 1 per cent —is captured by green plants and fixed through photosynthesis, yet this fraction operates all of the world's great ecosystems, such as tropical forests, marshlands, and deserts. About three-fourths of this energy is thought to be fixed on land and one-fourth in the seas. Tropical rainforests capture about 30 times as much solar energy as do deserts or tundras. Most of this photosynthetic energy is in ceaseless flux, flowing from one organism into another until eventually it is lost into outer space as radiated heat.

Of the total solar energy captured by a green plant through photosynthesis (i.e., its *gross production*), about one-half is devoted to living processes such as respiration, and about one-half is stored in the form of plant tissues, which represent *net production*. This net production is, worldwide, annually equivalent to roughly 172 thousand million metric tons of dry organic matter produced on land, and 60 thousand million tons produced in the seas (Rodin *et al.*, 1975). For purposes of broad comparisons, this amounts to approximately 970 million million (970×10^{12}) kilowatt hours of energy annually. Worldwide energy consumption by humans in 1967—from coal, oil, gas, water power, and nuclear sources—totaled about 45 million million kilowatt hours.

The green plants that fix solar energy in the form of potential energy in roots, stems, leaves, fruits, and the like are called *producers*. The herbivorous animals that eat the green plants to obtain energy and nutrients are called *primary consumers*. They, in turn, may be eaten by carnivores or *secondary consumers*. The energy of sunlight thus flows in progressively diluted amounts through successive organisms: from the producer (green plant), to the primary consumer (herbivore), to the secondary consumer (carnivore), and perhaps into another animal or two, until it reaches the end of the *food chain*. The sum of all of the interlacing food chains in a community makes up a *food web*.

Eventually these organisms will all die and will be reduced by *decomposer* organisms, such as bacteria and fungi, to the simple substances out of which they were originally constituted: carbon dioxide, water, nitrogen, phosphorus, and, in lesser amounts, other compounds and elements. The nutrient substances in all organisms are thus continually recycled between living organisms and the inorganic environment.

The efficiency of energy transfer from one organism in a chain to another is very low, largely due to heat losses. Roughly, only about 10 or 20 per cent of the net energy available in a green plant is assimilated by a primary consumer and made available as net production for consumption by a secondary consumer. The same reduction in energy occurs between the secondary and tertiary consumer, and so on to the end of the chain. That is, there is an 80 to 90 per cent loss in energy transfer between successive levels in a food chain. This means that as one advances along a food chain from producer to final consumer (e.g., grass→ meadow mouse→ weasel→ owl) the numbers of animals at successive levels are drastically reduced. This is why vegetarian finches can be so numerous in the same region where carnivorous hawks are scarce, and why vegetarians can subsist on smaller territories than carnivores.

Rarely will there be more than four levels in a food chain. This description represents a highly simplified version of the extremely complex phenomena of energy flow and nutrient cycling in natural environments. Consider, for example, the complexities posed by a robin that eats both berries and worms, and is thus simultaneously a primary and a secondary consumer. Parasites and bacteria of disease within the robin add secondary-tertiary consumers and decomposers, all in one organism. More complete discussions of this subject will be found in the references given at the end of this chapter. Figure 19–1 from Varley (1970) gives in a simplified form quantitative estimates of the energy flow in an English woodland community in which oak leaves are eaten by caterpillars, which are eaten by birds, which are eaten by weasels, and so on. The nutrient cycling and energy flow through birds in most habitats is surprisingly small. The energy consumption of all birds in a New Hampshire deciduous forest

amounted to 73,858 kilocalories per hectare per year, or only 0.17 per cent of its net productivity (Holmes and Sturges, 1975); and in western prairies of the United States, birds consumed about 20,000 kcal per hectare (Wiens, 1973).

The Substrate

Although some birds may be air- or water-borne for days on end, they must sooner or later come to earth, if for nothing else than to breed. The physical character of the ground on which a bird typically walks is commonly attested to by the bird's feet and legs (Fig. 19–2). Birds adapted to relatively firm, flat, open country typically have long, powerful legs, with short toes that are often reduced in number—a characteristic paralleled by the hoofed mammals of similar habitats. Among the long-legged, short-toed runners are ostriches, emus, rheas, coursers, bustards, tinamous, and a ground-dwelling bird of prey, the Secretary Bird, *Sagittarius serpentarius*, of Africa, whose toes are only about one-fifth as long as those of a comparably large hawk, but whose legs are perhaps three times as long.

Birds that do much scratching of the ground in

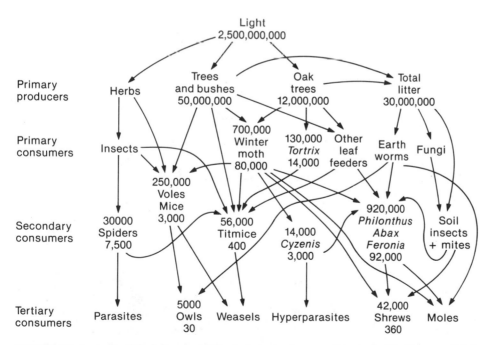

FIGURE 19–1 A simplified diagram of food chains and energy flow in an English woodland community. Estimates of energy consumption and production in kilocalories per hectare per year are inserted above and below each link for which figures are available. From Varley, 1970.

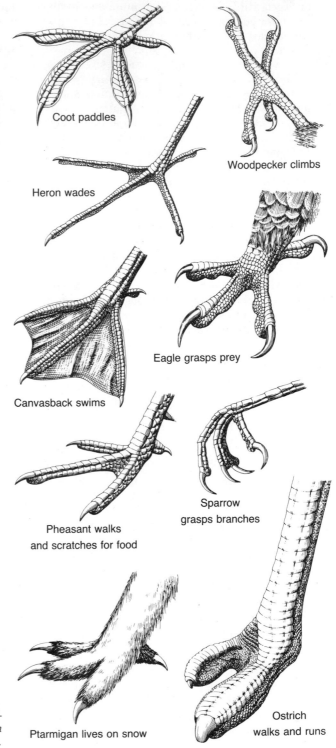

FIGURE 19–2 Examples of adaptations of feet to different substrates. From Wilson, *Birds: Readings from Scientific American.* Copyright 1980. All rights reserved.

seeking food (Galliformes) or in building mounts (Megapodiidae, Menuridae) have powerful legs and feet with heavy, blunt claws. Birds that frequent marshy ground are apt to have long legs and compensating long necks (herons, bitterns, cranes, storks), or lobate-webbed toes (grebes, coots, phalaropes, finfeet) suitable both for walking on mud or for swimming. Another adaptation for a soft substrate is found in the jaçanas, shore birds whose extremely long toes enable them to walk on floating aquatic vegetation such as water-lily pads, and thus exploit a habitat that attracts few, if any, avian competitors.

As a substrate, snow presents serious problems of both temperature and support, which many birds avoid by migrating. Several species of the grouse family solve the problem of locomotion by growing fringe-like scales along the side of each toe for the winter season only; in the spring these "snowshoe" fringes are molted. The feet of the Scottish Ptarmigan, *Lagopus mutus*, are relatively unfeathered in summer, but in winter each toe is so heavily feathered that the bird walks on feathers—an adaptation that provides both support and insulation. In addition to dense feathering on its feet, the arctic Willow Grouse, *Lagopus lagopus*, has claws which are 17 mm long in winter but only 9 mm in summer. The "snowshoe" effect of such feet was studied by Salomonsen (1972). He found that a square centimeter of this enlarged grouse foot surface needed to support only 14 or 15 g of body weight, whereas in the Hungarian Partridge, *Perdix perdix*, a bird lacking such winter adaptations, each square centimeter of foot had to support 40 to 41 g of body weight. The feet of Common Ravens, *Corvus corax*, show still another kind of winter adaptation. The cornified papillae on the soles of arctic Ravens' feet contain keratin, a poor heat conductor, and are about six times as thick as those of subtropical Ravens. The remarkable change in the plumage of several species of ptarmigan from white in winter to brown in summer was discussed earlier. Another type of adaptation to snow is found in the antarctic Adelie Penguin, *Pygoscelis adeliae*, whose nostrils are feathered over to keep snow out of its air passages. Other species of this genus living in warmer climates have exposed, unfeathered nostrils (Murphy, 1936).

The topography of the substrate is of ecological significance to many species, particularly during the nesting season. Many swallows, kingfishers, bee-eaters, South American ovenbirds, motmots, todies, and jacamars require vertical banks of soil into which they dig their nest burrows. The Peregrine Falcon,

Falco peregrinus, prefers as a nest site a ledge on a vertical cliff. Most gulls normally nest on shores that slope gradually to the water's edge. If overpopulation forces some gulls to nest on steep shores, they lose many young by drowning, because the latter rush to the water when disturbed, and are unable to clamber back up the steep slopes. In a somewhat contrary fashion, reproductive success of Puffins, *Fratercula arctica*, is higher when nests are situated on or near steep slopes and cliffs. Such sites provide antipredator advantages, as described on page 288.

Studies by Wilson (1959) have shown a relationship between the mineral content of soils and the ecological distribution of birds. In New York, the Hungarian Partridge, *Perdix perdix*, is restricted to soils of limestone origin. In two comparable field areas, the presence or absence of calcium was the only variable discovered that could explain the heavy population of Ring-necked Pheasants, *Phasianus colchicus*, in a limestone region and their scarcity in a noncalcareous area (Dale, 1955). Birds lacking calcium in their diet are unable to produce more than a very few eggs. Population densities and breeding success are higher in Red Grouse, *Lagopus lagopus*, on Scottish moors overlying basic rocks than on moors over granite or acid rocks. This is probably because heather growing on basic rocks is more nutritious, containing more potassium, phosphorus, cobalt, and copper than heather growing elsewhere (Moss, 1969).

Adaptations apply not only to finished structures and their functions, but also to the way in which structures grow. As described earlier, the precocial young of ground-nesting cranes leave the nest and follow their parents shortly after hatching. For over two months the young depend entirely on their legs for locomotion. Accordingly, their wings develop slowly and their legs very rapidly: the legs achieve their most rapid growth between the 8th and 16th days of life. Storks have a body architecture similar to that of cranes, but they raise their young not on the ground but high in roof- or tree-top nests. This ecological difference in nest sites correlates with a striking difference in leg and wing development of the young. Sturdy, rapidly growing legs would be not only useless in the young stork's nest habitat but positively dangerous. Accordingly, its legs develop slowly, growing most rapidly between the 20th and 38th days of life. Its wing feathers, however, sprout much earlier than those of the crane, and the young stork makes its first venture from the home nest into the world by means of its wings and not its legs (Heinroth, 1938). The primary wing feathers of the

FIGURE 19–3 Ecological adaptation in the growth of legs and wings. A, A young crane 2 days old is able to run, and at 32 days its legs have almost finished growing. Its wing feathers, however, have scarcely sprouted. B, A stork at 1^1/$_2$ days is blind and helpless. At 29 days its wing feathers are well along toward maturity, but its legs are still poorly developed. Adapted from Heinroth, 1938.

Paradise Flycatcher, *Terpsiphone atrocaudata*, grow rapidly and are so well developed at 10 days of age that the young are able to fly expertly, although the rest of the body is still covered with down (Jahn, 1939). Such a temporal bias in feather development is probably related to the bird's insect-catching mode of feeding.

Water is frequently a substrate for many water birds, and it may become a medium for a short while when a bird dives into it. Many of the adaptations of birds to an aquatic life are obvious and well known: webbed feet, powerful leg muscles, large uropygial glands, and dense waterproof plumage. In many ducks the oily flank feathers form pockets on each side into which the folded wings fit, dry and ready for instant action. Expert divers, such as the loon, have marrow-filled bones rather than the usual hollow bones—an adaptation promoting more efficient underwater swimming. Different species of penguins are known to swallow stones, presumably as ballast that would serve the same end. Penguins spend more time in water than any other birds, and they are well adapted to such an existence. Their bodies are smoothly streamlined and covered with small, hair-like feathers. Their wings have developed into flattened, powerful flippers adapted to skillful and rapid maneuvering in water. Their bodies are thickly covered with blubber for insulation against heat loss. Their eyes have flattened corneas adapted to vision both in water and in air.

Different kinds of water birds show distinct preferences for different kinds of aquatic habitats. Some species are exclusively fresh water birds, others salt water, and still others are able to live in either type of habitat. Some populations of Song Sparrows, *Melospiza melodia*, are adapted to live in salt marsh habitat on both coasts of North America (Marshall, 1948). The saltiness of sea water, when swallowed, causes the dehydration of body tissues and, in most animals, eventual death. This obstacle to the invasion of the sea has been overcome by petrels and other sea birds through the development of nasal or salt glands, which are many times more efficient than the kidneys in removing salt from body fluids. The remarkable ecological plasticity of a single species in adapting to this problem is demonstrated by the Mallard, *Anas platyrhynchos*, which in most parts of the world is a fresh water dweller and possesses ordinary-sized nasal glands. In Greenland, however, there is a race of Mallards that has become semimarine and whose salt glands are roughly 10 times as large as those of fresh water Mallards. This one ecological adaptation in water birds makes possible an enormous extension in the range of aquatic habitats open to them. It is very likely that similar structural and functional adaptations, as yet undiscovered, underlie such preferences and capabilities as those of the Snow Petrel, *Pagodroma nivea*, for cold water; the tropic birds, Phaëthontidae, for clear, saline, moderately warm water; the Laughing Gull, *Larus atricilla*, for warm water; and the Brown Pelican, *Pelecanus occidentalis*, for silt-free salt water (Murphy, 1936).

Climate and Weather

Climate, more than any other physical factor, determines whether or not a given species will live in a given region. Climate is largely a matter of rainfall and temperature, and its effects may be felt by the bird directly, or indirectly through the influence of climate on vegetation and other environmental features. In addition, humidity, wind, length of day, exposure to direct sunlight (e.g., north or south sides of a mountain), and slope and drainage of the soil play influential roles in the lives of birds. The twice-a-year global surge of migration shows to what extent temperature change alone requires adaptive adjustments in millions of birds. At all stages of their life histories, from eggs to adults, birds exhibit a tremendous variety of adaptations to temperature. Only a few representative examples of these adaptations can be given here.

Within vital limits, *change* in temperature works a greater hardship on an organism than a steady hot or a steady cold temperature. The Emperor Penguin, *Aptenodytes forsteri*, goes through the most exhausting stage of its life cycle, the breeding period, during the antarctic winter when temperatures range as low as $-60°C$ ($-76°F$). The Snow Petrel, *Pagodroma nivea*, nests atop mountains only 10 or 11 degrees in latitude from the south pole. The remarkably lethal effect of a 5° increase in water temperature has already been discussed in reference to El Niño. In 1982 a particularly intense warming-cycle occurred, resulting in widespread fish and sea bird die-offs in the Peru Current (Hughes, 1985) and the complete abandonment of Christmas Island by Blue-footed Boobies, *Sula dactylatra*, and Great Frigate Birds, *Fregata minor* (Barber and Chavez, 1983). Likewise, extreme cold can kill. After the severe winter of 1978–1979, the breeding population of five species of tits (Paridae) in a German larch forest was reduced from 50 per cent for Great Tits, *Parus major*, to 100 per cent for Willow Tits, *Parus atricapillus* (Winkel, 1981).

Cold weather can limit a species' activities and distribution indirectly through the freezing of water. Birds of northern lakes and streams must be able to migrate south as soon as ice covers the water where they make their living. The Belted Kingfisher, *Ceryle alcyon*, which fishes for its food by diving under water, is forced to fly south in the autumn ahead of the ice, and it migrates back north in the spring as soon as thaws open its feeding territories. Sea birds do not have to face this sudden metamorphosis of their habitat. As winter closes in, the sea never freezes solidly over wide areas, but leaves open waterways between ice floes through which sea birds may travel equatorward without flying. These facts may explain why there are flightless sea birds, but no flightless fresh water birds except in ice-free climates (Allen, 1925).

Through years of natural selection, temperature fluctuations have brought about adaptive changes in the structure, physiology, and behavior of birds. Feathers probably evolved primarily as heat-conserving devices and were later adapted for flight. A further adaptation to heat conservation is the increase in the number of feathers that many birds of cooler latitudes wear in winter. Various ecological adaptations in feather growth and molting patterns to cold weather, short growing seasons, and migration were considered in Chapter 3. Layers of body fat are also used for insulation against cold. Fat is particularly well developed in birds inhabiting cold waters, and, of course, birds of polar regions. Since large bodies have relatively less heat-radiation surface than small bodies (p. 139), large birds can tolerate cold climates better than small birds can. This explains why birds of high latitudes have, as a rule, larger bodies than their relatives in warm climates.

An ingenious adaptation that keeps the feet of ducks and geese from freezing while they stand on ice consists of a direct connection between the arteries and veins of the feet. This arrangement, found also in the beaks of some cold-climate birds, bypasses much of the capillary circulation, thus allowing a more rapid circulation and therefore a more rapid replacement of lost heat. The role of the *rete mirabile* in controlling heat loss in legs was described earlier in Chapter 7. In warm weather, air sacs promote the rapid cooling of the body as well as efficient respiration.

Chief among the physiological adaptations to temperature is thermoregulation, or warm-bloodedness, which enables birds to remain active day or night, winter or summer, in habitats spread over the surface of the entire earth. The complete absence of cold-blooded reptiles from high latitudes suggests the great ecological value of this adaptation in birds and mammals. Birds are able to maintain relatively constant body temperatures throughout a wide range of environmental temperatures, cooling themselves by sleeking their feathers and panting when external temperatures exceed their own, and generating more body heat and fluffing their feathers when cold air temperatures cause greater heat loss. As animals of incomparably high metabolism, birds maintain resting body temperatures ranging from 37.8°C to 44.6°C and averaging about 40.5°C (man's body temperature is 37.0°C). This temperature presses close to the lethal maximum body temperature of 46.7°C, but is quite far from the lethal minimum of 21.7°C (Bourlière, 1950). The efficiency of temperature regulation is seen in the fact that, within vital limits, for every rise or fall of 10°C in environmental temperature, a small bird's body temperature rises or falls only 0.1°C.

As in all animals, there are extremes in environmental temperatures beyond which birds cannot survive. These limits vary with species, health, age, and other factors. The Hoary Redpoll, *Carduelis hornemanni*, of North American arctic latitudes, is able to survive winter air temperatures as low as −67°C. This astonishing hardiness in a passerine is thought by Brooks (1968) to depend on the following adaptations: a rapid and large intake of food just before nightfall; large esophageal storage pouches to hold food; the selection of high-calorie food (birch seeds); increased digestive efficiency; probably unusually warm plumage; and an ability to remain active at low light intensities, thus lengthening the time available for feeding. Even closely related species may exhibit widely different temperature tolerances. The European Yellowhammer, *Emberiza citrinella*, is a nonmigratory species able to endure winter temperatures as low as −36°C. The Ortolan Bunting, *Emberiza hortulana*, which breeds in the same parts of Europe, tolerates a minimum of only −16°C, and consequently must emigrate south for the winter (Wallgren, 1954). Such factors as fat storage, metabolic efficiency, preferred diet, habitat choice, roosting habits, and thickness of plumage—all amenable to selective adaptation—determine a species' temperature tolerance and its limits of geographical dispersal. The lengths of cold winter nights, during which birds must fast, and of short winter days, during which they must eat, are also important considerations in the ecological distribution of cold climate species (pp. 104–105).

As a means of stretching their energy resources over cool periods of enforced fasting, a very few species have adopted a form of dormancy similar to mammalian hibernation. These periods of lowered body temperature (which greatly reduce the rate of heat loss and therefore postpone starvation) range from overnight in some hummingbirds, through a week to 9 days in young swifts, to a few months in some goatsuckers. The Andean Hillstar Hummingbird, *Oreotrochilus estella*, may drop its body temperature during night torpor to as low as 5°C, corresponding with ambient temperature in winter roosts. This feature enables it to live circumannually at altitudes up to 3850 meters or more in the Andes (Carpenter, 1976). To combat unusually high temperatures, many species begin panting when their body temperatures reach a certain threshold temperature. In the California Quail, *Lophortyx californicus*, this temperature is 43.5°C. An ability to tolerate a body temperature of as much as 4°C above normal is probably an adaptation to life in hot climates (Bartholomew and Dawson, 1958).

Physiological adaptation to chilling is found to a surprising degree in the embryos of incubating eggs, very likely an adaptation to the food-hunting absences of the incubating parent (p. 339). At the other extreme, the eggs of the Ostrich, *Struthio camelus*, are able to withstand high desert temperatures that would be deadly to eggs of other species. The rate at which the hatched young develop is also adapted in some species to thermal requirements. The large Emperor Penguin, *Aptenodytes forsteri*, hatches its chick in the coldest period of the antarctic winter and broods it assiduously until warmer weather arrives. The far smaller Adelie Penguin, *Pygoscelis adeliae*, living at the same latitude, hatches its chicks in the summer. These young grow and mature with astonishing speed and reach relative independence, along with the more slowly developing young Emperors, by the time the next winter arrives.

In the European Garden Warbler, *Sylvia borin*, and Blackcap, *S. atricapilla*, the later in the season a young bird hatches, the more rapidly it passes through its various developmental stages toward maturity, as if to catch up with earlier hatched birds of its species. This seasonal acceleration in development is especially pronounced in the Garden Warbler, a long-distance migrant, and is interpreted by Berthold and his associates (1970) as a genetically controlled mechanism to ensure that late-hatching birds reach migration-capability before winter closes in. Photoperiod experiments with birds showed that day-length was not the governing stimulus. In the same way, late-summer-hatched ducklings of the Redhead, *Aythya americana*, attain flight more quickly than those hatched earlier (Smart, 1965). Canada Geese, *Branta canadensis*, reared at 60° N. latitude, fledge and learn to fly in about two-thirds the time that is required at 45° N. latitude. This accelerated development may be due, however, to the longer feeding day available in arctic latitudes (Hansen and Nelson, 1964).

Numerous adaptations to temperature extremes are found in the behavior of birds. To avoid intolerable cold, many species migrate to warmer climates for the winter. Diurnal birds are almost all inactive at night, and this fact, coupled with a slightly lowered body temperature, greatly reduces their heat loss in cold weather. Many species form roosting or sleeping aggregations, some small species huddling together so tightly as to form feathered balls (Fig. 7–11). In the most bitter antarctic weather, incubating male Emperor Penguins huddle together for protection. Weight loss per day among these birds is only about one-half that of penguins kept isolated (Prevost and Bourlière, 1957). Since these birds fast during the incubation period, the metabolic economy of this behavior is doubly adaptive in that it conserves heat and also stretches energy reserves over the period of continuous, uninterrupted incubation that is essential in polar climates.

When sleeping during cold weather, most birds fluff their feathers, thus creating a thicker, warmer layer of insulating air. Most birds also tuck their beaks under their scapular feathers, thus providing themselves with prewarmed air to breathe and reducing the body surface area much like a human who hugs himself to keep warm. The Snow Bunting, *Plectrophenax nivalis*, some titmice, and several grouse escape the most extreme cold of winter by plunging into snow banks where the temperature is relatively mild. In extremely cold weather in Alaska, for example, the air temperature above the snow surface may be − 50°C, but only 60 cm below the surface it may be a relatively warm − 5°C. A northern European Hazel Hen, *Tetrastes bonasia*, spending the night in a snow burrow 25 cm deep, warms the burrow to nearly 0°C in 30 minutes when the air temperature outside is − 48°C (Andreev, 1977), and a Capercaillie, *Tetrao urogallus*, may warm its burrow air-temperature to 11°C when the surrounding snow temperature is as low as − 11.5°C (Marjakangas *et al.*, 1984). Many other species seek the thermal protection of dense thickets, natural cavities, and other places where the immedi-

ate temperature and wind are less severe than in the open.

Atmospheric conditions in these localized environments are known as *microclimates*, and they are widely exploited by birds living under extreme climatic conditions. Microclimates often figure in the selection of nesting sites. At altitudes of 1500 to 1700 meters in the Alps, birds build their early season nests on southern exposures, but by June, as the season advances, many birds change to northern exposures (Heilfurth, 1936). Desert birds commonly build their nests in the shade of bushes or on the east side of rocks and bushes where they will avoid the sun during the hottest part of the day. In very hot weather birds largely restrict their activity to the morning and evening hours when the air is cooler. In caring for their eggs and young, parent birds will ordinarily brood them more attentively in cold weather, and in hot weather protect them from the sun by spreading their wings over them. In these and numberless other ways, bird adaptations serve as homeostatic agents to reduce the impact of temperature extremes in environments that would otherwise be intolerable.

The influence of precipitation and humidity on birds is largely indirect, through their control of the growth and distribution of vegetation and food. There are, however, some direct effects of ecological consequence. Prolonged cold rains are much more destructive of bird life than prolonged dry cold, because the insulating value of plumage deteriorates as it becomes wet. Deep snows cover the food of many species of birds and may cause widespread hardship or death. Seasonal snowstorms may cause adult mountain White-crowned Sparrows, *Zonotrichia leucophrys*, to abandon their nests, resulting in chilling of eggs or death of nestlings (Morton *et al.*, 1972; Morton, 1978). As described earlier, the reproductive cycle in many arid land species is stimulated by rainfall, long before the rain could have any effect on vegetation or food supply. While it is difficult to separate other characteristics of a habitat from humidity and rainfall, it seems likely that for some desert species dryness in itself is an ecological requirement. Certainly the eggs of a desert lark require less moisture for successful incubation than the eggs of a grebe. The California Quail, *Lophortyx californicus*, introduced into New Zealand, favors regions where rainfall averages less than 150 cm annually (Williams, 1952).

Some idea of the adaptive success of desert birds in conserving body water is seen in the following comparative statistics from Hadley (1972). The Poor-will, *Phalaenoptilus nuttallii*, at an air temperature of 35°C, lost 0.86 mg of evaporative water per sq cm of body surface per hour; the Painted Quail, *Excalfactoria chinensis*, 1.5 mg at 25°C; a naked man sitting in the sun at an air temperature of 35°C, lost 22.32 mg per sq cm per hour. Other water-conserving adaptations in birds were previously considered in Chapter 6.

Since climate is made up predominantly of temperature and precipitation, attempts have been made by game managers to increase the success of game bird introductions by planting an exotic species in a region whose climate matches that of its optimum native habitat. Two such climates are quickly compared by scrutinizing climographs of the two regions: graphs in which mean monthly temperatures are plotted against the ordinate, and total monthly rainfall against the abscissa. Introductions of the Hungarian Partridge, *Perdix perdix*, have been attempted in various parts of the United States. In Missouri and California, where the climographs show a wide deviation from that of the bird's optimum European habitat, introductions failed. But in Montana, where the two climographs showed greater agreement, especially during the crucial breeding season months, the introductions succeeded (Twomey, 1936) (Fig. 19–4).

Wind intensifies the ecological effects of temperature and precipitation. A driving rain or a blizzard is much more dangerous to birds than a gentle rain or snowfall. Regions with prevailingly steady winds present birds with distinctly different environments as compared with less windy regions with the same climographs. Wind inhibits singing and other activities of birds. In strong winds, many birds fly near the earth where wind velocity is lessened. Other species, on the contrary, use winds for soaring, particularly in regions where hills or mountains create updrafts. Wind also serves as a vehicle for bird distribution. Birds are more likely to disperse in directions that prevailing winds carry them than in contrary directions. The circumglobal Blue Petrel, *Halobaena caerulea*, is limited in its oceanic wanderings by the northern edge of the belt of the Roaring Forties to the north and by pack ice to the south (Murphy, 1936).

Nest sites are often chosen with reference to prevailing winds. Many of the larger procellariiform sea birds locate their nests on cliffs or slopes on the windward sides of islands where taking off in flight is facilitated by updrafts of air. Woodpeckers living in exposed regions, such as mountain heights, are likely to dig their nest holes on the sides of trees away from prevailing winds. Many smaller birds of exposed regions

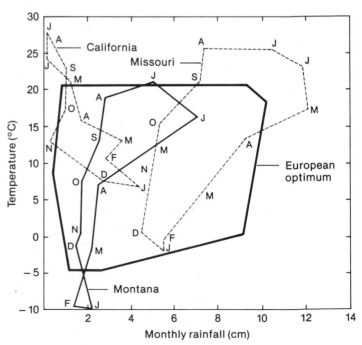

FIGURE 19–4 The Hungarian Partridge, transplanted from central Europe to America, has successfully established itself in Montana, but has failed conspicuously in California and Missouri, whose climates do not agree closely with the climate of the species' ancestral home. After Twomey, 1936.

in deserts and steppes build their nests on the lee side of rocks, shrubs, or other windbreaks to avoid the full effect of wind storms. Desert larks and wheatears often erect walls of small stones beside their nests. This behavior probably represents an age-old adaptive response to wind.

Light
As an ecological factor, light influences birds mainly through its intensity and duration, and to a lesser extent through its wave length (color) and direction. Eyes of nocturnal birds are usually larger than those of diurnal species. Not only are the eyes of the nocturnal Galapagos Swallow-tailed Gull, *Creagrus furcatus*, much larger than is usual for gulls, but like those of other nocturnal animals, they shine when illuminated (Thorton, 1971). Whereas the retinas of nocturnal species are rich in rods, those of diurnal species are rich in color-sensitive cones. The ability of most birds to distinguish colors undoubtedly influences their search for and selection of food. The variously colored oil droplets in the retinal cells of many birds may well be an adaptation for the precise detection of small objects (insects, worms, berries, seeds) of the same or complementary colors.

The two great natural rhythms of light—daily and yearly—have profound effects on birds: on their anatomy, physiology, behavior, and distribution. The daily rhythm of light and darkness has impressed on birds a daily rhythm of activity and rest, inverted, of course, in nocturnal species. Diurnal birds are more active, and have higher temperatures and higher metabolism, by day than by night. The close correspondence of their activity with light, rather than with temperature or some other environmental factor, is seen, for example, in the close relationship between sunrise and awakening song (p. 229). The annual cycle of change in day length in all parts of the world except equatorial regions inaugurates the global cycle of the seasons with all its profound and widespread effects on birds and plants. The many deep-seated adaptive responses of photoperiodism—courtship, mating, territorial battles, nest-building, egg-laying, incubation, plumage molt, migration—are, in hordes of birds, set into motion by changes in day length. These have all been treated in more detail in other chapters. It is enough here to emphasize the fact of their ecological connection with sunlight. Without the eons of fine polishing by natural selection, some birds might today wear their dullest colors for their courtship dances, lay their eggs in subzero weather, or migrate north rather than south in autumn. For most species the great clock of changing day length sets off all these activities and keeps them synchronized with the best times of year for their fruitful functioning.

BIOLOGICAL FACTORS OF ENVIRONMENT
The basic difference between plants and animals is that only plants can manufacture food from inorganic raw materials. This means that all animals, birds included, depend ultimately on sunlight and the food-manufacturing photosynthesis of green plants for their food—"All flesh is grass." Plants are of ecological importance to birds, not only as sources of food but also for nesting materials, nesting sites, lookout posts, singing stations, and protective cover. Plants probably satisfy some psychological needs in birds as well, but food is of primary importance.

Adaptations to the food supply are most obvious in the structure of birds' beaks. While there is wide diversity in the shapes and sizes of beaks, adaptations seem to tend either toward a generalized bill suitable for eating a variety of foods, or toward a highly specialized bill suitable for eating foods of a restricted type. Species ecologically tolerant of a wide variety of foods are said to be *euryphagous*. Crows and gulls, with their straight, simple beaks, eat fish, birds, eggs, invertebrates, fruits, vegetables, seeds, carrion, and other foods. Birds limited to a restricted diet are said to be *stenophagous*, or, if they are restricted to a single type of food—a very rare occurrence—*monophagous*. The Everglade Kite, *Rostrhamus sociabilis*, is reputed to be monophagous, feeding only on the freshwater snail *Pomacea*. With their long beaks and brushy-tipped or tubular tongues, hummingbirds are rather closely restricted to nectar and insects as food. The long, sensitive beaks of snipe and woodcock limit their diets to the worms and other invertebrates they can find by probing soft earth. Crossbills find their peculiarly asymmetrical beaks most useful in extracting the seeds of pine and spruce cones. Many species of ducks employ their fringe-edged beaks for sifting seeds and small animals from muddy water, while the Broad-billed Whalebird, *Pachyptila vittata*, uses the whalebone-like fringes of its upper mandible for filtering plankton from sea water (Fig. 19–5). Other examples of adaptations in beaks are given in Chapter 6.

Adaptations to food are also found in internal organs such as the food-storing crop of pigeons, the seed-grinding gizzard of the turkey, and the cellulose-digesting caeca of the domestic fowl. In general, seed-

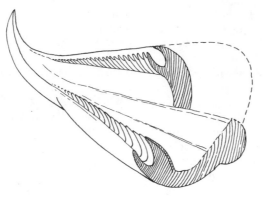

FIGURE 19–5 Maxillary lamellae of the Broad-billed Whalebird. There are about 150 of these fringe-like keratin plates along each margin of the upper jaw, and they grow as long as 3.5 mm tall at the inner end of the mouth. At right is a series of plates as seen from the midline of the mouth. After Murphy, 1936.

eating birds have muscular, grinding gizzards, while meat eaters have glandular stomachs that digest food almost entirely by enzymes. But individual birds may have further adaptations to changes in diet. The stomachs of some gulls are thin-walled and richly supplied with digestive glands while the gulls feed on fish, but in the autumn when they may move inland and feed on grain, their stomachs gradually change into muscular gizzards capable of crushing corn (Beebe, 1906). A corresponding increase in the musculature of the gizzard occurs in the Red-winged Blackbird, *Agelaius phoeniceus*, when it seasonally shifts its diet from insects to seeds (Allen, 1914). An ability to adjust to a different kind of diet at different times of year is a valuable adaptation for a bird living in a region with pronounced seasonal changes, for food availability may change from insects in spring and summer to fruits in the fall, and to seeds in winter. The same kind of digestive flexibility may occur in some long-distance migrants, but apparently the possibility remains to be investigated.

As regards their feeding habits there seem to be two general categories within each of the following groups of sea birds: penguins, boobies, gannets, gulls, terns, auks, murres, and puffins. Either they feed near shore (inshore feeders) or far out at sea (offshore feeders). Lack (1968) makes the interesting observation that offshore feeders, as compared with inshore, breed in larger and more widely spaced colonies, have smaller clutches, longer incubation and fledg-

ing periods, longer spells of incubation, longer periods between feeding visits to the young, and probably start breeding later in life and have a lower adult mortality.

Too narrow a specialization in diet can be dangerous to a species, because some natural catastrophe may destroy the one or several foods on which the bird depends. This nearly happened in 1931–33 when a blight killed most of the eel grass that was the preferred food of the Atlantic Brant, *Branta bernicla*. Apparently unable to turn to substitute foods, about 80 per cent of the Brant population disappeared (p. 102). Whatever a bird's feeding adaptations and habits may be, it must live in those regions where its preferred foods are found. It follows, therefore, that slight adaptive changes in the structure of a bird's beak, or in the chemistry of its digestive enzymes, might so change its diet that definite changes in its geographical distribution would result. These latter changes in turn might require new adaptations in limits of toleration for temperature extremes, rainfall, photoperiod, and other environmental conditions. Ecological adaptations may, as a consequence, reinforce, conflict with, or have no influence on one another.

Of the other needs supplied to birds by vegetation, one of the most important is a secure and effective nesting site. Some species show pronounced preferences for specific types of nest-site vegetation. Ring-necked Pheasants, *Phasianus colchicus*, choose hayfields, preferably older ones, for nesting. In its breeding ground the Rockhopper Penguin, *Eudyptes crestatus*, tolerates no vegetation larger than grass, while the Jackass Penguin, *Spheniscus demersus*, will nest either among tufts of tussock grass or in the midst of thickets of brush. Vesper Sparrows, *Pooecetes gramineus*, or Horned Larks, *Eremophila alpestris*, are likely to build their nests in a meadow with short grass, but if the grass is as high as half a meter or more, Bobolinks, *Dolichonyx oryzivorus*, are more likely to nest there.

Not only the height, but also the density and character of vegetation often determine its nesting population. Some species, like the Yellow-rumped Warbler, *Dendroica coronata*, typically nest in conifers, while others, like the vireos, nest in deciduous trees. Some, like the Wood Thrush, *Hylocichla mustelina*, nest in dense woods; others like the Robin, *Turdus migratorius*, in more open, park-like locations. Seasonal changes in the character of a stand of vegetation may affect its suitability as a habitat. The Ruffed Grouse, *Bonasa umbellus*, prefers coniferous woods as cover in winter, and deciduous woods in summer. It seems, as a general rule, that dense vegetation discourages the flocking habit in birds, while more open plant formations, such as prairies, steppes, or savannas, encourage flocking.

Birds are not only affected ecologically by vegetation, but, on occasion, they reciprocate and cause changes in the vegetation. The feces of blackbirds roosting in a pine plantation caused changes in phosphorus and potassium levels in the soil to such an extent that almost all of the trees died (Gilmore *et al.*, 1984). On the other hand, the Rockhopper Penguins, *Eudyptes crestatus*, on the South Atlantic island of Tristan da Cunha, fertilize with their droppings the tall tussock grass, which responds with vigorous growth and shelters the Penguins from wind, rain, and predators (Murphy, 1936). Herons and other fish-eating birds of Tampa Bay, Florida, used to be killed as competitors of human fishermen. After 12 years of protection, however, their rookeries grew so large that they dropped an estimated 50 tons of guano each day into the bay, with the result that mullet fishing became better than at any time in the last half century (Peterson, 1948)! This example of food ecology illustrates how minerals and living organisms participate in ongoing cycles of energy and materials. The manure of birds supplies the nutrients for microscopic plankton organisms in the water that provide food for the fishes that are eaten by the birds. Sunlight falling on the plankton provides the basic energy that runs this cycle, and the guano provides some of the chemicals that are assembled in energy-storing combinations as food. Similar food cycles exist all over the world, on land and in water, but ordinarily they do not occur in such a compact and obvious form.

A clear demonstration of the relationship between water fertility and duck breeding is found in the lakes and ponds of Minnesota (Leitch, 1964). There is a 25-cm excess of rain available for runoff in northeastern Minnesota and a 25-cm deficit in the southwest. As a consequence, there is a gradual increase in dissolved carbonates, sulfates, and chlorides in natural waters from northeast to southwest. The waters in the northeast have fewer than 200 parts per million dissolved solids while those in the southwest have between 1000 and 13,000 parts per million. The ponds and lakes in the northeast have approximately 22.5 kg of bottom organisms per hectare; those in the southwest, 225 kg per hectare (200 lbs per acre). Quite logically, the prairie ponds and "potholes" of the southwest are excellent for duck breeding, and those in the northeast are poor.

Other forms of bird-plant interdependence are seen in the pollination and distribution of plants by birds. Ornithophilous plants are those whose flowers are so constructed that they can be pollinated only by birds of certain special habits and with beaks of highly specific shapes that fit the flower's structure (p. 114). In some instances the mutual interdependence is very close. A mistletoe (*Loranthus*) of India will not fruit unless its flowers are pollinated by sunbirds of the genus *Cinnyris*. Experiments by Hainsworth and Wolf (1976) involving the effects of flower corolla-length and nectar composition on the feeding choices of hummingbirds suggest the kinds of selective pressures that promoted the co-evolution of hummingbirds and the ornithophilous plants associated with them. Birds that feed on and pollinate ornithophilous plants belong chiefly to the families Psittacidae (Lories), Nectariniidae, Meliphagidae, Dicaeidae, Drepanididae, and Trochilidae (Paton and Ford, 1977). The geographical dispersal of plants by birds is relatively common and does not require highly specialized adaptations either in the bird or the plant. The European Mistle Thrush, *Turdus viscivorus*, and New World Phainopepla, *Phainopepla nitens*, are well known for their roles in spreading the parasitic mistletoe. Feeding experiments have shown that certain shore birds can retain viable seeds several hundred hours before regurgitating and introducing them to the most remote oceanic island (Proctor, 1968). The large wingless seeds of Limber pine, *Pinus flexilis*, cannot be wind-dispersed and are carried by Clark's Nutcrackers, *Nucifraga columbiana*, to sites as far as 22 km away. The Nutcrackers bury the seeds, which germinate if not recovered (Lanner and Van der Wall, 1980). Recently, in west Greenland, at 67 degrees N. latitude, a new plant, *Spergularia canadensis* (Caryophyllaceae), appeared. Previously it was known to occur only in North America south of 60 degrees N. latitude. It was thought to have been brought to Greenland by migrating seed-eating Common Redpolls, *Carduelis flammea* (Bocher, 1978).

Part of every bird's biological environment is composed of other animals: mates, offspring, predators, prey, competitors, parasites, commensals, or symbionts. Many of the relationships between birds and these other animals have been considered earlier under such topics as courtship, care of the young, predation, and parasitism. In every instance, these relationships can be classified as being either beneficial, neutral, or harmful to the bird in question. Sometimes the relationships may be individually harmful but racially beneficial, as in the exhaustion of a parent bird feeding its brood of young.

Beneficial relationships are of two kinds: those beneficial only to the bird, and those mutually beneficial to both the bird and its partners. Among the latter type are many family and colony relationships that often demonstrate positive cooperation between individuals. The colonies may be composed of a single species or of several. Benefits to the members of a colony are obtained by the many eyes alert to detect danger, the strength in numbers for defense against predators, opportunities for mutual education, cooperative nest-building, helpers at the nest, the advantages of synchronized breeding, and, occasionally, cooperative food finding.

In many instances, mutually beneficial relations occur between birds and mammals. A typical example is furnished by the African tick birds of the genus *Buphagus*, which groom buffalo, rhinoceros, giraffes, and other mammals, removing their ticks and other vermin. The Ground Finch, *Geospiza fuliginosa*, of the Galapagos Islands, regularly grooms marine iguanas and giant tortoises of ticks (Amadon, 1967; MacFarland and MacFarland, 1972). The tortoises crane their necks and stretch high on their legs and remain motionless as long as 5 minutes while the finches extract ticks from the crevices in their skins. African

honeyguides *(Indicator* spp.) lead honey badgers to beehives. The badgers then raid the hive for its honey and larvae, and the honeyguides feed on the wax from the combs (Friedmann, 1955).

Other interspecific relationships that are primarily beneficial to the bird alone are, among others, predator-prey relations and brood parasite-host relations. In the case of bird nesting-associations with ants, bees, wasps, and termites, what is beneficial to the bird may be harmful to the insects. Other examples of unilateral benefits to birds include associations such as those of antbirds that feed on the insects stirred up by army ants; ptarmigan that follow caribou and benefit from the food they dig up in winter; the Rough-legged Hawk, *Buteo lagopus*, which follows the arctic fox to feed on the mice it stirs up; and the great variety of birds, such as egrets, bee-eaters, anis, starlings, and cowbirds, that follow herds of antelopes, cattle, and other ungulates, using them as beaters to raise up insects. Work by Grubb (1976) in Georgia showed that a Cattle Egret, *Bubulcus ibis*, associated with a cow, captured, on average, two insects per minute and required only 13.7 steps for each capture, whereas an egret feeding alone caught only 0.6 insects per minute and took 76.5 steps between captures. On a cost-benefit basis, an egret following a cow is approximately 18 times more efficient than one feeding alone. Ducks frequently nest in or near colonies of gulls or terns, and profit from the latter species' aggressive attacks on nest robbers such as crows. An exceptional instance of the adaptive recogition of a habitat free of predators (in this instance, arctic foxes) was demonstrated in northeast Greenland when, over a period of 20 years, a colony of nesting Common Eiders, *Somateria mollissima*, gradually increased from two nests to roughly 1300, around an area where 60 to 80 dog-sled huskies were normally tethered (Meltofte, 1978). A variety of small birds build their nests near or in the edges of the larger nests of hawks and eagles, where they obtain immunity from predation by other birds but seem to suffer few, if any, attacks by their host.

Among the more passive interspecific relationships are those of birds that inhabit the nest burrows of other species. The Elf Owl, *Micrathene whitneyi*, nests in saguaro cactus cavities excavated by woodpeckers; the Peruvian Inca Tern, *Larosterna inca*, sometimes nests in a little vestibule cavity that it scratches out in the occupied burrow of a penguin; the Flesh-footed Shearwater, *Puffinus carneipes*, occasionally shares a burrow with the lizard-like *Tuatara sphenodon* of

New Zealand. The Minera, *Geositta cunicularia*, of South America is reputed to nest only in the burrows of the Vizcacha, *Lagostromus trichodactylus*. The Isabelline Wheatear, *Oenanthe isabellina*, uses rodent burrows for nesting sites. A census of these birds in the Karakum Desert, USSR, revealed 9.8 wheatears per 10-km transect in areas where there were 4.7 rodent colonies per hectare, but only 1.1 wheatears per 10-km transect in areas with two rodent colonies per hectare (Babaev, 1967).

Competition and Niches

Competition is inevitable wherever birds congregate, particularly if the different individuals require for their survival the same exhaustible elements from the environment. The keenest competition occurs between birds of the same species because of the identity of their requirements. For similar reasons, closely related species are likely to be more competitive than distant relatives.

The population grows as a result of the breeding effort, and the competition for life-needs soon intensifies. Since a given habitat can support only a certain number of birds, only the fittest members of the crop of juveniles acquire territories. The rest are forced to disperse from the birth area or become relegated to the rank of floaters. Since no two species are alike, competition for limited resources usually results in one species increasing its numbers at the expense of another.

Competition may be by direct aggression, as when two or more individuals fight for such resources as territory, food, nesting sites, or peck-order dominance, or it may be by exploitation of a limited resource. If two birds, or two different species, compete for the same food, and one harvests it more efficiently than the other, that species will lower the resource level to a point where it can survive but its competitor cannot (Diamond, 1978). Diesselhorst (1971) cites an instance in which Black Weavers, *Ploceus nigerrimus*, eliminated nest-building Spectacled Weavers, *P. ocularis*, by aggressively stealing their palm-leaf nesting materials.

The outcome of competition depends on the type of competition involved, the resource being contested, the genetic or phenotypic variance in character of the competitors, priority in time of arrival of the individual or species in the field of competition, and the length of time over which the contest endures. Ordinarily, competition results in the loser's being expelled into a different geographic or ecological area, or in his death,

or even, ultimately, species extinction. The examples that follow will clarify these points.

In North America the Black-capped Chickadee, *Parus atricapillus*, covers an extensive range where it has no closely related competitors and where it inhabits a variety of wooded environments: wet and dry woods, deciduous and coniferous woods, mountain and valley woods. In Europe, however, where this species (known there as the Willow Tit) lives among six different species of small titmice of the same genus, its range is largely restricted to swampy forests, while the other titmice occupy the other types of forests that the chickadee inhabits in America. The reverse of this situation occurs among wrens. In the New World where there are 63 species of wrens, the Winter Wren, *Troglodytes troglodytes*, is restricted to breeding in boreal forests. This same species is the only wren that breeds in the Old World, and there it occupies a great variety of breeding environments. Similarly, Diamond (1978) found that doves on Pacific Islands also expanded the range of habitats occupied in the absence of competitors (Fig. 19–6). Savannah Sparrows, *Passerculus sandwichensis*, usually occupy grasslands on the mainland, but, in the absence of competition on San Benito Island off Baja California, this species occupies a variety of habitats, including scrub desert, rocky shores, floating kelp, and driftwood (Boswall, 1978).

The pressures of intraspecific competition may also be diminished if each sex occupies a different microhabitat. For example, female Downy Woodpeckers, *Picoides pubescens*, tend to feed on large branches and tree trunks, whereas males tend to feed on small branches and twigs. When males were experimentally removed, females began occupying branches, indicating that they had been avoiding this microhabitat used by the socially dominant males (Peters and Grubb, 1983). Intraspecific competition may also be relieved if different individuals (independent of age or sex classes) tend to feed on different-sized foods. Populations of Hook-billed Kites, *Chondrohierax uncinatus*, in Mexico, consist of large- and small-billed individuals which predominantly feed on large and small snails, respectively (Smith and Temple, 1982).

The reality of interspecific competition was clearly demonstrated by Williams and Batzli (1979). From a guild (i.e., the same "profession") of several bark-foraging birds in a small Illinois woodlot, all Red-headed Woodpeckers, *Melanerpes erythrocephalus*, were experimentally removed. Red-bellied Wood-

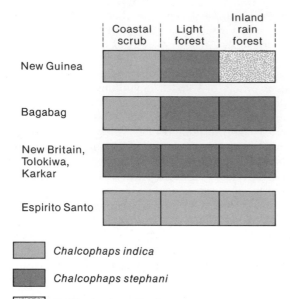

Chalcophaps indica

Chalcophaps stephani

Gallicolumba rufigula

FIGURE 19–6 The effect of the presence or absence of competitors on the distribution of three similar species of ground doves along a habitat gradient ranging from coastal scrub, through light forest, to inland rain forest, on six Pacific islands. Each species occupies a restricted habitat range in the presence of its competitors, but occupies the whole gradient in their absence. After Diamond, 1978.

peckers, *Melanerpes carolinensis*, and White-breasted Nuthatches, *Sitta carolinensis*, entered the experimental area, but not the control area where the Redheads remained. In similar fashion, interspecific territoriality was demonstrated on the Scottish Isle of Eigg when Great Tits, *Parus major*, moved into the wooded territory of Chaffinches, *Fringilla coelebs*, after the latter had been experimentally removed (Reed, 1982).

What is assumed to be interspecific competition may not be the case, since a species may be genetically preadapted to a given niche. Hand-raised Blue Tits, *Parus caeruleus*, and Coal Tits, *P. ater*, were reared in aviaries devoid of plants. When introduced into an aviary containing branches of Scotch pine and oak, Blue Tits perched more on oak branches, and Coal Tits on pine branches (Partridge, 1974).

Interspecific competition is summarized in "Grinnel's Axiom" or "Gause's Rule" (Gause, 1934), or in the "Gause-Volterra Principle," which states that two species cannot coexist indefinitely in the same habitat if they require the same things from the environment; sooner or later one will eliminate the other. In studying California titmice (*Parus* spp.), Grinnell (1904) observed that "two species of approximately the same food habits are not likely long to remain evenly balanced in numbers in the same region. One will crowd out the other" (Udvardy, 1959). This is also known as the principle of competitive exclusion (Hardin, 1960), and it accounts for the fact that closely related species are usually separated either ecologically

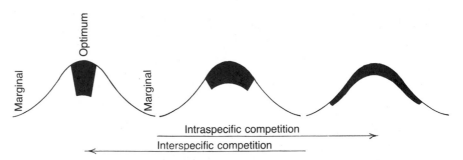

FIGURE 19–7 The effect of competition on habitat selection. The peak of each curve represents optimum habitat conditions for any particular species; the descending arms, the less desirable, marginal conditions. Strong intraspecific competition causes a population to spread out into marginal habitats, while strong interspecific competition causes each species to retreat to its optimum habitat where it is best able to meet the competition of other species. After Svärdson, 1949.

or geographically. Competitive exclusion is demonstrated by removal experiments (see above).

The habitat of any given bird may be considered from three different standpoints: Where is it? What is it? Why does this particular species live there? The first or spatial standpoint is a simple question of geography. The second or structural standpoint is more complex and concerns such things as the geology, topography, chemistry, meteorology, botany, and zoology of the habitat. The third or functional aspect is the most complex of all and concerns the intimate give-and-take between the bird and its total environment. It attempts to determine why species A and not species B is located there. Considered from this dynamic, functional perspective, an environment becomes a *niche*. As Odum (1971) puts it, a bird's habitat is its address; its niche is its profession. Niches are not necessarily restricted to fixed geographic habitats such as marshes, deserts, or forests. Some species depend on small "insular" habitats such as widely scattered flowering shrubs or trees, or on unstable habitats such as shifting sandbars in rivers, or drying ponds in steppes.

While many species of birds demand highly specific types of niches and are intolerant of even slight deviations from the standard form, others show considerable tolerance for deviations from the optimum. The Ring-necked Pheasant, *Phasianus colchicus*, introduced into Hawaii in 1865, now ranges from sea level to altitudes of 3700 meters; from subtropical to occasionally freezing temperatures; from lands with less than 24 cm rainfall per year to those with over 750 cm: from recent lava flows to deep loam soils. It is associated with all sorts of vegetation: forests, grasslands, deserts, and cultivated fields (Schwartz and Schwartz, 1951).

More precisely, a niche may be defined as the sum total of all environmental factors important to a bird's survival and to its ability to reproduce itself (Hutchinson, 1957). The size of a niche depends on the ecological tolerances of its inhabitants. A niche may have many dimensions: e.g., light, precipitation, temperature, food, and so on, and these usually occur in gradients or axes of two different types. The first type is composed of physicochemical gradients such as light, temperature, soil pH, or rainfall. These, of course, the animal must tolerate. And they are inexhaustible: no matter how many individuals inhabit a niche, these ingredients will ordinarily remain constant. They are particularly important to animals living in arctic or boreal zones. The second type of axis is an exhaustible

resource used by the animal, such as food, nest sites, nesting materials, and so on. While the presence of a species in a niche depends on its tolerance to certain ranges of both kinds of axes, competition is always directed toward the exhaustible resources. Niches may be illustrated graphically with each factor represented along the axes of one, two, three, or more dimensions. For example, the first axis in the niche of a salt marsh Song Sparrow, *Melospiza melodia*, might be the temperature range within which it can survive and reproduce; the second axis might represent the salinity range in water that it can tolerate; the third, the size of the insects it can eat. Some of these axial gradients play greater roles than others in determining a bird's survival in a particular habitat. In practice, two or three axes will often suffice to describe the differences in niche requirements separating two species. Two species may be in competition when one or more of their axes overlap.

Ecological niche-partitioning may occur when different species, coexisting in the same habitat and having the same general food requirements, either feed on different items or on different sizes of food. Several hummingbird species may coexist when flowers are abundant, as each species feeds selectively on particular blossoms. More aggression ensues when flowers are few (Leck, 1971). One would expect Cormorants, *Phalacrocorax carbo*, and Shags, *P. aristotelis*, which nest on the same cliff and feed in the same waters, to be keen competitors. On the contrary, a study by Lack (1945) showed that the Shags ate mainly free-swimming fish and eels, while the Cormorants fed on bottom-dwelling flatfish and crustaceans. A compact example of food-niche segregation is illustrated by four species of vultures scavenging, perhaps not in perfect amity, on one carcass in Spain (König, 1974, 1976). The Griffon Vulture, *Gyps fulvus*, with its nearly naked head and neck, pulls soft parts from deep inside the cadaver (see Fig. 6–5). The Black Vulture, *Aegypius monachus*, tears apart skin and firm muscle tissue. The Egyptian Vulture, *Neophron percnopterus*, pecks at small particles of food lying about the corpse or on bones. The Bearded Vulture (Lammergeier, or "Bone-breaker"), *Gypaëtus barbatus*, consumes cleaned bones or breaks the larger ones to get at the marrow (p. 111).

In New Guinea there are eight species of fruit pigeons of the genera *Ptilinopus* and *Ducula* that coexist in the same habitats but avoid competition by partitioning the food axis of their niches. The pigeons swallow fruits whole. The largest pigeons, weighing up to 800 g, swallow fruits about 40 mm in diameter; the smallest pigeons, weighing about 50 g, swallow fruits about 7 mm in diameter. Small pigeons can perch and feed on branches that would break under the larger pigeons; larger pigeons aggressively displace smaller pigeons that attempt to perch on the larger branches. In studying these birds, Diamond (1978) was concerned with how different or how widely separated species must be in order to coexist along this food-resource axis. He found that coexistence depended more on the variance in food size a given species was capable of eating than in variance in the quantity of food available. Further, the pigeons seemed to confirm MacArthur's (1972) principle that when species differ chiefly in size, the larger must be about two times as heavy as the smaller to coexist in the same habitat. Stated differently, if two species eat fruits of different sizes, the mean difference in niche-breadth allowing coexistence would approximate the standard of deviation of the fruit sizes. This principle introduces the idea of *species packing*, i.e., the number of species that a given niche axis can contain. Apparently, for at least some species, there are precise, mathematical limits to the number of species that can be accommodated or packed along a given resource axis.

Niches can be separated in time as well as in space. On the Galapagos Islands both the Galapagos Hawk, *Buteo galapagoensis*, and the Short-eared Owl, *Asio flammeus*, are native. On islands where the hawk is absent, the owl is active day and night, but on other islands it restricts its activity entirely to nighttime when it is safe from attacks by the hawk (deVries, 1973). Another form of temporal segregation occurs on coastal New Guinea at Merauke where the harrier *Circus spilonotus*, is present in the wet season, and *C. approximans*, in the dry (Diamond, 1972). Acting over a longer period, time may segregate similar species on a "first come, first served" basis. Diamond believes, for example, that two congeneric mannikins, *Lonchura spectabilis*, and *L. caniceps*, occupy extensive but sharply separated montane grasslands in New Guinea because the first immigrant in a given patch of habitat became so entrenched that the other species was unable to invade it and become a competitor.

Behavioral niche-partitioning occurs, for example, when different species use different foraging techniques to capture the same kinds of food in the same habitat. Insects may be captured by foliage gleaning, bark gleaning, wood probing, or by sallying into the air. On the Galapagos Islands, Blue-footed Boobies, *Sula nebouxii*, feed in shallow water, and the smaller male feeds closer to shore than the female. Masked Boobies, *Sula dactylatra*, feed about a mile offshore, and the Red-footed Booby, *Sula sula*, feeds several miles out at sea (Thornton, 1971). In part, the niche separation between Shags and Cormorants, mentioned above, is brought about by their difference in behavior.

Is there a relationship between niche-segregation and bird species diversity? Why, for example, do so many, and so varied, species of birds live in a deciduous forest? One might expect that a given patch of woodland composed of a wide variety of trees would support a greater diversity of bird species than an equal area composed of one species of trees, but this does not necessarily follow. Several investigators (MacArthur and MacArthur, 1961; Orians, 1971; Recher, 1969) have concluded that in both temperate and tropical woodlands the vegetation profile, or the degree of diversity in the heights and configurations of trees, correlates more closely with the number of resident bird species than does the degree of diversity of tree species—i.e., spatial heterogeneity is more influential than tree species heterogeneity. A forest of various sized and shaped trees of one species, they believe, will contain a greater variety of birds than a structurally uniform stand of trees of various different species. This relationship stems from the fact that many birds select their nest sites on the basis of the density of foliage and its height above the ground, and that birds' chief food during the breeding season consists of insects gleaned from the leaves.

But other students have arrived at somewhat different conclusions. In her statistical analysis of niches for different woodland species, James (1971) listed canopy cover, canopy height, and tree species diversity as the most influential variables in determining a specific bird's niche. In Bermuda, Emlen (1977) found that bird species diversity was best correlated with total volume of standing vegetation per square kilometer. Bird species diversity on 20 montane islands in the Great Basin of California was positively correlated with habitat diversity (Johnson, 1975).

In the Peruvian Andes, Terborgh (1977), following a transect from lowland rainforests at 500 meters elevation, through montane rainforests, and cloud rainforests, to elfin forests at 3500 meters elevation, discovered that bird species diversity was also a function of food preference. Insect-eaters decreased 5.2-fold from the bottom to the top of the altitudinal gradient; fruit-eaters, 2.3-fold; and nectar-eaters, not at all. Terborgh judged that while foliage profiles played an important role in bird diversity, other environmental ingredients were also influential: competitive interactions with other birds, declining food productivity at higher elevations, and the greater patchiness of montane forests. He concluded that bird species diversity "is a complex community property that is responsive to many types of influences beyond simply the structure of the habitat." Bird species diversity is further influenced by the size of the area inhabited and its distance from similar areas, and also by latitude. These factors will be considered in the chapter on bird geography.

CARRYING CAPACITY AND LIMITING FACTORS

Man's first dim awareness of the ecological interplay between animals and their environments probably arose as a consequence of failing supplies of game. As human populations increased and the hunting problem became more acute, man began to apply tribal taboos, and later game laws, to conserve game. Perhaps the earliest such law on record is that from Moses:

> If a bird's nest chance to be before thee in the way, in any tree or on the ground, whether they be young ones, or eggs, and the dam sitting upon the young, or upon the eggs, thou shalt not take the dam with the young: but thou shalt in any wise let the dam go, and take the young to thee; that it may be well with thee, and that thou mayest prolong thy days.—Deuteronomy 22:6.

Here, quite obviously, is a law intended to conserve breeding stock.

When the principal game bird of Britain, the Red Grouse, *Lagopus lagopus*, began to decrease in numbers, the traditional and obvious control measures were invoked; restrictions in hunting, predator control, the establishment of reservations and refuges, game farming—all to no avail. The grouse population steadily decreased. Finally, an intensive, ecological study of the problem was undertaken to seek out actual limiting factors—that is, the crucial niche axes—that affected the population. As a result of the information obtained, several unorthodox control measures were applied: patches of heather were burned; many older, breeding grouse were removed; predators were controlled; grit provided; wet places drained. As a consequence of these measures, the grouse population

increased 30-fold, chiefly, perhaps, because of the first two controls (Leopold, 1933). This historically important experiment in game management revealed that the *critical limiting factors* that held down the grouse population were not the apparent and obvious ones such as over-hunting, or starvation, or even heavy predation. More subtle limiting factors were involved, but they were the more significant ones. Burning the heather provided variety in habitats and created open sunning-yards where young chicks could dry themselves after rain and fog and thus reduce mortality from respiratory diseases. In addition, the new crop of heather was more nutritious and, being lower, more easily browsed than the old. Removing the old breeders opened up their territories to younger, more productive birds. Scattering grit, a dietary necessity, was required because the native grit was too deeply covered with years of peaty accumulations of organic matter. All these and other limiting factors exercised a quantitative control of the number of grouse a given area of land could support—that is, its *carrying capacity*. Every habitat thus has a carrying capacity which, of necessity, varies with the different species of birds living there.

Among all the limiting environmental factors that surround and influence birds—food, vegetation, precipitation, temperature, sunlight, soil, secure nest sites, predators, diseases, and others—certain ones will have a more direct effect on population increase than others. As a rule, biotic factors (e.g., vegetation, food, predators, diseases) are density-dependent, while climatic factors (e.g., temperature, rainfall, sunlight) are density-independent. The effect of a food shortage will vary, depending on the number of birds competing for food, but a late spring freeze will affect all birds in a population regardless of their number. The critical factors that limit population density may be very different for different species of birds. These facts underlie the principle proposed by Liebig in 1840 as the "law of the minimum," later revised by Shelford in 1911 as the "law of toleration." These laws hold that the survival of any organism, plant or animal, depends on the presence in minimum (or maximum) amount of that environmental factor that immediately exercises a controlling or limiting effect on the life of the organism, despite the favorable condition of other environmental factors. For example, a well-fed grouse, disease-free, supplied with a safe nesting site, exposed to optimum climate, and protected from predators, will still weaken and die if the grit available to it falls below a certain critical minimum. Similarly, too much of certain environmental elements—high temperature, sunlight, rain, calcium in the soil, industrial pollutants, pesticides—may operate as limiting factors. These influences may reduce a population either by causing the death of individuals or by reducing their reproductive potential, for example, by causing infertility of eggs. And probably for every species there is one developmental stage in its life cycle during which it is most sensitive, or least adaptable, to a given limiting factor. A further complication involves a possible synergism between environmental factors, i.e., the way they reinforce or counteract one another. The combined effect of certain pollutants in the environment may be much greater than the sum of their individual effects (p. 571).

The complex interrelations of factors often make it exceedingly difficult to identify a single environmental agent or material as *the* limiting factor. The minimum lethal air temperature for a bird varies, for example, with duration of exposure, season of the year, age of the bird, condition of plumage, amount of fat stored in its tissues, amount of food in the digestive tract, the presence or absence of parasites, and other factors. So one cannot say without qualification that a given air temperature is the lethal minimum for a given species. Nevertheless, the principle has enough validity to be useful in ecological studies.

MAJOR TYPES OF HABITATS

Although there is an almost limitless variety of natural habitats, the more prominent kinds can all be grouped into about a dozen major classes. These broad categories, such as evergreen forests, deserts, or the seas, are easily recognized even by the uninitiated. The chief distinctions between them are based on the kind and amount of vegetation they possess, and vegetation is largely determined by temperature and rainfall, although soil, day length, altitude, wind, and other influences play roles of varying importance. Animals, of course, are important components of any habitat, but plants are considered the key to habitat type because they do not move; they are generally the dominant form of life, and they are the food source on which animals depend for their energy. Plant distribution, therefore, normally determines animal distribution.

Different attempts have been made to identify and describe with some precision the major habitat types

in North America. In the late 19th century, C. Hart Merriam proposed a series of great transcontinental *life zones* based primarily on latitudinal temperature differences. These life zones, a series of broad bands, each running roughly east and west across the continent, he called, in sequence from north to south, the Arctic, Hudsonian, Canadian, Transition, Upper Austral, Lower Austral, and Tropical Zones. Because of pronounced differences in precipitation between the eastern and western halves of the United States, Merriam called the Transition Zone in the moist east the Alleghanian Zone; and the Upper and Lower Austral zones in the drier west were named the Upper and Lower Sonoran Zones. This division of habitats, in spite of its defects, won wide acceptance from naturalists, and life zones are still occasionally used in descriptive ecological work.

More recently, Clements and Shelford (1939) proposed a system of *biomes* in which each major habitat was defined according to its dominant form of vegetation. These great areas of vegetation are believed to be climaxes, or stable forms of vegetation, climatically determined, and capable of maintaining themselves indefinitely if undisturbed. Examples of major biomes in North America are Tundra, Coniferous Forest, Deciduous Forest, Grassland, and Desert. Should a biome be disturbed or destroyed by man or by some natural catastrophe, it would eventually revert to its original state if left to itself. With time, the disturbed area would pass through a series of characteristic vegetation types, or successions, until it again reached the climax form, at which stage it would again become stabilized. Since these and other schemes of habitat classifications depend largely on individual judgment, their application to natural situations is never perfect.

One biome does not stop at a sharp boundary where another begins. Instead, a region or transition known as an *ecotone* occurs between them, as, for example, between the coniferous forest and grassland biomes. Smaller subdivisions of biomes and ecotones also occur, all the way down to such units as individual thickets or small woodland pools.

A given species of bird may be restricted to a minor subdivision of a biome; it may range throughout a biome; or it may be distributed over several biomes. Even closely related species vary markedly in this regard. The Song Sparrow, *Melospiza melodia*, breeds throughout North America, but the Ipswich Sparrow, *Passerculus sandwichensis princeps*, breeds only on Sable Island, Nova Scotia. The Sharp-tailed Sparrow,

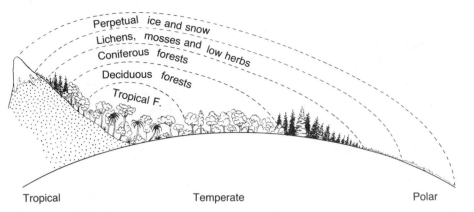

FIGURE 19–8 Starting in the tropics, one may travel across the major biomes of the earth either by climbing about 4 km vertically on a mountainside, or by traveling about 7000 km poleward at sea level.

Ammospiza caudacuta, breeds across Canada and the United States from British Columbia and southern Mackenzie to the St. Lawrence valley and North Carolina, but its close relative, the Cape Sable Sparrow, *A. maritima mirabilis*, is restricted to a coastal fringe in Florida about 95 km long.

Every major habitat presents special conditions of life and, usually, peculiar problems of existence for birds living there. Birds occupying a given habitat, as a rule, are adapted to exploit these conditions and to meet these problems, at least sufficiently well to meet local competition of other species. In the brief survey of major habitats that follows, only a few of the special conditions and problems relating to each can be described, as well as a few of the avian adaptations that relate to them.

POLAR REGIONS. These regions of eternal ice and snow present drastic problems of low temperatures, long winters, no land plants, and practically no cover. Very few birds are able to tolerate such extreme conditions. The antarctic penguins have evolved the most effective adaptations to this harsh environment. They combat the low temperatures by accumulating great reserves of insulating body fat, huddling together in storms, and occasionally burrowing in the snow. They have a remarkably strong urge to brood eggs and young, and have developed warm feet and special brood pouches for this function. All of their food comes from the sea, and penguins are fitted to capture fish, prawns, and other sea animals by such adaptations as eyes accommodated for underwater vision, fish-holding tongues, streamlined bodies, reduced feathers, and powerful flippers instead of wings.

TUNDRA. The vast, circumpolar belt of mosses, sedges, grasses, lichens, flowering perennials and dwarf shrubby alders, willows, and conifers is called the tundra. It extends across the northern parts of North America, Europe, and Asia, just south of the perpetual snows and north of the tree line. A short distance below the surface the ground is permanently frozen. In the brief summers of 24-hour days and no nights, the ground becomes wet and boggy; wild flowers and insects appear in profusion; and many ponds and streams arise. Because of the impossibility of hibernation, predatory reptiles are completely absent in the tundra, and the widespread interspersion of land and water gives nesting birds added security from the few predatory mammals present. Great numbers of shore birds and waterfowl breed in the tundra in summer, as well as a lesser number of hawks, ptarmigan, cranes, owls, larks, pipits, and finches. Practically all of these birds escape the rigors of the tundra winter by migrating south in the fall. As an adaptation to the brief summer season, birds exhibit precisely timed and strikingly accelerated reproductive cycles: quick nest-building, short incubation periods, rapid molting, and rapid development of precocial young.

Were it otherwise, the closing in of winter would cause the deaths of myriads of young birds unable to migrate or to care for themselves. The abundant insects, coupled with the long northern days, enable most tundra species to raise larger broods than their relatives of temperate or equatorial regions.

Ptarmigan are year-round residents of the tundra. They change their plumage from an inconspicuous white in winter to an inconspicuous brown in summer. They can sustain themselves on a vegetable diet in winter, and they survive the bitterest cold by burrowing under the snow. By means of fringed scales or heavy feathers on their feet, they are able to walk on soft snow. The Snowy Owl, *Nyctea scandiaca*, also has heavily feathered feet, and it feeds throughout the year on the usually abundant lemming. Resident birds of the tundra and other cold habitats commonly have larger and more compact bodies than their temperate and tropical zone relatives. This conserves heat because a large body has proportionately less heat-losing surface than a similar small one.

ALPINE REGIONS. In many ways, the alpine communities resemble the tundra. Vegetation is dwarf and scrubby, nights are cold, and the growing season is short. However, compared with the tundra, drainage is much better; wind is usually stronger; the sun's rays, including the ultraviolet, are more direct and intense; day length is "normal," and the air is more rarefied. As in the tundra, ptarmigan change colors seasonally. Smaller birds avoid the wind by building their nests in the shelter of rocks, crevices, and vegetation. One would expect birds that live in the thin air of high altitudes to have relatively larger wings than lowland species. This is at least true of African White-eyes of the genus *Zosterops*, whose wing length increases with altitude regardless of temperature (Moreau, 1964). Birds of high altitudes also have relatively larger hearts (p. 120), which help to support the greater exertion characteristic of mountaintop living.

As described in Chapter 7, American Robins, *Turdus migratorius*, living at altitudes of 2900 meters have hearts 12 per cent larger and lungs 41 per cent larger than Robins living at sea level (Dunson, 1965). Adaptations to the rarefied air of high altitudes are quickly achieved in some species. Seven female Pintail ducks, *Anas acuta*, were kept by Cohen (1965) in an atmospheric chamber at one-half sea-level pressure (equivalent to 7000 meters elevation). Within 20 days the birds responded to the low concentration of oxygen (hypoxia) by increasing total red blood cell volume and circulating hemoglobin by about 75 per cent. Hummingbirds under simulated high altitude conditions, consumed oxygen while in flight at the highest rate known for homeotherms (Berger, 1974). Other probable responses of birds to high altitudes include body fat deposits in Rosy Finches, *Leucosticte tephrocotis*, smaller-sized clutches in larks, swallows, and wheatears, and deeper pigmentation in the skin of naked nestlings of larks and other birds with nests open to the highly actinic rays of the sun.

Among birds living at extreme altitudes are a hummingbird, the Andean Hill Star, *Oreotrochilus chimborazo*, which lives all year at 4500 meters in Peru, and the Twite, *Carduelis flavirostris*, which is found in Tibet also at 4500 meters. An Alpine Chough, *Pyrrhocorax graculus*, has been observed at the extraordinary altitude of 8229 meters in Tibet (Vaurie, 1972).

CONIFEROUS FORESTS. These are the great evergreen forests of spruce, fir, and larch, found chiefly in Canada, northern Europe, and Siberia, with occasional extensions southward along mountains, and isolated fragments in a few other parts of the world. The northern coniferous forest, or *taiga*, may have winter temperatures as severe as those of the tundra lying immediately to its north, but the summer growing season is warmer and longer. In the growth of organic matter the coniferous forest is 50- to 100-fold more productive than the tundra (Deevey, 1960) and therefore is better able to support a large year-round population of birds and other animals. Since most conifers retain their leaves during the winter, they provide good shelter against the wind, and also prevent the windpacking of ground snow that characterizes the tundra. The trunks and stems of conifers and other woody plants expose seeds, berries, nuts, and fruits above the snow where they are available for winter feeding. Insects are abundant during summers in the taiga; reptiles are scarce, but mammalian predators are more common than in the tundra. About 80 per cent of the birds that breed in the taiga in the summer migrate south for the winter. Among these are eagles, hawks, flycatchers, swallows, goatsuckers, thrushes, kinglets, vireos, warblers, and finches. Among the year-round residents are owls, grouse, woodpeckers, ravens, jays, nutcrackers, titmice, nuthatches, creepers, grosbeaks, and crossbills. The majority of these birds are either insect- or seed- and nut-eaters, and various adaptations associated with these diets have been described earlier (p. 113).

DECIDUOUS FORESTS. The typical forests of temperate Eurasia and North America are deciduous forests whose trees bear thin, flat leaves which are shed during winter. This biome demands a moderate amount of rainfall—75 to 150 cm per year. The larger deciduous forest regions are in the eastern United States, western Europe, eastern China, and Japan, with smaller areas in southern Africa, eastern Australia, New Zealand, and Patagonia. Deciduous forests contain a greater variety of trees, shrubs, and herbaceous plants than the coniferous forests. This provides a more varied vegetable diet for resident birds, and a greater abundance of caterpillars and insects. The commoner trees in the deciduous forests of North America are the oak, hickory, beech, maple, elm, ash, sycamore, and poplar. Many of the trees and shrubs of this biome produce hard-shelled nuts—oak, hickory, beech, hazel—and many of the birds living here possess structural and functional adaptations for exploiting this nutritious source of food. A few species, such as jays, nutcrackers, woodpeckers, and nuthatches, have learned to hide stores of seeds or nuts for consumption in winter and the following spring when other foods are scarce. Other species inspect crevices in bark for insects and their larvae, while woodpeckers dig them out of dead wood with their sharp beaks and extensible tongues. For permanent residents the diet may change from berries and insects in summer to nuts and seeds in winter. Other typical birds of the North American deciduous forest include hawks, quail, turkeys, doves, owls, tyrant flycatchers, crows, titmice, creepers, wrens, thrashers, thrushes, vireos, warblers, blackbirds, and finches. Reflecting the comparative abundance of food in different forest biomes, the population density of Chaffinches, *Fringilla coelebs*, varies from 49 to 145 pairs per sq km in various deciduous woods, to 20 to 120 pairs in spruce woods, and 12 to 29 pairs in pine woods (Newton, 1972). Of the species of birds that breed in the deciduous forests of North America, about three-fourths migrate south for the winter.

TROPICAL RAIN FORESTS. The rain forests of the tropics constitute the largest terrestrial biome on earth, covering some 20 million sq km. The air is constantly warm and moist, and the rainfall exceeds 120 cm, averaging 225 cm per year. Tropical rain forests produce more organic material than any other type of biome: an estimated 24 thousand million tons of carbon per year as against 970 million tons for the deserts of the world and only 68 million tons for all tundra regions (Deevey, 1960). The main tropical rain forests occur in the lowlands of Central and South America, Africa, and the East Indies. The luxuriance of the rain forest vegetation is important to the animal residents, but probably more significant is its astonishing variety. Whereas a dozen species of trees might be the maximum found in a deciduous forest, 400 to 500 species of trees and some 800 of woody plants have been identified in the Cameroon rain forest (Hesse *et al.*, 1937). Moreover, this tremendous variety of plant life is thoroughly intermixed: the trees do not normally grow in stands of a single species as is commonly the case with other forests. This fact places a premium on mobility of rainforest animals. However, seasonal migration, as practiced by temperate zone birds, is essentially nonexistent. Probably because of the great variety of plants, food, and other necessities for birds that it produces, the rain forest harbors a greater variety of birds, showing possibly closer niche-packing, than any other biome.

The phenomenon of habitat stratification is well demonstrated in the rain forest. A series of different microclimates is arranged in horizontal strata as one descends from the sun-drenched forest canopy to the ground. The bright green crowns of the tall forest trees constitute the microhabitat where the sun shines most intensely and the frequent tropical rainstorms beat most furiously; and it is the location of most flowers, fruits, and animals. The birds of this roof stratum include many brilliantly colored forms, such as parrots, macaws, toucans, trogons, and cotingas or chatterers, as well as several predatory hawks, eagles, and vultures. Midway to the ground is an understory of small trees, large shrubs, and lianas bedecked with orchids, bromeliads, and other epiphytes. Here in subdued light and warm, humid, quiet air occur wrens, pigeons, curassows, woodpeckers, and other species. On the forest floor relatively little vegetation grows because of the perpetual gloom. Air currents are negligible, and although the beating rains on the canopy produce no violence at this stratum, the humidity of the air is almost constantly at the saturation point. On this level are found the dull-colored partridge, rails, tinamous, antbirds, and in the Orient, colorful pittas.

By adapting themselves to these different strata the different species of birds are able to avoid some of the competition for food, nesting sites, and territories.

Because of the scattering of the many types of plants, birds must keep moving, often in mixed flocks, to obtain their preferred food. In most rain forests one can find trees budding, flowering, fruiting, and shedding their leaves at almost any time of year. The presence of flowers and fruits the year around makes possible the existence of stenophagous species, such as the fruit-eating pigeons or nectar-sipping hummingbirds. These latter often have bills whose shapes and lengths are remarkably adapted to the corollas of the flowers they visit. Seed eaters are uncommon.

Most birds of the tropical rain forests have loud calls that enable them to keep together when separated by dense foliage. Many tropical birds are active mainly in the morning and evening, and rest during the heat of the day. Breeding seasons are usually prolonged, and body fat deposits are rare. Many species build sleeve-like hangnests that are less easily robbed than are open nests by the abundant monkeys, toucans, snakes, lizards, and other predators. Open nests are usually small and inconspicuous, and clutches of eggs are smaller than those of temperate zone relatives. This is perhaps an adjustment to the constantly available food, the shorter day length, and the absence of the dangers of migration. Rounded, rather than long tapered wings are the rule.

GRASSLANDS. As a major habitat type, grasslands are distributed throughout the temperate and tropical zones. They vary from the tall-grass, park-like savannas of Africa and the pampas of Argentina through the prairies of midwestern United States to the almost desert-like, short-grass steppes of Asia. The dominance of grass in all these habitats is the result of rainfall that is insufficient to grow trees, but copious enough to prevent deserts. Rain, 30 to 100 cm per year, often comes in pronounced seasonal cycles that cause luxuriant vegetation and food in the spring and summer, followed by searing dry seasons that force many animals to emigrate. Grass fires may be a recurrent problem for animals. Grasslands are usually flat or slightly undulating open plains, lacking trees or rocks that might provide shelter against the sun, rain, or winter storms. Having very little incentive to leave the ground except for annual migrations, many birds, even though they may fly well, have adopted the cursorial habit: tinamous, bustards, quail, sand-grouse,

the Secretary Bird, *Sagittarius serpentarius*, the Road Runner, *Geococcyx californianus*, larks, and others. In the emus, rheas, and ostriches, adaptations for running have been so adequate that their ability to fly has been lost. These large ratites or keel-less birds have long, powerful legs with reduced toes, long necks, and vestigial wings and feathers. The flocking habit, as seen in ostriches, certain grouse, longspurs, and buntings, is a common grassland adaptation. As a response to the lack of cover, several species either nest on the open ground (eagles, larks) or use underground burrows (Burrowing Owl, *Athene cunicularia*). In Tibet, three finches of the genus *Montifringilla* share burrows with native hares. Because of the absence of singing posts, many grassland birds give their territory and courtship songs while on the wing. Insect eaters and seed eaters predominate, and predatory birds are abundant.

DESERTS. Deserts cover about 28 per cent of the earth's land surface. Very little rainfall, very low humidity, and little or no vegetation are characteristic of deserts everywhere. The lack of cloud cover makes for extreme temperature fluctuations, often extending from freezing temperatures by night to intolerably hot temperatures by day. Air temperatures of 58°C (136°F) in the shade have been recorded in the Libyan desert, and soil surface temperatures were much higher. The desert substrate may be rock, shingle, or sand; the last type is the poorest in animal life. Drinking water is extremely rare and even nonexistent in many deserts. Accordingly, birds must get their water from early morning dew, or from insects, seeds, or other available foods. As described in Chapter 6, some desert birds can survive without water for weeks and even months. The water they require for body tissues and vital functions is obtained as metabolic water from the breakdown, for example, of nothing more fluid than a diet of dry seeds.

As in the grasslands, many birds are cursorial in habit. Plumage commonly matches the sandy or stony ground; in the Sahara, 24 of 47 species showed homochromy, or ground-matching coloration (Bourlière, 1950). As protection against both the intense sun and the violent winds, birds seek the shelter of shrubs and rocks for nesting and resting sites. They generally confine their activities to the cooler times of day. In Patagonian deserts, many ground-dwelling birds possess horny opercula over their nostrils, presumably as an adaptation against inhalation of dust and sand. Earth burrows do not seem to be commonly used by desert birds, but cavities in cacti are occupied by some owls and woodpeckers. In both desert and steppe biomes, the reproductive cycle of many birds is closely correlated with the season of rains; and when there is no rain at all, the birds ordinarily fail to breed. Other desert birds—more than one-fourth of all Australian birds—are, like the human aborigines, opportunistic nomads, following the sporadic, unpredictable rains. Characteristic birds of desert regions include ratites, hawks, doves, owls, woodpeckers, wrens, thrashers, shrikes, larks, and finches.

FRESHWATER LAKES, PONDS AND STREAMS. Freshwater habitats are spread over the terrestrial world, particularly in regions of abundant rainfall. Lakes vary in size from the smallest ponds to the Great Lakes of North America. As a rule, the smaller, shallower lakes are much richer in plant and bird life than the larger, deeper ones. Summarizing eight studies of lake productivity in southern Finland, von Haartman (1971) finds that vegetation-rich (eutrophic) lakes support 75 pairs of water birds per square kilometer, whereas vegetation-poor (oligotrophic) lakes support only nine pairs. Water birds typically have larger-than-average uropygial glands and dense, oily plumage. Many of them have strong, short legs with webbed toes. Diving birds, such as grebes and loons, often have dorsoventrally flattened bodies, and legs placed toward the rear. Their bones are generally less pneumatic than those of surface-feeding waterfowl. The dippers (Cinclidae) are passerine birds that plunge into streams, propel themselves under water with their wings like auklets, and scramble along the bottom in search of food. In addition to the structural adaptations of large oil glands and nasal flaps to keep water from their nostrils, they possess two striking physiological adaptations. Upon diving, the Dipper, *Cinclus mexicanus*, depresses its heart rate 55 to 69 per cent and increases the store of oxygen in its blood (Murrish, 1970).

In taking off from the water in flight, birds of small bodies of water, for example, "puddle ducks," rise directly into the air, while those of larger lakes and streams patter along the surface for some distance before they are airborne. Ducks adapted to feeding in shallow water tip their bodies vertically while remaining afloat on the surface and reach what submerged foods they can with their extended head and neck.

Deep-water ducks, on the other hand, dive for their food. Many kinds of feeding adaptations are found in the beaks of water birds: sifting devices for seeds and small animals, gaffhooks, spears, serrated forceps for fish, and others. These are described in more detail in Chapter 6.

SHORES AND MARSHES. These are the amphibious environments, part land and part water, where vegetation and food are unusually abundant but problems of space, substrate, and seasonal food shortage affect bird residents. Shores and many marshes are fringe habitats, linear and narrow, so that gregariousness is forced on the birds that inhabit them. Long, wading legs and long, agile necks are common features of birds of these regions. The long legs of such birds as cranes, herons, and storks enable them to enjoy the advantages of both land and shallow water, while the shorter legs and webbed toes of coots and grebes fit them to exploit muddy shores as well as open water. Toes may be webbed fully, partly, or not at all, correlating in part with the swimming habits of the species, and with the firmness of its typical feeding ground substrate. The exceptionally long toes of the jaçana have already been mentioned. Certain of the sandpipers are typically found along muddy shores, others along sandy shores. Among shore birds, long, probing bills predominate, and many of them are heavily supplied with sensitive touch corpuscles that enable them to discriminate food particles from others in muddy water or soft earth. Correlated with the abrupt cessation of food supplies when, in colder habitats, the lakes and streams freeze over in autumn, is the strong migratory impulse found in most shore birds. Birds of sea coasts often show a daily behavior cycle, oscillating between land-feeding when the tide is in and shore-feeding when it is out, exposing many marine invertebrates.

THE SEAS. The oceans of the world make up a continuous, highly uniform habitat. Although they vary markedly in temperature and food resources, they are all flat, wet, salty, windy, and vast. The problems they present to birds are simple but formidable: no places to perch or to build nests, no cover for hiding, no fresh drinking water, a diet limited to sea food. The use by birds of a method of reproduction involving nests and eggs has prevented them from adapting completely to a marine existence. Although the seas make up five-sevenths of the earth's surface, they harbor only about 3 per cent (about 250 species) of the world's birds. Sea birds can be divided into inshore species, which generally sleep ashore and find their food on shore or no farther at sea than 5 to 10 km, and the open-sea or pelagic birds that find their food and their rest throughout the wide-ranging oceans. Some of the pelagic birds, such as albatrosses and penguins, may roam the seas for months without resting on land.

Sea birds show many of the same adaptations to an aquatic life as do fresh water birds: webbed feet, oily feathers, fish-eating adaptations in their beaks, and others. One peculiar adaptation in the beaks of Procellariiformes (albatrosses, shearwaters, fulmars, and petrels) is a pair of tubular projections covering their nostrils and apparently protecting them from salt spray on stormy days. More important are the nasal or salt glands in the heads of sea birds that make it possible to drink sea water (p. 107). Without them it would be impossible for sea birds to explore the seas very far from land and fresh water. Long, narrow wings, most efficient for soaring, are found in many sea birds like the albatrosses, which spend much time on the wing. The distribution of sea birds follows closely the distribution of fish and other sea foods; these, in turn, reflect the concentration of nitrates, phosphates, and other nutrient chemicals in surface sea water. These nutrients are richest near the mouths of great rivers and in regions where deep ocean currents rise to the surface, e.g., along shores, or boundaries where warm and cold ocean currents meet.

ALTITUDE AND LATITUDE

On a vegetation map of North America, the biomes of high (northern) latitudes extend southward along mountain ranges, and the biomes of low tropical latitudes extend northward along valleys. This is due to the fact that temperature is one of the main controlling factors of vegetation, and that a low mean annual temperature, whether at sea level in Alaska or on a mountain top in the equatorial Andes, will result in the growth of similar types of vegetation in both habitats. One can therefore cross a series of biomes extending from the tropical rainforest to the tundra either by journeying from the equator to the Arctic Circle, about 7000 km north, or by ascending an equatorial mountain from sea level to an altitude of about 4 km. For every 150 meters one climbs upwards, the air temperature drops approximately 1°C. On either trip one would pass over the same successive biomes,

with differences in details caused by such factors as soil, wind, rainfall, exposure to sunlight, past geological history, and above all, the different annual cycles. Palm trees, for example, thrive in the temperate zone of tropical mountains, but not in the latitudinal temperate zone. In the former the temperature never drops to freezing, but in the latter it may fall many degrees below freezing. These differences in the climates of altitudinal and latitudinal temperate zones depend largely on the angle with which the sun's rays strike the earth (roughly vertical at the equator, but inclined an annual mean of 50 degrees to the horizon at 40 degrees N. latitude), and length of day (always 12 hours at the equator and varying roughly between 8 hours in winter to 16 hours in summer at 40 degrees N. latitude) (see Fig. 19–8).

An important advantage in comparing a series of biomes altitudinally (vertically) rather than latitudinally (horizontally) is the economy of time and effort required, and the opportunity for close comparisons of vegetation and animals of neighboring zones. An altitudinal study of the birds in the life zones (Merriam's) of Yosemite Park by Grinnell and Storer (1924) showed that species found in the Canadian zone high up in the Sierra Nevada of Yosemite were also found near sea level in the same Canadian zone in Canada. The study also revealed that some species are multizoned, tolerating the varied living conditions of five different successive zones, while others, much less tolerant, were restricted to a single zone (Fig. 19–9). A study by Miller (1951) of the ecological distribution of California birds showed that there was a greater variety of birds in zones of lower altitude, as one would expect, than in the zones of higher altitude where living conditions are more rigorous. The study also revealed that of the 143 species of birds found, 54 species were restricted to one life zone, 89 species lived in two zones, 88 in three zones, 23 in four zones, 5 in five zones, and 1 species spread its residence over six zones. Such studies as these reveal the great difference among species in environmental tolerance.

HABITAT INTERSPERSION AND EDGE EFFECT

Every ornithologist soon learns that the best way to see a large variety of birds in a short time is to visit *ecotones* or margins between different types of habitats. A study by Lay (1938) in Texas revealed that the margins of clearings had 95 per cent more birds, representing

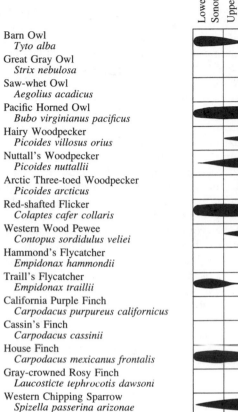

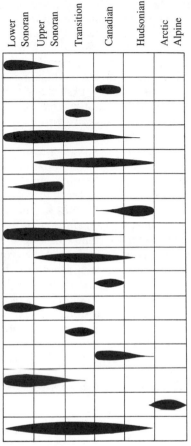

FIGURE 19–9 The vertical distribution of various birds in the life zones of Yosemite National Park. Some species are multizonal, while others are restricted to a single zone. The peculiar two-zoned distribution of Traill's Flycatcher shows that, for this species at least, habitat selection is more than a simple temperature response. After Grinnell and Storer, 1924.

41 per cent more species, than comparable areas of the interiors of adjacent woodland. Birds commonly require more things for survival and reproduction than can be had from a single meadow or wood lot, but if their needs can be met by using both habitats, chances of successful living are increased. The meadow may furnish insects for food and grass for lining nests; the woods, twigs for a nest platform and cover for escape from enemies. This is why so many birds live on the edges of habitats.

Treefalls create a mosaic of gaps in tropical forests in Panama. Floral composition in these gaps is different from that in forest interiors so that an edge effect is created. Sixty-six species were found occupying these gaps, versus 53 in the forests (Schemske and Brokaw, 1981). For a given species there may be differences in abundance between edge and forest interior. For example, Acadian Flycatchers, *Empidonax virescens*, and Ovenbirds, *Seiurus aurocapillus*, are more abundant in the interior of a hardwood-pine forest in Tennessee than at the forest edge. Conversely, Summer Tanagers, *Piranga rubra*, Cardinals, *Cardinalis cardinalis*, and Towhees, *Pipilo erythrophthalmus*, are more abundant at the forest edge (Kroodsma, 1984).

One reason why modern agriculture is so detrimental to wild birds is that it depends on large areas of homogeneous vegetation with little edge. Even in birds, mobility can become a problem if the necessities of life are spread too far apart. The chief reason for high population densities in parks and gardens (p. 387) is their rich interspersion of trees, shrubs, lawns, walks, flower beds, pools, and even buildings. Ecotones may also attract enemies of birds. In a forest and field ecotone in Michigan, nest density of 21 species of passerines increased with nearness to the edge, but so did mortality of the nestlings, largely because of predators and brood parasitism by the Brown-headed Cowbird, *Molothrus ater* (Gates and Gysel, 1978).

SUCCESSION

If one should fence off a square kilometer of flat, polished granite, and allow only natural forces and agents to affect it for many centuries, eventually a climax form of vegetation would establish itself on it: a beech-maple forest, or a tundra, or a prairie, depending on the climate. Left undisturbed, the granite would have been successively covered with lichens, mosses, herbs, shrubs, trees, and finally beech and maple trees with their associated plants, assuming that this experiment took place in the climate now characteristic of northern Indiana. If, at the same time and in the same general location, one fenced off a square-kilometer freshwater pond and left it undisturbed long enough, it too would eventually become a beech-maple forest, via submerged vegetation followed by floating plants, emergent marsh plants, shrubs, and finally trees, each growing on the accumulated organic debris of the previous generations. Such successions as these are constantly going on in nature wherever hurricanes, fires, glaciers, volcanic eruptions, landslides, epi-

demic plant diseases, or human agencies have disturbed the established climax succession. The time required for a complete series of succession from bare ground to climax varies in different parts of the world. The rapidity with which tropical woodland invades deserted plantations in humid regions is well known.

Disturbances of the natural habitat are not necessarily a bad thing for birds. Kirtland's Warbler, *Dendroica kirtlandii*, breeds only in stands of young jack-pine that range between 1 and 6 meters high. The pines exist in this state for only about 15 years after a region has been logged or swept by fire. Controlled burning can initiate habitat succession that is beneficial both to Kirtland's Warbler and to Bobwhite Quail, *Colinus virginianus*. The successional stages in a typical forest have been classified by May (1982) into the following categories: (a) herbaceous; (b) herb, shrub, and sapling; (c) young forest; (d) older forest. The density of breeding birds was found to be lowest in (a), rose sharply in (b), and reached a maximum in (d). The numbers of species and of feeding guilds showed a similar pattern of increase.

In a succession study involving tidal salt-marsh reclamation in England, a series of dykes was built that resulted in polders 3, 12, 19, and 30 years of age (Glue, 1971). Breeding bird censuses revealed that each reclaimed section of marsh was dominated

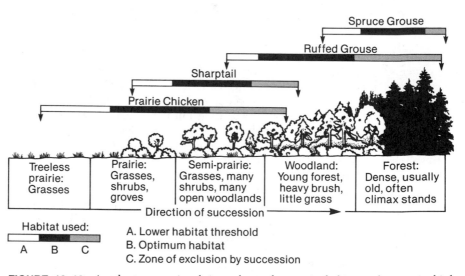

FIGURE 19–10 As plant succession brings about changes in habitats, changes in bird population naturally follow. The bars indicate those habitats used during the year by four different species of Wisconsin grouse. After Grange, 1948.

by a different species: the three-year-old salt marsh and brackish pools succession was dominated by the Meadow Pipit, *Anthus pratensis;* the 12-year-old soft mud pans, by the Reed Bunting, *Emberiza schoeniclus;* the 19-year-old hard mud pans, by the Yellow Wagtail, *Motacilla flava;* and the 30-year-old grassland, by the Skylark, *Alauda arvensis.* Birds themselves may initiate successional changes that affect other birds. An increase in the population of Horned Grebes, *Podiceps auritus,* on the Baltic island of Gotland was attributed by Högström (1970) to the improvement of their habitat by the fertilization of water and soil by Black-headed Gulls, *Larus ridibundus.*

If climate changes, as it does, the climax succession must also change. Lands that are now desert in the Near East were once thriving centers of human populations, luxuriant with vegetation and "flowing with milk and honey" when more rainfall was characteristic of the region. Man himself has been responsible for the development of extensive deserts in China, Saudi Arabia, Mexico, Africa, and elsewhere through his misuse of the land.

In regions of active plant succession, the bird population slowly changes its character to conform to changes in plants and other animals associated with given forms of plants—insects, worms, snails, mice, and other birds. Woodpeckers, for example, not only require trees for their particular way of life, but wood-tunneling beetle larvae, acorns, and other such associates of trees. Some of these successional changes may be invisible to human eyes, but they may still involve critical limiting factors that determine whether a given species of bird will live in a given habitat. Occasionally a species will persist nesting in a given location even though succession changes the vegetation enough to repel newcomers. If offspring reared in this changed habitat become imprinted on it and return there to breed, they may thereby expand the breeding range of the species (Orians, 1971).

Sometimes the interrelations between birds and their environments are obvious and even dramatic. Elton (1927) tells how an Indian mynah bird and a Mexican lantana plant were introduced into Hawaii with the following results. Feeding on the lantana berries, the mynah and other birds caused the rapid spread of the plant. The mynahs also increased enormously in numbers. Although the lantana became a serious weed, the mynah was welcomed because it ate the army-worm caterpillars that periodically ravaged young sugarcane plantations. To stop the spread of the lantana, an agromyzid fly that ate its seeds was introduced. This control succeeded so well that the lantana decreased greatly in abundance, and as a result the mynahs also decreased to such an extent that there was a resumption of severe outbreaks of army-worms in sugar plantations. As the lantanas decreased, other species of introduced shrubs took their place and became serious weeds.

All of these multitudinous crisscrossing threads of relationship among birds, plants, insects, climate, and other environmental factors make up Darwin's web of life, a tremendously intricate, dynamic system, and one, for the moment, in equilibrium. But let one filament of the web be cut—let one tree die, or a new bird move in—and new adjustments are in order.

■ SUGGESTED READINGS

A valuable source book in ecology, although now somewhat dated, is *Principles of Animal Ecology* by Allee *et al.* Excellent contemporary texts are Odum's *Fundamentals of Ecology,* which emphasizes quantitative, analytical ecology; Kendeigh's *Animal Ecology,* which places much emphasis on birds; and Ricklefs' *Ecology,* which emphasizes genetic, evolutionary, and mathematical aspects of the subject. For the ecology of birds of specific habitats see Buxton's *Animal Life in Deserts,* Haviland's *Forest, Steppe and Tundra,* and Schmidt-Nielsen's *Desert Animals.* Specifically devoted to birds are Lack's *Ecological Adaptations for Breeding in Birds,* and *Ecological Isolation in Birds.* Much of Dorst's *The Life of Birds* is devoted to descriptive avian ecology. The photosynthetic production of plants in numerous biomes is quantitatively treated by Reichle *et al.* in *Productivity of World Ecosystems.* An excellent series of essays on various topics in ecology may be found in J.R. Krebs and N.B. Davies's *Behavioural Ecology.*

The Geography of Birds

20

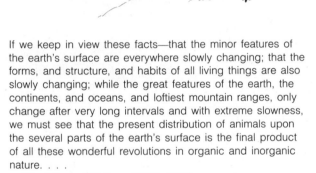

If we keep in view these facts—that the minor features of the earth's surface are everywhere slowly changing; that the forms, and structure, and habits of all living things are also slowly changing; while the great features of the earth, the continents, and oceans, and loftiest mountain ranges, only change after very long intervals and with extreme slowness, we must see that the present distribution of animals upon the several parts of the earth's surface is the final product of all these wonderful revolutions in organic and inorganic nature. . . .

—Alfred Russel Wallace (1823–1913),
The Geographical Distribution of Animals

When birds took to the air, some 150 million years before the Wright brothers, or before Icarus, for that matter, they obtained a highway to every possible habitat on the earth's surface. Today, birds are at home in polar regions and in the tropics, in forests and deserts, on mountains and prairies, and on the oceans and distant islands. Yet, when one considers the superb mobility of birds and the eons of time they have had to populate the globe, it is surprising how few cosmopolitan species there are. Some shore and sea birds—sandpipers and plovers, petrels and gulls —are worldwide in their distribution. Representatives of grebes, cormorants, herons, ducks, owls, ospreys, falcons, rails, doves, kingfishers, and swallows are at home on every continent. Ravens have inherited the entire earth except, for some obscure reason, South America. But these are exceptional. One usually sees, especially among land birds, a picture of curiously limited and seemingly haphazard distribution.

Concerning the apparently incongruous distribution of birds in the British Isles, David Lack (1969) facetiously remarked, "Saint Patrick explicitly banned snakes from Ireland, but the present state of the island suggests that his prejudices extended to woodpeckers, nightingales, and many other animals and plants. In all, only three-fifths of the land and fresh-water birds that breed regularly in Britain breed regularly in Ireland . . . ." Similar anomalies can be seen almost anywhere one looks. Why, for example, should the birds of England and Japan be more alike, though 11,000 kilometers apart, than the birds of Africa and Madagascar, separated by a mere 400 kilometers? Why does South America have more than 300 species of hummingbirds while Africa, with very similar habitats, has not a single one? Why does the North American Turkey, *Meleagris gallopavo*, occur naturally nowhere else in the world?

In sharp contrast to species with wide distribution are those with very limited ranges. Oceanic islands commonly have endemic species with highly restricted distribution. The Atitlan Grebe, *Podilymbus gigas*, is known to occur only on Lake Atitlan in Guatemala. Kirtland's Warbler, *Dendroica kirtlandii*, nests in only a few counties of north central Michigan, and the Ipswich Sparrow, *Passerculus sandwichensis princeps*, nests only on Sable Island, Nova Scotia (about 50 sq km). Several South American hummingbirds have strikingly limited ranges. The Racket-tailed Hummingbird, *Loddigesia mirabilis*, is known to breed only in a tiny mountain valley in Peru, and *Oreotrochilus chimborazo*, just below snowline on several volcanoes in Ecuador. The Seychelles Warbler, *Bebrornis sechellensis*, consists of a population of about 30 birds on Cousin Island that has an area of about 24 hectares. Such a drastically limited geographical range commonly foreshadows extinction. Hall and Moreau (1962) listed nearly 100 species of African birds that are each restricted to ranges measuring less than 400 kilometers in diameter. The populations of these species are generally small, and in a few instances they consist of a few hundred birds.

DESCRIPTIVE BIRD GEOGRAPHY

There are two ways of looking at bird distribution. The first and simplest is a description of the static, geographical location of the various species of birds. The second viewpoint recognizes the fact that bird distribution is a dynamic, constantly changing affair; that no species stays in one place forever. Intellectually a more lively approach, this second way of looking at bird geography attempts to explain why certain species live where they do. This can be done in two ways, one of which is ecological. Water birds obviously do not live in deserts, nor tree creepers in treeless steppes. But ecology does not explain why there are many hummingbirds in South America but none in similar rainforests in Africa; nor why the Red-breasted Nuthatch, *Sitta canadensis*, is found in conifer forests all over North America, but in the Old World only in eastern Siberia, China, and on the mountaintops of Corsica in the Mediterranean. Even more puzzling is the curious distribution of wrens and larks. In the New World there are 63 species of wrens (Troglodytidae); in the Old World, only one. In the Old World there are 75 species of larks (Alaudidae); in the New World, only one native species. The explanation of riddles such as these depends on historical principles involving not only ecology but also geology and evolution.

Considering first the static, descriptive picture of bird distribution, one is impressed with the fact that the vast majority of birds live on only 29 per cent of the earth's surface—the land surface. Of the approximately 9100 species of birds only about 290 are marine species; the remainder make their living on land or in fresh waters. Of the earth's land surface, about two-thirds lies north of the equator, and about two-thirds lies in the eastern hemisphere. As a habitat for birds, land provides much greater diversity than does the sea. Its surface may be ice, rock, sand, or fertile soil. It may be flat or deeply ribbed with high mountains. Its temperature may vary four or five times as much as that of the seas. The vegetation that clothes it, and that ultimately provides all food for the birds, varies primarily according to temperature and precipitation, and it is exceedingly varied in quantity, quality, and distribution. The directions of the various mountain chains have profound effects on bird distribution, migration, and even evolution. As a consequence of these variables, the study of the terrestrial distribution of birds is a much larger and more complex problem than the study of aquatic distribution.

Apparently, there is no place on earth so remote or so isolated as to be completely deprived of bird life. A United States Air Force research party, drifting on a huge cake of ice within 240 km (150 miles) of the north pole, between the latitudes of 88°01' and 88°30' N, sighted birds described as "sea gulls" on eight different occasions between June 12 and August 15, 1952. The following year, birds were observed 15 times between 88°15' and 86°10' N. latitude, ten of the observations being of small white birds that were probably Snow Buntings, *Plectrophenax nivalis*. One of the larger birds was collected, and proved to be an immature Kittiwake, *Rissa tridactyla* (Paynter, 1955). At the other end of the globe breeding colonies of Antarctic Petrels, *Thalassoica antarctica*, Snow Petrels, *Pagodroma nivea*, and South Polar Skuas, *Catharacta maccormicki*, have been found at 79° S. latitude and 28° W. longitude, about 250 km from the ocast of the Weddell Sea (Brook and Beck, 1972); Great Skuas, *Catharacta skua*, and Wilson's Petrels, *Oceanites oceanicus*, freely cross Anatarctica and have been seen at the South Pole (Dorst, 1974).

Relatively few breeding species are able to withstand the rigors of polar environments, although flocks of a given species may include tremendous numbers of individuals, as in the case of penguins, auks, and other sea birds. Breeding in the ecologically monotonous arctic tundras are many water birds and also a few land birds, such as ptarmigan, birds of prey, and a few finches. As one moves from polar regions toward the equator, the variety of bird life gradually increases until it reaches its maximum in tropical regions. It is difficult to know exactly why bird species are so abundant in the humid tropics. At least in part it is due to the stability and diversity of habitats and niches, to the rich food resources, and to the absence of stress from cold and freezing weather. But much of the avifaunal richness of the tropics stems from its geological and evolutionary history and this topic will be considered later. Table 20–1 illustrates this pole-to-equator geographical gradient in the numbers of terrestrial bird species. A similar decline in species diversity occurs along altitudinal gradients. On Mt. Karimui in Papua New Guinea, about 100 species of birds occur between sea level and 1000 meters altitude, 45 species at 1800 meters, and 12 species at 2450 meters (Diamond, 1972).

A closer look at the world distribution of birds reveals other geographical patterns. Whereas one species may spread over a broad, continuous geographical area, another may show a spotty, discontinuous type of distribution with unoccupied gaps between "islands" of residence. In certain regions a given species or family may be dominant; in another region it may be scarce. Certain regions may be characterized by the residence of many species or families peculiar to them, while other regions may possess not a single peculiar species.

Any widely traveled naturalist quickly becomes aware of the fact that the different continental land masses are populated with different kinds of birds. Careful scrutiny soon shows, however, that the more distinct changes in faunal differences do not necessarily occur precisely at continental boundaries. For example, in Asia no parrots live north of the Himalayan mountains, but they are common south of them. The sharpest change in character between North and South American birds comes not at the Isthmus of Panama but at the northern limit of the tropics

TABLE 20–1.
THE NUMBER OF SPECIES OF BREEDING BIRDS AT DIFFERENT LATITUDES

LOCATION	MEAN DEGREES NORTH LATITUDE	NUMBER OF SPECIES	SOURCE
Ellesmere Island	82°	14	Irving, 1972
Spitzbergen	78°	28	Ogilvie and Taylor, 1967
Greenland	70°	57	Salomonsen, 1950, 1951
Labrador	55°	88	Peterson, 1963
Newfoundland	49°	128	Peterson, 1963
Maine	45°	176	Peterson, 1963
Pennsylvania	41°	185	Peterson, 1963
Georgia	33°	160	Peterson, 1963
Texas	31°	300	Peterson, 1963
Mexico	23°	750	Blake, 1953
Costa Rica	10°	758	Slud, 1964
Panama	8°	1100	Griscom, 1945
Colombia	5°	1556	de Schauensee, 1970

in Mexico. African and Eurasian avifaunas are separated more by the Sahara Desert than by the Mediterranean Sea.

Pondering such facts as these, P. L. Sclater in 1858 proposed dividing the land surface of the earth into six great avian realms, each separated from the others by the peculiarity of its bird fauna. In 1876 Alfred Russel Wallace published his classic work, *The Geographical Distribution of Animals*. In it he proposed a modification of Sclater's zoogeographical regions to fit animals generally. Although no such partitioning of the world can have universal validity, these faunal regions have enough natural reality that they are still widely used by zoogeographers. Figure 20–1 shows these six regions in a polar projection of the earth. The endpapers of this book show these faunal regions and some of the distinctive features of the bird faunas of each region. In reading the following material it will be helpful to refer to these maps.

The following list of geographical regions of the world with their corresponding avian characteristics must be considered somewhat tentative, especially since the classification of many families is still fluid and subject to change. Nonetheless, the broad outlines of bird distribution as presented here are agreed upon generally by students of bird geography.

The **Palearctic** region is by far the largest and probably contains the site of the origin of all birds. It includes Europe, Africa north of the Sahara, and all of Asia except its southerly projections (see maps). It possesses great diversity in climates, physiography, and types of ecological habitats. Its major mountain ranges run east and west, a fact that has an important bearing on bird distribution and evolution. Despite its large size, the Palearctic is relatively poor in bird variety, a fact probably due to the cold climate of its northern portions. The great majority of its birds are migratory and many are insect-eaters. According to Meinertzhagen (1964), of the region's estimated 1026 species of breeding birds, about 55 per cent are passerines and 20 per cent are water birds. Of its 69 families, only one is unique—the Prunellidae or hedge sparrows.

FIGURE 20–1 Wallace's zoogeographical regions of the world as seen on a polar projection map. Note that South America, Africa, and Australia are relatively isolated from the main land masses of the world.

The warbler family, Sylviidae, is predominant in the number of its species.

The Palearctic shares many birds with the adjoining zoogeographical realms. Of its 329 genera, 35 per cent are also found in the Nearctic, 34 per cent in the Ethiopian, and 38 per cent in the Oriental regions. Of the Palearctic species, 12.5 per cent are shared with the Nearctic, 15 per cent with the Ethiopian, and 22 per cent with the Oriental regions. As in all regions, some birds are ecologically tolerant and widespread; some are not. The Common Raven, *Corvus corax*, and the Horned Lark, *Eremophila alpestris*, breed from the arctic to the tropics, and from an altitude of 5500 meters to sea level. This region shares with the Nearctic regions of the New World (the two areas together are called the Holarctic) 48 families of birds of which the following are found in no other regions: loons (Gaviidae), grouse (Tetraonidae), auks (Alcidae) and waxwings (Bombycillidae). Other characteristic birds of the Holarctic include hawks, owls, woodpeckers, swallows, thrushes, kinglets, titmice, creepers, crows, jays, and many northern-breeding shore and water birds. Additional birds characteristic of the Palearctic, but not exclusive to it, include larks, Old World flycatchers, Old World warblers, pipits, weaverbirds, and starlings.

The **Nearctic** region includes North America north of the tropics, and Greenland. It contains an estimated 750 species of breeding birds. It is similar to the Palearctic in its physical and biotic features and in its restrictive, cold northern climates. It differs most significantly from the Palearctic in the north-and-south direction of its mountain ranges and valleys. As in the Palearctic, most birds are migratory. There are many water and shore birds, and many insect eaters. The predominant families, in numbers of species, are the wood warblers, Parulidae, and blackbirds, Icteridae.

In spite of the fact that 61 families of birds breed in the Nearctic, it too is poor in avian variety and can boast no endemic family. However, the turkey family, Meleagrididae, is nearly restricted to the Nearctic area. One reason that the Nearctic shares so many bird groups with the Palearctic is that many Nearctic birds have been derived from the latter region.

Ernst Mayr (1946, 1972) has extensively analyzed the Nearctic avifauna and deduces three elements in its make-up. The first is an indigenous element so old that its origins are presently untraceable. To this element he ascribes five passerine families: wrens (Troglodytidae), mockingbirds (Mimidae), vireos (Vireonidae), American wood warblers (Parulidae), and buntings and sparrows (Fringillidae). Other possible indigenous families of ancient lineage include waterfowl (Anatidae), New World vultures (Cathartidae), hawks (Accipitridae), grouse (Tetraonidae), quail (Phasianidae), turkeys (Meleagrididae), swifts (Apodidae), motmots (Momotidae), woodpeckers (Picidae), swallows (Hirundinidae), dippers (Cinclidae), gnatcatchers (Sylviidae), and waxwings (Bombycillidae).

The second element includes immigrants from the Old World, probably by way of the Bering Strait land bridge: cranes (Gruidae), pigeons and doves (Columbidae), cuckoos (Cuculidae), owls (Strigidae), crows (Corvidae), and thrushes (Turdidae). Other likely immigrants from the Old World include parrots (Psittacidae), trogons (Trogonidae), barbets (Capitonidae), kingfishers (Alcedinidae), larks (Alaudidae), titmice (Paridae), creepers (Certhiidae), and shrikes (Laniidae).

The third element includes two major families from South America: hummingbirds (Trochilidae), and tyrant flycatchers (Tyrannidae); and possibly, other smaller groups as well—immigrant toucans (Ramphastidae), cotingas (Cotingidae), manakins (Pipridae), blackbirds (Icteridae), and tanagers (Thraupinae).

The **Neotropical** region embraces South America, Central America, the lowlands of Mexico, and the West Indies. It differs greatly from all other continental land masses in that about 70 per cent of its surface is low, well watered, and tropical. About 32 per cent of the South American continent is covered with tropical rain forest (as against about 9 per cent of Africa and 4.5 per cent of Australia), and about 38 per cent is covered with savanna. There is very little desert habitat in the Neotropical realm and very little temperate zone vegetation. There are no major latitudinal barriers to bird dispersal. This is by far the richest of all realms in bird life, both in numbers and in variety. The number of species per 260,000 sq km (100,000 sq mi) is double that for Africa (39.7 *vs* 20.0) (Keast, 1972). Yet, more primitive species live here than in any other region. About 3300 distinct species are known to occur in the Neotropics, of which 2930 species in 95 families occur in South America. This represents about one-third of all known species. Of these families, 31 are peculiar to the region. This is more than twice as many endemic families as are found in any other geographical division. The dominant Neotropical families are the hummingbirds (Trochilidae); miners, spinetails, and foliage-gleaners (Furnariidae); antbirds (Formicariidae); tyrant flycatchers (Tyrannidae); and tanagers (Thraupinae).

Among the 31 families that are exclusively Neotropical or nearly so are the following:

Rheidae—Rheas
Tinamidae—Tinamous
Anhimidae—Screamers
Cracidae—Curassows
Opisthocomidae—Hoatzin
Psophiidae—Trumpeters
Eurypygidae—Sunbitterns
Cariamidae—Cariamas
Thinocoridae—Seedsnipes
Steatornithidae—Oilbirds
Nyctibiidae—Potoos
Trochilidae—Hummingbirds
Todidae—Todies
Momotidae—Motmots
Galbulidae—Jacamars
Bucconidae—Puffbirds
Ramphastidae—Toucans
Dendrocolaptidae—Woodcreepers
Furnariidae—Ovenbirds, Miners
Rhinocryptidae—Tapaculos
Formicariidae—Antbirds
Conopophagidae—Antpittas
Pipridae—Manakins
Cotingidae—Cotingas
Phytotomidae—Plant-cutters

According to Mayr (1964) there is a very ancient, indigenous South American avifauna composed largely of these 25 families. In addition, there is a large element composed of somewhat more recent immigrants from the Nearctic: pigeons (Columbidae), jays (Corvidae), wrens (Troglodytidae), thrushes (Turdidae), vireos (Vireonidae), wood warblers (Parulidae), blackbirds (Icteridae), tanagers (Thraupinae), cardinals (Cardinalinae), and finches (Fringillidae). To these should be added the sea birds and the pantropical parrots and trogons.

Whereas in every other part of the world the songbirds or oscines predominate in numbers, in the Neotropical region they are in the minority, and the more primitive suboscines or clamatores are the dominant forms. These birds have simple syringes and they fall into two groups, depending on the anatomy of the syrinx. The first group is composed of those families with a tracheal syrinx, and includes the tapaculos, antpipits, antbirds, ovenbirds, and woodcreepers. Birds of the second group, possessing a bronchotracheal syrinx, include the tyrant flycatchers, manakins, cotingas, and plant-cutters. Together these two groups include over 1000 species of birds, or nearly one-eighth of all known species.

The **Ethiopian** region is made up of Africa south of the Sahara, southern Arabia, and Madagascar. This part of the world lacks the climatic variety displayed by the three previous regions. Continent-wide, Africa receives only about one-half as much rain as does South America. Consequently, it has a much larger proportion of desert, grassland, and savanna habitats. This leads, on the one hand, to an avifauna rich in terrestrial and cursorial birds and in seed-eaters, and, on the other hand, to a scarcity (about 7 per cent) of water birds. Of the estimated 1556 species of birds in the Ethiopian region, about 62 per cent are passerines and about 24 per cent are seed-eaters (Keast, 1972).

The predominant types of birds are the weaverbirds, shrikes, and the larks. Among the distinctive terrestrial birds are the Ostrich (Struthionidae), Hammerhead (Scopidae), Shoe-billed Stork (Balaenicipitidae), Secretary Bird (Sagittariidae), quail and francolins (Phasianidae), guinea fowl (Numididae), bustards (Otididae), coursers (Glareolidae), sand grouse (Pteroclidae), larks (Alaudidae), and pipits (Motacillidae). Among the distinctive arboreal birds are hawks, kites, and eagles (Accipitridae), falcons (Falconidae), pigeons and doves (Columbidae), bee-eaters (Meropidae), wood-hoopoes (Upupidae), mousebirds (Coliidae), barbets (Capitonidae), honey-guides (Indicatoridae), shrikes (Laniidae), helmet-shrikes (Prionopidae), starlings and tick-birds (Sturnidae), bald crows (Corvidae), thrushes (Turdidae), finches and buntings (Fringillidae), grass finches (Estrildidae), and buffalo weavers and widow birds (Ploceidae).

As one might expect, birds of the Ethiopian region show strong affinities with those of the Oriental region: 30 per cent of the Ethiopian genera occur also in the Oriental region. However, agreement at the species level drops to 2 per cent, which suggests a long period of independent evolution in the two areas (Moreau, 1952). About one-third of all Palearctic bird species, especially the insect-eaters, migrate to Africa and spend the northern winter there, mainly south of the Sahara Desert. Seventy-three families of land and fresh-water birds breed in the Ethiopian region, but only six of these families are peculiar, or nearly so, to the region: Ostrich (Struthionidae), Hammerhead (Scopidae), Secretary Bird (Sagittariidae), Mousebirds (Cotiidae), Touracos (Musophagidae), and Helmet Shrikes (Prionopidae).

The **Oriental** region includes India, Burma, Indo-China, Malaysia, Sumatra, Java, Borneo, and the Philippines. The Himalayan mountains present a formidable barrier separating the Oriental and Palearctic regions, and water separates the Oriental and Australian regions. The climate of the Oriental region is predominantly tropical or subtropical and its annual rainfall averages over 1500 mm. The vegetation is mainly tropical and subtropical rain forest, with progressively smaller areas of monsoon forest, dry tropical forest, scrub, savanna, and desert. There are an estimated 961 resident species of birds in this region. Of the 66 families of land and fresh water birds resident in this region, only one, the Irenidae, containing the leafbirds and fairy bluebirds, is endemic, and the Eurylaimidae, or broadbills, is nearly so. All the other families are also represented in one or more of the other geographical regions. Birds of the Oriental region resemble those of tropical Africa more than those of any other region, but they also show affinities with those of the Palearctic and Australian realms.

Characteristic birds of the Oriental region include many pheasants and pigeons, some owls, parrots, woodpeckers, pittas, babblers, corvids, shrikes, hoopoes, barbets, honey-guides, sunbirds, and several finch families. Less common are megapodes, fruit pigeons, frogmouths, wood swallows, flowerpeckers, honey-eaters, lories, and cockatoos. Many migratory northern birds spend the winter in the Oriental region.

As one progresses eastward in the East Indies from Java toward Papua New Guinea the number of typical Oriental region birds decreases, and coincidentally, the number of Australian region birds increases. In fact, certain dominant Oriental birds such as the barbets range as far east as the island of Bali and then suddenly break off completely and are not found at all on the island of Lombok, only about 35 km away. In a reciprocal fashion, the distribution of many Australian birds such as honey-eaters and cockatoos extends westward only as far as Lombok. Wallace was so impressed with the rather abrupt change in faunas in this region that he suggested as a boundary between Oriental and Australian faunas a line of demarcation (Wallace's Line) separating Bali, Borneo, and the Philippines to the west and north, from Lombok, Celebes, and Papua New Guinea to the east and south. Subsequent studies have shown that a more reasonable line that would strike a balance between a preponderance of Oriental animals to the west and Australian to the east would run roughly north and south between Celebes and the Moluccas—a line approximately 600 km east of Wallace's Line and known as Weber's Line. In any case, somewhere in this region the two faunas are fairly effectively separated from each other.

The **Australian** region is limited to Australia, Tasmania, New Zealand, the Moluccas, Papua New Guinea, and the smaller islands of this general area of the East Indies and Polynesia. The entire region is a composite of varied ingredients: dry, relatively flat Australia, half tropical and half temperate; humid, tropical, mountainous Papua New Guinea; isolated, temperate, mountainous New Zealand; and numerous isolated, oceanic islands, both tropical and temperate. Forty per cent of the region lies within the tropics. Of the entire Australian region only 15 per cent receives more than 100 mm of rainfall per year (as against 71 per cent of the Neotropical region), and 57 per cent receives less than 50 mm per year (as against 15 per cent of the Neotropical) (Keast, 1972). As a consequence of this relative dryness only 4.5 per cent of the region is covered with tropical rainforest (mostly in New Guinea), 20 per cent is woodland, 23 per cent is savanna, grassland, and scrub, and 30 per cent is desert. About 70 per cent of Australia is covered with desert and this sterile habitat harbors only 17 species of birds, or about 3 per cent of the total Australian avifauna.

Based on their studies on DNA/DNA hybridization, Sibley and Ahlquist (1985) concluded that one group of birds—the parvorder Corvi—radiated in parallel with marsupials and eucalypts in Australia. The Corvi gave rise to three groups (superfamilies). One superfamily, Menuridea, consists of the Lyrebirds, Scrub-birds, Bower-birds, and Australian Treecreepers (*Climacteris* spp.). A second superfamily, Meliphagoidea, is comprised of such birds as fairy wrens, honey-eaters, and thornbills. The third superfamily, Corvoidea, includes such birds as birds of paradise, wood swallows, crows and jays (Corvidea), the Australian robins (Eopsaltridae), and log-runners.

In the late Tertiary, Australia drifted closer to Asia; the Corvi dispersed into Eurasia and radiated even further. A reciprocal invasion of Australia and Papua New Guinea by the Muscicapidae then occurred. Thus, the avifauna of Australia can be said to consist of a large endemic radiation plus elements from the Oriental region.

Notable absentees in the Australian region are flamingos, Old World vultures, pheasants, skimmers, sand-grouse, trogons, barbets, woodpeckers, broadbills, and finches and buntings (Serventy, 1964). Very few Palearctic migrants winter in the Australian region; essentially no migratory passerines reach Australia proper. Parrots, doves, and honey-eaters are perhaps the dominant groups of the region. In and around New Guinea there are more kingfishers than in all the rest of the world combined.

Keast estimates that there are 906 species of birds in the Australian region, and 531 of these in Australia alone. Of the 64 families of terrestrial and freshwater birds represented in the region, the following 13 are essentially unique:

Casuariidae—Cassowaries
Dromiceidae—Emus
Apterygidae—Kiwis
Megapodiidae—Megapodes
Aegothelidae—Owlet Nightjars
Xenicidae—New Zealand Wrens
Menuridae—Lyrebirds
Atrichornithidae—Scrub-birds
Cracticidae—Bell Magpies
Grallinidae—Magpie Larks
Paradisaeidae—Birds of Paradise
Ptilonorhynchidae—Bower-birds
Meliphagidae—Honey-eaters

In addition to these exclusive families, the Australian region has many characteristic pigeons, parrots, cockatoos, frogmouths, wood swallows, cuckoo shrikes, and fairy wrens.

Bird geography may be studied equally well from this standpoint of geographical regions, or from the standpoint of the distribution of families of birds. Such a study, conducted by Barden (1941), revealed that of 144 families (the widely ranging sea birds and several other families were not considered), 33 or almost one-quarter were each represented in every one of Wallace's six geographical regions. Table 20–2, based on Barden and on Bartholomew's *Atlas of Zoogeography* (1911) gives, in order of decreasing ubiquity, examples of the geographical range of various families.

Oceanic Islands constitute a special case of bird geography. As a rule, oceanic islands support fewer species of land and freshwater birds than equal areas of comparable mainland. A given species, however,

is likely to have denser populations, and to occupy broader niches (i.e., be less specialized) than the same species on the mainland. The two most obvious influences on bird distribution are an island's size and its isolation from the nearest source of potential immigrants. The larger the island, the easier the target it makes for possible immigrants and, usually, the greater its ecological diversity to support new colonists. The remoter the island, the more difficult it will be to reach. Gilpin and Diamond (1976), studying the birds on 52 islands in the Solomon archipelago, calculated that about 98 per cent of the variation in numbers of species per island depended on island area and remoteness. Diamond (1973) estimated that a 10-fold increase in an island's area would increase its species diversity nearly twofold.

Combining the two influences of island size (i.e., ecological diversity) and isolation, MacArthur and Wilson (1967) proposed their equilibrium theory of island populations. This holds that the isolation of an island influences the rate of its colonization, whereas its size influences the rate of extinction of those species already there. The smallest islands with the smallest populations obviously run the highest random fluctuation risks of being wiped out by disease, starvation, hurricanes, and the like. Immigration rates increase, of course, with island size and proximity to the mainland. Through the millenia an equilibrium is established between immigration rates and extinction rates so that even though island populations tend to be in a constant state of flux, numbers of species tend to be the same. Obviously, the shifts in balance between immigration and extinction play an intimate role in the evolution of island birds—a topic considered in Chapter 23. Implicit in the MacArthur-Wilson theory is that competition with a superior immigrant causes the extinction ("turnover") of an island resident. Some authors have cautioned that these ideas, stimulating as they are, may not be indicative of the real picture. Extinction of insular species may be precipitated by factors other than competition. Changes in climate toward the end of the Pleistocene, 10- to 12,000 years ago, caused the extinctions of many species on Caribbean Islands, so that species once widespread (e.g., Burrowing Owls, *Athene cunicularia*) are now restricted and disjunct in distribution (Pregill and Olson, 1981).

The endemic Grayson's Dove, *Zenaida graysoni*, is now extinct on Socorro Island on which the mainland Mourning Dove, *Z. macroura*, now abounds. This

TABLE 20–2.

THE GEOGRAPHICAL DISTRIBUTION OF SELECTED BIRD FAMILIES*

FAMILY	NEOTROPICAL	NEARCTIC	PALEARCTIC	ETHIOPIAN	ORIENTAL	AUSTRALIAN
Anatidae—Ducks, Geese, Swans	X	X	X	X	X	X
Accipitridae—Hawks, Eagles	X	X	X	X	X	X
Charadriidae—Plovers, Turnstones	X	X	X	X	X	X
Cuculidae—Cuckoos	X	X	X	X	X	X
Strigidae—Owls	X	X	X	X	X	X
Caprimulgidae—Goatsuckers	X	X	X	X	X	X
Apodidae—Swifts	X	X	X	X	X	X
Alcedinidae—Kingfishers	X	X	X	X	X	X
Corvidae—Crows, Magpies, Jays	X	X	X	X	X	X
Gruidae—Cranes		X	X	X	X	X
Burhinidae—Thick-knees	X		X	X	X	X
Laniidae—Shrikes		X	X	X	X	X
Paridae—Titmice		X	X	X	X	X
Fringillidae—Finches, Grosbeaks	X	X	X	X	X	
Otididae—Bustards			X	X	X	X
Trogonidae—Trogons	X	X		X	X	
Muscicapidae—Old World Flycatchers			X	X	X	X
Troglodytidae—Wrens	X	X	X		X	
Upupidae—Hoopoes			X	X	X	
Bucerotidae—Hornbills				X	X	X
Capitonidae—Barbets	X			X	X	
Pittidae—Pittas				X	X	X
Gaviidae—Loons, Divers		X	X			
Tetraonidae—Grouse		X	X			
Podargidae—Frogmouths					X	X
Trochilidae—Hummingbirds	X	X				
Indicatoridae—Honey-guides				X	X	
Tyrannidae—Tyrant flycatchers	X	X				
Dicaeidae—Flowerpeckers					X	X
Thraunidae—Tanagers	X	X				
Struthionidae—Ostriches				X		
Rheidae—Rheas	X					
Dromiceidae—Emus						X
Coliidae—Colies				X		
Ramphastidae—Toucans	X					
Pipridae—Manakins	X					
Menuridae—Lyre birds						X
Ptilonorhynchidae—Bower-birds						X
Irenidae—Fairy bluebirds					X	
Prunellidae—Hedge sparrows			X			
Chamaeidae—Wren-tits		X				

*After Barden, 1941, and Bartholomew, 1911.

would appear to be a case of island turnover. However, Jehl and Parkes (1983) have persuasively argued that this is not the case. Grayson's Dove is larger and more aggressive than the Mourning Dove, so that extinction resulting from competition is unlikely. The island is now replete with feral cats which have probably caused the demise of this naturally tame island native.

The activities of prehistoric humans may also have precipitated the extinction of species (Olson and James, 1982). All these factors indicate that census of species numbers on islands gives us a somewhat distorted view of turnover rates if we are not aware of the other causes of extinctions in addition to competition.

Other influences that determine colonization of an island are its ecological diversity, food resources, climate, the direction and strength of prevailing winds, the presence or absence of competitors and predators, and the genetic dispersal powers of potential immigrants—some species being much more prone to exploratory wandering than others. Birds that can fly 100 km over land may be very hesitant to fly 1 km across water. The first species to repopulate islands recently defaunated by volcanic explosions, Diamond (1974) characterizes as "supertramps"—species that have high dispersal ability and reproductive potential but low competitive ability.

As evidence that isolation is not necessarily the primary factor in determining the native avifauna of islands, Lack (1969b, 1970) notes that the most isolated islands in the world, Tristan da Cunha and Gough Islands in the South Atlantic, each have no more than four species of land birds, and these birds are derived from South America, 3200 km to the windward. Nevertheless, within two years' time nine vagrant species were recorded on Tristan, and within three weeks, six species on Gough. Ecological diversity rather than isolation is considered responsible for the bird diversity in the Galapagos Islands (Harris, 1973) and the California Channel Islands (Power, 1976).

Competitive exclusion also helps determine the character of island faunas. Lack (1976) suggests that the number of resident species on oceanic islands is small "not because few species have been able to reach them, but because nearly all those that have done so have been excluded through competition with the existing residents." This phenomenon probably accounts for the fact that the European Goldcrest, *Regulus regulus*, occupies the Azores, while the

Firecrest, *R. ignicapillus*, holds sway in the Canary Islands and Madeira. Fifteen species of hummingbirds occur in the West Indies, but nearly every island has no more than one small and two larger species—a situation that cannot be ascribed to chance immigration but rather to the probable fact that all potential hummingbird niches on each island have been filled (Lack, 1976).

DYNAMIC BIRD GEOGRAPHY

A restless world of heaving earthquakes, wandering shorelines, shifting climates, and changing coats of vegetation can scarcely be expected to have sedentary tenants. Consequently, as the environment changes, a species must respond either by adapting to the changed conditions or by moving into a more suitable range—or by dying.

Birds must be reasonably well adapted to their habitats or they will lose out in the relentless, competitive struggle for existence which envelops all organisms. But ecological fitness is not the only explanation of their distribution. The successful transplantation of exotic species, such as the Starling, *Sturnus vulgaris*, to North America, or the Skylark, *Alauda arvensis*, to New Zealand, proves that other factors must be considered. Very often, similar ecological niches in different parts of the world are occupied by quite unrelated species of birds. The nectar-feeding niche occupied by the hummingbirds (Trochilidae) in South America is filled by the sunbirds (Nectariniidae) in Africa, and the honey-eaters (Meliphagidae) in Australia. The insect-eating niche of the tyrant flycatchers (Tyrannidae) of the New World is occupied by the flycatchers (Muscicapidae) in the Old World. Of the 398 species of Old World flycatchers, not a single one is represented in the New World, very probably because potential colonizers were excluded by competition with the native tyrannid flycatchers. Penguins (Sphenisciformes) in the antarctic are represented by the auks (Alcidae) in the arctic. Complete perfection in ecological fitness probably never occurs in nature, partly because the environment never stands still, and partly because birds continue to evolve. Mutations occur spontaneously, and if they are adaptive, they are selected for and perpetuated in the population. Individuals possessing genes enabling them to adapt to the environmental changes are the survivors, and thus pass those genes on to subsequent generations. A bird's physiological idiosyncrasies undoubtedly play a major role in its capacity for emigration. The striking effect on dispersal of differences in incubation constancy between the Crested Mynah, *Sturnus cristatellus*, and the Starling, *S. vulgaris*, was mentioned earlier (p. 341).

Animal distribution, then, is the result of the interplay of two great dynamic agents: the perpetually changing environment and the continually evolving bird. Furthermore, the very geological and climatic changes that shift and isolate existing species over the face of the earth become agents of natural selection through which new species evolve and old species die out. While ecological fitness furnishes the clue to many of the problems of animal distribution, especially on the local level, the historic forces of geology, climatology, and evolution must be invoked to explain many of the large-scale patterns of bird distribution.

Now and then nature performs an experiment through which man can observe in a relatively short time some of these forces at work. In 1883 the East Indian island of Krakatoa was overwhelmed by a tremendous volcanic eruption. Heat and volcanic ash sterilized the island of all life. By 1933, 50 years later, Krakatoa had about two-thirds the number of insect species and about 100 per cent of the bird species found on comparable nearby islands (MacArthur and Connel, 1966). The nearest land from which these plants and animals might have come was 18.5 km away.

Invasion of new territory is only a part of the story; withdrawals from previously occupied range probably occur just as often. A given species' geographical distribution, seen as a time-lapse motion picture—that is, the dynamic view of a species' range—would through the years appear something like a gigantic ameba extending and retracting its lobes as it slowly crawls over the surface of the earth.

A species may change its geographical range as a result of its own activities, or it may be shoved about by environmental forces. Evidence shows that the present map of global distribution is the result of the long-continued operation of both.

ACTIVE BIRD DISPERSAL

As used here, the word *dispersal* means both the outward spread of birds from an established area of residence and also the withdrawal from such an area. Mobility itself is one of the chief forces promoting the wide and rapid geographical dispersal of birds. No other class of animal can match birds in the speed,

ease, and efficiency of their flight—flight that may make accessible even the remotest areas.

Birds, however, vary greatly in the exercise of their powers of flight. In California, both adults and young of the salt marsh Song Sparrow, *Melospiza melodia*, are remarkably sedentary. Thirty-four banded nestlings established their first breeding territories an average distance of only 185 meters from their natal homes (Johnston, 1956). Among 9587 Great Tits, *Parus major*, banded in Denmark, 278 were later recovered, and only 9 of these (3 per cent) had moved more than 5 km from the banding site (Hansen, 1978). At the other extreme are such earth-girdling species as the albatrosses. A nestling Wandering Albatross, *Diomedea exulans*, banded at Kerguelen Island, was found dead one year later, 13,000 km westward in Chile. A South Polar Skua, *Catharacta maccormicki*, banded as a nestling in Antarctica, was later recovered in Greenland (Salomonsen, 1976). An American Black Duck, *Anas rubripes*, banded in Virginia in 1969, was recovered in Korea in 1977 (Banks, 1985). Such birds as these may be the pioneers that open up new range for a species (Grinnell, 1922).

Population pressure is another powerful force that may cause changes in the range of a species. Particularly in species that show territorial behavior, the birds living on the periphery of a species' range may be forced into new areas (p. 242). Expanding numbers may have been the force behind the spectacular recent increase in the range of the Fulmar, *Fulmarus glacialis*, in the British Isles. Previously nesting in Iceland and on the isolated isle of St. Kilda, the species began breeding on mainland cliffs of Ireland and Great Britain around 1880, and today, with a population of about 200,000 birds, the bird's colonies ring the British Isles. Contrariwise, in a species whose numbers are declining, there will be a recession from parts of the originally occupied range.

Withdrawal from a formerly occupied range is commonly forced on a species either by competition with a more successful species or by some change, either in the environment or in the species itself, that may reduce the bird's ecological fitness for that particular area. The Common Murre, or Guillemot, *Uria aalge*, has been shrinking in numbers (as much as 40 per cent in 15 years), and losing geographical range in northern Norway, probably as a result of food shortages and predation on eggs and young by gulls and corvids (Tschanz, 1978). Perhaps because of a gradual change to a warmer and more humid climate in the past century, the Red-backed Shrike, *Lanius collurio*, has given up much of its former range in Britain and Europe (Fig. 20–2). Birds live constantly under such threats of dispossession in their home ranges. Consequently, the most successful species are those acted upon by a mechanism that may promote dispersal into new and untried ranges.

Irruption is one such mechanism. As discussed earlier, certain species resident in the north temperate boreal zones burst out of their traditional range in years of high population and low food supplies to invade new areas. During periods of low temperatures and heavy snows in 1970 and 1981, Great Bustards, *Otis tarda*, from East Germany invaded western Europe and England (Hummell, 1983). Siberian Nutcrackers, *Nucifraga caryocatactes*, have invaded Germany

15 times between 1896 and 1933, each time when the pine-seed crop failed in their home range. A biennial rhythm in unusually heavy fall migrations of Purple Finches, *Carpodacus purpureus*, from central Canada to the southern United States coincided, over 18 years, in every instance but one, with years of poor seed production in Canada (Kennard, 1977). Similar irruptions have been recorded for the sand-grouse, titmice, crossbills, and numerous other species.

In these irruptions one sees a temporary surmounting of the normal barriers of a species' range through an increase in population pressure. Invasions provide a mechanism for the sampling of new ranges, and although they do not normally result in establishing new permanent homes for the species, even rare and sporadic successes can be of enormous significance in the long-range history of a species. As a result of irruptions from the continent, the Red Crossbill, *Loxia curvirostra*, established itself in Ireland in the 19th century and in England in the 20th (Thomson, 1926).

Population pressure also promotes range expansion through what seems to be an inborn tendency in the young of many species to strike out and explore the world in all directions. Every new generation puts some added strain on the traditional habitat for food, nesting sites, and territory; and the younger

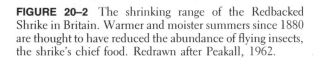

FIGURE 20–2 The shrinking range of the Redbacked Shrike in Britain. Warmer and moister summers since 1880 are thought to have reduced the abundance of flying insects, the shrike's chief food. Redrawn after Peakall, 1962.

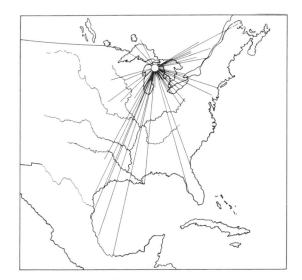

FIGURE 20–3 Young Herring Gulls banded as nestlings at Beaver Islands, Michigan, dispersed during their first year of life as shown by the radiating lines. Older gulls are much more likely to remain sedentary. After Lincoln, 1950.

birds find themselves in unequal competition with the entrenched older ones. The wanderlust of first-year birds is apparently an adaptive device whereby natural selection has met this contingency. For example, nestling Great Tits, *Parus major*, were banded in a study area in Germany covering 120 square km. Few birds settled near their birthplace, and most immigrants came from outside the study area (Bäumer-März and Schmidt, 1985). Tits gather into flocks about one or two weeks after fledging; a rapid explosive dispersal out of the breeding area then ensues (Goodbody, 1952).

Similar results were obtained by Gross (1940) from a banding study of over 23,000 Herring Gulls, *Larus argentatus*, on Kent Island, New Brunswick, where young birds disperse at the end of the breeding season. Subsequent recoveries of 773 (3.3 per cent) of these gulls revealed that the younger birds flew greater distances to their recovery points than the older ones. First-year birds were recovered an average distance of 1380 km from Kent Island; second-year birds, 695 km; third-year birds, 465 km; fourth-year and older birds, 467 km (Fig. 20–4). The young of numerous other species of birds show a similar nomadism. Very probably the yearling birds of a given species are ecologically more adaptable than the older ones, and hence are better equipped to travel, withstand different climates, eat strange foods, and, probably, exploit new range. Older birds, once they have nested in a given location, are more likely to be tied to it by site fidelity and peck-order seniority.

Seasonal migration is quite different from youthful nomadism, but at times it undoubtedly encourages the extension of range. Long-distance migrants are in a better position to discover new habitats, and they are naturally more tolerant of diversity in the environment than are sedentary species. In the mountains of Colombia, such winter visitors from North America as the Yellow-billed Cuckoo, *Coccyzus americanus*, the Rose-breasted Grosbeak, *Pheucticus ludovicianus*, and several warblers have been observed ranging freely through the temperate, subtropical, and tropical life zones, whereas the permanent residents are more rigidly confined in zonal boundaries (Chapman, 1917). Numerous wood warblers (Parulidae) breed in the boreal and temperate zones of the New World and migrate to the tropics for the winter (Keast, 1980). A number of species belonging to the genera *Parula*, *Geothlypis*, *Dendroica*, *Vermivora*, and others have developed both migratory and resident races, the latter living permanently, for example, in Central America, South America, or the islands of the Caribbean. Similar origins account for small African populations of White Storks, *Ciconia ciconia*, that have become non-

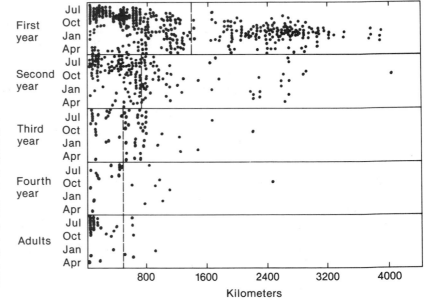

FIGURE 20–4 The dispersal tendency of young Herring Gulls is shown by this scatter diagram based on the recapture of 773 out of 23,434 birds banded as nestlings in New Brunswick. Each dot represents a recaptured bird plotted against its age, the season of the year recaptured, and the distance from its natal colony to the place of recapture. The mean distance of the recovery points from the colony for each age group is indicated by the broken line. Adapted from Gross, 1940.

migratory. Allen's Hummingbirds, *Selasphorus sasin*, have also developed migratory and sedentary populations in California.

Migration may, on the contrary, restrict the dispersal of a species. Darlington (1957) makes the point that much bird migration is an orderly, north and south, annual movement that, as an instinctive routine, prevents certain species from the random scattering in other directions which might otherwise increase opportunities for dispersal.

PASSIVE BIRD DISPERSAL

As airborne creatures, birds are at times passively displaced from their usual haunts by prevailing winds and hurricanes. For birds, as for airplanes, the flight eastward across the Atlantic is much easier than the return westward flight in the teeth of the prevailing westerly air currents. Fifty-two species of North American land and freshwater birds have been recorded in western Europe, usually during the spring or fall migration periods and often during or at the end of periods of strong westerly winds.

Ornithological literature abounds in references to land birds observed from ships far at sea. A record of such sightings, kept over a period of 22 years by naturalists aboard research vessels of the Woods Hole Oceanographic Institution, listed 54 species of land birds. Most of these birds were observed within 320 km of the nearest land, but several of them were seen as far out as 640 km (Scholander, 1955). As a rule, these birds were observed during migration periods, particularly in the fall, and generally at times when there were strong offshore winds. Many of the birds alighted on the ships, exhausted. A list by Fitter (1969) of all known American vagrant species identified in Britain totaled 49 species, including 3 of gulls and terns, 5 freshwater birds, 17 shore and marsh birds, 1 cuckoo, 1 goatsucker, and 22 passerines. Since then 3 more passerine species have been identified, making a total of 52 species. Summarizing records from 100 steamship crossings of the north Atlantic made between 1951 and 1962, Durand (1972) identified 58 North American and 9 European land bird species that had alighted on ships far at sea.

Numerous other observations have shown that various land bird species have been assisted in crossing oceans by ships. Until about three decades ago many ornithologists believed that most if not all records of American species found wild in Europe arrived there with such aid. Recent evidence, however, suggests that probably the majority of such vagrants cross the ocean under their own power assisted only by winds. For one thing, European arrivals of American species often coincide with strong westerly winds or even hurricanes, and most commonly during periods of migration. In late September, 1968, for example, an unprecedented invasion of vagrant American birds reached Ireland and Britain accompanied by record numbers of Monarch Butterflies, *Danaus plexippus*, on the heels of gale-force westerly winds (Burton and French, 1969).

Perhaps the best evidence of independent transocean flight by birds comes from weather ship observations. In mid-September, 1959, during a period of strong offshore winds, two Ruddy Turnstones, *Arenaria interpres*, two Kestrels, *Falco tinnunculus*, and 16 passerine birds of eight species appeared near or on a weather ship anchored in the Atlantic 640 km west of Ireland and 1300 km south of Iceland (McLean and Williamson, 1960). Between 1950 and 1961 a weather ship anchored midway between Norway and Iceland collected 654 migratory birds of 58 species, including the tiny Goldcrest, *Regulus regulus* (Hjelmtveit, 1969).

Although east-to-west crossing of the Atlantic by birds is less common because of prevailing westerly winds, Tuck (1971) records the capture in Newfoundland of 15 species of land and freshwater birds that had been banded in either Greenland, Iceland, or Europe. On occasion there are mass flights of European birds to North America. In December, 1927, hundreds of Lapwings, *Vanellus vanellus*, suddenly appeared in Labrador and Newfoundland, and a similar but smaller flight occurred in 1966 (Bagg, 1967). Both flights were preceded by cold weather in Europe, which caused a mass movement of Lapwings toward Ireland, followed by an intense low pressure area across the Atlantic that caused easterly gales.

There is persuasive evidence that wind-driven birds may at times enlarge a species' range. The island of South Georgia is 1600 km due east and leeward of Tierra del Fuego, in the belt of the Roaring Forties. There are two endemic land birds there: a pipit, *Anthus antarcticus*, and a teal, *Anas georgica*, whose nearest relatives live in Argentina directly to the windward (Murphy, 1936). Thirteen species of passerines introduced into New Zealand and Australia have established themselves on nine islands lying between 160 and 880 km off shore. It is thought that strong winds were largely responsible for their dispersal (Williams, 1953). In the past century ten Australian species have established themselves in New Zealand, 1600 km leeward in the Roaring Forties (Lack, 1969).

Wind storms are notorious transporters of birds. There are 21 records of the American Purple Gallinule, *Porphyrula martinica*, occurring in the southwest Cape Province of Africa. Most of these are juveniles apparently caught by strong westerly winds as they begin their northward migration in Argentina, and carried downwind to South Africa (Silbernagl, 1982).

At least one historic range expansion can be attributed to a specific storm. On the night of January 19, 1937, a flock of European thrushes known as Fieldfares, *Turdus pilaris*, was migrating from Norway toward England, probably to escape a sudden cold spell. They were caught in midcourse by strong southeasterly gales and carried to Jan Mayen and the northeastern coast of Greenland. At both places, the birds were observed and collected on January 20. By the end of January, the birds on Greenland had moved to its southern tip, where they have since established a sedentary population (Salomonsen, 1951). There seem to be strong potentialities in this species for invading and colonizing the North American continent. It has already been sighted on four separate occasions in North America (Threlfall *et al.*, 1973) and is currently expanding its range in central Europe.

Suggestive evidence that drought may cause range expansion in some species is seen in the fact that 230 Pintail Ducks, *Anas acuta*, banded in North America, were recovered in Siberia, particularly during years when drought conditions caused the birds to overfly their traditional United States and Canadian breeding grounds (Henny, 1973).

There is little evidence that water currents assist in the distribution of birds, but the possibility should not be ruled out. Penguins are typically antarctic birds, but the Galapagos Penguin, *Spheniscus mendiculus*, lives on the equator at the northern extremity of the cold, north-flowing Peru Current. In this case, the temperature of the water and its food resources were probably more responsible for this penguin's unorthodox distribution than the current itself.

Man himself has been an important agent in the passive dispersal of birds. The imported European House Sparrow, *Passer domesticus*, Starling, *Sturnus vulgaris*, and Chinese Ring-necked Pheasant, *Phasianus colchicus*, have succeeded phenomenally in establishing themselves in North America. Around 1865 four species of European finches were introduced into New Zealand. Within 30 years they had occupied the entire country and had even spread to several islands as far as 800 km away (Newton, 1972). Wild populations of several species of tropical parrots, doves, various finches, and other exotic species breed in the warm climates of Florida and southern California (Owre, 1973; Hardy, 1973). In the Hawaiian Islands, some 50 introduced species have established themselves, unfortunately at the expense of about two-thirds of the original native species.

FIGURE 20–5 The European Fieldfare, a close relative of the American Robin, was, in a 1937 storm, driven from Norway to Greenland, and has since established a colony there. It seems very possible that in future years this species may invade the North American Continent. Photo by E. Hosking.

BARRIERS AND ROUTES OF DISPERSAL

In spite of their superior powers of flight, warm-bloodedness, and high reproductive potential, certain birds do not exist everywhere on earth that conditions are ecologically tolerable for them. Obstacles of different sorts and of variable powers of exclusion prevent a species from expanding its range into ecologically suitable regions. These barriers may be physical, spatial, biological, psychological, or temporal. A mountain can be a barrier to a plains bird, an ocean to a land bird, a continent to a sea bird, and a prairie to a forest bird. A high mountain is a stronger barrier than a low one, a broad ocean stronger than a narrow strait. Wallace's zoogeographical regions are all surrounded by major physical barriers such as the oceans, the Sahara desert, or the Himalayan mountains.

Physical Barriers

High mountains are obvious barriers to bird dispersal. Their crests frequently offer obstacles in the form of cold, thin air, little oxygen, snow and ice, or rocky soil, little food and cover, and vertical height itself, which requires great exertion to overcome. High mountain ranges commonly mark the distributional limits of bird populations.

As physical barriers, deserts offer such obstacles to dispersal as rapid and extreme temperature fluctuations, lack of water, a high rate of evaporation, lack of food and cover, and dust storms. Such a sterile complex of ecological factors offers little attraction to a species to enlarge its range. Even flying across a desert may be hazardous. White Storks, *Ciconia ciconia*, migrating in large flocks from Europe to their winter quarters in Africa, sometimes die by the hun-

dreds in the Sinai or Egyptian deserts bordering the Red Sea, especially during wind storms (Schüz *et al.*, 1963).

Since very few barriers are absolute, their effects on bird distribution are selective and retarding rather than completely prohibitive. Only the more vigorous, mobile, adventurous, lucky, or ecologically tolerant species or individuals are able to surmount certain barriers to achieve an enlarged range. Thus, barriers commonly act as filters. Since some of the more difficult barriers may be overcome only on rare occasions when all contributing conditions are "right," it may be thousands of years before a given species may leap a given barrier and appropriate a new range. This is illustrated by the fact that few land birds reside on isolated oceanic islands, but even more significantly by the fact that no habitable oceanic island is so remote that some species of land bird cannot reach and colonize it (Darlington, 1957).

Of the great physical barriers, the oceans are probably the most effective in restricting land birds within their home ranges. Until man intervened, it was the Atlantic Ocean that kept the House Sparrow, *Passer domesticus*, from the Americas. One hundred individuals were released in New York in 1852 and 1853 (Robbins, 1973). Many other releases followed. Like other successful invaders of new territory, the House Sparrow increased its numbers and its range slowly when first introduced, but later increased markedly the pace of its invasion. By 1910 it had reached all the states except central Nevada, southern California, and the Northern Rockies.

During a period of forty years the Collared Turtle Dove, *Streptopelia decaocto*, expanded its range

FIGURE 20–6 The Collared Turtle Dove, a species that has recently been expanding its range in Europe with unmatched speed. Photo by E. Hosking.

with explosive speed, spreading some 2400 km northwest across Europe from Yugoslavia to Britain, the Outer Hebrides, and Sweden, sometimes advancing as much as 240 km per year. The species seems spectacularly well adapted for exploiting its new range, and breeds within a year or two of its arrival in new regions. It has been breeding in Britain since 1955 (Hudson, 1972). Few, if any, other species have ever been known to take over new range as rapidly. It is not known what barriers restrained this species from its dramatic dispersal until so recently. Mayr (1951) suggests that this expansion in range was initiated by genetic changes in peripheral populations of the species.

Another species which has displayed phenomenal extension of its range is the Old World Cattle Egret, *Bubulcus ibis*. Perhaps assisted by the northeast trade winds, this bird crossed the Atlantic from Africa to Central or South America late in the 19th century. In Florida its first known breeding was by seven pairs in 1952. Ten years later this "seeding" became a colony of about 4000 birds. It arrived in Texas in 1954, and by 1979 there were 300,000 breeding pairs (Telfair *et al.*, 1984). By 1972 this species became the most plentiful egret in North America, nesting in all but ten of the continental United States and also in Ontario. It has been observed in Newfoundland, Alberta, and even in the Northwest Territories (Crosby, 1972; Fogarty and Hetrick, 1973; Kuyt, 1972). Currently the Cattle Egret is also spreading in South America, Africa, Australia, and New Zealand.

In its native home the Cattle Egret is a commensal of African elephants, buffaloes, and antelopes. As it invaded other continents it found a ready-made, unoccupied niche as a commensal of domestic cattle.

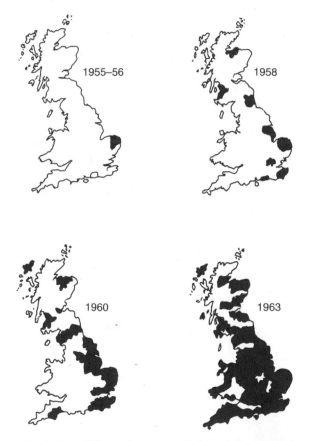

FIGURE 20–7 The explosive spread of the Collared Turtle Dove in Britain. In only 30 years this species expanded its range by a radius of 2400 kilometers. By 1970 it was found breeding in all counties in England. Redrawn after Hudson, 1965.

FIGURE 20–8 The Cattle Egret is currently invading and establishing new range in North America. The Cattle Egret and the Fieldfare are the only Old World species known to have established themselves in the New World within historic times without man's aid. Photo by E. Hosking.

Unlike other herons and egrets, it specializes in terrestrial insects, especially orthoptera. It is apparently the only egret that breeds when only one year old and that has two breeding cycles per year, at least in the tropics (Lowe-McConnell, 1967). The Cattle Egret and the Fieldfare are the only species of birds known historically to have established themselves in the New World as breeding birds without human help.

Without question, the surmounting of barriers and the invasion of new ranges will continue indefinitely as long as birds evolve and environments change. For certain species one can anticipate possible future changes in residence. The larks (Alaudidae) of North America, dwellers of plains and open country, have begun to filter into South America via the Andean paramos. If some of them can cross either the great Brazilian rain forests or the high Andes mountains, and reach the extensive pampas of Argentina, they should find an ecological paradise awaiting them (Barden, 1941). The Brown Pelican, *Pelecanus occidentalis*, of the Caribbean is a sea bird that does not venture far from shore. It feeds on fish that it sights from the air and then seizes in plunging dives. The muddy waters pouring out of the mouths of the Amazon make this method of feeding fruitless, and have prevented the extension of the pelican's range southward.

Even to a sea bird the ocean presents barriers to dispersal. The story of "El Niño," related earlier, reveals the striking effect that temperature may have on bird distribution. There are warm-water species and cold-water species. The Royal Tern, *Thalasseus maximus*, a warm-water bird, is bottled up in the Pacific Ocean within 30 degrees of latitude, between the cool south-flowing California Current and the chilly north-flowing Peru Current. But in the Atlantic, thanks to the warm Gulf Stream and Brazil Current, its range covers 70 degrees of latitude, extending from Florida to Argentina (Fig. 20–9). The White-chinned Petrel, *Procellaria aequinoctialis*, on the other hand, is tied to cold surface waters and is sandwiched between the antarctic pack ice and the Equatorial and Brazil Currents. Similarly, the antarctic Snow Petrel, *Pagodroma nivea*, is a bird of ice floes and cold water. The 12°C surface water isotherm marks both the northern limits of its range and that of the opossum-shrimp, its most abundant food source (Murphy, 1936).

Although some species seem to be directly limited by temperature boundaries, temperature is probably of greater importance indirectly, through controlling the food supplies of a given range. According to Stresemann (1927–1934) the decisive factors in determining the geographical distribution of birds are usually climate and the nature of the soil, these two elements accounting for minerals, food, vegetation, cover, nesting sites, and other requirements.

As described earlier, the interplay between birds and climate involves many and diverse ecological relationships, most of which are important for distribution. The effects of winds have already been mentioned. Of other climatic factors important in distribution, only a few examples can be given in the space available.

A hundred years ago in southern Sweden, the Hooded Crow, *Corvus cornix*, and the European Blackbird, *Turdus merula*, were migratory harbingers of spring. Today they are both all-year residents. This change in status is due to the warmer springs and milder winters that this Scandinavian peninsula has undergone since the 1880's. A parallel shift in range has occurred in certain sea birds. The Herring Gull, *Larus argentatus*, ranged northward to Iceland by 1927, to Bear Island by 1932, and to Spitzbergen by 1950. This warming trend, paralleled in many parts of the north temperate zone, is attributed to changes in the Azores high pressure area and the Icelandic low pressure mass, which permit an increased northward flow of warm air. As a consequence, there has been in recent years a striking northward expansion of range by resident and migratory birds (Hustich, 1952).

In arctic Siberia, not only have some 50 species of birds extended their ranges northward, but in certain areas the northern limit of trees has advanced on to the tundra 200 to 700 meters per year (Uspenskii, 1969). In 1900 only 33 species of birds were known to be year-around residents of the Connecticut Valley in New England. Today 70 species may be found there throughout the year; among them are 13 new permanent residents including the Cardinal, *Cardinalis cardinalis*, Carolina Wren, *Thryothorus ludovicianus*, and the Song Sparrow, *Melospiza melodia* (Boyd and Nunneley, 1964). During the last century in Finland, about one-third of the bird species increased in numbers and/or extended their ranges northward, while about one-fourth exhibited the opposite tendencies. The amelioration of climate has undoubtedly influenced these changes, but von Haartman (1978) points out that while the mean isotherms have advanced locally about 1¹/₂ degrees northward in latitude in the past century, many species of birds have extended their ranges northward much more: the Lapwing, *Vanellus vanellus*, by at least .8 degrees; the

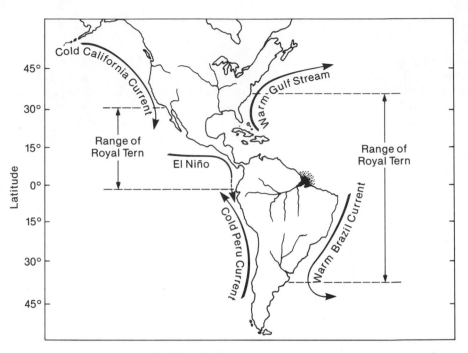

FIGURE 20–9 Warm-water birds like the Royal Tern are restricted to a narrow range along the Pacific shores of the Americas because of the flow of cold ocean currents toward the equator. The outward-flowing warm currents in the Atlantic more than double the width of this species' range. The turbid water at the mouth of the Amazon restricts the Brown Pelican to the Caribbean Sea. Periodic invasions of warm water by "El Niño" along the west coast of South America have disastrous effects on fish and water birds. Adapted from Murphy, 1936.

European Blackbird, *Turdus merula*, by 6 degrees. He believes that such faunal changes in range are more likely the result of man's cultural impact on the environment (e.g., deforestation, farming, urbanization) and the increase in the number of bird observers than in the warming trend in climate (Fig. 20–10).

Directly or indirectly the low temperatures of arctic or subarctic winters are the chief restraining agent in limiting the northward range expansion of many birds. There is of course great variation in the cold-hardiness of different species. Rock Ptarmigan, *Lagopus mutus*, for example, survive the winter in northern Greenland (83 degrees N. latitude) by burrowing under the snow and feeding on twigs and buds, even during the long arctic nights and when temperatures may drop to −40°C. On the other hand, of the tiny Goldcrests, *Regulus regulus*, that winter in northern Norway (70 degrees N. latitude), only about

10 per cent survive until spring. In some species cold-hardiness seems to be a physiological trait sensitive to natural selection. In less than a century the introduced House Sparrow, *Passer domesticus*, has evolved local populations with significantly different cold-hardiness. Fully acclimated specimens from Florida and Arizona (28 and 32 degrees N. latitude) survived to mean lethal temperatures of −23.5° and −25.2°C, while sparrows from Winnipeg and Churchill, Manitoba (50 and 59 degrees N. latitude) survived to −30.8° and −31.4°C (Blem, 1973).

Sometimes it is the length of winter rather than its coldness that limits range. Several species of grouse (Tetraonidae) feed on leaves, seeds, insects, and grit in late summer and early fall. In late fall when the snow arrives they change their diet to buds, twigs, or conifer needles that are ground by the grit. Should early fall or late spring snows prolong the duration

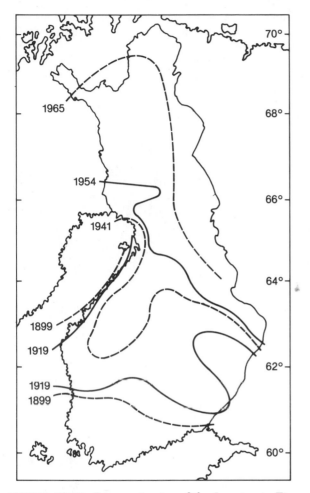

FIGURE 20–10 Range extension of the Lapwing in Finland, probably as a consequence of warming climate and human modification of the environment. After Kalela, 1949, and von Haartman, 1978.

of snow groundcover, the birds may perish from starvation in late winter even though food is abundant, because their supply of grit is worn out and additional grit is not yet available (Udvardy, 1969).

The effects of temperature on geographical range are not necessarily limited to adult birds. Newly hatched ducklings of 10 different species were exposed to near-freezing temperatures for varying time intervals. It was found that some ducklings, e.g., Eiders, *Somateria mollissima*, maintained normal high body temperatures for as long as 18 hours, whereas in other

species, e.g., Mallards, *Anas platyrhynchos*, temperature regulation collapsed instantly. It was found that there was an almost perfect correlation between the cold-hardiness of the young and the geographical range of the species (Koskimies and Lahti, 1964). The length of subarctic summers also affects geographical distribution. In northern Finland, 60 per cent of the summers are now long enough to allow cygnets of the Whooper Swan, *Cygnus cygnus*, to grow to fledging stage before the lakes and ponds freeze over. Between 1950 and 1970 the breeding population increased 20- to 30-fold (Haapanen *et al.*, 1973).

In arid regions, such as South Africa, rainfall may occur in spotty patterns. Many breeding birds will shift their residence from year to year to coincide with the more humid regions where vegetation and food are more abundant. Among the birds of Nepal, Ripley (1950a) found that many species existed in two races, an eastern and a western Himalayan race, respectively, and that the line separating these close relatives was probably the isohyet near the 87th degree of longitude, east of which the annual rainfall is about 190 cm and west of which it is only 125 cm.

Even fog may determine the geographical range of a species. Investigations by Hawksley (1957) of breeding colonies of Arctic Terns, *Sterna paradisaea*, in the Bay of Fundy, showed clearly that young chicks gained weight on clear days but generally lost weight on foggy days, which restricted the ability of adult birds to provide food. The absence of this species as a breeding bird on the Bering Sea shores of Siberia is probably due to the frequent fogs there.

Weather disasters may remove a bird from part of its established range. Persistent cold rains in 1903 practically eliminated Purple Martins, *Progne subis*, from northern New Jersey to southern New Hampshire, and this lost range was not recaptured for many years (Griscom, 1941). Severe winters of 1976–77 and 1977–78 drastically reduced populations of Ring-necked Pheasants, *Phasianus colchicus*, in central Illinois (Warner and David, 1982).

Length of day, which becomes progressively more variable as one moves away from the equator, probably influences the geographical breeding ranges of many species. The assumption of breeding plumage and behavior, and the timing of the sex cycle in many species are closely related to day length. This relationship is a hereditary matter originating in an adaptation to a specific latitude and seasonal light pattern. The possession of relatively fixed photoperiodic breeding responses very likely prevents the successful spread of some species into latitudes with pronouncedly different daylight patterns.

There are still other ways in which day length affects bird geography. As winter sets in, in temperate and polar latitudes, temperature decreases and day length shortens. At colder temperatures birds need more food to compensate for increased heat loss, but they have shorter days in which to gather it. At −18°C, for example, the Yellow-billed Magpie, *Pica nuttalli*, must spend 100 per cent of its time in feeding to keep warm (Verbeek, 1972). Either shorter days or

FIGURE 20–11 An Emperor Penguin and its chick. This largest and most southerly of penguins lays, incubates, and hatches its single egg in the frigid depths of the antarctic winter—a physiological feat without parallel in the animal kingdom. Photo by Expéditions Polaires Françaises.

colder temperatures will effectively restrict its range. The Nightjar, *Caprimulgus europaeus*, is a twilight-niche feeder. It feeds, ordinarily, neither by broad daylight nor after dark. In Norway the northern limit of its range is 64 degrees to 65 degrees N. latitude. Beyond that boundary the midsummer "midnight sun" keeps the nights too bright for this bird's special way of feeding (Lehtonen, 1972).

A widespread, continuous population of a given kind of bird may, through many years of natural selection by physical factors, be broken up into geographical populations of graduated types. This is particularly true for the climatic factors of temperature and humidity. Species and races of birds that live in colder climates commonly have larger bodies than their relatives living in warmer climates. For example, Dark-eyed Juncos, *Junco hyemalis*, at higher latitudes (Michigan and Indiana) were significantly heavier than birds from more southern localities, for example, Alabama and Mississippi (Nolan and Ketterson, 1983). This principle is known as Bergmann's Rule, and it has been found by Rensch (1936) to hold true for about 90 per cent of the geographical races of nonmigratory Palearctic species (Fig. 20–12). Showing rather remarkable evolutionary plasticity, northern populations of the introduced House Sparrow, *Passer domesticus*, in North America, are now larger than southern populations (Johnston and Selander, 1971).

FIGURE 20–12 Hairy Woodpeckers of Canada *(left)* and Costa Rica *(right)* illustrate Bergmann's Rule that warm-blooded animals living in the cooler, higher latitudes are likely to have larger bodies that conserve heat more effectively than small bodies.

For the Chickadee (or Willow Tit), *Parus atricapillus*, Rensch (1947) found that the geographical distribution of body size paralleled very closely the mean January isotherms of western Europe, larger bodies correlating with lower temperatures. January is the most critical month of the year for the survival of small species such as these. Presumably birds with relatively smaller bodies are the first to die of exposure to cold in bitter winter months, leaving the increasingly larger types to populate the increasingly colder zones. Bergmann's Rule does not apply with equal faithfulness to migratory species, since they are not subject to this extreme form of selection.

Geographical gradients, or clines, are found in other attributes of widely distributed species. Allen's Rule states that birds that live in colder regions generally have relatively shorter beaks, legs, and wings than their nearest relatives in warmer regions. Such projections lose heat more rapidly than the main bulk of a bird's body. A difference of 1 per cent in wing-length corresponds to a difference of 2 degrees N. latitude in Redpolls, *Carduelis flammea*, of just over 1 degree in Puffins, *Fratercula arctica*, and a little over 0.5 degree in European Wrens, *Troglodytes troglodytes* (Huxley, 1942).

Pigmentation likewise shows widespread geographical gradients. Gloger's Rule states that races of birds (and mammals) that live in warm and humid regions have darker pigmentation than races of the same species that live in cooler or drier regions (Fig. 20–13).

The clutch-size rule, referred to earlier, points out that races of species living at higher latitudes lay more eggs per clutch than races at lower latitudes (p. 326). Clutch size is significantly higher in the introduced populations of House Finches, *Carpodacus mexicanus*, in eastern North America than in the ancestral western populations (Wooton, 1986). In this case the differences are not a consequence of latitude, but possibly the result of higher predation rates and population densities (and reduced food supply) in the west.

Races of birds living in cooler climates are more likely to be migratory than warm-climate relatives. For many species this is a simple matter of race-survival. Recalling the fact that many birds have more feathers in winter than in summer (p. 35), it seems logical to expect that sedentary cold-climate races should have a greater number of feathers per bird than their warm-climate relatives, but this point, apparently, has not been established. However, Irving (1960) found that the tips of contour feathers of winter residents of arctic

FIGURE 20–13 Gloger's Rule as illustrated by the Goshawk. At the left is the nearly white northerly race, *Accipiter gentilis buteoides*, from western Siberia; at the right, the heavily pigmented race, *A. g. melanoleucus*, from East Africa. After Kleinschmidt, 1958.

Alaska are fluffier than those of summer residents. In cold weather, when a bird erects its feathers, the soft tips interlock to imprison an insulating layer of warm air, whereas the stiff-tipped feathers of migrants readily separate and lose the insulating air.

Biological Barriers

Since green plants are the ultimate source of all food, the ranges of many birds depend almost entirely on the distribution of plants. In addition, birds commonly require nesting sites, nesting materials, song posts, and protective cover that are provided by plants. Of course the prey of insectivorous and predatory birds depend also upon plants.

Certain bird species are closely tied to one or a few plant species. Some hummingbirds have bills of such a shape that they can sip nectar only from particular flowers. Certain crossbills, nutcrackers, titmice, and kinglets are closely restricted to evergreen forests: for instance, the European Nutcracker, *Nucifraga caryocatactes*, prefers conifer forests in which the Arolla pine occurs. The African vulture, *Gypohierax angolensis*, lives mainly on the fruits of the oil palm *Elaeis*, and consequently its geographical range coincides with that of the palm. The Palm Swift, *Cypsiurus parvus*, is chiefly limited to the range of the fan palm, *Barassus flabelliformis*, on whose leaves the bird cements its nest (Stresemann, 1927–1934). Narrow dependencies such as these are invitations to extinction.

Sea birds, no less than land birds, are limited in their distribution by food resources. In the seas, however, food distribution depends on somewhat different environmental controls than is the case on land. Also, sea birds commonly use feeding techniques quite different from those of land birds. As an example, several species of albatrosses of southern seas circle the globe in their constant search for food. No land birds make comparable feeding travels.

The basic food producers in the oceans are microscopic plankton organisms, primarily algae. The carbon compounds (food) that are photosynthetically fixed by plankton pass through a chain of consumers that commonly ends in birds. Therefore, sea birds are likely to be abundant where plankton is abundant. In the south Atlantic Ocean, plankton organisms vary from about 5000 per liter in equatorial waters to about 100,000 per liter at 55 degrees S. latitude. Sea birds are more than 7 times as abundant at 55 degrees S. latitude as at the equator (Hentschel, *in* Murphy, 1936) (Fig. 20–14).

The deep, cold waters of the seas constitute an enormous reservoir of plant nutrients, particularly nitrogen and phosphorus. But they are of no avail for photosynthetic food production except near the surface where sunlight can penetrate. Once the illuminated surface layers of water are exhausted of their nutrients by photosynthesizing plankton, they are not immediately recharged with nutrients from the dead and decaying bodies of plankton organisms or their consumers. The majority of them drift to the bottom of the sea where their fertilizing minerals may be unavailable for centuries. This is the great difference between cycles of nutrients in the sea and on land where the nitrogen, phosphorus, and other minerals become almost immediately available for re-use.

The photosynthetic fixation of carbon compounds by plankton varies from less than 100 to more than 500 milligrams of carbon per square meter of surface water per day (Ashmole, 1971). Large areas of tropic and subtropic seas are very low in food productivity despite the intense solar radiation they receive. The sun's rays heat and stratify the surface water to such an extent that it is effectively prevented from mixing with

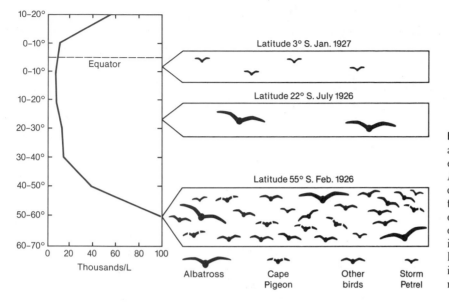

FIGURE 20–14 The relative abundance of sea birds at different latitudes in the pelagic South Atlantic as related to the abundance of plankton. The graph at the left shows the mean abundance of plankton in the upper 50 meters of the ocean in thousands of organisms per liter (after Hentschel). Relative bird abundance, at right, is after Spiess. Both diagrams are modified after Murphy, 1936.

the lower, colder, nutrient-rich water. In such regions sea birds are scarce. But high and continuous nutrient productivity will occur along shores of continents and islands where prevailing winds create surface currents that stimulate up-welling of deep waters, or in the open seas where great currents of waters with different temperatures or salinities meet and create deep turbulence. The circumglobal "antarctic convergence" at about 60 degrees S. latitude is such a region. It is exceedingly rich in bird life. Off the southeast coast of the Arabian peninsula the southwest monsoon winds from May to September cause a sharply seasonal up-welling of deep water that results in a seasonal abundance of plankton and sea birds (Bailey, 1966). The rest of the year the seas of this area are essentially a watery desert.

Such facts as these help explain the drastic effects of "El Niño" in the highly productive Peru Current off the west coast of South America. In normal years prevailing winds cause the up-welling of cold, nutrient-rich waters along most of this South American coast. Year-in and year-out, these waters support the photosynthetic fixation of up to 200 mg of carbon per square meter of surface water per day, as compared with about 15 mg in adjacent warm waters. Most of this carbon becomes stored in an annual production of 15 to 20 million metric tons of Anchovies, *Engraulis ringens* (Idyll, 1973). The diet of the guano

birds of Peru—the Guanay Cormorant, *Phalacrocorax bougainvillei*, Peruvian Booby, *Sula variegata*, and Brown Pelican, *Pelecanus occidentalis*—is 80 to 95 per cent Anchovies, and the birds annually eat about 2.5 million tons of them. In 1956–57 the guano bird population exceeded 27 million birds. In 1972 a devastating cycle of El Niño displaced for 11 months the normally cold (22°C) Peru Current with surface water of 30.3°C. The Anchovy population declined drastically (partly, also because of excessive commercial fishing) and the guano birds starved until their total population was reduced to about one million birds (Idyll, 1973).

Plants may either promote or hinder the dispersal of a species. Early in this century extensive fires in the western part of the upper peninsula of Michigan burned many openings in the solid growths of forest. The grass and shubbery that soon clothed these openings provided a suitable habitat for Sharp-tailed Grouse, *Pedioecetes phasianellus*, with the result that this species quickly expanded into this new range (Baumgartner, 1939). Changes in the general character of the vegetation of a given region almost inevitably are followed by changes in bird distribution. In Fiji, the distribution of the Red-vented Bulbul, *Pycnonotus cafer*, coincides with the distribution of three weed species that provide its main food supply (Watling, 1978). Birds, in turn, often affect the distribution of plants through the dispersal of their seeds. And, as was discussed in Chapter 6, certain plants, especially those of high-altitude-cloud forests, depend on hummingbirds that feed on their flowers for much of their reproductive success and distribution.

Other animals likewise may affect bird distribution. As discussed in the previous chapter, many species of birds are associated with other animals as commensals, symbionts, competitors, brood-parasites, predators, nest-cavity users, and so on. A tick bird species probably adjusts its geographical range to that of the ungulates it grooms; a cuckoo, to the range of its preferred host; a falcon, to the range of its preferred prey; an antbird, to the range of army ants that stir up the insects it feeds on. The House Sparrow, *Passer domesticus*, which has been a resoundingly successful introduction in North and South America, has been slow in colonizing Africa because its niche there is already occupied by other Ploceidae (Sick, 1968). There are no ducks whatever on Lake Edward in the Congo, and geese seem reluctant to swim there, probably because of the presence of large carnivorous fish (Lippens, 1938).

Other Barriers

In addition to the physical and biological barriers to bird dispersal, there seem to be psychological barriers.

Some land birds are capable fliers but are nevertheless stopped by moderate water gaps. The Central and South American cracids, puffbirds, toucans, ovenbirds, and manakins have failed to populate the West Indies; and the quail, trogons, and cotingas have reached only one or two islands (Darlington, 1957). This fact may result simply from the reluctance of these birds to fly across water. Races of South American barbets and manakins are similarly isolated from each other by the broad tributaries of the Amazon. Of the 325 species of Papua New Guinea lowland birds, 134 have failed to reach and populate any nearby oceanic island, despite the fact that many of them, like the Harpy Eagle, *Harpyosis novaeguineae*, and Spinetailed Swift, *Chaetura novaeguineae*, fly scores of miles each day in search of food. Diamond (1976) suggests that these "strong overland fliers apparently have insuperable psychological barriers to crossing water gaps." In contrast, the New Guinea Reef Egret, *Egretta sacra*, and Scrub Fowl, *Megapodius freycinet* "have colonized virtually every tropical island within hundreds of miles of New Guinea. . . ."

Theoretically, the number of land bird species found on an oceanic island varies inversely with the island's remoteness from the nearest mainland center of dispersal. There are variables other than remoteness, of course, including latitude, wind direction, size of island, habitat niches available, and extinction caused by man. Although the Hawaiian Islands are some 3200 km from the nearest mainland, Mayr (1943) estimated that they were invaded and colonized by land or freshwater birds on 14 separate occasions. Ten of these colonizations gave rise to endemic species (Olson and James, 1982). These include several spectacular extinct forms known only from recent fossil discoveries, e.g., an eagle, at least one *Accipiter* hawk, two flightless ibises, seven geese, including a large flightless species, *Tambetochen chauliodons*, and others. Olson and James believe that habitat modification and predation by prehistoric Polynesians caused the demise of many avian species including those listed above. The fossil record also tells us that many Hawaiian species with restricted ranges today once were more widespread. The past history of a species must therefore be taken into account in any attempts to explain current avian geography.

HISTORICAL BIRD GEOGRAPHY

A ceaseless ebb and flow probably characterizes the range pattern of nearly every species of bird, particularly when it is considered over a long period of time. In the sometimes gradual, sometimes sporadic, shifting of bird ranges over the surface of the earth, changes of different sorts may occur. A species may extend its range or withdraw from a range previously occupied. It may increase its numbers or its dominance within its range, or decrease them. It may decrease its numbers to the vanishing point and become extinct. Finally, a species may evolve into a different race or species, especially when it exists in small, isolated populations. Considering the constant interaction between evolving birds and changing environments, it is inevitable that geographical ranges change.

An exact study of historical bird geography is a practical impossibility because so many complex and unpredictable variables are involved: evolution, geologic change, climatic change, shifting vegetation, weather disasters, epizootics, rapid dispersal, slow dispersal, ecological adaptation, competition between species, barriers. The fossil record of ancient distributions of birds may sometimes throw light on historical ranges of species (Olson, 1982).

Since birds often invade a new range with explosive

rapidity—as measured in geologic time—their routes of dispersal cannot readily be reconstructed. The fossil record does show, however, that certain species once lived in areas outside their present range: for example, parrots in Nebraska, England, and Germany, and flamingoes in South Dakota. It also suggests that "certain groups of birds have always been confined to certain places: elephant birds to Madagascar, moas to New Zealand, penguins to southern and auks to northern parts of the world" (Darlington, 1957).

Beyond such facts as these, the past history of bird distribution must be reconstructed by the use of inferences based on present distribution (remembering that the absence of certain bird groups may be as significant as the presence of others), and on the geological history of changes in glaciation, climate, sea level, land bridges, and water gaps between land masses.

One of the most striking facts revealed by Wallace's map of zoogeography is the correlation between the degree of uniqueness of birds of certain continents and islands and the duration of isolation of these pieces of land from the rest of the world. The northern hemisphere is roughly the land hemisphere of the world, and possibly the site of the origin of birds. The outposts of the great Eurasian and North American land masses, best seen on a polar projection map (see Fig. 20–1), are New Zealand, Australia, Africa, Madagascar, and South America. Each of these regions is noted for the large number of resident species that are found nowhere else on earth today. This fact is a logical consequence of the long-continued isolation of these regions either by climatic barriers, like that of the Sahara desert, or by broad water barriers.

There is currently much interest in the principle of continental drift and the role it may have had in the distribution of birds. Continental drift, or plate tectonics, refers to the fact that the earth's continents are not rigidly fixed in place but slowly move with reference to one another, a few centimeters a year, as a consequence of convection currents of molten minerals in the earth's underlying mantle. Whether continental drift has had a significant impact on the geographical distribution of birds is a controversial question. Proponents of the theory postulate that a single supercontinent, Pangea, existed at the end of the Permian period, about 220 million years ago. It then began to break up into the earth's continents, and huge crustal segments slowly drifted apart, carrying with them, perhaps, early ancestors of modern birds. South America and Africa, for example, began severing connec-

tions in the late Jurassic and early Cretaceous periods about 130 million years ago. It is proposed that the phylogenetic resemblance of South American Rheas (Rheidae) and African Ostriches (*Struthio* spp.) is due to the former connection of these continents. Fossils of reptiles and fishes do suggest that these two continents were probably joined in the early Mesozoic. Diatrymids, cathartid vultures, some Galliformes, and others perhaps dispersed while North America and Eurasia were still united, possibly as late as 65 million years ago (Cracraft, 1973, 1976).

Opponents of the drift-dispersal theory hold that continental drift took place too long ago—practically at the time of *Archaeopteryx*—to play a significant role in the evolution and distribution of modern birds and that the present distribution of birds can be otherwise accounted for (Darlington, 1957; Mayr, 1952, 1972).

A look at the history of American land mammals illustrates the early isolation of certain regions. Geologists believe that for perhaps 50 to 70 million years North and South America were separated by a water gap. The two continents were united by the Isthmus of Panama only a few million years ago. Until they were united, North America had 27 families of land mammals, South America 29 families, but only one or two of these families were common to both continents. After the rising isthmus joined the two continents, they exchanged mammals freely, so that by the Pleistocene epoch, two million years ago, they had 22 families in common (Simpson, 1940). Since many birds can and do fly across broad water gaps, they are not as hemmed in geographically as are mammals. Nevertheless, the distribution of birds in North and South America shows distinct similarities to that of mammals.

An interesting case study in the historical geography of North American birds has been made by Mayr (1946, 1964, 1972). Not only have North and South America been separated by water until recently, but North America and Asia were at several different times joined by a land bridge 2000 km broad across Bering Strait during the Tertiary period (10 to 75 million years ago), during which time modern bird families were evolving. During much of the Tertiary, this land bridge enjoyed a mild climate, which may have permitted even tropical species, such as parrots, to cross over into North America from the Old World.

As described earlier in this chapter, Mayr suggests several agencies to account for the origins of birds in the New World. Certain families in both North and

South America have existed in isolation for so long and have changed so much in character through evolution that their origins are indeterminable until a more complete fossil record becomes available. These are the "indigenous elements" of each continent. Evidence strongly suggests that the majority of North American families were immigrants from the Old World, probably by way of the Bering Strait land bridge during periods of mild climate and lowered sea levels. Some of these families eventually filtered down to South America in geologically recent times when the Panama land bridge was thrust up. A smaller number of South American families migrated to North America by the same route.

For millions of years during the early Tertiary, large parts of both North and South America enjoyed tropical climates in which many species of tropical birds evolved. When the Panama bridge arose, these two diverse, tropical avifaunas were allowed to interchange. This mixture probably accounts for the superlative diversity of birds in tropical South America today. They have of course largely disappeared from North America because of its loss of a tropical climate.

Mayr makes clear, however, that bird distribution in the New World is more than the pouring of populations of birds from one continent into another through funnels of land bridges or by island-hopping. There is evidence, for example, that during the height of Pleistocene glaciation, drastic climatic changes caused wide extinction of tropical species in Central America. Later, when the climate ameliorated, the empty niches were filled largely by immigrants from South America. Today in the tropical forests of Central America about 38 per cent of the bird population is of South American origin and 13 per cent of North American, while in the more arid habitats only 8.5 per cent of the occupants are birds of South American origin and 49 per cent are of northern origin (Mayr, 1972). Zoogeography is still a highly speculative science with much yet to be learned.

The study of long-term changes in sea level presents evidence of drastic, worldwide fluctuations. Between 17,000 and 6000 years ago, the melting of North American and Scandinavian glaciers poured sufficient water into the seas to raise the general level 100 to 200 meters, with upward surges of as much as 10 meters per century (Fairbridge, 1960; Diamond, 1976). Such changes as these clearly must have flooded broad expanses of low-lying lands, and created new water gaps between formerly contiguous areas.

The islands of Sumatra, Java, and Borneo are separated from each other and from the mainland of Malaya by water less than 50 meters deep. When the seas were lower during periods of Pleistocene glaciation, these lands were all united into a single mass. As a consequence, the birds of all three islands today show considerable similarity, particularly the woodpeckers, barbets, cuckoos, and babbling thrushes. In the chain of islands east of Java, a deep but narrow strait (Wallace's Line) separates Bali from Lombok. Apparently this water gap is an ancient one that persisted even during periods of low sea levels. Consequently, birds to the west of this strait in Bali, Java, Sumatra, and Borneo are more likely to have Asiatic affinities, while those of Lombok and islands to the east have progressively stronger Australian and Papuan relationships. Here the trogons, hornbills, barbets, woodpeckers, and broadbills of the western islands disappear, and the cockatoos, lorikeets, whistlers, and honey-eaters of the eastern islands take over.

Birds living on long and widely isolated oceanic islands illustrate a special aspect of zoogeographical adjustment. Mayr (1954) points out that although island birds constitute less than 20 per cent of all species, yet of all the birds that have become extinct in historic times, over 90 per cent were island species. Apparently, the introduced species, which evolved on larger land areas amidst more rigorous competition, dominated and supplanted the island forms because of their superior evolutionary experience. Darwin (1859) was aware of this principle: "Widely-ranging species abounding in individuals which have already triumphed over many competitors in their own widely-extended homes, will have the best chance of seizing on new places when they spread into new countries." This principle also accounts for the one-way traffic of immigrant colonists from ancient centers of bird origin into new regions with smaller and less highly adapted populations.

During the Tertiary and Quaternary periods, earth climates fluctuated greatly. Periods of glaciation not only concentrated much of the earth's water in enormous continental ice masses and thus lowered the sea level, but also caused many north temperate zone birds to shift their ranges southward toward the tropics. Other species were undoubtedly extinguished on a catastrophic scale. Later, as the climate moderated and glaciers melted, birds could either extend their ranges northward again, or upward on mountain ranges into corresponding life zones. That both types of range extension occurred is seen today in the discontinuous distribution of many species. The Azure-winged Magpie, *Cyanopica cyanus*, has one subspecies that lives in the mountains of Spain and Portugal, while another subspecies, its nearest relative, lives in brushy thickets along rivers of southeastern Siberia (Stresemann, 1927–1934). In other words, certain species were left "high and dry" in the southern mountains as the glaciers withdrew to the north, pulling the lowland representatives north with them. Today, in the mountains of Guatemala, there are isolated populations of North American ravens, crossbills, flickers, creepers, bluebirds, and juncos left behind, in effect, by receding glaciers (Griscom, 1945). Such isolated populations are known as *relicts*.

Glaciation affects distribution in still other ways. Successive glacial periods during the Pleistocene epoch resulted in alternating wet and dry climatic periods. And these caused the repeated shrinkage, or rupturing, and later expansion and joining, of the low Amazon rain forests (Haffer, 1970, 1974). During the dry periods forest birds were restricted to isolated, humid, grassland-surrounded "refugia" where the forests survived. Some of these isolated bird populations probably evolved into new species; other probably became extinct. Haffer believes that the present distribution of several Amazon bird groups supports this hypothesis. This subject will be considered further in Chapter 23.

Discontinuous distribution may also result when a species surmounts a barrier that confines it. Aside from marine and seasonally coastal species, there are only two species of birds represented by breeding populations in both South America and Africa that are restricted to these two continents. These are two species of tree ducks of the genus *Dendrocygna*. Such a pattern of distribution can be explained only by assuming that these birds flew or were storm-driven across the south Atlantic (Friedmann, 1947). This assumption is supported by the observation of three Fulvous Tree Ducks, *Dendrocygna bicolor*, swimming in the Atlantic Ocean about 2000 km east of Florida (Watson, 1967).

A range of a species may also be fragmented into two or more isolated regions by movements of the earth's crust. Evidence from studies of fossil pollen shows that the Andes in Colombia have risen at least 2000 meters, probably in the late Pliocene or early Pleistocene (Haffer, 1968). This geologically recent upthrust of the Andes mountains has split apart numerous populations of tropical birds in Colombia and Ecuador, so that today their descendants occupy ranges on opposite sides of the mountains. With sufficiently long isolation from each other, such populations will eventually evolve into different races or species. The ornate Cock-of-the-Rock, *Rupicola peruviana*, is, for example, represented on the Pacific side of the Andes by the subspecies *R. p. sanguinolenta*, and on the Amazon side by *R. p. aurea* (Chapman, 1917). The freshwater fishes in the streams of the opposite sides of the Andes show similar affinities.

Although mountain ranges may isolate species, mountain passes may provide bridges to join them. Chapman describes a pass in the Andes at San Antonio, Colombia, where the tropical zone of the Cauca Valley practically reaches the divide. Here one can see a large reservoir of tropical bird species ready to spill over into a new and enlarged range as soon as the mountains sink a few hundred meters or climatic changes cause the tropical zone to rise an equal amount.

An interesting form of altitudinal competitive exclusion occurs among many mountain-dwelling species in Papua New Guinea. Diamond (1972, 1973) found that species of birds with no close relatives in the area generally exhibited gradual changes in abundance with changes in altitude. But many other birds occur in sequences of two or more closely related species that abruptly replace each other at certain altitudes. For example, *Crateroscelis murina*, and *C. robusta* are similar warblers with similar feeding habits. With increasing altitude *C. murina* becomes increasingly abundant until it suddenly disappears at 1643 meters altitude. At this height *C. robusta* suddenly appears at nearly its peak population density and then becomes progressively less abundant with increasing altitude. The two species are territorially intolerant of each other. All told, Diamond found 45 such pairs of altitudinally abruptly segregated close relatives. Among certain birds of paradise "the displaying adult males may be compressed into the top 180 meters of the altitudinal range. . . ." When a given species has no close relatives as competitors, it expands its ecological-geographical range altitudinally, as is shown in the distribution of two species of congeneric honey-eaters on different mountains (Fig. 20–15).

Today, the environmental force that has the harshest and most far-reaching impact on bird distribution is man. In a very few instances man's cultural interference with the natural environment has encour-

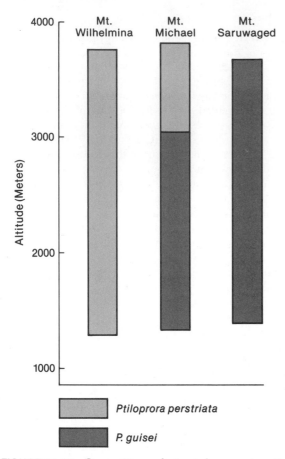

Ptiloprora perstriata

P. guisei

FIGURE 20–15 Competitive exclusion in honey-eaters. On Mt. Michael, where both *Ptiloprora perstriata* and *P. guisei* exist, the birds are altitudinally segregated through interspecific competition. But on mountains where only one of the species exists, each expands into the other's altitudinal range. Redrawn after Diamond, 1973.

aged the spread of species. His parks and lawns have increased the numbers and dominance of the American Robin, *Turdus migratorius*; his barns, for a time, the Barn Swallow, *Hirundo rustica*; his chimneys, the Chimney Swift, *Chaetura pelagica*. The range of the Bobolink, *Dolichonyx orizivorus*, has spread westward across North America simultaneously with the west-

ward establishment of farms by agricultural pioneers. Some 141 species of birds have been documented as occupants of urban habitat in Papua New Guinea (Bell, 1986).

But in the main, man has been a force for restriction and extermination. Natural prairies, woods, ponds and streams, once teeming with bird life, have been converted into relatively sterile agricultural or urbanized environments, ecologically unsuitable for most native birds. In addition to the vast areas of virgin prairies and forests converted into cities and farms, additional land has been requisitioned for such arteries of transportation and communication as highways, railroads, electric power and telephone lines, and gas and oil pipelines. The amount of land in the United States devoted to rights-of-way for utility corporations alone has been estimated by Egler (1958) to be greater than the area of all six New England states! A multitude of somber problems has accompanied man, the newcomer on the avian world scene. They will be treated more in detail in Chapter 24.

■ SUGGESTED READINGS

Although published in 1876, Wallace's *The Geographical Distribution of Animals* still remains the classic in animal geography. For charts showing the world distribution of all the principal living families of birds, see Bartholomew's *Atlas of Zoogeography*. A splendid pictorial book with scores of bird distribution maps is Fisher and Peterson's *The World of Birds*. Two excellent modern texts on the principles of animal geography are Darlington's *Zoogeography: the Geographical Distribution of Animals*, and Udvardy's *Dynamic Zoogeography*. An extensive treatment of European distribution is given in Voous' *Atlas of European Birds*. Brief articles on distribution and on the several zoogeographical regions are given in Thomson's *A New Dictionary of Birds*. For contemporary discussions of some of the causal mechanisms in bird geography see the articles by Mayr, Olrog, Serventy, Hall, and Keast in *Proceedings of the XVth International Ornithological Congress*, and by Cracraft, Diamond, and Salomonsen in *Proceedings of the 16th International Ornithological Congress*. Vuilleumier's chapter in Farner and King's *Avian Biology* V and Simberloff's chapter in Brush and Clark's *Perspectives in Ornithology* provide excellent summaries of current thought in avian geography.

They hover, swoop, and flutter, soar and glide;
On beating wings they climb to heaven's
height;
On unseen currents of the air they ride
Secure in space, in mastery of flight.
—Joel Peters (1905–   )
O, for the Wings of a Bird!

When native school children of Zambia were asked, given a free choice, what they most wanted to be, nearly half of the boys wanted to be birds, and about one-quarter of the girls wished that they, too, might become birds (Powdermaker, 1957). This resounding vote of confidence in the avian way of life probably revealed a discontent with, and a desire to escape from, the lacks and tribulations of the youngsters' earthbound lives. The reptilian forerunners of birds, through the evolution of wings and feathers, achieved the children's goal of avoiding many problems that plague ground dwellers. Greater opportunities to secure food, particularly flying insects, and easy escape from hungry, earth-borne predators, very likely provided primitive flying birds with enough selective survival to bring about rapid evolutionary refinements in flight. These advantages plus many others made possible through the conquest of air—rapid travel across great expanses and over otherwise impenetrable

barriers, safer living and breeding quarters, seasonal migration—have made birds the lively, successful vertebrates that we know today.

ORIGIN OF FLIGHT

Cold-blooded, earth-bound reptiles and warm-blooded, flying birds seem to have little in common, but all evidence indicates that modern birds descended from reptiles. Fossil forms linking reptiles to birds have been too few to show exactly how the front leg of a lizard-like reptile became modified into the wing of a bird.

Two theories have been advanced to explain how this was brought about. Nopsca (1907) assumed that the reptilian proavian ran on the ground, flapping its front legs in the air, much as does a running and flapping barnyard chicken. The reptilian scales on the rear margins of the front limbs gradually lengthened and

467

broadened into avian feathers, as natural selection, through many millenia, favored those individuals with longer, slightly airborne strides and, therefore, speedier locomotion. In time, according to this theory, the wings evolved sufficiently to bear the entire weight of the bird in the air. Ostrom (1976a, 1979) proposes a special modification of Nopsca's theory on the terrestrial origin of flight. From their skeletal architecture he argues that the bipedal ancestors of *Archaeopteryx*, the first known bird, had forelimbs adapted more for grasping prey than for flying, and that "wing" feathers were originally enlarged to serve as "insect nets" rather than to enhance locomotion. The arboreal theory on the origin of flight, proposed by Marsh (1880), Osborn (1900), Abel (1911), Heilmann (1927), and others, assumed that the bipedal reptile took to clambering about in the trees, leaping from limb to limb, gliding from one tree to another, and eventually perfected flapping flight. The prominent claws on the "fingers" of the front limbs or wings of *Archaeopteryx*, and the backward-directed hind toe on each of its feet suggest that the bird was arboreal. Additional morphological and anatomical evidence indicating that *Archaeopteryx* was at least capable of flapping flight is reviewed in Chapter 23.

PHYSICAL PRINCIPLES OF FLIGHT

The flapping flight of birds is exceedingly complex and not well understood. However, birds' gliding flight is much like that of the airplane, and both are subject to the same laws of aerodynamics. The following discussion of aerodynamic principles is based chiefly on Storer's (1948) *The Flight of Birds*.

When a drop of water or other fluid falls through the air, it is shaped by the friction and pressures of the resisting air into a "teardrop" or streamlined form, a shape with less resistance or "drag" than an equal volume of some other shape. Forms with such contours—blunt and rounded in front and tapering, more or less, to a point in the rear—are found in the bodies of most birds. This is also true of fishes, which likewise face the problem of passing through a resisting, fluid medium. However, slipping through the air efficiently is not enough to make flight possible. The bird must also be supported by the air, and it must have "oars" or "propellers" to gain a purchase on the air to thrust its body forward. These are the functions of the bird's wings. Each wing, moreover, like a bird's body, is streamlined in cross-section.

When the leading edge of a streamlined form, such as a wing, cleaves the air "head on," it thrusts the air upward and downward so that the resultant movement of the air reduces its pressure equally on both the upper and lower surfaces of the wing. But if the contours on the two surfaces of a wing differ from each other, the air pressures against them will be unequal, because one air stream must travel farther and, therefore, faster than the other, and this will reduce its pressure on the wing surface. If, for example, the lower surface of the wing is flattened or made concave, and thus shortened in breadth, the air on that side will not be thrust aside as much as before, and therefore its pressure against the underside of the wing will be greater than that on the longer, convex, upper surface.

This difference in wing contours will create an increased pressure against the under surface and a partial vacuum over the upper surface, or a net vertical lifting force at right angles to the surface of the wing. It is this lifting force that makes flight possible. Furthermore, if the leading edge of the wing is tilted upward with reference to its direction of forward motion (the angle of tilt being known as the "angle of attack"), lift will be further increased. This results not so much from increased pressure on the lower surface of the wing as from decreased pressure on its upper surface. In a typical airplane wing the upper surface provides about three-quarters of the lift.

An inclined flat plane, moving through still air, will create some lift, and also some drag. If the plane is "cambered," or curved, so that it is concave below and convex above, it will create more lift. If it is both cambered and streamlined, it creates the greatest lift with the least drag (Fig. 21–1). Drag is, of course, the backward force opposing the wing's motion through the resistant air. Both lift and drag vary according to the wing's area, shape, surface texture, velocity, and angle of attack.

The total drag in a flying bird involves more than wings. Drag varies with air density and with total surface area, shape, weight, and speed of the flying bird. Schmidt-Nielsen (1972) describes three kinds of drag in a flying animal: (1) friction drag, caused when viscous air is dragged along by the moving body and creates a gradient of shearing forces concentric with the animal's body surface; (2) pressure drag, which is the energy dissipated in the turbulent wake of a moving body; and (3) induced drag, which represents the force required to move wings so as to create the lift that supports the body's weight in midair.

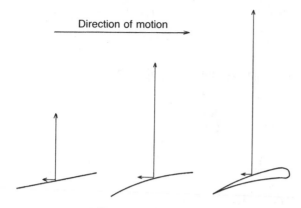

Direction of motion

FIGURE 21-1 Efficiencies of an inclined plane, a cambered plane, and an airfoil, moving through the air. Vertical arrows represent lift; horizontal arrows, drag. The cambered airfoil has by far the greatest lift-to-drag ratio. After Sutton, 1955.

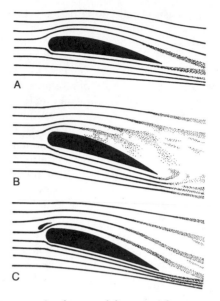

A

B

C

FIGURE 21-2 Smoke tunnel diagrams of a wing, or airfoil, at different angles of attack, moving through the air toward the left. A, In normal flight at a low angle of attack, the air streams smoothly over the upper surface of the wing and creates lift. B, At too steep an angle of attack, air passing over the wing becomes turbulent, lift disappears, and a stall develops. C, A wing-slot may prevent stalling turbulence by directing a layer of rapidly moving air close to the upper surface of the wing.

As long as air flows smoothly over the longer, upper surface of a wing, it creates lift. Within limits, the greater the angle of attack, the greater the lift, and also the greater the drag. At a given speed, a wing operates at maximum efficiency at that angle of attack that gives the highest lift-to-drag ratio. But when the angle of attack reaches about 15 degrees to the direction of the wing's motion through the air, it becomes too steep. Then the air stream begins to separate from the wing's upper surface and becomes turbulent, and lift disappears. For a given wing, this point at which lift vanishes is known as its stalling angle (Fig. 21-2).

Since lift also varies with speed, a given load can be carried at high speed with a small angle of attack, or at a lower speed with a larger angle of attack. Nevertheless, velocity may be reduced to a point below which the smooth air flow is disrupted, and stalling occurs regardless of the angle of attack. This speed is called the stalling speed. The stalling angle of a given wing can be increased somewhat, and hence the stalling speed can be lowered, if the air flow over the upper surface of the wing can be prevented from breaking away and creating turbulence. Wing slots achieve this effect. For example, an open slot along the front edge of a wing can be so designed as to direct, over the top of the wing, a stream of rapidly moving air, which will prevent break-away turbulence at the normal stalling angle of attack. The alula of a bird's wing acts in this way. Flaps hanging diagonally downward from the trailing edge of an airplane wing have the same func-

tion. Such accessories as these enable airplanes and birds to take off and land at steeper angles of attack, and therefore at slower speeds, than would otherwise be possible.

When the air on the underside of a wing slips out from under the trailing edge, it tends to swirl upward into the low pressure area above the wing and thus create a sheet of eddies, which disrupt the smooth flow of air across the wing's upper trailing edge. This lift-destroying, drag-creating turbulence is particularly strong and extensive near the wing tips, where it is called tip vortex. One way of minimizing these disturbances is to lengthen the wings so that the tip vortices are widely separated, since this makes a proportionately larger area of wing between them where the air can flow smoothly. Long wings are thus more efficient than short, stubby wings, but, naturally, not as strong. Long wings are also more difficult to flap than short wings. The ratio of the length to the width of the wing is known as its aspect ratio; long narrow wings have a high aspect ratio. Sailplanes and albatrosses may have aspect ratios as great as 18:1. The extreme efficiency of such long wings is seen in their high lift-to-drag ratio, which may be as high as 40:1. That is, for every 40 g of vertical lift there is only 1 g of resistance to forward motion (Jameson, 1960). Small wings are less efficient than large wings of the same design, because a larger proportion of the small wing's surface will be in the inefficient zones of edge and tip turbulence. Therefore, proportionately large wings are more necessary to small airplanes and small birds than to larger ones.

Another characteristic that influences a wing's performance is wing area in relation to the weight that must be carried. This area-to-weight ratio is known as *wing-loading* or span-loading. In general, the larger this ratio (i.e., the larger the wings in proportion to the load carried), the less power needed to sustain flight. A light sailplane with large wings needs much less power to keep aloft than a heavy, small-winged, fighter plane.

BIRDS AS FLYING MACHINES

In Chapter 1 it was pointed out that, in order to fly, a bird must solve two basic problems: the reduction of weight and the increase of power. Even with reduced weights, birds need increased power to support their bodies in the air, to overcome the air friction against their bodies and wings, to power their wing and tail muscles, and to support their increased circulation and breathing rates while flying. Most of the anatomical and physiological differences that set birds apart from other vertebrates seem to be adaptations devoted to the solution of these two problems. These adaptations were considered in more detail earlier. It is enough here to list the most important ones.

Weight-Reducing Adaptations

Thin, hollow bones.
Extremely light feathers.
Elimination of most skin glands.
Elimination of teeth and heavy jaws.
Elimination of tail vertebrae and some digits.
Extensive bone fusion, especially in the pectoral and pelvic girdles and vertebral column.
A system of branching air-sacs.
Oviparous rather than viviparous reproduction.
The atrophy of gonads between breeding seasons.

The eating of concentrated foods.
Rapid and efficient digestion.
The excretion of uric acid instead of urine.

Power-Increasing Adaptations

Warm-bloodedness.
Heat-conserving plumage.
An energy-rich diet.
Rapid and efficient digestion.
High glucose content of the blood.
A four-chambered heart that provides double circulation.
Rapid and high-pressure circulation.
A highly efficient respiratory system.
Breathing movements often synchronized with wing beats.
A high rate of metabolism.

Flight Metabolism

The long-distance, nonstop flights of birds over oceans and sterile deserts confirm the remarkable efficiency of bird flight. Millions of small Palearctic passerines cross the Sahara Desert in migration flights twice each year. Thousands of Lesser Golden Plovers, *Pluvialis dominica*, fly nonstop about 4000 km from the Aleutian Islands to the Hawaiian Islands with no opportunity to feed en route.

A variety of attempts have been made to determine the efficiency of bird flight in terms of energy consumption or carbon dioxide production of birds flying in closed chambers, or of birds trained to fly in wind tunnels while wearing gas-sampling head masks. Another technique involves the measurement of fat (the chief fuel used on migratory flights) in birds' bodies before and after long migration flights. Still another technique measures the loss of isotopes of oxygen and hydrogen used as metabolic tracers. So many uncontrolled variables creep into some of these experiments—size, age, and health of bird; air temperature; head- or tail-winds; fair or stormy weather; long or short flights; high or low altitudes; high or low flight speeds—that it is scarcely surprising that conclusions regarding energy requirements for flight in birds of even the same size may vary by 100-fold (Tucker, 1971).

However, results from some of the more tightly controlled experiments indicate that a flying bird consumes energy at a rate 6 to 20 times its basal or resting rate (Tucker, 1971; Lasiewski, 1972; Teal, 1969).

Summarizing results from several experimental studies, Berger and Hart (1974) found that heart rates in flight are about 2.4 times those at rest for small birds (10 to 20 g) and 3 times those for large birds (500 to 1000 g). During flight, respiration rates range from 3 to 19 times those at rest. From his experiments with Budgerigars, *Melopsittacus undulatus*, which were tethered to gas-sampling masks and tubes while flying in wind tunnels, Tucker (1969) made the interesting discovery that the birds expended minimum energy while flying at 35 km per hour, and that, per gram of body weight, they expended 3 calories of energy per kilometer flown. At speeds both faster and slower than 35 km per hour, the birds consumed more energy, although at higher speeds they could fly a greater distance on a given quantity of fuel. Assuming that body fat is the fuel used for long-distance flights, a Budgerigar expends about 0.03 per cent of its body weight for each kilometer traveled; a gull, about 0.015 per cent. Since many migratory birds store fat amounting to 25 to 50 per cent of their body weight before embarking on long flights, it is apparent that continuous flights of 1000 or more kilometers are quite possible. Compared with walking or running mammals of the same size, a flying bird is 10 to 25 or more times as efficient in energy consumption per kilometer traveled.

Since air speed is crucially important to a flying bird, especially on long flights, some means of determining velocity would be decidedly helpful. Gewecke and Woike (1978) have discovered that the breast feathers of a Eurasian Siskin, *Carduelis spinus*, with their associated mechanoreceptors (Herbst corpuscles) act as air-current sense organs. If, for example, the breast feathers are immobilized, or the breast skin anesthetized, a flying Siskin will increase the frequency of its wing-beats.

Another facet of flight metabolism concerns heat dissipation, since vigorously contracting muscles generate excess heat. As described earlier, heat may be eliminated by radiation or convection from the body surface, or by the evaporation of water from respiratory organs. Radio-telemetered European Starlings, *Sturnus vulgaris*, in 30-minute flights, maintained temperatures of 42.7 to 44.0°C—2 to 4 degrees above resting temperatures (Torre-Bueno, 1976). At these high temperatures, muscles presumably act more powerfully and efficiently, but the excess heat must be eliminated. During sustained flight of starlings in an air tunnel, Torre-Bueno (1978) found that evaporative heat loss ranged from 5 per cent of the metabolic rate at

− 5°C to 19 per cent at 29°C. Radiation and convection from skin and feathers accounted for the rest of the heat loss. At all temperatures above 7°C, flying starlings began dehydrating—i.e., water loss from breathing exceeded metabolic water produced from metabolized fats. Therefore, body dehydration may become a significant limiting factor in long migratory flights. The fact that warm air temperatures may increase respiratory water loss several-fold over that at cool temperatures may explain in part the flights of many migrants at night and in the cold air of high altitudes.

In addition to these structural, sensory, and metabolic adaptations, there are others that conserve power by promoting clean aerodynamic design in the bird's body; this design permits the bird to pass through the air smoothly, with little friction and a high degree of stability. The streamlining of the body, wings, and even the individual primary wing feathers contributes greatly toward this end. Most organs and the heaviest muscles are centrally placed between and beneath the wings, providing an automatic stability that requires little muscular correction. For example, the weight of one muscle that flexes the wrist is kept near the body because, although it originates above the elbow, its tendon is inserted beyond the wrist where the action takes place. The supracoracoideus muscle (p. 66), which elevates the wing, is not situated (as is the corresponding muscle in a reptile or a mammal) beside the backbone. Instead, it is inserted ventrally on the sternum and produces its lifting action through a pulley-like slip high in the shoulder girdle. Similarly, many of the muscles of the legs are concentrated near the body and carry out their functions by means of long, string-like tendons that activate the extremities. Such central placement of the heavy muscles of the wings and legs greatly reduces their moment of inertia and, therefore, the work required to move them.

Various skeletal adaptations related to flight have already been mentioned. Probably one of the earliest modifications was the development of bipedal locomotion, which released the forelimbs for flight. This required a shortening of the long reptilian body axis so that the hind feet could be placed under the center of gravity. The long and heavy reptilian tail was reduced to a tiny vestige, the pygostyle. The pelvic girdle was lengthened and fused to the vertebral column so that it protects and supports the visceral organs, provides mechanically advantageous sites for the origin of heavy leg muscles, and gives strength and rigidity for withstanding the shocks of leaps and landings.

The sternum, as remodeled, now provides similar protection and support on the underside of the body, and its keel furnishes ample room and suitable leverage for attachment of the breast muscles. The powerful, wing-flapping contractions of these muscles might crush the rib cage were it not for the development of stout coracoid bones, which brace the wings and shoulder girdles against this compressive force. The neck in birds has become highly mobile, taking over several functions of arms and hands. These and many less profound modifications in the skeleton have kept pace with the most fundamental adaptation of all, the development of the forelimbs into wings (See Fig. 4–2).

ADAPTIVE STRUCTURE OF WINGS

Relieved of their ancestral job of walking on the ground, the forelimbs of birds have become strikingly modified for supporting the body in air. Their attachment to the body has been moved dorsally and posteriorly so as to be located nearly over the body's center of gravity. The new motions required for flight have resulted in several major changes. The humerus is oriented in its shoulder socket so that the wing's chief movement is up and down rather than forward and backward, although it is still capable of considerable rotation. At rest, the wing can be folded into a compact Z against the body, out of harm's way. The elbow joint has become a hinge joint that permits movement of the forearm only in the plane of the wing. This stiffens the entire extended wing into a plane, which resists twisting when exposed to air pressures. The total length of the wing has been increased to accommodate enough flight feathers to support the bird in the air. However, in different birds the various segments of the wing are not lengthened proportionately. In the hummingbird, the bones of the hand are longer than those of the upper arm and forearm together; in the frigatebird, all three segments of the arm skeleton are about equal in length; in the albatross, the humerus, or upper arm, is the longest segment. These varying proportions in wing length are associated with different modes of flight. The ulna of the forearm, in all living birds but the kiwis, is now a stout bone on whose posterior margin the secondary wing feathers are attached.

The hand, no longer concerned with walking, climbing, or manipulating objects, has been drastically changed. It supports the large primary feathers, whose chief functions are to propel and maneuver the bird in flight. The almost fleshless hand has been simplified through a reduction in the number of its bones and a fusion and flattening of those remaining. The carpal bones of the wrist have been reduced from five to two; the digits, from five to three. The second digit supports the alula and is independently movable. The third and fourth digits, which support the primary feathers, are partly fused together, and their segments or phalanges flattened, fused, and reduced in number. The hand bones, in short, are nearly reduced to a rigid, flat paddle on whose rear margin are supported the all-important flight primaries.

Stretching along the front edge of the wing, from the shoulder to the wrist, is a triangular elastic membrane, the patagium. This tough membrane acts as a gliding plane, and also partly controls the movements of the larger wing feathers. A similar flat membrane extends along the rear of the wing from the shoulder to the elbow.

As seen earlier, on page 98, when any object increases in size, its surface increases as the square, and its mass as the cube, of its linear dimensions. If a large bird were, so to speak, no more than a photographic enlargement of a small bird, its wings would not be large enough to function properly. As the bird doubled in size "photographically," its wings would increase in area four times, but the weight of the body they must support would increase eight times. This means that large, heavy birds should have proportionately larger wings, or more efficient wings, than small birds. Because of the great moment of inertia in large wings, and the problem of making them strong enough for flight without making them too heavy, flying birds much larger than a condor are probably a physical impossibility.

One might suppose that in an ideal bird, regardless of its size, there should be a fairly constant wing-loading ratio. This is far from true. Wing-loading among the different species of birds may vary by a factor of as much as 19. This is not surprising when one considers the variable size and efficiency of wings, the variation in wing-beat frequency and amplitude, the different types of flight (flapping, soaring, hovering), and the periodic variations in the weight of a given bird. Table 21–1, taken from Poole (1938), gives wing-loading data for representative North American species. Wing areas are based on the two outstretched wings but not the tail.

The table clearly shows that the heavier birds have a higher wing-loading (fewer square centimeters of sur-

TABLE 21–1.
WING-LOADING IN REPRESENTATIVE NORTH AMERICAN BIRDS*

SPECIES	WEIGHT (G)	WING AREA (SQ CM)	WING AREA PER G
Ruby-throated Hummingbird, *Archilochus colubris*	3.0	12.4	4.2
House Wren, *Troglodytes aedon*	11.0	48.4	4.4
Black-capped Chickadee, *Parus atricapillus*	12.5	76.0	6.1
Barn Swallow, *Hirundo rustica*	17.0	118.5	7.0
Chimney Swift, *Chaetura pelagica*	17.3	104.0	6.0
Song Sparrow, *Melospiza melodia*	22.0	86.5	3.9
Leach's Petrel, *Oceanodroma leucorhoa*	26.5	251.0	9.5
Purple Martin, *Progne subis*	43.0	185.5	4.3
Red-winged Blackbird, *Agelaius phoeniceus*	70.0	245.0	3.5
European Starling, *Sturnus vulgaris*	84.0	190.3	2.2
Mourning Dove, *Zenaida macroura*	130.0	357.0	2.4
Pied-billed Grebe, *Podilymbus podiceps*	343.5	291.0	0.8
Common Barn Owl, *Tyto alba*	505.0	1683.0	3.4
American Crow, *Corvus brachyrhynchos*	552.0	1344.0	2.4
Herring Gull, *Larus argentatus*	850.0	2006.0	2.4
Peregrine Falcon, *Falco peregrinus*	1222.5	1342.0	1.1
Mallard, *Anas platyrhynchos*	1408.0	1029.0	0.7
Great Blue Heron, *Ardea herodias*	1905.0	4436.0	2.3
Common Loon, *Gavia immer*	2425.0	1358.0	0.6
Golden Eagle, *Aquila chrysaetos*	4664.0	6520.0	1.4
Canada Goose, *Branta canadensis*	5662.0	2820.0	0.5
Mute Swan, *Cygnus olor*	11602.0	6808.0	0.6

*After Poole, 1938.

face per gram weight) than the lighter ones. This probably means that the larger birds approach the absolute limit of wing-loading. Because of the anatomical and physiological extravagance of extremely large wings, large birds cannot afford the luxury of the generous margin of safety that small birds can. The large wings of small birds also compensate for the relatively large proportions of their wing areas devoted to inefficient margins. For these two reasons, small birds tend to have large wings and light wing loads. If a Mute Swan, *Cygnus olor*, had the same wing-loading as a Barn Swallow, *Hirundo rustica*, its two wings would have 81,000 sq cm of surface instead of an actual 6800. If these giant wings were designed with the same aspect ratio as those of the swallow (figured at the conservative ratio of 6:1), each wing would be seven meters long, and the bird's total wing span would be over 14 meters (46 feet)! The structural, mechanical, and physiological absurdity of such wings is evident.

FUNCTIONS OF WINGS

Wings of birds vary in shape as well as in size, and both characteristics are intimately related to function, habitat, and niche. Wings have been classified into four major types by Savile (1957). *Elliptical* wings are found on birds adapted to forested or shrubby habitats where birds must maneuver in close quarters. Such wings have a low aspect ratio and a somewhat reduced tip vortex. Associated with the slow flight that is often advantageous in such habitats, many elliptical wings have a high degree of slotting, especially in the form of separated primaries. Elliptical wings are found in most gallinaceous species as well as in *Archaeopteryx*, many doves, woodpeckers, and passerines, especially the Corvidae. *High speed* wings are characteristic of birds that feed on the wing or make long migrations. Such wings have a low camber (flattish profile) and a fairly high aspect ratio. They taper to a rather slender tip, without slots, and they show the sweep-back and "fairing" to the body of modern fighter-plane wings. Such wings are found in shore birds, swifts, hummingbirds, falcons, and swallows. Savile's third type is the *high aspect-ratio* wing, most commonly found in such soaring sea birds as albatrosses and frigate-birds. These long and narrow wings rarely have wing-tip slotting. The *slotted high-lift* wing is the fourth type. It possesses a moderate aspect ratio, deep camber, and marked slotting. Characteristic species are terrestrial soaring birds, such as vultures and hawks, owls, and predators that carry heavy loads.

To illustrate one way in which different types of wings suit different ways of living, Storer (1948) compares two birds of roughly equal weight but different ways of life. A Great Blue Heron, *Ardea herodias*, uses its long legs to wade in marshes where it stalks its prey. In alighting, it must take care to land slowly and carefully to avoid damaging its legs. The wings of the heron are deeply cambered and, for such a large bird, generous in size, with a square-centimeter-per-gram value of 2.3. The Common Loon, *Gavia immer*, on the contrary, is a heavy diving bird of open lakes. Large wings would only be a mechanical embarrassment under water. However, with its extremely small wings, giving it a wing-loading value of 0.6, the loon must fly rapidly to sustain itself in air. Hovering to a gentle landing is out of the question. In alighting,

the bird must dissipate the momentum of its flight in a long, splashing glide over the surface of the water. If two birds of the same weights have different wing-loadings, it is likely that their wing-beat frequencies will differ. The frequency for a 1-kg gull with a wing area of about 2000 sq cm is about 3 per second; for a 1-kg duck with a wing area of 1000 sq cm, it is about 8 per second.

To perform its two basic functions, propulsion and lift, the typical wing must drive the body through the air and must also support it there. Its construction is adapted to these requirements. It is thickest at its leading edge, tapering to a thin edge at the rear. At the wing's leading edge, bones give it rigidity and strength, while covert feathers, standing perpendicularly to the wing's surface, give the wing a rounded and smooth streamlining. The upper surface of the wing is convex in cross-section; the lower surface is concave, like the cambered wing of an airplane. In flight, this shape causes a partial vacuum above the wing and pressure below, creating lift. Such cambering is especially pronounced in slow fliers. Since the trailing edge of a wing is more flexible than the leading edge, it bends upward under air pressure from below, resulting (as in a propeller blade) in *pitch*, which drives the bird forward with each downward wing beat.

Flight feathers are also streamlined in cross-section,

FIGURE 21–3 A Red-throated Loon, *Gavia stellata*, because of its heavy wing-loading, cannot become airborne until it has gained a speed of about 36 km per hour by skittering across the water, using its feet. From Norberg and Norberg, 1971.

particularly those primaries that meet the air on edge. The stiff quill or rachis of each feather lies not in its middle but toward the leading edge. The leading vane is thicker and narrower, and the trailing vane thinner and wider.

Feathers overlap each other somewhat like the shingles on a roof, so that the firm, leading edge of one feather lies above the more flexible trailing edge of the feather in front. This arrangement provides a surface relatively impermeable to air on the downstroke, but one that may open like a venetian blind on the upstroke to let air slip through (see Fig. 21–6).

Unlike the wing of an airplane, the wing of a bird is both a wing and a propeller. The hand, with its large primaries, does most of the propelling; the forearm, with its secondaries, provides most of the lift. In flight, the hand is the most active member, moving through a slanting oval, or a figure 8, while the upper arm and forearm move very little. As a rule, propelling force is developed only during the approximately vertical movements of the wing.

The relative importance of the hand and its primaries was revealed by an experiment of Chapeau (Stresemann, 1927–1934) in which the removal of only a small portion of the tips of its primaries prevented a dove from flying. On the contrary, removing enough of the arm or secondary feathers to reduce the area of the entire wing by 55 per cent (from 547 sq cm to 248 sq cm) still did not keep the bird from flying! This experiment illustrates the extreme importance of the outer ends of the primaries in propelling the bird.

Secondary feathers are characteristically smaller than primaries; otherwise the wing would be too bulky to fold compactly. The number of flight feathers carried by a wing varies. Most birds have ten primaries on each hand, the outermost usually reduced in size. Finches, wood warblers, swallows, honey-guides, and others have only nine primaries, while herons, ducks, geese, gulls, and terns have 11, and grebes, storks, and flamingos usually have 12. Greater variability occurs in the number of secondaries carried on the forearm. Most passerines have nine secondaries, but swifts and hummingbirds have only six or seven. Large soaring birds with elongated forearms may have about 20 secondaries, and the Wandering Albatross, *Diomedea exulans*, has 32.

In addition to the primaries, the hand also supports a few small feathers on the thumb or first digit. These constitute the alula or "bastard wing" and function to produce a wing slot during slow, labored flight (Fig. 21–4). Wind tunnel experiments on four different species of birds by Nachtigall and Kempf (1971) showed that the projecting alula increased lift, particularly at high angles of attack (30 to 50 degrees).

Because of its stiff but elastic construction, a feather under stress does not have the same shape it has while at rest. The relative size and rigidity of different parts of the barbs, which make up the vanes, and of the rachis and quill stiffening the center, provide a structure that is remarkably responsive to exploiting the air. Air striking an isolated or projecting feather from different directions and with different velocities will cause it to bend up or down or sideways, or to twist on its axis. Ordinarily, the broader, more flexible vane on the trailing edge of a feather bends up more easily than the leading vane when subjected to air pressure from below. This twists the leading edge of the feather downward and produces a propeller pitch, which results in the feather's forward motion with each downstroke.

Barbs of a contour feather are normally inserted on the rachis at an angle slanting outward. Feather resistance against an air stream moving in a direction from the quill toward the tip will cause the linked barbs to bend inward toward the rachis, thus narrowing the vanes. Air moving either crosswise on the feather, or from the tip toward the quill, will cause the barbs, particularly those of the wider, trailing edge, to swing outward and widen the vanes. Barbs are so constructed that they do not easily bend at right angles to the plane of the feather vanes. This mechanism usefully insures that feather vanes will be their widest when wings are outstretched, as in soaring flight or when birds are alighting, and, on the contrary, the vanes will be their narrowest and least air-resistant when wings are partly folded as in rapid flight.

Birds with high-lift, slotted wings have primaries that separate from each other at their outer ends when the wing is under pressure from below, and in this way produce a series of slots. Each projecting feather twists under pressure and acts like an individual propeller blade, or like an efficient, narrow wing capable of flight at a high angle of attack. Hawks, owls, ravens, and many other birds have primaries whose vanes are narrowed or notched along their outer ends, making feather separation and slot formation easier when required.

Not all the work of flying is done by the wings. Since wings are generally attached slightly in front of a bird's center of gravity, the body tends to trail

FIGURE 21–4 An American Robin in slow flight. Note the projecting alula. Photo by G. R. Austing.

downward. In flight, the lowered tail is lifted by the horizontal air stream in a compensating fashion. This arrangement constantly provides the tail with a firm "grip" on the air stream for quick maneuvering actions (Stresemann, 1927–1934).

Various actions of the tail—raising, lowering, twisting, opening, and closing—aid the wings in supporting, balancing, steering, and braking the body in flight. It is significant that long-tailed pheasants typically inhabit wooded areas where agile maneuvering among the trees is required. Loons, on the contrary, with their small wings and stubby tails, have practically no maneuvering ability and must take off, fly, and land on water in practically straight paths. As a rule, rapid fliers have short tails. Some aquatic birds, such as auks, loons, and grebes, have such stubby tails that their webbed feet are substituted for steering. Stresemann points out that broad-tailed predatory birds, such as eagles and hawks, carry their prey in their talons, posteriorly, under the tail, whereas birds like the pelican, cormorant, or albatross, with well-developed wings but meager tails, carry their booty in the beak or esophagus.

GLIDING AND SOARING FLIGHT

In all likelihood, gliding was the original form of bird flight, and it is still the simplest. It requires no propelling energy from the bird itself. It is the form of flight used by an airplane coasting to a landing.

Gliding has evolved independently in several vertebrate groups, including frogs (Wallace's Flying Frog, *Rhacophorus nigropalmatus*), lizards (the flying lizard, *Draco volans*), mammals (the flying squirrels, *Glaucomys* spp., and the Australian Sugar Glider, *Acrobates*), and birds. The two chief forces acting upon a bird in flight are the pull of gravity on its mass and the resistance of air to its passage. A gliding bird, coasting downward, is simply using its weight to overcome the air resistance to its forward motion. The most proficient gliders can glide forward 15 to 20 meters while descending only 1 meter.

A soaring bird is one that maintains or even increases its altitude without flapping its wings. It can do this either by gliding in rising currents of air (static soaring), or by exploiting adjacent air currents of different velocities (dynamic soaring). The three chief requirements for successful soaring are large size, light wing-loading, and maneuverability. Large size gives a bird sufficient momentum to carry it through small, erratic air currents without loss of stability or control. Stability and maneuverability, however, are antagonistic qualities, and perfection in both cannot be found in a single bird. The relatively small breast muscles of soaring birds prove that this is an economical method of flying. Some species, such as the vultures and albatrosses, have become so highly adapted to soaring that they fly almost exclusively by this means.

As soaring flight evolved, there developed two distinct types of wings, which function in different ways and are adapted to different habitats. On the one hand are the slotted, high-lift wings of land birds such as condors. These are adapted to slow speed, high altitude, static soaring. On the other hand are the high-aspect ratio wings, represented by those of the albatrosses. These are adapted to high speed, low altitude, dynamic soaring. The line between the two types is not always hard and fast, and some hawks, for example, practice both types of soaring.

Soaring land birds, static soarers, keep aloft mainly by seeking out and "riding" rising air currents. Of these there are two sorts: *slope* or *obstruction currents* and convection currents or *thermals*. Obstruction currents are updrafts caused when a steady or prevailing wind strikes and rises over such objects as hills, buildings, or ships at sea. A wind coming down the lee side of a mountain may strike the plain and rebound in a series of "standing waves," each with an updraft component.

Thermals are updrafts caused by the uneven heating of air near the surface of the earth. The air over cities or bare fields heats more quickly than that over forests or bodies of water. Since warm air expands and is therefore lighter, it rises above cooler air. As a result of his observations of African eagles, vultures, storks, and other soaring birds from a motorglider, Pennycuick (1972) found that they exploited three kinds of thermals: (1) columnar thermals or "dust-devils," (2) vortex rings consisting of a rising central core surrounded by an annulus of sinking air, and (3) thermal "streets" caused by rows of buildings, rocks, or similar structures that heat up quickly in the morning sun. He concluded that a large bird such as a stork could reduce its fuel consumption by a factor of 23 by soaring rather than flying under its own power. For a small warbler this factor reduces to 2.4.

Thermal updrafts near thunderclouds may rise at a rate of 6 meters per second (26 km per hr). In such forceful updrafts a White-backed Vulture, *Pseudogyps africanus*, may travel crosscountry at a speed of 15 meters per second (21.6 km per hr); the motorglider, at 97 km per hr. Common House Martins, *Delichon urbica*, indulge in "thunder flights" during summer storms, rising in lively flocks in the vigorous updrafts, possibly to feed on the up-welling insect plankton (Voipio, 1970). Thermals may also occur when an advancing cold front slides under and lifts a mass of warm air. As a rule, thermals rise to greater heights than obstruction currents. Although owls have slotted, high-lift wings, they are not soaring birds, because thermals and dependable winds rarely occur at night when owls are active.

Maneuverability is particularly important to land soarers since the thermals on which they rise are often small and undependable pillars of warm air. In order to circle in tight spirals within these pillars, the birds need ample tails and short, broad wings—short wings for low inertia and quick, sensitive response to capricious air currents; broad slotted wings for high lift capacity. It is essential, of course, that the rate of gliding descent, or *sinking speed*, be no greater than the rate of air rise in the thermal.

Observations from distances as close as 5 meters, made from a motorless sailplane, showed that the Turkey Vulture, *Cathartes aura*, has a minimum sinking speed of 0.61 meter per second (2.28 km per hr), and the Black Vulture, *Coragyps atratus*, 0.79 meter per second (2.83 km per hr) (Raspet, 1950). These different sinking speeds of the two species of vultures depend, of course, on differences in wing-loading; the differences, although apparently slight, have striking geographical consequences. The range of the more buoyant Turkey Vulture extends as far into the cool north as southern Canada, but the aerodynamically heavier Black Vulture is restricted to the southern half of the United States and to tropical regions where the warm sun generates many and vigorous thermals.

A bird soaring in tight spirals cannot fly at high speeds. To fly slowly and yet provide enough lift to avoid sinking requires a high angle of attack, and to achieve this without stalling is the land soarer's problem. The efficiency of deeply slotted, high-

FIGURE 21–5 The White-backed Vulture, *Pseudogyps africanus*, has the broad, slotted wings of a static soarer. Photo by E. Hosking.

camber wings makes possible the combination of low speed and high lift. Each separately extended primary feather acts as a narrow high-aspect wing set at a very high angle of attack. Such wing construction greatly reduces tip vortices, but the high angle of attack increases drag. However, since drag increases as the square of air speed, it is not a very great problem for slowly flying birds. In that accomplished soarer, the California Condor, *Gymnogyps californianus*, 40 per cent of the wing span is occupied by slots (Storer, 1948). The deeply slotted wings of a condor, however, would be useless for oceanic flying, because wet feathers could not be smoothly or quickly manipulated.

To suit the demands of the moment, a soaring bird can alter the shape and expanse of its wings by fanning or folding its feathers, by changing the wingspread, camber, sweepback, or angle of attack. Such changes allow the bird to select the optimum lift-to-drag ratio for any given speed.

Updrafts occur even over large bodies of water. The rising currents of air formed on the windward sides of waves and swells are used by shearwaters and other small species for water-level soaring and gliding. Gulls commonly soar on the rising obstruction currents caused by ships or seaside cliffs. The ease of soaring and gliding contrasted with the work of wing-flapping seems to have made some gulls lazy. At Woods Hole, Massachusetts, there is a windward embankment which creates updrafts and is used by Herring Gulls, *Larus argentatus*, to glide directly windward to an island two-thirds kilometer off shore where they roost (Woodcock, 1940).

Thermal updrafts occur over water only when the air temperature is colder than the underlying water temperature—a condition likely to occur in the North Atlantic in winter. Under such conditions, gulls may be seen far from land, soaring high in the air. But should a warm front bring in a mass of air warmer than the sea water, convection stops, and the gulls can soar only on obstruction currents or fly by the laborious flapping of their wings. In relatively calm, cold air, these sea thermals are thought to occur in roughly hexagonal, vertical cells, with the updraft in the center and the downdraft around the edges, or vice versa. Under these conditions, gulls soar in circles around the edges of the updrafts and glide from one thermal cell to another.

However, with surface winds of about 25 to 40 km per hr, these cells are laid down parallel to the sea's surface, and thus change from columnar cells to linear "strip" or "roll-convection" cells. At these wind velocities, gulls cease to soar in circles and begin soaring in straight lines along the upwelling air currents of adjoining strip cells. At wind velocities of over 45 km per hr, all soaring stops, probably because the pattern of strip cells is destroyed (Woodcock, 1942).

On the seas, thermals are too unreliable, and wave-deflected updrafts too small, irregular, and too near the surface, to sustain a large bird in soaring flight.

Fortunately, over wide stretches of the ocean, another and more reliable source of energy is available. Steady winds, such as the trade winds or winds of the "Roaring Forties," blow across the ocean surface. Their friction with the waves causes the lower levels of air to move more slowly than those higher up. As a result, wind velocity gradients are set up: wind just above the waves may be moving at a speed of 32 km per hr, whereas wind 20 meters above the surface may be moving at twice that speed.

Soaring sea birds, such as the albatrosses, have learned to exploit this velocity gradient. By gliding sharply downward *with* the wind, from upper, high speed levels to lower, low speed levels, the bird acquires momentum that it applies, after wheeling head-on *against* the wind, toward climbing back up. It is much like a man hopping off a moving bus and using his bus-acquired momentum to run up a slope at the edge of the road. As the climbing bird applies its momentum against the wind, the ever-increasing velocity of the wind, as the bird continues rising, acts, by increasing the bird's air speed, to extend the drive of its momentum until it reaches its original height, circles leeward, and repeats the cycle. This type of flight is known as dynamic soaring. Albatrosses fly almost exclusively by this means, and therefore are most commonly found in the great oceanic wind belts.

The special requirements for dynamic soaring are fairly heavy body weight for maneuvering momentum, light wing-loading, low drag, and a relatively high but variable speed. Long, narrow wings with high aspect ratio contribute to these requirements. The Wandering Albatross, *Diomedea exulans*, with an aspect ratio of about 18 to 1, and a lift-to-drag ratio of about 40 to 1, is reputed to have the greatest wingspread of any living bird—as much as 3.65 meters (12 feet). Although such wings are adapted to rapid flight, they are somewhat less maneuverable than the wide, slotted wings of land soarers, but this slight lack is compensated for by the reliability of sea winds.

FLAPPING FLIGHT

Perhaps one of the most remarkable things about flapping flight is the fact that an inexperienced fledgling can leave the nest and fly successfully on its maiden attempt, although it uses a form of flight so complex as to defy precise analysis. The young of some birds, such as hornbills, swifts, swallows, and wrens, are raised in small, crowded cavities, where flapping

their spread wings is out of the question. In spite of this, they are able to fly considerable distances on their first flights. On its first attempt, a young African Bank Martin, *Psalidoprocne holomelaena*, flew continuously for at least 6 minutes (Moreau, 1940). The young of many burrowing petrels fly with great skill on their initial attempts. Without previous practice, young whale birds and diving petrels leave their burrows and fly as far as 10 km on their maiden flights (Murphy, 1936).

On the other hand, the young of many species of birds practice flapping their wings before trying true flight. The role of this behavior in the development of flight is unclear, since controlled experiments indicate that chicks with wings immobilized by bandages from day one to day 13, showed normal flapping rate and flight distances compared to controls not so immobilized (Provine, 1981). Pigeons, *Columba livia*, raised in cages so small as to prevent flapping, flew as well as did controls that were permitted practice-flapping (Grohmann, 1939).

The difficulties of understanding flapping flight are apparent when one considers the variables involved. A beating wing is flexible and yields under pressure. Its shape, expanse, aspect ratio, camber, sweepback, and even the positions of individual feathers may change strikingly. Not only are the frequencies and amplitudes of wing-beats subject to change, but different parts of the wing change in velocity and angle of attack even during a single beat, or one part of a wing may be at a stalling angle while the rest is not. Further, the forces of inertia and air pressure change as a bird accelerates in taking off, or brakes in alighting. In addition, during prolonged flights, such variables as air temperature, altitude (air density), and change in body weight and profile as a result of fuel consumption, will all have an effect on flapping flight. All told, this is a formidable list of variables. Nevertheless, some of the principles of flapping flight have been discovered, largely through high-speed photography.

The popular notion that wing-flapping is a sort of swimmer's breast stroke pushing downward to support the bird in the air, and pushing backward to propel it forward, is not true. When a small bird takes off, the wing moves downward and *forward* on the downstroke. Since the trailing edge of the wing is less rigid than the leading edge, it bends upward under air pressure and forms the entire wing into a propeller that pulls the bird forward through the air, the wing feathers biting the air with their under surfaces. On the return stroke, upward and *backward*, the wing ordinarily does little or no propelling. It is partly folded against the body, and the hand primaries twist open, much like jalousies, so that the rising wing meets a minimum of air resistance (Fig. 21–6). The upstroke is largely a passive recovery stroke. The wing is then extended and makes another downstroke.

In larger birds, with their slower wing action, the time consumed by the upstroke is too long and too precious to waste in neutral "idling." The heavier wing-loading and the greater body inertia to be overcome in taking off require that both wing strokes accomplish some work. In rising, the wing bends slightly at the wrist and elbow, and the whole arm (humerus) rotates backward at the shoulder joint to such a degree that the all-important primary feathers now push against the air with their upper surfaces and drive the bird forward. At the end of the upstroke the upward flexed arm is rotated forward and extended, producing a rapid backward snap of the primaries, restoring them to the proper, more elevated position for the downstroke. The downstroke, as in smaller birds, is a downward and forward pulling action. The entire excursion of the wing tip in a completed down and up beat may be in the form of an oval or a figure 8, slanted downward

FIGURE 21–6 A Black-capped Chickadee in slow flight. During the upstroke of the wing, the primary feathers twist and open in a venetian-blind fashion that allows them to slip through the air with little effort. Photo by H. C. Johnson.

and forward in the direction of flight. During the hard work of taking off, the wings will beat more vigorously in greater arcs, and in ovals more inclined toward the horizontal. Not only are the wing slots likely to be opened at this time, but the alula is commonly extended; and it reduces air turbulence over the unslotted secondaries, creating even more lift. Some birds are said to be unable to take off if their alulas are removed. In taking off, the somewhat vertical position of the body enables the wings to beat in a more nearly horizontal plane and thus produces a more nearly vertical thrust for leaving the ground. From analyses of high-speed motion pictures of birds in hovering flight, Dathe and Oehme (1978) determined that Common Black-headed Gulls, *Larus ridibundus*, and Collared Doves, *Streptopelia decaocto*, produced lift both in the downstroke and the upstroke, but the Jackdaws, *Corvus monedula*, and European Starlings, *Sturnus vulgaris*, produced lift only on the downstroke.

The rate at which birds flap their wings varies inversely with their size. Large vultures flap their wings about once each second. Medium-sized birds, including small hawks, shore birds, doves, and crows, beat their wings two or three times a second (Blake, 1947). Chickadees beat with a frequency of about 30 times per second, an average-sized hummingbird about 40 times, and the smallest hummingbird perhaps 80 times per second (Greenewalt, 1960).

Studies on ten species of birds by Berger and others (1969) revealed that in certain species, for example, doves and crows, wing-beats were synchronized with respiration cycles on a one-to-one basis. In other species, such as ducks and game birds, there were from three to five wing-beats for each respiration cycle. As a rule, the beginning of inspiration was linked with the end of the wing upstroke. Aulie (1971) found that wing-beat and heart rate were synchronized in the Budgerigar, *Melopsittacus undulatus*, whereas in the seed-eater, *Sporophila sp.*, the heart beats once for every two wing-beats. By analyzing data obtained from muscle structure and from motion pictures of domestic pigeons and Collared Doves in level flight, Oehme and Kitzler (1975) concluded that their major flight muscles (pectoralis) possessed a power output, per unit weight, 10 to 20 times as great as that of mammalian muscles, with the probable exception of flight muscles of bats.

Large birds, whether they flap, glide, or soar, face special launching problems because of their heavy wing-loading and the high moment of inertia in their large wings. Unaided, their wings are incapable of ini-

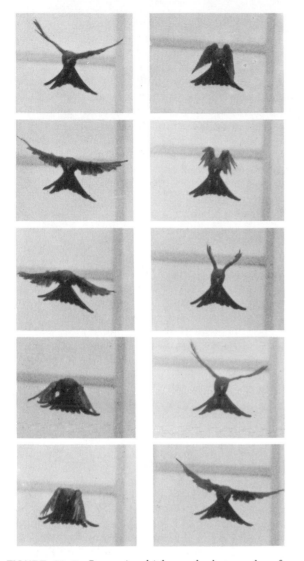

FIGURE 21–7 Successive high-speed photographs of a Black Drongo, *Dicrurus macrocercus*, approaching camera in slow, braking flight. Pictures were taken at intervals of $^1/_{80}$ second and proceed from the upper left downward. Note how the wings spread widely on the downstroke but move close to the body on the upstroke. Photos courtesy of Hans Oehme, 1965.

tiating flight. Some birds, like the herons and storks, combine a launching thrust of their long legs with the potential energy afforded by their elevated nesting sites. Others build up momentum by running windward on the ground like a condor, or pattering along the surface of the water like a coot or a goose. Albatrosses typically nest on the windward sides of islands where a short downhill run against the wind will see them airborne. In emergencies, albatrosses, vultures, and other soarers may regurgitate recently eaten food before taking off. In South America, vultures and condors are trapped by baiting enclosures that are too small to provide a runway long enough for the birds to take off. Petrels are so named because, like St. Peter, they "walk" on the water. The smaller petrels head into the wind and glide over the waves very near the surface, not by flapping their wings but by occasionally pushing against the water with one or both feet.

After a bird levels off in full flight, the wings do not beat as deeply as before, and the strokes are more nearly vertical. The arms become almost stationary, held out horizontally, and the hands do most of the flapping, bending at the wrists. It is now that the two functions of the wing become most obvious. The arm, with its secondary feathers, acts like the wing of an airplane and furnishes lift; the hand, with its primaries, acts like a propeller and provides forward drive, now, however, mainly on the downstroke.

There is great variation in the ways of flying. Some birds fly easily, from the first wing-beat, while others labor greatly to gain the air. Some fly with straightforward flapping, while others alternate short periods of flapping with gliding. Woodpeckers have an undulating flight, flapping a few times on the upcurve and diving on the downcurve with the wings tightly closed against the body. A number of birds, (e.g., many hummingbirds) in their bizarre courtship flights, gyrate through the air as though intoxicated. Some birds loop-the-loop and somersault; others zoom back and forth in arcs like a swinging pendulum, and still others essay breath-taking power dives, pulling up abruptly before the object of their attention (Wells and Baptista, 1979). A few birds, notably the hummingbirds, are able to fly backward a short distance, or to hover in a fixed spot in midair.

Hovering flight makes such special demands on the wings that, in the hummingbird, drastic modifications in wing structure have occurred. In its bony structure the wing is almost all hand; the upper arm and forearm are extremely short. The elbow and wrist joints are practically rigid; they make a permanently bent, inflexible framework that can be moved only at the shoulder, but there very freely and in almost any direction (Greenewalt, 1960). In proportion to the body, a hummingbird's wings are average in size; they are thin, flat, and pointed, with no slotting. The entire

FIGURE 21–8   While hovering or flying backward, a hummingbird applies power and lift on both up- and down-strokes of the wings. The extreme rotation of the wings at the shoulder sockets makes possible the use of their upper surfaces to provide lift on the upward or recovery stroke. After Greenewalt, 1960.

wing is essentially nothing but a variable pitch propeller.

In hovering flight, the hummingbird's body is slanted upward at about a 45-degree angle with the horizontal, so that the plane of the wing-beat is approximately horizontal. As the wing moves "up" (backward) and "down" (forward), its dorsal and ventral surfaces, respectively, face alternately downward. An extraordinary amount of rotation in the humerus makes this possible. The wing acts like an oscillating helicopter blade with a steep angle of attack, forcing the air downward with both up and down strokes. Each stroke, then, is a power stroke, which accounts for the fact that in hummingbirds the wing elevator muscles are about one-half as large as the depressor muscles, whereas in a non-hovering bird like the American Robin, *Turdus migratorius*, they are only one-ninth as large (Savile, 1950).

The chief reason hovering flight requires unusual amounts of energy is that the bird has no kinetic energy in the form of forward momentum through the air to provide lift against the pull of gravity. Compared to rotating helicopter blades, the twisting, vibrating wings of a hummingbird are grossly inefficient and require great power. The comparatively enormous breast muscles and keel of the bird show this. Breast muscles may account for 30 per cent of a hummingbird's total weight as against about 1 per cent in man. In strongly flying swallows, breast muscles may account for about 20 per cent of body weight; in

less competent fliers such as grebes and rails, about 15 per cent. One happy circumstance that helps to alleviate the inefficiencies of an alternating, helical flight mechanism in hummingbirds is the fact that each wing-beat creates a trailing current of air against which every successive reverse stroke moves, giving it the advantage of working in a higher air velocity, and hence with greater lift. In other words, with each stroke, the bird receives from the air some kinetic energy that was created by the preceding antagonistic stroke (Oemichen, 1950). Whatever the high metabolic cost of such flight, it provides superlative maneuverability.

There are probably energy-saving advantages in formation flying, although many questions remain to be answered. Many species of birds fly in flocks of different sizes and patterns, some of them so regularly organized (Heppner, 1974) that one is immediately tempted to look for adaptive advantages in their arrangements. Pelicans and numerous other species fly in long files with regular spacing between individual birds. They fly in such disciplined ranks that they even beat their wings in unison—the Germans call it "Resonanzflug"—like a chorus line. If the leader stops flapping and begins to glide, the followers glide also. It seems probable that each bird beats its wings against the rising, swirling wake of the bird immediately in front. In the V-formations of ducks and geese, each bird seems to rest its inner wing-tip on the rising vortex left by the bird in front. For such birds flying in V-formations, Lissaman and Schollenberger (1970) calculate that, theoretically, a flock of 25 or more birds could extend their flight range by 70 per cent as compared with a lone bird, because each bird flies in the upwash of its neighbor and therefore needs less power for lift. This may explain why wing-beat frequency was higher in American White Pelicans, *Pelecanus erythrorhynchos*, flying alone, than in pelicans flying in V-formations (O'Malley and Evans, 1982). A reduction in flight power demand for each wing and for the whole formation is attained as each bird in a formation flies in an upwash generated by all other wings in the group. Birds in the inner part of the formation have the advantage of the highest flight power reduction which decreases towards the apex and side edges of the formation (Hummel, 1983). The function of massed flights of hundreds and even thousands of shore birds (e.g., Dunlins, *Calidris alpina*), pigeons, and other species, is at present unknown. At one moment every Dunlin in a flock presents its back to an observer, and within 196 milliseconds, every bird, without exception, suddenly presents its underside (Davis, 1980). Just what stimulus-response mechanism orders this almost unbelievable synchrony and precision of maneuver is unknown.

For the bird in flight, landing is both more difficult and more dangerous than taking off. The great carnage that occurs, under certain atmospheric conditions, at radio and television towers, high buildings and monuments, and airport ceilometers, emphasizes the fact that abrupt landings can be as destructive to birds as to airplanes. Birds instinctively realize this, and they employ a variety of devices and actions to brake the momentum of their descent to earth.

Although some birds land vertically, helicopter fashion, the majority land obliquely, generally against the wind, and run on the ground or scud over the surface of the water to dissipate harmlessly the kinetic energy of their momentum. Water birds need not land as skillfully as land birds, because the damaging consequences of a miscalculation on water are relatively minor. Some birds, like woodpeckers, fly below their landing target and use up their kinetic

energy in a final upswing. Just before landing on the ground, a bird usually creates as much resistance against the air stream as it can. It erects its body nearly at right angles to the direction of flight. It opens its alula and creates other slots by spreading the wings and tail. In the last meter or so of flight, wings are vigorously fanned against the air. Water birds spread their extended webbed feet against the on-rushing air. Among land birds, the legs are stretched forward to absorb the shock of first impact with the ground, and in larger species, the head may continue to move earthward so as to lengthen slightly the period of abrupt deceleration.

FLYING SPEEDS, DISTANCES, AND HEIGHTS

Numerous tables have been compiled of the flying speeds of different birds, but often the data are unreliable because they have been gathered by imprecise methods, or the observer has not taken into account such modifying factors as assisting or hindering winds,

FIGURE 21–9 A male European Whinchat alighting. In landing, many birds spread their tails and fan the air vigorously with their wings to arrest their momentum. Photo by E. Hosking.

age and health of bird, state of its plumage, and conditions of escape or pursuit. A rough estimate of the top speeds of many roadside birds can be made by clocking them with an automobile speedometer. On a calm day it is a rare bird that can match the cruising speed of a modern automobile. Most passerines probably fly under 80 km per hour (50 mph).

Among the more accurate determinations of flying speeds are those obtained by using wind tunnels in which air velocity can be precisely controlled. Schnitzler (1972) observed that a White-crowned Sparrow, *Zonotrichia leucophrys*, flew with an average speed of 5.15 meters per second in still air. To maintain constant ground position at various wind-tunnel air speeds, the bird varied its air speeds from a maximum of 17.6 meters per second (63 km per hr) to a minimum of −0.9 meters per second—that is, the bird actually flew backwards to maintain position. From their observations of a Lagger Falcon, *Falco jugger*, trained to glide in a wind tunnel, Tucker and Parrott (1970) found that the bird varied its wing configurations in a way that achieved approximate maximum lift-to-drag ratios over its range of gliding speeds. They estimated that the maximum terminal speed of the falcon in a vertical dive would be about 100 meters per second (360 km per hr). Similar steep dives or "stoops" of a Peregrine Falcon, *Falco peregrinus*, that wintered on a tower of the Cologne Cathedral and preyed on urban pigeons, were calculated at 70 to 90 meters per second (252 to 324 km per hr) (Mebs, 1972).

Air speeds of birds can be determined by following them in airplanes, but the presence of the airplane itself is a disturbing factor. Ducks so pursued have flown at air speeds of about 90 km per hour. Using a Doppler radar-unit, Schnell and Hellack (1978) determined flight speeds of 13 species of birds flying at low altitudes in level flight. The birds ranged in size from pelicans to small sparrows, and their mean air speeds varied between 24 and 60 km per hour. A radio-tracking study of an adult Cape Griffin Vulture, *Gyps coprotheres*, indicated flight speeds of between 26 to 69 km per hour (Boshoff *et al.*, 1984).

One surprising aspect of bird flight is that the size of a bird has little if any relation to the speed of its flight. A hummingbird, a starling (20 times as heavy), and a goose (1000 times as heavy) all have maximum flight velocities of 22 to 25 meters per second (79 to 90 km per hr) (Hertel, 1963). For long migration flights, most birds, regardless of size, fly between 43 and 68 km per hour.

The speed with which a bird can maneuver while in flight probably depends on its size, the rate of its wing-beats, and the size, shape, and movability of its wing and tail feathers. In any case, it can be astonishingly rapid. A Black-capped Chickadee, *Parus atricapillus*, was photographed in flight with a series of four stroboscopic flashes, each operating 30 milliseconds later than the one preceding. The first flash apparently frightened the bird, so that by the second flash, 30 milliseconds later, the bird had already begun to take evasive action (Greenewalt, 1955).

There is little point in a bird's flying at high altitudes unless it can reap some advantage. The higher a bird flies, the more energy it requires to climb there, the colder it gets, the less air there is to support its wings, and the less oxygen there is to breathe. For most species, the rigorous demands of merely making a living preclude the physiological expense of high-altitude flight. However, for long-distance migrants crossing warm regions, high-altitude flying is essential to avoid dehydration. In the warmer air of low altitudes the amount of water lost through evaporative cooling exceeds the metabolic water available from the breakdown of fatty fuels. But in cold air, not only is more metabolic water available from fats generating extra heat and muscle energy, but more water

FIGURE 21–10 A Black-capped Chickadee launches itself with both its legs and its wings. If frightened in midflight, this tiny bird can begin to change its course within .03 second. Photo by H. C. Johnson.

is captured in each outgoing breath as water vapor is condensed in the cooler nasal passages. In addition, although less oxygen is available at high altitudes, there is also less aerodynamic drag to be overcome. Pennycuick (1969) differentiates between two optimum flying speeds: one speed that consumes minimum power, and the other that provides maximum range. Long-distance migrants are thought to employ the latter. It is, of course, highly unlikely that migrants travel at their energy-wasteful maximum air speeds. A few species, such as the hawks and vultures, fly at high altitudes in their sharp-eyed search for food. A very few species live in high mountains, probably driven there years ago by competition with lowland relatives. The Alpine Chough, *Coracia graculus*, has been found at altitudes around 8200 meters (27,000 feet) on Mt. Everest (Gilliard, 1958). In western Africa a commercial aircraft reportedly collided with a Rüppell's Griffon, *Gyps rueppellii*, at an altitude of 11,275 meters (37,000 feet) (Laybourne, 1974).

In a series of observations from airplanes, Mitchell (1955) found that very few birds flew at altitudes greater than 150 meters. On the average, he encountered birds above this height only once in every 70 hours of daylight flying. Other observations of birds at high altitudes by Carr-Lewty (1943) showed Pink-footed Geese, *Anser brachyrhynchos*, flying at 2100 meters; Mallards, *Anas platyrhynchos*, 1900 meters; Common Swifts, *Apus apus*, 1000 meters; Wood Pigeons, *Columba palumbus*, 950 meters; and Greater Golden Plovers, *Pluvialis apricaria*, European Starlings, *Sturnus vulgaris*, Barn Swallows, *Hirundo rustica*, and Rooks, *Corvus frugilegus*, all flying above 610 meters.

Radar scrutiny of nocturnal migrants showed that most passerine winter visitors in Britain fly below 1500 meters, although occasionally small passerines were seen flying at 4200 meters, and rarely at 6400 meters (21,000 feet). By day, the birds tended to fly somewhat lower (Lack, 1960a). During the autumn migration period, tracking-radar observations of large passerines or small shore birds flying by the island of Antigua in the West Indies showed that they traveled at a mean altitude of 2700 meters, with some birds seen as high as 6500 meters and a few recorded below 1500 meters (Hilditch *et al.*, 1973).

Records of long-distance, sustained flying, under natural conditions, are practically impossible to obtain. However, numerous records of transatlantic flights by land birds incapable of resting on the water testify to remarkable flight endurance, even though the birds may have been aided by strong tailwinds.

Some sea birds like the Sooty Tern, *Sterna fuscata*, seem unable to spend much time resting on the water without becoming waterlogged and unable to resume flight. Accordingly, their flight must be nearly perpetual except during the breeding season, since that is the only time they are ordinarily found on land. The spectacular overseas migrations of many shore birds, as well as the flights of small passerines across the Gulf of Mexico, are probably made without stopping. The tiny Andean Hillstar hummingbird, *Oreotrochilus estella*, of South America, lives at an altitude of about 3900 meters and is reputed to make daily flights of 120 kilometers and back in order to feed on its preferred flowers (Langner, 1973). Without question, birds have evolved a state of perfection in flight not yet achieved by man.

FLIGHTLESSNESS

Physiologically, flying is an expensive business, and when the advantages of flight no longer compensate for its cost, wings may atrophy or disappear. At least 13 orders of birds are known to contain flightless members (Raikow, 1985). As a rule, flightlessness is associated with geographical isolation and the relative absence of terrestrial predators. These factors probably accounted for the flightlessness of the now extinct Great Auk, *Pinguinis impennis*, in arctic regions, and explain that of penguins in the antarctic. Similar protective isolation occurs on oceanic islands where the majority of flightless grebes, cormorants, and rails exist. In fact, strong wings may be a handicap to island birds by increasing their chances of being blown to sea and drowned during hurricanes. In New Zealand, where land predators were unknown until man introduced them, there is even a flightless Kakapo (Owl Parrot), *Strigops habroptilus*, and only recently the flightless Stephen Island Wren, *Xenicus lyalli*, became extinct. The Dodo and solitaires (*Raphus* spp.) of the Mascarene Islands were clumsy, flightless birds, and they, too, became extinct when man introduced predators. Three species of duck have become flightless—the Flightless Steamer Ducks, *Tachyeres* spp., of Patagonia. Their powerful wings are too small for flight, but they use them (along with their hind legs) side-paddle fashion for swimming on the surface of the water (Humphrey and Livezey, 1982).

Avoidance of predators can be secured by adopting nocturnal living, secretive habits, or fleet-footedness. The flightless Kagu, *Rhinochetus jubatus*, of New Caledonia, the famous wingless kiwis (*Apteryx* spp.)

of New Zealand, and many rails hide by day and forage by night. The large, flightless ratites—ostriches, rheas, emus, and cassowaries—probably survive in the face of moderate predation because of their large size, keen vision, rapid running, and aggressiveness. A New Guinea cassowary can easily disembowel a man or sever his arm with its powerful kicking legs (Gilliard, 1958). Absence of terrestrial predators *permits* the evolution of flightlessness. But why is it favored? Flight muscles are energetically expensive to produce embryologically and maintain. A reduction of these muscle masses then would mean a considerable savings in energy which may be used for other functions such as reproduction, more rapid development, and larger legs (Feduccia, 1980; Diamond, 1981).

WALKING, SWIMMING, AND DIVING

Unlike most terrestrial vertebrates, birds are built for two forms of locomotion—walking and flying—and many species, such as the shore birds, have excellent equipment for both forms. They can run and they can fly—in either case skillfully and rapidly. Rails and gallinules are exceptional in being both good runners and good swimmers. However, as birds became molded to fit different niches, natural selection often specialized and refined the legs at the expense of the wings, or vice versa. This again seems to be a matter of anatomical and physiological economy. Why maintain splendid wings if the legs can do an adequate job? This principle may well explain why many good runners fly poorly or not at all. Contrariwise, some of the best fliers, such as the hummingbirds, swifts, and swallows, are all but helpless on their feet. Among the strong runners are the small plovers and sandpipers, many of the Galliformes, bustards, tinamous, and the ratites.

Sometimes the specialization of legs for one habitat diminishes their usefulness for another. The duck with its widespread legs and webbed feet can only waddle ploddingly when on shore; loons and penguins, with their legs placed far back on the body, can swim and dive magnificently, but they can walk only slowly and awkwardly on land.

Most nonpasserine birds seem to be walkers and runners, not hoppers. Hopping (when both legs act in synchrony) is mainly found among the perching birds, which means primarily the passerines. Certainly the majority of modern birds are perching birds, that is, tree dwellers, and the quickest, most effective way to get about in a tree is to hop from one branch to another. The typical perching foot has four toes of medium length, three in front and an opposing single toe behind. The toe segments are about the right lengths to allow the toes to bend snugly around small branches. The tendons that flex the toes slip in grooves and sheaths, which are located outside (anterior to) the knee joint and outside (posterior to) the ankle joint. This guarantees that the tendon will be tightened and the toes flexed around their perch by the weight of the bird as it bends its legs in settling down to roost.

A remarkable device, which automatically maintains a sleeping bird's grip on its perch, is found in the tendons that flex the toes. In the feet, these tendons are located under the toe bones. On the lower surface of each tendon are located hundreds of tiny, firm, hobnail-like projections. When the bird perches on a branch, its weight forces these projections to mesh, ratchet-wise, with hard ribs embossed on the inside surface of the adjacent tendon-sheath (Fig. 21–11). As long as the weight of the bird is opposed under its toes by the branch, the tendons will remain locked in their sheaths and the toes will retain their grip.

Many birds have become specialized as tree-trunk or rock-wall creepers: nuthatches, creepers, woodcreepers, woodpeckers, wallcreepers, wood-hoopoes, some ovenbirds, and others. These birds typically have short legs with strong, clinging toes, and, in the case of woodpeckers, generally two toes in front and two behind. The creepers, woodcreepers, and woodpeckers have stiff-quilled tail feathers, with acute tips that are used as a prop or third leg to brace the body against the pecking exertions of the head. The newly hatched Hoatzin, *Opisthocomus hoazin*, of South America, is equipped with functional claws on the ends of the second and third digits of the wings. These it uses competently for scrambling about in trees.

Parrots use their beaks as well as their feet in climbing, and, reciprocally, they use their feet in eating. Many species of birds may use their feet in manipulating food items or nest material (Clark, 1973). The preferential or exclusive use of one foot in this activity has been called "handedness" or "footedness," and is known in falconiformes, owls, parrots, tits, *Parus* spp., estrildid finches, Red Crossbills, *Loxia curvirostra*, and Yellow-faced Grassquits, *Tiaris olivacea* (Baptista, 1976). Handedness may be a manifestation of hemispheral dominance in the avian brain, and the preferential use of one limb often increases its size relative to the other limb (Smith, 1972).

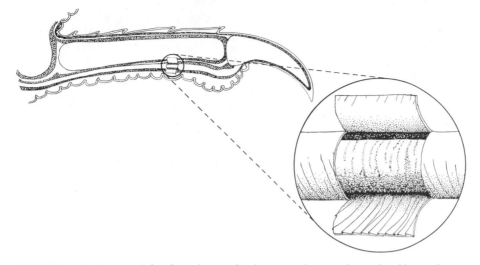

FIGURE 21–11 A pigeon's hind toe, longitudinal section, showing the ratchet-like mechanism that holds a sleeping bird on its perch. The weight of the bird, pressing through its toe bones on the underlying tendons, locks them in place in their sheaths when numerous projections on each tendon engage the sharp ridges on the adjacent wall of the sheath.

Walking and running birds, characteristic of steppes and similar open places, usually have long, powerful legs and very short toes. The toes are sometimes reduced in number: the ostrich has only two. Doves, gallinaceous birds, and rails normally accompany each walking step with a sudden forward jerk of the head. The head then remains fixed in space as the body moves forward, until the end of the step. Then the head is thrust forward before the next step is taken. This permits sharper vision than would otherwise be possible, for a steadily moving head cannot see objects—especially other *moving* objects—as clearly as can a head fixed in space (Stresemann, 1927–1934).

Motion picture analyses by Bangert (1960) of the head and leg movements of a walking domestic Chicken, *Gallus gallus*, showed that the two movements are strictly coordinated and in a rhythm of about 2.2 to 3 cycles per second. If a bird is passively carried forward, legs free from the ground, so that its retinal images will shift, the head will bob to and fro, along with perfectly coordinated leg movements.

A ground-living bird usually has a light but strong pelvic girdle to absorb the shocks of jumping and alighting. Such birds living on soft, yielding earth often have very long toes like the jacanas, or webbed toes like the ducks. Grouse grow special fringe-like scales on the edges of each toe in the wintertime only, when they function to support the bird on soft snow. Sandgrouse have short, thick toes adapted to walking on loose sand. Birds of marshes and shores have long, wading legs. These and many similar modifications show how natural selection has fitted birds' legs for locomotion in a variety of media and substrates.

Running speeds of birds cannot compare with flying speeds. However, an Emu, *Dromaius novaehollandiae*, can run 50 km per hour; the Ostrich, *Struthio camelus*, can run 50 km per hr for 30 minutes without obvious exertion and 70 km per hr for short bursts (Grzimek, 1972); a Ring-necked Pheasant, *Phasianus colchicus*, 34 km per hr; and a Greater Roadrunner, *Geococcyx californianus*, 32 km per hr. Even a two-day-old Piping Plover, *Charadrius melodus*, can run 6.5 km per hr for a distance of 25 meters. According to Tucker (1970), a 10-g bird flying 1 km uses less than 1 per cent of the amount of energy required by a 10-g mouse running the same distance.

A penguin is not ordinarily considered a land bird. It can travel on land or ice only by walking stiffly upright on its two short legs and tail prop, or by tobogganing on its belly. Nevertheless, penguins have been seen in Antarctica 110 km from the nearest sea and over 400 km from the nearest rookery. The track of one small penguin (Adelie?) was followed for 2 km. The bird walked upright for 2 meters and then tobogganed the rest of the way. In the 2-km path, the bird travelled a straight line, not deviating more than 2 degrees from a nearly due-south compass bearing (Sladen and Ostenso, 1960).

According to Boyd (1964) there are over 390 species of birds from 21 families and in nine orders that swim habitually. About two-fifths of the regular swimmers seek food by diving; the others feed from the surface, tip up for feeding just under the surface, or plunge into the water from the air. The adaptations required for diving are very different from those needed for flight. For propulsion under water, legs should be located near the rear (like the engine and propeller of a boat), where the leg muscles interfere least with the streamlining of the body, and where they can best control steering. Legs are so articulated as to thrust easily backward, and their bones are flattened laterally so that they slip through the water with little resistance (Figs. 21–12 and 21–13).

The energy cost of surface paddling for ducks is lowest at speeds of about 0.5 meter per second and amounts to 5.77 kilocalories per kilogram per kilometer. The oxygen consumption at this normal speed is 4.1 times the resting rate (Prange and Schmidt-Nielsen, 1970). Swimming birds have either webbed feet (e.g., ducks, loons, avocets) or lobed toes (grebes, phalaropes, coots). Because of their hollow bones, air sacs, and light, air-imprisoning feathers, birds have very low specific gravities. Diving birds, however, are heavier than surface swimmers or good fliers. Penguins, loons, grebes, cormorants, and auks all have relatively solid bones and heavy bodies. Penguins' stomachs commonly contain stones, eaten, perhaps, as ballast to make their bodies heavier and more easily controlled under water.

The body of the typical flying bird is streamlined, broad and blunt in front, tapering to a slender tail. The typical diver's body, on the other hand, is long and cylindrical, and its center of gravity is placed toward the rear. Streamlining is achieved in part by a narrowing of the pelvic girdle (Raikow, 1970). Water is a denser medium than air, and requires a different, more spindle-shaped form of streamlining. As insulation against the heat-absorbing water, many aquatic species have thick layers of subcutaneous fat and thick oily plumage on their ventral sides. Most water birds have large oil glands. Ducks and geese and other surface swimmers have a kind of pocket of waterproof flank feathers, which stick up on the sides of the body and in which the elbows and wrists of the wings may be kept dry while the bird paddles on the surface.

Because of their buoyancy, some water birds have difficulty in getting below the surface to feed. Pelicans, gannets, terns, and others enter the water from the air in plunging dives. Gannets, which may dive from heights of 100 meters, lack external nose openings. This is probably an adaptation that protects the respiratory system from the effects of sudden impact with the water. Grebes seem to be able to squeeze air not only from their plumage but also from the air sacs of the body, and slowly sink into the water as their specific gravity increases. The plumage of surface-feeding waterbirds is more wettable than that of divers. The former group gain about 6 per cent of their body weight when wet and the latter about 1 per cent. Anhinga feathers are more wettable, thus increasing their specific gravity as an adaptation to maneuver under water (Mahoney, 1984).

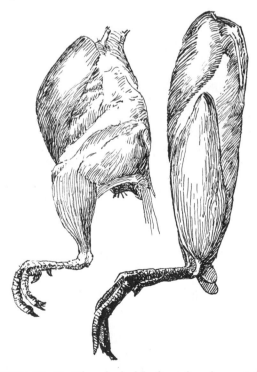

FIGURE 21–12   The plucked bodies of a pheasant *(left)* and a loon *(right)*, showing the adaptations for underwater swimming in the latter. After G. Heilmann, 1927.

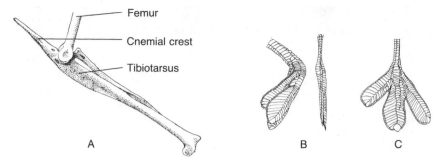

FIGURE 21–13 Structural adaptations for swimming and diving. A, Knee-joint and tibiotarsus of a Red-throated Diver (loon), showing the enormous development of the cnemial crest, on which are attached the powerful extensor muscles of the leg. After Pycraft, 1910. B, Side and front views of the lobate-webbed foot of the Great Crested Grebe, folding during forward movement in the water. C, Front view of the expanded foot during backward movement. Note the unusually thin silhouette of the shank and foot, which offer little more resistance to the water than a knife-edge. B and C after Barruel, 1954.

Geese, swans, and "surface (dabbling) ducks" feed in shallow waters, and tip their bodies head-down and tail-up to reach food on the bottom. "Diving ducks" submerge completely with quick, arching dives from the water's surface, and feed at various depths in the water. They have heavier wing-loading than surface, or "puddle," ducks and are unable to take flight without first "running" on the water surface. Diving ducks also have relatively larger "paddles" compared to dabbling ducks (Raikow, 1973). Once under water, they propel themselves with their feet, and hold their wings tightly against the body, although White- winged Scoters, *Melanitta fusca*, swim with their alulae extended.

Among species that use their wings to swim under water are penguins, shearwaters, diving petrels, cormorants, alcids, Oldsquaws, *Clangula hyemalis*, and dippers, *Cinclus* spp. In good divers, large wings can be a handicap because of their friction with the water. As a consequence, the better divers have small, thin, but muscular wings. This means heavy wing-loading and poor flying ability. In all penguins, flight has completely disappeared, and the wings have become highly specialized, powerful flippers, capable of exceptional performance in swimming. The wing, or flipper, bones have become broadly flattened and are fused at the wrist and elbow joints. The flippers may beat together or alternately. One may brake while the other beats forward, producing an extremely sharp turn. The surprising leaps that penguins can make

from the water to ice shelves at least 2 meters high testify to the speeds they can develop under water. Average speeds of Emperor Penguins, *Aptenodytes forsteri*, swimming under ice appear to be between 5.4 and 9.6 km per hour (Kooyman *et al.*, 1971).

Probably the majority of dives made by birds are under 5 meters in depth. Records of deep descents of diving alcids were obtained by recovery of birds caught in stationary sea-bottom gillnets off Newfoundland. Common Murres, *Uria aalge*, Razorbills, *Alca torda*, Atlantic Puffins, *Fratercula arctica*, and Black Guillemots, *Cepphus grylle*, may dive to depths of at least 180, 120, 60 and 50 m, respectively. There appears to be a direct correlation between size and diving ability (Piatt and Nettleship, 1985). Emperor Penguins and King Penguins, *Aptenodytes patagonicus*, equipped with depth recorders and observed from underwater chambers, achieved dives whose maximum depth was 265 m and 240 m, respectively (Kooyman *et al.*, 1971; Kooyman and Davis, 1982). Under natural conditions, diving birds rarely stay under water more than a minute, although loons have been known to remain submerged for 15 minutes. While the average dive of Emperor Penguins lasts less than 1 minute, they have been known to remain submerged for over 18 minutes. Experimentally, ducks have remained alive under water for 16 minutes.

The respiration physiology of diving birds is poorly understood. Oxygen for tissues cannot come from

FIGURE 21–14 The trunk skeleton of a guillemot, showing the dorsoventral compression of the body that is characteristic of diving birds. Protection against water pressure in deep dives is provided by the unusually long overlapping rib projections, or uncinate processes. After Pycraft, 1910.

the respiratory system, since the birds cease breathing while under water. Most of the needed oxygen comes from oxymyoglobin of the muscles and, of course, oxyhemoglobin of the blood. Both of these substances are more concentrated in diving birds than in nondivers. For example, the diving Tufted Duck, *Aythya fuligula*, may store 41.5 ml/kg of usable oxygen in the body as compared to 29.0 ml/kg for the surface-feeding Mallard, *Anas platyrhynchos*, (Keijer and Butler, 1982). These muscle and blood sources of oxygen are apparently consumed before anaerobic respiration sets in and creates an "oxygen debt." Oxygen stores in penguins are estimated by Kooyman (1975) to be roughly 60 per cent more per kg than in man. In addition, the blood is highly buffered and is therefore less sensitive to carbon dioxide accumulations. During prolonged dives ducks reduce their total energy metabolism by more than 90 per cent (Pickwell, 1968). This substantial reduction is in part revealed in the heart rate, which is reduced to as little as 5 to 8 per cent of stable predive rates. Peripheral temperatures, especially of the feet, drop precipitously, and organ systems are able to tolerate "total or near total reduction in the rate of metabolism for some

period of time." Still other adaptations to underwater respiration await discovery.

With their many adaptations for movement in water, on earth, and in the air, birds are, in Walt Whitman's phrase, born

. . . to match the gale,
To cope with heaven and earth and sea
and hurricane.

■ SUGGESTED READINGS

An excellent, lucid, and brief account of bird flight is found in Storer, *The Flight of Birds*. Rüppell's *Bird Flight* is a brief, well illustrated introduction to the subject. Greenewalt's *Hummingbirds* has a section on hovering flight with many superb illustrations. For the more technical aspects of aerodynamics, see Sherwood's *Aerodynamics*, or the paperbound book by Sutton, *The Science of Flight*. Also recommended are the articles by Pennycuick on Flight, Feduccia on Flightlessness, and Storer on Swimming and Diving, in *A Dictionary of Birds* by Campbell and Lack. In Farner and King's *Avian Biology* are excellent chapters by Berger and Hart on flight physiology (Vol. 4) and Pennycuick on flight mechanics (Vol. 5).

Migration and Orientation

22

They know the tundra of Siberian coasts
  And tropic marshes by the Indian seas;
They know the clouds and night and starry hosts
  From Crux to Pleiades.
—Frederick Peterson (1859–1938)
  Wild Geese

Vertebrate animals, unlike green plants, cannot root themselves in one place and make their own food. They must move about and search for it, and as soon as they have exhausted the food in one place they must move on to another. With their unparalleled mobility, birds have explored almost the whole earth, searching not only for food but also for territory, nesting sites, and other necessities for survival.

Because of their high metabolism, birds must have abundant and unfailing sources of rich food. It commonly happens that, in regions with seasonal climatic changes, ecological changes occur that may require the birds to move away if they are to survive. In the temperate zones the deep chill of winter closes lakes and streams to waterfowl, and so changes the land and its vegetation that most land birds must also leave for warmer regions. In the tropics, a dry season may so reduce the plants, berries, insects, worms, and other food, that the birds must emigrate to some more favorable place. These more or less regular, extensive, seasonal movements of birds between their breeding regions and their "wintering" regions are known as migration. This is simply a naturally selected tendency of birds to live the year through in places where conditions are optimum for them. Birds also make other rather extensive movements, such as the opportunistic nomadism of crossbills, or the sporadic irruptions of Sand Grouse, *Syrrhaptes paradoxus*, but only the periodic to-and-fro movements between nesting and winter quarters are considered to be true migration. Dispersal, as distinguished from migration, involves more or less random movements, often aperiodic, that have no directional bias.

Birds are by no means the only migratory animals. Among other periodic wanderers are some butterflies, squids, fishes, salamanders, reindeer, and bats. Some of these, like some birds, show remarkable capacities to "home" or to return to their birth places—for example, salmon and eels.

Some birds, of course, do not migrate at all. These

are the sedentary or resident species that inhabit a given locality throughout the year. In the United States the House Sparrow, *Passer domesticus*, is sedentary, and in Britain, the Starling, *Sturnus vulgaris*. Even in the coldest climates, some birds remain sedentary. Year-round residents in the arctic and northern part of the temperate zone of North America include the Willow Ptarmigan, *Lagopus lagopus*, Snowy Owl, *Nyctea scandiaca*, Northern Three-toed Woodpecker, *Picoïdes tridactylus*, Common Raven, *Corvus corax*, Gray Jay, *Perisoreus canadensis*, Redbreasted Nuthatch, *Sitta canadensis*, Boreal Chickadee, *Parus hudsonicus*, and some Snow Buntings, *Plectrophenax nivalis*.

Certain species are neither completely settled residents nor regular migrants. The House Sparrow, for example, is migratory in China, and the Starling in central Europe. In the United States the Barn Owl, *Tyto alba*, tends to be migratory in the northern part of its range and sedentary in the southern part (Stewart, 1952); and in central Ohio, Song Sparrows, *Melospiza melodia*, showed sex differences in migration: about one-half of the males and over two-thirds of the females migrated in the fall, while the rest of the population remained resident for the winter. Certain individual birds migrated one year but remained sedentary the next (Nice, 1937). Although male Blackbirds, *Turdus merula*, in southwest Germany, tended to winter in the breeding area, females changed from being sedentary to migratory in successive years (Schwabl, 1983). Among flocks of White-crowned Sparrows, *Zonotrichia leucophrys*, wintering in the southwestern United States, are four subspecies, each with different migration habits. *Z. l. nuttali* is a sedentary race, breeding along the California coast; *Z. l. pugetensis* is a short-distance migrant, breeding near the coasts of Oregon, Washington, and British Columbia; *Z. l. oriantha* is a long-distance migrant, breeding in the alpine zones of western mountains; and *Z. l. gambelli* is a long-distance migrant, breeding chiefly in Alaska and northwestern Canada (King, 1974).

The migration-stimulating effect of seasonal change in climate is strikingly demonstrated even within a single species. The Red-eyed Vireo, *Vireo olivaceus*, is thought by Zimmer (1938) to have three different subspecies: one remains a year-round resident in central South America, the second migrates from North America, and the third from Argentina, on seasonal visits to the locale of the first.

There is little doubt that heredity plays an important role in determining migratory behavior. In some species, particularly long-distance migrants, the instinct to migrate is firmly implanted, while in others it is weakly expressed, some individuals of a single brood migrating while others remain resident.

METHODS OF STUDYING MIGRATION

Even the most casual observer knows that vast numbers of birds depart from their northern homes at the end of the breeding season and fly to milder regions where they spend the winter. Great difficulties arise, however, when one seeks to learn more about migration. For example: Where does a specific, identified bird spend its winter? Does it migrate by day or night? At high altitudes or low? In one uninterrupted flight or in several broken flights? In a relatively direct path or otherwise? Singly or in groups? In bad weather or in good? What keeps a bird on its normal path? To answer such questions several different methods of observation have been developed recently and, as a result, much more is known today about bird migration than was known 25 years ago.

One direct approach to the first question is to identify thousands of birds at their breeding (or wintering) grounds with individually numbered aluminum leg bands and then hope to recover some of these birds in their winter (or breeding) quarters. In North America about 600,000 birds are banded each year. While this method is commendable for its simplicity, it requires great effort and yields minimum results. For example, of 226,516 White-crowned Sparrows banded in North America during some 40 years, only 198 (0.087 per cent) were recovered at any appreciable distance from their banding sites (Cortopassi and Mewaldt, 1965). Fortunately, recoveries are more numerous among game birds. In Britain each year some 1500 birdbanders band over 500,000 birds of all species. Of these, about 12,000 (2.4 per cent) are subsequently recovered (Spencer, 1977).

Another method of studying migration involves "moon watching." On four nights in October, 1952, observers at 265 scattered stations throughout the United States observed silhouettes of migrating birds as seen against the full moon. They recorded the time of night and direction of each bird's passage across the moon's disc (Lowery and Newman, 1966). This cooperative study provided valuable information about spatial grouping of nocturnal migrants, the time of peak night migration, the direction of migration paths and their relation to winds, weather fronts, and physiographic features such as rivers. An even more direct observation of nocturnal migrants can be made from a light airplane, equipped with extra landing lights and flying at different elevations and at different hours of the night among migrating birds (Bellrose, 1971). Even observing and tabulating the numerous corpses of nocturnal migrants killed after flying into high television towers provides information on geographical pathways taken, altitude of flight, fat (fuel) content per bird, the species participating, and their relative numbers.

A highly sophisticated, expensive, but productive method of observing the migration of an individual migrating bird is to capture it, equip it with a tiny (2.5 g) radio transmitter, and then, after its release, follow it with mobile receivers either on a truck or in an airplane. A Gray-cheeked Thrush, *Catharus minimus*, for example, was followed at night by an airplane for eight hours during a northerly migration flight of about 650 km from Illinois to northern Wisconsin (Cochran *et al.*, 1967).

Since 1940 the most productive method for the study of bird migration has been radar observation. Since World War II, radar instruments have increased in power and sophistication to such an extent that they are now able to detect the volume, direction, speed, height, and even wing-beats of migrating birds, over land or sea, by day or night, and as far away as about 100 km. Currently, radar is capable of detecting birds at heights of up to 6400 meters (21,000 feet), and a single gull-sized bird at a range of 80 km (Eastwood, 1967). Species cannot be identified by radar, and often the radar echoes of small groups of birds cannot be distinguished from individuals. Even so, this new technique has revealed much new information about migration. Another recent method of studying migration consists of confining migration-ready birds in circular, experimental cages equipped with devices to record the time of day, the compass direction, and the degree of migration-restlessness of each bird. Examples of some of the discoveries obtained by using these new techniques are given in the sections that follow.

ADVANTAGES OF MIGRATION

On the long and arduous journeys from their homes or breeding territories to their winter quarters, birds expend much energy and may encounter great dangers. The mean weight loss for five species of migrants after crossing the Sahara Desert was found by Ash (1969) to be between 26 and 44 per cent. The average small passerine crossing the North Sea between the British Isles and the continent loses about one-fifth of its weight. The windrows of dead birds that sometimes litter the beaches in various parts of the world bear testimony to the toll exacted of over-water migrants by adverse weather. In addition to the hazards of migration, a migrating bird normally gives up its hard-won territory once each year and must spend considerable energy to win a new territory again the following year. Bird species tend to migrate when the costs and hazards of migration are low relative to the gain in habitat suitability in their wintering grounds (Ketterson and Nolan, 1983).

Comparing the energy costs of migration (including the cost of adding premigratory fat) on a round trip of 2200 km, crossing 10 degrees of latitude, as against the stresses of 90 days of overwintering in the colder latitude, Kendeigh *et al.* (1977) estimate that small birds (10 g) profit more by migration than large birds (100 g), and that nonpasserines profit more than passerines.

The most obvious and perhaps the most important advantage provided by migration is the securing of a better climate for living. By flying hundreds and sometimes thousands of kilometers, birds can trade the bitter cold and long nights of northern winters for the gentler warmth and sunlight of southern climates. Conversely, in the summer they can escape the humid heat of the south and enjoy the long cool days of the northlands.

If all the birds that migrate should instead remain in their winter quarters the year around, they and the normally resident birds would make great demands on the food supply, and many birds might have difficulty in nourishing their young while rich food resources in northern habitats remained untouched. However, by alternately exploiting two different habitats for food, more birds are able to exist syntopically (occupying an identical microhabitat). Very probably some species migrate for this one advantage. The Pennant-winged Nightjar, *Semeïophorus vexillarius*, breeds during the rainy season (September to November) in southern Africa from Angola to Damaraland, and from southern Tanzania to the Transvaal. Then, in February, it follows the rains across the equator to "winter" quarters in Nigeria, Sudan, and Uganda. This periodic migra-

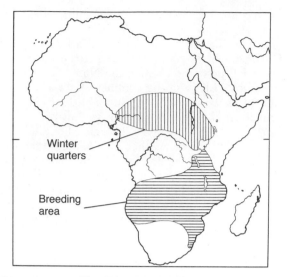

FIGURE 22–1 The Pennant-winged Nightjar breeds in southern Africa during the rainy season, and then migrates across the Congo basin to its winter quarters in the Sudan in time to enjoy the insect-rich rainy season there. The Nacunda Nightjar in South America makes a corresponding migration between Venezuela and Argentina. After Chapin, from Stresemann, 1927–1934.

tion places the bird in each habitat at the most favorable time of the year for feeding on its preferred food, flying termites (Stresemann, 1927–1934). Particularly in far northern habitats, insects are very abundant in summer, and they provide the rich protein food so necessary for young growing birds. But when cold weather arrives the insect population drops suddenly and drastically. This is why temperate and arctic zone insect-eaters must be early and long-distance migrators, while seed-eaters exhibit decidedly less migratory activity.

By changing their habitats twice each year, birds probably obtain benefits comparable to those provided by ecological interspersion or "edge effect" that are so valuable to static populations. Certainly, geographical wandering makes possible greater variety in birds' diets, and probably more nourishing diets as well. This kind of interspersion occurs both in space and time: one diet in winter quarters, another in the summer breeding range. It may well be that certain elements in the summer diet, such as vitamins, minerals, and trace elements, are particularly valuable to the growing young. As deer sometimes travel considerable dis-

tances to find a salt lick, so, perhaps, some migrating birds unconsciously seek special dietary needs, which may be exposed and available in recently glaciated northern regions but buried under organic debris in warmer climates. Contrariwise, it seems probable that some species acquire in their winter quarters stores of certain nutrients and minerals necessary for reproduction in their breeding range. By using emission spectroscopy of chemical elements in the ashes of feathers of wild geese, Hanson and Jones (1976) were able to match the birds with their localities of origin. The biochemistry of the feathers, of course, reflected the geochemistry of the environment in which they grew.

Especially in the far north, the long summer days provide birds with longer "working hours" to gather more food to feed more young and to raise them more quickly. In northern Alaska (69 degrees N. latitude) a female American Robin, *Turdus migratorius*, fed its brood approximately 21 hours each day (Karplus, 1952). As seen in Chapters 15 and 20, the farther north from the tropics a species breeds, the larger its brood generally is. The enlarged broods are more easily raised with the long days and abundant insects of northern summers. The longer breeding season of more southerly latitudes makes the raising of several broods possible. Because the Arctic breeding season is much shorter than that in lower latitudes, birds have to raise as many young as possible in the short time available. Development of the nestlings is also accelerated (Morton, 1976).

Migration furnishes another advantage in the vastly increased amount of space that it makes available to each breeding pair of birds. With more territory available, less time need be spent fighting to establish territory or nesting sites, and there is more space for the undisturbed rearing of the young. Population pressures are thus reduced at the most competitive time of the year. It may be that for some species, territorial aggressiveness is an ingredient in the forces responsible for migration.

Physiologically, migration makes possible a sort of external homeostasis that enables birds, simply by traveling, to avoid some of the harsher extremes in the physical and biotic environment. It seems possible that parasites and infectious microorganisms are reduced in abundance in far northern habitats as a result of long, cold winters, shallow soil, frozen subsoil, and relatively small animal populations. In a survey of birds of the Urals, Makarenko (1958) discovered that nematode parasites were more frequent among nonmigratory than among migratory birds. Migration also permits an economy of physiological adjustments to changing seasons—once the cost of travel itself is paid! Moreau (1972) estimates that the roughly 5000 million Palearctic migrants that spend their winter in Africa enjoy an approximately 40 per cent smaller daily energy budget than they did in their summer breeding range. An estimated 17 per cent saving comes from the nonbreeding existence and shorter African days, and another 16 per cent from the warmer temperature. It may be, too, that just as trout eggs develop best in cool waters, certain birds that breed in cool climates require a low thermoneutral environmental temperature for the most efficient development of their young. Both incubated and unincubated eggs of the Mallard Duck, *Anas platyrhynchos*, showed increased hatchability when exposed to periods of cold temperatures (Batt and Cornwell, 1972).

Finally, migration provides certain genetic and evolutionary benefits. For one thing, it affords a rigorous, three-pronged type of natural selection unknown to sedentary species. A long-distance migrant, especially, is subjected to different kinds of natural selection in its breeding quarters, its winter quarters, and while on the journey between the two habitats. Certainly, migratory species, subjected as they are to greater ecological diversity, have a greater range of adaptability than resident species (Morton, 1980). This is illustrated by the fact, referred to in Chapter 20, that North American migrants, while in their Colombian winter quarters, tolerate a greater variety of altitudinal life zones than do native sedentary species. In addition, races of birds toughened, that is, selected, by the hardships of migration, should be better able to meet the arduous requirements of the breeding season than resident races.

As discussed earlier, migration undoubtedly promotes the geographical dispersal of birds. Very likely some indigenous populations of land birds found on oceanic islands are descendants of birds that lost their way while migrating. The Arctic Tern, *Sterna paradisaea*, makes a migratory trip from the arctic to the antarctic each year. In the antarctic exists another, relatively sedentary, species, the Antarctic Tern, *Sterna vittata*, which closely resembles the Arctic Tern. It seems quite possible that years ago a few Arctic Terns remained behind in the antarctic winter quarters and began to breed there, and eventually evolved into the Antarctic Tern (Murphy, 1936). Of the 160 European species that overwinter in Africa, 32 species

have established isolated breeding populations south of the Sahara (Moreau, 1966), and 12 other species occasionally nest there without establishing permanent breeding populations (Leck, 1980).

In contrast to isolation, migration also permits the interbreeding or hereditary mixture of populations residing in different geographical ranges. A special instance of this occurs in ducks: male ducks in their winter quarters will often pair with and follow females from birthplaces other than those of the males. This explains why Mallards, *Anas platyrhynchos*, exist as a single race throughout Eurasia rather than as a number of isolated subspecies. In contrast, geese (e.g., Canada geese) pair for life, migrate in family parties, and return to breed near their birthplace. This probably explains the many races of Canada geese.

ORIGINS OF MIGRATION

Considering the great benefits of migration and the ease and perfection of bird flight, it would be one of the great mysteries of nature if birds did *not* migrate. Yet, when the average person looks at the streaming flocks of birds during a heavy migration movement, he is likely to pause and wonder, "When, and in what way, did this mysterious behavior begin?" Unfortunately, such a question cannot be answered simply and easily. Migration exists in great variety and complexity. It probably arose in different ways, at different times, and in different groups of birds. In fact, it is still arising—and disappearing—in different species today.

The Serin, *Serinus serinus*, formerly a sedentary Mediterranean bird, has in the past century extended its range throughout continental Europe as far north as the North and Baltic Seas. Around the Mediterranean Sea the species is still sedentary, but in the new range the bird has become migratory, probably in response to the more rigorous climate (Dorst, 1956). A contrary situation occurred in Allen's Hummingbird, *Selasphorus sasin*, which breeds in coastal California and winters in Mexico. On the Channel Islands off the sunny coast of southern California and on the immediate mainland the birds are sedentary (Grinnell, 1929; Wells and Baptista, 1979).

Like most animals, including man, birds are inclined to be creatures of comfortable habit, and they will change their habits only under great necessity. Without question, ecological stresses have been the most frequent cause of the migratory habit. Birds are driven out of their breeding areas by periodic climatic cycles, which bring on food shortages that are aggravated by the annual increase in bird population. Although the original impetus to migrate may be entirely external, the ceaseless repetition of seasons of famine, brought on by arctic frost or tropic drought, may eventually, through natural selection, winnow the sensitive protoplasm of a species until the urge to migrate at a particular time of year becomes innate, and comes into play at the appropriate season even in the absence of the original impelling threat of starvation. In this way climatic change has been an ultimate cause of migration, while each year the changing seasons commonly act as the proximate cause that sets off the actual migration journey.

There can be little doubt of the innate, genetic nature of migration behavior in many species. In an ingenious series of cross-fostering experiments, Harris (1970) interchanged eggs between British colonies of the nonmigratory Herring Gull, *Larus argentatus*, and the migratory Lesser Black-backed Gull, *L. fuscus*. About 900 young were reared by the "wrong" parents. Subsequent recoveries of the banded young revealed that many of the cross-fostered *argentatus* gulls migrated to France and Spain, perhaps following their foster parents. The cross-fostered *fuscus* gulls also migrated despite the fact that their foster parents remained in Britain. In their case migration was a matter of heredity and not education.

In those tropical regions that have humid, warm climates throughout the year, birds may wander at random after the breeding season, but they do not travel widely. The chief exceptions are the fruit eaters and flower feeders, particularly the fruit doves, lories, and hummingbirds (Stresemann, 1927–1934). Such birds are unable to find enough food in a limited area, and therefore must keep wandering. In the tropical rainforest there is a tremendous variety of plant life, but the plants are widely scattered, and rarely occur in solid stands of one species, as do plants of a temperate-zone forest. Consequently, birds that feed on only a few plants are required to travel more extensively than birds with more catholic appetites. The examination of the stomach contents of two species of rainforest doves in Colombia revealed only five species of plants eaten by the Band-tailed Pigeon, *Columba fasciata*, a wandering species, whereas 19 species of plants and two of molluscs were eaten by the Eared Dove, *Zenaida auriculata*, a sedentary species (Borrero, 1953). Food is apparently the key to the difference in mobility in these two birds.

Seasonal lack of food clearly affects migration among the Bramblings, *Fringilla montifringilla*. In southern Finland, up to 20 times more birds remain over winter when crops of Rowan berries, *Sorbus aucuparia*, are good than when they are poor. Furthermore, instead of shuttling back and forth within a rather restricted migration path, as does the Chaffinch, *F. coelebs*, the Brambling may show strikingly different migration routes, being found in Britain in one winter, and migrating to Switzerland, Czechoslovakia, or Italy the next—wherever the beech-mast crop is abundant (Newton, 1972). The Lesser Snow Goose, *Chen caerulescens*, was formerly a coastal migrant, but now only uses inland routes. This may have resulted from an increase in numbers and resulting feeding competition (McLaren and McLaren, 1982).

Most birds and most humans live in the northern hemisphere, which is roughly the land hemisphere of the world. It is here that the great seasonal surges of periodic north- and south-bound migration are best developed. Although corresponding migratory movements occur in the southern hemisphere, they are very meager in comparison because of the relatively small amount of land available for breeding areas in the south temperate zone. Although hundreds of species of northern hemisphere land birds migrate to lands in the southern hemisphere, no southern hemisphere land bird is known to migrate to the Nearctic or Palearctic regions. Moreau (1972) conjectures that some 4,000,000,000 Palearctic birds spend the northern hemisphere winter in Africa south of the Sahara. Ulfstrand (1973) estimates that these migrants, while wintering in Africa, make up 8.2 per cent of the total bird population. Worldwide, about 95 per cent of the land birds breeding between 40 and 50 degrees N. latitude fly south for the winter (Keast, 1976). The migratory movements, therefore, that most people know, are the preponderant south-in-winter and north-in-summer migrations of the northern hemisphere. The apparent simplicity of these movements has encouraged the appearance of several theories to explain their origin, despite the fact that many other types of migration occur.

One theory holds that birds originated in northern latitudes where Pleistocene (or earlier) glaciation forced them to seek greener pastures to the south—a tradition perpetuated today with each autumnal migration. A contrary theory maintains that birds originated in the tropics, and that population pressures, especially during breeding seasons with magnified food requirements, forced them northward each spring to exploit the food-rich lands, originally released by melting glaciers, and today, by the annual northward retreat of winter (Smith, 1980).

Still another theory suggests that migratory pathways originated in the drift of continents northward from Antarctica, and that migrations today are an attempt of various species to return to their ancestral homelands, now much removed from their ancient, southerly locations. The chief objection to this theory is that continental drift of sufficient magnitude to account for such migration occurred many millions of years before most modern species of birds, or their migration routes, had evolved. Besides, there are many migration paths whose patterns cannot easily be accounted for by such a theory.

While it is unquestionably true that Pleistocene glaciation influenced migratory patterns in the past, it is even more certain that other influences are operating today to lengthen or shorten migration routes, to change their directions, or to eliminate them entirely in some birds and originate altogether new paths in others. The fact of multiple origins of migration is more easily comprehended when one realizes that the geographical range of any animal is not the static area that a map of its distribution indicates. As Darlington (1957) makes clear, animal ranges are "inherently complex and unstable" and usually "change slightly with the animals' activities." At different times of the year animals may engage in different successive activities such as feeding, breeding, dispersal after breeding, and wintering. The niceties of ecological adjustment will often require different ranges for each of these different activities, although all of them are contained within a "sedentary" species' gross range. If an animal can travel far and fast, these subranges "can and sometimes do move far apart. This is what has happened in the case of strongly migratory birds."

Most bird migration today takes birds from a summer breeding area to a winter resting area. However, some species migrate between three different areas, each with a different function, a fact that gives force to the above argument. Several species of waterfowl, for example, make "molt migrations" by which they leave their breeding areas for a sheltered marshy place where they molt their flight feathers and grow new ones; then they fly to their winter quarters. Many thousands of European Eiders, *Somateria mollissima*, and Shelducks, *Tadorna tadorna*, congregate in July and August in the North Sea estuaries of the Elbe and Weser Rivers for their molting period.

Many Canada Geese, *Branta canadensis*, after breeding in the United States and Canada, regularly migrate northward to the Northwest Territories where they undergo molt prior to migrating south for the winter (Kuyt, 1966).

To sum up, whenever birds inhabit unstable, unpredictable environments, those that migrate and therefore survive to exploit more productive niches, will probably produce more young than those that remain sedentary. As a consequence, natural selection will favor migration (Koepcke, 1963; Alerstam and Enckell, 1979).

PATTERNS OF MIGRATION

Migration routes are probably determined more by climatic change and its power to vary the food supply than by any other environmental factor. This is why the majority of migration paths have a north and south direction. As a rule, the more rigorous the climate, especially in temperature extremes, the greater the percentage of birds that migrate. In Canada, the percentage of migrants is higher than in the United States. Still fewer of the Mexican birds migrate, and practically no birds migrate in the rain forests of the Amazon. About one-third of the breeding birds of Europe spend their winters in Africa south of the Sahara. Because the relatively small land areas south of the equator have much more uniform climates than those of the north temperate and boreal zones, very few southern hemisphere birds are strong migrants. In southern Africa there are about 20 species that move toward the equator to spend the winter.

A migratory species often winters in a region ecologically similar to its breeding area. Birds that breed in forests seek forested winter areas; desert breeders will seek deserts. This is one explanation of the fact that shore birds breeding in the far north make such tremendous migrations to similar habitats in the far south. However, the longer the migratory journey, the more costly it is biologically. This may explain the year-round residence in arctic regions of several varieties of birds. The hardships of living in an austere habitat are weighed by natural selection against the hardships of long migration trips. As long as food is available, ptarmigan and a few other residents can tolerate the intense cold. Among arctic sea birds the Puffin, *Fratercula arctica*, and Black Guillemot, *Cepphus grylle*, regularly stay through the winter in Baffin Bay at about 77 degrees N. latitude, and, in mild winters, even at Amsterdam Island, north of Spitzbergen, at 80 degrees N. latitude, feeding on littoral fish and invertebrates that they manage to catch, remarkably enough, in the darkness of the long winter nights. It is not the arctic cold, however, but the darkness that compels the Northern Fulmars, *Fulmarus glacialis*, to migrate south in mid-November when the arctic night deepens. The first birds return in late January with the reappearance of twilight (Stresemann, 1927–1934).

Winter quarters are not always equator-ward from breeding areas. If better feeding opportunities are to be found elsewhere, birds may migrate even in directions contrary to the seasonal trend. Thus, some albatrosses and guillemots migrate poleward after their breeding season to feed in cooler, more nutritious waters. Gulls whose breeding grounds are near Orlov, Russia (north of the Black Sea), winter to the west and southwest along the shores of the western Mediterranean. Banding recoveries of Redhead Ducks, *Aythya americana*, and Blue-winged Teal, *Anas discors*, which breed in the prairie states as far west as Utah, show that many of them migrate to the Atlantic Coast in the fall. Likewise, Evening Grosbeaks, *Coccothraustes vespertinus*, and Purple Finches, *Carpodacus purpureus*, banded in northern Michigan, have been recovered in numerous Atlantic coast states (Lincoln, 1950). The Rose-colored Starling, *Sturnus roseus*, which breeds in the Caucasian and Turkish steppes, spends its winter 3000 kilometers to the east in India; and Snow Buntings, *Plectrophenax nivalis*, which nest in northeast Greenland, have been recovered in northeast Russia.

Birds breeding in warmer climates are frequently rain followers, and rains do not always follow the sun in its annual movements north and south. In Madagascar there are five species of breeding birds that "winter" in Africa; one of them, the Broad-billed Roller *Eurystomus glaucurus*, migrates as far westward as Congo (Rand, 1936). A much shorter east-west migration is that of the Mourning Chat, *Oenanthe lugens*, which breeds between the Nile and the Red Sea, and spends its winters on the fringes of the Sahara (Hartley, 1949).

The geographical location of continents, mountain ranges, great rivers, deserts, peninsulas, and islands creates opportunities and problems for migrating birds, and consequently helps determine their migration paths. From this standpoint, the Old World and New World differ strikingly. In the New World the two continents are roughly north and south of each other. The great sierras run north and south, and there

is a land connection, plus many islands, between the two continents. In North America, where migration predominates, the great Mississippi Valley runs north and south.

In the Old World, on the contrary, mountain ranges and great desert systems run east to west, creating barriers to north and south seasonal migration. Southern land masses suitable for wintering areas are displaced to the west (Africa) and east (East Indies and Australia) of the center of the great Eurasian breeding range of most Old World birds. Accordingly, many Eurasian birds are unable to pursue a north-south migration route to their winter quarters. Although many species that breed in east Asia winter in the Indo-Malaysian regions, some of them winter far to the west in Africa. For example, the Spotted Flycatcher, *Muscicapa striata neumanni*, which breeds east of Lake Baikal in Siberia, the Common Swift, *Apus apus pekinensis*, of northeastern China, and the Red-footed Falcon, *Falco vespertinus amurensis*, of southeastern Siberia, all winter in Africa, flying there by way of India, Pakistan, and Saudi Arabia (Stresemann, 1927–1934). Moreau (1972) notes that at least 58 Palearctic species breeding between 45 and 90 degrees E. longitude, and 15 species breeding east of 90 degrees E. longitude (e.g., in Mongolia), regularly winter in Africa, whose central meridian is roughly 20 degrees E. longitude.

Some birds, reluctant to cross deserts or large expanses of water, take long, circuitous routes to reach their winter quarters. Numerous central European species, for example, go around either end of the Mediterranean rather than fly across it to their African winter quarters. Some species trace an angular course around the eastern end of the sea in order to follow the Nile southward and thus avoid the sterile Sahara. Storks and other soaring birds must remain over land if they are to use terrestrial thermal currents for their long soaring flights. Even so, at least 25 species of small birds, chiefly wood warblers (Parulidae), have been observed or captured over the Atlantic Ocean, as far as 2100 km from the mainland, in their fall migration from northeastern North America to the West Indies and South America (McClintock *et al.*, 1978).

Most species, in their migrations, sweep across the land in broad paths or "fronts," and usually take a rather direct route from breeding to wintering areas. Whether a species travels in a broad or a narrow path depends chiefly on the geographic extent of its breeding and wintering quarters and on the geographical and ecological nature of the intervening route. The Ross Goose, *Chen rossii*, breeds in a very restricted area in northern Canada near Queen Maude Gulf and on Southampton Island, and it winters in a small area in California; consequently it has a narrow migration path. Shore birds generally follow coasts of one sort or another to find food as they migrate, or else they migrate nonstop between summer and winter homes. Soaring birds, such as hawks and eagles, frequently migrate along very narrow corridors defined by mountains on whose flanks updrafts provide the motive force for their travels. The famous bird sanctuary at Hawk Mountain, Pennsylvania, is on such a migration highway. On September 14, 1978, over 21,000 Broadwinged Hawks, *Buteo platypterus*, flew over Hawk Mountain in their autumn migration (Nagy, 1979).

Broad-front migrants are occasionally funneled into narrow paths by the nature of the territory over which they pass. The Scarlet Tanager, *Piranga olivacea*, nests from the Dakotas to the Atlantic Coast in a belt about 3000 km wide. In fall migration it flies to the Gulf Coast, where its path narrows until it is about 1000 km wide. From here it flies across the Gulf of Mexico to Central America where, in Costa Rica, its path narrows further to a mere 160 km (Lincoln, 1950). Near the southeast shore of the Baltic is a long, narrow, offshore island called the Kurische Nehrung. In the autumn, land birds, migrating southwesterly from the east Baltic countries and the U.S.S.R., avoid the open water and become concentrated on this long, sandy ribbon of land. Sometimes as many as one-half million birds a day fly by. Until the end of World War II, the famous German ornithological research station, Vogelwarte Rossitten, was located here. In their autumn migration from Europe to Africa small passerines, unable to soar, generally cross the Mediterranean and the Sahara Desert on a broad front. Larger soaring species, such as storks, hawks, and cranes, travel on a broad front in Europe until they reach the Mediterranean, over which there are no thermal updrafts. Here the birds concentrate into narrow paths following the most direct route over land that will provide them with rising thermals. During 114 days from mid-July until early November, four observers counted 207,000 European White Storks, *Ciconia ciconia*, soaring across the Bosphorus at Istanbul (Porter and Willis, 1968). The energy-saving advantages of soaring migration are considerable. A soaring Herring Gull, *Larus argentatus*, consumes oxygen at less than one-third the rate

of that of a gull in flapping flight (Baudinette and Schmidt-Nielsen, 1974). During migration the European Crane, *Grus grus*, typically soars at about 50 km per hour, ground speed, while over land, but flaps its wings when crossing the Baltic, and then flies at about 77 km per hr ground speed, albeit with much less efficiency than in its flight over land (Alerstam, 1975).

Such migration-diverting objects as sea shores, mountain ranges, large rivers, desert rims, or forest edges are called guiding lines, or in German, *Leitlinie*. It was long thought that most migrants flew in narrow paths along such guiding lines as, for example, the Mississippi River, but recent studies of migration, using observation by airplanes and radar or observation of bird silhouettes against the full moon by telescope, have shown that the great majority of birds fly in broad fronts and are only rarely channeled into narrow paths by environmental guiding lines. Four such broad-front migration bands or "flyways" have been described by Lincoln for North America: the Atlantic, Mississippi, Central (or plains), and Pacific flyways. While there is a general consistency in the use of specific flyways by certain bird populations, birds are by no means rigidly restricted to them. Ducks breeding in the Canadian prairie provinces may migrate along all four flyways.

There is good evidence that *Leitlinie* may at times be used by birds to make corrections in their migratory paths when they have been displaced by strong sidewinds at night or when flying over featureless water. Migrating waterfowl may use large rivers such as the Mississippi as guidelines as long as the river parallels their chosen line of flight (Bellrose, 1967). Nocturnal fall migrants in New York fly a course paralleling the Hudson River when encountering strong winds. They apparently concentrate near the river as partial compensation for wind-drift (Bingham *et al.*, 1982). Several Semipalmated Plovers, *Charadrius semipalmatus*, banded at a beach at Manomet, Massachusetts, were captured again at this stopover area for from one to eight subsequent years (Smith and Houghton, 1984). Many birds seem to exhibit a homing fidelity for breeding grounds, migration route, and winter quarters. Passerines seem more likely to migrate on broad fronts, rarely showing route-fidelity.

There are certain deviations from the normal back-and-forth shuttling of most migration flights that throw further light on the origin of migration routes. The Lesser Golden Plover, *Pluvialis dominica*, breeds from northern Alaska across arctic Canada to southern Baffin Island, and winters in southern South America. When migrating north in the spring, this species travels by way of Central America, Mexico, the Mississippi Valley, and on through Canadian prairies west of Hudson Bay. However, to migrate south in the autumn, the majority of birds fly eastward to Labrador, Nova Scotia, and upper New England, where they strike out across the Atlantic Ocean, nonstop, for the West Indies and South America. Thus the annual circuit describes a loop migration, or *Schleifenzug* (Fig. 22–2). This particular loop migration is explained by Cooke (1915) as follows: Birds follow that route between summer and winter areas that provides the shortest path with adequate living conditions and food. The long Atlantic shore lines of Labrador, Nova Scotia, and New England provide both an ecologically suitable highway and a rich source of late summer

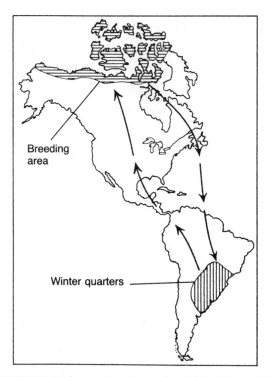

FIGURE 22–2 Loop migration in the Golden Plover. This species migrates south in the autumn via the Maritime Provinces and the Atlantic Ocean, but it returns north in the spring by a midcontinental route because ecological conditions there are more suitable for migration at that time. After Lincoln, 1950.

berries, which the birds use to fatten themselves for the long trip across the open sea. To return by this same path would be fatal, because spring arrives late and foggy on the Labrador coast, and the still-frosty soil could offer no food to the exhausted oceanic fliers. But the trip up the Isthmus of Panama and the grasslands of mid-America is congenial, warm, and already producing food for the hungry migrants.

Another example of *Schleifenzug* is that of the Arctic Loon, *Gavia arctica*, which breeds in the northern U.S.S.R. and migrates south for the winter via the Black Sea to southern Europe. In spring, however, it makes an early return to its home by first flying northwest to the Baltic, and then northeast, and finally east to its breeding grounds. The loon is a large bird with heavy wing-loading, and it can alight and take off only from water; therefore it can migrate only where open water is available for landing, feeding, and taking off. Thanks to the Gulf Stream, the waters of the maritime Baltic countries thaw out earlier than those of inland U.S.S.R., thus encouraging this loop migration (Bodenstein and Schüz, 1944). Still other loop migrations are thought to be the result of strong, prevailing side-winds during one migration trip but not during the return trip, thus causing a lateral wind drift during the one flight but not during the other. This is possibly the reason for the numerous loop migration paths described by various European species migrating through North Africa.

Many details of migration patterns have been made clear by the remarkable radar studies of Richardson (1974, 1976) in Puerto Rico (Fig. 22–3). Using long-range surveillance radar for 148 days and 130 nights, Richardson discovered the following facts. More birds, smaller birds (probably mostly passerines), and smaller groups, flew by night than by day. In the spring, most birds, mainly passerines, took off from Puerto Rico, ½ to 1 hour after sunset, and flew, mostly west, northwest, or north, toward the Bahamas and the United States. In the autumn, hordes of birds arrived in Puerto Rico from the northwest, north, and northeast, apparently having followed a curved, clockwise path, 2000 to 3000 km long, from southeastern Canada and New England. Those that continued on to South America flew at least 4000 km nonstop. In both spring and fall, flight tracks took advantage of wind direction, a fact resulting in an annual, transoceanic loop migration for many species. Cloud cover, spring and fall, had no apparent effect on the density, mean track, or mean heading of the migrants. The birds showed no evidence of seri-

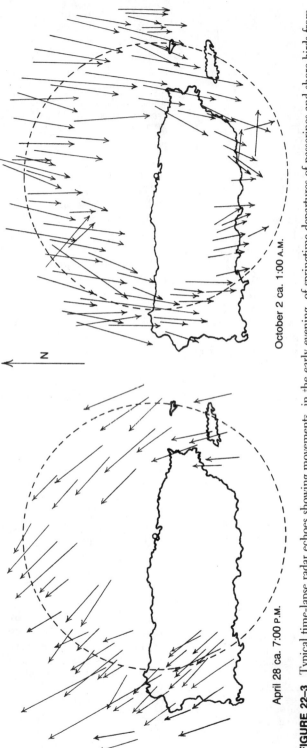
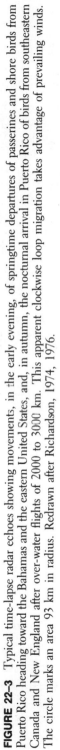

October 2 ca. 1:00 A.M.

April 28 ca. 7:00 P.M.

FIGURE 22–3 Typical time-lapse radar echoes showing movements, in the early evening, of springtime departures of passerines and shore birds from Puerto Rico heading toward the Bahamas and the eastern United States, and, in autumn, the nocturnal arrival in Puerto Rico of birds from southeastern Canada and New England after over-water flights of 2000 to 3000 km. This apparent clockwise loop migration takes advantage of prevailing winds. The circle marks an area 93 km in radius. Redrawn after Richardson, 1974, 1976.

ous disorientation on cloudy nights. Many birds flew at altitudes of 3 to 6 km. By night, birds generally flew on straight, broad-front courses without apparent response to topographic features, but by day, concentrations of birds often occurred along coast lines. Some migrants showed at least partial compensation for wind-drift.

"Leap-frog" migration aptly describes the pattern taken when races of the same species occupy two or more breeding areas (and also wintering areas) in the axis of migratory flight. There are six different races of the Fox Sparrow, *Passerella iliaca*, that breed and winter along the North American Pacific coast (Fig. 22–4). One race both breeds and winters in the Puget Sound area. The races that breed progressively farther north along the coast, up to the outermost of the Aleutian Islands, winter progressively farther south in Oregon and California; each race on its southern migration passes over winter areas already preempted by the race that breeds south of it (Swarth, 1920). Three races of the Common Ringed Plover, *Charadrius hiaticula*, and races of numerous other species show this same leap-frog migration. This type of migration is probably an adjustment to mutual intolerance engendered by

the sharpness of intraspecific competition. Many of the species and races that breed farthest north, winter farthest south.

Some migration paths seem to be remarkably crooked, illogical, out-of-the-way routes for connecting breeding and wintering areas. Careful studies show that these are not paths of biological necessity, but are the fruits of tradition, much as the crooked streets of Boston trace their lineage to ancestral cowpaths. It is, of course, impossible to trace the historic causes behind all present-day migration routes, but there is little question that past climatic changes, particularly glaciation, were responsible for the patterns many of them now display. In geologically recent times, birds of the boreal and north temperate zones have been shoved about, or exterminated, by extensive glaciation. The climatic changes wrought by glaciation caused widespread changes in the earth's vegetation and, therefore, in the birds' habitats. What is now the Sahara Desert was formerly a verdant winter refuge for birds from southern Europe. It takes very little imagination to realize the dislocations in migration routes that would be occasioned by the gradual drying up of the Sahara region, or by the slow shifting apart of any breeding and wintering areas.

As glaciers receded at the end of the last ice age, new northern and alpine habitats were opened up. These unoccupied habitats were ecological vacuums into which various species extended their ranges, pushed from behind by population pressure. Range extensions were not always in the direction of the chief migration axis. But when the summer residents of these new habitats were ready to migrate "south" in the fall, it was only natural for them to join their relatives and fly with them along the ancestral pathway to their traditional winter quarters, rather than to strike out by themselves on a direct and shorter new path. As a species gradually extended its new range, more and more birds followed the illogical ancestral path to winter quarters. In some cases this happened because younger birds learned the way from their elders; in others the habit of following the crooked path became innate. This hypothesis clearly fits numerous present-day migration paths, including the following examples.

Numerous Black Brant, *Branta bernicla*, banded on their breeding grounds in north central Siberia, and banded Snow Geese, *Chen caerulescens*, from Wrangel Island, Siberia, have been recovered in their winter quarters in western Canada and the United States.

The Pectoral Sandpiper, *Calidris melanotes*, breeds

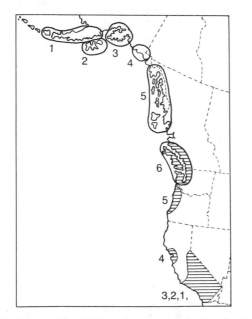

FIGURE 22–4 "Leap-frog" migration of the Pacific Coast Fox Sparrow. Breeding areas are encircled; winter quarters, shaded. The numbers refer to the different sub-species using the areas. After Swarth, from Lincoln, 1950.

all along the arctic coast of North America. It has extended its range across Bering Straight into northern Siberia as far westward as the Taimyr Peninsula. When the Siberian population of these birds migrates to winter quarters, most of them do not fly directly south to an Asiatic refuge as one might expect. On the contrary, they first fly east across the Bering Strait to Alaska, where they join the American birds and, with them, travel through North and South America to their winter quarters in Peru, Chile, and Argentina, including Patagonia (Fig. 22–5). In a reciprocal fashion, the Old World Common Ringed Plover, *Charadrius hiaticula*, has spread to Greenland, and to Ellesmere and Baffin Islands of the New World. The American population, instead of flying south along the shores of North America to its winter quarters, flies to Greenland, Iceland, and then down to its winter quarters along the shores of Britain, western Europe, and northwestern Africa.

Even more striking is the migration route followed by the Arctic Warbler, *Phylloscopus borealis*, of the Old World. This tiny bird presumably originated in northeastern Siberia. It has spread in two directions: eastward to northern Alaska, and westward across the entire breadth of the Siberian tundras to Finland and Norway. Its winter quarters are in southeastern Chi-

na, Indonesia, the Philippines, Borneo, and adjacent islands. The Alaskan warblers cross Bering Strait and migrate south with the Siberian birds. But more astonishing is the long trip taken by the Norwegian and Finnish birds. When migrating to their winter quarters, they first fly east some 3000 or 4000 km across Siberia, and then go south with their relatives to the tropics. The total journey from Norway to southeastern Asia is probably about 12,500 km—a long trip for a bird only 12 cm in length (Stresemann, 1927–1934)!

The Northern Wheatear, *Oenanthe oenanthe*, which breeds across the whole of northern Eurasia, has extended its range into North America both from the west, via Siberia into northern Alaska, and from the east, via Greenland into Labrador, northern Quebec, and the arctic islands to the north. Both of these American populations winter in tropical Africa, the birds from Alaska crossing Asia to reach Cameroon and Tanzania, and those from Labrador and Quebec flying across Greenland and Iceland to Europe and on to Morocco, Senegal, and Sierra Leone (Meinertzhagen, 1954).

Not all migratory birds are slaves to tradition. DeSante and Ainley (1980) recorded some 223 species of land and freshwater birds on the Farallon Islands, which lie about 32 km off the coast of San Francisco,

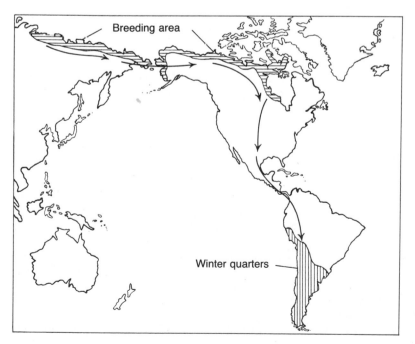

FIGURE 22–5 Many birds follow ancestral pathways to winter quarters. The Pectoral Sandpiper has extended its range from North America to north central Siberia. Each year, this Siberian population retraces its historic expansion path to join its American relatives when they migrate to South America for the winter. Contrariwise, an Alaskan race of the Asiatic Willow Warbler migrates each year through Siberia to spend the winter in the Philippines and southeast Asia.

California. Included on this list is a large number of eastern North American species, e.g., Blackpoll Warblers, *Dendroica striata*. Twenty-four of these warblers were placed in orientation cages by DeSante (*in* Diamond, 1982) who found them orienting in a "mirror image" direction of their normal compass bearing, which would place them somewhere over the California coast.

Some birds reap the advantages of long-distance migration with relatively little effort by migrating vertically instead of horizontally. Aristotle was aware of vertical, or altitudinal, migration:

Among birds the more feeble of them descend to the plains in winter when it is cold because there they find the air more temperate; in summer when the plains are burning, they return to the heights
—(Dorst, 1956).

This economical form of migration can be practiced only by a limited number of birds, because of the obvious limitations of alpine space. In the Rocky Mountains of Colorado, the following birds breed in the alpine or subalpine zones in summer and descend to lower levels in winter: Pine Grosbeak, *Pinicola enucleator*; Black-capped Rosy Finch, *Leucosticte australis*; Dark-eyed Junco, *Junco hyemalis*. A type of inverted altitudinal migration is found in the Blue Grouse, *Dendragapus obscurus*, which winters at high altitudes in the mountains of Idaho, where it feeds on needles and tree buds. The birds descend 300 meters or so in spring to feed on the earlier developing leaves and flowers, and the females with broods descend still farther in summer to feed their young on valley insects and berries (Marshall, 1946).

Akin to seasonal vertical migration are the daily movements of the Ethiopian White-collared Pigeons, *Columba albitorques*. They rise 2500 to 4400 meters each morning to feed on the plateaus, and descend each afternoon to roost in valley caves (Boswell and Demment, 1970).

Oceanic birds also show migration movements correlated with the seasons. These are most sharply evident when sea birds come ashore to breed. Because of the lack of fixed observation posts in the open sea, students know much less about the movements of sea birds than about those of land birds. The surface of the ocean, which is the only part that birds inhabit, is a much more uniform environment than the surface of the land. It is flat, constantly wet, relatively uniform in its physical and chemical properties, and it varies in temperature throughout the world only about one-

fourth or one-fifth as much as do land or air temperatures. As a result, sea-bird migrations are probably less complex and varied than those of land birds. Sea birds have less need for wide movements to escape drastic seasonal changes in temperature. Nevertheless, some sea birds make remarkably long migration journeys. For example, the Short-tailed Shearwater, *Puffinus tenuirostris*, breeds on the coastal islands of southern Australia and winters throughout the Pacific Ocean as far north as the Aleutian Islands. Its migration route in the Pacific describes a huge clockwise circle that enables it to take advantage of prevailing winds both in its spring and fall travels. The Ruddy Turnstone, *Arenaria interpres*, of the Pribilof Islands makes a similar clockwise oceanic migration (Thompson, 1973). More restricted, at least in latitude, and perhaps more typical of sea birds, are the seasonal movements of the Wandering Albatross, *Diomedea exulans*, as reported by Dixon (*in* Murphy, 1936). Observations during 2002 days in the south Atlantic, between 20 and 60 degrees S. latitude, showed that this species was practically confined between 30 and 60 degrees S. latitude. In the winter, 96.9 per cent of the birds seen were between 30 and 50 degrees, and only 0.1 per cent south of 50 degrees S. latitude, whereas in summer, 73 per cent were between 30 and 50 degrees, and 26.5 per cent south of 50 degrees. Albatrosses banded at South Georgia in the south Atlantic have often been recovered on the opposite side of the globe in New South Wales, Australia. As with land birds, food is an extremely important factor in determining migration movements of sea birds. Ocean waters rich in fish and other foods are more heavily populated with birds than waters poor in food, whether the birds are migrating or in their winter or breeding areas.

MECHANICS OF MIGRATION

Migratory behavior is probably about as old as birds themselves. Even so, birds have not evolved into two distinct types, migratory and nonmigratory. The many partial migrants living today—species in which some individuals migrate and others do not—suggest that migratory behavior is an evolutionary expedient, appearing or disappearing according to selective environmental pressures. Nevertheless, strongly migratory species or subspecies usually have longer, more pointed wings than their less migratory or sedentary relatives (Gatter, 1979). For example, the Northern Wheatear, *Oenanthe oenanthe*, which migrates from Eurasia to Africa, has primary wing feathers that extend more than twice as far beyond the secondaries as do those of the sedentary *Oenanthe phillipsi* of Somalia (Kipp, 1958). Similar differences in wings occur between migratory and sedentary species of the same genus in genera of Old World warblers, shrikes, wrynecks, rollers, and nightingales (Fig. 22–6). Kipp believes that these wing adaptations arose before the last glacial period. Migratory European Blackcaps, *Sylvia atricapilla*, have longer wings than the sedentary African races. F_1 hybrids produced in aviaries had wings intermediate in length between the two races, indicating genetic control for this character (Berthold and Querner, 1982).

The great distances covered by migrating birds are frequently cited as one of the wonders of migration. Many shore birds fly thousands of kilometers without stopping for rest or food. Many small passerines cross the Gulf of Mexico from the Gulf Coast of the United States to Yucatán or Tabasco, in one long flight. Radar observations now support the concept that the tiny Blackpoll Warbler, *Dendroica striata*, in its autumnal migration flies nonstop from New England either to the Lesser Antilles or to Venezuela (Nisbet *et al.*, 1963). With or without resting en route, many migrants make remarkably long journeys between breeding and wintering regions. One of the longest migration trips is that of the Arctic Tern, *Sterna paradisaea*, which nests in circumglobal arctic regions and winters on shores in the subantarctic. From the North American arctic, this species crosses the Atlantic to follow the cooler, more nutritious waters of the west shores of Europe and Africa to the tip of South Africa, whence it crosses the south Atlantic to Antarctica or Patagonia—a trip of perhaps 18,000 km. Shore birds that nest in the arctic make migrations nearly as long. The Bristle-thighed Curlew, *Numenius tahitiensis*, of Alaska flies as much as 10,000 km to reach its winter quarters in Polynesian islands.

The long land journeys of some passerines have already been mentioned. Remarkable as these trips are, they are no more surprising than the flights some birds perform in their daily activities. The European Common Swift, *Apus apus*, is estimated to fly at least 900 km per day gathering insects for its young. This is probably an extreme, but it is likely that many species fly at least 100 km per day in feeding activity. This same rate of daily travel would carry most of them fast enough to complete their normal migration flights with time to spare.

Direct measurements of the mean ground speeds of finches migrating against headwinds of about 7 km per hour were made by Gatter (1979). He found that the speed of the partially migrant Chaffinch, *Fringilla coelebs*, was 37.1 km per hr; that of the longer-winged, long-distance migrant, the Brambling, *Fringilla montifringilla*, was 41.4 km per hr.

One dependable method of learning migration speeds is to observe the migratory movements of large flocks that include so many birds that their progress across the country can be traced. In one October migration flight, about 500,000 ducks from eastern Saskatchewan and Manitoba flew, on a broad front several hundred kilometers in width, to their wintering areas in Louisiana in a little over one day. A heavy flight like this is called a "grand passage." Careful analysis of this flight by Bellrose and Sieh (1960) indicated that the ducks leaving Saskatchewan made a continuous flight of some 2400 kilometers to Louisiana at an average ground speed of about 65 to 80 km per hour, or 1600 to 1900 km per day. A flock of Blue Geese and Snow Geese, *Chen caerulescens*, color phases of the same species, travelled, apparently without stopping, from James Bay, Canada, to Louisiana, a distance of 2700 km, in 60 hours, or at the rate of 46 km per hour or 1100 km per day (Cooch, 1955).

Another way of determining the flight speeds of

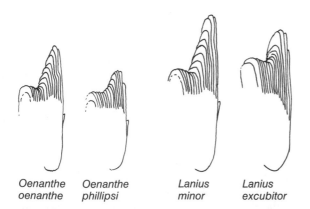

| Oenanthe | Oenanthe | Lanius | Lanius |
| oenanthe | phillipsi | minor | excubitor |

FIGURE 22–6 Migratory birds frequently have longer, more efficient wings than their nonmigratory relatives. The Wheatear, *Oenanthe oenanthe*, migrates from Europe to central Africa; the species *Oenanthe phillipsi*, of Somalia, is nonmigratory. The lesser Gray Shrike, *Lanius minor*, migrates from Europe to southern Africa; the Great Gray Shrike, *Lanius excubitor*, of Europe, is nonmigratory. After Kipp, 1958.

FIGURE 22–7 The champion long-distance migrant is probably the Arctic Tern, which may travel as far as 36,000 km each year in migration flights. Photo by E. Hosking.

birds under seminatural conditions is to displace a marked, breeding bird from its nest and see how long it takes to return. This technique does not reveal migration speeds, but at least it shows what a bird can do in natural flight. Table 22–1 gives flight speeds of various displaced birds that returned in the shortest times under such conditions. The birds listed in the table are all strong fliers. It is apparent that large size is not a requisite for flying either rapidly or far.

The most accurate and dependable measurements of migration speeds are those obtained through radar observations. Radar beams, although their reach is limited, can measure velocities of wild birds while flying under natural conditions. Most radar determinations place the ground speed of migrants at between 30 and 70 km per hour. Tracking radars detected fall migrants over the Atlantic near Bermuda traveling at a mean air speed of 67 km per hour and at an average height of 1734 meters (Ireland and Williams, 1972). Collective results from a network of radars suggest that

trans-ocean migrants require about 80 hours to fly from Nova Scotia and Cape Cod to South America (Williams *et al.*, 1978). Bruderer (1972) found that air speeds of migrants increased 10 per cent for each kilometer increase in flying altitude. This increased speed may reflect both a decrease in air resistance to flight and a preference of rapid fliers for high altitudes.

Performances of marked birds made during actual migration flights are given in Table 22–2. There is, of course, no guarantee that a banded bird will begin migrating immediately upon its release, that it will fly directly to its recovery point, or that it will be recovered (if at all) immediately upon its arrival there. While these are maximum records, they very likely do not represent maximum potentialities of the species named.

One feature of the migratory movements of many species, as yet unexplained, is the greater speed of spring migration compared with that of autumn. It is as though birds were impelled much more strongly to

TABLE 22–1.
FLIGHT SPEEDS OF EXPERIMENTALLY DISPLACED BIRDS RETURNING TO THEIR NESTS

SPECIES	DISTANCE DISPLACED (km)	ELAPSED TIME OF FLIGHT (Days)	KILOMETERS PER DAY	AUTHORITY
Laysan Albatross, *Diomedea immutabilis*	5150	10	510	Kenyon and Rice, 1958
Manx Shearwater, *Puffinus puffinus*	5300	12.5	420	Mazzeo, 1953
White Stork, *Ciconia ciconia*	2260	11.9	190	Wodzicki, 1953
Herring Gull, *Larus argentatus*	1400	4.1	340	Griffin, 1943
Noddy Tern, *Anoüs stolidus*	1739	5.0	350	Watson, 1910
Homing Pigeon, *Columba livia*	1620	1.5	1080	Allen, 1925
Alpine Swift, *Apus melba*	1620	3.0	540	Bourlière, 1950

reach their breeding areas than they are to reach their winter quarters. The Bar-tailed Godwit, *Limosa lapponica*, which breeds in Alaska and eastern Siberia, requires only one to one and a half months to migrate north from New Zealand in the spring, but it takes two to three months to make the return trip in the fall. European Wood Warblers, *Phylloscopus sibilatrix*, travel northward from their African winter quar-

ters in the spring at an average rate of about 180 km per day, but return in the fall at the more leisurely pace of 90 km per day (Stresemann, 1955). Many other species show this same tendency.

Some birds migrate at a fairly uniform speed, if one can judge by the times of first spring arrivals at different latitudes. As a rule, early spring migrants advance northward more slowly than late migrants. The Black-

TABLE 22–2.
FLIGHT SPEEDS OF BANDED WILD BIRDS DURING MIGRATION, MADE BETWEEN THEIR RELEASE AT ONE POINT AND THEIR EARLY RECOVERY AT A DISTANT POINT

SPECIES	DISTANCE COVERED (km)	ELAPSED TIME OF FLIGHT	KILOMETERS PER DAY	AUTHORITY
Manx Shearwater, *Puffinus puffinus*	9500	17 days	650	Ashmole, 1971
Mallard, *Anas platyrhychos*	890	2 days	445	Cooke, 1940
Blue-winged Teal, *Anas discors*	4800	35 days	140	Sharp, 1972
Peregrine Falcon, *Falco peregrinus*	1600	21 days	76	Cooke, 1946
Lesser Kestrel, *Falco naumanni*	8785	61 days	144	Preston, 1976
Ruddy Turnstone, *Arenaria interpres*	4655	4 days	1045	Sharp, 1972
Lesser Yellowlegs, *Tringa flavipes*	3100	7 days	444	Cooke, 1938
Red Knot, *Calidris canutus*	5640	8 days	705	Spencer, 1964
Ruff, *Philomachus pugnax*	6500	32 days	203	Lebedova, 1965
Arctic Tern, *Sterna paradisaea*	14,000	114 days	123	Lincoln, 1950
Barn Swallow, *Hirundo rustica*	8800	35 days	250	Rowan, 1968
Great Tit, *Parus major*	1200	21 days	57	Dhondt, 1966
Redwing, *Turdus iliascus*	2400	4 days	600	Barriéty, 1967
Yellow-rumped Warbler, *Dendroica coronata*	725	2 days	362	Novy, 1963
White-crowned Sparrow, *Zonotrichia leucophrys*	500	12 hrs	1000	Cotopassi, 1965

and-White Warbler, *Mniotilta varia*, is a slow but early migrant, advancing northward across the United States at an average rate of about 32 km per day. A more rapid pace is set by the Gray-cheeked Thrush, *Catharus minimus*, which migrates from Louisiana to Alaska at an average rate of about 210 km per day. In this case, however, the bird does not travel at a uniform speed but flies about four times as far each day in Alaska as it does earlier in the spring near the Gulf Coast (Lincoln, 1950).

Birds generally migrate at moderate to low altitudes. In the "grand passage" of ducks mentioned above, the birds flew between 460 and 850 meters high by day and dropped to a minimum of about 150 meters by night. The Blue and Snow Geese migrating from James Bay to Louisiana were observed at night flying at an altitude of 2400 meters.

Combined radar and visual observations indicated that various hawks migrating in South Texas flew at an altitude averaging 600 m and used thermal convection currents for lift (Kerlinger and Gauthreaux, 1985). Combined radar and telescopic observations by Gauthreaux (1972) of spring passerine migrants flying northward across the Gulf of Mexico indicated that at night while over the Gulf the birds flew singly at altitudes of 244 to 488 meters. At daylight, however, as the birds approached the Louisiana coast, they rose to heights of 1220 to 1524 meters and aggregated into compact flocks averaging about 20 birds each. Flocks often appeared to be composed of a single species, and in some cases, of a single sex: e.g., Northern Oriole, *Icterus galbula*, Scarlet Tanager, *Piranga olivacea*, and Rose-breasted Grosbeak, *Pheucticus ludovicianus*. At the peak of migration in early April, about 31,000 birds per mile of Gulf-coast front arrived each day. Radar studies of migrating passerines and shorebirds in Northwest Germany indicated that more than half the echoes were between 600 and 900 m. A few echoes were recorded at 3000 m, with a maximum of 5500 m (Jellmann, 1979).

Summarizing several years of radar observation of birds migrating between Britain and the European continent, Lack (1963) concluded that some migration goes on every month of the year; that birds fly higher with tailwinds and lower against headwinds; and that they generally travel in straight tracks on broad fronts rather than in narrow paths or ribbons. Lack also concluded that more migration occurs in clear than in cloudy or rainy weather, and that the peak in numbers of nocturnal migrants occurs before mid-

night. Contrary to Gauthreaux's study, Lack found that birds often begin migrating in dense flocks but arrive at their destination in much smaller flocks or singly. He observed that by day, migrants fly low over land, mostly below radar range, but as they put out to sea they rise higher into radar range (Fig. 22–8); and that by night, birds over land and sea fly at the same altitudes. This indicates that birds instinctively realize the dangers of flying at low levels over land in the dark.

Many species that show the aggressive intolerance of territorialism during the breeding season lose their mutual antagonism and become gregarious during migration periods. Preparatory to migration, many species gather in flocks, and often in flocks of several species. By traveling in flocks, young birds may learn the migration routes and successful migration behavior from older, experienced birds. These advantages do not apply, however, to species in which old and young migrate at different times of year. Flocks migrating at night are kept together by frequent chirps and calls that a listener on the ground can easily hear during a migration wave.

WEATHER AND MIGRATION

Weather influences migration in at least four different ways. It controls the advance of the seasons, or the phenology of natural events, such as the date of the appearance of the first violet or of the arrival of the first robin. Second, it affects the migrating bird in flight, helping it, hindering it, or even at times crushing it to earth. Third, weather may be the stimulus that initiates the migration journey in a bird physiologically prepared for it. Last, operating over long periods of time, the environmental effects of climate often become the ultimate determinant in establishing hereditary tendencies toward migration. A possible example of this influence is seen even within a single species, the Dominican Gull, *Larus dominicanus*. Populations of this gull in temperate New Zealand are sedentary, but those living in the rigorous climates of the Falkland and South Shetland Islands are migratory (Fordham, 1968).

The first, or phenological, influence of weather is essentially ecological. Water birds gain nothing migrating north in spring if the lakes and streams are still frozen, nor can warblers and flycatchers survive if there are no insects awaiting them. There are early springs and late springs, and many species tailor

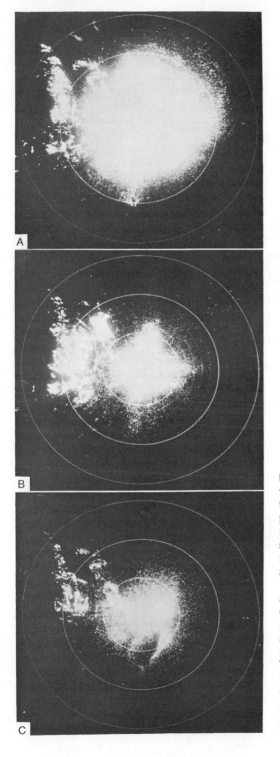

FIGURE 22–8 Three successive photographs of the same migration wave, as seen by radar from East Anglia, England. In each photo, the bright points of light represent radar echoes from birds. The larger patches of light to the left are chiefly high ground. North is at the top in each photo; the English Channel to the right, or East. The radar station itself is located in the center of the concentric circles, and the range from the station is shown by faint circles at 10-mile intervals, and heavier rings at 50-mile (80 km) intervals.

A, shows the migration wave near the peak of nocturnal departures at 10:00 P.M. on March 11, 1957, with bird echoes extending northeastward from the station about 130 miles. By 4:00 A.M. the next morning, B, the nocturnal movement had dwindled to a moderate size, and by 6:00 A.M. the same morning, C, nocturnal migration had nearly ceased and a moderate-sized morning migration had begun. The sharp boundary crossing the 50-mile circle (in C) southeast of the radar station represents the seacoast. In the daytime the birds fly low over land, below radar range, and therefore show no echoes. But on reaching the coast and putting out to sea (right), they rise higher into radar range. After Lack, 1959. Photos courtesy of the Royal Radar Establishment. Crown Copyright.

their migration schedules to fit the season. Records by Saunders (1959) of the first arrivals among 50 different species in southern Connecticut showed that in late, cold springs, migrants on the average arrived later than in early, warm springs. In 1917, for example, the average of all arrival dates for 50 species was 5.38 days late, while in 1938 it was 6.15 days early, as compared with the 40-year mean arrival dates for all species (Fig. 22–9). A similar, 48-year study by Wedemeyer (1973) in Montana, revealed greater seasonal variations in spring arrival dates. For 99 species the variations between extremely early and extremely late dates ranged from 16 to 94 days and averaged 36.6 days. Ranges in dates were greatest during late March and April and least in late May and June.

As spring advances northward each year, accompanying and successive waves of ecological phenomena also march northward: the thawing of ice and snow, the first green blades, the emergence of insects and worms, the flowering of trees. Some birds migrate northward immediately on the heels of minimum spring conditions that allow their survival. The Canada Goose, *Branta canadensis*, pushes northward roughly in step with the advance of the 2°C isotherm; and in Europe, the Willow Warbler, *Phylloscopus*

trochilus, migrates northward on a broad front which roughly parallels the progression of the 9°C isotherm. Other species, such as the Gray-cheeked Thrush, *Catharus minimus*, apparently wait in their winter quarters until ecological conditions are suitable in the breeding zone; then they migrate north with a rush.

The ecology of changing seasons also influences the geography of migration pathways for some species. As mentioned above, the Lesser Golden Plover, *Pluvialis dominica*, cannot retrace its fall migration route in the spring because the cold Labrador Current delays the arrival of spring along the northeast coast of North America. Consequently, the birds fly up the center of the continent where conditions are more favorable. Swifts are extremely sensitive to weather conditions and frequently make migration-like massed flights to avoid cold, wet weather. Because they feed exclusively on airborne insects, they require relatively fair weather if they are to avoid starvation.

Many studies have been made of the more direct influences of weather on birds while they are migrating. Attempts to correlate migration movements with barometric pressure, wind direction, precipitation, and temperature have resulted in conflicting conclusions. There is a general consensus, however, that

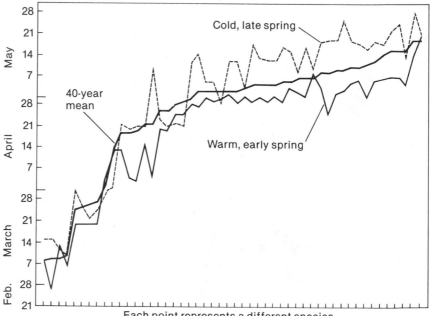

FIGURE 22–9 The arrival times of the first spring migrants vary with the weather. Each point on the thick line represents the 40-year mean of the first spring arrival-date for a given species in southeastern Connecticut. The species arriving earliest are shown at the left; those arriving latest, at the right. The solid light line shows the dates of the first arrivals of these same species during a warm early spring (1938) when most species arrived earlier than usual; the dotted line denotes arrivals during a late spring (1917). May, 1917, was especially cold and raw, and during this month the first arrivals averaged eight days late. After Saunders, 1959.

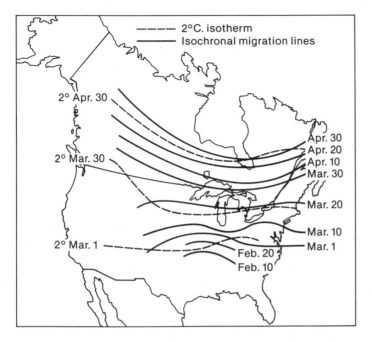

- - - - - 2°C. isotherm
————— Isochronal migration lines

2° Apr. 30

2° Mar. 30

2° Mar. 1

Apr. 30
Apr. 20
Apr. 10
Mar. 30

Mar. 20

Mar. 10
Mar. 1

Feb. 20
Feb. 10

FIGURE 22–10 In its spring migration, the Canada Goose advances northward roughly in step with the advance of the 2°C isotherm. After Lincoln, 1950.

in the north temperate zone, migration is intimately related to the eastward advancing high- and low barometric pressure centers. The high pressure center, or anticyclone, is composed of drier, cooler air flowing in a clockwise direction; the low pressure center, or cyclone, contains warmer, moister air flowing counterclockwise. Results of a six-year radar study in Switzerland indicated that the heaviest diurnal autumn migration of birds occurred in high-pressure zones with tailwinds or weak headwinds (Hilgerloh, 1981).

There seems to be little doubt that adverse weather may impede migration and cause concentrations of birds to pile up. Storms, strong headwinds, or fog ordinarily keep birds grounded. With the return of favorable weather, these birds move on in "waves" or "rushes." In southern Sweden in the autumn of 1952, ten days of strong winds held up the migration of great numbers of birds. When favorable conditions returned, the birds moved south like an avalanche (Mathiasson, 1957). At other times adverse weather or contrary winds may completely reverse the direction of migration. A coordinated study by Atlantic coast banding stations revealed that tens of thousands of passerines will fly northward in the autumn when winds are from the south (Duvall, 1966). Such reverse migra-

tion may continue for several hours or even several days, and involve flocks of many species of birds.

Numerous independent studies by radar of migrating birds have shown that winds are the most important influence in determining the time and direction of migration. During 64 nights in the spring and 38 in the autumn, Bruderer (1975) tracked with radar the paths of an average 150 migrating birds each night in Switzerland. He found that birds tended to migrate on those nights when, and at those altitudes where, winds were most favorable. Radar observation of birds migrating across the North Sea revealed that they will descend to pass beneath a rain or hail cloud and then rise again (Eastwood, 1967).

The direction of wind may be incidental to its directional effects on migration. Mueller and Berger (1961) kept daily counts of south-migrating hawks on the west shore of Lake Michigan at Cedar Grove, Wisconsin, on 256 autumn days during six years. Their data involved 29,061 hawks, over one-half of which were Broad-winged Hawks, *Buteo platypterus*. Over 92 per cent of those migrating appeared on days characterized by moderate westerly winds (15 to 35 km per hour) and the recent passage of a cold front. The winds concentrated the soaring migrants along the lake shore through the creation of updrafts.

Birds in migration may be forced off their course by side-winds. This lateral displacement is called wind-drift and it can have disastrous consequences for a land bird blown far to sea. The question immediately arises: can a migrating bird compensate for wind-drift? Radar observations have produced conflicting evidence on this point. Steidinger (1968), observing land birds migrating in mid-Switzerland, found little or no compensation for wind-drift. However, in radar observations of shore birds in fall migration in Britain, Evans (1968) reported that they compensated for wind-drift. In their radar observations of nocturnal migrants in mid-United States, Bellrose and Graber (1963) reported that the birds corrected for lateral drift but never quite completely, and the stronger the side-wind the greater the drifting off course. Schmidt-Koenig (1965) reported that daytime migrants compensated for wind-drift when flying over land but not when flying over water. Alerstam and Ulfstrand (1972) in Sweden made radar observations by day in autumn of Chaffinches, *Fringilla coelebs*. When flying at high altitudes the birds showed no compensation for wind-drift, but when flying low enough to be seen from the ground with binoculars, they not only corrected for lateral drift but were strongly influenced by topography and coastlines.

Current information about wind-drift compensation makes it impossible to draw any broad generalizations on the subject. As more facts accumulate it may be learned that correction for lateral wind displacement will vary according to species, to season, to geographical location, to altitude, and to other factors.

For many species, weather conditions may determine whether the birds will begin their migrations or not. But once under way, birds are deterred from migration only by extremely adverse weather. And weather as such cannot initiate migration in a bird physiologically unprepared for it.

TIMING OF MIGRATION

The ultimate stimuli of seasonal changes in such things as day length, temperature, and rainfall, operating over millenia, have, at least for temperate zone birds, set their internal clocks so that they are ready to migrate and breed at the optimal time of year. Once a bird is internally prepared for migration, various proximate external stimuli may trigger the beginning of the migratory journey. Among these stimuli weather

plays a dominant role. Although wind direction is vitally important, temperature and rainfall also act as controlling factors. In the spring, cold, wet weather may directly delay the departure of a species; in the fall such weather may hasten it. In the drier parts of Africa and Australia, certain species are rain followers, migrating to those regions where rain creates favorable conditions for raising young. Food failures may act as stimuli to initiate the migration of irruptive species like crossbills, while food abundance, on the other hand, may cause normally migratory species to become sedentary. The once migratory Kite, *Milvus milvus*, of southern Scandinavia, has recently become a year-around resident because of the availability of garbage and slaughter-house offal in the winter, and the Lesser Redpoll, *Carduelis flammea*, will winter in southern England if there is a good crop of bird seeds; otherwise it emigrates to the continent (Evans, 1969).

The calendar-like regularity of the spring and fall migrations of many species is without doubt the consequence of the accumulated effects of millenia of climatic cycles impressed on the sensitive living cells of birds. There is great variation in the precision with which different species follow the calendar. The famous swallows of San Juan Capistrano Mission in California are not as precise in their timing as popular accounts might lead one to believe. Nevertheless, some species arrive and depart on their migrations with striking calendar regularity. One of the physiological preparations for migration is the accumulation of body fat as fuel for the journey. The timing of pre-migratory fattening in some species is remarkably precise. Among White-crowned Sparrows, *Zonotrichia leucophrys*, kept for eight years in outdoor cages, the standard error for the mean date of the onset of spring-time fattening was plus or minus one day (King, 1972). Many pelagic sea birds, such as terns, boobies, and shearwaters, arrive at their breeding islands approximately the same week every year. The Short-tailed Shearwater, *Puffinus tenuirostris*, which winters in the Pacific up to the Aleutian Islands, arrives at its south Australian breeding grounds during the same 11 days every year (Marshall and Serventy, 1959).

Most species, however, show more diversity in their times of migration. Snow Geese, *Chen caerulescens*, have been known to arrive at their Delaware wildlife refuge as early as September and as late as November. A four-year study of migrating shore birds by New Jersey ornithologists showed that various species required as much as two to three months to pass through. As a rule, spring migration was of much shorter duration than fall migration, and young birds seemed to travel more leisurely than adults (Urner and Storer, 1949).

The apparently steady passage of migrant birds through a given region is not always the simple phenomenon it seems. The Yellow Warbler, *Dendroica petechia*, arrives in southern Arizona in mid-March, in northeastern Arizona in late April, and in Canada in May or early June. However, these dates alone give a misleading impression, for they do not apply to the same birds. The birds arriving in Arizona in mid-March are the resident race *sonorana*; more northerly races reach Arizona in late April; and the most northerly race, *rubiginosa*, does not enter Arizona until May, at which time it flies over the southern races on its way to Canada (Phillips, 1951).

Birds have their preferred times of day for migrating. The majority of small birds, including most passerines, feed by day and migrate by night, when they are safer from predators and when the air is generally more stable. Small birds, with their livelier metabolism, exhaust their energy stores in flight more rapidly than large birds do. Therefore, they must replace them quickly and effectively, and this can best be done by daylight. Furthermore, by migrating at night, they are less likely to be distracted by their surroundings and are better able to concentrate on covering distance. Numerous studies have shown that nocturnal migration is at its maximum between 1 and 3 hours after sunset, and has its minimum in the hours just before sunrise. Larger birds, such as hawks, eagles, storks, herons, and crows, migrate by day and rest by night, as do also the insect-feeding swifts and swallows which sweep up their food as they migrate. Loons, ducks and geese, some shore birds, and auks may migrate either by day or by night.

The time of year during which a migrant departs for winter quarters is related to its time of breeding. The European Swift, *Apus apus*, finishes nesting in Italy at the end of July, and leaves then for its winter quarters. In northern Germany it ceases breeding and departs around the first of August, and in mid-Finland, in late August and early September (Stresemann, 1927–1934). Conversely, the time consumed in flight by some long-distance migrants limits the time they have available for breeding. The Arctic Warbler, *Phylloscopus borealis*, mentioned earlier, winters in the Malay archipelago and arrives at its Norwegian breeding area, some 12,500 km away, between June 18 and 25. It immediately starts breeding, and the young leave the nest at the beginning of August. Shortly thereafter they fly away with their parents for winter quarters. Stresemann points out that this species could not migrate much farther and have any time left for breeding! The fact that so many shore birds and waterfowl make such long migrations and have such short breeding seasons in the far north probably explains why some of them become paired in winter quarters rather than in the breeding area.

The times at which migration occurs may vary not only between different races of a given species but also between old and young, or between males and females of the same species. In many species, particularly among the passerines, the males often arrive at the breeding area before the females. In the northern United States, for example, the male Red-winged Blackbird, *Agelaius phoeniceus*, arrives in spring from one to five weeks before the female; in England the male Skylark, *Alauda arvensis*, arrives on its territory one month before the female. Separation of sexes in migration need not be complete. In western North America, spring flocks of migrating Lapland Longspurs, *Calcarius lapponicus*, generally change from 90 per cent males among earliest migrants to about 25 per cent males in the latest flocks (West *et al.*, 1968). By arriving early in this way, the males have time to establish territory, so that when the females arrive, they can both proceed with breeding activities.

Apparently, most young birds migrate to winter quarters with the adults, but in some species, such as the Lesser Golden Plover, *Pluvialis dominica*, the adults migrate earlier in the fall than do the young, while in others, like the European Common Swift, *Apus apus*, the young often leave for winter quarters first. Immature Yellow-bellied Flycatchers, *Empidonax flaviventris*, leave for their wintering grounds about 24 days later than the adults (Hussell, 1982). An extraordinary migratory performance is characteristic of the Hudsonian Godwit, *Limosa haemastica*. The adults migrate down the west shores of Hudson and James Bays in late July and then are reported no more until they arrive in southern Chile and Argentina. Juveniles leave a month later. Presumably both groups make the 4800-km flight nonstop.

In some species, but certainly not in all, the time of migration seems to be related to the time of molting. In the European Red-backed Shrike, *Lanius collurio*, the adults have no postbreeding molt and are able to migrate south before the young have matured enough to accompany them. The White Wagtail, *Motacilla alba*, on the other hand, does have a complete post-breeding molt, and its young migrate from a week to 10 days before the adults. Other adaptive modifications involving molt and migration are described in Chapter 3.

Males tend to winter farther north than females, and adults tend to winter farther north than juveniles in a number of avian species (Ketterson and Noland, 1976; Morton, 1984). Several hypotheses have been advanced to explain this phenomenon (Ketterson and Nolan, 1983):

1. A premium is placed on birds returning to their breeding grounds early to claim or reclaim territories, so that members of the sex that defends territories winter nearer their breeding grounds (Myers, 1981).
2. The rigors of the winter climate cause the smaller-bodied sex to migrate farther south. For example, Ketterson and King (1977) found that the larger male White-crowned Sparrow, *Zonotrichia leucophrys*, can fast for longer stretches than can females, which winter farther south. The larger female Snowy Owl, *Nyctea scandiaca*, winters farther north than the smaller male (Kerlinger and Lein, 1986).
3. Intersexual competition during the nonbreeding season results in the subordinate sex segregating itself from the dominant one (Gauthreaux, 1982) as a mechanism to relieve competition.
4. One sex suffers a higher risk of mortality and therefore attenuates its migratory distance.

Ketterson and Nolan (1983) examine the pros and cons of each of these theories and conclude that no single-factor hypothesis adequately explains this phenomenon.

In some passerines, the males may remain sedentary while the females and young migrate to winter quarters. The European Bullfinch, *Pyrrhula pyrrhula*, is only partially migratory, many birds being year-round residents. A five-year tally of autumn-migrating Bullfinches near Stuttgart, Germany, revealed a ratio of 34 per cent males to 66 per cent females (Gatter, 1976). Winter residents there were approximately 70 per cent males to 30 per cent females. Gatter suggested that the temperature level needed to release migratory behavior was lower for the males than for the females. Very commonly, among partially migratory species, migratory activity decreases with increasing

age (Schüz *et al.*, 1971). Among many of the shore birds and waders, such as gulls, terns, herons, and storks, the young disperse widely after leaving the nest. They may wander as vagabonds for from two to five years before maturing sufficiently to become regular, breeding migrants. All told, a considerable variety of influences bear on the timing of migration. It is usually very difficult, if not impossible, to say why a given species migrates when it does.

PREPARATION FOR MIGRATION

It is obvious that warmer weather, longer days, abundant food, or any other of the numerous stimuli that release the departure of migrating birds do not cause migration irrespective of the condition of the birds themselves. This is apparent from the simple fact that of two races of birds living under identical environmental conditions, one may be stimulated to migrate when spring comes and the other may not. For example, both migratory and nonmigratory races of the Dark-eyed Junco, *Junco hyemalis*, live in northern California. When both types of birds were simultaneously subjected to increases in day length, either natural or artificial, only the migratory race became restless and departed on its northward migration. The nonmigratory race, subjected to identical conditions, remained sedentary (Wolfson, 1942).

A popular misconception holds that migration is caused by a shortage of food. The supposition is that hunger drives birds to regions with more abundant food. Although this idea does help to explain the long-range evolutionary origin of migration, it rarely accounts for present-day migratory movements. Baron von Pernau, aware of this fallacy, more than two centuries ago wrote:

> It is a very strange opinion if some believe the birds would emigrate, driven by hunger only. Instead they are usually very fat when about to leave us.
> —(Stresemann, 1947).

It is clear that certain conditions predispose certain birds to migrate, and certain external conditions act on these birds as stimuli that release migratory behavior. In all migratory species, this predisposition seems to be a cyclic mechanism closely tied to reproduction, and it operates to ensure migration to the breeding area at the best time of year for raising the young. The mechanism is sensitive to a variety of external factors, such as length of day, temperature, rainfall, food abundance, and behavioral interactions, which normally act as proximate agents that determine the time of migration either by stimulation or by inhibition.

Because of the intimate relation between migration and breeding, and particularly because, in early experiments, photoperiodism seemed to stimulate both sexual activity and migration, it was assumed that there was a causal connection between the two. This is not now thought to be the case. Arguing against such a connection is the fact that many sendentary species pass through their reproductive cycle without the slightest tendency to migrate. In addition, experiments by Putzig (1937) and others have shown that castrated birds are nevertheless able to migrate. And castrated Golden-crowned Sparrows, *Zonotrichia coronata*, were shown by Morton and Mewaldt (1962) to acquire both fat deposits and migratory restlessness similar to normal control birds. Photostimulated Bramblings, *Fringilla montifringilla*, prevented from accumulating fat by being denied sufficient food, nevertheless developed migratory restlessness (Lofts *et al.*, 1963).

Experiments by Berthold (1974, 1977b) have shown that readiness for migration is, at least in some species, determined by an innate circannual rhythm that neatly provides a temporal adaptation to a periodic environment by the optimal scheduling of such activities as reproduction, migration, and molting (see page 163). Migratory readiness is manifested in two stages that he calls *Zugdisposition*, or the tendency of birds to feed sufficiently to accumulate fat reserves, and *Zugaktivität*, of the actual flying of wild birds or, in caged birds, the nocturnal hopping and fluttering (*Zugunruhe*) in the direction of normal migration. The muscle mass of the Gray Catbird, *Dumatella carolinensis*, increases by some 85 per cent during premigratory fattening in the fall. This is due to muscle fiber hypertrophy in preparation for the long flight (Marsh, 1984).

Although there may be no causal connection between migration and fat accumulation, the function of fat as a fuel for migratory flight is clear. In preparation for their arduous journeys many migrants accumulate generous deposits of peritoneal and subcutaneous fat shortly before embarkation—as much as 50 per cent of total body weight in Bobolinks, *Dolichonyx oryzivorus*. This fatty fuel is primarily in the form of triglycerides, whereas structural body fats are chiefly phospholipids (Ricklefs, 1974). South-

migrating White-crowned Sparrows, *Z. leucophrys*, gain 2.5 g in 4.5 days or 0.6 g/day during their stopover in Maine. Stopovers are used to increase fat reserves depleted during their southward journey. Lean birds stay longer than heavier birds (Cherry, 1982). There is a reversed correlation between air temperature and quantity of fat in birds wintering in Washington when temperatures are below 3°C (King and Farner, 1966).

It is significant that the amount of fat accumulation in three races of the African Red-billed Quelea *Quelea quelea*, is correlated with the distance each race migrates (Ward and Jones, 1977). By measuring the body weights of over 2000 passerines of nine species that had been attracted to a Lake Ontario lighthouse during nocturnal migration, Hussell and Lambert (1980) determined that a typical bird lost 0.91 cent of its body weight per hour while migrating.

More convincing evidence of the role of fat fuel for migration comes from studies of the weight migratory birds before and after they cross a large and hostile environment where feeding would be impossible. Ash (1960) determined the mean gross weight four different species of Palearctic warblers in Nigeria on the south margin of the Sahara Desert and, at about the same time, in Morocco on the north edge. After crossing the desert the mean body-weight loss for the four species was 31, 34, 41, and 44 per cent. Presumably, most of this weight loss was in fat. Transocean crossings require comparable fuel reserves. A banded Ruddy Turnstone, *Arenaria interpres*, flew from Alaska to the Hawaiian Islands in four days, averaging over 1000 km per day. It weighed 174.0 g when banded and 154.8 when recaptured—a weight loss of 19.2 g (Thompson, 1973). This is equivalent to flying 100 km on $^1/_2$ g of fat. Figuring fuel consumption more conservatively than this, Tucker (1974) constructed a flight-range chart correlating distance with fat consumption (Fig. 22–11). It clearly indicates the superior efficiency of flight in larger birds. After examining the bodies of 20 exhausted or dead birds of nine species that had just crossed the western Sahara desert in spring migration, Haas and Beck (1979) found that the smaller species had failed because of exhausted fat stores, and the larger birds, apparently from dehydration. Underscoring the efficiency of air travel over land travel is the fact that an Emperor Penguin, *Aptenodytes forsteri*, needs about 1.5 kg of fat to "walk" 200 km from its rookery to the sea (Le Maho, 1977).

In New England migrating Blackpoll Warblers,

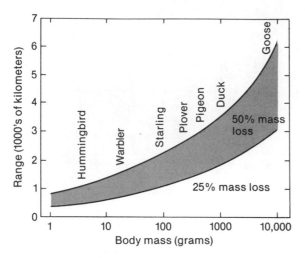

FIGURE 22–11 Flight ranges of various-sized birds, assuming that fat is the fuel used for flight. The two curves indicate the different amounts of fuel consumed expressed as proportions of initial body mass. Fifty per cent of body mass consumption is probably the maximum limit for long-distance migrants. Modified after Tucker, 1974.

Dendroica striata, rapidly accumulate body fat in the early fall and then strike off across the open Atlantic for the West Indies or South America. Ordinarily they pass Bermuda without stopping, but a few birds, attracted and killed by a Bermuda lighthouse, were found to have lost about 17 per cent of their original body weight. At this rate of consumption, it was estimated that these birds possessed sufficient fuel reserves on leaving New England to fly for 105 to 120 hours, or long enough to fly nonstop all the way to South America (Nisbet *et al.*, 1963). Little if any of the weight loss was attributed to water loss.

As a result of his experiments with hovering hummingbirds in metabolic chambers, Lasiewski (1962) concluded that a male Ruby-throated Hummingbird, *Archilochus colubris*, weighing about 4.5 g, of which 2 g was fat, could fly nonstop for 26 hours, consuming the fat at the rate of 0.69 calories per hour. At an average speed of 40 km per hour, the bird's flying range would be about 1050 km—easily enough to span the Gulf of Mexico.

Zugunruhe is a twice-yearly cycle of nocturnal restlessness that occurs in migratory birds during the migration seasons. Automatically recorded activity graphs of captive migrants show a daily rhythm in which the bird sleeps for a short period—15 minutes to 2 hours—after sunset, then awakens and hops and flutters in its cage with increasing vigor until shortly before midnight, after which the activity gradually dies down. This activity in captive birds closely parallels in time the nocturnal activity of freely flying migrants. In the springtime, warm weather will increase *Zugunruhe* in captive migrants and cold weather will depress it, whereas in autumn the exact reverse occurs. Among migratory *Zonotrichia* in the spring, most nighttime restlessness occurred prior to midnight, but in the fall it occurred between midnight and dawn (Mewaldt *et al.*, 1964). Nonmigratory birds like the House Sparrow, *Passer domesticus*, show no night-restlessness at any season of the year.

The relation of day length to *Zugunruhe* was demonstrated by Lofts and Marshall (1960) on captive Bramblings, *Fringilla montifringilla*, which were kept throughout their normal springtime migration period under artificially shortened days as though it were winter. Neither *Zugunruhe* nor gonad enlargement occurred. At the end of the normal migratory period, the light dosage was increased to 14½ hours daily. The birds then developed *Zugunruhe* and the males responded with fat deposition and enlargement of testes. Experiments with castrates indicated that *Zugunruhe* and migration could occur in the absence of sex hormones.

A special feature of nocturnal restlessness in certain species is its compass orientation. Migratory White-crowned Sparrows, *Zonotrichia leucophrys*, kept in circular cages under the open California sky, showed a southerly orientation at night during the fall migration period in August and September, and a pronounced northerly orientation in *Zugunruhe* during the spring migration period in April and early May (Mewaldt and Rose, 1960). The birds demonstrated their tendencies to orient by perching on any one of eight activity-sensitive, registering perches symmetrically arranged around the periphery of a cage.

There is more to *Zugunruhe* than an eagerness in a bird to begin migration at the proper time of year. Gwinner (1968) found that young warblers kept under constant photoperiods from September to June in Germany molted at the same time as control birds kept under natural conditions, and showed similar patterns of *Zugunruhe* and weight variation. Birds flown to their Congo winter quarters at the height of migratory restlessness persisted in their *Zugunruhe*, despite their early arrival, as long as did the control birds kept in

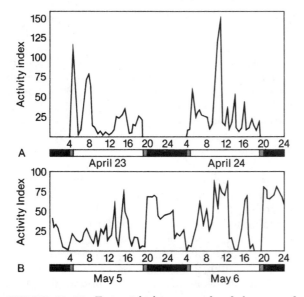

FIGURE 22–12 Forty-eight-hour records of the rate of activity of a first-year male White-crowned Sparrow. A, During the molting period just prior to the onset of *Zugunruhe*. B, After the onset of *Zugunruhe*. The black bar represents periods of darkness; the shaded bar, twilight; the open bar, sunrise to sunset. Redrawn after Farner, 1955.

Germany. Thus, stimuli from the winter habitat had no inhibiting effect on migratory restlessness. These results suggested that *Zugunruhe*, in these birds, is based on an endogenous rhythm independent of proximate environmental stimuli.

To test this concept further, Gwinner compared *Zugunruhe* in two northern Palearctic warblers of similar size and appearance: the Willow Warbler, *Phylloscopus trochilus*, which winters in southern Africa, and the Chiffchaff, *P. collybita*, which winters in southern Europe and northern Africa. The former spends three to four months on its long, autumn migration trip, and the latter, only one month. Hand-raised nestlings, when grown, were placed individually in *Zugunruhe*-registering cages. It was found that, whether kept under natural light conditions or under an artificial photoperiod of 12 hours light to 12 hours dark, all birds developed migratory restlessness and fat deposition corresponding to their conspecific wild relatives. For each species, both the onset and the end of *Zugunruhe* approximately coincided in time with the onset and end of migration in their wild relatives.

In other words, the distance that the young birds of each species travel on their first autumn migration is determined by an innate program that provides just sufficient migratory activity for them to reach their traditional winter quarters. This temporal budget of *Zugunruhe*, coupled with the typical directional orientation of inexperienced young birds, goes far to explain the striking homing ability of untutored young to find their proper winter quarters. Even different races of birds from the same species may show endogenous *Zugunruhe* rhythms adapted to their migration needs. Willow Warbler nestlings from near the Arctic Circle and others from southern Germany were both reared under constant photoperiods. The results showed that the earlier start and the longer duration of migration restlessness of the northern race were due to an endogenous rhythm and not to the longer northern daylight (Gwinner *et al.*, 1972).

Numerous experiments and much speculation have been expended on research into the basic causes of migration. As yet, there is no general agreement on the solution of the problem. For birds living in the north temperate zone, it seems likely that the predisposition to migrate hinges on a neuroendocrinological mechanism, which controls the timing and sequence of events in the annual cycle somewhat as follows: After a winter period of long nights and reproductive rest, the pituitary gland recovers from its refractory period (p. 163). The longer days and warm weather of springtime make possible increased feeding and less heat loss, and therefore a surplus energy balance. The increased day length also controls, perhaps through the pineal gland, the daily rhythms of peak flow of the two hormones, corticosterone and prolactin. The temporal interaction of these two hormones has been shown by Meier and his colleagues (1973) to influence in a most astonishing way a whole constellation of migratory preparations: gonad development, fat deposition, *Zugunruhe*, and compass orientation (see p. 167). Once in this condition, a bird is easily induced to migrate by some external stimulus such as favorable weather.

This admittedly hypothetical explanation does not fit birds passing the winter, for example, in equatorial regions where day length does not change appreciably throughout the year. Without question, different migration-stimulating or -inhibiting mechanisms exist for birds living under different environmental circumstances. Marshall (1961) makes the point that the neuroendocrinological machinery responsible for reproduction and migration is, week after week, under the influence of antagonistic sets of accelerators and inhibitors. Accelerators may be such influences as warmth, sunlight, territory, adequate food, the nest site, nest materials, and behavioral interactions. Inhibitors may include cold, inclement weather, hunger, fear, and lack of a mate, of nesting materials, or of the traditional nest site. For example, hand-raised Garden Warblers, *Sylvia borin*, subjected to simulated bad weather (drizzles all afternoon and complete night darkness), ceased their migratory restlessness activity (Schneider *et al.*, 1981).

ORIENTATION AND NAVIGATION

Probably the knottiest problem in all ornithology concerns how a bird finds its way home. Year in, year out, millions of migrating birds successfully travel hundreds and even thousands of kilometers, back and forth between their breeding and their wintering grounds, sometimes over what to man are featureless stretches of land or water, and very often at night. To do this successfully, a bird must know where it is; it must know the direction of its goal; it must be able to maintain its course, or navigate, in that direction; and then it must stop when it reaches its destination. Banded long-distance migrants have returned in successive years not only to the same pin-point nesting site, but to the same restricted wintering area. White-crowned Sparrows, *Zonotrichia leucophrys*, imprint on their wintering area during their first winter visit and return there year after year (Ralph and Mewaldt, 1975). Many individually marked birds have been experimentally displaced thousands of kilometers into completely strange areas, and yet have returned to their home nest sites, often with astonishing speed. For example, a Manx Shearwater, *Puffinus puffinus*, carried by airplane to Boston, over 5100 km from its breeding island off Wales, was back in its nesting burrow 12½ days later (Mazzeo, 1953).

Such spectacular homing is not restricted to marine birds. Hundreds of *Zonotrichia* sparrows were transported 2900 km from their winter quarters in San Jose, California, to Louisiana and released. The next winter 15 of them were recovered in San Jose. These, with additional sparrows, were then displaced 3800 km to Maryland. The following winter, among the birds that returned to San Jose, were six that had also returned from Louisiana the year before (Mewaldt, 1964).

Homing ability is widespread in the animal kingdom. It is found in limpets, squids, lobsters, ants, bees, and many fishes, salamanders, box turtles, bats, mice, and many other mammals. Although the capacity for homing is best developed in birds, not all birds have equal facility in finding their way. Some wrens, titmice, doves, and sparrows are unable to return to their nests when experimentally displaced only a few kilometers. Some breeds of pigeons are known to perform better at homing than others. To test for the hereditary nature of homing ability, 10- to 18-week-old feral and homing Pigeons were trained to home by releasing them at increasing distances from the loft beginning at 0.9 km and ending at 5 km. By the end of the training period only 19 per cent of the feral Pigeons remained whereas 85 per cent of the Homers were still at the training lofts. Both groups performed equally well at orienting and navigating, however, indicating that selection in homers is not for improved homing performance but rather for increased site attachment (Edrich and Keeton, 1977).

Whatever senses a bird may use to find its way, it still must recognize "home" when it arrives there. This recognition may be hereditary, or, more likely, acquired early in life, perhaps through imprinting. The experiments of Löhrl (1959) with displaced young flycatchers demonstrated the effectiveness of early imprinting on establishing a bird's sense of home (see p. 256).

Theories to explain a bird's ability to find its way in the absence of familiar landmarks can be grouped in three general categories: (1) using only its own body and no external cues for orientation; (2) using a single environmental cue for simple directional or compass orientation; (3) using two or more environmental cues simultaneously for undertaking true navigation.

The first, or no-cue, type of orientation may depend on simply flying in a random or patterned search for its goal. This method of home-finding undoubtedly exists, but it would be hopelessly inefficient and exhausting for a long-distance migrant. Another highly hypothetical type of orientation requiring no environmental cues is a form of inertial guidance in which the bird stores in its memory kinesthetic and semicircular canal impressions of every twist and turn of an outgoing journey, and recalls them, in reverse order, for guidance on the return trip. Numerous displacement experiments with birds either anesthetized or constantly spun on turn-tables on the outgoing trip, showed no differences in homing ability between experimental and control birds. Even pigeons with their semicircular canals bisected so that they could no longer perform head-stabilizing movements, nevertheless performed as well as normal, control birds in homing experiments (Wallraff, 1965). Moreover, a theory of this sort cannot account for the orientation on the maiden trip of a young, unaccompanied, inexperienced bird.

The second type of orientation theory, requiring a single environmental cue, or different cues individually used in chronological succession, assumes that the bird knows innately the direction of its goal, or home, and that it uses a single cue from the environment to select and maintain that direction during migration: e.g., the sun, the stars, wind direction, odors, terrestrial ecological gradients, or the earth's geomagnetic field.

The third theory involves true navigation, or the ability of a bird in unfamiliar surroundings to find its goal without using known landmarks. This theory, as described by Matthews (1964), postulates that the bird senses and reacts to quantitative differences between stimuli available at the unfamiliar site and those at home. These stimuli ordinarily involve a grid of at least two physical factors, X and Y, that vary quantitatively and regularly across the earth's surface, with the X gradient arranged at an angle to the Y gradient. The values of the X and Y gradients in an unknown area must then be compared by the bird to their values at home. Before the bird can head for home it must, of course, somehow relate the grid pattern to the surface of the earth just as a mariner must coordinate the north on his chart with the earth's magnetic north before he can plot his course to the desired destination. The bird may then simply fly in the compass direction of home (directional orientation), or it may constantly compare X and Y values with those at home and make corrections en route to its destination (true navigation). In either case the bird must have access to something akin to an internal map, one acquired either through heredity or through experience with previous travel.

Theorizing that birds have special senses or special refinements of known senses is not unreasonable if one considers some of the known sensory abilities of other animals: the exceedingly acute sense of taste in butterflies; the sharp olfactory sense of bloodhounds and moths; the ability of bees and birds to see ultravi-

olet radiation and to discriminate the plane of polarized light (Parrish *et al.*, 1984; Able, 1982); the ability of pit vipers to perceive infrared rays; the ability of certain sharks and electric fish to communicate and detect prey by means of extremely weak electric currents; the well-known capacity of bats and of at least three species of birds (Humboldt Penguin, *Spheniscus humboldti*, Oilbird, *Steatornis caripensis*, and Cave Swiftlet, *Collocalia brevirostris*) to avoid obstacles in the dark by echolocation (Griffin, 1953; Novick, 1959; Konishi and Knudsen, 1979).

Several navigation hypotheses have been proposed that involve exceptional cues from the environment. Ising (1946) suggested that because of their motions over the surface of the earth, birds add to or subtract from the effects of the earth's centrifugal force, and that they are able to feel the gravitational differences in weight that result. In opposition to this, Griffin (1955) points out that a bird flying east along the equator at 65 km per hour would find its weight increased by only 1 part in 2000 when it turned and flew west—hardly a detectable difference. Another force suggested as a possible clue to orientation is the Coriolis force. This is the lateral force generated in moving bodies that cross lines of latitude on the earth's surface: the force that causes the clockwise deflections of winds in the northern hemisphere. The almost infinitesimal value of Coriolis forces in a flying bird have discouraged wide acceptance of this means of orientation. A hypothesis advanced by Yeagley (1947, 1951) suggests that birds respond both to terrestrial magnetism and to Coriolis force, and that geographical gradients of the two forces create two sets of coordinates, something like the longitude-latitude grid, which birds unconsciously refer to while navigating. This theory has aroused much criticism and little support among ornithologists (Keeton, 1974).

For simple directional orientation some birds are thought to use wind direction as a clue (Vleugel, 1962; Bellrose, 1967). A migrant orienting visually by day may, when darkness comes, shift cues, for example, from the setting sun, and maintain a constant flight angle relative to the wind, probably with occasional reference to nocturnally visible landscape features or ocean wave patterns. It is possible that even smells may be used as orienting cues by those few birds that have good olfactory organs. An albatross, for example, may orient toward home upon encountering appropriately scented downwinds from its odorous breeding colony

(Fisher, 1971). Experiments with homing pigeons in Italy by Papi *et al.* (1974) seem to show that they associate different odors with winds from different directions, and that when displaced to strange sites, they use olfactory cues to establish the home direction. Pigeons had Xylocaine administered to their olfactory membranes, and were rendered unable to smell. They performed poorly in orientation and homing ability tests when compared to controls (Ioale, 1983; Meschini, 1983). Pigeons with their olfactory nerves sectioned also performed poorly at these tasks compared with controls (Walraff, 1981). This has been interpreted as evidence supporting the olfactory cue hypothesis. Bobolinks, *Dolichonyx oryzivorous*, were shown to orient to magnetic fields (Beason and Nichols, 1984). The authors also discovered iron-rich material around the olfactory nerve and the nasal cavity. They concluded that the Bobolink's innervation functions as a magnetic sensor so that local anesthesia or sectioning of the olfactory nerve may alter the magnetic perception of birds. The role of olfaction in navigation is thus still a controversial subject.

Still other formerly unsuspected environmental cues may be available to a migrating bird. The colored oil drops in its retinas may reveal significant landscape details to a daytime migrant. By means of conditioning pigeons to electrical shocks, it has been discovered that they can detect polarized and ultraviolet light (Kreithen and Eisner, 1978), and also the movement of an object as slow as 15 degrees per hour—the rate of the sun's movement in the sky (Meyer, 1964). Other conditioning experiments have shown that pigeons can detect extremely low-frequency sounds (infrasounds), as low as 0.05 hertz (or one cycle per 20 seconds). Below 10 Hz, Pigeons are at least 50 decibels more sensitive than humans (Kreithen and Quine, 1979). Homing pigeons can detect small shifts in sound frequency at 1, 2, 5, 10, and 20 Hz. Their acute sensitivity to these shifts makes it appear feasible that natural Doppler shifts may be detected (Quine and Kreithen, 1981). Very significantly, such natural infrasounds, generated by jet streams, winds over mountain ranges, thunderstorms, and ocean breakers, carry great distances and can be detected many hundreds of kilometers from their sources.

Sense of Direction

The idea that birds possess an innate "sense of direction" cannot be as easily dismissed as some of the above theories. Rüppell (1944) captured 896 Carrion (Hooded) Crows, *Corvus corone*, during their spring migration through Rossitten on the southeastern shore of the Baltic, and transported them by train to Flensburg (750 km to the west), Essen (1025 km westsouthwest), and Frankfurt (1010 km southwest), where they were released. Of these birds 176 were recaptured: 20 of them in their presumptive or traditional summer or winter ranges, but 156 of them in new and displaced summer and winter ranges, outside of and to the west of their normal ranges. The released birds had migrated in the traditional northeasterly direction, in a path parallel to their normal migration route but displaced to the west. This caused them to settle for the summer in Denmark and Sweden rather than in the east Baltic countries and Russia. The few displaced birds that returned to the normal breeding areas were probably older, experienced birds that had migrated there previously. Young White Storks, *Ciconia ciconia*, from Rossitten, which normally migrate in a southeasterly direction toward the eastern end of the Mediterranean Sea, were transported to western Germany and released after the native Storks of this region had already departed for winter quarters. The young Storks flew in a southerly or southeasterly direction, as they would have flown from Rossitten,

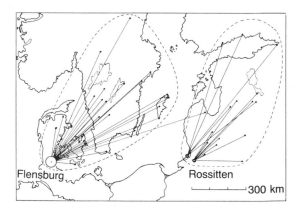

FIGURE 22–13 Rüppell's experiment demonstrating that migrating Hooded Crows orient by means of a sense of direction. When transported from Rossitten westward to Flensburg and then released, the birds migrated in a direction parallel to their normal migration route, and were subsequently recaptured in new areas to the west of their normal range. After Rüppell, 1944.

rather than in the southwesterly direction characteristic of west German White Storks (Schüz, 1949).

The significance of such innate, compass-direction migration was clearly demonstrated when Perdeck (1958) caught and banded over 11,000 Starlings, *Sturnus vulgaris*, near The Hague during fall migration and released them in Switzerland 560 to 670 km to the southeast. From 354 recoveries he discovered that the juvenile birds maintained their ancestral direction of migration (west-southwest) while most of the adults changed direction and flew northwest toward their normal winter range. Recoveries in subsequent years showed that the experimental displacement had, for the juveniles, established a permanent winter range that was completely new for this population. These results strongly indicate that, for inexperienced young birds, this innate directional migration, coupled with a specific duration of migratory restlessness and subsequent habitat imprinting (Ralph and Mewaldt, 1975) specifies the location of their winter quarters.

Tests on Mallard ducklings, *Anas platyrhynchos*, by Schadt and Southern (1972) showed that an unlearned, northwesterly directional preference appeared between the fifth and fourteenth days of age. Similar tests by Southern (1969) on young Ring-billed Gulls, *Larus delawarensis*, 2 to 20 days old, revealed an innate tendency to select a southeasterly heading. This is the direction the immature gulls must fly on their first fall migration to reach their traditional winter quarters.

Related to this directional orientation is the goalless, or "nonsense," orientation of many waterfowl. Experiments by Matthews (1961) and Matthews and Cook (1977) showed that Mallard ducks displaced from familiar haunts on clear days tended strongly to fly in a specific direction (commonly northwest) regardless of sex, age, topography of release point, time of day or of season, of wind direction, or of distance or direction from home. Landscape features, especially bodies of water, seemed to influence direction of flight. In birds whose circadian clocks had been shifted by artificial day-night regimes it was demonstrated that the sun was used for directional orientation in preference even to learned landmarks. The direction of goalless orientation varies between species and even among different populations of the same species. Mallards in southeast England and in Sweden, upon displacement from familiar surroundings, characteristically fly in a southeasterly direction. The ori-

gin, physiological mechanism, and function of this goalless orientation are as yet completely obscure.

A combination of goalless and goal-directed orientation seems to occur in inexperienced homing domestic pigeons, *Columba livia* (Wallraff, 1967; 1970). Such pigeons released in unfamiliar areas usually depart in directions that apparently are the result of two components: the direction of the home loft, and a preferred compass direction. The compass direction is specific for pigeons of a given loft and can be different for closely related populations of lofts only 20 or 30 km apart. In general, the greater the release-distance from the home loft, the more the initial flight direction is oriented toward the loft.

An innate sense of direction may be essential to orientation, especially for inexperienced young birds migrating without parental guidance. In experiments by Berthold et al. (1972) and Berthold (1973a), six different species of European warblers of the genus *Sylvia* were kept, during the migratory season, under both natural and artificially controlled photoperiodic conditions and their nocturnal *Zugunruhe* measured. There was a direct correlation between the amount of nocturnal restlessness exhibited by a given species and the length of its normal migration journey; long-distance migrants like the Garden Warbler, *S. borin*, produced more *Zugunruhe* than a short-distance migrant like the Blackcap, *S. atricapilla*. F$_1$ hybrids between migratory and sedentary subspecies of Blackcaps exhibited an amount of *Zugunruhe* intermediate between the parental types, indicating that this activity is under genetic control (Berthold and Querner, 1981).

Birds that follow the crooked paths of "dog-leg" migration routes would need an innate sequence of different and precisely timed directions in order to arrive successfully at their winter quarters. Evidence from experiments by Gwinner and Wiltschko (1978) indicate that this is the case, at least in the Garden Warbler, *S. borin*. This species, in its fall migration from north and central Europe, changes direction counterclockwise upon reaching northern Africa. Birds tested in Europe for orientation under constant photoperiodic conditions showed a counterclockwise shift in their mean direction of nocturnal orientation at about October 1, the time when free-flying birds should have reached Africa and shifted their migration paths similarly. The development of new migration paths in historic times, and the obvious dangers of using a petrified sense of direction in a constantly changing world, argue against its exclusive use as a means of bird orientation.

Visual Orientation

It would be strange indeed if birds did not use their superior eyesight to aid them in navigation. Visual orientation is not incompatible with a sense of direction, and is usefully supplementary to it. For example, a species may fly in a given direction until it reaches a seashore, and then follow that for the next leg of the journey. In some species, visual cues may be all that are required to guide birds on their migration travels. In the case of birds following traditional migratory routes, which the young birds learn from their elders, the birds need only to remember the varied visual landmarks over which they fly: shores, islands, mountains, river valleys, prairies, deserts, forests. Young geese very likely learn their migration routes in this way. Stresemann (1927–1934) remarks that young geese are so dependent on the leadership of older birds that the young, in spite of a lively *Zugunruhe*, will remain and spend the winter at the breeding area if older birds are restrained from migrating by wing-clipping. It may be, too, that birds flying in flocks orient themselves better than birds flying individually, because of the possibilities of pooling navigational resources. Flocks of five homing pigeons, followed by helicopter, flew as a group with one or two leaders (Wagner, 1982). This may explain why pigeons released singly showed greater indecision and poorer homing ability than those released in flocks (Hitchcock, 1955).

While there is much evidence that daytime migrants rely on such landscape features as shorelines, large rivers, and mountains for orientation, the majority of radar studies show that nocturnal migrants ignore the landscape and depend on other cues for orientation (Emlen, 1975). Even so, after a night's migration, a given bird will be able to observe its surroundings by daylight and possibly make adaptive adjustments in its next night's migration path as a result.

Strong, long-distance fliers probably combine several means of orientation to reach their goals: sense of direction, random search, topographic memory, and others. When one method proves inadequate, others may supplement it. The Northern Gannet, *Sula bassana*, is a highly marine species, rarely seen inland. When 17 marked gannets were displaced 342 km west-southwest from their nests on Bonaventure Island and released deep inland in Maine, they were observed by airplane to perform more or less random movements until they encountered the familiar Atlantic coast line, after which they made a more rapid and direct return home (Griffin and Hock, 1949).

In addition to topographic clues, birds may use ecological and meteorological clues for orientation. A migrating bird may well possess an inherited reaction to certain gross landmarks such as mountain ranges or forests, and also to certain finer details of landscape. The bird may prefer to fly over scrub desert rather than rock and sand desert, or over long-grass prairie rather than short-grass plains, or over oak-hickory forests rather than beech-maple woods. Still finer distinctions may be perceived by a bird whose eyes, supplied with colored oil droplets in many species, may discern otherwise invisible features of the landscape, as aerial cameras using special filters can detect features invisible to ordinary cameras.

The march of spring northward across the land provides seasonal phenological clues to the migrating bird. One species may fly northward from a condition of early spring into late winter, while a more slowly moving species may progress northward at the same rate as spring itself. In the north temperate zone in April and May, spring advances northward about 2 degrees of latitude every week. Therefore, the difference that one week of spring sunshine would make in the appearance of a flower garden might be evident to a bird, and hence a clue to its direction of flight, if it flew due north or south 2 degrees in latitude or approximately 210 km—not a strenuous day's flight for many migrants. Any bird flying even 100 km in the "wrong" direction would soon see discrepancies between the landscape beneath it and the learned or innate image of the correctly unfolding migration path. And even without phenological clues, the correlation between increasing day length and increasing latitude during the northern hemisphere breeding season could serve to inform any bird with even a modestly accurate circadian clock of its approximate geographical latitude.

Although migrating birds do not ordinarily fly at great speeds or great heights, their flight makes possible a type of orientation clue denied to earth-borne travelers: the juxtaposition of landscape features both in space (through altitude) and in time (through rapid movement). On a clear day, a bird flying at an altitude of only 600 meters can see the landscape within a radius of about 100 km. Thus, birds can grasp larger configurations of the landscape and make meaning out of topographic and ecological juxtapositions in somewhat the same way that a person can compare two color samples when they are side by side, but not when one is at the shop and the other at home. Ornithologists have perhaps been too prone to tackle the visual orientation problem analytically, looking for isolated clues when, possibly, birds recognize beneath them a terrestrial physiognomy much as we recognize the photograph of a friend, but would be at a loss to identify it if it were cut up and scrambled into jigsaw-puzzle pieces. If one person can recognize a friend among a thousand strangers, why should not a bird be able to recognize a "friendly" habitat or migration path among many similar-looking (to man) patches of the earth's surface? This does not deny the significance of prominent individual features of the landscape or face (mountain or wart) that might serve as visual clues to identity, but it helps to explain how a bird or a human may *unconsciously* use subtle differences in landscapes or faces to tell them apart. As Pascal remarked, "If the nose of Cleopatra had been shorter, the whole face of the earth would have been changed."

The advantages of an aerial inspection of the landscape are clearly revealed in the fact that "within just a few hours of flying time [an aerial archaeologist] discovered more Roman military sites in the North [Britain] than had come to be known in the previous two hundred years, during which several seasoned antiquarians and archaeologists had spent lifetimes in their searches" (Deuel, 1969).

Visual orientation by features of the landscape does not seem particularly helpful to migrants crossing large stretches of the ocean. But even here there are a few clues to location or direction, some visual and others not: wind and wave direction (particularly in zones with relatively steady prevailing winds); islands, reefs, atolls; cloud formations over oceanic islands and over many cold waters such as the California Current; temperature and humidity differences in the air; temperature, color, and turbidity differences in the great surface currents of the seas and at the mouths of great rivers; fog belts where cold and warm currents come together, as on the Newfoundland Bank; the presence or absence of organisms visible from the air, such as dolphins, flying fishes, jellyfishes, whales, or, at night, glowing dinoflagellates, or accumulations of floating algae, as in the Sargasso Sea. These do not make an impressive list of road signs in the vast reaches of the oceans, but they may nevertheless be useful clues to some wanderers.

One difficulty in the use of ecological and pheno-logical clues for navigation is that they change with the seasons, and the changes do not always come in precisely predictable cycles. For example, an early autumn snow storm can greatly change the appearance of the land. This may be one of the reasons the great majority of fall migrants move southward well ahead of the average time of the first snowfall. Nor are ecological clues generally available in the midst of the great expanses of desert or sea. It is difficult to see how they could be of any use to the New Zealand Bronzed Cuckoo, *Chalcites lucidus*, whose young, one month after their parasitic parents have preceded them, migrate about 1900 km westward across the ocean to Australia, and then another 1600 km northward over water to the Solomon and Bismarck Islands, where they join their parents. This migration, one of the wonders of bird life, surely demonstrates the hereditary nature of some migration flights, but reveals nothing of the means by which the young birds navigate successfully over relatively featureless water to a tiny wintering area.

Landmarks are learned rapidly in certain species. Matthews (1963) discovered that homing pigeons could learn landmarks within about 16 km of their home loft after only three to five releases in the vicinity. Surprisingly, when the birds were released 37 to 56 km away from home, they did not orient properly, but at distances of more than 80 km, they again oriented towards home. Matthews attributes these paradoxical results to two different types of orientation: visual orientation, or "piloting," by landmarks near the home loft, and orientation by reference to a bicoordinate grid system, or true navigation, at distances over 80 km, with no initial orientation at all in the intermediate zone. The essential point here seems to be that the X and Y physical gradients of the grid, whatever they may be, do not acquire sufficient quantitative differences between the release point and the home loft until a distance of about 80 km separates the two regions. Supplementary reference to the sun, to wind direction, or to some other physical cue, presumably carries the birds successfully over the intermediate region. Results similar to Matthews' with pigeons were obtained by Ilenko (1974) with European Robins, *Erithacus rubecula*. When displaced 80 km from their nest sites, 82 per cent of the birds oriented toward home in their confining cages; when displaced 8 km, none of them did. The distance effect on orientation found by Matthews could not be confirmed by Keeton (1974) using American pigeons.

Other investigators have obtained results suggesting that vision may not be essential even for short-distance homing. St. Paul (1962) found that pigeons could home successfully from displacements of about 20 km by night. Success was much poorer in cold winter weather than in summer. Inexperienced pigeons transported in a distorted magnetic field or in total darkness to a release site exhibited disorientation, but improved with repeated trials (Wiltschko and Wiltschko, 1985). The authors concluded that pigeons collect information during the outward journey and learn to use the local map with experience.

Magnetic Orientation

For over a century some ornithologists have hypothesized that birds may "home" or find their ways in migration by somehow referring to the earth's magnetic field. Experimental proof establishing the capacity of birds to sense geomagnetism was first provided by the pioneering studies of Merkel *et al.* (1964). Today a lively increase in experimental studies has enlarged and refined our understanding of this intriguing problem.

That birds should have the ability to sense the earth's magnetism is not so remarkable when one considers the variety of organisms that have demonstrated the capacity to sense magnetic fields. Sharks, catfish, and other electric fish find their way, locate prey, and even communicate by means of animal-generated electric fields (Bullock, 1973; Moller and Serrier, 1986). The cave salamander, *Eurycea lucifuga*, can be trained to orient either to the earth's magnetic field or to an artificial substitute field (Phillips, 1977). Bees have demonstrated a capacity to orient to the earth's magnetic field (Gould *et al.*, 1978), and even some bacteria orient their swimming in response to natural or artificial fields (Anonymous, 1978).

If European Robins, *Erithacus rubecula*, undergoing spring *Zugunruhe* are kept in circular experimental cages devoid of directional cues or of vision of the sky, they nevertheless confine their activity to the northeast sector of the cage—their natural migration direction. However, if they are enclosed in a steel chamber, this directional orientation disappears (Fromme, 1961). Merkel and Wiltschko (1965) found that this directional choice could be significantly altered by subjecting the birds to an artificial magnetic field generated with Helmholtz coils.

For centuries man has known of and used the ability of the domestic Pigeon, *Columba livia*, to return to its home loft. He has bred certain strains that excel in this capacity (Edrich and Keeton, 1977). Today such "homing pigeons" have been used in a variety of studies of magnetic field sensitivity, and they provide convincing proof of its reality. In a typical experiment pigeons are released singly from an unfamiliar location and observed with binoculars until they vanish from sight. Their "vanishing bearing" with reference to that of home is then recorded.

A group of young pigeons were reared in a light-tight room in a natural photoperiod. They could not see the morning sun, but were allowed to exercise in the afternoons in an outdoor aviary. Controls were given access to an outdoor aviary all day. Experimentals were trained to home only on afternoons, whereas controls also homed in the morning. Both groups were released from a distant site on a sunny morning when they were four months old. Half the controls and half the experimentals carried bar magnets on their backs. Controls and experimentals without magnets oriented correctly. Controls with magnets also performed correctly; they were accustomed to using the morning sun to home by. Experimentals with magnets did not orient. The magnets distorted their magnetic compass, and as they had never seen the morning sun, they could not use the sun compass. Thus, exposure to the morning sun enables the sun compass to develop, but in its absence pigeons use the magnetic compass, apparently their first source of compass information (Wiltschko *et al.*, 1981).

In another experiment Helmholtz coils were fitted to nest-boxes which deflected the magnetic field either 90 degrees to the right or 90 degrees to the left. A third group of boxes was left unmanipulated as controls. Pied Flycatchers, *Ficedula hypoleuca*, were hatched in these boxes, removed at 13 days of age, and raised in the laboratory with windows facing south and west. At the age of two months the birds were placed in orientation cages. Controls oriented approximately west whereas the other two groups oriented in the deflected directions, 90 degrees to the right or left (Alerstam and Hogstedt, 1983). These data suggest that the magnetic compass serves as the primary reference system in relation to which a secondary (e.g., celestial) cue is calibrated.

Walcott and Green (1974) attached batteries and coils to pigeons so that the polarity of the magnetic field created in the bird's head had its north pole pointing downward as does the earth's magnetic field in the northern hemisphere. On overcast days these pigeons flew toward home. But when the field was reversed so that the north pole pointed upward, most birds flew away from home.

These results, as well as those of Wiltschko and Wiltschko (1972) with European Robins, *Erithacus rubecula*, subjected to artifical magnetic fields, indicated that the birds were not responding to the north-south compass polarity (azimuth) of the field but to its dip vector or the angle it makes with the perpendicular. Single European Robins were placed in special L-shaped cages fitted with magnetic coils which could decrease the intensity of the horizontal component of the magnetic field. Birds reacted to the induced changes in magnetism, suggesting that they may refer to changes in magnetic inclination to obtain information about latitude (Fiore *et al.*, 1984).

The sensitivity of birds to magnetic fields is further revealed in their reactions to abnormal magnetic environments. Two- to 20-day-old chicks of the Ring-billed Gull, *Larus delawarensis*, confined in large orientation pens by Southern (1969), showed significant innate tendencies to move in a southeasterly direction, the normal migratory direction, except during times when intense solar storms caused disturbances in the earth's magnetic field. At such times the birds headed in random directions. An analysis by Schreiber and Rossi (1976) of the returns of 1200 homing pigeons flown in 18 Italian races revealed that a significantly smaller percentage of birds returned home by their first night on days with high sunspot activity (and presumably its disturbing effect on the earth's magnetic field). When the United States Navy's large underground antenna (Seafarer-Sanguine) for radio communication with submarines was energized, it caused disorientation of both Ring-billed Gull chicks (Southern, 1975) and low-flying nocturnal migrants (Larkin and Sutherland, 1977). Pollution of the earth's magnetic field is apparently a new hazard facing wildlife.

Natural magnetic anomalies in the earth's field also affect bird orientation. Near Kursk, in the U.S.S.R., is a region where the magnetic field is more than twice the normal strength and is inclined more than 60 degrees from the normal dip of that latitude. When placed in circular activity-recording cages in cloudy weather, birds of five different species showed none of the typical migratory orientation exhibited in normal surroundings. However, they all doubled or trebled their physical activity while subjected to this aberrant magnetism (Shumakov, 1967). Similarly, homing pigeons released at a magnetic anomaly near Iron Mountain, Rhode Island, became greatly disoriented (Walcott, 1978).

Whether birds "read" and respond to the azimuth, the dip, or the intensity of the earth's magnetic topography, they need an appropriate sense organ for the job. Already, bees have been shown to possess about one million crystals of a magnetic substance, so arranged in front of the abdomen as to create a horizontal magnetic field (Gould *et al.*, 1978). Preliminary research on this problem by Walcott *et al.* (1979) indicates that homing pigeons possess about 100 million needle-shaped, iron-containing (magnetite) objects, each about 0.1 micron long, on the anterior brain next to the skull. Presti and Pettigrew (1980) have detected "permanently magnetic material" in the neck muscles of both pigeons and migratory White-crowned Sparrows, *Zonotrichia leucophrys*. The authors tentatively propose that stretch-sensitive muscle spindles in the birds' necks may act as sensory receptors of magnetism when magnetic granules are subject to torque exerted by the earth's magnetism. Iron-rich material has also been located around the olfactory nerve, between the eyes near the olfactory bulb, and in the nasal cavity of Bobolinks, *Dolichonyx oryzivorous* (Beason and Nichols, 1984). The authors conclude that the olfactory nerve might be involved with the magnetic compass. In contrast, electrophysiological studies indicated that 70 per cent of the cells in the basal optic root nucleus of the pigeon responded to gradual changes in strength in both the vertical and horizontal component of the magnetic field (Semm *et al.*, 1984). Further investigations of this problem will be eagerly awaited by ornithologists.

Celestial Navigation

When prehistoric man traveled to gather food, he did not have the help of such civilized crutches as highways, road signs, maps, or compasses. It seems reasonable to assume that, at least on long journeys, he depended on celestial cues for a sense of place and direction. If, for example, he traveled far to the north, he undoubtedly depended on the sun's position to maintain his course: the sun at his right shoulder at sunrise, behind him at noon, and at his left at sunset. Very long northerly trips probably made him aware of the sun's lower altitude at its noontime zenith. And possibly, extensive travels east or west made him aware that such movements caused the sun to rise and set earlier or later. Artificial travel aids have, for civilized man, all but extinguished reliance on these cues to orientation. But wild animals, especially those that travel widely and rapidly, are highly sensitive to celestial cues. They use them in several ways.

Early experiments by Kramer (1957; 1961) and his students revealed how certain species of birds rely on the sun for orientation. Birds kept in circular cages during the migration season often show their restlessness by facing and fluttering in a direction conforming to that taken by migrating birds of the same species. Kramer placed Starlings, *Sturnus vulgaris*, in a cage exposed to sunlight and observed that the birds fluttered on their circular perch in the normal migration direction as long as the sun shone, but on cloudy days their activity was disoriented. When the apparent position of the sun was shifted by means of mirrors, the orientation of the birds shifted correspondingly. Since the direction of orientation of the Starlings remained constant for hours, this indicated that they made allowance for the sun's changing position. That is, the birds compensated for the apparent movement of the sun by reference to their physiological clocks.

These internal clocks can be reset or shifted by subjecting birds to an artificial light and dark cycle out of phase with natural day and night. Hoffmann (1954) reset the clocks of Starlings by subjecting them for several days to a light-dark cycle six hours advanced over the natural outdoor cycle. That is, "dawn" for the birds coincided with natural midnight. As a result, when the birds with reset clocks were exposed to natural daylight, their directional fluttering was shifted about 90 degrees in a counterclockwise direction (Fig. 22–14). Since these birds respond to the azimuth, or compass bearing, of the sun and not to its height above the horizon, this type of orientation is called *sun azimuth orientation*. In similar experiments with pigeons whose clocks had been shifted, Keeton (1969) discovered that when the birds were released in sunlight, their flight bearings were shifted in the direction predicted by the sun azimuth theory, but when released under total overcast, the bearings of both clock-shifted and non-shifted control birds were homeward-oriented. Keeton concluded that the birds used the sun as a compass when it was available, but could call on other means of orientation in the absence of either the sun or familiar landmarks. The actual results of an experiment by Emlen and Keeton on clock-shifted pigeons are shown in Figure 22–15. The pigeons were obviously using the sun as their compass. Sun azimuth orientation has been demonstrated not only in many species of birds but also in box tortoises (Gould, 1967), lizards (Fischer, 1961), frogs (Ferguson, 1967), and insects. And when the clocks of these animals are reset, they respond like the birds by shifting their orientation accordingly.

The pioneer work on orientation in bees by von Frisch (1956) showed that they not only use the sun in navigation, but that scout bees inform workers in the

FIGURE 22–14 If, for a week or two, a bird is subjected to an artificial light-dark cycle six hours advanced over the natural day-night cycle so that it becomes accustomed to arising at midnight and going to sleep at noon, its directional orientation will be shifted 90° counterclockwise, when it is again exposed to the natural day-night cycle. Thus, a bird that under normal conditions would migrate north (or away from the noon sun) will, when its clock is reset six hours, migrate west (or away from the natural rising sun), as if still using the sun as a noontime guidepost.

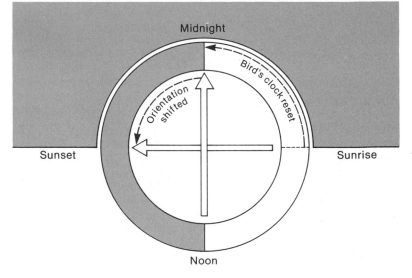

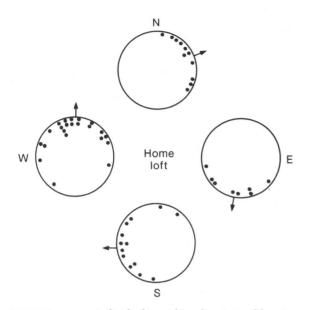

FIGURE 22–15 Individual vanishing bearings of homing pigeons whose clocks had been advanced 6 hours, and were then released between 30 and 80 km north, east, south, and west of their home loft. Arrows indicate mean directions taken, showing an approximately 90° counterclockwise shift away from the normal homing direction. Redrawn after Emlen, 1975.

dark hive just how far to travel and in what direction, with reference to the sun's position, to find a source of nectar.

Hand-reared Savannah Sparrows, *Passerculus sandwichensis*, may orient with respect to sunset only in the presence of a directional magnetic field. Viewing the setting sun is not sufficient (Bingham, 1983). Moore (1985) found, however, that this behavior changes with age. He subjected Savannah Sparrows to two "cue-conflict" experiments. In one case, sparrows saw a sunset shifted 90 degrees north of its westerly position followed by a view of stars. They shifted their migratory activity with reference to the setting sun. In a second experiment birds were exposed to a sunset shifted 90 degrees north of its westerly position, and the magnetic north shifted 90 degrees west with Helmholtz coils. Again birds oriented with reference to the sun (Moore, 1985). Savannah Sparrows with experience may use solar cues to select a migratory direction independently of geomagnetic or celestial information. Moreover, Able (1982) manipulated

the *e*-vector of polarized light during sunset and found that White-throated Sparrows, *Zonotrichia albicollis*, shifted their activity in predictable directions. Thus polarized light may be used as yet another cue in migratory orientation.

Wiltschko (1981) reviewed the literature on sun orientation and concluded that there is no evidence in favor of birds using the sun as a navigational cue. However, many avian species use a sun compass to maintain a selected direction.

The obstinately innate nature of the sun azimuth orientation mechanism in birds was well demonstrated in homing experiments with Adelie Penguins, *Pygoscelis adeliae*. Penguins transported by air from their coastal rookeries to release points in the interior of Antarctica consistently pursued straight courses north-northeast toward the coast (and survival), provided the sun was shining (Emlen and Penney, 1964). Since the Antarctic sun moves through the sky in a counterclockwise direction at 15 degrees per hour, the penguins could use it for a compass only by shifting their path with a compensating deviation of 15 degrees per hour in a clockwise direction. However, when penguins were flown to North Dakota (and later to the Arctic) and released on snowy fields under clear skies, their southern hemisphere clockworks persisted in "compensating" clockwise for the motion of the sun, which in the northern hemisphere moves clockwise. As a result, the birds did not compensate for the sun's motion but instead compounded it and shifted their compass orientation 30 degrees per hour, describing large circles rather than straight paths (Penney and Riker, 1969). Transequatorial displacement of homing pigeons resulted in similarly false orientation (Schmidt-Koenig, 1963).

Sun azimuth orientation is of little use to a displaced bird that is unaware of its location with reference to home, just as a compass is useless to a man lost in the wilderness if he has no knowledge of the direction of his goal. Two sets of grid coordinates, such as degrees of latitude and longitude, are required to determine with any geographical precision the location of an object on the face of the earth. Numerous experimental attempts have been made to determine what bicoordinate cues are used by displaced birds that enable them to return home, even to tiny islands in immense oceans.

Thus far no proposed system of bicoordinate navigation has received general acceptance by ornithologists. Yeagley's hypothesis based on geomagnetism and the

Coriolis force has never been confirmed. Matthews (1968) proposed a sun-arc hypothesis in which birds determined their comparative latitude by the height of the sun's noon zenith, and their comparative longitude by comparing the time of the sun's zenith in an unfamiliar region with its time at home. Although conditioning experiments with pigeons have shown that they can detect movements as slow as 15 degrees per hour—the rate of the sun's motion through the sky (Meyer, 1964)—no convincing experimental evidence has yet been demonstrated that birds use the sun in any other way than as a compass.

For nocturnal migrants the starry skies constitute the most obvious source of reference for long-distance orientation and navigation. Sauer (1957) placed several species of migratory-restless European warblers in circular activity cages and exposed them to clear night skies. The birds persistently oriented in the direction taken by normal wild migrants for that place and time of year. On cloudy nights the birds showed no preferred directional orientation. The cages were next placed under the dome of a planetarium. With the star images set in time and latitude for the local region (Freiburg, Germany), the birds oriented as they had under the natural sky. By shifting the artificial planetarium sky, Sauer could control the orientation of the restless warblers. One glimpse of the sky was sufficient, he believed, to tell a bird where it was and where it should go.

Since Sauer's pioneer experiments, other investigators have confirmed that many species of birds orient at least in the appropriate direction for seasonal migration by reference to natural or artificial night skies. But questions still remain as to whether and how the birds use stars for bicoordinate navigation—that is, to tell where they are and where home is. When Gauthreaux (1969) exposed White-throated Sparrows, *Zonotrichia albicollis*, during spring migration to planetarium star patterns, the birds oriented northerly when the north star was projected to the north, as normal, and southerly when the north star was projected abnormally to the south. Autumn star patterns presented to birds in springtime did not influence their vernal northerly orientation. Emlen (1967; 1970) found that Indigo Buntings, *Passerina cyanea*, also reversed their spring *Zugunruhe* orientation when the planetarium sky was rotated 180 degrees horizontally. When the birds were exposed to planetarium skies advanced or retarded 3, 6, or 12 hours from local time, the birds still maintained their normal seasonal orien-

tation. That is, they did not interpret the earlier- or later-rising stars as shifts in the birds' longitude. Through photoperiod manipulation birds were brought into autumnal *Zugunruhe*. Such birds exposed to planetarium spring skies oriented southward. From these results Emlen argued that these birds did not rely on a bicoordinate system of celestial navigation, nor did they possess a genetic star map that enabled them, through time-compensation, to orient longitudinally. Rather, he believed, the birds used *Gestalt* configurations of stars as directional cues—especially those stars that rotated about the north star. That they oriented with reference to the stellar axis of rotation rather than to specific stars was borne out when the birds still oriented normally when presented with a fictitious sky in which the stars rotated about Betelgeuse of Orion as a pole star. Emlen (1969) also found that Indigo Buntings hand-raised in visual isolation from stars and later tested in a planetarium exhibited random orientation for many nights. Early visual experience appears to be necessary for this species to develop normal migratory orientation capabilities.

In contrast to Emlen's theory that night migrants use the axis of stellar rotation as an orienting cue, Sauer (1971) found that three species of *Sylvia* warblers oriented properly when under a stationary planetarium sky, and Wiltschko and Wiltschko (1974) reported the same for European Robins, *Erithacus rubecula*, that were able to observe only a sector of the night sky restricted to 90 degrees centered on the zenith (excluding Polaris).

The recognition by birds of seasonal star patterns as cues for migrating north in the spring and south in the fall seems unlikely in light of the fact that stars and seasons reverse their relationships every 13,000 years as a result of the earth's axial precession (Agron, 1962). One would guess that evolutionary adaptation to the simple and more stable axis of celestial rotation would take precedence over adaptation to the more changeable patterns of star distribution, especially the distribution of the bright but roving planets.

There is also the possibility that stars may be used only as a substitute compass after a migrating bird has determined its proper bearing by other means. From experiments with 14 night-migrating Savannah Sparrows, *Passerculus sandwichensis*, Moore (1978; 1985) concluded that the birds were oriented primarily by the position of the setting sun, and maintained that orientation after dark by reference to the stars. Wiltschko and Wiltschko (1975), from experiments with *Sylvia* warblers, concluded also that the stars are secondary sources for maintaining orientation after information from the earth's magnetic field has been transferred to them.

Much still remains to be learned about the neurophysiological machinery involved in orientation and navigation. As was described in Chapter 8, preparation for reproduction, *Zugunruhe*, and the contrary directional tendencies in spring and fall depended on the temporal interaction of two hormones, corticosterone and prolactin, for at least one migratory species, (Meier, 1973). The steps through which these hormones control migration direction are still largely unknown, as are the receptor and effector mechanisms responsive to geomagnetism. There can be no doubt that different species of birds rely on different physiological mechanisms to carry out their astonishing feats of orientation and navigation.

To sum up, present evidence suggests that birds rely on a variety of orientation and navigation systems, one or more of which may be used as back-up or redundant systems when a given means of orientation fails them, e.g., the sun compass is useless at night or in overcast weather. Currently, it seems likely that a typical migrant depends, at least in part, on its ability to determine position and/or direction by reference to celestial bodies or to the earth's magnetism, or by sharp-sighted scanning of the landscape; by an innate timing mechanism; possibly by an inherited recognition of certain features of the earth's surface; and by the capacity to learn migration routes by experience and remember them from year to year. Perhaps to a lesser extent migrants may employ their abilities to perceive odors, or very slight changes in barometric pressure, infrasounds (through their Doppler effects), ultraviolet and infrared radiations, and the plane of polarization of light from the sky. But whatever the mechanisms may be, it is certain that they are constantly being sharpened and refined by natural selection. The longer the migration journey and the smaller its end targets, the more rigorous the selection and the more disciplined the bird's performance must be. Successful migration pays high rewards, but even slight errors in navigation may cost the bird its life.

■ SUGGESTED READINGS

Largely descriptive of migration patterns are Dorst's *The Migrations of Birds*, and Moreau's *The Palaearctic-African Bird Migration Systems.* Among the more analytical studies of orientation and navigation are Baker's *The Evolutionary Ecology of Animal Migration*; S. A. Gauthreaux Jr.'s *Animal Migration, Orientation and Navigation.* Berthold's chapter on Migration: Control and metabolic physiology, and Emlen's chapter on Migration: Orientation and navigation, *in* Farner and King's *Avian Biology*, Vol. 5; Keeton's chapter on The orientational and navigational basis of homing in birds, *in* Lehrman *et al.*, *Advances in the Study of Behavior*, Vol. 5; Matthews' *Bird Navigation*; Schmidt-Koenig and Keeton's *Animal Migration, Navigation and Homing*; Schüz's *Grundriss der Vogelzugkunde*, and A. Keast and E. S. Morton (eds.) *Migrant Birds in the Neotropics: Ecology, Behavior, Distribution and Conservation.*

The Origin and Evolution of Birds
23

Is truth or fable really the more strange?
  In Egypt's myth the Phoenix rose from fire;
In Nature's buried book the reptiles range
    By evolution's tricks
    Through Archaeopteryx
Until some sport from out the tree-tops springs
Air-borne; and its descendants all aspire,
    On stronger pinions, like the Eagle, higher!
—Joel Peters (1905–   )
  The Sun-Birds

In the broad sweep of time between the present and that earlier era when the planet Earth had cooled sufficiently to cradle its first primitive organism, man has existed for only the briefest moment. If Mother Nature had kept a diary of the happenings on Earth since life first appeared, and if, unlike human diarists, she had written only one page every ten thousand years, her diary today would comprise over 300 volumes, each with 1000 pages. Somewhere near page 700 in the most recent volume she would have recorded the first appearance of man. Only on the very last page would she have made the observation that man had finally achieved the wit to record his own history. Small wonder it is that much of the history of life on earth is hidden from man, the naked, bipedal newcomer.

Man, however, has been able to reconstruct the broad outlines of much of the history of life on earth from the scraps of nature's diary he has found in the form of fossils and living vestiges. These show incontrovertibly that living organisms have evolved, and that birds, for example, have evolved from reptiles.

It is universally agreed among contemporary authorities that birds originated in the diapsid reptiles and that their derivation involved relatively modest modifications of basic reptilian anatomy. Consequently, not only fossil birds but also many features of living birds testify to that relationship. Some of these reptilian affinities of birds have been mentioned earlier. The following summary, based on details from Heilmann (1927), shows how impressive the resemblances are.

Many of the similarities between birds and reptiles are found in the skeleton. Both birds and reptiles have skulls that hinge on similar neck bones by means of a single condyle, or ball-and-socket arrangement. Their lower jaws are made of several bones, and articulate with the skull by means of a movable quadrate bone. The lateral brain case (pleurosphenoid bone) is much extended in each. There is a single sound-transmitting bone, or columella, in the middle ear. Ribs in birds and certain reptiles (*Sphenodon*) have overlapping tabs, or uncinate processes. The ankle joint in all birds and in several reptiles is intertarsal. Likewise, certain bones in each may be hollow, or pneumatic.

Birds possess scales on their legs that are remarkably similar to reptilian scales. The pleural cavity in birds is very similar to that in crocodiles, and avian air sacs resemble those of turtles and chameleons. The brain of a bird is more like that of a crocodile than that of a mammal. Eyes of birds and of many lizards contain a pecten. Both birds and reptiles have nucleated red corpuscles. Precipitin tests show that the blood proteins of birds are similar to those of turtles and crocodiles. Birds and reptiles lay similar eggs, and the hatching young of each often possess egg-teeth. In their embryonic development, birds and reptiles show many close parallels. On rare occasions vestigial or atavistic claws appear on the wings of such birds as ducks, hawks, cranes, rails, and crows. In the nestling Hoatzin, *Opisthocomus hoazin*, the ends of the second and third fingers have claws, which the young birds use in clambering about trees.

The differences between birds and reptiles are nevertheless pronounced. Birds are warm-blooded and have a double circulation, pulmonary and systemic, based on a four-chambered heart. Reptiles are cold-blooded and have a single circulation put in motion by a three-chambered heart, although in crocodiles the heart approaches the four-chambered condition. Birds, and only birds, have feathers. It seems probable that feathers evolved to conserve heat and make endothermy possible, before they were used for flight.

ARCHAEOPTERYX

Some 160 million years ago, a flying reptile fell into the water of a tropical lagoon, drowned, settled into the fine silt at the bottom, and eventually became fossilized. If paleontologists had tried to draw up specifications for a missing link between birds and reptiles, they could scarcely have improved on this fossil, called *Archaeopteryx lithographica*, which was uncovered in 1861 in a lithographic limestone quarry in Bavaria. The discovery of this Jurassic fossil was one of the most fortunate finds in the history of paleontology. It has settled once and for all any question regarding the reptilian ancestry of birds. Since the original discovery, four other fossils of *Archaeopteryx* have come to light, all of them in the limestone beds near Solnhofen, Bavaria. Two of them, found in 1857 and 1951, were first misidentified as reptiles and only later correctly identified as specimens of *Archaeopteryx* (Ostrom, 1970; Kleinschmidt, 1974). The other two were discovered in 1877 and 1974. Fossil feathers of late Jurassic and early Cretaceous deposits have been discovered in Australia, the U.S.S.R., and Lebanon, indicating that birds had a worldwide distribution not long after *Archaeopteryx* (Talent et al., 1966; Rautian, 1978; Schlee, 1973). The Lebanon feather, preserved in amber, is essentially identical with modern contour feathers, clearly showing barbs, barbules, and hooklets.

Archaeopteryx possessed a shoulder girdle, pelvis, and legs roughly similar to those of modern birds. It was similar in standard external measurements (culmen, wing, tail, tarsus) to present day chachalacas, *Ortalis spp.*, Hoatzin, *Opisthocomus hoazin*, and touracos, *Crinifer spp.* (Brodkorb, 1971). Its feathers were, as far as can be known from fossil imprints, exactly like those of modern birds. On the other hand, most of its skeletal structures were decidedly reptilian—so much so that, were it not for its feathers, it would be classified as a reptile. It had a skull with toothed jaws, a long bony tail, clawed fingers, and abdominal ribs, none of which are bird-like characters. Since this reptile-bird did not fit into any existing scheme of classification, a new taxonomic group was created for its solitary use: the subclass Archaeornithes of the Class Aves. All other birds belong to the subclass Neornithes. The following lists of details are developed from Heilmann (1927), Brodkorb (1971), Feduccia (1980), Martin (1983), and Olson (1985).

Archaeopteryx was a lightly-built, crow-sized bird

FIGURE 23–1 A cast of the 1861 fossil of *Archaeopteryx*. About one-fourth actual size. Photo courtesy of the American Museum of Natural History.

that may have weighed about 200 g. Like the other small dinosaurs of its day, its skull articulated with the neck by a single ball-like condyle. Both jaws carried socket teeth, another reptilian character. The spinal column was made of 50 biconcave vertebrae; the long flexible neck was supported by 10 vertebrae. It possessed 11 pairs of thoracic ribs, which, unlike those of modern birds, were unjointed, lacked uncinate processes and did not attach to the sternum. It had a reptile-like tail; in modern birds caudal vertebrae are reduced in number and fused to form the pygostyle from which the tail feathers emerge. A single pair of feathers emerged from each caudal vertebra, providing an air foil with the same area as the wings.

The long bipedal hindlimb had four diverging toes, the first of which, the hallux, was turned backwards as in modern birds. The bird probably ran along branches in the manner of touracos (Brodkorb, 1971).

The outline of the wings in the Berlin specimen revealed that they were "elliptical" in shape with a low aspect ratio, typical of birds adapted to moving through restricted openings in vegetation (e.g., *Accipiter* hawks). The wing had 9 primaries (modern birds have 9 to 12) and 14 secondaries (modern birds have 7 to 32).

There has been some debate as to whether or not *Archaeopteryx* could fly. Typical of flying birds, the rachis of its wing feathers supported two vanes which were unequal in width, i.e., the vanes were asymmetrical; in flightless birds the vanes are symmetrical (Feduccia and Tordoff, 1979). Some paleontologists have argued that the apparent absence of a keel indicates that *Archaeopteryx* could not fly. The principal muscle of avian flight is the pectoralis. This muscle, Olson and Feduccia (1979) remind us, does not necessarily attach to the keel but in some groups, to the furcula and coracoclavicular membrane. The furcula in *Archaeopteryx* appears hypertrophied compared to those in modern birds, probably providing a place of anchorage for powerful muscles. During

FIGURE 23–2 A restoration of *Archaeopteryx* based on the 1877 fossil. The body parts are arranged in the same position as the corresponding distorted bones of the fossil. Photo courtesy of the American Museum of Natural History.

flight the down stroke is provided by the pectoralis, and the recovery stroke by the dorsal levators. The keel is the attachment site for the supracoracoides. A pigeon or crow with a severed supracoracoides is no longer able to take off from the ground, but is still capable of sustained flight (Sy, 1936). The supracoracoides is thus not really necessary for flight itself. (Bats fly and have no keel.) Olson and Feduccia suggested that the supracoracoides is a feature added to an avian architecture already adapted for flight. The acute angle of the scapula is also indicative of a bird with flying abilities—flightless birds have lost this acute angle.

Although there is no question that *Archaeopteryx* is a link between birds and reptiles, Steadman (1983) provides evidence suggesting that it may have been contemporaneous with more advanced avian forms.

Forerunners of *Archaeopteryx*

No fossil series has yet been discovered that links *Archaeopteryx* to the ancestral reptiles from which it must have emerged. The thousands of reptiles of the Mesozoic period were divided into five large subclasses, one of which, the Diapsida, was further divided into superorders. The first of these, the Lepidosauria, gave rise to three orders that included, among others, the lizards and snakes. The other superorder, Archosauria, or ruling reptiles, gave rise to five orders, only one of which survives today—the Crocodilia. The other four orders were the Thecodontia (socket teeth) or ancestral archosaurs; the Saurischia (reptile pelvis), which gave rise to gigantic dinosaurs like *Tyrannosaurus* and *Brontosaurus*; the Pterosauria (winged reptile) archosaurs, which solved the problem of flight with huge bat-like membrane-wings supported chiefly by an enormous finger, some wings having a spread of 7.5 meters (25 feet); and the order Ornithischia (bird pelvis), which included herbivorous dinosaurs such as *Stegosaurus* and *Triceratops*.

Students of avian paleontology currently agree that birds and dinosaurs are close relatives, descended from a common ancestor. They do not agree, however, as to the exact path of descent followed by *Archaeopteryx*. Some students, following Broom (1913) and especially Heilmann (1927), believe that both dinosaurs and birds arose from the small, bipedal, thecodont reptiles of the suborder Pseudosuchia. Several genera of these primitive Triassic reptiles have been discovered that show structural similarities to *Archaeopteryx*. Among them are *Euparkeria*, discovered in southern Africa, *Ornithosuchus* in Scotland, and *Saltoposuchus* in southern Germany. Similarities between these pseudosuchians and *Archaeopteryx* are particularly striking in the skulls (Fig. 23–3), but skeletal similarities also exist in the fore- and hindlimbs, pectoral and pelvic girdles, ribs, and tail vertebrae. The epidermal scales of *Euparkeria* suggest a possible first step toward feathers in that they were about twice as long as wide and had a central axis from which fine striations ran out on the two sides. After an exhaustive study of numerous reptilian fossils, Ostrom (1976b) makes a strong case for the descent of *Archaeopteryx* from the saurischian dinosaurs of the suborder Theropoda, with the late Jurassic *Ornitholestes* of North America as a possible ancestral type. Basing their conclusions largely on the nature of the teeth and on the ossification of the tibia and tarsal bones, Martin *et al.* (1980) still maintain that *Archaeopteryx* descends from the pseudosuchians. No fossil intermediate between Triassic reptiles and *Archaeopteryx* has come to light, but Heilmann attempted the reconstruction of a hypothetical "proavian" as shown in Figure 23–4. Resolution of this fascinating problem must await further study and, one hopes, the discovery of one or more significant missing links.

Fossil Record Since *Archaeopteryx*

Compared with the fossil record of other vertebrates, that of birds is disappointingly incomplete and fragmentary. Birds are light-weight, fragile animals, easily destroyed or decomposed. When a bird dies and floats on some body of water, it is subject to destruction by aquatic predators and scavengers, and its buoyancy usually prevents its settling into bottom sediments to become fossilized as was *Archaeopteryx*. According-

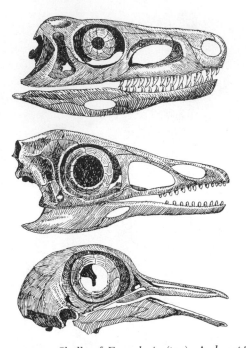

FIGURE 23–3 Skulls of *Euparkeria* (top), *Archaeopteryx* (middle), and a modern pigeon (bottom), showing the intermediate character of *Archaeopteryx*. *Euparkeria*, and *Ornithosuchus* (the latter shown in reconstruction at the beginning of this chapter) are representatives of the Pseudosuchian reptiles, which are thought to be the ancestors of the first birds. After Heilmann, 1927.

ly, there are many large gaps in the fossil history of birds, and even the relationships between many avian orders are still obscure because of these breaks in the continuity of the record.

There are, fortunately, certain localities where fossils of birds are found in relative abundance. These include such paleontologic sites as caves, dried-up lakes, diatomaceous earth strata, bogs, rock quarries, tar pits, and kitchen middens. In the United States, the most prolific source of avian fossils is California, which contains 44 different fossil localities, from which remains of over 220 species of birds have been taken. By far the richest of these sites is in Los Angeles at Rancho La Brea, where several asphalt deposits ooze to the surface. Here, in the past, these tar pits acted as giant sheets of flypaper to trap unwary birds and mammals coming for drinks of water. Their struggles to free themselves from the sticky asphalt attracted predatory hawks, vultures, and even saber-toothed tigers, which

FIGURE 23–4 A hypothetical ancestor of birds, intermediate between the Pseudosuchian reptiles and *Archaeopteryx*, as conceived by Gerhard Heilmann. From Heilmann, 1927.

in turn were trapped in the tar. From one La Brea pit alone 30,000 fossil bird bones representing 81 species were removed (Howard, 1955). In certain Pleistocene deposits in Wyoming, there is an average of 215 bird bones per cubic meter of material in an outcrop 30 to 100 cm thick and nearly a kilometer long. At Fossil Lake, Oregon, fossil fragments of 66 varieties of birds have been discovered, 16 of which are extinct. Unfor-

tunately, most of these rich deposits contain relatively recent fossils which tell little, if anything, about the descent of birds from reptiles. Table 23–1 attempts to give a brief overall picture of the fossil record of birds. Inasmuch as there are occasional disagreements as to the dating of fossil discoveries and the judging of prehistoric climates, the data given in this table must be considered somewhat tentative.

TABLE 23-1.
A TENTATIVE SUMMARY OF THE FOSSIL HISTORY OF BIRDS*

GEOLOGICAL ERA	GEOLOGICAL PERIOD AND EPOCH		YEARS SINCE BEGINNING OF PERIOD	IMPORTANT EVENTS IN THE HISTORY OF BIRDS	NUMBER OF NEW ORDERS APPEARING		
					Reptiles	Birds	Mammals
CENOZOIC Age of Birds and Mammals	QUATERNARY	Recent	11,000	Modern birds, 8900 species. Geographic races differentiated. Passeriformes dominant. Extinction of many island birds.			
		Pleistocene	2,000,000	All modern orders and families of birds represented. About 40% of living genera of nonpasserine, and 11% of passerine birds appear. About 900 fossils of living bird species described. A period of repeated glaciation, changing sea levels, great dispersals, and widespread extinction.			1
	TERTIARY	Pliocene	13,000,000	Bird species probably reached their maximum numbers. First appearance of moas, tinamous. Climate mild and dry. Mountains rise in western North and South America and join continents at Panama. Extensive grasslands and deserts.			1
		Miocene	25,000,000	The majority of modern bird families and genera probably in existence. North temperate zone initially very warm, gradually cools. Many species adapt to dry habitats. Some families become extinct.			
		Oligocene	36,000,000	First appearance of shearwaters, petrels, gannets, boobies, turkeys, *Phororhacos*, parrots, pigeons, goatsuckers, swifts, kingfishers, Old World warblers, sparrows. Climates gradually drying. Spread of great forests.		4	3
		Eocene	65,000,000	Probable period of major bird evolution. Twenty or more modern orders now in existence. First appearance of penguins, *Aepyornis*, ostriches, rheas, albatrosses, herons, storks, ducks, hawks, eagles,		16	21

	Period				Characteristics	Years Ago
MESOZOIC Age of Reptiles	CRETACEOUS		8		Grouse, pheasants, cranes, bustards, *Diatryma*, gulls, terns, auks, cuckoos, true owls, trogons, woodpeckers, titmice, starlings. Erosion of mountains. Tropical climate in north temperate zone. Rise of modern mammals.	130,000,000
	CRETACEOUS				Rise and extinction of the toothed birds, *Hesperornis* and *Ichthyornis*. Earliest fossils of birds resembling loons, grebes, cormorants, pelicans, flamingos, ibises, rails, and sandpipers. Extensive inland swamps and seas in early Cretaceous; mountain building, in late. Separation of South America and Africa by continental drift. Dinosaurs reach peak and become extinct. Rise of angiosperm plants.	
	JURASSIC	2	1	4	*Archaeopteryx*, the first bird. Flying reptiles. Large dinosaurs dominant. Primitive mammals. Cycads and conifers abundant.	180,000,000
	TRIASSIC	2	2	10	Pseudosuchia, possible ancestors of *Archaeopteryx*. First dinosaurs and egg-laying mammals. Gymnosperms dominant vegetation. Widespread deserts.	220,000,000
PALEOZOIC Age of Amphibians, Fishes, and Invertebrates	PERMIAN			5	Expansion of reptiles and decline of amphibians. Widespread glaciation. Arid climates.	270,000,000
	CARBONIFEROUS			2	Age of amphibians. First reptiles. Great coal swamps.	350,000,000
	DEVONIAN				Age of fishes. First amphibians. First trees. Much glaciation and arid climates.	400,000,000
	SILURIAN				Rise of fishes. First insects. First land plants.	425,000,000
	ORDOVICIAN				First vertebrates. Much land submergence.	500,000,000
	CAMBRIAN				Appearance of most kinds of invertebrates. Trilobites and brachiopods dominant. Mild climates.	600,000,000

*Adapted from Brodkorb (1971); Colbert (1955); Fisher and Peterson (1964); Howard (1955); Rensch (1959); Storer (1960); and Wetmore (1952).

For all its gaps, the fossil record does furnish much important information about the history of birds. For example, it reveals that in the past certain birds lived outside their present ranges (parrots and hornbills in central Europe), and suggests that some groups have been permanently confined to specific places (moas to New Zealand and penguins to southern parts of the world) (Darlington, 1957). It should be made clear, however, that authorities sometimes differ widely in their interpretation of fossils, and that much present information is provisional in nature, awaiting further discoveries.

CRETACEOUS BIRDS. Thirty million years after *Archaeopteryx* appeared, birds had evolved into essentially the modern form. The oldest fossil bird known next to *Archaeopteryx* is *Ambiortus dementjevi* from Mongolia. It was a truly volent bird possessing coracoids, scapulae, furcula, and a keel along with feather impressions (Olson, 1985). In the Cretaceous shales of Kansas, Montana, and Texas, six species of a gull-like bird of the genus *Ichthyornis* have been uncovered. In almost every detail these were modern birds. They had a deeply keeled sternum and modern bird wings, indicating strong powers of flight. The vertebrae were still biconcave as in reptiles, but the long tail of *Archaeopteryx* had disappeared. Although the brain was smaller than that in contemporary birds, it was definitely larger than in reptiles of comparable size, and it had prominent optic lobes. A fragmentary lower jaw of *Ichthyornis* shows that it had teeth individually set in sockets (Gingerich, 1972). Numerous fish bones associated with the fossils indicate that the bird was a fish-eater.

Three families of large, flightless diving birds have also been found in the Cretaceous. These include *Hesperornis* and related forms, some of them nearly 2 meters long (Fig. 23–5). They were loon-like birds

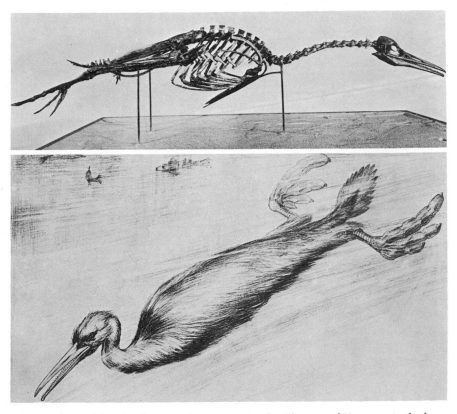

FIGURE 23–5 A fossil skeleton, and reconstruction by Gleeson, of *Hesperornis*, the large diving bird of the Cretaceous Period. Photos courtesy of the American Museum of Natural History.

with vestigial wings, stubby tails, and powerful short legs placed far back on the body. They had conical teeth, set in grooves, in both upper and lower jaws. The cerebrum and optic lobes of the brain were smaller than those of Recent birds, and the olfactory lobes were larger and more reptilian. The ribs had uncinate processes as in contemporary birds, and the sternum had no keel. The neck was long and flexible. In addition to *Ichthyornis* and *Hesperornis*, fragments of 18 other species of Cretaceous birds resembling grebes, cormorants, rails, sandpipers, pelicans, flamingos, and herons, all without teeth, have been found.

ECOCENE BIRDS. The beginning of the Tertiary probably saw the major blossoming, or peak radiation, of bird types. Certainly, by the beginning of the Eocene, birds had acquired their basic forms as fliers, swimmers, and runners; and by the end of the period at least 80 per cent of the modern orders of birds had made their appearance. Major groups that had not yet appeared in the fossil record included the cassowaries, moas, kiwis, parrots, and colies. The disappearance of the dinosaurs had opened many niches into which hordes of birds now moved. During the Eocene a few genera of Recent birds had already become established: *Milvus, Aquila, Haliaëtus, Phoenicopterus, Charadrius, Totanus, Numenius,* and others. In the Paris basin there have been discovered 40 species of 25 genera of birds, including owls, vultures, secretary-birds, herons, partridge, sand grouse, and trogons. The presence of trogons, which are now practically confined to the tropics, indicates that 60 million years ago France had a tropical climate.

Among early Tertiary fossils are three species having such generalized and intermediate characters that they throw light on the taxonomic relationships of several modern troups. *Romainvillia*, from France, links the ducks and geese. *Eostega*, from Hungary, is intermediate between boobies and cormorants. *Telmabates*, from Argentina, points toward a common ancestry for flamingos, ducks, and geese (Howard, 1955).

Fossil penguins dating back to the Eocene have been found in New Zealand. Apparently they became differentiated into streamlined, submarine swimmers very early in bird history. They remain today the most highly specialized of all birds. Fossils of the largest penguins indicate that they were as tall as a man, standing $1^{1}/_{2}$ meters high and weighing approximately 120 kg.

Probably the most awe-inspiring bird to inhabit the North American Eocene landscape was the large, flightless *Diatryma*. This stout-legged giant stood over 2 meters high, was built somewhat like a stocky ostrich, and had a horse-sized head with an enormous hooked beak. In spite of its structural relationship to the cranes and rails, the bird was probably predatory. With some similar ground-dwelling relatives (*Gastornis* and *Remiornis*), it moved into the ecological vacuum left by the dying predatory dinosaurs. It has been argued that the disappearance of these giant cursorial birds logically coincided with the appearance of the predatory placental mammals of the Oligocene. The argument gains force from the fact that in South America, which was geographically isolated from North America and its predatory mammals until the late Pliocene, flightless heavy-bodied relatives of *Diatryma* (*Phororhacos, Brontornis,* and *Andrewsornis*) all survived in Patagonia until the end of the Pliocene, some 60 million years later. Although this may be true, the fact that rheas in South America and ostriches in Africa survive even today in the midst of predatory mammals suggests that they may have died out from other causes.

OLIGOCENE BIRDS. After the millions of years available for bird evolution in the Eocene, evolutionary inventiveness on a broad scale began to run out. In the Oligocene, fossils first appeared that represent about a dozen families of existing birds. These include shearwaters, boobies, limpkins, stilts, turkeys, pigeons, parrots, kingfishers, Old World warblers, and sparrows (Storer, 1960; Brodkorb, 1971).

Several genera of Recent birds also appeared in the Oligocene period: *Sula, Pelecanus, Puffinus, Podiceps, Colymbus, Anas, Lanius,* and *Motacilla*. In South America, separated from North America's predatory mammals by a water gap, large flightless birds of prey persisted in the forms of *Brontornis, Andrewsornis,* and *Phororhacos*, of which the last belonged to a group that contained three families and a dozen or more genera. *Phororhacos*, nearly as tall as a man, had rudimentary wings and a powerfully beaked head 65 cm long. Anatomic evidence suggests that the modern cariamas of South America are lineal descendants of *Phororhacos* and its allies.

MIOCENE BIRDS. By the end of the Miocene period, some 15 million years ago, probably the majority of avian families and many genera of contemporary birds were in existence. Fossils representing 20 living families first appeared during the Miocene. Among

FIGURE 23–6 Cast of a fossil skeleton of *Diatryma*, and its reconstruction. This predatory, flightless bird was taller than a man, and lived in North America during the Eocene Period. Photos courtesy of the American Museum of Natural History.

them were grebes, storm petrels, pelicans, falcons, and doves. Among the passerines appeared crows, thrushes, wagtails, shrikes and wood warblers (Brodkorb, 1971). *Phororhacos* was still living in Patagonia, and a new order of giants, the moas, arose in New Zealand. An enormous flying sea bird, *Osteodontornis*, with a wingspread of 4.5 meters (15 feet) was found in the Miocene of California. Although it possessed tooth-like bony projections on each jaw, it seems to have been an early relative of the petrels and pelicans (Howard, 1957). Some of the complexities of tracing bird origins can be inferred from the fact that Miocene fossils excavated in South Dakota have included birds characteristic today not only of North America but also of South America and the Old World.

PLIOCENE BIRDS. The Pliocene, covering a span of about 12 million years, ended only two million years ago. Very probably, by its end, essentially all modern bird genera were established. Among the fossil forms that made their first appearance in this period were

tinamous, larks, swallows, nuthatches, creepers, and pipits.

PLEISTOCENE BIRDS. The Pleistocene, covering the last two million years of fossil history, was a period of drastic climatic change and widespread glaciation. Among the consequences of these global changes were great shifts in the geographical ranges of birds, and wholesale extermination of many species. Basing his calculations on admittedly highly speculative premises, Brodkorb (1971) estimated that between 24,000 and 32,000 new species of birds evolved in each of the four Tertiary epochs, plus 21,000 more during the Pleistocene, for a very rough total of 150,000 species so far during the earth's history. Of these birds, only about 10,600 survived into the late Pleiostocene, and of these, about 9000 survive as living species today. If this is true, then the Pleistocene was a period of decline and extinction, with evolution limited mainly to the development of geographical races. From intensive studies of Pleistocene birds, Howard (1955) con-

FIGURE 23–7 The flightless giant *Phororhacos* lived in Patagonia during the Miocene. Drawing by C. R. Knight. Photo courtesy of the American Museum of Natural History.

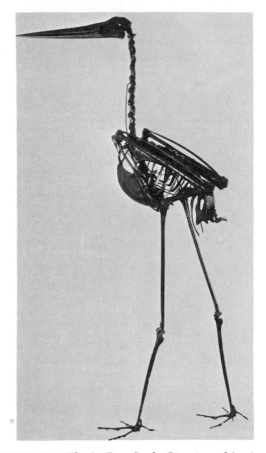

FIGURE 23–8 The La Brea Stork, *Ciconia maltha*, is one of the 180 species of fossil birds found in the Pleistocene deposits of California. After Howard, 1955. Photo courtesy of the Los Angeles County Museum.

cludes that " . . . bird life as we know it today was essentially established some twenty-five to fifty thousand years ago."

Even though the Pleistocene was a great dying-off period, it was also a period of expansion for some of the large flightless species, including the emus, cassowaries, moas, and elephant birds. In isolated New Zealand, there were between 14 and 20 species of moas belonging to the two families Dinornithidae and Anomalopterygidae. They varied in size from $1/2$ meter to over 3 meters in height, and were constructed on the pattern of an enlarged kiwi, a bird to which they were related. Their legs were large and powerful, the largest species having tibiotarsal bones 1 meter in length. The moas, like the contemporary ostrich and emu, were ratites with a flat, keelless sternum. Instead of vestigial wings, however, the moas had no wings at all. Great numbers of moa skeletons, and also their eggshells, feathers, and bits of skin and flesh, have been found in caves, bogs, and Maori camp grounds in New Zealand. One bog alone yielded 140 skeletons of four genera of moas. These birds appeared to be herbivorous.

On sheltered Madagascar lived another flightless group of Pleistocene giants, the elephant birds, or Aepyornithiformes. Although not as tall as the largest moa, the larger species of *Aepyornis* weighed more (up to 450 kg, or 1000 pounds) and laid larger eggs (7.5 liters, or 2 gallons, in capacity) than any other bird known, living or fossil. Like the moas, the elephant birds varied considerably in size, but all of them were stockily built with enormous legs and broad, heavy pelvic girdles. Their eggs had such stout shells that Madagascans once used them for water containers and mixing bowls.

There were also aerial giants among the Pleistocene birds. The condors, *Teratornis incredibilis*, from LaBrea, California, and *Argentavis magnificens* from Argentina (Howard, 1952; Campbell and Tonnie, 1980), had wingspans of 5 and 6 to 8 m, respectively (Campbell and Tonnie, 1983).

RECENT BIRDS. The Recent period, which embraces the last 11,000 years, continues the Pleistocene motif in bird history. The major evolutionary innovations had long since occurred, and only minor specific and subspecific variations were left to be incorporated in the family tree of birds. Extinction of species continued until the number of birds throughout the world was reduced to its present level of about 9000 species. The dominant birds are now the Passeriformes, which outnumber all other orders combined.

Each of the glacial epochs of the Pleistocene undoubtedly caused widespread avian extinction, but the extreme specialization of some birds may have contributed greatly to the disappearance of many

species. Particularly in isolated or sheltered habitats, evolutionary adaptations tend to develop in ways that may become harmful under changed conditions. The well-known inability of oceanic island birds to survive the competition and predation of rats, pigs, and other introduced human satellites is a case in point. Gigantism seems to have been one of these excessive developments, for the Pleistocene giants were prominent among those birds that suffered the evolutionary penalties of overspecialization.

Man himself has been one of the chief agents of bird extinction in the Recent period. He has exterminated species both indirectly by modifying their habitats through his cultural changes (including the introduction of competitors, predators, and diseases), and directly by slaughtering them for food, feathers, or "sport." The majestic moas survived in New Zealand until Polynesians (later to become Maoris) immigrated there and began hunting them around 1350 A.D. Excavations of Maori kitchen middens have revealed burned fragments of the bones of six genera and perhaps twelve species of moas along with the bones and shells of other food animals (Cracraft, 1976; Millener, 1982). Included in the midden debris were also awls,

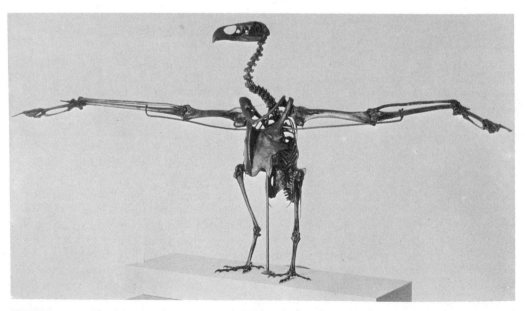

FIGURE 23–9 The Merriam Teratorn, *Teratornis merriami*, from the Pleistocene La Brea tar pits of California, was a vulture-like bird with a wing spread of about 3 meters. After Howard, 1955. Photo courtesy of the Los Angeles County Museum.

fishhooks, and sartorial ornaments fashioned of moa bones (Buick, 1937). Radiocarbon dating of skeletons indicate that they were extinct by 1800 or nearly so. In her later years, a Mrs. Alice McKenzie claimed that as a child of seven, around 1880, she saw a large bird of dark bluish plumage standing close to her. It was 1 m high, had large protruding eyes, a broad beak, and powerful scaly legs. A Moa?

SPECIATION—THE MECHANISM OF EVOLUTION

During his five years' voyage around the world in HMS *Beagle*, Charles Darwin was particularly impressed by the numerous and often bizarre fossils he saw in South America, and with the variety and peculiarity of organisms, especially birds, he found on the Galapagos Islands. In his *Journal of Researches*, Darwin (1845) wrote, with reference to these islands, the following prophetic words:

The remaining land birds form a most singular group of finches, related to each other in the structure of their beaks, short tails, form of body and plumage. All these species are peculiar to this archipelago . . . . Seeing this gradation and diversity of structure in one small, intimately related group of birds, one might easily fancy that from an original paucity of birds in this archipelago, one species had been taken and modified for different ends.

Darwin's observation that species might not be forever changeless and immutable provided the germinal idea that grew into his book *On the Origin of Species by Means of Natural Selection* that shook the intellec-

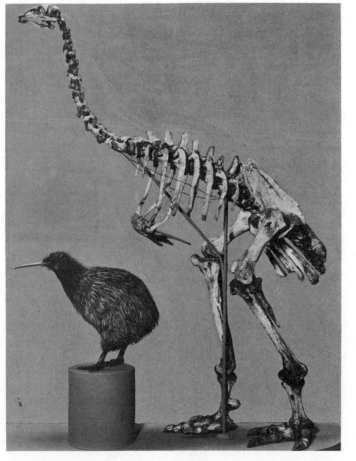

FIGURE 23–10 The skeleton of a small New Zealand Moa, compared with its living relative, the Kiwi. Moas have become extinct within historic times. Photo courtesy of the American Museum of Natural History.

tual world. In its essence, Darwin's theory of evolution proposed that since all organisms vary, and since more individuals are born than can possibly survive, there is a competitive struggle for existence in which the fittest survive to propagate their kind. This principle of the survival of the fittest and the elimination of the less fit, or the principle of natural selection, has dominated biological thought ever since Darwin's day. But, like the organisms to which it applies, the principle itself has had to undergo some minor evolutionary changes.

Genetic Variation

Darwin noted that all organisms possess more or less variability with which natural selection may cooperate to produce new races and species. He was impressed by the great reservoir of variability in the Rock Dove, *Columba livia*, that enabled man to breed domestic pouters, fantails, trumpeters, homers, tumblers, runts, web-footed, and even featherless pigeons Darwin (1875). The heritability of some of these characteristics has since been demonstrated by hybridization experiments (Entrikin and Erway, 1972; Baptista and Abs, 1983). Darwin recognized the similarities between the processes of domestication and speciation. In the domestication process, man isolates a certain genetic trait and artificially selects for it. In the speciation process, a segment of a gene pool is isolated geographically, and acted upon by natural selection.

Darwin assumed that both acquired and genetic variations could be passed on from generation to generation. It was not until the rediscovery of Mendel's laws in 1900 that a clear distinction began to be drawn between genetic variations, or mutations, which are heritable, and acquired variations, or somatic modifications, which are not heritable. Today it is well established that only genetic variations—that is, those originating in the genes and chromosomes of the germ plasm—play a role in evolutionary change.

Mutations such as crests, frizzled feathers, feathered tarsi, albinism, and piedness, have been artificially selected for in some or all of the following domesticated birds: ducks, chickens, pigeons, canaries, *Serinus canaria*, Bengalese Finches, *Lonchura striata*, and Zebra Finches, *Poephila guttata* (Buckley, 1982). Genetic transformations for egg and feather color in chickens have been obtained by the irradiation of male gametes (Pandy and Patchell, 1982). Breeds with special voice qualities are known for such domestic birds as Japanese Quail, *Coturnix coturnix*, chicken, canary (chopper, roller, waterschlager

breeds), pigeon (laugher, trumpeter breeds) (Baptista and Abs, 1983).

Man has produced large and miniature breeds of all the domesticated species listed above. The heritability of morphometric traits has also been demonstrated in the wild. Smith and Dhondt (1981) cross-fostered 140 broods of Song Sparrows, *Melospiza melodia*, and left 47 broods to be raised by their own parents as controls. The young strongly resembled their true parents and showed no resemblance to their foster parents.

Many of these variant characters are due to mutations, or permanent changes, in single genes. The rolling behavior of some domestic pigeon breeds is such a mutation (Entrikin and Erway, 1972). Others, such as body size, are due to mutations in a number of genes that affect the same trait. As a rule, a single-gene mutation affects various structures and functions in the body, but it gets its name from its most apparent influence. For example, the mutation "frizzle," which changes normal feathers into ruffled curly feathers, also causes changes in metabolism and in the sizes of the thyroid gland, heart, gizzard, and other organs. Most of the known genetic changes in domestic birds are nonadaptive, or even definitely harmful; several result in death and are therefore known as lethal mutations. Even though most mutations seem to produce nonadaptive characters, adaptive mutations will occur in sufficient numbers over a long period of time to provide the raw materials for evolutionary change.

Genes achieve their effects by producing highly specific protein molecules (enzymes) that catalyze biochemical reactions in cells of growing organisms. These reactions, in turn, determine such genetic traits as feather color, stereotyped nesting behavior, courtship patterns, and methods of feeding and drinking. Some genes may block the action of other genes, or may regulate the performance of entire groups of genes. The animal that results from the entire genotype, or whole gene package, is then subject to natural selection.

Speciation of Birds

Darwin believed that an adaptive variation in a single individual could, through sheer adaptive merit, assure that individual's survival, the eventual ascendancy of the adapted type in future generations, and thus, in time, the appearance of a new species. Recent work in genetics and evolution has shown that the production of new races and species in nature rarely, if ever, occurs so simply and directly. Instead of working on single hereditary factors in individuals, natural selection works on combinations of genes in populations of organisms. For example, when Turkeys, *Meleagris gallopavo*, artificially selected for meat production, are released in the wild, there is a drop in their general fitness. "The commercial qualitites of the domestic birds had been bought at the price of abandoning qualities favoring survival in the wild" (Leopold, *in* Mayr, 1970).

Evolution in a population usually depends on the accumulation of small differences (micromutations) in a number of genetic traits, especially those that may improve a species' means of obtaining food, shelter, mates, and other necessities of life, or of evading predators, resisting disease, and heightening physiological resistance to environmental extremes. Individuals with such advantageous variations will in general leave more offspring than individuals lacking them, and thus will eventually replace the latter in the population. If, for example, in a particular species, those individuals (A) that possess certain beneficial variations leave 100 offspring as against 90 offspring left by those individuals (B) that lack the variations, group A is said to possess a selection coefficient of + 10 per cent with respect to B. Therefore, B will rapidly vanish from the population. In large populations a selection coefficient of + 1 per cent is sufficient to cause a particular variety of bird or animal to spread through the population in a very few generations (Cain, 1964). Natural selection thus depends on the differential reproductive success of individuals brought on by their genetic differences.

The total genetic endowment, or *gene pool*, of a given population is ordinarily very stable, provided the population is large and mating is random. Regardless of the frequencies of various genes, dominant or recessive, their ratios will remain the same indefinitely —the so-called Hardy-Weinberg equilibrium—unless disturbed either by mutations or by differential rates of reproduction brought on by natural selection, as just described.

In order to understand how species evolve, one must first know what a species is. A working definition of species is that they are "groups of actually or potentially interbreeding natural populations, which are reproductively isolated from other such groups" (Mayr, 1942). We have seen earlier (Chapter 2) that two species can sometimes hybridize and produce viable young. According to Mayr (1970) the three chief attributes of a species are: adaptation to its

physical environment; ability to coexist with potential competitors; and ability to maintain reproductive isolation from other species. The advantages of coping with one's environment and competitors are self-evident. Reproductive isolation is a protective device that guards against the disintegration of a species' inventory of genes that through years of natural selection have become a well-integrated and interrelated system for meeting the problems of a particular niche.

In simplest outline, the evolution of a new species of a sexually reproducing animal involves the following steps:

1. Geographical or ecological isolation of one or more fragments of the population.

2. Genetic changes in the isolated populations, due to mutation, selection, and other forces, especially changes leading toward

3. The development of mechanisms for reproductive isolation and ecological divergence.

4. Development of sufficient genetic change to make both the isolated and parent populations sterile should they again meet, or at least to cause hybrids between them to be inferior to either in competition. In short, spatial isolation permits genetic divergence to continue until reproductive isolation is achieved. The separate elements in the process are worth examining in more detail. The discussion that follows is based largely on Mayr (1942, 1949, 1970).

Isolation and Evolution

The chief evolutionary significance of isolation is that it prevents the flow of genes—the trading back and forth of hereditary factors—between the parent population and the incipient new-species population. Any mutations, new combinations of genes, or other hereditary novelties will be confined to the group in which they occur as long as it is reproductively isolated from the parent population. Genes provide the raw materials for creating differences between species.

By virtue of its superior gene combinations, each species fits a given ecological niche, where it thrives better than its competitors. Isolating devices protect these superior gene aggregations. "In order to survive," says Mayr, "each species must be supreme master in its own niche." If, for example, the genes determining brown fur were introduced into a race of Arctic Foxes, they would produce offspring quickly doomed to extinction in competition with their normal white relatives.

There are two major types of isolating mechanisms.

First are those that prevent mating between potential mates. Either they do not meet because of isolation in space (geographical) or in time (asynchronous breeding cycles, for example); or they meet and do not mate because of behavioral incompatibilities (e.g., diverse courtship patterns); or they attempt copulation but are unsuccessful because of mechanical, anatomical incompatibilities. Second, even though the birds mate, there may be isolating mechanisms that reduce the full success of interspecific crosses through the failure of sperm to fertilize the egg, or the failure of the fertilized egg to produce a viable embryo, chick, or adult. Further, hybrid adults, although healthy, may be sterile. In crosses between the Eastern and Western Meadowlarks, *Sturnella magna*, and *S. neglecta*, hatching and fledging success were both low, and hybrids that matured were apparently sterile (Szijj, 1966). As a further barrier to interspecific crossing, the females customarily responded only to the courtship songs of males of their own species.

Among the various forms of reproductive isolation, the geographical form is by far the most common in promoting speciation among animals. As a consequence, areas of the world that are rich in geographical barriers are also correspondingly rich in species. Species that are to all appearances uniform over large areas of a continent are commonly strongly differentiated on islands, even those but a short distance from the mainland. This is illustrated by the House Finch, *Carpodacus mexicanus*, on islands off the coast of California (Power, 1979). Gene flow is relatively unimpeded between segments of the continental population, but may be strongly restricted between island populations. A variety of influences promote speciation on islands. The island population may be founded by a small group that carries only a small but not necessarily representative fragment of the gene pool of the ancestral mainland population. Further, mutations in the two populations will be different, as will natural selection pressures generated by different climates, foods, predation, diseases, and the like. For example, island populations of birds tend to have longer bills and tarsi than do mainland relatives. This is true of birds on the Tres Marias Islands off the coast of Mexico, the Queen Charlotte Islands off the coast of British Columbia, and the Channel Islands off the Santa Barbara coast of California (Power, 1980).

Greater bill size may reflect the absence of competitors, enabling island birds to take a wider range of food sizes. Longer tarsi may reflect an adaptation to use a greater variety of perches.

Rate of racial differentiation may be affected by the dispersal habits of a species. Blue Tits, *Parus caeruleus*, tend to disperse short distances from their birthplace, whereas Pied Flycatchers, *Ficedula hypoleuca*, tend to be long-distance dispersers. The relatively sedentary Blue Tits are represented by 14 races as compared to three in the Flycatchers (Berndt and Sternberg, 1968). In the warm and food-rich tropics birds are more sedentary, less migratory, and probably less subject to extinction, than are birds of the higher latitudes. These facts probably contribute to the relative abundance of tropical species (Diamond, 1973).

The most instructive example of the influence of geographical isolation on species-formation may be seen when a small population of a continental species somehow finds its way to an isolated oceanic archipelago. This is precisely what happened to Darwin's finches (Family Fringillidae, subfamily Geospizinae) when a small "seed" population of them, probably storm driven, bridged the 950-km water barrier between their ancestral home in South America and the volcanic Galapagos Islands. This group of more than a dozen islands opened up to the birds a whole series of unoccupied ecological niches. The islands of the archipelago are so far apart that only on rare occasions could the finches disperse from one island to another. In time, however, most of the islands became occupied, and the various populations, isolated from each other, evolved into 14 distinct species. Undoubtedly descendants of a common finch-like ancestor, these birds now are adapted to an astonishing variety of niches (Fig. 23–11). Some are still ground finches with heavy bills, feeding on seeds; some are tree dwellers specializing on insect food; one of the tree finches looks and behaves like a warbler, while another uses its bill like a woodpecker to chisel out wood-boring insects. Lacking a long tongue, however, this woodpecker-finch has developed a remarkable substitute: it breaks off a cactus spine and uses it as a probe to extract its insect prey (Lack, 1947; Milliken and Bowman, 1967)!

This branching out of one type of bird into a diversity of ecological niches is known as *adaptive radiation*. It has occurred many times in the evolutionary history of birds and other animals. Another striking example of adaptive radiation is seen in the sicklebills, or Drepanididae, of the Hawaiian Islands. These birds, related to carduelid finches from the American mainland, have evolved, on the various islands of the archipelago, into nectar, fruit, seed, nut,

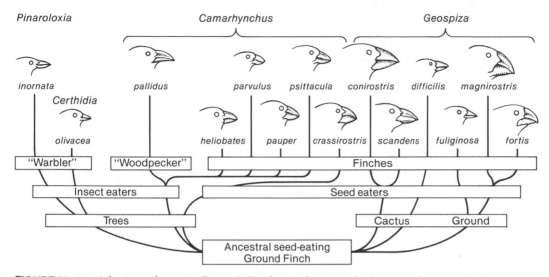

FIGURE 23–11 Adaptive radiation in Darwin's Finches. Isolation on the various Galapagos Islands provided evolutionary opportunities for the development of new varieties of birds from an ancestral seed-eating ground finch. Modified after Simpson *et al.*, from Lack, 1947.

caterpillar and insect eaters (Amadon, 1950; Sibley and Ahlquist, 1982). All told, there are 22 species and 45 subspecies of the Hawaiian Drepanididae, all probably descended from not more than one or two species of ancestral immigrants (Mayr, 1970). Similarly, the 4300 known species of Hawaiian insects are probable descendants of only about 250 species of original immigrants (Zimmerman, 1948). Other bird species demonstrating extensive adaptive radiation are the hummingbirds, tyrant flycatchers, and ovenbirds of Central and South America, the parrots in Australia, and the Icteridae of the Americas.

When a population of birds gains a foothold on a single isolated island, however, continuing adaptive radiation does not occur even though a variety of unoccupied niches is available. That is because a single small island does not provide conditions for geographical isolation to occur within itself. The Geospizinae years ago made their way to Cocos Island, about 900 km north-northeast of the Galapagos Islands, yet in spite of the presence of considerable ecological variety in habitats the bird has remained a single species because of the impossibility of geographical isolation. A single island of an archipelago may, however, support two or more species descended from a common ancestor as a consequence of successive invasions widely spaced in time. Tiny Norfolk Island, about 1250 km from Australia, supports three species of

White-eyes: *Zosterops norfolkensis, Z. tenuirostris,* and *Z. albogularis.* Their only close relative is *Z. lateralis,* which lives on the Australian mainland. Stresemann (*in* Mayr, 1942) suggested that their presence on Norfolk Island resulted from three widely spaced waves of immigration. The first immigrants had evolved into a distinct and populous species by the time the second invasion of one or two pairs arrived, and the same process was repeated for the third wave. The two species of stilts on New Zealand are thought to represent two separate invasions, the first giving rise to the endemic Black Stilt, *Himantopus novaezelandiae,* and the second to the Pied Stilt, *H. h. leucocephalus,* (Pierce, 1984). Alpine birds, living on isolated mountain peaks on continents, show this same phenomenon of double and multiple invasions.

One of the most effective barriers to gene flow in the recent history of birds has been Pleistocene glaciation. As the tremendous masses of ice advanced and retreated, the accompanying changes in sea level, climate, and terrestrial vegetation created widespread barriers that promoted speciation over the entire earth. In the Amazon basin, for example, the wet and dry cycles that accompanied glacial cycles caused radical changes in the distribution of forests and grasslands. During dry periods certain isolated areas remained as humid tropical forests, and these areas, isolated geographically and ecologically by grassland barriers from each other, became refuges and regions of active speciation for forest-dwelling animals (Haffer, 1969, 1970, 1974). Later, when wet cycles allowed the ruptured forests to rejoin, ruptured bird populations were similarly rejoined geographically, but were then composed of new species that had become reproductively isolated from their parental forms. Today, the geographical distribution of closely related bird species in the Amazon valley, as well as that of certain butterflies, lizards, tapirs, monkeys, and even plant families, accords with this interpretation of Pleistocene speciation. Parallel studies of Pleistocene speciation in North America (Rand, 1948; Mengel, 1970), Australia (Serventy, 1971), and Africa (Moreau, 1966); come to the same conclusion: that fragmentation of large continental habitats, brought on by glaciation-induced climatic changes, provided the geographical isolation essential to the speciation of many birds. Paleobotanists have postulated that closely related species of North American spruces (*Picea* spp.) probably evolved in ice-free refugia during the Pleistocene (Halliday and Brown, 1943). One refugium was southeast of the icesheet, two or more southwest in the Rockies, and one or more in the Yukon-Bering Sea region. The extant subspecies of White-crowned Sparrows, *Zonotrichia leucophrys*, warblers, and other birds probably evolved while isolated in these ice-free "islands" (Hubbard, 1969).

In Europe and Asia a special feature of Pleistocene evolution resulted from the east-to-west positioning of mountain ranges as barriers to southerly dispersal. The southward advance of glaciers forced many European species, for example, into refuges both in the southeastern and southwestern parts of Europe, or in northeastern or northwestern Africa. There they remained, isolated from each other for several thousands of years. After their postglacial return to central Europe, many birds that had been presumably monospecific before their fragmented southern residence revealed distinct eastern and western species or subspecies. Thus, the Nightingale, *Luscinia megarhynchos*, of western Europe, and the Thrush Nightingale, *L. luscinia*, of eastern Europe, are distinct *sibling species*, thought to have arisen from a single preglacial parental species during their period of enforced geographical separation. Although today their ranges overlap, the two species do not interbreed. The Carrion and Hooded Crows, *Corvus c. corone*, and *C. c. cornix*, the Icterine and Olivaceous Warblers, *Hippolais icterina*, and *H. pallida*, and numerous other related pairs of European bird species owe their kinship and geographical distribution to this same east-west glacial separation (Rensch, 1959). When such related populations of birds live in completely separated geographical areas, they are said to be *allopatric*; when they reside together in the same geographical area, as when their ranges overlap, they are said to be *sympatric*.

As described on page 465, the upthrust of the Andes mountains about 10 million years ago split the populations of many tropical South American birds, so that today many species in the Pacific lowlands of Colombia are geographically isolated from their parent populations on the eastern side of the mountains. Haffer (1967) analyzed 332 species of Pacific lowland birds and found that over one-half of them had evolved into new subspecies, new species, or, in 12 instances, even new genera.

Pleistocene climatic changes have also influenced speciation of alpine birds by forcing them into altitudinal displacements. Radiocarbon dating of pollen profiles from east African mountains shows that during the chilly periods of glacial maxima, the cool montane forests descended, along with their avifauna, and spread out over valleys as much as 1200 meters lower in altitude. With the retreat of glaciers and the advent of warmer climates, the montane forests again climbed to higher and fragmented "island" peaks and their montane bird populations became separated from one another. Moreau (1966) attributes the striking similarity between montane species in east and west Africa to such slow, vertical oscillations in climate and vegetation. Vuilleumier (1969), explaining the speciation of birds in the high Andes, suggests that gene flow between incipient species across what are today valley barriers, was highest during periods of glacial maxima and lowest during their minima.

A very different form of reproductive isolation depends on the appearance or morphology of the bird itself. The lively breeding season plumages of many male birds are undoubtedly species-specific identification badges (as well as aphrodisiacs) that function to restrict breeding to birds of the same species. Snow Geese, *Chen caerulescens*, occur in two color phases, pure white or pale blue-gray. In wild, mixed populations, about 90 per cent of the birds choose mates of a color similar to that of the family in which they were raised, regardless of their own color (Cooke, 1978). Here, early learning determines mate choice, and thus affects the genetic pool of the population. The

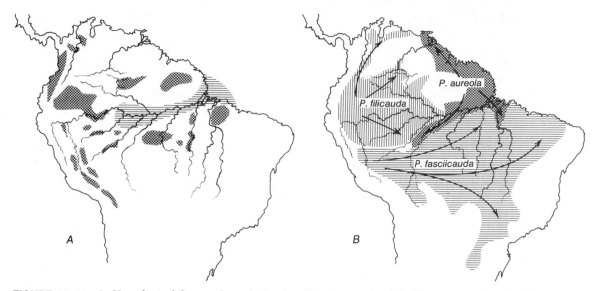

FIGURE 23–12 A, Hypothetical forest refuges during dry climatic periods of the Pleistocene. The shaded regions represent fragments of the Amazonian forest separated by savannas and grasslands; the horizontal shading represents an ocean bay caused by a 50-meter rise in sea level during interglacial periods. B, The approximate distribution today of three species of manakins of the genus *Pipra*. Arrows indicate the dispersal paths of the birds from their presumed Pleistocene refuges. Adapted from Haffer, 1969, 1970.

Western Grebe, *Aechmophorus occidentalis*, occurs in two color phases or morphs, light and dark. In mixed wild flocks observed in Utah by Ratti (1979) purely random mating would have resulted in 33 per cent mixed light-dark pairs. Instead, only 1.2 per cent of the pairs were mixed. Differences in "advertising" calls have been described for each morph, and males selectively approach tape-recorded playback of their own call versus the other morph's call (Neuchterlein, 1981). This type of non-random, selective mating is known as *assortative mating* and is a potent mechanism to ensure gene pool stability. Some ornithologists recognize these grebes as two species, A. *occidentalis* and A. *clarkii*.

The grebes represent an example of ethological isolation, showing how behavior patterns may erect barriers to gene flow between species. Such behavior may involve song, social behavior, feeding habits, and the like. Of special importance are the behavior patterns concerned with courtship and mating. All of the elaborate displays, dances, flights, songs, calls, and colors involved in courtship have an isolating function in that they tend to confine interbreeding to birds of a specific behavioral as well as morphologic type. If two

birds pair and the responses of one sex to the courtship advances of the other are not "correct," copulation is unlikely to occur. Related species living in the same geographical locality (sympatric) are thus restrained from hybridizing and losing their genetic identity.

To judge by the energy that most birds devote to courtship, one of the keenest forms of competition is that for mates. Darwin considered sexual selection an important form of natural selection. Certainly it can be a crucial form of selection, for the transmission of an individual's genetic peculiarities to the next generation depends on his securing a mate. As Darwin observed, this form of selection may be either intersexual (the power of a male to charm a female), or intrasexual (the power of one male to conquer another). Sexual selection is unquestionably responsible for much of the richness and variety of bird coloration, especially in males, and for the elaborate displays, ritualized fighting, and profuse singing of many species. It is often difficult to decide whether a certain color pattern is used chiefly in courtship, and hence is attributable to sexual selection, or whether it is used principally in species recognition or in warning, thus evolving from other selection pressures. But the fact that the most colorful plumage of the year appears in many species just before the breeding season, and that courting males usually expose their brightest adornments to competitors or to their intended mates, indicates that sexual selection is a highly effective form of natural selection.

The bower-birds provide convincing evidence of the efficiency of sexual selection. In these birds the display function has been transferred from their plumage to inanimate eye-catching objects spread out in their bowers. In one study, female Satin Bower-birds, *Ptilonorhynchos violaceus*, tended to choose older males to mate with. Older males built more elaborate bowers and sang more mimetic songs than younger males. These behaviors probably provided females with information about their individual fitness, thus influencing mate choice (Loffredo and Borgia, 1986).

That birds really pay attention to the appearance of other individuals of their kind, and make choices on that basis, has been demonstrated by captive studies on breeding preferences. The Timor (nominate) subspecies of the Zebra Finch, *Poephila g. guttata*, differs in appearance from the Australian subspecies, *P. g. castanotis*. Eight males and eight females of each race were placed in an aviary and pair formation recorded. No mixed pairs were formed; each form paired with its own kind, which suggests that these two taxa have achieved full species status (Böhner *et al.*, 1984). Australian Zebra Finches imprint sexually on the characteristics of their parents (see Chapter 10). In these finches pairing preferences are influenced by learning.

Social behavior combined with mating habits may also affect the isolation of populations. Numerous species of ducks with circumpolar Holarctic distribution have few, if any, subspecies because in winter quarters a male and a female from widely separated breeding ranges may pair. A drake from Maine, for example, may pair with a duck from North Dakota at their common winter quarters in Florida, and follow the female back to her Dakota breeding grounds. Such a breeding pattern results in a thorough panmixia, or continuing exchange of genes, on a broad geographic scale. As a result of this panmictic phenomenon, one species of Mallard, *Anas platyrhynchos*, lives in a broad band encircling the earth in the northern hemisphere. Geese, on the contrary, have a type of family organization that promotes inbreeding and hereditary isolation. Parents and young remain together as a family unit, migrating together both to their winter quarters and back the next spring to their breeding area. As a consequence Canada Geese of the genus *Branta* have broken up into 11 pronounced geographical races in North America (Lack, 1974).

Song can also be an important isolating mechanism as evidenced by field and laboratory experiments of viduine finches (Payne, 1973a, 1973b). Each viduine species selects a particular estrildid finch host's nest to lay in. Males learn the song of the host species and use it in courtship. Playback experiments using different kinds of estrildid songs indicated that females approached more to playback of their own host's song than to an alien host song. Moreover, female *Vidua chalybeata* develop larger ovarian follicles when stimulated with their own host species song than with a foreign host song (Payne, 1983).

However, the relative importance of song versus appearance as species-isolating mechanisms in wild populations is still poorly known for most species. In Indigo and Lazuli Buntings, *Passerina cyanea* and *P. amoena*, appearance seems to take precedence over song. In Nebraska, these buntings sometimes learn each other's song syllables and respond to each other's songs in playback experiments. Yet few hybrid pairs are established (Emlen *et al.*, 1975). A male's advertising song may attract a female; if he doesn't look like the correct species, she probably won't pair.

Psychological isolation, a special form of ethological isolation, is not a common barrier mechanism but it does occur in a few species. The numerous species of manakins of Central and South America are very reluctant to leave the undergrowth and lower canopy of tropical forest. Threfore the broad rivers of the Amazon valley have become effective barriers to their dispersal and have accounted for the speciation of several different populations. The preference of larks for soils that match the color of their feathers (p. 54) is another example of psychological isolation that probably promotes speciation. Still another is the innate tendency of some species to disperse widely, whereas others travel very little. Mayr (1970) tells of a small White-eye, *Zosterops lateralis*, that crossed the 2000-km gap between Tasmania and New Zealand around 1850 to become the most common native passerine there. In the next few decades "it successfully colonized all the outlying islands of the stormy New Zealand seas." In contrast, a close relative White-eye, *Z. rendovae*, of the Solomon Islands, "refuses to cross barriers only a few kilometers wide . . . even though its flying equipment is essentially the same. . . ."

Temporal isolation, like geographical separation,

can effectively prevent the mixing of genes between two related populations of birds. A subspecies of the Australian Roller, *Eurystomus orientalis pacificus*, winters with other subspecific relatives in the tropical East Indies. The birds do not interbreed because the Australian race is never in breeding condition while in its winter quarters (Mayr, 1942). The maturation of gonads in two races of White-crowned Sparrows, *Zonotrichia leucophrys*, is not synchronized in the same photoperiod, so that although they occupy the same winter quarters in California, a reproductive barrier exists between them.

Temporal events operate in still other ways to affect avian evolution. Apparently the incubation calendar of the Horned Lark, *Eremophila alpestris*, is fixed partly by the weather and partly by predators. In the northern United States in April, many young of the Horned Lark succumb to late snows, cold rains, or food shortages. But late hatching is no cure for the trouble since "predaceous enemies cause a greater and greater loss as the season advances into June and July. The optimum season for the welfare of the young is shown to be May" (Bent, 1942). The fledglings are thus caught in a selective vise, one jaw of which brings physical pressures, the other, biological pressures. In woodlands near Oxford, England, the Long-tailed Tit, *Aegithalos caudatus*, faces comparable seasonal hazards. In the early spring, as leaves are emerging and food is abundant, many nests fail because they are easily found by predators. Later in spring when vegetation is thicker, nests are safer from predators, but food for the young is less abundant (Lack, 1958). It is reasonable to assume that the evolutionary adjustments made by many species are compromises between two or more selective pressures.

As seen in the foregoing examples, natural selection may operate through physical agents, biological agents, or both. It is often difficult to distinguish the effects of one from the other. Biological agents include food, predators, competition, mates, parasites and diseases, nesting materials, vegetation cover, and other selective factors.

The chapter on ecology outlined the many ways in which physical environmental factors influence birds. Insofar as a given factor determines whether an individual bird survives or perishes, that factor has a selective influence on the evolution of the species. Numerous examples were given earlier of *limiting* factors that acted in such a fashion: temperature, humidity, precipitation, sunlight, wind, air density, soil miner-

als, water salinity, day-length, and others. It will be enough here to give one more example, pointing out in this case some of the evolutionary consequences of a change in nesting substrate.

Most gulls nest on flat ground, but the Kittiwake, *Rissa tridactyla*, has abandoned this orthodox nesting behavior and shifted to nesting on narrow ledges of steep, seaside cliffs. Presumably this is a naturally selected antipredator arrangement. With the change in nesting sites has come an array of correlated adaptations in structure, physiology, and behavior. Among some of those reported in an interesting paper on this species by Cullen (1957) are the following. The claws of the Kittiwake, longer and sharper than those of most gulls, give it a more secure footing on its narrow ledge. Using mud, it builds a nest more deeply cupped than that of other gulls. Correlated with the safety of the cliff-edge location, the clutch has been reduced from three to two eggs. The adults do not carry away eggshells as do other gulls—a confirmation of the fact that such behavior is a naturally selected device to prevent advertising the nest site to predators. The adults give the alarm call less frequently than other gulls, and remain on the nest when predators do happen to approach, or make only weak attempts to fight the predators off. The young Kittiwakes with their conspicuous, black neckbands, have lost the cryptic coloration of other young gulls, and do not run from the nest when attacked. They sit close in the nest and face the cliff much of the time. Since the young are not in the habit of wandering about the colony as is commonly the case in other species of gulls, there is no need for the parents to know them individually. Whereas in other gulls the parents learn to recognize their own young in a few days, Kittiwakes are unable to do so until the nestlings are at least four weeks old. Although some of these changes may not be hereditary, others unmistakably are, and all of the changes are clearly related to the nature of the nest site and to the accompanying lack of predation pressure.

Natural selection, acting through climatic factors, often selects heritable variations, which gradually change in magnitude or intensity over an extensive geographical area. Northern populations of Wrentits, *Chamaea fasciata*, are heavier than southern populations, and birds from the humid California coast are darker than birds from the drier interior (Bowers, 1960). Species in which genetic characters vary in parallel with selective factors in the environment are said to form *clines* (Huxley, 1939). The gradual variation of species in such characteristics as body size, color, and size of clutch, in parallel with such factors as temperature, humidity, and altitude are the basis for the geographical rules of Bergmann, Gloger, Allen, and others, which were described earlier.

Continuously varying populations in a long, unbroken chain show increasingly pronounced changes in character as one moves geographically outward from the center of the species' range. Often the terminal populations in such a series of interbreeding groups have become so different from each other that they cannot interbreed, although they are connected by a series of freely interbreeding populations. This form of "speciation by distance" is perfectly demonstrated by chains of gradually varying subspecies of the Herring Gull, *Larus argentatus*, which form a ring in the northern hemisphere whose center of origin was presumably in eastern Siberia, and whose end-links overlap in western Europe (Fig. 23–13). Here the extremely variant races coexist but do not interbreed (Mayr, 1942). Such rings of gradually varying races whose terminal points overlap but are intersterile are called *Rassenkreise*, or "circles of races." A situation

similar to that of the Herring Gull has been found in titmice, kingfishers, white-eyes, babblers, warblers, wheatears, sparrows, and other species in various parts of the world.

In the light of these examples, races or subspecies appear to be incipient species that may become full species if the gradually increasing divergences in character pass a certain threshold that confers reproductive isolation on the most extreme forms. Species that show two or more subspecific variant forms are called *polytypic* species. The varying characters may correlate with selective environmental factors, or they may not. At times a given character change continues in a given direction through a geographical series of populations, which suggests that it, or some underlying related trait, may have adaptive value. In other cases the characters seem to vary geographically in a haphazard fashion, correlated with no known selective influence of the environment. One of the most impressive examples of a polytypic species is the Song Sparrow, *Melospiza melodia*, of North America, of which 31 subspecies are listed in the A.O.U. *Checklist of North American Birds* (1957). These vary in size from the large Alaskan

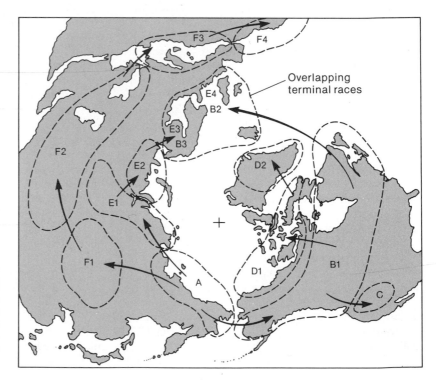

FIGURE 23–13 The circumpolar ranges of the different subspecies of the Herring Gull show an overlap in western Europe of the terminal links in a chain of intergrading races. The terminal races coexist without interbreeding, like genuine species. After Mayr, 1942.

races *(M. m. sanaka)* weighing about 52 g to small (e.g., *M. m. saltonis)* weighing about 18 g. Those from dry areas (e.g., *M. m. saltonis)* are light-colored, and those from moist coastal California are dark (e.g., *M. m. morphna)*. Moreover, several subspecies are endemic to specific marshes around the San Francisco Bay Area of California (Marshall, 1948). Many tropical birds show polytypic variations of this sort, particularly in their coloration (Fig. 23–14). Mayr estimates that at least 80 per cent of the birds of Papua New Guinea, and 70 per cent of all Palearctic passerine species, are polytypic.

When sharply discontinuous, or nonblending, genetic variations occur within a population, the phenomenon is known as *polymorphism* and the different forms are known as morphs. Although the various morphs may differ strikingly in appearance, they are all completely interfertile and may occur in the same brood. Perhaps the best-studied example of polymorphism is that in White-throated Sparrows, *Zonotrichia albicollis*. It can occur as either a white-crowned or tan-crowned morph. White-crowned birds have an extra (supernumerary) chromosome not found in tan-crowns (Thorneycroft, 1976). White males tend to be larger than tan males, and the two morphs differ in the structure of the 2nd and 3rd chromosomes (Rising and Shields, 1980). White females dominate tan females

in spring, but the reverse is true in winter (Watt *et al.*, 1984). The two morphs mate disassortatively, i.e., tan females pair with white males and white females with tan males (Lowther, 1961). Other examples of polymorphism are found in the strikingly varied plumage colors in African shrikes of the genus *Chlorophoneus*, or the tan versus grey morphs of many owls, or in the neck plumages of the male ruffs, *Philomachus pugnax*, or in the egg-color morphs of certain Cuckoos, e.g., *Cuculus canorus*.

Somewhat like the gradually blending variations mentioned earlier, discontinuously variable morphs often occur in geographical clines. The Reddish Egret, *Egretta rufescens*, occurs in two color phases—white and a dark gray-brown. In Texas about 99 per cent of the birds are dark; in the West Indies, about 50 per cent; and in the Bahamas, only 10 per cent.

THE ROLES OF COMPETITION AND PREDATION

Finding enough food for survival and avoiding predators are two of the most pressing problems that any bird faces. Therefore, food, competition, and predation are powerful biotic forces of selective importance in any bird's evolutionary history. Some of the many ways in which food brings selective pressures on

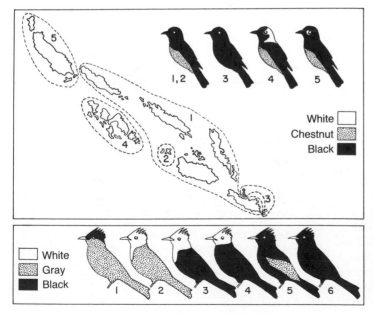

FIGURE 23–14 *Upper*, Geographic variation in the polytypic Flycatcher, *Monarcha castaneoventris*, of the Solomon Islands. The subspecies show four principal color patterns. 1, *Monarcha castaneoventris castaneoventris*, 2, *M. c. obscurior*, 3, *M. c ugiensis*, 4, *M. c. richardsii*, 5, *M. c. erythrosticta*. *Lower*, Similar discontinuous variation is shown in races of the Asiatic bulbul, *Microscelis leucocephalus*, 1, lives in India and Burma, 2, in Yunnan, 3, in Szechuan, 4, in southern China, 5, in Taiwan, and 6, in Hainan. After Mayr, 1942.

a species have been discussed earlier in the chapter on ecology. It was pointed out there that two species with the same ways of life, especially identical feeding habits, cannot live indefinitely in the same habitat, because eventually one of them will prove to be more efficient in the habitat to the other and eventually outnumber and even eliminate it (Gause, 1934). There are, however, other possible solutions to the problem. One species may become extinct, or the two species may divide the area geographically, one living in one part and the other occupying the remainder; or they may divide the habitat ecologically, feeding throughout the area but each on a different food. Different cormorants, Pelagic, *Phalacrocorax pelagicus*, Brandt's, *P. penicillatus*, and Double-crested, *P. auritis*, solve the problem of coexistence by specializing on different foods (Ainley and Anderson, 1981). In a detailed study of the food habits of 172 species of birds in Tanzania, Moreau (1948) discovered that in 94 per cent of the cases in which two or more sympatric species belonged to the same genus, the different species were ecologically isolated by diet, habitat, or both, and thus avoided competing with each other. Wherever related species overlapped in range, they were either different in size of body or beak, or they used different methods of seeking food. Fruit pigeons in Papua New Guinea avoid feeding competition by a simple arrangement—larger pigeons eat larger fruits on the bigger branches while smaller pigeons consistently eat the smaller fruits among the twigs (Diamond, 1973).

The avoidance of feeding competition may be more than a matter of the selection of certain foods and the rejection of others. At times it unquestionably involves physiological adaptations as, in an extreme case, the unique ability of honey-guides to digest beeswax. Sympatric species of flamingos show a structural difference in bills which enables them to feed side by side without competing. The Greater Flamingo, *Phoenicopterus ruber*, has a shallow-keeled bill, adapted to filtering small molluscs, crustacea, and plankton from bottom mud, while the Lesser Flamingo, *P. minor*, has a deep-keeled bill, adapted to filtering microscopic plankton from surface water.

Competitive exclusion has obviously been a key mechanism in promoting isolation and evolution in birds. As Mayr (1948) remarks, ". . . there is no more formidable zoogeographical barrier for a subspecies than the range of another subspecies." However, when the ranges of two closely related birds overlap (i.e., the birds are sympatric) and some critical resource such as

food is in such short supply that it creates competition, the situation is likely to lead to evolutionary changes in one or both species that will lessen this competition. Such changes are then known as *character divergence* or character displacement.

This happened to Darwin's finches on the Galapagos Islands. When two related species meet and compete for food on the same small island, they can both survive only if they evolve a measure of ecological isolation. In both the Tree finches (*Camarhynchus* spp.) and the Ground finches (*Geopiza* spp.), the bills of two species inhabiting the same island differ more strongly than those of the two species living on separate islands (Fig. 23–15). Competition has promoted increased specialization in their feeding tools; this in turn has brought partial ecological isolation that has reduced competition between the birds (Lack, 1947; Abbott *et al.*, 1977). Competitive exclusion must not be taken for granted, however (see Chapter 19), as two species may occupy two different habitats because they are genetically disposed to do so, or because one species is excluding the other. Only removal experiments (p. 424) can test for either possibility.

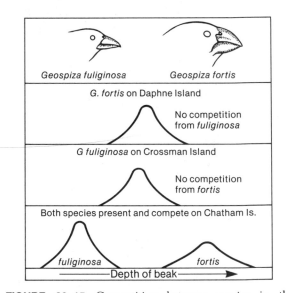

FIGURE 23–15 Competition between species in the same habitat promotes evolutionary divergence. Species of *Geospiza* living on the same island show greater divergence in beak form than the same species living on isolated islands where they need not compete. After Simpson *et al.*, from Lack, 1947.

It is difficult to measure the selective pressures of predators upon prey species, but without question such pressures exist and have evolutionary consequences on such characters as coloration, sharpness of senses, "freezing" behavior, quickness and agility in flight. The remarkably close resemblance of the plumage of some larks to the soils they habitually frequent, and of the mimetic resemblance of the frogmouths (Podargidae) to dead stubs of trees, must have been brought about largely by the millennia of unceasing elimination of the more conspicuous variants by predators. The evolution of nocturnal habits by gulls in the Galapagos probably resulted from kleptoparasitism and nest predation by frigate birds (Snow and Nelson, 1984).

A special form of predation is the brood parasitism of the European Cuckoo, *Cuculus canorus*. In southern England and other parts of Europe the Cuckoo lays, in nests of a variety of birds, eggs that do not closely resemble those of the host species. In the Khasi hills of India, however, the local race of Cuckoos lays eggs that are remarkably similar in size and coloration to those of its host or fosterer species. Baker (1942) observed the fate of Cuckoos' eggs deposited in 1642 nests of normal fosterers (where mimicry was very close) and in 298 nests of abnormal fosterers (where egg color agreement was poor). The incidence of nest desertion or other forms of Cuckoo-egg rejection by host species was only 8 per cent among the normal fosterers, but 24 per cent among the abnormal fosterers. This selective rejection of Cuckoos' eggs by fostering species depends both on the closeness of egg mimicry and the discrimination of the host species. Here is a clear example of natural selection in operation. The Cuckoo, in attempting to extend its parasitic sway over new species, meets drastic selection in the higher percentage of rejections that it encounters.

A tiny warbler or sparrow will continue to feed and protect a parasitic nestling Cuckoo even when it has grown to be ten or more times as large as its foster parent. The nestlings of most races of Cuckoos do not resemble the young of their hosts, even though the eggs of both species often agree closely in appearance. Once the feeding instinct has been awakened in the foster parent, the cuckoo nestling's cry and open mouth prove irresistible, despite the lack of resemblance between the young Cuckoo and the fosterer's own nestlings. A few species, however, seem to be intelligent enough to distinguish the differences between the interloper and their own young. This dilemma has resulted in shifting the burden of adaptation back upon the Cuckoo. In Spain, both the eggs and the young of the Great Spotted Cuckoo, *Clamator glandarius*, resemble closely those of its host, the Magpie, *Pica pica*. A logical assumption is that the Magpie over many years exercised a form of intelligent selection by ejecting or abandoning young Cuckoos that looked "strange," and that natural selection reacted on the young Cuckoos by tailoring their appearance to match that of the host young (Cott, 1940).

The phenomenal precision with which natural selection can direct adaptive character divergence in birds is seen nowhere more convincingly than in the evolution of certain African brood parasites of the subfamily Viduinae of the family Ploceidae. Nicolai (1964, 1967), and Payne (1983) have shown that species formation among the parasitic widowbirds (Viduinae) parallels very closely that of their hosts, the waxbills (Estrildidae). The concurrent evolution of parasite and host depends in part upon imprinting and vocal mimicry of the host species by the parasite. For example, the Indigo-bird, *Vidua chalybeata* (viduine parasite), of Cameroon, is able to sing the entire vocal repertory of its estrildid host, the Red-billed Fire-finch, *Lagonosticta senegala*—its song, distance call, contact notes, nest call, and begging notes of the nestlings. Geographical subspecies of the parasite similarly mimic precisely the corresponding dialects of races of the host species living in given localities.

The significance of precise mimicry of the host's songs and calls lies in the fact that the female Indigo-bird is stimulated to breed and lay eggs by observing the breeding preparations of a pair of host Fire-finches, and that she prefers to copulate with a male partner whose calls most completely and accurately mimic the notes of that particular host species (Payne, 1973). Moreover, two closely related and similar looking Indigo-birds may occur sympatrically, each one parasitizing a different Fire-finch host. Mimicking the correct Fire-finch song ensures that females will make the correct mate choice. For example, *Vidua funerea* may be green or blue, depending on the region, but mimics the song of its host, *L. rubricata*. *Vidua chalybeata* may vary from green to blue in color but mimics the song of its host *L. senegala* (Fig. 23–16). The parasite egg, laid in the host's nest, hatches into a nestling that closely resembles the host's young in size, shape, color, down pattern, peculiar begging movements, and even in mouth palate colors and markings. Unlike most hosts of the European Cuckoo, *Cuculus canorus*, the waxbills refuse to feed young that do not

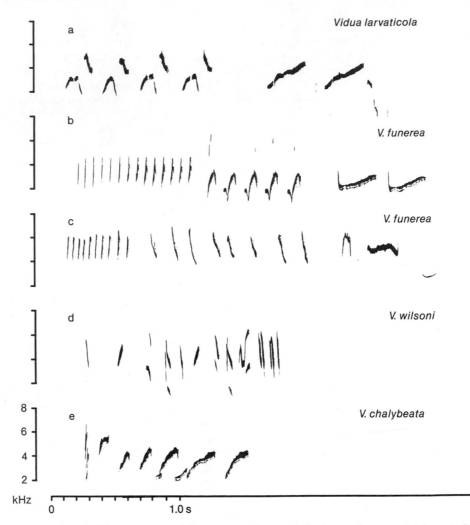

FIGURE 23–16 Mimetic songs of four species of parasitic finches in the genus *Vidua*. A, Song of *Vidua larvaticola* learned from its host, the Masked Firefinch, *Lagonosticta larvata*; B, and C, songs of V. *funerea* from Cameroon and Nigeria, respectively, learned from their host, the Dark Firefinch, *L. rubricata*; D, Song of V. *wilsoni* learned from its host, the Bar-breasted Firefinch, *L. rufopicta*. E, Song of V. *chalybeata* learned from its host, the Senegal Firefinch, *L. senegala*. Viduine finches are difficult to tell apart by appearance, but are easily distinguished by their mimetic songs. From Payne, 1986.

closely resemble their own young. Host resemblance has in this way been forced on the parasitic widowbirds by the selectivity of the host waxbills.

Parasites and diseases have brought heavy selective pressures on birds through the years, and birds have probably responded with the evolutionary development of various mechanisms of resistance and immunity. But diseases are also important because of the role they play in aggravating population fluctuations. At the peak of a population cycle, intraspecific competition generates keen selective pressures, but at the low ebb of population following an epizootic, so few animals remain that competitive pressures are negligible (Woolfenden and Fitzpatrick, 1984).

Everyone is aware that in human affairs a sampling of public opinion is unreliable when it is based on a small sample. To attempt to learn who the next president will be by asking a dozen or a hundred persons is not a dependable procedure. Likewise, when a large population is reduced by disease to a small remnant, the remnant is very unlikely to have precisely the same genetic composition as the original population. In small samples, the genetic character may shift pronouncedly in one direction or another. This shift is known as *genetic drift*. Variations in the gene pools of such samples are therefore primarily a matter of chance and of selective pressures. For example, a drought decimated 85 per cent of the 1500 color-marked *Geospiza fortis* on Isla Daphne Major in the Galapagos. Survivors tended to be birds with the largest bills, as only they were able to crack the large and hard seeds that predominated in the drought (Boag and Grant, 1981).

Should a small population become isolated, as on an oceanic island, where there are numerous unfilled niches and reduced competition, its genetic changes, less rigorously disciplined by selection than on the mainland, may produce forms quite deviant from those of the ancestral species. Relaxed selection pressure allows more genetic variability. The large flightless dodos (Raphidae) of the Mascarene Islands were no doubt well adapted to their particular habitat. As long as no predators existed on the islands the lack of wings and possession of heavy bodies carried no serious penalties. But once man introduced predators, the species quickly became extinct. This has been a common fate of birds isolated on oceanic islands.

KIN SELECTION

Natural selection acts on individuals, preserving only those that are best able to cope with the problems of existence. A question then arises when one considers altruistic behavior such as that of nonbreeding helpers at the nest (p. 360). How can such behavior, that appears to diminish the fitness of an individual, possibly evolve through natural selection? One possibility is that although the altruist's behavior reduces his own chances of contributing his socially helpful genes to the next generation, it increases those for the group, particularly if they are close relatives. By increasing the reproductive success of his relatives the altruist may propagate his own kind of genes (including tendencies toward altruism) more effectively than if he were

to concentrate his efforts "selfishly" on his own reproduction. This form of group selection is known as *kin selection* (Hamilton, 1963, 1970). Care of one's own young is not considered altruism since such care is an attribute of the individual's fitness.

However, recent studies have revealed that some helper systems uncommonly but regularly contain unrelated individuals (Emlen and Vehrencamp, 1983). For example, 8 to 11 per cent of the helper population of Green Woodhoopoes, *Phoeniculus purpureus*, studied in Kenya, consisted of individuals unrelated to the breeding pair (Ligon, 1983). The detailed studies of Woolfenden (1973, 1975) and his students on Florida Scrub Jays, *Aphelocoma coerulescens*, revealed that helpers were relatives of breeders and that pairs with helpers fledged more young than pairs without helpers. Following a crash in the population due to an epidemic, helpers moved into vacant territories and became breeders (Woolfenden and Fitzpatrick, 1984). Rowley (1983) studied the helper system in the Splendid Wren, *Malurus splendens*, of Australia and found that the population usually suffered from a shortage of females. A season of good rains was followed by a production of twice as many surviving juveniles as usual. For the first time in nine years there were more females than males, so that new territories that were set up were occupied by single pairs: helpers now became breeders. Thus it appears to be the strategy of some birds living in crowded environments to function as helpers and wait for the first available vacancy to establish their own territories. Long-lived birds, e.g., in tropical countries, are more likely to develop helper systems because they can "afford" to spend time waiting to be a breeder. R-selected species cannot wait, and must breed at the earliest opportunity or it may be too late. What appears at first sight to be altruism may be explained in terms of selection on the individual level after all.

■ SUGGESTED READINGS

The standard reference on the ancestry of birds is the excellent and well-illustrated book by Heilmann, *The Origin of Birds.* The adaptive radiation of birds is well summarized in Storer's chapter in Marshall's *Biology and Comparative Physiology of Birds*, and a survey of fossil finds is given by Wetmore in Wolfson's *Recent Studies in Avian Biology.* A brief but well-illustrated booklet dealing chiefly with La Brea Pleistocene birds is Howard's *Fossil Birds.* Brodkorb's definitive *Catalogue of Fossil Birds* is a useful reference. For interesting and thorough discussions on the principles and mechanisms of evolution in the lives of birds today, read Mayr's *Populations, Species, and Evolution*, and Rensch's *Evolution Above the Species Level*. Lack's small book, *Darwin's Finches*, gives a convincing account of the evolutionary history of the birds which first turned Darwin's thoughts toward the possibility of an evolutionary process. For an illuminating, recent study of avian evolution, see Haffer's *Avian Speciation in Tropical South America*. The latest review of our knowledge of fossil birds can be found in Storrs Olson's "The fossil record of birds," *in Avian Biology*, Vol. 8, 1985).

Once birds were gods to men,
  Magical, strange, and wise;
Now men are devils to them
  On earth, on sea, in skies.
—Joel Peters (1905–   ),
  Who cares?

Birds have affected man, and man has affected birds, for countless thousands of years. The Stone Age cave drawings of birds in France, Spain, Africa, and elsewhere reveal that man was interested in birds at least 22,000 years ago, and probably long before that.

It is impossible to know whether man's original interest in birds centered on their food value, their apparently magical powers of flight, their mysterious appearance and disappearance in migratory movements, or their bright colors, striking behavior, and intriguing songs. Myths, stories, paintings, and sculptures from the most ancient times suggest that all these ideas were interwoven. Primitive peoples from the earliest times to very recent years have ascribed supernatural powers to a great variety of birds. Some tribes

have believed that men's souls reside in birds, and that when a specific bird dies, its individual human counterpart also dies. In *The Golden Bough* Frazer (1922) related numerous instances in which primitive people ate birds or parts of birds not for nourishment but as one of the rites of homeopathic or sympathetic magic. Japanese Ainu believed that if one ate the still-warm heart of a freshly killed water ouzel, he would become wise and fluent. Primitive people of northern India believed that eating the eyeballs of an owl would give one the ability, like an owl, to see in the dark. Certain tribes in Morocco would not eat the heart of a chicken for fear that it would make them timid, and in certain Turkish tribes, children who were slow in learning to speak were fed the tongues of birds.

FOOD AND OTHER PRODUCTS FROM BIRDS

It seems probable, in any case, that man the omnivore has relished birds and their eggs as food for as long as he has existed. Archeological evidence suggests that the Red Jungle Fowl, *Gallus gallus*, was domesticated in India about 3200 B.C. and in China and Egypt about 1500 B.C. (Wood-Gush, 1964). There is even reason to believe that the Jungle Fowl, as an immigrant from Asia, was domesticated in Ecuador and Peru in pre-Columbian times.

Certainly in the modern world birds play an important economic role as a cheap and easily raised source of protein food. Today there are over 60 breeds of domestic Chickens, *Gallus gallus*, raised for meat, eggs, or ornament. In the United States alone, over 2000 million chickens are raised for food each year. This means that the average person eats about 18 kilograms of chicken per year. Also consumed each year in the United States are about 100 million Turkeys, *Meleagris gallopavo* (domesticated in pre-Columbian Mexico), and about 11 million ducks. Lesser numbers of geese, pheasants, Guinea Fowl, *Numida meleagris*, Japanese quail, partridge, and domestic Pigeons, *Columba livia*, are raised either for food or for hunting. The pigeon, descendant of the Rock Dove, was domesticated in Egypt as early as 3100 B.C. as a sacred bird associated with fertility, and has been widely bred ever since.

In addition to meat and eggs, birds produce other products of value to man. Early in this century hundreds of thousands of Ostriches, *Struthio camelus*, were raised each year in southern Africa for their ornamental plumes. Today about 25,000 are raised annually as a source of leather. Various species of wild birds have become extinct or nearly so because of man's desire for ornamental feathers. Primitive tribes in all parts of the world have at various times used the colorful feathers of birds for personal adornment, religious or magical rites, and even as money for trading. In 1902, over 1300 kg of plumes from the Great Egret, *Casmerodius albus*, were sold in London to decorate ladies' hats. They were worth more than their weight in gold. Today, under protection, the once decimated egrets have regained their former abundance. Iceland today exports annually about 4000 kg of down from the nests of Common Eider Ducks, *Somateria mollissima*. The ducks, although semiwild, are protected, and their down is collected on a conservative, sustained-yield basis.

More valuable than their feathers is the guano of certain sea birds. On desert islands off the west coasts of South America and Africa, cormorants, boobies, pelicans, and other birds drop between 200,000 and 300,000 metric tons of dry guano per year. One Peruvian Cormorant, *Phalacrocorax bougainvillei*, deposits on land about 1 kg of dry guano per month. Rich in nitrogen (*ca.* 16 per cent) and phosphorus (*ca.* 10 per cent), guano constitutes a valuable organic fertilizer. Ancient guano deposits, some as deep as 55 meters on Peruvian islands, were recklessly exploited in the 19th century when many millions of tons were shipped to the United States, Europe, and elsewhere. To stave off complete collapse of the guano industry, conservation and restoration measures were effected in 1901. Today, guano is no longer "mined" but is harvested as an annually renewable resource. It accumulates annually in layers about 8 cm deep.

Of less importance are several other products of birds put to use by man. The stomach oil of the Short-tailed Shearwater, *Puffinus tenuirostris*, was formerly used by Tasmanians as a beverage, a medicine (rich in vitamins), a lubricant, and as fuel for lamps (Murphy, 1936). The cave-dwelling Oilbird, *Steatornis caripensis*, has long been slaughtered by South American Indians for its body fat, which they render and use for cooking or as an oil in lamps. The fat, chiefly obtained from nestlings, will keep for months without becoming rancid. The edible nest of cave swiftlets, *Collocalia* spp., has long been highly prized as the chief ingredient of Chinese bird's-nest soup and other gourmet delicacies. Twenty to 30 tons of the nests, largely composed of a mucilaginous protein from the birds' salivary glands, are exported from Borneo and other Indo-Malaysian islands each year (Smythies, 1968). White nests of high quality may sell for as much as 120 dollars per kg.

OTHER MATERIAL BENEFITS TO MAN

Birds are undeniably of great economic benefit to man through their insatiable appetites for insects, rodents, and other organisms that are harmful to agricultural crops. But these avian benefits are rarely as simple and direct as they at first seem. Although orioles are avid eaters of the cotton boll weevil, they also eat or damage small fruits such as blueberries and grapes. Red-winged Blackbirds, *Agelaius phoeniceus*, consume hordes of destructive caterpillars and insects, but they also pull up sprouting corn and eat rice, corn, and other mature crops. Despite their great value as producers of guano, the Peruvian guano birds are in direct competition with the lucrative Peruvian fish-meal industry, because both guano and fishmeal are primarily derived from the same limited annual crop of marine anchovies. Although a given species of bird may feed exclusively on insects, it is not necessarily true that it is therefore beneficial to man. The species may feed indiscriminately on beneficial insects such as those that prey on harmful insects, or on crop-pollinating bees; or it may feed chiefly on easy-to-catch diseased or parasitized insects and thus increase the relative proportion of healthy but harmful insects. Similar logic can be advanced in appraising a species that eats weed seeds: is a seed-eating species valuable because it destroys weed seeds, or is it harmful because it distributes the seeds and thereby extends the weed's range? Thus, it is often difficult to describe a given species of bird as beneficial or detrimental to man until all the facts are known—until its complete economic niche is described. And this has rarely if ever been done.

The preponderance of evidence available indicates that the majority of birds are, in the main, either neutral or beneficial to man's interests. Numerous field studies have shown the clearly beneficial effects of certain birds on specific crops. For example, Northern Flickers, *Colaptes auratus*, in two winters removed 64 per cent and 82 per cent, respectively, of overwintering populations of corn-borer larvae from a Mississippi corn field (Black *et al.*, 1970). A study conducted in Washington and Idaho tested the efficacy of birds in controlling budworm density by comparing trees protected in enclosures versus trees that were exposed. Ten to 15 times as many moths were produced in the enclosed trees as compared to exposed trees (Campbell *et al.*, 1984). In European forests where birdhouses had been erected to attract increased populations of resident birds, insect damage to trees was significantly lower than in control forests with normal bird populations.

Not only do birds eat insects whose larvae may live as parasites on and in domestic livestock, but certain birds such as starlings and tick-birds relieve cattle and wild animals of their ticks and other vermin. A variety of owls specialize in preying on mice and other small rodents (Earhart and Johnson, 1970). In a five-year study of the Long-eared Owl, *Asio otus*, by Lundin (1960), 2667 prey animals were identified from the regurgitated owl pellets. The vast majority of animals eaten were mice and other small rodents; only 1.3 per cent were birds. Of further benefit to man are scavengers such as vultures, gulls, crows, and other birds that feed on carrion, offal, garbage, and other organic pollutants of the environment.

Less appreciated, perhaps, is the beneficent role birds play as pollinators and distributors of flowering plants. Ecological balance in nature would be seriously disturbed if these bird functions were to cease. Particularly in the tropics, certain birds are important pollinators of flowering trees and shrubs: hummingbirds (Trochilidae), honeyeaters (Meliphagidae), sunbirds (Nectariniidae), flowerpeckers (Dicaeidae), white-eyes (Zosteropidae), bulbuls (Pycnonotidae), honey-creepers (Drepanididae), and parrots (Psittacidae). Most of these birds are, at least in part, nectar feeders. In Australia there are about 80 species of birds, mainly honeyeaters and parrots, that feed on

FIGURE 24–1 Kestrel, *Falco tinnunculus*, is an especially effective predator on small rodents that damage agricultural crops. Photo by E. Hosking.

FIGURE 24–2 The skulls of small rodents, mainly voles, *Microtus pennsylvanicus*, found in pellets collected from a communal roost of Long-eared Owls in eastern Pennsylvania. A few undissected pellets are seen in the center of the circles and at the outer edges. Photo by G. R. Austing.

nectar. Ducks (Anatidae), pigeons and fruit pigeons (Columbidae), woodpeckers (Picidae), thrushes (Turdidae), nuthatches (Sittidae), and crows and jays (Corvidae) are exceptionally effective dispersers of seeds. On occasion, seeds that pass through a bird's digestive tract germinate more freely than those that come directly from the plant. In the 18th century the Dutch were frustrated in their effort to sustain a monopoly of the nutmeg trade on Amboina in the East Indies because fruit pigeons carried the undigested nutmeg seeds to other islands.

As described elsewhere in this book, birds have provided the experimental data for many notable sci-

entific discoveries in physiology, sensory perception, endocrinology, reproduction, population dynamics, behavior, vocalizations, social parasitism, flight, geographical distribution, migration, taxonomy, speciation, and other fields. The experimental study of bird malaria has contributed to a better understanding of human malaria. The pharmaceutical manufacture of vaccines for yellow fever and smallpox is accomplished through the inoculation of hens' eggs undergoing incubation. Finally, some minor contributions of birds to man's economy are found in such activities as fishing with cormorants in Japan and China, the observation of seabird aggregations as indicators of schools of fish, or the carrying of messages by pigeons or frigate-birds.

BIRDS IN ESTHETICS, THE SUPERNATURAL, AND RECREATION

Living in a forbidding Ice Age environment, Paleolithic man was by necessity a hunter rather than a farmer. He was understandably preoccupied with the animals that swarmed over the steppes and tundras around him: bison, horses, reindeer, aurochs, mammoths, tigers, cave bears, and others. He preyed on them; some of them preyed on him. Quite naturally his cave art predominantly featured animals. Among these were painted or engraved various birds: ostriches, auks, grouse, Snowy Owls, swans, ducks, bustards, eagles, passerines, and others (Armstrong, 1958). Surprisingly, some of the cave figures were of men with birds' heads. Quite obviously these prehistoric drawings represented more than the whim of an ignorant savage to mimic on cave walls what he observed outside. Armstrong, with persuasive logic, suggests that these caves were not homes but grottoes reserved for rituals, and that the bird-headed humans possessed magicoreligious significance.

Certainly it is true that, since the days of Ice Age art, birds have figured prominently in man's magic, religion, totemism, superstition, augury, folklore, and art. Space limitations will allow only brief mention of a few examples of the place of birds in man's supernatural beliefs.

In ancient Egypt a hawk-headed man represented Horus, the supreme sun god, and an ibis-headed man represented Thoth, the moon god of wisdom and magic. In ancient Greece, Athena, the goddess of wisdom, was symbolized by an owl. In the Christian religion the dove symbolizes the Holy Spirit, the European Goldfinch, *Carduelis carduelis*, represents Christ, and the cock, which reprimanded St. Peter, signifies watchfulness and vigilance. The emblem of Roman legions was the eagle of Jove. Even today an eagle is often used as the symbol of power or war, and the dove as a symbol of love and peace. In medieval times the swan figured prominently in numerous swan-maiden legends.

For many centuries, notably in Roman times, birds have figured prominently in augury—the art of observing and interpreting signs of approval or disapproval sent by the gods regarding certain human undertakings (Durant, 1944).

Among the Romans not a bird
Without a prophecy was heard;
Fortunes of empire often hung
On the magician magpie's tongue,
And every crow was to the state
A sure interpreter of fate.
—Charles Churchill
(1731–1764), *The Ghost*

In the folklore of many nations, as well as in Shakespeare, the Common Raven, *Corvus corax*, is commonly associated with ominous events or death. In an Irish tract dated around 1100 A.D., 28 different prognostications could be deduced from the behavior of Ravens; each of nine different calls had a different significance (O'Ruadháin, 1955).

Among the Sea Dayaks of southeast Borneo the art of augury persists vigorously even today. Seven species of birds—a piculet, a kingfisher, two trogons, a jay, a woodpecker, and a shama—are considered sacred sons-in-law of the supreme gods.

Divination of a bird's behavior requires the learning of a highly complex code of signs. Alarm cries of birds are associated with danger or harm; the omen of a bird flying from left to right across the observer's line of vision carries less force than that of one flying from right to left; a bird flying in the same direction as that taken by the observer represents an auspicious augury, as opposed to that for a bird flying in the opposite direction (Freeman, 1968).

Remnants of belief in birds' supernatural powers persist even in contemporary western civilization. In the living memory of people in western Eire a goose was killed at Martinmas and its blood sprinkled on the four corners of the house as a rite invoking St. Martin to bless the house and its dwellers.

Birds are frequently used as heraldic emblems. The pelican as a symbol of Christ has long been favored by ecclesiastics. The mother pelican is usually shown on her nest, pecking at her breast so that she can feed her young with her own blood. This example of the pelican in heraldry is found in the coat of arms of Richard Foxe, Bishop of Winchester, who founded Corpus Christi College, Oxford, in 1515. The emblem is still used by the College (Brook-Little, 1964).

Probably the chief recreational activity associated with birds is the systematic killing of them through hunting. Hunting arose primarily as a means of providing man with food, and still serves that function among primitive peoples. Among more "civilized" peoples it has become primarily an outdoor sport and a profitable business in which food is of distinctly secondary importance to the excitement of the chase. One measure of the magnitude of bird hunting as a business can be obtained from the fact that each autumn about 50,000 hunters invade the single state of South Dakota to hunt the Ring-necked Pheasant, *Phasianus colchicus*, and while there spend about 12.5 million dollars for that privilege (Linduska, 1966). These figures do not include the total costs of travel, guns, ammunition, or special clothing. In 1986 there were in the United States about 16 million licensed hunters, the majority of whom probably hunted game birds. To sustain such heavy human predation on birds and to keep the sport of hunting alive, various hunters' organizations, sporting goods associations, and government bureaus have felt obliged to finance game bird research and management as well as land acquisition for breeding areas and hunting grounds.

The chief game birds of North America fall into two groups: upland species and waterfowl. The upland game birds include pheasants, quail, partridge, grouse, turkeys, and doves. Waterfowl include ducks, geese, and so-called shorebirds: woodcock, snipe, plovers, rails, and coots.

A very different form of hunting is the ancient sport of falconry, in which falcons, hawks, or eagles are trained to attack and capture not only game birds but rabbits, foxes, gazelles, and other quarry. The oldest records of falconry go back to about 2000 B.C. in China; in Japan, India, and the Middle East, they reach back to about 600 B.C. From medieval times to the 17th century, falconry was pursued in Europe with a passion enjoyed by no other sport. From the reign of William the Conqueror to that of Elizabeth I the flying of falcons was regulated by stringent laws. "Falcons and hawks were allotted to degrees and orders of men according to rank and station—for instance, to royalty the gyrfalcons [*Falco rusticolus*], to an earl the peregrine [*F. peregrinus*], to a yeoman the goshawk [*Accipiter gentilis*], to a priest a sparrow hawk [*A. nisus*], and to knave or servant the useless kestrel [*F. tinnunculus*]" (Radcliffe, 1898). Even today falconry is still practiced as a sport in central and southeast Asia, and, to a lesser extent, in Europe and the United States. The proper training of a falcon requires great skill and patience. Falconry poses no threat to the falcon's prey species, but rather to the falcons themselves. Unscrupulous individuals have been known to rob eyries of wild falcons at a time when the population of falcons is in a serious decline in the United States, Europe, and elsewhere.

Today, increasing numbers of persons obtain esthetic and recreational satisfactions from birds by observing them undisturbed in their natural habitats. Here, with binoculars, onlookers can see the colors of a bird, hear its songs, and take note of its interesting behavior—and then leave it unharmed for others to enjoy later. In their search for new and strange species of birds, observers often travel to foreign countries. Current ornithological and natural history journals carry advertisements of group tours to almost every part of the world, designed to broaden the ornithological horizons of "bird watchers."

Rather than traveling to find birds, other people attract them to their homes with bird houses, bird baths, and feeding shelves, or keep them in cages or aviaries. As increasing human populations cause the shrinking of natural habitats throughout the world, more and more people attempt to maintain in this way some limited contact with the natural world. The striking increase in recent years of memberships in bird clubs, conservation groups, ecology classes and the like bears witness to this urge for man to keep alive some shred of contact with his natural origins.

BIRDS IN CONFLICT WITH MAN

Despite their small size and apparent fragility, birds can occasionally cause man to suffer great economic hardship. This is most likely to occur when man, through his manipulation of the natural environment, brings about changes conducive to explosive growth in the populations of certain birds. Chief among these changes is the so-called monoculture of agricultural crops—the practice of growing a single food plant over a wide expanse of land. Although this practice sharply reduces the populations of most native birds, a very few species seem to be preadapted to the changed environment, therefore increasing their numbers greatly and becoming agricultural pests. Here again, however, the damage that a bird inflicts on a certain crop during a brief season must be weighed against its value as a destroyer of noxious insects the remainder of the year. The chief crop-damaging birds include the waterfowl (Anatidae), pigeons and doves (Columbidae), parrots and parakeets (Psittacidae), crows and jays (Corvidae), robins (Turdidae), starlings (Sturnidae), blackbirds (Icteridae), weaver finches (Ploceidae), and finches (Fringillidae).

Following a nation-wide survey by agricultural and wildlife officials, it was estimated that in the United States birds destroyed over six million bushels of corn (maize) in 1970, or roughly 0.2 per cent of the total crop (Stone *et al.*, 1972). In the southern United States huge winter roosts of blackbirds, each containing over a million birds, are often located within foraging distance (*ca.* 60 km) of rice fields where the birds find approximately 50 per cent of their food. In Arkansas, blackbird consumption of rice, however, was several times less than the normal wastage that occurred in combine harvesting (Meanley, 1971).

Probably the most spectacular instance of bird damage to crops is that caused by the small Red-billed Quelea, *Quelea quelea*, of southern Africa. A ploceid, this bird normally lives in uninhabited thornbush country. With as many as 400 nests in one tree or bush, this species breeds in huge colonies which may cover a few hectares or many square kilometers. The young are first fed insects, and then grass and grain seeds in the milky stage. Disney (1964) gives examples of the damage to grain crops caused by these birds, and of man's attempts to control this damage: In

Tanzania in 1890 the birds caused famine and forced native farmers to shift crops from millet to sorghum; in Kenya (1952) 100,000 bags of wheat were lost; in Transvaal (1953) native crops completely failed, and many farmers had to abandon sorghum as a crop. One million birds are considered capable of destroying 60 tons of grain per day. Suggested reasons for the burgeoning success of the Quelea are the protection against predators provided by colonialism; its high mobility, which enables it to breed hundreds of kilometers from an originally unsuccessful attempt; its persistence in breeding after initial failure; the absence of numerically effective predators; and the early age (nine months) at which the young begin breeding. The development of water holes and grain farming on marginal lands, owing to a rising human population, has obviated the former necessity of the birds to migrate to rivers and lakes in regions with limited food supplies. Disney reports that men have attacked colonies and roosts with poison aerial sprays, disease

FIGURE 24–3 A typical view of the vast numbers of blackbirds from one roost that prey on the Arkansas rice crop. Photo courtesy of B. Meanley and the U.S. Fish and Wildlife Service.

germs, axes, flame-throwers, and explosives. In Tanzania in 1956 and 1957 some 72 million nestlings and over 6 million adults were destroyed in addition to another 5 million adults that had migrated to Kenya. The Quelea problem still awaits a successful solution.

Related to the ecological dislocations caused by plant monoculture are the problems caused by the feeding of dense concentrations of cattle in feed lots. At one such lot, grain consumed by Starlings, *Sturnus vulgaris*, amounted to 1 1/2 to 2 tons (6.4 to 12.4 per cent) per feeding season over three years. The erection of plastic netting ended further depredations (Feare and Swannack, 1978).

Birds, of course, damage other crops as well as grains. The eight million waterfowl that visit refuges in Oregon and California each year destroy lettuce worth several million dollars. In the Niagara Peninsula of Canada, bird damage of fruit commonly causes a 25 per cent loss of the sweet cherry crop annually; damage to grapes is usually less than 10 per cent of the crop (Brown, 1974). When the spraying of forest budworms destroyed their natural food supply, American Robins, *Turdus migratorius*, invaded Canadian blueberry farms. Berry growers killed thousands of Robins in 1972, but were denied permits to shoot them the following year (Sayre, 1973). The Monk or Quaker Parakeet, *Myiopsitta monachus*, recently introduced into the United States from South America, eats fruits, grains, and other crops. It may become a North American agricultural pest as was the now extinct Carolina Parakeet, *Conuropsis carolinensis*. In one Argentine province alone from 1958 to 1960, bounties were paid on 427,206 Monk Parakeets in an attempt to reduce agricultural damage (Bull and Riccuiti, 1974). Currently, eradication programs may have stopped its growth and spread in the United States (Niedermeyer and Hickey, 1977).

On rare occasions birds prey on domestic animals. In Europe members of the crow family have been known to prey on sheep and young deer, especially the sick and the weak. Hawks, eagles, and other falconiform birds are commonly accused by farmers of killing chickens, lambs, and other small domestic animals, but quantitative studies of alleged predation are rare. Such a study, however, was made of the remains of prey in 41 nests of the Golden Eagle, *Aquila chrysaetos*, in sheep- and goat-raising areas of Texas and New Mexico (Mollhagen *et al.*, 1972). Rabbits, squirrels, and prairie dogs comprised about 90 per cent of all prey individuals found in the nests. Remains of at least 66 sheep and goats, mainly young, were found

in two-thirds of the nests, but it could not be determined whether they represented fresh kills or carrion. A similar study of the nests of 28 Bald Eagles, *Haliaetus leucocephalus*, in the Aleutian Islands revealed no prey of economic importance to man.

Related to agricultural predation is the predation of birds on wild animals that man also depends on for food. Certain raptors prey on grouse, pheasant, and quail; the Osprey, *Pandion haliaetus*, feeds on game fish; kingfishers steal fingerlings from fish hatcheries; and guano birds eat millions of tons of anchovies each year. These and similar forms of avian predation place birds in competition with man. Very probably, in most instances the cost to man of such competition is more than balanced by the value of the bird in question as a factor in the ecological equilibrium of the total environment.

Another conflict between birds and man arises when large concentrations of birds roost in cities, and foul buildings, monuments, and parks with their excrement. The average domestic Pigeon, *Columba livia*, voids about 2.5 kg of feces annually. On the ledges of some buildings in Paris, pigeon droppings have accumulated to the depth of 1 meter. Since the beginning of this century Starlings, *Sturnus vulgaris*, have been roosting on building ledges in cities, perhaps because they are safe there from predators, or because they obtain thermal advantages. Even Pied Wagtails, *Motacilla alba*, have adopted urban dormitories in London, Frankfurt, Rome, Hong Kong, and elsewhere. The urban birds are objectionable not only because of their noise, but because their droppings are odorous, corrosive of masonry, and a source of human histoplasmosis.

A more modern hazard caused by bird aggregations occurs at or near airports. Modern, high-speed jet engines are highly vulnerable to damage upon striking a bird. Two major passenger plane disasters in the United States were caused by birds, and in Britain about one serious bird strike occurs in 20,000 flights (Murton and Wright, 1968). It is estimated that bird strikes cost England's Royal Air Force about £1 million per year. As a general rule, airplane-bird strikes occur at altitudes under 1800 meters.

Possibly the greatest and most poorly understood harm inflicted by birds on man occurs when they become carriers or transmitters (vectors) of human diseases. In addition to encephalitis, as described in Chapter 18, birds also are vectors for chlamydiosis (formerly called ornithosis). Chlamydiosis is an airborne respiratory disease caused by a bacterium.

Both birds and humans may become infected merely by breathing contaminated dust from droppings of infected birds. Domestic pigeons in the United States show a high incidence of chlamydiosis infection. Salmonella food poisoning in man is occasionally contracted by eating contaminated poultry or egg products such as custards. Birds commonly harbor blood-sucking ticks and other arthropods that may be vectors of serious virus and other diseases in man. There are about 150 different arboviruses (i.e., arthropod-borne viruses) known worldwide, of which about 50 are known to produce disease in man and domestic animals. Eight of these latter arboviruses have been isolated from wild birds (Stamm, 1966). Hoogstraal and Kaiser (1961) found that among 8379 Eurasian birds examined during their migrations through Egypt into Africa, 6 per cent of them carried ticks, two species of which were vectors of causal organisms of such human diseases as Crimean hemorrhagic fever, Q fever, tularemia, tick typhus, and brucellosis. Fifteen different strains of viruses, chiefly of influenza, have been isolated from various pet birds imported into Japan (Nishikawa *et al.*, 1977), and Chagas' disease (p. 305) is being spread in Brazil by House Sparrows, *Passer domesticus*, introduced in 1906 to combat mosquitoes (Smith, 1973). The causal agent of a fungal ringworm disease in man has been isolated from grouse. Pigeon droppings have been known to harbor a virulent skin fungus that also infects man's lungs and central nervous system, causing a form of meningitis. It is estimated that in a large city like New York the average person inhales 3 μg of pigeon fecal dust daily (Murton, 1971). Healthy pigs have contracted viral gastroenteritis after ingesting the droppings of experimentally infected starlings. The large and complex subject of birds as vectors of diseases of man and domestic animals requires much further study.

THE CONTROL OF PROBLEM BIRDS

Wherever large concentrations of birds have created problems for man, he has attempted various control measures. Even primitive tribes erect scarecrows and noisemaking devices in food patches to repel unwanted birds. In urban regions gunfire may be dangerous to other organisms and to property. In open regions gunfire merely disperses the birds, and they then congregate elsewhere or at the original site when gunfire ceases. In 1932 when the Emu, *Dromaius novaehollandiae*, threatened to become a serious pest to wheat growers in Australia, the government dispatched a machine gun detachment from the Royal Australian Artillery to engage with the enemy (Serventy, 1964). Loud explosive sounds are effective repellents with some species, but they cover a limited area and are unacceptable near human communities. Many species quickly become habituated to loud noises—even the nearby roar of jet planes—when the noise proves to be harmless. However, the broadcast of species-specific tape-recorded alarm calls is often very effective. In grain fields and vineyards the loud broadcast of a species' alarm calls can effectively reduce the unwanted bird population by 80 per cent in four days (Boudreau, 1968). An effective, automatic rotating sound source in a vineyard may cover an area 600 meters in diameter. Broadcast distress calls are not as effective as alarm calls. The response threshold to alarm calls varies with bird species, social organization, activity, season, habituation, and other variables.

Another form of bird repellent for agricultural crops requires making the food unpalatable. Several species of blackbirds (Icteridae) have learned to uproot sprouting corn to get at the germinating seed. Corn plots treated before planting with methylcarbamate will lose 70 per cent fewer plants than untreated plots (West and Dunks, 1969).

The most effective and lasting control of problem birds is achieved by reducing their numbers, either directly by destroying them with toxic substances or indirectly by controlling their reproductive success. At a New Zealand airfield, gulls from a nearby breeding colony were causing frequent and dangerous air-strikes. The gulls were baited with bread dosed with alpha-chloralose (200 mg per bait). This caused a reduction in their population of over 85 per cent, and a subsequent reduction of over 92 per cent in air-strikes (Caithness, 1968). A less drastic solution of some avian overpopulation problems can be achieved by using antifecundity chemical compounds. The estrogen mestranol incorporated in synthetic grit will inhibit ovulation in pigeons without inhibiting courtship behavior; offspring from eggs which do hatch may become sterile through mestranol transferred either through crop milk or in the grit itself (Sturtevant, 1970). Field trials in Germany of the chemosterilant Busulfan resulted in the sterilization of feral pigeons that lasted several months (Geisthardt, 1977).

The special problem caused by too many birds near airports can sometimes be solved by habitat management. Food, cover, and water can often be controlled to discourage resident birds. Grass mowing discourages some birds from nesting and also reduces the insect population that attracts birds. Orange or yellow runway lights attract fewer insects than white lights. Hills and sand dunes that create updrafts attractive to gulls, albatrosses, and other soaring birds can be leveled with bulldozers. Radar detection of bird flocks in the air can be used to alert pilots to potential danger. A highly unusual method of eliminating air-strikes was employed at a British Navy airfield. A team of falconers, flying two falcons per year, sharply reduced airfield bird populations and, as a consequence, lowered air-strike damage from 600,000 dollars per year to zero (Kuhring, 1969).

MAN'S DESTRUCTION OF BIRDS

The earth that harbors both birds and man is a finite sphere. During the past two centuries man's numbers and his domination of the natural environment have increased dramatically, and his destructive impact on birds has increased proportionately. His environment-modifying tools are powerful, pervasive, and destructive: gang-plows, power reapers, ditchdiggers, earthmovers, paving machines, chainsaws, and power sprayers. Forest fires, soil erosion, and water and air pollution contribute further to the destruction of natural habitats. Livestock and introduced pests compound the deterioration. That byword for extinction, the flightless Dodo, *Raphus* spp., was sent on its way by the pigs, cats, and rats introduced by man on the Mascarene Islands. As man's tools gnaw away at the earth's green covering, its area shrinks, habitats deteriorate, and birds are forced into extinction.

Man destroys birds by both direct and indirect means. Direct destruction of birds occurs when they are killed for food, feathers, oil, "sport," or museum specimens. It has been estimated that today the annual toll of birds from human predation amounts to "many hundreds of millions" (Elliott, 1970). In the late 19th century some 20 million birds were imported annually into Great Britain solely to supply the demand for hat trimmings: parrots, hummingbirds, birds of paradise, waterfowl, and others (Anon., 1979). A century or more ago birds were generally considered an inexhaustible natural resource and were exploited accordingly. The fallacy of this viewpoint should have become apparent in 1844 when, as a result of wholesale slaughter by whaling-boat crews, the Great Auk, *Pinguinus impennis*, became extinct. But the slaugh-

ter continued. In 1857 about one-half million penguins in the Falkland Islands were killed for their oil. Around 1900, Japanese poachers nearly wiped out the albatrosses on one of the Bonin Islands, killing over 5 million birds for their feathers. In the first decades of this century about one-half million Short-tailed Shearwaters, *Puffinus tenuirostris*, were killed annually in Australia for their flesh and oil, and in the 1920's in New Zealand, about 250,000 Sooty Shearwaters, *P. griseus*, were killed each year. Many similar examples of mass slaughter could be given. Even songbirds are slaughtered for food. The Swedish physician Axel Munthe (1957) wrote that on the island of Capri, songbirds were used as caged, singing decoys to lure migrating passerines into nets. From one mountainside alone, 15,000 netted birds were shipped annually to Paris to be eaten in smart restaurants. Eventually, after many setbacks, Munthe was able to buy the mountainside and turn it into a bird refuge. Even today, millions of songbirds are being killed each year for food in Italy, Greece, Spain, Thailand, Japan, and other countries.

The wholesale collection of birds' eggs brings similar adverse pressures on many species. Colonial sea birds are peculiarly susceptible to egg collecting by man. In 1901 about 630,000 eggs of the Jackass Penguin, *Spheniscus demersus*, were collected on south African islands; in 1946, in the Seychelles Islands, 1,996,400 tern eggs were collected (Fisher and Peterson, 1964). In a few rare instances the decline in the population of an exploited bird suggested a cause-and-effect relationship to the egg collectors and they responded by instituting conservation measures. Hundreds of thousands of ducks breed on the shores of Lake Myvatn in northern Iceland. Farmers whose land adjoins the lake are careful to leave four eggs in each nest each year, removing the remainder for their own use. They have followed this custom for the past 700 years. Scott (1953) reported no decline in the egg crop in the past 50 years.

High mortality generally occurs in any species that is legally designated a game bird. In the United States during years with open hunting seasons, the mortality rates for Canvasback Ducks, *Aythya valisneria*, and Redhead Ducks, *A. americana*, average 50 per cent or more, but in years when hunting is prohibited because of low populations, their mortality rates decline by 14 to 21 per cent (Geis and Crissey, 1969). For the American Black Duck, *Anas rubripes*, it has been estimated that one-half of the total deaths of adult birds are caused by hunting, and that this mortality is largely in addition to, rather than instead of, nonhunting mortality. There are, of course, many birds killed illegally in the name of sport. Between the years 1941 and 1946, pilots flying from the Alpine, Texas, airport killed between 675 and 1008 Golden Eagles, *Aquila chrysaetos*, each year by shooting them from airplanes (Spofford, 1964). Since 1961 aerial eagle hunting has been prohibited by federal law.

Too often birds are killed by gunners simply because they offer a moving target. As described earlier, sometimes migrating birds follow natural "aerial highways" in their annual journeys. Before the conservation ethic was as widely appreciated as it is today, hunters used to gather along these migration routes for target practice. Peterson (1948) gave a vivid description of the annual slaughter of Sharp-shinned Hawks, *Accipiter striatus*, in the fall of 1935 on Cape May, New Jersey. He saw some 800 hawks try to fly across a road one morning, and by noon there were 254 dead on the pavement. He remarked, "It would have done no good to explain predation, ecology, and natural balance to these folk. Having lived at Cape May all their lives, they had a distorted idea of the abundance of hawks."

Sometimes birds are killed as "innocent bystanders" when man is preying on other animals. About 500,000 Thick-billed Murres, *Uria lomvia*, are killed each year in fish nets, and these, added to the 750,000 harvested annually for food, clearly threaten the survival of the species (Tull *et al.*, 1972). Along the coast of Texas, commercial fisherman set thousands of trot-lines in shallow water, each with scores of baited hooks. Many ducks and cormorants are caught on the hooks and drown. McMahan and Fritz (1967) estimated that each winter over 21,000 Redhead Ducks, *Aythya americana*, are hooked on the lines. Of these over 3000 drown, and over 7000 others that break loose or are removed from the hooks have poor chances of survival.

By all odds the harshest impact of man on birds has been through his modification of the natural environment into farms, plantations, cities, highways, and industries, with all their supporting technologies. Modern agriculture, in particular, has supplanted native habitats over much of the earth. Primeval forests have given way to millions of hectares of monocultured corn; virgin prairies, with all their rich diversity of plants, have been plowed under and planted with endless vistas of wheat. After the harvest season, little remains but a relatively sterile, unprotective stubble. As mentioned earlier, while some of man's modifications of the natural environment have created useful habitats for some birds (lawns and parks for robins, chimneys for swifts) the overall effect has been detrimental to birds. Monocultural environments provide variety neither in food nor in habitat that most species require. Occasionally the foods made available, although edible and nutritious, may be harmful. For example, Canada Geese, *Branta canadensis*, that eat large quantities of soybeans will die because the beans swell, block the esophagus, and cause starvation.

Contemporary industrial civilization requires many physical structures and implements that create unnatural hazards for birds: towers, high-tension lines, highways and automobiles, farm machinery, and the like. Television and radio towers with their supporting guy-wires create death-traps for many nocturnal migrants, especially during the overcast or foggy nights which accompany arriving cold fronts. At the foot of one 205-meter television tower in Florida, Stoddard (1962) made an effort to pick up all bird casualties that occurred between October 1, 1955, and June 30, 1961. Although many bird carcasses were carried off by predators, Stoddard managed to collect and identify 15,200 individual birds of 150 different species. Perhaps the most concentrated slaughter occurred at a 330-meter tower at Eau Claire, Wisconsin, on two nights, September 18 to 20, 1963, when some 30,000 migrating birds were killed. Among the broken bodies were 1121 Red-eyed Vireos, *Vireo olivaceus*, and over 900 each of four species of warblers (Parulidae) (Kemper, 1964). Even greater destruction has occurred at airport ceilometers—intensely brilliant vertical searchlights that measure the altitude of clouds. At the Warner Robins Air Force Base in Georgia on the night of October 8, 1954, an estimated 50,000 birds of 53 species were killed (Johnston and Haines, 1957). The migrating birds, apparently confused by the brilliant beam, dashed themselves against the ground. The use of ceilometers is now controlled with the hope of protecting birds during peak migration periods. Skyscrapers and tall monuments, illuminated at night, also exact their toll of migrating birds. On September 15, 1964, 497 birds of 32 species were picked up from the streets and rooftops surrounding the Empire State Building in New York City (Vosburgh, 1966).

High-tension cables, towers, and telephone wires also take heavy tolls of nocturnal migrants, but since the obstacles are so widely spread, often over thick ground vegetation, the dead birds are not as conspicuous as those near television towers. An analysis of the known causes of mortality of 294 banded, wild White Storks, *Ciconia ciconia*, in Germany by Riegel and Winkel (1971) gave the following statistics: 226 (77 per cent) were killed by flying into wires; 16 died from human persecution; 8 hit moving vehicles; 7 struck factory chimneys; 11 were killed by lightning; 16 died in battles with rivals; and 9 were found dangling from crotches of trees. High-tension power lines are a major mortality factor for swans, cranes, eagles, and other large birds (Braun *et al.*, 1978; Boeker and Nickerson, 1975). In wildlife refuges, equipping power lines with aircraft markers has greatly reduced bird mortality. The electric wires sometimes used to fence in cattle also kill birds. On one farm alone, 200 birds were electrocuted by such fences in one year (Stewart, 1973).

Farm machinery is a major indirect threat to bird survival, mainly because it so drastically changes the natural environment. But sometimes it acts all too directly in killing birds that do manage to live in an agricultural habitat. In several studies of the nesting success of Ring-necked Pheasants, *Phasianus colchicus*, in hay and grain fields in the United States, it was revealed that farm machinery, particularly mowing and reaping machines, destroyed between 50 and 75 per cent of the nests, whereas predators destroyed between 18 and 31 per cent. In Oregon two farmers estimated that they had destroyed numerous waterfowl nests and had killed between 400 and 600 young shore birds, ducks, and cranes during a two-week period of mowing (Braun *et al.*, 1978). In Britain the virtual disappearance of the Corncrake, *Crex crex*, in the last 50 years has been attributed to the destruction of eggs and young by mowing machines.

In a special study of road deaths carried out by 176 observers in England, it was estimated that $2^{1}/_{2}$ million birds were killed annually on the highways (Hodson and Snow, 1965). It was concluded that road mortality was insignificant as a factor in reducing bird population. A similar study placed road deaths in Denmark at an estimated rate of 3 million birds per year.

A new and unexpected threat to bird survival has been traced to the sonic booms of supersonic aircraft. On the Dry Tortugas of the Caribbean the Sooty Terns, *Sterna fuscata*, had been rearing between 20,000 and 25,000 young each year for the past 50 years. In 1969 about 50,000 pairs settled on the islands

to breed, but that year they reared only about 240 chicks. The only discernible unusual circumstances that particular year were a lusher growth of vegetation and frequent sonic booms from jet aircraft. Upon hearing the booms the birds would rise in "panic flights." This probably interrupted their normal incubation rhythm and caused mass nest desertion (Austin *et al.*, 1972). In like fashion, the few surviving Whooping Cranes, *Grus americana*, wintering on the Aransas, Texas, refuge, were threatened by United States Air Force planes that flew over the refuge and bombed an adjacent island (Sayre, 1974). These examples show how easily the thin thread of survival can be unintentionally snapped by uninformed human actions.

MAN'S POLLUTION OF THE ENVIRONMENT

The excretions of industrial civilization, intensified by the human population explosion, have reached a magnitude that threatens the survival not only of birds and other wild organisms but of man himself. Western "civilized" man, in his pursuit of well-being, has poured countless millions of tons of noxious substances into the air, the waters, and the land on which he lives. Immense though it is, the earth's atmosphere is showing signs of impending disaster. It is becoming burdened with ash, dust, metallic particles, radioactive fallout, and noxious chemical compounds and gases. Automobiles and factory chimneys are estimated to pour millions of tons of carbon monoxide alone into the world's atmosphere every year. Air pollution, notably from pulp mills, was found to cause a significant reduction in the nesting density, colony size, and occupancy of House Martins, *Delichon urbica*, in Czechoslovakia (Neuman *et al.*, 1985).

One malignant and only recently recognized consequence of air pollution from the burning of fossil fuels is acid rain. Ice that originated as snow in Greenland about 180 years ago has a hydrogen ion concentration (pH) of between 6 and 7.6—essentially neutral in reaction. But many freshwater streams and lakes in the eastern United States, eastern Canada, and western Europe have, as a result of acid precipitation, pH readings of 5 and even 4. The average annual pH of precipitation in these regions ranges from 4 to 4.5. Such extreme changes in the nature of surface waters affect the entire ecological environment: producers, consumers, and decomposers. Fish populations in acid lakes are either wiped out or greatly reduced, as are the dependent fish-eating birds. Bacteria of decomposition are adversely affected. The chief culprits are emissions of sulfur and nitrogen oxides. The urgency of the problem is revealed in the fact that between the years 1950 and 1975, atmospheric oxides of nitrogen trebled in the United States, and oxides of sulfur more than doubled in western Europe (Likens *et al.*, 1979).

Approximately 100 square km of the bottoms of Back Bay, Virginia, and Currituck Sound, North Carolina, are smothered under tons of silt representing human, agricultural, and industrial waste, and erosion from upstream. The resulting turbidity allows only 5 per cent of the sunlight to penetrate as much as 1.5 meters, and this for only two months of the year. Submerged vegetation, the source of food for all aquatic life, is accordingly reduced to less than one-fifth its normal abundance. Active and abandoned coal mines are estimated to pour about 3.5 million tons of acids into the streams of the United States each year, damaging 6500 km of these streams as waterfowl habitat (McCallum, 1964). The river Rhine in Germany has become so heavily charged with industrial wastes that Dutch agriculture, at its mouth, is on the verge of ruin. The pollutants include organic chemicals, mercury compounds, carcinogens, and, especially, refuse from the French potassium mines. In spite of an international commission set up three decades ago to deal with Rhine pollution (Quigg, 1974), serious problems remain. From similar neglect Lake Erie became essentially an inland dead sea. However, emergency measures to rehabilitate the lake seem to be making it again a livable habitat.

An especially malignant form of pollution results from the deposits of spent lead pellets left by hunters in the bottoms of ponds and other waterfowl hunting grounds. The typical hunter shoots between 5 and 30 shotgun shells for every duck or goose he fells, and each 12-gauge shell contains over 250 lead pellets of number six shot. In lakes with firm bottoms there may be as many as 290,000 accessible pellets per hectare; 150,000 per hectare is a common density. Ducks and geese ingest these toxic pellets along with the seeds they eat. Examination of 36,000 gizzards from 20 species of waterfowl showed lead shot in 6.7 per cent of all ducks and 1 per cent of all geese (Bellrose, 1964). The toxicity of the pellets varies with the species of waterfowl, the number of pellets eaten, and whether or not they are pulverized by grit or coarse seeds in the gizzard. A single pulverized pellet is a fatal dose to most waterfowl. Waterfowl also carry lead pellets in their body tissues as the result of non-fatal gunshot wounds. Fluoroscopic examination of 745 Mississippi Flyway Canada Geese, *Branta canadensis*, showed that 44 per cent of them carried lead pellets; similar burdens were found to occur in wild geese in Britain, each bird carrying from 1 to 23 pellets (Elder, 1955).

Lead poisoning was responsible for the deaths of over 500 Tundra Swans, *Cygnus columbianus*, in a North Carolina refuge in 1974, and of about 1500 waterfowl at Rice Lake, Illinois, in 1972 (Anderson, 1975). For terrestrial species exposed to air contaminated with automobile exhausts, leaded gasoline poses a hazard. Laughing Doves, *Streptopelia senegalensis*, living in the city of Cape Town had a mean content of lead in their bones of 84.3 parts per million (ppm), whereas in rural doves it was 13.1 ppm (Siegfried *et al.*, 1972). It is difficult to say how general lead-poison mortality may be among game birds. Statistical results from the experimental feeding of measured doses of lead shot to wild Mallards, *Anas platyrhynchos*, convinced Bellrose that about 4 per cent of them died annually from lead poisoning. By extrapolation this means that probably millions of waterfowl in North America die each year from this cause. Currently, nontoxic steel shot is expected to replace lead shot in the near future.

While lead poisoning has deleterious effects on the brain and other vital organs, perhaps its most lethal influence is on the enzyme delta-aminolevulinic acid dehydratase—the enzyme essential for the synthesis of hemoglobin. Ducks dosed with only one lead shot suffered an 85 per cent inhibition of delta-ALAD activity for at least four weeks (Finley *et al.*, 1976).

THE EFFECTS OF PESTICIDES ON BIRDS

The year 1942 marked the beginning of a chemical revolution in agriculture—the introduction of a whole new class of synthetic pesticides. The first of these was DDT (dichlorodiphenyltrichloroethane), which was soon followed by hundreds of others. Initially the new pesticides were so successful that their use increased dramatically. Whereas in the years between 1951 and 1966 world food production increased 34 per cent, and the use of farm tractors 63 per cent, the use of pesticides increased 300 per cent (Woodwell, 1974). Part of this increase occurred also in the traditional or old-type pesticides such as the botanical poisons (e.g., rotenone, pyrethrins), or inorganic salts (such as mercury and arsenic). In Japan the rice crop was increased 33 per cent, mainly through the yearly application of 140,000 tons of phenylmercury compounds—but at the cost of the extinction of the Japanese race of the White Stork, *Ciconia ciconia* (Elliott, 1970). Pesticides need not kill wildlife directly to have ill effects. Diets containing 3 ppm of methyl mercury were fed to 13 pairs of American Black Ducks, *Anas rubripes*, during two breeding seasons. These birds produced 16 ducklings surviving one week, whereas 13 control pairs produced 73 ducklings (Finley and Stendell, 1978). The mercury compounds caused both eggshell thinning and, in the ducklings, brain lesions.

The old pesticides soon began to give way to the new synthetic compounds which were "broad spectrum" poisons—i.e., each killed a large variety of agricultural pests—and also seemed to be less toxic to vertebrates. Chemically the new pesticides fall into two major groups: (1) the organophosphorus compounds, such as parathion and malathion. They are highly toxic but quickly break down into harmless compounds; (2) the chlorinated hydrocarbons, such as DDT, aldrin, dieldrin, and heptachlor. These do not readily break down and may persist for years as toxic compounds in the water, soil, or ground litter. They are nearly insoluble in water but tend to adsorb on organic matter and become concentrated in the body fat of vertebrates.

The long-lasting residual toxicity of chlorinated hydrocarbons proved to be extremely useful in killing the anopheline mosquito, for example, thereby removing the threat of malarial fever from hundreds of millions of humans. The tremendous successes of the new pesticides produced an initial optimism that was soon seen to be unjustified. The very virtues that made such pesticides so successful—their broad toxicity and chemical stability—proved to be *too* effective. They not only killed their target organisms but also, as Rachel Carson dramatized in *Silent Spring* (1962), poisoned and killed whole food chains and entire ecosystems. As an example, the use of DDT in the watershed of Lake George, New York, caused the complete reproductive failure of trout in that lake (Graham, 1967). Thanks to their persistence and the ease with which they are distributed by air and water currents, the chlorinated hydrocarbons have produced one of the world's most serious pollution problems. DDT is probably the most widely distributed synthetic chemical in the world. It is even found in the bodies of penguins in the antarctic and of auks in the arctic.

The hazard of indiscriminate application of broad

spectrum pesticides was convincingly demonstrated when the United States Department of Agriculture attempted to eradicate the imported fire ant, *Solenopsis*, from several southern states. Dieldrin and heptachlor were applied from the air to thousands of hectares at the rate of about 2.2 kg per hectare, with the result that millions of birds, mammals, other vertebrates, and invertebrates perished. In areas where detailed estimates were made, over 80 per cent of the resident birds were destroyed, but the fire ant was not eradicated (Hickey, 1966). A survey by Presst and Ratcliffe (1972) showed that residues of chlorinated hydrocarbons occur in at least 154 different species from 47 families of European birds. As a rule, the geographic distribution of affected birds correlates with the degree of pesticide use in different parts of Europe, but Berthold (1973b) ascribes the 66 to 75 per cent decline in European populations of several warblers, flycatchers, martins, and other species between 1968 and 1973 to pesticide contamination picked up either in African winter quarters or along routes of migration.

A further hazard to wildlife resides in the capacity of chlorinated hydrocarbons to become "biologically magnified"—that is, to become progressively more and more concentrated in the body tissues of animals living at successively higher levels in a food chain. An instructive example of this principle was set in motion when Clear Lake, California, was sprayed with DDD (a close relative of DDT) to combat a plague of gnats (Hunt and Bischoff, 1960). In 1949 the lake water was treated with 0.014 ppm of DDD, as it was again in 1954 and 1957. Following each of the last two applications about 100 corpses of Western Grebes, *Aechmophorus occidentalis*, were found. Whereas there were over 1000 pairs of grebes before the insecticide program was begun, during the 1959 breeding season there were only 25 pairs living. No nests or young grebes were found, suggesting that the surviving birds were sublethally affected. The dead birds contained 1000 ppm of DDD in their body fat. It was later discovered that the microscopic plankton in the lake had concentrated the pesticide 265 times, the small fish that ate the plankton, 500 times, and the fish-eating grebes, 80,000 times!

In a Long Island marsh that had been sprayed for 20 years with DDT to control mosquitoes, it was found that the bottom mud contained 14 kg of DDT residues per hectare. Although the marsh water contained only 0.00005 ppm of residues, successive organisms in the local food chain concentrated the pesticide as follows: plankton, 0.04 ppm; minnows (two species), 0.23 and 0.94 ppm; pickerel, 1.33 ppm; gull, 6.0 ppm; cormorant, 26.4 ppm (Woodwell *et al.*, 1967).

In sufficiently high doses the immediate effect on birds of DDT and related pesticides is death. Acting on the nervous system, the pesticide causes typical muscular tremors and convulsions that end fatally. In sublethal doses, however, the toxins may cause long-term effects just as disastrous to a species as immediate death. Among these effects are aggressive nervous behavior; disrupted breeding behavior; egg eating; the laying of sterile eggs, or of eggs with thin shells, or no eggs at all; the death of embryos in the egg; and the premature death of both young and adults. The thinning of eggshells shows a strong positive correlation with DDT dosage in body tissues. DDT or its metabolites, in the bird's oviduct, cause the inhibition of the enzyme carbonic anhydrase, which, in turn, controls the transport of the calcium necessary for eggshell formation (Peakall, 1971). Eggshells that are reduced in thickness 20 per cent or more usually fail to support the weight of the incubating bird and are crushed (Fig. 24–4). The reproductive failures of the Peregrine Falcon, *Falco peregrinus*, in the eastern United States, the Brown Pelican, *Pelecanus*

FIGURE 24–4 A sampling of broken and intact pelican and cormorant eggs from Anacapa Island, California. In each species the eggshell thinning was caused by DDT contamination of seawater. Photo courtesy of D. W. Anderson and the U.S. Fish and Wildlife Service.

occidentalis, in the western United States, the Golden Eagle, *Aquila chrysaetos*, in Scotland, and many other species as well are now well documented as caused by eggshell thinning due to ingested chlorinated hydrocarbons. Experimental feeding of DDT and similar pesticides to captive birds produces the same results: thin eggshells and reproductive failure. In 1972, one hundred eggs of the White Pelican, *P. erythrorhynchos*, were collected from a colony in Great Salt Lake, Utah, and analyzed for insecticide residues by Knopf and Street (1974). Residues of DDE were roughly four times as abundant as those of either DDD or dieldrin. There was a significant negative correlation between eggshell thickness and DDE concentrations in the egg yolks (Fig. 24–5). Further, eggshell porosity is affected. DDE-contaminated Common Terns, *Sterna hirundo*, produce eggs with 44 per cent fewer breathing pores, resulting in the suffocation of the developing embryo within (Fox, 1976). The eggs of Black-crowned Night Herons, *Nycticorax nycticorax*, from Washington, Oregon, and Nevada, were found to be contaminated with DDE, causing nesting failure. Moreover, there was a north-south contamination gradient, with southern colonies more heavily affected, suggesting that pesticides were being picked up on their wintering grounds (Henny *et al.*, 1984).

Pesticides in a bird's body threaten its survival in still other ways. A bird may store in its body fat concentrations of DDT that would be overwhelmingly fatal if they were in its nervous system. Herring Gulls, *Larus argentatus*, can tolerate, in apparent good health, an average of 2441 ppm of DDT, or its metabolites DDE and DDD, in their body fat, but only 20 ppm in their brains. However, if the birds are denied food, they begin living on body fat. The stored pesticides are then released into the blood and begin accumulating in the nervous system until lethal levels are reached and the bird dies (Hickey, 1966). Twenty Brown-headed Cowbirds, *Molothrus ater*, were fed for 13 days on food dosed with 10 ppm DDT. After two days on full rations of untreated food, their food intake was reduced to 43 per cent of normal. Seven of the birds died within four days. On the contrary, control birds previously fed DDT-dosed food, survived indefinitely when put on full rations of undosed food (Van Velzen *et al.*, 1972). The authors concluded that stored DDT presents hazards to birds which utilize fat during periods of stress caused by reproduction, cold weather, injury, disease, limited food supply, or migration. This undoubtedly explains why female and

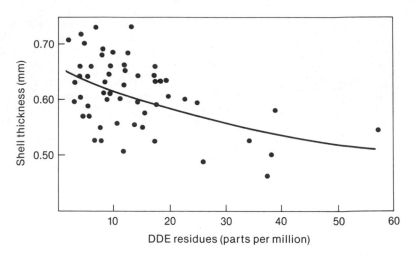

FIGURE 24–5 The relationship between eggshell thickness and DDE residues in the yolks of White Pelican eggs (after Knopf and Street, 1974).

not male Common Eider Ducks, *Somateria mollissima*, were found dying on the intertidal shores of the river Rhine in Holland between 1964 and 1967. Chlorinated hydrocarbon pesticides, originating in a Rotterdam factory, had been stored in the body fats of the ducks but were released to other tissues in lethal quantities only in the females during incubation (Koeman, 1975). When the factory ceased dumping pesticide wastes in the river in 1968, the populations of ducks and other sea birds began recovering.

If 10-day-old Mallard Ducklings, *Anas platyrhynchos*, are given food dosed with 25 to 100 ppm of polychlorinated biphenyls for ten days, they suffer no apparent intoxication. If, however, they are then inoculated with duck hepatitis virus, they will suffer more than twice the mortality of control ducklings that were not fed the chlorinated hydrocarbon (Friend and Trainer, 1970). Comparable experiments using DDT and hepatitis virus produced even greater mortality among the experimental ducks as compared with virus-only controls (Friend and Trainer, 1974).

There are, finally, two other aspects of pesticide pollution that pose still other ominous threats for birds and man. Some pesticides or their breakdown products have been shown to cause cancer or teratomorphic abnormalities in laboratory animals. Certain pesticides behave synergistically in combination. For example, the combined effects of the organophosphates malathion and EPN are ten times greater than the sum of their individual effects (Buckley and Springer, 1964).

These and the foregoing facts suggest strongly that man should proceed much more cautiously and intelligently with the use of chemical pesticides, new and old. He should immediately explore alternatives that are ecologically more wholesome. As examples of some of these alternatives, Odum (1971) suggests the use of:

1. Natural predators of harmful insects, e.g., lady beetles
2. Parasites of pests, such as chalcid wasps
3. Pathogens, such as viruses or bacteria, that are specific for a given pest
4. Radiation sterilization of male pest insects to reduce female fecundity
5. Hormonal stimulants that block a pest's development
6. Pheromones to lure insects and regulate their behavior
7. Rapidly degradable chemical pesticides
8. Rotation and diversification of crops
9. Genetic selection of disease- and pest-free plants.

There is some slender hope that birds may evolve pesticide-resistant populations as have many insects. By breeding the survivors in a series of five generations of European Quail, *Coturnix coturnix*, exposed to a fungal carcinogenic toxin, Marks and Syatt (1979) produced birds whose mortality was only about one-tenth that of unselected control birds.

Oil is yet another effluent of man's technology that cuts short the lives of many birds. Since World War II the phenomenal increase in the use of oil in industry and transportation has also increased opportunities for oil pollution of the environment. Oil spills from ship-wrecks are known to have caused the demise of hundreds of thousands of water birds (Bourne, 1970). It is estimated that about three million tons of oil are spilled into the seas annually, partly because of wrecks and accidents, and partly because ship owners are tempted to discard oil waste and tank ballast cheaply. Similar oil pollution exists in fresh waters. Oil pollution in the Detroit River was estimated to kill about 12,000 ducks during approximately three weeks in 1960.

Even on land, oil takes its toll. In the San Joaquin Valley of California there are about 1700 oil sump basins where waste effluents from oil wells are dumped. Birds mistake these basins for ponds. It is estimated that 150,000 birds perish in these deceptive ponds each year (*Audubon*, 1973).

Oiled birds may transfer oil from their plumage or feet to their eggs upon returning to their nests. Oil was applied to eggs of Great Black-backed Gulls, *Larus marinus*, and Herring Gulls, *L. argentatus*. A significant reduction in hatchability followed when eggs were oiled four to eight days after laying (Lewis and Malecki, 1984).

Mortality in oiled aquatic birds results both from chilling as a consequence of oil-matted feathers, and from toxic effects of oil ingested while the birds attempt to preen their oiled plumage. A duck that swallows oil may develop gastrointestinal irritation, lipid pneumonia, fatty degeneration of the liver, adrenal and salt gland hypertrophy, and other disorders. Swallowed oil may also inhibit egg-laying (Hartung and Hunt, 1966). Thus far, efforts to rehabilitate oiled sea birds have been prohibitively expensive and almost completely ineffectual (Clark and Kennedy, 1968).

HUMAN SATELLITES AS PREDATORS AND COMPETITORS

Man's faulty stewardship of the natural environment threatens birds and other wildlife in still another way. Wherever man goes he takes with him animal satellites whose impact on native birds is almost universally harmful. These animals include cats, rats, rabbits, dogs, pigs, goats, sheep, and also other birds. The small island of Herokapare off New Zealand contained 400,000 birds before the advent of cats around 1931. Twelve years later only a few thousand birds were left, and six species had disappeared entirely (Richdale, 1943). In 1970 on Christmas Island in the equatorial Pacific, one colony of Sooty Terns, *Sterna fus-*

FIGURE 24–6   A Murre (Guillemot), *Uria aalge*, so badly oiled that its death is almost inevitable. Photo by E. Hosking.

cata, laid an estimated 600,000 eggs between May and August. Native Gilbertese collected and ate about 250,000 of the eggs. Great Frigate Birds, *Fregata minor*, preyed on the young chicks, and domestic cats killed and ate between 50 and 75 adult terns each night (Schreiber and Ashmole, 1970). Rats are even worse predators on birds than cats. On Lord Howe Island, Australia, eight species of birds have become extinct since its discovery, five of them because of rats deposited there in a shipwreck in 1918 (Hindwood, 1940). On Green Island of the Kure Atoll, rats destroyed over 50 per cent of the nests of Laysan Albatrosses, *Diomedea immutabilis*, both by eating the eggs and young and by attacking the adults. At night the rats attacked incubating birds, gnawing off feathers and exposing skin and flesh on their backs—sometimes as many as 20 rats on one bird. They would create holes up to 15 cm in diameter, exposing the bird's lungs. Generally, the birds died the next day (Kepler, 1967). Rabbits introduced on Laysan Island so denuded the vegetation that four of the five endemic bird species were extinguished. On Inaccessible Island in the south

Atlantic, introduced pigs have completely wiped out a penguin colony. According to Mayr (1947) "more kinds of birds have become extinct on the islands of the Pacific than in all the rest of the world put together."

EXTINCT AND ENDANGERED SPECIES OF BIRDS

In the light of man's predation on birds and his mistreatment of the natural environment, it is not surprising that many species of birds have become extinct within historic times. Extinction, however, is not only a recent phenomenon, brought on by man's activities. It is also a natural phenomenon that had been occurring long before man appeared on the scene. In the prehistoric past, extinction of species was roughly balanced by the evolution of new species. Today, unfortunately, the rate of extinction has been accelerated by man's actions while there has been no counterbalancing increase in the rate of speciation.

It is impossible to say how many species of birds have existed on earth, chiefly because the fossil record for birds is far from complete. Many species may never have been fossilized, or, if fossilized, their fossil remains have never been found.

One may assume that man's scientific interest in birds began about 1600 A.D. Bird species known to be living since then may be considered neospecies; those known to have existed only prior to 1600, paleospecies. Most paleospecies are represented only by fos-

sils. An extensive survey of fossil species was undertaken by Fisher and Peterson (1964). They found that of the 202 families of birds recognized by them, 153 (74.5 per cent) were represented, in part or in whole, by fossils. All told, they listed a total of 1382 fossil species of birds. Among the families listed are: Turkey (Meleagrididae), 8 fossil and 2 living species; Pelican (Pelecanidae), 15 fossil and 6 living species; Hawks and Eagles (Accipitridae), 97 fossil and 208 living species; Cranes (Gruidae), 31 fossil and 14 living species. But representation of families by fossils is extraordinarily uneven. In some families the fossil representatives outnumber the living species, and in other families the reverse is true. Table 24-1 gives a few representative examples that illustrate this phenomenon.

Several explanations help to account for the extreme inequality of family representation in the fossil record. The following kinds of birds are much more likely to become fossilized than others: aquatic species (in bottom sediments) rather than terrestrial species; large birds, with their stouter bones, rather than small birds with fragile bones; cave-nesting species (e.g., vultures), rather than open-country nesters; temperate and polar zone species rather than tropical species that decompose faster. Moreover, the geographic distribution of paleontologists and their searchings determine the kinds of avian fossils they uncover.

According to Fisher and Peterson, 85 species of birds, representing 27 families, have become extinct

TABLE 24-1.
FOSSIL REPRESENTATION IN SELECTED FAMILIES OF BIRDS*

FAMILY	NUMBER OF FOSSIL SPECIES	NUMBER OF LIVING SPECIES
Moas (Dinornithidae)	27	0
Ostriches (Struthionidae)	7	1
Loons (Gaviidae)	12	4
Penguins (Sphenicidae)	36	15
Cormorants (Phalacrocoracidae)	31	26
Grouse (Tetraonidae)	18	17
Rails (Rallidae)	67	119
Parrots (Psittacidae)	19	317
Hummingbirds (Trochilidae)	1	320

*After Fisher and Peterson, 1964.

since 1600—that is, roughly, the period since human pressures began to build up against bird survival. According to Greenway (1958) there are 87 such species, plus an additional 19 forms that are probably but not surely extinct. In contrast, only 36 species of mammals have become extinct in the same period. Among the recently extinct bird species of North America are the Passenger Pigeon, *Ectopistes migratorius*, the Heath Hen, *Tympanuchus cupido*, (a race of the Greater Prairie Chicken), and, possibly, the Ivory-billed Woodpecker, *Campephilus principalis*. On islands of the North Atlantic the Great Auk, *Pinguinus impennis*, became extinct in 1844, the victim of meat and feather hunters.

Careful study of fossils collected in the Hawaiian Islands in the 1970s revealed some 39 extinct avian species (Olson and James, 1982). Notable among these were seven species of geese (several flightless), two flightless ibises, and seven flightless rails. Their demise is attributed to hunting and extensive clearing of lowlands for agriculture by the early Polynesians. It is estimated that only 25 per cent of the original avian taxa are still living (Scott and Kepler, 1985). In the Hawaiian Islands, 28 endemic species are now in danger of extinction (Wallace, 1971). Additional causes of this disastrous rate of extinction are the introduction of exotic fauna, rats, cats, pigs, cattle, goats, and sheep, plus imported game animals such as pronghorn antelopes, deer, mouflon sheep, and numerous species of game birds. Berger (1970, 1972) notes that 76 species of game birds and about 60 of nongame birds have been introduced into Hawaii. These immigrants not only compete with indigenous birds for food and habitats but also introduce foreign parasites and diseases to which the local birds have little or no resistance (p. 399).

What of the future? Species now in danger of extinction should claim the immediate and most urgent attention of concerned naturalists. Extinction of any species is terribly, irretrievably final. Bird conservationists are in general agreement that the following North American birds are in danger of impending extinction:

1. California Condor, *Gymnogyps californianus*. Only 19 to 22 individuals were censused in the wild in 1983 (Snyder and Johnson, 1985). All remaining wild individuals were subsequently taken into captivity for captive breeding. As of 1987, there are 27 captives (Kiff, personal communication).

2. Everglade Kite, *Rostrhamus sociabilis*. In 1963 its total population was estimated at about 58 birds, but in 1984 the number had risen to 668. Periodic droughts severely reduce the population.

3. Attwater's Greater Prairie Chicken, *Tympanuchus cupido attwateri*. In 1966 there were an estimated 750 birds; a survey in Texas in 1986 produced 880 birds.

4. Whooping Crane, *Grus americana*. In 1987, thanks to conservation efforts, about 128 to 132 were censused in the wild, and 46 are in captive breeding programs (Gable, personal communication).

5. Eskimo Curlew, *Numenius borealis*. An extremely rare species. In 1965 there were five records of birds having been sighted, and in 1970, two records.

6. Ivory-billed Woodpecker, *Campephilus principalis*. There are no living birds known for certain, in the United States, but the Cuban subspecies was sighted in 1986 (Horn and Short, personal communication).

7. Kirtland's Warbler, *Dendroica kirtlandii*. About 1000 birds survived in 1961, but only a mean of 207 in annual counts between 1971 and 1981 (Kelly, 1982).

8. Bachman's Warbler, *Vermivora bachmanii*. Last seen (one male) in 1975.

Species on the brink of extinction are not unique to North America. Alert to this problem of imminent extinction, conservationists around the world have banded together to study the problems of endangered species of animals and to organize action to save them from disappearance. The most pressing problem is to identify those species threatened with immediate extinction. The International Union for Conservation of Nature and Natural Resources has organized committees for this purpose, and they publish a continuously up-dated *Red Data Book* (Vincent, 1970; King, 1981). This book names endangered species, gives their geographical location, their present chances of survival, and other information relevant to their conservation. Volume Two of the *Red Data Book* is devoted to birds. It classifies threatened species under three degrees of urgency:

1. Endangered species—those in immediate danger of extinction whose survival is unlikely without prompt protective measures.
2. Rare species—those not in immediate danger but

present in such small numbers or in such restricted or specialized habitats that they could quickly disappear.

3. Depleted species—those whose numbers are still adequate for survival but whose depleted or declining populations are a cause for serious concern.

The U.S. Fish and Wildlife Service listed 218 endangered North American species in 1987. Predictably, the majority of endangered species live on oceanic islands or within tightly circumscribed habitats, e.g., the Atitlán Grebe, *Podilymbus gigas*, found only on Lake Atitlán in Guatemala.

THE PRESERVATION AND RESTORATION OF ENDANGERED SPECIES

The momentum inherent in human affairs makes it inevitable that additional species of birds will disappear before the current discouraging trend toward extinction is halted. Nonetheless, there is cause for cautious optimism in several contemporary trends toward the preservation of threatened species.

One of the simplest and most obvious ways to preserve a species is to protect it in its native habitat. In 1960 there were only about 200 Atitlán Grebes, *Podilymbus gigas*, in Lake Atitlán, Guatemala. Then, because of the introduction of large-mouthed bass, the Grebe population dwindled to 80 in 1965. Through ecological research, the establishment of a sheltered refuge, political lobbying, and educational programs among the local Indians, the trend toward extinction was reversed, and in 1969 the Grebe population had been restored to about 130 individuals (LaBastille, 1972). In 1935 there were surviving in North America only 35 known Trumpeter Swans, *Cygnus buccinator*. They were given increased protection, especially in suitable refuges, and by 1969 their numbers had increased to over 5000. Through protection, the remnants of a once heavily exploited population of King Penguins, *Aptenodytes patagonicus*, grew from 3400 birds in 1930 to 218,000 birds in 1980 (Rounsevell and Copson, 1982).

Sometimes obviously simple changes in a threatened species' situation will save it from disaster. In 1903 rabbits were introduced on Laysan Island, Hawaii. They overran the island and nearly denuded it of vegetation. The population of the endemic Laysan Duck, *Anas laysanensis*, dropped to seven individu-

als by 1912. Around 1924 the rabbits were eliminated and by 1961 the duck population had increased to 688 (Warner, 1963). The natural population now fluctuates between 400 and 600 (Ripley, 1985).

An ingenious stratagem is helping to stave off extinction of the imperiled Imperial Eagle, *Aquila heliaca*, in Spain (Fig. 24–7). Infant mortality is being reduced by transferring the third hatchling (which ordinarily dies) from its home nest to another active eagle nest that contains only one chick or infertile eggs. Fledging success at nine nests has been thus increased 43 per cent (Meyburg and Heydt, 1973). In a variation of this theme, eggs of the scarce Peregrine Falcon, *Falco peregrinus*, have been transferred into the nests of the more numerous and more prolific Prairie Falcon, *F. mexicanus*. Since the female Peregrine is an indeterminate layer, she will promptly lay three or four more eggs to replace those removed. This use of foster parents has made possible an impressive recovery of Peregrine numbers in regions where both species exist and where pesticide residues are sufficiently low in the environment (Cade and Hardaswick, 1985).

Between 1975 and 1982, 127 eggs of Whooping Cranes, *Grus americana*, were transferred from breeding grounds in northern Alberta to nests in a thriving colony of Sandhill Cranes, *G. canadensis*, in Idaho. Twenty-three young were seen on their wintering grounds (Drewien and Kuyt, 1979; Johnsgard, 1983). It remains to be seen if these cross-fostered young will reproduce successfully on reaching breeding age.

A somewhat similar stop-gap solution to the declining population of the Kirtland's Warbler, *Dendroica kirtlandii*, involved another species. In recent years between 60 and 70 per cent of the Warbler nests have been parasitized by the Brown-headed Cowbird, *Molothrus ater*. Probably as the result of trapping and removing 2400 Cowbirds in 1972 and 3000 in 1973, the Warbler population increased in those years from about 400 to about 432 (Mayfield, 1973). Since Kirtland's Warbler nests in young Jack Pine successions, the controlled burning of habitat is another technique designed to encourage the bird's survival. The Warbler population has not increased substantially despite all the measures taken so that more research is needed to identify factors limiting their population levels (Kelly and DeCapita, 1982). Chapter 19 on avian ecology has considered a few of the many environmental influences that affect birds' survival.

FIGURE 24–7 The Spanish Imperial Eagle, an endangered species of which about 50 pairs are known to exist today. Photo by E. Hosking.

Often, when a species' natural habitat has been largely preempted by humans for their own purposes, certain suitable portions may be set aside as bird sanctuaries or refuges to ensure the species' survival. This is in the main a recent development in human history, but it is heartening to know that as early as 677 A.D., the simple ascetic monk, St. Cuthbert, established what was perhaps the world's first bird sanctuary on Lindisfarne Island, Northumberland (Fisher and Peterson, 1964). As of 1987 the United States National Wildlife Refuge System had established 450 refuges, covering over 36.4 million hectares, where all forms of wildlife may be protected.

There are occasions when refuges are, at least at the time, impractical solutions to the problems of bird survival. In such instances the breeding of captive birds may be undertaken. The population of Peregrine Falcons in the United States declined drastically as a result of eggshell thinning induced by the ubiquitous, persistent DDT in the environment, especially in the eastern and far western states. Now that the use of DDT is banned in the United States, its gradual diminution in the environment makes possible the recovery of Peregrine populations. In an effort to restore the species to its former habitats, Tom Cade and his colleagues at Cornell University have established an impressive breeding program involving various combinations of natural and artificial breeding techniques—imprinting, artificial insemination and incubation, foster parents, "hacking," nest platforms, and the like. The fledged young are then released to live independent natural lives. A few of these Peregrines have established eyries on tall office buildings in New York, Washington, Norfolk, and Baltimore, where they live on feral pigeons, House Sparrows, and Starlings. In 1980 two pairs of wild Peregrines (originally from Cornell) fledged their own young for the first time since the 1950's east of the Mississippi River (Cade and Dague, 1980). Encouraging as these results are, Peregrines, as widely peregrinating birds, still suffer heavy reproductive failures, even in arctic regions, because of the use of chlorinated hydrocarbons in tropical American countries.

In 1951, the population of the Hawaiian Mountain Goose or Nene, *Branta sandvicensis*, was reduced to about 33 individuals. That year, the British Wildfowl Trust at Slimbridge, Gloucestershire, acquired a male and two females. Descendants of these three birds now number over 2500 living in Hawaii and various other parts of the world (Kear, 1985). The species now seems safe from extinction.

As in the case of refuges, captive breeding also has its ancient antecedents. In 16th century Mexico, Montezuma maintained a huge royal aviary. As described by one of Cortés' companions, the aviary contained a great variety of birds, "from the Royal Eagle . . . down to tiny birds of many-colored plumage . . . from which they take rich plumage which they use in their green feather work. All the birds . . . breed in these houses . . . and two hundred men and women to attend them . . ." (von Hagen, 1961).

Most encouraging is the fact that most zoos throughout the world carry on breeding programs. According to the 1979 *International Zoo Yearbook* census, at least 801 bird species have been bred in zoos (Conway, 1983). Of the species of birds listed in the *Red Data Book*, 62 were reported to be in captivity between 1964 and 1970, and about one-half of these were successfully bred (Bridgewater, 1972).

Legislation is another all-important tool in man's efforts to protect wildlife. It may be as broad as the Migratory Bird Treaty Acts between the United States, Mexico, and Canada, or as narrowly defined as laws banning lead ammunition, or limiting hunting seasons, or prohibiting sonic booms over refuges. The effectiveness of laws in protecting wildlife is illustrated by the recovery of species recently decimated by insecticide poisoning. Numerous governments now ban the use of DDT and similar wide-spectrum pes-

ticides. During the 1960s when DDT was in common use, a chemical firm in Los Angeles dumped an estimated 450 kg (1000 lb) of waste DDT into the ocean daily (McNulty, 1971). As a consequence, Brown Pelicans, *Pelecanus occidentalis*, and other sea birds breeding on coastal islands suffered devastating reproductive failures. On the nearby Channel Islands, Pelican eggshells were 50 per cent thinner than in years before the pollution, and they of course became crushed during incubation. In 1969, from 1272 breeding attempts, the Pelicans reared a maximum of four young, and in 1970, only one young was raised in the entire colony. In 1970 the chemical firm ceased dumping DDT into the sea. Since then, breeding success in the colony has increased strikingly and populations continue to grow (Anderson and Gress, 1983). A similar success story has been reported for Bald Eagles, *Haliaetus leucocephalus*, in northwestern Ontario (Grier, 1982). Golden Eagles, *Aquila chrysaetos*, Peregrine Falcons, and other raptors have appeared in the British Isles following the cessation of use of DDT, dieldrin, and similar products.

Even today laws are still needed to protect rare species from "plume hunters." In Oklahoma, 30 persons were arrested in 1974 for illegally killing thousands of birds whose feathers were used in making "antique Indian artifacts." One Indian war bonnet required the feathers of as many as 10 eagles; a single eagle carcass was reported to sell for $125 on the black market (Sayre, 1974).

Research is yet another approach to solving the problems of vanishing birds, particularly research into the agents and forces that seem most influential in forcing species into extinction. In this area, high priority should be given to investigation of alternatives to broad-spectrum pesticides. Several biologically acceptable substitutes for harmful chemicals were mentioned earlier, and a number of them have already been successfully applied. A new forest spray, Zectran, to control spruce budworms has been used in the field without apparent harm to forest bird populations (Pillmore *et al.*, 1971). As a natural biological control, hills of ants have been transplanted into Russian forests where the ants successfully checked the numbers of insect pests (Zavednyuk, 1968). Undoubtedly, scores of biological controls of insect pests already exist. They need only to be discovered and applied.

Finally, and most importantly, the safe future of birds on earth depends on human attitudes. It is here especially that there is considerable cause for optimism. In industrial countries where pollution and other pressures on bird life are extensive, educational standards are fortunately high. When vast expanses of black asphalt, seen through brown smog, separate man from the green earth and blue skies of his earlier days, a self-correcting feedback generally comes into play. The less man is able to see of unspoiled nature the more likely he is to appreciate it and to yearn for a cabin in the wilderness. This, in part, accounts for the recent blossoming of interest in all sorts of natural history activities and organizations: nature study camps, nature trails, ecology clubs, nature photography, natural history lecture series, wildlife television programs, nature tours, and so on. Some of this interest has crystallized in national and international organizations devoted to the protection of birds and other wildlife—for example, in the United States, the National Audubon Society, the Nature Conservancy, the Sierra Club, and the Wilderness Society; in Britain, the Wildfowl Trust; and internationally, the International Union for Conservation of Nature and Natural Resources, and the World Wildlife Fund. These organizations are supplemented by thousands of other smaller associations more concerned with local bird life. Members of all of these groups combine an appreciation of birds with an active concern for their preservation. They have won increasing popular support for saving and restoring wildlife. There is much hope for the survival of many endangered birds and the continued human enjoyment of bird life for years to come.

■ SUGGESTED READINGS

Interesting and well-illustrated books on numerous relationships between birds and man are found in two government publications: Stefferud and Nelson, *Birds in Our Lives*, and Linduska and Nelson, *Waterfowl Tomorrow*. A scholarly and fascinating treatment of bird folklore and mythology is found in Armstrong's *The Folklore of Birds*. The problems that birds create for man are well considered in Murton's *Man and Birds*, and also in Murton and Wright's *The Problem of Birds as Pests*. Fisher and Peterson's *The World of Birds* presents a detailed catalog of fossil birds of the world and a list of endangered species. For a more extensive treatment of endangered species see the books by Greenway, *Extinct and Vanishing Birds of the World*; Vincent and King, *Red Data Book*; and Fisher *et al.*, *The Red Book: Wildlife in Danger*. Informative progress reports on efforts to save endangered species in various parts of the world may be found in Stanley A. Temple (ed), *Bird Conservation*, Vols. I and II. The role zoos play in preserving and breeding endangered birds is described in the various volumes of the *International Zoo Yearbook*. For continuing articles devoted to the vigorous promotion of bird and wildlife conservation, read *Audubon* magazine.

REFERENCES

Abbott, I., L. K. Abbott, and P. R. Grant. 1977. Comparative ecology of Galapagos ground finches (*Geospiza* Gould): Evaluation of the importance of floristic diversity and interspecific competition. *Ecological Monograph*, 47:151–184.

Abel, O. 1911. *Die Vorfahren der Vögel und ihre Lebensweise*. Vienna.

Able, K. P. 1982. Skylight polarization patterns at dusk influence migratory orientation in birds. *Nature*, 299:550–551.

Abramson. M. 1979. Vigilance as a factor influencing flock formation among Curlews, *Numenius arquata*. *Ibis*, 121:213–216.

Abs, M. 1980. Zur Bioakustik des Stimmbruchs bei Vögeln. *Zoologische Jahrbücher, Abteilurg für allgemeine Zoologie und Physiologie dertiere*, 84:289–382.

Ackermann, A. 1967. Quantitative Untersuchungenan körnerfressenden Singvögeln. *Journal für Ornithologie*, 108:430–473.

Agron, S. L. 1962. Evolution of bird navigation and the earth's axial precession. *Evolution*, 16:524–527.

Ainley, G., and D. W. Anderson. 1981. Feeding ecology of marine cormorants in southwestern North America. *Condor*, 83:120–131.

Ainley, D. G., and R. P. Schlatter. 1972. Chick-raising ability in Adelie Penguins. *Auk*, 89:559–566.

Akester, A. A., D. E. Pomeroy, and M. D. Purton. 1973. Subcutaneous air pouches in the Marabou Stork (*Leptoptilos crumeniferus*). *Proceedings of the Zoological Society of London*, 170:493–499.

Alatalo, R. V. 1981. Interspecific competition in Tits *Parus* spp. and the Goldcrest *Regulus regulus*: Foraging shifts in multispecific flocks. *Oikos*, 37:335–344.

Alatalo, R. V., A. Carlson, A. Lundberg, and S. Ulfstrand. 1981. The conflict between male polygamy and female monogamy: the case of the pied flycatcher *Ficedula hypoleuca*. *American Naturalist*, 117:738–753.

Alatalo, R. V., A. Lundberg, and K. Stilbrandt. 1982. Why do Pied Flycatcher females mate with already-mated males? *Animal Behavior*, 30:585–593.

Albert, E. N., M. C. Seremet, and R. E. Daniel. 1978. Thrombocyte (platelet) adherence and endothelial alteration in young pigeons. *Scanning Electron Microscopy*, 1978:855–860.

Alcock, J. 1969. Observational learning by Fork-tailed Flycatchers (*Muscivora tyrannus*). *Animal Behavior*, 17:652–657.

Alcock, J. 1979. The origins of tool-using by Egyptian Vultures, *Neophron percnopterus*. *Ibis*, 112:542.

Alcock, J. 1975. *Animal Behavior: An Evolutionary Approach*. Sinauer Associates, Inc., Sunderland, Massachusetts.

Aldrich, J. W., and J. S. Weske. 1978. Origin and evolution of the eastern house finch population. *Auk*, 95:528–536.

Aldrich, T. W., and W. G. Rawling, 1983. Effects of experience and body weight on incubation behavior of Canada Geese. *Auk*, 100:670–679.

Alerstam, T. 1975. Crane, *Grus grus*, migration over sea and land. *Ibis*. 117:489–495.

Alerstam, T., and P. H. Enckell. 1979. Unpredictable habitats and evolution of bird migration. *Oikos*, 33:228–232.

Alerstam, T., and G. Hogstedt. 1983. The role of geomagnetic field in the development of birds' compass sense. *Nature*, 306:463–465.

Alerstam, T., and S. Ulfstrand. 1972. Radar and field observations of diurnal migration in south Sweden, autumn, 1971. *Ornis Scandinavica*, 3:99–139.

Alexander, W. B. 1948. The index of Heron population, 1947. *British Birds*, 41:146–148.

Alexander, W. B. 1954. *Birds of the Ocean*. G. P. Putnam's Sons, New York.

Alexander, W. B., and R. S. R. Fitter. 1955. American land birds in western Europe. *British Birds*, 48:1–14.

Ali, S., and S. D. Ripley. 1968. *Handbook of the Birds of India and Pakistan, Together with Those of Nepal, Sikkim, Bhutan and Ceylon*. Vol. 1. Oxford University Press, London.

Allard, H. A. 1930. The first morning song of some birds of Washington, D.C.: its relation to light. *American Naturalist*, 64:436–439.

Allee, W. C. 1936. Analytical studies of group behavior in birds. *Wilson Bulletin*, 48:145–151.

Allee, W. C., and N. Collias. 1938. Effect of injections of testosterone propionate on small flocks of hens. *Anatomical Record*, 72(4) plus *Supplement*: 60.

Allee, W. C., A. E. Emerson, O. Park, T. Park, and K. Schmidt. 1949. *Principles of Animal Ecology*. W. B. Saunders Company, Philadelphia.

Allee, W. C., and R. H. Masure. 1936. A comparison of maze behavior in paired and isolated Shell Parakeets. *Journal of Comparative Psychology*, 22:131–155.

Allen, A. A. 1914. The Red-winged Blackbird: a study in the ecology of a cat-tail marsh. *Proceedings of the Linnaean Society of New York*, 24–25:43–128.

Allen, D. L. 1954. *Our Wildlife Legacy*. Funk and Wagnalls Company, New York.

Allen, G. M. 1925. *Birds and Their Attributes*. Marshall Jones Company, Boston.

Altevogt, R., and T. A. Davis. 1980. Urbanization in nest building of Indian House Crows (*Corvus splendens* Vieillot) *Journal of the Bombay Natural History Society*, 76:283–290.

Alvarez, H. 1975. The social system of the Green Jay in Colombia. *Living Bird*, 14:5–44.

Amadon, D. 1950. The Hawaiian Honeycreepers (Aves, Drepanididae). *Bulletin of the American Museum of Natural History*, 95.

Amadon, D. 1967. Galapagos finches grooming marine iguanas. *Condor*, 69:311.

American Ornithologists' Union Committee. 1957. *Check-list of North American Birds*. Fifth edition. American Ornithologists' Union.

Anderson, D. W., and F. Gress. 1983. Status of a northern population of California Brown Pelicans. *Condor*, 85:79–88.

Anderson, G. L., and E. J. Braun. 1985. Postrenal modification of urine in birds. *American Journal of Physiology*, 17:93–98.

Anderson, S. H. 1970. Water balance of the Oregon Junco. *Auk*, 87:161–163.

Anderson, W. L. 1975. Lead poisoning in waterfowl at Rice Lake, Illinois. *Journal of Wildlife Management*, 39:264–270.

Andersson, M. 1982. Female choice selects for extreme tail length in a widowbird. *Nature*, 299:818–820.

Andreev, A. V. 1977. [Temperature conditions in snow burrows of the Hazel Hen, *Tetrastes bonasia kolymenis* But.] *Ekologiya*, 5:93–95.

Andrew, R. J. 1961. The displays given by passerines in courtship and reproductive fighting: a review. *Ibis*, 103a:315–348; 549–579.

Andrew, R. J. 1962. Evolution of intelligence and vocal mimicking. *Science*, 137:585–589.

Anonymous. 1957. Mourning Dove investigations 1948–1956. *Technical Bulletin No. 1, Southeastern Association of Game and Fish Commissioners*. Columbia, South Carolina.

Anonymous, 1978. Microbial magnets. Science and the Citizen. *Scientific American*, 238(3):72–74.

Anonymous, 1979. Eighty years ago in Sports Afield. *Sports Afield Outdoor Almanac*.

Ar, A., and Y. Yom-Tov. 1978. The evolution of parental care in birds. *Evolution*, 32:655–669.

Ar, A., *et al*. 1974. The avian egg: water vapor conductance, shell thickness, and functional pore area. *Condor*, 76:153–158.

Arheimer, O. 1978. Laying, incubation, and hatching in the Redwing, *Turdus iliacus*, in subalpine birch forest at Ammaranäs, Swedish Lapland. *Vår Fågelvärld*, 37:297–312.

Armstrong, E. A. 1942. *Bird Display*. Cambridge University Press, Cambridge.

Armstrong, E. A. 1949. *Bird Life*. Lindsay Drummond, London.

Armstrong, E. A. 1950. The nature and function of displacement activities. *Physiological Mechanisms in Animal Behavior. Symposia of the Society for Experimental Biology*, Vol. 4. Academic Press, New York.

Armstrong, E. A. 1953. The history, behavior and breeding biology of the St. Kilda Wren, *Auk*, 70:127–150.

Armstrong, E. A. 1963. *A Study of Bird Song*. Oxford University Press, London.

Armstrong, E. A. 1965. *Bird Display and Behaviour*. Dover Publications, New York.

Aschoff, J. 1967. Circadian rhythms in birds. *Proceedings of the XIVth International Ornithological Congress*, 81–105. Blackwell Scientific Publications, Oxford.

Ash, J. S. 1964. Observations in Hampshire and Dorset during the 1963 cold spell. *British Birds*, 57:221–241.

Ash, J. S. 1969. Spring weights of trans-Saharan migrants in Morocco. *Ibis*, 111:1–10.

Ashmole, N. P. 1971. Sea bird ecology and the marine environment. *In* Farner, D. S., and J. R. King (eds.). *Avian Biology*, Vol. 1. Academic Press, New York.

Ashmole, N. P., and T. S. Humberto. 1968. Prolonged parental care in Royal Terns and other birds. *Auk*, 85:90–100.

Audubon, 1973. Oil sump cleanup stuck in tar while thousands of animals die. *Audubon*, 75(3):114–115.

Aulie, A. 1971. Coordination between the activity of the heart and the flight muscles during flight in small birds. *Comparative Biochemistry and Physiology*, 38A:91–97.

Austin, G. T. 1970. Experimental hypothermia in a Lesser Nighthawk. *Auk*, 87:372–374.

Austin, O. L. 1940. Some aspects of individual distribution in the Cape Cod tern colonies. *Bird-Banding*, 11:155–169.

Austin, O. L. 1946. The status of the Cape Cod terns in 1944: a behaviour study. *Bird-Banding*, 17:10–27.

Austin, O. L. 1947. A Study of the mating of the Common Tern *(Sterna h. hirundo)*. *Bird-Banding*, 18:1–16.

Austin, O. L. 1951. The mourning dove on Cape Cod. *Bird-Banding*, 22:149–174.

Austin, O. L., and O. L. Austin, Jr. 1956. Some demographic aspects of the Cape Cod population of Common Terns *(Sterna hirundo)*. *Bird-Banding*. 27:55–66.

Austin, O. L., Jr., and A. Singer. 1961. *Birds of the World*. Golden Press, New York.

Austin, O. L., Jr., and A. Singer. 1971. *Families of Birds*. Golden Press, New York.

Austin, O. L., *et al.* 1972. Mass hatching failure in Dry Tortugas Sooty Terns, *Proceedings of the XVth International Ornithological Congress*, 627. E. J. Brill, Leiden.

Avery, M. L. 1980. Diet and breeding seasonality among a population of sharp-tailed munias, *Lonchura striata*, in Malaysia. *Auk*, 97:160–166.

Avery, M. L. 1983. Social transmission of an acquired dietary aversion among captive House Finches. Abstract No. 353 of *Presented Posters and Papers, 101st Meeting*. American Ornithologists' Union.

Avery, M. L., and J. R. Krebs. 1984. Temperature and foraging success of Great Tits, *Parus major*, hunting for spiders. *Ibis*, 126:33–38.

Aymar, G. C. 1935. *Bird Flight*. Dodd, Mead and Company, New York.

Babaev, K. 1967. [Dancing Wheatear abundance and population density of *Rhombomys opimus* in Turkmeniya.] *Ornitologiya*, 8:333. From *Bird-Banding*, 39:142.

Baerends, G. P., and R. H. Drent (eds.). 1970. The Herring Gull and its egg. *Behaviour, Suppl*, 17:1–312.

Baeyens, G. 1981. Magpie breeding success and carrion crow interference. *Ardea*, 69:125–139.

Bagg, A. M. 1967. Factors affecting the occurrence of the Eurasian Lapwing in eastern North America. *The Living Bird*, 6:87–121.

Bailey, E. D., and J. A. Baker. 1982. Recognition characteristics in covey dialects of Bobwhite Quail. *Condor*, 84:317–320.

Bailey, E. P., and G. H. Davenport. 1972. Die-off of Common Murres on the Alaska peninsula and Unimak Island. *Condor*, 74:215–219.

Bailey, R. E. 1952. The incubation patch of passerine birds. *Condor*, 54:121–136.

Bailey, R. 1966. The sea-birds of the southeast coast of Arabia. *Ibis*, 108:224–264.

Baird, J., and I. C. T. Nisbet. 1960. Northward fall migration on the Atlantic coast, and its relation to offshore drift. *Auk*, 77:119–149.

Baker, E. C. S. 1923. Cuckoos' eggs and evolution. *Proceedings of the Zoological Society of London*, 277–294.

Baker, E. C. S. 1942. *Cuckoo Problems*. H. F. & G. Witherby, Ltd, London.

Baker, J. R. 1938. The relation between latitude and breeding seasons in birds. *Proceedings of the Zoological Society of London*, Series A, 108:557–582.

Baker, M. C. 1982. Vocal dialect recognition and population genetic consequences. *American Zoologist*, 22:561–569.

Baker, R. R. 1978. *The Evolutionary Ecology of Animal Migration*. Holmes and Meier, Publishers, New York.

Balda, R. P. 1980. Recovery of cached seeds by a captive *Nucifraga caryocatactes*. *Zeitschrift für Tierpsychologie*, 52:331–346.

Balda, R. P., and G. C. Bateman. 1972. The breeding biologyy of the Piñon Jay. *Living Bird*, 11:5–42.

Baldwin, P. H., and D. E. Oehlerts. 1964. *Studies in Biological Literature and Communications*. No. 4 Biological Abstracts, Philadelphia.

Baldwin, P. S., H. C. Oberholser, and L. G. Worley. 1931. Measurements of birds. *Scientific Publications of the Cleveland Museum of Natural History*, 2:1–165.

Balen, J. H. van, and A. J. Cave. 1972. Survival and weight loss of nestling Great Tits *(Parus major)* in relation to brood size and air temperature. *Proceedings of the XVth International Ornithological Congress*, 628. E. J. Brill, Leiden.

Balfour, E., and C. J. Cadbury. 1979. Polygyny, spacing and sex ratio among Hen Harriers, *Circus cyaneus*, in Orkney, Scotland. *Ornis Scandinavica*, 10:133–141.

A.4 References

Ballman, E. 1976. The contribution of fossil birds to avian classification. *In* Frith, H. J., and J. H. Calaby (eds). *Proceedings of the 16th International Ornithological Congress.* Australian Academy of Science, Canberra.

Balph, M. H. 1977. Winter social behavior of dark-eyed juncos: communication, social organization, and ecological implications. *Animal Behaviour, 25*:859–884.

Balthazart, J. 1976. Daily variations of behavioural activities and of plasma testosterone levels in the domestic duck *Anas platyrhynchos. Journal of Zoology* (London), *180*:155–173.

Bambridge, R. 1962. Early experience and sexual behaviour in the domestic chicken. *Science, 136*:259–260.

Bancroft, W. D., *et al.* 1923. Blue feathers. *Auk, 40*:275–300.

Bang, B. G. 1960. Anatomical evidence for olfactory function in some species of birds. *Nature, 188*:547–549.

Bang, B. G., and S. Cobb. 1968. The size of the olfactory bulb in 108 species of birds. *Auk, 85*:55–61.

Bang, B. G., and B. M. Wenzel. 1985. Nasal cavity and olfactory system. *In* King, A. S., and J. McLelland (eds). *Form and Function in Birds*, 3:195–225. Academic Press, New York.

Bangert, H. 1960. Untersuchungen zur Koordination der Kopf-und Beinbewegungen beim Haushuln. *Zeitschrift für Tierpsychologie, 17*:143–164.

Banks, H. C. 1985. American Black Duck record from Korea. *Journal of Field Ornithology, 56*:277.

Baptista, L. F. 1973. On courtship displays and the taxonomic positions of the Grey-headed Silverbill. *Avicultural Magazine, 79*:148–154.

Baptista, L. F. 1974. The effects of songs of wintering White-crowned Sparrows on song development in sedentary populations of the species. *Zeitschrift für Tierpsychologie, 34*:147–171.

Baptista, L. F. 1975. Song dialects and demes in sedentary populations of the White-crowned Sparrow. *University of California Publications in Zoology, 105*:1–53.

Baptista, L. F. 1976. Handedness, holding and its possible taxonomic significance in Grassquits, *Tiaris* spp. *Ibis, 118*:218–222.

Baptista, L. F. 1977. Geographic variation in song and dialects of the Puget Sound White-crowned Sparrow. *Condor, 79*:356–370.

Baptista, L. F. 1978. Territorial, courtship and duet songs of the Cuban Grassquit *(Tiaris canora). Journal für Ornithologie, 119*:91–101.

Baptista, L. F. 1981. Behavior genetics studies with birds. *In* Risser, A. C., Jr., L. F. Baptista, S. R. Wylie, and N. B. Gale (eds). *Proceedings 1st International Birds in Captivity Symposium*, 217–249. Seattle, Washington (1978). International Foundation for Conservation of Birds, Hollywood, Calif.

Baptista, L. F. 1983. Commentary: Bird song learning: theme and variations. *In* Brush, A. H., and G. A. Clark, Jr. (eds). *Perspectives in Ornithology*, 499–506. Cambridge University Press.

Baptista, L. F. 1985. The functional significance of song sharing in the White-crowned Sparrow. *Canadian Journal of Zoology, 63*:1741–1752.

Baptista, L. F., and M. Abs. 1983. Vocalizations. *In* Abs M. (ed.), *Physiology and Behavior of the Pigeon*, 309–325. Academic Press, New York and London.

Baptista, L. F., W. I. Boarman, and P. Kandianidis. 1983. Behavior and taxonomic status of Grayson's Dove. *Auk, 100*:907–919.

Baptista, L. F., and R. B. Johnson. 1982. Song variation in insular and mainland California Brown Creepers. *Journal für Ornithologie, 123*:131–144.

Baptista, L. F., and J. R. King. 1980. Geographical variation in song and song dialects of montane White-crowned Sparrows. *Condor, 82*:267–284.

Baptista, L. F., and M. L. Morton. 1982. Song dialects and mate selection in montane White-crowned Sparrows. *Auk, 99*:537–547.

Baptista, L. F., and L. Petrinovich. 1984. Social interaction, sensitive phases and the song template hypothesis in the White-crowned Sparrow. *Animal Behaviour, 32*:172–181.

Baptista, L. F., and L. Petrinovich. 1986. Song development in the White-crowned Sparrow: social factors and sex differences. *Animal Behaviour, 35*:1359–1371.

Barash, D. P. 1976. Mobbing behavior by crows: The effect of the "Crow-in-distress" model. *Condor, 78*:120.

Barber, R. T., and F. P. Chavez. 1983. Biological consequences of El Niño. *Science, 222*:1203–1210.

Barden, A. A. 1941. Distribution of the families of birds. *Auk, 58*:543–557.

Barfield, R. J. 1966. Stimulation of gonadotropin secretion in the female Ring Dove by limited exposure to a mate. *American Zoologist, 6*:518.

Barfield, R. J. 1967. Activation of sexual and aggressive behavior by androgen implants in the brain of the male Ring Dove. *American Zoologist, 7*:800.

Barfield, R. J. 1971. Activity of sexual aggressive behavior by androgen implantation in the male Ring Dove brain. *Endocrinology*, 89:1470–1476.

Barlow, G. W. 1977. Modal action patterns. *In* Sebeok, T. A. (ed). *How Animals Communicate*, 98–133. Indiana University Press.

Barnas, G., and W. Rautenberg. 1985. Cardiorespiratory responses in the pigeon to thermal stimulation of the spinal cord in the opposite direction to external stimulation. *Physiological Zoology*, 58:232–235.

Barrington, D. 1773. Experiments and observations on the singing of birds. *Philosophical Transactions of the Royal Society*, 63:249–291.

Barrowclough, G. F., N. K. Johnson, and R. M. Zink.1985. On the nature of genic variation in birds. *In* R. F. Johnston (ed), *Current Ornithology*, 2:135–154.

Barruel, P. 1954. *Birds of the World*. Oxford University Press, New York.

Barry, T. W. 1962. Effect of late season on Atlantic Brant reproduction. *Journal of Wildlife Management*, 26(1):19–26.

Barry, T. W. 1964. Brant, Ross' Goose, and Emperor Goose. *In* Linduska, J. P., and A. L. Nelson (eds). *Waterfowl Tomorrow*, 145–154. U.S. Fish and Wildlife Service, Washington, D.C.

Barry, T. W. 1968. Observations on natural mortality and native use of Eider Ducks along the Beaufort Sea Coast. *Canadian Field Naturalist*, 82:140–144.

Barth, E. K. 1949. Kroppstemperatur hos fugler og pattedyr. *Fauna och Flora*, 4/5:163–177.

Bartholomew, G. A. 1949. The effect of light intensity and day length on reproduction in the English Sparrow. *Bulletin of the Museum of Comparative Zoology at Harvard College*, 101:433–476.

Bartholomew, G. A. 1972. The water economy of seed-eating birds that survive without drinking. *In* Voous, K. H. (ed). *Proceedings of the 15th International Ornithological Congress*, 237–254. E. J. Brill, Leiden.

Bartholomew, G. A., and W. R. Dawson. 1954. Body temperature and water requirements in the Mourning Dove. *Ecology*, 35:181–187.

Bartholomew, G. A., and W. R. Dawson. 1958. Body temperatures in California and Gambel's Quail. *Auk*, 75:150–156.

Bartholomew, G. A., T. R. Howell, and T. J. Cade. 1957. Torpidity in the White-throated Swift, Anna Hummingbird, and the Poorwill. *Condor*, 59:145–155.

Bartholomew, G. A., and R. E. MacMillen. 1961. Water economy of the California Quail and its use of sea water. *Auk*, 78:505–514.

Bartholomew, G. A., and T. J. Cade. 1963. The water economy of land birds. *Auk*, 80:504–539.

Bartholomew, G. A., and T. R. Howell. 1964. Egg and chick displacement experiments: Albatrosses. *Animal Behaviour*, 12:549–559.

Bartholomew, G. A., C. M. Vleck, and T. L. Bucher. 1983. Energy metabolism and nocturnal hypothermia in two tropical passerine frugivores, *Manacus vitellinus* and *Pipra mentalis*. *Physiological Zoology*, 56:370–379.

Bartholomew, G. A., F. N. White, and T. R. Howell. 1976. The thermal significance of the nest of the Sociable Weaver, *Philetairus socius*: summer observations. *Ibis*, 118:402–410.

Bartholomew, J. G., W. E. Clarke, and P. H. Grimshaw. 1911. *Atlas of Zoogeography*. John Bartholomew and Company, Edinburgh.

Bartonek, J. C., and J. J. Hickey. 1969. Selective feeding by juvenile diving ducks in summer. *Auk*, 86:443–457.

Bastock, M. 1967. *Courtship: An Ethological Study*. Heinemann Education, London.

Basteson, P. 1979. Sexual imprinting and optimal outbreeding. *Nature*, 273:659–660.

Bateson, P. 1980. Optimal outbreeding and the development of sexual preferences in Japanese Quail. *Zeitschrift für Tierpsychologie*, 53:231–244.

Bateson, P. P. G., G. Horn, and S. P. R. Rose. 1969. Effects of an imprinting procedure on regional incorporation of triated lysine into protein of chick brain. *Nature*, 223:534–535.

Batt, B. D. J., and G. W. Conwell. 1972. The effects of cold on Mallard embryos. *Journal of Wildlife Management*, 36:745–751.

Baudinette, R. V., and K. Schmidt-Nielsen. 1974. Energy cost of gliding flight in Herring Gulls. *Nature (London)*, 248:83–84.

Baudvin, H. 1979. Taille des pontes et reussite des nichées chez la Chouette Effraie, *Tyto alba* en Bourgogne. *Alauda*, 47:13–16.

Bäumer-März, C., and K. H. Schmidt. 1985 Bruterfolg und Dispersion regulieren die Bestände der Kohlmeise *(Parus major)*. *Die Vogelwarte*, 33:1–7.

Baumgartner, F. M. 1939. Studies on the distribution and habits of the Sharptail Grouse in Michigan. *Transactions of the North American Wildlife Conference*, 4:485–489.

Bayer, E. 1929. Beiträge zur Zweikomponententheorie des Hungers. *Zeitschrift für Psychologie,* 112:1–54.

Beason, R. C., and J. E. Nichols. 1984. Magnetic orientation and magnetically sensitive material in a transequatorial migratory bird. *Nature,* 309:151–153.

Bech, C. 1982. Regional blood flow changes in response to thermal stimulation of the brain and spinal cord in the Pekin duck *(Anas platyrhynchos). Journal of Comparative Physiology B. Biochemical/Systemic Environmental Physiology,* 147:71–78.

Beck, J. R., and D. W. Brown. 1972. The breeding biology of the Black-backed Storm Petrel, *Fregetta tropica. Ibis,* 113:73–90.

Becker, P. H. 1982. The coding of species-specific characteristics in bird songs. *In* Kroodsma, D. E., and E. H. Miller (eds). *Acoustic Communication in Birds,* Vol. 1. Academic Press, New York, 214–252.

Becker, P. H., G. Thielcke, and K. Wüstenburg. 1980. Der Tonhöhenverlauf ist entscheiden für das Gesangserkennen beim mitteleuropäischen Zilpzalp *(Phylloscopus collybita). Journal für Ornithologie,* 121:229–244.

Beckerton, P. R., and A. L. A. Middleton. 1982. Effects of dietary protein levels on ruffed grouse *(Bonasa umbellus)* reproduction. *Journal Wildlife Management* 46:569–579.

Beebe, C. W. 1906. *The Bird, Its Form and Function.* Henry Holt and Company, New York.

Beebe, C. W. 1908. Preliminary report on an investigation of the seasonal changes of color in birds. *American Naturalist,* 42:34–56.

Beecher, W. J. 1951. Adaptations for food-getting in the American blackbirds. *Auk,* 68:411–440.

Beehler, B. 1983. Lek behavior of the Lesser Bird of Paradise. *Auk,* 100:992–995.

Beer, C. G. 1966. Incubation and nest-building behaviour of Black-headed Gulls. *Behaviour,* 26:189–214.

Beletsky, D. L. 1983. Aggressive and Pair-bond maintenance songs of female Red-winged Blackbirds *(Agelaius phoeniceus). Zeitschrift für Tierpsychologie,* 62:47–54.

Bell, H. L. 1986. Occupation of urban habitats by birds in Papua New Guinea. *Proceedings of the Western Foundation of Vertebrate Zoology,* 3:1–48.

Bellairs, A. d'A. 1964. Skeleton. *In* Thomson, A. L. (ed.). *A New Dictionary of Birds.* McGraw-Hill Book Company, New York.

Bellrose, F. C. 1964. Spent shot and lead poisoning. *In* Linduska, J. P., and A. L. Nelson (eds). *Waterfowl Tomorrow.* U.S. Fish and Wildlife Service, Washington, D.C.

Bellrose, F. C. 1967. Orientation in waterfowl migration. *In* Storm, R. M. (ed). *Animal Orientation and Navigation.* Oregon State University Press, Corvallis, Oregon.

Bellrose, F. C. 1967. Radar in orientation research. *Proceedings of the XIVth International Ornithological Congress,* 281–309. Blackwell Scientific Publications, Oxford.

Bellrose, F. C. 1971. The distribution of nocturnal migration in the air space. *Auk,* 88:397–424.

Bellrose, F. C., and J. G. Sieh. 1960. Massed waterfowl flights in the Mississippi flyway, 1956 and 1957. *Wilson Bulletin,* 72:29–59.

Bellrose, F. C., and R. R. Graber. 1963. A radar study of the flight directions of nocturnal migrants. *Proceedings of the XIIIth International Ornithological Congress,* 362–389. American Ornithologists' Union.

Bengtsson, H., and O. Ryden. 1983. Parental feeding rate in relation to begging behavior in asynchronously hatched broods of Great Tits, *Parus major. Behavioral Ecology and Sociobiology,* 12:243–251.

Bennett, G. F., E. M. White, and N. A. Williams. 1977. Additional observations on the blood parasites of Uganda birds. *Journal of Wildlife Diseases,* 13:251–257.

Bennett, T. 1974. Peripheral and autonomic nervous systems. *In* Farner, D. S., and J. R. King (eds). *Avian Biology,* Vol. 4. Academic Press, New York.

Benoit, J. 1950. Organes uro-génitaux; reproduction-charactères sexuels et hormones determinisme du cycle sexuel seasonier. *In* Grassé, P. (ed.). *Traité de Zoologie.* Tome XV, *Oiseaux.* Masson et Cie, Paris.

Benoit, J., and L. Ott. 1944. External and internal factors in sexual activity. Effect of irradiation with different wavelengths on the mechanisms of photo-stimulation of the hypophysis and on testicular growth in the immune duck. *Yale Journal of Biology and Medicine,* 17:27–46.

Benoit, J. M. 1978. Chronobiologic study of the domestic duck: II. *Chronobiologia,* 5:158–168.

Benson, C. W. 1948. Geographic voice variation in African birds. *Ibis,* 90:48–71.

Benson, C. W. 1963. The breeding seasons of birds in the Rhodesias and Nyasaland. *Proceedings of the XIIIth International Ornithological Congress,* 2:623–639. American Ornithologists' Union.

Bent, A. C. 1923. Life histories of North American wild fowl. *U.S. National Museum Bulletin 126.* Washington, D.C.

Bent, A. C. 1932. Life histories of North American gallinaceous birds. *U.S. National Museum Bulletin 162.* Washington, D.C.

Bent, A. C. 1937. Life histories of North American birds of prey. *U.S. National Museum Bulletin 167.* Washington, D.C.

Bent, A. C. 1940. Life histories of North American cuckoos, goatsuckers, hummingbirds and their allies. *U.S. National Museum Bulletin 176.* Washington, D.C.

Bent, A. C. 1942. Life histories of North American flycatchers, swallows, larks, and their allies. *U.S. National Museum Bulletin 179.* Washington, D.C.

Bent, A. C. 1948. Life histories of North American nuthatches, wrens, thrashers, and their allies. *U.S. National Museum Bulletin 195.* Washington, D.C.

Bent, A. C. 1949. Life histories of North American thrushes, kinglets and their allies. *U.S. National Museum Bulletin 196.* Washington, D.C.

Bent, A. C. 1953. Life histories of North American wood warblers. *U.S. National Museum Bulletin 203.* Washington, D.C.

Bent, A. C. 1958. Life histories of North American blackbirds, orioles, tanagers, and their allies. *U.S. National Museum Bulletin 211.* Washington, D.C.

Bercovitz, A. B., *et al.* 1972. Light induced alterations in growth pattern of the avian eye. *Vision Research,* 12:1253–1259.

Berger, A. J. 1953. On the locomotor anatomy of the Blue Coua, *Coua caerulea. Auk,* 70:49–82.

Berger, A. J. 1970. The present status of birds in Hawaii. *Pacific Science,* 24:29–42.

Berger, A. J. 1972. Hawaiian birds. *Wilson Bulletin,* 84:212–222.

Berger, M. 1974. Energiewechsel von Kolibris beim Schwirrflug unter Höhenbedingungen. *Journal für Ornithologie,* 115:273–288.

Berger, M., and J. S. Hart. 1974. Physiology and energetics of flight. *In* Farner, D. S., and J. R. King (eds). *Avian Biology,* Vol. 4. Academic Press, New York.

Berger, M., O. Z. Roy, and J. S. Hart. 1970. The coordination between respiration and wing-beats in birds. *Zeitschrift für vergleichende Physiologie,* 66:190– 200.

Berger, M., *et al.* 1970. Respiration, oxygen consumption and heart rate in some birds during rest and flight. *Zeitschrift für vergleichende Physiologie,* 66:201– 214.

Berger, P. J., N. C. Negus, E. H. Sanders, and P. D. Gardner. 1981. Chemical triggering of reproduction in *Microtus montanus. Science,* 214:69–70.

Bergman, G. 1946. Der Steinwälzer, *Arenaria i. interpres* (L.) in seiner Beziehung zur Umwelt. *Acta Zoologica Fennica,* 47:1–144.

Berkhoudt, H. 1985. Structure and function of avian taste reception. *In* King, A. S., and J. McLelland (eds). *Form and Function in Birds,* 3:463–496. Academic Press, New York.

Berndt, R. 1973. Subfamily: true flycatchers. *In* Grzimek, B. (ed). *Grzimek's Animal Life Encyclopedia.* Vol. 9. Van Nostrand Reinhold Company, New York.

Berndt, R., and H. Sternberg. 1963. Ist die Mortalitätsrate adulter *Ficedula hypoleuca* wirklich unabhängig vom Lebensalter? *Proceedings of the XIIIth International Ornithological Congress,* 2:675–684. American Ornithologists' Union.

Berndt, R., and H. Sternberg. 1968. Terms, studies and experiments on the problems of bird dispersion. *Ibis,* 110:256–269.

Berndt, R., and W. Winkel. 1975. Gibt es beim Trauerschnäpper *Ficedula hypoleuca* eine Prägung auf den Biotop des Geburtsortes? *Journal für Ornithologie,* 116:195– 201.

Berndt, R., and W. Winkel. 1979. Verfrachtungs-Experiment zur Frage der Geburtsortsprägung beim Trauerschnäpper *(Ficedula hypoleuca). Journal für Ornithologie,* 120:41–53.

Bernstein, M. H. 1971a. Cutaneous water loss in small birds. *Condor,* 73:268–269.

Bernstein, M. H. 1971b. Vascular adjustments and their control, influencing heat loss in pigeon feet. *American Zoologist,* 11:671.

Berthold, P. 1973a. Relationships between migratory restlessness and migration distance in six *Sylvia* species. *Ibis,* 115:594–599.

Berthold, P. 1973b. Über starker Rückgang der Dorngrasmücke *Sylvia communis* und anderer Singvogelarten im westlichen Europa. *Journal für Ornithologie,* 114:348–360.

Berthold, P. 1974. Circannuale Periodik bei Grasmüken *(Sylvia). Journal für Ornithologie,* 115:215–272.

Berthold, P. 1976. Animalische und vegetabilische Ernährung omnivorer Singvogelarten: Nahrungsbevorzugung, Jahresperiodik der Nahrungswahl, physiologische und ökologische Bedeutung. *Journal für Ornithologie*, 117:145–209.

Berthold, P. 1977a. Über die künstliche Aufzucht nestjunger Amseln *(Turdus merula)* mit Beeren des Efeus *(Hedera helix)*. *Die Vogelwarte*, 29:110–113.

Berthold, P. 1977b. Endogene Steuerung des Vogelzuges. *Die Vogelwarte*, 29 *(Sonderheft)*:4–15.

Berthold, P. 1977c. Steuerung der Jugendentwicklung bei verschiedenen Populationen derselben Art: Untersuchungen an südfinnischen und südwestdeutschen Gartengrasmücken. *Die Vogelwarte*, 29:38–44.

Berthold, P. 1979. Über die photoperiodische Synchronisation circannualer Rhythmen bei Grasmücken *(Sylvia)*. *Die Vogelwarte*, 30:7–10.

Berthold, P., and H. Berthold. 1971. Über Jahreszeitliche Änderungen der Kleingefiederquantität in Beziehung zum Winterquartier bei *Sylvia atricapilla* und *S. borin*. *Die Vogelwarte*, 26:160–164.

Berthold, P., E. Gwinner, and H. Klein. 1970. Vergleichende Untersuchung der Jugendentwicklung eines ausgeprägten Zugvogels *Sylvia borin*, und eines weniger ausgeprägten Zugvogels, *S. atricapilla*. *Die Vogelwarte*, 25:297–331.

Berthold, P., E. Gwinner, and H. Klein. 1972. Circannuale Periodik bei Grasmucken. *Journal für Ornithologie*, 113:179–190.

Berthold, P., E. Gwinner, H. Klein, and P. Westrich. 1972. Beziehungen zwischen Zugunruhe und Zugablauf bei Garten- und Mönchgrasmücke *(Sylvia borin* und *S. atricapilla)*. *Zeitschrift für Tierpsychologie*, 30:26–35.

Berthold, P., and U.Querner. 1981. Genetic basis of migratory behavior in European Warblers. *Science*, 212:77–79.

Berthold, P., and U. Querner. 1982. Genetic basis of molt, wing length and body weight in a migratory bird species, *Sylvia atricapilla*. *Experientia*, 38:801–802.

Biaggi, V. 1955. The Puerto Rican Honeycreeper, *Coereba flaveola portoricensis* (Bryant). *University of Puerto Rico Agricultural Station, Special Bulletin*.

Biebach, H. 1979. Energetik des Breitens beim Star *(Sturnus vulgaris)*. *Journal für Ornithologie*, 120:121–138.

Bierens de Haan, J. A. 1926. Die Balz des Argusfasans, *Biologische Zentralblatt*, 46:428–435.

Bigelow, H. B., and W. W. Welsh. 1924. Fishes of the Gulf of Maine. *Bulletin, U.S. Bureau of Fisheries*, 40(1):527.

Bijlsma, R. G. 1984. On the breeding association between Wood Pigeons, *Columba palumbus*, and Hobbies, *Falco subbuteo*. *Limosa*, 57:133–139.

Bildstein, K. L. 1980. Corn cob manipulation in Northern Harriers. *Wilson Bulletin*, 92:128–130.

Binggeli, R., and W. Pauli. 1969. The pigeon retina: quantitative aspects of the optic nerve and ganglion cell layer. *Journal of Comparative Neurology*, 137:1–18.

Bingham, V. P. 1983. Importance of earth magnetism for the sunset orientation of migratory naive Savannah Sparrows *(Passerculus sandwichensis)*. *Monitore Zoologico Italiano Firenze*, 17:395–400.

Bingham, V. P., K. P. Able, and P. Kerlinger. 1982. Wind drift, compensation, and the use of landmarks by nocturnal bird migrants. *Animal Behaviour*, 30:49–53.

Binkley, S., E. Kluth, and M. Menaker. 1971. Pineal function in sparrows: circadian rhythms and body temperature. *Science*, 174:311–314.

Binkley, S., S. E. MacBride, D. C. Klein, and C. L. Ralph. 1973. Pineal enzymes: regulation of avian melatonin synthesis. *Science*, 181:273–275.

Binkley, S., J. B. Richman, and K. B. Reilly. 1978. The pineal gland: a biological clock in vitro. *Science*, 202:1198–1201.

Bird, R. D., and L. B. Smith. 1964. The food habits of the Red-winged Blackbird, *Agelaius phoeniceus*, in Manitoba. *Canadian Field Naturalist*, 78:179–186.

Birkhead, T. R., S. D. Johnson, and D. N. Nettleship. 1985. Extra-pair matings and mate guarding in the common murre *Uria aalge*. *Animal Behaviour*, 33:608–619.

Birkhead, T. R., and D. N. Nettleship. 1984. Alloparental care in the Common Murre *(Uria aalge)*. *Canadian Journal of Zoology*, 62:2121–2124.

Bissonnette, T. H. 1932. Light and diet as factors in relation to sexual photoperiodicity. *Nature*, 129:613.

Bissonnette, T. H. 1937. Photoperiodicity in birds. *Wilson Bulletin*, 49:241–270.

Bissonnette, T. H., and A. G. Csech. 1936. Fertile eggs from pheasants in January by "night lighting." *Bird-Banding*, 7:108–111.

Black, E. R., *et al.* 1970. The role of birds in reducing overwintering populations of southwestern corn borer *Diatraea grandiosella*, (Lepidoptera: Crambidae), in Mississippi. *Annals of the Entomological Society of America*, 63:701–706.

Blake, C. H. 1947. Wing-flapping rates in birds. *Auk*, 64:619–620.

Blake, C. H. 1956. Weight changes in birds. *Bird-Banding*, 27:16–22.

Blake, C. H. 1958. Respiration rates. *Bird-Banding*, 29:38–40.

Blake, E. R. 1953. *Birds of Mexico.* University of Chicago Press, Chicago.

Blake, E. R. 1977. *Manual of Neotropical Birds*, Vol. 1. The University of Chicago Press, Chicago and London.

Blanchard, B. D. 1936. Continuity of behavior in the Nuttall White-crowned Sparrow. *Condor*, 38:145–150.

Blanchard, B. D. 1941. The White-crowned Sparrows *(Zonotrichia leucophrys)* of the Pacific Seaboard; Environment and annual cycle. *University of California Publications in Zoology*, 46:1–178.

Blank, J. L., and V. Nolan, Jr. 1983. Offspring sex ratio in Red-winged Blackbirds is dependent on maternal age. *Proceedings of the National Academy of Science, U.S.A.*, 80:6141–6145.

Blem, C. R. 1973. Geographic variation in the bioenergetics of the House Sparrow. *In* Kendeigh, S. C. (ed). A Symposium on the House Sparrow . . . and European Tree Sparrow . . . in North America. *Ornithological Monographs No. 14.* American Ornithologists Union.

Blem, C. R. 1980. The energetics of migration. *In* Gauthreaux, S. A., Jr. (ed). *Animal Migration, Orientation, and Navigation.* Academic Press, New York, 174–224.

Bloze, G. 1968. Anordnung und Bau der Herbstschen Körperschen in Limicolenschnabeln im Zusammenhang mit der Nahrungsfindung. *Zoologischer Anzeiger*, 181:313–355.

Bluhm, C. K. 1985. Mate preferences and mating patterns of Canvasbacks *(Aythya valisneria).* *In* Gowaty, P. A., and D. W. Mock (eds). *Avian Monogamy, Ornithological Monographs* 37:45–56.

Blume, D. 1973. Ausdrucksformen unserer Vögel. *Die Neue Brehm-Bücherei*, 342. A. Ziemsen Press, Wittenberg, Lutherstadt.

Boag, D. A. 1982. How dominance status of adult Japanese Quail *(Coturnix coturnix)* influences the viability and dominance status of their offspring. *Canadian Journal of Zoology*, 60:1885–1891.

Boag, P., and P. R. Grant. 1981. Intense natural selection in a population of Darwin's Finches (Geospizinae) in the Galapagos Islands. *Science*, 214:82–85.

Bocher, J. 1978. *Spergularia canadensis* (Caryophyllaceae) in Greenland. *Botanisk Tidsskrift*, 72:245–248.

Bock, W. J. 1961. Salivary glands in gray jays *(Perisoreus).* *Auk*, 78:355–365.

Bock, W. J. 1974. The avian skeletomuscular system. *In* Farner, D. S., and J. R. King (eds). *Avian Biology*, Vol. 4, 119–257. Academic Press, New York.

Bock, W. J. 1976. Recent advances and the future of avian classification. *In* Frith, H. J., and J. H. Halaby (eds). *Proceedings of the 16th International Ornithological Congress.* Australian Academy of Science, Canberra.

Bock, W. J., and J. Farrand, Jr. 1980. The number of species and genera of recent birds: a contribution to comparative systematics. *American Museum Novitates*, 2703:1–29.

Bock, W. J., and J. Morony. 1972. Snap-closing jaw ligaments in flycatchers. *American Zoologist*, 12:729–730.

Bock, W. J., *et al.* 1973. Morphology of the sublingual pouch and tongue musculature in Clark's Nutcracker. *Auk*, 90:491–519.

Bodenstein, G., and E. Schüz. 1944. Vom Schleifenzug des Prachttauchers *(Colymbus arcticus).* *Ornithologische Monatsberichte*, 52:98–105.

Boeker, E. L., and P. R. Nickerson. 1975. Raptor electrocutions. *Wildlife Society Bulletin*, 3:79–81.

Boersma, P. D., *et al.* 1980. The breeding biology of the Fork-tailed Storm-Petrel *(Oceanodroma furcata).* *Auk*, 97:268–282.

Böhner, J., F. Cooke, and K. Immelmann. 1984. Verhaltensbedingte Isolation zwischen den beiden Rassen des Zebrafinken *(Taeniopygia guttata).* *Journal für Ornithologie*, 125:473-477.

Bond, R. M. 1942. Development of young Goshawks. *Wilson Bulletin*, 54:81–88.

Booth, D. T. 1984. Thermoregulation in neonate mallee fowl *Leipoa ocellata.* *Physiological Zoology*, 57:251–261.

Borchelt, P. L., and L. Duncan. 1974. Dustbathing and feather lipid in Bobwhite *(Colinus virginianus).* *Condor*, 76:471–472.

Borgia, G. 1985. Bower quality, number of decorations and mating success of male satin bowerbirds *(Ptilonorhynchos violaceus)*: an experimental analysis. *Animal Behaviour*, 33:266–271.

Borodulina, T. L. 1966. On the morphology of filoplumes. *In* Kleinenberg, S. E. (ed). *Flight Mechanisms and the Orientation of Birds.* Nauka Publishing House, Moscow. From *Bird-Banding*, 38:169–170.

Borrero, H. J. 1953. Notas preliminares sobre habitos alimenticios de palomas silvestres Colombianos. *Caldasia*, 6(27):75–80.

Borrero H. J. I. 1970. A photographic study of the Potoo in Colombia. *Living Bird*, 9:257–263.

Borror, D. J. 1967. Songs of the Yellowthroat. *Living Bird*, 1967:141–161.

Borror, D. J., and C. R. Reese. 1956. Vocal gymnastics in Woodthrush songs. *Ohio Journal of Science*, 56:177–182.

Boshoff, A. F., A. S. Robertson, and P. M. Norton. 1984. A radiotracking study of an adult Cape Griffin Vulture, *Gyps coprotheres* in the SW Cape Province. *South African Wildlife Research*, 14:73–78.

Boswall, J. 1977. Tool-using by birds and related behaviors. *Avicultural Magazine*, 83:88–97; 146–159; 220–228. 1978, 84:162–166.

Boswall, J. 1978. The birds of the San Benito Islands, Lower California, Mexico. *Journal Bristol Ornithological Club*, 11:23–30.

Boswall, J. 1983. Tool-using and related behaviours in birds: more notes. *Avicultural Magazine*, 89:94–108.

Boswell, J., and M. Demment. 1970. The daily altitude movement of the White-collared Pigeon, *Columba albitorques*, in the High Simien, Ethiopia. *Bulletin of the British Ornithologists' Club*, 90:105–107.

Boudreau, G. W. 1968. Alarm sounds and responses of birds and their application in controlling problem species. *The Living Bird*, 1968:27–46.

Bourgeois, C. E., and W. Threlfall. 1982. Metazoan parasites of 3 species of scoter (Anatidae). *Canadian Journal of Zoology*, 60:2253–2257.

Bourne, W. R. P. 1970. Special review, after the *Torrey Canyon* disaster. *Ibis*, 112:120–125.

Bowers, D. E. 1960. Correlation of variation in the wrentit with environmental gradients. *Condor*, 62:91–120.

Bowmaker, J. K., and A. Knowles. 1977. The visual pigments and oil droplets of the chicken retina. *Vision Research*, 17:755–764.

Bowman, R. I. 1961. Morphological differentiation and adaptation in the Galapagos Finches. *University of California Publications in Zoology*, 58:1–302.

Bowman, R. I. 1979. Adaptive morphology of song dialects in Darwin's Finches. *Journal für Ornithologie*, 120:353–389.

Bowman, R. I. 1983. The evolution of song in Darwin's Finches. *In* Bowman, R. I., M. Berson, and A. E. Leviton (eds). *Patterns of Evolution in Galapagos Organisms*, 237–537. Pacific Division, AAAS, San Francisco, California.

Bowman, R. I., and S. L. Billeb. 1965. Blood-eating in a Galapagos finch. *The Living Bird*, 1965:29–44.

Boxall, P. C., and M. R. Lein. 1982. Territoriality and habitat selection of female snowy owls *(Nyctea scandiaca)* in winter. *Canadian Journal of Zoology*, 60:2344–2350.

Boyd, A. W., and A. L. Thomson. 1936. Recoveries of marked swallows within the British Isles. *British Birds*, 30:278–287.

Boyd, E. M. 1951. The external parasites of birds; a review. *Wilson Bulletin*, 63:363–369.

Boyd, E. M. 1958. Birds and some human diseases. *Bird-Banding*, 29:34–38.

Boyd, E. M., and A. E. Fry. 1971. Metazoan parasites of the Eastern Belted Kingfisher, *Megaceryle alcyon*. *Journal of Parasitology*, 57:150–156.

Boyd, E. M., and S. A. Nunneley. 1964. Banding records substantiating the changed status of 10 species of birds since 1900 in the Connecticut Valley. *Bird-Banding*, 35:1–8.

Boyd, H., and E. Fabricius. 1965. Observations on the incidence of following of visual and auditory stimuli in naive Mallard duckling *(Anas platyrhynchos)*. *Behaviour*, 25:1–15.

Brackenbury, J. H. 1978. Respiratory mechanics of sound production in chickens and geese. *Journal of Experimental Biology*, 72:229–250.

Bradley, C. C. 1978. Play behaviour in Northern Ravens. *Passenger Pigeon*, 40:493–495.

Bradley, O. C., and T. Grahame. 1950. *The Structure of the Fowl*. J. B. Lippincott Company, Philadelphia.

Braun, C. E., *et al.* 1978. Management of national wildlife refuges in the United States: its impact on birds. *Wilson Bulletin*, 90:309–321.

Braun, H. 1952. Über das Unterscheidungsvermögen unbenannter Anzahlen bei Papageien. *Zeitschrift für Tierpsychologie*, 9(1):40–91.

Brauner. J. 1952. Reactions of Poor-wills to light and temperature. *Condor*, 54:152–159.

Bray, O. E., J. J. Kennelly, and J. L. Guarino. 1975. Fertility of eggs produced on territories of vasectomized red-winged blackbirds. *Wilson Bulletin*, 87:187–195.

Brenner, F. J. 1966. Energy and nutrient requirements of the Red-winged Blackbird. *Wilson Bulletin*, 78:111–120.

Brereton, J. L. 1963. Evolution within the Psittaciformes. *Proceedings of the XIIIth International Ornithological Congress*, 499–517. American Ornithologists' Union.

Bretz, W. L., and K. Schmidt-Nielsen. 1971. Bird respiration: flow patterns in the duck lung. *Journal of Experimental Biology*, 54:103–118.

Bretz, W. L., and K. Schmidt-Nielsen. 1972. The movement of gas in the respiratory system of the duck. *Journal of Experimental Biology*, 56:57–65.

Bridgeman, B. 1972. Visual receptive fields sensitive to absolute and relative motion during tracking. *Science*, 178:1106–1108.

Brisbin, I. L. 1969. Behavioral differentiation of wildness in two strains of Red Junglefowl. *American Zoologist*, 9:1072.

Brockway, B. F. 1965. Stimulation of ovarian development and egg laying by male courtship vocalization in Budgerigars *(Melopsittacus undulatus). Animal Behaviour,* 13:507–512.

Brockway, B. F. 1967. The influence of vocal behavior on the performer's testicular activity in Budgerigars *(Melopsittacus undulatus). Wilson Bulletin,* 79:328–334.

Brodkorb, P. 1955. Number of feathers and weights of various systems in a Bald Eagle. *Wilson Bulletin,* 67:142.

Brodkorb, P. 1960–1971. Catalogue of fossil birds. *Bulletin of the Florida State Museum,* Gainesville, Florida.

Brodkorb, P. 1971. Origin and evolution of birds. *In,* Farner, D. S., and J. R. King (eds). *Avian Biology,* Vol. 1, 19–54. Academic Press, New York.

Broley, C. L. 1947. Migration and nesting of Florida Bald Eagles. *Wilson Bulletin,* 59:3–20.

Brook, D., and J. R. Beck. 1972. Antarctic Petrels, Snow Petrels, and South Polar Skuas breeding in the Theron Mountains. *Ibis, 114:*569.

Brooke-Little, J. P. 1964. Heraldic birds. *In* Thomson, A. L. (ed). *A New Dictionary of Birds.* McGraw-Hill Book Company, New York.

Brooks, R. J., and J. B. Falls. 1975. Individual recognition by song in the White-throated Sparrow. III. *Canadian Journal of Zoology,* 53:1749–1761.

Brooks, W. S. 1968. Comparative adaptations of the Alaska redpolls to the arctic environment. *Wilson Bulletin,* 80:253–280.

Broom, R. 1913. On the South African pseudosuchian *Euparkeria* and allied genera. *Proceedings of the Zoological Society of London, 1913:*619–633.

Brosset, A. 1971. L' "imprinting" chez les Columbides—Etude des modifications comportementales au cours du vieillissement. *Zeitschrift für Tierpsychologie,* 29:279–300.

Brosset, A. 1976. Observations sur le parasitisme de la reproduction du Coucou Emeraude, *Chrysococcyx cupreus* au Gabon. *Oiseau, 46:*201–208.

Brosset, A. 1978. Social organization and nest-building in the forest weaver birds of the genus *Malimbus* (Ploceinae). *Ibis, 120:*27–37.

Broun, M., and B. V. Goodwin 1943. Flight speeds of hawks and crows. *Auk,* 60:487–492.

Brower, L. P. 1969. Ecological chemistry. *Scientific American,* 220(2):22–29.

Brower, L. P., B. S. Alpert, and S. C. Glazier. 1970. Observational learning in the feeding behavior of Blue Jays *(Cyanocitta cristata). American Zoologist,* 10:475–476.

Brown, C. H. 1982. Ventriloquial and locatable vocalizations in birds. *Zeitschrift für Tierpsychologie,* 59:338–350.

Brown, C. R. 1985. Energetic cost of molt in macaroni penguins *(Eudyptes chrysolophus)* and rockhopper penguins *(E. chrysocome). Journal of Comparative Physiology, B. Biochemical Systemic Environmental Physiology,* 155:515–520.

Brown, G. 1937. Aggressive display of birds before a looking glass. *British Birds,* 31:137–138.

Brown, J. L. 1959. Method of head-scratching in the wrentit and other species. *Condor,* 61:53.

Brown, J. L. 1964. The evolution of diversity in avian territorial systems. *Wilson Bulletin,* 76:160–169.

Brown, J. L. 1969. The control of avian vocalization by the central nervous system. *In* Hinde, R. A. (ed). *Bird Vocalizations,* 79–96. Cambridge University Press.

Brown, L. H. 1966. Observations on some Kenya eagles. *Ibis, 108:*531–572.

Brown, L. H., and D. Amadon, 1968. *Eagles, Hawks and Falcons of the World.* 2 vols. McGraw-Hill Book Company, New York.

Brown, R. G. B. 1963. The behavior of the Willow Warbler, *Phylloscopus trochilus,* in continuous daylight. *Ibis,* 105:63–75.

Brown, R. G. B. 1974. Bird damage to fruit crops in the Niagara Peninsula. Canadian Wildlife Service Reports, *Series* no. 27:1–57.

Brown, R. G., P. R. Sweeney, and E. T. Moran, Jr. 1982. Collagen levels in tissues of silenium deficient ducks. *Comparative Biochemistry and Physiology,* 71:383–390.

Bruderer, B. 1972. Radar studies in spring migration in northern Switzerland. *Proceedings of the XVth International Ornithological Congress,* 635. E. J. Brill, Leiden.

Bruderer, B. 1975. Zeitliche und räumliche Unterschiede in der Richtung und Richtungsstreuung des Vogelzuges im Schweizerischen Mittelland. *Ornithologische Beobachter,* 72:169–179.

Bruijns, M. F. 1959. Stookolievogels op de Nederlandse kust. *De Levende Natuur,* 62:172–178.

Brüll, H. 1972. Goshawks and falcons. *In* Grzimek, B. (ed). *Grzimek's Animal Life Encyclopedia,* 7:339. Van Nostrand Reinhold Company, New York.

Brunning, D. F. 1974. Social structure and reproductive behaviour in the Greater Rhea. *Living Bird,* 13:251–294.

Brush, A. H. 1978. Avian Pigmentation. *In* Brush, A. H. (ed), *Chemical Zoology.* Academic Press, New York.

Brush, A. H., and D. M. Power. 1976. House finch pigmentation: carotenoid metabolism and the effect of diet. *Auk*, 93:725–739.

Brush, A. H., and H. Seifried. 1968. Pigmentation and feather structure in genetic variants of the Gouldian Finch, *Poephila gouldiae*. *Auk*, 85:416–430.

Bryant, D. M. 1983. Heat stress in tropical birds: Behavioral thermoregulation during flight. *Ibis*, 125:313–323.

Buckley, J. L., and P. F. Springer. 1964. Insecticides. *In* Linduska, J. P., and A. L. Nelson (eds). *Waterfowl Tomorrow*. U.S. Fish and Wildlife Service, Washington, D.C.

Buckley, P. A. 1982. Avian Genetics. *In* Petrak, M. (ed), *Diseases of Cage and Aviary Birds*, 2nd Edition, 21–110. Lea and Febiger, Philadelphia.

Buckley, P. A., and F. G. Buckley. 1977. Hexagonal packing of royal tern nests. *Auk*, 94:36–43.

Budgell, P. 1971. Behavioural thermoregulation in the Barbary Dove (*Streptopelia risoria*). *Animal Behaviour*, 19:524–531.

Buick, T. L. 1937. *The Moa-Hunters of New Zealand, Sportsmen of the Stone Age*. Thomas Avery and Sons, Ltd., New Plymouth, New Zealand.

Buitron, D., and G. L. Nuechterlein. 1985. Experiments on olfactory detection of food caches by Black-billed Magpies. *Condor*, 87:92–95.

Bull, J., and E. R. Ricciuti. 1974. Polly want an apple? *Audubon*, 76(3):48–54.

Bullock, T. J. 1973. Seeing the world through a new sense: electroreception in fish. *American Scientist*, 61:316–325.

Burger, A. E. 1979. Breeding biology, moult and survival of Lesser Sheathbills *Chionis minor* at Marion Island. *Ardea*, 67:1–14.

Burger, A. E., and R. P. Miller. 1980. Seasonal changes of sexual and territorial behaviour and plasma testosterone levels in male lesser sheathbills (*Chionis minor*). *Zeitschrift für Tierpsychologie*, 52:397–406.

Burger, J. W. 1939. Some aspects of the roles of light intensity and the daily length of exposure to light in the sexual photoperiodic activation of the male Starling. *Journal of Experimental Zoology*, 81:333–340.

Burger, J. W. 1949. A review of experimental investigations on seasonal reproduction in birds. *Wilson Bulletin*, 61:211–230.

Burley, N. 1981. Mate choice by multiple criteria in a monogamous species. *American Naturalist*, 117:515–528.

Burley, N., and N. Moran. 1979. The significance of age and reproductive experience in the mate preferences of feral Pigeons, *Columba livia*. *Animal Behavior*, 27:686–698.

Burrows, W. H., and S. J. Marsden. 1938. Artificial breeding of Turkeys. *Poultry Science*, 17:408.

Burton, J. F., and R. A. French. 1969. Monarch Butterflies coinciding with American passerines in Britain and Ireland in 1968. *British Birds*, 62:493–494.

Busse, K. 1977. Prägungsbedingte akustische Artkennungsfähigkeit der Küken der Flusseeschwalben und Küstenseeschwalben (*Sterna hirundo* L. und *S. paradisaea* Pont.). *Zeitschrift für Tierpsychologie*, 44:154–161.

Bussman, J. 1953. Beitrag zur kenntnis der Brutbiologie des Kleibers (*Sitta europaea caesia*). *Der Ornithologische Beobachter*, 40:57–67.

Butler, R. G., and S. J. Butler. 1982. Territoriality and behavioral correlates of reproductive success of Great Blacked-backed Gulls. *Auk*, 99:58–66.

Butterfield, P. A. 1970. *In* Crook, J. H. (ed). *Social Behaviour in Birds and Mammals*. Academic Press, New York.

Butz, E. L. (ed). 1976. The Face of Rural America. *1976 Yearbook of the U.S. Department of Agriculture*, U.S. Government Printing Office, Washington, D.C.

Buxton, P. A. 1923. *Animal Life in Deserts*. E. Arnold & Company, London.

Cade, T. J. 1953. Behavior of a young Gyrfalcon. *Wilson Bulletin*, 65:26–31.

Cade, T. J. 1973. Sunbathing as a thermoregulatory aid in birds. *Condor*, 75:106–108.

Cade, T. J. 1982. *The Falcons of the World*. Collins, London.

Cade, T. J., and P. R. Dague (eds). 1980. *The Peregrine Fund Newsletter*, No. 8. Cornell University Laboratory of Ornithology. Ithaca, New York.

Cade, T. J., and J. A. Dybas, Jr. 1962. Water economy in the Budgerygah. *Auk*, 79:345–364.

Cade, T. J., and V. J. Hardaswick. 1985. Summary of peregrin falcon production and re-introduction by the Peregrine Fund in the United States, 1873–1984. *Avicultural Magazine*, 91:79–92.

Cade, T. J., and G. L. Maclean. 1967. Transport of water by adult Sandgrouse to their young. *Condor*, 69:323–343.

Cain, A. J. 1964. Natural selection. *In* Thomson, A. L. (ed). *A New Dictionary of Birds*. McGraw-Hill Book Company, New York.

Caithness, T. A. 1968. Poisoning gulls with alpha-chloralose near a New Zealand airfield. *Journal of Wildlife Management*, 32:279–286.

Calder, W. A. 1968. Respiratory and heart rates of birds at rest. *Condor*, 70:358–365.

Calder, W. A. 1970. Respiration during song in the Canary *Serinus canarius*. *Comparative Biochemistry and Physiology*, 32:251–258.

Calder, W. A., and J. R. King. 1974. Thermal and caloric relations of birds. *In* Farner, D. S., and J. R. King (eds). *Avian Biology*, Vol. 4. Academic Press, New York.

Calisher, C. H. *et al.* 1971. Identification of two South American strains of Eastern equine encephalitis virus from migrating birds. . . . *American Journal of Epidemiology*, 94:172–178.

Campbell, K. E., and E. P. Tonni. 1980. A new genus of teratorn from the Huayquerian of Argentina (Aves: Teratonithidae). *Contributions in Science*, 330:59–68, Los Angeles County Natural History Museum.

Campbell, R. W., T. R. Torgersen, and N. Srivastava. 1984. A suggested role for predaceous birds and ants in the population dynamics of the western spruce budworm. *Forest Science*, 29:779–790.

Caraco, T. 1982. Flock size and the behavioral sequences in juncos (*Junco phaeonotus*). *Condor*, 84:101–105.

Carey, C. 1983. Structure and function of avian eggs. *In* Johnston, R. F. (ed). *Current Ornithology*, Vol. 1, 69–103. Plenum Press, New York.

Carey C., and M. L. Morton. 1976. Aspects of circulatory physiology of montane and lowland birds. *Comparative Biochemistry and Physiology*, 54A:61–74.

Carey, C., H. Rahn, and P. Parisi. 1980. Calories, water, lipid and yolk in avian eggs. *Condor*, 82:335–343.

Carey, C., *et al.* 1980. Physiology of the avian egg. *American Zoologist*, 20:325–484.

Carlson, C. W. 1960. Aortic rupture. *Turkey Producer*, January, 1960.

Carpenter, F. L., and R. E. MacMillen. 1976. Energetic cost of feeding territories in an Hawaiian honeycreeper. *Oecologia Berlin*, 26:213–223.

Carr-Lewty, R. A. 1943. Reactions of birds to aircraft. *British Birds*, 36:151–152.

Carson, R. 1962. *Silent Spring*. Houghton Mifflin Company, Boston.

Cartwright, B. W. 1944. The "crash" decline in Sharp-tailed Grouse and Hungarian Partridge in Western Canada and the role of the predator. *Transactions of the North American Wildlife Conference*, 9:324–329.

Catchpole, C. K. 1973. The functions of advertising song in the Sedge Warbler (*Acrocephalus schoenobaenus*) and the Reed Warbler (*A. scirpaceus*). *Behaviour*, 46:300–320.

Catchpole, C. K. 1978. Interspecific territorialism and competition in *Acrocephalus* Warblers as revealed by playback experiments in areas of sympatry and allopatry. *Animal Behaviour*, 26:1072–1080.

Catchpole, C. K., J. Dittani, and B. Leisler. 1984. Differential responses to male song repertoires in female songbirds implanted with oestradiol. *Nature*, 312:563–565.

Catchpole, C. K., B. Leisler, and H. Winkler. 1985. Polygyny in the great reed warbler, *Acrocephalus arundinaceus*: a possible case of deception. *Behavioral Ecology and Sociobiology*, 16:285–291.

Catlett, R. H., *et al.* 1978. The effect of flying and not flying on myoglobin content of heart muscle of the pigeon *Columba livia domestica*. *Comparative Biochemistry and Physiology A*, 59:401–402.

Chamberlain, D. R. 1968. Syringeal anatomy in the common Crow. *Auk*, 85:244–252.

Chance, E. P. 1940. *The Truth about the Cuckoo*. Country Life, London.

Chapin, J. P., and L. W. Wing. 1959. The wideawake calendar, 1953 to 1958. *Auk*, 76:153–158.

Chapman, F. M. 1908. *Camps and Cruises of an Ornithologist*. D. Appleton and Company, New York.

Chapman, F. M. 1917. The distribution of bird life in Colombia. *Bulletin of the American Museum of Natural History*, 36.

Chapman, F. M. 1926. *Handbook of Birds of Eastern North America*. D. Appleton and Company, New York.

Chapman, F. M. 1940. The post-glacial history of *Zonotrichia capensis*. *Bulletin of the American Museum of Natural History*, 77:381–438.

Chapuis, C. 1971. Un exemple de l'influence du milieu sur les émissions vocales des oiseaux: l'évolution des chants en forêt equatoriale. *La Terre et la Vie*, 25:183–202.

Cheke, A. S. 1969. Mechanism and consequences of hybridization in sparrows. *Passer. Nature*, 222:179–180.

Cheke, R. A. 1971. Temperature rhythms in African montane sunbirds. *Ibis*, 113:500–506.

Chen, D. M., J. C. Collins, and T. H. Goldsmith. 1984. The ultraviolet receptor of bird retinas. *Science*, 225:337–340.

Cheng, K. M., J. T. Burns, and F. McKinney. 1982. Forced copulation in captive mallards (*Anas platyrhynchos*): II. Temporal Factors. *Animal Behaviour*, 30:695–699.

Cheng, K. M., J. T. Burns, and F. McKinney. 1983. Forced copulation in captive mallards: III. Sperm competition. *Auk*, 100:302–310.

Cheng, M. F., and J. Balthazart. 1982. Role of nest-building activity in gonadotropin secretions and the reproductive success of ring doves *(Streptopelia risoria)*. *Journal Comparative Physiology and Psychology*, 96:307–324.

Chernin, E. 1952. The epizootiology of *Leucocytozoon simondi* infections in domestic ducks in northern Michigan. *American Journal of Hygiene*, 56:39–57; 101–118.

Cherry, J. D. 1982. Fat deposition and length of stopover in migrant White-crowned Sparrows. *Auk*, 99:725–734.

Chesness, R. A., M. M. Nelson, and W. A. Longley, 1968. The effect of predator removal on pheasant reproductive success. *Journal of Wildlife Management*, 32:683–697.

Childe, G. V. 1941. *Man Makes Himself*. Watts and Co., London.

Chura, N. J. 1961. Food availability and preferences of juvenile Mallards. *Transactions of the North American Wildlife Conference*, 26:121–134.

Clapp, R. B., M. K. Klimkiewicz, and A. G. Futcher. 1983. Longevity records of North American birds: Columbide through Paridae. *Journal of Field Ornithology*, 54:123–137.

Clapp, R. B., M. K. Klimkiewicz, and J. H. Kennard. 1982. Longevity records of North American birds: Gaviidae through Alcidae. *Journal of Field Ornithology*, 53:81–208.

Clark, G. A. 1961. Occurrence and timing of egg teeth in birds. *Wilson Bulletin*, 73:268–278.

Clark, G. A., Jr. 1973. Holding food with the feet in passerines. *Bird-Banding*, 44:91–99.

Clark, R. B., and R. J. Kennedy. 1968. Rehabilitation of oiled sea birds. A *Report to the Advisory Committee on Oil Pollution of the Sea*. University of Newcastle upon Tyne.

Clarke, L. F., H. Rahn, and M. D. Martin. 1942. Sage Grouse studies II. *Wyoming Game and Fish Department Bulletin*, No. 2:13–17.

Clemens, S. L. (Mark Twain). 1935. Life on the Mississippi. *In The Family Mark Twain*. Harper and Brothers, New York.

Clements, F. E., and V. E. Shelford. 1939. *Bio-ecology*. John Wiley and Sons, New York.

Cobb, S. 1960. Observations on the comparative anatomy of the avian brain. *Perspectives in Biology and Medicine*, 3:383–408.

Cochran, W. W., G. G. Montgomery, and R. R. Graber. 1967. Migratory flights of *Hylocichla* thrushes in spring: a radio-telemetry study. *The Living Bird*, 6:213–225.

Cody, M. L. 1966. A general theory of clutch size. *Evolution*, 20:174–184.

Cody, M. L. 1971. Ecological aspects of reproduction. *In* Farner, D. S., and King, J. R. (eds). *Avian Biology*, Vol. 1. Academic Press, New York.

Cohen, R. R. 1965. Avian adjustment of blood oxygen capacity in and after chronic hypoxia stress, as seen in the Pintail Duck *(Anas acuta)*. *American Zoologist*, 5:207–208.

Cohen, R. R. 1969. Total and relative erythrocyte levels of Pintail Ducks *(Anas acuta)* in chronic depression hypoxia. *Physiological Zoology*, 42:108–119.

Colbert, E. H. 1955. *Evolution of the Vertebrates*. John Wiley and Sons, New York.

Collias, E. C. 1983. Inheritance of egg color in Village Weaverbird *(Ploceus cucullatus)*. *Abstracts of Presented Posters and Papers*, No. 53. 101st Meeting of American Ornithologists Union, New York.

Collias, E. C., and N. E. Collias. 1964. The development of nest-building behavior in a weaverbird. *Auk*, 81:42–52.

Collias, N. E. 1952. The development of social behavior in birds. *Auk*, 69:127–159.

Collias, N. E. 1960. An ecological and functional classification of animal sounds. *In* Lanyon, W. E., and W. N. Tavolga, *Animal Sounds and Communication*. American Institute of biological Sciences, Arlington, Virginia.

Collias, N. E. 1964. The evolution of nests and nest-building in birds. *American Zoologist*, 4:175–190.

Collias, N. E., and E. C. Collias. 1968. Anna's Hummingbirds trained to select different colors in feeding. *Condor*, 70:273–274.

Collias, N. E., and E. C. Collias. 1969a. Size of breeding colony related to attraction of mates in a tropical passerine bird. *Ecology*, 50:481–488.

Collias, N. E., and E. C. Collias. 1969b. Some observations on behavioral energetics of two related subspecies of birds. *American Zoologist*, 9:1064.

Collias, N. E., and E. C. Collias. 1984. *Nest Building and Bird Behavior*. Princeton University Press, Princeton, New Jersey.

Collias, N. E., *et al.* 1961. Nest-building and breeding behavior of castrated Village Weaverbirds *(Textor cucullatus)*. *American Zoologist*, 1:349.

Colquhoun, M. K. 1939. The vocal activity of Blackbirds at a winter roost. *British Birds*, 33:44–47.

Comar, C. L., and J. C. Driggers. 1949. Secretion of radioactive calcium in the hen's egg. *Science*, 109:282.

Cone, C. D., Jr. 1962. Thermal soaring of birds. *American Scientist*, 50:180–209.

Conover, M. R., and G. L. Hunt, Jr. 1984. Experimental evidence that female-female pairs in gulls result from a shortage of breeding males. *Condor*, 86:472–476.

Contino, F. 1968. Observations on the nesting of *Sporophila obscura* in association with wasps. *Auk*, 85:137–138.

Conway, W. 1983. Captive birds and conservation. *In* Brush, A. H., and G. A. Clark, Jr. (eds). *Prespectives in Ornithology*, 23–36. Cambridge University Press.

Cooch, G. 1955. Observations on the autumn migration of Blue Geese, *Wilson Bulletin*, 67:171–174.

Cooke, F. 1978. Early learning and its effect on population structure. Studies on a wild population of Snow Geese. *Zeitschrift für Tierpsychologie*, 46:344– 358.

Cooke, W. W. 1915. Bird migration. *U.S. Department of Agriculture Bulletin*, 185:1–47.

Cooper, W. T., and J. M. Forshaw. 1977. *The Birds of Paradise and Bower-Birds*. David R. Godine, Australia.

Corti, U. A., R. Melcher, and T. Tinner. 1949. Beiträge zur Biologie der Blaumerle, *Monticola solitarius* (L.). *Archives swisses d'Ornithologie*, 2:185–212.

Cortopassi, A. J., and L. R. Mewaldt. 1965. The circumannual distribution of White-crowned Sparrows. *Bird-Banding*, 36:141–169.

Cott, H. B. 1940. *Adaptive Coloration in Animals*. Methuen, London.

Cott, H. B. 1964. Coloration, adaptive. *In* Thomson, A. L. (ed). A *New Dictionary of Birds*, McGraw-Hill Book Company, New York.

Coues, E. 1903. *Key to North American Birds*. Page and Company, Boston.

Coulson, J. C. 1956. Mortality and egg production of the Meadow Pipit with special reference to altitude. *Bird Study*, 3:119–132.

Coulson, J. C. 1960. A study of mortality of the Starling based on ringing records. *Journal of Animal Ecology*, 29:251–271.

Coulson, J. C. 1966. The influence of the pair bond and age on the breeding biology of the Kittiwake Gull, *Rissa tridactyla*. *Journal of Animal Ecology*, 35:269–279.

Coulson, J. C. 1968. Differences in the quality of birds nesting in the centre and on the edges of a colony. *Nature*, 217:478–479.

Coulson, J. C., and J. Horobin. 1976. The influence of age on the breeding biology and survival of the Arctic Tern, *Sterna paradisaea*. *Journal of Zoology* (London), 178:247–260.

Coulson, J. C., and E. White. 1959. The post-fledging mortality of the Kittiwake. *Bird Study*, 6:97–102.

Coulson, J. C., *et al*. 1968. Mortality of Shags and other sea birds caused by a paralytic shellfish poison. *Nature*, 220:23–24.

Coutlee, E. L. 1971. Vocalizations in the genus *Spinus*. *Animal Behaviour*, 19:556–565.

Cowie, R. J., J. R. Krebs, and D. F. Sherry. 1981. Food storage by Marsh Tits *(Parus palustris)*. *Animal Behaviour*, 29:1252–1259.

Cox, P. R. 1944. A statistical investigation into bird song. *British Birds*, 38:3–9.

Cracraft, J. 1973. Continental drift, paleoclimatology, and the evolution and biogeography of birds. *Journal of Zoology (London)*, 169:455–545.

Cracraft, J. 1976. Avian evolution on southern continents. *In* Frith, H. J., and J. H. Calaby (eds). *Proceedings of the 16th International Ornithological Congress*. Australian Academy of Science, Canberra.

Cracraft, J. 1976. The species of Moas. *Smithsonian Contributions to Palaeobiology*, 27:189–205.

Craig, J. L., A. M. Stewart, and J. L. Brown. 1982. Subordinates must wait. *Zeitschrift für Tierpsychologie*, 60:275–280.

Craig, J. V., L. L. Ortman, and A. M. Guhl. 1965. Genetic selection for social dominance ability in chickens. *Animal Behaviour*, 13:114–131.

Craig, R. B. 1978. An analysis of the predatory behavior of the loggerhead shrike. *Auk*, 95:221–234.

Craig, W. 1908. The voices of pigeons regarded as a means of social control. *American Journal of Sociology*, 14:86–100.

Craig, W. 1918. Appetites and aversions as constituents of instincts. *Biological Bulletin*, Woods Hole, 34(2):91–107.

Cram, E. B. 1927. Bird parasites of the nematode suborders Strongylata, Ascaridata, and Spirurata. *U.S. National Museum Bulletin 140*.

Cramp, S., and K. E. L. Simmons (eds). 1977. *Handbook of the Birds of Europe, the Middle East, and North Africa*, Vol. 1. Oxford University Press, Oxford.

Crawford, E. C., Jr., and K. Schmidt-Nielson. 1967. Temperature regulation and evaporative cooling in the Ostrich. *American Journal of Physiology*, 212:347–353.

Crescitelli, F., *et al*. (eds). 1977. The visual system in Vertebrates. *Handbook of Sensory Physiology*, Vol. VII-5. Springer-Verlag, New York.

Creutz, G. 1949. Die Entwicklung zweier Populationen des Trauerschnäppers, *Muscicapa h. hypoleuca* (Pall.), nach Herkunft und Alter. *Beiträge zur Vogelkunde*, 27–53. Akademische Verlagsgesellschaft, Leipzig.

Crew, F. A. E. 1923. Studies in intersexuality II: Sex reversal in the fowl. *Proceedings of the Royal Society*, Series B, 95:256–278.

Christensen, P. V. 1975. [Size and nestling production of a Swallow, *Hirundo rustica*, population in North Zealand.] *Dansk Ornithologisk Forenings Tidsskrift*, 69:19–29.

Crome, F. H. J. 1986. Australian waterfowl do not necessarily breed on a rising water level. *Australian Wildlife Research*, 13:461–480.

Cronin, E. W., and P. S. Sherman. 1976. A resource-based mating system: the Orange-rumped Honeyguide. *Living Bird*, 15:5–32.

Crook, J. H. 1963. A comparative analysis of nest structure in the weaver birds (Ploceinae). *Ibis*, 105:238–262.

Crook, J. H., and P. A. Butterfield. 1968. Effects of testosterone propionate and luteinizing hormone on agonistic and nest-building behavior of *Quelea quelea*. *Animal Behaviour*, 16:370–384.

Crook, J. H., and P. A. Butterfield. 1970. *In* Crook, J. H. (ed). *Social Behaviour in Birds and Mammals*. Academic Press, New York.

Crosby, G. T. 1972. Spread of the Cattle Egret in the western hemisphere. *Bird-Banding*, 43:205–212.

Crossner, K. A. 1977. Natural selection and clutch size in the European Starling. *Ecology*, 58:885–892.

Croze, H. 1970. Searching image in carrion crows. *Zeitschrift für Tierpsychologie*, Supplement 5.

Cruden, R. W. 1972. Pollinators in high-elevation ecosystems: relative effectiveness of birds and bees. *Science*, 176:1439–1440.

Cruickshank, A. D. 1956. Nesting heights of some woodland warblers in Maine. *Wilson Bulletin*, 68:157.

Cuisin, M. 1983. Note sur certaines adaptations du Pic Noir *(Dryocopus martius* (L.)) et sa niche ecologique dans deux biocenoses. *Oiseau Revue Francaise Ornithologie*, 53:63–77.

Cullen, E. 1957. Adaptations in the Kittiwake to cliff-nesting. *Ibis*, 99:275–302.

Cullen, E., and J. M. Cullen. 1962. The pecking response of young Kittiwakes and a Black-headed Gull foster chick. *Bird Study*, 9:1–6.

Curio, E. 1959. Beiträge zur Populationsoekologie des Trauerschnäppers *(Ficedula hypoleuca*, Pallas). *Zoologischer Jahrbücher Abteilung für Systematik*, 87(3):185–230.

Curio, E. 1970. Kaspar-häuser-versuche zum Feinderkennen junger Trauerschnäpper *(Ficedula hypoleuca hypoleuca*, Pall.) *Journal für Ornithologie*, 111:438–455.

Curio, E., U. Ernst, and W. Veith. 1978. The adaptive significance of avian mobbing. II. *Zeitschrift für Tierpsychologie*, 48:184–202.

Curio, E., and P. Kramer. 1964. Vom Mangrovefinken *(Cactospiza heliobates* Snodgrass and Heller). *Zeitschrift für Tierpsychologie*, 21:223–234.

Cushing, J. E., and A. O. Ramsey. 1949. The non-heritable aspects of family unity in birds. *Condor*, 51:82–87.

Daan, S., and J. Tinbergen. 1979. Young Guillemots *(Uria lomvia)* leaving their Arctic breeding cliffs: a daily rhythm in numbers and risk. *Ardea*, 67:96–100.

Daanje, A. 1941. Über das Verhaltung des Haussperlings *(Passer d. domesticus*, L.). *Ardea*, 30:1–42.

Dale, F. M. 1955. The role of calcium in reproduction of the Ring-necked Pheasant. *Journal of Wildlife Management*, 19:325–331.

Dambach, C. A. 1941. The effect of land-use adjustments on wildlife populations in the Ohio Valley region. *Transactions of the North American Wildlife Conference*, 5:331–337.

Daniel, M. J. 1963. Observations of chick mortality in a colony of Black-billed Gulls in the thermal area of Whakarewarewa. *Notornis*, 10:277–278; 285–286; 386–392.

Darley, J. A. 1983. Territorial behavior of the female brown-headed cowbird *(Molothrus ater)*. *Canadian Journal of Zoology*, 61:65–69.

Darling, F. F. 1938. *Bird Flocks and the Breeding Cycle*. Cambridge University Press, Cambridge.

Darling, F. F. 1952. Social behavior and survival. *Auk*. 69:183–191.

Darlington, P. J. 1957. *Zoogeography: the Geographical Distribution of Animals*. John Wiley and Sons, New York.

Darwin, C. 1845. *Journal of Researchers into the Geology and Natural History of the Various Countries Visited During the Voyage of H.M.S. Beagle Round the World*. 1902 edition. P. F. Collier and Son, New York.

Darwin, C. 1859. *On the Origin of Species by Means of Natural Selection*. Various editions.

Darwin, C. 1871. *The Descent of Man and Selection in Relation to Sex*. J. Murray, London.

Darwin, C. R. 1875. *The Variation of Animals and Plants Under Domestication*, 2nd Edition. J. Murray, London.

Dathe, H. 1955. Uber die Schreckmauser. *Journal für Ornithologie*, 96:5–14.

Dathe, H. H., and H. Oehme. 1978. Typen des Rüttelfluges der Vögel. *Biologisches Zentralblatt*, 97:299–306.

Dauber, D. V. 1944. Spontaneous arteriosclerosis in chickens. *Archives of Pathology*, 38:46.

Dave, B., C. Walcott, and W. H. Dury. 1959. The form and duration of the display actions of the goldeneye *(Bucephala clangula)*. *Behavior*, 14:265–281.

Davies, N. B. 1976. Food, flocking, and territorial behaviour of the Pied Wagtail *(Motacilla alba yarrelli* Gould) in winter. *Journal of Animal Ecology*, 45:235–253.

Davies, N. B. 1978a. Parental meanness and offspring independence. *Ibis*, 120:509–516.

Davies, N. B. 1978b. Ecological questions about territorial behaviour. *In* Krebs, J. R., and N. B. Davies (eds). *Behavioural Ecology: An Evolutionary Approach*. Sinauer Associates, Inc., Boston, Massachusetts.

Davies, N. B. 1985. Cooperation and conflict among dunnocks, *Prunella modularis*, in a variable mating system. *Animal Behaviour*, 33:628–648.

Davies, D. E. 1942a. Number of eggs laid by Herring Gulls. *Auk*, 59:549–554.

Davies, D. E. 1942b. The phylogeny of social nesting habits in the Crotophaginae. *Quarterly Review of Biology*, 17:115–134.

Davis, D. E. 1955. Breeding biology of birds. *In* Wolfson, A. (ed). *Recent Studies in Avian Biology*. University of Illinois Press, Urbana, Illinois.

Davis, D. E. 1957. Aggressive behavior in castrated Starlings. *Science*, 126:253.

Davis, D. E. 1961. Principles for population control by gametocides. *Transactions of the North American Wildlife Conference*, 26:160–167.

Davis, D. E. 1963. The hormonal control of aggressive behavior. *Proceedings of the XIIIth International Ornithological Congress*, 2:994–1003. American Ornithologists' Union.

Davis, D. E., and L. V. Domm. 1941. The sexual behavior of hormonally treated domestic fowl. *Proceedings of the Society for Experimental Biology and Medicine*, 48:667-669.

Davis, J. M. 1980. The coordinated aerobatics of Dunlin *(Calidris alpina)* flocks. *Animal Behaviour*, 28:668–673.

Davis, P. 1957. The breeding of the Storm Petrel. *British Birds*, 50:85–101.

Davis, T. A., and R. A. Ackerman. 1985. Adaptations of Black Tern *(Chlidonias niger)* eggs for water loss in a moist nest. *Auk*, 102:640–643.

Davis, T. A., M. Platter-Reiger, and R. A. Ackerman. 1984. Incubation water loss by Pied-billed Grebe eggs: adaptation to a hot, wet nest. *Physiological Zoology*, 57:384–391.

Davis, W. J. 1982. Territory size in *Megaceryle alcyon* along a stream habitat. *Auk*, 99:353–362.

Dawson, W. R. 1958. Relation of oxygen consumption and evaporative water loss to temperature in the Cardinal. *Physiological Zoology*, 31:37–48.

Dawson, W. R. 1976. *In* Frith, H. J., and J. H. Calaby (eds). *Proceedings of the 16th International Ornithological Congress*. Australian Academy of Science, Canberra.

Dawson, W. R., and G. A. Bartholomew. 1968. *In* Brown, G. W. (ed). *Desert Biology*. Academic Press, New York.

Deadhikari H., and S. P. Bhattacharyya. 1985. The influence of gonadotropins in the physiology of the uropygial gland of domestic pigeons. *Zoologischer Anzeiger*, 214:61–69.

deBoer, L. E. M., and R. van Bocxstaele. 1981. Somatic chromosomes of the congo peafowl *(Afropavo congensis)* and their bearing on the species affinities. *Condor*, 83:204–208.

Deevey, E. S. 1960. The human population. *Scientific American*, 203:(3):195–204.

De Guirtchitch, G. 1937. Chronique Ornithologique Tunisienne pour l'Année 1936. *L'Oiseau et la Revue Francaise d'Ornithologie*, 7:450–472.

Deighton, T., and J. C. D. Hutchinson. 1940. Studies on the metabolism of the fowls. II. The effect of activity on metabolism. *Journal of Agricultural Science*, 30:141–157.

Delacour, J. 1946. Les Timaliinés. *L'Oiseau et la Revue Francaise d'Ornithologie*, 41:7–36.

Delacour, J. 1954. *The Waterfowl of the World*. 4 vols. Country Life, Ltd., London.

Delacour, J., and E. Mayr. 1945. The family Anatidae. *Wilson Bulletin*, 57:3–55.

Delius, J. D. 1983. Learning. *In* Abs, M. (ed). *Physiology and Behavior of the Pigeon*, 327–355. Academic Press, New York.

Delius, J. D., R. J. Perchard, and J. Emmerton. 1976. Polarized light discrimination by pigeons and an electroretinographic correlate. *Journal of Comparative Physiological Psychology*, 90:560–571.

del Toro, A. 1971. On the biology of the American finfoot in southern Mexico. *Living Bird*, 1971:79–88.

DeMan, E., and H. V. S. Peeke. 1982. Dietary ferulic acid, biochanin A, and the inhibition of reproductive behavior in Japanese Quail *(Coturnix coturnix)*. *Pharmacology, Biochemistry & Behavior, 17:405–411.*

Dement'ev, G. P., *et al.* 1966. *Birds of the Soviet Union.* 6 vols. Smithsonian Institution, Washington, D.C.

Dennison, M. D., H. A. Robertson, and D. Crouchley. 1984. Breeding in the Chatham Island Warbler *(Gerygone albofrontata). Notornis, 31:97–105.*

DeSante, D. F., and D. G. Ainley. 1980. The avifauna of the South Farallon Islands, California. *Studies in Avian Biology, 4.*

de Schauensee, R. M. 1970. *A Guide to the Birds of South America.* Livingston Publishing Company, Wynnewood, Pennsylvania.

Detert, H., and H. H. Bergmann. 1984. Regenrufdialekte von Buchfinken *(Fringilla coelebs,* L.): Untersuchungen an einer Population von *Mischrufern. Ökologie der Vögel, 6:101–118.*

Deuel, L. 1969. *Flights into Yesterday—The Story of Aerial Archaeology.* St. Martin's Press, New York.

de Vries, T. 1973. The Galapagos Hawk. Ph.D. thesis. University of Amsterdam.

Dewsbury, D. A. 1978. *Comparative Animal Behavior.* McGraw-Hill Book Company, New York.

Diamond, A. W., and R. W. Smith. 1973. Returns and survival of banded warblers wintering in Jamaica. *Bird-Banding, 44:221–224.*

Diamond, J. M. 1972. *Avifauna of the Eastern Highlands of New Guinea.* Publications of the Nuttall Ornithology Club, No. 12. Cambridge, Massachusetts.

Diamond, J. M. 1973. Distributional ecology of New Guinea birds. *Science, 179:759–769.*

Diamond, J. M. 1974. Colonization of volcanic islands by birds: the supertramp strategy. *Science, 184:803–806.*

Diamond, J. M. 1976. Relaxation and differential extinction on land-bridge islands: applications to natural preserves. *In* Frith, H. J., and J. H. Calaby (eds). *Proceedings of the 16th International Ornithological Congress.* Australian Academy of Science, Canberra.

Diamond, J. M. 1978. Niche shifts and the rediscovery of interspecific competition. *American Scientist, 66:322–331.*

Diamond, J. M. 1981. Flightlessness and fear of flying in island species. *Nature,* London, *293:507–508.*

Diamond, J. M. 1982a. Mirror-image navigational errors in migrating birds. *Nature, 295:277–278.*

Diamond, J. M. 1982b. Evolution of bowerbirds' bowers: animal origins of the aesthetic sense. *Nature, 297:99–102.*

Diamond, J. M. 1983. Taxonomy by nucleotides. *Nature, 305:17–18.*

Diamond, J. M., and A. G. Marshall. 1972. Niche shifts in New Hebridean birds. *Emu, 77:61–72.*

Dice, L. R. 1945. Minimum intensities of illumination under which owls can find dead prey by sight. *American Naturalist, 79:385–416.*

Diesselhorst, G. 1950. Erkennen des Geschlechts und Paarbildung bei der Goldammer *(Emberiza citrinella* L.). *Ornithologische Berichte, 3:69–112.*

Diesselhorst, G. 1971. Ein Fall interspezifischer Kompetition bei Webervögeln *(Ploceus). Journal für Ornithologie, 112:227–229.*

Dilger, W. C. 1955. Ruptured heart in the Cardinal, *Richmondena cardinalis. Auk, 72:85.*

Dilger, W. C. 1956. Nest-building movements performed by a juvenile Olive-backed Thrush. *Wilson Bulletin, 68:157–158.*

Dilger, W. C. 1962. The behavior of lovebirds. *Scientific American, 206(1):88–98.*

Dinnendahl, L. 1954. Nächtlicher Zug und Windrichtung auf Helgoland. *Die Vogelwarte, 17:188–194.*

Dinsmore, J. J. 1973. Foraging success of Cattle Egrets, *Bubulcus ibis. American Midland Naturalist, 89:242–246.*

Disney, H. J. de S. 1964. Quelea control. *In* Thomson, A. L. (ed). *A New Dictionary of Birds.* McGraw-Hill Book Company, New York.

Dixon, C. 1902. *Birds' Nests.* Frederick A. Stokes Company, New York.

Dobzhansky, T. 1950. Evolution in the tropics. *American Scientist, 38:209–221.*

Dobzhansky, T. 1967. Review of *On Aggression* by K. Lorenz. *Animal Behaviour, 15:392–393.*

Dolnik, T. V. 1974. [Circadian and circannual periodicity in Chaffinches *(Fringilla coelebs)* trained to turn a light on and off.] *Zoologicheskii Zhurnal, 53:888–897.*

Dolnik, V. R., and V. M. Gavrilov. 1979. Bioenergetics of molt in the Chaffinch *(Fringilla coelebs). Auk, 96:253–264.*

Domm, L. V. 1955. Recent advances in knowledge concerning the role of hormones in the sex differentiation of birds. *In* Wolfson, A. (ed). *Recent Studies in Avian Biology.* University of Illinois Press, Urbana, Illinois.

Domm, L. V., and E. Taber. 1946. Endocrine factors controlling erythrocyte concentration in the blood of the domestic fowl. *Physiological Zoology, 19:258–281.*

Donner, K. O. 1951. The visual acuity of some passerine birds. *Acta Zoologica Fennica, 66:1–40.*

Donner, K. O. 1953. The spectral sensitivity of the pigeon's retinal elements. *Journal of Physiology* (London), *122:524–537.*

Dooling, R. J. 1982. Auditory perception in birds. *In* Kroodsma, D. E., and E. H. Miller (eds). *Acoustic Communication in Birds*, Vol. 1, 95–130. Academic Press, New York.

Dorrscheidt, G. J. 1978. A reversible method for sound analysis adapted to bioacoustical experiments. *Journal of Bio-Medical Computing*, 9:127–145.

Dorst, J. 1962. *The Migrations of Birds*. Houghton Mifflin Company, Boston.

Dorst, J. 1974. *The Life of Birds*. Columbia University Press, New York.

Dorward, P. K. 1970. Response patterns of cutaneous mechanoreceptors in the domestic duck. *Comparative Biochemical Physiology*, 35:729–735.

Doty, H. A. 1972. Hatchability tests with eggs from captive Wood Ducks. *Poultry Science*, 51:849–853.

Doty, H. A., and F. B. Lee. 1974. Homing to nest baskets by wild female Mallards. *Journal of Wildlife Management*, 38:714–719.

Dow, D. D. 1977. Indiscriminate interspecific aggression leading to almost sole occupancy of space by a single species of bird. *Emu*, 77:115–121.

Dowsett-Lemaire, F. 1979. The imitative range of the song of the Marsh Warbler, *Acrocephalus palustris*, with special reference to imitations of African birds. *Ibis*, 121:453–468.

Drent, R. H. 1970. Functional aspects of incubation in the Herring Gull. *Behaviour*, Suppl. 17:1–132.

Drent, R. H. 1972. Adaptive aspects of the physiology of incubation. *Proceedings of the XVth International Ornithological Congress*, 255–280. E. J. Brill, Leiden.

Drent, R. 1975. Incubation. *In* Farner, D. S., and J. R. King (eds). *Avian Biology*, Vol. 5. Academic Press, New York.

Drewien, R. C., and E. Kuyt. 1979. Teamwork helps the Whooping Crane. *National Geographic*, 155:680–693.

Drobney, R. D. 1984. Effect of diet on visceral morphology of breeding woods ducks. *Auk*, 101:93–98.

Drost, R., and G. Hartmann. 1949. Hohes Alter einer Population des Austerfischers. *Die Vogelwarte*, 2:102–104.

DuBois, A. 1923. The Short-Eared Owl as a foster-mother. *Auk*, 40:383–393.

DuBois, A. D. 1936. Habits and nest life of the Desert Horned Lark. *Condor*, 38:49–56.

Dücker, G., and I. Schulze. 1977. Color vision and color preferences in Japanese Quail (*Coturnix c. japonica*) with colorless oil droplets. *Journal of Comparative Physiological Psychology*, 91:1110–1117.

Deubbert, H. F. 1966. Island nesting of the Cadwall in North Dakota. *Wilson Bulletin*, 78:12–25.

Duffy, D. C. 1983. The ecology of tick parasitism on densely nesting Peruvian seabirds. *Ecology*, 64:110–119.

Duke, G. E., *et al.* 1968. Chromium-51 in food metabolizability and passage rate studies with the Ring-necked Pheasant. *Poultry Science*, 47:1356–1364.

Duncan, C. J. 1960. Preference tests and the sense of taste in the feral pigeon (*Columba livia* Gmelin). *Animal Behaviour*, 8:54–60.

Duncan, C. J. 1964. Taste. *In* Thomson, A. L. (ed). *A New Dictionary of Birds*. McGraw-Hill Book Company, New York.

Duncan, I. J. H., and D. G. M. Wood-Gush. 1971. Frustration and aggression in the Domestic Fowl. *Animal Behaviour*, 19:500–504.

Duncker, H. -R. 1971. *The lung air sac system of birds*. Ergebnisse Anatomische Entwicklungs-Geschichte 45, Heft 6:1–171. Springer Verlag, Berlin, New York.

Dunn, E. H. 1975. The timing of endothermy in the development of altricial birds. *Condor*, 77:288–293.

Durand, A. L. 1972. Land birds over the North Atlantic. *British Birds*, 65:428–442.

Durant, W. 1944. *The Story of Civilisation III: Caesar and Christ*. Simon & Schuster, New York.

Durrer, H., and W. Villiger. 1970. Schillerfarben der Stare (Sturnidae). *Journal für Ornithologie*, 111:133–153.

Durrer, H. and W. Villiger. 1975. Schillerstruktur des Kongofaus (*Afropavo congensis*, Chapin 1936). *Journal für Ornithologie*, 116:94–102.

Duvall, A. J. 1966. Where do they go? *In* Stefferud, A. (ed). *Birds in Our Lives*. U.S. Fish and Wildlife Service, Washington, D.C.

Dwight, J., Jr. 1900. Sequence and plumage of moults of the Passerine birds of New York. *Annals of the New York Academy of Science*, 13:73–360.

Dyrcz, A. 1983. Breeding ecology of the Clay-colored Robin (*Turdus grayi*) in lowland Panama. *Ibis*, 125:287–304.

Earhart, C. M., and N. K. Johnson. 1970. Size dimorphism and food habits of North American owls. *Condor*, 72:251–264.

Eastwood, E. 1967. *Radar Ornithology*. Methuen and Company, Ltd., London.

Eastwood, E., and G. C. Rider. 1965. Some radar measurements of the altitude of bird flight. *British Birds*, 58:393–426.

Ederstrom, H. E., and S. J. Brumleve. 1964. Temperature gradients in the legs of cold-acclimatized pheasants. *American Journal of Physiology*, 207:457–459.

Edminster, F. C. 1939. The effect of predator control on Ruffed Grouse populations in New York. *Journal of Wildlife Management*, 3:345–352.

Edrich, W., and W. T. Keeton. 1977. A comparison of homing behavior in feral and homing pigeons. *Zeitschrift für Tierpsychologie*, 44:389–401.

Egler, F. E. 1958. Science, industry, and the abuse of rights of way. *Science*, 127:573–580.

Ehrlich, P. R., D. S. Dobkin, and D. Wheye. 1986. Commentary: The adaptive significance of anting. *Auk*, 103:835.

Eibl-Eibesfeldt, I. 1970. *Ethology: The Biology of Behavior*. Holt, Rinehart and Winston, New York.

Einarsen, A. S. 1942. Specific results from Ring-necked Pheasant studies in the Pacific Northwest. *Transactions of the North American Wildlife Conference*, 7:130–145.

Eisner, E. 1960. The relationship of hormones to the reproductive behaviour of birds, referring especially to parental behaviour: A review. *Animal Behaviour*, 8:155–179.

Eisner, E. 1969. The effect of hormone treatment upon the duration of incubation in the Bengalese Finch. *Behaviour*, 33:262–276.

Eklund, C. R., and F. E. Charlton. 1959. Measuring the temperatures of incubating penguin eggs. *American Scientist*, 47:80–86.

Elder, W. H. 1954. The oil gland of birds. *Wilson Bulletin*, 66:6–31.

Elder, W. H. 1955. Fluoroscopic measures of shooting pressure on Pink-footed and Grey Lag Geese. *The Wildfowl Trust 7th Annual Report, London, 1953–54*, 123–126.

Elder, W. H. 1964. Chemical inhibitors of ovulation in the pigeon. *Journal of Wildlife Management*, 28:556–575.

Elder, W. H., and M. W. Weller. 1954. Duration of fertility in the domestic Mallard hen after isolation from the drake. *Journal of Wildlife Management*, 18:495–502.

Eliassen, E. 1957. Right ventricle pressures and heartrate in diving birds. *Nature*, 180:512–513.

Eliassen, E. 1960. University of Bergen, *Yearbook*. Mathematical and Natural Sciences Series, No. 12.

Elliott, H. F. I. 1970. Outlook for bird conservation. *Bird Study*, 17:81–94.

Elton, C. 1927. *Animal Ecology*. Sidgwick and Jackson, Ltd., London.

El-Wailly, A. 1966. Energy requirements for egg-laying and incubation in the Zebra Finch, *Taeniopygia castanotis*. *Condor*, 68:582–594.

Emeis, W. 1942. Über den ungünstigen Verlauf des Brutgeschäfts der schleswigholsteinischen Störche (*Ciconia ciconia*) im Sommer 1941. *Beiträger zur Fortpflanzungbiologie der Vögel*, 18:153–155.

Emlen, J. T. 1937. Morning awakening time of a Mockingbird. *Bird-Banding*, 8:81–82.

Emlen, J. T. 1940. Sex and age ratios in survival of the California Quail. *Journal of Wildlife Management*, 4:92–99.

Emlen, J. T. 1941. An experimental analysis of the breeding cycle of the Tricolored Red-wing. *Condor*, 43:209–219.

Emlen, J. T. 1942. Notes on a nesting colony of Western Crows. *Bird-Banding*, 13:143–154.

Emlen, J. T. 1954. Territory, nest building and pair formation in the Cliff Swallow. *Auk*, 71:16–35.

Emlen, J. T. 1955. The study of behavior in birds. *In* Wolfson, A. (ed). *Recent Studies in Avian Biology*. University of Illinois Press, Urbana, Illinois.

Emlen, J. T. 1965. Notes and news. *Auk*, 82:319–320.

Emlen, J. T. 1977. Land bird communities of Grand Bahama Island: the structure and dynamics of an avifauna. *Ornithological Monographs No. 24*. American Ornithologists Union.

Emlen, J. T., and F. W. Lorenz. 1942. Pairing responses of free-living valley quail to sex-hormone pellet implants. *Auk*, 59:369–378.

Emlen, J. T., and D. E. Miller. 1969. Pace-setting mechanisms of the nesting-cycle in the Ring-billed Gull. *Behaviour*, 33:237–261.

Emlen, J. T., and R. L. Penney. 1964. Distance navigation in the Adelie Penguin. *Ibis*, 106:417–431.

Emlen, J. T., *et al.* 1966. Predator-induced parental neglect in a Ring-billed Gull colony. *Auk*, 83:677–679.

Emlen, S. T. 1967. Migratory orientation in the Indigo Bunting, *Passerina cyanea*. *Auk*, 84:309–345.

Emlen, S. T. 1969. The development of migratory orientation in young Indigo Buntings. *Living Bird*, 8:113–126.

Emlen, S. T. 1970. Celestial rotation: its importance in the development of migratory orientation. *Science*, 170:1198–1201.

Emlen, S. T. 1972. An experimental analysis of the parameters of bird song eliciting species recognition. *Behaviour*, 41:130–171.

Emlen, S. T. 1975. Migration: Orientation and Navigation. *In* Farner, D. S., and J. R. King (eds). *Avian Biology*, Vol. 5. Academic Press, New York.

Emlen, S. T., and L. W. Oring. 1977. Ecology, sexual selection and the evolution of mating systems. *Science*, 197:215–223.

Emlen, S. T., J. D. Rising, and W. L. Thompson. 1975. A behavioral and morphological study of sympatry in the Indigo and Lazuli Buntings of the Great Plains. *Wilson Bulletin*, 87:145–179.

Emlen, S. T., and S. L. Vehrencamp. 1983. Cooperative breeding strategies among birds. *In* Brush, A. H., and G. A. Clark, Jr. (eds). *Perspectives in Ornithology*, Cambridge University Press.

Enemar, A., and Sjöstrand. 1972. Effects of the introduction of Pied Flycatchers, *Ficedula hypoleuca*, on the composition of a passerine bird community. *Ornis Scandinavica*, 3:79–89.

Enoksson, B., and S. G. Nilsson. 1983. Territory size and population density in relation to food supply in the Nuthatch *Sitta europaea (Aves)*. *Journal of Animal Ecology*, 52:927–935.

Entrikin, R. K., and L. C. Erway. 1972. A genetic investigation of roller and tumbler pigeons. *Journal of Heredity*, 63:351–354.

Erickson, M. M. 1938. Territory, annual cycle, and numbers in a population of wren-tits (*Chamaea fasciata*). *University of California Publications in Zoology*, 42:247–334.

Errington, P. L. 1945. Some contributions of a 15-year local study of the Northern Bobwhite to a knowledge of population phenomena. *Ecological Monographs*, 15:1–34.

Ettinger, A. O., and J. R. King. 1981. Consumption of green wheat enhances photostimulated ovarian growth in White-crowned Sparrows. *Auk*, 98:832–834.

Eutermoser, A. 1961. Schlagen Beizfalken bevorzugt kranke Krähen? *Vogelwelt*, 82:101–104.

Evans, L. T. 1936. Territorial behavior of normal and castrated females of *Anolis carolinensis*. *Journal of Genetic Psychology*, 49:49–60.

Evans, P. R. 1968. Autumn movements and orientation of waders in northeast England and southern Scotland studied by radar. *Bird Study*, 15:53–64.

Evans, P. R. 1969. Ecological aspects of migration and premigratory fat deposition in the Lesser Redpoll, *Carduelis flammea cabaret*. *Condor*, 71:316–330.

Evans, R. M. 1970. Parental recognition and the "mew" call in Black-bellied Gulls (*Larus bulleri*). *Auk*, 87:503–513.

Ewald, P. W., and S. Rohwer. 1982. Effects of supplemental feeding on timing of breeding, clutch-size and polygyny in Red-winged Blackbirds, (*Agelaius phoeniceus*). *Journal of Animal Ecology*, 51:429–450.

Fabricius, E. 1959. What makes plumage waterproof? *Report of the Wildfowl Trust*, 10:105–113.

Fabricius, E. 1964. Crucial periods in the development of the following response in young nidifugous birds. *Zeitschrift für Tierpsychologie*, 21:326–337.

Fairbridge, R. W. 1960. The changing level of the sea. *Scientific American*, 202(5):70–79.

Fallis, A. M., and D. O. Trainer, 1964. Blood parasites. *In* Linduska, J. P., and A. L. Nelson (eds). *Waterfowl Tomorrow*. U.S. Fish and Wildlife Service, Washington, D.C.

Falls, J. B. 1963. Properties of bird song eliciting responses from territorial males. *Proceedings of the XIIIth International Ornithological Congress*, 1:259–271. American Ornithologists' Union.

Falls, J. B. 1969. Functions of territorial song in the White-throated Sparrow. *In* Hinde, R. A. (ed). *Bird Vocalizations: Their Relation to Current Problems in Biology and Psychology*. Cambridge University Press, Cambridge.

Falls, J. B. 1982. Individual recognition by sound in birds. *In* Kroodsma, D. E., and E. H. Miller (eds). *Acoustic Communication in Birds*, Vol. 2, 237–278. Academic Press, New York.

Falls, J. B. 1985. Song matching in western meadowlarks. *Canadian Journal of Zoology*, 63:2520–2524.

Falls, J. B., and M. K. McNicholl. 1979. Neighbor-stranger discrimination by song in male Blue Grouse. *Canadian Journal of Zoology*, 57:457–462.

Falls, J. B., and L. Szijj. 1959. Reactions of Eastern and Western meadowlarks in Ontario to each others vocalizations. *Anatomical Record*, 134:560.

Farabaugh, S. M. 1982. The ecological and social significance of duetting. *In* Kroodsma, D. E., and E. H. Miller (eds). *Acoustic Communication in Birds*, Vol. 2, 85–124. Academic Press, New York.

Faraci, F. M., *et al.* 1984. Oxygen delivery to the heart and brain during hypoxia: Pekin duck vs. bar-headed goose. *American Journal of Physiology*, 247:169–175.

Farmer, J. N., and R. P. Breitenbach. 1966. A comparison of the course of *Plasmodium lophurae* infections in control and hormonally bursectomized chickens. *American Zoologist*, 6:307–308.

Farner, D. S. 1945. The return of Robins to their birthplaces. *Bird-Banding,* 16:81–99.

Farner, D. S. 1955. The annual stimulus for migration: experimental and physiologic aspects. *In* Wolfson A. (ed). *Recent Studies in Avian Biology.* University of Illinois Press, Urbana, Illinois.

Farner, D. S. 1964. The photoperiodic control of reproductive cycles in birds. *American Scientist,* 52:137–156.

Farner, D. S. 1970. Some glimpses of comparative avian physiology. *Federation Proceedings,* 29:1649–1663.

Farner, D. S. (ed). 1973. *Breeding Biology of Birds.* National Academy of Sciences. Washington, D.C.

Farner, D. S., R. S. Donham, K. S. Matt, P. W. Mattocks, Jr., M. C. Moore, and J. C. Wingfield. 1983. The nature of photorefractoriness. *In* Mikami, S. *et al.* (eds). *Avian endocrinology: Environmental and Ecological Perspectives,* 249–266.

Farner, D. S., and J. R. King (eds). 1971–1975. *Avian Biology.* 5 vols. Academic Press, New York.

Feare, C. J. 1976. Desertion and abnormal development in a colony of Sooty Terns, *Sterna fuscata,* infested by virus-infected ticks. *Ibis,* 188:112–115.

Feare, C. J., and K. P. Swannack. 1978. Starling damage and its prevention at an open-fronted calf yard. *Animal Production,* 26:259–266.

Feduccia, A. 1980. *The Age of Birds.* Harvard University Press, Cambridge, Massachusetts.

Feduccia, A., and H. B. Tordoff. 1979. Feathers of *Archaeopteryx:* asymmetric vanes indicate aerodynamic function. *Science,* 203:1021–1022.

Ferguson, D. E. 1967. Sun compass orientation of the Northern Cricket Frog, *Acris crepitans. Animal Behaviour,* 15:45–53.

Ferrell, R., and L. F. Baptista. 1982. Diurnal rhythms in the vocalizations of budgerigars. *Condor,* 84:123–124.

Ferster, C. B., and B. F. Skinner. 1957. *Schedules of Reinforcement.* Appleton-Century-Crofts, New York.

Fiala, K. L. 1979. A laparotomy technique for nestling birds. *Bird-banding,* 50:366–367.

Ficken, M. S. 1977. Avian play. *Auk,* 94:573–582.

Finley, M. T., M. P. Dieter, and L. N. Locke. 1976. Delta-aminolevulinic acid dehydratase: inhibition in ducks dosed with lead shot. *Environmental Research,* 12:243–249.

Finley, M. T., and R. C. Stendell. 1978. Survival and reproductive success of Black Ducks fed methyl mercury. *Environmental Pollution,* 16:51–64.

Finnlund, M., R. Hisso, J. Koirusaari, E. Merila, and I. Nuuja. 1985. Eggshells of Arctic Terns from Finland: Effects of incubation and geography. *Condor,* 87:79–86.

Fiore, L., *et al.* 1984. Effects of intensity changes in the horizontal geomagnetic field component on the oriented behavior of caged robins, (*Erithacus rubecula*). *Monitore Zoologico Italiano Firenze,* 18:337–346.

Fischdick, G., V. Hahn, and K. Immelmann. 1984. Die Sozialisation beim Rosenköpfchen *Agapornis roseicollis. Journal für Ornithologie,* 125:307–319.

Fischer, A. B. 1969. Laboruntersuchungen und Freilandbeobachtungen zum Sehvermögen und Verhalten von Altweltgeiern. *Zoologischer Jahrbücher Abteilung für Systematik,* 96:81–132.

Fischer, K. 1961. Untersuchungen zur Sonnenkompassorientierung und Laufaktivität von Smaragdeidechsen (*Lacerta viridis* Laur.). *Zeitschrift für Tierpsychologie,* 18:450–470.

Fisher, H. I. 1958. The "hatching muscle" in the chick. *Auk,* 75:391–399.

Fisher, H. I. 1962. The hatching muscle in Franklin's Gull. *Wilson Bulletin,* 74:166–172.

Fisher, H. I. 1971a. Experiments on homing in Laysan Albatrosses, *Diomedea immutabilis. Condor,* 73:389–400.

Fisher, H. I. 1971b. The Laysan Albatross: its incubation, hatching, and associated behaviors. *Living Bird,* 1971: 19–78.

Fisher, H. I. 1972. The nutrition of birds. *In* Farner, D. S., and J. R. King (eds). *Avian Biology.* Vol. 2. Academic Press, New York.

Fisher, H. I. 1975. Longevity in the Laysan Albatross, *Diomedea immutabilis. Bird-Banding,* 46:1–6.

Fisher, H. I., and M. L. Fisher. 1969. The visits of the Laysan Albatrosses to the breeding colony. *Micronesica,* 5:173–221.

Fisher, J. 1951. *Watching Birds.* Penguin Books, Harmondsworth, Middlesex.

Fisher, J. 1952. *The Fulmar.* Collins, London.

Fisher, J., and R. A. Hinde. 1949. The opening of milk bottles by birds. *British Birds,* 42:347–357.

Fisher, J., and R. M. Lockley. 1954. *Sea-Birds.* Houghton Mifflin Company, Boston.

Fisher, J., N. Simon and J. Vincent. 1969. *The Red Book: Wildlife in Danger.* Collins, London.

Fisler, G. F. 1977. Interspecific hierarchy at an artificial food source. *Animal Behavior,* 25:240–244.

Fitter, R. 1969. *Book of British Birds.* Drive Publications, London.

Flegg, J. J. M., and C. J. Cox. 1975. Mortality in the Black-headed Gull. *British Birds,* 68:437–449.

Flegg, J. J. M., and C. J. Cox. 1977. Morphometric studies of a population of Blue and Great Tits. *Ringing and Migration*, 1:135–140.

Fleuster, W. 1973. Versuche zur Reaktion freilebender Vögel auf Klangattrapen verschiedener Büchfinkalarme. *Journal für Ornithologie*, 114:417–428.

Flower, S. S. 1938. Further notes on the duration of life in animals. IV. Birds. *Proceedings of the Zoological Society of London*, Series A, 108:195–235.

Flower, W. U. 1969. Over 60 Wrens roosting together in one nest box. *British Birds*, 62:157–158.

Fogarty, M. J., and W. M. Hetrick. 1973. Summer foods of Cattle Egrets in north central Florida. *Auk*, 90:268–280.

Fordham, R. A. 1968. Dispersion and dispersal of the Dominican Gull in Wellington, N. Z. *Proceedings of the New Zealand Ecological Society*, 15:40–50.

Fordham, R. A. 1970. Mortality and population change of Dominican Gulls in Wellington, New Zealand. *Journal of Animal Ecology*, 39:13–27.

Forrester, D. J., *et al.* 1977. An epizootic of waterfowl associated with a red tide episode in Florida. *Journal of Wildlife Diseases*, 13:160–167.

Foster, M. S. 1969. Synchronized life cycles in the Orange-crowned Warbler and its Mallophagan parasites. *Ecology*, 50:315–323.

Foster, M. S. 1974. A model to explain molt-breeding overlap and clutch size in some tropical birds. *Evolution*, 8:182–190.

Foster, M. S. 1975. The overlap of molting and breeding in some tropical birds. *Condor*, 77:304–314.

Fox, G. A. 1976. Eggshell quality: its ecological and physiological significance in a DDE-contaminated Common Tern population. *Wilson Bulletin*, 88:459–477.

Frank, F. 1941. Besondere Nistweise der Feldsperlings in Bessarabien. *Ornithologische Monatsberichte*, 52:156–157.

Frankel, A. I., and T. S. Baskett. 1961. The effect of pairing on cooing of penned Mourning Doves. *Journal of Wildlife Management*, 25:372–384.

Franks, E. C. 1967. The responses of incubating Ringed Turtle Doves (*Streptopelia risoria*) to manipulated egg temperatures. *Condor*, 69:268–276.

Fraser, J. G. 1922. *The Golden Bough*. Macmillan Publishing Company, New York.

Frederickson, L. H. 1969. An experimental study of clutch size of the American Coot. *Auk*, 86:541–550.

Frederikson, K. A. 1940. Über das Brüten der Lachmöwe, *Larus ridibundus* L., auf Felseninseln im Schärenhof und die Ursachen dazu. *Ornis Fennica*, 17:62–63.

Freeman, H. D. 1968. Iban augury. *In* Smythies, B. E. (ed) *The Birds of Borneo*. Oliver and Boyd, Edinburgh and London.

Fretwell, S. 1968. Habitat distribution and survival in the field sparrow (*Spizella pusilla*). *Bird-Banding*, 39:293–306.

Fretwell, S. 1969. Dominance behavior and winter habitat distribution in juncos (*Junco hyemalis*). *Bird-Banding*, 40:1–25.

Friedmann, H. 1929. *The Cowbirds*. Charles C Thomas Publishers, Springfield, Illinois.

Friedmann, H. 1947. Geographic variations of the Black-bellied, Fulvous and White-faced Tree Ducks. *Condor*, 49:189–195.

Friedmann, H. 1955. The Honey-guides. *U.S. National Museum Bulletin 208*. Washington, D.C.

Friedmann, H. 1968. The evolutionary history of the avian genus *Chrysococcyx*. *U.S. National Museum Bulletin*, 265:1–137. Washington, D.C.

Friedmann, H., and L. F. Kiff. 1985. The parasitic cowbirds and their hosts. *Proceedings Western Foundation of Vertebrate Zoology*, 2:227–302.

Friend, M., and D. O. Trainer. 1970. Polychlorinated biphenyl: interaction with duck hepatitis virus. *Science*, 170:1314–1316.

Friend, M., and D. O. Trainer. 1974. Experimental DDT-duck hepatitis virus interaction studies. *Journal of Wildlife Management*, 38:887–895.

Frings, H., and M. Frings. 1957. Recorded calls of the Eastern Crow as attractants and repellents. *Journal of Wildlife Management*, 21:91.

Frings, H., and J. Jumber. 1954. Preliminary studies on the use of specific sound to repel Starlings (*Sturnus vulgaris*) from objectionable roosts. *Science*, 119:318–319.

Frisch, K. von. 1956. The "language" and the orientation of the bees. *Proceedings of the American Philosophical Society*, 100:515–519.

Frith, H. J. 1956. Temperature regulation in the nesting mounds of the Mallee-Fowl, *Leipoa ocellata* Gould. Commonwealth Scientific and Industrial Research Organization, *Wildlife Research*, 1:79–95.

Frith, H. J. 1957a. Clutch size in the Goldfinch. *Emu*, 57:287–288.

Frith, H. J. 1957b. Experiments on the control of temperature in the mound of the Mallee-Fowl, *Leipoa ocellata* Gould (Megapodiidae). Commonwealth Scientific and Industrial Research Organization, *Wildlife Research*, 2:101–110.

Frith, H. J. 1961. Ecology of wild ducks in inland Australia. *The Wildfowl Trust 12th Annual Report, 1959–1960*, 81–91.

Frith, H. J., and R. A. Tilt. 1959. Breeding of the Zebra Finch in the Murrumbridgee irrigation area, New South Wales. *Emu*, 59:289–295.

Fromme, H. G. 1961. Untersuchungen über das Orientierungsvermögen nächtliche ziehender Kleinvogel (*Erithacus rubecula* und *Sylvia communis*). *Zeitschrift für Tierpsychologie*, 18:205–219.

Fry, C. H. 1972. The biology of African Bee-eaters. *Living Bird*, 1972:75–112.

Fry, D. M., and C. K. Toone. 1981. DDT-induced feminization of gull embryos. *Science*, 213:922–924.

Gabrielsen, G. W. 1985. Free and forced diving in ducks: Habituation of the initial dive response. *Acta Physiologica Scandinavica*, 123:67–72.

Gabrielson, I. F. 1948. A study of the home life of the Brown Thrasher. *In* Bent, A. C. *U.S. National Museum Bulletin*, 195:361–362.

Gallet, E. 1950. *The Flamingos of the Camargue*. Oxford University Press, New York.

Galushin, V. 1974. Synchronous fluctuations in populations of some raptors and their prey. *Ibis*, 116:127–134.

Gasaway, W. C. 1976. Volatile fatty acid and metabolizable energy derived from cecal fermentation in the Willow Ptarmigan. *Comparative Biochemistry and Physiology* A, 53:115–121.

Gatehouse, S. N., and B. J. Markham. 1970. Respiratory metabolism in three species of raptors. *Auk*, 87:738–741.

Gates, J. E., and L. W. Gysel. 1978. Avian nest dispersion and fledging success in field-forest ecotones. *Ecology*, 59:871–883.

Gatter, W. 1976. Über den Wegzug des Gimpels, *P. pyrrhula*: Geschlechteverhaltnis und Einfluss von Witterungsfaktoren. *Die Vogelwarte*, 28:165–170.

Gatter, W. 1979. Unterschiedliche Zuggeschwindigkeit nähe verwandter Vogelarten. *Journal für Ornithologie*, 120:221–225.

Gaunt, A. S. 1971. Pressure and air flow during distress calls of *Sturnus vulgaris*. *American Zoologist*, 11:706.

Gaunt, A. S. 1983. On sonograms, harmonics and assumptions. *Condor*, 85:259–261.

Gaunt, A. S. 1983. An hypothesis concerning the relationship of syringeal structure to vocal abilities. *Auk*, 100:853–862.

Gaunt, A. S., and S. L. L. Gaunt. 1985a. Electromyographic studies of the syrinx in parrots (Aves, Psittacidae). *Zoomorphology*, 105:1–11.

Gaunt, A. S., and S. L. L. Gaunt. 1985b. Syringeal structure and avian phonation. *Current Ornithology*, 2:213–245.

Gaunt, A. S., S. L. L. Gaunt, and R. M. Casey. 1982. Syringeal mechanics reassessed: Evidence from *Streptopelia*. *Auk*, 99:474–494.

Gaunt, A. S., S. L. L. Gaunt, and D. W. Hector. 1976. Mechanics of the syrinx in *Gallus gallus*: I. *Condor*, 78:208–223.

Gause, G. F. 1934. Experimental studies on the struggle for existence in *Paramecium caudatum, Paramecium aurelia*, and *Stylonichia mytilus*. *Zoologischeskii Zhurnal*, 12.

Gause, G. F. 1934. *The Struggle for Existence*. Williams and Wilkins Company, Baltimore.

Gauthreaux, S. A. 1969. Stellar orientation of migratory restlessness in the White-throated Sparrow, *Zonotrichia albicollis*. *American Zoologist*, 9:1065.

Gauthreaux, S. A. 1972. Behavioral responses of migrating birds to daylight and darkness: a radar and direct visual study. *Wilson Bulletin*, 84:136–148.

Gauthreaux, S. A., Jr. 1982. The ecology and evolution of avian migration systems. *In* Farner, D. S., and J. R. King (eds). *Avian Biology*, 6:93–167. Academic Press, New York.

Gavin, T. A., and E. K. Bollinger. 1985. Multiple paternity in a territorial passerine: The Bobolink. *Auk*, 102:550–555.

Geis, A. D., and W. F. Crissey. 1969. Effect of restrictive hunting regulations on Canvasback and Redhead harvest rates and survival. *Journal of Wildlife Management*, 33:860–866.

Geisthardt, G. 1977. [On the possibilities of reducing the population of wild domestic pigeons.] *Zeitschrift für Angewandte Zoologie*, 64:27–36.

Genovese, R. F., and M. P. Browne. 1978. Sickness-induced learning in chicks. *Behavioral Biology*, 24:68–76.

George, J. C., and A. J. Berger. 1965. *Avian Myology*. Academic Press, New York.

George, J. C., and R. M. Naik. 1960a. Some observations on the distribution of the blood capillaries in the pigeon breast muscle. *Auk*, 77:224–226.

George, J. C., and R. M. Naik. 1960b. Intramuscular fat store in the pectoralis of birds. *Auk*, 77:216–217.

George, U. 1970. Beobachtungen an *Pterocles senegallus* und *P. coronatus* in der Nordwest-Sahara. *Journal für Ornithologie*, 111:175–188.

Gerstell, R. 1942. The place of winter feeding in practical wildlife management. *In* Pennsylvania Game Commission, *Research Bulletin No. 2*. Harrisburg, Pennsylvania.

Gewecke, M., and M. Woike. 1978. Breast feathers as an air-current sense organ for the control of flight behaviour in a songbird (*Carduelis spinus*). *Zeitschrift für Tierpsychologie*, 47:293–298.

Gibb, J. 1947. Sun-bathing in birds. *British Birds*, 40:172–174.

Gibb, J. 1958. Predation by tits and squirrels on the eucosmid *Enarmonia conicolana* (Heyl.). *Journal of Animal Ecology*, 27:375–396.

Gibb, J., and C. Gibb. 1951. Waxwings in the winter of 1949–50. *British Birds*, 44:158–163.

Gilbert, A. B. 1971. *In* Bell, D. J., and B. M. Freeman (eds), *Physiology and Biochemistry of the Domestic Fowl*, Vol. 3. Academic Press, New York.

Gill, F. B., and L. L. Wolf. 1979. Nectar loss by Golden-winged Sunbirds to competitors. *Auk*, 96:448–461.

Gilliard, E. T. 1956. Bower ornamentation versus plumage characteristics in bower-birds. *Auk*, 73:450–451.

Gilpin, M. E., and J. M. Diamond. 1976. Calculation of immigration and extinction curves from the species-area-distance relation. *Proceedings of the National Academy of Sciences of the U.S.A.*, 73:4130–4134.

Gingerich, P. D. 1972. A new partial mandible of *Ichthyornis*. *Condor*, 74:471–473.

Giroux, J. F. 1981. Interspecific parasitism by Redhead (*Aythya americana*) on islands in southeastern Alberta, Canada. *Canadian Journal of Zoology*, 59:2053–2057.

Gish, S. L., and E. S. Morton. 1981. Structural adaptations to local habitat acoustics in Carolina Wren songs. *Zeitschrift für Tierpsychologie*, 56:74–84.

Glase, J. C. 1973. Ecology of social organization in the Black-capped Chickadee. *Living Bird*, 12:235–267.

Glover, F. A. 1956. Nesting and production of the Blue-winged Teal (*Anas discors* Linnaeus) in northwest Iowa. *Journal of Wildlife Management*, 20:28–46.

Glutz von Blotzheim, U. N., K. Bauer, and E. Bezzel. 1971. *Handbuch der Vögel Mitteleuropas*. 6 vol. Akademische Verlagsgesellschaft, Frankfurt.

Goethe, F. 1941. Beobachtungen am Neusiedlersee und in dem Gebiet der Salzlachen. *Journal für Ornithologie*, 89:268–281.

Goethe, F. 1953. Soziale Hierarchie im Aufzuchtschwarm der Silbermöwen. *Zeitschrift für Tierpsychologie*, 10:44–50.

Goforth, W. R., and T. S. Baskett. 1971. Social organization of penned Mourning Doves. *Auk*, 88:528–542.

Goldby, F. 1964. Nervous system. *In* Thomson. A. L. (ed). *A New Dictionary of Birds*. McGraw-Hill Book Company, New York.

Goldsmith, K. M., and T. H. Goldsmith. 1982. Sense of smell in the Black-chinned Hummingbird. *Condor*, 84:237–238.

Goldsmith, T. H. 1980. Hummingbirds see near ultraviolet light. *Science*, 207:786–788.

Goodall, J. D., R. A. Philippi, B. and A. W. Johnson. 1945. Nesting habits of the Peruvian Gray Gull. *Auk*, 62:450–451.

Goodbody, I. M. 1952. The post-fledging dispersal of juvenile titmice. *British Birds*, 45:279–285.

Goodpasture, K. A. 1955. Recovery of a Chickadee population from the 1951 ice storm. *Migrant*, 26(2):21–23.

Goodridge, A. G., and E. G. Ball. 1967. The effect of prolactin on lipogenesis in the pigeon: *in vivo* studies; *in vitro* studies. *Biochemistry*, 6:1676–1682; 2335–2343.

Goodwin, D. 1956. Observations on the voice and some displays of certain pigeons. *Avicultural Magazine*, 62:17–33; 62–70.

Goodwin, D. 1956. Care of the body surface—preening, bathing, dusting, and anting. *In* Hutson, H. P. W. (ed). *The Ornithologists' Guide*. Philosophical Library, New York.

Goodwin, D. 1966. The bowing display of pigeons in reference to phylogeny. *Auk*, 83:117–123.

Goodwin, D. 1967. *Pigeons and Doves of the World*. British Museum (Natural History), London.

Goodwin, D. 1982. *Estrildid finches of the world*, British Museum, London.

Goodwin, D. 1983. *Pigeons and Doves of the World*. Comstock Press, Ithaca, New York.

Gosney, S., and R. A. Hinde. 1976. Changes in the sensitivity of female budgerigars to male vocalizations. *Journal of Zoology* (London) 179:407–410.

Goss-Custard, J. D., and W. J. Sutherland. 1984. Feeding specializations in oystercatchers, *Haematopus ostralegus*. *Avian Behavior*, 32:299–300.

Gossett, R. L., M. F. Gossette, and W. Riddell. 1966. Comparisons of successive discrimination reversal performances among closely and remotely related avian species. *Animal Behaviour*, 14:560–564.

Gotmark, F., and M. Andersson. 1984. Colonial breeding reduces nest predation in the common gull (*Larus canus*). *Animal Behaviour*, 32:485–492.

Gottlieb, G. 1963. A naturalistic study of imprinting in Wood Ducklings (*Aix sponsa*). *Journal of Comparative and Physiological Psychology*, 56:86–91.

Gottlieb, G. 1966. Species identification by avian neonates: contributory effect of perinatal auditory stimulation. *Animal Behaviour*, 14:282–290.

Gottlieb, G. 1971a. *Development of Species Identification in Birds*. University of Chicago Press, Chicago.

Gottlieb, G. 1971b. Ontogenesis of sensory function in birds and mammals. *In* Tobach, E. *et al.* (eds). *The Biopsychology of Development*. Academic Press, New York.

Gould, E. 1957. Orientation in box turtles, *Terrapene c. carolina* (L). *Biological Bulletin*, 112:336–348.

Gould, J. L., J. L. Kirschvink, and K. S. Deffeyes. 1978. Bees have magnetic remanence. *Science*, 102:1026–1028.

Govardovskii, V. I., and L. V. Zueva. 1977. Visual pigments of chicken and pigeon. *Vision Research*, 17:537–543.

Gowaty, P. A. 1983. Male parental care and apparent monogamy among Eastern Bluebirds (*Sialia sialis*). *American Naturalist* 121:149–157.

Gowaty, P. A. 1985. Multiple parentage and apparent monogamy in birds. *In* Gowaty, P. A., and D. W. Mock (eds). *Avian Monogamy. Ornithological Monogographs*, 37:11–21.

Gowaty, P. A., and A. A. Karlin. 1984. Multiple paternity and maternity in single broods of apparently monogamous Eastern Bluebirds. *Behavioral Ecology and Sociobiology*, 15:91–95.

Gower, C. 1936. The cause of blue color as found in the Bluebird and the Blue Jay. *Auk*, 53:178–185.

Graber, R. R., and J. W. Graber. 1962. Weight characteristics of birds killed in nocturnal migration. *Wilson Bulletin*, 74:74–78.

Graham, F. 1967. The uncertain defenders. *Audubon*, 69(3):28–37.

Grammeltvedt, R. 1978. Attrophy of a breast muscle with a single fibre type (m. pectoralis) in fasting Willow Grouse, *Lagopus lagopus* (L). *Journal of Experimental Zoology*, 205:195–204.

Grange, W. B. 1948. *Wisconsin Grouse Problems*. Wisconsin Conservation Department, Madison, Wisconsin.

Grant, B. R. 1984. The significance of song variation in a population of Darwin's finches. *Behaviour*, 89:90–116.

Grant, G. S. 1982. Avian incubation: egg temperature, nest humidity and behavioral thermoregulation in a hot environment. *Ornithological Monographs*, 30:1–75.

Grant, G. S., T. N. Pettit, H. Rahn, G. C. Whittow, and C. V. Paganelli. 1982. Regulation of water loss from Bonin Petrel (*Pterodroma hypoleuca*) eggs. *Auk*, 99:236–242.

Grant, P. R., B. R. Grant, J. N. M. Smith, I. Abbott, and L. K. Abbott. 1976. Darwin's finches: population variation and natural selection. *Proceedings of the National Academy of Sciences of the U.S.A.*, 73:257–261.

Grassé, P. P. 1950. Organization des sociétés d'oiseaux. *Traité de Zoologie*. Tome XV, *Oiseaux*. Masson et Cie, Paris.

Grau, C. R. 1976. Ring structure of avian egg yolk. *Poultry Science*, 55:1418–1422.

Grau, C. R., and L. B. Astheimer. 1982. Patterns of yolk growth in seabirds. *Pacific Seabird Group, Bulletin*, 9(2):65 (Abstract).

Graves, H. B. 1970a. Development of social preferences in selected lines of *Gallus* chicks. *American Zoologist*, 10:289.

Graves, H. B. 1970b. Comparative ethology of imprinting: field and laboratory studies of Wild Turkeys, Jungle Fowl, and Domestic Fowl. *American Zoologist*, 10:483.

Graves, H. B., and P. B. Siegel. 1966. Qualitative inheritance of the approach response of chicks. *American Zoologist*, 6:568.

Green, C. 1949. The Black-shouldered Kite in Masira (Oman). *Ibis*, 91:450–464.

Green, C. 1972. Use of tool by Orange-winged Sitella. *Emu*, 72:185–186.

Green, R., W. J. Carr, and M. Green. 1968. The hawk-goose phenomenon: Further confirmation and a search for the releaser. *Journal of Psychology*, 69:271–276.

Greene, H. W. 1986. Natural History and evolutionary biology. *In* Feder, M. E., and O. V. Lauder (eds). Predator-Prey Relations: Perspectives and Approaches from the Study of Lower Vertebrates, 99–108. University of Chicago Press.

Greenewalt, C. H. 1955. The flight of the Black-capped Chickadee and the White-breasted Nuthatch. *Auk*, 72:1–5.

Greenewalt, C. H. 1960. *Hummingbirds*. Doubleday and Company, Inc., Garden City, New York.

Greenewalt, C. H. 1968. *Bird Song: Acoustics and Physiology*. Smithsonian Institution Press, Washington, D.C.

Greenlaw, J. 1977. Taxonomic distribution, origin and evolution of bilateral scratching in ground-feeding birds. *Condor*, 79:426–439.

Greenlaw, J. S., and W. Post. 1985. Evolution of monogamy in seaside sparrows, *Ammodramus maritimus*: tests of hypothesis. *Animal Behaviour*, 33:373–383.

Greenway, J. C. 1958. *Extinct and Vanishing Birds of the World*. American Committee for International Wildlife Protection, Special Publication No. 13, New York.

Greenwood, P. J. 1980. Mating systems, philopatry and dispersal in birds and mammals. *Animal Behaviour*, 28:1140–1162.

Greenwood, P. J., and P. H. Harvey. 1982. The natal and breeding dispersal of birds. *Annual Reviews of Ecology and Systematics*, 13:1–21.

Greenwood, P. J., P. H. Harvey, and C. M. Perrins. 1978. Inbreeding and dispersal in the Great Tit. *Nature*, 271:52–54.

Greenwood, R. J. 1969. Mallard hatching from an egg cracked by freezing. *Auk*, 86:752–754.

Gress, F., *et al.* 1973. Reproductive failures of Double-Crested Cormorants in southern California and Baja California. *Wilson Bulletin*, 85:197–208.

Grieg-Smith, P. W. 1978. The formation, structure, and function of mixed species insectivorous bird flocks in West African savanna woodland. *Ibis*, 120:284–297.

Grier, J. W. 1982, Ban of DDT and subsequent recovery of reproduction in Bald Eagles. *Science*, 218:1232–1235.

Griffin, D. R. 1953. Acoustic orientation in the Oil Bird, *Steatornis*. *Proceedings of the National Academy of Science*, 39:884–893.

Griffin, D. R. 1955. Bird navigation. *In* Wolfson, A. (ed). *Recent Studies in Avian Biology*. University of Illinois Press, Urbana, Illinois.

Griffin, D. R. 1964. *Bird Migration*. Natural History Press, Garden City, New York.

Griffin, D. R., and R. J. Hock. 1949. Airplane observations of homing birds. *Ecology*, 30:176–198.

Grimes, L. 1965. Antiphonal singing in *Laniarius barbarus* and the auditory reaction time *Ibis*, 107:101–104.

Grimes, L. G. 1976. Cooperative breeding in African birds. *In* Frith, H. J., and J. H. Calaby (eds). *Proceedings of the 16th International Ornithological Congress*. Australian Academy of Science, Canberra.

Grinnell, J. 1904. The origin and distribution of the chestnut-backed chickadee. *Auk*, 21:364–382.

Grinnell, J. 1922. The role of the "accidental." *Auk*, 39:373–389.

Grinnell, J. 1929. A new race of hummingbird from southern California. *Condor*, 31:226–227.

Grinnell, J., and T. I. Storer. 1924. *Animal Life in the Yosemite*. University of California Press. Berkeley.

Griscom, L. 1937. A monographic study of the Red Crossbill. *Proceedings of the Boston Society of Natural History*, 41:77–210.

Griscom, L. 1941. The recovery of birds from disaster. *Audubon Magazine*, 43:191–196.

Griscom, L. 1945. *Modern Bird Study*. Harvard University Press, Cambridge.

Grohmann, J. 1939. Modifikation oder Funktionsreifung? *Zeitschrift für Tierpsychologie*, 2:132–144.

Groskin, H. 1950. Banding 4469 Purple Finches at Ardmore, Pa. *Bird-Banding*, 21:93–99.

Gross, A. O. 1940. The migration of Kent Island Herring Gulls. *Bird-Banding*, 11:129–155.

Gross, A. O. 1965a. The incidence of albinism in American birds. *Bird-Banding*, 36:67–71.

Gross, A. O. 1965b. Melanism in North American birds. *Bird-Banding*, 36:240–242.

Gross, A. O. 1968. Albinistic eggs (white eggs) of some North American Birds. *Bird-Banding*, 39:1–6.

Grubb, T. C., Jr. 1972. Smell and foraging in shearwaters and petrels. *Nature*, 237:404–405.

Grubb, T. C., Jr. 1974. Olfactory navigation to the nesting burrow in Leach's Petrel (*Oceanodroma leucorrhoa*). *Animal Behaviour*, 22:192–202.

Grubb, T. C., Jr. 1976. Adaptiveness of foraging in the Cattle Egret. *Wilson Bulletin*, 88:145–148.

Grzimek, B. 1972. Physical capacities. *In* Grzimek, B. (ed). *Grzimek's Animal Life Encyclopedia*, 7:95–96. Van Nostrand Reinhold Company, New York.

Grzimek, B. (ed). 1972–1973. *Grzimek's Animal Life Encyclopedia*. Vols. 7–9, *Birds*. Van Nostrand Reinhold Company, New York.

Guhl, A. M. 1956. The social order of chickens. *Scientific American*, 194(2):42–46.

Guhl, A. M. 1968. Social inertia and social stability in chickens. *Animal Behaviour*, 16:219–232.

Guiton, P. 1978. Visuo-motor plasticity in the chick: adaptive readjustments with distorted visual input. *Biology of Behaviour*, 3:63–70.

Gurney, M. E. 1981. Hormonal control of cell form and number in Zebra finch song system. *Journal of Neuroscience*, 6:658–673.

Gurney, M. E., and M. Konishi, 1980. Hormone induced sexual differentiation of brain and behavior in Zebra finches. *Science,* 208:1380–1383.

Guttierez, R. J., R. M. Zink, and S. Y. Yang. 1983. Genic variation, systematic and biogeographic relationships of some galliform birds. *Auk, 100*:33–47.

Güttinger, H. R. 1970. Zur Evolution von Verhaltensweisen und Lautäusserungen bei Prachtfinken (Estrildidae). *Zeitschrift für Tierpsychologie,* 27:1011–1075.

Güttinger, H. R. 1976. Zur Systematischen Stellung der Gattungen *Amadina, Lepidopygia,* und *Lonchura* (Aves, Estrildidae). *Bonner Zoologische Beiträge,* 27:218–244.

Güttinger, H. R. 1979. The integration of learnt and genetically programmed behaviour: a study of hierachical organization in songs of canaries, greenfinches and their hybrids. *Zeitschrift für Tierpsychologie,* 49: 285–303.

Güttinger, H. R., and J. Nicolai. 1973. Struktur und Funktion der Rufe bei Prachtfinken (Estrildidae). *Zeitschrift für Tierpsychologie,* 33:319–334.

Güttinger, H. R., E. Pröve, K. Wiechel, and A. Pesch. 1984. Hormonelle Korrelate zur Gesangsentwicklung der Kanarienvögel. *Journal für Ornithologie,* 125:245–247.

Güttinger, H. R., I. Wolffgramm, and F. Thimm. 1978. The relationship between species specific song programs and individual learning in song birds. *Behaviour,* 65:241–262.

Gwinner, E. 1964. Untersuchungen über das Ausdrucks- und Sozialverhalten des Kolkraben (*Corvus corax corax* L.). *Zeitschrift für Tierpsychologie,* 21:657–748.

Gwinner, E. 1965. Über den Einfluss des Hungers und anderen Factoren auf die Versteck-Aktiv-ität des Kolkraben (*Corvus corax*). *Die Vogelwarte,* 23:1–4.

Gwinner, E. 1968. Artspezifische Muster der Zugunruhe bei Laubsängern und ihre mögliche Bedeutung für die Beendigung des Zuges im Winterquartier. *Zeitschrift für Tierpsychologie,* 25:843–853.

Gwinner, E. 1979. Jugendentwicklung südfinnischer und süddeutscher Gartengrasmücken *(Sylvia borin)* unter derselben Bedingungen. *Die Vogelwarte,* 30:41–43.

Gwinner, E., and L.-O. Ericksson. 1977. Circadiane Rhythmik und photoperiodische Zeitmessung beim Star *(Sturnus vulgaris). Journal für Ornithologie,* 118:60–67.

Gwinner, E., and W. Wiltschko. 1978. Endogenously controlled changes in migratory direction of the Garden Warbler *(Sylvia borin). Journal of Comparative Physiology* A, 125:267–273.

Gwinner, E., *et al.* 1972. Untersuchungen zur Jahresperiodik von Laubsängern. *Journal für Ornithologie,* 113:1–8.

Haapanen, A., M. Helminen, and H. K. Suomalainen. 1973. Population growth and breeding biology of the Whooper Swan, *Cygnus c. cygnus,* in Finland in 1950–1970. *Riistatiet Julk,* 33:40–60.

Haartman, L. von. 1949. Der Trauerfliegenschnäpper. I. *Acta Zoologica Fennica,* 56:1–104.

Haartman, L. von. 1953. Was reizt den Trauerfliegenschnäpper *(Muscicapa hypoleuca)* zu füttern? *Die Vogelwarte,* 16:157–164.

Haartman, L. von. 1954. Der Trauerfliegenschnäpper. III. Die Nahrungsbiologie. *Acta Zoologica Fennica,* 83:1–96.

Haartman, L. von. 1955. Clutch size in polygamous species. *Acta XI Congressus Internationalis Ornithologici,* 450–453.

Haartman, L. von. 1956. Einfluss der Temperatur auf den Brutrhythmus. *Ornis Fennica,* 33:100–107.

Haartman, L. von. 1960. The Ortstreue of the Pied Flycatcher. *Proceedings of the XIIth International Ornithological Congress,* 266–278. Helsinki.

Haartman, L. von. 1969. Nest-site and evolution of polygamy in European passerine birds. *Ornis Fennica,* 46:1–12.

Haartman, L. von. 1971. Population dynamics. *In* Farner, D. S., and J. R. King (eds). *Avian Biology.* Vol. 1. Academic Press, New York.

Haartman, L. von. 1978. Changes in the bird fauna in Finland and their causes. *Fennia,* 150:25–32.

Haas, W., and P. Beck. 1979. Zum Frühjahrzug paläarktischer Vögel über die westliche Sahara. *Journal für Ornithologie,* 120:237–246.

Hadley, N. F. 1972. Desert species and adaptation. *American Scientist,* 60:338–347.

Haffer, J. 1967. Speciation in Colombian forest birds west of the Andes. *American Museum Novitates,* No. 2294.

Haffer, J. 1968. Über die Entstehung der nördlichen Anden und das vermütliche Alter colombianischer Vogelarten. *Journal für Ornithologie,* 109:67–69.

Haffer, J. 1969. Speciation in Amazon birds. *Science*, 165:131–137.

Haffer, J. 1970. Art-Entstehung bei einigen Waldvögeln Amazoniens. *Journal für Ornithologie*, 111:285–331.

Haffer, J. 1974. Avian speciation in tropical South America. *Publications of the Nuttall Ornithology Club No. 14.*

Hafner, D. J., and K. E. Petersen. 1985. Song dialects and gene flow in the White-crowned Sparrow, *Zonotrichia leucophrys nuttalli*. *Evolution*, 39:687–694.

Hagen, V. W. von. 1961. *The Aztec: Man and Tribe*. Mentor Books, New York.

Hailman, J. P. 1966. Four color preferences of the Laughing Gull, *Larus atricilla*. *American Zoologist*, 6:568.

Hailman, J. P. 1967. The ontogeny of an instinct: The pecking response in chicks of the Laughing Gull (*Larus atricilla* L.) and related species. *Behavior*, Supplement 15.

Hails, C. J., and D. M. Bryant. 1979. Reproductive energetics of a free-living bird. *Journal of Animal Ecology*, 48:471–482.

Hainsworth, F. R., and L. L. Wolf. 1972. Energetics of nectar extraction in a small, high-altitude hummingbird, *Selasphorus flammula*. *Journal of Comparative Physiology* A, 80:377.

Hainsworth, F. R., and L. L. Wolf. 1976. Nectar characteristics and food selection by hummingbirds. *Oecologia*, 25:101–113.

Hair, J. D., and D. J. Forrester. 1970. The helminth parasites of the Starling (*Sturnus vulgaris*). *American Midland Naturalist*, 83:555–564.

Hale, E. B. 1956–1957. Breed recognition in the social interactions of domestic fowl. *Behaviour*, 10:240–254.

Hall, P. B., and R. E. Moreau. 1962. A study of the rare birds of Africa. *Bulletin of the British Museum (Natural History)*, Zoology 8(7):313–378.

Hall, T. S. 1951. *Source Book in Animal Biology*. McGraw-Hill Book Company, New York.

Halliday, W. E. D., and A. W. A. Brown. 1943. The distribution of some important forest trees in Canada: *Ecology*, 24:353–373.

Hamed, D. M., and S. M. Evans. 1984. Social influences on foraging behaviour of the red-cheeked cordon bleu *Uraeginthus bengalus* Estrildidae. *Ibis*, 126:156–167.

Hamilton, W. D. 1963. The evolution of altruistic behavior. *American Naturalist*, 97:354–356.

Hamilton, W. D. 1970. Selfish and spiteful behaviour in an evolutionary model. *Nature*, 228:1218–1220.

Hamilton, W. J. 1943. Nesting of the Eastern Bluebird. *Auk*, 60:91–94.

Hamilton, W. J. 1965. Sun-oriented display of the Anna's Hummingbird. *Wilson Bulletin*, 77:38–44.

Hamilton, W. J., and M. C. Hammond. 1960. Oriented overland spring migration of pinioned Canada Geese. *Wilson Bulletin*, 72:385–391.

Hamilton, W. J., and G. H. Orians. 1965. The evolution of brood parasitism in altricial birds. *Condor*, 67:361–382.

Hann, H. W. 1953. *The Biology of Birds*. Ulrich Book Store, Ann Arbor, Michigan.

Hannon, S. J. 1984. Factors limiting polygyny in the Willow Ptarmigan. *Animal Behaviour*, 32:153–161.

Hannon, S. J., R. L. Mumme, W. D. Koenig, and F. A. Pitelka, 1983. Removal and replacement of a cooperatively breeding bird, the Acorn Woodpecker: a test of the habitat saturation hypothesis. Abstract No. 93. *101st American Orthologists Union*, 38. New York. This has been updated in Hannon, *et al.*, 1985 below.

Hannon, S. J., R. L. Mumme, W. D. Koenig, and F. A. Pitelka, 1985. Replacement of breeders and within-group conflict in the cooperatively breeding Acorn Woodpecker. *Behavioral Ecology and Social Biology*, 17:303–312.

Hansen, H. A., and H. K. Nelson. 1964. Honkers large and small. *In* Linduska, J. P., and A. L. Nelson (eds). *Waterfowl Tomorrow*. U.S. Fish and Wildlife Service, Washington, D.C.

Hansen, K. 1978. [Migration and dispersal in Danish Great Tits.] *Dansk Ornithologisk Forenings Tidsskrift*, 72:97–104.

Hanson, H. C. 1963. The dynamics of condition factors in Canada Geese and their relation to seasonal stresses. *Arctic Institute of North America Technical Bulletin*, 12:1–68.

Hanson, H. C., and R. L. Jones. 1976. *The Biogeochemistry of Blue, Snow, and Ross' Geese*. Southern Illinois University Press, Edwardsville, Illinois.

Hanson, H. C., and R. H. Smith. 1950. Canada Geese of the Mississippi flyway. *Bulletin of the Illinois Natural History Survey*, 25:67–210.

Hardin, G. 1960. The competitive exclusion principle. *Science*, 131:1292–1297.

Hardy, J. W. 1963. Epigamic and reproductive behavior of the Orange-fronted parakeet. *Condor*, 65:169–199.

Hardy, J. W. 1973. Feral exotic birds in southern California. *Wilson Bulletin*, 85:506–512.

Harper, F. 1938. The Chuck-will's-widow in the Okefenokee region. *The Oriole*, 3:9–13.

Harrington, B. A. 1974. Colony visitation behavior and breeding ages of Sooty Terns (*Sterna fuscata*). *Bird-Banding*, 45:115–144.

Harris, M. P. 1969. The biology of Storm Petrels in the Galapagos Islands. *Proceedings of the California Academy of Science*, 37:95–165.

Harris, M. P. 1969. Effect of laying date on chick production in Oystercatchers and Herring Gulls. *British Birds*, 62:70–75.

Harris, M. P. 1970. Abnormal migration and hybridization of *Larus argentatus* and *L. fuscus* after interspecies fostering experiments. *Ibis*, 112:488–498.

Harris, M. P. 1973. The Galapagos avifauna. *Condor*, 75:265–278.

Harrison, C. J. O. 1960. The food of some urban Tawny Owls. *Bird Study*, 7:236–240.

Harrison, C. J. O. 1965. Allopreening as agonistic behaviour. *Behaviour*, 24:161–209.

Hart, J. S., and M. Berger. 1972. Energetics, water economy, and temperature regulation during flight. *Proceedings of the XVth International Ornithological Congress*, 189–199. E. J. Brill, Leiden.

Hart, J. S., and O. Z. Roy. 1966. Respiratory and cardiac responses to flight in pigeons. *Physiological Zoology*, 39:291–306.

Hartley, P. H. T. 1949. The biology of the Mourning Chat in winter quarters. *Ibis*, 91:393–413.

Hartman, F. A. 1961. Locomotor mechanisms of birds. *Smithsonian Miscellaneous Collections*, 143(1):1–99.

Hartung, R., and G. S. Hunt, 1966. Toxicity of some oils to waterfowl. *Journal of Wildlife Management*, 30:564–570.

Harvey, S., H. Klandorf, and Y. Pinchasov. 1983. Visual and metabolic stimuli cause adrenocortical suppression in fasted chickens during refeeding. *Neuroendocrinology*, 37:59–63.

Hassler, R. 1968. An account of a Ruffed Grouse with a tractor. *Migrant*, 39(1):17.

Hatch, D. E. 1970. Energy conserving and heat dissipating mechanisms of the Turkey Vulture. *Auk*, 87:111–124.

Hatch, S. A. 1983. Mechanism and ecological significance of sperm storage in the Northern Fulmar with reference to its occurrence in other birds. *Auk*, 100:593–600.

Haukioja, E., and J. Reponen. 1969. On the molt of the House Sparrow, *Passer domesticus*. *Annual Reports of the Ornithological Society of Pori*, 1968:49–51.

Haverschmidt, F. 1970. Notes on the Snail Kite in Surinam. *Auk*, 87:580–584.

Haviland, M. 1926. *Forest, Steppe, and Tundra*. Cambridge University Press, Cambridge.

Hawksley, O. 1957. Ecology of a breeding population of Arctic Terns. *Bird-Banding*, 28:57–92.

Haymes, G. T., and H. Blokpoel. 1980. The influence of age on the breeding biology of Ringbilled Gulls. *Wilson Bulletin*, 92:221–228.

Hazelhoff, E. H. 1951. Structure and function of the lung of birds. *Poultry Science*, 30:3–10.

Hazelwood, R. L. 1972. The intermediary metabolism of birds. *In* Farner, D. S., and J. R. King (eds). *Avian Biology*, Vol. 2. Academic Press, New York.

Headstrom, R. 1961. *Birds' Nests: A Field Guide*. Ives Washburn, New York.

Headstrom, R. 1951. *Birds' Nests of the West: A Field Guide*. Ives Washburn, New York.

Heaton, M. B. 1972. Prenatal auditory discrimination in the Wood Duck (*Aix sponsa*). *Animal Behaviour*, 20:421–424.

Heaton, M. B. 1975. Behavioral adaptation in neonatal chicks following embryonic vestibular system rearrangement. *Journal of Experimental Zoology*, 194:495–509.

Hediger, H. 1964. *Wild Animals in Captivity*. Dover, New York.

Hegner, R. E., S. T. Emlen, and N. J. Demong. 1982. Spatial organization of the White-fronted Bee-eater. *Nature*, 298:264–266.

Heilfurth, F. 1936. Beiträge zur Fortpflanzungsökologie der Hochgebirgsvögel. *Beiträge zur Fortpflanzungsbiologie der Vogel*, 12:98–105.

Heilmann, G. 1927. *The Origin of Birds*. D. Appleton and Company, New York.

Heinroth, O. 1911. Beiträge zur Biologie, namentlich Ethologie und Psychologie der Anatiden. *Verhandlung des V Internationalen Ornithologen Kongresses*, 589–702. Berlin. Deutsche Ornithologische Gesellschaft.

Heinroth, O. 1922. Die Beziehungen zwischen Vogelgewicht, Eigewicht, Gelegegewicht und Brutdauer. *Journal für Ornithologie*, 70:172–285.

Heinroth, O. 1938. *Aus dem Leben der Vögel*. Julius Springer, Berlin.

Heinroth, O., and K. Heinroth. 1958. *The Birds*. University of Michigan Press, Ann Arbor, Michigan.

Heinroth, O., and M. Heinroth. 1924–1933. *Die Vögel Mitteleuropas*. 4 vols. Hugo Bermühler Verlag, Berlin, Lichterfelde.

Heinz, G. 1973. Responses of Ring-necked Pheasant chicks (*Phasianus colchicus*) to conspecific calls. *Animal Behaviour*, 21:1–9.

Helb, H.-W. 1973. Analyze der artisolierenden Parameter in Gesang des Fitis (*Phylloscopus t. trochilus*) mit Untersuchungen zur Objektivierung der analytischen Methode. *Journal für Ornithologie*, 114:145–206.

Henny, C. J. 1969. Geographical variation in mortality rates of the Barn Owl (*Tyto alba* spp.) *Bird-Banding*, 40:277–290.

Henry, C. J. 1973. Drought-displaced movement of North American Pintails into Siberia. *Journal of Wildlife Management*, 37:23–29.

Henny, C. J., L. J. Blus, A. J. Krynitsky, and C. M. Bunck. 1984. Current impact of DDE on Black-crowned Night-herons in the intermountain west. *Journal of Wildlife Management*, 48:1–13.

Hensel, R. J., and W. A. Troyer. 1964. Nesting studies of the Bald eagle in Alaska. *Condor*, 66:282–286.

Hensley, M. M., and J. B. Cope. 1951. Further data on removal and repopulation of the breeding birds in a spruce-fir forest community. *Auk*, 68:483–493.

Heppner, F. 1970. The metabolic significance of differential absorption of radiant energy by black and white birds. *Condor*, 72:50–59.

Heppner, F. H. 1974. Avian flight formations. *Bird-Banding*, 45:160–169.

Herman, C. M. 1938. Epidemiology of malaria in Eastern Redwings (*Agelaius p. phoeniceus*). *American Journal of Hygiene*, 28:232–241.

Herman, C. M. 1944. The blood protozoa of North American birds. *Bird-Banding*, 15:89–112.

Herman, C. M. 1955. Diseases of birds. *In* Wolfson, A. (ed). *Recent Studies in Avian Biology.* University of Illinois Press, Urbana, Illinois.

Herman, C. M., and E. E. Wehr. 1954. The occurrence of gizzard worms in Canada Geese. *Journal of Wildlife Management*, 18:509–513.

Hernández-Peón, R., H. Scherrer, and M. Jouvet. 1956. Modification of electric activity in cochlear nucleus during "attention" in unanesthetized cats. *Science*, 123:331–332.

Herrera, C. M. 1979. Ecological aspects of heterospecific flock formation in a Mediterranean passerine bird community. *Oikos*, 33:85–96.

Herrera, C. M. 1984. Adaptation to frugivory of Mediterranean avian seed dispersers. *Ecology*, 65:609–617.

Herrick, E. H., and J. O. Harris. 1957. Singing female Canaries. *Science*, 125:1299–1300.

Herrick, F. H. 1924. The daily life of the American Eagle: late phase. *Auk*, 41:517–541.

Hertel, H. 1963. *Struktur-Form-Bewegung.* Krauskopf-Verlag, Mainz.

Hess, E. H. 1956. Space perception in the chick. *Scientific American*, 195(1):71–80.

Hess, E. H. 1958. "Imprinting" in animals. *Scientific American*, 198(3):81–90.

Hess, E. H. 1959. Imprinting. *Science*, 130:133–141.

Hess, E. H. 1970. Imprinting. *Encyclopedia Americana.* Vol. 14:829–830. Americana Corporation, New York.

Hess, E. H. 1972. "Imprinting" in a natural laboratory. *Scientific American*, 227(2):24–31.

Hess, E. H., and S. B. Petrovich (eds). 1977. *Benchmark Papers in Animal Behavior, 5: Imprinting.* Hutchinson and Ross, Stroudsburg, Pa.

Hess, E. H., and H. H. Schaefer, 1959. Innate behavior patterns as indicators of the "critical period." *Zeitschrift für Tierpsychologie*, 16:155–160.

Hess, G. 1951. *The Bird: Its Life and Structure.* Greenberg, New York.

Hesse, R., W. C. Allee, and K. P. Schmidt. 1937. *Ecological Animal Geography.* John Wiley and Sons, New York.

Hetrick, W., and G. McCaskie. 1965. Unusual behavior of a White-tailed Tropic-bird in California. *Condor*, 67:186–187.

Hickey, J. J. 1942. Eastern population of the Duck Hawk. *Auk*, 59:176–204.

Hickey, J. J. 1943. *A Guide to Bird Watching.* Oxford University Press, New York.

Hickey, J. J. 1955. An elevated nest of a Barn Swallow. *Wilson Bulletin*, 67:135.

Hickey, J. J. 1966. Birds and pesticides. *In* Strefferud, A., and A. L. Nelson (eds). *Birds in Our Lives.* U.S. Fish and Wildlife Service, Washington, D.C.

Hickey, J. J. (ed). 1969. *Peregrine Falcon Populations: Their Biology and Decline.* University of Wisconsin Press, Madison, Wisconsin.

Hildén, O. 1975. Breeding system of Temminck's Stint, *Calidris temminckii. Ornis Fennica*, 52:117–144.

Hildén, O. 1977. Mass irruption of Long-tailed Tits, *Aegithalos caudatus*, in northern Europe in 1973. *Ornis Fennica*, 54:47–65.

Hildtich, C. D. M., T. C. Williams, and I. C. T. Nisbet. 1973. Autumnal bird migration over Antigua, W. I. *Bird-Banding*, 44:171–179.

Hilgerloh, G. 1981. Die Wetterabhängigkeit von Zugintensität, Zughöhe und Richtungsstreuung bei tagziehenden Vögeln im Schweizerischen Mittelland. *Ornithologische Beobachter*, 78:245–263.

Hill, D. A. 1984. Factors affecting success in the Mallard and Tufted Duck. *Ornis Scandinavica*, 15:115–122.

Hill, D. C., *et al.* 1968. Metabolizable energy of aspen buds for captive Ruffed Grouse. *Journal of Wildlife Management*, 32:854–858.

Hill, R. A. 1976. Host-parasite relationships of the Brown-headed Cowbird in a prairie habitat of west-central Kansas. *Wilson Bulletin*, 88:555–565.

Hills, S. 1980. Incubation capacity as a limiting factor of a shorebird clutch size. *American Zoologist*, 20:774.

Hinchliffe, J. R., and M. K. Hecht. 1984. Homology of the bird wing skeleton. *Evolutionary Biology*, 18:21–39.

Hinde, R. A. 1956. The biological significance of territories of birds. *Ibis*, 98:340–369.

Hinde, R. A. 1958. The nest-building behaviour of domesticated Canaries. *Proceedings of the Zoological Society of London*, 131:1–48.

Hinde, R. A. 1964. Pair formation. *In* Thomson, A. L. (ed). *A New Dictionary of Birds*. McGraw-Hill Book Company, New York.

Hinde, R. A. (ed). 1969. *Bird Vocalizations: Their Relation to Current Problems in Biology and Psychology*. Cambridge University Press, New York.

Hindwood, K. A. 1959. The nesting of birds in the nests of social insects. *Emu*, 59:1–36

Hindwood, K. A. 1940. The birds of Lord Howe Island. *Emu*, 40:1–86.

Hindwood, K. A. 1955. Bird-wasp nesting associations. *Emu*, 55:263–274.

Hirsch, K. V., and C. R. Grau. 1981. Yolk formation and oviposition in captive emus. *Condor*, 83:381–382.

Hitchcock, H. B. 1955. Homing flights and orientation of pigeons. *Auk*, 72:355–373.

Hjelmtveit, I. 1969. On migratory birds at ocean weather station. M. Årbok. Univers. Bergen, Naturv. Serie, 1969, No. 4.

Hodson, N. L., and D. W. Snow. 1965. The road deaths inquiry, 1960–61. *Bird Study*, 12:90–99.

Hoffman, H. S. 1968. Effect of companion on distress calls of an imprinted duckling. *Behaviour*, 30:175–191.

Hoffman, K. 1954. Versuche zu der im Richtungsfinden der Vögel enthaltenen Zeitschätzung. *Zeitschrift für Tierpsychologie*, 11:453–475.

Hoffmann, A. 1949. Über die Brutpflege des polyandrischen Wasserfasans *Hydrophasianus chirurgus* (Scop.). *Zoologische Jahrbücher*, 78:367–403.

Hofstad, M. S., *et al.* 1972. *Diseases of Poultry*. Sixth edition. Iowa State University Press, Ames, Iowa.

Hoglund, N., and K. Borg. 1955. Über die Grunde für die Frequenzvariation beim Auerwild. *Zeitschrift für Jagdwissenschaft*, 1:59–62.

Högström, S. 1970. [Horned Grebe *(Podiceps auritus)* on the Baltic Island of Gotland]. *Vår Fågel-wärld*, 29:60–66. From *Auk*, 88:462.

Höhn, E. O. 1970. Gonadal hormone concentrations in Northern Phalaropes in relation to nuptial plumage. *Canadian Journal of Zoology*, 48:400–401.

Höhn, E. O., and S. C. Cheng. 1967. Gonadal hormones in Wilson's phalarope *(Steganopus tricolor)* and other birds in relation to plumage and sex behavior. *General and Comparative Endocrinology*, 8:1–11.

Holcomb, L. 1967. Goldfinch accept young after long or short incubation. *Wilson Bulletin*, 79:348.

Holm, E. R., and M. L. Scott. 1954. Studies on the nutrition of wild waterfowl. *New York Fish and Game Journal*, 1:171–187.

Holmes, R. T. 1970. Differences in population density, territoriality, and food supply of Dunlin on arctic and subarctic tundra. *In* Watson, A. (ed). *Animal Populations in Relation to their Food Resources*. Blackwell Scientific Publications, Oxford.

Holmes, R. T., and F. W. Sturges. 1975. Bird community dynamics and energetics in a northern hardwoods ecosystem. *Journal of Animal Ecology*, 44:175–200.

Holst, E. von., and U. von Saint Paul. 1962. Electrically controlled behavior. *Scientific American*, 206(3):50–59.

Holst, E. von., and U. von Saint Paul. 1963. On the functional organization of drives. *Animal Behaviour*, 11:1–20.

Holzapfel, M. 1939. Analyse des Sperrens und Pickens in der Entwicklung des Stars. *Journal für Ornithologie*, 87:525–553.

Homberg, L. 1957. Fiskande Kråkor. *Fauna och Flora*, 5:182–185.

Homberger, D. G. 1981. Functional morphology and evolution of the feeding apparatus in parrots, with special reference to the pasquet's parrot, *Psittrichas fulgidus* (Lesson). *In* Pasquier, R. F. (ed). Conservation of New World Parrots. *International Council for Bird Preservation Technical Publication No. 1*: 471–485.

Hoogerwerk, A. 1937. Uit het leven der witte ibissen *Threskiornis aethiopicus melanocephalus*. *Limosa*, 10:137–146.

Hoogstraal, H., and M. N. Kaiser, 1961. Ticks from European-Asiatic birds migrating through Egypt into Africa. *Science*, 133:277–278.

Horn, G., B. J. McCabe, and J. Cipolla-neto. 1983. Imprinting in the domestic chick: The role of each side of the hyperstriatum centrale in acquisition and retention. *Experimental Brain Research*, 53:91–98.

Horton-Smith, C. 1938. *The Flight of Birds*. H. F. & G. Witherby Ltd., London.

Horvath, L. 1955. Red-footed Falcons in Ohat-Woods near Hortobagy. *Acta Zoologica Academiae Hungaricae*, 1: 245–287.

Höst, P. 1942. Effect of light on moults and sequences of plumage in the Willow Ptarmigan. *Auk*, 59:388–403.

Howard, H. E. 1920. *Territory in Bird Life*. John Murray (Publishers) Ltd., London.

Howard, H. E. 1935. Territory and food. *British Birds*, 28:285–287.

Howard, H. 1952. The prehistoric avifauna of Smith Creek Cave, Nevada, with a description of a new gigantic raptor. *Bulletin of the Southern California Academy of Science*, 51:50–54.

Howard, H. 1955. Fossil birds with especial reference to the birds of Rancho LaBrea. *Los Angeles County Museum Science Series No. 17*.

Howard, H. 1957. A gigantic "toothed" marine bird from the Miocene of California. *Santa Barbara Museum of Natural History Bulletin No. 1*.

Howard, R. D. 1974. The influence of sexual selection and interspecific competion on Mockingbird song *(Mimus polyglottos)*. *Evolution*, 28:428–438.

Howe, H. F. 1977. Nestling sex ratio adjustment among common grackles. *Science*, 198:744–746.

Howell, J. C., and G. M. Heinzmann. 1965. Comparison of nesting sites of Bald Eagles in central Florida from 1930 to 1965. *Auk*, 84:602–603.

Howell, T. R. 1978. Breeding biology of the Egyptian Plover, *Pluvianus aegyptius*. *Nigerian Ornithological Society Bulletin*, 14:51–53.

Howell, T. R., B. Araya, and W. R. Millie. 1974. Breeding biology of the Gray gull, *Larus modestus*. *University of California Publications in Zoology*, 104:57.

Hubbard, J. P. 1969. The relationships and evolution of the *Dendroica coronata* complex. *Auk*, 86:393–432.

Hudson, R. 1972. Collared doves in Britain and Ireland. *British Birds*, 65:139–154.

Hudson, W. H. 1920. *The Birds of La Plata*. London.

Huggins, R. A. 1941. Egg temperature in wild birds under natural conditions. *Ecology*, 22:148–157.

Hughes, M. R. 1970. Some observations on ion and water balance in the Puffin, *Fratercula arctica*. *Canadian Journal of Zoology*, 48:479–482.

Hughes, R. A. 1985. Notes on the effects of El Niño on the seabirds of the Mollendo district, southwest Peru, in 1983. *Ibis*, 127:385–388.

Hummel, D. 1983a. The invasion of Western Europe by the Great bustard *(Otis tarda)* in the winter season 1978/1979. *Die Vogelwarte*, 104:41–53.

Hummel, D. 1983b. Aerodynamic aspects of formation flight in birds. *Journal of Theoretical Biology*, 104:321–348.

Humphrey, P. S., and B. C. Livezey. 1982. Flightlessness in flying Steamer-Ducks. *Auk*, 99:368–372

Humphrey, P. S., and K. C. Parkes. 1959. An approach to the study of molts and plumages. *Auk*, 76:1–31.

Hundley, M. H. 1963. Notes on methods of feeding and the use of tools in the Geospizinae. *Auk*, 80:372–373.

Hunt, J. H. 1971. A field study of the Wrenthrush *Zeledonia coronata*. *Auk*, 88:1–20.

Hurwitz, S., and A. Bar. 1969. Intestinal calcium absorption in the laying fowl and its importance in calcium homeostasis. *American Journal of Clinical Nutrition*, 22:391.

Hussell, D. J. T. 1969. Weight loss of birds during nocturnal migration. *Auk*, 86:75–83.

Hussell, D. J. T. 1982. The timing of fall migration in Yellow-bellied Flycatchers. *Journal of Field Ornithology*, 53: 1–6.

Hussell, D. J. T., and A. B. Lambert. 1980. New estimates of weight loss in birds during nocturnal migration. *Auk*, 97:547–558.

Hustich, I. (ed). 1952. The recent climatic fluctuation in Finland and its consequences. *Fennia*, 75:1–128.

Hutchinson, G. E. 1978. *An Introduction to Population Ecology*. Yale University Press, New Haven, Connecticut.

Hutchinson, L. V., and B. M. Wenzel. 1980. Olfactory guidance in foraging by Procellariiformes. *Condor*, 82: 314–319.

Hutt, F. B. 1949. *Genetics of the Fowl*. McGraw-Hill Book Company, New York.

Huxley, J. 1914. The courtship habits of the great crested grebe *(Podiceps cristatus)*. *Proceedings of the Zoological Society of London*, 2:491–562.

Huxley, J. 1934. A natural experiment on the territorial instinct. *British Birds*, 27:270–277.

Huxley, J. S. 1939. A discussion on subspecies and varieties. *Proceedings of the Linnaean Society, London*, 151: 105–106.

Huxley, J. S. 1942. *Evolution: The Modern Synthesis*. Harper and Brothers, New York.

Huxley, J. S. 1949. Wren feeding young on fish. *British Birds*, 42:185–186.

Ickes, R. A., and M. S. Ficken. 1970. An investigation of territorial behavior in the American Redstart utilizing recorded songs. *Wilson Bulletin*, 82:167–176.

Idyll, C. P. 1973. The Anchovy crisis. *Scientific American*, 288(6):22–29.

Ilenko, E. 1947. [Orientation of the Robin.] *Vestnik Zoologii*, 1974(2):78–79.

Imber, M. J. 1973. The food of Grey-faced Petrels (*Pterodroma macroptera gouldii*. Hutton) with special reference to diurnal vertical migration of their prey. *Journal of Animal Ecology*, 42:645–662.

Immelmann, K. 1962. Beiträge zur einer vergleichenden Biologie australischer Prachtfinken (Spermestidae). *Zoologischer Jahrbücher Abteilung für Systematik*, 90:1– 196.

Immelmann, K. 1963. Drought adaptations in Australian desert birds. *Proceedings of the XIIIth International Ornithological Congress*, 2:649–657. American Ornithologists' Union.

Immelmann, K. 1966. Beobachtungen an Schwalbenstaren. *Journal für Ornithologie*, 107:37-69.

Immelmann, K. 1969. Über den Einfluss frühkindlichen Erfahrungen auf die geschlechtliche Objectfixierung bei Estrildiden. *Zeitschrift für Tierpsychologie*, 26:677-691.

Immelmann, K. 1969. Song development in the *Zebra Finch* and other estrildid finches. *In* Hinde, R. A. (ed). *Bird Vocalizations: Their Relation to Current Problems in Biology and Psychology*. Cambridge University Press, Cambridge.

Immelmann, K. 1970. Ecological significance of imprinting and early learning. *Annual Review Ecology and Systematics*, 6:15–37.

Immelmann, K. 1971. Ecological aspects of periodic reproduction. *In* Farner, D. S., and J. R. King (eds). *Avian Biology*. Vol. 1. Academic Press, New York.

Immelmann, K. 1972a. The influence of early experience upon the development of social behavior in estrildine finches. *Proceedings of the XVth International Ornithological Congress*, 316–338. E. J. Brill, Leiden.

Immelmann, K. 1972b. Sexual and other long-term aspects of imprinting in birds and other species. *In* Lehrman, D. S., *et al.* (eds). *Advances in the Study of Behavior*. Academic Press, New York.

Immelmann, K. 1976. The use of behavioural characters in avian classification. *In* Frith, H. J., and J. H. Calaby (eds). *Proceedings of the 16th International Ornithological Congress*. Australian Academy of Science, Canberra.

Immelmann, K. 1982. *Australian Finches*. Angus and Robertson, Sydney.

Immelmann, K., H. H. Kalberlah, P. Rausch, and A. Stahnke. 1978. Sexuelle Prägung als Möglicher Faktor innerartlicher Isolation beim Zebrafinken. *Journal für Ornithologie*, 119:197–212.

Immelmann, K., A. Piltz, and R. Sossinka. 1977. Experimentelle Untersuchungen zur Bedeutung der Rachenzeichnung junger Zebrafinken. *Zeitschrift für Tierpsychologie*, 45:210–218.

Immelmann, K., and S. J. Suomi. 1981. Sensitive phases in development. *In* Immelmann, K., G. W. Barlow, L. Petrinovich, and M. Main (eds). *Behavioral development*. The Bielefeld Interdisciplinary Project. Cambridge University Press, 395–431.

Inglis, I. R., and J. Lazarus. 1981. Vigilance and flock size in Brent Geese (*Branta bernicla bernicla*): The edge effect. *Zeitschrift für Tierpsychologie*, 57:193– 200.

Ingold, P. 1980. Anpassung der Eier und des Brutverhaltens von Trottellummen (*Uria aalge aalge* Pont.) an das Brüten auf Felssimsen. *Zeitschrift für Tierpsychologie*, 53:341–388.

Ingram, C. 1962. Cannibalism by nestling Short-eared Owls. *Auk*, 79:715.

Ioale, P. 1983. Effect of Anaesthesia of the nasal muscosae on the homing behaviour of pigeons. *Zeitschrift für Tierpsychologie*, 61:102–110.

Ireland, L. C., and T. C. Williams. 1972. Bird migration over Bermuda detected by tracking radars. *American Zoologist*, 12:663.

Irving, L. 1960. Birds of Anaktuvuk Pass, Kobuk, and Old Crow. A study in arctic adaptation. *U.S. National Museum Bulletin 217*.

Irving, L. 1972. *Arctic LIfe of Birds and Mammals, Including Man*. Springer-Verlag, Berlin.

Ising, G. 1946. Die physikalische Möglichkeit eines tierischen Orientierungssinnes auf Basis der Erdrotation. *Ark. Matematik, Astronomi, och Fysik*, 32A(18):1–23.

Jacob, J. 1977. Die systematische Stellung der Dampfschiffenten (*Tachyeres*) innerhalb der Ordnung Anseriformes. *Journal für Ornithologie*, 118:52–59.

Jacob, J. 1978. Uropygial gland secretions and feather waxes. *In* Brush, A. H. (ed). *Chemical Zoology*, 10:165–211. Academic Press, New York.

Jacob, J., and H. Hoerschelmann. 1982. Chemotaxonomische Untersuchungen zur Systematik der Rohrennasen (Procellariformes). *Journal für Ornithologie*, 123:63–84.

Jacob, J., and V. Ziswiler. 1982. The uropygial gland. *In* Farner, D. S., J. R. King, and K. E. Parkes (eds). *Avian Biology* 6:199–324.

Jaeger, E. C. 1949. Further observations on the hibernation of the Poor-will. *Condor*, 51:105–109.

Jahn, H. 1939. Zur Biologie des japanischen Paradiesfliegenschnäppers *Terpsiphone a. atrocaudata* (Eyton). *Journal für Ornithologie*, 87:216–223.

James, F. C. 1970. Geographic size variation in birds and its relationship to climate. *Ecology*, 51:365–390.

James, F. C. 1971. Ordinations of habitat relationships among breeding birds. *Wilson Bulletin*, 83:215–236.

James, F. C. 1983. Environmental component of morphological differentiation in birds. *Science*, 221:184–185.

James, W. 1890. *Principles of Psychology*. Henry Holt and Company, New York.

Jameson, W. 1960. Flight of the Albatross. *Natural History*, 69(4):62–69.

Jannson, C., J. Ekman, and A. von Bromssen. 1982. Winter mortality and food supply in tits *Parus spp*. *Oikos*, 37: 313–322.

Järvi, T., T. Radesäter, and S. Jakobsson. 1978. Aggressive responses of two hole-nesting passerines, *Parus major* and *Ficedula hypoleuca*, to the play-back of sympatric species song. *Ornis Fennica*, 55:154–157.

Jehl, J. R., Jr., and K. C. Parkes. 1983. "Replacements" of landbird species on Socorro Island, Mexico. *Auk*, 100: 551–559.

Jellmann, J. 1979. Flughöhen ziehender Vögel in Nordwest Deutschland nach Radarmessungen. *Vogelwarte*, 30: 118–134.

Jenkins, D. 1957. The breeding of the Red-legged Partridge. *Bird Study*, 4:97–100.

Jenkins, P. F. 1978. Cultural transmission of song patterns and dialect development in a free-living bird population. *Animal Behaviour*, 25:50–78.

Jenkins, P. F., and A. J. Baker. 1984. Mechanisms of song differentiation in introduced populations of Chaffinches *Fringilla coelebs* in New Zealand. *Ibis*, 126:510–524.

Jenni, D. A. 1969. A study of the ecology of four species of herons during the breeding season at Lake Alice, Florida. *Ecological Monographs*, 39:245–270.

Jenni, D. A., and G. Collier. 1972. Polyandry in the American Jacana (*Jacana spinosa*). *Auk*, 89:743–765.

Jesperson, P. 1924. On the frequency of birds over the high Atlantic Ocean. *Nature*, 114:281–283.

Jesperson, P. 1929. On the frequency of birds over the high Atlantic Ocean. *Verhandlungen des VI Internationalen Ornithologen Kongresses, Kopenhagen, 1926*. Berlin.

Johns, J. E. 1969. Field studies of Wilson's Phalarope. *Auk*, 86:660–670.

Johnsgard, P. A. 1965. *Handbook of Waterfowl Behavior*. Cornell University Press, Ithaca, New York.

Johnsgard, P. A. 1967. *Animal Behavior*. William C. Brown Company, Publishers, Dubuque, Iowa.

Johnsgard, P. A. 1983. *Cranes of the World*. Indiana University Press, Bloomington, Indiana.

Johnsgard, P. A., and J. Kear. 1968. A review of parental carrying of young by waterfowl. *Living Bird*, 1968:89–102.

Johnson, N. F. 1971. Effects of levels of dietary protein on Wood Duck growth. *Journal of Wildlife Management*, 35: 798–802.

Johnson, N. K., and C. B. Johnson. 1985. Speciation in Sapsuckers (*Sphyrapicus*): II. Sympatry, hybridization, and mate preference in *S. ruber daggetti* and *S. nuchalis*. *Auk*, 102:1–15.

Johnson, O. W., and J. N. Mugaas. 1970. Quantitative and organizational features of the avian renal medulla. *Condor*, 72:288–292.

Johnson, R. A. 1941. Nesting behavior of the Atlantic Murre. *Auk*, 58:153–163.

Johnson, R. A. 1969. Hatching behavior of the Bobwhite. *Wilson Bulletin*, 81:79–86.

Johnson, R. E. 1965. Reproductive activities of rosy finches, with special reference to Montana. *Auk*, 82:190–205.

Johnson, R. F., Jr., and N. F. Sloan. 1978. White Pelican production and survival of young at Chase Lake National Wildlife Refuge, North Dakota. *Wilson Bulletin*, 90:346–352.

Johnston, D. W. 1956. The annual reproductive cycle of the California Gull. *Condor*, 58:134–162.

Johnston, D. W., and T. P. Haines. 1957. Analysis of mass bird mortality in October, 1954. *Auk*, 74:447–458.

Johnston, R. F. 1956a. Population structure in salt marsh Song Sparrows. Part I. *Condor*, 58:24–44.

Johnston, R. F. 1956b. Population structure in salt marsh Song Sparrows. Part II. *Condor*, 58:254–272.

Johnston, R. F. 1962. A review of courtship feeding in birds. *Kansas Ornithological Society Bulletin*, 13:25–32.

Johnston, R. F. 1972. Other pigeons. *In* Grzimek, B. (ed). *Grzimek's Animal Life Encyclopedia*, 8:278–279. Van Nostrand Reinhold Company, New York.

Johnston, R. F., and R. K. Selander. 1971. Evolution in the House Sparrow. *Evolution*, 25:1–28.

Johnston, T. D., and G. Gottlieb. 1981. Development of visual species identification in ducklings: what is the role of imprinting? *Animal Behaviour*, 29:1082–1099.

Jones, D. R., and K. Johansen. The blood vascular system of birds. 1972. *In* Farner, D. S., and J. R. King (eds). *Avian Biology*, Vol. 2. Academic Press, New York.

Jones, L. R. 1986. The effect of photoperiods and temperature on testicular growth in captive Black-billed Magpies. *Condor*, 88:91–93.

Jones, P. J., and P. Ward. 1976. The level of reserve protein as the proximate factor controlling the timing of breeding and clutch-size in the Red-billed Quelea, *Quelea quelea*. *Ibis*, 110:547–574.

Jones, P. J., and P. Ward. 1979. A physiological basis for colony desertion by Red-billed Queleas (*Quelea quelea*). *Journal of Zoology* (London), 189:1–19.

Jones, R. E. 1969. Epidermal hyperplasia in the incubation patch of the California Quail *(Lophortyx californicus)* in relation to pituitary prolactin content. *General and Comparative Endocrinology*, 12:498–502.

Jones, R. E. 1971. The incubation patch of birds. *Biological Reviews*, 46:315–339.

Jordan, J. S. 1953a. Consumption of cereal grains by migratory waterfowl. *Journal of Wildlife Management*, 17:120–123.

Jordan. J. S. 1953b. Effects of starvation on wild Mallards. *Journal of Wildlife Management*, 17:304–311.

Juhn, M. 1963. An examination of some interpretations of molt with added data from progesterone and thyroxine. *Wilson Bulletin*, 75:191–197.

Juhn, M., G. H. Faulkner, and R. G. Gustavson. 1931. The correlation of rates of growth and hormone threshold in the feathers of fowls. *Journal of Experimental Zoology*, 58:69–111.

Jull, M. A. 1952. *Poultry Breeding*. John Wiley and Sons, New York.

Jutro, P. R. 1975. Territorial defense of people by Laughing Gulls. *Living Bird*, 14:157–161.

Kahl, M. P., Jr. 1963. Thermoregulation in the Wood Stork with special reference to the role of the legs. *Physiological Zoology*, 36:141–151.

Kahl, M. P. 1964. Food ecology of the Wood Stork (*Mycteria americana*), in Florida. *Ecological Monographs*, 34:97–117.

Kahl, M. P. 1967. Observations on the behaviour of the Hammerkop *Scopus umbretta* in Uganda. *Ibis*, 109:25–32.

Kahl, M. P., Jr., and L. J. Peacock. 1963. The bill-snap reflex: a feeding mechanism in the American Wood Stork. *Nature*, 199:505–506.

Kalela, O. 1949. Changes in geographic ranges in the avifauna of northern and central Europe in relation to recent changes in climate. *Bird-Banding*, 20:77–103.

Kalmbach, E. R. 1958. Quoted in Bent, A. C. Life histories of North American blackbirds, orioles, tanagers and their allies. *U.S. National Museum Bulletin 211*. Washington, D.C.

Kalmus, H. 1956. Sun navigation of *Aptis mellifica* L. in the southern hemisphere. *Journal of Experimental Biology*, 33:554–565.

Kaltenhäuser, D. 1971. Über Evolutionsvorgänge in der Schwimmentenbalz. *Zeitschrift für Tierpsychologie*, 29:481–540.

Kanwisher, J. W., *et al.* 1978. Radiotelemetry of heart rates from free-flying gulls. *Auk*, 92:288–293.

Karplus, M. 1952. Bird activity in the continuous daylight of arctic summer. *Ecology*, 33:129–134.

Kasal, C. A., M. Menaker, and J. R. Perez-Palo. 1979. Circadian clock in culture: N-acetyltransferase activity in chick pineal gland oscillates in vitro. *Science*, 203:656–658.

Kato, M., and T. Oishi. 1967. Radioluminous paints as activator of photoreceptor systems. *Proceedings of the Japanese Academy of Science*, 43:220–223.

Kear, J. 1964. Colour preferences in anatinae. *Ibis*, 106:361–369.

Kear, J. 1967. Experiments with young nidifugous birds on a visual cliff. *The Wildfowl Trust 18th Annual Report*, 122–124.

Kear, J., and R. K. Murton. 1976. The origins of Australian waterfowl as indicated by their photoresponses. *In* Frith, H. J., and J. H. Calaby (eds). *Proceedings of the 16th International Ornithological Congress*. Australian Academy of Science, Canberra.

Keast, A. 1972. Faunal elements and evolutionary patterns: some comparisons between the continental avifaunas of Africa, South America, and Australia. *Proceedings of the XVth International Ornithological Congress*, 594–622. E. J. Brill, Leiden.

Keast, A. 1976. Biological attributes and continental characteristics. *In* Frith, H. J., and J. H. Calaby (eds). *Proceedings of the 16th International Ornithological Congress*. Australian Academy of Science, Canberra.

Keast, A. 1980. Spatial relationships between migratory parulid warblers and their ecological counterparts in the Neotropics. *In* Keast, A., and E. S. Morton (eds). *Migrant Birds in the Neotropics.* 109–130. Smithsonian Institution Press, Washington, D.C.

Keast, J. A., and A. J. Marshall. 1954. The influence of drought and rainfall on reproduction in Australian desert birds. *Proceedings of the Zoological Society of London, 124*:493–499.

Keeton, W. T. 1969. Orientation by pigeons: is the sun necessary? *Science, 165*:922–928.

Keeton, W. T. 1974. The orientational and navigational basis of homing in birds. *In* Lehrman, D. S., and J. S. Rosenblatt. *Advances in the Study of Behavior*, Vol. 5. Academic Press, New York.

Keijer, E., and P. Butler. 1982. Volumes of the respiratory and circulatory systems in tufted ducks *(Aythya fuligula)* and mallard ducks *(Anas platyrhynchos)*. *Journal of Experimental Biology, 10*:213–220.

Keith, S. 1978. Review of D. Ripley's *Rails of the World. Wilson Bulletin, 90*:322.

Kelley, D. B. 1978. Neuroanatomical correlates of hormone sensitive behaviors in frogs and birds. *American Zoologist, 18*:477–488.

Kellogg, F. E., and J. P. Calpin. 1971. A checklist of parasites and diseases reported from the Bobwhite Quail. *Avian Diseases, 15*:704–715.

Kelly, S. T., and M. E. DeCapita. 1982. Cowbird control and its effect on Kirtland's Warbler reproductive success. *Wilson Bulletin, 94*:363–365.

Kelso, L., and M. M. Nice. 1963. A Russian contribution to anting and feather mites. *Wilson Bulletin, 75*:23–26.

Kelty, M. P., and S. I. Lustick. 1977. Energetics of the Starling *(Sturnus vulgaris)* in a pine woods. *Ecology, 58*: 1181–1185.

Kemp, A. C. 1978. A review of the Hornbills: Biology and radiation. *Living Bird, 17*:105–136.

Kemper, C. A. 1964. A tower for T.V.: 30,000 dead birds. *Audubon, 66*(2):86–90.

Kendeigh, S. C. 1940. Factors affecting length of incubation. *Auk, 57*:499–513.

Kendeigh, S. C. 1941. Territorial and mating behavior of the House Wren. *Illinois Biological Monographs, 18*(3):1–120. University of Illinois Press, Urbana, Illinois.

Kendeigh, S. C. 1945. Resistance to hunger in birds. *Journal of Wildlife Management, 9*:217–226.

Kendeigh, S. C. 1949. Effect of temperature and season on energy resources of the English Sparrow. *Auk, 66*: 113–127.

Kendeigh, S. C. 1952. Parental care and its evolution in birds. *Illinois Biological Monographs, 22*:1–356. University of Illinois Press, Urbana, Illinois.

Kendeigh, S. C. 1961. *Animal Ecology.* Prentice-Hall, Inc., Englewood Cliffs, New Jersey.

Kendeigh, S. C. 1969. Energy responses of birds to their thermal requirements. *Wilson Bulletin, 81*:441–449.

Kendeigh, S. C., *et al.* 1977. *In* Pinowski, J., and S. C. Kendeigh (eds). *Granivorous Birds in Ecosystems.* Cambridge Uiversity Press, Cambridge.

Kennard, J. H. 1977. Biennial rhythm in Purple Finch migration. *Bird-Banding, 48*:155–157.

Kennedy, P. L., and D. R. Johnson. 1986. Prey-size selection in nesting male and female Cooper's Hawks. *Wilson Bulletin, 98*:110–115.

Kennedy, R. J. 1969. Sunbathing behaviour of birds. *British Birds, 62*:249–258.

Kennedy, R. J. 1972. The probable function of flexules. *Ibis, 114*:265–266.

Kenyon, K. W. 1942. Hunting strategy of Pigeon Hawks. *Auk, 59*:443–444.

Kepler, C. B. 1967. Polynesian rat predation on nesting Laysan Albatrosses and other Pacific seabirds. *Auk, 84*: 426–430.

Kerlinger, P., and M. R. Lein. 1986. Differences in winter range among age-sex classes of Snowy Owls *Nyctea scandiaca* in North America. *Ornis Scandinavica, 17*:1–7.

Kern, M. D. 1979. Seasonal changes in the reproductive system of the female White-crowned Sparrow, *Zonotrichia leucophrys gambelii*, in captivity and in the field. *Cell Tissue Research, 202*:379–398.

Kern, M. D., and L. Coruzzi. 1979. The structure of the Canary's incubation patch. *Journal of Morphology, 162*: 425–452.

Kern, M. D., and J. R. King. 1972. Testosterone-induced singing in female White-crowned Sparrows. *Condor, 74*: 204–209.

Kern, M. D., and C. Van Riper. 1984. Altitudinal variations in nests of the Hawaiian honeycreeper *Hemignathus virens virens. Condor, 86*:443–454.

Kessel, B. 1976. Winter patterns of Black-capped Chickadees in interior Alaska. *Wilson Bulletin, 88*:36–61.

Ketterson, E. D., and J. R. King. 1977. Metabolic and behavioral responses to fasting in the White-crowned Sparrow *(Zonotrichia leucophrys gambelii)*. *Physiological Zoology, 50*:115–129.

Ketterson, E. D., and V. Nolan, Jr. 1976. Geographic variation and its climatic correlates in the sex ratio of eastern-wintering dark-eyed juncos *(Junco hyemalis hyemalis)*. *Ecology*, 57:679–693.

Ketterson, E. D., and V. Nolan, Jr. 1979. Seasonal, annual and geographic variation in sex ratio of wintering populations of Dark-eyed Juncos *(Junco hyemalis)*. *Auk*, 96:532–536.

Ketterson, E. D., and V. Nolan, Jr. 1983. The evolution of differential bird migration. *In* Johnston, R. F. (ed). *Current Ornithology*, 1:357–393.

Kikkawa, J. 1980. Winter survival in relation to dominance among silvereyes, *Zosterops lateralis chlorocephala* of Heron Island, Great Barrier Reef, Australia. *Ibis*, 22:37–446.

Kilham, L. 1958. Territorial behavior of wintering Red-headed Woodpeckers. *Wilson Bulletin*, 70:347–358.

Kilham, L. 1971. Use of blister beetle in bill-sweeping by White-breasted Nuthatch. *Auk*, 88:175–176.

King, A. S., and D. C. Payne. 1962. The maximum capacities of the lungs and air sacs of *Gallus domesticus. Journal of Anatomy*, 96:495–503.

King, B. 1969. Hooded Crows dropping and transferring objects from bill to foot in flight. *British Birds*, 62:201.

King, B. 1969. Swallow banding in Bangkok, Thailand. *Bird-banding*, 40:95–104.

King, J. E., and R. L. Pyle. 1966. Some birds like fish. *In* Stefferud, A. (ed). *Birds in Our Lives*, 230–239. U.S. Fish and Wildlife Service, Washington, D.C.

King, J. R. 1961. The bioenergetics of vernal premigratory fat deposition in the White-crowned Sparrow. *Condor*, 63:128–142.

King, J. R. 1968. Cycles of fat deposition and molt in White-crowned Sparrows in constant environmental conditions. *Comparative Biochemistry and Physiology*, 24:827–837.

King, J. R. 1972. Adaptive periodic fat storage by birds. *Proceeding of the XVth International Ornithological Congress*, 200–217. E. J. Brill, Leiden.

King, J. R. 1972. Variation in the song of the Rufous-collared Sparrow, *Zonotrichia capensis*, in northwestern Argentina. *Zeitschrift für Tierpsychologie*, 30:344–373.

King, J. R. 1973. *In* Farner, D. S. (ed). *Breeding Biology of Birds*. National Academy of Sciences, Washington, D.C.

King, J. R. 1974. *In* Paynter, R. A., Jr. (ed). *Avian Energetics*. Publications of the Nuttall Ornithology Club, No. 15. Cambridge, Massachusetts.

King, J. R. 1976. The annual cycle and its control in subequatorial Rufous-collared Sparrows. *In* Frith, H. J., and J. H. Calaby (eds). *Proceedings of the 16th International Ornithological Congress*. Australian Academy of Science, Canberra.

King, J. R., and D. S. Farner. 1966. The adaptive role of winter fattening in the White-crowned Sparrow with comments on its regulation. *American Naturalist*, 100:403–418.

King, M. J. 1965. Disruptions in the pecking order of cockerels concomitant with degrees of accessibility to food. *Animal Behaviour*, 13:504–506.

King, W. B. 1979. *Red Data Book*, Vol. 2. Second edition. International Union for the Conservation of Nature and Natural Resources. Morges, Switzerland.

Kinsky, F. C. 1971. The consistent presence of paired ovaries in the Kiwi *(Apteryx)* with some discussion of this condition in other birds. *Journal für Ornithologie*, 112:334–357.

Kipp, F. A. 1958. Zur Geschichte des Vogelzuges auf der Grundlage der Flügelanpassungen. *Die Vogelwarte*, 19:233–242.

Kiriline, L. de. 1954. The voluble singer of the tree-tops. *Audubon Magazine*, 56:109–111.

Kirkman, F. B. 1937. *Bird Behaviour*. Thomas Nelson & Sons, London.

Kirkpatrick, C. M., and A. S. Leopold. 1952. The role of darkness in sexual activity of the quail. *Science*, 116:280–281.

Kleinschmidt, A. 1974. Ein neuer-fünfter-vollständiger Archaeopteryx-Fund. *Die Vogelwarte*, 27:224.

Kleinschmidt, O. 1958. *Rawbvögel und Eulen der Heimat*. Ziemsen Verlag, Wittenberg.

Klimkiewicz, M. K., R. B. Clapp, and A. G. Futcher. 1983. Longevity records of North American birds: Remizidae through Parulinae. *Journal of Field Ornithology*, 54:287–294.

Kling, J. W., and J. Stevenson-Hinde. 1977. Development of song and reinforcing effects of song in female chaffinches. *Animal Behaviour*, 25:215–220.

Klint, T. 1973. Prakdrakten som "sexuall utlosare" hos grasand. *Zoologisk Revy*, 35:11–21.

Klint, T. 1978. Significance of mother and sibling experience for mating preferences in the Mallard *(Anas platyrhynchos)*. *Zeitschrift für Tierpsychologie*, 47:50–60.

Klomp, H. 1970. The determination of clutch-size in birds. *Ardea*, 58:1–124.

Klopfer, P. H. 1956. Goose-behavior by a White Leghorn chick. *Wilson Bulletin*, 68:68–69.

Klopfer, P. H. 1957. Imitative learning in ducks. *American Naturalist*, 91:61–63.

Klopfer, P. H. 1959a. Imprinting. *Science, 130*:730.

Klopfer, P. H. 1959b. Social interactions in discrimination learning with special reference to feeding behavior in birds. *Behaviour,* 14:282–299.

Klopfer, P. H., and J. P. Hailman. 1965. Habitat selection in birds. *In* Lehrman, D.S. *et al. (eds). Advances in the Study of Behavior,* 1:279–303. Academic Press, New York.

Kluijver, H. N. 1935. Waarnemingen over de Levenswijze van den Spreeuw *(Sturnus v. vulgaris* L.) met Behulp van Geringde Individuen. *Ardea,* 24:133–166.

Kluijver, H. N. 1950. Daily routines of the Great Tit, *Parus major,* L. *Ardea,* 38:99–135.

Kluijver, H. N. 1951. The population ecology of the Great Tit, *Parus m. major* L. *Ardea,* 39:1–135.

Kluijver, H. N. 1955. Das Verhalten des Drosselrohrsängers, *Acrocephalus arundinaceus* (L.). *Ardea,* 43(1/3):1–50.

Kluyver (Kluijver), H. N. 1971. Reduced clutches increase adult survival of tits. *In* den Boer, P. J., and G. R. Gradwell (eds). *Dynamics of Populations.* Centre for Agricultural Publishing and Documentations. Wageningen, Netherlands. From *Science,* 177:507.

Kluijver, H. N., *et al.* 1940. De levenswijse van den Winterkoning, *Troglodytes tr. troglodytes* (L.). *Limosa,* 13:1–51.

Knapton, R. W., and J. R. Krebs. 1974. Settlement patterns, territory size, and breeding density in the Song Sparrow *(Melospiza melodia). Canadian Journal of Zoology,* 52:1413–1420.

Kneutgen, J. 1969. "Musikalische" Formen in Gesang der Schamadrossel *(Kittacincla macroura* Gm.) und ihre Funktionen. *Journal für Ornithologie,* 110:245–285.

Kneutgen, J. 1970. Ein Kranken Vogel gelangt durch ein Missverständnis an die Spitze der Rangordnen. *Zeitschrift für Tierpsychologie,* 27:840–841.

Knopf, F. L., and J. C. Street. 1974. Insecticide residues in White Pelican eggs from Utah. *Wilson Bulletin,* 86: 428–434.

Koch, H. J., and A. F. de Bont. 1944. influence de la mue sur l'intensite de metabolisme chez le pinson, *Fringilla coelebs coelebs.* L. *Annales de la Societé Royale Zoologique Belgique,* 75:81–86.

Koehler, O. 1951a. The ability of birds to "count." *Bulletin of Animal Behaviour,* 9:41–45.

Koehler, O. 1956. Thinking without words. *Proceedings of the XIVth International Congress of Zoology,* 75–88. Danish Science Press, Ltd., Copenhagen.

Koeman, J. H. 1975. The toxological importance of chemical pollution for marine birds in the Netherlands. *Die Vogelwarte,* 28:145–150.

Koenig, W. D. 1981. Space competition in the Acorn Woodpecker: Power struggles in a cooperative breeder. *Animal Behaviour,* 29:396–409.

Koenig, W. D. 1984. Geographic variation in clutch size in the Northern Flicker *(Colaptes auratus):* support of Ashmole's hypothesis. *Auk,* 101:698–706.

Koenig, W. D., and F. A. Pitelka. 1979. Relatedness and inbreeding avoidance: counterploys in the communally nesting acorn woodpecker. *Science,* 206:1103–1105.

Koepcke, H.-W. 1963. Probleme des Vogelzuges in Peru. *Proceedings of the XIIIth International Ornithological Congress,* 1:396–411. American Ornithologists' Union.

Koepcke, M. 1963. Anpassungen und geographische Isolation bei Vögeln der peruanischen Küstenlomas. *Proceedings of the XIIIth International Ornithological Congress,* 2:1195–1213. American Ornithologists' Union.

Koepcke, M. 1972. Über die Resistenzformen der Vogelnester in einem begrenzten Gebiet des tropischen Regenwaldes in Peru. *Journal für Ornithologie,* 113:138–160.

Kok, O. B. 1972. Breeding success and territorial behavior of male boat-tailed grackles. *Auk,* 89:528–540.

König, C. 1974. Zum Verhalten spanischer Geier an Kadavern. *Journal für Ornithologie,* 115:289–320.

König, C. 1976. Inter- und intraspezifische Nahrungskonkurrenz bei Altweltgeiern (Aegypiinae). *Journal für Ornithologie,* 117:297–316.

König, C. 1982. Zur systematischen Stellung der Neuweltgeier (Cathartidae). *Journal für Ornithologie,* 123:259–267.

Konishi, M. 1969. Time resolution by single auditory neurons in birds. *Nature,* 222:566–567.

Konishi, M. 1973. How the owl tracks its prey. *American Scientist,* 61:414–424.

Konishi, M. 1974. Hearing and vocalizations in songbirds. *In* Goodman, I. J., and M. W. Schein (eds). *Birds: Brain and Behavior.* Academic Press, New York.

Konishi, M., and E. I. Knudsen. 1979. The Oilbird: Hearing and echolocation. *Science,* 204:425–427.

Konishi, M., and F. Nottebohm. 1969. Experimental studies in the ontogeny of avian vocalizations. *In* Hinde, R. A. (ed). *Bird Vocalizations: Their Relation to Current Problems in Biology and Psychology.* Cambridge University Press, Cambridge.

Kooyman, G. L. 1975. *In* Stonehouse, B. (ed). *The Biology of Penguins.* University Park Press, Baltimore.

Kooyman, G. L., and R. W. Davis. 1982. Diving depths and energy requirements of King Penguins (*Aptenodytes patagonicus*). *Science*, 217:726–727.

Kooyman, G. L., *et al.* 1971. Diving behavior of the Emperor Penguin, *Aptenodytes forsteri*. *Auk*, 88:775–795.

Körner, H. K. 1966. Zur Eirollbewegung der Küstenseeschwalbe (*Sterna macrura* Naum.) *Zeitschrift für Tierpsychologie*, 23:315–323.

Korschgen, C. E. 1977. Breeding stress of female Eiders in Maine. *Journal of Wildlife Management*, 41:360–373.

Kortlandt, A. 1942. Levensloop, samenstelling in structuur der Nederlandse aalscholverbevolking. *Ardea*, 31:175–280.

Kosin, I. L. 1942. Observations on effect of esterified androgen on sex eminence of the chick. *Endocrinology*, 30: 767–772.

Koskimies, J. 1950. The life of the Swift, *Micropus apus* (L.). in relation to the weather. *Annales Academiae Scientiarum Fennicae, Series A, IV. Biologica*, 12:1–151.

Koskimies, J., and L. Lahti. 1964. Cold-hardiness of the newly hatched young in relation to ecology and distribution in ten species of European ducks. *Auk*, 81:281–307.

Kosten, P. A. 1982. Egg retrieval by Clapper Rails. *Journal of Field Ornithology*, 53:274–275.

Kramer, G. 1957. Experiments on bird orientation and their interpretation. *Ibis*, 99:196–227.

Kramer, G. 1961. Long distance orientation. *In* Marshall, A. J. (ed). *Biology and Comparative Physiology of Birds.* Academic Press, New York.

Kramer, G., J. G. Pratt, and U. von St. Paul. 1956. Directional differences in homing pigeons. *Science*, 123:329–330.

Kramer, H. 1972. Family: Herons. *In* Grzimek, B. (ed). *Grzimek's Animal Life Encyclopedia*, Vol. 7. Van Nostrand Reinhold Company, New York.

Krause, H. 1965. Nesting of a pair of Canada Warblers. *Living Bird*, 1965:5–11.

Krebs, J. R. 1970. Territory and breeding density in the Great Tit, *Parus major*. *Ecology*, 52:2–22.

Krebs, J. R. 1977. *In* Stonehouse, B., and C. Perrins (eds). *Evolutionary Ecology.* University Park Press, Baltimore.

Krebs, J. R. 1978. Optimal foraging: decision rules for predators. *In* Krebs, J. R., and N. B. Davies (eds). *Behavioural Ecology: An Evolutionary Approach*, 23–63. Sinauer, Sunderland, Massachusetts.

Krebs, J. R., R. Aschroft, and K. van Orsdol. 1981. Song matching in the great-tit *Parus major* L. *Animal Behaviour*, 29:918–923.

Krebs, J. R., J. T. Erichsen, T. Webber, M. I. Charnov, and E. L. Charnov. 1977. Optimal prey selection in the Great Tit (*Parus major*). *Animal Behaviour*, 25:30–38.

Krebs, J. R., and D. E. Kroodsma. 1980. Repertoires and geographical variation in bird song. *Advances in the Study of Behavior*, 11:143–177.

Krebs, J. R., D. W. Stephens, and W. J. Sutherland. 1983. Perspectives in optimal foraging. *In* Brush, A. H., and G. A. Clark, Jr. (eds). *Perspectives in Ornithology*, 165–216. Cambridge University Press.

Kreithen, M. L., and T. Eisner. 1978. Ultraviolet light detection by the homing pigeon. *Nature*, 272:347–348.

Kreithen, M. L., and D. B. Quine. 1979. Infrasound detection by the homing pigeon: a behavioral audiogram. *Journal of Comparative Physiology*, A. 129:1–4.

Kroodsma, D. E. 1974. Song learning, dialects, and dispersal in the Bewick's Wren. *Zeitschrift für Tierpsychologie*, 35: 352–380.

Kroodsma, D. E. 1976. Reproductive development in a female songbird: Differential stimulation by quality of male song. *Science*, 192:574–575.

Kroodsma, D. E. 1977. A re-evaluation of song development in the Song Sparrow. *Animal Behaviour*, 25:390–399.

Kroodsma, D. E. 1979. Vocal dueling among male Marsh Wrens: evidence for ritualized expressions of dominance/subordinance. *Auk*, 96:506–515.

Kroodsma, D. E. 1984. Songs of the Alder Flycatcher (*Empidonax alnorum*) and Willow Flycatcher (*Empidonax traillii*) are innate. *Auk*, 101:13–24.

Kroodsma, D. E., M. C. Baker, L. F. Baptista, and L. Petrinovich. 1985. Vocal "dialects" in Nuttall's White-crowned Sparrow. *In* Johnston, R. F. (ed). *Current Ornithology*, Vol. 2. Plenum Press, New York.

Kroodsma, D. E., and J. R. Baylis. 1982. Appendix: A world survey of evidence for vocal learning in birds. *In* Kroodsma, D. E., and E. H. Miller (eds). *Acoustic Communication in Birds*, Vol. 2. Academic Press, New York.

Kroodsma, D. E., and L. D. Parker. 1977. Vocal virtuosity in Brown Thrasher. *Auk*, 94:783–785.

Kroodsma, D. E., and R. Pickert. 1980. Environmentally dependent sensitive periods for avian vocal learning. *Nature*, 288:477–479.

Kroodsma, D. E., R. Pickert. 1984a. Sensitive phases for song learning: effects of social interaction and individual variation. *Animal Behaviour*, 32:389–394.

Kroodsma, D. E., and R. Pickert. 1984b. Repertoire size, auditory templates and selective vocal learning in songbirds. *Animal Behaviour*, 32:395–399.

Kroodsma, R. L. 1974. Species recognition behavior of territorial male Rose-breasted and Black-headed Grosbeaks *(Pheucticus)*. *Auk*, 91:54–64.

Krüger, K., R. Prinzinger, and K. L.-Schuchmann. 1982. Torpor and metabolism in hummingbirds. *Comparative Biochemistry and Physiology*, 73A:679–689.

Kruijt, J. P., I. Bossema, and G. J. Lammers. 1982. Effects of early experience and male activity on male choice in mallard females *(Anas platyrhynchos)*. *Behaviour*, 80:32–43.

Kruijt, J. P., C. J. Ten Cate, and G. B. Meeuwissen. 1983. The influence of siblings on the development of sexual preference of male zebra finches. *Development Psychobiology*, 16:233–239.

Krushinskaya, N. L. 1970. [On the memory problem.] *Priroda*, 1970(9):75–78. From *Bird-Banding*, 42:142.

Kruuk, H. 1964. Predators and anti-predator behaviour of the Black-headed Gull, *Larus ridibundus* L. *Behaviour*, Suppl. 2:1–130.

Küchler, W. 1935. Jahreszyklische Veränderung im histologhischen Bau der Vogelschilddrüse. *Journal für Ornithologie*, 83:414–461.

Kuhk, R. 1960. Ein $31^{1}/2$ jahriger Grosser Brachvogel *(Numenius arquata)*. *Die Vogelwarte*, 20:233.

Kuhring, M. S. (ed). 1969. *Proceedings of the World Conference on Bird Hazards to Aircraft*. National Research Council, Ottawa.

Kunkel, P. 1969. Zur Rückwirkung der Nestform auf Verhalten und Auslöserausbildung bei den Prachtfinken (Estrildidae). *Zeitschrift für Tierpsychologie*, 26:277– 283.

Kunkel, P. 1974. Mating systems of tropical birds: the effects of weakness or absence of external reproduction-timing factors, with special reference to prolonged pair bonds. *Zeitschrift für Tierpsychologie*, 34:265–307.

Kusuhara, S., and K. Ishida. 1984. Influence of sexual hormones on the secretory rate in uropygial glands of cockerels. *Japanese Journal of Zootechnical Science*, 55:760–764.

Kuyt, E. 1966. Further observation on large Canada Geese moulting on the Thelon River, N. W. Territories. *Canadian Field Naturalist*, 80:63–69.

Kuyt, E. 1972. First record of the Cattle Egret in the Northwest Territories. *Canadian Field Naturalist*, 86:83–84.

LaBastille, A. 1972. How fares the Poc? *Audubon*, 74(2):36–43.

Lack, D. 1939. The display of the Blackcock. *British Birds*, 32:290–303.

Lack, D. 1940a. Courtship feeding in birds. *Auk*, 57:169–178.

Lack, D. 1940b. Observations on captive Robins. *British Birds*, 33:262–270.

Lack, D. 1943. *The Life of the Robin*. H. F. & G. Witherby, London.

Lack, D. 1945. The ecology of closely related species with special reference to Cormorant *(Phalacrocorax carbo)* and Shag *(Phalacrocorax aristotelis)*. *Journal of Animal Ecology*, 14:12–16.

Lack, D. 1947. *Darwin's Finches*. Cambridge University Press, Cambridge.

Lack, D. 1948. The significance of clutch-size. III. *Ibis*, 90:25–45.

Lack, D. 1954. *The Natural Regulation of Animal Numbers*. Oxford University Press, London.

Lack, D. 1955. British tits *(Parus* spp.) in nesting boxes. *Ardea*, 43:50–84.

Lack, D. 1958. Balancing one nesting hazard against another. *Bird Study*, 5:1–19.

Lack, D. 1959. Watching migration by radar. *British Birds*, 52:258–267.

Lack, D. 1960a. The height of bird migration. *British Birds*, 53:5–10.

Lack, D. 1960b. The influence of weather on passerine migration. *Auk*, 77:171–209.

Lack, D. 1960c. Migration across the North Sea studied by radar. Part II. *Ibis*, 102:26–57.

Lack, D. 1963. Migration across the southern North Sea studied by radar. Part V. *Ibis*, 105:461–492.

Lack, D. 1966. *Population Studies of Birds*. Clarendon Press, Oxford.

Lack, D. 1967. Interrelationships in breeding adaptations as shown by marine birds. *Proceedings of the XIVth International Ornithological Congress*, 3–42. Blackwell Scientific Publications, Oxford.

Lack, D. 1968. *Ecological Adaptations for Breeding in Birds*. Methuen & Co., Ltd., London.

Lack, D. 1969. Island birds. *Biotropica*, 2:29–31.

Lack, D. 1970. The numbers of bird species on islands. *Bird Study*, 16:193–209.

Lack, D. 1971. *Ecological Isolation in Birds*. Harvard University Press, Cambridge.

Lack, D. 1974. *Evolution Illustrated by Waterfowl*. Blackwell Scientific Publications, London.

Lack, D. 1976. *Island Biology, Illustrated by the Land Birds of Jamaica*. University of California Press, Berkeley.

Lack, D., and H. Arn. 1947. Die Bedeutung der Gelegegrösse beim Alpensegler. *Ornithologische Beobachter, 44:* 188–210.

Lack, D., and E. Lack. 1952. The breeding behaviour of the Swift. *British Birds, 45*:186–215.

Lack, D., and E. Lack. 1958. The nesting of the Long-tailed Tit. *Bird Study, 5*:1–19.

Lance, A. N. 1978. Territories and the food plant of individual Red Grouse II. *Journal of Animal Ecology, 47*:307–314.

Lange, H. 1948. Sløruglens *(Tyto alba guttata (Brehm))* Føde, belyst gennem Undersøgelser a Gylp. *Dansk Ornithologisk Forenings Tidsskrift, 42*:50–84.

Lanner, R. M., and S. B. Van der Wall. 1980. Dispersal of timber pine seed by Clark's Nutcrackers. *Journal of Forestry, 78*:637–639.

Lanyon, W. E. 1958. The motivation of sunbathing in birds. *Wilson Bulletin, 70*:280.

Lanyon, W. E., and W. N. Tavolga. 1960. *Animal Sounds and Communication*. American Institute of Biological Sciences, Washington, D.C.

Larkin, R. P., and P. J. Sutherland. 1977. Migrating birds respond to Project Seafarer's electromagnetic field. *Science, 195*:777–779.

Lasiewski, R. C. 1962. The energetics of migrating hummingbirds. *Condor, 64*:324.

Lasiewski, R. C. 1963. Oxygen consumption of torpid, resting, active, and flying hummingbirds. *Physiological Zoology, 36*:122–140.

Lasiewski, R. C. 1972. Respiration of flying birds. *In* Farner, D. S., and J. R. King (eds). *Avian Biology*, Vol. 2. Academic Press, New York.

Lasiewski, R. C., and R. J. Lasiewski. 1967. Physiological responses of the Blue-throated and Rivoli's Hummingbirds. *Auk, 84*:34–48.

Laskey, A. R. 1943. The nesting of Bluebirds banded as nestlings. *Bird-Banding, 14*:39–43.

Laskey, A. R. 1946. Snake depredations of bird nests. *Wilson Bulletin, 58*:217–218.

Laskey, A. R. 1948. Some nesting data on the Carolina Wren at Nashville, Tennessee. *Bird-Banding, 19*:101–121.

Laskey, A. R. 1950. A courting Carolina Wren building over nestlings. *Bird-Banding, 21*:1–6.

Lavern, H. 1940. Beiträge zur Biologie des Sandregenspfeiters *(Charadrius hiaticula* L.). *Journal für Ornithologie, 88:* 184–287.

Lawick-Goodall, J. van. 1968. Tool-using bird: the Egyptian Vulture. *National Geographic, 133*:630–641.

Lawrence, L. de K. 1948. Comparative study of the nesting behavior of Chestnut-sided and Nashville Warblers. *Auk, 65*:204–219.

Lawton, M. F., and C. F. Guidon. 1981. Flock composition, breeding success and learning in the brown jay *(Psilorhinus morio)*. *Condor, 83*:27–33.

Lay, D. W. 1938. How valuable are woodland clearings to birdlife? *Wilson Bulletin, 50*:254–256.

Laybourne, R. C. 1974. Collision between a vulture and an aircraft at an altitude of 37,000 feet. *Wilson Bulletin, 86:* 461–462.

Leck, C. F. 1980. Establishment of new population centers with changes in migration patterns. *Journal of Field Ornithology, 51*:168–173.

LeCroy, M., A. Kulupi, and W. S. Peckover. 1980. Goldie's bird of paradise: display, natural history and traditional relationships of people to the birds. *Wilson Bulletin, 92*:289–301.

LeFebvre, E. A. 1964. The use of D_2O^{18} for measuring energy metabolism in *Columba Livia* at rest and in flight. *Auk, 81*:403–416.

Lehrman, D. S., R. A. Hinde, and E. Shaw (eds). 1965. *Advances in the Study of Behaviour*, Vol. 2. Academic Press, New York.

Lehrman, D. S., *et al.* (eds). 1972. *Advances in the Study of Behavior*, Vol. 5. Academic Press, New York.

Lehtonen, L. 1972. Family: Nightjars. *In* Grzimek, B. (ed). *Grzimek's Animal Life Encyclopedia*, Vol. 9:429–436. Van Nostrand Reinhold Company, New York.

Leiber, A. 1907. Vergleichende Anatomie der Spechtzunge. *Zoologica*, Stuttgart, 20:1–79.

Leitch, W. G. 1964. Water. *In* Linduska J. P., and A. L. Nelson (eds). *Waterfowl Tomorrow*. U.S. Fish and Wildlife Service, Washington, D.C.

Lemon, R. E. 1967. The response of cardinals to songs of different dialects. *Animal Behaviour, 15*;538–545.

Leopold, A. 1933. *Game Management*. Charles Scribner's Sons, New York.

Leopold, A. S. 1953. Intestinal morphology of gallinaceous birds in relation to food habits. *Journal of Wildlife Management*, 17:197–203.

Leopold, A., and A. E. Eynon. 1961. Avian daybreak and evening song in relation to time and light intensity. *Condor*, 63:209–293.

Leopold, A. S., *et al.* 1976. Phytoestrogens: adverse effects on reproduction in California Quail. *Science*, 191:98–100.

Lepkovsky, S., and M. Yasuda. 1967. Adipsia in chickens. *Physiology and Behavior*, 2:45–47.

Leppelsack, H. J. 1983. Analysis of song in the auditory pathway of song birds. *In* Ewert, J. P., R. R. Capranica, and D. J. Ingle (eds). *Advances in Vertebrate Neuroethology*, 788–799. Plenum Press, New York.

Le Resche, R. E., and W. J. L. Sladen. 1970. Establishment of pair and breeding site bonds by young known-age Adelie Penguins, *Pygoscelis adeliae*. *Animal Behaviour*, 18:517–526.

Levine, S., and G. W. Lewis. 1959. Critical period for effects of infantile experience on maturation of stress response. *Science*, 129:42–43.

Lewis, F. T. 1944. The Passenger Pigeon as observed by the Rev. Cotton Mather. *Auk*, 61:587–592.

Lewis, R. W. 1969. Studies on the stomach oils of marine animals. *Comparative Biochemistry and Physiology*, 31: 725–731.

Lewis, S. J., and R. A. Malecki. 1984. Effects of egg oiling on larid productivity and population dynamics. *Auk*, 101: 584–592.

Ligon, J. D. 1967. Relationships of the cathartid vultures. *University Michigan Occasional Papers*, 651. Ann Arbor, Michigan.

Ligon, J. D. 1970. Still more responses of the Poor-will to low temperatures. *Condor*, 72:496–498.

Ligon, J. D. 1983. Commentary: Cooperative breeding strategies among birds. *In* Brush, A. H., and G. A. Clark, Jr. (eds). *Perspectives in Ornithology*, 120–127. Cambridge University Press.

Likens, G. E., *et al.* 1979. Acid rain. *Scientific American*, 241(4):43–51.

Lill, A. 1968. An analysis of sexual isolation in the domestic fowl. *Behaviour*, 30:107–126.

Lill, A. 1974a. Sexual behavior of the lek-forming White-bearded Manakin *(Manacus m. trinitatis* Hartert). *Zeitschrift für Tierpsychologie*, 36:1–36.

Lill, A. 1974b. The evolution of clutch size and male "chauvinism" in the White-bearded Manakin. *Living Bird*, 13: 211–231.

Lillie, F. R. 1940. Physiology of the development of the feather. *Physiological Zoology*, 13:143–175.

Lillie, F. R., and M. Juhn. 1938. Physiology of development of the feather. *Physiological Zoology*, 11:434–448.

Lincoln, F. C. 1928. The migration of young North American Herring Gulls. *Auk*, 45:49–59.

Lincoln, F. C. 1950. *Migration of Birds*. U.S. Fish and Wildlife Service, Circular No. 16, Washington, D.C.

Lindeboom, H. J. 1984. The nitrogen pathway in a penguin rookery. *Ecology*, 65:269–277.

Lindsay, G. 1967. Prairie Chickens died under crusted snow. *Passenger Pigeon*, 29:25–28.

Linduska, J. P. 1966. Hunting is a positive thing. *In* Stefferud, A. (ed). *Birds in Our Lives*. U.S. Fish and Wildlife Service, Washington, D.C.

Linduska, J. P., and A. L. Nelson (eds). 1964. *Waterfowl Tomorrow*. U.S. Fish and Wildlife Service, Washington, D.C.

Lippens, L. 1939. Les oiseaux aquatiques du Kivu. *Le Gerfaut*, 28, Fasc. Spécial: 1–104.

Lissaman, P. B. S., and C. A. Schollenberger. 1970. Formation flight of birds. *Science*, 168:1003–1005.

Liversidge, R. 1972. Family: Hammerheads. *In* Grzimek, B. (ed). *Grzimek's Animal Life Encyclopedia*. Vol. 7. Van Nostrand Reinhold Company, New York.

Lloyd, C. S. 1979. Factors affecting breeding of Razorbills, *Alca torda*, on Stokholm. *Ibis*, 121:165–176.

Lockie, J. D. 1955. The breeding habits and food of Short-eared Owls after a vole plague. *Bird Study*, 2:53–69.

Loefer, J. B., and J. A. Pattern. 1941. Starlings at a Blackbird Roost. *Auk*, 58:584–586.

Loffredo, C. A., and G. Borgia. 1986. Male courtship vocalization as cues for mate choice in the Satin Bowerbird *(Ptilonorhynchus violaceus)*. *Auk*, 103:189–195.

Lofts, B., and A. J. Marshall. 1960. The experimental regulation of Zugunruhe and the sexual cycle in the Brambling, *Fringilla montifringilla*. *Ibis*, 102:209–214.

Lofts, B., and B. K. Murton. 1973. Reproduction in birds. *In* Farner, D. S., and J. R. King (eds). *Avian Biology*, Vol. 3. Academic Press, New York.

Lofts, B., *et al.* 1963. The experimental demonstration of premigration activity in the absence of fat deposition in birds. *Ibis*, 105:99–105.

Lögler, P. 1959. Versuche zur Frage des "Zahlvermögens" an einem Graupapagei und Vergleichsversuche an Menschen. *Zeitschrift für Tierpsychologie*, 16:179–217.

Löhrl, H. 1951. Balz und Paarbildung beim Halsbandfliegenschnäpper. *Journal für Ornithologie*, 93:46–60.

Löhrl, H. 1959. Zur Frage des Zeitpunktes einer Prägung auf die Heimatregion beim Halsbandschnäpper *(Ficedula albicollis)*. *Journal für Ornithologie*, 100:132–140.

Löhrl, H. 1957. *Der Kleiber*. Ziemsen Verlag, Wittenberg.

Löhrl, H. 1973. Einfluss der Brutraumfläche auf die Gelegegrösse der Kohlmeise *(Parus major)*. *Journal für Ornithologie*, 114:339–347.

Löhrl, H. 1978. Beiträge zur Ethologie und Gewichtsentwicklung beim Wendhals, *Jynx torquilla*. *Ornithologische Beobachter*, 75:193–201.

Löhrl, H. 1980. Weitere Versuche zur Frage "Brutraum und Gelegegrösse" bei der Kohlmeise. *Parus major. Journal für Ornithologie*, 121:403–405.

Löhrl, H. 1983. Zur Feindabwehr der Wacholderdrossel *(Turdus pilaris)*. *Journal für Ornithologie*, 124:271–280.

Lord, R. D., *et al.* 1962. Radiotelemetry of the respiration of a flying duck. *Science*, 137:39–40.

Lorenz, K. Z. 1935. Der Kumpan in der Umwelt des Vogels. *Journal für Ornithologie*, 83:137–213.

Lorenz, K. Z. 1937. The companion in the bird's world. *Auk*, 54:245–273.

Lorenz, K. Z. 1938. A contribution to the comparative sociology of colonial nesting birds. *Proceedings of the VIIIth International Ornithological Congress, Oxford, July 1934*. Oxford University Press, London.

Lorenz, K. Z. 1941. Vergleichende Bewegungsstudien an Anatiden. *Journal für Ornithologie*, 89:194–293.

Lorenz, K. Z. 1950. The comparative method of studying innate behavior patterns. *Physiological Mechanisms in Animal Behaviour. Symposia for Experimental Biology*. Vol. 4. Academic Press, New York.

Lorenz, K. Z. 1958. The evolution of behavior. *Scientific American*, 199(6):67–78.

Lorenz, K. Z. 1970. *Studies in Animal and Human Behavior*, Vol. 1. Harvard University Press, Cambridge.

Lott, D. F., and S. Comerford. 1968. Hormonal initiation of parental behavior in inexperienced Ring Doves. *Zeitschrift für Tierpsychologie*, 25:71–75.

Lovell, H. B. 1958. Baiting of fish by a Green Heron. *Wilson Bulletin*, 70:280–281.

Low, S. H. 1957. Banding with mist nets. *Bird-Banding*, 28:115–128.

Lowe-McConnell, R. H. 1967. Biology of the immigrant Cattle Egret, *Ardeola ibis*, in Guyana, South America. *Ibis*, 109:168–179.

Lowery, G. H., and R. J. Newman. 1955. Direct studies of nocturnal bird migration. *In* Wolfson, A. (ed). *Recent Studies in Avian Biology*. University of Illinois Press, Urbana, Illinois.

Lowery, G. H., and R. J. Newman. 1966. A continent-wide view of bird migration of four nights in October. *Auk*, 83:547–586.

Lowther, J. K. 1961. Polymorphism in the White-throated Sparrow, *Zonotrichia albicollis* (Gmelin). *Canadian Journal of Zoology*, 39:281–292.

Lucas, A. M. 1979. Integumentum commune. *In* Baumel, J. J., A. S. King, A. M. Lucas, J. F. Breazile, and H. E. Evans (eds). *Nomina Anatomica Avium*. Academic Press, New York.

Lucas, A. M., and P. A. Stettenheim. 1972. *Avian Anatomy: Integument*, 2 vols. Agriculture Handbook 362. U.S. Government Printing Office, Washington, D.C.

Lumsden, H. G. 1976. Choice of nest boxes by Starlings. *Wilson Bulletin*, 88:665–666.

Lundin, A. 1960. En undersökning av hornugglans *(Asio otus)* föda. *Vår Fågelvärld*, 19:43–50. From *Biological Abstracts*, 35:4972.

Lydekker, R. 1901. *The New Natural History*. Merrill and Baker, New York.

Lyon, D. L. 1976. A montane hummingbird territorial system in Oaxaca, Mexico. *Wilson Bulletin*, 88:220–299.

MacArthur, R. H. 1972. *Geographical Ecology*. Harper and Row, Publishers, New York.

MacArthur, R. H., and J. H. Connell, 1966. *The Biology of Populations*. J. Wiley and Sons, New York.

MacArthur, R. H., and J. W. MacArthur. 1961. On bird species diversity. *Ecology*, 42:594–598.

MacArthur, R. H., and E. O. Wilson. 1967. *The Theory of Island Biogeography*. Princeton University Press, Princeton, New Jersey.

MacDonald, D. W., and D. G. Henderson. 1977. Aspects of the behaviour and ecology of mixed-species bird flocks in Kashmir. *Ibis*, 119:481–493.

MacDonald, M. A. 1977. Adult mortality and fidelity to mate and nest-site in a group of marked Fulmars. *Bird Study*, 24:165–168.

MacFarland, C., and J. MacFarland. 1972. Goliaths of the Galapagos. *National Geographic*, 142:632–649.

Mackworth-Praed, C. W., and C. H. B. Grant. 1957–1960. *African Handbook of Birds: Birds of Eastern and Northeastern Africa.* 2 Vols. Longmans, Green and Company, London, New York.

Mackworth-Praed, C. W., and C. H. B. Grant. 1961. *African Handbook of Birds: Birds of the Southern Third of Africa.* 2 Vols. Longmans, Green and Company, London, New York.

Maclatchey, A. R. 1937. Contribution à l'étude des oiseaux du Gabon Meridional (Suite). *L'Oiseau et la Revue Française d'Ornithologie,* 7:60–80.

Maclean, G. L. 1968. Field studies on the Sandgrouse of the Kalahari Desert. *Living Bird,* 1968:209–235.

Maclean, G. L. 1976. Adaptations of Sandgrouse for life in arid lands. *In* Frith, H. J., and J. H. Calaby (eds). *Proceedings of the 16th International Ornithological Congress.* Australian Academy of Science, Canberra.

Maclean, S. F., Jr. 1974. Lemming bones as a source of calcium for arctic sandpipers. *Ibis,* 116:552–557.

MacLellan, C. R. 1961. Woodpecker control of the codling moth in Nova Scotia orchards. *Atlantic Naturalist,* 16(1): 17–25.

MacRoberts, B. R., and M. H. MacRoberts. 1972. Social stimulation of reproduction in Herring and Lesser Black-backed Gulls. *Ibis,* 114:495–506.

MacRoberts, M. H., and B. R. MacRoberts. 1976. Social organization and behavior of the Acorn Woodpecker in central coastal California. *Ornithological Monographs No. 21.* American Ornithologists Union.

Magee, M. J. 1938. Estimated sex ratio of the Eastern Purple Finch. . . . *Bird-Banding,* 9:199.

Magee, M. J. 1940. Notes on the returns of the Eastern Purple Finch *(Carpodacus purpureus purpureus)* and their sex ratio. *Bird-Banding,* 11:110–111.

Mahoney, S. A. 1984. Plumage wetability of aquatic birds. *Auk,* 101:181–184.

Makarenko, V. F. 1958. [Change in the helminth fauna of birds in relation to ecological conditions.] Moscow Izdat. Akad. Nauk S.S.S.R., pp. 211–215. From *Wildlife Review,* 104:51.

Makatsch, W. 1950. *Der Vogel und sein Nest.* Akademische Verlagsgesellschaft, Leipzig.

Mangold, E. 1929. Die Verdauung des Geflügels. In *Handbuch der Ernährung und des Stoffwechsels der landwirtschaften Nutziere. Vol. 2.* Julius Springer, Berlin.

Mangold, O. 1946. Die Nasc der segelnden Vögel ein Organ des Strömungssinnes? *Die Naturwissenschaften,* 33: 19–23.

Manuwal, D. A. 1974. Effects of territoriality on breeding in a population of Cassin's Auklet. *Ecology,* 55:1399–1406.

Marchand, C.-R., and L. Gomot. 1976. Abortive spermatogenesis in the hybrid duck (cross male *Anas platyrhynchos* by female *Cairina moschata*): ultrastructural study. *Bulletin de l'Association des Anatomistes,* 60:613–622.

Marder J., I. Gavrieli-Levin, and Räber H. 1986. Cutaneous evaporation in heat-stressed spotted sand-grouse. *Condor,* 88:99–100.

Margoliash, D. 1983. Acoustic parameters underlying the responses of song-specific neurons in the White-crowned Sparrow. *Journal of Neuroscience,* 3:1038–1057.

Marks, H. L., P. B. Siegel, and C. Y. Kramer. 1960. Effect of comb and wattle removal on the social organization of mixed flocks of chickens. *Animal Behaviour,* 8:192–196.

Marks, H. L., and R. D. Syatt. 1979. Genetic resistance to aflatoxin in Japanese Quail. *Science,* 206:1329–1330.

Marler, P. 1955. Studies of fighting in Chaffinches. *British Journal of Animal Behaviour,* 3:137–146.

Marler, P. 1956a. Studies of fighting in chaffinches. Proximity as a cause of aggression. *British Journal of Animal Behaviour,* 3:111–117.

Marler, P. 1956b. Behaviour of the Chaffinch, *Fringilla coelebs, Behaviour,* Suppl. 5:1–184.

Marler, P. 1964. Aggression. *In* Thomson, A. L. (ed). *A New Dictionary of Birds.* McGraw-Hill Book Company, New York.

Marler, P. 1967. Comparative study of song development in sparrows. *Proceedings of the XIVth International Ornithological Congress,* 231–244. Blackwell Scientific Publications, Oxford.

Marler, P. 1969. Tonal quality of bird sounds. *In* Hinde, R. A. (ed). *Bird Vocalizations: Their Relations to Current Problems in Biology and Psychology.* Cambridge University Press, Cambridge.

Marler, P. 1970. A comparative approach to vocal learning: Song development in the White-crowned Sparrows. *Journal of Comparative Physiology,* 71:1–25.

Marler, P. R., and W. J. Hamilton. 1966. *Mechanisms of Animal Behavior.* John Wiley and Sons, New York.

Marler, P., and S. Peters. 1977. Selective vocal learning in a sparrow. *Science,* 198:519–521.

Marler, P., and M. Tamura. 1962. Song "dialects" in three populations of White-crowned Sparrows. *Condor, 64:* 368–377.

Marler, P., and M. Tamura. 1964. Culturally transmitted patterns of vocal behavior in sparrows. *Science, 146:* 1483–1486.

Marsh, C. O. 1880. *Odontornithes: a Monograph on the Extinct Toothed Birds of North America*. U.S. Government Printing Office, Washington, D.C.

Marsh, R. I., and W. R. Dawson. 1982. Substrate metabolism in seasonally acclimatized American Goldfinches. *American Journal of Physiology*, 24:563–569.

Marsh, R. L. 1984. Adaptations of the gray catbird *Dumatella carolinensis* to long-distance migration: Flight muscle hypertrophy associated with clevated body mass. *Physiological Zoology*, 57:105–117.

Marshall, A. J. 1950. The function of vocal mimicry in birds. *Emu*, 50:5–16.

Marshall, A. J. 1954. *Bower-Birds*. Oxford University Press, London.

Marshall, A. J. (ed). 1960–1961. *Biology and Comparative Physiology of Birds*. 2 vols. Academic Press, New York and London.

Marshall, A. J., and J. D. Roberts. 1959. The breeding biology of equatorial vertebrates. *Proceedings of the Zoological Society of London*, 132:617–625.

Marshall, A. J., and D. L. Serventy. 1956a. The breeding cycle of the Short-tailed Shearwater *Puffinus tenuirostris* (Temminck) in relation to trans-equatorial migration and its enviornment. *Proceedings of the Zoological Society of London*, 127:489–510.

Marshall, A. J., and D. L. Serventy. 1956b. Moult adaptation in relation to long-distance migration in petrels. *Nature*, 177:943.

Marshall, A. J., and D. L. Serventy. 1959. The experimental demonstration of an internal rhythm of reproduction in transequatorial migrant, the Short-tailed Shearwater, *Puffinus tenuirostris*. *Nature*, 184:1704–1705.

Marshall, H. 1947. Longevity of the American Herring Gull. *Auk*, 64:188–198.

Marshall, J. T. 1955. Hibernation in captive goatsuckers. *Condor*, 57:129–134.

Marshall, W. H. 1946. Cover preferences, seasonal movements, and food habits of Richardson's Grouse and Ruffed Grouse in southern Idaho. *Wilson Bulletin*, 58:42–52.

Marti, C. D. 1973. Food consumption and pellet formation rates in four owl species. *Wilson Bulletin*, 85:178–181.

Martin, D. J. 1973. A spectrographic analysis of Burrowing Owl vocalizations. *Auk*, 90:564–578.

Martin, L. D. 1983. The origin and early radiation of birds. *In* Brush, A. H., and G. A. Clark, Jr. (eds). *Perspectives in Ornithology*, 291–338. Cambridge University Press.

Martin, L. D., *et al.* 1980. The origin of birds: structure of the tarsus and teeth. *Auk*, 97:86–93.

Martin, R. C., and K. B. Melvin. 1964. Fear responses of Bobwhite Quail (*Colinus virginianus*) to a model and a live Red-tailed hawk (*Buteo jamaicensis*). *Psychologische Forschung*, 27:323–336.

Mason, A. G. 1945. The display of the Corn-Crake. *British Birds*, 38:351–352.

Mason, E. A. 1944. Parasitism by Protocalliphora and management of cavity-nesting birds. *Journal of Wildlife Management*, 8:232–247.

Mathews, F. S. 1921. *Field Book of Wild Birds and Their Music*, G. P. Putnam's Sons, New York.

Mathiasson, A. 1957. Fågelsträcket vid Falsterbo. *Vår Fågelvärld*, 16:90–104.

Matthews, G. V. T. 1954. Some aspects of incubation in the Manx Shearwater, *Procellaria puffinus*, with particular reference to chilling resistance in the embryo. *Ibis*, 96:432–440.

Matthews, G. V. T. 1961. "Nonsense" orientation in Mallard, *Anas platyrhynchos*, and its relation to experiments on bird navigation. *Ibis*. 103:211–230.

Matthews, G. V. T. 1963. The orientation of pigeons as affected by the learning of landmarks and by the distance of displacement. *Animal Behaviour*, 11:310–317.

Matthews, G. V. T. 1964. Navigation. *In* Thomson, A. L. (ed). A *New Dictionary of Birds*. McGraw-Hill Book Company, New York.

Matthews, G. V. T. 1968. *Bird Navigation*. Cambridge University Press, Cambridge.

Matthews, G. V. T., and W. A. Cook. 1977. The role of landscape features in the "nonsense" orientation of Mallard. *Animal Behaviour*, 25:508–517.

Matthijsen, E., and A. A. Dhont. 1983. Die Ansiedlung junger Kleiber (*Sitta europaea*) in Spätsommer und Herbst. *Journal für Ornithologie*, 124:281–290.

Mayaud, N. 1950. *In* Grassé, P. (ed). *Traité de Zoologie*. Tome XV, *Oiseaux*. Masson et Cie, Paris.

Mayfield, H. F. 1952. Nesting-height preference of the Eastern Kingbird. *Wilson Bulletin*, 64:160.

Mayfield, H. F. 1973. Census of Kirtland's Warbler in 1972. *Auk*, 90:684–685.

Mayhew, W. W. 1955. Spring rainfall in relation to Mallard production in the Sacramento Valley, California. *Journal of Wildlife Management*, 19:36–47.

Mayr, E. 1935. Bernard Altum and the territory theory. *Proceedings of the Linnaean Society of New York*, 45–46:1–15.

Mayr, E. 1939. The sex ratio in wild birds. *American Naturalist*, 73:156–179.

Mayr, E. 1942. *Systematics and the Origin of Species*. Columbia University Press, New York.

Mayr, E. 1943. The zoogeographic position of the Hawaiian Islands. *Condor*, 45:45–48.

Mayr, E. 1945. Bird conservation problems in the Southwest Pacific. *Audubon Magazine*, 47:279–282.

Mayr, E. 1946a. History of North American bird fauna. *Wilson Bulletin*, 58:3–41.

Mayr, E. 1946b. The number of species of birds. *Auk*, 63:64–69.

Mayr, E. 1948. The bearing of the new systematics on genetical problems. The nature of species. *In* Demerec, M. (ed). *Advances in Genetics*, Vol. 2. Academic Press, New York.

Mayr, E. 1949. Speciation and systematics. *In* Jepsen, G. L., E. Mayr, and G. G. Simpson, *Genetics, Paleontology and Evolution*. Princeton University Press, Princeton.

Mayr, E. 1951. Speciation in birds. in *Proceedings of the Xth International Ornithological Congress, Uppsala, June 1950*, 91–131. Almqvist and Wiksells, Uppsala.

Mayr, E. (ed). 1952. The problem of land connections across the South Atlantic, with special reference to the Mesozoic. *Bulletin of the American Museum of Natural History*, 99:81–258.

Mayr, E. 1954. Change of genetic environment and evolution. *In* Huxley, J., A. C. Hardy, and E. B. Ford (eds). *Evolution as a Process*. G. Allen and Unwin, London.

Mayr, E. 1957. New species of birds described from 1941 to 1955. *Journal für Ornithologie*, 98:22–35.

Mayr, E. 1958. *In* Roe, A., and G. G. Simpson (eds). *Behavior and Evolution*. Yale University Press, New Haven, Connecticut.

Mayr, E. 1964. Inference concerning the Tertiary American bird faunas. *Proceedings of the National Academy of Sciences*, 51:280–288. United States.

Mayr, E. 1970. *Population, Species, and Evolution*. Harvard University Press, Cambridge.

Mayr, E. 1971. New species of birds described from 1956 to 1965. *Journal für Ornithologie*, 112:302–316.

Mayr, E. 1972. Continental drift and the history of the Australian bird fauna. *Emu*, 72:26–28.

Mayr, E. 1976. the value of various taxonomic characters in avian classification. *In* Frith, H. J., and J. H. Calaby (eds). *Proceedings of the 16th International Ornithological Congress*. Australian Academy of Science, Canberra.

Mayr, E. 1983a. *In* Brush, A. H., and G. A. Clark (eds). Introduction to *Perspectives in Ornithology*, 1–21. Cambridge University Press.

Mayr, E. 1983b. The joy of birds. *Natural History* 92:2.

Mayr, E., and D. Amadon. 1951. A classification of recent birds. *American Museum Novitates*, 1496:1–42.

Mayr, E., E. G. Linsley, and R. L. Usinger. 1953. *Methods and Principles of Systematic Zoology*. McGraw-Hill Book Company, New York.

Mayr, E., and M. Mayr. 1954. The tail molt of small owls. *Auk*, 71:172–178.

Mayr, E., and F. Vuilleumier. 1983. New species of birds described from 1966 to 1975. *Journal für Ornithologie*, 124:217–232.

Mayr, I. 1970. Olkugelverteilung in der Retina männlicher und weiblicher Feuerweber *(Euplectes orix franciscanus)*. *Journal für Ornithologie*, 111:30–37.

Mazzeo, R. 1953. Homing of the Manx Shearwater. *Auk*, 70:200–201.

McBee, R. H., and G. C. West. 1969. Caecal fermentation in the Willow Ptarmigan. *Condor*, 71:54–58.

McCabe, R. A., and H. F. Deutsch. 1952. The relationships of certain birds as indicated by their egg white proteins. *Auk*, 69:1–18.

McCabe, B. J., G. Horn, and P. P. G. Bateson. 1981. Effects of restricted lesions on the chick forebrain on the acquisition of filial preferences during imprintings. *Brain Research*, 205(1):29–38.

McCallum, G. E. 1964. Clean water and enough of it. *In* Linduska, J. P., and A. L. Nelson (eds). *Waterfowl Tomorrow*. U.S. Fish and Wildlife Service, Washington, D.C.

McClintock, C. P., T. C. Williams, and J. M. Teal. 1978. Autumnal bird migration observed from ships in the western North Atlantic Ocean. *Bird-Banding*, 49:262–277.

McClure, H. E. 1945. Reaction of the Mourning Dove to colored eggs. *Auk*, 62:270–272.

McDonald, M. E. 1969. *Catalogue of Helminths of Waterfowl (Anatidae)*. U.S. Bureau of Sport Fisheries and Wildlife, Washington, D.C.

McFarland, D. J., and E. Baher. 1968. Factors affecting feather posture in the barbary dove. *Animal Behaviour*, 16:171–177.

McGahan, J. 1968. Ecology of the Golden Eagle. *Auk*, 85:1–12.

McGillivray, W. B. 1983. Nestling feeding rates and body size of adult House Sparrows. *Canadian Journal of Zoology*, 62:381–385.

McGillivray, W. B. 1983. Intraseasonal reproductive costs for the house sparrow *(Passer domesticus)*. *Auk*, 100:25–32.

McGowan, J. D. 1973. Fall and winter foods of Ruffed Grouse in interior Alaska. *Auk*, 90:636–640.

McIlhenny, E. A. 1940. Effect of excessive cold on birds in southern Louisiana. *Auk*, 57:408–410.

McKean, J. L. 1960. Movements of Cattle Egrets. *Emu*, 60:202.

McKernan, D. L., and V. B. Scheffer. 1942. Unusual numbers of dead birds on the Washington coast. *Condor*, 44:264–266.

McKinney, F. 1965. The comfort movements of Anatidae. *Behaviour*, 25:120–220.

McKinney, F., S. R. Derrickson, and P. Mineau. 1983. Forced copulation in waterfowl. *Behaviour*, 86:250–294.

McLaren, P. L., and M. A. MacLaren. 1982. Migration and summer distribution of Lesser Snow Geese *(Chen. c. caerulescens)* in interior Keewatin. *Wilson Bulletin*, 94:494–504.

McLaughlin, R. L., and R. D. Montgomerie. 1985. Brood division by Lapland Longspurs. *Auk*, 102:686–695.

McLean, I., and K. Williamson. 1960. Migrants at station "Juliett" in September 1959. *British Birds*, 53:215–219.

McMahan, C. A., and R. L. Fritz. 1967. Mortality to ducks from trotlines in Lower Laguna Madre, Texas. *Journal of Wildlife Management*, 31:783–787.

McNulty, F. 1971. The silent shore. *Audubon*, 73(6):4–11.

Meanley, B. 1971. Blackbirds and the southern rice crop. U.S. Fish and Wildlife Service, *Resource Publication 100*, Washington, D.C.

Meanley, B., and J. S. Webb. 1965. Nationwide population estimates of blackbirds and Starlings. *Atlantic Naturalist*, 20:189–191.

Mebs, T. 1964. Zur Biologie und Populationsdynamik des Mäusebussards *(Buteo buteo)*. *Journal für Ornithologie*, 105:247–306.

Mebs, T. 1972. Family: Falcons. *In* Grzimek, B. (ed). *Grzimek's Animal Life Encyclopedia*, Vol. 7. Van Nostrand Reinhold Company, New York.

Meier, A. H. 1973. Daily hormone rhythms in the White-throated Sparrow. *American Scientist*, 61:184–187.

Meier, A. H. 1976. Chronoendocrinology of the White-throated Sparrow. *In* Frith, H. J., and J. H. Calaby (eds). *Proceedings of the 16th International Ornithological Congress*. Australian Academy of Science, Canberra.

Meier, A. H., and A. C. Russo. 1985. Circadian organization of the avian annual cycle. *In* Johnston, R. F. (ed). *Current Ornithology*, 303–343. New York and London, Plenum Press.

Meijknecht, J. T. V. 1941. Farbsehen und Helligkeitsunterscheidung beim Steinkauz *(Athene noctua vidalii* E. A. Brehm). *Ardea*, 30:129–174.

Meinertzhagen, R. 1954. *Birds of Arabia*. Oliver and Boyd, London.

Meinertzhagen, R. 1964. Palearctic region. *In* Thomson, A. L. (ed). *A New Dictionary of Birds*. McGraw-Hill Book Company, New York.

Meise, W. 1954. Über Zucht, Eintritt der Geschlectsreife, Zwischen und Weiterzug der Wachtel *(C. coturnix)*. *Die Vogelwarte*, 17:211–215.

Meise, W. 1972. Family: Cuckoos and Coucals. *In* Grzimek, B. (ed). *Grzimek's Animal Life Encyclopedia*, Vol. 8. Van Nostrand Reinhold Company, New York.

Melemis, S. M., and J. B. Falls. 1982. The defense function: a measure of territorial behavior. *Canadian Journal of Zoology*, 60:495–501.

Meltofte, H. 1978. A breeding association between Eiders and tethered huskies in Northwest Greenland. *Wildfowl*, 29:45–54.

Menaker, M., R. Roberts, J. Elliott, and H. Underwood. 1970. Extraretinal light perception in the sparrow. III. *Proceedings of the National Academy of Science*, 67:320–325.

Mengel, R. M. 1970. The North American central plains as an isolating agent in bird speciation. *In* Pleistocene and recent environments of the Central Great Plains. *University of Kansas Special Publication*, 3:279–340.

Menne, M., and E. Curio. 1978. Investigations into the symmetry concept in the Great Tit *(Parus major)*. *Zeitschrift für Tierpsychologie*, 47:299–322.

Menon, G. K. 1984. Glandular functions of avian integument: An overview. *Journal of the Yamashina Institute of Ornithology*, 16:1–12.

Menon, G. K., S. K. Aggaarwal, and A. M. Lucas. 1981. Evidence for the holocrine nature of lipoid secretion by avian epidermal cells: a histochemical and fine structural study of rictus and the uropygial gland. *Journal of Morphology*, 167:185–199.

Menon, G. K., L. F. Baptista, P. M. Elias, and M. Bouvier. 1986. Fine structural basis of the cutaneous water barrier in nestling Zebra Finches. *Ibis*, in press.

Merkel, F. W., H. G. Fromme, and W. Wiltschko. 1964. Nachtvisuelles Orientierungsvermögen bei nächtlich zugunruhigen Rotkehlchen. *Die Vogelwarte*, 22:168–173.

Merkel, F. W., and W. Wiltschko. 1965. Magnetismus und Richtungsfinden zugunruhiger Rotkehlchen *(Erithacus rubecula). Die Vogelwarte,* 23:71–77.

Merrill, G. W. 1961. Loss of 1000 Lesser Sandhill Cranes. *Auk,* 78:641–642.

Merton, D. V., R. B. Morris, and I. A. E. Atkinson. 1984. Lek behaviour in a parrot: the Kakapo *Strigops haproptilus* of New Zealand. *Ibis,* 126:277–283.

Meschini, E. 1983. Pigeon navigation: some experiments on the importance of olfactory cues at short distances from the loft. *Journal of Comparative Physiology, A. Series Neural Behavioral Physiology,* 150:493–498.

Messmer, E., and I. Messmer. 1956. Die Entwicklung der Lautäusserungen und einiger Verhaltungsweisen der Amsel *(Turdus merula merula* L.) unter natürlichen Bedingungen und nach Einzelaufzucht in schalldichten Räumen. *Zeitschrift für Tierpsychologie,* 13:341– 344.

Metzgar, L. H. 1967. An experimental comparison of Screech Owl predation on resident and transient White-footed Mice *(Peromyscus leucopus). Journal of Mammalogy,* 48:387–391.

Mewaldt, L. R. 1964. California 'crowned' sparrows return from Maryland. *Western Bird-Bander,* 39:1–2.

Mewaldt, L. R. 1964. Effects of bird removal on a winter population of sparrows. *Bird-Banding,* 35:184–195.

Mewaldt, L. R. 1976. Winter philopatry in White-crowned Sparrows *(Zonotrichia leucophrys). North American Bird Bander,* 1:14–20.

Mewaldt, L. R., S. S. Kibby, and M. L. Morton. 1968. Comparative biology of Pacific coastal White-crowned Sparrows. *Condor,* 70:14–30.

Mewaldt, L. R., and J. R. King. 1978. Latitudinal variation of postnuptial molt in Pacific coast White-crowned Sparrows. *Auk,* 95:168–174.

Mewaldt, L. R., M. L. Morton, and I. L. Brown. 1964. Orientation of migratory restlessness in *Zonotrichia. Condor,* 66:377–417.

Mewaldt, L. R., and R. G. Rose. 1960. Orientation of migratory restlessness in the White-crowned Sparrow. *Science,* 131:105–106.

Meyer, M. E., 1964. Discriminative basis for astronavigation in birds. *Journal of Comparative and Physiological Psychology,* 58:403–406.

Meyer, O. 1930. Untersuchungen an den Eiern von *Megapodius eremita. Ornithologische Monatsberichte,* 38:1–5.

Mickey, F. W. 1943. Breeding habits of McCown's Longspur. *Auk,* 60:181–209.

Mildenberger, H. 1940. Beobachtungen über Fitis- Weiden- und Waldlaubsänger im Rheinland. *Journal für Ornithologie,* 88:537–549.

Miles, F. A. 1970. Centrifugal effects in the avian retina. *Science,* 170:992–995.

Millay, E. St. V. 1922. A *Few Figs from Thistles.* Harper and Brothers, New York.

Millener, P. R. 1982. And then were twelve: the taxonomic status of *Anomalopteryx oweni* (Aves: Dinornithidae). *Notornis,* 29:165–168.

Miller, A. H. 1951. An analysis of the distribution of the birds of California. *University of California Publications in Zoology,* 50:531–644.

Miller, A. H. 1954. The occurrence and maintenance of the refractory period in crowned sparrows. *Condor,* 56:13–20.

Miller, A. H. 1961. Bimodal occurrence of breeding in an equatorial sparrow. *Proceedings of the National Academy of Science,* 48:396–400.

Miller, A. H. 1961. Molt cycles in equatorial Andean sparrows. *Condor,* 63:143–161.

Miller, D. E. 1972. Parental acceptance of young as a function of incubation time in the Ring-billed Gull. *Condor,* 74:482–484.

Miller, E. H. 1985. Parental behavior in the Least Sandpiper *(Calidris minutilla). Canadian Journal of Zoology,* 63: 1593–1601.

Miller, M. R. 1975. Gut morphology of Mallards in relation to diet quality. *Journal of Wildlife Management,* 39: 168–173.

Miller, R. C. 1958. Morning and evening song of Robins in different latitudes. *Condor,* 60:105–107.

Miller, F. L., and E. Broughton. 1971. An unusual display of territorial aggressiveness by Sandhill Cranes *(Grus canadensis* L.). *Canadian Field Naturalist,* 85:66–67.

Miller, S. C., B. M. Bowman, and R. L. Myers. 1984. Morphological and ultrastructural aspects of the activation of avian medullary bone osteoclasts by parathyroid hormone. *Anatomical Record,* 258:223–232.

Miller, S. J., and D. W. Inouye. 1983. Roles of the wing whistle in the territorial behaviour of male broad-tailed hummingbirds *(Selasphorus platycercus). Animal Behaviour,* 31:689–700.

Millikan, G. C., and R. I. Bowman. 1967. Observations on Galapagos tool-using finches in captivity. *Living Bird,* 6: 23–41.

Millington, S. J., and T. D. Price. 1985. Song inheritance and mating patterns in Darwin's finches. *Auk, 102:* 342–346.

Mills, J. A. 1973. The influence of age and pair-bond on the breeding biology of the Red-billed Gull *(Larus novaehollandiae scopulinus). Journal of Animal Ecology,* 42:147–162.

Mills, S. H., and J. E. Heath. 1970. Thermoresponsiveness of the preoptic region of the brain in House Sparrows. *Science,* 168:1008–1009.

Milne, H., and A. Reed. 1974. Annual production of fledged young from the Eider colonies of the St. Lawrence estuary. *Canadian Field Naturalist,* 88:163–169.

Minton, C. D. T. 1968. Pairing and breeding of Mute Swans. *Wildfowl,* 19:41–60.

Mirsky, E. N. 1976. Song divergence in hummingbird and junco populations on Guadalupe Island. *Condor,* 78:230–235.

Mitchell, K. D. G. 1955. Aircraft observations of birds in flight. *British Birds,* 48:59–70.

Mitchell, W. G., and E. C. Turner. 1969. Arthropod parasites on the Starling, *Sturnus vulgaris,* in southwest Virginia. *Journal of Economic Entomology,* 62:195–197.

Mock, D. W. 1983. On the study of avian mating systems. *In* Brush, A. H., and G. A. Clark, Jr. (eds). *Perspectives in Ornithology,* 55–84. Cambridge University Press, Cambridge.

Mock, D. W. 1984. Siblicidal aggression and resource monopolization in birds. *Science,* 225:731–733.

Mock, D. W. 1985. Siblicidal brood reduction: The prey-size hypothesis. *American Naturalist,* 125:327–343.

Moermond, T. C. 1983. Suction-drinking in tanagers Thraupidae and its relation to frugivory. *Ibis,* 125:545–549.

Moermond, T. C., and J. S. Denslow. 1985. Neotropical avian frugivores: patterns of behavior, morphology, and nutrition, with consequences for fruit selection: *In Neotropical Ornithology,* Ornithological Monograph, 36:865–897.

Moffat, C. B. 1903. The spring rivalry of birds. *The Irish Naturalist,* 12:152–166.

Moffett, G. M., Jr. 1970. A study of nesting torrent ducks in the Andes. *Living Bird,* 9:5–27.

Moffit, J., and C. Cottam. 1941. The eel-grass blight and its effect on Brant. *U.S. Fish and Wildlife Service Leaflet* 204:1–26.

Moller, A. P. 1984. On the use of feathers in birds' nests: predictions and tests. *Ornis Scandinavica,* 15:38–42.

Moller, P., and J. Serrier. 1986. Species recognition in mormyrid weakly electric fish. *Animal Behaviour,* 34:333–339.

Mollhagen, T. R., *et al.* 1972. Prey remains in Golden Eagle nests: Texas and New Mexico. *Journal of Wildlife Management,* 36:784–792.

Montevecchi, W. A., and J. Wells. 1984. Fledging success of northern gannets from different nest sites. *Bird Behavior,* 5:90–95.

Moore, E. L., and H. Mueller. 1982. Cardiac response of domestic chickens to hawk and goose models. *Behavioral Processes,* 7:255–258.

Moore, F. R. 1976. The dynamics of seasonal distribution of Great Lakes Herring Gulls. *Bird-Banding,* 47:141–159.

Moore, F. R. 1978. Sunset and the orientation of a nocturnal migrant bird. *Nature,* 274:154–156.

Moore, F. R. 1985. Integration of environmental stimuli in the migratory orientation of the Savannah sparrow *(Passerculus sandwichensis). Animal Behaviour,* 33:657–663.

Moore, M. C. 1982. Hormonal response of free-living male White-crowned Sparrows to experimental manipulation of female sexual behavior. *Hormones and Behavior,* 16:323–329.

Moore, M. C. 1983. Effect of female sexual displays on the endocrine physiology and behaviour of male White-crowned Sparrows, *Zonotrichia leucophrys. Journal of Zoology* (London), 199:137–148.

Moreau, R. E. 1936. Bird-insect nesting associations. *Ibis,* 6:460–471.

Moreau, R. E. 1940. Numerical data of African birds' behaviour at the nest. II. *Ibis,* 14th Series, 4:234–248.

Moreau, R. E. 1948. Ecological isolation in a rich tropical avifauna. *Journal of Animal Ecology,* 17:113–126.

Moreau, R. E. 1950. The breeding seasons of African birds. *Ibis,* 92:223–267.

Moreau, R. E. 1952. Africa since the Mesozoic: with particular reference to certain biological problems. *Proceedings of the Zoological Society of London,* 121:869–913.

Moreau, R. E. 1964. Breeding season. *In* Thomson, A. L. (ed). *A New Dictionary of Birds.* McGraw-Hill Book Company, New York.

Moreau, R. E. 1964. White-eye. *In* Thomson, A. L. (ed). *A New Dictionary of Birds.* McGraw-Hill Book Company, New York.

Moreau, R. E. 1966. *The Bird Faunas of Africa and its Islands.* Academic Press, London.

Moreau, R. E. 1969. The recurrence in winter quarters (Ortstreue) of trans-Saharan migrants. *Bird Study,* 16:108–110.

Moreau, R. E. 1972. *The Palearctic-African Bird Migration Systems.* Academic Press, New York.

Moreau, R. E., and W. M. Moreau. 1938. The comparative breeding ecology of two species of Euplectes (Bishop Birds) in Usambara. *Journal of Animal Ecology,* 7:314–327.

Moreau, R. E., and W. M. Moreau. 1940. Incubation and fledging periods of African birds. *Auk*, 57:313–325.

Morejohn, G. V., and R. E. Genelly, 1961. Plumage differentiation of normal and sex-anomalous Ring-necked Pheasants in response to synthetic hormone implants. *Condor*, 63:101–110.

Morel, G., and M.-Y. Morel. 1974. Recherches ecologiques sur une savane sahélienne du Ferlo septentrional Sénégal. . . . *Terre et la Vie*, 28:95–123.

Morel, M.-Y. 1964. Natalité et mortalité dans une population naturelle d'un passereau tropicale, le *Lagonosticta senegale*. *La Terre et la Vie*, 3:436–451.

Morel, M.-Y. 1967. Les oiseaux tropicaux élèvent-ils autant de jeunes qu'ils peuvent en nourrir? *La Terre et la Vie*, 114:77–82.

Morgan, P. A., and P. E. House. 1973. Avoidance conditioning of Jackdaws (*Corvus monedula*) to distress calls. *Animal Behaviour*, 21:481–491.

Morike, K. D. 1953. Der Leier-Überschlag der Monchgrasmucke. *Ornithologische Mitteilungen*, 5:90–95.

Morrell, J. M., and J. R. G. Turner. 1970. Experiments in mimicry: I. The response of wild birds to artificial prey. *Behaviour*, 36:116–130.

Morris, D. 1956. The feather postures of birds and the problem of the origin of social signals. *Behaviour*, 9:75–113.

Morris, R. D., and M. J. Biodochka. 1982. Mate guarding in Herring Gulls. *Colonial Waterbirds*, 5:124–130.

Morris, R. D., and J. E. Black. 1980. Radiotelemetry and Herring Gull foraging patterns. *Journal of Field Ornithology*, 51:110–118.

Morris, R. L., and C. J. Erickson. 1971. Pair bond maintenance in the Ring Dove (*Streptopelia risoria*). *Animal Behaviour*, 19:398–406.

Morrissey, R. B., and W. E. Donaldson. 1977. Rapid accumulation of cholesterol in serum, liver and aorta of Japanese Quail. *Poultry Science*, 56:2003–2008.

Morse, D. H. 1968. The use of tools by Brown-headed Nuthatches. *Wilson Bulletin*, 80:220–224.

Morse, D. H. 1970. Territorial and courtship songs of birds. *Nature*, 226:659–661.

Morton, E. S. 1970. Ecological sources of selection on avian sounds. Ph.D. dissertation, Yale University, New Haven, Connecticut.

Morton E. S. 1971. Nest predation affecting the breeding season of the clay-colored Robin, a tropical song bird. *Science*, 171:920–921.

Morton, E. S. 1976. Vocal mimicry in the Thick-billed Euphonia. *Wilson Bulletin*, 88:485–487.

Morton, E. S. 1980. The importance of migrant birds to the advancement of evolutionary theory. *In* Keast, A., and E. S. Morton (eds). *Migrant Birds in the Neotropics: Ecology, Behavior, Distribution and Conservation*, 555–557. Smithsonian Institute Press, Washington, D. C.

Morton, E. S. 1982. Grading, discreteness, redundancy and motivational structural rules. *In* Kroodsma, D. E., and E. H. Miller (eds). *Acoustic Communication in Birds*, Vol. 1. Academic Press, New York.

Morton, E. S., and M. D. Shalter. 1977. Vocal response to predators in pair bonded Carolina Wrens. *Condor*, 79:222–227.

Morton, M. L. 1967. The effects of insulation on the diurnal feeding patterns of White-crowned Sparrows (*Zonotrichia leucophrys gambelii*). *Ecology*, 48:690–694.

Morton, M. L. 1976. Adaptive strategies of *Zonotrichia* breeding at high latitude or high altitude. *In* Frith, H. J., and J. H. Calaby (eds). *Proceedings of the 16th International Ornithological Congress*, Australian Academy of Science, Canberra.

Morton, M. L. 1979. Fecal sac ingestion in the Mountain White-crowned Sparrow. *Condor*, 81:72–77.

Morton, M. L. 1984. Sex and age ratios in wintering White-crowned Sparrows. *Condor*, 86:85–87.

Morton, M. L., J. L. Horstmann, and J. M. Osborn. 1972. Reproductive cycle and nesting success of the Mountain White-crowned Sparrow (*Zonotrichia leucophrys oriantha*) in the central Sierra Nevada. *Condor*, 74:152–163.

Morton, M. L., and L. R. Mewaldt, 1962. Some effects of castration on migratory sparrow (*Zonotrichia atricapilla*). *Physiological Zoology*, 35:237–247.

Morton, M. L., M. E. Pereyra, and L. F. Baptista. 1985. Photoperiodically induced ovarian growth in the White-crowned Sparrow (*Zonotrichia leucophrys gambelii*) and its augmentation by song. *Comparative Biochemistry and Physiology*, 80A:93–97.

Moss, R. 1969. A comparison of Red Grouse (*Lagopus l. scoticus*) stocks with the production and nutritive value of heather (*Calluna vulgaris*). *Journal of Animal Ecology*, 38:103–122.

Moss, R. 1972. Effects of captivity on gut lengths in Red Grouse. *Journal of Wildlife Management*, 36:99–104.

Moss, R., and I. Lockie. 1979. Infrasonic components in the song of the Capercaillie, *Tetrao urogallus*. *Ibis*, 121:95–97.

Moss R., A. Watson, R. Rothery, and W. Glenmi. 1982. Inheritance of dominance and aggressiveness in captive red grouse *Lagopus l. scoticus*. *Aggressive Behavior*, 8:1–18.

Motai, T. 1973. Male behavior and polygamy in *Cisticola juncindis*. *Miscellaneous Reports of the Yamashina's Institute for Ornithology and Zoology*, 7:87–103.

Mountfort, G. R. 1935. Manifestations visibles du développement sexual des oiseaux. *L'Oiseau et la Revue Française d'Ornithologie*, 5:494–505.

Mountfort, G. 1957. *The Hawfinch*. William Collins, Sons & Co. Ltd., London.

Moynihan, M. 1958. Notes on the behavior of some North American gulls. *Behaviour*, 13:112–130.

Mueller, H. C. 1970. Stimuli eliciting bathing behavior in hand-reared hawks. *Auk*, 87:810–811.

Mueller, H. C., and D. D. Berger. 1961. Weather and fall migration of hawks at Cedar Grove, Wisconsin. *Wilson Bulletin*, 73:171–192.

Mueller, H. C., and K. Meyer. 1985. The evolution of reversed sexual dimorphism in size: A comparative analysis of the falconiformes of the Western Palearctic. *In* Johnston, R. F. (ed). *Current Ornithology*, 2:65–101.

Mueller, H. C., and P. C. Parker. 1980. Naive ducklings show different cardiac response to hawk and to goose models. *Behaviour*, 74:101–113.

Mühl, K. 1954. Krähen und Elstern fressen Schafwolle. *Ornithologische Mitteillungen*, 6:236.

Mühlthalen, F. 1952. Beobachtungen am Bergfinken Schlafplatz bei Thun, 1950–51. *Der Ornithologische Beobachter*, 49:173–192.

Mulligan, J. A. 1966. Singing behavior and its development in the Song Sparrow, *Melospiza melodia*. *University of California Publications in Zoology*, 81:1–76. University of California Press, Berkeley.

Mumme, R. L., W. D. Koenig, and F. A. Pitelka. 1983. Mate guarding in the Acorn Woodpecker: within-group reproductive competition in a cooperative breeder. *Animal Behaviour*, 31:1094–1106.

Mundinger, P. C. 1979. Call learning in the Carduelinae: ethological and systematic considerations. *Systematic Zoology*, 28:270–283.

Mundinger, P. C. 1982. Microgeographic and macrogeographic variation in acquired vocalizations in birds. *In* Kroodsma, D. E., and E. Miller (eds). *Acoustic Communication in Birds*, Vol. 2. Academic Press, New York.

Munn, C. A., and J. W. Terborgh. 1979. Multi-species territoriality in neotropical foraging flocks. *Condor*, 81:338–347.

Munro, J. A. 1938. The Northern Bald Eagle in British Columbia. *Wilson Bulletin*, 50:28–35.

Munro, D. A. 1954. Prairie Falcon "playing." *Auk*, 71:333–334.

Munthe, A. 1957. *The Story of San Michele*. E. P. Dutton and Co., New York.

Murphy, M. T. 1983. Clutch size in the eastern kingbird: Factors affecting nestling survival. *Auk*, 100:326–334.

Murphy, R. C. 1936. *Oceanic Birds of South America*. Macmillan Publishing Company, New York.

Murrish, D. 1970. Responses to diving in the Dipper, *Cinclus mexicanus*. *Comparative Biochemistry and Physiology*, 34:853–858.

Murrish, D. E. 1973. Respiratory heat and water exchange in penguins. *Respiratory Physiology*, 19:262–270.

Murton, R. K. 1968. Breeding, migration and survival of Turtle Doves. *British Birds*, 61:193–212.

Murton, R. K. 1971. *Man and Birds*. William Collins, Sons & Co. Ltd. London.

Murton, R. K., A. J. Isaacson, and N. J. Westwood. 1966. The relationships between wood-pigeons and their clover food supply and the mechanism of population control. *Journal of Applied Ecology*, 3:55–96.

Murton, R. K., A. J. Isaacson, and N. J. Westwood. 1983. The food and growth of nestling Wood Pigeons in relation to the breeding season. *Proceedings of the Zoological Society of London*, 141:747–781.

Murton, R. K., B. Lofts, and N. J. Westwood. 1970. Manipulation of photo-refractoriness in the House Sparrow, *Passer domesticus*, by circadian light regimes. *General and Comparative Endocrinology*, 14:107–113.

Murton, R. K., and N. J. Westwood. 1977. *Avian Breeding Cycles*. Oxford University Press, Oxford.

Murton, R. K., and E. N. Wright (eds). 1968. The problems of birds as pests. *Institute of Biology Symposium No. 17*. Academic Press, London, New York.

Musselman, T. E. 1941. Bluebird mortality in 1940. *Auk*, 58:409–410.

Musselman, T. E. 1946. Some interesting nest habits of the Eastern Bluebird *(Sialia sialis sialis)*. *Bird-Banding*, 17:60–63.

Myers, *In* Mackworth-Praed, C. W., and C. H. B. Grant. 1960. *African Handbook of Birds: Birds of Eastern and Northeastern Africa*. Vol. 2, p. 867. Longmans, Green and Company, London.

Myers, J. P. 1981. A test of three hypotheses for latitudinal segregation of the sexes in wintering birds, *Canadian Journal of Zoology*, 59:1527–1534.

Myers, J. P., P. G. Conners, and F. A. Pitelka. 1979. Territoriality in non-breeding shorebirds. *Studies in Avian Biology*, 2:231–246.

Nachtigall, W., and B. Kempf. 1971. Vergleichenden Untersuchungen zur flugbiologischen Funktion des Daumenfittichs (alula spuria) bei Vögeln. *Zeitschrift für Vergleichende Physiologie*, 71:326–341.

Nagy, A. 1979. Miracle Day. *Hawk Mountain News*, 50:25–29.

Nekipelov, N. V. 1969. [Importance of bright male coloration to sexual selection in Mallard ducks.] *Izvestiya vost. siber. Otdel. Geograph. Soc. USSR*, 66:93–97. From *Bird-Banding*, 41:259.

Nelder, J. A. 1962. A statistical examination of the Hastings Rarities. *British Birds*, 55:281–384.

Nelson, J. B. 1969. The breeding ecology of the Red-footed Booby in the Galapagos. *Journal of Animal Ecology*, 38: 181–198.

Nero, R. W., and J. T. Emlen. 1951. An experimental study of territorial behavior in breeding Red-winged Blackbirds. *Condor*, 53:105–116.

Nesbitt, S. A. 1978. Sandhill Cranes in Florida flying with their legs drawn up. *Florida Field Naturalist*, 6:17–18.

Nestler, R. B. 1946. Mechanical value of grit for Bob-white Quail. *Journal of Wildlife Management*, 10:137–142.

Nestler, R. B., *et al.* 1944. Winter protein requirements of Bob-white Quail. *Journal of Wildlife Management*, 8: 218–222.

Nestler, R. B., and A. L. Nelson. 1945. Inbreeding among pen-reared Quail. *Auk*, 62:217–222.

Neub, M. 1979. Brutbiologische Konsequenzen des asynchronen Schlüpfens bei Kohlmeise *(Parus major)* und Blaumeise *(Parus caeruleus)*. *Journal für Ornithologie*, 120:196–214.

Neumann, G.-H. 1962. Das visuelle Lernvermögen eines Emus. *Journal für Ornithologie*, 103:153–165.

Newman, J. R., E. Novakova, and J. T. McClave. 1985. The influence of industrial air emission on the nesting ecology of the House Martin *Delichon urbica* in Czechoslovakia. *Bird Conservation*, 31:229–248.

Newton, A. 1896. *A Dictionary of Birds*. A. and C. Black, London.

Newton, I. 1970. Irruptions of Crossbills in Europe. *In* Watson, A. (ed). *Animal Populations in Relation to Their Food Resources*. Blackwell Scientific Publications, Oxford.

Newton, I. 1972. *Finches*. Collins, London.

Newton, I. 1979. *Population Ecology of Raptors*. Buteo Books, Vermillion, South Dakota.

Nice, M. M. 1937. Studies in the life history of the Song Sparrow. I. A population study of the Song Sparrow. *Transactions of the Linnaean Society of New York*, 4:1–247.

Nice, M. M. 1938a. The biological significance of bird weights. *Bird-Banding*, 9:1–11.

Nice, M. M. 1938b. What determines the time of a Song Sparrow's awakening song? IXe Congres Ornithologique International, Rouen, 249–255.

Nice, M. M. 1939. The Social Kumpan and the Song Sparrow. *Auk*, 56:255–262.

Nice, M. M. 1941. The role of territory in bird life. *American Midland Naturalist*, 26:441–487.

Nice, M. M. 1943. Studies in the life history of the Song Sparrow. II. *Transactions of the Linnaean Society of New York*, 6:1–328.

Nice, M. M. 1949. [Review]. *Bird-Banding*, 20:192.

Nice, M. M. 1953. The earliest mention of territory. *Condor*, 55:316–317.

Nice, M. M. 1954. Incubation periods throughout the ages. *Centaurus*, 3:311–359.

Nice, M. M. 1957. Nesting success in altricial birds. *Auk*, 74:305–231.

Nice, M. M. 1962. Development of behavior in precocial birds. *Transactions of the Linnaean Society of New York*, 8: 1–211.

Nicod, L. 1952. Augmentation des Hirondelles de cheminèe. *Nos Oiseaux*, 21:168–170.

Nicolai, J. 1956. Zur Biologie und Ethologie des Gimpels *(Pyrrhula pyrrhula* L.). *Zeitschrift für Tierpsychologie*, 13: 93–132.

Nicolai, J. 1959. Familientradition in der Gesangsentwicklung des Gimpels *(Pyrrhula pyrrhula* L.). *Journal für Ornithologie*, 100:39–46.

Nicolai, J. 1964. Brutparasitismus der Viduinae als ethologische Problem. *Zeitschrift für Tierpsychologie*, 21:129–204.

Nicolai, J. 1967. Rassen und Artbildung in der Viduinengattung *Hypochera*: I. . . . *Journal für Ornithologie*, 108: 309–319.

Nicolai, J. 1973. Das Lernprogramm in der Gesangsbildung der Strohwitwe *Tetraenura fischeri* Reichenow. *Zeitschrift für Tierpsychologie*, 32:113– 138.

Niedermeyer, W. J., and J. J. Hickey. 1977. The Monk Parakeet in the United States. 1970–1975. *American Birds*, 31:273–278.

Niethammer, G. 1937–1942. *Handbuch der Deutschen Vogelkunde*. Akademische Verlagsgesellschaft. Leipzig.

Niethammer, G. 1970. Clutch size of introduced Passeriformes in New Zealand. *Notornis*, 17:214–222.

Niethammer, G. 1972. Waders and gull-like birds. *In* Grzimek, B. (ed). *Grzimek's Animal Life Encyclopedia*, Vol. 8. Van Nostrand Reinhold Company, New York.

Nisbet, I. C. T. 1977. *In* Stonehouse, B., and C. M. Perrins (eds). *Evolutionary Ecology*. Macmillan, London.

Nisbet, I. C. T. 1978. Population models for Common Terns in Massachusetts. *Bird-Banding*, 49:50–58.

Nisbet, I. C. T. 1983. Paralytic shellfish poisoning: Effects on breeding terns. *Condor*, 85:338–345.

Nisbet, I. C. T., W. H. Drury, and J. Bird. 1963. Weight-loss during migration. Part I. *Bird-banding*, 34:107–138.

Nishikawa, F., *et al.* 1977. Isolation of influenza type A viruses from imported pet birds. *Japanese Journal of Medical Science and Biology*, 30:31–36.

Noble, G. K. 1936. Courtship and sexual selection of the Flicker. *Auk*, 53:269–282.

Noble, G. K. 1939. The role of dominance in the social life of birds. *Auk*, 56:263–273.

Noble, G. K., M. Wurm, and A. Schmidt. 1938. Social behavior of the Black-crowned Night Heron. *Auk*, 55:7–40.

Nolan, V., Jr. 1978. The ecology and behavior of the Prairie Warbler, *Dendroica discolor*. *Ornithological monographs* No. 26. American Ornithologists Union.

Nolan, V., Jr., and E. D. Ketterson. 1983. An analysis of body mass, wing length and visible fat deposits of dark-eyed juncos *(Junco hyemalis)* wintering at different latitudes. *Wilson Bulletin*, 95:603–620.

Nopcsa, 1907. Ideas of the origin of flight. *Proceedings of the Zoological Society of London*, 1907:223–236.

Norberg, R. Å. 1977. Occurrence and independent evolution of bilateral ear asymmetry in owls and implications in owl taxonomy. *Philosophical Transactions of the Royal Society of London, B. Biological Science*, 280:375–408.

Norberg, R. A., and V. M. Norberg, 1971. Take-off, landing, and flight speed during fishing flights of *Gavia stellata* (Pont.). *Ornis Scandinavica*, 2:55–67.

Nottebohm, F. 1968. Auditory experiences and song development in the Chaffinch *Fringilla coelebs*. *Ibis*, 110:549–568.

Nottebohm, F. 1969a. The "critical period" for song learning. *Ibis*, 111:386–387.

Nottebohm, F. 1969b. The song of the chingolo, *Zonotrichia capensis*, in Argentina: description and evaluation of a system of dialects. *Condor*, 71:299–315.

Nottebohm, F. 1971. Neural lateralization of vocal control in a passerine bird. *Journal of Experimental Zoology*, 117:229–262.

Nottebohm, F. 1975. Vocal behavior in birds. *In* Farner, D. S., and J. R. King (eds). *Avian Biology*, Vol. 5. Academic Press, New York.

Nottebohm, F. 1976. Continental patterns of song variability in *Zonotrichia capensis*. . . . *Proceedings of the 16th International Ornithological Congress*. Australian Academy of Science, Canberra.

Nottebohm, F. 1984. Birdsong as a model in which to study brain processes related to learning. *Condor*, 86:227–236.

Nottebohm, F., and A. P. Arnold. 1976. Sexual dimorphism in vocal control areas of the songbird brain. *Science*, 194:211–213.

Nottebohm, F., and M. E. Nottebohm. 1971. Vocalizations and breeding behaviour of surgically deafened Ring Doves *(Streptopelia risoria)*. *Animal Behaviour*, 19:313–327.

Nottebohm, F., and M. E. Nottebohm. 1978. Relationship between song repertoire and age in the Canary, *Serinus canarius*. *Zeitschrift für Tierpsychologie*, 46:298–305.

Novick, A. 1959. Acoustic orientation in the Cave Swiftlet. *Biological Bulletin*, 117:497–503.

Nuechterlein, G. L. 1981. Courtship behavior and reproductive isolation between Western grebe color morphs. *Auk*, 98:335–349.

Nur, N. 1984. The consequences of brood size for breeding blue Tits: I. Adult survival, weight change and the cost of reproduction. *Journal of Animal Ecology*, 53:479–496.

Nyström, M. 1972. On the quantification of pecking responses in young gulls *(Larus argentatus)*. *Zeitschrift für Tierpsychologie*, 30:36–44.

O'Connell, M. E. 1981. Social interactions and androgen levels in birds: Female charcteristics assoicated with increased plasma androgen levels in the male ring dove *(Streptopelia risoria)*. *General Comparative Endocrinology*, 44:454–463.

O'Conner, R. J. 1978. Brood reduction in birds: selection for fratricide, infanticide or suicide? *Animal Behavior*, 26:79–96.

Odlaug, T. O. 1955. *Laboratory Anatomy of the Fetal Pig*. W. C. Brown Company, Dubuque, Iowa.

Odum, E. P. 1941a. Variation in the heart rate of birds. *Ecological Monographs*, 11:299–326.

Odum, E. P. 1941b. Winter homing behavior of the Chickadee. *Bird-Banding*, 12:113–119.

Odum, E. P. 1943. Some physiological variations in the Black-capped Chickadee. *Wilson Bulletin*, 55:178–191.

Odum, E. P. 1958. The fat deposition picture in the White-throated Sparrow in comparison with that in long-range migrants. *Bird-Banding*, 29:105–108.

Odum, E. P. 1971. *Fundamentals of Ecology*. W. B. Saunders Company, Philadelphia.

Odum, E. P., and J. D. Perkinson. 1951. Relation of lipid metabolism to migration in birds. *Physiological Zoology*, 24:216–230.

Oehme, H. 1959. Untersuchungen über Flug und Flügelbau von Kleinvögeln. *Journal für Ornithologie*, 100:363–396.

Oehme, H. 1963. Flug und Flügel von Star und Amsel. *Biologisches Zentralblatt*, 82:413–454.

Oehme, H., and U. Kitzler. 1975. Die Bestimmung der Muskelleistung beim Kraftflug der Vögel aus kinmatischen und morphologischen Daten. . . . *Zoologische Jahrbücher Abeilung Für Allgemeinee Zoologie und Physiologie der Tiere*, 79:425–458.

Oemichen, E. 1950. Le vol des oiseaux. *In* Grassé, P. P. (ed). *Traité de Zoologie*. Tome XV, *Oiseaux*. Masson et Cie, Paris.

Ogilvie, C. M. 1951. The building of a rookery. *British Birds*, 44:1–5.

Ohmart, R. D., and R. C. Lasiewski. 1971. Roadrunners: energy conservation by hypothermia and absorption of sunlight. *Science*, 172:67–69.

Ohta, M., W. Wada, and K. Homma. 1984. Induction of rapid testicular growth in Japanese Quail by phasic electrical stimulation of the hypothalamic photosensitive area. *Journal of Comparative Physiology*. A. *Sensory Neural and Behavioral Physiology*, 154:583–590.

Olney, P. J. S. (ed). 1979. *International Zoo Year-book*. Zoological Society of London.

Olrog, C. C. 1965. Diferencias en el ciclo sexual de algunas aves. *Hornero*, 10:269–272.

Olsen, M. W. 1960. Performance record of a parthenogenetic Turkey male. *Science*, 132:1661.

Olson, S. L. 1982. Fossil vertebrates from the Bahamas. *Smithsonian Contributions to Paleobiology*, No. 48.

Olson, S. L. 1985. The fossil record of birds. *In* Farner, D. S., J. R. King, and K. C. Parkes (eds). *Avian Biology*, 8: 79–238.

Olson, S. L., and A. Feduccia. 1979. Flight capability and the pectoral girdle of *Archaeopteryx*. *Nature*, 278:247–248.

Olson, S. L., and H. F. James. 1982. Prodromus of the fossil avifauna of the Hawaiian Islands. *Smithsonian Contributions to Zoology*, 365:1–59.

Olson, S. L., and H. F. James. 1982. Fossil birds from the Hawaiian Islands: evidence for wholesale extinction by man before western contact. *Science*, 217:633–635.

O'Malley, J. B., and R. M. Evans. 1982. Structure and behavior of white pelican (*Pelecanus erythrorhynchos*) formation flocks. *Canadian Journal of Zoology*, 60:1388–1396.

O'Neill, E. J. 1947. Waterfowl grounded at the Muleshoe National Wildlife Refuge, Texas. *Auk*, 64:457.

Oniki, Y. 1970. Nesting behavior of Reddish Hermits (*Phaethornis ruber*) and occurrence of wasp cells in nests. *Auk*, 87:720–728.

Oniki, Y. 1984. Why robin eggs are blue and birds build nests: statistical tests for Amazon birds. *Neotropical Ornithology*, 36:536–545.

Oomen, H. C. J. 1972. Exhibition of primitive blowing instruments made from bird bones. In *Proceedings of the XVth International Ornithological Congress*. E. J. Brill, Leiden.

Oppenheim, R. W. 1968. Light responsivity in chick and duck embryos just prior to hatching. *Animal Behavior*, 16: 276–280.

Orians, G. H. 1969. On the evolution of mating systems in birds and mammals. *American Naturalist*, 103:589–603.

Orians, G. 1971. Ecological aspects of behavior. *In* Farner, D. S., and J. R. King (eds). *Avian Biology*, Vol. 1, Academic Press, New York.

Orr, R. T. 1939. Observations on the nesting of the Allen hummingbird. *Condor*, 41:17–24.

O'Ruadháin, M. 1955. The Raven in Irish bird-lore. In *Acta XI Congressus Internationalis Ornithologici*. Basel, Switzerland.

Ostrom, J. H. 1970. Archaeopteryx: notice of a "new" specimen. *Science*, 170:537–538.

Ostrom, J. H. 1976a. Some hypothetical anatomical stages in the evolution of avian flight. *In* Olson, S. L. (ed). *Smithsonian Contributions to Paleobiology* No. 27.

Ostrom, J. H. 1976b. *Archaeopteryx* and the origin of birds. *Biological Journal of the Linnean Society*, 8:91–182.

Ostrom, J. H. 1979. Bird flight: How did it begin? *American Scientist*, 67:46–56.

Owre, O. T. 1973. A consideration of the exotic avifauna of southeastern Florida. *Wilson Bulletin*, 85:491–500.

Palmer, R. S. (ed). 1962. *Handbook of North American Birds*. Yale University Press, New Haven, Conn.

Pandey, K. K., and M. R. Patchell. 1982. Genetic transformation in chicken by the use of irradiated male gametes. *Molecular and General Genetics*, 186:305–308.

Papi, F. 1975. Experiments on the sense of time in *Talitrus saltator* (Montagu) (Crustacea, amphipoda). *Experientia* (Basel), *11*:230–233.

Papi, F:, *et al.* 1974. Olfactory navigation in pigeons: The effect of treatment with odorous air currents. *Journal of Comparative Physiology A*, *94*:187–194.

Parkes, K. C. 1975. Special review. *Auk*, *92*:818–830.

Parrish, J. W., J. A. Ptcek, and K. L. Will. 1984. The detection of near-UV light by nonmigratory and migratory birds. *Auk*, *101*:53–58.

Parry, V. 1973. The auxiliary social system and its effect on territory and breeding in Kookaburras. *Emu*, *73*:81–100.

Parslow, J. L. F. 1969. The migration of passerine night migrants across the English Channel studied by radar, *Ibis*, *111*:48–79.

Parsons, J. 1970. Relationship between egg size and post-hatching chick mortality in the Herring Gull *(Larus argentatus)*. *Nature*, *228*:1221–1222.

Parsons, J. 1971. Cannibalism in Herring Gulls. *British Birds*, *64*:528–537.

Parsons, J. 1975. Asynchronous hatching and chick mortality in Herring Gulls, *Larus argentatus*. *Ibis*, *117*:517–520.

Parsons, J., and L. F. Baptista. 1980. Crown color and dominance in the White-crowned Sparrow. *Auk*, *97*:807–815.

Partridge, L. 1974. Habitat selection in titmice. *Nature*, *247*:573–577.

Partridge, L. 1978. Habitat selection. *In* Krebs, J. R., and N. B. Davies (eds). *Behavioural Ecology*, 351–376. Sinauer, Sunderland, Massachusetts.

Pastore, N. 1954. Discrimination learning in the Canary. *Journal of Comparative and Physiological Psychology*, *47*: 288–289; 389–390.

Pastore, N. 1961. Number sense and "counting" ability in the Canary. *Zeitschrift für Tierpsychologie*, *18*:561–573.

Paton, D. D., and H. A. Ford. 1977. Pollination by birds of native plants in South Australia. *Emu*, *77*:73–85.

Payne, K., and R. Payne. 1985. Large scale changes over 19 years in songs of Humpback Whales in Bermuda. *Zeitschrift für Tierpsychologie*, *68*:89– 176.

Payne, R. B. 1969. Nest parasitism and display of Chestnut Sparrows in a colony of Gray-capped Social Weavers. *Ibis*, *111*:300–307.

Payne, R. B. 1969. Overlap of breeding and molting schedules in a collection of African birds. *Condor*, *71*:140–145.

Payne, R. B. 1972. Mechanisms and control of molt. *In* Farner, D. S., and J. R. King (eds). *Avian Biology*, Vol. 2. Academic Press, New York.

Payne, R. B. 1973a. Individual laying histories and the clutch size and numbers of eggs of parasitic cuckoos. *Condor*, *75*:414–438.

Payne, R. B. 1973b. Vocal mimicry of the Paradise Whydahs *(Vidua)* and response of female whydahs to the songs of their hosts *(Pytilia)* and their mimics. *Animal Behaviour*, *21*:762–771.

Payne, R. B. 1973c. Behavior, mimetic songs and song dialects, and relationships of the parasitic indigobirds *(Vidua)* of Africa. *Ornithological Monographs*, *11*.

Payne, R. B. 1977. The ecology of brood parasitism in birds. *American Review of Ecology and Systematics*, *8*:1–28.

Payne, R. B. 1978. Local dialects in the wing-flaps of Flappet Larks, *Mirafra rufocinnamomea*. *Ibis*, *120*:204–207.

Payne, R. B. 1981a. Song learning and social interaction in indigo buntings. *Animal Behaviour*, *29*:688–697.

Payne, R. B. 1981b. Population structure and social behavior: Models for testing the ecological significance of song dialects in birds. *In* Alexander, R. D., and D. W. Tinkle (eds). *Natural Selection and Social Behavior: Recent Research and New Theory*. Chiron Press, New York.

Payne, R. B. 1982a. Ecological consequences of song matching: breeding success and intraspecific song mimicry in Indigo Buntings. *Ecology*, *63*:401–411.

Payne, R. B. 1982b. Species limits in the Indigobirds (Ploceidae, *Vidua*) of West Africa: Mouth mimicry, song mimicry, and description of new species. *Miscellaneous Publications, Museum of Zoology, University of Michigan*, *162*:1–96.

Payne, R. B. 1983a. The social context of song mimicry: song-matching dialects in indigo buntings *(Passerina cyanea)*. *Animal Behaviour*, *31*:788–805.

Payne, R. B. 1983b. Bird songs, sexual selection, and female mating strategies. *In* Wasser, S. K. (ed). *Social Behavior of Female Vertebrates*. Academic Press, New York.

Payne, R. B. 1984. Sexual selection, lek and arena behavior, and sexual size dimorphism in birds. *Ornithological Monograph*, *33*.

Payne, R. B. 1985. Bird songs and avian systematics. *In* Johnston, R. F. (ed). *Current Ornithology*, *3*:87–126.

Payne, R. B. and K. D. Groschupf. 1984. Sexual selection and interspecific competition: a field experiment on territorial behavior of nonparental finches *(Vidua spp.)*. *Auk*, *101*:140–145.

Payne, R. B., and K. Payne. 1977. Social organization and mating success in local song populations of Village Indigobirds, *Vidua chalybeata*. *Zeitschrift für Tierpsychologie*, 45:113–173.

Payne, R. B., L. L. Payne, and I. Rowley. 1985. Splendid Wren *Malurus splendens* response to Cuckoos: an experimental test of social organization in a communal bird. *Behaviour*, 94:108–127.

Payne, R. S. 1961. The acoustical localization of prey in the Barn Owl *(Tyto alba)*. *American Zoologist*, 1:379.

Paynter, R. A. 1955. Birds in the upper arctic. *Auk*, 72:79–80.

Paynter, R. A., Jr. (ed). 1974. *Avian Energetics*. Publications of the Nuttall Ornithology Club, No. 15. Cambridge, Massachusetts.

Pearson, O. P. 1950. The metabolism of hummingbirds. *Condor*, 52:145–152.

Pearson, O. P. 1953a. The metabolism of hummingbirds. *Scientific American*, 188(1):69–72.

Pearson, O. P. 1953b. Use of caves by hummingbirds and other species at high altitudes in Peru. *Condor*, 55:17–20.

Pearson, O. P. 1979. Spacing and orientation among feeding Golden-crowned Sparrows. *Condon*, 81:278–285.

Pearson, R. 1972. *The Avian Brain*. Academic Press, London.

Peiponen, V. A. 1963. Experimentelle Untersuchungen über das Farbsehen beim Blaukehlchen, *Luscinia svecica* (L.) und Rotkehlchen, *Erithacus rubecula* (L.). *Annales Zoologici Societatis Zoologicae-Botanicae Fennicae, Vanamo, Finland*, 24(8):1–49.

Penney, R. L., and D. K. Riker. 1969. Adelie Penguin orientation under the northern sun. *Antarctic Journal*, 4:116–117.

Pennycuick, C. J. 1969. The mechanics of bird migration. *Ibis*, 111:525–556.

Pennycuick, C. J. 1972. Soaring behavior and performance of some East African birds, observed from a motor-glider. *Ibis*, 114:178–218.

Pennycuick, C. J. 1975. Mechanics of flight. *In* Farner, D. S., and J. R. King (eds). *Avian Biology*, Vol. 5. Academic Press, New York.

Pepperberg, I. M. 1983. Cognition in the African Gray Parrot: Preliminary evidence for auditory-vocal comprehension of the class concept. *Animal Learning Behavior*, 11:179–185.

Perdeck, A. C. 1958. Two types of orientation in migrating Starlings, *Sturnus vulgaris* L., and Chaffinches, *Fringilla coelebs* L., as revealed by displacement experiments. *Ardea*, 46:1–37.

Perrins, C. M., and P. J Jones. 1974. Inheritance of clutch size in the Great Tit *(Parus major* L.). *Condor*, 76:225–229.

Peters, H. S. 1936. A list of external parasites from birds of the eastern part of the United States. *Bird-Banding*, 7:9–27.

Peters, J. L. 1931–1987. *Check-list of Birds of the World*. 16 vols. Harvard University Press, Cambridge.

Peters, W. D., and T. C. Grubb, Jr. 1983. An experimental analysis of sex-specific foraging in the Downy Woodpecker. *Ecology*, 64:1437–1443.

Petersen, A., and H. Young. 1950. A nesting study of the Bronzed Grackle. *Auk*, 67:466–476.

Petersen, E. 1953. Orienteringsforsøg med. Haettemåge *(Larus r. ridibundus* L.) og stormmåge *(Larus c. canus* L.) i vinterkvarteret. *Dansk Ornithologisk Forenings Tidsskrift*, 47:153–178.

Peterson, D. F., and M. R. Fedde. 1969. Receptors sensitive to CO_2 in lungs of chickens. *Science*, 162:1499–1501.

Peterson, R. T. 1941. *A Field Guide to the Western Birds*. Houghton Mifflin Company, Boston.

Peterson, R. T. 1963. *The Birds*. Time, Inc. New York.

Petrak, M. L. (ed). 1969. *Diseases of Cage and Aviary Birds*. Lea & Febiger, Philadelphia.

Patrides, G. A. 1944. Sex ratios in ducks. *Auk*, 61:564–571.

Petrinovich, L. 1985. Factors influencing song development in the White-crowned Sparrow *(Zonotrichia leucophrys)*. *Journal of Comparative Psychology*, 99:15–29.

Petrinovich, L., and T. L. Patterson. 1979. Field studies of habituation: I. Effects of reproduction condition, number of trials, and different relay intervals on responses of the White-crowned Sparrow. *Journal of Comparative Physiology and Psychology*, 93:337–350.

Petrinovich, L., and T. L. Patterson. 1981. The responses of White-crowned Sparrows to songs of different dialects and subspecies. *Zeitschrift für Tierpsychologie*, 57:1–14.

Petrinovich, L., and T. L. Patterson. 1982. The White-crowned Sparrow: stability, recruitment, and population structure in the nuttall subspecies (1975–1980). *Auk*, 99:1–14.

Petrinovich, L. and H. V. S. Peeke. 1973. Habituation to territorial song in the White-crowned Sparrow *(Zonotrichia leucophrys)*. *Behavioral Biology*, 8:743–748.

Pettigrew, J. D., and M. Konishi. 1976. Neurons selective for orientation and binocular disparity in the visual wulst of the Barn Owl *(Tyto alba)*. *Science*, 193:675–678.

Pettingill, O. S. 1942. The birds of a Bull's Horn Acacia. *Wilson Bulletin*, 54:89–96.

Pettingill, O. S. 1970. *Ornithology in Laboratory and Field.* Burgess Publishing Company, Minneapolis.

Pfeifer, S. 1963. Dichte und Dynamik von Brutpopulationen zweier deutscher Waldgebiete 1949–61. *Proceedings of the XIIIth International Ornithological Congress,* 2:754–765. American Ornithologists' Union.

Phillips, A. R. 1951. Complexities of migration: a review. *Wilson Bulletin,* 63:129–136.

Phillips, C. L. 1887. Egg-laying extraordinary in *Colaptes auratus. Auk,* 4:346.

Phillips, J. B. 1977. Use of the earth's magnetic field by orienting cave salamanders *(Eurycea lucifuga). Journal of Comparative Physiology,* A, 121:273–288.

Phillips, R. E., and F. McKinney. 1962. The role of testosterone in the display of some ducks. *Animal Behaviour,* 10:244–246.

Phillips, R. E., and O. M. Youngren. 1971. Brain stimulation and species-typical behavior: activities evoked by electrical stimulation of the brain of chickens *(Gallus gallus). Animal Behaviour,* 19:757–779.

Pianka, E. R. 1970. On r- and K-selection. *American Naturalist,* 104:592–597.

Piatt, J. F., and D. N. Nettleship. 1985. Diving depths of four alcids. *Auk,* 102:293–297.

Pickwell, G. V. 1968. Energy metabolism in ducks during submergence asphyxia: assessment by direct method. *Comparative Biochemistry and Physiology,* 27:455–485.

Picman, J. 1980. Responses of red-winged blackbirds to nests of long-billed marsh wrens. *Canadian Journal of Zoology,* 58:1821–1827.

Picman, J. 1980. Impact of marsh wrens on reproductive strategy of red-winged blackbirds. *Canadian Journal of Zoology,* 58:337–350.

Picman, J. 1982. Impact of red-winged blackbirds on singing activities of long-billed marsh wrens. *Canadian Journal of Zoology,* 60:1683–1689.

Picman, J. 1983. Aggression by red-winged blackbirds towards marsh wrens. *Canadian Journal of Zoology,* 61:1896–1899.

Pienkowski, M. W., and P. R. Evans. 1982. Breeding behavior, productivity and survival of colonial and noncolonial shellducks, *Tadorna tadorna. Ornis Scandinavica,* 13:101–116.

Pierce, R. J. 1984. Plumage, morphology and hybridization of New Zealand stilts *Himantopus* spp. *Notornis,* 31:106–130.

Pillmore, R. E., *et al.* 1971. Forest spraying of Zectran and its safety to wildlife. *Journal of Forestry,* 69:721–727.

Pincus, G., and T. F. Hopkins. 1958. The effects of various estrogens and steroid substances on sex differentiation in the fowl. *Endocrinology,* 62:112–118.

Pinowski, J., and S. C. Kendeigh (eds). 1977. *Granivorous Birds in Ecosystems.* Cambridge University Press, Cambridge.

Pitelka, F. A., and T. W. Schoener. 1968. Sizes of feeding territories among birds. *Ecology,* 49:123–141.

Pitelka, F. A., P. Q. Tomich, and G. W. Treichel. 1955. Ecological relations of jaegers and owls as lemming predators near Barrow, Alaska. *Ecological Monographs,* 25:85–117.

Pitts, T. D. 1977. Do eastern Bluebirds and House Sparrows prefer nest boxes with white or black interiors? *Bird-Banding,* 48:75–76.

Pitz, G. F., and R. B. Ross. 1961. Imprinting as a function of arousal. *Journal of Comparative and Physiological Psychology,* 54:602–604.

Platz, F. 1964. 'Zeremonielles Futtern' bei der Kolbenente, *Netta rufina. Journal für Ornithologie,* 105:190–196.

Pohl-Apel, G., and R. Sossinka. 1984. Hormonal determination of song capacity in females of the Zebra Finch: Critical phase of treatment. *Zeitschrift für Tierpsychologie,* 64:330–336.

Polyak, S. 1957. *The Vertebrate Visual System.* University of Chicago Press, Chicago.

Poole, E. L. 1938. Weights and wing areas in North American birds. *Auk,* 55:511–517.

Porter, R., and I. Willis. 1968. The autumn migration of soaring birds at the Bosphorus. *Ibis,* 110:520–536.

Porter, R. D., and S. W. Wiemeyer. 1970. Propagation of captive American Kestrels. *Journal of Wildlife Management,* 34:594–604.

Portielje, A. F. J. 1921. Zur Ethologie bzw. Psychologie von *Botaurus stellaris. Ardea,* 15:1–15.

Portmann, A. 1950. Le developpement postembryonnaire. *In* Grassé, P. (ed). *Traité de Zoologie.* Tome XV, *Oiseaux.* Masson et Cie, Paris.

Portmann, A., and W. Stingelin. 1961. The central nervous system. *In* Marshall, A. J. (ed). *Biology and Comparative Physiology of Birds.* Academic Press, New York.

Post, W., and F. Enders. 1970. The occurrence of Mallophaga on two bird species occupying the same habitat. *Ibis,* 112:539–540.

Potter, P. E. 1972. Territorial behavior in Savannah Sparrows in Southwest Michigan. *Wilson Bulletin,* 84:48–59.

Poulson, H. 1950. Morphological and ethological notes on a hybrid between a domestic duck and a domestic goose. *Behaviour,* 3:99–104.

Poulsen, H. 1951. Inheritance and learning in the song of the Chaffinch (*Fringilla coelebs* L.). *Behaviour*, 3:216–228.

Poulsen, H. 1953. A study of incubation responses and some other behaviour patterns in birds. *Videnskabelige Meddelelser fra Dansk Naturhistorisk Forening*, 115:131.

Powdermaker, H. 1957. Science and the citizen. *Scientific American*, 196(1):60–62.

Power, D. M. 1976. Avifauna richness on the California Channel Islands. *Condor*, 78:394–398.

Power, D. M. 1979. Evolution in peripheral isolated populations: *Carpodacus* finches on the California islands. *Evolution*, 33:834–847.

Prange, H. D., and K. Schmidt-Nielsen. 1970. The metabolic cost of swimming in ducks. *Journal of Experimental Biology*, 53:763–777.

Pratt, A. 1933. *The Lore of the Lyrebird*. Robertson and Mullens, Melbourne.

Pregill, G. K., and S. L. Olsen. 1981. Zoogeography of West Indian vertebrates in relation to Pleistocene climate cycles. *Annual Review of Ecology and Systematics*, 12:75–98.

Presst, I., and D. A. Ratcliffe. 1972. Effects of organochlorine insecticides on European birdlife. *Proceedings of the XVth International Ornithological Congress*, 486–513. E. J. Brill, Leiden.

Presti, D., and J. D. Pettigrew. 1980. Ferromagnetic coupling to muscle receptors as a basis for geomagnetic field sensitivity in animals. *Nature*, 285:99–100.

Preston, F. W. 1962. A nesting of Amazonian terns and skimmers. *Wilson Bulletin*, 74:286–287.

Preston, F. W. 1969. Shapes of birds' eggs: extant North American families. *Auk*, 86:246–264.

Prevost, J. 1953. Formation des couples, ponte et incubation chez le Manchot Empereur. *Alauda*, 21:141–156.

Prevost, J., and F. Bourlière. 1957. Vie Sociale et thermoregulation chez le Manchot Empereur *Aptenodytes forsteri*. *Alauda*, 25:167–173.

Price, P. H. 1979. Developmental determinants of structure in Zebra Finch song. *Journal of Comparative Physiology*, 93:260–277.

Prinzinger, R. R. Goppel, A. Lorenz, and E. Kulzer. 1981. Body temperature and metabolism in the Redbacked Mousebird (*Colius castanotus*) during fasting and torpor. *Comparative Biochemistry and Physiology*, 69A:689–692.

Proctor, V. W. 1968. Long distance dispersal of seeds by retention in digestive tract of birds. *Science*, 160:321–322.

Prosser, C. L., et al. 1950. *Comparative Animal Physiology*. W. B. Saunders Company, Philadelphia.

Pröve, E. 1974. Der Einfluss von Kastration und Testosteronsubstitution auf das Sexualverhalten männlicher Zebrafinken *Taeniopygia guttata castanotis* Gould. *Journal für Ornithologie*, 115:338–347.

Pröve, E. 1985. Steroid hormones as a physiological basis of sexual imprinting in male Zebra Finches (*Taeniopygia guttata castanotis*). In Follett, B. K., S. Ishii, and A. Chandola (eds). *The Endocrine System and the Environment*. Springer-Verlag, Berlin, 235–245.

Provine, R. R. 1981. Development of wing-flapping and flight in normal and flap-deprived domestic chicks. *Developmental Psychobiology*, 14:279–292.

Pruett-Jones, M. A., and S. G. Pruett-Jones. 1985. Food caching in the tropical frugivore, MacGregor's Bowerbird (*Amblyornis macgregoriae*). *Auk*, 102:334–341.

Pullainen, E. 1983. Seasonal changes in the gut-length of the willow grouse (*Lagopus lagopus*) in Finnish Lapland. *Annales Zoologicae Fennicae*, 20:53–56.

Pulliam, R., et al. 1972. On the evaluation of sociality, with particular reference to *Tiaris olivacea*. *Wilson Bulletin*, 84:77–89.

Putnam, L. S. 1949. The life history of the Cedar Waxwing. *Wilson Bulletin*, 61:141–182.

Putzig, P. 1937. Von der Beziehung des Zugablaufs zum Inkretdrüsensystem. *Vogelzug*, 8:116–130.

Putzig, P. 1938. Der Frühwehzug des Kiebitzes (*Vanellus vanellus* L.) unter Berücksichtigung anderen Limicolen. *Journal für Ornithologie*, 86:123–163.

Pynnönen, A. 1939. Beiträge zur Kenntnis der Biologie finnischer Spechte. *Annales Zoologici Societatis Zoologicae-Botanicae-Fennicae Vanamo*, 7:1–166.

Pynnönen, A. 1950. Om Järpens Levnadsvanor. *Suonem Riista* [Helsinki], 5:7–27.

Quaintance, C. W. 1938. Content, meaning and possible origin of male song in the Brown Towhee. *Condor*, 40:97–101.

Quay, W. B. 1985. Cloacal sperm in spring migrants: occurrence and interpretation. *Condor*, 87:273–280.

Querengässer, A. 1973. Über das Einemsen von Singvögeln und die Reifung dieses Verhaltens. *Journal für Ornithologie*, 114:96–117.

Quigg, P. W. 1974. One earth. *Audubon*, 76(2):104–105.

Quine, D. B., and M. L. Kreithen. 1981. Frequency shift discrimination: Can homing pigeons locate infrasounds by Doppler shifts? *Journal of Comparative Physiology*, A. *Series Neural Behavioral Physiology*, 141:153–156.

Quinlan, S. E., and W. A. Lehnhausen, 1982. Arctic Fox, *Alopex lagopus*, predation on nesting common Eiders, *Somateria molissima*, at Icy Cape, Alaska. *Canadian Field Naturalist*, 97:62–65.

Quinney, T. E. 1983. Tree swallows cross a polygyny threshold. *Auk*, 100:750–754.

Quiring, D. P. 1950. *Functional Anatomy of the Vertebrates*. McGraw-Hill Book Company, New York.

Räber, H. 1950. Das Verhalten gefangenen Waldohreulen *(Asio otus otus)* und Waldkäuze *(Strix aluco aluco)* sur Beute. *Behaviour*, 2:1–95.

Rabinowitch, V. E. 1966. The role of early experience in the development of food habits in gull chicks. *American Zoologist*, 6:309.

Rabinowitch, V. E. 1968. The role of experience in the development of food preferences in gull chicks. *Animal Behaviour*, 16:425–428.

Rabinowitch, V. E. 1969. The role of experience in the development and retention of seed preferences in Zebra Finches. *Behaviour*, 33:222–236.

Radcliffe, E. D. 1898. Falconry. *The Encyclopaedia Britannica*, Ninth edition. Adam and Charles Black, Edinburgh.

Raikow, R. J. 1970. Evolution of diving adaptations in the stifftail ducks. *University of California Publications in Zoology*, 94:1–52.

Raikow, R. J. 1973. Locomotor mechanisms in North American ducks. *Wilson Bulletin*, 85:295–307.

Raikow, R. J. 1985a. Locomotor system. *In* Farner, D. S., and J. R. King (eds). *Avian Biology*, Vol. 1. Academic Press, New York.

Raikow, R. J. 1985b. Locomotor system. *In* King, A. S., and J. McLelland (eds). *Form and Function in Birds*, 3:57–147.

Raitasuo, K. 1964. Social behavior of the Mallard, *Anas platyrhynchos*, in the course of the annual cycle. *Papers on Game Research* (Helsinki), 24:1–72.

Raitt, R. J., Jr. 1961. Plumage development and molts of California Quail. *Condor*, 63:294–303.

Ralph, C. J., and L. R. Mewaldt. 1975. Time of site fixation upon the wintering ground in sparrows. *Auk*, 92:698–705.

Ralph, C. L., *et al.* 1967. Studies of the melanogenic response of regenerating feathers in the weaver bird. *Journal of Experimental Zoology*, 166:283–288.

Ramsay, A. O. 1953. Variations in the development of broodiness in fowl. *Behaviour*, 5:51–57.

Rand, A. L. 1936. The distribution and habits of Madagascar birds. *Bulletin of the American Museum of Natural History*, 72:143–499.

Rand, A. L. 1943. Some irrelevant behavior in birds. *Auk*, 60:168–171.

Rand, A. L. 1948. Glaciation, an isolating factor in speciation. *Evolution*, 2:314–321.

Rankin, M. N., and D. H. Rankin. 1940. Additional notes on the roosting habits of the Tree Creeper. *British Birds*, 34:56–60.

Raspet, A. 1950 Pereformance measurements of a soaring bird. *Aeronautical Engineering Review*, 9(12):1–4.

Ratti, J. T. 1979. Reproductive separation and isolating mechanisms between sympatric dark- and light-phase Western Grebes. *Auk*, 96:573–586.

Rautenberg, A. 1972. Interaction of central and peripheral thermodetectors in body temperature regulation in the pigeon. *Proceedings of the XVth International Ornithological Congress*, 680–681. E. J. Brill, Leiden.

Rautian, A. S. 1978. [A unique bird feather from Jurassic lake deposits in the Kara-tau ridge (Kazakh SSR, USSR).] *Paleontologicheskii Zhurnal*, 4:106–114.

Raveling, D. G. 1979. The annual cycle of body composition of Canada Geese with special reference to control of reproduction. *Auk*, 96:234–252.

Rawles, M. E. 1960. The integumentary system. *In* Marshall, A. J. (ed). *Biology and Comparative Physiology of Birds*. Academic Press, New York.

Rawles, M. E. 1963. Tissue interactions in scale and feather development as studied in dermalepidermal recombinations. *Journal of Embryology and Experimental Morphology*, 11:765–789.

Recher, H. F. 1969. Birds species diversity and habitat diversity in Australia and North America. *American Naturalist*, 103:75–80.

Redding, E. 1978. Der Ausdrucksflug der Bekassine (Capella gallinago gallinago). Journal für Ornithologie, 119:357–387.

Reed, T. M. 1982. Interspecific territoriality of the Chaffinch (Fringilla coelebs) and Great Tit (Parus major) on islands and the mainland of Scotland—Playback and removal experiments. Animal Behavior, 30:171–181.

Regal, P. J. 1975. The evolutionary origin of feathers. The Quarterly Review of Biology, 50:35–66.

Rehder, N. B., and D. M. Bird. 1983. Annual profiles of blood packed cell volumes of captive American kestrels. Canadian Journal of Zoology, 61:2550–2555.

Reichholf, H., and J. Reichholf, 1973. "Honigtan" der Bracaatinga-Schildlaus als Winternahrung von Kolibris (Trochilidae) in Süd-Brasilien. Bonner Zoologische Beiträge, 24(1/2):7–14.

Reichle, D. E., J. F. Franklin, and D. W. Goodall (eds). 1975. Productivity of World Ecosystems. National Academy of Sciences, Washington, D.C.

Reinecke, K. J. 1979. Feeding ecology and development of juvenile Black Ducks in Maine. Auk, 96:737–745.

Reinert, J. 1965. Takt- und Rhythmusunterscheidung bei Dohlen. Zeitschrift für Tierpsychologie, 22:623–671.

Reinikainen, A. 1937. The irregular migrations of the Crossbill Loxia c. curvirostra, and their relation to the cone-crop of the conifers. Ornis Fennica, 14:55–64.

Rensch, B. 1925a. Experimentelle Untersuchungen über den Geschmackssinn der Vögel. Journal für Ornithologie, 73:1–8.

Rensch, B. 1931. Der Einfluss des Tropenklimas auf den Vogel. Proceedings of the VIIth International Ornithological Congress, Amsterdam, 1930, 197–205. Amsterdam.

Rensch, B. 1947. Neuere Probleme der Abstammungslehre. Ferdinand Enke Verlag, Stuttgart.

Rensch, B. 1959. Evolution Above the Species Level. Columbia University Press, New York.

Reynolds, C. M. 1978. The heronries census: 1972–1977 population changes and a review. Bird Study, 26:7–12.

Reynolds, J. 1965. 'Feeding hygiene' in birds. British Birds, 58:384–385.

Rice, W. R. 1982. Acoustical location of prey by the Marsh Hawk: Adaptation to concealed prey. Auk, 99:403–413.

Richards, D. G. 1979. Recognition of neighbors by associative learning in Rufous-sided Towhees. Auk, 96:688–693.

Richardson, F. 1965. Breeding and feeding habits of the Black Wheatear, Oenanthe leucura, in southern Spain. Ibis, 107:1–16.

Richardson, W. J. 1974. Spring migration over Puerto Rico and the western Atlantic: a radar study. Ibis, 116:172–193.

Richardson, W. J. 1976. Autumn migration over Puerto Rico and the western Atlantic: a radar study. Ibis, 118:309–332.

Richdale, L. E. 1941. A brief summary of the history of the Yellow-eyed Penguin. Emu, 40:265–387.

Richdale, L. E. 1943. The Kuaka or Diving Petrel, Pelecanoides urinatrix (Gmelin). Emu, 43:24–48.

Richdale, L. E. 1943. Whero, island of sea birds. Wild Life in New Zealand, 3. Dunedin, New Zealand.

Richdale, L. E. 1949. A study of a group of penguins of known age. Otago Daily Times, Dunedin, New Zealand.

Richdale, L. E. 1950. Biological Monographs No. 4. Dunedin, New Zealand.

Richdale, L. E. 1951. Sexual Behavior in Penguins. University of Kansas Press, Lawrence, Kansas.

Richdale, L. E. 1952. The post-egg period in albatrosses. Biological Monographs, No. 4. Dunedin, New Zealand.

Richdale, L. E. 1954. Breeding efficiency in Yellow-eyed Penguins. Ibis, 96:207–224.

Richdale, L. E. 1957. A Population Study of Penguins. Oxford University Press, Oxford.

Richdale, L. E., and J. Warham. 1973. Survival, pair bond retention and nest-site tenacity in Buller's Mollymawk. Ibis, 115:257–263.

Richter, R. 1939. Weitere Beobachtungen an einer gemischter Kolonie von Larus fuscus graellsi Brehm und Larus argentatus Pontopp. Journal für Ornithologie, 87:75–86.

Ricklefs, R. E. 1968a. Patterns of growth in birds. Ibis, 110:419–451.

Ricklefs, R. E. 1968b. Weight recession in nestling birds. Auk, 85:30–35.

Ricklefs, R. E. 1969. An analysis of nestling mortality in birds. Smithsonian Contributions to Zoology, No. 9.

Ricklefs, R. E. 1973. Ecology. Chiron Press, Inc., Newton, Massachusetts.

Ricklefs, R. E. 1974. In Paynter, R. A., Jr. (ed). Avian Energetics. Publications of the Nuttall Ornithology Club, No. 15. Cambridge, Massachusetts.

Ricklefs, R. E. 1976. Growth rates of birds in the humid New World tropics. Ibis, 118:179–207.

Ricklefs, R. E., and F. R. Hainsworth. 1969. Temperature regulation in nestling Cactus Wrens: the nest environment. Condor, 71:32–37.

Riddle, O. 1927. Metabolic changes in the body of female pigeons at ovulation. Proceedings of the American Philosophical Society, 66:497–509.

Riddle, O. 1938. The changing organism. Cooperation in Research. The Carnegie Institution of Washington, Publication No. 501, 259–273.

Ridgway, R., and H. Friedmann. 1901–1950. *Birds of North and Middle America*. U.S. National Museum Bulletins, Washington, D.C.

Ridpath, M. G. 1972. The Tasmanian Native Hen, *Tribonyx mortierii*. Commonwealth Scientific and Industrial Research Organization. *Wildlife Research*, 17:53–90; 91–118.

Riegel, M., and W. Winkel. 1971. Über Todesursachen beim Weissstorch *(C. ciconia)* an Hand von Ringfundangaben. *Die Vogelwarte*, 26:128–135.

Riley, G. M. 1937. Experimental studies on spermatogenesis in the House Sparrow, *Passer Domesticus* (L.). *Anatomical Record*, 67:327.

Riley, G. M. 1940. Light versus activity as a regulator of the sexual cycle of the House Sparrow. *Wilson Bulletin*, 52:73–86.

Ringer, R. K., and K. Rood. 1959. Hemodynamic changes associated with aging in the Broad Breasted Bronze Turkey. *Poultry Science*, 38:395–397.

Ripley, S. D. 1950a. Birds from Nepal, 1947–49. *Journal of Bombay Natural History Society*, 49:355–417.

Ripley, S. D. 1950b. Strange courtship of birds of paradise. *National Geographic Magazine*, 97:247–278.

Ripley, S. D. 1985. The Laysan teal – recent history and future? *Avicultural Magazine*, 91:76–78.

Rising, J. D. 1983. The Great Plains hybrid zones. *In* Johnston, R. F. (ed). *Current Ornithology*, 1:131–157.

Rising, J. D., and G. F. Shields. 1980. Chromosomal and morphological correlates in two New World Sparrows (Emberizidae). *Evolution*, 34:654–662.

Rissi, P. J. 1968. Adaptation and negative aftereffect to lateral optical displacement in newly hatched chicks. *Science*, 160:430–432.

Ritchinson, G. 1983. The function of singing in female black-headed Grosbeaks *(Pheucticus melanocephalus)*: family group maintainance. *Auk*, 100:105–116.

Ritter, W. E., and S. B. Benson. 1934. "Is the poor bird demented?" *Auk*, 51:169–179.

Rivière, B. B. 1940. House martins rebuilding broken nest and feeding young. *British Birds*, 34:87–88.

Robbins, C. S. 1973. House Sparrow in North America. *In* Kendeigh, S. C. (ed). *American Ornithologists Union Monograph*, 14:3–9.

Roberts, B. 1934. Notes on the birds of central and south-east Iceland with special reference to food habits. *Ibis*, 4:252.

Roberts, N. L. 1942. Breeding activities of three common species. *Emu*, 41:185–194.

Roberts, N. L. 1955. A survey of the habit of nest-appropriation. *Emu*, 55:110–126; 173–184.

Roberts, T. S. 1932. *The Birds of Minnesota*. University of Minnesota Press, Minneapolis.

Robertson, W. B. 1964. The terns of the Dry Tortugas. *Bulletin of the Florida State Museum*, 8:1–94.

Robinson, A. 1945. The application of 'territory and the breeding cycle' to some Australian birds. *Emu*, 45:100–109.

Rochon-Duvigneaud, A. 1950. Les yeux et la vision. *In* Grassé, P. (ed). *Traité de Zoologie*. Tome XV, *Oiseaux*. Masson et Cie, Paris.

Rodbard, A. 1950. Weight and body temperature. *Science*, 111:465–466.

Rodbard, A. 1953. Warm-bloodedness. *Scientific Monthly*, 77:137–142.

Rodin, L. E., *et al.* 1975. *In* Reichle, D. E., *et al.* (eds). *Productivity of World Ecosystems*. National Academy of Sciences, Washington, D.C.

Rohwer, S. A., and F. C. Rohwer. 1978. Status signaling in Harris' Sparrows: Experimental deceptions achieved. *Animal Behaviour*, 26:1012–1022.

Rohwer, S., and P. W. Ewald. 1981. The cost of dominance and advantage of subordination in a badge signaling system. *Evolution*, 35:441–454.

Rohwer, S., and J. C. Wingfield. 1981. A field study of social dominance, plasma levels of luteinizing hormone, and steroid hormones in wintering Harris Sparrows. *Zeitschrift für Tierpsychologie*, 57:173–183.

Rollin, N. 1958. Late season singing of the Yellow-hammer. *British Birds*, 51:290–303.

Romanowski, E. 1979. Der Gesang von Sumpf- und Weidenmeise *(Parus palustris* und *P. montanus)*—reaktionsauslösende Parameter. *Die Vogelwarte*, 30:48–65.

Romer, A. S. 1970. *The Vertebrate Body*. W. B. Saunders Company, Philadelphia.

Roseberry, J. L., and W. D. Klimstra. 1970. The nesting ecology and reproductive performance of the Eastern Meadowlark, *Sturnella magna*. *Wilson Bulletin*, 82:243–267.

Rössler, E. 1976. Übertragung von Verhaltensweisen durch Transplantation von Anlagen neuroanatomischer Strukturen bei Amphilienlarven. *Zeitschrift für Tierpsychologie*, 41:244–265.

Rothschild, M. 1962. Development of paddling and other movements in young Black-headed Gulls. *British Birds*, 55:114–117.

Rothschild, M., and T. Clay. 1957. *Fleas, Flukes, and Cuckoos*. William Collins Sons & Co., Ltd., London.

Rothstein, S. I. 1972. Eggshell thickness and its variation in the Cedar Waxwing. *Wilson Bulletin*, 84:469–474.

Rothstein, S. I. 1975. Mechanisms of avian egg-recognition: Do birds know their own eggs? *Animal Behaviour*, 23: 268–278.

Rothstein, S. I. 1977. Cowbird parasitism and egg recognition of the Northern Oriole. *Wilson Bulletin*, 89:21–32.

Rothstein, S. I. 1978. Mechanisms of avian egg-recognition: additional evidence for learned components. *Animal Behaviour*, 26:671–677.

Rounsevell, D. E., and G.R. Copson. 1982. Growth rate and recovery of a king penguin, *Aptenodytes patagonicus*, population after exploitation. *Australian Wildlife Research*, 9:519–526.

Rowan, W. 1921. Observations on the breeding habits of the Merlin. *British Birds*, 15:122–129.

Rowan, W. 1929. Experiments in bird migration. I. Manipulation of the reproductive cycle: Seasonal histological changes in the gonads. *Proceedings of the Boston Society of Natural History*, 39:151–208.

Rowan, W. 1938. Light and seasonal reproduction in animals. *Biological Review*, 13:374–402.

Rowley, I. 1976. Cooperative breeding in Australian birds. *In* Frith, H. J., and J. H. Calaby (eds). *Proceedings of the 16th International Ornithological Congress*. Australian Academy of Science, Canberra.

Rowley, I. 1978. Communal activities among White-winged Choughs, *Crocorax melanorhamphus*. *Ibis*, 120:178–197.

Rowley, I. 1981. The communal way of life in the Splendid Wren, *Malurus splendens*. *Zeitschrift für Tierpsychologie*, 55:228–267.

Rowley, I. 1983. Commentary: Cooperative breeding strategies among birds. *In* A. H., Brush, and G. A. Clark, Jr. (eds). *Perspectives in Ornithology*, 127–133. Cambridge University Press.

Rubin, C. T., and L. E. Lanyon. 1984. Regulation of bone formation by applied dynamic loads. *Journal of Bone and Joint Surgery* (American), 66:397–402.

Ruge, K. 1974. Europäische Schwalbenkatastrophe im Oktober 1974. *Die Vogelwarte*, 27:299–300.

Ruiter, C. J. S. 1941. Waarneminingen omtrent de levenswijze de Gekraagde Roodstaart, *Phoenicurus ph. phoenicurus* (L.). *Ardea*, 41:175–214.

Ruiter, I. de. 1952. Some experiments on the camouflage of stick caterpillars. *Behaviour*, 4:222–232.

Runfeldt, S., and J. C. Wingfield. 1985. Experimentally prolonged sexual activity in female sparrows delays termination of reproductive activity in their untreated mates. *Animal Behavior*, 33:403–410.

Rüppell, G. 1977. *Bird Flight*. Van Nostrand and Reinhold, New York.

Rüppell, W. 1944. Versuche über Heimfinden ziehender Nebelkrähen nach Verfrachtung. *Journal für Ornithologie*, 92:106–132.

Rutledge, J. T., and M. J. Angle. 1977. Persistence of circadian activity rhythms in pinealectomized European Starlings (*Sturnus vulgaris*). *Journal of Experimental Zoology*, 202:333–338.

Ryden, O. 1978. The significance of antecedent auditory experiences on later reactions to the"seeet" alarm-call in Great Tit nestling *Parus major*. *Zeitschrift für Tierpsychologie*, 47:396–409.

Ryden, O., and H. Bergtsson. 1980. Differential begging and locomotony behaviour by early and late hatched nestlings affecting the distribution of food in asynchronously hatched broods of altricial birds. *Zeitschrift für Tierpsychologie*, 53:209–224.

Ryser, F. A., and P. R. Morrison. 1954. Cold resistance in the young Ring-necked Pheasant. *Auk*, 71:253–266.

Saalfield, E. von. 1936. Untersuchungen über das Hackeln bei Tauben. *Zeitschrift für Vergleichende Physiologie*, 23:727–743.

Sadovnikova, M. P. 1923. A study of the behavior of birds in the maze. *Journal of Comparative Psychology*, 3:123–139.

Saint-Paul, U. von. 1954. Nachweis der Sonnenorientierung bei nächtlich ziehenden Vögeln. *Behaviour*, 6:1–7.

Saint-Paul, U. von. 1962. Das Nachtfliegen von Brieftauben. *Journal für Ornithologie*, 103:337–343.

Sage, B. L. 1963. The incidence of albinism and melanism in British birds. *British Birds*, 56:409–416.

Salinukul, N., and G. K. Menon. 1983. Histochemical profile of steroid dehydrogenase in uropygial gland of euthyroidic and hypothyroidic pigeon. *Indian Journal of Experimental Biology* 21:208–211.

Salomonsen, F. 1938. Notes on the moults of the Rock Ptarmigan (*Lagopus mutus*). IXe *Congrès Ornithologique International, Rouen, 9–13 Mai, 1938*, 295–310. Secretariat du Congrés, Rouen.

Salomonsen, F. 1947. Maagekoloniere paa Hirshomene. *Dansk Ornitologisk Forenings Tidsskrift*, 41:174–186.

Salomonsen, F. 1951. The immigration and breeding of the Fieldfare (*Turdus pilaris* L.) in Greenland. *Proceedings of the Xth International Ornithological Congress, Uppsala, June, 1950*, 515–526. Almqvist & Wiksells, Uppsala.

Salomonsen, F. 1972. Zoogeographical and ecological problems in arctic birds. *Proceedings of the XVth International Ornithological Congress*. E. J. Brill, Leiden.

Salomonsen, F. 1976. The South Polar Skua *Stercorarius maccormicki* Saunders, in Greenland. *Dansk Forenings Tidsskrift*, 70(³/4):81–89.

Sanders, E. H., P. D. Gardner, P. J. Berger, and N. C. Negus. 1981. 6-Methoxybenzoxazolinone: A plant derivative that stimulates reproduction in *Microtus montanus*. *Science*, 214:67–69.

Sapin-Jaloustre, J. 1952. Découverte et description de la rookery de Manchot Empereur *(Aptenodytes forsteri)* de Pointe Géologie (Terra Adélie). *L'Oiseau*, 22:153–184.

Sargent, T. D. 1965. Experience and nest-building in the Zebra Finch. *Auk*, 82:48–61.

Sauer, E. G. F. 1954. Die Entwicklung der Lautäusserungen vom Ei ab schalldicht gehaltener Dorngrasmücken (*Sylvia c. communis*, Latham) im Vergleich mit später isolierten und mit wildlebenden Artgenossen. *Zeitschrift für Tierpsychologie*, 11:10–93.

Sauer, E. G. F. 1957. Die Sternenorientierung nächtlich ziehender Grasmücken *(Sylvia atricapilla, borin,* und *curruca)*. *Zeitschrift für Tierpsychologie*, 14:29– 70.

Sauer, E. G. F. 1971. Celestial rotation and stellar orientation in migrating warblers. *Science*, 173:459–460.

Sauer, E. G. F. 1972. The impact of unusual rains and food on the social organization of the South African Ostrich. *Proceedings of the XVth International Ornithological Congress*, 684–685. E. J. Brill, Leiden.

Saunders, A. A. 1947. The seasons of bird song: the beginning of song in spring. *Auk*, 64:97–107.

Saunders, A. A. 1948. The seasons of bird song: the cessation of song after the nesting season. *Auk*, 65:19–30.

Saunders, A. A. 1951. *A Guide to Bird Songs*. Doubleday and Company, Inc. New York.

Saunders, A. A. 1959. Forty years of spring migration in southern Connecticut. *Wilson Bulletin*, 71:208–219.

Savile, D. B. O. 1950. The flight mechanism of swifts and hummingbirds. *Auk*, 67:499–504.

Savile, D. B. O. 1957. Adaptive evolution in the avian wing. *Evolution*, 11:212–224.

Savory, C. J. 1983. Selection of heather age and chemical composition by red grouse in relation to physiological state, season and time of day. *Ornis Scandinavica*, 14:135–143.

Sayre, R. 1973. Econotes. *Audubon*, 75(6):132–133.

Sayre, R. 1974. Econotes. *Audubon*, 76(2):100–103.

Schadt, J. C., and W. E. Southern. 1972. Effects of solar cues on basic directional preferences of young Mallards. *Bird-Banding*, 43:47–53.

Scheer, G. 1951. Über die Zeitliche Differenz zwischen Erwachen und Gesangbeginn. *Die Vogelwarte*, 16:13–15.

Schenkel, R. 1958. Zur Deutung der Balzleistungen einiger Phasianiden und Tetraoniden, *Ornithologische Boebachter*, 55:65–95.

Schindler, J., P. Berthold, and F. Bairlein. 1981. Über den Einfluss simulierter Wetterbedingungen auf das endogene Zugzeitprogramm der Gartengrasmücke *(Sylvia borin)*. *Vogelwarte*, 31:14–32.

Schjelderup-Ebbe, T. 1923. Weiters Beiträge zur Social- und Individual-Psychologie des Haushühns. *Zeitschrift für Psychologie*, 132:289–303.

Schjelderup-Ebbe, T. 1935. Social behavior in birds. *In* Murchison, C. *(ed)*. *A Handbook of Social Psychology*. Clark University Press, Worcester, Massachusetts.

Schlee, D. 1973. Harzkonservierte fossile Vogelfedern ausder untersten Kreide. *Journal für Ornithologie*, 114:207–219.

Schleidt, W. M. 1961. Reaktionen von Truthühnern auf fliegende Raubvögel und Versuche zur Analyse ihrer AAM's. *Zeitschrift für Tierpsychologie*, 18:534– 560.

Schleidt, W. M., M. Schleidt, and M. Magg. 1960. Störung des Mutter-Kind-Beziehung durch Gehörverlust. *Behaviour*, 16:254–260.

Schleidt, W. M., and M. D. Shalter. 1972. Cloacal foam glands in the quail *Coturnix coturnix*. *Ibis*, 114:558.

Schleidt, W. M., and M. D. Shalter. 1973. Stereotype of a fixed action pattern during ontogeny in *Coturnix coturnix coturnix*. *Zeitschrift für Tierpsychology*, 33:35–37.

Schluter, D. 1984. Feeding correlates of breeding and social organization in two Galapagos Finches. *Auk*, 101:59–68.

Schmekel, L. 1960. Datenüber das Gewicht des Vogeldottersackes vom Schlüpftag bis zum Schwinden. *Revue Swisse de Zoologie*, 68:103–110.

Schmidt, K. H., and J. Steinbach. 1985. Jahreszeitliche Änderung der Gelegegrösse bei der Kohlmeise *(Parus major)*. *Journal für Ornithologie*, 126:163– 174.

Schmidt, W. J. 1952. Neuere Unterschungen über Schillerfarben. *Journal für Ornithologie*, 93:130–135.

Schmidt-Koenig, K. 1963. Sun compass orientation of pigeons upon equatorial and trans-equatorial displacement. *Biological Bulletin*, 124:311–321.

Schmidt-Koenig, K. 1965. Current problems in bird orientation. *In* Lehrman and Hinde (eds). *Advances in the Study of Behaviour*, 217–278. Academic Press, New York.

Schmidt-Koenig, K., and W. T. Keeton (eds). 1978. *Animal Migration, Navigation and Homing*. Springer-Verlag, New York.

Schmidt-Koenig, K., and H. J. Schlichte. 1972. Homing in pigeons with impaired vision. *Proceedings of the National Academy of Science*, 69:2446–2447.

Schmidt-Nielsen, K. 1959. Salt glands. *Scientific American*, 200(1):109–116.

Schmidt-Nielsen, K. 1960. *Animal Physiology*. Prentice-Hall, Inc., Englewood Cliffs, New Jersey.

Schmidt-Nielsen, K. 1960. The salt secreting gland of marine birds. *Circulation*, 21:955–967.

Schmidt-Nielsen, K. 1964. *Desert Animals: Physiological Problems of Heat and Water*. Oxford University Press, London.

Schmidt-Nielsen, K. 1971. How birds breathe. *Scientific American*, 225(6):72–79.

Schmidt-Nielsen, K. 1972. Locomotion: energy cost of swimming, flying, and running. *Science*, 177:222–228.

Schmidt-Nielsen, K., *et al*. 1970. Countercurrent heat exchange in the respiratory passagers: Effect on water and heat balance. *Respiration Physiology*, 9:263–276.

Schmidt-Nielsen, K., and Y. T. Kim. 1964. The effect of salt intake on the size and function of the salt gland of ducks. *Auk*, 81:160–172.

Schneider, K. J. 1984. Dominance, predation and optimal foraging in the White-throated sparrow. *Ecology*, 65:1820–1827.

Schnell, G. D., and J. J. Hellack. 1978. Flight speeds of Brown Pelicans, Chimney Swifts, and other birds. *Bird-Banding*, 49:108–112.

Schnitzler, H.-V. 1972. Windkanalversuche zur Abhängigkeit der Flugschwindigkeit einer Weisscheitelammer (*Zonotrichia leucophrys*). *Journal für Ornithologie*, 113: 21–28.

Schoener, T. W. 1968. Sizes of feeding territories among birds. *Ecology*, 49:123–141.

Scholander, S. I. 1955. Land birds over the western North Atlantic. *Auk*, 72:225–239.

Scholz, A. T., *et l*. 1978. Homing of Rainbow Trout in Lake Michigan: a comparison of three procedures for imprinting and stocking. *Transactions of the American Fisheries Society*, 107:439–443.

Schorger, A. W. 1937. The great Wisconsin Passenger Pigeon nesting of 1871. *Proceedings of the Linnaean Society of New York*, 48:1–26.

Schorger, A. w. 1947. The deep diving of the Loon and the Old Squaw and its mechanism. *Wilson Bulletin*, 59: 151–159.

Schorger, A. w. 1960. The crushing of *Carya* nuts in the gizzard of the Turkey, *Auk*, 77:337–340.

Schrantz, F. G. 1943. Nest life of the eastern Yellow Warbler. *Auk*, 90:367–387.

Schreiber, B., and O. Rossi. 1976(77). Correlation between race arrivals of homing pigeons and solar activity. *Bollettino di Zoologia*, 43:317–320.

Schreiber, R. W., and N. P. Ashmole. 1970. Sea-bird breeding seasons on Christmas Island, Pacific Ocean. *Ibis*, 112: 363–394.

Schreiber, R. W. and E. A. Schreiber. 1984. Central Pacific seabirds and the El Niño southern oscillations 1982 to 1983. *Science*, 225:713–716.

Schubert, M. 1971. Untersuchungen über die Reaktionauslösenden Signalstrukturen des Fitisgesangs, *Phylloscopus t. trochilus* (L.), und des Verhalten gegenüber arteigenen Rufen. *Behaviour*, 38:250–288.

Schuchmann, K.-L. 1978. Allopatrische Artbildung bei der Kilibrigattung *Trochilus*. *Ardea*, 66:156–172.

Schuchmann, K.-L. 1983. Analyse und Ontogenese des Sperrverhaltens bei Trochiliden. *Journal für Ornithologie*, 124: 65–74.

Schuchmann, K.-L., and H. Jakob. 1981. Energy expenditure of an incubating tropical hummingbird under laboratory conditions. *Le Gerfaut*, 71:227–233.

Schuchmann, K.-L., K. Krüger, and R. Prinzinger. 1983. Torpor in hummingbirds. *Bonner Zoologische Beiträge*, 34: 273–277.

Schuchmann, K.-L., and D. Schmidt-Marloh. 1979. Temperature regulation in non-torpid hummingbirds. *Ibis*, 121: 354–356.

Schuler, W. 1974. Die Schuzwirkung künstlicher Bates'scher Mimikry abhängig von Modellänlichkeit und Beuteangebot. *Zeitschrift für Tierpsychologie*, 36:71–127.

Schultz, F. 1965. Sexuelle Prägung bei Anatiden. *Zeitschrift für Tierpsychologie*, 22:50–103.

Schutz, F. 1974. Der Einfluss von Testosteron auf die Partneerwahl bei geprägt aufgezoenen Stockentenweibchen. *Verhandlungen der Deutschen Zoologischen Gesellschaft*, 67:339–344.

Schüz, E. 1936. The White Stork as a subject of research. *Bird-Banding*, 7:99–107.

Shuüz, E. 1938. Über Biologie und Ökologie des Weiszen Storches (*Cicomia c. cicomia*). *Proceedings of the VIIIth International Ornithological Congress, Oxford, July, 1934*, 577–591. Oxford University Press, London.

Schüz, E. 1943. Über die Jungen aufzucht des Weiszen Storches. *Zeitschrift für Morphologie und Oekologie der Tierre*, 40:181–237.

Schüz, E. 1944. Nest-Erwerb und Nest-Besitz beim Weiszen Storch (*C. cicomia*). *Zeitschrift für Tierpsychologie*, 6: 1–25.

Schüz, E. 1949a. Reifung, Ansiedlung und Bestandwechsel beim Weiszen Storch (*C. cicomia*). *Ornithologie als biologische Wissenschaft*. Carl Winter Universitätsverlag, Heidelberg.

Schüz, E. 1949b. Die Spät-auflassung ostpreussicher Jungstörche in West-Deutschland durch die Vogelwarte Rossitten, 1933. *Die Vogelwarte*, 2:63–78.

Schüz, E., M. Casement, and H. Seilkopf. 1963. Weisser Storch: weitere Fälle von Suesgolf-querung und Sinai-zug. *Die Vogelwarte*, 22:26–30.

Schüz, E., *et al.* 1971. *Grundriss der Vogelzugkunde*. Second edition. Verlag Paul Parey, Berlin.

Schwabel, H. 1983. Ausprägung und Bedeutung des Teilzugverhaltens einer Südwestdeutschen Population der Amsel *Turdus merula*. *Journal für Ornithologie*, 124:101–116.

Schwartz, C. W., and E. R. Schwartz. 1951. An ecological reconnaissance of the pheasants of Hawaii. *Auk*, 68: 281–314.

Schwartzkopff, J. 1949. Über den Zusammenhang von Gehör und Vibrationssinn bei Vögeln. *Experientia*, 5:159–161.

Schwartzkopff, J. 1955a. On the hearing of birds. *Auk*, 72:340–347.

Schwartzkopff, J. 1955b. Schallsinnesorgan bei Vögeln. In *Acta XI Congressus Internationalis Ornithologici*. Berhauser Verlag, Basel and Stuttgart.

Schwartzkopff, J. 1963. Morphological and physiological properties of the auditory system in birds. *Proceedings of XIIIth International Ornithological Congress*, 2:1059–1068. American Ornithologists' Union.

Schwartzkopff, J. 1973. Mechanoreception. In Farner, D. S., and J. R. King (eds). *Avian Biology*, Vol. 3. Academic Press, New York.

Sclater, P. L. 1858. On the general geographical distribution of the members of the Class Aves. *Journal of the Proceedings of the Linnaean Society, London, Zoology*, 2:130–145.

Scott, J. M., and C. B. Kepler. 1985. Distribution and abundance of Hawaiian native birds: a status report. In Temple, S. A. (ed). *Bird Conservation*, 2:43–70.

Scott, P. 1953. *Severn Wildfowl Trust Annual Report 1951–1952*. Myvatn, 1951.

Scott, V. E., *et al.* 1977. Cavity nesting birds of North American forests. *U.S. Department of Agriculture, Handbook 511*.

Scott, W. E. 1938. Old Squaws taken in gill nets. *Auk*, 55:668.

Searcy, W. A., and K. Yasukawa. 1983. Sexual selection and Red-winged Blackbirds. *American Scientist*, 71:166–174.

Seitz, A., 1940–1941. Die Paarbildung bei einigen Cichliden. *Zeitschrift für Tierpsychologie*, 4:40–84; 5:74–101.

Selander, R. K. 1966. Sexual dimorphism and differential niche utilization in birds. *Condor*, 68:113–151.

Selander, R. K., and R. F. Johnston. 1967. Evolution in the house sparrow: I. Intrapopulation variation in North America. *Condor*, 69:217–258.

Selander, R. K., and L. L. Kuich. 1963. Hormonal control and development of the incubation patch in icterids, with notes on behavior of Cowbirds. *Condor*, 65:73–90.

Semm, P., D. Nohr, C. Demaine, and W. Wiltschko. 1984. Neural basis of the magnetic compass: interactions of visual, magnetic, and vestibular inputs in the pigeon's brain. *Journal of Comparative Physiology, A. Series Neural Behavioral Physiology*, 155:283–288.

Sengupta, S. 1981. Adaptive significance of the use of margosa leaves in nests of House Sparrows. *Emu*, 81:114–115.

Seppa, J. 1969. The Cuckoo's ability to find a nest where it can lay an egg. *Ornis Fennica*, 46:78–79.

Serventy, D. L. 1938. Notes on Cormorants. *Emu*, 38:357–371.

Serventy, D. L. 1957. Duration of immaturity in the Short-tailed Shearwater, *Puffinus tenuirostris* (Temminck). Commonwealth Scientific and Industrial Research Organization, *Wildlife Research*, 2:60–62.

Serventy, D. L. 1963. Egg-laying timetable of the Slender-billed Shearwater, *Puffinus tenuirostris*. *Proceedings of the XIIIth International Ornithological Congress*, 1:338–343. American Ornithologists' Union.

Serventy, D. L. 1964. Australasian region. In Thomson, A. L. (ed). *A New Dictionary of Birds*. McGraw-Hill Book Company, New York.

Serventy, D. L. 1964. Emu. In Thomson, A. L. (ed). *A New Dictionary of Birds*. McGraw-Hill Book Company, New York.

Serventy, D. L. 1967. Aspects of the population ecology of the Short-tailed Shearwater, *Puffinus tenuirostris*. *Proceedings of the XIVth International Ornithological Congress*, 165–190. Blackwell Scientific Publications, Oxford.

Serventy, D. L. 1971. Biology of desert birds. In Farner, D. S., and J. R. King (eds). *Avian Biology*, Vol. 1. Academic Press, New York.

Serventy, D. L. 1973. Family: Magpie-larks. In Grzimek, B. (ed). *Grzimek's Animal Life Encyclopedia*, Vol. 9. Van Nostrand Reinhold Company, New York.

Serventy, D. L., and H. M. Whittell. 1950. A *Handbook of the Birds of Western Australia*. Second edition. Patterson Press, Perth, Australia.

Seuter, F. 1970. Ist eine endozoische Verbreitung der Tollkirsche durch Amsel und Stat möglich? *Zoologischer Jahrbücher Abteilung für Physiologie*, 75:342– 359.

Shaffner, C. S. 1954. Progesterone induced molt. *Science*, 120:345–346.

Sharpe, R. S., and P. A. Johnsgard. 1966. Inheritance of behavioral characters in F2 Mallard × Pintail (*Anas platyrhynchos*, L.–*Anas acuta*, L.) hybrids. *Behaviour*, 27:259–272.

Shaw, P. 1984. The social behaviour of the Pin-tailed Whydah (*Vidua macroura*) in northern Ghana. *Ibis*, 126: 463–473.

Shelford, V. E. 1945.The relation of Snowy Owl migration to the abundance of the Collared Lemming. *Auk*, 62: 592–596.

Shelford, V. E., and L. Martin, 1946. Reactions of young birds to atmospheric humidity. *Journal of Wildlife Management*, 10:66–68.

Shellswell, G. B., S. G. Gosney, and R. A. Hinde. 1975. Photoperiodic control of Budgerigar reproduction: circadian changes in sensitivity. *Journal of Zoology* (London), 175:53–60.

Shepard, J. 1971. Factors influencing non-random mating in the Ruff (*Philomachus pugnax*). *American Zoologist*, 11:624.

Sherman, A. R. 1910. At the sign of the Northern Flicker. *Wilson Bulletin*, 22:135–166.

Sherwood, A. W. 1946. *Aerodynamics*. McGraw-Hill Book Company, New York.

Shields, G. F. 1982. Comparative avian cytogenetics: A review. *Condor*, 84:45–58.

Shields, G. F. 1983. Bird chromosomes. *In* Johnston, R. F. (ed). *Current Ornithology*, 1:189–209.

Shoemaker, H. H. 1939. Social hierarchy in flocks in the Canary. *Auk*, 56:381–406.

Shoemaker, V. H. 1972. *In* Farner, D. S., and J. R. King (eds). *Avian Biology*, Vol. 2. Academic Press, New York.

Short, L. 1961. Interspecies flocking of birds of montane forest in Oaxaca, Mexico. *Wilson Bulletin*, 73:341–347.

Shufeldt, R. W. 1909. Osteology of birds. *Bulletin of the New York State Museum*, 130:5–381.

Shumakov, M. E. 1967. [An investigation of the migratory orientation of passerine birds.] *Vestnik Leningradskogo Universiteta*, Biology Series, 1967 (3):106–118. From *Bird-Banding*, 38:328.

Sibley, C. G. 1968. The relationships of the "wren-thrush," *Zeledonia coronata* Ridgway. *Postilla*, 125:1–12.

Sibley, C. G. 1970. A comparative study of the egg-white proteins of passerine birds. *Peabody Museum of Natural History Yale University, Bulletin No. 32*.

Sibley, C. G. 1972. A comparative study of the egg white proteins of non-passerine birds. *Peabody Museum of Natural History. Yale University Bulletin*, 39:1–276.

Sibley, C. G., and J. E. Ahlquist. 1972. A comparative study of the egg-white proteins of non-passerine birds. *Peabody Museum of Natural History Yale University, Bulletin No. 39*.

Sibley, C. G., and J. E. Ahlquist. 1981. The phylogeny and relationships of the ratite birds as indicated by DNA-DNA hybridization. *In* Scudder, G. G. E., and J. L. Reveal (eds). *Evolution Today. Proceedings of the 2nd International Congress in Systematics and Evolutionary Biology*, 301–335. Hunt Institute Botanic Document. Carnegie-Mellon University, Pittsburgh, Pennsylvania.

Sibley, C. G., and J. E. Ahlquist. 1982a. The relationships of the Wrentit as indicated by DNA-DNA hybridization. *Condor*, 84:40–44.

Sibley, C. G., and J. E. Ahlquist. 1982b. The relationships of the Hawaiian Honeycreepers (Drepaninini) as indicated by DNA-DNA hybridization. *Auk*, 99:130–140.

Sibley, C. G., and J. E. Ahlquist. 1983. Phylogeny and classification of birds based on the data of DNA-DNA hybridization. *In* Johnston, R. F. (ed). *Current Ornithology*, 1:245–292.

Sibley, C. G., and J. E. Ahlquist. 1985a. The phylogeny and classification of the Australo-Papuan passerine birds. *Emu*, 85:1–14.

Sibley, C. G., and J. E. Ahlquist. 1985b. The relationships of some groups of African birds, based on comparisons of the genetic material, DNA. *In* Schuchmann, K. L. (ed). *Proceedings International Symposium on African Vertebrates*, 115–161. Museum Alexander Koenig, Bonn.

Sibley, C. G., and J. E. Ahlquist. 1986. Reconstructing bird phylogeny by comparing DNA's. *Scientific American*, 254:82–92.

Sick, H. 1939. Über die Dialektbildung beim "Regenfur" des Buchfinken. *Journal für Ornithologie*, 87:568–592.

Sick, H. 1968. Über in Südamerika eingefuhrte Vogelarten. *Bonner Zoologische Beiträge*, 19:298–306.

Sick, H. 1973. Family: antbirds. *In* Grzimek, B. (ed). *Grzimek's Animal Life Encyclopedia*, 9:139. Van Nostrand Reinhold Company, New York.

Sick, H. 1981. Zum Problem der Elimination der Nestgeschwister beim Lerchenkuckuck *Tapera naevia*. *Journal für Ornithologie*, 122:437–438.

Siegel, H. S., and P. B. Siegel. 1961. The relationship of social competition with endocrine weights and activity in male chickens. *Animal Behaviour*, 9:151–158.

Siegfried, W. R. 1968. Breeding season, clutch and brood sizes in Verreaux's Eagle. *Ostrich*, 39:139–145.

Siegfried, W. R., *et al.* 1972. Lead concentrations in the bones of city and country doves. *South African Journal of Science*, 68:229–230.

Sielmann, H. 1958. *Das Jahr mit den Spechten*. Verlag Ullstein, Berlin.

Silbernagl, H. P. 1982. Seasonal and spatial distribution of the American purple gallinule in South Africa. *Ostrich*, 53:236–240.

Sillman, A. J. 1973. Avian vision. *In* Farner, D. S., and J. R. King (eds). *Avian Biology*, Vol 3. Academic Press, New York.

Simmons, K. E. L. 1963. Some behaviour characters of the babblers (Timaliidae). *Avicultural Magazine*, 69:183–193.

Simmons, K. E. L. 1985. Anting. *In* Campbell, B. and E. Lack (eds). *A Dictionary of Birds*, 19. Buteo Books, South Dakota.

Simpson, G. G. 1940. Mammals and land bridges. *Journal of the Washington Academy of Science*, 30:137–163.

Simpson, G. G. 1976. *Penguins. Past and Present, Here and There*. Yale University Press, New Haven, Connecticut and London.

Simpson, G. G., C. S. Pittendrigh, and L. H. Tiffany. 1957. *Life—An Introduction to Biology*. Harcourt, Brace and Company, New York.

Sivak, J. G., and M. Millodot. 1977. Optical performance of the penguin eye in air and water. *Journal of Comparative Physiology A*, 119:241–248.

Skead, C. J. 1971. Use of tools by the Egyptian Vulture. *Ostrich*, 42:226.

Skowron, C., and M. Kern. 1980. The insulation in nests of selected North American songbirds. *Auk*, 97:816–824.

Skutch, A. F. 1931. The life history of Rieffer's Hummingbird (*Amazilia tzacatl tzacatl*) in Panama and Honduras. *Auk*, 48:481–500.

Skutch, A. F. 1940. Social and sleeping habits of Central American wrens. *Auk*, 57:293–312.

Skutch, A. F. 1942. Life history of the Mexican Trogon. *Auk*, 59:341–363.

Skutch, A. F. 1944a. The life-history of the Prong-billed Barbet. *Auk*, 61:61–88.

Skutch, A. F. 1944b. Life history of the Quetzal. *Condor*, 46:213–235.

Skutch, A. F. 1945. Incubation and nestling periods of Central American birds. *Auk*, 62:8–37.

Skutch, A. F. 1946a. Life history of the Costa Rican Tityra. *Auk*, 63:327–362.

Skutch, A. F. 1946b. The parental devotion of birds. *Scientific Monthly*, 62:364–374.

Skutch, A. F. 1949. Do tropical birds rear as many young as they can nourish? *Ibis*, 91:430–458.

Skutch, A. F. 1952. On the hour of laying and hatching of birds' eggs. *Ibis*, 94:49–61.

Skutch, A. F. 1954, 1960, 1969. *Life Histories of Central American Birds*. 3 vols. Cooper Ornithological Society, Los Angeles, California.

Skutch, A. F. 1957. The incubation patterns of birds. *Ibis*, 99:69–93.

Skutch, A. F. 1959. Life history of the Groove-billed Ani. *Auk*, 76:281–317.

Skutch, A. F. 1961. Helpers among birds. *Condor*, 63:198–226.

Skutch, A. F. 1962. The constancy of incubation. *Wilson Bulletin*, 74:115–152.

Skutch, A. F. 1964. Life history of the Blue-diademed Motmot, *Momotus momota*. *Ibis*, 106:321–332.

Skutch, A. F. 1966. Life history notes on three tropical American cuckoos. *Wilson Bulletin*, 78:139–165.

Skutch, A. F. 1967. Adaptive limitation of the reproductive rate of birds. *Ibis*, 109:579–599.

Skutch, A. F. 1970. Life History of the Common Potoo. *Living Bird*, 9:265–280.

Skutch, A. F. 1976. *Parent Birds and Their Young*. University of Texas Press, Austin, Texas.

Sladen, W. J. L., and N. A. Ostenso. 1960. Penguin tracks far inland in the Antarctic. *Auk*, 77:466–469.

Sladen, W. J. L., and W. L. N. Tickell. 1958. Antarctic bird-banding by the Falkland Islands Dependencies Survey, 1945–1957. *Bird-Banding*, 29:1–26.

Slagsvold, T. 1980. Egg predation in woodland in relation to the presence and density of breeding fieldfares (*Turdus pilaris*). *Ornis Scandinavica*, 11:92–98.

Slud, P. 1957. The song and dance of the Long-tailed Manakin, *Chiroxiphia linearis*. *Auk*, 74:335–339.

Smart, G. 1965. Development and maturation of primary feathers of Redhead ducklings. *Journal of Wildlife Management*, 29:533–536.

Smith, A. G., and H. R. Webster. 1955. Effects of hail storms on waterfowl population in Alberta, Canada—1953. *Journal of Wildlife Management*, 19:368–374.

Smith, C. C., and O. J. Reichman. 1984. The evolution of food caching by birds and mammals. *Annual Review of Ecology and Systematics*, 15:329–351.

Smith, D. G. 1976. An experimental analysis of the function of red-winged blackbird song. *Behaviour*, 56:136–156.

Smith, D. G. 1979. Male singing ability and territory integrity in Red-winged Blackbirds *(Agelaius phoeniceus)*. *Behaviour*, 68:193–206.

Smith, D. J. 1972. Role of the epaulets in the Red-winged Blackbird, *Agelaius phoeniceus*, social system. *Behaviour*, 41:251–268.

Smith, G. A. 1972. "Handedness" in parrots. *Ibis*, 114:109–110.

Smith, G. A. 1975. Systematics of Parrots. *Ibis*, 117:18–68.

Smith, G. A. 1979. *Lovebirds and Related Parrots*. TFH Publications, Inc., New Jersey.

Smith, G. T. C. 1969. A high altitude hummingbird on the volcano Cotopaxi. *Ibis*, 111:17–22.

Smith, J. N. M., and A. A. Dhont. 1981. Experimental confirmation of heritable morphological variation in a natural population of song sparrows *(Melospiza melodia)*. *Evolution*, 14:1155–1158.

Smith, J. N. M., and J. R. Merkt. 1980. Development and stability of single-parent family units in the song sparrow. *Canadian Journal of Zoology*, 58:1869–1875.

Smith, J. N. M., Y. Yom-Tov, and R. Moses. 1982. Polygyny, male parental care, and sex ratio in Song Sparrows: An experimental study. *Auk*, 99:555–564.

Smith, L. H. 1968. *The Lyrebird*. Lansdowne Press, Melbourne, Australia.

Smith, N. G. 1968. The advantage of being parasitized. *Nature*, 219:690–694.

Smith, N. G. 1980. Hawk and vulture migrations in the neotropics. *In* Keast, A., and E. S. Morton (eds). *Migrant Birds in the Tropics, Ecology, Behaviour, Distribution and Conservation*. Smithsonian Institute Press, Washington. D. C.

Smith, N. J. H. 1973. House Sparrows *(Passer domesticus)* in the Amazon. *Condor*, 75:242–243.

Smith, P. W., and N. T. Houghton. 1984. Fidelity of Semipalmated Plovers to a migration stopover area. *Journal of Field Ornithology*, 55:247–249.

Smith, R. L. 1966. *Ecology and Field Biology*. Harper and Row, New York.

Smith, S. 1942. Field observations on the breeding biology of the Yellow Wagtail. *British Birds*, 35:186–189.

Smith, S. M. 1972. Roosting aggregations of Bushtits in response to cold temperatures. *Condor*, 74:478–479.

Smith, S. M. 1975. Innate recognition of Coral Snake pattern by a possible avian predator. *Science*, 187:759–760.

Smith, S. M. 1977. The behaviour and vocalization of young Turquoise-browed Motmots. *Biotropica*, 9:127–130.

Smith, S. M. 1978. The "underworld" in a territorial sparrow: adaptive strategy for floaters. *The American Naturalist*, 112:571–582.

Smith, S. M. 1980. Henpecked males: the general pattern of monogamy? *Journal of Field Ornithology*, 51:55–64.

Smith, S. M. 1983. The ontogeny of avian behavior. *In* Farner, D. S., and J. R. King (eds). *Avian Biology*, 7:85–160.

Smith, T. B., and S. A. Temple. 1982. Feeding habits and bill polymorphism in hook-billed kites. *Auk*, 99:197–207.

Smythies, B. E. 1953. *The Birds of Burma*. Oliver and Boyd, Ltd., London Edinburgh.

Smythies, B. E. 1968. *The Birds of Borneo*. Oliver and Boyd, Ltd., London Edinburgh.

Smythies, B. E. 1973. Family: Flowerpeckers, *In* Grzimek, B. (ed). *Grzimek's Animal Life Encyclopedia*. Van Nostrand Reinhold Company, New York.

Snow, B. K. 1970. A field study of the Bearded Bellbird in Trinidad. *Ibis*, 111:299–329.

Snow, D. W. 1955. The abnormal breeding of birds in the winter 1953–54. *British Birds*, 48:121–126.

Snow, D. W. 1956. Courtship ritual: the dance of the manakins. *Animal Kingdom*. 59:86–91.

Snow, D. W. 1961. The displays of the manakins *Pipra pipra*, and *Tyranneutes virescens*. *Ibis*, 103a:110–113.

Snow, D. W. 1978. The nest as a factor determining clutch size in tropical birds. *Journal für Ornithologie*, 119:227–230.

Snow, D. W., and J. B. Nelson. 1984. Evolution and adaptations of Galapagos sea birds. *Biological Journal of the Linnean Society*, 21:137–156.

Snyder, N. F. R., and E. V. Johnson. 1985. Photographic censusing of the 1982–1983 California Condor population. *Condor*, 87:1–13.

Sonnemann, P., and S. Sjolander. 1977. Effects of cross-fostering on the sexual imprinting of the female zebra finch. *Taeniopygia guttata*. *Zeitschrift für Tierpsychologie*, 45:337–348.

Sossinka, R. 1972. Langfristiges Durstvermögen wilder und domestizierter Zebrafinken *(Taeniopygia guttata castanotis* Gould). *Journal für Ornithologie*, 113:418–426.

Sossinka, R. 1975. Quantitative Untersuchungen zur sexuellen Reifung des Zebrafinken. *Taeniopygia castanotis* Gould. *Verhandlungen der Deutschen Zoologischen Gesellschaft*, 67:344–347.

Southern, H. N. 1970. The natural control of a population of Tawny Owls *(Strix aluco)*. *Journal of Zoology*, 162:197–285.

Southern, W. E. 1959. Homing of Purple Martins. *Wilson Bulletin, 71*:254–261.

Southern, W. F. 1969. Orientation behavior of Ring-billed Gull chicks and fledglings. *Condor, 71*:418–425.

Southern, W. E. 1972. Magnets disrupt the orientation of juvenile Ring-billed Gulls. *BioScience, 22*:476–479.

Southern, W. E. 1975. Orientation of gull chicks exposed to Project Sanguine's electromagnetic field. *Science, 189*: 143–145.

Spalding, D. 1873. Instinct: with original observations on young animals. *Macmillan's Magazine, 27*:282–293.

Spearman, R. I. C. 1980. The avian skin in relation to surface ecology. *Proceedings of the Royal Society of Edinburgh, 79*:57–74.

Spearman, R. I. C., and J. A. Hardy. 1985. Integument. *In* Kingard, A. S., and J. McLelland (eds). *Form and Function in Birds, 3*:1–56.

Spellerberg, I. F. 1975. *In* Stonehouse, B. (ed). *The Biology of Penguins*. University Park Press, Baltimore.

Spencer, R. 1977. The role of the amateur in bird research in Britain and Ireland. *Die Vogelwarte, 29 Sonderheft*: 54–56.

Spilhaus, A. (ed). 1966. *Waste Management and Control*. Publication 1400. National Academy of Sciences, Washington, D.C.

Spitzer, G. 1972. Jahreszeitliche Aspekte der Biologie der Bartmeise *(Panurus biarmicus)*. *Journal für Ornithologie, 113*:241–275.

Spofford, W. R 1964. The Golden Eagle in the Trans-Pecos and Edwards Plateau of Texas. *Audubon Conservation Report, 1*:1–47.

Sprot, G. D. 1937. Migratory behavior of some Glaucous-winged Gulls in the Strait of Georgia, British columbia. *Condor, 39*:238–242.

Sprunt, A. 1955. The spread of the Cattle Egret. *Smithsonian Report for 1954*, 259–276.

Srb, A. M., and R. D. Owen. 1952. *General Genetics*. Freeman and Company, San Francisco.

Staaland, H. 1967. Temperature sensitivity of the avian salt gland. *Comparative Biochemistry and Physiology, 23*: 991–993.

Stager, K. E. 1941. A group of bat-eating duck hawks. *Condor, 43*:137–139.

Stager, K. E. 1964. The role of olfactation in food location in the Turkey Vulture *(Cathartes aura)*. *Los Angeles County Museum Contributions in Science, 81*:1–63.

Stager, K. E. 1967. Avian olfaction. *American Zoologist, 7*:415–419.

Stamm, D. D. 1966. Relationships of birds and arboviruses. *Auk, 83*:84–97.

Steadman, D. W. 1983. Commentary on the origin and early radiation of birds. *In* brush, A. H., and G. A. Clark (eds). *Perspectives in Ornithology*, 338–344. Cambridge University Press, London.

Stefferud, A., and A. L. Nelson (eds). 1966. *Birds in Our Lives*. U. S. Fish and Wildlife Service, Washington, D.C.

Steidinger, P. 1968. Radar Beobachtungen über die Richtung und deren Streuung beim nächtlichen Vogelzug im Schweizerischen Mittelland. *Ornithologische Beobachter, 65*:197–226.

Steinbacher, J. 1936. Zur Frage der Geschlechtsreife von Kleinvögeln. *Beträge zur Fortpflanzungsbiologie der Vögel, 12*:139–144.

Steinbacher, J. 1951. *Vogelzug und Vogelzugforschung*. Kramer, Frankfurt am Mainz.

Steiner, H. 1960. Klassifikation der Prachtfinken, Spermestidae, auf Grund der Rachenzeichnungen ihrer Nestlinge. *Journal für Ornithologie, 101*:92– 112.

Steinfatt, O. 1938. Das Brutleben der Sumpfmeise und einige Vergleiche mit dem Brutleben der anderen einheimischen Meisen. *Beträge Fortpflanzungsbiologie der Vögel, 14*:84–89; 137–144.

Stenger, J., and J. B. Falls. 1959. The utilized territory of the Ovenbird. *Wilson Bulletin, 71*:125–140.

Stephens, M. L. 1984. Interspecific aggressive behavior of the polyandrous northern jacana *(Jacana spinosa)*. *Auk, 101*: 508–518.

Stettenheim, P. 1972. The integument of birds. *In* Farner, D. S. and J. R. King (eds). *Avian Biology*, Vol. 2. Academic Press, New York.

Stettenheim, P. 1973. The bristles of birds. *Living Bird, 12*:201–234.

Stettenheim, P., *et al.* 1963. The arrangement and action of feather muscles in chickens. *Proceedings of the XIIIth International Ornithological Congress*, 918–924. American Ornithologists' Union.

Stewart, P. A. 1952. Dispersal, breeding behavior, and longevity of banded Barn Owls in North America. *Auk, 69*: 227–245.

Stewart, P. A. 1973. Electrocution of birds by an electric fence. *Wilson Bulletin, 85*:476–477.

Stewart, P. A. 1973. Estimating numbers in a roosting aggregation of blackbirds and Starlings. *Auk, 90*:353–358.

Stewart, R. E. 1956. Ecological study of Ruffed Grouse broods in Virginia. *Auk, 73*:33–41.

Stewart, R. M., *et al.* 1977. Breeding ecology of the Wilson's Warbler in the high Sierra Nevada, California. *Living Bird*, 16:83–102.

Stiles, F. G. 1973. Food supply and the annual cycles of the Anna Hummingbird. *University of California Publications in Zoology*, 97:1–116.

Stiles, F. G. 1976. Taste preferences, color preferences, and flower choice in hummingbirds. *Condor*, 78:10–26.

Stiles, F. G. 1985. Seasonal patterns and coevolution in the hummingbird-flower community of a Costa Rican subtropical forest. *In Neotropical Ornithology*. American Ornithologists' Monograph, 36:757–785.

Stirling, I. J., and J. F. Bendell. 1966. Census of Blue Grouse with recorded calls of a female. *Journal of Wildlife Management*, 30:184–187.

Stoddard, H. L. 1962. Bird casualties at a Leon County, Florida, television tower. *Tall Timbers Research Station Bulletin*, No. 1.

Stokes, A. W. 1974. *Benchmark Papers in Animal Behaviour: Territory*. Dowden, Hutchinson and Ross, Inc., Stroudsburg, Pennsylvania.

Stokes, A. W., and H. W. Williams. 1971. Courtship feeding in gallinaceous birds. *Auk*, 88:543–559.

Stone, C. P., *et al.* 1972. Bird damage to corn in the United States in 1970. *Wilson Bulletin*, 85:101–105.

Stonehouse, B. 1956. The King Penguin of South Georgia. *Nature*, 178:1424–1426.

Stonehouse, B., and C. M. Perrins (eds). 1977. *Evolutionary Ecology*. University Park Press, Baltimore.

Stoner, D. 1945. Temperature and growth studies of the Northern Cliff Swallow. *Auk*, 62:207–216.

Storer, J. H. 1948. *The Flight of Birds*. Cranbrook Institute of Science, Bloomfield Hills, Michigan.

Storer, R. 1963. The courtship and mating behavior and the phylogeny of grebes. *Proceedings XIII International Ornithological Congress*, 562–569. American Ornithologists Union.

Storer, R. W. 1960. Adaptive radiation in birds. *In Marshall, A. J. (ed)*. *Biology and Comparative Physiology of Birds*. Academic Press, New York.

Storer, R. W. 1966. Sexual dimorphism and food habits in three North American accipiters. *Auk*, 83:423–436.

Storer, R. W. 1971. Adaptive radiation of birds. *In Farner, D. S., and J. R. King (eds)*. *Avian Biology*, Vol. 1. Academic Press, New York.

Storer, T. I. 1943. *General Zoology*. McGraw-Hill Book Company, New York.

Stoudt, J. H. 1982. Habitat use and productivity of Canvasbacks (*Aythya valisneria*) in southcoast Manitoba, Canada. *U. S. Fish and Wildlife Service Special Science Report. Wildlife*, 248:1–31.

Stout, I. J., and G. W. Cornwell. 1976. Nonhunting mortality of fledged North American waterfowl. *Journal of Wildlife Management*, 40:681–693.

Streicher, E., D. B. Backel, and W. Fleischmann. 1950. Effects of extreme cold on the fasting pigeon. *American Journal of Physiology*, 161:300.

Stresemann, E. 1927–1934. *In Kükenthal, W., and T. Krumbach. Handbuch der Zoologie. Sauropsida: Aves*. W. de Gruyter and Co., Berlin and Leipzig.

Stresemann, E. 1940. Zeitpunkt und Verlauf der Mauser bei einigen Entenarten. *Journal für Ornithologie*, 88:288–333.

Stresemann, E. 1947. Baron von Pernau, pioneer student of bird behavior. *Auk*, 64:35–52.

Stresemann, E. 1955. Die Wanderungen des Waldlaubsängers (*Phylloscopus sibilatrix*). *Journal für Ornithologie*, 96: 153–167.

Stresemann, E., and V. Stresemann. 1966. Die Mauser der Vögel. *Journal für Ornithologie*, Sonderheft 107:1–445.

Studd, M., *et al.* 1983. Group size and predator surveillance in foraging house sparrows (*Passer domesticus*). *Canadian Journal of Zoology*, 61:226–231.

Sturkie, P. D. 1965. *Avian Physiology*. Comstock Publishing Associates, Ithaca, New York.

Sturkie, P. D. (ed). 1976. *Avian Physiology*. Third edition. Springer-Verlag, New York.

Sturtevant, J. 1970. Pigeon control by chemosterilization: Population model from laboratory results, *Science*, 170: 322–324.

Sullivan, G. C., and Rubinoff. 1983. Avoidance of venomous sea snakes by naive Herons and Egrets. *Auk*, 100: 195–198.

Summers, D. D. B., and F. J. S. Jones. 1976. The importance of protein in the selection of fruit buds by Bullfinches. *Experimental Horticulture*, 28:47–50.

Summers-Smith, D. 1956. Mortality of the House Sparrow. *Bird Study*, 3:265–270.

Suomalainen, H. 1937. The effect of temperature on the sexual activity of non-migratory birds, stimulated by artificial lighting. *Ornis Fennica*, 14:108–112.

Sutter, E. 1951. Growth and differentiation of the brain in nidifugous and nidicolous birds. *Proceedings of the Xth International Ornithological Congress. Uppsala, June 1950*, 636–643. Almqvist and Wiksells, Uppsala.

Sutter, E. 1957. Radar als Hilfsmittel der Vogelzugforschung. *Ornithologische Beobachter*, 54:70–96.

Sutton, G. M. 1951. *Mexican Birds* University of Oklahoma Press, Norman, Oklahoma.

Sutton, O. G. 1955, *The Science of Flight*. Penguin Books, Baltimore.

Svärdson, G. 1957. The 'invasion' type of bird migration. *British Birds*, 50:314–343.

Svärdson, G. 1958. Biotop och häckning hos skrattmäsen *(Larus ridibundus)*. *Vår Fagelvärld*, 17:1–23.

Swank, W. C. 1944. Germination of seeds after ingestion by Ring-necked Pheasants. *Journal of Wildlife Management*, 8:223–231.

Swarth, H. S. 1920. Revision of the avian genus *Passerella* with special reference to the distribution and migration of the races in California. *University of California Publications in Zoology*, 21:75–224.

Swennen, C. 1968. Nest protection of Eiderducks and Shovellers by means of faeces. *Ardea*, 56:248–258.

Swennen, C. 1974. Observations on the effect of ejection of stomach oil by the Fulmar, *Fulmarus glacialis*, on other birds. *Ardea*, 62:111–117.

Swingland, I. R. 1977. The social and spatial organization of winter communal roosting in Rooks *(Corvus frugilegus)*. *Journal of Zoology* (London), 182:509–528.

Swinton, W. E. 1958. *Fossil Birds*. British Museum (Natural History), London.

Sy, M. 1936. Funktionell-anatomische Untersuchungen am Vogelflügel. *Journal für Ornithologie*, 84:199–296.

Szijj, L. J. 1966. Hybridization and the nature of the isolating mechanism in sympatric populations of Meadowlarks *(Sturnella)* in Ontario. *Zeitschrift für Tierpsychologie*, 23:677–690.

Szaro, R. C., and P. H. Albers. 1977. *In* Wolfe, D. A. (ed). *Fate and Effects of Petroleum Hydrocarbons in Marine Ecosystems and Organisms*. Pergamon Press, New York.

Talent, J. A., *et al.* 1966. Early Cretaceous feathers from Victoria. *Emu*, 66:81–86.

Tanner, J. T. 1941. Three years with the Ivory-billed Woodpecker, America's rarest bird. *Audubon Magazine*, 43:5–14.

Tanner, J. T. 1942. The Ivory-billed Woodpecker. *Research Report No. 1 of the National Audubon Society*, New York.

Taylor, R. H. 1979. Predation on Sooty Terns at Raoul Island by rats and cats. *Notornis*, 26:199–202.

Taylor, T. G. 1970. How an eggshell is made. *Scientific American*, 222(3):88–95.

Teal, J. M. 1969. Direct measurement of carbon dioxide production during flight in small birds. *Zoologica*, 54:17–23.

Telfair, R., C. Raymond II, and L. E. Marcy. 1984. Cattle egrets (*Ardeola ibis* equals *Bubulcus ibis*) in Texas. *Texas Journal of Science*, 35:303–314.

Terborgh, J. 1977. Bird species diversity on an Andean elevational gradient. *Ecology*, 58:1007–1019.

Terhivuo, J. 1983. Why does the Wryneck, *Jynx torquilla*, bring strange items to the nest? *Ornis Fennica*, 60:51–57.

Terres, J. K. 1980. *The Audubon Society Encyclopedia of North American Birds*. Alfred A. Knopf, New York.

Test, F. H. 1969. Relation of wing and tail color of the woodpeckers *Colaptes auratus* and *C. cafer* to their food. *Condor*, 71:206–211.

Theurich, M., G. Langner, and H. Scheich. 1984. Infrasound responses in the midbrain of the Guinea Fowl. *Neuroscience Letters*, 49:81–86.

Thibault, J. C., and D. T. Holyoak. 1978. Vocal and olfactory displays in the Petrel Genera *Bulweria* and *Pterodroma*. *Ardea*, 66:53–56.

Thielcke, G. 1969a. Geographic variations in bird vocalizations. *In* Hinde, R. A. (ed). *Bird Vocalizations: Their Relation to Current Problems in Biology and Psychology*. Cambridge University Press, Cambridge.

Thielcke, G. 1969b. Die Reaktion von Tannen- und Kohlmeise *(Parus ater, Parus major)* auf den Gesang nahverwandten Formen. *Journal für Ornithologie*, 110:148–157.

Thielcke, G. 1970a. Die sozialen Funktionen der Vogelstimmen. *Vogelwarte*, 25:204–279.

Thielcke, G. 1970b. Lernen von Gesang als möglicher Schrittmacher der Evolution. *Zeitschrift für Zoologische Systematik und Evolutionsforschung*, 8:309–320.

Thielcke, G. 1972. Learning song patterns as pacemaker of evolution. *Proceedings of the XVth International Ornithological Congress*, 694. E. J. Brill, Leiden.

Thielcke, G. 1972. Waldaumläufer *(Certhis familiaris)* ahmen Artfremdes Signal nach und reagieren darauf. *Journal für Ornithologie*, 113:287–296.

Thielcke, G. 1973a. Uniformierung des Gesangs der Tannenmeise *(Parus ater)* durch Lernen. *Journal für Ornithologie*, 114:443–454.

Thielcke, G. 1973b. On the origin of divergence of learned signals (songs) in isolated populations. *Ibis,* 115:511–516.

Thielcke-Poltz, H., and G. Thielcke. 1960. Akustisches Lernen verschieden alter schallisolierter Amseln (*Turdus merula* L.) und die Entwicklung erlernter Motive ohne und mit künstlichem Einfluss von Testosteron. *Zeitschrift für Tierpsychologie,* 17:211– 244.

Thiollay, J. M. 1976. [Quantitative food requirements of some tropical birds.] *Terre et la Vie,* 30:229–245.

Thompson, D. W. 1942. *On Growth and Form.* Cambridge University Press, Cambridge.

Thompson, M. C. 1973. Migratory patterns of the Ruddy Turnstone in the central Pacific region. *Living Bird,* 12: 5–23.

Thompson, R. D., *et al.* 1968. Differential heart rate response of Starlings to sound stimuli of biological origin. *Journal of Wildlife Management,* 32:888–893.

Thompson, W. C. 1960. Agonistic behavior in the House Finch: I. Annual cycle and display patterns. *Condor,* 62: 245–271.

Thompson, W. R., and R. A. Dubonski. 1964. Imprinting and the "law of effort." *Animal Behaviour,* 12:213–218.

Thomson, A. L. 1926. *Problems of Bird Migration.* Houghton Mifflin Company, Boston.

Thomson, A. L. 1936. *Bird Migration.* H. F. and G. Witherby, Ltd., London.

Thomson, A. L. (ed). 1964. A *New Dictionary of Birds.* McGraw-Hill Book Company, New York.

Thomson, A. L. 1964. Articles: Crocodile Bird; Numbers; Oil Gland; Systematics. *In* Thomson, A. L. (ed). A *New Dictionary of Birds.* McGraw-Hill Book Company, New York.

Thomson, J. A. 1923. *The Biology of Birds.* Macmillan Publishing Company, New York.

Thorndike, L. 1929. A *History of Magic and Experimental Science,* Macmillan Publishing Company, New York.

Thorneycroft, H. B. 1976. A genetic study of the White-throated Sparrow, *Zonotrichia albicollis* (Gmelin). *Evolution,* 29:611–621.

Thorpe, W. H. 1956a. The language of birds. *Scientific American,* 195(4):128–138.

Thorpe, W. H. 1956b. *Learning and Instinct in Animals.* Harvard University Press, Cambridge.

Thorpe, W. H. 1958. The learning of song patterns by birds, with especial reference to the song of the chaffinch, *Fringilla coelebs. Ibis,* 100:535–570.

Thorpe, W. H. 1961. *Bird Song: The Biology of Vocal Communication and Expression in Birds.* Cambridge University Press, Cambridge.

Thorpe, W. H. 1964. Imprinting. *In* Thomson, A. L. (ed). A *New Dictionary of Birds.* McGraw-Hill Book Company, New York.

Thorpe, W. H. 1964. Learning. *In* Thomson, A. L. (ed). A *New Dictionary of Birds.* McGraw-Hill Book Company, New York.

Thorpe, W. H. 1964. Singing. *In* Thomson, A. L. (ed). A *New Dictionary of Birds.* McGraw-Hill Book Company, New York.

Thorpe, W. H. 1969. *In* Hinde, R. A. (ed). *Bird Vocalizations: Their Relation to Current Problems in Biology and Psychology.* Cambridge University Press, Cambridge.

Thorpe, W. H. 1972. Duetting and antiphonal song in birds. *Behaviour, Supplement* 17:1–197.

Threlfall, W., *et al.* 1973. A new record of the Fieldfare. *(Turdus pilaris)* in Canada. *Canadian Field Naturalist,* 87:311.

Tiedemann, M. 1943. Ornithologische Beobachtungen aus dem Hornsund-Gebiet auf West Spitzbergen. *Journal für Ornithologie,* 91:239–267.

Tiemeier, O. W. 1941. Repaired bone injuries in birds. *Auk,* 58:350–359. *Time Magazine,* June 23, 1958, p. 66.

Tinbergen, N. 1930a. On the analysis of social organization among vertebrates, with special reference to birds. *American Midland Naturalist,* 21:210–234.

Tinbergen, N. 1939b. The behavior of the Snow Bunting in spring. *Transactions of the Linnaean Society of New York,* 5:1–95.

Tinbergen, N. 1948. Social releasers and the experimental method required for their study. *Wilson Bulletin,* 60:6–51.

Tinbergen, N. 1951. *The Study of Instinct.* Clarendon Press, Oxford.

Tinbergen, N. 1952. "Derived" activities; their causation, biological significance, orign and emancipation during evolution. *Quarterly Review Biology,* 27:1–32.

Tinbergen, N. 1953a. *The Herring Gull's World.* William Collins, Sons & Co., London.

Tinbergen, N. 1953b. *Social Behavior in Animals.* John Wiley and Sons, New York.

Tinbergen, N. 1957. Defense by color. *Scientific American,* 194(4):48–54.

Tinbergen, N. 1959. Comparative studies of the behaviour of gulls (Laridae): a progress report. *Behaviour,* 15:1–70.

Tinbergen, N. 1960. The dynamics of insect and bird populations in pine woods. *Archives Neerlandaises de Zoologie,* 13:259–473.

Tinbergen, N. 1963a. On aims and methods of ethology. *Zeitschrift für Tierpsychologie,* 20:410–433.

Tinbergen, N. 1963b. The shell menace. *Natural History,* 72:28–35.

Tinbergen, N., M. Impekoven, and D. Franck. 1967. An experiment in spacing out as a defense against predation. *Behaviour,* 28:307–321.

Tinbergen, N., and D. J. Koenen. 1939. Über die auslösenden und die richtunggedbenden Reizsituationen der Sperrbewegung von jungen Drosseln. *Zeitschrift für Tierpsychologie,* 3:37–60.

Tinbergen, N., and A. C. Perdeck. 1951. On the stimulus situation releasing the begging response in the newly hatched Herring Gull chick (*Larus argentatus* Pont.). *Behaviour,* 3:1–39.

Titman, R. D., and J. K. Lowther. 1975. The breeding behavior of a crowded population of mallards. *Canadian Journal of Zoology,* 53:1270–1283.

Todt, D., *et al.* 1979. Conditions affecting song acquisition in Nightingales (*Luscinia megarhynchos* L.). *Zeitschrift für Tierpsychologie,* 51:23–35.

Tolman, C. W. 1967. The feeding behavior of domestic chicks as a function of the rate of pecking by a surrogate companion. *Behaviour,* 29:57–62.

Tolman, C. W., and G. F. Wilson. 1965. Social feeding in domestic chicks. *Animal Behaviour,* 13:134–142.

Tomback, D. F., D. B. Thompson, and M. C. Baker, 1983. Dialect discrimination by White-crowned Sparrows: reaction to near and distant dialects. *Auk,* 100:452–460.

Tompa, F. S. 1963. Behavioral response of the Song Sparrow to different environmental conditions. *Proceedings of the XIIIth International Ornithological Congress,* 729–739. American Ornithologists' Union.

Tonosaki, K., and T. Shibuya. 1985. Olfactory receptor cell responses of a pigeon to some odors. *Comparative Biochemistry, Physiology and Comparative Physiology,* 81:329–334.

Torre-Bueno, J. R. 1976. Temperature regulation and heat dissipation during flight in birds. *Journal of Experimental Biology,* 65:471–482.

Torre-Bueno, J. R. 1978. Evaporative cooling and water balance during flight in birds. *Journal of Experimental Biology,* 75:231–236.

Trail, P. W. 1985. Courtship disruption modifies mate choice in a lek-breeding bird. *Science,* 227:778–780.

Trautman, M. B. 1947. Courtship behavior of the Black Duck. *Wilson Bulletin,* 59:26–35.

Traylor, M. A., and J. W. Fitzpatrick. 1982. A survey of the tyrant flycatchers. *Living Bird,* 19:7–45.

Tretzel, E. 1965. Imitation und variation von Schäferpfiffen durch Haubenlerchen (*Galerida c. cristata* L.). Ein Beispiel für specielle Spottmotiv-Prädisposition. *Zeitschrift für Tierpsychologie,* 22:784–809.

Trivers, R. L. 1972. Parental investment and sexual selection. *In* Campbell, B. (ed). *Sexual Selection and the Descent of Man 1871–1971.* Aldine, Chicago, Illinois, 136–179.

Tschanz, B., 1959. Zur Brutbiologie der Trottellumen (*Uria aalge aalge* Pont.). *Behaviour,* 14:1–100.

Tschanz, B., 1968. Trottellumen: die Entstehen der-persönlichen Beziehungen zwischen Jungvogel und Eltern. *Zeitschrift für Tierpsychologie,* 4:1– 103.

Tschanz, B. 1978. Unterschungen zur Entwicklung des Trottellummenbestandes auf Vedöy (Röst, Lofoten). *Journal für Ornithology,* 119:133–145.

Tuck, L. M. 1971. The occurence of Greenland and European birds in Newfoundland. *Bird-Banding,* 42:184–209.

Tucker, B. W. 1943. *In* Witherby, H. F., F. C. R. Jourdain, N. Ticehurst, and B. W. Tucker. *The Handbook of British Birds.* H. F. and G. Witherby, Ltd., London.

Tucker, V. A. 1968a. Respiratory physiology of House Sparrows in relation to high altitude flight. *Journal of Experimental Biology,* 48:55–66.

Tucker, V. A. 1968b. Respiratory exchange and evaporative water loss in the flying Budgerigar. *Journal of Experimental Biology,* 48:67–87.

Tucker, V. A. 1969. The energetics of bird flight. *Scientific American,* 220(5):70–78.

Tucker, V. A. 1970. Energetic cost of locomotion in animal. *Comparative Biochemistry and Physiology,* 34:841–846.

Tucker, V. A. 1971. Flight energies in birds. *American Zoologist,* 11:115–124.

Tucker, V. A. 1974. *In* Paynter, R. A., Jr. (ed). *Avian Energetics.* Publications of the Nuttall Ornithology Club, No. 15. Cambridge, Massachusetts.

Tucker, V. A., and G. C. Parrott, 1970. Aerodynamics of gliding flight in a falcon and other birds. *Journal of Experimental Biology,* 52:345–367.

Tull, C. E., *et al.* 1972. Mortality of Thick-billed Murres in the West Greenland salmon fishery. *Nature,* 237:42–44.

Tûrcek, F. J. 1952. An ecological analysis of the bird and mammalian population of a primeval forest on the Polona-mountain (Slovakia). *Bulletin international de l'Academie tcheque des Sciences*, 53:1–25.

Tûrcek, F. 1961. *Oekologische Beziehungen der Vögel and Gehölze*. Verlag der Slowakischen Akademie der Wissenschaften, Bratislava.

Tûrcek, F. J. 1966. On plumage quantity in birds. *Ekologia Polska*, Series A, *14*(32):617–634.

Turney, T. H. 1982. The associations of visual concepts and imitative vocalizations in the myna *(Gracula religiosa)*. *Bulletin of the Psychonomic Society*, 19:59–62.

Twomey, A. C. 1936. Climographic studies of certain introduced and migratory birds. *Ecology*, 17:122–132.

Udvardy, M. D. F. 1953. Contributions to the knowledge of the body temperature of birds. *Zoologiska Bidrag Fran Uppsala*, 30:25–42.

Udvardy, M. D. F. 1957. An evaluation of quantitative studies in birds. *Cold Spring Harbor Symposia on Quantitative Biology*, 22:301–311.

Udvardy, M. D. F. 1959. Notes on the ecological concepts of habitat, biotope, and niche. *Ecology*, 40:725–728.

Udvardy, M. D. F. 1969. *Dynamic Zoogeography*. Van Nostrand Reinhold Company, New York.

Udvardy, M. D. F. 1972. Plumage cycles in adult North American birds. *Aquila Ser Zoologica*, 13:56–60.

Udvardy, M. D. F. 1983. The role of the feet in behavioral thermoregulation of hummingbirds. *Condor*, 85:281–285.

Ulfstrand, S. 1973. Proportions of Palaearctic birds in some east African habitats. *Die Vogelwarte*, 27:137–141.

Urner, C. A., and R. W. Storer. 1949. The distribution and abundance of shorebirds on the north and central New Jersey coast, 1928–1938. *Auk*, 66:177–194.

Uspenskii, S. M. 1969. [*Life at High Altitudes as Exemplified by Birds.*] Mysl Publishing House, Moscow. From *Bird-Banding*, 41:161–163.

Vanderschaegen, P. V. 1972. A partial list of parasites of the Ruffed Grouse. *Passenger Pigeon*, 34:70–73.

Vander Wall, S. B. 1982. An experimental analysis of cache recovery in Clark's Nutcracker *(Nucifraga columbiana)*. *Animal Behavior*, 30:84–94.

van Heezik, Y. M., A. F. C. Gerritsen, and C. Swennen. 1983. The influence of chemoreception on the foraging behaviour of two species of sandpiper, *Calidris alba* (Pallas) and *Calidris alpina* (L.). *Netherlands Journal of Sea Research* 17:47–56.

van Rhijn, J. G. 1973. Behavioural dimorphism in male ruffs, *Philomachus pugnax* (L.). *Behaviour*, 47:10–227.

van Rhijn, J. G. 1977. Processes in feathers caused by bathing in water. *Ardea*, 65:126–147.

van Tets, G. F. 1965. A comparative study of some social communication patterns in the pelecaniformes. *American Ornithologists Union, Monograph No. 2*.

Van Tyne, J., and A. J. Berger. 1959. *Fundamentals of Ornithology*. John Wiley and Sons, New York.

Van Velzen, A. C., *et al.* 1972. Lethal mobilization of DDT by Cowbirds. *Journal of Wildlife Management*, 36:733–739.

Varley, G. C. 1970. The concept of energy flow applied to a woodland community. *In* Watson, A. (ed.). *Animal Populations in Relation to their Food Resources*, 389–405. Blackwell Scientific Publications, Oxford and Edinburgh.

Vaurie, C. 1959. *The Birds of the Palearctic Fauna*. 2 vols. H. F. and G. Witherby, Ltd. London.

Vaurie, C. 1972. *Tibet and its Birds*. H. F. and G. Witherby, Ltd., London.

Verbeek, N. 1972. Daily and annual time budget of the Yellow-billed Magpie, *Auk*, 89:567–582.

Verbeek, N. A. M. 1971. Hummingbirds feeding on sand. *Condor*, 73:112–113.

Verbeek, N. A. M. 1977. Age differences in the digging frequency of Herring Gulls on a dump. *Condor*, 79:123–125.

Verner, J. 1964. Evolution of polygamy in the Long-billed Marsh Wren. *Evolution*, 18:252–261.

Verner, J. 1977. On the adaptive significance of territoriality. *American Naturalist*, 111:769–775.

Verner, J., and G. H. Engelsen. 1970. Territories, multiple nest-building, and polygamy in the Long-billed Marsh Wren. *Auk*, 87:557–567.

Verner, J., and M. F. Willson. 1966. The influence of habitats on mating systems of North American passerine birds. *Ecology*, 47:143–147.

Verner, J., and M. F. Willson. 1969. Mating systems, sexual dimorphism and the role of male North American passerine birds in the nesting cycle. *Ornithological Monograph*, 9:1–76.

Veselovsky, A. 1978. Beobachtungen zur Biologie und Verhalten des grossen Kragenlaubenvogels (*Chlamydera nuchalis*). *Journal für Ornithologie*, 19:74–90.

Vevers, G. 1964. Colour. *In* Thomson, A. L. (ed). *A New Dictionary of Birds*. McGraw-Hill Book Company, New York.

Victoria, J. K. 1972. Clutch characteristics and egg discriminative ability of the African Village Weaverbird *Ploceus cucullatus*. *Ibis*, 114:367–376.

Vince, M. A. 1966. Potential stimulation produced by avian embryos. *Animal Behaviour*, 14:34–40.

Vince, M. A. 1969. Embryonic communication and synchronized hatching. *In* Hinde, R. A. (ed). *Bird Vocalizations*. Cambridge University Press, Cambridge.

Vince, M. A. 1972. Communication between quail embryos and the synchronization of hatching. *Proceedings of the XVth International Ornithological Congress*, 357–362. E. J. Brill, Leiden.

Vince, M. A., E. Okleford, and M. Reader. 1984. The synchronization of hatching in quail embryos: aspects of development affected by a retarding stimulus. *Journal of Experimental Zoology*, 229:273–282.

Vincent, J. 1970. *Red Data Book*. Vol. 2, *Aves*. International Union for Conservation of Nature and Natural Resources, Morges, Switzerland.

Vleck, C. M. 1981. Energetic cost of incubation in the Zebra Finch (*Poephila guttata*). *Condor*, 83:229–237.

Vleck, D., C. M. Vleck, and R. S. Seymour. 1984. Energetics of embryonic development in the megapode birds, Mallee Fowl, *Leipoa ocellata* and the brush turkey *Alectura lathami*. *Physiological Zoology*, 57:444–456.

Vleck, C. M., and J. Priedkalns. 1985. Reproduction in Zebra Finches: Hormone levels and effect of dehydration. *Condor*, 87:37–46.

Vleugel, D. A. 1951. A case of Herring Gulls learning by experience to feed after explosions by mines. *British Birds*, 44:180.

Vleugel, D. A. 1962. Über nächtlichen Zug von Drosseln und ihre Orientierung. *Die Vogelwarte*, 21:307–313.

Vogt, W. 1941. Food detection by vultures and condors. *Auk*, 58:571.

Voipio, P. 1970. On 'thunder flights' of the House Martin *Delichon urbica*. *Ornis Fennica*, 47:15–19.

Voous, K. H. 1957. The birds of Aruba, Curaçao and Bonaire. *Studies on the Fauna of Curaçao and other Caribbean Islands*, Vol. 7. The Hague.

Voous, K. H. 1960. *Atlas of European Birds*. Thomas Nelson and Sons, Edinburgh.

Vosburgh, F. G. 1948. Easter egg chickens. *National Geographic Magazine*, 94:377–387.

Vosburgh, J. 1966. Deathtraps in the flyways. *In* Stefferud, A., and A. L. Nelson (eds). *Birds in Our Lives*. U.S. Fish and Wildlife Service, Washington, D.C.

Vuilleumier, F. 1969. Pleistocene speciation of birds living in the high Andes. *Nature*, 223:1179–1180.

Vuilleumier, F. 1972. Flocking vs. non-flocking in Andean forest birds. *Proceedings of the XVth International Ornithological Congress*, 702. E. J. Brill, Leiden.

Vuilleumier, F. 1975. Zoogeography. *In* Farner, D. S., and J. R. King (eds). *Avian Biology*, Vol. 5. Academic Press, New York.

Wagner, G. 1982. Leadership in flocks of homing pigeons. *Revue Suisse de Zoologie Geneve*, 89:297–306.

Wagner, H. O. 1945. Notes on the life history of the Mexican Violet-ear. *Wilson Bulletin*, 57:165–187.

Wagner, H. O. 1957. The molting periods of Mexican hummingbirds. *Auk*, 74:251–257.

Wagner, H. O. 1957. Variation in clutch size at different latitudes. *Auk*, 74:243–250.

Walcott, C. 1978. Anomalies in the earth's magnetic field increase the scatter of pigeons' vanishing bearings. *In* Schmidt-Koenig, K., and W. T. Keeton (eds). *Animal Migration, Navigation, and Homing*. Springer-Verlag, Berlin, New York.

Walcott, C., *et al.* 1974. Orientation of homing pigeons is altered by a change in the direction of an applied magnetic field. *Science*, 184:180–182.

Walcott, C., *et al.* 1979. Pigeons have magnets. *Science*, 205;1027–1028.

Walkinshaw, L. H. 1939. Additional information on the Prothonotary Warbler. *Jack-Pine Warbler*, 17:64–71.

Walkinshaw, L. H. 1963. Some life history studies of the Stanley Crane. *Proceedings of the XIIIth International Ornithological Congress*, 1:344–353. American Ornithologists' Union.

Wallace, A. R. 1876. *The Geographical Distribution of Animals*. Harper and Brothers, New York.

Wallace, A. R. 1889. *Travels on the Amazon and Rio Negro*. Ward, Lock & Co., Ltd., London.

Wallace, G. J. 1955. *An Introduction to Ornithology*. Macmillan Publishing Company, New York.

Wallace, G. J. (chr). 1971. Report of the Committee on Conservation. *Auk*, 88:902–910.

Wallgren, H. 1954. Energy metabolism of two species of the genus *Emberiza* as correlated with distribution and migration. *Acta Zoologica Fennica*, 84:1–110.

Wallraff, H. 1965. Über das Heimfindenvermögen von Brieftauben mit durchtrenten Bogengängen. *Zeitschrift für Vergleichende Physiologie*, 50:313–330.

Wallraff, H. G. 1967. The status of our knowledge about pigeon homing. *Proceedings of the XIVth International Ornithological Congress*, 331–358. Blackwell Scientific Publications, Oxford.

Wallfraff, H. G. 1970. Über die Flugrichtungen verfrachteter Brieftauben in Abhängigkeit vom Heimatort vom Ort der Freilassung. *Zeitschrift für Tierpsychologie*, 27:303–351.

Wallraff, H. G. 1981. The olfactory component of pigeon navigation: steps of analysis. *Journal of Comparative Physiology, A. Series Neural Behavioral Physiology*, 13:411–422.

Walls, G. L. 1942. *The Vertebrate Eye and Its Adaptive Radiation*. Cranbrook Institute of Science, Bloomfield Hills, Michigan.

Walsberg, G. E. 1975 Digestive adaptations of *Phainopepla nitens* associated with the eating of mistletoe berries. *Condor*, 77:169–174.

Walsberg, G. E. 1977. Ecology and energetics of contrasting social systems in *Phainopepla nitens* (Aves: Ptilogonatidae). *University of California Publications in Zoology*, 108:1–63.

Walter, H. 1968. Zur Abhängigkeit des Eleonorenfalken *(Falco eleonorae)* vom mediterranen Vogelzug. *Journal für Ornithologie*, 109:323–365.

Walter, H. 1979. *Eleanora's Falcon: Adaptation to Prey and Habitat in a Social Raptor*. University of Chicago Press.

Walters, J. R. 1980. The evolution of parental behavior and clutch size in shorebirds. *In* Burger, J., and B. L. Olla, (eds). *Behavior of Marine Animals*, Vol. 1. Shorebirds: breeding behavior and populations. Plenum Press, New York, 243–287.

Ward, P. 1963. Lipid levels of birds preparing to cross the Sahara. *Ibis*, 105:109–11.

Ward, P., and P. J. Jones. 1977. Premigratory fattening in three races of the Red-billed Quelea, *Quelea quelea* (Aves: Ploceidae) and intra-tropical migrant. *Journal of Zoology* (London), 181:43–56.

Warham, J., G. J. Wilson, and B. R. Keeley. 1982. The annual cycle of the sooty shearwater *Puffinus griseus* at the Snares Islands, New Zealand. *Notornis*, 29:269–292.

Warner, R. E. 1963. Recent history and ecology of the Laysan Duck. *Condor*, 65:3–23.

Warner, R. E. 1968. The role of introduced diseases in the extinction of the endemic Hawaiian avifauna. *Condor*, 70:101–120.

Warner, R. E., and L. M. David. 1982. Woody habitat and severe winter mortality of ring-necked pheasants in central Illinois, U.S.A. *Journal of Wildlife Management*, 46:923–932.

Waser, M. S., and P. Marler. 1977. Song learning in canaries. *Journal of Comparative Physiology Psychology*, 91:1–7.

Wasserman, F. E. 1977. Mate attraction function of song in the White-throated Sparrow. *Condor*, 79:125–127.

Wasserman, F. E. 1983. Territories of Rufous-sided Towhees contain more than minimal food resources. *Wilson Bulletin*, 95:664–667.

Watling, D. 1978. Observations on the naturalized distribution of the Red-vented Bulbul in the Pacific, with special reference to the Fiji Islands, *Notornis*, 25:109–117.

Watson, A. (ed). 1970. *Animal Populations in Relation to their Food Resources*. Blackwell Scientific Publications, Oxford.

Watson, A., and R. Moss. 1972. A current model of population dynamics in Red Grouse. *Proceedings of the XVth International Ornithological Congress*, 134–149. E. J. Brill, Leiden.

Watson, A., R. Moss, and R. Parr. 1984. Effects of food enrichment on numbers and spacing behavior of Red Grouse *(Lagopus lagopus scoticus)*. *Journal of Animal Ecology*, 53:663–678.

Watson, G. E. 1967. Fulvous Tree Duck observed in the southern Sargasso Sea. *Auk*, 84:424.

Watson, J. B. 1908. The behavior of Noddy and Sooty Terns. *Papers from the Tortugas Laboratory of the Carnegie Institution of Washington*, 2:187–255.

Watt, D. J., C. J. Ralph, and C. T. Atkinson. 1984. The role of plumage polymorphism in dominance relationships of the White-throated Sparrow. *Auk*, 101:110–120.

Weatherhead, P. J. 1979. Ecological correlates of monogamy in tundra-breeding Savannah Sparrows. *Auk*, 96:391–401.

Weathers, W. W., and G. K. Snyder. 1974. Functional acclimation of Japanese Quail to simulated high altitude. *Journal of Comparative Physiology B*, 93:127–138.

Wedemeyer, W. 1973. The spring migration pattern at Fortine, Montana. *Condor*, 75:400–413.

Weeden, J. S. 1965. Territorial behavior of the Tree Sparrow. *Condor*, 67:193–209.

Weidmann, U. 1956. Observations and experiments on egg-laying in the black-headed Gull (*Larus ridibundus* L.). *British Journal of Animal Behaviour*, 4:150–161.

Weir, D. G. 1973. Status and habits of *Megapodius pritchardii*. *Wilson Bulletin*, 85:79–82.

Weller, M. W. 1958. Observations on the incubation behavior of a Common Nighthawk. *Auk*, 75:48–59.

Weller, M. W. 1968. Brood parasitism of the Black-headed Duck. *Living Bird*, 1968:169–207.

Wells, S., and L. F. Baptista. 1978. Displays and morphology of an Anna × Allen Hummingbird hybrid. *Wilson Bulletin*, 91:524–532.

Wells, S., and L. F. Baptista. 1979. Breeding of Allen's Hummingbird *(Selasphorus sasin sedentarius)* on the southern California mainland. *Western Birds*, 10:83–85.

Welty, J. C. 1934. Experiments in group behavior of fishes. *Physiological Zoology*, 7:85–128.

Welty, J. C. 1955. Birds as flying machines. *Scientific American*, 192(3):88–96.

Welty, J. C. 1957. The geography of birds. *Scientific American*, 197(1):118–128.

Wendland, V. 1958. Zum Problem des vorzeitigen Sterbens von jungen Greifvögeln und Eulen. *Die Vogelwarte*, 19:186–191.

Wenzel, B. M. 1972. Olfactory sensation in the Kiwi and other birds. *Annals of the New York Academy of Sciences*, 188:183–193.

West, G. C. 1965. Shivering and heat production in wild birds. *Physiological Zoology*, 38:111–120.

West, G. C., L. J. Peyton, and L. Irving. 1968. Analysis of spring migration of Lapland Longspurs to Alaska. *Auk*, 85:639–653.

West, R. R., and J. H. Dunks. 1969. Repelling Boat-tailed Grackles from sprouting corn with a carbamate compound. *Texas Journal of Science*, 21:231–233.

Westerskov, K. 1956. Incubation temperatures of the pheasant, *Phasianus colchicus*. *Emu*, 56:405–420.

Westerskov, K. 1963. Ecological factors affecting distribution of a nesting Royal Albatross population. *Proceedings of the XIIIth International Ornithological Congress*, 795–811. American Ornithologists' Union.

Wetherbee, D. K., *et al.* 1964. Antifecundity effects of Sudan Black B and transovarian vital staining in population control. *Experimental Station Bulletin 543*. College of Agriculture, University of Massachusetts, Amherst.

Wetmore, A. 1919. Lead poisoning in waterfowl. *U.S. Department of Agriculture Bulletin*, No. 793.

Wetmore, A. 1936. The number of contour feathers in Passeriform and related birds. *Auk*, 53:159–169.

Wetmore, A. 1951. A revised classification for the birds of the world. *Smithsonian Miscellaneous Collections*, 117(4).

Wetmore, A. 1952. Recent additions to our knowledge of prehistoric birds. *Proceedings of the Xth International Ornithological Congress, Uppsala, June 1950*. Almqvist and Wiksells, Uppsala.

Wetmore, A. 1955. Paleontology. *In* Wolfson, A. (ed). *Recent Studies in Avian Biology*. University of Illinois Press, Urbana, Illinois.

Wetmore, A. 1956. A checklist of the fossil and prehistoric birds of North America and the West Indies. *Smithsonian Miscellaneous Collections*, 131(5).

Wheelwright, N. T., and P. D. Boersma. 1979. Egg chilling and the thermal environment of the Fork-tailed Storm Petrel *(Oceanodroma furcata)* nest. *Physiological Zoology*, 52:231–239.

Whistler, H. 1941. Differences in moult in closely allied forms. *Ibis*, 1941:173–174.

Whitaker, L. M. 1957. A resumé of anting, with particular reference to a captive Orchard Oriole. *Wilson Bulletin*, 69:195–262.

White, F. N., *et al.* 1975. The thermal significance of the nest of the Sociable Weaver *Philetarius socius*: winter observations. *Ibis*, 117:171–179.

White, F. N., and J. L. Kinney. 1974. Avian incubation. *Science*, 186:107–115.

White, S. J. 1971. Selective responsiveness by the Gannet *(Sula bassana)* to played-back calls. *Animal Behaviour*, 19:125–131.

Whitman, C. O. 1899. Animal behavior. *Biological Lectures of the Marine Biological Laboratory, Woods Hole, Massachusetts, 1898*, 285–338.

Whitman, C. O. 1919. The behavior of pigeons. *Posthumous Works*, Vol. III. Publication 257, No. 3, Carnegie Institute, Washington, D. C.

Wiens, J. A. 1973. Pattern and process in grassland bird communities. *Ecological Monographs*, 43:237–270.

Wiese, J. H., *et al.* 1977. Large scale mortality of nestling ardeids caused by nematode infection. *Journal of Wildlife Diseases*, 13:376–382.

Wiley, R. H. 1973. Territoriality and non-random mating in Sage Grouse, *Centrocercus urophasianus*. *Animal Behavior Monographs*, 6:85–169.

William, P. L. 1963. Further notes on the African Finfoot, *Podica senegalensis* (Vieillot). *Bulletin British Ornithological Club*, 83:127–132.

Williams, A. J. 1975. Guillemot fledging and predation on Bear Island. *Ornis Scandinavica*, 6:117–124.

Williams, G. G. 1950. Weather and spring migration. *Auk*, 67:52–65.

Williams, G. R. 1952. The California Quail in New Zealand. *Journal of Wildlife Management*, 16:460–483.

Williams, G. R. 1953. The dispersal from New Zealand and Australia of some introduced passerines. *Ibis*, 95:676–692.

Williams, J. B., and G. O. Batzli, 1979. Competition among bark-foraging birds in central Illinois: experimental evidence. *Condor*, 81:122–132.

Williams, T. C., *et al.* 1978. Estimated flight time for transatlantic autumnal migrants. *American Birds*, 32:275–280.

Williamson, K. 1949. The distraction display of the Arctic Skua. *Ibis*, 91:307–313.

Williamson, K. 1959. The September drift movements of 1956 and 1958. *British Birds*, 52:334–377.

Willier, B. H. 1952. Cells, feathers, and colors. *Bios*, 22:109–125.

Willier, B. H. and M. E. Rawles. 1940. The control of feather color pattern by melanophores grafted from one embryo to another of a different breed of fowl. *Physiological Zoology*, 13:177–199.

Willis, E. O. 1966. Notes on a display and nest of the Club-winged Manakin. *Auk*, 83:475–476.

Willis, E. O., and Y. Oniki. 1978. Birds and army ants. *Annual Review of Ecology and Systematics*, 9:243–63.

Willoughby, E. J. 1966. Water requirements of the Ground Dove, *Condor*, 68:243–248.

Willson, M. F. 1971. Seed selection in some North American finches. *Condor*, 73:415–429.

Wilson, E. O. 1975. *Sociobiology: the New Synthesis*. Harvard University Press, Cambridge.

Wilson, E. O. 1978. *On Human Nature*. Harvard University Press, Cambridge, Massachusetts.

Wilson, J. E. 1959. The status of the Hungarian Partridge in New York. *Kingbird*, 9:54–57.

Wiltschko, R. 1981. Die Sonnenorientierung der Vögel. *Journal für Ornithologie*, 121:1–43; 122:1–22.

Wiltschko, R., N. Nohr, and W. Wiltschko. 1981. Pigeons with a deficient sun compass use the magnetic compass. *Science*, 214:343–345.

Wiltschko, R., and W. Wiltschko. 1985. Pigeon homing: change in navigational strategy during ontogeny. *Animal Behaviour*, 33:583–590.

Wiltschko, W. 1980. The relative importance and integration of different directional cues during ontogeny. *In* Nohring, R. (ed). *Acta XVII International Ornithology Congress*, 561–565. Deutsche Ornithologen-Gesellschaft, Berlin.

Wiltschko, W., H. Hock, and F. W. Merkel. 1971. Outdoor experiments with migrating Robins in artificial magnetic fields. *Zeitschrift für Tierpsychologie*, 29:409–415.

Wiltschko, W., and R. Wiltschko. 1972. Magnetic compass orientation of European Robins. *Science*, 176:62–64.

Wiltschko, W., and R. Wiltschko. 1974. Bird orientation under different sky sectors. *Zeitschrift für Tierpsychologie*, 35: 536–542.

Wiltschko, W., and R. Wiltschko. 1975. The interaction of stars and magnetic field in the orientation system of night migrating birds. *Zeitschrift für Tierpsychologie*, 37:337–355: 39:265–282.

Wingfield, J. C. 1984. Androgens and mating systems: Testosterone induced polygyny in normally monogamous birds. *Auk*, 4:665–671.

Wingfield, J. C. 1984. Environmental and endocrine control of reproduction in the Song Sparrow, *Melospiza melodia*. II. Agonistic interactions as environmental information stimulating secretion of testosterone. *General and Comparative Endocrinology*, 56:417–424.

Winkel, W. 1969. Experimentelle Untersuchungen an Zuckervögeln (Coerebidae) im Funktionskreis der Nahrungssuche. . . . *Zeitschrift für Tierpsychologie*, 26:573–608.

Winkel, W., and R. Berndt. 1972. Beobachtungen und Experimente zur Dauer der Huderperiode beim Trauerschnäpper (*Ficedula hypoleuca*). *Journal für Ornithologie*, 113:9–20.

Winkler, H. 1982. Das Jagdverhalten des Glockenreihers *Egretta ardesiaca*. *Journal für Ornithologie*, 123:307–314.

Winterbottom, J. M. 1950. Related species in mixed bird parties in Northern Rhodesia. *Ostrich*, 21:77–83.

Witherby, H. F., F. C. R. Jourdain, N. F. Ticehurst, and B. W. Tucker. 1943. *The Handbook of British Birds*. H. F. and G. Witherby, Ltd., London.

Witschi, W. 1935. Seasonal sex characters in birds and their hormonal control. *Wilson Bulletin*, 47:177–188.

Wittenberger, J. F. 1983. A contextual analysis of two song variants in the Bobolink. *Condor*, 85:172–184.

Wolf, L. L. 1975. "Prostitution" behavior in a tropical hummingbird. *Condor*, 77:140–144.

Wolfson, A. 1942. Regulation of spring migration in Juncos. *Condor*, 44:237–263.

Wolfson, A. 1945. The role of the pituitary, fat deposition and body weight in bird migration. *Condor*, 47:95–127.

Wolfson, A. 1954. Sperm storage at lower-than-body temperature outside the body cavity of some passerine birds. *Science*, 120:68–71.

Wolfson, A. (ed). 1955. *Recent Studies in Avian Biology*. University of Illinois Press, Urbana, Illinois.

Wolstenholme, G. E. W., and J. Knight (eds). 1970. The pineal gland. Ciba Foundation Symposium, London. *Science*, 175:618–619.

Wong, M. 1983. Effect of unlimited food availability on the breeding biology of wild Eurasian tree sparrows in West Malaysia. *Wilson Bulletin,* 95:287–294.

Wood, C. A., and F. M. Fyfe. 1943. *The Art of Falconry.* Stanford University Press, Stanford, California.

Wood, H. B. 1950. Growth bars in feathers. *Auk,* 67:486–491.

Wood, S. F., and C. M. Herman. 1943. The occurrence of blood parasites in birds from southwestern United States. *Journal of Parasitology,* 29:187–196.

Woodbury, R. A., and W. F. Hamilton. 1937. Blood pressure studies in small animals. *American Journal of Physiology,* 119:663.

Woodcock, A. H. 1940. Observations on Herring Gull soaring. *Auk,* 57:219–224.

Woodcock, A. H. 1942. Soaring over the open sea. *Scientific Monthly,* 55:226–232.

Wood-Gush, D. G. M. 1960. A study of sex drive of the two strains of cockerel through three generations. *Animal Behaviour,* 8:43–53.

Wood-Gush, D. G. M. 1964. Domestication. *In* Thomson, A. L. (ed). *A New Dictionary of Birds.* McGraw-Hill Book Company, New York.

Woodwell, G. M. 1970. The energy cycle of the biosphere. *Scientific American,* 223(3):64–74.

Woodwell, G. M. 1974. Success, succession, and Adam Smith. *BioScience,* 24:81–87.

Woodwell, G. M., C. F. Wurster, and P. A. Isaacson. 1967. DDT residues in an east coast estuary: a case of biological concentration of a persistent insecticide. *Science,* 156:821–824.

Woolfenden, G. E. 1973. Nesting and survival in a population of Florida Scrub Jays. *Living Bird,* 12:25–49.

Woolfenden, G. E. 1975. Florida Scrub Jay helpers at the nest. *Auk,* 92:1–15.

Woolfenden, G. E. 1978. Growth and survival of young Florida Scrub Jays. *Wilson Bulletin,* 90:1–18.

Woolfenden, G. E., and J. W. Fitzpatrick. 1984. *The Florida Scrub Jay.* Princeton University Press, New Jersey.

Woller, R. D., and J. C. Coulson. 1977. Factors affecting the age of first breeding of the Kittiwake, *Rissa tridactyla. Ibis,* 119:339–349.

Wooton, J. T. 1986. Clutch-size differences in western and introduced eastern populations of houses finches: patterns and hypotheses. *Wilson Bulletin,* 98:459–462.

Wrenn, M. S. 1978. The Gouldian Finch: A complete list of mating expectations. *Avicultural Magazine,* 84:224–231.

Wunderle, J. M. 1981. An analysis of a morph ratio cline in the Bananaquit *(Coereba flaveola)* on Grenada, W. I. *Evolution,* 35:333–344.

Würdinger, I. 1970. Erzeugung, Ontogenie und Funktion der Lautäusserungen bei vier Gänsearten. *Zeitschrift für Tierpsychologie,* 27:257–302.

Wynne-Edwards, V. C. 1930. On the waking time of the nightjar *(Caprimulgus e. europaeus). Journal of Experimental Biology,* 7:241–247.

Wynne-Edwards, V. C. 1939. Intermittent breeding of the Fulmar *Fulmaris glacialis* (L). *Proceedings of the Zoological Society of London,* Series A, 109:127–132.

Wynne-Edwards, V. C. 1948. Yeagley's theory of bird navigation. *Ibis,* 90:606–611.

Wynne-Edwards, V. C. 1962. *Animal Dispersion in Relation to Social Behaviour.* Oliver & Boyd, Edinburgh and London.

Yamashina, M. 1939. A social breeding habit among Timaliine birds. *IXme Congrès Ornithologique International, Rouen, 9–13 Mai, 1938,* 453–456. Secretariat du Congres, Rouen.

Yamashina, Y. 1952. Classification of the Anatidae based on cytogenetics. *Papers from the Coordinating Committee on Research Genetics,* 3:1–34.

Yang, S. Y., and R. K. Selander. 1968. Hybridization in the Grackle, *Quiscalus quiscula* in Louisiana. *Systematic Zoology,* 17:107–143.

Yasukawa, K. 1981. Song repertoires in the Red-winged Blackbird *(Agelaius phoeniceus):* a test of the Beau Geste hypothesis. *Animal Behaviour,* 29:114–125.

Yeagley, H. L. 1947. A preliminary study of a physical basis of bird navigation. *Journal of Applied Physics,* 18:1035–1063.

Yeagley, H. L. 1951. A preliminary study of a physical basis of bird navigation, Part II. *Journal of Applied Physics,* 22:746–760.

Yeatter, R. E. 1934. The Hungarian Partridge in the Great Lakes Region. University of Michigan School of Forestry and Conservation, *Bulletin No. 5.*

Yocum, C. F. 1943. The Hungarian Partridge (*Perdix perdix* Linn.) in the Palouse Region, Washington. *Ecological Monographs*, 13:167–201.

Young, E. C. 1963. The breeding behaviour of the South Polar Skua, *Catharacta maccormicki*. *Ibis*, 105:203–233.

Young, H. 1951. Territorial behavior of the Eastern Robin. *Proceedings of the Linnaean Society of New York*, 58–62: 1–37.

Young, H. 1955. Breeding behavior and nesting of the Eastern Robin. *American Midland Naturalist*, 53:329–352.

Yunick, R. P. 1984. An assessment of the irruptive status of the Boreal Chickadee in New York State. *Journal of Field Ornithology*, 55:31–37.

Ytreberg, N. J. 1956. Contribution to the breeding biology of the Black-headed Gull (*Larus ridibundus*) in Norway. *Nytt Magasin for Zoologi*, 4:5–106.

Zablotskaya, M. M. 1984. Systematics and Cardueline songs. Academy of Sciences of the U.S.S.R., *Scientific Centre of Biological Research Newsletter*, 3–21.

Zach, R. 1978. Selection and dropping of whelks by Northwestern Crows. *Behaviour*, 67:134–148.

Zagniborodova, E. N., and G. S. Belskaya. 1965. [The fleas of burrow-dwelling birds and their possible role in plague epizootology in Turkmeniya.] *Izvestiya Akad. Nauk Turkmenskoi SSR, Seriya Biol.*, 3:69–74. From *Bird-Banding*, 38:332.

Zahavi, A. 1979. Why shouting? *American Naturalist*, 113:155–156.

Zar, J. H. 1969. The use of allometric model for avian standard metabolism-body weight relationships. *Comparative Biochemistry and Physiology*, 29:227–234.

Zavednyuk, V. F. 1968. [Ant dispersal for forest protection.] *Lesmoe khazyaistvo*, 21:(8):71–72. From *Bird-Banding*, 40:342.

Zerba, E., and M. L. Morton. 1983. Dynamics of incubation in mountain White-crowned Sparrows. *Condor*, 85: 1–11.

Zeuthen, E. 1942. The ventilation of the respiratory tract in birds. *Kgl. Danske Videnskabernes Selskab, Bilogiske Meddelelser*, 17:1–51.

Zigmond, R. E., F. Nottebohm, and D. W. Pfaff. 1973. Androgen-concentrating cells in the midbrain of a songbird. *Science*, 179:1005–1007.

Zigmond, R. E., *et al.* 1977. An autoradiographic study of ^{3}H-androgen retention in the brain of a songbird (*Fringilla coelebs*). *Abstracts of the Society for Neuroscience*, 360.

Zimmer, J. T. 1938. Notes on migrations of South American birds. *Auk*, 55:405–410.

Zimmerman, E. C. 1948. *Insects of Hawaii*, Vol. 1. University of Hawaii Press, Honolulu.

Zimmerman, J. L. 1971. The territory and its density dependent effect in *Spiza Americana*. *Auk*, 88:591–612.

Zink, G. 1959. Zeitliche Faktoren in Brutablauf der Kohlmeise (*Parus major*). *Die Vogelwarte*, 20:128–134.

Zink, G. 1967. Populationsdynamik des Weissen-Storchs, *Ciconia ciconia*, in Mitteleuropa. *Proceedings of the XIVth International Ornithologic Congress*, 191–215. Blackwell Scientific Publications, Oxford.

Zink, R. M. 1983. Evolutionary and systematic significance of temporal variation in the Fox Sparrow. *Systematic Zoology*, 32:223–238.

Zink, R. M. 1985. Genetical population structure and song dialects in birds. *The Behavioral and Brain Sciences*, 8: 118–119.

Zinkl, J. G., *et al.* 1977. Avian cholera in waterfowl and Common Crows in Phelps County, Nebraska, in the spring, 1975. *Journal of Wildlife Diseases*, 13:194–198.

Zippelius, H.-M. 1972. Die Karawanenbildung bei Feld- und Hausspitzmaus. *Zeitschrift für Tierpsychologie*, 30: 305–320.

Ziswiler, V. 1965. Zur Kenntnis des Samenoffnens und der Struktur des horneren Gaumens bei Kornerfressenden Oscines. *Journal für Ornithologie*, 106:1–48.

Ziswiler, V., and D. S. Farner. 1972. Digestion and the digestive system. *In* Farner, D. S., and J. R. King (eds). *Avian Biology*, Vol. 2. Academic Press, New York.

Zuckerman, S. 1957. Hormones. *Scientific American*, 196(3):76–87.

Zwickel, F. C., and J. F. Bendell. 1972. Blue Grouse, habitat and populations. *Proceedings of the XVth International Ornithological Congress*, 150–169. E. J. Brill, Leiden.

INDEX

(Page numbers in italics refer to figures in the text)

The Geographic

PALEARCTIC

ISOLATION BY SEA AND COLD CLIMATE
SHARPLY ACCENTED BREEDING SEASONS
BIRDS POOR IN VARIETY
MOST SPECIES MIGRATORY
MANY PRECOCIAL SPECIES
MANY INSECT EATERS. MANY WATER BIRDS

ORIENTAL

ISOLATION BY SEA AND MOUNTAINS
MANY TROPICAL SPECIES
MANY FRUIT EATERS
FEW SEED EATERS

ETHIOPIAN

ISOLATION BY SEA AND DESERT
MANY GROUND DWELLERS
RICH IN PASSERINE SPECIES
BIRDS SHOW ORIENTAL AFFINITIES

AUSTRA

LONG-TIME ISOLA
MANY GROUND D
MANY NOMADIC
MANY NECTAR